Handbook on Applications of
ULTRASOUND
Sonochemistry for Sustainability

Handbook on Applications of
ULTRASOUND
Sonochemistry for Sustainability

EDITED BY
DONG CHEN | SANJAY K. SHARMA | ACKMEZ MUDHOO

CRC Press
Taylor & Francis Group
Boca Raton London New York

CRC Press is an imprint of the
Taylor & Francis Group, an **informa** business

MATLAB® is a trademark of The MathWorks, Inc. and is used with permission. The MathWorks does not warrant the accuracy of the text or exercises in this book. This book's use or discussion of MATLAB® software or related products does not constitute endorsement or sponsorship by The MathWorks of a particular pedagogical approach or particular use of the MATLAB® software.

First published in paperback 2024

First published 2012
by CRC Press
2385 NW Executive Center Drive, Suite 320, Boca Raton FL 33431

and by CRC Press
4 Park Square, Milton Park, Abingdon, Oxon, OX14 4RN

CRC Press is an imprint of Taylor & Francis Group, LLC

© 2012, 2024 Taylor & Francis Group, LLC

Reasonable efforts have been made to publish reliable data and information, but the author and publisher cannot assume responsibility for the validity of all materials or the consequences of their use. The authors and publishers have attempted to trace the copyright holders of all material reproduced in this publication and apologize to copyright holders if permission to publish in this form has not been obtained. If any copyright material has not been acknowledged please write and let us know so we may rectify in any future reprint.

Except as permitted under U.S. Copyright Law, no part of this book may be reprinted, reproduced, transmitted, or utilized in any form by any electronic, mechanical, or other means, now known or hereafter invented, including photocopying, microfilming, and recording, or in any information storage or retrieval system, without written permission from the publishers.

For permission to photocopy or use material electronically from this work, access www.copyright.com or contact the Copyright Clearance Center, Inc. (CCC), 222 Rosewood Drive, Danvers, MA 01923, 978-750-8400. For works that are not available on CCC please contact mpkbookspermissions@tandf.co.uk

Trademark notice: Product or corporate names may be trademarks or registered trademarks and are used only for identification and explanation without intent to infringe.

Publisher's Note
The publisher has gone to great lengths to ensure the quality of this reprint but points out that some imperfections in the original copies may be apparent.

ISBN: 978-1-4398-4206-5 (hbk)
ISBN: 978-1-03-291768-9 (pbk)
ISBN: 978-0-429-10574-6 (ebk)

DOI: 10.1201/b11012

Visit the Taylor & Francis Web site at
http://www.taylorandfrancis.com

and the CRC Press Web site at
http://www.crcpress.com

To my lovely parents, wife Wei, and daughters Elena and Elisa

Dong Chen

In memory of my grandfather, Pt. Vishwambhar Dayal Sharma

Sanjay K. Sharma

For Yana, Teena, Assad, mum and dad

Ackmez Mudhoo

Contents

Foreword xi
Preface xiii
Acknowledgments xv
Editors xvii
Contributors xix

Chapter 1 Emerging Ubiquity of Green Chemistry in Engineering and Technology 1

Pavel Pazdera

Chapter 2 Introduction to Sonochemistry: A Historical and Conceptual Overview 23

Giancarlo Cravotto and Pedro Cintas

Chapter 3 Aspects of Ultrasound and Materials Science 41

Andrew Cobley, Timothy J. Mason, Larisa Paniwnyk, and Veronica Saez

Chapter 4 Ultrasound-Assisted Particle Engineering 75

Anant Paradkar and Ravindra Dhumal

Chapter 5 Applications of Sonochemistry in Pharmaceutical Sciences 97

Robina Farooq

Chapter 6 Ultrasound-Assisted Synthesis of Nanomaterials 105

Siamak Dadras, Mohammad Javad Torkamany, and Jamshid Sabbaghzadeh

Chapter 7 Ultrasound for Fruit and Vegetable Quality Evaluation 129

Amos Mizrach

Chapter 8 Ultrasound in Food Technology 163

Taner Baysal and Aslihan Demirdoven

Chapter 9 Use of Ultrasound in Coordination and Organometallic Chemistry 183

Boris Ildusovich Kharisov, Oxana Vasilievna Kharissova, and Ubaldo Ortiz-Méndez

Chapter 10 Ultrasound in Synthetic Applications and Organic Chemistry 213

Murlidhar S. Shingare and Bapurao B. Shingate

Chapter 11 Ultrasound in Synthetic Applications and Organic Chemistry 263

    Rodrigo Cella

Chapter 12 Ultrasound Applications in Synthetic Organic Chemistry 281

    Mohammad Majid Mojtahedi and Mohammad Saeed Abaee

Chapter 13 Ultrasound-Assisted Anaerobic Digestion of Sludge 323

    Ackmez Mudhoo and Sanjay K. Sharma

Chapter 14 Ultrasound Application in Analyses of Organic Pollutants in Environment 345

    Senar Ozcan, Ali Tor, and Mehmet Emin Aydin

Chapter 15 Applications of Ultrasound in Water and Wastewater Treatment 373

    Dong Chen

Chapter 16 Ultrasound and Sonochemistry in the Treatment of Contaminated Soils
by Persistent Organic Pollutants 407

    *Reena Amatya Shrestha, Ackmez Mudhoo, Thuy-Duong Pham,
and Mika Sillanpää*

Chapter 17 Role of Heterogeneous Catalysis in the Sonocatalytic Degradation of Organic
Pollutants in Wastewater 419

    Juan A. Melero, Fernando Martínez, Raul Molina, and Yolanda Segura

Chapter 18 Degradation of Organic Pollutants Using Ultrasound 447

    *Kandasamy Thangavadivel, Mallavarapu Megharaj, Ackmez Mudhoo,
and Ravi Naidu*

Chapter 19 Applications of Ultrasound to Polymer Synthesis 475

    Boon Mian Teo, Franz Grieser, and Muthupandian Ashokkumar

Chapter 20 Mechanistic Aspects of Ultrasound-Enhanced Physical and Chemical Processes 501

    *Vijayanand S. Moholkar, Thirugnanasambandam Sivasankar,
and Venkata Swamy Nalajala*

Chapter 21 Ultrasound-Assisted Industrial Synthesis and Processes 535

    *Cezar Augusto Bizzi, Edson Irineu Müller, Érico Marlon de Moraes Flores,
Fábio Andrei Duarte, Mauro Korn, Matheus Augusto Gonçalves Nunes,
Paola de Azevedo Mello, and Valderi Luiz Dressler*

Contents

Chapter 22 Development of Sonochemical Reactor 581

Keiji Yasuda and Shinobu Koda

Chapter 23 Ultrasound for Better Reactor Design: How Chemical Engineering Tools Can Help Sonoreactor Characterization and Scale-Up 599

Jean-Yves Hihn, Marie-Laure Doche, Audrey Mandroyan, Loic Hallez, and Bruno G. Pollet

Chapter 24 Sonoelectrochemistry: From Theory to Applications 623

Bruno G. Pollet and Jean-Yves Hihn

Chapter 25 Combined Ultrasound–Microwave Technologies 659

Pedro Cintas, Giancarlo Cravotto, and Antonio Canals

Chapter 26 Integrating Ultrasound with Other Green Technologies: Toward Sustainable Chemistry 675

Julien Estager

Index 697

Foreword

In the eyes of nature, we cannot achieve a sustainable future by the linear extension of existing technologies. This observation drives the quest for new ways to practice chemistry. Over the past 20 years, an approach to chemistry and engineering defined by the 12 principles of green chemistry and green engineering offers us a proven and systematic way to address sustainability from a first design basis. We know there are more than 85,000 chemicals used in commerce around the world and the vast majority has never been tested for human health and environmental impacts. We know there are trillions of dollars of capital invested in existing chemical manufacturing plants that must be considered in any transition plan. There are no silver bullets to addressing the systems already in place. However, a steady and ever-expanding application of the principles of green chemistry and engineering over time will allow us to make progress in replacing our dependence on petroleum-derived fuels and feedstocks.

Chemists today can go to the literature and find tools to synthesize just about any compound one can conceive based on established transformations, many of which have been around for more than a hundred years. Yet we know man does not practice chemistry the way nature does chemistry and therein lies our hope for the future. We must continue to invent and develop the tools of green chemistry to allow a transition to bio-inspired chemistry.

This book is one such commitment to building the new green chemistry toolbox. It is focused primarily on transformations aided by the use of sonochemistry (acoustic cavitation). The ability to deliver rapid, high-density energy to a system facilitates new possibilities. While the concepts of sonochemistry have been known for more than 80 years, in-depth understanding of this phenomenon continues to evolve. Recently, the technique has begun to see applications ranging from nanoparticle formation to carbohydrate synthesis to waste destruction. Thus, sonochemistry appears to be applicable to a wide variety of chemical transformations and has the potential to influence yields, dramatically reduce reaction times, and increase throughput. All these elements are consistent with the principles of green chemistry and engineering.

As with any new technology, a gradual development and adoption process sets in. Initial efforts focus on understanding and scoping the new tools and we now see synthetic sonochemistry technology finding an ever-increasing variety of applications.

This book represents a wonderful effort to bring together the latest developments in the field and should prove useful to all practicing scientists who have an interest in exploring new methods to input energy into a reaction process.

P. Robert Peoples, PhD
American Chemical Society
Green Chemistry Institute®
Washington, District of Columbia

Preface

Following the establishment of the *12 Principles of Green Chemistry* (Anastas and Warner, 1998), there has been a steady growth in our understanding of what green chemistry means. Green chemistry is a relatively young science in its own respect. Interest in this discipline is growing rapidly and is transgressing several cascading research areas in science, engineering, and technology (Sharma and Mudhoo, 2010). The understanding of the principles that backbone green chemistry has spurred many outstanding efforts to implement chemical processes and innovative technologies that are incrementally taking modern society toward safer and more sustainable practices and products that embody and foster environmental stewardship.

Sonochemistry is a branch of chemical research dealing with the chemical effects and applications of ultrasonic waves, i.e., longitudinal sound waves with frequencies above 20 kHz that lie beyond the upper limit of human hearing—although the range of ultrasonic frequencies can be extended up to 100 MHz (Cravotto and Cintas, 2006). Sonochemistry shares with sustainable chemistry such aims as the use of less hazardous chemicals and solvents, a reduced energy consumption, and an increased product selectivity (Cravotto and Cintas, 2006). In this regard, ultrasonic heating and irradiation are in many instances complementary techniques for driving chemical reactions with a higher efficiency and effectiveness. Ultrasound, an efficient and virtually innocuous means of activation in synthetic chemistry, has been employed for decades with varied successes (Cravotto and Cintas, 2006). Not only can this high-energy input enhance mechanical effects in heterogeneous processes, but it is also known to induce new reactivities leading to the formation of unexpected chemical species. Sonochemistry is unique in its remarkable phenomenon of cavitation, currently the subject of intense research, and has already produced interesting results.

Imaging techniques using echolocation, such as SONAR systems for target detection or echography in health care, represent perhaps the best-known use of ultrasound. Chemical applications extend to such varied areas as organic and organometallic chemistry, materials science, aerogels, water and wastewater treatment, food chemistry, and medicinal research (Cravotto and Cintas, 2006). The writing of this handbook has been undertaken because it was earnestly felt to bring forward an updated pool of the latest research and development findings that reasonably encompass a fair number of most relevant aspects linked to and linking green chemistry practices to environmental sustainability through the uses and applications of ultrasound-mediated and ultrasound-assisted biological, biochemical, chemical, and physical processes. In this handbook, a rich panoply of novel research findings and applications of ultrasonic radiation and sonochemistry have been presented. Several chapters have been presented in the following areas: medical applications, drug and gene delivery, nanotechnology, food technology, synthetic applications and organic chemistry, anaerobic digestion, pollutant degradation, polymer chemistry, industrial syntheses and processes, reactor design, electrochemical systems, and combined ultrasound–microwave technologies.

We sincerely hope this handbook provides a robust pool of knowledge on the green applications of sonochemistry. We also feel it provides up-to-date information on some selected fields of applied research of ultrasound where the principles of green chemistry are being embraced by the scientific, engineering, and technological communities for safeguarding and improving the quality of the environment and human life, at large. We also want to share that Professor Sanjay K. Sharma and A. Mudhoo have recently edited a book, *Green Chemistry for Environmental Sustainability* (CRC Press, Taylor & Francis Group, 2010), which is an up-to-date and humble contribution to the literature on green chemistry.

For MATLAB® and Simulink® product information, please contact:

The MathWorks, Inc.
3 Apple Hill Drive
Natick, MA, 01760-2098 USA
Tel: 508-647-7000
Fax: 508-647-7001
E-mail: info@mathworks.com
Web: www.mathworks.com

REFERENCES

Anastas, P.T. and Warner, J.C. 1998. *Green Chemistry: Theory and Practice.* New York: Oxford University Press.

Cravotto, G. and Cintas, P. 2006. Power ultrasound in organic synthesis: Moving cavitational chemistry from academia to innovative and large-scale applications. *Chemical Society Reviews*, 35:180–96.

Sharma, S.K. and Mudhoo, A. 2010. *Green Chemistry for Environmental Sustainability.* Boca Raton, FL: CRC Press, Taylor & Francis Group.

Dong Chen
Sanjay K. Sharma
Ackmez Mudhoo

Acknowledgments

This bold undertaking has provided us with a unique opportunity to renew some old friendships and hopefully weave some new ones in pursuit of gathering and distilling the expertise required for editing and compiling this handbook on the applications of sonochemistry for sustainability. Without any reservation, we heartily thank our esteemed contributors for the way they have graciously responded with characteristic good humor and patience to our deadlines. We also appreciate their constructive criticisms and suggestions, which have enhanced the content of this work. We hope they effortlessly feel that the final result does ample justice to their painstaking efforts deployed in preparing their respective chapter(s). We are equally appreciative toward other colleagues and fellow researchers who volunteered their help in reviewing the scientific contents of the manuscripts.

Professor Sanjay K. Sharma especially expresses his heartfelt gratitude to his respected parents, Dr. M.P. Sharma and Smt. Parmeshwari Devi. He also extends his regards to Professor R.K. Bansal, who has been a source of inspiration to him, and to Dr. V.K. Agrawal, chairman of the Institute of Engineering and Technology, Alwar (India), for his encouraging words.

Ackmez Mudhoo expresses his appreciation for the faith his parents, Azad A. Mudhoo and Ruxana B. Mudhoo, his brother Assad, sister-in-law Teena, and lovely niece Yanna have placed in him throughout the writing and compilation of this handbook. He is also thankful to Professor Konrad Morgan (vice-chancellor and chairman of Senate of the University of Mauritius, Réduit, Mauritius), Professor Romeela Mohee (dean, Faculty of Engineering, University of Mauritius, Réduit, Mauritius), Professor Pavel Pazdera (Masaryk University, Brno, Czech Republic), Professor Muthupandian Ashokkumar (Particulate Fluids Processing Centre, University of Melbourne, Australia), Professor Giancarlo Cravotto (Università di Torino, Torino, Italy) and Dr. Vinod K. Garg (Guru Jambheshwar University of Science and Technology, Hisar, Haryana, India) for their presence, encouragement, and support.

Note: This work was supported by New Faculty Starting Fund from Indiana University–Purdue University Fort Wayne.

Editors

Dr. Dong Chen is an assistant professor at Indiana University-Purdue University Fort Wayne, Indiana. He has been doing important fundamental research in sonochemistry, environmental chemistry, and water and wastewater treatment technologies. He has served as the principal and coprincipal investigator of several funded scientific research projects worth over $1 million. Dr. Chen's work enjoys a high reputation in the international scientific community and has been widely cited by his peers. His research in the area of ultrasonic control of membrane fouling was reported by many scientific news media, including *Nature*. He has more than 30 journal, book, and conference publications and 2 U.S. patents. He has been serving as a grant reviewer for the U.S. Department of Agriculture and National Institute for Water Resources research programs. In addition, Dr. Chen is a routine reviewer for 12 scientific journals and books. He received his PhD in civil (environmental) engineering with a minor in geological science from The Ohio State University, Columbus, Ohio, in 2005. Besides his research experience, Dr. Chen had been working as a full-time engineering consultant for more than two years. He is a licensed professional engineer in the state of Ohio.

Professor (Dr.) Sanjay K. Sharma is a well-known author and editor of many books, research journals, and hundreds of articles over the last 20 years. One of his books, *Green Chemistry for Environmental Sustainability*, has been recently published by CRC Taylor & Francis Group, LLC, Boca Raton, Florida. He has also been appointed as series editor by Springer's London for their prestigious book series *Green Chemistry for Sustainability*.

Dr. Sharma completed his postgraduation in 1995 and received his PhD from the University of Rajasthan, Jaipur, in 1999. His PhD thesis covered the field of synthetic organophosphorus chemistry and computational chemistry. In 1999, he started working additionally in the field of environmental chemistry and green chemistry and produced very good research papers and books during his 12 year long stay at the Institute of Engineering and Technology, Alwar, Rajasthan, India.

His work in the field of green corrosion inhibitors is well recognized and has been well received by the international research community. He is also actively involved in raising environmental awareness, especially with regard to rainwater harvesting.

Presently, Dr. Sharma works as a professor of chemistry at Jaipur Engineering College and Research Centre, JECRC Foundation, Jaipur, Rajasthan, India—one of the best engineering colleges in North India—where he teaches engineering chemistry and environmental engineering courses to BTech students and pursues his research interests. He has delivered many guest lectures on different topics of applied chemistry in various reputed institutions. His students appreciate his teaching skills and hold him in high esteem.

Dr. Sharma is a member of the American Chemical Society (United States), the International Society for Environmental Information Sciences (Canada), and Green Chemistry Network (Royal Society of Chemists, United Kingdom). He is also a life member of various international professional societies, including the International Society of Analytical Scientists, the Indian Council of

Chemists, the International Congress of Chemistry and Environment, and the Indian Chemical Society.

Dr. Sharma has 9 books on chemistry and over 40 research papers of national and international repute to his credit, which is sufficient evidence of his fair track record as a researcher. He also works as editor in chief for three international research journals, *RASAYAN Journal of Chemistry; the International Journal of Chemical, Environmental and Pharmaceutical Research*; and *the International Journal of Water Treatment and Green Chemistry*, and serves as a reviewer for many other international journals, including the prestigious *Green Chemistry Letters and Reviews*.

Ackmez Mudhoo obtained his BEng (Hons) in chemical and environmental engineering from the University of Mauritius in 2004. His research interests encompass the bioremediation of solid wastes and wastewaters by composting, anaerobic digestion, phytoremediation, and biosorption. Ackmez has 48 international journal publications (original research papers, critical reviews, and book chapters) and 5 conference papers to his credit, and an additional 7 research and review papers in the pipeline in his early career. Ackmez also serves as peer reviewer for *Waste Management*; the *International Journal of Environment and Waste Management*; the *Journal of Hazardous Materials*; the *Journal of Environmental Informatics*; *Environmental Engineering Science*; *RASAYAN Journal of Chemistry*; *Ecological Engineering*; *Green Chemistry Letters and Reviews*; *Chemical Engineering Journal*; and *Water Research*. He is also the editor in chief for the *International Journal of Process Wastes Treatment* and the *International Journal of Wastewater Treatment and Green Chemistry*, and serves as handling editor for the *International Journal of Environment and Waste Management* and the *International Journal of Environmental Engineering*. Ackmez also reckons professional experience as consultant chemical process engineer for China International Water & Electric Corp. (CWE, Mauritius) from February 2006 to March 2008. He is also the coeditor of *Green Chemistry for Environmental Sustainability* (Publisher: CRC Press, Taylor & Francis Group, LLC, Boca Raton, Florida, 454pp, ISBN: 978-1-4398-2473-3). He is presently a lecturer in the Department of Chemical and Environmental Engineering at the University of Mauritius.

Contributors

Mohammad Saeed Abaee
Faculty of Organic Chemistry and Natural Products
Chemistry and Chemical Engineering Research Center of Iran
Tehran, Iran

Muthupandian Ashokkumar
School of Chemistry
University of Melbourne
Parkville, Victoria, Australia

Mehmet Emin Aydin
Environmental Engineering Department
Selcuk University
Konya, Turkey

Taner Baysal
Faculty of Engineering
Department of Food Engineering
Ege University
Izmir, Turkey

Cezar Augusto Bizzi
Department of Chemistry
Federal University of Santa Maria
Santa Maria, Brazil

Antonio Canals
Department of Analytical Chemistry
Nutrition and Food Science and Institute of Materials
University of Alicante
Alicante, Spain

Rodrigo Cella
Oxiteno S/A Industry and Trade
Santo Andre, Brazil

Dong Chen
Department of Engineering
Indiana University–Purdue University
Fort Wayne, Indiana

Pedro Cintas
Faculty of Science
Department of Organic and Inorganic Chemistry
University of Extremadura
Badajoz, Spain

Andrew Cobley
Faculty of Health and Life Sciences
The Sonochemistry Centre
Coventry University
Coventry, United Kingdom

Giancarlo Cravotto
Department of Science and Technology of Drug
University of Turin
Torino, Italy

Siamak Dadras
Laser Materials Processing Laboratory
Iranian National Centre for Laser Science and Technology
Tehran, Iran

Aslihan Demirdoven
Faculty of Engineering and Natural Sciences
Department of Food Engineering
Gaziosmanpaşa University
Tokat, Turkey

Ravindra Dhumal
Centre for Pharmaceutical Engineering Science
University of Bradford
West Yorkshire, United Kingdom

Marie-Laure Doche
National Centre for Scientific Research
University of Franche-Comte
Besançon, France

Valderi Luiz Dressler
Department of Chemistry
Federal University of Santa Maria
Santa Maria, Brazil

Fábio Andrei Duarte
School of Chemical and Food
Federal University of Rio Grande
Rio Grande do Sul, Brazil

Julien Estager
The QUILL Research Centre
The Queen's University of Belfast
Belfast, United Kingdom

Robina Farooq
Department of Chemical Engineering
COMSATS Institute of Information
 Technology
Lahore, Pakistan

Érico Marlon de Moraes Flores
Department of Chemistry
Federal University of Santa Maria
Santa Maria, Brazil

Franz Grieser
School of Chemistry
University of Melbourne
Parkville, Victoria, Australia

Loic Hallez
National Centre for Scientific Research
University of Franche-Comte
Besançon, France

Jean-Yves Hihn
National Centre for Scientific Research
University of Franche-Comte
Besançon, France

Boris Ildusovich Kharisov
School of Chemical Sciences
Autonomous University of Nuevo Leon
Monterrey, Mexico

Oxana Vasilievna Kharissova
School of Physico-Mathematical Sciences
Autonomous University of Nuevo Leon
Monterrey, Mexico

Shinobu Koda
Department of Molecular Design
 and Engineering
Graduation School of Engineering
Nagoya University
Aichi, Japan

Mauro Korn
Department of Mathematical Sciences and Earth
University of Bahia
Salvador, Brazil

Audrey Mandroyan
Centre National de la Recherche Scientifique
Université de Franche-Comté
Besançon, France

Fernando Martínez
Department of Chemical and Environmental
 Technology
Universidad Rey Juan Carlos
Madrid, Spain

Timothy J. Mason
Faculty of Health and Life Sciences
The Sonochemistry Centre
Coventry University
Coventry, United Kingdom

Mallavarapu Megharaj
Centre for Environmental Risk Assessment
 and Remediation
University of South Australia
Adelaide, South Australia, Australia

and

Cooperative Research Centre for
 Contamination Assessment and Remediation
 of the Environment
Salisbury, South Australia, Australia

Juan A. Melero
Department of Chemical and Environmental
 Technology
Universidad Rey Juan Carlos
Madrid, Spain

Paola de Azevedo Mello
Department of Chemistry
Federal University of Santa Maria
Santa Maria, Brazil

Amos Mizrach
Agricultural Research Organization
The Volcani Center
The Institute of Agricultural Engineering
Bet Dagan, Israel

Contributors

Vijayanand S. Moholkar
Department of Chemical Engineering
Indian Institute of Technology Guwahati
Guwahati, India

Mohammad Majid Mojtahedi
Faculty of Organic Chemistry and Natural Products
Chemistry and Chemical Engineering Research Center of Iran
Tehran, Iran

Raul Molina
Department of Chemical and Environmental Technology
King Juan Carlos University
Madrid, Spain

Ackmez Mudhoo
Faculty of Engineering
Department of Chemical and Environmental Engineering
University of Mauritius
Reduit, Mauritius

Edson Irineu Müller
Department of Chemistry
Federal University of Santa Maria
Santa Maria, Brazil

Ravi Naidu
Centre for Environmental Risk Assessment and Remediation
University of South Australia
Adelaide, South Australia, Australia

and

Cooperative Research Centre for Contamination Assessment and Remediation of the Environment
Salisbury, South Australia, Australia

Venkata Swamy Nalajala
Department of Chemical Engineering
Indian Institute of Technology Guwahati
Guwahati, India

Matheus Augusto Gonçalves Nunes
Department of Chemistry
Federal University of Santa Maria
Santa Maria, Brazil

Ubaldo Ortiz-Méndez
School of Mechanical and Electrical Engineering
Autonomous University of Nuevo Leon
Monterrey, México

Senar Ozcan
Environmental Engineering Department
Selcuk University
Konya, Turkey

Larisa Paniwnyk
Faculty of Health and Life Sciences
The Sonochemistry Centre
Coventry University
Coventry, United Kingdom

Anant Paradkar
Centre for Pharmaceutical Engineering Science
University of Bradford
West Yorkshire, United Kingdom

Pavel Pazdera
Faculty of Sciences
Department of Chemistry
Centre for Syntheses at Sustainable Conditions and Their Management
Masaryk University
Brno, Czech Republic

Thuy-Duong Pham
Laboratory of Applied Environmental Chemistry
Department of Environmental Sciences
University of Eastern Finland
Mikkeli, Finland

Bruno G. Pollet
PEM Fuel Cell Research Group
Centre for Hydrogen and Fuel Cell Research
School of Chemical Engineering
The University of Birmingham
Edgbaston, United Kingdom

Jamshid Sabbaghzadeh
Iranian National Centre for Laser Science and Technology
Tehran, Iran

Veronica Saez
Faculty of Health and Life Sciences
The Sonochemistry Center
Coventry University
Coventry, United Kingdom

Yolanda Segura
Department of Chemical and Environmental Technology
Universidad Rey Juan Carlos
Madrid, Spain

Sanjay K. Sharma
Jaipur Engineering College & Research Centre
Jaipur, Rajasthan, India

Murlidhar S. Shingare
Department of Chemistry
Dr. Babasaheb Ambedkar Marathwada University
Aurangabad, India

Bapurao B. Shingate
Department of Chemistry
Dr. Babasaheb Ambedkar Marathwada University
Aurangabad, India

Reena Amatya Shrestha
Department of Civil and Environmental Engineering
Lehigh University
Bethlehem, Pennsylvania

Mika Sillanpää
Laboratory of Applied Environmental Chemistry
Department of Environmental Sciences
University of Eastern Finland

and

Faculty of Technology
Lappeenranta University of Technology
Mikkeli, Finland

Thirugnanasambandam Sivasankar
Department of Chemical Engineering
National Institute of Technology
Tiruchirappalli, India

Boon Mian Teo
School of Chemistry
University of Melbourne
Parkville, Victoria, Australia

Kandasamy Thangavadivel
Centre for Environmental Risk Assessment and Remediation
University of South Australia
Adelaide, South Australia, Australia

and

Cooperative Research Centre for Contamination Assessment and Remediation of the Environment
Salisbury, South Australia, Australia

Ali Tor
Environmental Engineering Department
Selcuk University
Konya, Turkey

Mohammad Javad Torkamany
Laser Materials Processing Laboratory
Iranian National Centre for Laser Science and Technology
Tehran, Iran

Keiji Yasuda
Department of Chemical Engineering
Graduate School of Engineering
Nagoya University
Aichi, Japan

1 Emerging Ubiquity of Green Chemistry in Engineering and Technology

Pavel Pazdera

CONTENTS

Causes of a "Chemistry Curse"..1
Wastes as Real Effect of the Anthropogenic Activities...4
Anthropogenic Wastes and Their Impacts on Nature and Development of Humankind...................4
Chemical Wastes...7
Chemical Wastes and Synthetic Chemistry Metrics..9
Sustainable Development: Starting Point and Prevention of an Ulterior Being of Human Civilization...12
Sustainable Development, Chemistry, and Its Engineering and Technological Application............14
Principles and Goals of Green Chemistry..16
Methodology and Methods of Green Chemistry..17
References and Literature Resources...19

CAUSES OF A "CHEMISTRY CURSE"

The terms chemistry, chemical industry, and chemicals evoke in the minds of people very negative associations and oftentimes excite fears. The reasons for these negative associations are observed from the days of antiquity up to the present due to negative or rather bad human experiences in a number of phenomena connected with chemistry and its conquest.

At the present time, these negative associations are also evoked and provoked by the media or by some ecological activists. It may also be said that such negative associations with regard to chemistry and its symptoms are based on rational or irrational fictions and feelings.

In ancient times and in the Middle Ages when alchemy reigned as the predecessor of present chemistry, people connected alchemy and alchemist activities with evils, ghosts, demons, ghouls, devils, and Satan in spite of all positive things that it brought to human civilization such as new practical and applicable knowledge, findings and observations, chemical elements, compounds, and new medical products. On the other hand, new poisons and toxic drugs, caustic agents, black gunpowder, and other negativities were invented and used (Armitage, 2010; Brown, 2006).

The nineteenth and the twentieth centuries brought to mankind a great deal of new discoveries as well as practical and applicable knowledge. Also, during this time chemistry gave humankind new materials, medical products, phytoeffectors, new technologies, and synthetic procedures (phytoeffectors may be defined as substances or their mixtures that enable vegetables to develop prosperously from germination to harvest time and they contain not only pesticides but also stimulants for growth and immunity).

Chemistry helped to boost the quality of human life and its development though it had its own disadvantages. Moreover, chemistry and its products may be misappropriated knowingly or unpredictably. Several examples of "badly" applied chemistry can be demonstrated. Phosgene (carbonyl dichloride, dichloromethane) (*Merck Index*, 11th edition, 7310), yperite (mustard gas, 1,5-dichloro-3-thiapentane or bis(2-chloroethyl) sulfide) (Cook, 1999), new brilliant explosives (e.g., 2,4,6-trinitrophenol—TNP, 2,4,6-trinitrotoluene—TNT) (Urbanski, 1964; Ferro and Morrow Jr., 2005), and other chemical products were developed and used for killing soldiers during World War I. Phosgene as well as diphosgene (trichloromethyl chloroformate) or triphosgene (bis(trichloromethyl) carbonate), safer to handle than phosgene (Kurita and Iwakura, 1979), was used for the synthesis of functional derivatives of carbonic acid. For example, carbamates have been applied as pesticides or plastics (polyurethane).

At the time of World War II, many men, mainly Jews, were poisoned with Cyclon B (i.e., hydrogen cyanide) (Zabecki, 1999) in gas chambers of Nazi concentration camps. Widespread "biological" application of hydrogen cyanide was initially limited to the fumigation of valuable tree crops, namely, citrus fruit, spreading in 1887 from California to Spain and other countries (Baur, 1984). Hydrogen cyanide is still in production in the Czech Republic in the Synthetic Factory Draslovka Kolín Co. in the city of Kolín under the trademark name Uragan D2 and is sold for eradicating insects and small animals. Other fumigants such as methyl bromide, cyanogens (dicyan), and carbonyl sulfide are also used (Messenger and Braun, 2000). Since 2000 BC, the human population has utilized substances or their mixtures with phytoeffector action to protect and intensify their crops. Some of the phytoeffectors, which were later called pesticides, were at first applied as natural compounds, e.g., sulfur or extracts from tobacco. Over time, the preparation of synthetic molecules and substances began.

Two types of pesticides played a key role in the middle of the twentieth century: the very unfortunate case of DDT (1,1,1-trichloro-2,2-di(4-chlorophenyl)ethane) and the 1:1 mixture of isooctyl 2,4-dichlorophenoxyacetate and 2,4,5-trichlorophenoxyacetate contaminated by 2,3,7,8-tetrachlordibenzo-p-dioxine (TCDD), well known as Agent Orange. DDT is one of the world's most famous synthetic pesticides with a long, unique, and controversial history. The Swiss chemist Paul Hermann Müller was awarded the Nobel Prize in Physiology or Medicine in 1948 "for his discovery of the high efficiency of DDT as a contact poison against several arthropods" (http: //nobelprize.org/nobel_prizes/medicine/laureates/1948/). After World War II, DDT was used as an agricultural miraculous insecticide, and soon its production and widespread use erupted. Its decline occurred after the discovery that DDT is a persistent organic pollutant and is extremely hydrophobic and strongly absorbed by soil. Further, DDT and its metabolites are toxic to a wide range of animals in addition to insects, including marine animals such as crayfish, daphnids, sea shrimp, and many species of fish. They are less toxic to mammals but may be moderately toxic to some amphibian species, especially in the larval stage. Most significantly, they are a reproductive toxicant for certain marine and continental bird species. The search for further negative effects of DDT has been pursued relentlessly by the World Health Organization and by other international and national specialized agencies. It is a terrible fact that some poor African countries used DDT till lately (U.S. Environmental Protection Agency, 1975).

Agent Orange is the code name for one of the herbicides and defoliants used by the U.S. Army in its herbicidal warfare program during the Second Vietnam War (from 1961 to 1971) (Stellman, 2003).

Dioxin, namely, 2,3,7,8-tetrachlordibenzo-p-dioxin, is the synthetic by-product of the reaction of two phenoxyl herbicides mentioned earlier: iso-octyl ester of 2,4-dichlorophenoxyacetic acid and 2,4,5-trichlorophenoxyacetic acid. Agent Orange is not a chemical agent against militants. It was applied to aid the sapper agent in destroying the landscape blanketed by full-grown plants and forests which served as a natural cover for the native warriors of Viet Cong in Vietnam, Laos, and Cambodia. However, all negative impacts of Agent Orange were observed both on the warriors of Viet Cong and the U.S. military. Vietnamese scientists have been conducting epidemiological research on the impact of dioxin on human health since the late 1960s. According to the Vietnam

Red Cross, as many as 3 million Vietnamese people were affected by Agent Orange including at least 150,000 children born with congenital defects. The question as to whether or not the exposure to dioxin affected the health of the Vietnamese was debated since the time of the war when the first animal studies were released showing that 2,3,7,8-tetrachlorodibenzo-p-dioxine caused cancer and congenital defects in rodents. Studies of U.S. veterans who served in the south during the war compared to those who did not go south found increased rates of cancer and nerve, digestive, skin, and respiratory disorders among the former. Among the cancers, veterans from the south had higher rates of throat cancer, acute/chronic leukemia, Hodgkin's lymphoma and non-Hodgkin's lymphoma, prostate cancer, lung cancer, soft tissue sarcoma, and liver cancer (U.S. Environmental Protection Agency—Dioxin Web site).

The second very unfortunate case connected with 2,3,7,8-tetrachlordibenzo-p-dioxine took place on July 10, 1976, in a small chemical manufacturing plant approximately 15 km north of Milan in the Lombardy region in Italy and is known as Seveso Disaster (De Marchi et al., 1972). The factory of the company Industrie Chimiche Meda Società Azionaria (ICMESA) produced 2,4,5-trichlorophenol from 1,2,4,5-tetrachlorobenzene as an intermediate product for 2,4,5-trichlorophenoxyacetic acid. The reaction of 1,2,4,5-tetrachlorobenzene with sodium hydroxide should have proceeded under controlled temperature. However, excessive pressure put the operation out of control and 6 tons of reaction material was dispersed over an area of 18 km². This material also included about 1 kg of 2,3,7,8-tetrachlorodibenzodioxin, which is normally seen only in trace amounts of less than 1 ppm. However, in the higher-temperature conditions associated with the runaway reaction, the production of the mentioned dioxin apparently reached 100 ppm or more. Industrial safety regulations called the Seveso Directive passed the European Community in 1982 imposing more stringent and harsher industrial regulations. The Seveso Directive was updated in 1999, amended again in 2005, and is currently referred to as the Seveso II Directive (or COMAH Regulations in the United Kingdom). Similar directives were also established by government agencies of other countries (e.g., U.S. Environmental Protection Agency—EPA).

Another similar disaster caused by erroneous handling of technological equipment occurred during the night of December 2–3, 1984, in Bhopal, Madhya Pradesh, India. This industrial catastrophe is generally well known as the Bhopal Disaster or the Bhopal Gas Tragedy (Broughton, 2005). In a pesticide plant of Union Carbide India Limited (UCIL), a tank containing about 40 tons of methyl isocyanate (MIC) as an intermediate in the manufacturing the pesticide Carbaryl (1-naphthyl N-methylcarbamate, trademark Sevin) entered into contact with large amounts of water. This caused very rapid decomposition combined with an acute rise in temperature and pressure (exothermic reaction under carbon dioxide formation), and leading to the worst industrial disaster to date. An explosion resulted in a large volume of mixed toxic gases, including MIC, which contaminated the region of Bhopal. The official death toll was initially recorded at around 5000. Many sources indicated that 18,000 had died within 2 weeks, and it is estimated that around 8000 have died since then as victims of gas-poisoning-related diseases that cropped up (Browning, 1993). The decisive factors that contributed to the disaster include the chemical plant's poorly chosen location, the use of hazardous ingredient chemicals such as MIC instead of less dangerous ones, storage of these chemicals in large tanks instead of several smaller storage tanks, poor maintenance and control of equipment at the chemical plant, failure of several safety systems, which were not in operation at the time (Carbaryl is not at present registered for pesticide use because it is classified as a likely human carcinogen).

Both these disasters led to tighter specifications of safety in the chemical industry and technology application sectors. The tragedy of Thalidomide (Moghe et al., 2008) is also a very well-known chemical disaster because of literary and cinematographic elaborations. Thalidomide ((RS)-2-(2,6-dioxopiperidin-3-yl)-1H-isoindole-1,3(2H)-dione) was introduced as a sedative drug in the late 1950s in West Germany and consequently in the rest of the world except the United States. In the late 1950s and early 1960s, more than 10,000 children in 46 countries were born with deformities such as phocomelia as a consequence of thalidomide use.

In 1962, the U.S. Congress enacted laws requiring tests for safety during pregnancy before a drug could receive approval for sale in the United States. Other countries enacted similar legislation, and thalidomide was not prescribed or sold for decades. The later examples of a "sad" and "bad" applied chemistry, risk of some chemicals, chemical technologies, industry, and pollution of the environment by different wastes contribute to the explanation of the execration of chemistry. On the other hand, chemistry and its products are employed by people daily, and, without them, the nature and contemporary characteristics of human life and quality would become generally very difficult.

WASTES AS REAL EFFECT OF THE ANTHROPOGENIC ACTIVITIES

Often, we hear views that regard chemicals as something bad, in the form of a food and/or a drink. If we define the term "chemicals" as elements, their compounds, and compound mixtures (Hill et al., 2005), it reveals the absurdity of these views, because, by strict definition, drinking water, air, bread, ham and eggs, steak, housefly, dog, many others, and, even man could be "chemicals." Finally, we can define life as a highly sophisticated self-organized set of transformations and processes of some chemicals into other chemicals which take place inside a living organism from the time of their conception to the moment of their fatality (Koshland, 2002). In the course of their life cycle, wastes are formed (containing water, inorganic, and organic substances), and products of aerobic and/or anaerobic respiration (carbon dioxide and oxygen), dead matter of tabernacles, and other residues are produced.

Processes of waste formation and their circulation are inevitable for life. Because other organisms consume and derive benefit from such wastes or these wastes stay in the environment as sedimentary rocks (limestone, flintstone, guano, coal bed, and other rock), one may find some regions with high concentrations of these wastes (Blatt et al., 1980).

All highly sophisticated self-organized transformation processes of wastes, and their use and reuse in the environment, including the biosphere, proceed therefore in closed and well-balanced cycles (Beckett, 1981). Any organism needs chemical substances such as food and products other than chemical substances such as wastes and all occur throughout the complex food chains and food webs. A simple example would be green plants that are food for fauna, as they yield oxygen for the animal kingdom and plants. Vice versa, animals produce excrements and expired carbon dioxide as a food for flora (Polis and Winemiller, 1996).

ANTHROPOGENIC WASTES AND THEIR IMPACTS ON NATURE AND DEVELOPMENT OF HUMANKIND

Wastes produced by the human society profoundly underlie different processes because of the different qualities and quantities of the wastes that are produced at the end of these complex processes. The quantity of wastes increases due to a rise in population, and the quality of waste is more diverse because of the ever-increasing industrial activities and their complexity. Both are affected by a change in lifestyle of humans and by the growth in their consumption patterns. Over time, the development of humankind leads to the release of toxic chemicals into the environment. These toxic chemicals as compounds of heavy metals or radioactive chemicals get concentrated in the environment. The concentration of toxic or hazardous substances has a very negative effect on the environment including the biosphere. This is because the natural environmental cycles cannot assimilate and degrade these pollutants any further in a sustainable and effective manner. Processes of waste accumulation are therefore non-sustainable. At the time of the industrial and postindustrial society, the term "waste" took on a new dimension (Pongrácz and Pohjola, 2004).

Waste may be defined as mass (or energy) that is formed during the manufacturing process of a product with a utility value and which remains when this product loses its utility value. This waste continues to be a waste until it is either reintegrated into an environment or it is changed into a new product(s) with a new utility value.

A copy of the newspaper *The Times* can help as an example for the above-mentioned definition. It is evident that during the preparation process of a copy of *The Times* a different but broad spectrum of wastes is produced. However, this is not a problem for most newspaper readers since they look forward to learn about current news. The newspaper has for them an actual utility value equal to the price of £1, which for a majority of readers acquires zero-value after a read-through and turns into waste ending up in a container. It is in principle "bad" when this waste is terminated as municipal waste because its current utility value is negative. On the other hand, this waste paper may be reused as firing material (small but still with a positive utility value) or preferably recycled as fresh raw material (a net more positive utility value). Archiving newspapers in libraries, archives, or in record offices is an unusual way of using read-through newspapers. The utility value of this read-through copy of *The Times* is positive and, over a period of time, it increases.

In this context, it points toward a product of life-cycle assessment (LCA). LCA (Cooper and Fava, 2006) may be defined as a method to assess the environmental aspects and potential impacts associated with a product based on compiling an inventory of relevant energy, material inputs, and environmental releases, on evaluating the potential environmental impacts associated with identified inputs and releases including potential hazards, and on interpreting the results help to come up with a more informed decision.

The possibility of accessing the sources of base materials and energy is the second most important factor in the development of humankind. These sources can be classified in terms of their availability as permanent (e.g., base materials such as water, salty sea and ocean water, nitrogen, oxygen, and other air gases and energy such as solar and geothermal energy and kinetic or potential energy of flowing water and agitate air) or as renewable (e.g., green algae and plants, charcoal, and nuclear energy) biomass. The permanent and renewable sources are therefore classified as sustainable whereas the nonrenewable sources are categorized as non-sustainable (Lancaster, 2002). In the sequel, a number of human activities connected with their development, namely in the time of industrial and postindustrial society, have always resulted in hazardous consequences. As examples, clearing of forest and tropical rain forest (Moran, 1993) for obtaining agricultural farms, land for civil constructions, or for obtaining roundwood as raw material (Hartman, 1992); the use of cyanide for gold mining (Ali, 2006); or the use of sulfuric acid for uranium mining result in the pollution of waters, production of greenhouse and other hazardous atmospheric gases. Furthermore, the production and exploitation of generally toxic and hazardous products and the application of methods for their production lead to environmental pollution.

The possibility of hazard is the third important factor in the development of humankind (Ericson III, 2005). By nature, hazard involves something that could potentially be harmful to a person's life, health, property, or the environment (MacCollum, 2006). One key concept in identifying any hazard is the presence of stored energy that, when released, can cause damage. Stored energy can occur in many forms: chemical, mechanical, thermal, radioactive, or electrical. Another class of hazard does not involve the release of stored energy. Rather, it involves the presence of hazardous situations. Examples include confined or limited egress, oxygen-depleted atmospheres, awkward positions, repetitive motions, and low-hanging or protruding objects. Hazard and vulnerability interact together to create risk.

The cause of the 2008–2009 worldwide financial and economic crisis, which had arisen as a result of derivatives of dangerous financial products or phosgene substitution by dimethyl carbonate—which is produced during a hazardous high-pressure synthesis from methanol and carbon dioxide—constitutes the possibility of hazard. Hazard cannot be completely eliminated but can be avoided by expert handling of safety devices, and can thus be minimized. It is necessary to eliminate hazardous situations or states which are non-sustainable. It stands to reason that the LCA for a paper copy of the *The Times* and for the above-mentioned herbicide Agent Orange will be totally different. The paper version of the *The Times* therefore continues to be issued whereas Agent Orange is forbidden for use.

The quantity of waste during a manufacturing process of a target product can be defined as the mass difference between the amount of raw materials and the target product. The value is commensurable to the efficiency of raw material exploitation, production costs, and the final price of products.

Waste is directly linked to human development, both technologically and socially. The composition of different wastes has varied over time and location, with industrial development and innovation. Examples of this include plastics and nuclear technology. Anthropogenic waste types defined by modern systems of waste management (Rhyner, 1995) are municipal solid waste (MSW), construction and demolition waste (C&DW), institutional waste, commercial waste, and industrial waste (IC&I), medical waste (also known as clinical waste), hazardous waste, radioactive waste, and electronic waste.

Wastes that are formed during anthropogenic activities may be localized in an integral environment, i.e., on the surface of the Earth in soils, in the water of rivers, seas, oceans and subterranean waters, and in the atmosphere. Chemical substances which are contained in anthropogenic wastes may be transformed in the environment by natural environmental processes, e.g., during biodegradation in soil, water, and air by the enzymatic action of living organisms such as bacteria, yeasts, and green plants under aerobic and/or anaerobic conditions (Diaz, 2008).

So long as wastes are not transformed in the environment into environment-friendly and acceptable derivatives, they function as pollutants with all negative results, e.g., as general toxic residues and agents depleting the ozone layer. It is generally known that chemical compounds with strong chemical bonds such as polyaromatic hydrocarbons (PAHs), chlorinated and/or fluorinated hydrocarbons such as chlorofluorocarbons (CFCs, freons), chloromethanes, and the above-mentioned dioxins are dangerous for the environment (Williams et al., 2000).

Wastes can also be classified according to their state of matter such as solid, liquid, and/or gaseous or their material uniformity such as homogeneous or heterogeneous; industrial waste can be categorized by the location of their genesis, e.g., "at source" or "end of pipe" wastes. The best situation is to have wastes that are not generated. Prevention of waste production and prevention of pollution are principal approaches of waste management (Clark and Macquarrie, 2002). Disposal of "end of pipe" wastes includes all anthropogenic waste types and may be described by the following sequence of procedures:

1. Reuse of original product material as fresh resources of material or energy
2. Biodegradation by composting or usage in aerobic/anaerobic sewerage plants
3. Waste incineration with cleaning of combustion products
4. Controlled waste dumping
5. Waste incineration without cleaning of combustion products
6. Noncontrolled waste dumping

Waste separation generally prevents its disposal. The environmental advantage of this sequence is decreasing, and is hence rather analogous to material merit and utility value. The hazard of the presented procedures escalates especially for events 4–6. Cost-efficiency is variable and depends on waste characteristics, costs of procedure, and environmental assessment (Doble and Kruthiventi, 2007).

Moreover, some companies move polluting manufacturing processes from developed countries to the poorer developing countries (least-developed countries), for reasons that need no comment, in an effort to save costs thereafter. This solution does not prevent the consequences. On the other hand, it creates for these companies problems with flexibility of production and transport. For the prevention of hazardous or risky situations, disasters, and accidents, the monitoring and analysis of all hazardous factors (Mannan, 2005) are important. The monitoring and analysis of the movement of chemical substances include all waste types in the environment, a toxicity analysis of current and new chemicals, a capacity assessment of material and energy sources, other hazardous factors, and environmental costs. Consequently, a rise in the quality and quantity of wastes, their local and temporal concentrations, and the self-regulated processes in the environment may be inhibited and the entire system may be deflected from balance.

A number of persons (environmentalists and politicians) have realized this "fatal" problem of environmental pollution. And for the perpetuity of human civilization, these people have recently started seeking solutions. The approaches to address these impending environmental health threats are discussed in the following section.

CHEMICAL WASTES

As mentioned earlier, the existence of contemporary human civilization without chemistry, chemicals, and the pharmaceutical industry, and without chemical processes as applicable in the processing industry and energy industry is hardly conceivable. However, human activities connected with chemistry, chemical engineering processes, and the chemical industry yield very negative impacts on the environment and create all sorts of hazards.

Polyacrylonitrile (PAN) prepared by free-radical vinyl polymerization of monomer acrylonitrile (vinyl cyanide) is a resinous, fibrous, or rubbery organic polymer (Morgan, 2005). The synthetic retro-projection of PAN manufacturing is illustrated in Scheme 1.1. It is generally known as a material for use in the textile industry, often in combination with natural fibers such as cotton and wool. PAN fibers are also the chemical precursors of high-quality carbon fibers. They are chemically modified to make carbon fibers found in plenty of both high-tech and common daily applications such as primary and secondary structures of civil and military aircraft, missiles, solid propellant rocket motors, pressure vessels, fishing rods, tennis rackets, badminton rackets, and high-tech bicycles. Almost all PAN resins are copolymers made from mixtures of monomers with acrylonitrile as the main component. It is a component repeat unit in several important copolymers, such as styrene–acrylonitrile (SAN) and acrylonitrile butadiene styrene (ABS) plastic. Evidently, PAN and its copolymers are very important for the existence of contemporary human life. On the other hand, PAN is a very stable material which is however problematically integrated into the environment because of its incomplete combustion that produces hydrogen cyanide and nitrogen oxides.

Monomer acrylonitrile (Dalin et al., 1971) is manufactured by the catalytic ammoxidation of propylene or by the catalytic additive reaction of hydrogen cyanide with acetylene (Goodrich process). Hydrogen cyanide is produced mainly by the Andrussov oxidation process in which methane and ammonia react in the presence of oxygen at about 1200°C over a platinum catalyst. Methane, propylene, and acetylene are petrochemical products, and therefore nonrenewable raw materials. However, ammonia can be characterized as a renewable raw material and oxygen as a permanent one. All the chemicals used for acrylonitrile manufacturing including acrylonitrile itself are highly

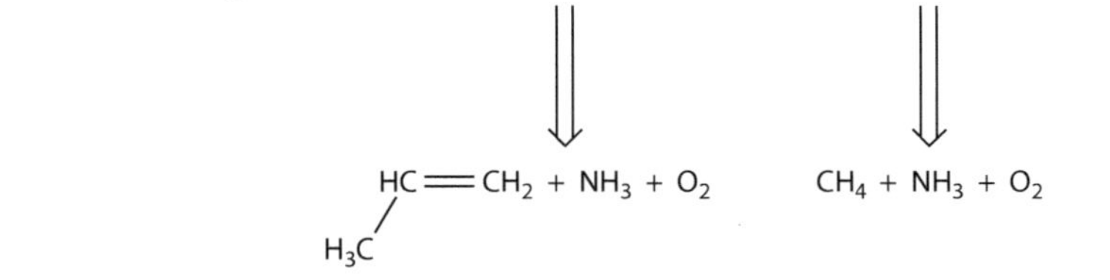

SCHEME 1.1 Synthetic retro-projection of PAN manufacturing.

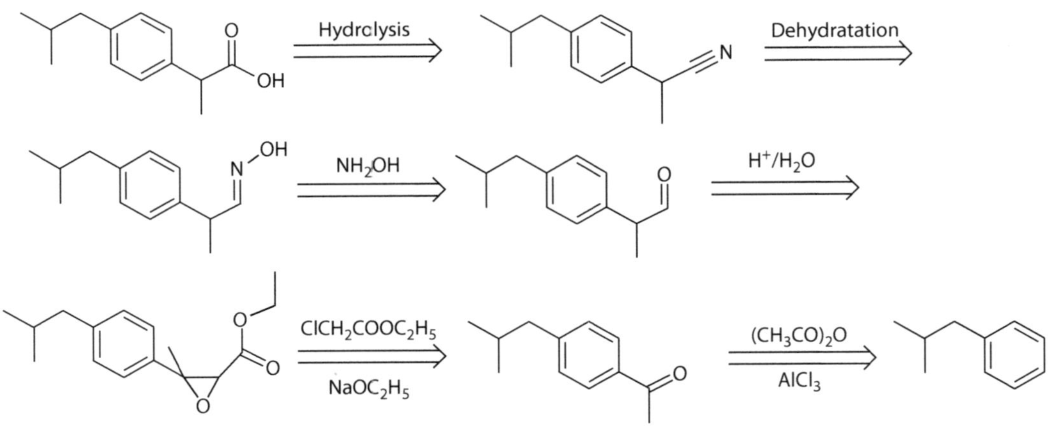

SCHEME 1.2 Synthetic retro-projection of ibuprofen manufacturing by Boots method.

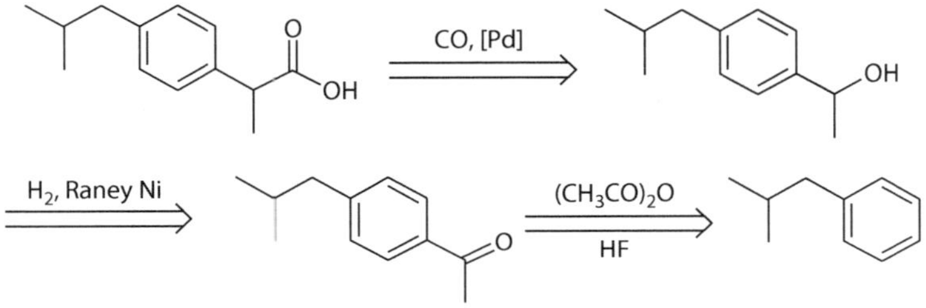

SCHEME 1.3 Synthetic retro-projection of ibuprofen manufacturing by Hoechst method.

hazardous and explosive in a mixture with air, flammable, and generally toxic. These technological processes are also hazardous because they involve high pressures and high temperatures.

Ibuprofen (now outdated in the nomenclature—iso-butyl-propanoic-phenolic acid, Brufen, Nurofen, and Advil, systematic chemical IUPAC name (*RS*)-2-(4-(2-methylpropyl)phenyl)propanoic acid) is one of three mass-applied pharmaceuticals (the other two being aspirin–acetylsalicylic acid—ASA, and Paracetamol–*N*-acetyl-4-aminophenol—APAP) with analgesic, antiphlogistic (anti-inflammatory), and antipyretic effects (Rainsford, 1999). Scheme 1.2 depicts the synthetic retro-projection production of ibuprofen by the Boots method and Scheme 1.3 shows the synthetic retro-projection of ibuprofen manufacturing by the Hoechst method.

Ibuprofen is manufactured currently (procedure of Hoechst company, Presidential [USA] Green Chemistry Challenge: Greener Synthetic Pathways Award in 1997) by a three-step synthesis starting with the Friedel–Crafts para-acetylation of iso-butylbenzene by acetanhydride in the presence of hydrogen fluoride as the Lewis acid catalyst. This synthetic step produces acetic acid and hydrogen fluoride as waste, but acetic acid can be used as a raw material for other chemical syntheses and hydrogen fluoride can be reused. Just using hydrogen fluoride is very innovative because Friedel–Crafts acetylation of aromatics using aluminium chloride as a Lewis acid are commonly applied.

On the other hand, there is no doubt that both acetic acid and increasingly hydrogen fluoride are hazardous substances. In the next step, 4-acetylated iso-butylbenzene is hydrogenated on Raney nickel as a heterogeneous catalyst under pressure to give the corresponding alcohol, which in a third step undergoes palladium-catalyzed carbonylation by carbon monoxide.

Once again, all the above reagents, intermediate products, and catalysts are somehow hazardous. In terms of the source inputs, nickel and palladium are non-renewable, and hydrogen and carbon monoxide can be marked as partly renewable raw materials. Classification in both cases depends on the used energy and carbon monoxide on carbon source. A sustainable method for carbon monoxide manufacturing is hence based either on the reaction of carbon (i.e., charcoal) with carbon dioxide, controlled oxidation of charcoal, or cellulose pyrolysis.

Evaluations of synthetic processes are necessary for an assessment of their environmental impacts, the utilization of material and energy inputs, waste production, and costs involved. We therefore need to define the content of "a synthetic process" or of a chemical synthesis. Chemical synthesis may be defined as a process that starts from the time of location of reagent(s), possible solvent(s), catalyst(s), and auxiliaries in a reactor up to the point of ending their reaction, separation, and purification process of reaction product(s), finally culminating in the adjustment of product(s). The first and last parts of the definition are important because they help to explore the requirement of energy and work and their costs for all processes. On the other hand, they help to obtain a real outlook on some sensational so-called solvent-free syntheses. These cases are reported very often, but, in a further view, they are not solvent-free syntheses. Respective reactions proceed indeed without a solvent, but a mixture of reagents have been mixed in solvents (for deposition on solid support) and/or the final (solid) product has been purified by using a solvent. The advantages of one-pot, multicomponent, and domino reactions excel in the context of the synthetic process defined earlier because several synthetic or reaction steps are located in the same reactor without separation, purification, transport, and adjustment of the intermediary product.

CHEMICAL WASTES AND SYNTHETIC CHEMISTRY METRICS

A chemical one-step synthesis may be described as in Scheme 1.4.

An ideal goal of synthetic processes is the full conversion of starting reactants into a pure final target product C without the application of other substances (catalyst, solvent, auxiliaries) and without energy claims. In addition, this process should be conducted without the formation of regio- and/or stereoisomers of the target product C', C", by-product(s) D, nonconverted educts %A, %B, catalyst, solvents, and auxiliaries, which are a breeding ground for waste formation because they are not embedded in the product, and additional energy and labor are needed for their separation. However, ideal synthetic procedures occur very sporadically. Hence, trade-offs must be sought and considered. Designing synthetic processes helps optimize these compromises.

Current chemistry, engineering, and technology use a number of indexes (synthetic chemistry metrics) to assess the efficiency of synthetic processes (Lapkin and Constable, 2008):

1. Conversion X of reactant A or B (one of two is marked as key or limiting reactant) is the degree of its utilization for any product formation in a given moment. This index (a value of 0–1 or 0%–100%, theoretical maximum $X = 1$ or $X = 100\%$) shows how many key reactants are utilized for product formation and signifies how much of it is left as waste for waste management. Conversion may be increased by the arrangement of synthetic procedure

$$A + B \xrightarrow[\text{Energy and work}]{\text{Catalyst, solvent, auxiliaries}} C + C' + C'' + D + \%A + \%B$$

SCHEME 1.4 General scheme of one-step chemical synthetic process. (A, B: Starting reactants, educts; C: target product; C', C": isomers, both regio and stereo, of the target product (undesirable); D: by-product (undesirable); %A, %B: nonconverted educts; catalyst: acid–base, metal complex, homogeneous or heterogeneous solvent for reaction and/or for purification; auxiliaries: sorbent for purification, surfactant, inert gas; energy for heating, cooling, stirring, high pressure, vacuum, transport and work of staff.)

conditions, e.g., by a change of solvent, and a possible pressure- or temperature-induced displacement of chemical/dynamic equilibrium.

2. Yield of the target product C is the amount of product obtained as a result of a chemical reaction. The relationship between yield Y and conversion X is given by the multiplication operation $Y = X \times S$, where S is a selectivity of reaction for the target product, all calculated on molar basis. Otherwise, a relationship is found between the mass of the target product and its theoretical calculated value, ranging from 0 to 1 or 0% to 100%, with a theoretical maximum for $Y = 1$ or $Y(\%) = 100\%$. Yield as well as conversion and reaction selectivity may be increased by a change of synthetic procedure conditions.

3. Effective mass yield is defined as the ratio between the weight of the target product and the mass of all non-benign materials incident in the course of its synthesis (i.e., regio- and/or stereoisomers of the target product, by-product(s), catalyst, solvents, and auxiliaries). The weakness of this metric is the requirement for a further definition of a benign substance. It is assumed that these have no environmental risk associated with them, e.g., water, low-concentration brine, inert gases, dilute ethanol, autoclaved cell mass may be referred to as benign. This definition is very subjective because environmental data may be incomplete.

4. Environmental factor (E-factor): Roger Sheldon's E-factor (Sheldon, 1992, 1994, 1997a,b, 2000, 2007, 2008) can be viewed as complex and thorough and on the other hand as simple when required. Assumptions on solvent and other factors can be made or a total analysis can be performed. The E-factor calculation is defined by the ratio of the mass of waste per unit of product. This value is very simple to understand and to use. E-factor ignores recyclable factors such as recycled solvents and reused catalysts, which obviously increase the accuracy but ignore the energy involved in the recovery. The main difficulty with E-factors is the need to define system boundaries before reliable and meaningful calculations can be performed and these differ from assessor to assessor. This limitation is the main drawback of all metrics with the exception of the extremely complex LCA of a product. By incorporating yield, stoichiometry, and solvent usage, the E-factor is an excellent metric. Crucially, E-factors can be combined to assess multistep reactions step by step or in one calculation. Sheldon demonstrated in his documents that the chemical industry sector with a bulky tonnage of annual production achieving relatively small values of E-factor, e.g., petrochemical industry and industry, produced bulk chemicals having an E-factor of 0.1–5 at an annual production of 10^4–10^8 tons. On the other hand, chemical companies producing fine chemicals and pharmaceuticals had for their annual production of 10^4–10 tons an E-factor incomparably higher at 5–100. These data hence demonstrate that oil companies produce a lot less waste than pharmaceuticals as a percentage of material processed. This reflects the fact that the profit margins in the oil industry require them to minimize waste and find alternative uses for products which would "normally" be discarded as waste, or it may be transformed by catalytic processes to reusable fundamental products (methane, ethane, ethylene, and hydro-crafting of "heavy" natural resin paraffines). By contrast, the pharmaceutical sector is more focused on molecule manufacturing and quality. The actually high profit margins within the pharmaceutical sector mean that there is less concern about the comparatively large amounts of waste that are produced (especially considering the volumes used) although it has to be noted that, despite the percentage of waste and E-factor being high, the pharmaceutical sector produces much lower tonnage of waste per product unit than any other sector. As mentioned earlier, by obtaining the relevant data for the calculation of a conversion, yield, effective mass yield, and E-factor, a synthetic or process experiment may be performed. Synthetic or process experiments are not necessary for the calculation of a further 4–6 chemistry metrics because of the chemical synthesis and synthetic process modeling discussed below.

5. Atom economy (atom efficiency—AE) is designed in a different way from all the above metrics. It can be designed as a method by which organic chemists would plan on "cleaner" synthetic processes very simply. The essential definition of atom economy is based on how much of the reactant remains in the final product. For a single-step procedure of the above synthesis, atom economy may be calculated as

$$AE = \frac{\text{molecular weight of C}}{\text{molecular weight of } (A+B)}$$

For a general multistep reaction scheme

$$A + B \rightarrow C \text{ followed } C + D \rightarrow E \text{ and next } E + F \rightarrow G$$

the mathematical definition of atom economy is described as

$$AE = \frac{\text{molecular weight of G}}{\text{molecular weight of } (A+B+D+F)}$$

The weakness of AE is that the used catalyst, solvents, and auxiliaries are ignored as they are not incorporated into the final product. AE also ignores the possibility for reuse of secondary reaction products by recycling. AE calculation is very simple and useful as a low atom economy at the design stage of a reaction prior to entering the laboratory can drive a cleaner synthetic strategy to be formulated. The range of AE is 0–1 or 0%–100%, for a theoretical maximum $Y = 1$ or $Y(\%) = 100\%$ achievable for isomerization reactions, rearrangement, additive reactions including Diels–Alder and similar cyclo-additions if these proceed without more isomer formation. In these reaction types, only one product is formed. On the other hand, in the Wittig reaction or Mitsunobu reaction, a very poor AE is obtained because heavy triphenylphosphine oxide is formed. Other examples of poor AE may be the nucleophilic substitution reactions of halogens on saturated carbon atom under S_N2 reaction conditions. Alkyl iodides are very reactive at these alkylating reactions.
6. Carbon efficiency (CE) is another metric derived from AE, but is used only for the efficiency of carbon atoms involved in the synthetic process. The mathematical representation is according to AE, but in lieu of mass of all atoms the mass of carbon atoms is considered. This metric is a good simplification for use in the pharmaceutical industry (and in sustainable petrochemistry too) as it takes into account the stoichiometry of reactants and products. Furthermore, this metric is of interest to the pharmaceutical industry where the development of carbon skeletons is important for the work. The goal is to conserve all carbon atoms in the matrix of the compound. A value of CE less than 1 is connected with cracking and decarboxylation reactions. Bulky decarboxylation reactions observed in petrochemistry might be the source of massive volumes of greenhouse carbon dioxide. Hence, this index shows that the potential transformation of plant oils into high alkanes or alkenes as renewable products will not be based on hydrolysis and decarboxylation reactions, but on reduction through the catalytic hydrogenation of functionalized carboxyl groups in plant oils.
7. Reaction mass efficiency is also another metric derived from AE, but in the calculation, the mass of product and educts is used in lieu of molecular weight. The difference between AE and reaction mass efficiency values may arise if educts react not in a real reaction but on a stoichiometric basis (excess of non-key reactant). In this case, a value for reaction mass efficiency less than the atom economy is obtained.
8. The EcoScale (Van Aken et al., 2006) is a recently developed metric tool for the evaluation of the effectiveness of a synthetic reaction. It is characterized by simplicity and general

applicability. Like the yield-based scale, the EcoScale gives a score from 0 to 100, but also takes into account cost, safety, technical setup, energy, source, and purification aspects. It is obtained by assigning a value of 100 to an ideal reaction. The proposed approach is based on assigning a range of penalty points to these parameters:

$$\text{EcoScale} = 100 - \text{sum of individual penalties}$$

This semiquantitative analysis can easily be modified by other synthetic chemists who may feel that different relative penalty points should be assigned to some parameters. It is a powerful tool to compare several preparations of the same product based on safety, economical, source, and environmental features.

SUSTAINABLE DEVELOPMENT: STARTING POINT AND PREVENTION OF AN ULTERIOR BEING OF HUMAN CIVILIZATION

Since the late 1960s and early 1970s some people from academia, civil society, diplomacy, and industry have begun to be aware that the current life style and development of humankind is not further sustainable. A massive pollution of the environment in consequence of industrial activities and manufacturing, and overexploitation of natural raw material and energy sources have taken place.

Hence, in April 1968, The Club of Rome (King and Schneider, 1993) was founded as a global think tank that would deal with a variety of international political issues. It has raised considerable public attention in 1972 with its report book *The Limits to Growth* (Donella et al., 1972), which used the World3 model to simulate the consequences of interactions between the Earth's systems and human systems. The club stated that its mission is "to act as a global catalyst for change through the identification and analysis of the crucial problems facing humanity and the communication of such problems to the most important public and private decision makers as well as to the general public."

On July 9, 1970, citing rising concerns over environmental protection and conservation, U.S. president Richard Nixon transmitted Reorganization Plan No. 3 to the United States Congress by executive order, creating the EPA as a single, independent agency from a number of smaller arms of different federal agencies. Prior to the establishment of the EPA, the federal U.S. government was not structured to comprehensively regulate environmental pollutants. The United Nations (UN) Conference on the Human Environment (also known as the Stockholm Conference) was an international conference convened under the UN auspices held in Stockholm, Sweden, from June 5 to 16, 1972.

It was the UN's first major conference on international environmental issues, and marked a turning point in the development of international environmental politics. The meeting agreed upon a declaration containing 26 principles concerning the environment and development, an Action Plan with 109 recommendations, and a Resolution. In 1973 the European Union (EU) created the Environmental and Consumer Protection Directorate, and composed the first Environmental Action Program.

In 1987, "Our Common Future" (Our Common Future (1987), Oxford: Oxford University Press), also known as the Brundtland Report, from the UN World Commission on Environment and Development (WCED) was published. The publication of "Our Common Future" and the work of the WCED laid the groundwork for the convening of the 1992 Earth Summit and the adoption of Agenda 21, the Rio Declaration, and the establishment of the Commission on Sustainable Development.

An oft-quoted definition of sustainable development (SD) is defined in the report: "Development that meets the needs of the present without compromising the ability of future generations to meet their own needs."

In addition, key contributions of "Our Common Future" to the concept of SD include the recognition that the many crises facing the planet are interlocking crises that are elements of a single crisis of the whole and of the vital need for the active participation of all sectors of society in consultation and decisions relating to SD.

Agenda 21 (http://www.un.org/esa/dsd/agenda21/) is a program run by the UN related to SD and was the planet's first summit to discuss global warming–related issues. Agenda 21 clearly identified information, integration, and participation as key building blocks to help countries to achieve development that recognizes these interdependent pillars. It emphasizes that in SD everyone is a user and provider of information. It stresses the need to change from old sector-centered ways of doing business to new approaches that involve cross-sectoral coordination and the integration of environmental and social concerns into all development processes. Furthermore, Agenda 21 emphasizes that broad public participation in decision making is a fundamental prerequisite for achieving SD. It is a comprehensive blueprint of action to be taken globally, nationally, and locally by organizations of the UN, governments, and major groups in every area in which humans directly affect the environment.

Such increased interest and research collaboration arguably paved the way for further understanding of global warming, which has led to such agreements as the Kyoto Protocol. This is a protocol to the United Nations Framework Convention on Climate Change (UNFCCC or FCCC), aimed at fighting global warming. The UNFCCC is an international environmental treaty with the goal of achieving "stabilization of greenhouse gas concentrations in the atmosphere at a level that would prevent dangerous anthropogenic interference with the climate system." The Protocol was initially adopted on December 11, 1997 in Kyoto, Japan, and entered into force on February 16, 2005. As of November 2009, 187 states have signed and ratified the protocol, but without the United States and other countries.

On September 8, 2000, following a 3-day Millennium Summit of world leaders at the headquarters of the United Nations, the General Assembly adopted the Millennium Declaration. A follow-up outcome of the resolution passed the General Assembly on December 14, 2000 to guide its implementation. Progress on implementation of the Declaration was reviewed at the 2005 World Summit of leaders.

The World Summit on Sustainable Development (WSSD) or Earth Summit 2002 took place in Johannesburg, South Africa, from August 26 to September 4, 2002. It was convened to discuss SD by the UN. WSSD gathered a number of leaders from business and nongovernmental organizations, 10 years after the first Earth Summit in Rio de Janeiro. (It was therefore also informally nicknamed "Rio+10".)

The 2009 UN Climate Change Conference, commonly known as the Copenhagen Summit, was held in Copenhagen, Denmark, between December 7 and 18.

The conference included the 15th Conference of the Parties (COP 15) in the UNFCCC and the 5th Meeting of the Parties (COP/MOP 5) in the Kyoto Protocol. According to the Bali Road Map, a framework for climate change mitigation beyond 2012 was to be agreed there. The conference was preceded by the Climate Change: Global Risks, Challenges, and Decisions scientific conference, which took place in March 2009 and was also held at the Bella Centre. The negotiations began to take a new format when in May 2009 UN Secretary General Ban Ki-moon attended the World Business Summit on Climate Change in Copenhagen, organized by the Copenhagen Climate Council (COC), where he requested COC councilors to attend New York's Climate Week at the Summit on Climate Change on September 22 and engage with heads of government on the topic of the climate problem.

What results from SD might be a starting point, solution, and prevention of an ulterior being of human civilization. It might optimize close relations between production, economy, human society, the biosphere, and the environment. Hence, SD is built on three pillars: environment, economy, and society.

SD is an optimal intersection of sets of all entities in the environment, economy (including production), and human society. SD could be viable (for production, economy, and the environment), socially and economically equitable, and socially and environmentally bearable in all and for all. Most countries have no problem in accepting these environmental pillars. However, the acceptance of these economical and social pillars may generate scruple or even a silent opposition in poorer developing countries or least-developed countries on the one hand, but also in developed countries on the other, namely, those with a neoliberal government on principle as a "Third Way."

SD also has some shortfalls at its start. The proportion between the size of the population and nonrenewable resources is incommensurable and this trend will grow even further (Cohen, 1995). The proportion between the standard of life of the people in developed countries and that in least-developed countries is incommensurable and this difference will not cease during the lifetime of one human generation. Every one of three richest men in the world is in possession of more property than the wealth of 48 most needy countries in the world, while the richest woman in France has an annual income that equals that of 15,700 of her fellow citizens who draw minimal wages (Keller, 2010).

These proportions show that the inequality between some people and some countries produces a difference that will probably rise further. Moreover, if the economy in developed countries does not show characteristics of growth in human consumption, the global economy will suffer problems. Albert Bartlett, in his contribution "The Laws of Sustainability" in the anthology *The Future of Sustainability* (Keiner, 2006), is very skeptical about the word "sustainability" in connection with the term "development" and in the context of an exponentially growing human population. He asserts in his first of 21 "Laws of Sustainability" that the term "Sustainable Growth" is an oxymoron. In the next laws, he declares that "One cannot sustain a world in which some regions have high standards of living while others have low standards of living" (5th Law), "The benefits of population growth and of growth in the rates of consumption of resources accrue to a few; the costs of population growth and growth in the rates of consumption of resources are borne by all of society" (9th Law), "Humans will always be dependent on agriculture" (16th Law), "If, for whatever reason, humans fail to stop population growth and growth in the rates of consumption of resources, Nature will stop these growths"(18th Law), "Starving people don't care about sustainability" (19th Law), "The addition of the word 'sustainable' to our vocabulary, to our reports, programs, and papers, to the names of our academic institutes and research programs, and to our community initiatives, is not sufficient to ensure that our society becomes sustainable" (20th Law), and he finishes with "Extinction is forever."

All these realities do not leave us feeling optimistic about the further SD of humankind. Hence, the activities of current political and economic elites, establishments, scientists, engineers, and technologists are focused on the environmental and economical pillars of SD as its profitable parts. On the other hand, a mutual relationship between the environment, economy, and production activities is evident.

Consequently, the concept of "sustainable development" may be understood as a set of sustainable endeavors aiming at the betterment and survival of humankind following the rational development and responsive growth that this concept entails.

SUSTAINABLE DEVELOPMENT, CHEMISTRY, AND ITS ENGINEERING AND TECHNOLOGICAL APPLICATION

Chemical production and its engineering and technological application are economically and socially very beneficial to humanity but are very disadvantageous for natural sources and for the rest of the environment. With the goal of harmonizing this disproportionate balance, the following remedial solutions have come into demand.

In 1990, the Pollution Prevention Act passed in the United States. This act helped to create a modus operandi for dealing with pollution in an original and innovative way. It aimed at avoiding problems before they actually happened. Shortly after the passage of the Pollution Prevention Act of 1990, the EPA's Office of Pollution Prevention and Toxics (OPPT) began to explore the idea of developing new or improving existing chemical products and processes to make them less hazardous to human health and the environment.

In 1991, the OPPT launched the model research grants program "Alternative Synthetic Pathways for Pollution Prevention." This program provided, for the first time, grants for research projects that included pollution prevention in the synthesis of chemicals. Since that time the Green Chemistry

Program has built collaborations with many partners to promote pollution prevention through green chemistry. Partnering organizations represent academia, industry, other government agencies, and nongovernmental organizations.

In the name of green chemistry, corresponding activities have started for the first time in all countries except Continental Europe. In the EU analogous activities may be present under the indices of sustainable chemistry.

The European Environment Agency (EEA), an agency devoted to establishing a network for the monitoring of the European environment, was founded by the EU Regulation (European Economic Committee (EEC) Regulation) 1210/1990, as amended by EEC Regulation 933/1999. Work started in earnest in 1994. The regulation also established the European environment information and observation network (Eionet).

Registration, Evaluation, Authorization and Restriction of Chemicals (REACH) is an EU Regulation of December 2006. REACH addresses the production and use of chemical substances and their potential impacts on both human health and the environment.

REACH has been described as the most complex legislation in the EU's history and the most important in 20 years. It is the strictest law to date regulating chemical substances and will affect industries throughout the world. REACH entered into force in June 2007, with a phased implementation over the next decade. When REACH is fully enforced, it will require all companies manufacturing or importing chemical substances into the EU in quantities of 1 ton or more per year to register these substances with a new European Chemicals Agency (ECHA) in Helsinki, Finland. Because REACH applies to some substances that are contained in objects ("articles" in REACH terminology), any company importing goods into Europe could be affected.

About 143,000 chemical substances marketed in the EU were preregistered by the December 1, 2008 deadline. Although preregistering is not mandatory, it allows potential registrants much more time before they have to fully register.

Supply of substances to the European market which have not been preregistered or registered is illegal (known in REACH as "no data, no market"). REACH legislation can therefore help to prevent a risk connected with the use of hazardous chemicals.

In contrast to the EPA, the EEA does not organize programs such as the EPA Green Chemistry Program. The European Technology Platform for Sustainable Chemistry (SusChem) is a European Technology Platform (ETP) initiative to improve the competitive position of the EU in the field of chemistry in three domains: industrial biotechnology, materials technology, and reaction and process design.

The program is a joint initiative (public–private partnership) of the European Commission, representing the European communities and the industry. The main objective of the program is to produce and implement a Strategic Research Agenda (SRA).

Sustainable chemistry, its engineering and technological applications are also focused on cleaner chemical processes and their designing using the "best available technologies" (BAT). The term "best available technology" is applied under regulations on limiting pollutant discharges with regard to the abatement strategy. Similar terms are "best available techniques," "best practicable means," and "best practicable environmental option." The term constitutes moving targets on practices, since developing societal values and advancing techniques may change what is currently regarded as "reasonably achievable," "best practicable," and "best available."

A literal understanding will connect it with a "spare no expense" doctrine which prescribes the acquisition of the best state-of-the-art technology available, without regard for traditional cost–benefit analysis. In practical usage, the cost aspect is also taken into account.

In principle, sustainable chemistry and technology platforms for sustainable chemistry (all European and national) and green chemistry include green engineering, technology, and production, and all have the same goals and principles and use analogous methodologies and methods.

Fundamental connections between SD and sustainable and green chemistry are explained and described by Eissen et al. (2002), Metzger (2004), and in the context of industries by Poliakoff and

License (2007). Two exclusive journals, *Green Chemistry* (RSC) since 1999 and *ChemSusChem* (Wiley Interscience) since 2008, publish the most up-to-date results and solutions of sustainable and green chemistry.

PRINCIPLES AND GOALS OF GREEN CHEMISTRY

The goals of green chemistry are focused on four of the current demands of humankind which are minimizing waste and pollution, efficient exploitation of material and energy sources, minimizing hazard, and minimizing costs as a result of the previous three. The key term for the set of green chemistry goals is "minimizing" but in the context of terms like "efficiently," "rationally," "really," and "preferably." This is because the declared goals may not be achieved promptly and absolutely. For example, it is generally known that freons (CFCs) used as propellants for spray cans were replaced by *n*-butane (hazardous but less than CFCs). Similarly, environmentally hazardous chlorinated hydrocarbons as tetrachloromethane, trichloroethylene, and the other chlorohydrocarbons were alternated by supercritical carbon dioxide as washing media. Principles for the achievement of the set goals may be grouped in the following way:

A. *Minimizing waste and pollution*
 1. Prevention of waste formation is preferred before waste disposal. It is better to handle waste "at source" than at "end of pipe."
B. *Efficient exploitation of material and energy sources*
 2. Syntheses, synthetic processes, must be designed with highest atom economy, i.e., with maximal incorporation of inputs into the product.
 3. Rational reduction for the use of solvents and other auxiliaries is preferred before their recycling and/or regeneration.
 4. Preference of catalytic reagents.
 5. Preference of (solid) supported catalysts.
 6. Multistep syntheses are preferred to one-pot procedures, ideally as multicomponent reactions (MCRs) and/or domino syntheses.
 7. Permanent and renewable material and energy sources should be practicable and applicable rather than nonrenewable wherever technically and economically possible.
 8. Rational reduction for the use of a raw material or feedstock derivative for the sake of its protection, activation, and other temporary modification.
C. *Minimizing general hazard*
 9. In synthetic processes, incoming and outgoing chemicals must be minimal and generally not hazardous.
 10. Life cycle assessment of chemical products.
 11. Development and use of precise analytical techniques and methods to allow for real-time, in-process monitoring and control prior to the formation of hazardous substances.
D. *Minimizing costs*
 12. This can result from a rational and efficient use of the above-presented principles.

On the other hand, it should be remembered that Professor Paul Anastas and Professor John C. Warner (Anastas and Warner, 1998) were the first to develop and formulate the 12 principles of green chemistry. Nevertheless, the above-mentioned principles, which are newer and more innovative, reformulate the essence of green chemistry and chemistry for SD emanating from newer understandings of chemists and engineers. The 12 principles of green chemistry as formulated by Anastas and Warner (1998) are given below for comparison:

1. *Prevention*: It is better to prevent waste than to treat or clean up waste after it has been created.
2. *Atom economy*: Synthetic methods should be designed to maximize the incorporation of all materials used in the process into the final product.

3. *Less hazardous chemical syntheses*: Wherever practicable, synthetic methods should be designed to use and generate substances that possess little or no toxicity to human health and the environment.
4. *Designing safer chemicals*: Chemical products should be designed to affect their desired function while minimizing their toxicity.
5. *Safer solvents and auxiliaries*: The use of auxiliary substances (e.g., solvents, separation agents, etc.) should be made unnecessary wherever possible and innocuous when used.
6. *Design for energy efficiency*: Energy requirements of chemical processes should be recognized for their environmental and economic impacts and should be minimized. If possible, synthetic methods should be conducted at ambient temperature and pressure.
7. *Use of renewable feedstocks*: A raw material or feedstock should be renewable rather than depleting whenever technically and economically practicable.
8. *Reduce derivatives*: Unnecessary derivatization (use of blocking groups, protection/deprotection, temporary modification of physical/chemical processes) should be minimized or avoided if possible, because such steps require additional reagents and can generate waste.
9. *Catalysis*: Catalytic reagents (as selective as possible) are superior to stoichiometric reagents.
10. *Design for degradation*: Chemical products should be designed so that at the end of their function they break down into innocuous degradation products and do not persist in the environment.
11. *Real-time analysis for pollution prevention*: Analytical methodologies need to be further developed to allow for real-time, in-process monitoring and control prior to the formation of hazardous substances.
12. *Inherently safer chemistry for accident prevention*: Substances and the form of a substance used in a chemical process should be chosen to minimize the potential for chemical accidents, including releases, explosions, and fires.

METHODOLOGY AND METHODS OF GREEN CHEMISTRY

Science and advances in engineering, technology, and technical production have given a sophisticated methodology and broad scale of methods to current green chemistry. A sophisticated approach to solve environmental problems, green chemistry, may be essentially characterized as a "trivial solution" to a nontrivial and complicated problem. Sophisticated approaches such as the field of green chemistry methodology are often very effective with minimal energy and technology demands and costs. The manufacturing of 2,4-dichlorobenzyl cyanide, an important fine chemical and intermediate for phytoeffector production, from 2,4-dichlorobenzyl chloride and sodium cyanide with a relatively low level of E-factor and minimum waste formation (Czech Patent 301063 (2009), Lučební závody Draslovka Co. Kolín) can serve as an example of a complete sophisticated solution. This process was industrially realized with high yield and purity in methanol as solvent and in the catalytic presence of sodium iodide (Scheme 1.5).

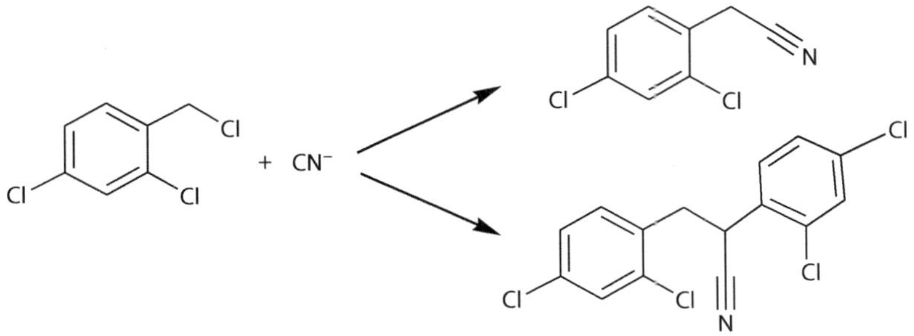

SCHEME 1.5 Preparation of 2,4-dichlorobenzyl cyanide or 2,3-bis(2,4-dichlorophenyl)propanenitrile from 2,4-dichlorobenzyl chloride and sodium cyanide.

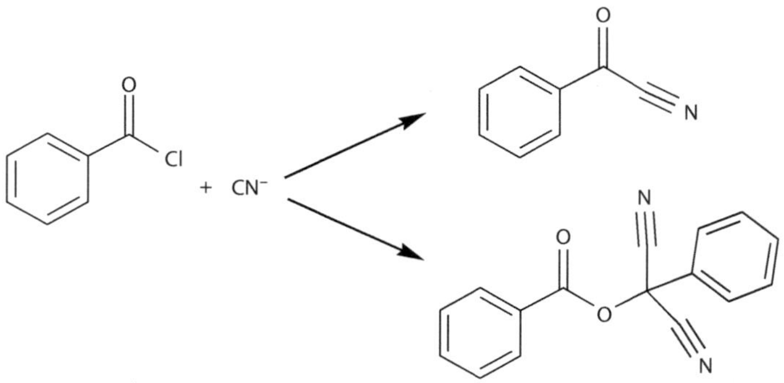

SCHEME 1.6 Preparation of benzoyl cyanide and 1,1-dicyanobenzyl-benzoate from benzoyl chloride and sodium cyanide.

On the other hand, the same reagents and catalyst yield in acetone solvent and in the presence of the catalytic addition of benzyl-trimethylammonium chloride only 2,3-bis(2,4-dichlorophenyl) propanenitrile in high yield and purity (Pazdera and Šimbera, 2011). In the latter study, a suitable selection of solvents as sophisticated resolution for cleaner synthesis was demonstrated for the preparation of benzoyl cyanide and 1,1-dicyanobenzyl-benzoate, which is formed by domino reaction (Scheme 1.6).

Further, similar examples of sophisticated approaches such as the field of green chemistry methodology might be found abundantly in the literature. Hence, several synthetic methods, approaches, and techniques can be presented as feasible and applicable routes for the realization of the green chemistry goals and the fulfillment of its principles. Examples of such synthetic methods are as follows:

1. Sophisticated approaches including regio- and stereoselective synthetic procedures (Vögtle et al., 2000)
2. One-pot, multicomponent (MCRs), and domino reactions (Zhu and Bienaymé, 2005; Tietze et al., 2006; Ishikawa et al., 2009)
3. Acid–base catalysis, catalysis by transition metal and their complexes, and enzymatic catalysis (Sheldon et al., 2007; Anastas, 2009)
4. Interphase catalysis, i.e., phase transfer catalysis (PTC)—both "classic and inverse, and micellar catalysis (Goldberg, 1992; Starks et al., 1994; Khan, 2007)
5. Synthetic applications of supported catalysts and auxiliaries because of their easy separation, regeneration, and reusability, and solid supported and combinatorial syntheses (Seneci and Pierfausto, 2000; Sherrington and Kybett, 2001)
6. Syntheses realized under nonclassic conditions:
   a. Syntheses in supercritical (sc) water and sc-carbon dioxide (both sc-fluids with a risk factor because of temperature and high pressure, respectively) and in ionic liquids as solvents (Leitner and Jessop, 1999; Brunner 2004)
   b. Microwaves (MW) as a low-energy and efficient alternative to classic heating (Kappe and Stadler, 2005)
   c. Ultrasound (US) as a low-energy and efficient alternative to classic stirring, shaking, and heating (Mason and Lorimer, 2002)
7. Combinations of the above methods and approaches, which often bear very effective and surprising results as a consequence of their synergism

The various applications of ultrasound power as a suitable and effective green method in science and engineering research, chemical and biochemical processing, and applied science and engineering technology are discussed in the subsequent chapters.

REFERENCES AND LITERATURE RESOURCES

Ali, S.H. 2006. Gold mining and the golden rule: A challenge for developed and developing countries. *Journal of Cleaner Production*, 14(3–4): 455–462.

Anastas, P.T. 2009. *Handbook of Green Chemistry—Green Catalysis*, Vols. 1–3. Wiley-VCH, Weinheim, Germany.

Anastas, P.T. and Warner, J.C. 1998. *Green Chemistry: Theory and Practice*. Oxford University Press, New York, p. 30.

Armitage, F.P. 2010. *A History of Chemistry*. BiblioBazaar LLC, Charleston, SC.

Baur, F.J. 1984. *Insect Management for Food Storage and Processing*. American Assessment of Cereal Chemists, St. Paul, MN, pp. 162–165.

Beckett, B.S. 1981. *Illustrated Human and Social Biology*. Oxford University Press, Oxford, U.K., p. 38.

Blatt, H., Middleton, G., and Murray, R. 1980. *Origin of Sedimentary Rocks*, Prentice-Hall, Englewood Cliffs, NJ.

Broughton, E. 2005. The Bhopal disaster and its aftermath: A review. *Environmental Health*, 4: 6.

Brown, J.C. 2006. *A History of Chemistry—from the Earliest Times Fill the Present Day*. Kessinger Publishing LLC, Whitefish, MT.

Browning, J. 1993. Union carbide: Disaster at Bhopal. In *Crisis Response: Inside Stories on Managing Image under Siege*, J.A. Gottschalk, Ed. Visible Ink Press, Detroit, MI. http://www.bhopal.com/pdfs/browning.pdf

Brunner, G. 2004. *Supercritical Fluids as Solvents and Reaction Media*. Elsevier, Amsterdam, the Netherlands.

Clark, J. and Macquarrie, D. 2002. *Handbook of Green Chemistry and Technology*. Blackwell science Ltd., Oxford, U.K.

Cohen, J.E. 1995. *How Many People Can the Earth Support?* W.W. Norton & Co., New York City, NY.

Cook, T. 1999. *No Place to Run: The Canadian Corps and Gas Warfare in the First World War*. UBC Press, Vancouver, British Columbia, Canada.

Cooper, J.S. and Fava, J. 2006. Life cycle assessment practitioner survey: Summary of results. *Journal of Industrial Ecology*, 10: 12.

Dalin, M.A., Igor, K.K., and Serebriakov, B.R. 1971. *Acrylonitrile*. Technomic Publication, Westport, CT.

De Marchi, B., Funtowicz, S., and Ravetz, J. 1972. 4 Seveso: A paradoxical classic disaster. United Nations University. http://www.unu.edu/unupress/unupbooks/uu21le/uu21le09.htm

Diaz, E. (Ed.). 2008. *Microbial Biodegradation: Genomics and Molecular Biology*. Caister Academic Press, Norfolk, U.K.

Doble, M. and Kruthiventi, A.K. 2007. *Green Chemistry and Engineering*. Academic Press Elsevier, Amsterdam, the Netherlands.

Donella, H., Meadows, D.L., Meadows, J.R., and Behrens III, W.W. 1972. *The Limits to Growth*. Universe Books, New York.

Eissen, M., Metzger, J.O. et al. 2002. 10 years after Rio-concepts on the contribution of chemistry to a sustainable development. *Angewandte Chemie International Edition*, 41: 414–436.

Ericson III, C.A. 2005. *Hazard Analysis Techniques for System Safety*. Wiley-Interscience, Hoboken, NJ.

Ferro, M. and Morrow Jr., J.H. 2005. *The Great War*. Taylor & Francis Group, Boca Raton, FL, p. 98.

Goldberg, Y. 1992. *Phase Transfer Catalysis: Selected Problems and Applications*. Gordon and Breach Science Publishers, Philadelphia, PA.

Hartman, H.L. 1992. *SME Mining Engineering Handbook*. Society for Mining, Metallurgy, and Exploration Inc., Littleton, CO, p. 3.

Hill, J.W., Petrucci, R.H., McCreary, T.W., and Perry, S.S. 2005. *General Chemistry*, 4th edn. Pearson Prentice-Hall, Upper Saddle River, NJ, p. 5.

Ishikawa, H., Suzuki, T., and Hayashi, Y. 2009. High-yielding synthesis of the anti-influenza neuramidase inhibitor (-)-oseltamivir by three "one–pot" operations. *Angewandte Chemie* (International ed. in English), 48(7): 1304–1307.

Kappe, C.O. and Stadler, A. 2005. *Microwaves in Organic and Medicinal Chemistry*. Wiley-VCH, Weinheim, Germany.

Keiner, M. (Ed.). 2006. *The Future of Sustainability*, Springer, Dordrecht, the Netherlands.

Keller, J. 2010. *Tři sociální světy—Sociální struktura postindustriální společnosti (Three Social Words—Social Structure of the Post-Industrial Society)*. SLON, Prague, p. 12.

Khan, M.N. 2007. *Micellar Catalysis*. CRC/Taylor & Francis, Boca Raton, FL.

King, A. and Schneider, B. 1993. *The First Global Revolution (Club of Rome)*. Orient Longman, Bombay, India.

Koshland Jr., D.E. 2002. The seven pillars of life. *Science*, 295: 2215–2216.

Kurita, K. and Iwakura, Y. 1979. Trichloromethyl chloroformate as a phosgene equivalent: 3-isocyanatopropanoyl chloride. *Organic Synthesis*, 59: 195; 6: 715. http://www.orgsyn.org/orgsyn/orgsyn/prepContent.asp?prep=cv6p0715

Lancaster, M. 2002. *Green Chemistry—An Introductory Text*, Chap. 6. Royal Society of Chemistry, London, U.K.

Lapkin, A. and Constable, D. 2008. *Green Chemistry Metrics. Measuring and Monitoring Sustainable Processes*. Wiley, Chichester, U.K.

Leitner, W. and Jessop, P.G. 1999. *Chemical Synthesis Using Supercritical Fluids*. Wiley-VCH, Weinheim, Germany.

MacCollum, D. 2006. *Construction Safety Engineering Principles: Designing and Managing Safer Job Sites*. McGraw-Hill Professional, New York.

Mannan, S. 2005. *Lee's Loss Prevention in the Process Industries: Hazard Identification, Assessment and Control*. Elsevier, Amsterdam, the Netherlands.

Mason, T.J. and Lorimer, J.P. 2002. *Applied Sonochemistry: The Uses of Power Ultrasound in Chemistry and Processing*. Wiley-VCH, Weinheim, Germany.

Messenger, B. and Braun, A. 2000. Alternatives to methyl bromide for the control of soil-borne diseases and pests in California. Pest Management Analysis and Planning Program. http://www.cdpr.ca.gov/docs/emon/methbrom/alt-anal/sept2000.pdf

Metzger, J.O. 2004. Guest editorial: Agenda 21 as a guide for green chemistry research and a sustainable future. *Green Chemistry*, 6: 15–16.

Moghe, V.V., Kulkarni, U., and Parmar, U.I. 2008. Thalidomide. *Bombay Hospital Journal*, 50(3): 446. http://www.bhj.org/journal/2008_5003_july/download/page-472-476.pdf

Moran, E.F. 1993. Deforestation and land use in the Brazilian Amazon. *Human Ecology*, 21(1): 1–21.

Morgan, P. 2005. *Carbon Fibers and Their Composites*. Taylor & Francis, Boca Raton, FL.

Pazdera, P. and Šimbera, J. 2011. "Cleaner synthesis" of Substituted Aryl Cyanides and 1,1-Dicyanobenzyl Benzoate. In *Organic Preparation and Proceedings International* (in press).

Poliakoff, M. and License, P. 2007. Sustainable technology—Green chemistry. *Nature*, 450: 810–812.

Polis, G.A. and Winemiller, K.O. 1996. *Food Webs: Integration of Patterns and Dynamics*. Chapman & Hall, New York.

Pongrácz, E. and Pohjola, V.J. 2004. Re-defining waste, the concept of ownership and the roles of waste management. *Resources Conservation & Recycling*, 40(2): 141–153.

Rainsford, K.D. 1999. *Ibuprofen*. Taylor & Francis, Boca Raton, FL.

Rhyner, C.R. 1995. *Waste Management and Resource Recovery*. CRC Press, Boca Raton, FL.

Selected Classic Papers from the History of Chemistry. Available at http://web.lemoync.edu/~giunta/papers.html.

Seneci, P. 2000. *Solid Phase Synthesis and Combinatorial Technologies*. Wiley-VCH, Weinheim, Germany.

Sheldon, R.A. 1992. Organic synthesis; past, present and future. *Chemical Industry (London)*, 23: 903–906.

Sheldon, R.A. 1994. Consider the environmental quotient. *Chemtech*, 24: 38–47.

Sheldon, R.A. 1997a. Catalysis and pollution prevention. *Chemical Industry (London)*, 1: 12–15.

Sheldon, R.A. 1997b. Catalysis: The key to waste minimization. *Journal of Chemical Technology and Biotechnology*, 68: 381–388.

Sheldon, R.A. 2000. Atom efficiency and catalysis in organic synthesis. *Pure and Applied Chemistry*, 72: 1233–1246.

Sheldon, R.A. 2007. The E factor: Fifteen years on. *Green Chemistry*, 9: 1273–1283.

Sheldon, R.A. 2008. E-factors, green chemistry and catalysis: An odyssey. *Chemical Communications*, 3352–3365.

Sheldon, R.A., Arends, I., and Hanefeld, U. 2007. *Green Chemistry and Catalysis*. Wiley-VCH, Weinheim, Germany.

Sherrington, D.C. and Kybett, A.P. 2001. *Supported Catalysts and Their Applications*. Royal Society of Chemistry, Great Britain, U.K.

Starks, C.M., Liotta, C.L., and Halpern, M. 1994. *Phase-Transfer Catalysis: Fundamentals, Applications, and Industrial Perspectives*. Springer Verlag, Berlin, Germany.

Stellman, J. et al. 2003. The extent and patterns of usage of agent orange and other herbicides in Vietnam. *Nature*, 422: 681–687.

Tietze, L.F., Brasche, G., and Gericke, K. 2006. *Domino Reactions in Organic Synthesis*. Wiley-VCH, Weinheim, Germany.

U.S. Environmental Protection Agency—Dioxin website. http://cfpub.epa.gov/ncea/CFM/nceaQFind.cfm?keyword=Dioxin

U.S. Environmental Protection Agency, DDT website. http://www.epa.gov/pesticides/factsheets/chemicals/ddt-brief-history-status.htm

Urbanski, T. 1964. *Chemistry and Technology of Explosives*, 1st edn. Pergamon Press, Oxford, Oxford, U.K., pp. 389–391.

Van Aken, K., Strekowski, L., and Patiny, L. 2006. EcoScale, a semi-quantitative tool to select an organic preparation based on economical and ecological parameters. *Beilstein Journal of Organic Chemistry*, (3): 2.

Vögtle, F., Stoddart, J.F., and Shibasaki, M. 2000. *Stimulating Concepts in Chemistry*. Wiley-VCH, Weinheim, Germany.

Williams, P.L., James, R.C., and Roberts, S.M. 2000. *Principles of Toxicology—Environmental and Industrial Applications*, 2nd edn. Wiley-Interscience, New York, p. 6.

Zabecki, D.T. 1999. *Zyklon B. World War II in Europe: An Encyclopedia*. Garland Pub., New York.

Zhu, J. and Bienaymé, H. 2005. *Multicomponent Reactions*. Wiley-VCH, Weinheim, Germany.

2 Introduction to Sonochemistry: *A Historical and Conceptual Overview*

Giancarlo Cravotto and Pedro Cintas

CONTENTS

Introduction 23
Origin of Acoustics: The Sound Spectrum 24
Bubble Formation in Liquids: Cavitation 26
Chemistry in High-Energy Cavities 27
    Sonoluminescence Studies 28
Sonochemical Activation 29
    Reactivity Rules in Ultrasonic Fields 32
Conclusions 37
Acknowledgments 37
References 38

INTRODUCTION

Although the use of ultrasound as the primary means of stimulating chemical reactions and processes has been known for many years, this safe form of irradiation has become increasingly popular during the last two decades along with the emergence of other stimulating techniques (e.g., microwaves, photochemistry, electrochemistry, or high pressure) in the search for more environmentally benign conditions. Large rate enhancements and selectivities are usually observed and represent the most important pluses.

    The interaction of sound waves with matter is far from being trivial, though a rather reductionist approach points to effects related to both piezo- and thermochemistries (*vide infra*). Sonochemistry is a branch of science that deals with the chemical and mechanical effects of ultrasound. This irradiation appears to be more general than other activation techniques as the system requires essentially an ultrasound source and a liquid (either aqueous or organic), which contrasts with specific requirements in electrochemistry (conducting media), microwaves (polar media or species/ions), photochemistry (presence of chromophores), or supercritical conditions (elevated pressure or temperature in closed systems). Ultrasound-assisted chemistry is generally associated with a series of key characteristics such as safety, energy savings, the use of ambient conditions, waste prevention, and improved mass transfer, among others. The greenness of sonochemistry, like other activation techniques, should however be assessed with care and from a critical point of view. If one uses either microwaves or ultrasound with toxic reagents and the subsequent separation and purification steps require the extensive use of volatile organic solvents, the whole protocol is not green at all. Clearly, the benefits associated with a safer technique disappear. This book concentrates on the wide range of ultrasonic applications with a focus on sustainable processes. This introductory chapter intends to highlight the historical basis and fundamentals of power ultrasound, the unique phenomenon of cavitation, and the different factors influencing the fate of a sonochemical reaction. This contribution specifically excludes discussion of experimental setup, reactor design, or transducer engineering, which will be treated in other chapters. It is hoped such considerations will be of benefit to the large readership which is interested in this multidisciplinary technology.

As sonochemists, we have incredulously observed that many entering the field have simply received one message: just switch on, thus overlooking (even worse ignoring) the science behind this sort of activation. Although this chapter in particular and the whole book in general will doubtless serve as tutorial in many aspects, our treatment cannot be comprehensive. The interested reader is also referred to some previous excellent monographs, which cover both theoretical fundamentals in detail and numerous domains of chemical application (Suslick, 1988; Mason, 1991; Luche, 1998; Mason and Lorimer, 2002; Mason and Peters, 2002).

ORIGIN OF ACOUSTICS: THE SOUND SPECTRUM

Studies and empirical observations aimed at elucidating the nature of sound date back to ancient times, although only from the seventeenth century on a series of key scientists such as Boyle, Hooke, or Newton were capable of formulating plausible hypotheses and theories (Hunt, 1978; Beyer, 1999).

Sound consists of pressure waves transmitted through a medium (gas, liquid or solid) as cycles of compression and expansion. Newton paid attention to sound propagation and was probably the first to describe the relationships between the speed of sound and measurable properties of the medium such as density and pressure. In his *Principia*, Newton postulated that the speed of sound is equal to the square root of the ratio between the elastic force and the density of the propagation medium. This assumption, however, gives rise to inaccurate estimates (lower than experimental values). Sound propagation is not an isothermal process, but rather adiabatic, a fact that could only be proved in the nineteenth century with the advent of thermodynamics. The reason is that propagation does not allow enough time for heat exchange to occur between the compression and rarefaction zones of the sound wave; approximately isothermal processes will only occur at the boundary layer to a surface (Carr Everbach, 2007). On the other hand, Augustin-Jean Fresnel (1788–1827) recognized the differences between the diffraction of sound and of light in terms of the differences in wavelength. Since sound wavelengths may be comparable in size to objects, he anticipated that diffraction and interference phenomena could even be more important in acoustics than in optics (Hunt, 1978).

Humans can sense these pressure waves through their ears if frequencies lie in the range of 10 Hz to approximately 18 kHz (the Hertz unit meaning one cycle of compression or rarefaction per second). These frequencies are similar to those of low-frequency radio waves; however, sound is markedly different to electromagnetic radiation and lacks a quantum nature. Although sound transmission causes some excitation in the medium, in the form of enhanced molecular motion, appreciable effects of interest in chemistry and other applied sciences require higher frequencies with a threshold intensity. Ultrasound has therefore frequencies beyond human hearing, i.e., above 18 kHz. For comparative purposes, one can speak of *the sound spectrum* covering a range of frequencies in which audible sound occurs in a narrow region between infrasound (just below 10 Hz) and ultrasound irradiation, the latter being the subject of interest to sonochemists. It is customary to divide ultrasound into two regions: *conventional power ultrasound*, up to 100 kHz, as most devices usually operate within such frequencies, and *diagnostic ultrasound*, from about 1–10 MHz (Figure 2.1). The latter, which possess much shorter wavelengths and hence better resolution in detecting phase changes, is used in medicine (*echography*) for fetal and soft-tissue imaging. Provided that the ultrasonic intensity is sufficiently low, the irradiation causes no harmful effects and represents the ideal choice in the early stages of pregnancy. Fetal echoes are obtainable many weeks before the fetus skeleton is visible by x-rays.

From a historical viewpoint, the dawn of ultrasonics can be traced to observations of animal orientation methods made at the end of the eighteenth century. Italian polymath Lazzaro Spallanzani (1729–1799) observed that bats, even blind animals, could fly freely in a dark room detecting their prey and avoiding obstacles. A Swiss zoologist, Charles Jurine (1751–1819) heard from this

Introduction to Sonochemistry

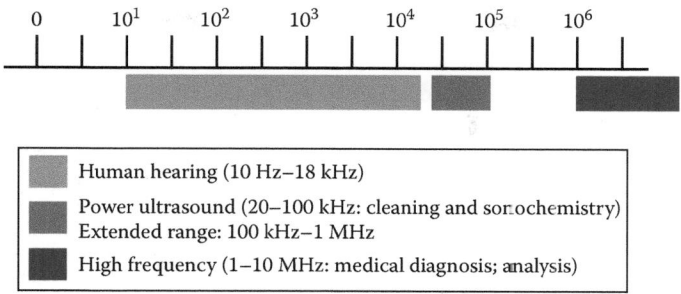

FIGURE 2.1 (See color insert.) Sound frequencies (scale in Hz).

observation through a letter submitted by Spallanzani to the Geneva Natural History Society. Jurine reported in 1794 that when the ear canals of bats were closed with wax, the animals collided with wires and other obstacles. Jurine corresponded with Spallanzani and both natural scientists agreed that bats use their ears to navigate, although they were unable to formulate a theory of such a biosonar. Moreover, the influential paleobiologist, Georges Cuvier (1769–1832), considered such experiments as flawed concluding that bats use touch organs for orientation. This conjecture along with the fact that bats use frequencies inaudible to human ears did much to hinder further research on bat orientation for over a century. It was not until the late 1930s when American biophysicist Donald Griffin and others showed that bats produce ultrasonic sounds and detect objects and insects that reflect such sounds, thus proving that the animals orient themselves by echolocation (Griffin, 1958). This seminal breakthrough was instrumental in discovering echolocation in dolphins a decade later (Whitlow, 1993; Simmons, 2007). Although echolocation lies beyond the scope of this article, these preliminary and historical foundations illustrate salient characteristics of high frequencies, which show strong reflection characteristics. The range of sounds employed by different species varies from 10 kHz to about 150 kHz. The animals emit whistle signals and broadband echolocation cycles lasting for 40–70 μs in the case of dolphins. The energy flux density (EFD) is small (from 0.001 to 0.033 J m^{-2}) and is related to the time-dependent acoustic pressure $p(t)$ by

$$\text{EFD} = \frac{1}{\rho c} \int p(t)^2 dt$$

where
  ρ is the density of the medium
  c is the speed of sound

The product ρc represents the impedance.

The present-day generation of ultrasound has its origin in the piezoelectric effect, discovered by Jacques and Pierre Curie in 1880 (Curie and Curie, 1880; Mould, 2007). These brothers showed that some crystals, especially quartz, generated electrical polarization under mechanical stress. The essential unit of ultrasonic devices is the transducer, composed of a piezoelectric material, in which the inverse effect is observed, i.e., a high alternating electrical potentially causes mechanical vibration of high frequency (ultrasound). The first practical application was an ultrasonic submarine detector developed in 1917, during World War I, by Paul Langevin, who had been one of Pierre Curie's students. The device consisted of thin quartz crystals glued between two steel plates, along with a hydrophone to detect the returning echo. This transducer emits a pulse of ultrasound, and by measuring the time taken to hear the echo of the sound waves from a submarine object, the distance to the object can be determined. Subsequent developments in underwater range finding resulted in the system now called SONAR (Sound Navigation and Ranging).

BUBBLE FORMATION IN LIQUIDS: CAVITATION

When a pressure wave of sufficient intensity propagates through a liquid, formation of vapor bubbles may occur. Such cavities result when a negative pressure exceeds the tensile strength of the liquid, which is the maximum stress that a substance can withstand from stretching without tearing. Although acoustic cavitation will be discussed later, early observations and descriptions of cavitational phenomena emerged from hydrodynamic experiments. Within the historical context of this chapter, it is appropriate to recreate such events, which are largely unknown and carve out a long journey toward understanding. It has been claimed (Young, 1999) that Isaac Newton was the first to observe cavitation in low-pressure regions in a thin layer of water between two rolling pieces of glass. He was examining rings formed between a convex lens and a plane glass surface. In his *Optiks* (1704), one reads: "When the water was between the glasses, if I pressed the upper glass variously at its edges to make the rings move nimbly from one place to another, a little white spot would immediately follow the center of them, which upon creeping in of the ambient water into that place would presently vanish." Newton then adds: "Its appearance was such as interjacent air would have caused, and it exhibited the same colors. But it was not air, for where any bubbles of air were in the water they would not vanish. The reflexion must have rather been caused by a subtiler medium, which could recede through the glasses at the creeping in of the water." Clearly, Newton could not account for bubble formation as a result of the reduced pressure and the subsequent "dissolution" in water; so he invokes a *subtiler* medium.

Leonhard Euler (1707–1783), who had studied mathematics with Johann and Daniel Bernoulli, became interested in turbines and reported in 1754 the case of inhomogeneities in liquids resulting from air cavities when the liquid is accelerated at high velocities under a negative pressure (Euler, 1754; Truesdell, 1955). Georges Stokes (1819–1903) had formulated a series of intriguing questions on fluid dynamics to his students in Cambridge, one of them directly related to cavity formation: "An infinite mass of homogeneous incompressible fluid acted upon by no forces is at rest, and a spherical portion of the fluid is suddenly annihilated; it is required to find the instantaneous alteration of pressure at any point of the mass, and the time in which the cavity will be filled up, the pressure at an infinite distance being supposed to remain constant." Besant solved the problem by using the equation of motion of the fluid (Besant, 1859). Several decades later, Lord Rayleigh (who had been commissioned by the Royal Navy to shed light into cavitation once the phenomenon was clearly identified) obtained an elegant solution for the bubble wall velocity and the time of collapse using energy considerations (Rayleigh, 1917).

An interesting historical debate, dealing with the origin of boiling points and effervescent bubbles, is also quoted as the prescientific foundation of cavitation. Désiré–Jean–Baptiste Gernez (1834–1910) reported that boiling could always be induced in superheated water by the insertion of a trapped pocket of air into the liquid. Thus, the presence of internal gases was the key enabling condition for boiling, and together with the balance of pressure, it constituted a sufficient condition as well (Gernez, 1867). In stark contrast, Charles Tomlinson (1808–1897) believed that the crucial factor in boiling was not gases, but small solid particles (Tomlinson, 1867, 1869). Tomlinson showed that metallic objects lost their vapor-liberating power if they were chemically cleaned to remove all specks of dust. Tomlinson declared: "A liquid at or near the boiling point is a supersaturated solution of its own vapor, constituted exactly like soda–water, Seltzer–water, champagne, and solutions of some soluble gases" (Chang, 2004). However, in the beginning of the twentieth century, some controversy still existed; while numerous researchers agreed with the influence of dissolved air in facilitating boiling, it was unclear whether the action was directly linked to the air itself, to particles of dust suspended in it, or to other impurities. The Tomlinson–Gernez debate may be irrelevant under our modern perspective. The pressure–temperature relationship is the same, no matter how the vapor is produced, i.e., steady common boiling, unstable superheated boiling, or explosion. Nevertheless, it has been claimed that Gernez, with his conjecture on the liberation of gas bubbles trapped in liquids or solids, pioneered the so-called *crevice model* of cavitation nucleation (Harvey et al., 1944, 1947; Atchley and Prosperetti, 1989).

However, it is widely recognized that the launch pad in cavitational studies comes from observations in 1895 by Sir John Thornycroft and Sidney Barnaby, who noticed the poor performance of a newly built destroyer, *HMS Daring* (Thornycroft and Barnaby, 1895). It was found that due to the rapid motion of the propeller blade in water, the trailing edge created sufficient negative pressure to create microbubbles. The sudden growth and collapse of those vapor cavities caused the extreme effects that pitted the metal surfaces exposed to the cavitating liquid. This seminal paper coined the term "cavitation" for the first time: "Cavitation, as Mr. Froude has suggested to the Authors that the phenomenon should be called, appears to manifest itself when the mean negative pressure exceeds about 6 3/4 pounds per square inch, or when the whole thrust exceeds 11 1/4 pounds per square inch." This Mr. Froude is actually Robert E. Froude, the third son and assistant of William Froude (1810–1879) who was a notable engineer and naval architect. Thorneycroft and Barnaby envisaged a solution to increase performance by modifying the propeller surface, thereby decreasing its angular velocity and therefore reducing bubble formation (Bremner, 1990).

CHEMISTRY IN HIGH-ENERGY CAVITIES

Pressure variations in a flowing stream or the propagation of pressure waves (ultrasound) may generate the same phenomenon, *cavitation*, which in both cases is brought about by tension in liquids. The chemical and physical effects of ultrasound cannot result from the direct interaction of sound waves with matter. On the one hand, ultrasound has wavelengths much larger than molecular dimensions and on the other, the average energy is insufficient to even modify ro-vibrational levels.

Through acoustic cavitation, i.e., rapid formation, growth, and violent collapse of bubbles, power ultrasound promotes and enhances chemical and physical changes. With ultrasound of sufficient intensity, rarefaction cycles exert negative pressure capable of exceeding attractive intermolecular forces, thereby creating cavities in the liquid. From the preceding historical perspective, it is convenient to bear in mind that cavitation is a nucleated process. Thus, the intriguing question "Where do the bubbles, which take part in cavitation come from?", has puzzled scientists over one century. Contrary to the classical debate, cavitation can be produced without the addition of gas bubbles. However, it is generally assumed that a liquid contains microscopic spaces, filled with gas or vapor, which act as cavitation nuclei. The nature of these may be varied, from gas-filled crevices in solid surfaces or suspended matter to gaseous spheres, or even quantum vortices in certain media (e.g., liquid helium).

Bubble dynamics is a complex issue, largely influenced by the local environment and applied intensity (Leighton, 1994; Brennen, 1995; Young, 1999), though some preliminary aspects suffice in this overview. In *stable cavitation*, bubbles oscillate gently around some equilibrium size and their mean life time may be longer than a cycle of the sound pressure. In contrast, transient cavities generally exist for less than one cycle and they will collapse violently releasing enough kinetic energy to drive chemical reactions. At high intensities, transient cavitation bubbles will produce high temperatures and pressures on collapse (*vide infra*), which account for most chemical and mechanical effects such as erosion, cleaning, bond cleavage, molecular aggregation, sonoluminescence, among others. The stable types of cavitation, occurring at lower acoustic intensities, should not be neglected and are just as important as transient bubbles. Surface oscillations and microstreaming stem from stable cavitation and, in addition, stable bubbles often evolve into transient ones over time due to mass or heat transfer, resulting in bubble growth.

A theory which is capable of rationalizing the nature of cavitation is a complicated affair, but numerous experimental data agree with a model called *hot spot*. This means that after cavity collapse, the surrounding liquid will quickly quench a short-lived, localized entity (hot spot) with temperatures in the range of 4500–5000 K and pressures exceeding 1000 atmospheres. Since this event occurs with a lifetime of a few microseconds and cooling rates of about 10^{10} K^{-1}, this quasi-adiabatic high-energy process has a profound influence on the physical properties of the cavitating liquid and dictates the chemical fate of volatile species in particular (Suslick, 1990; Flint and Suslick, 1991; Henglein, 1993).

$$M(CO)_x \xrightarrow{))))} M(CO)_{x-n} + nCO$$

$$M(CO)_{x-n} + nL \xrightarrow{))))} M(CO)_{x-n}L_n$$

M = Fe, Cr, Mo, W

L = PR$_3$ (R = alkyl, aryl)

SCHEME 2.1 Sonochemical ligand substitution of volatile metal carbonyls as rate probes.

Although from the viewpoint of most practitioners, the above estimates provide an intuitive picture of the high-energy scenario involved in cavitational collapse, it is worth noting that the temperatures of cavitation, which are also dependent on solvents, solutes present, and working conditions, constitute a major scientific challenge. Accurate temperatures are instrumental in elucidating the nature of cavitation and may offer clues to help us better understand chemical reactivity under sonication.

An indirect way to assess the temperatures reached during cavitation was devised by Suslick and associates through the use of competing unimolecular reactions whose temperature-dependent rates had been previously established (Suslick et al., 1986). Thus, by measuring the relative sonochemical rates for some ligand substitutions of volatile metal carbonyls (Scheme 2.1) and combining them with the known thermal behaviour of these reactions, the effective temperature of the hot spots was inferred to be *ca.* 5200 K in the inner gas-phase and approximately 1900 K at the interfacial shell.

SONOLUMINESCENCE STUDIES

The spectral analysis of sonoluminescence represents the most useful and convenient probe to ascertain the conditions in a cavitating bubble. Sonoluminescence is light emission by a sound wave of sufficient intensity. Like sonochemistry, sonoluminescence is rooted in cavitation. The energy of collapse is delivered to a small number of molecules, which are thereby excited or dissociated so as to emit light flashes when they return to their ground state, in a close analogy to chemiluminescence.

Historians of sonochemistry allude to observations of sonoluminescence-like phenomena as early as the late seventeenth century! An evaluation of such experiments, however, suggests situations rather related to tribo- or chemiluminescence. In 1769, Joseph Priestley reported a curious effect previously observed by Giambatista Beccaria (1716–1781), a Physics professor at the University of Turin: "Beccaria observed that hollow glass vessels, of a certain thickness, exhausted of air, gave a light when they were broken in the dark. By a beautiful train of experiments, he found, at length, that the luminous appearance was not occasioned by the breaking of the glass, but by the dashing of the external air against the inside, when it was broken" (Young, 1999). In 1933, Marinesco and Trillat found unexpected images when they were attempting to accelerate the process of developing photos by immersing them in water under ultrasound (Marinesco and Trillat, 1933). However, the genuine interpretation of sonoluminescence took place the year after when German scientists, Frenzel and Schultes, demonstrated in 1934 that the effect produced on photographic plates came from the energy-focusing power of cavitation clouds (Frenzel and Schultes, 1934). The phenomenon is now referred to as multi-bubble sonoluminescence (MBSL). The effect appeared initially as random and uncontrollable and, with a few exceptions, was not studied scientifically until much later. The spectra of MBSL in both aqueous and nonaqueous solutions now constitute a fruitful research domain and provide an alternative measure of the temperature generated by cavitation (Suslick et al., 1999).

Chemistry (in other words, the formation and evolution of high-energy species) during MBSL cannot be accurately determined as there are numerous, often unknown, parameters involved such as the number of active bubbles, the acoustic pressure on each bubble, and bubble size distribution. In the 1990s, Gaitan and Crum were able to detect sonoluminescence from a single bubble (SBSL). This phenomenon produces very short light flashes (for less than 50 ps), a regular light emission, and the appearance of light immediately before collapse (Gaitan, 1990;

Gaitan and Crum, 1992). Since then, the process has been extensively documented with several mechanistic hypotheses (Brenner et al., 2002).

Remarkably, SBSL points to higher cavitational temperatures than those predicted by an almost adiabatic compression. Calculations also predict that chemical reactions in an air bubble under SBSL conditions should afford H, OH, and HOO radicals, along with H_2O_2, O_3, and H_2 as the prevalent species, with minor amounts of nitrogen-containing compounds such as NO_x, NH_x, and HNO_x (Yasui, 1997). Given the tiny amount of gas within a single bubble ($<10^{-13}$ mol), its chemical activity is elusive. Didenko, Suslick, and associates, however, demonstrated the existence of molecular-excited states and chemical reactions during cavitation in both polar aprotic solvents and water at different frequencies and bulk temperatures (Didenko et al., 2000; Didenko and Suslick, 2002). Under such conditions, the efficiency of OH radical formation was comparable to that in MBSL, but the efficiency of light emission was much higher. Nitrite formation correlated well with the diffusion rate of N_2 in the bubble. The temperatures reached in SBSL extended over a wide range (from 1,600 to about 15,000 K) and were limited by the endothermic reactions of the polyatomic species inside the bubble. The latter rules out exceptionally high temperatures in solvents with significant vapor pressures. Evidence for a plasma-like condition arose from studies with Xe- or Ar-filled sonoluminescing bubbles in concentrated sulphuric acid (Flannigan and Suslick, 2005). The low volatility and high viscosity of sulphric acid relative to water ensured that the bubble consisted almost entirely of gas atoms. Moreover, the low vapor pressure of the medium also facilitates working at higher acoustic pressures (above 5 bar) than is possible with a bubble in water. Spectral lines correspond to excited ionic species that could only have been formed by collisions with energetic charged particles, and not via thermal excitation. Accordingly, gas-phase light emission from ions signals the formation of a plasma (Figure 2.2).

Despite the above-mentioned limitations of MBSL, practical imperatives force researchers to return to experiments in a multi-bubble acoustic field. Under appropriate conditions, relevant conclusions regarding cavitational phenomena may be attained. For instance, the observation of molecular and atomic ions during MBSL in 95% H_2SO_4 under Ar equally points to an inner ionized plasma core. Counterintuitively, strong Ar and SO lines are present at low acoustic intensities (14–22 W cm^{-2}) and disappear as intensity (30 W cm^{-2}) increases (Eddingsaas and Suslick, 2007a). Brighter sonoluminescence has also been observed in MBSL from 85% H_3PO_4 saturated with noble gases (at 20 kHz, Ti horn). The high viscosity and low vapor pressure of phosphoric acid guarantee again that the only volatile component inside the bubbles is water vapor (Xu et al., 2010). Interestingly, this study proved the existence of two distinct bubble populations that emerge from OH radical and PO radical emissions, respectively (Figure 2.3). Therefore, there should be two cavitating sites: (1) bubbles whose collapse is highly symmetric and occurs near the ultrasonic horn (*ca.* 9500 K, preferential OH radical emission) and (2) rapidly moving bubbles which show much less symmetric collapse involving injection of liquid nanodroplets into the gas phase (far from the horn, *ca.* 4000 K, PO radical emission).

Moreover, another study on the sonochemical disproportionation of carbon monoxide via vibrational excitation cannot be reconciled with the adiabatic heating model, which assumes a near-equilibrium gas inside the bubble. The result in which water sonolysis (20 kHz) in the presence of CO/Ar gas mixture produces a carbonaceous solid material enriched in ^{13}C would be consistent with a far-from-equilibrium nonthermal plasma (Nikitenko et al., 2009). Finally, the fact that a more intense mechanoluminescence can be induced by acoustic cavitation than by grinding points to the uniqueness of sonication in promoting more efficient interparticle collisions and fractures (Eddingsaas and Suslick, 2007b).

SONOCHEMICAL ACTIVATION

The preceding discussion reveals that sonochemistry should largely be interpreted in terms of cavitational chemistry (Cravotto and Cintas, 2006). The mechanisms of single-bubble and multi-bubble cavitation show that a collapsing bubble behaves as a particular microreactor where high-energy species (ions and radicals) and excited states may be involved in the reaction outcome.

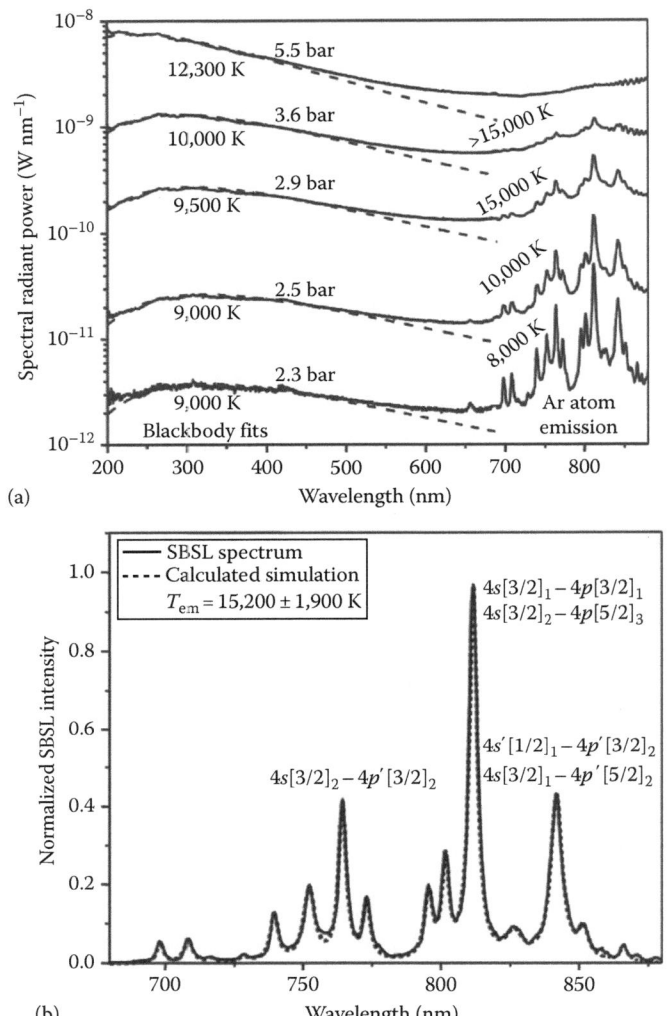

FIGURE 2.2 Emission from a single bubble (SBSL) in 85% aqueous H_2SO_4. (a) Solid lines are the observed SBSL emission spectra, while dashed lines correspond to calculated blackbody spectra. The applied acoustic pressures as well as temperatures of blackbody fits and Ar atom emission (see (b)) are next to their plots. (b) Ar atom line emission (solid line at 2.8 bar) compared to a calculated Ar atom emission spectrum at 15,200 K (dashed line). (Reprinted by permission from Macmillan Publishers Ltd. *Nature*, Flannigan, D.J. and Suslick, K.S., Plasma formation and temperature measurement during single-bubble cavitation, 434, 52–55, Copyright (2005).)

A simple, yet intuitive, picture that rationalizes how sonochemical reactions occur is provided by the shell model (Figure 2.4) highlighting three different temperature domains. Volatile molecules enter the microbubbles where the high temperatures and pressures generated during cavitation cause bond scission and short-lived species return to the liquid phase. Compounds of low volatility hardly enter the bubbles, although they may experience the effects arising from the pressure changes associated with the propagation of the wave or from bubble collapse (shock waves). These substances can also react with reactive species (chiefly radicals) generated by the sonolysis of the solvent.

This model applies well to homogeneous conditions. However, as seen before, the unexpected emission from nonvolatile species during MBSL is more consistent with nanodroplet injection (Xu et al., 2009, 2010) as depicted in Figure 2.3. In this model, interfacial instabilities, such as

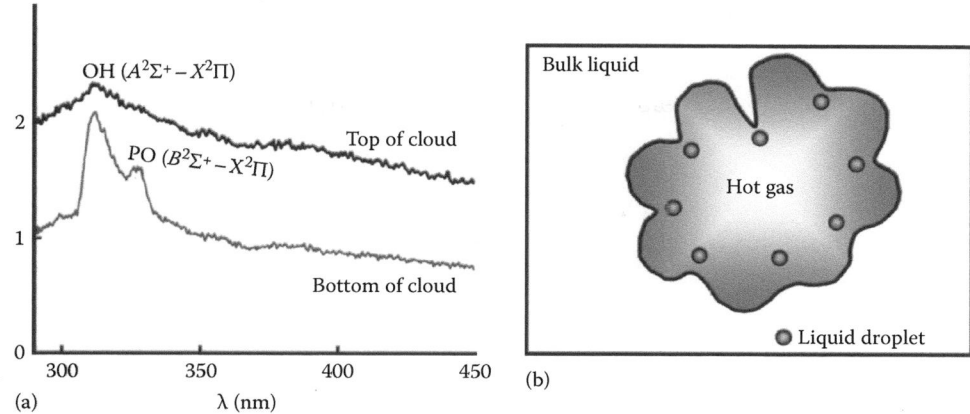

FIGURE 2.3 (a) MBSL spectra taken at the top and bottom of a cavitating bubble cloud (20 kHz, 25 W cm^{-2}, 298 K) from 85% H_3PO_4 saturated with He. (b) Schematic representation of the injected droplet model. (Xu, H., Glumac, N.G., and Suslick, K.S.: Temperature inhomogeneity during multi-bubble sonoluminescence. *Angew. Chem. Int. Ed.* 2010. 49. 1079–1082. Copyright Wiley-VCH Verlag GmbH & Co. KGaA. Reproduced with permission.)

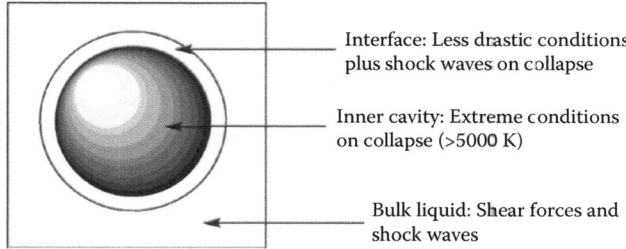

FIGURE 2.4 A simplified model for cavitation bubbles in a homogeneous medium.

capillary surface waves and microjet formation, will nebulize during collapse nanodroplets of liquid into the hot cavity of the bubble, with subsequent thermolysis or reduction of nonvolatile substances (e.g., ions from metal salts).

Cavity collapse under heterogeneous conditions, such as near a liquid–solid interface, is essentially different and other side effects appear (Mason, 1991; Mason and Lorimer, 2002). Collapse is now asymmetrical and an inrush of liquid from one side of the bubble gives rise to a violent liquid jet targeted at the surface. The net effects are surface cleaning, the destruction of boundary layers, and concomitant mass and heat transfer improvements. Suspended power, as is also mentioned above, contains defects or trapped gas that serves as cavitation nucleus. Bubble collapse on the surface of a particle forces it into rapid motion and collision with vicinal solid matter. Overall, such effects account for dispersion, erosion, and size reduction, which represent driving forces in the activation of solid reagents and catalysts. Likewise, in a heterogeneous liquid–liquid system, the powerful disruption of the interface will cause efficient mixing and fine emulsions. As a result, the presence of a catalyst is often unnecessary when phase-transfer reactions are conducted under sonication.

Historically, the chemical effects of ultrasound were first reported in the late 1920s using high frequencies, from 100 to 500 kHz (Richards and Loomis, 1927). The observed acceleration of certain chemical reactions was largely attributed to frequency effects. A paper unveiling true sonochemical effects on oxidation reactions appeared two years later (Schmitt et al., 1929). The authors noted: "If an aqueous solution containing dissolved oxygen is radiated, hydrogen peroxide

$$H_2O \longrightarrow H\bullet + HO\bullet$$

$$H\bullet + HO\bullet \longrightarrow H_2O$$

$$H\bullet + H\bullet \longrightarrow H_2$$

$$HO\bullet + HO\bullet \longrightarrow H_2O_2$$

$$H\bullet + O_2 \longrightarrow HOO\bullet$$

$$H\bullet + HOO\bullet \longrightarrow H_2O_2$$

$$HOO\bullet + HOO\bullet \longrightarrow H_2O_2 + O_2$$

$$H_2O + HO\bullet \longrightarrow H_2O_2 + H\bullet$$

SCHEME 2.2 Formation and recombination of radical species produced by water sonolysis.

or something analogous to it is formed." Moreover, they also observed: "It seems necessary to assume that every phase of bubble formation is realized in this process, and it must be concluded, therefore, that the mere formation of bubbles in the absence of ultrasonic vibrations is unable to effect the oxidation."

For decades, sonochemistry was explored in aqueous solutions, which highlighted the similarities between sonolysis and radiolysis. Cavitation produces solvent radicals; in the case of water, these species are H and OH radicals that recombine to produce hydrogen and hydrogen peroxide (Scheme 2.2). At low intensities, yields of the resulting radicals and molecules are scarce and less than those found in radiation chemistry. Detection of some intermediates proved to be difficult, and, thus, unequivocal confirmation for ultrasound-generated hydrogen radicals could only be obtained by electrochemical methods in recent decades (Birkin et al., 2001). While water saturated with oxygen generates H_2O_2, the presence of air gives rise to nitrous acid; further oxidation also leads to nitric acid. This sequence constitutes a mechanism for the fixation of atmospheric nitrogen under sonication (Bremner, 1990).

In practice, the synthetic utility of sonochemistry, especially the application of power ultrasound in organic solvents, was with a few exceptions (Porter and Young, 1938; Renaud, 1950), was ignored for nearly 60 years. In the early 1980s, it was shown that organometallic reagents derived from lithium, which are otherwise sensitive to air and moisture, could be easily prepared in cleaning baths (50 kHz, 60 W) without prior metal activation and in undried solvents. Remarkably, one-step Barbier reactions involving carbonyls and alkyl halides were complete in less than 1 h (often in only 10–15 min) and with excellent yields (Luche and Damiano, 1980). With magnesium, the induction period is suppressed or considerably reduced; sonication helps the removal of the superficial oxide layer, which is actually harder than the metal itself. In fact, ultrasound appears to accelerate all the steps of Grignard reagent formation (Tuulmets et al., 1995).

REACTIVITY RULES IN ULTRASONIC FIELDS

Can we predict the fate of a sonochemical reaction? or alternatively, can we ascertain whether a given reaction will be sensitive to ultrasound? Unfortunately, definitive answers cannot be formulated. This reflects our modest understanding of cavitational phenomena, which are indeed complex and often follow a nonlinear behavior. On the other hand, chemistry under sonication is markedly dependent on solvent effects and physical properties, which may significantly influence cavitation but are usually overlooked in conventional chemistry. The latter pays attention to more chemical concepts, such as acid-base strength, donor–acceptor character, polarity, orbital interactions, etc. To guide newcomers, Table 2.1 summarizes a series of key effects and their influence on bubble collapse, thereby altering its formation and energy. For some factors, conclusive statements cannot easily be formulated, which is true in the case of intensity and frequency

TABLE 2.1
Experimental Parameters Affecting Cavitation

Parameter	Influence on Cavitation
Frequency	At higher frequency, the rarefaction phase shortens. More power is required to make a liquid cavitate as the frequency increases
Solvent viscosity	Collapse produces shear forces in the bulk liquid; viscosity increases the resistance to shear
Surface tension	No simple relationship. Cavitation generates liquid–gas interfaces; addition of a surfactant facilitates cavitation
Vapor pressure	Cavitation is difficult in solvents of low vapor pressure. A more volatile solvent supports cavitation at lower acoustic energy
Bubbled gases	The energy on collapse increases for gases with a large polytropic ratio (C_p/C_v); monoatomic gases are preferred
Temperature	Any increase in temperature will raise the vapor pressure and cavitation will be easier, though a less violent collapse
Intensity	In general, an increase in intensity will also increase the sonochemical effects. A minimum intensity is required to reach the cavitation threshold
External pressure	Raising the external pressure will produce a larger intensity of cavitational collapse

(Mason and Lorimer, 2002; Cravotto and Cintas, 2006). The temporal evolution of the pressure $P(t)$ through the elastic medium can be expressed as follows:

$$P(t) = P_A \sin(2\pi ft + \theta)$$

where

P_A is the acoustic pressure amplitude
f is the frequency of the alternating pressure wave

As frequency increases, it is also necessary to increase the amplitude (or power) of irradiation to maintain the same amount of cavitational energy. Accordingly, it becomes more difficult to make a liquid cavitate at higher frequencies (such as the MHz region). The reason is that at high frequencies there are shorter cycles of compression and rarefaction, and cavitation requires enough time for molecules to be pulled apart.

Frequency effects can even be more difficult to rationalize as sound, unlike electromagnetic radiation, is not quantized. In fact similar chemical effects can be attained at different frequencies. In any event, chemical reactions induced by cavitation depend on the lifetimes of primary radicals relative to bubble lifetime. Frequency then influences the time it takes for a bubble to collapse; however, studies in this context are rather scarce and caution should be paid when applying homogeneous or heterogeneous conditions (Pétrier et al., 1992; Wu et al., 2009; Kojima et al., 2010).

In an attempt to rationalize the effects of sonication on reactivity, Luche suggested a tentative classification highlighting the fact that reactions which are truly sensitive to ultrasound will lead to alternative intermediates or products, via alternative mechanisms (stepwise versus concerted; radical versus polar) to conventional thermal activation (Luche, 1993):

Type I: homogeneous reactions following radical mechanisms or via coordinatively unsaturated species (e.g., organometallics): acceleration and different product distribution may be observed.

Type II: Heterogeneous reactions following a polar mechanism. In this case, appreciable mechanical effects such as emulsification or enhanced mass transfer should be observed. This type represents the so-called *false sonochemistry*, because the participation of radical species arising from bubble collapse hardly occurs.

Type III: Heterogeneous reactions following either polar or radical pathways; the latter will usually be favored under sonication.

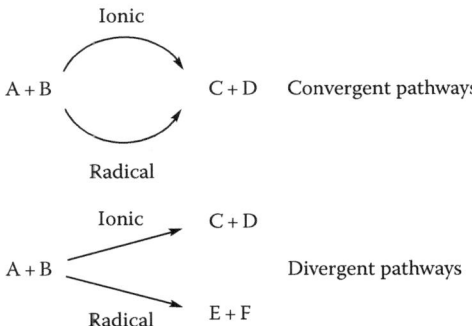

SCHEME 2.3 Schematic representation of different reaction pathways under sonication.

Along with rate acceleration and improved yields, the possibility of switching the conventional pathway or significantly altering the product ratio constitutes a conspicuous characteristic of some ultrasound-assisted reactions. For the sake of clarity, it is appropriate to use the terms *convergent* or *divergent*, according to the nomenclature of synthetic chemists, to describe mechanisms that can take place under sonication (Scheme 2.3). In a convergent process, the reagents can react by following either a radical or a polar pathway, each leading however to the same product. A representative case is provided by the Grignard reaction (and similar carbon–metal bond formation), in which a radical pathway through the intermediacy of radical ions will be preferentially induced by ultrasound, yet producing the same organometallic reagent derived from an ionic route. A divergent process is one in which the competing ionic and radical mechanisms lead to different products, the situation for *sonochemical switching* to occur.

The first observation of a different reaction route in the presence of ultrasound was reported by Ando and coworkers. A mixture of benzyl bromide, potassium cyanide, and alumina in an aromatic solvent (toluene) yields benzyl cyanide under sonication. However, mechanical stirring gives the Friedel–Crafts product (Ando et al., 1984). Ultrasound irradiation clearly switches the reaction pathway from aromatic electrophilic substitution to aliphatic nucleophilic substitution (Scheme 2.4).

A divergent result however does not necessarily mean that ultrasound invariably favors a single-electron transfer (SET) pathway at the expenses of an ionic mechanism. Further investigations on the above switching revealed that ultrasonic irradiation accelerates poisoning by KCN of the catalytic active sites of alumina, thereby impeding the Friedel–Crafts substitution (Ando and Kimura, 1991).

Much more definitive results about the divergence between polar and radical pathways were achieved by the Japanese group in the reactions of styrenes with lead tetraacetate (Ando et al., 1991). Transformations conducted at 50°C under stirring are sluggish and products derived from ionic intermediates are obtained in poor yields. In contrast, overall yields, still modest, largely increase under sonication affording products that can be rationalized in terms of cationic and radical intermediates (Scheme 2.5). Since an increase in temperature favors radical formation in the reactions of $Pb(OAc)_4$,

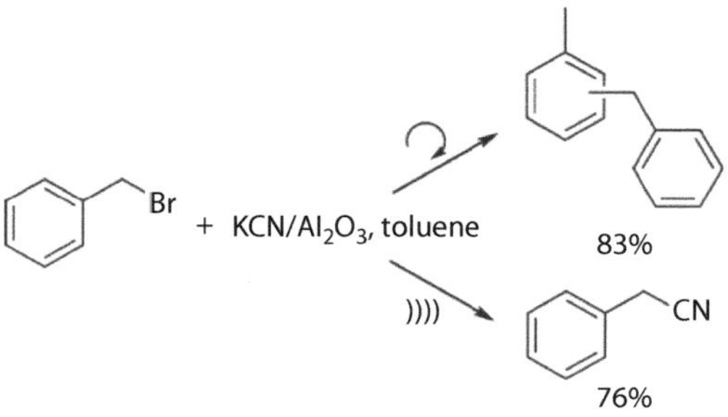

SCHEME 2.4 The original sonochemical switching.

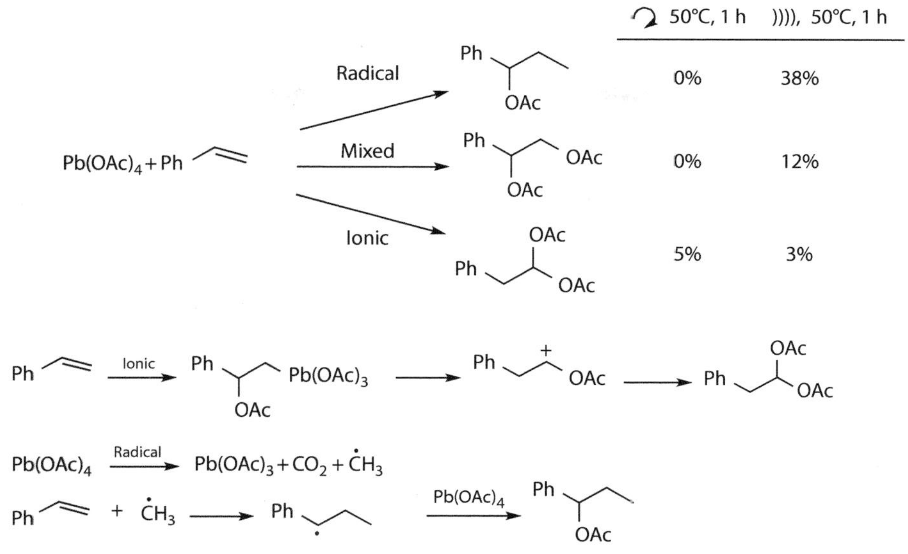

SCHEME 2.5 Divergent pathways observed in the reaction of styrene with Pb(OAc)$_4$.

the authors conducted the sonochemical reactions at the same temperature with minor experimental variations. The major product under sonication arises from radical addition, even though a competing polar mechanism seems to be operative as well.

The unequivocal participation of species derived from cavitation is beautifully illustrated by the type-I reaction of tin hydrides to alkenes initiated by AIBN (Nakamura et al., 1994). Thus, tin radical species react under ultrasonic irradiation in a manner hitherto unknown. *Hydroxystannation*, and not the expected hydrostannation (which proceeds via radicals as well), takes place when an aerated solution of R$_3$SnH and an alkene are irradiated between 0°C and 10°C. Ultrasound does generate tin radicals in the region of hot cavities, which then undergo further reactions in the bulk liquid with species derived from the cavitational collapse. Remarkably, ultrasound also gives rise to vinylstannanes when a mixture of a tin hydride and an alkyne are irradiated at −50°C, such reactions are more than 100 times faster than those without sonication (Scheme 2.6). The latter likewise evidences the paradoxical effect of temperature on certain ultrasound-assisted reactions. The anti-Arrhenius effect that a sonochemical reaction may indeed be accelerated by lowering the temperature witnesses again the crucial role exerted by physical properties on cavitation: larger effects due to a lower solvent vapor pressure.

Although the above examples show the ability of ultrasound to bias the reaction outcome and prove the validity of the mentioned empirical rules, a classical illustration of the uniqueness of sonication with respect to thermal and photochemical processes is provided by the decomposition of volatile metal carbonyls (Suslick et al., 1983, 1991). Sonolysis of, for instance, iron pentacarbonyl in hydrocarbon solvents in the presence of ligands (phosphines or phosphites) causes CO substitution yielding Fe(CO)$_4$L or Fe(CO)$_3$L$_2$; whereas sonication in the absence of ligands produces the cluster

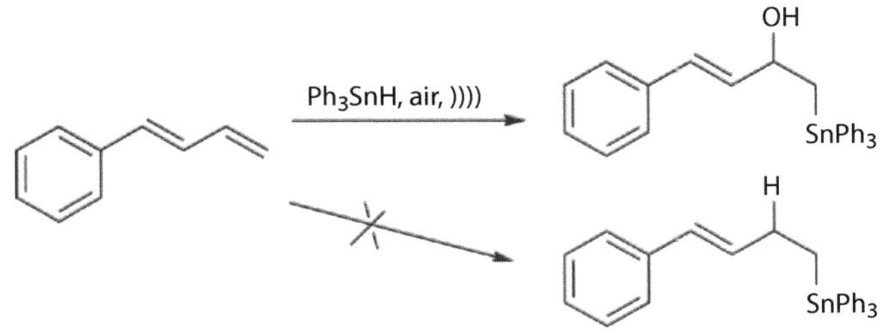

SCHEME 2.6 Hydroxystannation of alkenes under sonochemical activation.

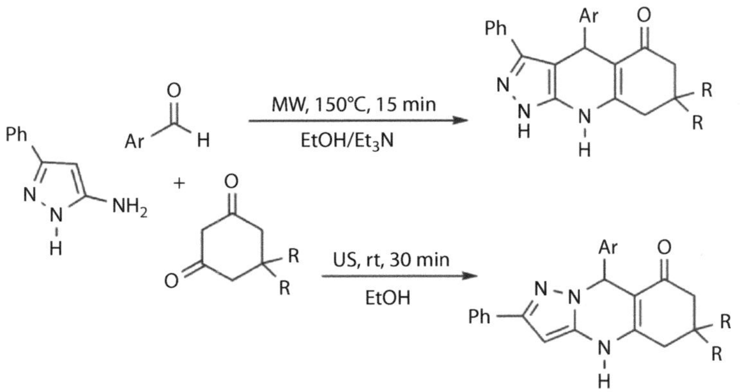

SCHEME 2.7 Comparative thermal results for a multicomponent reaction.

compound Fe$_3$(CO)$_{12}$. Such results are consistent with a dissociative mechanism in which coordinatively unsaturated species (playing a similar role to radicals) are produced by the cavitation process. The photochemical reaction leads to diiron nonacarbonyl, Fe$_2$(CO)$_9$. Interestingly, sonolysis in a high-boiling solvent (decalin) also gives rise to amorphous iron. In general, amorphous metallic glasses are difficult to obtain by thermal methods; however, cavitational collapse, due to its extreme cooling rates, produces nearly pure amorphous iron, which proved to be an efficient catalyst for hydrogenation (Fischer–Tropsch) and dehydrogenation reactions.

Likewise, distinctive pathways between ultrasound and other thermal methods, especially microwave irradiation, have recently been reported for a three-component coupling involving nitrogen heterocycles and carbonyl compounds (Chebanov et al., 2008). Compared with the microwaved reaction, ultrasound promotes cyclization at room temperature and is only slightly slower (Scheme 2.7).

In general, polar homogeneous reactions which are not switched to a free radical pathway will take place in solution experiencing the effects of shock waves and shear forces. This conclusion can also be extrapolated to concerted transformations, such as the venerable Diels–Alder cycloaddition, i.e., the reaction will significantly be accelerated only if an SET process can be induced (Tuulmets, 1997). Cycloadditions do indeed represent an open scenario in studies of sonochemical reactivity as these processes, which are usually sensitive to temperature and pressure, should be activated by cavitation. Unfortunately, this is not the general observation. Most substrates are not volatile enough and, in addition, numerous transformations are run under heterogeneous conditions, thus complicating the rationale. Ultrasound may however be a mechanistic probe capable of ruling out concerted pathways. Thus, the formal cycloaddition of masked *o*-benzoquinones and furans (these acting as dienophiles) is greatly improved under sonication (Scheme 2.8). The sonochemical study shows a significant dependence on acoustic energy, temperature, and solvent composition. These facts, together with the inertness to spin traps, are consistent with a stepwise addition and suggest that ultrasound does not cause switching to a radical pathway (Avalos et al., 2003). Effects are of mechanochemical origin in which ultrasonic agitation favors enhanced mass transfer and produces a nearly perfect mixing.

As a final illustration of the potentialities of sonication in synthetic chemistry, closely related to Type II reactions, the preparation of ionic liquids deserves attention. These substances, which may serve as both solvents and catalysts, have triggered a profound revolution in organic syntheses and materials science. From a sonochemical viewpoint, they are ideal solvents owing to their negligible vapor pressure, thus avoiding competing reactions with volatile substrates which may be susceptible to activation in the bubbles. The conventional preparation of these high-boiling and viscous liquids is tedious and usually involves a two-step protocol of nitrogen quaternization followed by anion metathesis of the resulting salt. Under ultrasound, the synthesis is dramatically accelerated as depicted in Scheme 2.9 (Levêque et al., 2002) and can even be conducted with neat liquid reagents, without solvent (Namboodiri and Varma, 2002).

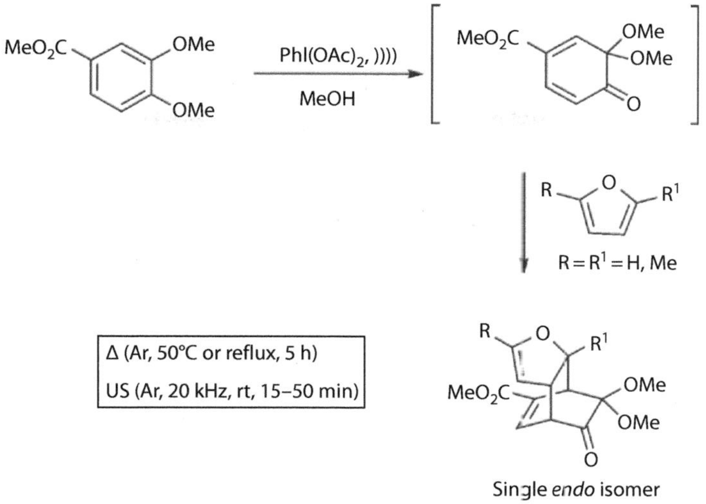

Introduction to Sonochemistry

SCHEME 2.8 Ultrasound-promoted coupling of quinones with furans.

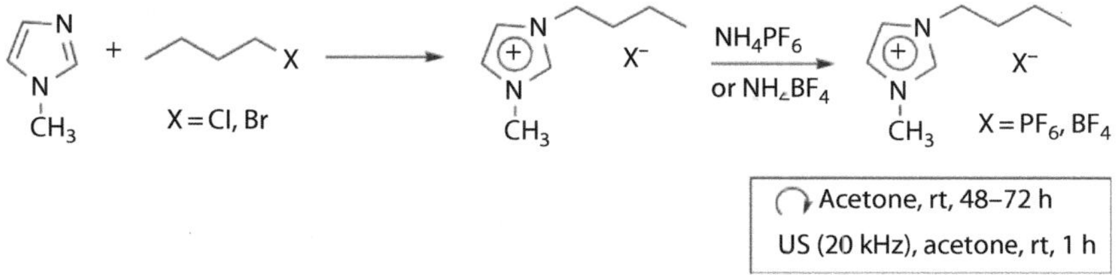

SCHEME 2.9 Ultrasound-accelerated synthesis of ionic liquids.

CONCLUSIONS

This preliminary survey has summarized the long journey toward our current understanding of cavitational phenomena and hence of sonochemistry. Further studies will be required to provide more satisfactory pictures and a complete rationale of cavitation as a unique activation phenomenon which resides at the interface between physics and chemistry. It suffices, however, to note a series of macroscopic characteristics that boost ultrasound as a mild, safe, and particularly green technology (Cintas and Luche, 1999; Mason and Cintas, 2002), namely, (1) accelerate chemical reactions, (2) permit the use of less forcing conditions, (3) make a process more economical by the use of standard reagents and solvents, (4) often reduce the number of synthetic steps required, (5) reduce any induction period with unwilling synthetic partners and initiate stubborn reactions, (6) enhance catalyst efficiency, and (7) enhance radical reactions at the expenses of polar ones. Clearly, the contributions in this book will illustrate these and other pluses, convincingly showing that the technique and its development deserve particular attention in applied science.

ACKNOWLEDGMENTS

We sincerely appreciate the financial support by the Spanish Ministry of Science and Innovation (Grant MAT2009-14695-C04-C01), the University of Turin and MIUR (PRIN, Prot. 2008M3Y5WX). We would also like to thank the European Union for supporting our research program on green technologies focused on ultrasound and microwaves, through the past COST Action D32/006/04.

REFERENCES

Ando, T., Bauchat, P., Foucaud, A. et al. 1991. Sonochemical switching from ionic to radical pathways in the reaction of styrene and trans-β-methylstyrene with lead tetraacetate. *Tetrahedron Letters*, 32: 6379–6382.

Ando, T. and Kimura, T. 1991. Ultrasonic organic synthesis involving non-metal solids. In *Advances in Sonochemistry*, Vol. 2, ed. T.J. Mason, pp. 211–252. London, U.K.: JAI Press Ltd.

Ando, T., Sumi, S., Kawate, T., Ichihara, J., and Hanafusa, T. 1984. Sonochemical switching of reaction pathways in solid–liquid two-phase reactions. *Journal of the Chemical Society, Chemical Communications*, 439–440.

Atchley, A.A. and Prosperetti, A. 1989. The crevice model of bubble nucleation. *Journal of the Acoustical Society of America*, 86: 1065–1084.

Avalos, M., Babiano, R., Cabello, N. et al. 2003. Thermal and sonochemical studies on the Diels–Alder cycloadditions of masked *o*-benzoquinones with furans: New insights into the reaction mechanism. *Journal of Organic Chemistry*, 68: 7193–7203.

Besant, W.H. 1859. *A Treatise on Hydrostatics and Hydrodynamics*, pp. 170–171. Cambridge, U.K.: Deighton, Bell and Co.

Beyer, R.T. 1999. *Sounds of Our Times. Two Hundred Years of Acoustics*. New York: Springer/AIP Press.

Birkin, P.R., Power, J.F., and Leighton, T.G. 2001. Electrochemical evidence of H· produced by ultrasound. *Chemical Communications*, 2230–2231.

Bremner, D. 1990. Historical introduction to sonochemistry. In: *Advances in Sonochemistry*, Vol. 1, ed. T.J. Mason, pp. 1–37. London, U.K.: JAI Press Ltd.

Brennen, C.E. 1995. *Cavitation and Bubble Dynamics*. Oxford, U.K.: Oxford University Press.

Brenner, M.P., Hilgenfeldt, S., and Lohse, D. 2002. Single-bubble sonoluminescence. *Reviews of Modern Physics*, 74: 425–484.

Carr Everback, E. 2007. Medical diagnostic ultrasound. *Physics Today*, 60(3): 44–48.

Chang, H. 2004. *Inventing Temperature. Measurement and Scientific Progress*. Oxford, U.K.: Oxford University Press.

Chebanov, A.A., Saraev, V.E., Desenko, S.M. et al. 2008. Tuning of chemo- and regioselectivities in multicomponent condensations of 5-aminopyrazoles, dimedone, and aldehydes. *Journal of Organic Chemistry*, 73: 5110–5118.

Cintas, P. and Luche, J.-L. 1999. Green chemistry. The sonochemical approach. *Green Chemistry*, 1: 115–125.

Cravotto, G. and Cintas, P. 2006. Power ultrasound in organic synthesis: Moving cavitational chemistry from academia to innovative and large-scale applications. *Chemical Society Reviews*, 35: 180–196.

Curie, J. and Curie, P. 1880. Développement, par pression, de l'électricité polaire dans les cristaux hémièdres à faces inclinées. *Comptes Rendus de l'Acadamie de Sciences*, 91: 294–295.

Didenko, Y.T., McNamara III, W.B., and Suslick, K.S. 2000. Molecular emission from single-bubble sonoluminescence. *Nature*, 407: 877–879.

Didenko, Y.T. and Suslick, K.S. 2002. The energy efficiency of formation of photons, radicals and ions during single-bubble cavitation. *Nature*, 418: 394–397.

Eddingsaas, N.C. and Suslick, K.S. 2007a. Evidence for a plasma core during multibubble sonoluminescence in sulfuric acid. *Journal of the American Chemical Society*, 129: 3838–3839.

Eddingsaas, N.C. and Suslick, K.S. 2007b. Intense mechanoluminescence and gas phase reactions from the sonication of an organic slurry. *Journal of the American Chemical Society*, 129: 6718–6719.

Euler, L. 1754. Classe de philosophie expérimentalle. *Histoire de l'Academie Royale des Sciences et Belles Lettres*, Memo R.10. Chap. 81: 227–295.

Flannigan, D.J. and Suslick, K.S. 2005. Plasma formation and temperature measurement during single-bubble cavitation. *Nature*, 434: 52–55.

Flint, E.B. and Suslick, K.S. 1991. The temperature of cavitation. *Science*, 253: 1397–1399.

Frenzel, H. and Schultes, H. 1934. Lumineszenz im ultraschall-beschickten wasser. *Zeitschrift fuer Physikalische Chemie-Abteilung B*, 27B: 421–424.

Gaitan, D.F. 1990. PhD dissertation. The University of Mississippi, Oxford, MS.

Gaitan, D.F., Crum, L.A., Church, C.C., and Roy, R.A. 1992. Sonoluminescence and bubble dynamics for a single, stable, cavitation bubble. *Journal of the Acoustical Society of America*, 91: 3166–3183.

Gernez, M. 1867. On the disengagement of gases from their saturated solutions. *Philosophical Magazine*, 33: 479–481.

Griffin, D.R. 1958. *Listening in the Dark: The Acoustic Orientation of Bats and Men*. New Haven, CT: Yale University Press.

Harvey, E.N., Barnes, K.K., McElroy, W.D. et al. 1944. Bubbles formation in animals, I: Physical factors. *Journal of Cellular and Comparative Physiology*, 24: 1–22.

Harvey, E.N., McElroy, W.D., and Whiteley, A.H. 1947. On cavity formation in water. *Journal of Applied Physics*, 18: 162–172.

Henglein, A. 1993. Contributions to various aspects of cavitation chemistry. In *Advances in Sonochemistry*, Vol. 3, pp.17–83, ed. Mason, T.J. London, U.K.: JAI Press Ltd.

Hunt, F.V. 1978. *Origins in Acoustics. The Science of Sound from Antiquity to the Age of Newton*. New Haven, CT: Yale University Press.

Kojima, Y., Yamaguchi, K., and Nishimiya, N. 2010. Effect of amplitude and frequency of ultrasonic irradiation on morphological characteristics control of calcium carbonate. *Ultrasonics Sonochemistry*, 17: 617–620.

Leighton, T.G. 1994. *The Acoustic Bubble*. San Diego, CA: Academic Press.

Levêque, J.-M., Luche, J.-L., Pétrier, C., Roux, R., and Bonrath, W. 2002. An improved preparation of ionic liquids by ultrasound. *Green Chemistry*, 4: 357–360.

Luche, J.-L. 1993. Sonochemistry. From experiment to theoretical considerations. In *Advances in Sonochemistry*, Vol. 3, pp. 85–124, ed. Mason, T.J. London, U.K.: JAI Press Ltd.

Luche, J.-L. 1998. *Synthetic Organic Sonochemistry*. New York: Plenum Press.

Luche, J.-L. and Damiano, J.C. 1980. Ultrasounds in organic syntheses. 1. Effect on the formation of lithium organometallic reagents. *Journal of the American Chemical Society*, 102: 7926–7927.

Marinesco, N. and Trillat, J.J. 1933. Action des ultrasons sur les plaques photographiques. *Proceedings of the Royal Academy of Science, Amsterdam*, 196: 858–860.

Mason, T.J. 1991. *Practical Sonochemistry. User's Guide to Applications in Chemistry and Chemical Engineering*. Chichester, U.K.: Ellis Horwood Publishers.

Mason, T.J. and Cintas, P. 2002. Sonochemistry. In *Handbook of Green Chemistry and Technology*, pp. 372–396, eds. J. Clark, and D. Macquarrie. Oxford, U.K.: Blackwell Science

Mason, T.J. and Lorimer, J.P. 2002. *Applied Sonochemistry. The Uses of Power Ultrasound in Chemistry and Processing*. Weinheim, Germany: Wiley-VCH.

Mason, T.J. and Peters, D. 2002. *Practical Sonochemistry. Power Ultrasound Uses and Applications*. Chichester, U.K.: Ellis Horwood Publishers.

Mould, R.F. 2007. Pierre Curie, 1859–1906. *Current Oncology*, 14: 74–82.

Nakamura, E., Imanishi, Y., and Machii, D. 1994. Sonochemical initiation of radical chain reactions. Hydrostannation and hydroxystannation of C–C multiple bonds. *Journal of Organic Chemistry*, 59: 8178–8186.

Namboodiri, V.V. and Varma, R.S. 2002. Solvent-free sonochemical preparation of ionic liquids. *Organic Letters*, 4: 3161–3163.

Nikitenko, S.I., Martinez, P., Chave, T., and Billy, I. 2009. Sonochemical disproportionation of carbon monoxide in water: Evidence for Treanor effect during multibubble cavitation. *Angewandte Chemie International Edition*, 48: 9529–9532.

Pétrier, C., Jeunet, A., Luche, J.-L., and Reverdy, G. 1992. Unexpected frequency effects on the rate of oxidative processes induced by ultrasound. *Journal of the American Chemical Society*, 114: 3148–3150.

Porter, C.W. and Young, L. 1938. A molecular rearrangement induced by ultrasonic waves. *Journal of the American Chemical Society*, 60: 1497–1500.

Rayleigh, Lord. 1917. On the pressure developed in a liquid during the collapse of a spherical cavity. *Philosophical Magazine*, 34: 94–98.

Renaud, P. 1950. Note de laboratoire sur l'application des ultrasons à la préparation de composes organométalliques. *Bulletin de la Société Chimique de France*, 17: 1044–1045.

Richards, W.T. and Loomis, A. 1927. The chemical effects of high frequency sound waves I. A preliminary survey. *Journal of the American Chemical Society*, 49: 3086–3100.

Schmitt, F.O., Johnson, C.H., and Olson, A.R. 1929. Oxidations promoted by ultrasonic radiation. *Journal of the American Chemical Society*, 51: 370–375.

Simmons, J.A. 2007. Echolocation in dolphins and bats. *Physics Today*, 60(9): 40–45.

Suslick, K.S. 1988. *Ultrasound. Its Chemical, Physical and Biological Effects*. New York: VCH Publishers.

Suslick, K.S. 1990. Sonochemistry. *Science*, 247: 1439–1445.

Suslick, K.S., Choe, S.-B., Cichowlas, A.A., and Grinstaff, M.W. 1991. Sonochemical synthesis of amorphous iron. *Nature*, 353: 414–416.

Suslick, K.S., Didenko, Y., Fang, M.M. et al. 1999. Acoustic cavitation and its chemical consequences. *Philosophical Transactions: Mathematical, Physical and Engineering Sciences*, 357(1751): 335–353.

Suslick, K.S., Goodale, J.W., Schubert, P.F., and Wang, H.H. 1983. Sonochemistry and sonocatalysis of metal carbonyls. *Journal of the American Chemical Society*, 105: 5781–5785.

Suslick, K.S., Hammerton, D.A., and Cline Jr., R.E. 1986. Sonochemical hot spot. *Journal of the American Chemical Society*, 108: 5641–5642.

Thornycroft, J.I. and Barnaby, S.W. 1895. Torpedo-boat destroyers. *Proceedings of the Institution of Civil Engineering*, 122: 51–69.

Tomlinson, C. 1867. On some phenomena connected with the adhesion of liquids to liquids. *Philosophical Magazine*, 33: 401–412.

Tomlinson, C. 1869. On the formation of bubbles of gas and of vapour in liquids. *Philosophical Magazine*, 38: 204–206.

Truesdell III, C.A. 1955. Lausanne: Auctoritate et Impensis Societatis Scientiarum Naturalium Helveticae. *Leonhardi Euleri Opera Omnia*, 196–239.

Tuulmets, A. 1997. Ultrasound and polar homogeneous reactions. *Ultrasonics Sonochemistry*, 4: 189–193.

Tuulmets, A., Kaubi, K., and Heinoja, K. 1995. Influence of sonication on Grignard reagent formation. *Ultrasonics Sonochemistry*, 2: S75–S78.

Whitlow, W.L. 1993. *The Sonar of Dolphins*. New York: Springer.

Wu, S., Leong, T., Kentish, S., and Ashokkumar, M. 2009. Frequency effects during acoustic cavitation in surfactant solution. *The Journal of Physical Chemistry B*, 113: 16568–16573.

Xu, H., Eddingsaas, N.C., and Suslick, K.S. 2009. Spatial separation of cavitating bubble populations: The nanodroplet injection model. *Journal of the American Chemical Society*, 131: 6060–6061.

Xu, H., Glumac, N.G., and Suslick, K.S. 2010. Temperature inhomogeneity during multibubble sonoluminescence. *Angewandte Chemie International Edition*, 49: 1079–1082.

Yasui, K. 1997. Chemical reactions in a sonoluminescing bubble. *Journal of the Physical Society of Japan*, 66: 2911–2920.

Young, F.R. 1999. *Cavitation*. London, U.K.: Imperial College Press.

3 Aspects of Ultrasound and Materials Science

*Andrew Cobley, Timothy J. Mason,
Larisa Paniwnyk, and Veronica Saez*

CONTENTS

Background: Ultrasound and Materials 41
Sonochemical Surface Modification of Materials 42
Effect of Ultrasound on the Electrochemical Metallization of Materials 46
    Electroplating 46
    Electroless Plating 49
        Effect of Ultrasound on Electroless Nickel Plating 50
        Effect of Ultrasound on Electroless Copper 51
Sonoelectrochemical Synthesis of Nanoparticles (Saez and Mason, 2009) 53
    Preparation of Nanoparticles by Pulsed Sonoelectrochemistry 55
        Metallic Nanopowders 55
        Alloy Nanopowders 58
        Semiconductor Nanopowders 58
        Conducting Polymer Nanoparticles 59
Ultrasound in Polymer Technology 59
Ultrasound and Environmental Protection 62
    Use of Ultrasound in the Degradation of Textile Dyes 62
    Treatment of Aromatic Hydrocarbons 65
    Treatment of Sewage Sludge (Mason and Lorimer, 2002; Mason, 2007) 66
    Surface Decontamination 66
Therapeutic Ultrasound 67
    Cancer Treatment 67
    Transdermal Drug Delivery 67
Conclusions 67
References 68

BACKGROUND: ULTRASOUND AND MATERIALS

If you are a materials scientist, you will probably have encountered ultrasound in the context of the pulse echo technique used in nondestructive testing. This field is often referred to as diagnostic ultrasound and it has expanded enormously from its origins of fault or crack detection to encompassing of medical imaging. However, there is an altogether different range of applications of ultrasound that have been used for modification of the physical and chemical properties of the materials and in processing. This latter field has its origins in two uses in materials technology—ultrasonic cleaning and welding.

Ultrasound is defined as sound with a frequency that is too high for the human ear to detect and is usually considered to be above 20 kHz and the first commercial use of pulse echo location was developed by Langevin in 1918. The remarkable effects of high power ultrasonic irradiation were initially reported a few years later in 1927 when two papers were published by Loomis entitled "The chemical effects of high frequency sound waves 1. A preliminary survey" (Richards and Loomis, 1927) and "Physical and biological effects of high frequency sound waves of great intensity" (Wood and Loomis, 1927). There were a number of reports of the effects of the so-called power ultrasound following these revelations and in the 1987 the advances were drawn together in two reviews, which established a field that became known as sonochemistry. (Lindley and Mason, 1987; Lorimer and Mason, 1987). It is the development and exploitation of sonochemistry and ultrasonic processing that is the concern of this chapter.

Sonochemistry effects change in materials as a result of acoustic cavitation, which is induced by the powerful ultrasound waves. Like any sound wave, ultrasound is propagated via a series of compression and rarefaction waves induced in the molecules of the medium through which it passes. At sufficiently high power, the rarefaction cycle may exceed the attractive forces of the molecules of the liquid and in this way the cavitation bubbles are formed. It is the fate of these cavities when they collapse in succeeding compression cycles that generates the energy for chemical and mechanical effects. The violent collapse of such bubbles generates large shear forces in the bulk medium that can be used for mixing and particle dispersion. In the collapsing bubble itself, high energies are generated, which can liberate very reactive species such as free radicals.

In a heterogeneous solid–liquid system the collapse of the cavitation bubble will have significant mechanical effects. Collapse near to a surface produces an unsymmetrical collapse of the bubble in which the surface interferes with the inrushing fluid from that side. The main inrush is from the side of the bubble distant from the surface and this generates a liquid jet, targeted at the surface, with speeds well in excess of $100\,m\,s^{-1}$. The effect is equivalent to high pressure jetting and is the reason why ultrasound is used for surface cleaning. The jetting effect can also activate solid catalysts and increase mass and heat transfer to the surface by disruption of the interfacial boundary layer.

Current major areas of study in sonochemistry research and development at the Sonochemistry Centre in Coventry University include

- Electrochemistry
- Environmental protection
- Food technology
- Materials technology
- Therapeutic ultrasound

SONOCHEMICAL SURFACE MODIFICATION OF MATERIALS

To ensure the adhesion of a coating to its substrate, it is essential to form a mixture of chemical and physical (or mechanical) bonds between them. To achieve this, the substrate is often roughened or textured in a process known as surface modification (or adhesion promotion).

The electronics and metal finishing industries have always had a requirement for adhesion promotion on a vast array of dielectric substrates and with the emergence of printed electronics the choice of substrate will increase still further as, theoretically, anything that can be printed could become an electronic device. The surface modification of polymers and plastics is important in the traditional manufacture of printed circuit boards (PCBs) (i.e., the desmear process) and molded interconnect devices (MIDs), but will become even more so for polymer electronics, printed electronics, radio frequency identification (RFID) technology, etc. The metallization of glass and ceramics is critical to the success of all WiFi equipment, flat panel displays, organic light emitting diodes (OLEDs), and light emitting polymers (LEPs) as well as solar panels. In addition, the etching of silicon and the newer semiconductor materials such as silicon-germanium, gallium arsenide, and indium phosphide is an essential step in the processing of these substrates.

Achieving good adhesion between these substrates and the conductive tracks or pixels, etc. (whether a metal, semiconductor, conductive ink, polymer, or paste) is perhaps the most critical stage of the manufacturing process since, if the adhesion is poor, the device will simply fail. Traditional "wet" manufacturing techniques for surface modification lend themselves most readily to high volume manufacturing but utilize hazardous chemistry (chromic acid, hydrofluoric acid, etc.), operate at high temperatures, and require copious rinsing. Alternative methods of surface modification are available (Cobley, 2007) but often require high capital investment, do not fit easily into existing manufacturing sequences, and have high energy needs.

Manufacturing facilities throughout the World are becoming subject to ever stricter environmental and health and safety legislation. For example, in Europe these traditional surface modification processes are affected by the control of major accident hazards (COMAH) regulations and the restriction of the use of hazardous substances (RoHS), while the chemistry used may be affected by the Registration Evaluation Authorization and Restriction of Chemicals (REACH) directive. Therefore, to avoid falling foul of existing and proposed environmental legislation, to reduce water usage and the carbon footprint of this essential manufacturing process, clean, green, and more sustainable techniques need to be evaluated. Sonochemical surface modification techniques fit these criteria. They have the potential to process a diverse range of substrates, employ fewer process stages, require less rinsing, utilize nonhazardous, benign aqueous solutions, and operate at low temperatures.

When ultrasound is applied to a liquid medium, a number of effects occur that can be used in the processing of materials:

- Physical—the abrasive, erosive action of the microjets hitting the surface of the material can remove contaminants or roughen and texture it, providing an ideal surface for subsequent coating
- Chemical—the generation of radical and other reactive species can oxidize the substrate while the localized high temperatures and pressures can break bonds on the surface of materials such as polymers and plastics

It has been reported that ultrasound can surface modify materials such as acrylonitrile butadiene styrene (ABS) (Zhao et al., 1998a) polyvinyl chloride (PVC) (Zhao et al., 1998b), polyethylene (Price et al., 1996a,b) as well as piezoelectrics such as lead zirconium titanate (Baumgartner, 1989).

Ultrasound can be used in conjunction with wet chemical treatments, e.g., persulfates (Price et al., 1996a) and other mild oxidizing agents (Price et al., 1996b), where it has been shown that under sonication, polyethylene materials can be surface modified and become hydrophilic, as determined by contact angle. More aggressive formulations were employed (e.g., tetrafluoroboric acid/nitric acid) to etch lead zirconium titanate (Baumgartner, 1989), and the application of ultrasound produced a linear increase in weight loss. The PCB industry has used ultrasonics to enhance the desmear process for many years and it is particularly useful in horizontal equipment (Kreisel and Dudik, 1987), where it has been shown to improve the topography, debris removal, and the adhesion of subsequent metallization in through holes.

An important evaluation of sonochemical surface modification of materials indicated that significant surface modification of substrates could be achieved using ultrasound in solutions as benign as water (Cobley and Mason, 2007). Figure 3.1 shows the weight loss results for an inert ceramic material, which was surface modified with (20 kHz ultrasonic horn, intensity 39.7 W cm^{-2}) and without ultrasound in conjunction with a number of chemical "etch" solutions (Solutions 1–15).

The results indicated that not only did the application of an acoustic field almost always led to enhanced weight loss but that one of the highest values was achieved in water (Solution 4).

Other workers (Zhao et al., 1998a) showed that by using an ultrasonic horn in water the adhesion of electroplated copper to ABS was always better compared to equivalent chromic acid etching times, while weight loss and roughness were higher when treatment times of more than 30 min were used (Figure 3.2).

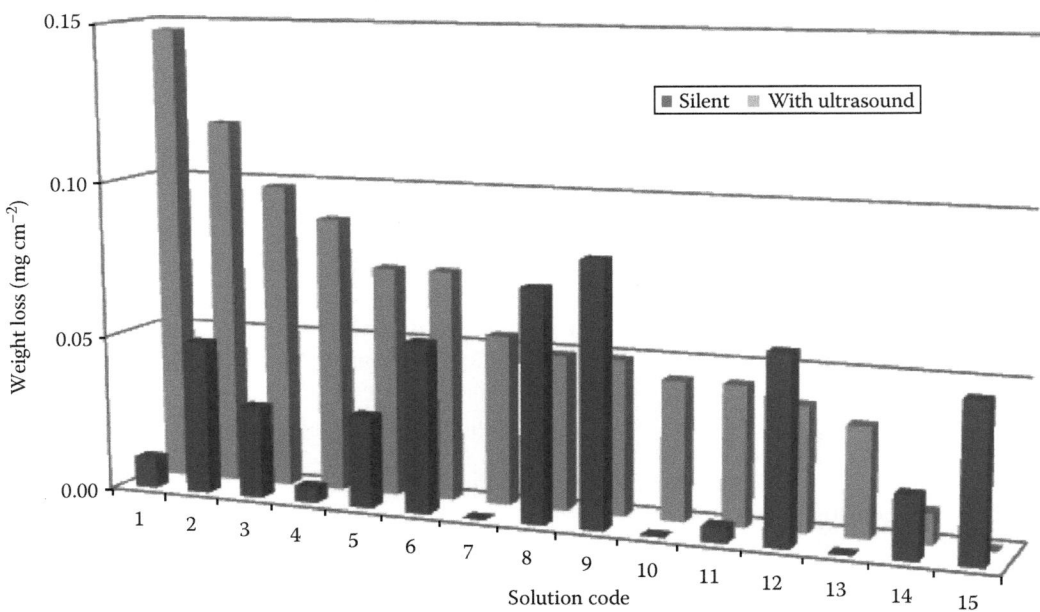

FIGURE 3.1 Effect of ultrasound on the weight loss of a ceramic material using various surface modification formulations.

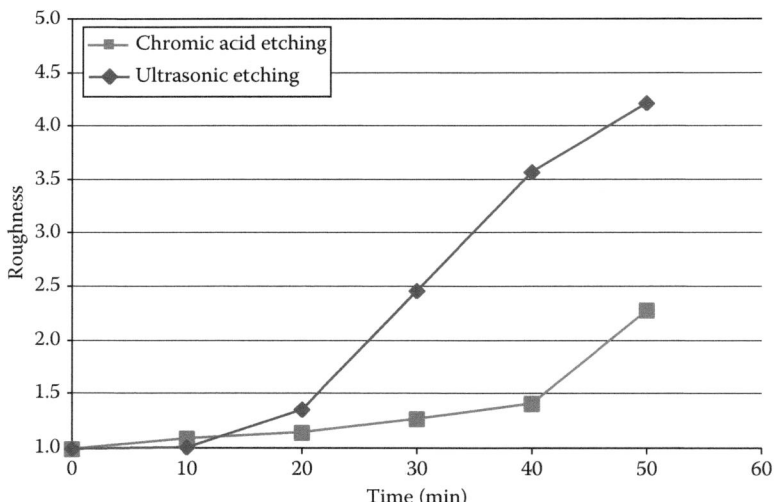

FIGURE 3.2 Surface roughness of ABS versus etching time. (Data taken from Zhao, Y. et al., *Plating Surf. Fin.*, 85(9), 98, 1998a.)

X-ray photoelectron microscopy (XPS) measurements also indicated a chemical change to the surface and these workers found similar results with PVC (Zhao et al., 1998b). Cobley and Mason carried out further work on sonochemical surface modification of materials in water to determine the main effects on the process. For example, the workers have shown that, when using an ultrasonic horn, the distance between the horn tip and the substrate surface is critical (Cobley and Mason, 2008a). They have also investigated various other process enhancements (Cobley and Mason, 2008b) such as adding a surfactant and lowering the solution temperature close to zero.

Aspects of Ultrasound and Materials Science

FIGURE 3.3 (a) Isola 370HR—no treatment—X500. (b) Isola 370HR—standard sonochemical process—X500 DI water, 15 min, 20 kHz, 4.8 W cm^{-2}, 40°C. (c) Isola 370HR—sonochemical process + surfactant—X500 DI water, 15 min, 20 kHz, 4.8 W cm^{-2}, 40°C. (d) Isola 370HR—sonochemical process at low temperature—X500 DI water, 15 min, 20 kHz, 4.8 W cm^{-2}, 9°C.

Comparison of the scanning electron micrographs (SEM) (Figure 3.3a through d) clearly indicates the significance of these factors on an epoxy laminate used in the manufacture of PCBs, namely, Isola 370HR.

Compared to the "as received" material the "standard" sonochemical treatment has cleaned the surface and removed any debris. However, the addition of surfactant or reducing the solution temperature has lead to much greater depth of epoxy removal, revealing the underlying glass reinforcing fibers as well as changing the structure of the epoxy to some extent.

The ultrasonic frequency used is also important and weight loss results on a Noryl (polyphenylene/polyester) material (Cobley et al., 2010) clearly indicated that *physical* surface modification of materials is favored at low frequencies (Figure 3.4).

However, XPS data from the same study suggested that the only chemical changes were occurring at higher frequencies (Figure 3.5), the percentage oxygen content peaking at 850 kHz. It was proposed that under these conditions oxidizing radical species are being formed, which chemically attack the surface of the material.

There have been several studies investigating the use of ultrasound to surface modify a range of materials. However, from a "green chemistry" perspective perhaps the most promising is that which has been carried out using simply water as the liquid medium. Although the effects are mostly physical in nature and can be somewhat localized, this work suggests that sonochemistry could provide a route to more green, energy efficient, environmentally friendly surface modification.

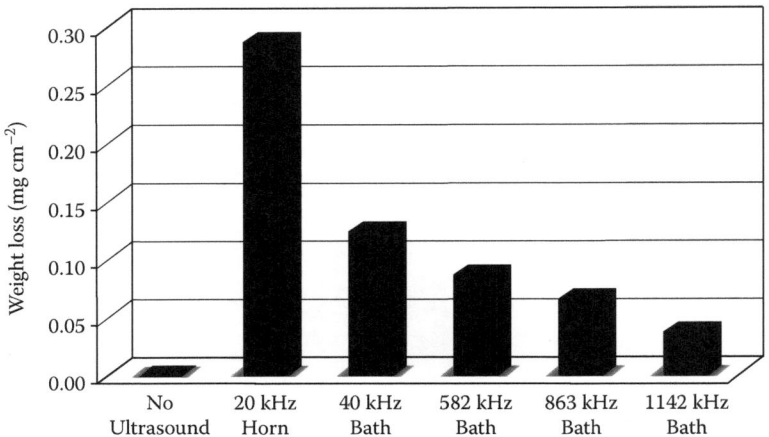

FIGURE 3.4 Effect of ultrasonic frequency on weight loss.

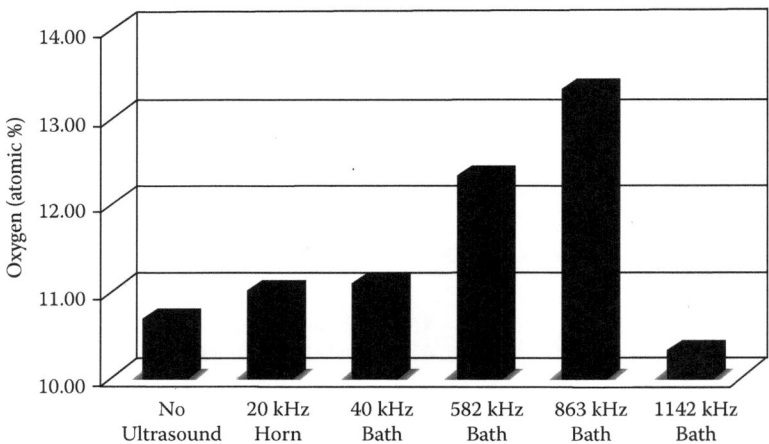

FIGURE 3.5 Effect of frequency on atomic percentage oxygen by XPS.

EFFECT OF ULTRASOUND ON THE ELECTROCHEMICAL METALLIZATION OF MATERIALS

When ultrasound is applied to a liquid electrolyte, a number of well-known effects occur that can influence an electrochemical process. In particular microjetting or microstreaming can improve mass transport, thin diffusion layers (such as the Nernst diffusion layer) (Compton et al., 1997a), and generate localized heating.

Such benefits have been used to enhance two widely used electrochemical metallization processes, namely, electroplating and electroless plating, which are employed in key industrial sectors such as electronics, aerospace, and automotive.

Electroplating

The effects of electroplating in an acoustic field have been studied for at least 70 years (Young and Kersten, 1936) and have been the subject of previous reviews (Lorimer and Mason, 1999), where a number of significant effects have been identified.

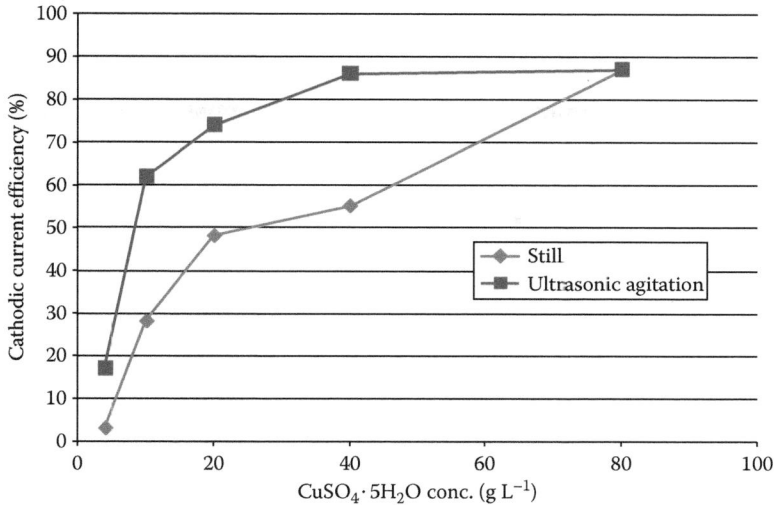

FIGURE 3.6 Effect of ultrasound on cathodic current efficiency for a copper electroplating solution. (From Walker, R., *Ultrason. Sonochem.*, 4, 39, 1997.)

One of the most important factors when electroplating is the current efficiency of the electrode particularly at the cathode. A high cathodic current efficiency is desirable so that most of the energy going into the electroplating cell is being used to deposit the metal and is not wasted on side reactions such as hydrogen evolution.

It has been demonstrated (Walker, 1997) that if copper electroplating was performed in an acoustic field, then the current efficiency could be significantly enhanced, particularly at low metal concentrations (Figure 3.6).

This confirms the findings from the more fundamental electrochemical studies (Compton et al., 1997b), i.e., that ultrasound is more effective on an electrochemical system when it is under mass transport control. Similar results were obtained on a silver electroplating solution (Wei et al., 1979) where ultrasound was found to enhance the cathode current efficiency at high current density.

Another useful effect is the ultrasonically induced microjetting at electrode surfaces, which will thin diffusion layers. This effect has been demonstrated in a study (Drake, 1980), where a copper electroplating solution was utilized and the limiting current and diffusion layer thickness were measured initially, using a rotating disk electrode (RDE) (Figure 3.7).

As would be expected, as the rotation speed of the RDE rose, the diffusion layer thickness was reduced and the limiting current density increased. Applying high frequency ultrasound showed a significant effect on both these responses compared to a static solution; but, at best, was only as effective as utilizing the RDE. However, when low frequency ultrasound was applied, a dramatic increase in the limiting current was observed with a concurrent fall in diffusion layer thickness.

This work also illustrates how the frequency of the sonication used can result in quite different effects. The use of "megasonics" in electroplating (i.e., the use of ultrasound at a frequency of 1 MHz) has been investigated (Kaufmann et al., 2009) and it was found that the throwing power of a copper electroplating bath could be enhanced. This was determined by measuring the depth of penetration of the electroplate into the "blind via" of a PCB.

Apart from its influence on the electrochemistry of a plating solution, ultrasound has been shown to affect the grain structure of the plated deposit. Generally, when electroplating occurs in an ultrasonic field the grain size will be smaller (Walker and Benn, 1971) and this will also have an influence on the physical properties of the deposit such as hardness The tendency for ultrasound to produce finer grained electrodeposits can also produce brighter deposits (Touyeras et al., 2005) and lower porosity (Barnes and Ward, 1977). This latter effect can be important for corrosion resistant

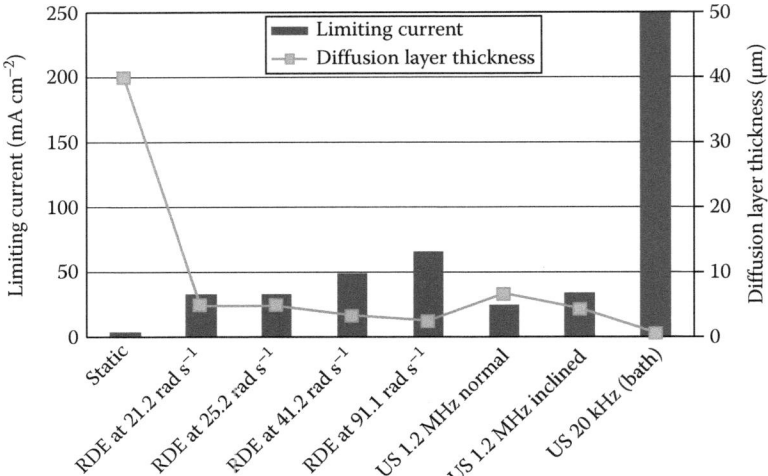

FIGURE 3.7 Effect of ultrasound on the limiting current density and diffusion layer thickness. (From Drake, M.P., *Trans. IMF*, 58(2), 67, 1980.)

coatings and multilayer coatings such as nickel–gold for connectors on PCBs. Not only does a fine grained, coherent deposit lead to better electrical conductivity but the low porosity is essential to prevent corrosion of the nickel undercoat.

The dispersive properties of ultrasound are useful in the deposition of composite coatings and enhanced deposit characteristics have been found for gold/PTFE (Rezrazi et al., 2005), nickel–cobalt/alumina (Chang et al., 2008), and nickel/ titanium nitride (Xia et al., 2009). Studies on a zinc–nickel alloy/alumina composite coating (Zheng and An, 2008) indicated that at a frequency of 25 kHz the amount of alumina in the coating increased with ultrasonic power until an optimum power was reached, as shown in Figure 3.8. Not surprisingly, the hardness of the coating followed a similar pattern.

It is suggested that this optimum ultrasonic power occurs due to the abrasive action of microjetting on the surface of the substrate occurring at the highest ultrasonic intensities, which effectively remove the alumina from the cathode, preventing incorporation into the coating.

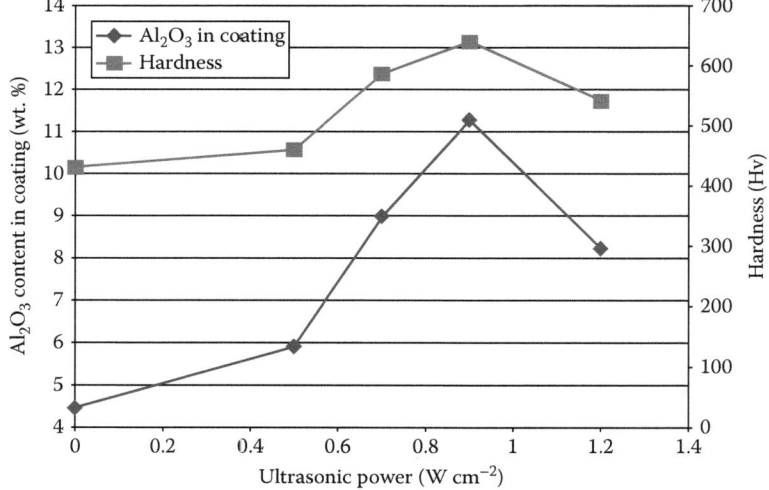

FIGURE 3.8 Effect of ultrasonic power on composite coating composition and hardness. (From Zheng, H.-Y. and An, M.-Z., *J. Alloys Comp.*, 459(1–2), 548, 2008.)

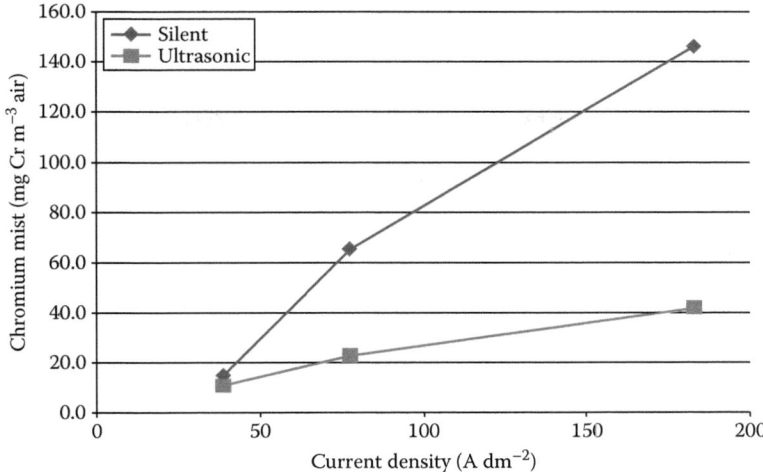

FIGURE 3.9 Effect of ultrasound on mist reduction above a chrome plating electrolyte. (From Mason, T.J. and Tiehm, A., Ultrasound in environmental protection, in *Advances in Sonochemistry*, Vol. 6., Eds. T.J. Mason and A. Tiehm, Elsevier Science B.V., Amsterdam, the Netherlands, 2001.)

A number of workers have studied the effect of chrome plating in an acoustic field and this has led to some quite unexpected findings (Mason et al., 2001). A significant problem with chrome plating is that a mist of chromic acid is produced above the plating solution due to the fact that chrome plating is extremely inefficient (typically 15%–25% cathode current efficiency), and, therefore, a copious amount of hydrogen gas is produced at the cathode. Normally, this is controlled by placing a surfactant type chemical in the electrolyte, which forms foam on top of the solution and acts as a "fume suppressant." However, many of the surfactants used tend to have negative environmental characteristics and many have been banned under environmental legislation. Despite this, because of their importance to the chrome plating industry, a derogation has been allowed for their use in chrome plating although a more satisfactory solution to this issue would be desirable. Studies have indicated (Mason et al., 2001) that if chrome plating is performed in an acoustic field the amount of chromic acid "fume" above the electrolyte could be dramatically reduced, as shown in Figure 3.9.

Although the effects of ultrasound on electroplating systems have been known for some time, it is perhaps important to revisit some of the effects in light of the demands required of the modern day electroplater. Ultrasound can reduce the carbon footprint of the process by improving the current efficiency and throwing power while enabling high speed plating by increasing the limiting current density. In addition, sonication is most effective at low metal concentrations and the reduction in porosity may allow for thinner metal coatings, therefore, improving the resource efficiency and cost-effectiveness of the process. There are requirements for new advanced coatings, and plating in an acoustic field is clearly beneficial for composite deposits. Finally, the work on chrome plating clearly shows some significant health and safety benefits to the use of ultrasound.

Electroless Plating

Electroless plating processes are also electrochemical in nature and, therefore, many of the effects of ultrasound on electroplating systems should also influence electroless plating. There have been studies on the utilization of ultrasound for electroless plating since the early 1960s (Kuzub and Mukhlya, 1963) and it will be seen that the incorporation of ultrasound in electroless plating processes has the potential to make them more efficient and sustainable.

Effect of Ultrasound on Electroless Nickel Plating

As is the case for most electroless plating, there are a number of variables employed in electroless nickel formulations such as the complexant type and concentration, reducing agent, and additives, while the solution itself can be operated at a range of pHs and temperatures. Mallory (Mallory, 1978) used five different chelating systems in his study and showed that in all but one case, ultrasound (49 kHz, 30 W) increased the deposition rate. He suggested that the influence of ultrasound was reduced if the stability of the nickel-chelate system was high. The effect of using succinic acid and citric acid as complexing agents has been reported (Matsuoka and Hayashi, 1986) and it was found that ultrasound (25 kHz, 150 W) had a much greater effect on the deposition rate from the citrate bath, particularly at relatively high pHs of 8–9 (80°C). Other workers (Abyaneh et al., 2007) demonstrated that by increasing the complexing agent concentration, the effect of ultrasound on the plating bath could be dramatically changed. Figure 3.10 illustrates how, at low sodium acetate concentrations (0.04 mol dm^{-3}), ultrasonic irradiation (35 kHz, 0.8 W cm^{-2}) caused an increase in plating rate at all nickel chloride concentrations tested. Similar results were seen when the sodium hypophosphite concentration was varied. However, at high sodium acetate concentrations (0.24 mol dm^{-3}) sonication had a negative effect on plating rates.

Studies using a range of ultrasonic frequencies indicated that, although each frequency investigated caused the plating rate to increase, compared to a still bath an optimum existed at 45 kHz (Kobayashi et al., 2000). Other investigations (Park et al., 2002) into the effect of ultrasonic frequency under conditions where the pH and stabilizer concentrations were varied suggested that, after statistical analysis of the data (design of experiment [DOX] methodology) at 40 kHz, the deposit thickness was increased by 15%. In another study (Yang et al., 1997), a 30% increase in plating rate was observed in the presence of ultrasound.

There have been many theories put forward as to why ultrasonic agitation increases the plating rate of electroless nickel solutions. It is generally accepted that improved mass transport (Mallory, 1978; Matsuoka and Hayashi, 1986) is a factor although this can also have an adverse effect if it brings stabilizers such as thiourea (Mallory, 1978; Abyaneh et al., 2007) and thallous nitrate (Matsuoka and Hayashi, 1986) to the reaction surface causing plating rates to drop. Microjetting at the substrate surface will reduce the diffusion layer thickness (Matsuoka and Hayashi, 1986), and

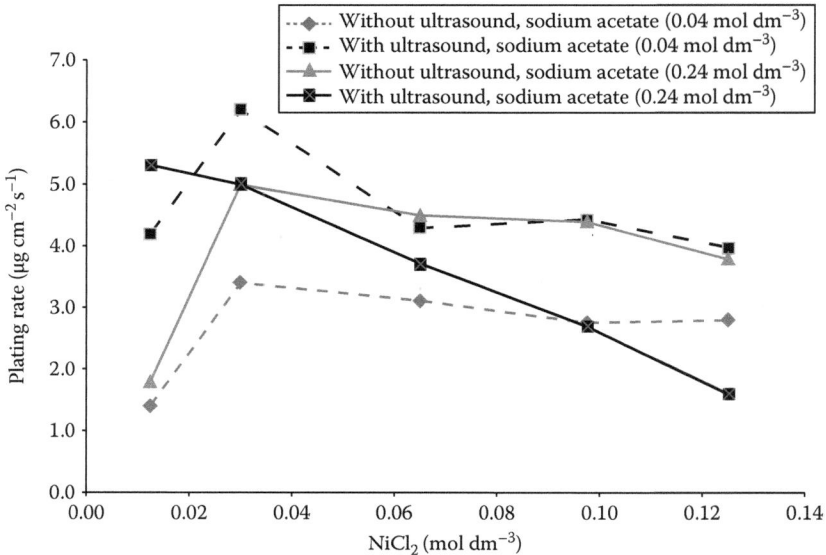

FIGURE 3.10 Effect of ultrasound on electroless nickel plating rates. (From Abyaneh, M.Y. et al., *J. Electrochem. Soc.*, 154, D467, 2007.)

cavitation near the surface may cause localized heating in this layer (Mallory, 1978) and perhaps increase the concentration of adsorbed hydrogen (Yang et al., 1997). It has also been proposed that ultrasound enhances the rate of crystal nucleation (Abyaneh et al., 2007) and produces more active sites (Wu et al., 2009) on which deposition can take place. A further theory is that the inactive form of hypophosphite can be converted to the active form sonochemically (Mallory, 1978).

The application of ultrasound to an electroless nickel plating bath generally reduces the amount of phosphorous in the subsequent deposit. This has been attributed to a number of causes including higher localized temperatures induced by cavitation at the solution/substrate interface (Mallory, 1978; Abyaneh et al., 2007), inhibition of phosphorus codeposition by stabilizers (Matsuoka and Hayashi, 1986), and changes in the local pH in the presence of an ultrasonic field (Yang et al., 1997). It has also been suggested (Mallory, 1985) that the reduction in phosphorus content could cause higher internal stress of a deposit and an increase in the "as plated" hardness (Matsuoka and Hayashi, 1986). However, after heat treatment a comparison of ultrasonic with conventional hardness of the electroless deposit revealed very little difference. Investigators have found (Park et al., 2002) that in the absence of ultrasound the electroless nickel deposit could be pitted due to the presence of hydrogen at the surface, where the electroless reaction occurs. This was alleviated by utilizing ultrasound at a frequency of 40 and 68 kHz, but the use of a lower frequency ultrasound (28 kHz) resulted in severe pitting on the surface due to the impact of high energy microjets produced by cavitation bubble collapse. These workers concluded that for their system the optimum ultrasonic frequency was 40 kHz as this also produced a deposit with enhanced hardness. This effect was again attributed not only to the lower phosphorus content but also to a change from amorphous to more microcrystalline deposit structure under the influence of an acoustic field (Yang et al., 1997). Another reported benefit of an electroless nickel deposit plated with ultrasound is increased fatigue life (Prasad et al., 1994). This was ascribed to a lower number of irregularities in its grain structure compared to a deposit plated from a conventional still bath.

Ultrasound is also known to be extremely good at dispersing micro and nanoparticles, and this property was utilized in the electroless nickel plating of polystyrene microspheres (Jiang et al., 2007). The polystyrene was subsequently removed by calcination to leave hollow nickel microspheres. These workers found that if the microspheres were dispersed in the electroless nickel bath using ultrasound (rather than mechanical agitation) not only was the plating time shorter but the coverage was enhanced and no agglomeration of plated spheres occurred.

Effect of Ultrasound on Electroless Copper

Electroless copper is extensively employed in electronic manufacturing where it is used to metallize dielectric materials. The use of relatively high frequency ultrasound (530 kHz) at a range of powers (5, 10, and 15 W) on a commercial electroless copper plating bath using an epoxy material as the substrate, which had been catalyzed using a palladium-tin colloidal solution has been reported (Touyeras et al., 2001). They found that the highest plating rates were achieved with the lowest ultrasonic power (5 W) if this was applied in the first 5 min of an overall 1 h plating process (i.e., 5 min ultrasonic, 55 min silent). This is in contrast with other studies (Touyeras et al., 2005) at a range of ultrasonic frequencies (300, 500, and 800 kHz) that had indicated the most enhanced mass transfer coefficients occurred at high powers for all three frequencies (with 500 kHz being the optimum). From these results it was concluded that ultrasound mainly affected the initiation stages of the electroless copper process by removing any remaining tin from the colloidal catalyst and enhancing the reduction of ionic palladium to the more catalytic palladium metal. At higher powers the catalyst was ultrasonically scrubbed from the surface of the epoxy substrate. The workers then took this study a stage further (Touyeras et al., 2003) and investigated the effect of ultrasound not only in the electroless copper solution but also in the colloidal palladium-tin catalyst bath. XPS data demonstrated that when the catalyst bath was sonicated a higher amount of palladium was found on the surface of the epoxy than without. Figure 3.11 shows the results of the plating rates obtained under silent conditions and when the catalyst and/or the electroless copper bath were sonicated.

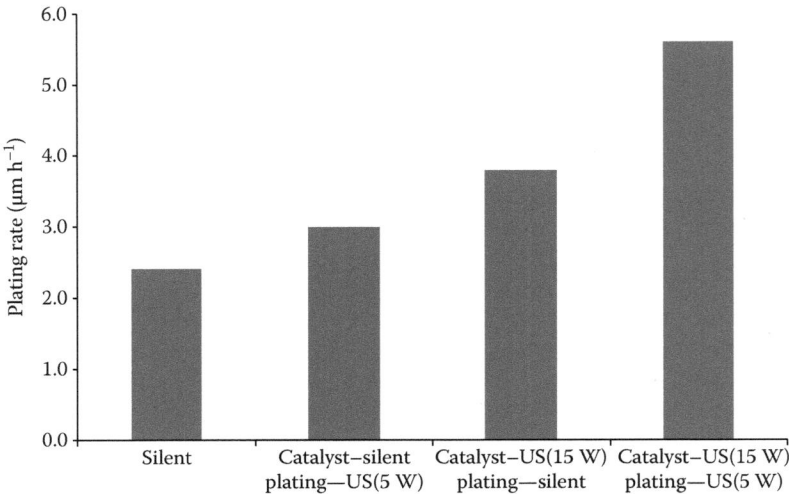

FIGURE 3.11 Effect of applying ultrasound in the catalyst and electroless copper solution on the plating rate. (From Touyeras, F. et al., *Ultrason. Sonochem.*, 10, 363, 2003.)

It can be seen that when ultrasound was used in both solutions the plating rate could be more than doubled.

Other workers (Zhao et al., 1995) have studied the effect of ultrasonic irradiation on electroless copper coating on ceramic. They reported that sonication produced a more uniform and well-bonded coating with a faster plating rate compared to the control process and that it was most effective in the early stages of plating particularly using a higher frequency of 1020 kHz rather than 28.2 kHz. The explanations for the improvements conformed to others in that they included enhanced mass transfer and cavitation generating localized heating in the diffusion layer. However, they also suggested the possibility that ultrasound could produce more fine-grained copper crystals in the initial stages of plating, which are more catalytic.

One of the benefits of electroless copper plating in electronics manufacturing is that it can be used to metalize etched trenches and/or the vias and through holes in materials used for electronic manufacturing. It has been suggested (Kim et al., 2006a) that ultrasonic agitation could enhance the filling of trenches in silicon dioxide materials. Other workers have proposed (Kou and Hung, 1999) that the degassing of small blind vias in PCBs by the application of ultrasound during electroless copper plating enabled more complete metallization than air agitation. In this way, subsequent electroplating produced better coverage of the via. In addition, similar studies (Matsushima et al., 2000) have demonstrated that ultrasonic agitation could improve the electroless copper coverage into vias with different aspect ratios, as shown in Table 3.1.

This enhancement was attributed to improved mass transport of the copper-EDTA complex into the vias coupled with a thinning of the diffusion layer and the removal of hydrogen gas from them.

A number of advantages were seen when ultrasound rather than mechanical agitation was used to plate a silane modified PET film (Lu, 2010). The deposit was smoother and had a better color (referred to as "bright" copper as opposed to "brown"). In addition, the adhesion was significantly improved (16.7 N cm^{-1} with ultrasound, 11.9 N cm^{-1} without) and this was thought to be due to ultrasonic degassing of hydrogen from between the substrate and coating. The electrical conductivity was also enhanced due to the smaller grain size in the deposit plated in an acoustic field.

Enhanced coverage of electroless copper plating on nanomaterials has been reported (Peng and Chen, 2009). They found that ultrasonic agitation (42 kHz, 100 W) aided the dispersion and plating of multiwalled carbon nanotubes (nanowires) in an electroless solution with improved adhesion of the plated copper.

TABLE 3.1
Experimental Values for Coverage of Blind via Holes of 0.09 mm Diameter

Hole Depth (mm)	Tm/To		Tm/Tb	
	Mechanical Agitation	Ultrasonic Agitation	Mechanical Agitation	Ultrasonic Agitation
0.06	0.63	1.10	0.42	1.00
0.12	0.11	0.32	0.00	0.08
0.18	0.00	0.16	0.00	0.00

Source: Matsushima, T. et al., *J. Electrochem.*, 68, 568, 2000.

Note: Ultrasonic frequency, 47 kHz; To, electroless copper thickness at opening of via; Tm, electroless copper thickness in middle of via; and Tb, electroless copper thickness in bottom of via.

In conclusion, a number of effects have been identified that are common to many of the studies. Perhaps the most obvious is that electroless plating in an acoustic field tends to enhance plating rates although, particularly for electroless nickel formulations, this is strongly affected by the chelating species used. For palladium catalyzed electroless copper deposition, the main effect of ultrasound on the plating rate occurred during the early stages of the process and may have more to do with enhancement of the catalytic activity of the palladium than to electrochemical effects. There is clear evidence that ultrasound can improve the coverage of electroless deposition and this is particularly important in the plating of small blind via holes in PCBs and when coating nanomaterials. In this latter case, the dispersive properties of ultrasound prevent agglomeration of nanopowders, which further aids uniform plating. Several workers have shown that ultrasound can affect the composition of the electroless deposit, most clearly by reducing the amount of phosphorus codeposited in electroless nickel plating. This, and changes to the grain structure and morphology of the deposit, can also have effects on the mechanical properties such as hardness and internal stress.

With industry now looking for higher plating speeds, reduced production times, and enhanced material properties, the results from this review have clearly shown that ultrasonic technology has the potential to fulfill these requirements.

SONOELECTROCHEMICAL SYNTHESIS OF NANOPARTICLES (SAEZ AND MASON, 2009)

There are a range of methods for producing metallic nanosized materials including radiation methods (Yeh et al., 1999), thermal decomposition (Kim et al., 2006b), vapor deposition (Ponce and Klabunde, 2005), reduction in microemulsions (Haram et al., 1996), and chemical reduction methods (Athawale et al., 2005). However, most of these techniques tend to be expensive and time-consuming. An alternative method, simple and cost-effective, is the use of sonoelectrochemistry (Zin et al., 2009). The following examines the work employed in the preparation of nanomaterials using the pulsed sonoelectrochemistry method. A device for the production of metal powders using pulsed sonoelectrochemical reduction has been described (Reisse et al., 1994; Durant et al., 1995). Figure 3.12 shows the experimental setup used. In these experiments, a titanium probe (20 kHz) acts both as a cathode and an ultrasound emitter. The electroactive part of the sonoelectrode is the planar circular surface at the bottom of the horn, and the immersed cylindrical part into the electrolyte is covered by an isolating plastic jacket. The ultrasound probe is connected to a generator and a potentiostat using a pulse driver.

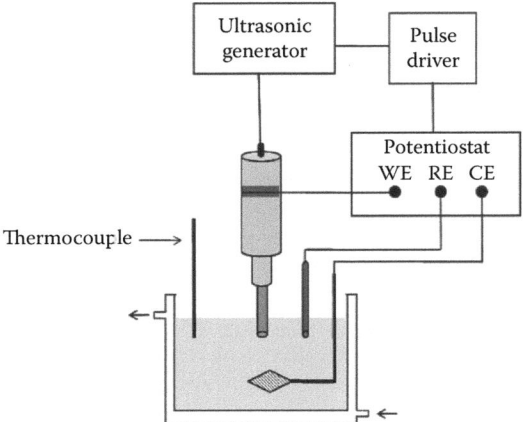

FIGURE 3.12 Sonoelectrochemistry setup used in the production of nanopowders (WE, working; RE, reference electrode; and CE, auxiliary electrode).

The first system used the simplest configuration of a two-electrode cell because the process is carried out under galvanostatic conditions. The drawback of this configuration is the presence of undesirable secondary reactions under galvanostatic control and to overcome this, an adaptation was made. The replacement of a two-electrode configuration (cathode and anode) by a three-electrode configuration (working, reference, and auxiliary electrodes) (Aqil et al., 2008) in the sonoelectrochemistry system was carried out with the aim of applying a controlled potential to the sonoelectrode to get a better control of the process.

The fundamental basis of the pulsed sonoelectrochemical technique for the production of nanopowders is massive nucleation (Rao et al., 2008). At the cathode, a pulse of current (or potential) reduces a number of cations, depositing a high density of metal nuclei on the sonoelectrode surface, and the titanium horn works only as an electrode during this time (T_{ON}). This short electrochemical pulse is immediately followed by a short pulse of high intensity ultrasound (T_{US}) that removes the metal particles from the cathode surface and replenishes the double layer with metal cations by stirring the solution. Sometimes, a rest time (T_{OFF}), without current or ultrasonic vibrations, follows the two previous pulses and it is useful to restore the initial conditions close to the sonoelectrode surface.

Figure 3.13 shows the distribution of the pulses with the time. Electrochemical and ultrasound pulses typically range between 100 and 500 ms and the rest time lasts no more than 1 s.

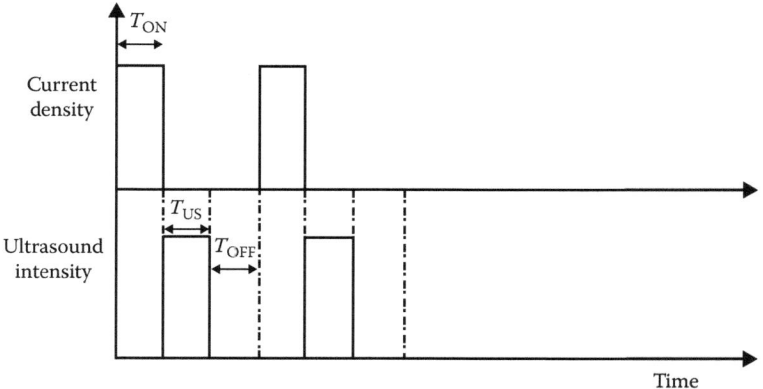

FIGURE 3.13 Distribution of the ultrasound and current pulses with the time (T_{ON}, current pulse time; T_{US}, ultrasound pulse time; and T_{OFF}, rest time).

In most commercial systems, the ultrasound horns are made from titanium alloy (Ti:Al:V 90:6:4). In air, titanium forms a surface oxide layer, consisting of a mixture of TiO_2, Ti_2O_3, and adsorbed oxygen. Under an oxidation process a passivated film can grow on the sonoelectrode surface and this acts as an insulator (Compton et al., 1996). This limitation restricts the use of the sonoelectrode to reduction processes.

This new electrochemical method has since been employed to produce numerous pure metals (Qiu et al., 2003; Aqil et al., 2008) or alloys nanopowders (Mancier et al., 2004; Dabala et al., 2008) and semiconductor nanoparticles (Qiu et al., 2005). More recently, conductive polymer nanoparticles (Atobe et al., 2009) have also been synthesized by pulsed sonoelectrochemistry. The metal powders obtained are in a finely divided state with high surface areas, an average particle size of 100 nm, and high chemical purity (Reisse et al., 1996). Table 3.2 summarizes the experimental conditions used in the synthesis of different materials by pulsed sonoelectrochemistry.

Preparation of Nanoparticles by Pulsed Sonoelectrochemistry

Metallic Nanopowders

Copper electrodeposition is a familiar industrial process where all electrodeposition parameters and electrolytes are well established, and thus copper was one of the first metals to be synthesized using pulsed sonoelectrochemical methods. The synthesis of a range of metallic copper nanostructures by pulsed sonoelectrochemistry has been reported (Reisse et al., 1994; Durant et al., 1995; Haas et al., 2006, 2008). Copper nanoparticles were synthesized (Haas et al., 2006) from an aqueous acidic solution of $CuSO_4$ using polyvinylpyrrolidone (PVP) as a stabilizer. By applying a range of current densities between 55 and 100 mA cm^{-2}, monodispersed spherical copper nanoparticles with a diameter range of 25–60 nm were observed. A reaction mechanism between copper ions and PVP was proposed. The first step was the formation of a coordinative bonding between PVP and copper ions, forming a Cu^{2+}-PVP complex. The formed complex was present in the solution and when the current pulse was applied, the Cu^{2+} was reduced to Cu^0 on the polymer, preventing the agglomeration of the metallic nanoparticles. Infrared (IR) studies showed that the PVP was coordinated with Cu through C–N and C=O bonds. When PVA (polyvinylacetate) was used as a stabilizing agent (Haas et al., 2008), copper with dendritic morphologies was obtained.

The production of platinum nanoparticles from aqueous chloroplatinic solutions under galvanostatic conditions has been reported (Zin et al., 2009). The platinum nanoparticles produced were spherical with an average size ranging from 10 to 20 nm. These aggregated into secondary structures with a mean size ranging between 100 and 200 nm. Tridimensional dendritic Pt nanostructures (Shen et al., 2008a) were synthesized when PVP was used as stabilizer.

Stable suspensions of gold nanoparticles in water have been reported (Aqil et al., 2008) using pulsed sonoelectrochemistry. Some polymers were added to the electrolyte to avoid the nanoparticle aggregation which is frequently observed in such sonoelectrochemical procedures (Durant et al., 1995). The gold electrodeposition was carried out by applying a potential in the range of −850 to −1300 mV vs. NHE, and, in the presence of α-methoxy-ω-hydroxyl polyethylene (MPEO), the nanoparticles aggregated and settled down in the electrochemical cell in a similar way to that observed without a stabilizer present. However, using a MPEO/PVP polymer mixture, a stable violet suspension was obtained at the end of the process without sedimentation. This sample showed a narrow size distribution centered on 12 nm, together with a few larger particles (30 nm). In the presence of polyethylene oxide (PEO), disulfide polymer, a very stable suspension of gold nanoparticles with a mean diameter of 35 nm was obtained. Table 3.3 summarizes the different strategies used to prepare stable suspensions of gold nanoparticles and the average size obtained.

Silver nanoparticles have also been synthesized using different electrolytes and stabilizers by the pulsed sonoelectrochemistry method (Zhu et al., 2000a; Liu et al., 2001; Socol et al., 2002; Jiang et al., 2004). Shaped silver nanoparticles including spheres, rods, and dendrites were prepared from

TABLE 3.2
Experimental Conditions for the Pulsed Sonoelectrochemistry Synthesis for Different Nanomaterials Using a 20 kHz Titanium Horn as Working Electrode

Species	Solution	I_{US} (W cm^{-2})	T_{US} (ms)	Electrochem. Conditions	T_{ON} (ms)	T_{OFF} (ms)	Experiment Duration	Size	Reference
Cu	0.16 mol L^{-1} CuSO$_4 \cdot$5H$_2$O, 1.84 mol L^{-1} H$_2$SO$_4$, pH 0.5	62	100–600	440–480 mA cm^{-2}	250–900	150–300	30 min	With PVP 29–34 nm Without PVP 200 nm aggregates	Haas et al. (2006)
Pt	0.1 mol L^{-1} K$_2$PtCl$_4$, 0.5 mol L^{-1} NaCl pH 1	62	300–500	50 mA cm^{-2}	200–500	—	1 h	10–20 nm (some aggregated 100 and 200 nm)	Zin et al. (2009)
Au	2.8 × 10^{-4} mol L^{-1} HAuCl$_4 \cdot n$H$_2$O 1 g L^{-1} MPEO pH 1	Not indicated	100	−850 to −1300 mV/NHE	10–50	100–200	5 h	5–35 nm	Aqil et al. (2008)
Mg	Grignard reagents (EtMgCl and BuMgCl), AlCl$_3$ in THF and DBDG	62	300	5 mA cm^{-2}	6 × 10^5	600	Not indicated	4.5 ± 0.5 nm	Jia et al. (2007)
CdSe	CdCl$_2$ 2.5H$_2$O, NTA, Na$_2$SeO$_3$ with PVP	Not indicated	Cont.	60–80 mA cm^{-2}	Cont.	—	2 h	80 nm diameter nanotubes	Shen et al. (2008)
Co$_{65}$Fe$_{35}$	Sulfate bath based on Aotani's formulation pH 3	62	300–500	8–380 mA cm^{-2}	300–500	Not used	90 min	3D structures 300 nm	Dabala et al. (2008)
PANI	0.5 mol L^{-1} aniline, 0.5 mol L^{-1} HCl	62	Not indicated	+1 V/Ag/AgCl (3M)	8 × 10^6	800	2 h	20–40 nm	Ganesan et al. (2008)
Cu$_2$O	0.45 mol L^{-1} CuSO$_4$ 5H$_2$O + 3.25 mol L^{-1} lactic acid pH 9.1	110	100–400	−0.65, −1.2 V/SSE	100–300	200–400	Not indicated	8 nm	Mancier et al. (2008)

TABLE 3.3
Summary of Experimental Conditions in Gold Nanoparticles Synthesis

Polymer	E (mV) versus ENH	T_{ON} (ms)	Size (nm)
MPEO	−850	50	Sediment
MPEO/PVP	−850	50	12
PEO disulfide	−1300	20	35

Source: Aqil, A. et al., *Ultrason. Sonochem.*, 15, 1055, 2008. With permission.

an aqueous solution of AgNO$_3$ in the presence of nitriloacetate (NTA) (Zhu et al., 2000a). It was found that the electrolyte composition and the reaction time can greatly affect the shape and growth of the nanoparticles. Without NTA, shaped Ag nanoparticles were not formed. If the concentration of NTA is very low (less than 0.1 g L^{-1}), only randomly shaped aggregates were obtained but if the concentration of NTA was increased to 1 g L^{-1} the formation of the shaped particles was favored. Using this stabilizer concentration and varying the concentration of silver ions in the electrolyte, silver nanoparticles were formed with different shapes. If a concentration of 0.6 g L^{-1} AgNO$_3$ was utilized, the nanoparticles were spherical and well-dispersed, with a size of 20 nm if the reaction time was kept below 5 min. Nanorod structures with a width of 10–20 nm in diameter were obtained with 2 g L^{-1} AgNO$_3$ after 12 min of reaction and when the concentration of AgNO$_3$ was increased to 20 g L^{-1}, while highly ordered dendritic nanostructures were observed for a 25 min reaction. The authors suggested that the excess of silver ions in the solution favors the aggregation and growth into the dendritic structure of Ag clusters. In subsequent work, Socol et al. (2002) used the same system, i.e., silver plus NTA and proposed a model based on a suspensive electrode formation to explain the growth of the dendritic structures. The model involves expulsion of the particles from the electrode surface as a result of the ultrasonic pulse into the solution, where they can remain suspended. These suspended nanoparticles are then moved about in the solution through the influence of the ultrasonic wave and as a result could hit the electrode and thus acquire a charge. These charged particles traveling into the bulk solution could reach the anode and initiate electrodeposition onto themselves and grow in size into a dendritic structure.

The synthesis of silver nanoparticles with a face-centered cubic structure in a saturated solution of silver citrate in the presence of PVP has also been reported (Jiang et al., 2004). Under the experimental conditions used, the Ag nanoparticles were prepared as spherical and monodisperse with an average size of 20–25 nm. In contrast, amorphous silver nanoparticles (Liu et al., 2001) of 20 nm size were prepared from an aqueous solution of AgBr in the presence of gelatin.

The synthesis of spherical nanoparticles of tungsten by pulsed sonoelectrochemistry has been reported (Lei et al., 2007). The average diameter of these nanoparticles was 30 nm and some aggregated particles were observed. The synthesis of highly dispersed palladium nanoparticles by pulsed sonoelectrochemistry methods with different sizes and shapes using a solution of PdCl$_2$ in the presence of cetyltrimethylammonium bromide (CTAB) has been achieved (Qiu et al., 2003). The nanoparticles were mostly spherical with an average size of 4–5 nm and, for a longer reaction time than 2.5 h, dendritic structured Pd was detected. The dendritic palladium was made up of numerous spherical Pd nanoparticles with a diameter of approximately 10 nm. Zinc (Durant et al., 1995; Reisse et al., 1996; Jia et al., 2007), nickel (Reisse et al., 1996; Jia et al., 2007), and cobalt (Reisse et al., 1996) nanoparticles have also been successfully synthesized by pulsed sonoelectrochemical methods.

Nanoparticles of very reactive metals with a high negative reduction potential, e.g., magnesium and aluminum can also be synthesized using pulsed sonoelectrochemistry. The synthesis of Mg nanoparticles (Haas and Gedanken, 2008) was carried out using two different electrolyte solutions based on Grignard reagents (EtMgCl and BuMgCl) in ethers. The ethers used were tetrahydrofuran (THF) and dibutyldiglyme (DBDG), and $AlCl_3$ was added to increase the conductivity of the electrolyte. Four nm-sized metallic magnesium particles were obtained for both electrolytes. The product efficiencies in THF and in DBDG were different at 41% and 33%, respectively, and this was ascribed to the difference in solution viscosity. MgO nanoparticles could also be found in the product due to the ease of oxidation of the metal.

In the case of aluminum nanoparticles (Mahendiran et al., 2009), a solution of $LiAlH_4$ and $AlCl_3$ in THF was used. TEM analysis showed the formation of aggregated particles in the range 10–20 nm, and energy-dispersive x-ray (EDX) analysis confirmed that the surface of the material is mainly composed of aluminum, but there were also small peaks observed for oxygen, which showed that the surface of the aluminum was partially oxidized. This was a similar result to that obtained for the magnesium nanoparticles. The current pulse used in the synthesis of aluminum and magnesium nanoparticles was 600 s, which is much longer than those commonly used in the other nanometal syntheses.

Alloy Nanopowders

The pulsed sonoelectrochemistry technique has been applied to the synthesis of binary and ternary alloyed nanopowders containing iron, cobalt, and nickel (Atobe and Nonaka, 1995; Mancier et al., 2004; Dabala et al., 2008). All these alloys were synthesized using an electrolyte bases on Aotani's formulation. Binary and ternary alloys were deposited galvanostatically at 8000 A m^{-2} and particles with a mean diameter of 100 nm were produced (Delplancke et al., 2000). In a recent report, a series of iron–cobalt alloy nanopowders were prepared by pulsed sonoelectrochemistry under different potentiostatic conditions (Mancier et al., 2004). Under these experimental conditions the composition of the alloys is essentially independent of the deposition potential and is only determined by the composition of the electrolytic bath. The smallest particles detected had a mean diameter of 7 nm and were strongly aggregated in 3D clusters.

It has been calculated (Dabala et al., 2008) that the cathode efficiency for the production of $Co_{65}Fe_{35}$ nanoparticles is the ratio of the produced mass of powders to the Faradaic yield. For applied currents lower than −20 mA the efficiency was high and remained over 50% but as the applied current increased 50-fold, the cathode efficiency decreased by 10-fold.

Semiconductor Nanopowders

Semiconductor materials are the foundation of modern electronics and are finding applications in photochemistry, dye-sensitized solar cells, and in the photocatalytic treatment of chemical waste (Hoffmann et al., 1995; Serpone and Khairutdinov, 1996).

Cu_2O nanopowders have been prepared in potentiostatic mode (Mancier et al., 2008). The work was based on a previous voltametric study that showed that, at an applied potential ranged between −0.65 and −1 V/SSE, it was possible to avoid the formation of a mixture of Cu_2O and Cu. The powders generated were analyzed by XRD and only Cu_2O peaks were detected indicating that neither metallic copper nor CuO were formed at these potentials. TEM micrographs showed numerous agglomerates of nanoparticles of a variety sizes but isolated particles were also found with diameter ranges between 7 and 20 nm.

The synthesis of CdSe with a tubular structure has been reported (Shen et al., 2008b). This synthesis was carried out by applying a constant current density in the range of 60–80 mA cm^{-2}, but in this case under continuous sonication. The CdSe nanotubes had an outer diameter of 80 nm and a wall thickness of 10 nm and were obtained through a roll-up mechanism. In the first stage, 2D nanosheets were formed on the electrode and these were then dislodged from the sonotrode surface. Due to the high surface energy of the edges of the nanosheets, these flexible and unstable nanosheets are thought to roll-up and form tubular nanostructures during the cavitation process.

PbTe nanorods (Qiu et al., 2005) were synthesized by pulsed sonoelectrochemistry methods using NTA as stabilizer. PbTe formation predominated when the Pb^{2+}/NTA ratio was high, even at a very low concentration of Te^{2-} ions. However, for low Pb^{2+}/NTA ratio, the thermodynamic solubility constant limited the precipitation of PbTe even at relatively high concentrations of Te^{2-} ions. With very low concentrations of Pb^{2+} (2 mmol L^{-1}), only spherical particles were observed and the increase in the concentration of Pb^{2+} ions caused the spheres to grow and rod-shaped morphologies to be formed. PbTe nanorods showed highly uniform nanorod morphology with an average diameter of 7 nm.

Other semiconductor materials such as PbSe (Zhu et al., 2000b), Bi_2Se_3 (Qiu et al., 2004), and MoS_2 (Mastai et al., 1999) have been successfully synthesized with different morphologies by using the pulsed method in aqueous solution at room temperature.

Conducting Polymer Nanoparticles

Polyaniline (PANI) and other conducting polymers such as polythiophene, polypyrrole, and poly(methylaniline) have great potential in numerous technological applications (Rupprecht, 1999). These materials can exist as bulk films or as dispersions; but a common problem with the latter is particle aggregation, which limits the range of applications. Conducting-polymer synthesis using pulsed sonoelectrochemistry has been reported (Ganesan et al., 2008; Atobe et al., 2009) despite the fact that oxidation of the horn surface can generate an insulating layer when the sonoelectrode is used in oxidation process (Compton et al., 1996). The synthesis of PANI nanomaterial has been reported (Ganesan et al., 2008) by oxidative polymerization using the sonoelectrode as anode. In this synthesis, a constant potential pulse of +1 V versus Ag/AgCl/3 mol L^{-1} NaCl was applied to the sonoelectrode in an aqueous solution of aniline in HCl. After 2 h, the formation of PANI nanostructures was confirmed by UV-Vis, and TEM micrographs showed particles with a diameter of 2–4 µm made up of very small nanoparticles of average size 20–40 nm.

The synthesis of nanopoly(methylaniline) by the use of a pulsed sonoelectrochemical method has been reported (Atobe et al., 2009), but in this case the anode was a platinum electrode placed face-to-face with the ultrasound emitting surface of a horn. This arrangement overcame the drawback related to the use of an ultrasound horn in oxidation process. Poly(methylaniline) synthesis was achieved at a constant potential pulse of 0.75 V vs. SCE to the platinum electrode in an aqueous solution of methylaniline in $HClO_4$. PNMA microspheres were obtained and their size distribution depended on the electric pulse width. Thus, for a pulse width of 40 s, the average size was 1.4 µm, whereas for 90 s the average size was 2.4 µm.

Pulsed sonoelectrochemistry techniques, which use 20 kHz ultrasound horn both as working electrode and ultrasound emitter, have been used to prepare nanopowders. The majority of nanomaterials produced by this method are pure metals. More recently, the syntheses have been extended to include the preparation of nanosized metallic alloys, metal oxide semiconductors, and conductive polymers. Factors that affect the process yield and particle size are the ultrasound pulse time and the current density. In general, decreasing temperature, shorter pulse duration, high current density, and high ultrasound intensity will lead to a reduction in crystal size. These parameters need to be optimized in order to maximize the nanoparticles production yield and to obtain the lowest size of the products depending of the applications.

It can be seen that a variety of nanomaterials have been prepared to date by pulsed sonoelectrochemistry. Although these nanoparticles were aggregated, the shape and size of the nanoparticles can be adjusted by controlling the various electrodeposition and ultrasound parameters and by using a suitable stabilizer.

ULTRASOUND IN POLYMER TECHNOLOGY

Ultrasound has been employed to affect polymers for many years. Both depolymerization and synthesis can be achieved when utilizing the correct conditions. Depolymerization is thought to be, in part, due to the rupture of the C–C polymer backbone chain by the intense shear forces released

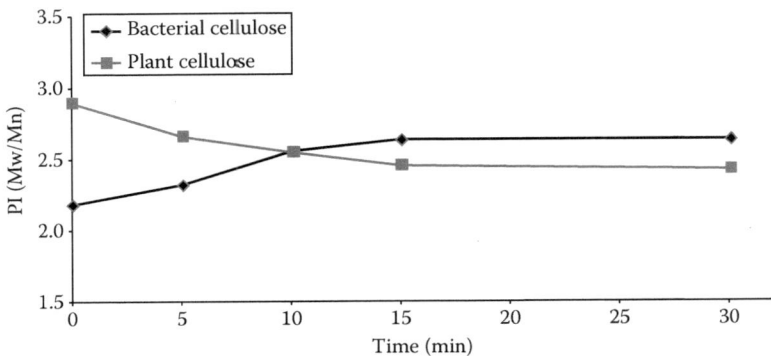

FIGURE 3.14 Polydispersity-index variation of plant and bacterial cellulose as a function of time of ultrasonication. (From Wonga, S.-S. et al., *Carbohydr. Polym.*, 77, 280, 2009.)

on cavitational collapse. Synthesis can be improved due to enhanced mixing during sonication as well as radical generation. Interest in the use of ultrasound allows for novel materials to be formed by employing an environmentally clean technology. Often ultrasound can be utilized to improve product yields and selectivity. It can reduce reaction temperatures while enabling less acidic or corrosive chemicals to be used and enhancing product recovery and final quality. Performing these reactions in an acoustic field can facilitate the use of aqueous solvents and, in some cases, reduces the need for pure chemicals or initiators; and is also considered to be a nontoxic and environmentally friendly technology. Recently, ultrasound has been employed to disperse nanoparticles within a polymeric base, producing new hydrogels and has also been used for controlled drug release from a polymer matrix.

It has been reported (Wonga et al., 2009) that ultrasound can be utilized to control the depolymerization of plant celluloses and bacterial celluloses by employing suitable sonication settings, namely, an ultrasonic frequency of 37 kHz and ultrasonic power of 150 W over time periods of 5, 10, 15, and 30 min (Figure 3.14).

By employing size exclusion chromatography, they determined that a decrease in polymer molecular weight was matched by a reduction in the polydispersity of the plant polymers but an increase in the bacterial polymers. This was thought to be due to lack of change in the crystallinity of plant polymers but a marked increase in crystallinity of the bacterial polymers.

Other workers (Jevtic et al., 2009) used ultrasound to achieve a controlled assembly of poly(D,L-lactide-coglycolide) (PLGA)/hydroxyapatite (Hap) core-shell nanospheres, which are desirable for use as implants that accelerate the reconstitution of damaged bones. They applied ultrasound at powers of 142.5 and 50 W to form biocomposite nanospheres of PLGA/Hap. They revealed a significant dependence of the morphology of the resultant composites upon parameters such as the intensity of the applied ultrasonic field, weight percentage ratio of ceramic/polymer, and the temperature of the medium. The optimal settings, which provided a significant improvement in final composite appeared to be a low treatment temp of 8°C with high ultrasonic power of 142.4 W. The final morphology of the PLGA/Hap particles synthesized under these conditions was highly regular sphere-like, with particles of very small dimensions (150–320 nm), extremely uniform particle size distribution with planar spatial self-organization. These characteristics indicate significant improvements in PLGA/Hap composite resulting from this ultrasonic procedure.

Many materials require specially engineered microstructures, e.g., those which are used in orthopedic applications or are required to meet specific thermal or mechanical constraints. Ultrasound (Torres-Sanchez and Corney, 2008) was used at 20, 25, and 30 kHz for 20 min in an off/on cycle of 2 min on/1 min off during the production of a polyurethane foam to see how it affected the foam cellular structure. They determined that the frequency and power of the applied ultrasound affected the volume and distribution of pores within the matrix and that porosity varied in direct proportion to both the acoustic pressure and frequency of the ultrasound signal. It was proposed that ultrasound could be employed to control the geometry of the foam and link this to specific material requirements.

Other workers (Kang et al., 2010) have studied the effect of ultrasound on the conformation and crystallization behavior of isotactic polypropylene and b-isotactic polypropylene. They showed that application of ultrasound at a frequency of 20 kHz frequency and powers of 0, 50, 100, 150, 200, 250 W for 3 min at 190°C decreased the helical conformation order and changed the crystalline structures of samples studied.

Ultrasound has also been used (Park et al., 2008) to induce a selective covering of carbon nanotubes with sol-gel sheaths. These carbon nanotubes can be used to produce novel advanced materials, which exhibit high mechanical strength and large electrical and thermal conductivities. However, poor dispersion and low interfacial bonding within polymer matrixes reduces their effectiveness. The authors produced sol–gel sheaths on the surface of multiwalled carbon nanotubes in water by employing a hydrophobicity-induced covering with the assistance of ultrasound. They employed a frequency of 24 kHz at 110 W to prevent agglomeration induced by the hydrophobicity of the nanotubes and phenyl-containing sols, leading to a selective construction of sol–gel sheaths on the nanotube surface. The phenyl groups of the resulting sol–gel sheaths were then successfully removed by air-oxidation to provide the nanotubes to be covered with amorphous silicon oxide sheaths.

A variety of different hydrogels were prepared (Cass et al., 2010) by utilizing ultrasound to aid polymerization of water soluble monomers, such as 2-hydroxyethyl methacrylate, poly (ethylene glycol) dimethacrylate, dextran methacrylate, and macromonomers. Ultrasound itself was used to create the initiating radicals in viscous aqueous monomer solutions by using glycerol, sorbitol, and glucose as additives. A 5% w/w solution of dextran methacrylate formed a hydrogel in 6.5 min in a 70% w/w solution of glycerol in water at 37°C with 20 kHz ultrasound, 56 W cm^{-2}. It is thought that this method could be used for processes where additional initiators are not wanted.

Work has also been performed (Kowalski et al., 2008) on the effect of ultrasonic irradiation during electropolymerization of polypyrrole on corrosion prevention of coated steel. They used ultrasound during the electropolymerization of polypyrrole (PPy) in acid phosphate solution containing molybdophosphate ions and pyrrole monomer. The corrosion of the steel coated by the film prepared under ultrasonic irradiation was tested in 3.5 wt % NaCl solution and compared to a non-ultrasonically prepared coating. They discovered that the film prepared under ultrasonic irradiation kept the steel in the passive state one and a half times longer as that prepared without ultrasonic irradiation. They suggest that the surface morphology of the ultrasonic film was changed during the electropolymerization process. Under sonication, a dense and compact polypyrrole layer was formed. The structure of the film obtained under ultrasonic irradiation was assumed to result from a change in the nucleation-growth mechanism.

Ultrasound has also been utilized to develop an efficient and green procedure for the synthesis of 1,8-dioxo-octahydroxanthenes derivatives using MCM-41-SO$_3$H as a nanoreactor and catalyst (Rostamizadeh et al., 2010). The reaction was carried out in water under ultrasonic irradiation at 25 kHz, using nanosized MCM-41-SO$_3$H. In this method, several types of aromatic aldehydes, containing electron-withdrawing groups as well as electron-donating groups, were rapidly converted to the corresponding 1,8-dioxo-octahydroxanthenes in good to excellent yields.

By using a combination of MCM-41-SO$_3$H and ultrasound allowed for a better intercalation of guest molecules (reactant) into the host nanoreactors. Ultrasound irradiation produced high yields, short reaction times with a simpler, and easy workup compared to the conventional methods reported in the literature.

An attempt to synthesis a novel chitosan-mesoporous silica hybrid composite has been reported (Depan et al., 2010) to design a drug delivery system based on ultrasound triggered stimuli-responsive smart release using Ibuprofen as a model drug. Ultrasound triggered release of Ibuprofen in a simulated body fluid (pH 7.4) was examined by utilizing an ultrasonic bath of frequency 33 kHz. The results indicated that ultrasound irradiation can be used as a noninvasive technique for drug release from polymeric materials. The authors suggested that the enhanced rate of drug released observed with ultrasound is due to the cavitation effect, which is achieved without causing any significant destruction of the polymer morphology.

ULTRASOUND AND ENVIRONMENTAL PROTECTION

Water used for drinking purposes can contain many different types of pollutants of a chemical and also of a biological nature and therefore must be treated prior to subsequent use, and the utilization of sonication to destroy such pollutants has attracted much attention in recent years. Ultrasound has been employed in many ways to deal with issues within the environment from treating chemicals in wastewater to enhancing aerobic and anaerobic sewage treatment. The production of free radicals by the sonication of water is thought to contribute to the degradation of chemicals, whether they are dyes from the textile industry or agrochemicals and pesticides from the agricultural industry. In addition, by using different frequencies of ultrasound, bacteria present in biofilms and wastewater may also be treated with the main component for effective kill rate being the direct physical effect of the ultrasound applied. The physical and chemical affects produced by ultrasound, can have a beneficial effect in wastewater treatment. Recent investigations involved the examination of the effects of ultrasound on biotechnology (enhancing enzyme activity, e.g., activated sludge, biodegradation) (Schlafer et al., 2000; Chu et al., 2001; Mason and Tiehm, 2001) agglomeration for use in coagulation and flocculation, filtration, and decontamination (destruction of pesticides, chlorinated hydrocarbons, PCBs) (Price et al., 1994; Naffrechoux et al., 2000; Peters, 2001).

USE OF ULTRASOUND IN THE DEGRADATION OF TEXTILE DYES

Recently, ultrasound has been used to degrade dyes as both an individual oxidation process and combined with others. Azo dyes are the largest group of colorants used in a variety of industries ranging from textile manufacture to paper production. The textile industry involves several processes, which produce contaminated wastewaters. This waste must be treated prior to subsequent use but the dye chemicals employed are often resistant to treatment whether it is physical, chemical, and/or biological. Ultrasound has the potential for use in environmental remediation due to the formation of highly concentrated oxidizing species and as a result several studies of its use for dye degradation have been undertaken.

For example, ultrasound has been employed (Onat et al., 2010) in the decoloration of textile azo dyes by using it in combination with microbial removal. They examined the decolorization of Reactive Red 2, Reactive Blue 4, and Basic Yellow 2 first by employing continuous ultrasonic irradiation at 20 kHz and second by employing microbial digestion. After the ultrasonic/microbial step-by-step applications, the highest decolorization yield achieved was 93% in the sample with Reactive Red 2 at 50 mg L^{-1} initial dye and 6 g L^{-1} cell concentrations. The lowest dye removal yield was 25% for about 100 mg L^{-1} initial Basic Yellow 2 and 4 g L^{-1} cell concentrations.

Ultrasound was utilized in combination with Fenton's reagent for the degradation of Acid Red 97 (Li and Song, 2010). By using ultrasound at frequency of 40 kHz, the degradation ratio was 99.9% for the initial concentration of 50 mg L^{-1} using 1.57 mmol L^{-1} H_2O_2, 0.054 mmol L^{-1} Fe^{2+} at 40°C within 140 min. They found that a combination of ultrasound and Fenton's reagent was more effective than individual sonolysis or Fenton reagent process treatments.

The degradation of C.I. Acid Orange 7 by ultrasound has also been investigated (Zhang et al., 2009). They studied the effect of ultrasonic power density (Figure 3.15), goethite addition, hydrogen peroxide concentration, and initial pH, among other factors.

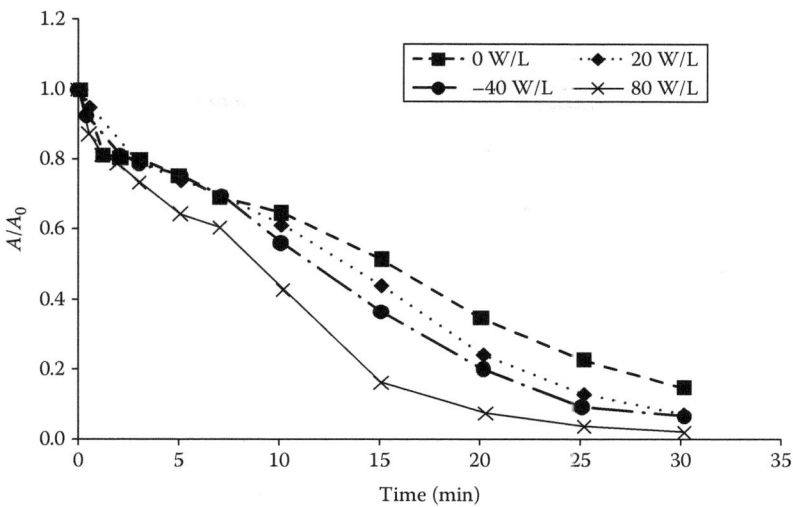

FIGURE 3.15 Effect of power density on the decolorization (C_0 = 79.5 mg L^{-1}, [FeOOH] = 0.2 g L^{-1}, [H_2O_2] = 7.77 mmol L^{-1}, pH$_0$ 3). (From Zhang, Z. and Zheng, H., *J. Hazard. Mater.*, 72, 1388, 2009.)

The results showed that the decolorization rate increased with ultrasonic power density, goethite addition, and hydrogen peroxide concentration, but decreased with the increase of initial dye concentration—an effect observed by many studies.

The influence of bicarbonate and carbonate ions on sonochemical degradation of Rhodamine B in aqueous phase has also been studied (Merouani et al., 2010a). They discovered that as a result of the generation of •OH radicals by cavitational collapse, the presence of carbonate radicals were formed as secondary products of water sonication when the water contained dissolved bicarbonate or carbonate ions. The results clearly demonstrated the significant destruction of Rhodamine B when sonicated in the presence of bicarbonate and carbonate, especially at lower dye concentrations. They continued their studies (Merouani et al., 2010b) by examining the effect of additives in the degradation process. Additives such as iron (elemental, bivalent and trivalent), carbon tetrachloride, hydrogen peroxide, *tert*-butyl alcohol, salt (Na_2SO_4), sucrose, and glucose were employed. The ultrasonic degradation of dye was enhanced by iron addition. The rate of degradation employing iron was of the following order Fe(II) > Fe(III) > Fe^0. Carbon Tetrachloride aided degradation through the formation of oxidant chlorine and it was found that there were optimum concentration requirements for the use of H_2O_2 and Na_2SO_4 as additives while glucose and sucrose also showed very promising results.

The optimization of decolorization of azo dye Acid Green 20 by ultrasound and H_2O_2 has been reported (Zhang and Zheng, 2009). The optimum operating conditions for Acid Green 20 decolorization were found to be 1.08 W mL^{-1} ultrasonic power density, 4.35 pH and 1.94 mmol L^{-1} of H_2O_2 concentration. The decolorization rate was found to be 96.3%.

An electrocatalytic process has also been used (Ai et al., 2010) to enhance the ultrasonic degradation of Rhodamine B. They optimized oxidation factors by examining parameters such as applied potentials, power of the ultrasound, initial pH of the solution, and initial concentration of dye. The eventual result was that Rhodamine B was decolorized completely within several minutes. They also found the process to be effective in the decolorization of other azo dyes, such as methylene blue, reactive brilliant red X-3B, and methyl orange.

Congo Red was degraded using ultrasound in another study by (Gopinath et al., 2010). The dye solution was subjected to ultrasound irradiation of 30 kHz at various initial concentrations and pH. The results showed that the initial dye concentration and pH of the dye solution influenced the percentage decolorization and low initial values resulted in high percentage decolorization.

Ultrasound has also been used (Wang et al., 2010) to enhance the degradation of dye pollutants by also using H_2O_2 activated by Fe_3O_4 magnetic nanoparticles as a peroxidase mimetic. They found that Fe_3O_4 could catalyze H_2O_2 by removing Rhodamine B in a wide pH range from 3.0 to 9.0. The degradation was significantly enhanced by employing ultrasound irradiation. At pH 5.0 and at a temperature of 55°C, the ultrasound-assisted H_2O_2-Fe_3O_4 catalysis removed about 95% of Rhodamine B (0.02 mmol L^{-1}) in 15 min with an apparent rate constant of 0.15 min^{-1} for the degradation of Rhodamine B, i.e., 6.5–37.6 times of that observed for a H_2O_2-Fe_3O_4 system and an ultrasonic ultrasound-H_2O_2 system, respectively.

Other workers (Eren and Ince, 2010) examined the degradation of two azo dyes at low and high frequency ultrasound to compare their reactivity and to assess the impacts of frequency, OH•, chemical structure, and soluble/nonsoluble additives. Low frequency ultrasound alone (20 kHz) was found to be totally ineffective for bleaching the dyes even after 2 h irradiation, while high frequency (861 kHz) provided significant color decay in 30 min contact. This study was a continuation of their previous work where they stated that azo dyes were water soluble and mostly hydrophilic. Hence, the degradation of dyes is mainly due to chemical oxidation by OH radicals in the bulk liquid (Ince and Tezcanli-Güyer, 2004). Reactive Black 5 was also used as a model dye by Ince and Tezcanlí (2001). The degradation was studied using individual sonolysis at 520 kHz, ozonation, and a combination of both methods. The degradation was monitored by the decay of absorbance at 596 nm, which was found to follow pseudo first-order kinetics. Under the same conditions and for identical contact times, color removal with ultrasound/O_3 was twice as fast as that with ozone alone, while no significant removal was observed in the control experiments with ultrasound alone. This finding was attributed to the relative shortness of the contact period. It is inferred that during simultaneous operation, the early oxidation reactions of the dye bleaching are governed by ozone and its decomposition products, whereas the destruction of the oxidation intermediates is achieved by the joint action of ultrasound and ozone.

Optimizing the degradation of Rhodamine B was carried (Sivakumar and Pandit, 2001) out using different ultrasonic equipment with varying ultrasonic power. It was found that the degradation rate increased with an increase in the power parameters up to a definite optimum value and then began to decrease with further increases in these power parameters. The occurrence of the optimum level can be explained by the formation of larger bubble clouds and the associated decoupling effect. The optimal values were not the same for all the ultrasonic equipment used in the experiments, which can be attributed to the difference in the acoustic field generated by each of them. It has been observed that for the optimization studies, it is better to consider power density (W mL^{-1}) parameter rather than power intensity (W cm^{-2}), because the former directly gives an idea about the amount of energy dissipated in the volume of the solution. The optimization method with power density and power intensity aims to indicate the energy requirements of the process and the effective mode of supply (intensity) or assists in proper utilization of energy for a given chemical reaction to avoid oversupply of energy.

The degradation of industrial azo dyes Acid Orange 5 and 52, Direct Blue 71, Reactive Black 5, Reactive Orange 16, and Reactive Orange 107 was performed by power ultrasound of 850 kHz at 60, 90, and 120 W (Rehorek et al., 2004). All investigated dyes have been decolorized and degraded within 3–15 h at 90 W and within 1–4 h at 120 W, respectively. The results also show that power ultrasound is able to mineralize azo dyes to nontoxic end products, confirmed by the respiratory inhibition test of *Pseudomonas putida*. The mass spectrometric results show that hydroxyl radicals attack azo dyes by simultaneous azo bond scission, oxidation of nitrogen atoms, and hydroxylation of aromatic ring structures.

Acid Orange 7 and Reactive Orange 16 were studied (Ince and TezcanliGüyer, 2004) at 300 kHz. The results showed that the decrease of absorbance at the visible band was much greater than in the UV band, which indicated the priority of OH radicals addition to the N=N or C–N bonds of the chromophore. Moreover, the bleaching of Acid Orange 7 was considerably faster than that of Reactive Orange 16, which has a higher molecular weight and more complex structure. The degradation was also pH-dependent. Acidic conditions accelerated the degradation by hydrophobic enrichment and of the molecules and the formation of excess $H_2O_2^-$, $•HO^{2-}$, and $•O^{2-}$.

TREATMENT OF AROMATIC HYDROCARBONS

Decontamination of chemicals such as polychlorinated biphenyls (PCBs) and other chlorine disinfection by-products (DBPs) found in water are currently of interest as more information is known about their adverse effect on the environment and health.

Ultrasound as one of the advanced oxidation processes has been used to treat chlorophenols and poly aromatic hydrocarbons (PAH) by many researchers, as described below.

The sonochemical degradation of polyaromatic hydrocarbons in aqueous solution has been studied (David, 2009) at two frequencies (20 and 506 kHz) and the greatest activity was observed at 506 kHz. It was suggested that this was due to sonodegradation occurring via pyrolysis of the hydrocarbons within the center of the cavitation bubbles themselves.

The degradation of polycyclic aromatic hydrocarbons in aqueous solutions has also been investigated (Psillakis et al., 2004). They investigated the effect of concentration, temperature, applied power, and ultrasound frequency on the sonochemical degradation of naphthalene, acenaphthylene, and phenanthrene. Using an ultrasonic horn at applied power of 45, 75, and 150 W and frequencies of 24 and 80 kHz, all the PAHs were susceptible to sonochemical treatment and, in most cases, complete degradation could be achieved in up to 120 min of treatment. Degradation was found to be reduced with increasing initial concentration and temperature and decreasing power and frequency as well as in the presence of an excess of dissolved salts. Addition of 1-butanol, a known hydroxyl radical scavenger, substantially suppressed degradation throughout the course of the reaction suggesting a radical attack as the main method of degradation.

The degradation of 4-chlorophenol in O_2-saturated solutions at different concentrations, temperatures, and frequencies has also been studied (Jiang et al., 2006). From their study, 500 μmol L^{-1} of 4-chlorophenol was completely destroyed after 300 min of sonication at 500 kHz with the power of 30 W and liquid temperature 20°C ± 1°C. The degradation reaction was pseudo-first-order reaction with a rate constant of 0.0012 ± 0.002 min^{-1}. Primary intermediates formed were hydroquinone and 4-chlorocatechol with final products of chloride ions, CO, CO_2, and HCO_2H. The intermediates indicated that hydroxyl radicals are involved in the degradation of 4-chlorophenol. The initial degradation rate increased with the increase of the concentration up to about 1000 μmol L^{-1}. They concluded that at high concentrations the degradation occurs predominantly at the liquid–gas bubble interface and at low concentrations, degradation occurs in the bulk solution.

The degradation of 4-chlorophenol at 516 kHz has been reported (Hamdaoui and Naffrechoux, 2008). A solution of 4-chlorophenol at a concentration of 0.78 mmol L^{-1} was completely destroyed after 70, 80, 100, and 130 min of sonication for the power 38.3, 31.1, 21.5, and 15.2 W, respectively. The primary intermediates were hydroquinone, 4-chlorocatechol and in a small quantity 4-chlororesorcinol. Final degradation products were formic acid (HCOOH), CO, CO_2, and chloride ions. The addition of *tert*-butanol as a radical scavenger found that the degradation was effectively quenched but not completely. As a result, they suggested that the main mechanism of 4-chlorophenol destruction is chemical oxidation by hydroxyl radicals.

The degradation of 2,3,5-trichlorophenol was studied (Tiehm and Neis, 2005) at 41, 206, 360, 618, 1068, and 3217 kHz. In their study, greatest decomposition was observed at 360 kHz. The degradation followed pseudo first-order rate with efficiency, decreasing in the presence of hydrophobic radical scavenger, *tert*-butanol as also observed by other authors.

Other workers (Peters, 2001) have investigated the degradation of volatile organic compounds (VOC) in natural ground water with a range of high frequency ultrasound of 206, 361, 620, and 1086 kHz. All VOC are almost completely destroyed within 60–120 min, and most after 30 min, and are within water requirement levels. There is minimal difference between the frequencies employed if the ultrasonic dose is kept the same; however, it must be noted that more power is needed for higher frequencies to keep the dose the same. The kinetics for degradation in ground water follows pseudo first-order behavior and the mechanism takes place inside the collapsing bubble by high temperature pyrolysis.

A sono-photochemical reactor was used (Gogate et al., 2002) for the degradation of formic acid as a model for organic oxidation in wastewater treatment. A combination of ultrasound at 20 kHz/30 kHz/50 kHz with ultraviolet light gave higher degradation of formic acid than either treatment individually. Sono-photochemical treatment was greatest when ultrasound was applied simultaneously rather that sequentially, with addition of catalysts of hydrogen peroxide or TiO_2 titanium oxide depending upon the concentration.

TREATMENT OF SEWAGE SLUDGE (MASON AND LORIMER, 2002; MASON, 2007)

Biological treatment produces large quantities of biomass or sludge. Although some of the sludge will be recycled in the treatment system, most of the excess will be removed and will continue onward to be treated further. Sludge will carry on degrading producing noxious gases and as a result it has to be stabilized before it is disposed of. This stabilization involves digestion to complete the breakdown process and dewatering to reduce its bulk. Ultrasound can improve both the digestion process by disintegrating the sludge and the dewatering process by facilitating the migration of moisture, and also enabling the filtration system to operate more efficiently and much longer periods without maintenance.

A review of ultrasound and its effect on sludge treatment has been undertaken by Pilli et al. (2010). They determined that sonication enhances sludge digestibility by disrupting the physical, chemical, and biological properties of the sludge. The degree of disintegration depends on the sonication parameters and also on sludge characteristics Some full-scale ultrasonic installations demonstrated that there is 50% increase in generation of biogas and an energy balance evaluation showed that the average ratio of the net energy gain to electric consumed by the ultrasound device is 2.5, indicating its viability for use on a large scale.

The chemical, physical, and biological properties of waste activated sludge have been monitored by treatment with ultrasound at 20 kHz (Chu et al., 2001), using different intensities and times, for sludge conditioning. Microbial flocs were disintegrated by ultrasound when a critical power was reached with particle size reducing from 100 µm to less than 10 µm at an intensity of 0.44 and 0.33 W mL^{-1}, with intensities lower than this smaller reduction were found.

Studies have also been performed (Xie et al., 2009) to investigate the use of ultrasound to improve the activity of anaerobic sludge. They utilized ultrasound at 35 kHz to sonicate the sludge and monitored the dehydrogenate activity and the amount of coenzyme F420 as a method of indicating the change of activity. They determined that the optimal ultrasonic intensity and irradiation period were 0.2 W cm^{-2} and 10 min, respectively, and that the biological activity was enhanced dramatically under the optimal conditions.

An assessment (Commenges-Bernole and Marguerie, 2009) of heavy metals fixation capacity on sonicated activated sludge determined that on sonication of 15 min and storage of three days after irradiation, the equilibrium capacity of the sludge increased by about 45%.

The effect of low power ultrasonic radiation on anaerobic biodegradability of sewage sludge has been reported (Liu et al., 2009). Soluble substances and microbial systems of sewage sludge were subjected to low power ultrasonic radiation and the optimal parameters were found to be an exposure time of 15 min with an ultrasonic intensity of 0.35 W cm^{-2} and ultrasonic power density of 0.25 W mL^{-1}. Under these conditions the anaerobic biodegradability of sewage sludge was increased by 67.6%.

SURFACE DECONTAMINATION

The primary cleaning action of sonication is the result of unsymmetrical collapse of cavitational bubbles near surfaces to produce liquid jets, which can remove bacteria and other contaminants from the surfaces. The objects to be cleaned can range from large crates used for food packaging and transportation to delicate surgical implements such as endoscopes. The advantage of using ultrasound is that it can reach crevices that are not easily accessed by conventional cleaning methods (Mason and Tiehm, 2001; Mason and Lorimer, 2002). This principal may also be beneficial for the cleaning of instruments for medical, surgical, and dental industries (Sierra and Boucher, 1971; Mason and Tiehm, 2001). The combination of ultrasound with heat and pressure treatments is also under investigation for the sterilization and/or pasteurization of food spoilage causing bacteria in the food industry (Ordonez et al., 1987; Raso et al., 1998a,b; Pagan et al., 1999a,b; Manas et al., 2000).

THERAPEUTIC ULTRASOUND

Cancer Treatment

The key to ultrasonic cancer treatment is the ability to focus low energy ultrasound beams to a small, high energy target in a tumor inside the body. The individual beams are harmless but the focus heats up and kills the cancerous cells in a procedure known as high-intensity focused ultrasound (HIFU). Patients lie over a small bath of water in which there are two concentric ultrasound transducers. One transmits a low-power diagnostic beam, allowing the doctor to visualize the tumor and guide the treatment; the other produces the focused beam.

HIFU is useful in treating a single tumor or part of a large tumor but cannot be used to treat widespread cancers. It is, therefore, not suitable for people with cancer that has spread to more than one place in their body. HIFU treatments are, therefore, most promising for the treatment of localized cancers such as prostrate (Chaussy and Thüroff, 2010; Uchida et al., 2010) liver (Ohto et al., 2009), pancreatic (Zaitsev et al., 1996), and breast (Ma et al., 2009). Further studies are investigating the possibility of utilizing HIFU for lithotripsy, i.e., the erosion and breakdown of kidney stones.

Other methods of employing sonication in cancer treatment include the enhancement of chemotherapeutic drugs. This might be achieved by employing acoustic waves to disrupt the blood–brain barrier (Kinoshita et al., 2006; Liu et al., 2010) or by using ultrasound to activate the release of drugs contained within a polymeric matrix (Husseini et al., 2000).

Transdermal Drug Delivery

There is much evidence to suggest that the application ultrasound to the skin (sonophoresis) can increase its permeability and thus enhance transdermal drug deliver (TDD). Studies using insulin (Tachibana, 1992), glucose, and mannitol (Merino et al., 2003) indicate that low frequency ultrasound is most effective and it has been concluded that cavitation is an important part of the mechanism. Other workers have shown that ultrasound can have a synergistic effect on other methods of TDD such as chemical (Tezel et al., 2002), iontophoresis (Le et al., 2000), and electroporation (Kost et al., 1996).

CONCLUSIONS

What conclusions can be drawn about future developments in this field? Undoubtedly, it is an expanding field of study that continues to thrive on outstanding laboratory results particularly in the fields of catalysis, nanoparticles, electrodeposition, and bioeffects. In the last few years, there has also been a major change in that many power ultrasonic equipment manufacturers have become involved in the construction of large scale ultrasonic apparatus for industry. There are now sonochemistry or sonoprocessing sessions at most major international conferences on acoustics and the impact factor of the Elsevier journal *Ultrasonics Sonochemistry*, the only journal devoted exclusively to sonochemistry, has increased year on year. Compared with the past, there is also far greater contact and cooperation between the scientific disciplines interested in the effects of cavitation which certainly indicates a very rosy future for the many and varied uses of ultrasound in materials technology.

REFERENCES

Abyaneh, M.Y., Sterritt, A., and Mason, T.J. 2007. Effects of ultrasonic irradiation on the kinetics of formation, structure, and hardness of electroless nickel deposits. *Journal of Electrochemical Society*, 154: D467–D472.

Ai, Z., Li, J., Zhang, L., and Lee, S. 2010. Rapid decolorization of azo dyes in aqueous solution by an ultrasound-assisted electrocatalytic oxidation process. *Ultrasonics Sonochemistry*, 17: 370–375.

Aqil, A., Serwas, H., Delplancke, J.L., Jerome, R., Jerome, C., and Canet, L. 2008. Preparation of stable suspensions of gold nanoparticles in water by sonoelectrochemistry. *Ultrasonics Sonochemistry*, 15: 1055–1061.

Athawale, A.A., Katre, P.P., Kumar, M., and Majumda, M.B. 2005. Synthesis of CTAB-IPA reduced copper nanoparticles. *Materials Chemistry and Physics*, 91: 507–512.

Atobe, M., Ishikawa, K., Asami, R., and Fuchigami, T. 2009. Size-controlled synthesis of conducting-polymer microspheres by pulsed sonoelectrochemical polymerization. *Angewandte Chemie International Edition*, 48: 6069–6072.

Atobe, M. and Nonaka, T. 1995. Ultrasonic effects on electroorganic processes. Electroreduction of benzaldehydes on ultrasound-vibrating electrodes. *Chemistry Letters*, 24: 669–670.

Barnes, C. and Ward, J.J.B. 1977. Use of ultrasonic agitation in gold plating for electronic applications. *Transactions of IMF*, 55(3): 101–103.

Baumgartner, C.E. 1989. Adhesion of electrolessly deposited nickel on lead zirconate titanate ceramic. *Journal of American Ceramic Society*, 72(6): 890–895.

Cass, P., Knower, W., Pereeia, E., Holmes, N.P., and Hughes, T. 2010. Preparation of hydrogels via ultrasonic polymerization. *Ultrasonics Sonochemistry*, 17: 326–332.

Chang, L.M., Guo, H.F., and An, M.-Z. 2008. Electrodeposition of Ni-Co/Al$_2$O$_3$ composite coating by pulse reverse method under ultrasonic condition. *Materials Letters*, 2(19): 3313–3315.

Chaussy, C. and Thüroff, S. 2010. High-intensity focused ultrasound in the management of prostate cancer. *Expert Review of Medical Devices*, 7(2): 209–217.

Chu, C.P., Chang, B.-V., Liao, G.S., Jean, D.S., and Lee, D.J. 2001. Observations on changes in ultrasonically treated waste-activated sludge. *Water Research*, 35(4): 1038–1046.

Cobley, A.J. 2007. Alternative surface modification processes in metal finishing and electronic manufacturing industries. *Transactions of IMF*, 85(6): 293–297.

Cobley, A.J. and Mason, T.J. 2007. The evaluation of sonochemical techniques for sustainable surface modification in electronic manufacturing. *Circuit World*, 33(3): 29–34.

Cobley, A.J. and Mason, T.J. 2008a. The sonochemical surface modification of materials for electronic manufacturing. The effect of ultrasonic source to sample distance. *Circuit World*, 34(3): 18–22.

Cobley, A.J. and Mason, T.J. 2008b. Sonochemistry—A sound approach to surface modification, in *Proceedings—2008 2nd Electronics Systemintegration Technology Conference, ESTC*, Greenwich, art. no. 4684433, pp. 685–690.

Cobley, A.J., Mason, T.J., Graves, J.G., and Morgan, D. 2010. New evidence for the inverse dependence of mechanical and chemical effects on the frequency of ultrasound. *Ultrasonics Sonochemistry*, 18: 226–230.

Commenges-Bernole, N. and Marguerie, J. 2009. Adsorption of heavy metals on sonicated activated sludge. *Ultrasonics Sonochemistry*, 16: 83–87.

Compton, R.G., Eklund, J.C., and Marken, F. 1997a. Sonoelectrochemical processes. A review. *Electroanalysis*, 9: 509–522.

Compton, R.G., Eklund, J.C., Marken, F., Rebbitt, T.O., Akkermans, R.P., and Walle, D.N. 1997b. Dual activation: Coupling ultrasound electrochemistry—An overview. *Electrochimica Acta*, 42: 2919–2927.

Compton, R.G., Eklund, J.C., Marken, F., and Waller, D.N. 1996. Electrode processes at the surfaces of sonotrodes. *Electrochimica Acta*, 41: 315–320.

Dabala, M., Pollet, B.G., Zin, V., Campadello, E., and Mason, T.J. 2008. Sonoelectrochemical (20 kHz) production of Co65Fe35 alloy nanoparticles from Aotani solutions. *Journal of Applied Electrochemistry*, 38: 395–402.

David, B. 2009. Sonochemical degradation of PAH in aqueous solution. Part I: Monocomponent PAH solution. *Ultrasonics Sonochemistry*, 16: 260–265.

Delplancke, J.L., Dille, J., Reisse, J., Long, G.J., Mohan, A., and Grandjean, F. 2000. Magnetic nanopowders: Ultrasound assisted electrochemical preparation and properties. *Chemistry of Materials*, 12: 946–955.

Depan, D., Saikia, L., and Singh, R.P. 2010. Ultrasound-triggered release of ibuprofen from a chitosan-mesoporous silica composite—Novel approach for controlled drug release. *Macromolecular Symposia*, 287: 80–88.

Drake, M.P. 1980. Electrodeposition in ultrasonic fields. *Transactions of IMF*, 58(2): 67–71.

Durant, A., Delplancke, J.L., Winand, R., and Reisse, J.A. 1995. A new procedure for the production of highly reactive metal powders by pulsed sonoelectrochemical reduction. *Tetrohedron Lett.*, 36: 4257–4260.

Eren, Z. and Ince, N.H. 2010. Sonolytic and sonocatalytic degradation of azo dyes by low and high frequency ultrasound. *Journal of Hazardous Materials*, 177: 1019–1024.

Ganesan, R., Shanmugam, S., and Gedanken, A. 2008. Pulsed sonoelectrochemical synthesis of polyaniline nanoparticles and their capacitance properties. *Synthetic Metals*, 158: 848–853.

Gogate, P.R., Mujumdar, S., and Pandt, A.B. 2002. A sonophotochemical reactor for the removal of formic acid from wastewater. *Industrial and Engineering Chemistry Research*, 41(14): 3370–3378.

Gopinath, K.P., Muthukumar, K., and Velan, M. 2010. Sonochemical degradation of Congo red: Optimization through response surface methodology. *Chemical Engineering Journal*, 157: 427–433.

Haas, I. and Gedanken, A. 2008. Synthesis of metallic magnesium nanoparticles by sonoelectrochemistry. *Chemical Communications*, 15: 1795–1797.

Haas, I., Shanmugam, S., and Gedanken, A. 2006. Pulsed sonoelectrochemical synthesis of size-controlled copper nanoparticles stabilized by poly(N-vinylpyrrolidone). *Journal of Physical Chemistry B*, 110: 16947–16952.

Haas, I., Shanmugam, S., and Gedanken, A. 2008. Synthesis of copper dendrite nanostructures by a sonoelectrochemical method. *Chemistry: A European Journal*, 14: 4696–4703.

Hamdaoui, O. and Naffrechoux, E. 2008. Sonochemical and photosonochemical degradation of 4-chlorophenol in aqueous media. *Ultrasonics Sonochemistry*, 15(6): 981–987.

Haram, S.K., Mahadeshwar, A.R., and Dixit, S.G. 1996. Synthesis and characterization of copper sulfide nanoparticles in Triton-X 100 water-in-oil microemulsions. *Journal of Physical Chemistry*, 100: 5868–5873.

Hoffmann, M.R., Martin, S., Choi, W., and Bahnemann, D.W. 1995. Environmental applications of semiconductor photocatalysis. *Chemical Reviews*, 95: 69–96.

Husseini, G.A., Myrup, G.D., Pitt, W.G., Christensen, D.A., and Rapoport, N.Y. 2000. Factors affecting acoustically triggered release of drugs from polymeric micelles. *Journal of Controlled Release*, 69: 43–52.

Ince, N.H. and Tezcanlí, G. 2001. Reactive dyestuff degradation by combined sonolysis and ozonation. *Dyes and Pigments*, 49(3): 145–153.

Ince, N.H. and Tezcanlí-Güyer, G. 2004. Impacts of pH and molecular structure on ultrasonic degradation of azo dyes. *Ultrasonics* 42(1–9): 591–596.

Jevtic, M., Radulovic, A., Ignjatovic, N., Mitric, M., and Uskokovic, D. 2009. Controlled assembly of poly(D,L-lactide-co-glycolide)/hydroxyapatite core-shell nanospheres under ultrasonic irradiation. *Acta Biomaterialia*, 5: 208–218.

Jia, F., Hu, Y., Tang, Y., and Zhang, L. 2007. A general nonaqueous sonoelectrochemical approach to nanoporous Zn and Ni particles. *Powder Technology*, 176: 130–136.

Jiang, J., Lu, H., Zhang, L., and Xu, N. 2007. Preparation of monodisperse Ni/PS spheres and hollow nickel spheres by ultrasonic electroless plating. *Surface and Coatings Technology*, 201: 7174–7179.

Jiang, L.P., Wang, A.N., Zhao, Y., Zhang, J.R., and Zhu, J.J.A. 2004. A novel route for the preparation of monodisperse silver nanoparticles via a pulsed sonoelectrochemical technique. *Inorg. Chem. Commun.* 7: 506–509.

Jiang, Y., Petrier, C., and Waite, T.D. 2006. Sonolysis of 4-chlorophenol in aqueous solution: Effects of substrate concentration, aqueous temperature and ultrasonic frequency. *Ultrasonics Sonochemistry*, 9(6): 317–323.

Kang, J., Chen, J., Cao, Y., and Li, H. 2010. Effects of ultrasound on the conformation and crystallization behaviour of isotactic polypropylene and b-isotactic polypropylene. *Polymer*, 51: 249–256.

Kaufmann, J., Desmulliez, M.P.Y., Tian, Y., Price, D., Hughes, M., Strussevich, N., Bailey, C., Liu, C., and Hutt, D. 2009. Megasonic agitation for enhanced electrodeposition of copper. *Microsystem Technology*, 15: 1245–1254.

Kim, Y.S., Kim, H.I., Cho, J.H., Seo, H.K., Dar, M.A., Shin, H.S., Ten Eyck, G.A., Lu, T.M., and Senkevich, J.J. 2006a. Electroless copper on refractory and noble metal substrates with an ultra-thin plasma-assisted atomic layer deposited palladium layer. *Electrochimica Acta*, 51: 2400–2406.

Kim, Y.H., Lee, D.K., Jo, B.G., Jeong, J.H., and Kang, Y.S. 2006b. Synthesis of oleate capped Cu nanoparticles by thermal decomposition. *Colloids and Surfaces A*, 284–285: 364–368.

Kinoshita, M., McDannold, N., Jolesz, F.A., and Hynynen, K. 2006. Noninvasive localized delivery of Herceptin to the mouse brain by MRI-guided focused ultrasound-induced blood-brain barrier disruption. *Proceedings of the National Academy of Sciences USA*, 103(31): 11719–11723.

Kobayashi, K., Chiba, A., and Minami, N. 2000. Effects of ultrasound on both electrolytic and electroless nickel depositions. *Ultrasonics*, 38: 676–681.

Kost, J., Pliquett, U., Mitragotri, S., Yamamoto, A., Langer, R., and Weaver, J. 1996. Synergistic effect of electric field and ultrasound on transdermal transport. *Pharmaceutical Research*, 13: 633–638.

Kou, S.-C. and Hung, A. 1999. Plating of small blind vias. *IEEE Transactions of Electronic, Packaging and Manufacturing Society*, 22: 202–208.

Kowalski, D., Ueda, M., and Ohtsuka, T. 2008. The effect of ultrasonic irradiation during electropolymerization of polypyrrole on corrosion prevention of the coated steel. *Corrosion Science*, 50: 286–291.

Kreisel, R. and Dudik, R. 1987. New process forces a solution. *Circuits Manufacturing*, 27(6): 18–20.

Kuzub, V.S. and Mukhlya, S.Y.Z. 1963. Nickel plating by chemical reduction in an ultrasonic field. *Priklad. Khim.*, 36: 2762–2764.

Le, L., Kost, J., and Mitragotri, S. 2000. Combined effect of low frequency ultrasound and iontophoresis: Applications for transdermal heparin delivery. *Pharmaceutical Research*, 17: 1151–1154.

Lei, H., Tang, Y.J., Wei, J.J., Li, J., Li, X.B., and Shi, H.L. 2007. Synthesis of tungsten nanoparticles by sonoelectrochemistry. *Ultrasonics Sonochemistry*, 14: 81–83.

Li, J.-T. and Song, Y.L. 2010. Degradation of AR 97 aqueous solution by combination of ultrasound and Fenton reagent. *Environmental Progress and Sustainable Energy* 29(1). doi 10.1002, April 2010.

Lindley, J. and Mason, T.J. 1987. Sonochemistry. Part 2—Synthetic applications. *Chemical Society Reviews*, 16: 275–311.

Liu, H.-L., Hua, M.-Y., Chen, P.-Y., Huang, C.-Y., Wang, J.-J., and Wei, K.-C. 2010. Focused ultrasound induced blood–brain barrier disruption to enhance chemotherapeutic drugs (BCNU) delivery for glioblastoma treatment. *AIP Conference Proceedings*, 1215: 176–181.

Liu, S., Huang, W., Chen, S., Avivi, S., and Gedanken, A. 2001. Synthesis of x-ray amorphous silver nanoparticles by the pulse sonoelectrochemical method. *Journal of Non-Crystalline Solids*, 283: 231–236.

Liu, C., Xiao, B., Dauta, A., Peng, G., Liu, S., and Hu, Z. 2009. Effect of low power ultrasonic radiation on anaerobic biodegradability of sewage sludge. *Bioresource Technology*, 24: 6217–6222.

Lorimer, J.P. and Mason, T.J. 1987. Sonochemistry. Part 1—The physical aspects. *Chemical Society Reviews*, 16: 239–274.

Lorimer, P. and Mason, T.J. 1999. The applications of ultrasound in electroplating. *Electrochemistry*, 67: 924–930.

Lu, Y. 2010. Improvement of copper plating adhesion on silane modified PET film by ultrasonic-assisted electroless deposition. *Applied Surface Science*, 256(11): 3554–3558.

Ma, C.C.-M., Chen, L., Freedman, G., and Bleicher, R. 2009. MR guided focused ultrasound for high risk and recurrent breast cancer. *IFMBE Proceedings*, 25(6): 140–143.

Mahendiran, C., Ganesan, R., and Gedanken, A. 2009. Sonoelectrochemical synthesis of metallic aluminum nanoparticles. *European Journal of Inorganic Chemistry*, 2009: 2050–2053.

Mallory, G.O. 1978. The effects of ultrasonic irradiation on electroless nickel plating. *Transactions of IMF*, 56: 81–86.

Mallory, G.O. 1985. Electroless nickel deposition in an ultrasonic field. *Plating and Surface Finishing*, 72: 64–68.

Manas, P., Pagan, R., and Raso, J. 2000. Predicting lethal effects of ultrasonic waves under pressure treatments on listeria monocytogenes ATCC 15313 by power measurements. *Journal of Food Science*, 65(4): 663–667.

Mancier, V., Daltin, A.L., and Leclercq, D. 2008. Synthesis and characterization of copper oxide (I) nanoparticles produced by pulsed sonoelectrochemistry. *Ultrasonics Sonochemistry*, 15: 157–163.

Mancier, V., Delplancke, J.L., Delwiche, J., Hubin-Franskin, M.J., Piquer, C., Rebbouh, L., and Grandjean, F. 2004. Morphologic, magnetic, and Mössbauer spectral properties of $Fe_{75}Co_{25}$ nanoparticles prepared by ultrasound-assisted electrochemistry. *Journal of Magnetism and Magnetic Materials*, 281: 27–35.

Mason, T.J. 2007. Developments in ultrasound-non-medical. *Progress Biophysics and Molecular Biology*, 93: 166–175.

Mason, T.J. and Lorimer, J.P. 2002. *Applied Sonochemistry: Uses of Power Ultrasound in Chemistry and Processing*, pp. 152–153. Weinheim, Germany: Wiley-VCH GmbH.

Mason, T.J., Lorimer, J.P., Saleem, S., and Paniwnyk, L. 2001. Controlling emissions from electroplating by the application of ultrasound. *Environmental Science and Technology*, 35: 3375–3377.

Mason, T.J. and Tiehm, A. 2001. Ultrasound in environmental protection. In *Advances in Sonochemistry*, Vol. 6, eds. T.J. Mason and A. Tiehm. Amsterdam, the Netherlands: Elsevier Science B.V.

Mastai, Y., Homyonfer, M., Gedanke, A., and Hodes, G. 1999. Room temperature sonoelectrochemical synthesis of molybdenum sulfide fullerene-like nanoparticles. *Advanced Materials*, 11: 1010–1013.

Matsuoka, M. and Hayashi, T. 1986. The influence of ultrasonic radiation on chemical nickel-phosphorus plating. *Metal Finishing*, 84: 27–31.

Matsushima, T., Habaki, H., and Kawasaki, J. 2000. Electroless copper deposition in a blind via hole of printed wiring board. Uniform copper deposition by axial temperature gradient. *Journal of Electrochemistry*, 68: 568–571.

Merino, G., Kalia, Y.N., Delgado-Charro, M.B., Potts, R.O., and Guy, R.H 2003. Frequency and thermal effects on the enhancement of transdermal transport by sonophoresis. *Journal of Controlled Release*, 88: 85–94.

Merouani, S., Hamdaoui, O., Saoudi, F., Chiha, M., and Pétrier, C. 2010a. Influence of bicarbonate and carbonate ions on sonochemical degradation of Rhodamine B in aqueous phase. *Journal of Hazardous Materials*, 175: 593–599.

Merouani, S., Hamdaoui, O., Saoudi, F., and Chiha, M. 2010b. Sonochemical degradation of Rhodamine B in aqueous phase: Effects of additives. *Chemical Engineering Journal*, 158: 550–557.

Naffrechoux, E., Chanoux, S., Petrier, C., and Suptil, J. 2000. Sonochemical and photochemical oxidation of organic matter. *Ultrasonics Sonochemistry*, 7: 255–259.

Ohto, M., Fukuda, H., Ito, R., Shinohara, Y., Sakamoto, A., and Karasawa, E. 2009. Contrast-enhanced three dimensional ultrasonography supporting HIFU treatment of small liver cancer. *AIP Conference Proceedings*, 1113: 86–90.

Onat, T.A., Gümüşdere, H.T., Güvenç, A., Dönmez, G., and Mehmetoğlu, A. 2010. Decolorization of textile azo dyes by ultrasonication and microbial removal. *Desalination*, 255(103): 154–158.

Ordonez, J.A., Aguilera, M., Garcia, M.L., and Sanz, B. 1987. Effect of combined ultrasonic and heat treatment (thermoultraonication) on the survival of a strain of *Staphyloccus aureus*. *Journal of Dairy Research*, 54: 61–67.

Pagan, R., Manas, P., Alvarez, I., and Condon, S. 1999a. Resistance of listeria monocytogenes to ultrasonic waves under pressure at sublethal (manosonication) and tethal (manothermosonication) temperatures. *Food Microbiology*, 16: 139–148.

Pagan, R., Manas, P., Palop, A., and Sala, F.J. 1999b. Resistance of heat-shocked cells of *Listeria monocytogenes* to mano-sonication and mano-thermo-sonication. *Letters in Applied Microbiology*, 28: 71–75.

Park, Y.S., Kim, T.H., Lee, M.H., and Kwon, S.C. 2002. Study on the effect of ultrasonic waves on the characteristics of electroless nickel deposits from an acid bath. *Surfaces and Coatings Technology*, 153: 245–251.

Park, K.C., Mahiko, T., Morimoto, S., Takeuchi, K., and Morinobu, E. 2008. Hydrophobicity-induced selective covering of carbon nanotubes with sol-gel sheaths achieved by ultrasound assistance. *Applied Surfaces Science*, 254: 7438–7445.

Peng, Y. and Chen, Q. 2009. Ultrasonic assisted fabrication of highly dispersed copper/multi-walled carbon nanotube nanowires. *Colloids and Surfaces A*, 342: 132–135.

Peters, D. 2001.Sonolytic degradation of volatile pollutants in natural ground water: Conclusions from a model study. *Ultrasonics Sonochemistry*, 8: 221–226.

Pilli, S., Bhunia, P., Yan, S., LeBlanc, R.J., Tyagi, R.D., and Surampalli, R.Y. 2010. Ultrasonic pretreatment of sludge: A review. *Ultrasonics Sonochemistry*, 18(1): 1–18.

Ponce, A.A. and Klabunde, K.J. 2005. Chemical and catalytic activity of copper nanoparticles prepared via metal vapor synthesis. *Journal of Molecular Catalysis A Chemistry*, 225: 1–6.

Prasad, P.B.S.N.V., Ahila, S., Vasudevan, R., and Seshari, S.K.J. 1994. Fatigue strength of nickel electrodeposits prepared in ultrasonically agitated bath. *Journal of Materials Science Letters*, 13: 15–16.

Price, G.J., Clifton, A.A., and Keen, F. 1996a. Ultrasonically enhanced persulfate oxidation of polyethylene surfaces. *Polymer*, 37(26): 5825–5829.

Price, G.J., Keene, F., and Clifton, A.A. 1996b. Sonochemically-assisted modification of polyethylene surfaces. *Macromolecules*, 29(17): 5664–5670.

Price, G.J., Mathhias, P., and Lenz, E.J. 1994. The use of high power ultrasound for the destruction of aromatic compounds in aqueous solution. *Transactions of the Institution of Chemical Engineers*, 72(B): 27–31.

Psillakis, E., Goula, G., Kalogerakis, N., and Mantzavinos, D. 2004. Degradation of polycyclic aromatic hydrocarbons in aqueous solutions by ultrasonic irradiation. *Journal of Hazardous Materials*, B108: 95–102.

Qiu, X., Burda, C., Fu, R., Pu, L., Chen, H., and Zhu, J. 2004. Heterostructured Bi_2Se_3 nanowires with periodic phase boundaries. *Journal of American Chemical Society*, 126: 16276–16277.

Qiu, X., Lou, Y., Samia, A.C.S., Devadoss, A., Burgess, J.D., Dayal, S., and Burda, C. 2005. PbTe nanorods by sonoelectrochemistry. *Angewandte Chemie International Edition*, 44: 5855–5857.

Qiu, X.F., Xu, J.Z., Zhu, J.M., Zhu, J.J., Xu, S., and Chen, H.Y. 2003. Controllable synthesis of palladium nanoparticles via a simple sonoelectrochemical method. *Journal of Materials Research*, 18: 1399–1404.

Rao, C.N.R., Muller, A., and Cheetan, A.K. 2008. *The Chemistry of Nanomaterials Synthesis, Properties an Applications*, Vol. 1, p. 151. Weinheim, Germany: Wiley-VCH Verlag GmbH & Co.

Raso, J., Pagan, R., Condon, S., and Sala, F. 1998a. Influence of temperature and pressure on the lethality of ultrasound. *Applied and Environmental Microbiology*, 64(2): 465–471.

Raso, J., Palop, A., Pagan, R., and Condon, S. 1998b. Inactivation of *Bacillus subtilis* spores by combining ultrasonic waves under pressure and mild heat treatment. *Journal of Applied Microbiology*, 85: 849–854.

Rehorek, A., Tauber, M., and Gübitz, G. 2004. Application of power ultrasound for azo dye degradation. *Ultrasonics Sonochemistry*, 11: 177–182.

Reisse, J., Caulier, T., Deckerkheer, C., Fabre, O., Vandercammen, J., Delplancke, J.L., and Winand, R. 1996. Quantitative sonochemistry. *Ultrasonics Sonochemistry*, 3: S147–S151.

Reisse, J., Francois, H., Vandercammen, J., Fabre, O., Kirsch-de Mesmaeker, A., Maerschalk, C., and Delplancke, J.L. 1994. Sonoelectrochemistry in aqueous electrolyte: A new type of sonoelectroreactor. *Electrochimica Acta*, 39: 37–39.

Rezrazi, M., Doche, M.L., Bercot, P., and Hihn, J.Y. 2005. Au-PTFE composite coatings elaborated under ultrasonic stirring. *Surfaces and Coatings Technology*, 192(1): 124–130.

Richards, W.T. and Loomis, A.L. 1927. The chemical effects of high frequency sound waves 1. A preliminary survey. *Journal of the American Chemical Society*, 49(12): 3086–3100.

Rostamizadeh, S., Amani, A.M., Mahdavinia, G.H., Amiri, G., and Sepehrian, H. 2010. Ultrasound promoted rapid and green synthesis of 1,8-dioxo-octahydroxanthenes derivatives using nanosized MCM-41-SO_3H as a nanoreactor, nanocatalyst in aqueous media. *Ultrasonics Sonochemistry*, 17: 306–309.

Rupprecht, L.1999. *Conductive Polymers and Plastics in Industrial Applications*, New York: William Andrew Publishing/Plastics Design Library.

Saez, V. and Mason, T.J. 2009. Sonoelectrochemical synthesis of nanoparticles. *Molecules*, 14: 4284–4299.

Schlafer, O.S., Sievers, M., Klotzbucher, H., and Onyeche, T.I. 2000. Improvement of biological activity by low energy assisted bioreactors. *Ultrasonics*, 38: 711–716.

Serpone, N. and Khairutdinov, R.F. 1996. Semiconductor nanoclusters. In: *Studies in Surface Science and Catalysis*, eds. P.V. Kamat and D. Meisel, p. 417. New York: Elsevier Science.

Shen, Q., Jiang, L., Miao, J., Hou, W., and Zhu, J.J. 2008b. Sonoelectrochemical synthesis of CdSe nanotubes. *Chemical Communications*, 14: 1683–1685.

Shen, Q., Jiang, L., Zhang, H., Min, Q., Hou, W., and Zhu, J.J. 2008a. Three-dimensional dendritic Pt nanostructures: Sonoelectrochemical synthesis and electrochemical applications. *Journal of Physical Chemistry C*, 112: 16385–16392.

Sierra, G. and Boucher, R.M.G. 1971. Ultrasonic synergistic effects in liquid-phase chemical sterilization. *Applied Microbiology*, 22: 160–164.

Sivakumar, M. and Pandit, A.B. 2001. Ultrasound enhanced degradation of Rhodamine B: Optimization with power density. *Ultrasonics Sonochemistry*, 8(3): 233–240.

Socol, Y., Abramson, O., Gedanken, A., Meshorer, Y., Berenstein, L., and Zaban, A. 2002. Suspensive electrode formation in pulsed sonoelectrochemical synthesis of silver nanoparticles. *Langmuir*, 18: 4736–4740.

Tachibana, K. 1992. Transdermal delivery of insulin to alloxan-diabetic rabbits by ultrasound exposure. *Pharmaceutical Research*, 9: 952–954.

Tezel, A., Sens, A., Tuchscherer, J., and Mitragotri, S. 2002. Synergistic effect of low frequency ultrasound and surfactants on skin permeability. *Journal of Pharmaceutical Science*, 91: 91–100.

Tiehm, A. and Neis, U. 2005. Ultrasonic dehalogenation and toxicity reduction of trichlorophenol. *Ultrasonics Sonochemistry*, 12(1–2): 121–125.

Torres-Sanchez, C. and Corney, J.R. 2008. Effects of ultrasound on polymeric foam porosity. *Ultrasonics Sonochemistry*, 15: 408–415.

Touyeras, F., Hihn, J.Y., Bourgoin, X., Jacques, B., Hallez, L., and Branger, V. 2005. Effects of ultrasonic irradiation on the properties of coatings obtained by electroless plating and electro plating. *Ultrasonics Sonochemistry*, 12: 13–19.

Touyeras, F., Hihn, J.Y., Delalande, S., Viennet, R., and Doche, M.L. 2003. Ultrasound influence on the activation step before electroless coating. *Ultrasonics Sonochemistry*, 10: 363–368.

Touyeras, F., Hihn, J.Y., Doche, M.L., and Roizard, X. 2001. Electroless copper coating of epoxide plates in an ultrasonic field. *Ultrasonics Sonochemistry*, 8: 285–290.

Uchida, T., Nakano, M., Shoj, S., Omata, T., Harano, Y., Nagata, Y., Usui, Y., and Terachi, T. 2010. Ten-year biochemical disease-free survival after high-intensity focused ultrasound (HIFU) for localized prostate cancer: Comparison with four different generation devices. *AIP Conference Proceedings*, 1215: 216–219.

Walker, R. 1997. Ultrasound improves electrolytic recovery of metals. *Ultrasonics Sonochemistry*, 4: 39–43.

Walker, R. and Benn, R.C. 1971. Microhardness, grain size and topography of copper electrodeposits. *Plating*, 58(5): 476–481.

Wang, N., Zhu, L., Wang, D., Wang, M., Lin, Z., and Tang, H. 2010. Sono-assisted preparation of highly-efficient peroxidase-like Fe_3O_4 magnetic nanoparticles for catalytic removal of organic pollutants with H_2O_2. *Ultrasonics Sonochemistry*, 17(3): 526–533.

Wei, T.Y., Wang, Y.Y., and Wan, C.C. 1979. Effect of ultrasonic agitation on silver plating. *Plating and Surfaces Finishing*, 66(3): 47–50.

Wonga, S.-S., Kasapis, S., and Tan, Y.M. 2009. Bacterial and plant cellulose modification using ultrasound irradiation. *Carbohydrate Polymers*, 77: 280–287.

Wood, R.W. and Loomis, A.L. 1927. Physical and biological effects of high frequency sound waves of great intensity. *Philosophical Magazine*, 4: 414.

Wu, Z., Ge, S., Zhang, M., Li, W., and Tao, K. 2009. Synthesis of nickel nanoparticles supported on metal oxides using electroless plating: Controlling the dispersion and size of nickel nanoparticles. *Journal of Colloid and Interface Science*, 330: 359–366.

Xia, F.-F., Wu, M.-H., Wang, F., Jia, Z.-Y. and Wang, A.-I. 2009. Nanocomposite Ni-TiN coatings prepared by ultrasonic electrodeposition. *Current Applied Physics*, 9(1): 44–47.

Xie, B., Liu, H., and Yan, Y. 2009. Improvement of the activity of anaerobic sludge by low-intensity ultrasound. *Journal of Environmental Management*, 90: 260–264.

Yang, L.-X., Hou, W.-T., and Wu, Y.-S. 1997. Influence of ultrasound on the properties of electroless plated nickel. *Transactions of IMF*, 75: 131–133.

Yeh, M.S., Yang, Y.S., Lee, Y.P., Lee, H.F., Yeh, Y.H., and Yeh, C.S. 1999. Formation and characteristics of Cu colloids from CuO powder by laser irradiation in 2-propanol. *Journal of Physical Chemistry, B*, 103: 6851–6857.

Young, W.T. and Kersten, H. 1936. An effect of ultrasonic radiation on electrodeposits *Journal of Chemical Physics*, 4: 426.

Zaitsev, A.V., Sanghvi, N.T., Ikenberry, S., Worzalla, J.F., Schultz, R.M., and Self, T.D. 1996. High intensity focused ultrasound (HIFU) treatment of human pancreatic cancer. *Proceedings of the IEEE Ultrasonic Symposium*, 2: 1295–1298.

Zhang, Z. and Zheng, H. 2009. Optimization for decolorization of azo dye acid green 20 by ultrasound and H_2O_2 using response surface methodology. *Journal of Hazardous Materials*, 72: 1388–1393.

Zhao, Y., Bao, C., Feng, R., and Chen, Z. 1995. Electroless coating of copper on ceramic in an ultrasonic field. *Ultrasonics Sonochemistry*, 2: S99–S103.

Zhao, Y., Bao, C., Feng, R., and Li, R. 1998a. A new method of etching ABS plastic for plating by ultrasound. *Plating and Surface Finishing*, 85(9): 98–100.

Zhao, Y., Bao, C., Feng, R., and Mason, T.J. 1998b. New etching method of PVC plastic for plating by ultrasound. *Journal of Applied Polymer Science*, 68(9): 1411–1416.

Zheng, H.-Y., and An, M.-Z. 2008. Electrodeposition of Zn-Ni-Al_2O_3 nanocomposite coatings under ultrasound conditions. *Journal of Alloys and Compounds*, 459(1–2): 548–552.

Zhu, J., Aruna, S.T., Koltypin, Y., and Gedanken, A. 2000b. A novel method for the preparation of lead selenide: Pulse sonoelectrochemical synthesis of lead selenide nanoparticles. *Chemistry of Materials*, 12: 143–147.

Zhu, J., Liu, S., Palchik, O., Koltypin, Y., and Gedanken, A. 2000a. Shape-controlled synthesis of silver nanoparticles by pulse sonoelectrochemical methods. *Langmuir*, 16: 6396–6399.

Zin, V., Pollet, B.G., and Dabala, M. 2009. Sonoelectrochemical (20 kHz) production of platinum nanoparticles from aqueous solution. *Electrochimica Acta*, 54: 7201–7206.

4 Ultrasound-Assisted Particle Engineering

Anant Paradkar and Ravindra Dhumal

CONTENTS

Introduction 75
Ultrasound-Assisted Particle Engineering Techniques 76
    Crystallization from Solution 76
        Sonocrystallization 76
        Solution Atomization and Crystallization by Sonication 77
    Particle Engineering from Slurry/Suspension 78
        Ultrasound-Mediated Amorphous to Crystalline Transition 78
        Sonicslurry® and Particle Rounding Technology 79
    Particle Engineering from Melt 79
        Melt Sonocrystallization 79
        Ultrasound-Assisted Polymer Extrusion 80
Applications in Pharmaceutical Particle Engineering 81
    Application in Inhalation Drug Delivery 81
    Application in Design of Particles with Enhanced Dissolution Rate 89
Scale-Up 93
Limitations 93
Conclusions 94
References 95

INTRODUCTION

Particle engineering in pharmaceuticals has evolved from conventional granulation and milling techniques to advanced techniques like supercritical fluid technology, jet milling, spray drying, high pressure homogenization, spray congealing, extrusion–spheronization, spherical crystallization, and crystallo-co-agglomeration. This engineering of solids has been aimed to alter and improve the primary physical properties of solids such as particle size, shape, crystal habit, crystal form, density, porosity, etc. as well as secondary properties like flowability, compressibility, compactibility, consolidation, dust generation, and air entrapment during processing. In the pharmaceutical industry, efficient production of small and/or substantially uniform particles is desired not only due to its impact on performance during processing and storage but also due to its impact after consumption by patients. Micronization of drugs is routinely carried out for increasing dissolution rate. Currently, nanosized particles are explored due to advancement in techniques available for nanoparticle formation. For inhalation drug delivery system, the particle size, shape, and surface properties determine its interaction with the container and amount of dose deposited in lungs, which is generally the desired site of action.

    The advances in the drug delivery system demand special characteristics of drug particles to suite appropriate dosage form and route of administration for optimum efficiency of drug molecules. This has encouraged pharmaceutical scientist to explore and adapt new and diverse processing technologies. Ultrasound (US)-assisted particle engineering is being explored for producing particles with altered size, crystallinity, morphology, and surface properties. The first application of US to crystallization predates by decades (Richards and Loomis, 1927), followed by considerable literature in late twentieth century, dealing with small-scale applications (Kapustin, 1963; Reshetnyak, 1975). While the concept of ultrasonic processing is not new, the ability to use it on an industrial scale is. Interest in the application of US to crystallization in the pharmaceutical and fine chemical sector has received further impetus in recent years with the advances in equipment (McCausland et al., 2001). Ultrasonic technology offers a clean and efficient tool to improve traditional existing processes and provide alternative for innovation.

ULTRASOUND-ASSISTED PARTICLE ENGINEERING TECHNIQUES

US treatment influences particle formation through cavitation and acoustic streaming (Young, 1989). Cavitation appears to be particularly effective as a means of inducing nucleation during crystallization and there is evidence of dramatic improvements in reproducibility obtained through such sononucleation. Crystal growth rate and pattern is also known to modify with application of US, which influences the properties of the particle so formed. US is not only utilized to modify particle formation process but also employed to modify already formed particles. These techniques may be classified as

1. Crystallization from solution
    a. Sonocrystallization
    b. Solution atomization and crystallization by sonication (SAXS)
2. Particle engineering from slurry/suspension
    a. US-mediated amorphous to crystalline transition (UMAX®)
    b. Sonic slurry and particle rounding technology (PRT)
3. Particle engineering from melt
    a. Melt Sonocrystallization
    b. US-assisted melt extrusion

CRYSTALLIZATION FROM SOLUTION

Sonocrystallization

Crystallization is one of the oldest unit operation and particle engineering tool used at some stage or the other in nearly all process industries as a method of production, purification, or recovery of solid materials. Crystallization is considered to progress through achievement of supersaturation, nucleation or production of microscopic crystals, and subsequent crystal growth. A solute will remain in solution until a sufficiently high level of supersaturation has been developed to induce spontaneous nucleation. The extent of this supersaturation is referred to as the "metastable zone width." Nucleation is classified as primary and secondary nucleation. "Primary nucleation" occurs when a crystal is nucleated in a solution containing no preexisting crystals, while nucleation induced by preexisting crystals is called "secondary nucleation," which results from either fragmentation of crystals or these crystals acting as templates for new crystals. When primary nucleation occurs in, the bulk of liquid in the absence of solid surface is called "homogeneous nucleation" and when a solid interface, whether a container wall or a preexisting crystal is involved, the process is called "heterogeneous nucleation." As soon as stable nuclei have been formed in a supersaturated or supercooled system, they begin to grow into crystals of visible size. A period of time usually elapses between achievement of supersaturation and the appearance of crystals. This lag time is called "induction time or period." This is influenced by the level of supersaturation, state of agitation, presence of impurities, and viscosity.

Though crystallization can be achieved by sublimation and from melts, crystallization from solution remains the most explored and industrially used technique. Classically, supersaturation in solution is achieved either by cooling, evaporation, or addition of the anti-solvent. The rate of achievement of supersaturation, nucleation, and crystal growth has a great influence on the properties of crystals. Various particle engineering techniques use different means of modifying the rates of nucleation, crystal growth, or both.

In continuation with this, US is sought as an alternative and effective way of modifying and achieving the rates of nucleation and crystal growth. Application of power US (frequencies from 10 to 100 kHz) to crystallization is known as sonocrystallization. The research has revealed that the application of US has many benefits on the crystallization process. These benefits include faster and uniform primary nucleation, relatively easy nucleation in materials that are difficult to nucleate otherwise, nucleation of particles at lower supersaturation levels, initiation of secondary nucleation in certain cases, reduction of agglomeration, production of smaller, purer, and uniform size crystals. These effects can be attributed to the highly spacially resolved regions of extreme excitation, temperature, and pressure created by bubble collapse and release of shock wave. When US is applied to liquid medium it exerts alternative cycles of compression and rarefaction within a liquid, creating bubbles during rarefaction stage. The bubbles survive repeated cycles of compression and rarefaction until a critical size is reached and collapse occurs, initiating a well-known phenomenon of cavitation. The precise mechanisms for US action on crystallization remain to be established but known to influence nucleation induction time and metastable zone width.

Influence on Induction Time

In fact, US can induce primary nucleation in nominally particle-free solutions and, noteworthy, at much lower supersaturation levels than would otherwise be the case. Another effect of US on nucleation is shortening the induction time between the establishment of supersaturation and the onset of nucleation and crystallization. However, these effects depend on particular medium and working conditions. In an experiment conducted by Lyczko et al. (2002), the induction time for potassium sulfate was examined as a function of conductivity and changed from 9000 to 1000 s when the crystallization was carried out in the absence and presence of US, respectively. In another experiment carried out by Guo et al. (2005), the US was found to significantly reduce the induction time during crystallization of roxithromycin, particularly at low supersaturations and this effect fades as the concentrations increase (Roxy). They also studied the effect of US on primary nucleation by measuring the induction time. Their results showed that US has a significant effect on reducing the induction time and suggested that diffusion acceleration is the major reason for this effect.

Influence on Metastable Zone Width

The ease or difficulty with which the crystallization process can be carried out depends on the understanding of metastable zone width (MZW). During cooling, crystallization for a given supersaturation level and cooling rate, the temperature drop below the solubility curve at which the solid starts to separate spontaneously is described as MZW. The MZW of roxithromycin was significantly reduced on application of US, as observed by Guo et al. (2005). Thompson and Doraiswamy (2000) have also reported significant increase in the solubility of sparingly soluble solids when subjected to US, which results in narrowing of MZW. They have proposed that this could be either due to hot spot regions produced from cavitating bubbles. The nucleation mechanism itself modifies as the apparent order of nucleation is decreased by US. Therefore, US can induce nucleation under conditions where spontaneous nucleation cannot occur in its absence. This avoids seeding and hence the introduction of foreign particles into the solution.

Solution Atomization and Crystallization by Sonication

Kaerger and Price (2004) extended the process of sonocrystallization and developed solution atomization and crystallization by sonication (SAXS) technique to produce active pharmaceutical ingredients with a narrow particle size distribution, centered around the optimum particle size for reproducible and maximized therapeutic efficacy. The SAXS process consists of three interdependent steps: (a) the production of aerosol droplets of the solute from a carrier solvent using a suitable aerosol generator, (b) the collection of the highly supersaturated droplets in a crystallization vessel containing a non-solvent of the drug, and (c) the application of ultrasonic waves to a crystallization vessel to controllably induce homogeneous nucleation and crystal growth. By combining these steps and controlling relevant parameters, high-purity micron-sized sphere-like crystalline particles are produced. During SAXS process, micron-sized aerosol of a drug solution lead to rapid vaporization of the solvent and production of highly supersaturated droplets of the solute molecules. These droplets are collected in a crystallization vessel separated by specific distance. To enhance the kinetically limiting diffusion within the droplets, ultrasonic waves are concomitantly applied to reduce MZW for nucleation and to minimize agglomeration, which improve the powder-handling properties of crystallized particles. Only on application of US does the droplets in the anti-solvent undergo nucleation and crystallization. Without US the particles solidify to semicrystalline and poorly defined particles. The major advantage of this technique relates to the use of any suitable aerosol generator, and that the whole process can be carried out under atmospheric pressure and ambient conditions. SAXS results in spherical crystalline particles within well-defined particle size range. It has the potential for batch and continuous processing at an industrial scale. The limitation of the method is the requirement of suitable organic solvent to dissolve the drug that undergoes evaporation at required rate.

Particle Engineering from Slurry/Suspension

Recently, apart from conventional solution crystallization, application of US to suspension/slurry is also being explored for modification of particle size, particle size distribution, morphology, crystal habit, and more importantly surface characteristics. Propagation of ultrasonic waves through liquid medium results in the formation of cavitation bubbles, and formation of a localized hot spot along with an extremely large temperature and pressure gradient takes place when a cavitational bubble implodes. When solid particles are present in the fluid system, the cavitational event might occur symmetrically or asymmetrically, depending upon the proximity and size of the solids. Particles that are close to each other move away from the cavitational event in a radial direction with high speed due to the formation of particle acceleration shock waves. The high force collisions between the particles results in surface erosion, breaking, or rearrangement.

Ultrasound-Mediated Amorphous to Crystalline Transition

The UMAX particle engineering technology is developed by Prosonix® to increase the crystallinity of solid material, which is less than 100% crystalline. The process comprises the steps of contacting the material with solvent in which it is insoluble and applying US in order to convert the amorphous, semiamorphous, or semicrystalline material into highly crystalline particles with defined physicochemical properties. The process is completed in three distinct stages (Figure 4.1):

1. Dissolving the material in appropriate solvent followed by spray drying to form power of correct size.
2. Dispersing the powder in non-solvent followed by ultrasonic processing to form slurry.
3. Isolation of the crystallized dry powder either by spray drying or supercritical CO_2 processing.

UMAX is used for the manufacture of optimal particles designed and formulated as inhalable drug products for respiratory disorders with the optimal performance attributes of size, shape, surface rugosity, low surface free energy, crystallinity, and stability. UMAX facilitates the manufacture of spheroidal and more regular shaped particles for both dry powder inhalation (DPI) and pressurized metered dose inhalation (pMDI) with significantly enhanced performance when compared with mechanically micronized material.

Ultrasound-Assisted Particle Engineering

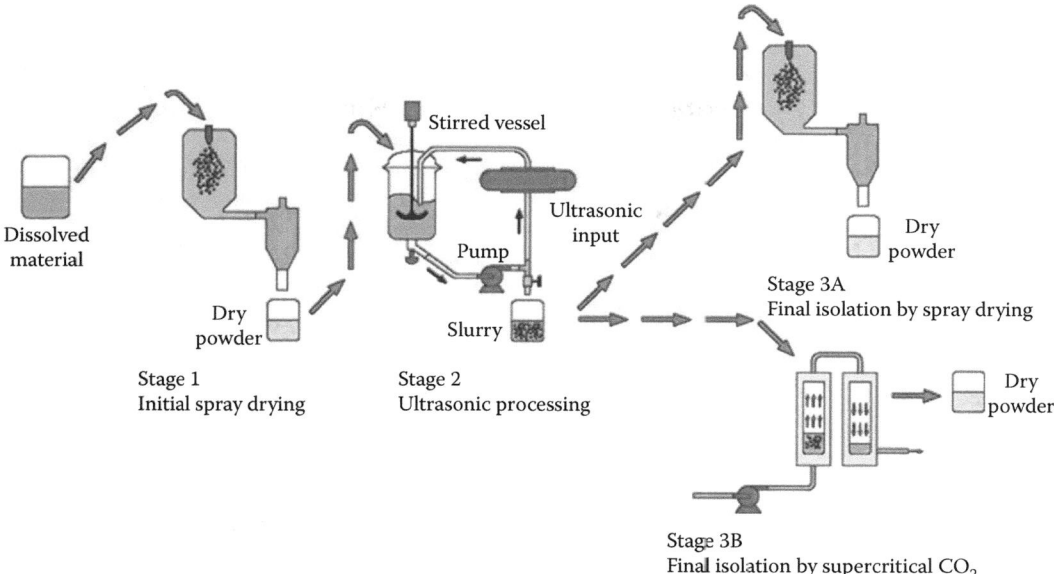

FIGURE 4.1 Representation of the UMAX process. (Reproduced from website: http://www.prosonix.co.uk/. With permission.)

Sonicslurry® and Particle Rounding Technology

Sonicslurry is a technique for crystallizing materials from the slurry by application of US (patent). In this technique the liquid is used as a medium to mediate the material transform. This technique has been used to transfer the metastable forms into stable crystalline forms. Childs et al. (2005) has applied the sonic slurry for preparing stable polymorphs and cocrystals.

PRT involves suspending the starting material in a liquid, which is a partial solvent, and applying ultrasonic vibrations to the suspension, whereby the starting material is shaped and ground to produce a final suspension of ground and rounded particles. PRT was invented by Armament Development Authority in Israel for shaping of explosives. In heterogeneous solid-liquid systems, the interface produces a perturbation in the sonic field, which induces an asymmetric collapse of the cavitational bubbles. At extended interfaces several times larger than the resonance cavitation size, the result is a microjet of liquid passing through the cavitation that impinges with the solid surface at velocities estimated around 100 m s^{-1}. This phenomenon is the origin of the erosion effect and used for ultrasonic shaping and surface modification. The technology is exclusively in-licensed by Prosionix, United Kingdom, and applied for shaping of drugs and excipients. The main benefits of PRT include: Improved packing density of powders, enhanced flow of powders, reduced electrostatic charges, manufacturing by direct compression of material without granulation, and higher filler loading in composite pastes. Well-defined rounded particles with improved flow of oxcarbazepine (antiepileptic API with rod like crystals), diltiazem HCl (antihypertensive API), and metformin (an oral type 2 antidiabetic drug, which do not flow at all) are developed. Prosonix have applied PRT to improve compressibility of fructose. Rounded particles of lactose are developed for improved inhalation delivery.

Particle Engineering from Melt

Melt Sonocrystallization

Melt Sonocrystallization is a particle engineering technique involving the application of ultrasonic energy to the soft or viscous molten mass, dispersed in immiscible liquid. In other words, solidification/crystallization from emulsified melt is carried out under the influence of ultrasonic energy. This technique was initially developed by Paradkar et al. for generation of sintered crystals (Maheshwari et al., 2005) and porous glassy beads (Paradkar et al., 2006). The extent of US energy received by the melt in the emulsified state determines properties of the resultant particles, which is dependent on US energy input and solidification rate of the melt. The rate of solidification depends on glass transition temperature (T_g) of the material and temperature of the medium. Application of US at temperatures above T_g has shown to favor crystallization, whereas processing below T_g results in amorphous state. The mechanical stress due to sonication results in sintered crystals or porous beads. The porous nature and potential for producing crystalline as well as amorphous particles offers flexibility to the technology and is looked upon for solubility enhancement of poorly soluble pharmaceuticals.

Ultrasound-Assisted Polymer Extrusion

Extrusion is the most extensively used and most promising method for polymer processing. Numerous efforts have been made to improve extrusion processing aimed at higher productivity and better product quality. Over the past several decades, a number of studies in the area of effects of US on polymers have been performed and reported. Dispersing of nanocomposites in polymer melts by extrusion is enhanced by application of US. Application of US field to polymer melt in the shaping zone has been approved to be a very efficient way to lower the resistance of the shaping channels or reduce the viscosity of polymer melts. The results showed that applying US disturbs the convergent flow of polymer melt in the entry zone and changes the stream patterns, which leads to lesser elastic tensile strains. It also improves the motion of the molecular chains, so elastic tensile strains can be recovered very quickly. Guo et al. demonstrated significant changes in the properties of the polymeric materials by the application of US to the polymer melt during extrusion process. Guo et al. (2003) applied ultrasonic oscillations in the direction parallel to the flow of polymer melt (Figure 4.2). Scientists at the University of Akron have applied longitudinal vibrations in the direction perpendicular to the flow direction using two horns mounted symmetrically on the slit die during twin screw extrusion of polymer nanocomposites containing carbon nanotubes and polypropylene clays (Isayev et al., 2009; Figure 4.3). Apart from dispersing nanocomposites, application of US to polymer has resulted in increase of crystallinity, reduction of structural defects, and enhancement of mechanical properties.

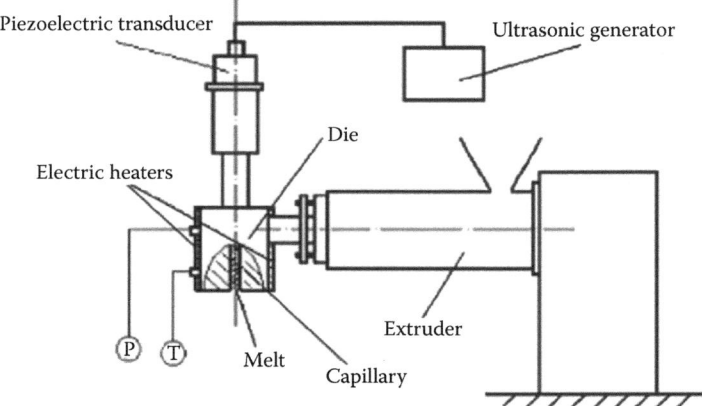

FIGURE 4.2 Schematic diagram of ultrasonic irradiation extrusion system as used by Guo et al. (2003). (Guo, S., Li, Y., Chen, G., and Li, H.: Ultrasonic improvement of rheological and processing behaviour of LLDPE during extrusion. *Polym. Int.* 2003. 52. 68–73. Copyright Wiley-VCH Verlag GmbH & Co. KGaA. Reproduced with permission.)

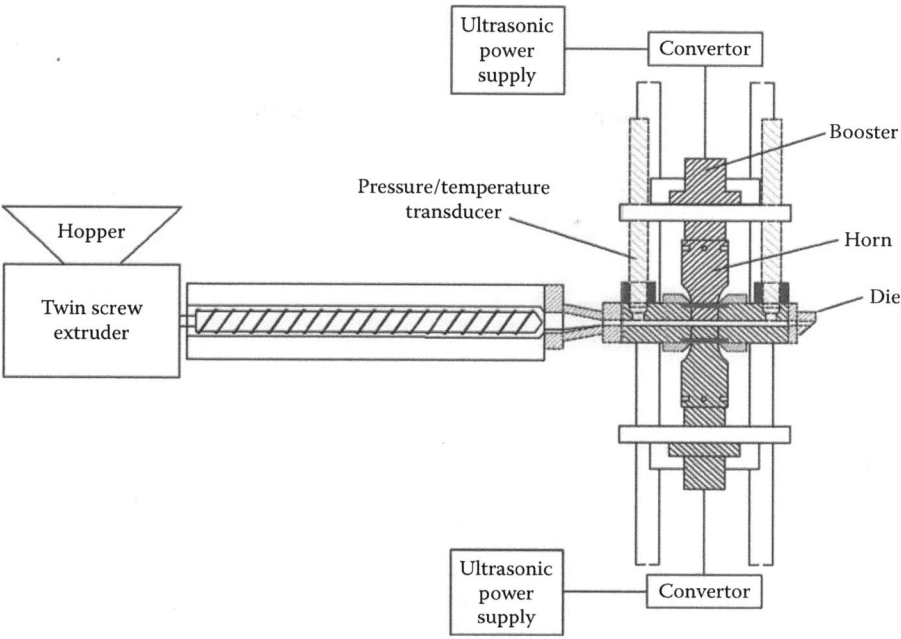

FIGURE 4.3 Schematic diagram of ultrasonic irradiation extrusion system as used by scientists at the University of Acron. (Reproduced from *Polymer*, 50, Isayev, A.I., Kumar, R., and Lewis, T.M., Ultrasound assisted twin screw extrusion of polymer–nanocomposites containing carbon nanotubes, 250–260, Copyright 2009, with permission from Elsevier.)

APPLICATIONS IN PHARMACEUTICAL PARTICLE ENGINEERING

As the crystal habit, size, and morphology are the function of crystallization process, slight modification in the rate of crystallization may have significant effect on these properties. Though US is found to have significant effect on the characteristics of the crystals (our papers or any), there are reports where authors have obtained similar crystals in presence and absence of US (Nishida, 2004). Therefore, judicious selection of processing conditions like solution concentration, anti-solvent, temperature, and US power has enabled scientists to reap the benefits of sonocrystallization. Recognizing the benefits US can bring to sonocrystallization, many researchers have studied sonocrystallization at laboratory and industrial scale, as reviewed by Ratsimba et al. (1999). The metastable or imperfectly formed particles are also treated with US after suspending or formation of slurry for application in the area of pharmaceutical particle engineering. The applications in inhalation drug delivery systems and solubility enhancement of poorly soluble APIs are discussed in detail.

APPLICATION IN INHALATION DRUG DELIVERY

Recently, inhalation therapy has sparked considerable biomedical interest in the development of novel particle technologies for respiratory drug formulation. Introduction of new potent medicines in various therapeutic areas such as asthma, chronic obstructive pulmonary disease (COPD), and various infectious diseases has necessitated an accurate and consistent dosing, using inhalation device. Pulmonary drug delivery by dry powder inhalers (DPIs), by virtue of its propellant-free nature, high-patient compliance, high dose carrying capacity, drug stability, and patent protection, has encouraged rapid development in recent past to realize full potential of lungs for local and systemic treatment of diseases. However, DPIs are complex in nature and their performance relies on many aspects including the design of inhaler, the powder formulation, and the airflow generated by the patient. DPI formulations require drug particles with aerodynamic particle size below 5 μm for deep lung deposition along with good flow properties to ensure accurate dose metering. Aerodynamic particle size is equal to the diameter of the sphere that has the same drag coefficient as a given particle. Airborne particles have irregular shapes, and their aerodynamic behavior is expressed in terms of the diameter of an idealized spherical particle. Therefore, particles having the same aerodynamic diameter may have different dimensions and shapes. The choice and design of processing technique is vital in manipulating macroscopic shape and mesoscopic surface topography of particles. These properties play a critical role in determining flow, deaggregation, and dispersion of fine particulates. In the last decade, performance of DPIs has improved significantly through the use of engineered drug particles and modified excipient systems.

The location in the respiratory tract where the drug is delivered depends upon its particle size. Drug particles of 2–5 μm are delivered to the lungs. Larger particles do not fully penetrate the respiratory tract, whereas particles smaller than 2 μm are exhaled. Most drug particles between 2 and 5 μm tend to agglomerate when they are not formulated with a carrier and will not, therefore, reach the lungs. To prevent this agglomeration, the drug is formulated with a carrier (generally lactose) to form an adhesive mixture. The materials are typically low-shear blended. This mixture uses the following principle: agglomeration of micronized drugs is caused by the hydrophobic character and large surface area of these particles. They are attracted to each other, resulting in agglomerate; however, blending with a carrier prevents this agglomeration. Thus, micronized drug particles are attached to the surface of a lactose carrier and detached when actuated through the device (Figure 4.4). Generally, such blends are formed when the lactose particles are at least five times larger than the drug particles. Additionally, there should be at least five times the amount of lactose compared with the drug. In these formulations the surface texture plays a crucial role and has a significant effect on the cohesive–adhesive balance, which governs the ease or difficulty by which the drug particles are released from the carrier particles under the force of inspiration. This understanding has lead to the development of particle engineering techniques to design drug microparticles and carrier macroparticles to maintain the cohesive–adhesive balance.

To achieve desired particle size jet-milling micronization and spray drying are routinely employed. However, micronization is energy intensive and time consuming and can introduce impurities into the product (Waltersson and Lundgren, 1985). The process also shows some disadvantages in practice such as inadequate control of particle size, undesired particle shape, surface charge modifications, decreased crystallinity, and possible chemical degradation (Briggner et al., 1994; Ticehurst et al., 2000). Spray drying yields amorphous product that is thermodynamically unstable and therefore show tendency to absorb moisture, agglomerate, and recrystallize on storage (Chawla et al., 1994). Inhalable particles can potentially be obtained by rapid precipitation from aqueous solutions using anti-solvents. However, due to dispersion in the nucleation rate and crystal growth, it is difficult to reproducibly generate particle size in the micron range for aerosol delivery. During conventional anti-solvent crystallization with mechanical agitation, poor mixing leads to heterogeneous growth of crystals and in turn variation in particle size and morphological features (Zeng et al., 2001). Application of US during crystallization is known to avoid this by

FIGURE 4.4 Microparticle loaded carriers are filled in capsules and are released after traveling certain distance enabling targeting of lower airways in lung.

maintaining reasonably uniform conditions throughout the crystallization vessel (Louhi-Kultanen et al., 2006). US irradiation is commonly known to induce acoustic streaming, microstreaming, and highly localized temperature and pressure within the fluid; these effects bring considerable benefits to crystallization process, such as rapid induction of primary nucleation, reduction of crystal size, inhibition of agglomeration, and manipulation of crystal size distribution (Luque de Castro and Priego-Capote, 2007). As the number of primary nuclei increases, amount of solute on each primary nucleus decreases, thus decreasing the size of final crystal (Louhi-Kultanen et al., 2006). Recently, ultrasonic radiation has been applied to control the precipitation process (Lancaster et al., 2000). The setup can simply comprise an US probe in a mechanically stirred reaction tank where the anti-solvent is mixed with the drug solution to precipitate the fine drug particles. The US frequency is crucial and 20–25 kHz (or higher) is reported to be suitable.

Tang et al. (2006) and Abbas et al. (2007) have applied US during anti-solvent crystallization of sodium chloride to obtain well-formed highly crystalline particles with cubic habit. Application of US resulted in reduced size and narrow particle size distribution. These particles were separated by spray drying. During spray drying, the cubic shape was altered to more rounded morphology (Figure 4.5). The particles obtained by spray drying, the sodium chloride solution was more round and agglomerated (Figure 4.6). The fine particle fraction (based on recovery) was improved from approximately 12% to 34% when dispersion air flow rate was increased from 60 to 120 L min^{-1}. Sodium chloride as an osmotic agent have been used widely for bronchial provocation tests to identify people with active asthma or exercised-induced asthma. The improved aerosol performance of the sonocrystallized particles is suitable for the bronchoprovocation test, making sonocrystallization a promising alternative for preparing sodium chloride dry powder inhalers for inhalation delivery. This method represents rapid and simple method of producing small sodium chloride crystal particles suitable for inhalation. The highly crystalline powder obtained from the ultrasonic precipitation is advantageous because it provides stability. Amorphous sodium chloride powder is physically unstable and in high humidity will recrystallize uncontrollably, forming solid bridges between particles that make them unsuitable for inhalation.

Kaerger and Price (2004) have developed SAXS technique for production of spherical particles with optimum size and morphology suitable for inhalation delivery. After the success with paracetamol as model drug, they applied the SAXS technique for designing the particles of

FIGURE 4.5 Scanning electron micrograph of the NaCl particles. (Reproduced with permission from Tang, P., Chan, H.K., Tam, E., Gruyter, N., and Chan, J., Preparation of NaCl powder suitable for inhalation, *Ind. Eng. Chem. Res.*, 45, 4188–4192, 2006. Copyright 2006 American Chemical Society.)

FIGURE 4.6 Scanning electron micrograph of the NaCl particles obtained by spray drying from solution alone. (Reproduced from *Chem. Eng. Sci.*, 62, Abbas, A, Srour, M., Tang, P., Chiou, H., Chanc, H.K., and Romagnolid, J.A., Sonocrystallisation of sodium chloride particles for inhalation, 2445–2253, Copyright (2007), with permission from Elsevier.)

budesonide for inhalation delivery. Budesonide is a synthetic steroid and potent anti-inflammatory used for treatment of asthma. Budesonide solution in methanol (4%) was sprayed into non-solvent. The highly supersaturated droplets were crystallized in 35–45 kHz US frequency range to obtain spherical agglomerates of the budesonide crystals (Figure 4.7). In vitro aerosolization properties of carrier free DPI formulation of micronized and SAXS produced budesonide particles were tested with Anderson cascade impactor after filling in turbuhaler reservoir device. The emitted dose and FPF was significantly higher for SAXS produced budesonide compared to micronized budesonide. The microparticles were also tested for settling behavior after dispersion in a model HFA–pMDI formulation. The SAXS microparticles flocculated instantaneously upon dispersion and remained freely dispersed in propellant for up to 45 min. This increase in stability was attributed to the reduction in surface free energy components and morphological characteristics of SAXS produced budesonide.

Traditionally, nonspherical particles have been avoided for potential problems of powder flow. However, the respirable fraction of drugs is reported to improve with the elongation ratio, as aerodynamic diameter shows a relatively small dependence on the length but is strongly correlated to the short dimension of the particles (Ikegami et al., 2002; Larhrib et al., 2003). Dhumal et al. (2009) have developed needle-shaped crystals with desirable size and uniform particle size distribution by application of US during anti-solvent crystallization of salbutamol sulphate (SS). Salbutamol sulfate

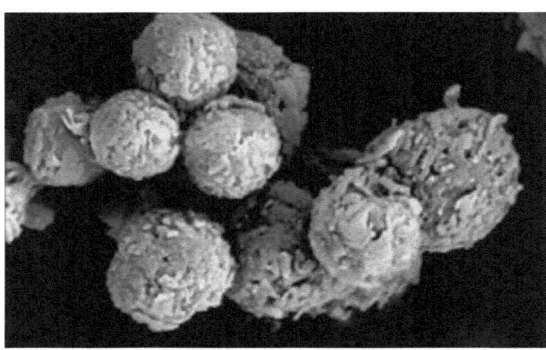

FIGURE 4.7 SAXS crystals of budesonide. (Reproduced from website: http://www.prosonix.co.uk/. With permission.)

is a short acting β-adrenergic receptor agonist used for relief of bronchospasm in conditions such as asthma and COPD. The sonocrystallized batches showed high crystallinity and significantly reduced particle size with narrow particle size distribution compared to particles obtained by anti-solvent crystallization. During sonocrystallization the intensified micromixing leads to enhanced mass transfer and diffusion between solvent and anti-solvent inducing instantaneous high levels of supersaturation, resulting in rapid nucleation. As the number of primary nuclei increases, the amount of solute depositing on each primary nucleus decreases, thus decreasing the size of final crystal. Whereas, anti-solvent crystallization in the absence of US exhibits poor micromixing process leading to accidental zones of supersaturation, heterogeneous growth of crystals, aggregation of particles, and, in turn, variation in the particle size. The crystalline particles were separated from suspension by spray drying and formulated with coarse lactose by blending, filled in capsule and studied for in vitro aerosol deposition in cascade impactor, using Rotahaler® at $28.3\,L\,min^{-1}$. The aerosolization performance was compared with micronized and amorphous (spray dried) salbutamol sulfate formulated with coarse lactose.

The SEM image of micronized SS shows the agglomerated particles, while amorphous SS obtained by spray drying SS solution shows uniform, spherical particles with some pitting and shrunken areas on the surface. Sonocrystallized SS before spray drying exhibited elongated, needle-shaped crystals. After spray drying, the suspension of these needle-shaped microcrystals, loose oval aggregates measuring around 3–7 μm in size were formed due to instantaneous drying of spherical droplets containing microcrystals. All the images are presented in Figure 4.8. Uniform supersaturated conditions achieved rapidly throughout the crystallization vessel with ultrasonic

FIGURE 4.8 (a) SEM photomicrographs of spray dried amorphous SS, (b) micronized SS, (c) anti-solvent precipitated SS, and (d) sonocrystallized SS. (Modified from *Int. J. Pharm.*, 368, Dhumal, R.S., Biradar, S.V., Paradkar, A.R., and York, P., Particle engineering using sonocrystallization: Salbutamol sulphate for pulmonary delivery, 129–137, Copyright (2009), with permission from Elsevier.)

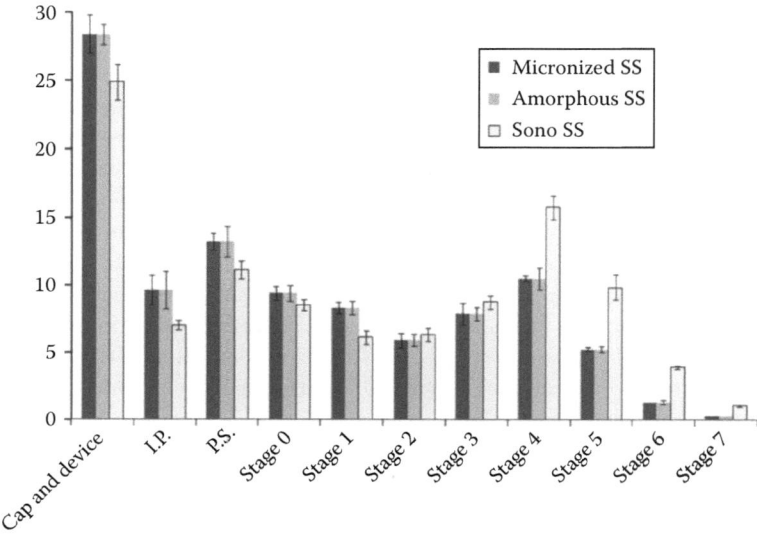

FIGURE 4.9 Drug deposition profiles of different salbutamol sulfate formulations (micronized, amorphous spray dried, and sonocrystallized) with lactose (pharmatose) as carrier in Andersen cascade impactor at 28.3 L min^{-1} via Rotahaler $n = 3$. (Reproduced from *Int. J. Pharm.*, 368, Dhumal, R.S., Biradar, S.V., Paradkar, A.R., and York, P., Particle engineering using sonocrystallization: Salbutamol sulphate for pulmonary delivery, 129–137, Copyright (2009), with permission from Elsevier.)

insonation resulted in rapid crystallization. Rate of crystallization is known to increase or decrease the growth of certain crystal faces (Guo et al., 2005). Faster crystallization rate in this case resulted in needle-shaped crystals.

The sonocrystallized SS showed improved inhalation performance over the micronized and the spray dried amorphous particles. The fine particle fraction (FPF) increased from 16.66% for micronized SS to 31.12% for amorphous SS and 44.21% for sonocrystallized SS, as seen in Figure 4.9. As particle size of all the powders was comparable, the enhanced aerosol behavior of sonocrystallized drug was attributed to the difference in shape and the surface properties of the particles. The reason for improved performance of spray dried sonocrystallized salbutamole sulphate (SD-SCSS) may be the reduced cohesive and adhesive forces in interactive mixture resulting in better flow and deagglomeration from the surface of carrier. The elongated, needle shape of individual crystals prevents close packing between the crystals, which effectively increases the interparticulate distance and lowers the van der Waals attractive forces giving loose aggregates. During mixing with lactose the loose aggregates, disaggregate into individual crystals by the shear forces from the collisions between the agglomerates and the carrier. Thus, individual crystals and not the aggregates adhere to the lactose surface as seen from SEM image of drug loaded lactose (Figure 4.10). In contrast, micronized drugs are reported to adhere to the carrier surface as aggregates owing to its activated surface and cohesive nature. The elongated shape of SD-SCSS crystals are known to reduce the effective areas of contact between the crystals and the carrier surface reducing the adhesive forces (Ikegami et al., 2002; Larhrib et al., 2003).

Though spray dried amorphous particles also showed better aerosolization behavior compared to micronized drug, it deteriorated on storage. This was due to the recrystallization of amorphous SS on storage with increase in moisture content and agglomeration, indicating instability and nonsuitability of amorphous systems for pulmonary delivery in the form of DPI. In contrast, microcrystals obtained by sonocrystallization showed good physical stability and no significant difference in the deposition profiles on storage.

FIGURE 4.10 SEM photomicrograph of the surface of SD-SCSS formulation with lactose. (Reproduced from *Int. J. Pharm.*, 368, Dhumal, R.S., Biradar, S.V., Paradkar, A.R., and York, P., Particle engineering using sonocrystallization: Salbutamol sulphate for pulmonary delivery, 129–137, Copyright (2009), with permission from Elsevier.)

Singh et al. (2002) had developed a method of producing crystalline particles suitable for inhalation delivery by anti-solvent crystallization in continuous flow cell in the presence of ultrasonic radiation. This method was applied to prepare particles suitable for inhalation delivery of compounds used for treating asthma, such as fluticasone propionate and salmeterol xinafoate. These particles were either filled in the aluminum canister with liquefied HFA134a for metered dose inhalers or formulated with milled lactose as DPI (Singh et al., 2002).

The most popular carrier used for inhalation drug delivery is lactose. Toxicology data of lactose after administration to the lungs are favorable. Moreover, lactose is readily available and well-established, having been used in DPIs for more than 20 years. When employed as carrier for dry powder aerosols, to overcome intrinsic cohesiveness of micronized drug particles intended for pulmonary delivery, crystal habit determines the surface interaction of lactose particles with adhered drugs (Steckel et al., 2004). Drug delivery from these interactive mixtures is known to be influenced by the physicochemical characteristics of lactose particles such as particle size, shape, and surface (Zeng et al., 2001; Larhrib et al., 2003). Particles with average size in the range 63–90 µm with smooth surfaces and high elongation ratios exhibit better dispersibility and deagglomeration of adhered drug, thus improving the FPF (Zeng et al., 2000). Therefore, there are attempts to either crystallize the lactose particles with these attributes or modify the already crystallized particles with these features.

Lactose is commercially produced by cooling crystallization, which shows slow crystallization rate extending up to 72 h. Bund and Pandit (2007a,b) have reported application of US during anti-solvent crystallization of lactose for achieving homogeneous supersaturation throughout crystallization vessel, resulting in rapid recovery of nonagglomerated fine lactose crystals (Bund and Pandit, 2007a,b). Though the crystallization time was reduced and recovery of lactose improved significantly the crystal size (5–10 µm) was too small for application as carrier for DPI application. Dhumal et al. (2008a) have achieved US assisted spontaneous in situ seeding during cooling crystallization of lactose. These microfine seed crystals were grown steadily in stagnant viscous glycerin solution to obtain lactose crystals with desirable size (63–90 µm), regular shape, smooth surface, and narrow PSD for application as carrier for DPI. Cooling crystallization and US assisted seeding followed by growth under stirring were employed as control experiments. Micronized SS was blended separately with 63–90 µm sieved fractions of pharmatose and crystallized lactose batches in a ratio of 1:67.5. The blend was filled in hard gelatin capsule and studied for in vitro aerosol

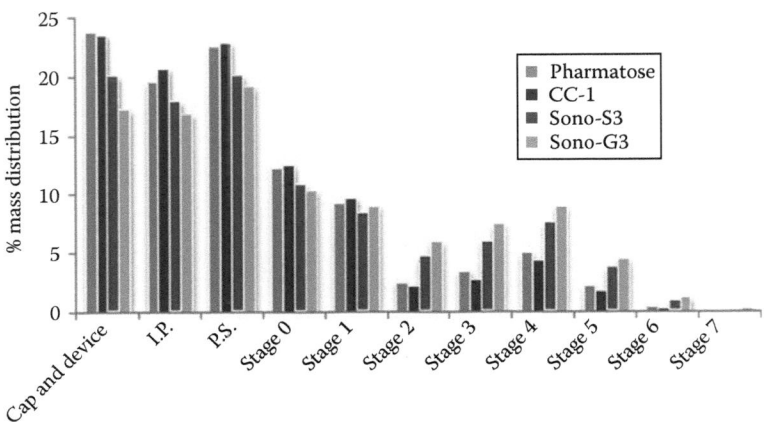

FIGURE 4.11 Drug deposition profiles of blends containing different lactose batches in Anderson cascade impactor at 28.3 L min^{-1} via Rotahaler. (Reproduced from *Int. J. Pharm.*, 368, Dhumal, R.S., Biradar, S.V., Paradkar, A.R., and York, P., Particle engineering using sonocrystallization: Salbutamol sulphate for pulmonary delivery, 129–137, Copyright (2009), with permission from Elsevier.)

deposition studies in a cascade impactor using Rotahaler at 28.3 L min^{-1}. The FPFs from pharmatose, lactose crystallized from ultrasonic seeding followed by stirring, and ultrasonic seeding followed by growth in glycerine were 13.01%, 22.82%, and 27.82%, respectively (Figure 4.11).

The improved deposition profiles from recrystallized lactose compared to pharmatose were attributed to the difference in the morphological features like elongation ratio and surface smoothness. This was explained by the fact that elongated particles have aerodynamic diameters almost independent of their length and is governed by the shorter dimension of the particle (Ikegami et al., 2002). Such particles exhibit smaller aerodynamic diameters than spherical particles of similar mass or volume (Hickey et al., 1992). More elongated particles may also be expected to travel a longer distance before collision occurs in comparison to less elongated particles of similar mass, as a result of lower relative aerodynamic diameters of the former. More drug particles can adhere to elongated carrier particles in comparison to spherical particles. Elongated particles are also known to experience drag forces of the air stream for longer period of time. This would be expected to result in a higher proportion of drug being detached from the carrier particle, leading to a higher FPF of the drug (Zeng et al., 2001).

In spite of only slight difference in the elongation ratio, lactose crystallized from ultrasonic seeding followed by growth in glycerine produced significantly higher FPF than lactose crystallized from ultrasonic seeding followed by stirring. Authors attributed this to the difference in the surface smoothness, which was determined by calculating the values of surface factor and the observation from SEM images. Thus, the microscopic asperities in sample obtained by stirring might have acted as adhering sites for drug, preventing the deagglomeration from its surface. However, growth of lactose crystals in viscous glycerine medium permits a steady diffusion of crystallizing molecules without any external turbulence providing a homogeneous environment in which the crystals grow to maturity without surface defects. Therefore, high elongation ratio and surface factor was considered responsible for highest FPF.

Apart from application during crystallization of lactose for design of crystals suitable as carrier for inhalation, US has also been utilized for rounding of lactose particles. This technique is called as PRT and applied to improve flowability and storage stability of alpha-lactose monohydrate without agglomeration (Figure 4.12). These rounded smooth lactose particles have shown improved performance when used as carrier for inhalation drug delivery. Prosonix have developed and further scaled the process of obtaining rounded lactose using PRT for inhalation delivery and propose to market the obtained lactose.

FIGURE 4.12 Ultrasonically rounded lactose crystals. (Reproduced from website: http://www.prosonix.co.uk/. With permission.)

APPLICATION IN DESIGN OF PARTICLES WITH ENHANCED DISSOLUTION RATE

Dhumal et al. (2008b) have obtained amorphous cefuroxime acetil (CA) nanoparticles by sonocrystallization and shown significantly faster dissolution as compared to unprocessed and micron-sized amorphous particles obtained by spray drying (Dhumal et al., 2008b). CA exists in crystalline as well as amorphous form, of these latter exhibits higher bioavailability owing to improved solubility. The most successful attempt for achieving amorphous CA from commercial point of view was through spray drying (Crisp et al., 1989). Application of US during anti-solvent crystallization of CA was found to be effective not only in generating nanoparticles (80 nm) but also in producing them as discreet, nonagglomerated particles (Figure 4.13d) compared to agglomerated particles with broad particle-size distribution (10–100 µm) in anti-solvent crystallization with mechanical stirring and absence of US (Figure 4.13c). When drug solution is added into the anti-solvent, it does not disperse immediately and produces local zones of excessive supersaturation. When anti-solvent process uses mechanical stirring, poor micromixing is unavoidable, which increases the precipitation rate locally leading to agglomeration and increase in particle size. In contrast, homogeneous micromixing during ultrasonication maintains reasonably uniform conditions throughout the vessel causing the formation of nanosized discreet particles. As the rate of precipitation during US assisted anti-solvent crystallization was too high, the drug precipitated in amorphous form. The particles were separated from the suspension by spray drying. CA solution was spray dried to yield smooth spherical amorphous particles (Figure 4.13b) with mean particle size of 10 µm amorphous CA as a control.

The dissolution rate of amorphous CA particles obtained by sonoprecipitation, anti-solvent crystallization, and spray drying of solution was found to significantly enhance over crystalline unprocessed CA with large particle size (Figure 4.13a). However, since all the processed particles were amorphous in nature the difference in dissolution was attributed to the particle size with sonoprecipitated discrete nanoparticles showing highest dissolution rate than anti-solvent processed agglomerates of CA particles nanoparticles, followed by spray dried CA microparticles (Figure 4.14). Similar trend was observed in the bioavailability tested in Wistar rats after oral administration with sonoprecipitaed CA showing maximum bioavailability. This dramatic improvement in rate as well as extent of in vitro drug release and oral bioavailability in Wistar rat was attributed to amorphization, increased surface area, and decreased diffusion layer thickness. This study demonstrates potential of sonoprecipitation technique to obtain discreet and amorphous nanoparticles with feasibility for scale-up.

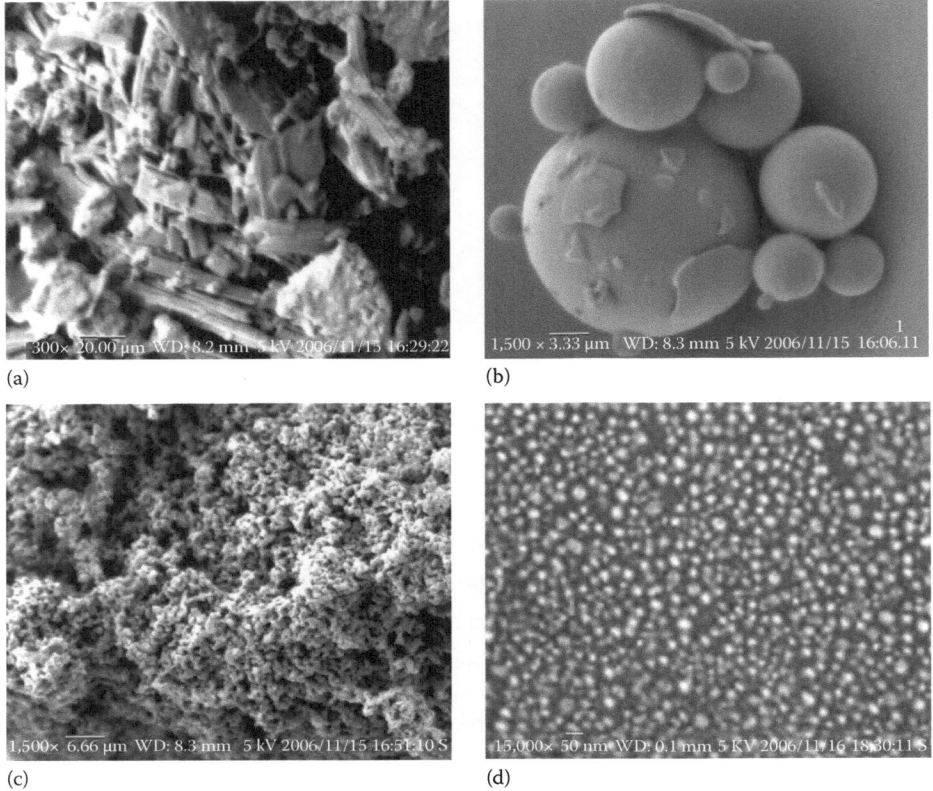

FIGURE 4.13 SEM photomicrographs of cefuroxime axetil (a) unprocessed, (b) spray dried, (c) anti-solvent crystallized, and (d) sonoprecipitated. (Reproduced from *Eur. J. Pharm. Biopharm.*, 70, Dhumal, R.S., Biradar, S.V., Yamamura, S., Paradkar, A.R., and York, P., Preparation of amorphous cefuroxime axetil nanoparticles by sonoprecipitation for enhancement of bioavailability, 109–115, Copyright (2008b), with permission from Elsevier.)

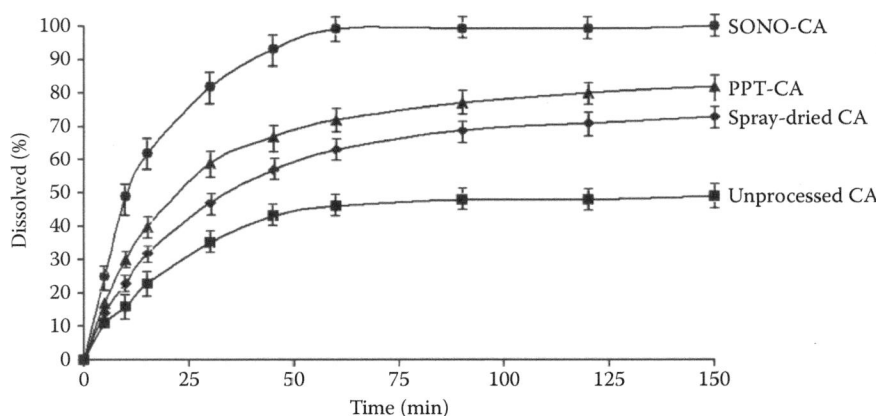

FIGURE 4.14 Dissolution profiles of cefuroxime axetil unprocessed, spray dried, anti-solvent crystallized, and sonoprecipitated. (With kind permission from Springer Science+Business Media: *Pharm. Res.*, Ultrasound assisted engineering of lactose crystals, 25, 2008a, 2835–2844, Dhumal, R.S., Biradar, S.V., Paradkar, A.R., and York, P.)

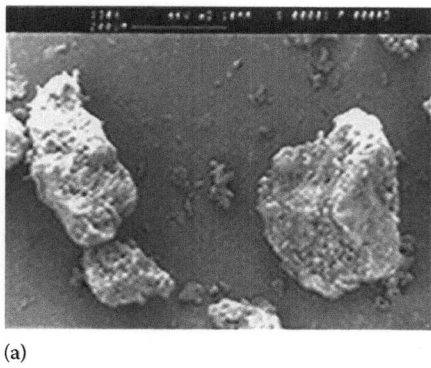 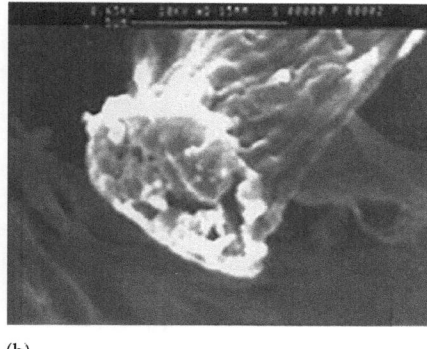

(a)   (b)

FIGURE 4.15 (a) SEM photographs of ibuprofen and (b) melt sonocrystallized ibuprofen. (Reproduced from *Eur. J. Pharm. Sci.*, 25, Maheshwari, M., Jahagidar, H., and Paradkar, A.R., Melt sonocrystallization of ibuprofen: Effect on crystal properties, 41–48, Copyright (2005), with permission from Elsevier.)

Apart from nanoparticles, development of porous particles is also known to improve solubility due to higher surface area. Paradkar et al., have developed porous amorphous celecoxib particles (Maheshwari et al., 2005) and porous crystalline ibuprofen particles (Paradkar et al., 2006), using melt sonocrystallization. Melt sonocrystallization is a particle engineering technique involving application of ultrasonic energy to the molten mass dispersed in suitable dispersion media. Insonation of ibuprofen melt dispersed in water was carried out at ambient temperature. Ibuprofen is a drug with a low glass transition temperature (T_g), and the melt remains in a low viscosity liquid state for a relatively longer period of time. As the ultrasonic energy could be applied to the melt for longer duration of time during crystallization, a product consisting of porous tubes and sintered crystals was obtained (Figure 4.15). Owing to high specific surface area product showed improved dissolution rate. Improvement in compressional properties and reduction in sticking were also observed due to the change in crystal habit.

Further, an attempt was made to study the effect of insonation on dispersed melt of celecoxib having relatively high viscosity and processed at a temperature below its glass transition temperature (T_g). Since, the process was carried out near T_g, the melt was viscous and solidified rapidly, hence received the insonation for shorter duration during solidification, resulting in porous amorphous form (Figure 4.16). These porous amorphous particles showed increase in the apparent solubility

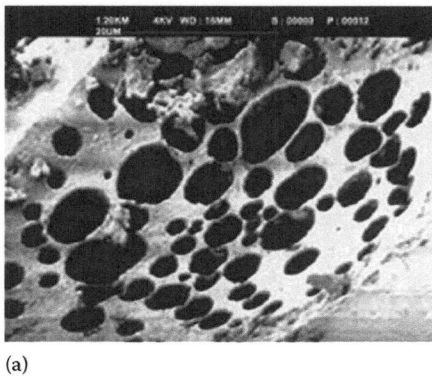 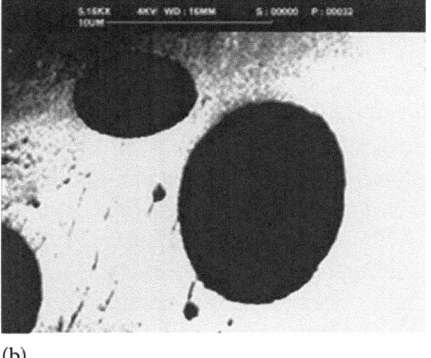

(a)   (b)

FIGURE 4.16 SEM photomicrographs of MSC celecoxib at (a) low and (b) high magnification. (With kind permission from Springer Science + Business Media: *Pharm. Res.*, Design and evaluation of celecoxib porous particles using melt sonocrystallization, 23, 2006, 1395–1400, Paradkar, A., Maheshwari, M., Kamble, R., Grimsey, I., and York, P.)

and exhibited a higher stability in the amorphous state compared with particles obtained by melt quenching. This technique has resulted in particles with improved physicochemical and biopharmaceutical properties and proved the possibility of generating crystalline as well as amorphous form. Technique has an advantage of producing particles with desired biopharmaceutical properties without the addition of excipients.

Ambrus et al. (2010) has compared the techniques of sonocrystallization for gembibrozil from melt and solution. Crystal habit and structure of the particles obtained by sonocrystallization from melt and solution differ significantly. They observed that the method of separation of the particles from the suspension also has significant effect on the particle size and morphology. Needle-shaped crystals were observed for the sample separated by lyophilization and spherical crystals for sample separated by spray drying (Ambrus et al., 2010).

Recently, US has been applied for the development of cocrystals. In forming a cocrystal the physicochemical properties of the API can be modified, while maintaining its intrinsic therapeutic activity. Cocrystal is made up of two or more components (API/guest(s)), usually in a stoichiometric ratio, each component being an atom, ionic compound, or molecule held together by noncovalent forces (Aakeroy, 1997). Cocrystal formation can not only improve solubility and bioavailability but also improve the physical properties of API including density, hygroscopicity, crystal morphology, loading volume, and compressibility (Blagden et al., 2007). Different pharmaceutical cocrystallization techniques have been reported such as solution crystallization, grinding/solvent drop grinding (Trask, 1997), melt crystallization (Berry et al., 2008), and cocrystallization using US (Childs et al., 2005; Bucar and MacGillivray, 2007; Friscic et al., 2009; Aher et al., 2010) and hot melt extrusion (Dhumal et al., 2010).

Childs et al. (2005) has used US for cocrystal formation from either solution or suspension, where US is thought to promote nucleation (sononucleation) during cocrystal formation. Slurry sonication experiments (Sonicslurry) were carried out by Friscic et al. (2009) to obtain cocrystals. An empirical parameter η (the volume of solvent in μL divided by the sample weight in mg) was used to understand the effect of solubility and saturation of cocrystal components on cocrystal formation. Pure cocrystal product was observed mainly at lower η values (<6 μL mg^{-1}), whereas, at higher η values (up to 12 μL mg^{-1}), either a mixture of cocrystal with a least soluble cocrystal component or only a least soluble cocrystal component was produced. However, the exact effect of US in this process remained unclear. Sonocrystallization is also used to obtain 2:2 resorcinol/*trans*-1,2-bis(4-pyridyl)ethylene cocrystal from cloudy suspension, where cavitation was thought to cause single crystal to single crystal reactivity in suspension (Bucar and MacGillivray, 2007). The ultrasonic radiations were believed to provide an environment that favored rapid solubilization of the cocrystal components and also provided a mechanism for rapid precipitation and formation of the small crystals. In this case, low-intensity ultrasonic radiations were applied using an ultrasonic cleaning bath to obtain nanostructured cocrystals, where reprecipitation alone failed.

Aher et al. (2010) has applied US to cocrystallize noncongruently soluble cocrystal component pair of caffeine and maleic acid in 2:1 ratio, which is otherwise difficult to obtain by conventional solution crystallization technique (Aher et al., 2010). Caffeine/maleic acid 2:1 cocrystal was prepared using US, where caffeine and maleic acid represent a noncongruently soluble pair in methanol. Simultaneous achievement of supersaturation of cocrystal components in solution is the basic requirement to obtain cocrystal. Molar concentrations of cocrystal components in solution were found to be an important parameter while designing experiments for noncongruently soluble cocrystal component pair. Pure caffeine/maleic acid 2:1 cocrystal was obtained by application of US to caffeine: maleic acid solution in 1:3.5 molar ratio in methanol. Solution crystallization, grinding, and solvent drop grinding techniques resulted in cocrystal product, which was accompanied with variable amount of caffeine. Only US assisted process offered pure caffeine/maleic acid 2:1 cocrystal.

SCALE-UP

While novel technologies can seem extremely promising during laboratory and early developmental phases, many fail on scale-up and on meeting the tighter controls required by the pharmaceutical industry and regulatory authorities. Although the application of ultrasonic energy to produce or to enhance a wide variety of processes has been explored since the middle of the twentieth century, only few ultrasonic processes have been established at the industrial level. However, during the last 10 years, the interest in ultrasonic processing has revived particularly in pharmaceutical sectors where the ultrasonic technology represent a clean and efficient tool to improve classical existing processes or an innovative alternative for the development of new processes. The possible major problem in the application of power US on industrial processing is the design and development of efficient ultrasonic systems (generators and reactors) capable of large scale successful operation specifically adapted to each individual process. The development of different cavitational reactors for liquid treatment in continuous flow is helping to introduce into industry the wide potential in the area of sonochemistry. Processes such as water and effluent treatment, soil remediation, etc. have been already implemented at semi-industrial and/or industrial stage.

Bristol Myers Squibb has successfully developed and scaled-up crystallization process for polymorph control of a proprietary substance to obtain uniform plate-type crystals. The US is applied to the crystal slurry just prior to filtration in a recirculation loop, where the particle size is reduced from 100–200 mm range to 20 mm for the purpose of enhancing physicochemical properties like solubility, improved bulk density, and flowability.

Prosonix has designed an industrial equipment to allow effective and focused distribution of acoustic energy into a liquid by using a number of low-power transducers bonded to the outside of a cylindrical duct. This equipment can be used as a recirculation or continuous flow cell. Prosonix scaled up SAXS process for single component as well as combination particles for asthma (Ruecroft, 2007). Prosinix also scaled up the PRT technology for rounding of lactose and propose to launch lactose obtained by this technique as a carrier for inhalation drug delivery. All these reports show great potential for use of US in pharmaceutical processes and we expect to see many more industrial applications in the near future.

LIMITATIONS

It is worth noting that, like any innovative processing technology, high power US is also not a standardized technology and therefore needs to be developed and scaled up for every new application on a case-to-case basis. Though US is used for long, the exact mechanism is not clear. A better understanding of the complex physicochemical mechanism of the US with relationship between the duration, intensity, and frequency of ultrasonic waves and particle formation would contribute to reinforce the applications in pharmaceutical and other industries. The hot spots formed in the liquid owing to the explosion of cavitational bubbles lead to the generation of free radicals, which undergo a number of chain reactions including the generation of hydrogen peroxide. As the energy released is high enough to cause a chemical reaction, the possible degradation of the actives should be taken into consideration. There are many reports describing the chemical degradation of actives when US was applied for enhancing the processes. Thongson et al. (2004) have applied US to improve the extraction rate and yield of ginger and turmeric. However, a slightly reduced antimicrobial activity was observed against few microorganisms. Zhao et al. (2006) has applied US for extraction of biologically active compound and observed degradation of few carotenoids. Gulseren et al. (2007) found that cavitation-generated hydrogen peroxide may alter the chemical structure of bovine serum albumin, oxidizing its free sulfhydryl groups, leading to the formation of sulfinic and sulfonic acid.

Another technical drawback associated with the use of power US is cavitation erosion. Cavitation bubbles collapse unsymmetrical when in close proximity to the emitting probe face (or another solid

FIGURE 4.17 Image showing the eroded ultrasonic probes. (Reproduced from website: http://www.prosonix.co.uk/. With permission.)

surface). A microjet forms, which impacts the solid surface, resulting in surface pitting, and the metal is released into the medium, especially titanium, which is a major component of the probe tip. While using a probe, relatively small surface area and high amplitude generates high surface intensity that easily decouples and promotes cavitation close to the tip of the probe with little penetration into the bulk of the liquid. This leads to rapid erosion of the probe and the shedding of probe material into the process liquid (see eroded probes images Figure 4.17). To overcome this problem, Prosonix have recently developed a Prosonitron technology, which provides ultrasonic power input through an array of bonded transducers mounted on the outside of a cylindrical pipe. This spreads the power input over the whole surface of the vessel allowing significant power input for a low local surface intensity, greatly reducing surface erosion and any associated contamination. As the waves travel radially into the liquid, the intensity of the waves increases, which carries the cavitation deep into the bulk of the liquid and in fact results in the strongest cavitation occurring at the center of the pipe.

CONCLUSIONS

Although the use of US in processing is well known, recently several novel and interesting US-based processes for particle production have been developed for controlling solid-state, morphology, and particle size of pharmaceuticals. Ultrasonics not only represents a rapid, efficient, reliable, green, and clean alternative to improve the quality of pharmaceuticals but also has the potential to develop new products with unique functionality. Though this chapter discusses in detail particle engineering for inhalation delivery and solubility enhancement of drugs, there are many other applications of US beyond the scope of this chapter. Well-characterized solid state, purity, and environmental acceptability are becoming more and more stringent requirements for regulatory authorities that reflect an approach that is in keeping with optimization of drug quality. In this regard, US should be considered to be in a prominent position in the drug development processes. As for the development of new technologies, a strict interdisciplinary approach must be sought, possibly by merging engineering, physicochemical, pharmaceutical technology, and biopharmaceutical expertise.

It is worthy to note that some processes have been already scaled up to industrial manufacturing. Nevertheless, the potential of this innovative technology still remains largely unexplored. With the obvious potential of US in designing pharmaceutical processes, it can be anticipated that technology needs to undergo considerable growth and advances over the coming years. The competitive energy costs and the low maintenance make US processes economically profitable. Advantages are also obtained for large-scale applications involving continuous flow-through US systems, as the high financial benefits and short payback period outweigh the costs of converting and maintaining the US processing equipment.

REFERENCES

Aakeroy, C. B. 1997. Crystal engineering: Strategies and architectures. *Acta Crystallographica*, B, 53: 569–586.

Abbas, A., Srour, M., Tang, P., Chiou, H., Chanc, H. K., and Romagnoli, J. A. 2007. Sonocrystallization of sodium chloride particles for inhalation. *Chemical Engineering Science*, 62: 2445–2453.

Aher, S., Dhumal, R. S., Mahadik, K. R., Paradkar, A. R., and York, P. 2010. Ultrasound assisted cocrystallization from solution (USSC) containing a non-congruently soluble cocrystal component pair: Caffeine/maleic acid. *European Journal of Pharmaceutical Science*, 41(5): 597–602.

Ambrus, R., Amirzadi, N. N., Sipos, P., and Szabó-Révész, P. 2010. Effect of sonocrystallization on the habit and structure of Gemfibrozil crystals. *Chemical Engineering and Technology*, 33: 827–832

Berry, D. J., Seaton, C. C., Clegg, W., Harrington, R. W., Coles, S. J., Horton, P. N., Hursthouse, M. B., Storey, R., Jones, W., Friscic, T., and Blagden, N. 2008. Applying hot-stage microscopy to cocrystal screening: A study of nicotinamide with seven active pharmaceutical ingredients. *Crystal Growth and Design*, 8: 1697–1712.

Blagden, N., de Matas, M., Gavan, P. T., and York, P. 2007. Crystal engineering of active pharmaceutical ingredients to improve solubility and dissolution rates. *Advances in Drug Delivery Reviews*, 59: 617–630.

Briggner, L. E., Buckton, G., Bystrom, K., and Darcy, P. 1994. The use of isothermal microcalorimetry in the study of changes in crystallinity induced during the processing of powders. *International Journal of Pharmaceutics*, 105: 125–135.

Bucar, D. and MacGillivray, L. R. 2007. Preparation and reactivity of nanocrystalline cocrystals formed via sonocrystallization. *Journal of American Chemical Society*, 129: 32–33.

Bund, R. K. and Pandit, A. B. 2007a. Sonocrystallization: Effect on lactose recovery and crystal habit. *Ultrasonics Sonochemistry*, 14: 143–152.

Bund, R. K. and Pandit, A. B. 2007b. Rapid lactose recovery from buffalo whey by use of 'anti-solvent, ethanol'. *Journal of Food Engineering*, 82: 333–341.

Chawla, A., Taylor, K. M. G., Newton, J. M., and Johnson, M. C. R. 1994. Production of spray dried salbutamol sulfate for use in dry powder aerosol formulation. *International Journal of Pharmaceutics*, 108: 233–240.

Childs, S. L., Mougin, P., and Stahly, B. 2005. Screening for solid forms by ultrasound crystallization and cocrystallization using ultrasound. WO/2005/089375.

Crisp, H. A., Clayton, J. C., Elliott, L. G., and Wilson, E. M. 1989. Preparation of a highly pure substantially amorphous form of cefuroxime axetil. US Patent No. 4,820,1833.

Dhumal, R. S., Biradar, S. V., Paradkar, A. R., and York, P. 2008a. Ultrasound assisted engineering of lactose crystals. *Pharmaceutical Research*, 25: 2835–2844.

Dhumal, R. S., Biradar, S. V., Paradkar, A. R., and York, P. 2009. Particle engineering using sonocrystallization: Salbutamol sulphate for pulmonary delivery. *International Journal of Pharmaceutics*, 368: 129–137.

Dhumal, R. S., Biradar, S. V., Yamamura, S., Paradkar, A. R., and York, P. 2008b. Preparation of amorphous cefuroxime axetil nanoparticles by sonoprecipitation for enhancement of bioavailability. *European Journal of Pharmaceutics and Biopharmaceutics*, 70: 109–115.

Dhumal, R. S., Kelly, A. L., York, P., Coates, P. D., and Paradkar, A. R. 2010. Cocrystallization and simultaneous agglomeration using hot melt extrusion. *Pharmaceutical Research*, 27(12): 2725–2733.

Friscic, T., Childs, S. L., Rizvic, S. A. A., and Jones, W. 2009. The role of solvent in mechanochemical and sonochemical cocrystal formation: A solubility-based approach for predicting cocrystallization outcome. *Crystal Engineering Communications*, 11: 418–426.

Gulseren, D. G., Bruce B. D., and Weiss, J. 2007. Structural and functional changes in ultrasonicated bovine serum albumin solutions. *Ultrasonics Sonochemistry*, 14: 173–183.

Guo, S., Li, Y., Chen, G., and Li, H. 2003. Ultrasonic improvement of rheological and processing behavior of LLDPE during extrusion. *Polymer International*, 52: 68–73.

Guo, Z., Zhang, M., Li, H., Wang, J., and Kougoulos, E. 2005. Effect of ultrasound on anti-solvent crystallization process. *Journal of Crystalline Growth*, 273: 555.

Hickey, A. J., Fults, K. A., and Pillai, R. S. 1992. Use of particle morphology to influence the delivery of drugs from dry powder aerosols. *Journal of Biopharmaceutical Science*, 3: 107–113.

Ikegami, K., Kawashima, Y., Takeuchi, H., Yamamoto, H., Isshiki, N., Momose, D., and Ouchi, K. 2002. Improved inhalation behavior of steroid KSR-592 in vitro with Jethaler® by polymorphic transformation to needle-like crystals (β-form). *Pharmaceutical Research*, 19: 1439–1445.

Isayev, A. I., Kumar, R., and Lewis, T. M. 2009. Ultrasound assisted twin screw extrusion of polymer-nanocomposites containing carbon nanotubes. *Polymer*, 50: 250–260.

Kaerger, J. S. and Price, R. 2004. Processing of spherical crystalline particles via a novel solution atomization and crystallization by sonication (SAXS) Technique. *Pharmaceutical Research*, 21: 372–381.

Kapustin, A. P. 1963. *The Effects of Ultrasound on the Kinetics of Crystallization*. U.S. Academy of Sciences Press, English Translation Consultants Bureau: New York.

Lancaster, R. W., Sigh, H., and Theophilus, A. L. 2000. Apparatus and process for preparing crystalline particles. WO/2000/038811.

Larhrib, H., Martin, G. P., Marriot, C., and Prime, D. 2003. The influence of carrier and drug morphology on the drug delivery from dry powder formulations. *International Journal of Pharmaceutics*, 257: 283–296.

Louhi-Kultanen, M., Karjalainen, M., Rantanen, J., Huhtanen, M., and Kallas, J. 2006. Crystallization of glycine with ultrasound. *International Journal of Pharmaceutics*, 320: 23–29.

Luque de Castro, M. D. and Priego-Capote, F. 2007. Ultrasound-assisted crystallization (sonocrystallization). *Ultrasonics Sonochemistry*, 14: 717–724.

Lyczko, N., Espitalier, F., Louisnard, O., and Schwartzentruber, J. 2002. Effect of ultrasound on the induction time and metastable zone width of potassium sulphate. *Chemical Engineering Journal*, 86: 233–241.

Maheshwari, M., Jahagidar, H., and Paradkar, A. R. 2005. Melt sonocrystallization of ibuprofen: Effect on crystal properties. *European Journal of Pharmaceutical Science*, 25: 41–48.

McCausland, L. J., Cains, P. W., and Martin, P. D. 2001. Use the power of sonocrystallization for improved properties. *Chemical Engineering Progress*, 97: 56.

Nishida, I. 2004. Precipitation of calcium carbonate by ultrasonic irradiation. *Ultrasonics Sonochemistry*, 11: 423–428.

Paradkar, A., Maheshwari, M., Kamble, R., Grimsey, I., and York, P. 2006. Design and evaluation of celecoxib porous particles using melt sonocrystallization. *Pharmaceutical Research*, 23: 1395–1400.

Ratsimba, B., Biscans, B., Delmas, H., and Jenck, J. 1999. Sonocrystallization: The end of empiricism? A review on the fundamental investigations and the industrial developments. *Kona*, 17: 38–48.

Reshetnyak, I. I. 1975. Effect of ultrasound on crystallization kinetics in small volumes of solutions. *Akust. Zh.*, 21: 99–103.

Richards, W. T. and Loomis, A. I. 1927. The chemical effects of high-frequency sound waves I. A preliminary survey. *Journal of American Chemical Society*, 49: 3086–3100.

Ruecroft, G. 2007. Sonocrystallization to rescue. *Innovative Pharmaceutical Technology*, 22: 74–76.

Singh, H., Theophilus, A., and Lancaster, R. W. 2002. Apparatus and process for preparing crystalline particles. US Patent No. 6482438.

Steckel, H., Markefka, P., Wierik, H., and Kammelar, R. 2004. Functionality testing of inhalation grade lactose. *European Journal of Pharmaceutical Science*, 57: 495–505.

Tang, P., Chan, H. K., Tam, E., Gruyter, N., and Chan, J. 2006. Preparation of NaCl powder suitable for inhalation. *Industrial Engineering and Chemistry Research*, 45: 4188–4192.

Thompson, L. H. and Doraiswamy, L. K. 2000. The rate enhancing effect of ultrasound by inducing supersaturation in a solid–liquid system. *Chemical Engineering Science*, 55: 3085–3090.

Thongson, C., Davidson, P. M., Mahakarnchanakul, W., and Weiss, J. 2004. Antimicrobial activity of ultrasound-assisted solvent-extracted spices. *Letters in Applied Microbiology*, 39: 401–406.

Ticehurst, M. D., Basford, P. A., Dallman, C. I., Lukas, T. M., Marshall, P. V., Nichols, G., and Smith, D. 2000. Characterization of the influence of micronization on the crystallinity and physical stability of revatropate hydrobromide. *International Journal of Pharmaceutics*, 193: 247–259.

Trask, A. V. 1997. An overview of pharmaceutical cocrystals as intellectual property. *Molecular Pharmaceutics*, 4: 301–309.

Waltersson, J. O. and Lundgren, P. 1985. The effect of mechanical comminution on drug stability. *Acta Pharmaceutica Suecica*, 22: 291–300.

Young, F. R. 1989. *Cavitation*, McGraw-Hill: New York.

Zeng, X. M., Martin, G. P., Marriott, C., and Pritchard, J. 2000. The influence of carrier morphology on drug delivery by dry powder inhalers. *International Journal of Pharmaceutics*, 200: 93–106.

Zeng, X. M., Martin, G. P., Marriott, C., and Pritchard, J. 2001. Crystallization of lactose from carbopol gels. *Pharmaceutical Research*, 17: 879–886.

Zhao, G., Chen, F., Wang, Z., Wu, J., and Hu, X. 2006. Different effects of microwave and ultrasound on the stability of (all-E)-astaxanthin. *Journal of Agricultural and Food Chemistry*, 54: 8346–8351.

5 Applications of Sonochemistry in Pharmaceutical Sciences

Robina Farooq

CONTENTS

Introduction 97
An Ultrasonic Driven Powder Transport System 97
Effect of Ultrasound on Synthesis of Drugs 98
Ultrasound-Induced Polymerization and Depolymerization 99
Role of Ultrasound in Extraction 99
Photo-Acoustic Evaluation of Pharmaceutical Tablets with Ultrasound 100
Effect of Sonocrystallization in Formation of Aerosols..100
Application of Ultrasound in Transdermal Drug Delivery 100
Effect of Ultrasound in Chemotherapy 101
Application of Ultrasound in Cell Therapy 101
Ultrasonic Irradiation of Toxic Effluent 101
Conclusion 102
References 102

INTRODUCTION

Pharmaceutics is a discipline of pharmacy that deals with all facets of the process to turn new chemical entity (NCE) into proper medication. Pharmaceutics deals with the formation of a pure substance into dosage form such as capsule, injection, suppository, cream, ointment, eye drop, inhaler, nasal spray, etc. Pharmaceuticals are indispensable to health systems; they can complement other types of health care services to reduce morbidity and mortality rates and enhance quality of life at the systems levels. Pharmaceuticals, if used appropriately, have the power to make our lives qualitatively better and longer. As pharmaceuticals have curative and therapeutic qualities, they cannot be considered simply ordinary commodities or even basic health stimuli for that matter (Floros and Liang, 1994; Mason et al., 1996; Mason, 1998; Ishtiaq et al., 2009).

This chapter is intended mainly to review the application of ultrasound in the field of pharmaceutics, from drug dispensation, formulation, to its delivery. It also identifies various related factors encompassing their diverse processes or methods. Various areas have been identified for their great potential for future development, e.g., crystallization, drying, extraction, filtration, homogenization, and synthesis.

AN ULTRASONIC DRIVEN POWDER TRANSPORT SYSTEM

The transport and dosage of granular materials are important components of process engineering. The most accurate mixing process in chemical, pharmaceutical, and food industries demands for an exact control of powder feeding. A novel powder-feeding device is developed at the Heinz Nixdorf Institute and is based on piezoelectrically excited traveling waves suggested by many workers on this principle (Kanbe et al., 1993; Takano and Tomikawa, 1998). An acrylic pipe, which is stimulated to oscillations in the form of traveling waves through piezoelectric impulses, is used to convey the powder. A piezoelectric actuator, as described in Mracek and Wallaschek (2005), is normally used for generating a progressive wave in an acrylic pipe. The acrylic pipe has specific damping properties that allow the excitation of a progressive flexural radial wave in a pipe, using only a single piezoelectric actuator (Mracek and Wallaschek, 2005).

An ultrasonic driven powder transport system was developed by Mracek and Wallaschek (2005), in which the wave stimulation and expansion pipes are constructed from relatively strongly absorbing materials. The movement of a progressive wave in the acrylic pipe and especially the desired movement of the surface points are essential for a powder transport. The developed ultrasonic driven powder system distinguishes itself from conventional dosage systems due to simple structure, nearly no wear and tear in operation, cost-effective, and easy to integrate into existing production plants.

The technology also allows miniaturization, operation in the low voltage area, and a very careful, exact dosage. The experimental investigations of this prototype for ultrasonic powder feeding showed that it is possible to feed small amounts of powder with extremely high quantitative accuracy. But it was also confirmed experimentally that the performance of this system, like most piezoelectric systems, showed a significant sensitivity concerning environmental conditions (e.g., temperature, aging, load influence of the outside, production inaccuracies).

Therefore, ultrasound-driven powder-feeding system can be used for very accurate measurements and movement of materials for the preparation of drugs.

EFFECT OF ULTRASOUND ON SYNTHESIS OF DRUGS

The advantages of ultrasound in chemical reactions, shorter reaction time and higher yields, can be used in industrial application in pharmaceutical or fine chemical industry. Sonochemistry can be used for fast reactions or in the synthesis of expensive products. Scale-up problems can be solved with the aim to develop a technically feasible process. Development of process technology (reactor design, process simulation) is necessary. The most important factor is the establishment of a commercialized process (Mason et al., 1997; Emery et al., 2005).

Many studies on the effect of ultrasound on the synthesis of drugs have been carried out and it is well documented (Mason et al., 1997; Emery et al., 2005). The interaction between ultrasound energy and the vibrational and electronic levels of molecules is an indirect phenomenon. In 1984, Ando and coworker published a paper about reaction of benzyl bromide and alumina supported potassium cyanide in toluene (Aquino et al., 1997). This is the example that ultrasound irradiation induces a particular reactivity leading to products differing in nature from those obtained conventionally and is called sonochemical switching, as shown in Scheme 5.1.

It has been found that, under the influence of ultrasound, this scheme can synthesize with 61% yield in a shorter reaction time. The application of ultrasound in chemical reactions for the reasons that using such types of nonclassical methods give an alternative method for the influence of selectivity and yield of reactions. This application of ultrasound in the dehydration process was tested

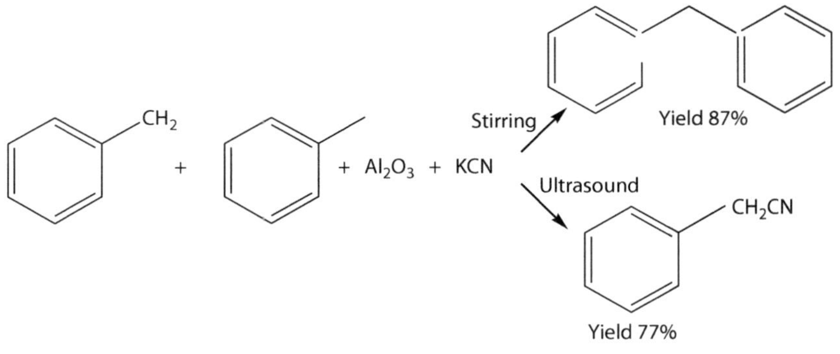

SCHEME 5.1

with other amides. It was found that in these reactions amides could be dehydrated to the corresponding nitriles in good yields and in a shorter reaction time (Aquino et al., 1998).

ULTRASOUND-INDUCED POLYMERIZATION AND DEPOLYMERIZATION

The chemical effects of ultrasound include the formation of radicals and the enhancement of reaction rates at ambient temperatures. The enhanced dissolution of a solid reactant or catalyst caused by renewal of the liquid at the solid–liquid interface illustrates a mechanical effect induced by ultrasound. The formation, growth, and collapse of a cavity occur in 0.1 ms. The implosions of these cavities generate temperatures and pressures of approximately 5000 K and 200 bar, respectively (Mason et al., 1997).

The possibility of cavitation at high pressures creates several interesting potentials in the development of ultrasound-induced chemical processes, such as, precipitation polymerization in CO_2, bulk ethylene polymerization, phase-transfer catalysis in liquid by using ultrasound. It is possible to carry out radical polymerizations without the addition of initiator or catalyst. However, the obtained yield for a bulk polymerization is rather low as a result of the increase in viscosity. A high viscosity hinders cavitations and consequently the production of radicals. In order to overcome this conversion limitation, precipitation polymerization can be a solution (Ando et al., 1984; Price, 1996).

Two types of precipitation systems are used in the high-pressure setup; systems in which the monomer acts as reacting species and as anti-solvent for the polymer, e.g., ethylene systems in which carbon dioxide is applied as anti-solvent, e.g., methyl methacrylate with carbon dioxide besides polymerization, depolymerization also occurs through irradiation with ultrasound. Under sufficiently strong flow conditions around a cavitation bubble, the solvent drag force causes extension of the polymer molecule. If the drag force on the stretched polymer molecule is larger than the bond strength of the polymer, scission will occur. This scission will preferably occur in the middle of the chain. Ultrasound-induced precipitation polymerization enables the production of well-defined polymers (MWD) without conversion limitations used in many processes like pharmaceutics (Zhu et al., 1993; Adeoya-Osiguwa et al., 2003).

ROLE OF ULTRASOUND IN EXTRACTION

A rapid, sensitive, and accurate ultrasonic extraction method has been developed for nicotine in pharmaceutical formulations. The results obtained in this study indicate that ultrasound is a reliable tool for the fast extraction of nicotine in pharmaceutical formulations. The ultrasonic extraction can shorten the extraction time from 24 h to <20 min as compared with the conventional cold extraction technique. Solvent consumption is six times lower in ultrasonic extraction in contrast to similar conventional extraction methods (Kanbe et al., 1993).

In order to extract nicotine from the pharmaceutical formulations, the ground chewing gum sample, or small pieces of transdermal patch was placed in 15 mL glass vial containing 10 mL heptane and sonicated for 20, 40, and 60 min at 37°C in the ultrasonic bath. The results indicated that the recovery of nicotine enhances with increased sonication time. The complete extraction of nicotine was obtained in <20 min from the transdermal patch and 60 min from the chewing gum. Ultrasonic extraction reduced the extraction time from 24 h to <20 min in comparison with the use of conventional cold extraction technique (Yuegang et al., 2004).

Ultrasound was also used to increase the extraction efficiency of carnosic acid from the herb *Rosmarinus officinalis* using butanone, ethyl acetate, and ethanol as solvents (Vinatoru et al., 1999). Similarly, specific examples of the benefits include the extraction of tea solids from dried leaves with water, using ultrasound and thus giving an improvement of almost 20% in yield at 60°C, approaching the efficiency of thermal extraction at 100°C (Mason and Zhao, 1994). A reduction in the maceration time from 8 h to 15 min has been reported in the extraction of the alkaloid reserpine from *Rawolfia serpentina* (Bose and Sen, 1961). The efficiency surged in extracting pharmacologically active compounds from Salvia Officinalis with some 60% of the target compounds extracted within 2 h at around ambient temperature (Salisova et al., 1997).

PHOTO-ACOUSTIC EVALUATION OF PHARMACEUTICAL TABLETS WITH ULTRASOUND

In photo-acoustics, the ultrasound is generated by means of pulsed laser illumination. Normally, photo-acoustical methods are used in applications where a touching ultrasonic transducer would damage the sample or it itself would be damaged. This can be the case with porous and hygroscopic systems, e.g., pharmaceutical tablets, where the use of a coupling liquid would be detrimental to the structure of the tablets. Despite its low efficiency in producing ultrasound, the so-called thermoelastic regime is attractive in many applications, because the phenomenon is nondestructive to the samples and the theory of such nondestructive testing is well-established (Berthelot, 1989; McDonald, 1989).

Variations in porosity, density, and sodium chloride content of microcrystalline cellulose tablets were found to be related to parameters extracted from the through-transmitted ultrasonic wave forms. By using the amplitudes and ultrasonic velocities of these wave forms, it was possible to obtain values of a transverse to longitudinal amplitude ratio, and also elastic parameters. The transverse to longitudinal amplitude ratio and the amplitudinal Poisson's ratio (Poisson's ratio is a measure of the simultaneous change in elongation and in cross-sectional area within the elastic range during a tensile or compressive test) were indicative of structural variations, e.g., changes in the porosity and the sodium chloride content of tablets. Changes in mechanical structure can affect the physical and biopharmaceutical properties, and in some cases even the chemical stability, of pharmaceutical tablets (Kitazawa et al., 1975).

EFFECT OF SONOCRYSTALLIZATION IN FORMATION OF AEROSOLS

Sodium chloride aerosols have been widely used as part of bronchial provocation tests to identify people with active asthma and exercise-induced asthma and those who wish to enter particular occupations (e.g., police, army) or sports (e.g., diving). It has become practical to prepare dry NaCl powder that can be delivered using dry powder inhalers (DPIs) instead of traditional nebulizers. Preparation of traditional nebulizers is energy intensive and time consuming and can induce impurities into the product. It also has several other disadvantages including inadequate control of particle size, undesired particle shape, surface charge modifications, decreased crystallinity, and possible chemical degradation (Waltersson and Lundgren, 1985; Parrott, 1990; Malcolmson and Embleton, 1998; Ticehurs et al., 2000; Chan and Chew, 2003; Rasenack et al., 2004; Shoyele and Cawthorne, 2006). These disadvantages of the micronization process ultimately jeopardize the NaCl powder performance. Sonocrystallization offers several advantages including production of smaller sized crystal as compared to conventional crystallization as well as cost-effectiveness of apparatus. The process can be run at ambient conditions and the reaction vessel involved is of simple geometry, making the cleaning process simple for the pharmaceutical requirements (Mullin and Nyvli, 1971).

Another important advantage is that crystal growth occurs at lower supersaturation levels, where initial growth is controlled; the size distribution of the products is narrower than uncontrolled conventional crystallization process. The limitations of conventional crystallization techniques in the processing pharmaceutical ingredients for a number of dosage forms typically require the need for micronization (Guo et al., 2006; Alza Cooperation).

APPLICATION OF ULTRASOUND IN TRANSDERMAL DRUG DELIVERY

Transdermal drug delivery offers several advantages over traditional drug delivery systems such as oral delivery and injections—the attractive attributes of transdermal drug delivery includes avoidance of first-pass metabolism, elimination of pain that is associated with injection, and the opportunity for the sustained release of drugs. However, the efficiency of transdermal transport of molecules is low because the stratum corneum of the human skin is an effective and selective barrier to chemical permeation (Scheuplein and Blank, 1971). Indeed, the low permeability of the stratum corneum is the key reason that only a small number of low molecular weight drugs are currently administered using this route. The mechanisms are developed for the transdermal transport of drugs at low frequency ultrasound (Prausnitz et al., 2004).

The in vitro permeation enhancement of several low molecular weight drugs under the same ultrasound conditions is reported (Bommannan et al., 1992a; Mitragotri et al., 1995; Lavon and Kost, 2004). It is hypothesized that since the absorption coefficient of the skin varies directly with the ultrasound frequency, high-frequency ultrasound energy would concentrate more in the epidermis, thus leading to higher enhancements. They found that a 20-min application of ultrasound (0.2 W cm^{-2}) at a frequency of 2 MHz did not significantly enhance the amount of salicylic acid penetrating the skin. However, 10 MHz ultrasound under the same conditions resulted in a 4-fold increase and 16 MHz ultrasound resulted in about a 2.5-fold increase in transdermal salicylic acid transport (Bommannan et al., 1992b; Ruecroft et al., 2005).

EFFECT OF ULTRASOUND IN CHEMOTHERAPY

Ultrasound has been used in many life science fields such as medical imaging and diagnostics, biological cell disruption, and fermentation processes. The application of power ultrasound (20–100 kHz, or even up to 2 MHz) to chemical processing has seen steady progress over the past 15 years or so, and a widely reported aspect of this is the use of sonochemistry to promote or modify chemical reactions (Beverley et al., 2004). Traditional preparations used for delivery of drugs include ointments, gels, creams, and medicinal plasters containing natural herbs and compounds. The development of the first pharmaceutical transdermal patch for motion sickness (Adewuyi, 2001) in the early 1980s heralded acceptance of the benefits and applicability of this method of administration for modern commercial drug products. The success of this approach is evidenced by the fact that there are currently more than 35 transdermal drug delivery (TDD) products approved in the United States for the treatment of a wide variety of conditions including: hypertension, angina, motion sickness, female menopause, male hypogonadism, severe pain, local pain control, nicotine dependence, and, recently, contraception and urinary incontinence. There are also several products in late-stage development that will further expand TDD usage into new therapeutic areas, including Parkinson's disease, attention deficit, and hyperactivity disorder as well as female sexual dysfunction (Hynynen et al., 2001).

APPLICATION OF ULTRASOUND IN CELL THERAPY

By combining focused ultrasound technology with the properties of magnetic resonance imaging, a system has been developed (ExAblate 2000) that enables precise targeting within tissues. In addition, temperature sensitive magnetic resonance (MR) sequences provide real-time feedback of focal rises in temperature to ensure safe delivery of an effective thermal dose. Although studies have been carried out in many different areas including breast, brain, and liver tumors, the largest body of work, to date, has taken place in women with symptomatic uterine fibroids (Adeoya-Osiguwa et al., 2003; Jolesz et al., 2004).

ULTRASONIC IRRADIATION OF TOXIC EFFLUENT

In recent years, considerable interest has been shown on the application of ultrasound for hazardous chemical destruction, including among others the degradation of chlorinated hydrocarbons, aromatic compounds, pesticides, explosives, dyes, and surfactants (Emery et al., 2005). The degradation of triphenylphosphine oxide (TPPO) in water, a toxic compound typically found in effluents from the pharmaceutical industry by means of ultrasonic irradiation at 20 kHz (Zhu et al., 1993).

There are many reports documenting the adverse effects, such as feminization of fish, of estrogen hormones in the environment. One of the major sources of these compounds is from municipal wastewater effluents. The occurrence of estrogen hormones in natural systems like surface water, soil, and sediment has become a subject of significant concern. There are many sources of estrogenic pollution, which include effluent from municipal and industrial wastewater treatment plants, livestock wastes, biosolids, septic tanks, and landfills. Estrogenic hormones have also been linked to lower sperm counts in adult males and an increase of cancer (Zhu et al., 1993; Emery et al., 2005).

The biological processes at municipal wastewater treatment plants cannot completely remove these compounds. The effect of ultrasound power density and power intensity on the destruction of various estrogen compounds that include 17α-estradiol, 17α-estradiol, estrone, estriol, equilin, 17α-dihydroequilin, 17α-ethinyl estradiol, and norgestrel were conducted in single component batch and flow-through reactors using 0.6, 2, and 4 kW ultrasound sources. The sonolysis process produced 80%–90% destruction of individual estrogens at initial concentration of 10 μg L^{-1} within 40–60 min. The estrogen degradation rates increase with increase in power intensity of ultrasound. The sonolysis process could be used for the effective destruction of estrogen compounds present in aqueous solutions. Sonication is a process wherein ultrasound waves are irradiated into a liquid medium to destroy the contaminants (Suri et al., 2007).

CONCLUSION

Ultrasound shows the main advantage shorter reaction/preparation time; reduction of the sample preparation time; usage of small amounts of material; efficient and minimum expenditure on solvents, reagents; and the increasing of the sample throughput; it is very useful for the isolation and purification of compounds of interest from the point of view of their pharmacological and other bioactive properties and pharmaceutical formulation. Ultrasonic techniques when compared with the conventional methods like extraction, crystallization, evaporation, sonication, and sonolysis TDD methods appears to be more effective. Sonolysis appeared to be more effective in the destruction of toxic materials in the pharmaceutical industrial effluents. In pharmaceutics, the use of ultrasound gives more economic process and improved environmental and health and safety considerations. Low-frequency sonophoresis has been shown to increase skin permeability to a variety of low- as well as high molecular weight drugs. Sonocrystallization can become a core technology in the pharmaceutical industry and it is expected to see more industrial application in the near future. A lower influence of the particle size on the release properties is obtained for the US tablets in comparison with traditional tablets. On the other hand, a lower variability in the pharmaceutical availability is expected for the systems obtained using ultrasound-assisted compression. It is expected that this ultrasound control system, after further studies, could be developed and applied for the optimization of pharmaceuticals freeze-drying cycles in industrial condition. Photo-acoustic evaluation using ultrasound not only shows the elastic properties of the tablet materials but also helps in evaluating its internal structure.

REFERENCES

Adeoya-Osiguwa, S.A., Markoulaki, S. et al. 2003. 17 β-Esteadiol and environmental estrogens significantly affect, mammalian sperm function. *Human Reproduction*, 18:100–107.

Adewuyi, Y.G. 2001. Sonochemistry: Environmental science and engineering applications. *Industrial Engineering and Chemistry Research*, 40:4681–4715.

Alza Corporation, Mountain View, CA. http://www.alza.com

Ando, T., Sumi, S., Kawata, T., Ichihara, J., and Hanafusa, T.J. 1984. *Journal of Chemical Society and Chemical Communications*, 3:436–439.

Aquino, F., Bonrath, W., Couturier, L., Pauling, H., and Villaine, S. 1997. *COST Workshop*, Santorini, Greece.

Aquino, F., Bonrath, W., and Pauling, H. 1998. Dehydration of 4-methyl-oxazole-5-carboxylic acid amide to 4-methyl-oxazole-5-carbonitrile. In Luche, J.L. et al. *Chemical Processes and Reactions under Extreme or Non-Classic Conditions*. European Commission: Luxembourg, Belgium, p. 123.

Berthelot, Y.H. 1989. Thermoacoustic generation of narrow-band signals with high repetition rate pulse lasers. *Journal of Acoustic Society of America*, 85:1173–1181.

Beverley, J. and Barrie, C.F. 2004. The transdermal revolution, research focus reviews. *Drug Discovery Today*, 9:697–703.

Bommannan, D., Menon, G.K., Okuyama, H. et al. 1992a. Sonophoresis: II. Examination of the mechanism(s) of ultrasound-enhanced transdermal drug delivery. *Pharmaceutical Research*, 9:47.

Bommannan, D., Okuyama, H., Stauffer, P. et al. 1992b. Sonophoresis: I. The use of high-frequency ultrasound to enhance transdermal drug delivery. *Pharmaceutical Research*, 9:559–564.

Bose, P.C. and Sen, T.C. 1961. Observations on the action and use of symplocos racemosa. *Indian Journal of Pharmacy*, 23:222.

Chan, H.K. and Chew, Y.N. 2003. Novel alternative methods for the delivery of drugs for the treatment of asthma. *Advanced Drug Delivery Reviews*, 55:793–805.

Emery, R.J., Papadebi, M., Treitas Dos Santos, L.M. et al. 2005. Extent of sonochemical degradation and change of toxicity of a pharmaceutical precursor (triphenylphosphine oxide) in water as a function of treatment conditions. *Environment International*, 31:207–211.

Floros, J.D. and Liang, H.H. 1994. Acoustically assisted diffusion through membranes. *Food Technology*, 48:79–84.

Guo, Z., Jones, A.G., and Li, N. 2006. The effect of ultrasound on the homogeneous nucleation of $BaSO_4$ during reactive crystallization. *Chemical Engineering Science*, 61:1617–1626.

Hynynen, K., Pomeroy, O., and Smith, D.N. 2001. MR imaging-guided focused ultrasound surgery of fibroadenomas in the breast a feasibility study. *Radiology*, 219:176–185.

Ishtiaq, F., Farooq, R., Farooq, U. et al. 2009. Application of ultrasound in pharmaceutics. *World Applied Science Journal*, 6:886–893.

Jolesz, F.A.K., Hynynen, N., McDonnold, D. et al. 2004. Noninvasive thermal ablation of hepatocellular carcinoma by using magnetic resonance imaging-guided focused ultrasound. *Gastroenterology*, 127:242–247.

Kanbe, N., Takano, T., Tomikawa, Y., and Adachi, K. 1993. Analysis of axisymmetric waves propagating along a hollow cylindrical ultrasonic transmission line. *Journal of Acoustic Society of America*, 93:6.

Kitazawa, S., Johno, I., Ito, Y., Teramura, S., and Okada, J. 1975. Effects of hardness on the disintegration time and dissolution rate of uncoated caffeine tablets. *Journal of Pharmacy and Pharmacology*, 27:765–770.

Lavon, I. and Kost, J. 2004. Ultrasound and transdermal drug delivery. *Drug Discovery Today*, 9:670–676.

Malcolmson, R.J. and Embleton, J.K. 1998. Dry powder formulations for pulmonary delivery. *Pharmaceutical Science and Technology Today*, 1:394–398.

Mason, T.J. 1998. Power ultrasound in food processing—The way forward. In Povey, M.J.W. and Mason, T.J. (Eds.), *Ultrasound in Food Processing*. Blackie Academic & Professional: Glasgow, U.K., pp. 104–124.

Mason, T.J., Luche, J.L., Eldick, R.V., and Hubbard, C.D. (Eds.). 1997. *Chemistry under Extreme or Non-Classical Conditions*. Wiley: New York, p. 317.

Mason, T.J., Paniwnyk, L., and Lorimer, J.P. 1996. The use of ultrasound in food technology. *Ultrasonics Sonochemistry*, 3:253–256.

Mason, T.J. and Zhao, Y. 1994. Enhanced extraction of tea solids using ultrasound. *Ultrasonics*, 32:375–377.

McDonald, F.A. 1989. Practical quantitative theory of photo acoustic pulse generation. *Applied Physical Letters*, 54:1504–1506.

Mitragotri, S., Edwards, D., Blankschtein, D., and Langer, R. 1995. A mechanistic study of ultrasonically enhanced transdermal drug delivery. *Journal of Pharmaceutical Sciences*, 84:697–706.

Mracek, M. and Wallaschek, J. 2005. A system for powder transport based on piezoelectrically excited ultrasonic progressive waves. *Materials Chemistry and Physics*, 90:378–380.

Mullin, J.W. and Nyvli, J. 1971. Programmed cooling of batched crystallizers. *Chemical Engineering Science*, 26:369–377.

Parrott, E.L. 1990. Communication. In Swarbrick, J., Boylan, J.C. (Eds.), *Encyclopedia of Pharmaceutical Technology*. Marcel Decker Inc.: New York, Vol. 3, pp. 101–121.

Prausnitz, M.R., Mitragotri, S., and Langer, R. 2004. Current status and future potential of transdermal drug delivery. *Nature Reviews in Drug Discovery*, 3:115–124.

Price, G.J. 1996. Ultrasonically enhanced polymer synthesis. *Ultrasonics Sonochemistry*, 3:229–238.

Rasenack, N., Steckel, H., and Muller, B.W. 2004. Preparation of microcrystals by in situ micronization. *Powder Technology*, 143–144:291–296.

Ruecroft, G., Hipkiss, D., Maxted, N., and Cains, P.W. 2005. *Journal of Organic Proceedings in Research and Development*, 9:923–932.

Salisova, M., Toma, S., and Mason, T.J. 1997. Ultrasound-assisted extraction of rutin and quercetin. *Ultrasonics Sonochemistry*, 4:131–134.

Scheuplein, R.J. and Blank, I.H. 1971. Permeability of the skin. *Physiology Reviews*, 51:702–747.
Shoyele, S.A. and Cawthorne, S. 2006. Particle engineering techniques for inhaled biopharmaceuticals. *Advanced Drug Delivery Reviews*, 58:1009–1029.
Suri, R.P.S., Nayak, M., Devaiah, U., and Helmig, E. 2007. Ultrasound assisted destruction of estrogen hormones in aqueous solution: Effect of power density, power intensity and reactor configuration. *Journal of Hazardous Materials*, 146:472–478.
Takano, T. and Tomikawa, Y. 1998. Excitation of a progressive wave in a loss ultrasonic transmission line and an application to a powder feeding device. *Smart Materials and Structures*, 7:417–421.
Ticehurst, M.D., Basford, P.A., Dallman, C.I. et al. 2000. Characterisation of the influence of micronisation on the crystallinity and physical stability of revatropate hydrobromide. *International Journal of Pharmacy*, 193:247–259.
Vinatoru, M., Toma, M., and Mason, T.J. 1999. Ultrasonically assisted extraction of bioactive principles from plants and their constituents. In Mason, T.J. (Ed.). *Advances in Sonochemistry*. JAI Press: London, U.K., Vol. 5, pp. 209–248.
Waltersson, J.O. and Lundgren, P. 1985. The effect of mechanical communication on drug stability. *Acta Pharmaceutica Suecica*, 22:291–300.
Yuegang, Z., Liliang, Z., Jingping, W., Fritz Johnathan, W., Suzanne, M., and Christopher, R. 2004. Ultrasonic extraction and capillary gas chromatography determination of nicotine in pharmaceutical formulations. *Analytica Chimica Acta*, 525:35–39.
Zhu, B.T., Roy, D., and Liehr, J.G. 1993. The carcinogenic activity of ethinyl estrogens in determined by both their hormonal characteristics and their conversion to catechol metabolites. *Endocrinology*, 132:577–583.

6 Ultrasound-Assisted Synthesis of Nanomaterials

Siamak Dadras, Mohammad Javad Torkamany, and Jamshid Sabbaghzadeh

CONTENTS

Introduction 105
Sonochemistry 107
    Sonochemical Reduction in Synthesis of Metallic Nanostructures 107
    Ultrasound-Assisted Sol–Gel in the Synthesis of Metal Oxide Nanostructures 108
    Sonochemical Decomposition in the Synthesis of Metal Chalcogenides 109
Ultrasound-Induced Deposition 110
Sonoelectrochemistry 111
    Preparation of Nanoparticles by Pulsed Sonoelectrochemistry 112
        Metallic Nanopowders 112
        Alloy Nanopowders 113
        Semiconductor Nanopowders 113
        Conducting Polymer Nanoparticles 113
    Parameters That Control the Formation of Nanoparticles 113
        Synthesis Cell Temperature 114
        Current Density 114
        Electrochemical Current Pulse Time 114
        Ultrasound Pulse Intensity 114
        Ultrasound Pulse Time 114
        Stabilizer 114
Ultrasound-Assisted Laser Ablation 115
Ultrasonic Spray Pyrolysis 116
    USP in the Synthesis of Nanostructured Materials 117
    Mesostructured Nanomaterials from Organic–Inorganic Hybrid Nanocomposites ... 117
    Nanosized Structures from Silica-Based Nanocomposites 118
    Macroporous Materials from Polymer-Based Nanocomposites 118
    Nanosized Structures from Metal Salt-Based Nanocomposites 118
    Nanostructured Semiconductors from Chemical Aerosols 119
Conclusions 120
References 121

INTRODUCTION

Nanostructured materials have broad applications in a variety of fields because of their unusual chemical, optical, electronic, and magnetic properties (Schmid, 1992; Burda et al., 2005). They are characterized by an extremely large specific surface area, and their properties are determined mainly by the behavior of their surface (Hodes, 2007). The applications of nanomaterials are well known in the fields of cosmetics and pharmaceutical products, coatings, electronics, energy, and catalysis.

One of the most important tasks in nanoscience is developing new versatile routes for the preparation of nanomaterials with tunable physical and chemical properties accompanied with a dominant control on their size and morphology. There are a range of methods of producing nanosized materials including sol–gel, chemical reduction, hydrothermal, chemical vapor deposition (CVD), thermal decomposition, laser ablation, physical vapor deposition (PVD), and electrochemical routes. Size, morphology, nanostructure, and phase composition of the products out of these methods could be varied with numerous indispensable factors described in detail elsewhere. However, in a number of cases, characteristics of synthesized nanomaterials can be influenced under the action of some additional external factors, such as ultrasound, microwave, and UV radiation. Among these factors, the ultrasound radiation has been widely exploited for the preparation of nanomaterials due to its applicability in most of the liquid-based chemical and physical synthetic routes. Currently, the ultrasound-assisted processes have been proved to be useful for generating various nanostructured materials. Utilization of ultrasound waves provides the capability of improving different features of the products out of the conventional synthetic routes or producing nanomaterials that are often unavailable by these routes.

The main physical phenomena associated with ultrasound in nanomaterials synthesis are cavitation and nebulization. Acoustic cavitation creates extreme conditions that serve as the origin of most sonochemical and sonophysical phenomena by providing an exclusive interaction between energy and matter with hot spots inside the cavitational bubbles in liquids or liquid–solid slurries. During sonication, ultrasonic longitudinal waves are radiated through the solution causing alternating high- and low-pressure regions in the liquid medium. Millions of microscopic bubbles form and grow in the low-pressure stage and subsequently collapse in the high-pressure stage. Hot spots that are localized regions of extremely high temperatures, as high as ~5000 K, and pressures of up to ~1000 bar can occur from the collapsing bubbles, and cooling rates can often exceed ~10^{10} K s^{-1}. The energy released from this process, known as cavitation, would lead to enhanced chemical or physical reaction rates in any of the liquid-based synthesis routes of nanomaterials.

Additionally, nebulization (the creation of mist by the ultrasound waves propagating through a liquid and impinging on a liquid–gas interface) is the basis for ultrasonic spray pyrolysis (USP) with subsequent reactions occurring in the heated droplets of the mist. These phase-separated microreactors created by USP facilitate the formation of a wide range of nanocomposites.

It could be suggested that the formation and growth of nanomaterials in the ultrasound-assisted routes will be promoted by all of the following processes:

1. Formation of extra centers of nucleation on cavitation bubbles or nebulization microdroplets
2. Increase of the particle growth rate due to rise of effective diffusion coefficients as well as continuous renewal of the surface growing particles
3. Breakage of particle aggregates due to shock waves and microjets of molecules

In this chapter, the most successful ultrasound-assisted methods for the synthesis of nanomaterials are discussed with the following approaches:

1. Sonochemical reduction, sol–gel, and decomposition which lead to the generation of metal, metal oxide, metal chalcogenide/carbide nanostructures respectively
2. Ultrasound-induced deposition
3. Sonoelectrochemistry
4. Ultrasound-assisted laser ablation in liquid media
5. USP

In the following sections, the above-mentioned methods are discussed in detail to provide a fundamental understanding of their basic principles and to illustrate the influential and exclusive features of ultrasound in nanostructured material synthesis.

SONOCHEMISTRY

Chemistry deals with the interaction between energy and matter, and chemical reactions require some forms of energy such as heat, light irradiation, electric potential, and so on to proceed. Dominant control on the chemical reactions is a key factor to accomplish the nanostructured material synthesis. However such a control is currently limited to the manipulations of various reaction parameters including time, energy input, and pressure. In comparison with traditional energy sources, ultrasonic irradiation provides rather exceptional reaction conditions (a short duration of extremely high temperatures and pressures in liquids) that cannot be achieved by other methods.

Acoustic cavitation, which is driven by high-intensity ultrasound irradiation, accounts for the chemical effects of ultrasound (Suslick and Doktycz, 1990). When liquids get sonicated, the alternating expansive and compressive acoustic waves create bubbles and make the bubbles oscillate. The oscillating bubbles can accumulate ultrasonic energy effectively while growing to a certain size, typically tens of microns. Under such a condition, a bubble can overgrow and subsequently collapse, releasing the concentrated energy stored in the bubble within a very short time (with a heating and cooling rate of $>10^{10}$ K s^{-1}). This cavitational implosion is very localized and transient with a temperature around 5000 K and a pressure around 1000 bar (Suslick, 1990).

The cavitation phenomenon occurs over a wide range of ultrasound frequencies, from tens of Hz to tens of MHz, and above this frequency interval, the intrinsic viscosity of liquids prevents cavitation from occurring. Most high-intensity ultrasonic horns operate at 20 or 40 kHz, most cleaning baths near 40 kHz, and there is specialized equipment available in the few hundred kHz to few MHz regime (Bang et al., 2010). Generally speaking, the physical effects of ultrasound such as emulsification and surface damage are more dominant at lower frequencies, whereas cavitational heating of collapsing bubbles occurs over the full frequency range (Bang et al., 2010). A powerful aspect of the sonochemical synthesis resides in its versatility; various forms of nanostructured metals, oxides, chalcogenides, and carbides can be prepared simply by changing reaction conditions as the following processes:

SONOCHEMICAL REDUCTION IN SYNTHESIS OF METALLIC NANOSTRUCTURES

Syntheses of metallic nanoparticles via sonochemical reduction have been investigated by a number of groups in decades 1990s and 2000s (Suslick et al., 1991, 1996a; Nagata et al., 1992; Yeung et al., 1993; Okitsu et al., 1996; Dhas et al., 1998; Mizukoshi et al.; 1999; Caruso et al., 2002; Su et al., 2003; Nemamcha et al., 2006; Okitsu et al., 2009). Reduction of metal salts through the ultrasound irradiation has considerable advantages over other conventional reduction methods; no chemical reducing agent is needed, the reaction times are relatively short, and generation of very small particles is possible (Dhas et al., 1998; Mizukoshi et al., 1999; Caruso et al., 2002; Su et al., 2003; Nemamcha et al., 2006). In this case, sonolysis of aqueous liquids leads to the generation of free H• and OH• radicals (Makino et al., 1982). Sonochemically generated H radicals are considered to act as reductants, as summarized below. Often, organic additives (e.g., 2-propanol or surfactants) are added to produce a secondary radical, which can considerably promote the reduction rate:

$$H_2O \rightarrow H\bullet + OH$$

$$RH + OH\bullet \text{ (or } H\bullet) \rightarrow R\bullet + H_2O \text{ (or } H_2)$$

$$Au(III) \text{ (Ag(I), Pt(II) or Pd(II))} + H\bullet \text{ (or } R\bullet) \rightarrow Au(0) \text{ (Ag(0), Pt(0) or Pd(0))}$$

$$nM(0) \rightarrow M_n \text{ (M = metallic nanostructure)}$$

A systematic set of studies have been done on sonochemical reduction process to understand the reduction mechanism and the effects of operating parameters (e.g., time, concentration, ultrasonic frequency, and different organic additives) on particle size and shape (Caruso et al., 2002; Okitsu et al., 2005, 2009; Vinodgopal et al., 2006; Anandan et al., 2008; Brotchie et al., 2008). An interesting observation of these studies is that the size of particles inversely depends on alcohol concentration and alkyl chain length (Caruso et al., 2002). Also, the rate of sonochemical reduction is strongly dependent on the ultrasonic frequency, at least within the range of the specific ultrasonic apparatus used in those studies, demonstrating that frequency can play a key role in controlling particle size (Okitsu et al., 2005).

The ultimate structure of the nanomaterials synthesized by sonochemical reduction technique can also be influenced by the type of the stabilizing agent utilized in the process. Stabilizers such as oleic acid or polyvinyl-pyrrolidone can trap the sonochemically decomposed metal nanoclusters before aggregation, resulting in colloidal nanoparticle formation (Suslick et al., 1996a). Depending on the presence of an organic or polymeric stabilizer, either agglomerates of nanoparticles or colloidal nanoparticles can be obtained (Suslick et al., 1991, 1996a).

Generally, the sonochemical reduction leads to the formation of spherical nanoparticles, and thus preparation of other metallic nanostructures via this method is less reported in literature. Among the shape-controlled structures, gold nanobelts and nanodecahedra and also silver nanoplates have been merely reported to be synthesized via the sonochemical reduction processes (Jiang et al., 2004b; Sánchez-Iglesias et al., 2006; Zhang et al., 2006; Pastoriza-Santos et al., 2007).

In addition to monometallic nanoparticles, preparation of colloidal bimetallic nanoparticles by the sonochemical reduction route has also been performed by a number of researchers. Bimetallic nanoparticles have been of great interest because of their increased use in catalysis and optoelectronic applications. They may be prepared by either sonochemical decomposition of two different metal salts (e.g., $(Fe(CO)_5)$ and $(Co(CO)_3(NO))$ leading to Fe–Co alloys) or a single-source precursor $(Pt_3Fe_3(CO)_{15}$ leading to FePt nanoparticles) (Bellissent et al., 1995; Suslick et al., 1996b; Rutledge et al., 2006). Composition of the resultant alloy is reported to be simply tuned by changing stoichiometric ratios of the two precursors in the gas phase via their respective vapor pressures (Bellissent et al., 1995).

Among the other bimetallic nanostructures, Au–Pd core-shell nanoparticles, various structures of Au–Ag nanocomposites, and Pt–Ru nanoparticles have been synthesized by different groups via sonochemical reduction in aqueous or alcohol solutions (Mizukoshi et al., 1997, 2000; Basnayake et al., 2006; Vinodgopal et al., 2006; Radziuk et al., 2008).

ULTRASOUND-ASSISTED SOL–GEL IN THE SYNTHESIS OF METAL OXIDE NANOSTRUCTURES

Utilization of sol–gel technique to prepare the metal oxide nanostructures has been widely reported in literature as one of the most successful routes in the synthesis of these materials. Sol–gel, also known as chemical solution deposition, is a wet-chemical technique used in the fields of materials science. Such a method is used primarily for the fabrication of materials (usually metal oxides) starting from a chemical solution which acts as the precursor for an integrated network (or gel) of either discrete particles or network polymers. Typical precursors are metal alkoxides in an alcoholic solvent, which undergo various forms of hydrolysis reactions. These reactions induce the substitution of OR groups linked to metal by metal–OH groups. These chemical species may react together to form metal–oxo (M–O–M) or metal–hydroxo (M–OH–M) bonds which lead to the formation of metal oxide network in solution. Thus, the sol evolves toward the formation of a gel-like diphasic system containing both liquid and solid phases whose morphologies range from discrete nanoparticles to continuous polymer networks (Brinker and Scherer, 1990). In the case of the colloid, a significant amount of fluid may need to be removed initially by sedimentation or centrifugation for the gel-like properties to be recognized. Removal of the remaining liquid (solvent) phase requires a drying process, which is typically accompanied by a thermal treatment in order to favor polycondensation and enhance structural stability (Brinker and Scherer, 1990).

Features of the nanosized structures synthesized by sol–gel route can be promoted by exploiting the ultrasonic irradiation during the hydrolysis process (Bang et al., 2010). The advantages of the ultrasound-assisted sol–gel approach over conventional routes in the synthesis of metal oxides have been recognized by many research groups. Shortening the synthesis time from several days to few hours due to faster hydrolysis process accompanied with more uniform particle size distribution, higher surface area, better thermal stability, and improved phase purity are the noticeable consequences of using ultrasound irradiation in sol–gel technique (Bang et al., 2010). The shortened reaction time can be attributed to an extremely high temperature at the interface between a collapsing bubble and the bulk solution, which was presumed to hasten the hydrolysis and condensation of the precursor (Bang et al., 2010). Examples of successful ultrasound-assisted sol–gel synthesis of metal oxide nanostructures include TiO_2, ZnO, MoO_3, In_2O_3, and SiO_2 (Enomoto et al., 1996; Qian et al., 2003; Rao et al., 2005; Krishnan et al., 2006; Dutta et al., 2008; Jung et al., 2008; Latt et al., 2008; Neppolian et al., 2008; Jafarzadeh et al., 2009; Xiong et al., 2009; Prasad et al., 2010).

SONOCHEMICAL DECOMPOSITION IN THE SYNTHESIS OF METAL CHALCOGENIDES

Metal chalcogenide nanoparticles with fine size and morphology and non-aggregated structures have recently gained considerable importance in the ongoing technological advancement. The conventional methods for preparing metal chalcogenides suffer from several limitations, such as high processing temperature, relatively high cost, nonstoichiometric compositions, and poor crystallinity (Zhu et al., 2005). In recent years, sonochemistry offers an advantageous alternate in the synthesis of these nanomaterials with eliminating or decreasing the inconveniences caused by these limitations. Several metal chalcogenides such as CdS, ZnS, PbS, MoS_2, Bi_2S_3, CdSe, ZnSe, PbSe, Bi_2Se_3, β-CuSe, Cu_3Se_2, Cu_7Te_4, Cu_4Te_3, GaSb, etc. (Mdleleni et al., 1998; Li et al., 2000, 2002, 2003; Zhu et al., 2000b, 2002; Wang et al., 2002; Xie et al., 2002; Zhou et al., 2003a,b; Qiu et al., 2004; Uzcanga et al., 2005) have been synthesized by sonochemical route.

A typical sonochemical preparation of these materials involves the ultrasonic irradiation of an aqueous solution of a metal salt and a chalcogen source (e.g., thiourea for sulfur or selenourea for selenium) in which the in-situ-generated H_2S or H_2Se by sonication reacts with sonochemically decomposed metal salts to produce metal chalcogenide nanoparticles. Using a structure directing agent, a variety of nanostructures such as nanorods, nanowires, or nanocubes can be prepared (Wang et al., 2002; Zhu et al., 2002; Zhou et al., 2003a,b; Qiu et al., 2004).

Sonochemically prepared metal chalcogenides would exhibit unusual morphologies compared to those synthesized by conventional methods. For instance, MoS_2 which is prepared via a conventional route in bulky platelike morphology can be synthesized in the form of a porous agglomeration of nanosized spherical particle clusters. Therefore, because of its nanoparticle nature, the sonochemically prepared MoS_2 has a higher surface area with much higher numbers of edges and defects on the surface than the conventionally obtained counterpart. This makes the layers in the agglomeration to bend, break, or distort to fit to the radius of curvature of the outer surface (Mdleleni et al., 1998).

Sonochemical methods have also been recognized to be very useful for the production of hollow metal chalcogenide nanostructures. In typical synthetic techniques, one utilizes sacrificial template materials such as preformed silica or polymer colloids, which make the synthetic process complicated and inefficient. However, in sonochemical synthesis, a facile and rapid synthetic route is provided to generate hollow inorganic spheres usually without the use of templates.

Synthesis of nanostructured metal chalcogenides via the sonochemical routes is reported to be preferred to that of traditional methods even in producing high-quality semiconductor nanoparticles, the so-called quantum dots. Several points such as the better control over growth rate of nanocrystals, significantly lower reaction temperatures, and high photoluminescence activities of the ultimate products can be highlighted as the advantages of sonochemical procedure over thermal synthesis routes (Murcia et al., 2006).

Sonochemical decomposition route is also applicable in the synthesis of refractory metal carbides such as molybdenum and tungsten carbides. Generally, conventional preparation of metal carbides by reacting a metal and carbon requires an extremely high temperature and also carbides themselves are intrinsically refractory. Thus, the synthesis of nanostructured metal carbides with a large surface area, associated with a high porosity, has still remained a substantial challenge in materials science. Alternatively, a facile sonochemical decomposition route has been developed to prepare nanostructured metal carbides such as molybdenum and tungsten carbides (Mo_2C and W_2C) (Hyeon et al., 1996; Oxley et al., 2004). In this procedure, amorphous oxycarbides are first obtained via sonochemical decomposition of molybdenum hexacarbonyl or tungsten hexacarbonyl in hexadecane. The oxygen is then eliminated by heat treatment under a gas composed of methane and hydrogen. This alternative technique not only reduces the synthesis temperature, but also enables the production of new carbide nanostructures. Nanosized molybdenum and tungsten carbides prepared by the sonochemical decomposition exhibit superior activity, selectivity, and stability for hydrodehalogenation of halogenated organic pollutants (Hyeon et al., 1996; Oxley et al., 2004).

ULTRASOUND-INDUCED DEPOSITION

The physical effects of sonication such as shock waves and microjets at the liquid–solid interface primarily induce unusual effects like surface damage and fragmentation of friable solid associated with rapid interparticle collisions (Suslick and Casadonte, 1987; Suslick and Doktycz, 1989a; Suslick et al., 1989b; Doktycz et al., 1990; Suslick and Price, 1999). Intense ultrasound irradiation is able to rapidly drive the low melting point metal particles (e.g., Zn or Sn) together to induce effective melting at the impact point (Doktycz et al., 1990; Prozorov et al., 2004). This physical phenomenon is the basis of the ultrasound-induced deposition route which is widely used in the synthesis of a myriad of nanomaterials whether on a substrate or another nanostructure.

Among the ultrasound-induced deposition reports, the in-situ-generated noble metal nanoparticles have been deposited on the surface of various substrates (e.g., silica, carbon, or polymer) exploiting the physical features of sonication (Pol et al., 2003a,b, 2005). This technique significantly reduces the reaction time accompanied with uniform coating of nanoparticles on substrates without tailoring surface properties (Pol et al., 2003a,b, 2005).

In addition to noble metal deposition on different substrates, the ultrasound-induced deposition route has been widely utilized for coating metal oxide nanoparticles with silica or metal sulfides (e.g., CdS). In a related report, deposition of silica over Fe_3O_4 nanoparticles has been described, in which the sonication significantly promotes the hydrolysis of TEOS in sol–gel process and improves the mass transport of silica sols to Fe_3O_4 nanoparticle surfaces to achieve homogeneously coated core-shell nanoparticles (Morel et al., 2008). Coating of silica over indium tin oxide (ITO) nanoparticles has also been investigated by the ultrasound-induced deposition route (Chen et al., 2008). This technique, additionally, has been exploited by a number of groups in the synthesis of metal oxide core-shell nanostructures (TiO_2, γ-Fe_2O_3/silane and SnO_2, ZnO/CdS) (Shafi et al., 2001, 2002; Gao et al., 2004, 2005).

Combining the physical and chemical effects of sonication, a bifunctional eggshell catalyst has also been prepared, in which the outer surface of ZSM-5 is decorated with nanometer-sized Mo_2C catalyst particles (Dantsin and Suslick, 2000). This facile route produces uniformly deposited Mo_2C nanoparticles on the outer surface of ZSM-5, greatly improving the poor dispersion of nanoparticles formed in supported catalysts obtained by conventional methods.

Ultrasonic irradiation promotes the formation of graphite intercalation compounds with remarkable efficiency in terms of reaction time. A facile ultrasound-induced synthesis of KC_8, a powerful reducing agent in organic reactions, has been developed in this approach, reducing the synthesis time from 1 to 2 days in a typical route to less than 5 min (Jones et al., 2004). Graphite intercalation reaction by means of ultrasound irradiation has also been reported in another study on the synthesis of Pt nanoparticle-intercalated graphite (Walter et al., 2001).

Among the attempts to synthesize shape-controlled nanomaterials, hollow MoS_2 and MoO_3 spherical nanostructures have been prepared via ultrasound-induced deposition on solid substrates (Dhas and Suslick, 2005). Sonication of a solution of molybdenum hexacarbonyl, sulfur, and silica nanospheres under an Ar flow yields an MoS_2/SiO_2 composite. A similar procedure in the presence of air and the absence of sulfur results in an MoO_3/SiO_2 composite. Hollow spheres of MoS_2 and MoO_3 can be produced by subsequent HF treatment of the composites to leach out the silica spheres. Because of its significantly enhanced number of edge defects and improved accessibility to both inner and outer surfaces, the catalytic activity of hollow MoS_2 is superior to that of sonochemically prepared nanostructures and conventionally synthesized micron-sized counterparts. Furthermore, the heat treatment can induce a strange phase transformation from the hollow nanospheres of MoO_3 into truncated cubic hollow crystals (Dhas et al., 2005).

Preparation of hollow nanostructures from the ultrasonically deposited nanocomposites has been further extended to the synthesis of porous Co_3O_4 nanotubes which are excellent electrode material in lithium batteries. Carbon nanotubes (CNTs) can be exploited as a sacrificial template in this technique and the ultrasonically deposited CoO_x on CNTs is calcinated under air to burn away the CNTs and convert the amorphous CoO_x into crystalline Co_3O_4 (Du et al., 2007). Hollow FePt nanospheres have also been prepared in this jargon by ultrasonically depositing FePt particles on silica spheres and subsequent HF treatment of the resultant core-shell to remove the silica cores (Wang et al., 2007).

SONOELECTROCHEMISTRY

The study of the ultrasound irradiation effects on electrochemical synthesis of nanomaterials was started at the beginning of the current century (Moriguchi, 1934) and since then, it was increasingly expanded due to the growing interest in developing novel synthesis approaches of nanomaterials (Walton, 2002). The cavitation through the ultrasound radiation in the electrolyte is recognized as the responsible phenomenon for unusual effects induced in the ultrasound-assisted electrochemical synthesis processes (Compton et al., 1997a). This phenomenon, in case of taking place close to the surface of the electrode, leads to the penetration of a high-velocity microjet of liquid inside the bubble perpendicular to the electrode surface (Mason, 1989). The collapse of the bubbles is also accompanied with shock waves and microstreaming if the ultrasound intensity is higher than the threshold (Klima et al., 1999; Birkin et al., 2005). So, the overall mass transport and the reaction rates will improve by decreasing the diffusion layer thickness (Compton et al., 1997b; Lorimer et al., 1998) associated with degassing and releasing the deposit of the electrode surface (Mason, 1989; Mason et al., 1990).

The well-defined arrangement for the electrochemically synthesis of nanomaterials in the presence of ultrasound waves is the introduction of an ultrasonic horn system directly into an electrochemistry cell. This allows the ultrasound waves to be directed onto the electrode surface and provides rather efficient power distribution. Different types of sonoelectrochemistry setups using ultrasound horns have been presented in literature. In the most common setup, the ultrasound horn is placed in front of the cathode surface at a known distance from the electrodes (Compton et al., 1997a). Alternatively, the ultrasound horn has itself been converted into the working electrode, the so-called sonotrode or sonoelectrode (Reisse et al., 1994; Compton et al., 1997a). The main idea of the sonoelectrochemical preparation of nanomaterials was subsequently inspired from this system involving consecutive pulses of electrolysis (for deposition) and ultrasound (to release the deposit).

The nucleation process through the pulsed sonoelectrochemical procedure in order to synthesize nanopowders occurs in distinct stages. Initially, a pulse of current (or potential) reduces a number of cations, and consequently a high density of metal nuclei deposit on the sonoelectrode surface, as long as the horn works only as an electrode. A high-intensity ultrasound pulse, applied immediately after the electrochemical pulse, makes the nanoparticles remove from the cathode surface, preparing it for the next stage of deposition. In some cases, a rest time without current or ultrasonic currents, coming after the two previous pulses, would be beneficial to remain the initial conditions around the sonoelectrode surface. Electrochemical and ultrasound pulses typically range between 100 and 500 ms, and the rest time lasts no more than 1 s.

The first system, in which a titanium horn works as both a cathode and an ultrasound probe (20 kHz), used a simple design of a two-electrode cell, and the process was performed under galvanostatic conditions. In this configuration, the ultrasound probe is connected to a generator and a potentiostat using a pulse driver. However, the synthesis process undergoes some undesirable secondary reactions in this case, and to overcome this drawback, a replacement was applied. A three-electrode configuration, involving the working, reference, and auxiliary electrodes, was utilized instead, to apply a controlled potential on the sonoelectrode to get an improved synthesis process. The most frequent case, exploited in the sonoelectrochemical synthesis process, has been the galvanostatic condition because of its simplicity to use for the large-scale production of nanoparticles.

PREPARATION OF NANOPARTICLES BY PULSED SONOELECTROCHEMISTRY

Sonoelectrochemistry has widely been employed to produce numerous pure metals, alloys, and semiconductor nanoparticles. More recently, conductive polymer nanoparticles have also been synthesized by this method. In the following, the attempts for sonoelectrochemical preparation of these nanostructures have been grouped in different categories based on their chemical nature.

Metallic Nanopowders

Copper was one of the first metals to be sonoelectrochemically synthesized in a diverse range of nanostructures (Reisse et al., 1994; Durant et al., 1995; Haas et al., 2006, 2008). Monodispersed spherical copper nanoparticles, the size of which (25–60 nm) is changed by the applied range of current density, have been synthesized from an aqueous acidic solution of $CuSO_4$ using polyvinylpyrrolidone (PVP) as a stabilizer (Haas et al., 2006). Copper with dendritic morphologies has also been obtained when PVA was used as the stabilizing agent (Haas et al., 2008).

Preparation of spherical platinum nanoparticles has been reported as well using aqueous chloroplatinic solutions under galvanostatic conditions (Zin et al., 2009). Tridimensional dendritic Pt nanostructures have also been synthesized when PVP was used as stabilizer (Shen et al., 2008b).

Stable suspensions of gold nanoparticles in water have been prepared using pulsed sonoelectrochemistry by adding some polymers to the electrolyte to avoid the nanoparticle aggregation (Aqil et al., 2008). In this report, a stable violet suspension, showing a narrow size distribution centered on 12 nm, together with a few larger particles (30 nm), was obtained using an α-methoxy-ω-hydroxyl polyethylene (MPEO)/PVP polymer mixture. In the presence of polyethylene oxide (PEO), however, a very stable suspension of gold nanoparticles with a mean diameter of 35 nm was prepared.

Sonoelectrochemical synthesis of silver nanoparticles has also been widely investigated using different electrolytes and stabilizers (Zhu et al., 2000c; Liu et al., 2001; Socol et al., 2002; Jiang et al., 2004a). In this approach, different shapes of silver nanoparticles including spheres, rods, and dendrites were prepared from an aqueous solution of $AgNO_3$ in the presence of nitriloacetate (NTA) (Zhu et al., 2000c). Composition of the electrolyte is argued to largely influence the shape and growth of the nanoparticles; formation of the shaped particles is favored by increasing the concentration of NTA. Additionally, varying the concentration of silver ions in the electrolyte, different shapes of silver nanoparticles are formed, purportedly because the excess of silver ions in the solution favors the aggregation and growth into the dendritic structure of Ag clusters (Zhu et al., 2000c). In a similar work, a silver and NTA system has been used, and a model based on a suspensive electrode formation is proposed to explain the growth of dendritic structures (Socol et al., 2002). Furthermore, synthesis of monodisperse spherical silver nanoparticles with an average size of 20–25 nm and a face-centered cubic structure has been reported in a saturated solution of silver citrate in the presence of PVP (Jiang et al., 2004a). In contrast, amorphous silver nanoparticles of 20 nm size have been prepared from an aqueous solution of AgBr in the presence of gelatin (Liu et al., 2001).

Among the other reports, concerning the sonoelectrochemical preparation of metallic nanostructures, synthesis of spherical tungsten nanoparticles with 30 nm diameter has been reported as well (Lei et al., 2007). Additionally, preparation of highly dispersed palladium nanoparticles with different sizes and shapes is performed in this respect using a solution of $PdCl_2$ in the presence of cetyltrimethylammonium bromide (CTAB) (Qiu et al., 2003).

Nanoparticles of very reactive metals with a high negative reduction potential, e.g., magnesium and aluminum, can also be prepared using pulsed sonoelectrochemistry. Zinc, nickel, and cobalt nanoparticles have also been successfully synthesized by pulsed sonoelectrochemical methods (Durant et al., 1995; Reisse et al., 1996; Jia et al., 2007).

Alloy Nanopowders

Applications of pulsed sonoelectrochemistry has been further extended for the preparation of binary and ternary alloyed nanopowders composed of iron, cobalt, and nickel (Delpancke et al., 2000; Mancier et al., 2004; Dabala et al., 2008). In a galvanostatic conditions, binary and ternary alloys have been deposited, and particles of 100 nm mean diameter were accordingly produced (Delpancke et al., 2000). In a recent report, a series of iron–cobalt alloy nanopowders have been prepared by pulsed sonoelectrochemistry under different potentiostatic conditions (Mancier et al., 2004). Under these experimental conditions, the composition of the alloys is essentially independent of the deposition potential and is only determined by the composition of the electrolyte utilized. The cathode efficiency for the production of $Co_{65}Fe_{35}$ nanoparticles has been calculated in another report versus the ratio of the produced mass of powders to the Faradaic yield (Dabala et al., 2008).

Semiconductor Nanopowders

Various semiconductor nanostructures have been reported to be synthesized via the sonoelectrochemical technique with different operating conditions. In this approach, Cu_2O nanopowders have been prepared in potentiostatic mode (Mancier et al., 2008). Synthesis of nanosized CdSe with a tubular structure has also been performed by applying a constant current density (Shen et al., 2008a). Additionally, PbTe nanorods have been prepared by pulsed sonoelectrochemical methods using NTA as stabilizer (Qiu et al., 2005). Other semiconductor materials such as PbSe, Bi_2Se_3, and MoS_2 have further been synthesized with different morphologies utilizing the pulsed method in aqueous solution at room temperature (Mastai et al., 1999a; Zhu et al., 2000a; Qiu et al., 2004).

Conducting Polymer Nanoparticles

Polyaniline (PANI) and other conducting polymers such as polythiophene, polypyrrole, and poly(methylaniline) can exist as dispersions. But a common problem with these materials is the particle aggregation which limits the range of applications. In this view, several investigations have been conducted to overcome this drawback exploiting different features of sonoelectrochemistry. For instance, synthesis of PANI nanomaterial by oxidative polymerization is performed using the sonoelectrode as the anode, to which a potentiostatic pulse is applied in an aqueous solution of aniline in HCl (Ganesan et al., 2008). Preparation of nanosized poly(methylaniline) has also been reported to be performed using a pulsed sonoelectrochemical method. However, in this case, the anode was a platinum electrode placed face to face with the ultrasound emitting surface of the horn. This synthesis was achieved by applying potentiostatic pulses in an aqueous solution of methylaniline in $HClO_4$. Under these operating conditions, the employed setup overcame the drawback concerning the use of an ultrasound horn in the oxidation process (Atobe et al., 2009).

PARAMETERS THAT CONTROL THE FORMATION OF NANOPARTICLES

The most promising operating parameters, applicable in sonoelectrochemical synthetic route, and their significant effects on size, morphology, crystallinity, and yield efficiency of the synthesized nanomaterials are briefly discussed below.

Synthesis Cell Temperature

Crystal growth is recognized to be slower at low temperatures, resulting in smaller crystal size (Glasstone, 1942). On the other hand, the yield of nanoparticles falls down with increasing cell temperature, originating from the increased rate of nanoparticle dissociation with the temperature enhancement (Sáez et al., 2009). Consequently, high synthesis efficiencies associated with small sizes of nanoparticles would merely be obtained by maintaining the sonoelectrochemistry cell at low temperatures.

Current Density

Current density has been found to affect the size of nanoparticles in two distinct directions (Zhu et al., 2000a). Although at low currents, the smaller size distribution of particles is expected, atomic diffusion processes have more time to occur in this situation, leading to the formation of particles with larger size distribution. Except than a few reports, where the current enhancement leads to a larger particle size (Mastai et al., 1999b; Jiang et al., 2004a), most studies illustrate that increasing the current density results in smaller nuclei and faster nucleation (Qiu et al., 2003; Haas et al., 2006; Shen et al., 2008b).

The current density can also affect the efficiency of the synthesis process in view of the fact that the secondary reactions, such as reduction of water, can take place at high current densities, leading to the low cathode efficiency and decreasing the yield of the nanoparticles.

Electrochemical Current Pulse Time

Since the deposit should be removed during the period of each sonic pulse, the electrochemical current pulse time (T_{EL}) should be short enough in order to have only primary nucleation and avoid further growth of nucleus. Consequently, to obtain a smaller crystal size associated with increased process efficiency, a short T_{EL} is required (Dabala et al., 2008).

Ultrasound Pulse Intensity

In the case that the intensity of the ultrasonic pulse is not high enough to entirely remove the deposit from the sonoelectrode surface, the residual nuclei on the surface would prevent the growth of the particles during the next electrochemical pulse (Durant et al., 1995). On the other hand, exceeding such required intensity will not affect crystal growth significantly and the extra energy will be wasted (Rao et al., 2008).

Moreover, an increase in the ultrasound pulse intensity can influence the morphology and size of the freshly formed nanoparticles in the solution by the formation of agglomerates due to rapid movement and further collisions of the particles (Suslick and Casadonte, 1987; Suslick, 1990).

Ultrasound Pulse Time

Although the ultrasonic pulse time (T_{US}) must be sufficient to remove the whole deposit from the sonoelectrode during electrodeposition, the excessively prolonging this period would not significantly affect the synthesis yield and size distribution of nanoparticles. This could be explained by the fact that there is an effective ultrasonic time in which all of the particles are expelled away from sonoelectrode surface and consequently their size and concentration do not change by extending the ultrasonic pulse duration.

Stabilizer

The initially formed nanoparticles via sonoelectrochemical route are so fine and can easily agglomerate in the solution. This can be induced by both the impact of the nanoparticles with the wall of electrochemical cell and also the acoustic streaming associated with the collisions of particles (Durant et al., 1995). In order to prevent such an inevitable agglomeration and control the shape and size of the nanoparticles, polymer stabilizers have been exploited in several reports (Zhu et al., 2000c; Jiang et al., 2004a; Aqil et al., 2008).

Nanomaterials that have been prepared to date by pulsed sonoelectrochemistry have been reviewed in this section. The shape and size of the nanoparticles can be adjusted in this technique by controlling the various electrodeposition and ultrasound parameters and using the suitable stabilizer.

ULTRASOUND-ASSISTED LASER ABLATION

Employing the ultrasound waves in laser-based generation of nanomaterials is presented as a newly emerged approach, complementing the well-defined ultrasound-assisted routes that are applicable in liquid phase. As mentioned earlier, the cavitation phenomenon due to ultrasonic irradiation in liquid media is responsible for the promotion of the nucleation processes, leading to noticeable changes in yield efficiency and qualitative features of the synthesized nanoparticles. The main idea of manipulating the laser-based synthesis of nanoparticles by simultaneous irradiation of the ultrasound waves has also been originated from this hypothesis, as one may improve the nucleation process in any liquid-based synthesis route of nanoparticles by means of the sonication. In order to realize how the cavitation phenomenon due to ultrasound irradiation affects the laser-based generation of nanoparticles, a brief review on the mechanism of nanoparticle synthesis through the laser ablation of the bulk target in liquid seems essential.

Synthesis of nanoparticles by the pulsed laser ablation method has been widely studied in recent years, starting with noble metals, e.g., gold and silver (Mafuné et al., 2000, 2001). The mechanism of nanoparticle formation would be deduced by investigating the reactions occurred through the plasma creation upon the laser ablation of the target in liquid. At the initial stage of interaction between the laser pulse and the target surface, the spot area of the target is heated up to its vaporization point leading to a hot plasma zone which heats the adjacent liquid layer as well. The plasma which is encapsulated in this layer, the so-called bubble, expands adiabatically at a supersonic velocity, creating a shock wave in front. This leads to a further increase of plasma temperature and an elevated vapor pressure of metal specious and liquid molecules up to the order of tens of atmospheres. In such a condition, the strong reaction of the active species, removed from the target, with liquid vapor leads to the fast agglomeration of these species into nanoparticles. These reactions originate from the successive collisions of the mentioned species due to the high vapor pressure inside the bubble (Jafarkhani et al., 2010).

Several investigations have been performed to study the effects of the operating parameters on the yield and morphology of the nanoparticles synthesized by laser ablation method (Neddersen et al., 1993; Sibbald et al., 1996). Besides the diversity of the laser parameters, i.e., the wavelength, energy density, and pulse repetition rate, the surrounding medium of the ablation process is quite noticeable in improving the findings. Qualitative and quantitative characteristics of nanoparticles are recognized to be controllable by performing the ablation process in different aqueous, organic, or surfactant-included environments (Mafuné et al., 2001). In addition to the composition of the synthesis medium, the sonication has also been reported to influence the characteristics of laser-synthesized nanoparticles (Dadras et al., 2009). In this report, effects of the sonication, associated with laser pulse energy, have been studied on the yield and size of the gold nanoparticles synthesized with the co-radiation of Nd:YAG laser pulses and ultrasound waves. Enhancement of the synthesis rate of nanoparticles in the diversity of laser pulse energies, applied in the presence of ultrasound waves, is concluded in this study. The behavior of electron temperature of the plasma-dominant species, calculated through the emission spectra of induced plasma in laser ablation of this metal, has also been investigated, to analyze the effects of sonication on the production yield. The results demonstrate that the plasma temperature, which can be directly attributed to the synthesis rate, has significantly enhanced by applying the ultrasound waves in all the irradiated laser pulse energies (Dadras et al., 2009).

The physical mechanism of the phenomenon occurring in the laser ablation process in the presence of ultrasound waves can be deduced by considering the influence of cavitation on the previously described laser-induced bubbles. As stated earlier, the creation of nanosized clusters by irradiating the laser beam on the target originates from an intense interaction of evaporated species with the surrounding liquid vapor (Dolgaev et al., 2002). Through a single laser pulse strike, the spot area of the target is heated up to its vaporization point, vaporizing the adjacent water layer as well. The thickness of this layer, $\sim(a\tau_p)^{1/2}$, could be estimated less than 150 nm, where a is the heat diffusion coefficient (10^{-3} cm^2 s^{-1} for water) and τ_p is the diffusion time (pulse duration) usually less than 250 ns. On the other hand, the plume temperature inside this bubble can be affected by applying the ultrasound waves. According to the cavitation theory, the sonication can cause creation, growth, and collapse of micrometer-size bubbles inside the liquid leading to localized high temperatures (Suslick et al., 1991; Gedanken, 2004). Therefore, applying the ultrasound radiation can result in the growth and collapse of the bubbles created during the laser pulse strike leading to the increase in the plume temperature (Dadras et al., 2009). The plasma with the increased temperature can play the role of another heat source on the target surface besides the laser beam. This results in more evaporation from the target and consequently the enhancement of the formation of nanoparticles (Dadras et al., 2009).

The improvement of Au nanoparticle synthesis rate has been demonstrated in both cases of applying the ultrasound waves and increasing the laser pulse energy. However, the nanoparticle size distribution has been reported to vary only by the laser pulse energy rather than the sonication (Dadras et al., 2009).

Although the generation of nanomaterials through the laser ablation in liquid has been widely investigated in literature and effects of numerous operating parameters have been studied in this method, the exploration on the influence of ultrasound waves on synthesis mechanism and product characteristics is quite incomplete. More comprehensive investigations on the effects of sonication parameters, e.g., intensity, frequency, and geometry, on the features of the products are required in developing the method of ultrasound-assisted laser synthesis of nanomaterials.

ULTRASONIC SPRAY PYROLYSIS

Spray pyrolysis has been widely used in industry for ultrafine nanoparticle preparation, because of its simplicity and scalability for mass production. In general, spray pyrolysis involves the thermal decomposition of solid or liquid particles (aerosols) generated by a nebulizer (e.g., pneumatic, ultrasonic, or electrostatic nebulizers) in a gas flow (Kodas and Hampden-Smith, 1999). Among the various nebulization techniques, utilization of ultrasonic nebulizers has been favored because of their exceptional energy efficiency in aerosol generation over other nebulization tools, affordability, and the intrinsically low velocity of the initially generated aerosols (Bang et al., 2010).

The role of the ultrasound waves in USP is to provide the phase isolation of one microdroplet reactor from another. In this technique, the high-frequency (~2 MHz) ultrasound nebulizes precursor solutions to produce the hot micron-sized droplets (500–1300 K temperature, 1 bar pressure and 10^4 K s^{-1} cooling rate) that act as isolated, individual micron-sized chemical reactors (Bang et al., 2010). The production of intended nanostructures is accomplished when these liquid droplets generated by ultrasonic nebulization are heated in a gas flow, and subsequently solid- or liquid-phase chemical reactions occur.

USP has several advantages over other conventional synthetic routes, among them one would mention the production of highly pure micron- or nanosized particles, incessant operation, and ease of composition control (Messing et al., 1993; Kodas and Hampden-Smith, 1999; Okuyama and Lenggoro, 2003). Additionally, the dominant control over chemical and physical compositions in the USP makes this technique specifically practical in the preparation of composite materials (Kodas and Hampden-Smith, 1999).

Ultrasonic nebulization is a result of capillary waves that are generated by ultrasonic vibrations at the liquid's surface and travel along the interface of liquid and the ambient gas. When the amplitude of the surface capillary waves is sufficiently high, the crests of these waves can break off, leading to the formation of liquid microdroplets (Bang et al., 2010). Since the wavelength of capillary waves is inversely proportional to ultrasound frequency, finer droplets can be produced by sonication with higher frequencies (Lang, 1962).

A usual USP equipment consists of a high-frequency ultrasonic transducer at the base of a precursor solution container and fitted with a gas flow (e.g., Ar, N_2, O_2, etc.) to drive the ultrasonically generated droplets into a tubular furnace. Typically, in the case of water solvent, spherical droplets with an initial diameter of ~5 μm are generated at a frequency of ~2 MHz. The microdroplets then undergo successive processes such as solvent evaporation, precipitation, decomposition, and densification after being carried into the furnace (Kodas and Hampden-Smith, 1999). During solvent evaporation, the droplets rapidly shrink, and further heating results in the supersaturation, at which the precipitation occurs, often at the surface of the droplet. In addition, decomposition of precursors may lead to porous or hollow particle formation, which may subsequently experience densification to generate solid particles. The formation of dense solid particles versus hollow shells is intimately related to solvent evaporation rate and solubility of precursors (Messing et al., 1993; Kodas and Hampden-Smith, 1999; Lenggoro et al., 2000). Finally, a collector such as bubbler, filter, or electrostatic precipitator is installed at the furnace outlet to accumulate the products at the end of the process (Kodas and Hampden-Smith, 1999).

USP in the Synthesis of Nanostructured Materials

During the last decade, USP has been exploited by several researchers as a powerful synthetic method for the preparation of nanostructured materials. This widespread utilization of the USP technique has originated from its exceptional ability to produce nanocomposites. In most cases, the nanocomposites are a combination of the desired materials (e.g., metal oxides, sulfides, or carbon) and sacrificial ones (e.g., surfactants, colloidal silica, polymers, or metal salts). The sacrificial materials are subsequently removed by several methods such as chemical etching, calcination, or dissolution, introducing different nanostructures into the ultimate products. In the following, a diversity of nanostructures, synthesized by USP, have been grouped into a number of categories based on the sacrificial materials employed.

Mesostructured Nanomaterials from Organic–Inorganic Hybrid Nanocomposites

Well-ordered mesoporous materials with their unique physical and chemical properties, which are quite different from their nonporous solid counterparts, have important applications to catalysis, sorption, gas sensing, optics, and photovoltaics (Davis, 2002; Soler-Illia et al., 2002; Boettcher et al., 2007; Kanatzidis, 2007). Conventional synthetic routes for mesostructured materials which are based on batch reactions, in which self-assembled organic–inorganic hybrid nanocomposites are produced with the assistance of pre-organized organic species, possess serious drawbacks. One would enumerate the time-consuming templating processes, difficulties to produce thin films, and limited utilization for producing patterned nanocomposites as the main disadvantages of typical routes for the synthesis of mesostructured materials (Brinker et al., 1999).

Recently, a facile route has been developed to prepare different mesostructured silica particles and film by combination of evaporation-induced self-assembly (EISA) process and nebulization techniques (Lu et al., 1999; Ji et al., 2006). In this report, homogeneous water/ethanol mixture solutions containing silica precursors and surfactants were ultrasonically nebulized into a furnace, and various surfactant-dependent forms of mesostructured silica spheres were produced by heat treatment of the obtained organic–inorganic hybrid nanocomposites (Lu et al., 1999; Ji et al., 2006). The aerosol (nebulization) procedure has been developed further to prepare other mesostructured metal oxides. Crystalline mesoporous γ-alumina with high thermal and chemical stability and also rare-earth-doped mesoporous titania microspheres have been reported to be prepared via this route (Boissière et al., 2006; Li et al., 2008).

Another advantage of aerosol synthetic route is the ability to incorporate metal species, organic dye molecules, or polymers within the mesostructured framework (Lu et al., 1999; Hampsey et al., 2005a). Noble metal/silica nanocomposites can be simply obtained via this method by nebulizing precursor solutions containing metal complexes (Lu et al., 1999; Hampsey et al., 2005a).

Nanosized Structures from Silica-Based Nanocomposites

Silica has been recognized quite useful in the USP preparation of nanostructured materials, as its nanoparticles can close-pack in the evaporating microdroplets and provide a nanostructured scaffold in real time. Silica nanoparticles have been utilized as a sacrificial agent to prepare porous MoS_2 via USP technique (Skrabalak and Suslick, 2005); SiO_2/MoS_2 nanocomposite is initially prepared from the decomposition of nebulized single-source MoS_2 precursor dissolved in colloidal silica nanoparticles suspension. Subsequently, silica nanostructures are selectively leached out of the nanocomposites by HF etching, resulting in the formation of porous MoS_2 network.

Similarly, nanostructured $ZnS:Ni^{2+}$ photocatalysts have been prepared via USP synthetic route. However, the ultimate structure of the products is reported to be quite temperature dependent, as mesoporous hollow microspheres were obtained at a low temperature after the removal of silica template, while they were damaged into nanoparticles at a high temperature (Bang et al., 2008).

Silica template has also been utilized for the USP synthesis of metal oxide nanostructures. Various forms of titania nanostructures, including porous, hollow, and ball-in-ball architectures have been synthesized from HF etching of the titania/silica nanocomposites, prepared by nebulization and decomposition of an aqueous solution containing a titanium complex and silica nanoparticles (Suh et al., 2006).

As another use of silica templating in USP, various kinds of porous carbon spheres have been synthesized from silica/sucrose nanocomposites, where the pore structures and macroscopic morphology of these spheres are simply controlled by changing the type of silica template and manipulating the template to sucrose ratio (Hampsey et al., 2005b; Hu et al., 2008).

Macroporous Materials from Polymer-Based Nanocomposites

Colloidal polymer particles have also been exploited as a sacrificial agent in USP synthesis of macroporous (pores with diameters >50 nm) nanostructures. Among the related reports, ordered silica macroporous spheres have been synthesized by ultrasonic nebulization of a suspension of silica and polystyrene latex and subsequent low- and high-temperature heating to create silica/polystyrene nanocomposites (solvent evaporation) and produce ordered macroporous silica spheres (pyrolysis) respectively (Iskandar et al., 2001, 2002). In a similar research, macroporous brookite titania spheres have been simply produced by replacing the colloidal silica nanoparticles with titania nanoparticles (Iskandar et al., 2007).

Furthermore, in an interesting research, polymer templating macroporosity and the surfactant self-assembly-based mesoporosity have been combined to create multiple-sized pores in silica spheres. This result is achieved by exploiting colloidal polystyrene particles and a surfactant as a structure-directing agent for macroporosity and mesoporosity, respectively (Fan et al., 2001).

Utilization of polymer templating in USP technique has been further extended for the preparation of macroporous silica spheres in a low-cost in situ-generated polymer template (Suh and Suslick, 2005). In this approach, the polymerization of an organic monomer (i.e., styrene) takes place at a low-temperature heating zone after nebulization of droplets containing silica nanoparticles to create a silica/polystyrene nanocomposite in real time. The in situ-generated polymer is then pyrolyzed out of the nanocomposites in a hotter zone to produce macroporous silica spheres, the morphology and surface area of which are easily controllable by changing the ratios of styrene and silica (Suh et al., 2005).

Nanosized Structures from Metal Salt-Based Nanocomposites

In the conventional USP synthetic processes, the nanocrystallites, created due to multiple nucleations at an early stage of the process, undergo the growth and aggregation into a larger micron-sized single particle, as the temperature in the droplet proceeds. This distinctive feature of the traditional USP procedure has limited the ability of this technique for the production of nanoparticles.

To overcome such an issue, a facile and rapid USP method has been proposed, in which the metal salts (e.g., chlorides or nitrates of Li, Na, K) are simply inserted into precursor solutions to inhibit the agglomeration of primarily created nanocrystallites (Xia et al., 2001a). In this "salt-assisted aerosol decomposition" approach, the added metal salts behave as hot liquid solvents, in which the nanocrystallites can dissolve and precipitate during USP process. Such dissolution/precipitation processes occurring in the molten droplets result in the production of individual nanoparticles trapped in a salt template. The molten salts and the nanoparticles then solidify, and the salt template is subsequently removed by several cycles of washing to obtain separated nanoparticles.

Plenty of nanoparticles, including Ni, Ag–Pd, NiO, CeO_2, ZnO, $LiCoO_2$, Y_2O_3–ZrO_2, $(Ba_{1-x}Sr_x)TiO_3$, CdS, and ZnS, have been prepared by salt-assisted aerosol decomposition method (Xia et al., 2001a,b; Itoh et al.; 2003, 2004; Panatarani et al., 2003; Lenggoro et al., 2004). Besides the versatility of this initiative technique, the molten salts serve as an effective liquid flux to improve mass transport and subsequently to enhance the crystallinity of the materials produced via this route (Xia et al., 2001a).

Porous architectures such as mesopore nanostructures can also be prepared by salt-assisted aerosol decomposition by a dominant control over the quantity of metal salts. In this respect, nanoporous metal oxides (e.g., Al_2O_3 and SiO_2) have been reported to be synthesized via this synthetic route (Kim et al., 2002, 2004). Intermediate concentrations of salts in the precursor solution can be used instead of excess amounts of metal salts; so the structural integrity of nanocrystallites in nanocomposite can remain after the removal of salts. Inexpensive and non-toxic in situ salt templating, good thermal stability of the salts at very high temperatures, and possibility of recycling the metal salts are the most advantageous points of using this strategy to generate porous structures (Bang et al., 2010).

Metal salt templating route has been further extended to prepare hollow nanonaterials such as mesoporous silica particles with NaCl cores. This synthesis is accomplished via an initial aerosol process followed by calcination to create a nanocomposite with cubic NaCl core. A subsequent washing procedure then removes NaCl to create a hollow cubic cavity (Jiang and Brinker, 2006). In a similar manner, in which ferric chloride ($FeCl_3$) was employed instead of sodium chloride (NaCl), nanocomposite with a ferric-rich core surrounded by a silica-rich shell is produced in the first stage of process. Hollow interiors are then created by subsequent calcination which results in a rattle-like nanostructure (i.e., hematite nanoparticles encapsulated in hollow silica spheres) (Zheng et al., 2007).

Furthermore, in a novel USP method, in-situ-generated metal salts have been utilized as the template materials for the preparation of porous carbons instead of pre-existing metal salts, exploited in the usual salt-assisted USP methods. Various nanostructured carbons have accordingly been synthesized from the decomposition of alkali halocarboxylates (Skrabalak and Suslick, 2006). This new approach is a one-step process with no need for expensive template materials. Additionally, a diversity of mesoporous, macroporous, or hollow nanostructures is achievable via this procedure, depending on the types of alkali halocarboxylates employed (Skrabalak et al., 2006).

Porous or hollow carbon spheres can also be produced by USP decomposition of aqueous solutions of substituted alkali benzoate salts (Skrabalak et al., 2007). The advantageous point of utilizing this salt is the capability of controlling size and morphology of the carbon spheres by changing the concentration and manipulating both cation and ring substituents of the precursor, respectively (Skrabalak et al., 2007).

Nanostructured Semiconductors from Chemical Aerosols

Notwithstanding the conspicuous achievements of the salt-assisted aerosol decomposition technique in the synthesis of nanoparticles, the production of high-quality semiconductor nanoparticles (or quantum dots) was still a controversial challenge. In this respect, USP has proposed a novel approach, in which the rapid heating and cooling of nebulized droplets, occurring throughout the USP process in a timescale of seconds, can quickly result in the nucleation required. This technique which is named "chemical aerosol flow synthesis," utilizes organic precursor solutions with high boiling points (e.g., octadecane) instead of their aqueous counterparts (Didenko and Suslick, 2005). To conduct this procedure, the viscosity of organic precursor solutions is initially reduced by diluting them with a low boiling point liquid (e.g., toluene). Nebulized droplets of the diluted organic solutions then lose the low boiling point solvent by passing through a heating zone and leave a concentrated precursor solution in the high boiling point solvent behind. Subsequently, chemical reactions occur in this organic liquid to produce highly crystalline nanoparticles, and the reactions are finally quenched in cold, solvent-filled bubblers (Didenko et al., 2005). Highly fluorescent cadmium-based quantum dots such as CdS, CdSe, CdTe, and also mixed chalcogenides have been reported to be successfully prepared via this method. The size-dependent photoluminescence of these nanoparticles can be easily tuned by changing the furnace temperature. Highly reproducible gram-scale synthesis with a production rate around 100 mg h^{-1} is achievable in this technique even with a laboratory-scale experimental setup.

CONCLUSIONS

Several ultrasound-assisted synthetic routes have been explored in the production of nanostructured materials, including sonochemistry, ultrasound-induced deposition, sonoelectrochemistry, ultrasound-assisted laser ablation in liquid media, and USP. Ultrasonic irradiation provides exceptional reaction conditions via acoustic cavitation phenomenon. Micron-sized bubbles created in this phenomenon can efficiently accumulate the diffuse energy of ultrasound waves, and upon their collapse, a large concentration of energy is released to heat up the contents of the bubble. These momentary, localized hot spots with extremely high temperatures and pressures primarily account for the chemical and physical effects of ultrasound.

The effectiveness of sonochemistry as an appropriate tool for the synthesis of nanomaterials resides in its versatility. With a simple manipulation in reaction conditions, a variety of nanostructured materials can be synthesized, including metals, alloys, oxides, chalcogenides, carbides, and nanostructured supported catalysts. The sonochemical method has been even further extended to the preparation of carbons, polymers, and biomaterials. Furthermore, by sonochemical decomposition of volatile organic precursors, associated with considerable enhancement of materials mass transport due to shock waves, one can achieve the synthesis of nanocomposites. Various hollow materials can be obtained after the removal of templating material which is initially used as one component of the nanocomposites.

Ultrasonic irradiation not only assists sonochemical synthesis of nanomaterials by promoting different chemical reactions, but also it can be exploited further in the deposition of nanomaterials as a synthetic route based on physical effects of sonication. Ultrasound-induced shock waves and microjets primarily result in some unusual effects at the liquid–solid interface. These effects, including surface damage and fragmentation of friable solids associated with rapid interparticle collisions, are the physical basis of ultrasound-induced nanoparticle deposition. In addition, low melting point metal particles can be rapidly driven together by the intense ultrasound irradiation to induce effective melting at their impact point. This physical phenomenon is responsible for the ultrasound-induced deposition which is appropriately applicable in the synthesis of various nanomaterials whether on a substrate or another nanostructure.

The cavitation through the ultrasound radiation is also recognized as the responsible phenomenon for unusual effects induced in the ultrasound-assisted electrochemical synthesis processes. This phenomenon, in case of taking place close to the surface of the electrode, results in the penetration of a high-velocity microjet of liquid inside the bubbles and consequent improvement of overall mass transport and the reaction rates by decreasing the diffusion layer thickness and releasing the deposit of the electrode surface. The technique based on this phenomenon, the so-called sonoelectrochemistry, has widely been employed to produce plenty of pure metals, alloys, semiconductor and conductive polymer nanoparticles.

Utilization of the ultrasound waves in laser-based generation of nanomaterials is presented as a newly emerged approach, complementing the well-defined ultrasound-assisted routes which are applicable in liquid media. Applying the ultrasound radiation can result in the growth and collapse of the bubbles created during the laser pulse strike, leading to the increase in the plume temperature. The plasma with the increased temperature can play the role of another heat source on the target surface besides the laser beam. This results in more evaporation from the target and consequently the enhancement of the formation of nanoparticles.

In USP, unlike the other ultrasound-assisted nanomaterials synthetic routes, sonication does not induce thermally driven chemical reactions in and of itself. Instead, the ultrasound serves as the nebulizer of precursor solutions to produce micron-sized droplets confining chemical reactions within their interior. Such isolated micron-sized reactors allow one to simply control the desired nanostructure composition and maintain the bulk chemical composition on the micron-size scale. With these advantages, USP has stood out among different techniques for the synthesis of multicomponent or composite materials.

The applications of ultrasound in the synthesis of nanomaterials are diverse, but there still remains much to investigate in ultrasound-assisted synthetic routes, surveyed in the current review. Considering the existing limitations and lack of research in some of these fields, future progress will require the development of more systematic investigations on numerous features of ultrasound-assisted synthesis of nanomaterials.

REFERENCES

Anandan, S., Grieser, F., and Ashokkumar, M. 2008. Sonochemical synthesis of Au–Ag core–shell bimetallic nanoparticles. *The Journal of Physical Chemistry, C*, 112: 15102–15105.

Aqil, A., Serwas, H., Delplancke, J. L. et al. 2008. Preparation of stable suspensions of gold nanoparticles in water by sonoelectrochemistry. *Ultrasonics Sonochemistry*, 15: 1055–1061.

Atobe, M., Ishikawa, K., Asami, R. et al. 2009. Size-controlled synthesis of conducting-polymer microspheres by pulsed sonoelectrochemical polymerization. *Angewandte Chemie International Edition*, 48: 6069–6072.

Bang, J. H., Helmich, R. J., and Suslick, K. S. 2008. Nanostructured ZnS:Ni^{2+} photocatalysts prepared by ultrasonic spray pyrolysis. *Advanced Materials*, 20: 2599–2603.

Bang, J. H. and Suslick, K. S. 2010. Applications of ultrasound to the synthesis of nanostructured materials. *Advanced Materials*, 22: 1039–1059.

Basnayake, R., Li, Z., Katar, S. et al. 2006. PtRu nanoparticle electrocatalyst with bulk alloy properties prepared through a sonochemical method. *Langmuir*, 22: 10446–10450.

Bellissent, R., Galli, G., Hyeon, Y. et al. 1995. Structural properties of amorphous bulk Fe, Co and Fe–Co binary alloys. *Physica Scripta*, T57: 79–83.

Birkin, P. R., Offin, D. G., Joseph, P. F. et al. 2005. Cavitation, shock waves and the invasive nature of sonoelectrochemistry. *The Journal of Physical Chemistry B*, 109: 16997–17005.

Boettcher, S. W., Fan, J., Tsung, C.-K. et al. 2007. Harnessing the sol–gel process for the assembly of non-silicate mesostructured oxide materials. *Accounts of Chemical Research*, 40: 784–789.

Boissière, C., Nicole, L., Gervais, C. et al. 2006. Nanocrystalline mesoporous gamma-alumina powders "UPMC1 Material" gathers thermal and chemical stability with high surface area. *Chemistry of Materials*, 18: 5238–5243.

Brinker, C. J., Lu, Y., Sellinger, A. et al. 1999. Evaporation-induced self-assembly: Nanostructures made easy. *Advanced Materials*, 11: 579–585.

Brinker, C. J. and Scherer, G. W. 1990. *Sol–Gel Science: The Physics and Chemistry of Sol–Gel Processing*. New York: Academic Press Inc.

Brotchie, A., Grieser, F., and Ashokkumar, M. 2008. Sonochemistry and sonoluminescence under dual frequency ultrasound irradiation in the presence of water soluble solutes. *The Journal of Physical Chemistry C*, 112: 10247–10250.

Burda, C., Chen, X., Narayanan, R. et al. 2005. Chemistry and properties of nanocrystals of different shapes. *Chemical Reviews*, 105: 1025–1102.

Caruso, R. A., Ashokkumar, M., and Grieser, F. 2002. Sonochemical formation of gold sols. *Langmuir*, 18: 7831–7836.

Chen, Q., Boothroyd, C., Tan, G. H. et al. 2008. Silica coating of nanoparticles by the sonogel process. *Langmuir*, 24: 650–663.

Compton, R. G., Eklund, J. C., and Marken, F. 1997a. Sonoelectrochemical processes. A review. *Electroanalysis*, 9: 509–522.

Compton, R. G., Eklund, J. C., Marken, F. et al. 1997b. Dual activation: Coupling ultrasound to electrochemistry. An overview. *Electrochimica Acta*, 42: 2919–2927.

Dabala, M., Pollet, B. G., Zin, V. et al. 2008. Sonoelectrochemical (20 kHz) production of $Co_{65}Fe_{35}$ alloy nanoparticles from Aotani solutions. *Journal of Applied Electrochemistry*, 38: 395–402.

Dadras, S., Jafarkhani, P., Torkamany, M. J. et al. 2009. Effects of ultrasound radiation on the synthesis of laser ablated gold nanoparticles. *Journal of Physics D: Applied Physics*, 42: 025405 (5pp.).

Dantsin, G. and Suslick, K. S. 2000. Sonochemical preparation of a nanostructured bifunctional catalyst. *Journal of the American Chemical Society*, 122: 5214–5215.

Davis, M. E. 2002. Ordered porous materials for emerging applications. *Nature*, 417: 813–821.

Delplancke, J. L., Dille, J., Reisse, J. et al. 2000. Magnetic nanopowders: Ultrasound assisted electrochemical preparation and properties. *Chemistry of Materials*, 12: 946–955.

Dhas, N. A., Raj, C. P., and Gedanken, A. 1998. Synthesis, characterization, and properties of metallic copper nanoparticles. *Chemistry of Materials*, 10: 1446–1452.

Dhas, N. A. and Suslick, K. S. 2005. Sonochemical preparation of hollow nanospheres and hollow nanocrystals. *Journal of the American Chemical Society*, 127: 2368–2369.

Didenko, Y. T. and Suslick, K. S. 2005. Chemical aerosol flow synthesis of semiconductor nanoparticles. *Journal of the American Chemical Society*, 127: 12196–12197.

Doktycz, S. J. and Suslick, K. S. 1990. Interparticle collisions driven by ultrasound. *Science*, 247: 1067–10679.

Dolgaev, S. I., Simakin, A. V., Voronov, V. V. et al. 2002. Nanoparticles produced by laser ablation of solids in liquid environment. *Applied Surface Science*, 186: 546–551.

Du, N., Zhang, H., Chen, B. D. et al. 2007. Porous Co_3O_4 nanotubes derived from $Co_4(CO)_{12}$ clusters on carbon nanotube templates: A highly efficient material for Li-battery applications. *Advanced Materials*, 19: 4505–4509.

Durant, A., Delplancke, J. L., Winand, R. et al. 1995. A new procedure for the production of highly reactive metal powders by pulsed sonoelectrochemical reduction. *Tetrahedron Letters*, 36: 4257–4260.

Dutta, D. P., Sudarsan, V., Srinivasu, P. et al. 2008. Indium oxide and europium/dysprosium doped indium oxide nanoparticles: Sonochemical synthesis, characterization, and photoluminescence studies. *The Journal of Physical Chemistry C*, 112: 6781–6785.

Enomoto, N., Koyano, T., and Nakagawa, Z. 1996. Effect of ultrasound on synthesis of spherical silica. *Ultrasonics Sonochemistry*, 3: S105–S109.

Fan, H., Van Swol, F., Lu, Y. et al. 2001. Multiphased assembly of nanoporous silica particles. *Journal of Non-Crystalline Solids*, 285: 71–78.

Ganesan, R., Shanmugam, S., and Gedanken, A. 2008. Pulsed sonoelectrochemical synthesis of polyaniline nanoparticles and their capacitance properties. *Synthetic Metals*, 158: 848–853.

Gao, T., Li, Q., and Wang, T. 2005. Sonochemical synthesis, optical properties, and electrical properties of core/shell-type ZnO nanorod/CdS nanoparticle composites. *Chemistry of Materials*, 17: 887–892.

Gao, T. and Wang, T. 2004. Sonochemical synthesis of SnO_2 nanobelt/CdS nanoparticle core/shell heterostructures. *Chemical Communications*, 22: 2558–2559.

Gedanken, A. 2004. Using sonochemistry for the fabrication of nanomaterials. *Ultrasonics Sonochemistry*, 11: 47–55.

Glasstone, S. 1942. *An Introduction to Electrochemistry*. New York: Van Nostrand Company.

Haas, I., Shanmugam, S., and Gedanken, A. 2006. Pulsed sonoelectrochemical synthesis of size-controlled copper nanoparticles stabilized by poly(N-vinylpyrrolidone). *The Journal of Physical Chemistry B*, 110: 16947–16452.

Haas, I., Shanmugam, S., and Gedanken, A. 2008. Synthesis of copper dendrite nanostructures by a sonoelectrochemical method. *Chemistry—A European Journal*, 14: 4696–4703.

Hampsey, J. E., Arsenault, S., Hu, Q. et al. 2005a. One-step synthesis of mesoporous metal–SiO_2 particles by an aerosol-assisted self-assembly process. *Chemistry of Materials*, 17: 2475–2480.

Hampsey, J. E., Hu, Q., Rice, L. et al. 2005b. A general approach towards hierarchical porous carbon particles. *Chemical Communications*, 28: 3606–3608.

Hodes, G. 2007. When small is different: Some recent advances in concept and applications of nanoscale phenomena. *Advanced Materials*, 19: 639–655.

Hu, Q., Lu, Y., and Meisner, G. P. 2008. Preparation of nanoporous carbon particles and their cryogenic hydrogen storage capacities. *The Journal of Physical Chemistry C*, 112: 1516–1523.

Hyeon, T., Fang, M., and Suslick, K. S. 1996. Nanostructured molybdenum carbide: Sonochemical synthesis and catalytic properties. *Journal of the American Chemical Society*, 118: 5492–5493.

Iskandar, F., Mikrajuddin, A., and Okuyama, K. 2001. In situ production of spherical silica particles containing self-organized mesopores. *Nano Letters*, 1: 231–234.

Iskandar, F., Mikrajuddin, A., and Okuyama, K. 2002. Controllability of pore size and porosity on self-organized porous silica particles. *Nano Letters*, 2: 389–392.

Iskandar, F., Nandiyanto, A. B. D., Yun, K. M. et al. 2007. Enhanced photocatalytic performance of brookite TiO_2 macroporous particles prepared by spray drying with colloidal templating. *Advanced Materials*, 19: 1408–1412.

Itoh, Y., Abdullah, M., and Okuyama, K. 2004. Direct preparation of nonagglomerated indium tin oxide nanoparticles using various spray pyrolysis methods. *Journal of Materials Research*, 19: 1077–1086.

Itoh, Y. Lenggoro, I. W., Okuyama, K. et al. 2003. Size tunable synthesis of highly crystalline $BaTiO_3$ nanoparticles using salt-assisted spray pyrolysis. *Journal of Nanoparticle Research*, 5: 191–198.

Jafarkhani, P., Dadras, S., Torkamany, M. J. et al. 2010. Synthesis of nanocrystalline titania in pure water by pulsed Nd:YAG Laser. *Applied Surface Science*, 256: 3817–3821.

Jafarzadeh, M., Rahman, I. A., and Sipaut, C. S. 2009. Synthesis of silica nanoparticles by modified sol–gel process: The effect of mixing modes of the reactants and drying techniques. *Journal of Sol–Gel Science and Technology*, 50: 328–336.

Ji, X., Hu, Q., Hampsey, J. E. et al. 2006. Synthesis and characterization of functionalized mesoporous silica by aerosol-assisted self-assembly. *Chemistry of Materials*, 18: 2265–2274.

Jia, F., Hu, Y., Tang, Y. et al. 2007. A general nonaqueous sonoelectrochemical approach to nanoporous Zn and Ni particles. *Powder Technology*, 176: 130–136.

Jiang, X. and Brinker, C. J. 2006. Aerosol-assisted self-assembly of single-crystal core/nanoporous shell particles as model controlled release capsules. *Journal of the American Chemical Society*, 128: 4512–4513.

Jiang, L. P., Wang, A. N., Zhao, Y. et al. 2004a. A novel route for the preparation of monodisperse silver nanoparticles via a pulsed sonoelectrochemical technique. *Inorganic Chemistry Communications*, 7: 506–509.

Jiang, L. P., Xu, S., Zhu, J. M. et al. 2004b. Ultrasonic-assisted synthesis of monodisperse single-crystalline silver nanoplates and gold nanorings. *Inorganic Chemistry*, 43: 5877–5883.

Jones, J. E., Cheshire, M. C., and Casadonte, D. J. 2004. Facile sonochemical synthesis of graphite intercalation compounds. *Organic Letters*, 6: 1915–1917.

Jung, S. H., Oh, E., Lee, K. H. et al. 2008. Sonochemical preparation of shape-selective ZnO nanostructures. *Crystal Growth & Design*, 8: 265–269.

Kanatzidis, M. G. 2007. Beyond silica: Nonoxidic mesostructured materials. *Advanced Materials*, 19: 1165–1183.

Kim, S. H., Liu, B. Y. H., and Zachariah, M. R. 2002. Synthesis of nanoporous metal oxide particles by a new inorganic matrix spray pyrolysis method. *Chemistry of Materials*, 14: 2889–2899.

Kim, S. H., Liu, B. Y. H., and Zachariah, M. R. 2004. Ultrahigh surface area nanoporous silica particles via an aero-sol–gel process. *Langmuir*, 20: 2523–2526.

Klima, J. and Bernard, C. 1999. Sonoassisted electrooxidative polymerisation of salicylic acid: Role of acoustic streaming and microjetting. *Journal of Electroanalytical Chemistry*, 462: 181–186.

Kodas, T. T. and Hampden-Smith, M. 1999. *Aerosol Processing of Materials*. New York: Wiley-VCH.

Krishnan, C. V., Chen, J., Burger, C. et al. 2006. Polymer-assisted growth of molybdenum oxide whiskers via a sonochemical process. *The Journal of Physical Chemistry B*, 110: 20182–20188.

Lang, R. J. 1962. Ultrasonic atomization of liquids. *The Journal of the Acoustical Society of America*, 34: 6–8.

Latt, K. K. and Kobayashi, T. 2008. TiO_2 nanosized powders controlling by ultrasound sol–gel reaction. *Ultrasonics Sonochemistry*, 15: 484–491.

Lei, H., Tang, Y. J., Wei, J. J. et al. 2007. Synthesis of tungsten nanoparticles by sonoelectrochemistry. *Ultrasonics Sonochemistry*, 14: 81–83.

Lenggoro, I. W., Hata, T., Iskandar, F. et al. 2000. An experimental and modeling investigation of particle production by spray pyrolysis using a laminar flow aerosol reactor. *Journal of Materials Research*, 15: 733–743.

Lenggoro, I. W., Itoh, Y., Okuyama, K. et al. 2004. Nanoparticles of a doped oxide phosphor prepared by direct-spray pyrolysis. *Journal of Materials Research*, 19: 3534–3539.

Li, L., Tsung, C.-K., Yang, Z. et al. 2008. Rare-earth-doped nanocrystalline titania microspheres emitting luminescence via energy transfer. *Advanced Materials*, 20: 903–908.

Li, B., Xie, Y., Huang, J. et al. 2000. Sonochemical synthesis of nanocrystalline copper tellurides Cu_7Te_4 and Cu_4Te_3 at room temperature. *Chemistry of Materials*, 12: 2614–2616.

Li, H. L., Zhu, Y. C., Chen, S. G. et al. 2003. A novel ultrasound-assisted approach to the synthesis of CdSe and CdS nanoparticles. *Journal of Solid State Chemistry*, 172: 102–110.

Li, H. L., Zhu, Y. C., Palchik, O. et al. 2002. Sonochemical preparation of GaSb nanoparticles. *Inorganic Chemistry*, 41: 637–639.

Liu, S., Huang, W., Chen, S. et al. 2001. Synthesis of X-ray amorphous silver nanoparticles by the pulse sonoelectrochemical method. *Journal of Non-Crystalline Solids*, 283: 231–236.

Lorimer, J. P., Pollet, B., Phull, S. S. et al. 1998. The effect upon limiting currents and potentials of coupling a rotating disc and cylindrical electrode with ultrasound. *Electrochimica Acta*, 43: 449–455.

Lu, Y., Fan, H., Stump, A. et al. 1999. Aerosol-assisted self-assembly of mesostructured spherical nanoparticles. *Nature*, 398: 223–226.

Mafuné, F., Kohno, J., Takeda, Y. et al. 2000. Formation and size control of silver nanoparticles by laser ablation in aqueous solution. *The Journal of Physical Chemistry B*, 104: 9111–9117.

Mafuné, F., Kohno, J., Takeda, Y. et al. 2001. Formation of gold nanoparticles by laser ablation in aqueous solution of surfactant. *The Journal of Physical Chemistry B*, 105: 5114–5120.

Makino, K., Mossoba, M. M., and Riesz, P. 1982. Chemical effects of ultrasound on aqueous solutions. Evidence for hydroxyl and hydrogen free radicals (•OH and •H) by spin trapping. *Journal of the American Chemical Society*, 104: 3537–3539.

Mancier, V., Daltin, A. L., and Leclercq, D. 2008. Synthesis and characterization of copper oxide (I) nanoparticles produced by pulsed sonoelectrochemistry. *Ultrasonics Sonochemistry*, 15: 157–163.

Mancier, V., Delplancke, J. L., Delwiche, J. et al. 2004. Morphologic, magnetic, and Mössbauer spectral properties of $Fe_{75}Co_{25}$ nanoparticles prepared by ultrasound-assisted electrochemistry. *Journal of Magnetism and Magnetic Materials*, 281: 27–35.

Mason, T. J. 1989. *Sonochemistry: The Uses of Ultrasound in Chemistry*. Cambridge: Royal Society of Chemistry.

Mason, T. J., Lorimer, J. P., and Walton, D. J. 1990. Sonoelectrochemistry. *Ultrasonics*, 28: 333–337.

Mastai, Y., Homyonfer, M., Gedanken, A. et al. 1999a. Room temperature sonoelectrochemical synthesis of molybdenum sulfide fullerene-like nanoparticles. *Advanced Materials*, 11: 1010–1013.

Mastai, Y., Polsky, R., Koltypin, Y. et al. 1999b. Pulsed sonoelectrochemical synthesis of cadmium selenide nanoparticles. *Journal of the American Chemical Society*, 121: 10047–10052.

Mdleleni, M. M., Hyeon, T., and Suslick, K. S. 1998. Sonochemical synthesis of nanostructured molybdenum sulfide. *Journal of the American Chemical Society*, 120: 6189–6190.

Messing, G. L., Zhang, S.-C., and Jayanthi, G. V. 1993. Ceramic powder synthesis by spray pyrolysis. *Journal of the American Ceramic Society*, 76: 2707–2726.

Mizukoshi, Y., Fujimoto, T., Nagata, Y. et al. 2000. Characterization and catalytic activity of core–shell structured gold/palladium bimetallic nanoparticles synthesized by the sonochemical method. *The Journal of Physical Chemistry B*, 104: 6028–6032.

Mizukoshi, Y., Okitsu, K., Maeda, Y. et al. 1997. Sonochemical preparation of bimetallic nanoparticles of gold/palladium in aqueous solution. *The Journal of Physical Chemistry B*, 101: 7033–7037.

Mizukoshi, Y., Oshima, R., Maeda, Y. et al. 1999. Preparation of platinum nanoparticles by sonochemical reduction of the Pt(II) ion. *Langmuir*, 15: 2733–2737.

Morel, A. L., Nikitenko, S. I., Gionnet, K. et al. 2008. Sonochemical approach to the synthesis of $Fe_3O_4SiO_2$ core–shell nanoparticles with tunable properties. *ACS Nano*, 2: 847–856.

Moriguchi, N. 1934. The effect of supersonic waves on chemical phenomena, (III). The effect on the concentration polarization. *Journal of the Chemical Society of Japan*, 55: 749–750.

Murcia, M. J., Shaw, D. L., Woodruff, H. et al. 2006. Facile sonochemical synthesis of highly luminescent ZnS-shelled CdSe quantum dots. *Chemistry of Materials*, 18: 2219–2225.

Nagata, Y., Watanabe, Y., Fujita, S. et al. 1992. Formation of colloidal silver in water by ultrasonic irradiation. *Journal of the Chemical Society, Chemical Communications*, 21: 620–622.

Neddersen, J., Chumanov, G., and Cotton, T. M. 1993. Laser ablation of metals: A new method for preparing SERS active colloids. *Applied Spectroscopy*, 47: 1959–2177.

Nemamcha, A., Rehspringer, J. L., and Khatmi, D. 2006. Synthesis of palladium nanoparticles by sonochemical reduction of palladium (II) nitrate in aqueous solution. *The Journal of Physical Chemistry B*, 110: 383–387.

Neppolian, B., Wang, Q., Jung, H. et al. 2008. Ultrasonic-assisted sol–gel method of preparation of TiO_2 nano-particles: Characterization, properties and 4-chlorophenol removal application, *Ultrasonics Sonochemistry*, 15: 649–658.

Okitsu, K., Ashokkumar, M., and Grieser, F. 2005. Sonochemical synthesis of gold nanoparticles: Effects of ultrasound frequency. *The Journal of Physical Chemistry B*, 109: 20673–20675.

Okitsu, K., Bandow, H., and Maeda, Y. 1996. Sonochemical preparation of ultrafine palladium particles. *Chemistry of Materials*, 8: 315–317.

Okitsu, K., Sharyo, K., and Nishimura, R. 2009. One-pot synthesis of gold nanorods by ultrasonic irradiation: The effect of pH on the shape of the gold nanorods and nanoparticles. *Langmuir*, 25: 7786–7790.

Okuyama, K. and Lenggoro, I. W. 2003. Preparation of nanoparticles via spray route. *Chemical Engineering Science*, 58: 537–547.

Oxley, J. D., Mdleleni, M. M., and Suslick, K. S. 2004. Hydrodehalogenation with sonochemically prepared Mo_2C and W_2C. *Catalysis Today*, 88: 139–151.

Panatarani, C., Lenggoro, I. W., and Okuyama, K. 2003. Synthesis of single crystalline ZnO nanoparticles by salt-assisted spray pyrolysis. *Journal of Nanoparticle Research*, 5: 47–53.

Pastoriza-Santos, I., Sánchez-Iglesias, A., García de Abajo, F. J. et al. 2007. Environmental optical sensitivity of gold nanodecahedra. *Advanced Functional Materials*, 17: 1443–1450.

Pol, V. G., Gedanken, A., and Calderon-Moreno, J. 2003a. Deposition of gold nanoparticles on silica spheres: A sonochemical approach. *Chemistry of Materials*, 15: 1111–1118.

Pol, V. G., Grisaru, H. Gedanken, A. et al. 2005. Coating noble metal nanocrystals (Ag, Au, Pd, and Pt) on polystyrene spheres via ultrasound irradiation. *Langmuir*, 21: 3635–3640.

Pol, V. G., Motiei, M., Gedanken, A. et al. 2003b. Sonochemical deposition of air-stable iron nanoparticles on monodispersed carbon spherules. *Chemistry of Materials*, 15: 1378–1384.

Prasad, K., Pinjari, D. V., Pandit, A. B. et al. 2010. Synthesis of titanium dioxide by ultrasound assisted sol–gel technique: Effect of amplitude (power density) variation. *Ultrasonics Sonochemistry*, 17: 697–703.

Prozorov, T., Prozorov, R., and Suslick, K. S. 2004. High velocity interparticle collisions driven by ultrasound. *Journal of the American Chemical Society*, 126: 13890–13891.

Qian, D., Jiang, J. Z., and Hansen, P. L. 2003. Preparation of ZnO nanocrystals via ultrasonic irradiation. *Chemical Communications*, 9: 1078–1079.

Qiu, X., Burda, C., Fu, R. et al. 2004. Heterostructured Bi_2Se_3 nanowires with periodic phase boundaries. *Journal of the American Chemical Society*, 126: 16276–16277.

Qiu, X., Lou, Y., Samia, A. C. S. et al. 2005. PbTe nanorods by sonoelectrochemistry. *Angewandte Chemie International Edition*, 44: 5855–5857.

Qiu, X. F., Xu, J. Z., Zhu, J. M. et al. 2003. Controllable synthesis of palladium nanoparticles via a simple sonoelectrochemical method. *Journal of Materials Research*, 18: 1399–1404.

Radziuk, D., Shchukin, D., and Möhwald, H. 2008. Sonochemical design of engineered gold–silver nanoparticles. *The Journal of Physical Chemistry* C, 112: 2462–2468.

Rao, K. S., El-Hami, K., Kodaki, T. et al. 2005. A novel method for synthesis of silica nanoparticles. *Journal of Colloid and Interface Science*, 289: 125–131.

Rao, C. N. R., Muller, A., and Cheetan, A. K. 2008. *The Chemistry of Nanomaterials Synthesis, Properties and Applications*, Vol. 1. Weinheim, Germany: Wiley-VCH Verlag GmbH & Co.

Reisse, J., Caulier, T., Deckerkheer, C. et al. 1996. Quantitative sonochemistry. *Ultrasonics Sonochemistry*, 3: S147–S151.

Reisse, J., Francois, H., Vandercammen, J. et al. 1994. Sonoelectrochemistry in aqueous electrolyte: A new type of sonoelectroreactor. *Electrochimica Acta*, 39: 37–39.

Rutledge, R. D., Morris, W. H., Wellons, M. S. et al. 2006. Formation of FePt nanoparticles having high coercivity. *Journal of the American Chemical Society*, 128: 14210–14211.

Sáez, V. and Mason, T. J. 2009. Sonoelectrochemical synthesis of nanoparticles. *Molecules*, 14: 4284–4299.

Sánchez-Iglesias, A., Pastoriza-Santos, I., Pérez-Juste, J. et al. 2006. Synthesis and optical properties of gold nanodecahedra with size control. *Advanced Materials*, 18: 2529–2534.

Schmid, G. 1992. Large clusters and colloids. Metals in the embryonic state. *Chemical Reviews*, 92: 1709–1727.

Shafi, K. V. P. M., Ulman, A., Dyal, A. et al. 2002. Magnetic enhancement of γ-Fe_2O_3 nanoparticles by sonochemical coating. *Chemistry of Materials*, 14: 1778–1787.

Shafi, K. V. P. M., Ulman, A., Yan, X. et al. 2001. Sonochemical preparation of silane-coated titania particles. *Langmuir*, 17: 1726–1730.

Shen, Q., Jiang, L., Miao, J. et al. 2008a. Sonoelectrochemical synthesis of CdSe nanotubes. *Chemical Communications*, 14: 1683–1685.

Shen, Q., Jiang, L., Zhang, H. et al. 2008b. Three-dimensional dendritic Pt nanostructures: Sonoelectrochemical synthesis and electrochemical applications. *The Journal of Physical Chemistry* C, 112: 16385–16392.

Sibbald, M. S., Chumanov, G., and Cotton, T. M. 1996. Reduction of cytochrome c by halide-modified, laser-ablated silver colloids. *The Journal of Physical Chemistry*, 100: 4672–4678.

Skrabalak, S. E. and Suslick, K. S. 2005. Porous MoS_2 synthesized by ultrasonic spray pyrolysis. *Journal of the American Chemical Society*, 127: 9990–9991.

Skrabalak, S. E. and Suslick, K. S. 2006. Porous carbon powders prepared by ultrasonic spray pyrolysis. *Journal of the American Chemical Society*, 128: 12642–12643.

Skrabalak, S. E. and Suslick, K. S. 2007. Carbon powders prepared by ultrasonic spray pyrolysis of substituted alkali benzoates. *The Journal of Physical Chemistry C*, 111: 17807–17811.
Socol, Y., Abramson, O., Gedanken, A. et al. 2002. Suspensive electrode formation in pulsed sonoelectrochemical synthesis of silver nanoparticles. *Langmuir*, 18: 4736–4740.
Soler-Illia, G. J. de, A. A., Sanchez, C., Lebeau, B. et al. 2002. Chemical strategies to design textured materials: From microporous and mesoporous oxides to nanonetworks and hierarchical structures. *Chemical Reviews*, 102: 4093–4138.
Su, C. H., Wu, P. L., and Yeh, C. S. 2003. Sonochemical synthesis of well-dispersed gold nanoparticles at the ice temperature. *The Journal of Physical Chemistry B*, 107: 14240–14243.
Suh, W. H., Jang, A. R., Suh, Y.-H. et al. 2006. Porous, hollow, and ball-in-ball metal oxide microspheres: Preparation, endocytosis, and cytotoxicity. *Advanced Materials*, 18: 1832–1837.
Suh, W. H. and Suslick, K. S. 2005. Magnetic and porous nanospheres from ultrasonic spray pyrolysis. *Journal of the American Chemical Society*, 127: 12007–12010.
Suslick, K. S. 1990. Sonochemistry. *Science* 247: 1439–1445.
Suslick, K. S. and Casadonte, D. J. 1987. Heterogeneous sonocatalysis with nickel powder. *Journal of the American Chemical Society*, 109: 3459–3461.
Suslick, K. S., Casadonte, D. J., and Doktycz, S. J. 1989b. Ultrasonic irradiation of copper powder. *Chemistry of Materials*, 1: 6–8.
Suslick, K. S., Choe, S. B., Cichowlas, A. A. et al. 1991. Sonochemical synthesis of amorphous iron. *Nature*, 353: 414–416.
Suslick, K. S. and Doktycz, S. J. 1989a. The sonochemistry of zinc powder. *Journal of the American Chemical Society*, 111: 2342–2344.
Suslick, K. S. and Doktycz, S. J. 1990. *Advances in Sonochemistry*, Vol. 1 (Ed.: Mason, T. J.). New York: JAI Press.
Suslick, K. S., Fang, M., and Hyeon, T. 1996a. Sonochemical synthesis of iron colloids. *Journal of the American Chemical Society*, 118: 11960–11961.
Suslick, K. S., Hyeon, T., and Fang, M. 1996b. Nanostructured materials generated by high-intensity ultrasound: Sonochemical synthesis and catalytic studies. *Chemistry of Materials*, 8: 2172–2179.
Suslick, K. S. and Price, G. J. 1999. Applications of ultrasound to materials chemistry. *Annual Review of Materials Science*, 29: 295–326.
Uzcanga, I., Bezverkhyy, I., Afanasiev, P. et al. 2005. Sonochemical preparation of MoS_2 in aqueous solution: Replication of the cavitation bubbles in an inorganic material morphology. *Chemistry of Materials*, 17: 3575–3577.
Vinodgopal, K., He, Y., Ashokkumar, M. et al. 2006. Sonochemically prepared platinum–ruthenium bimetallic nanoparticles. *The Journal of Physical Chemistry B*, 110: 3849–3852.
Walter, J., Nishioka, M., and Hara, S. 2001. Ultrathin platinum nanoparticles encapsulated in a graphite lattice prepared by a sonochemical approach. *Chemistry of Materials*, 13: 1828–1833.
Walton, D. J. 2002. Sonoelectrochemistry—The application of ultrasound to electrochemical systems. *ARKIVOC*, iii: 198–218.
Wang, J., Loh, K. P., Zhong, Y. L. et al. 2007. Bifunctional FePt core–shell and hollow spheres: Sonochemical preparation and self-assembly. *Chemistry of Materials*, 19: 2566–2572.
Wang, H., Zhu, J. J., Zhu, J. M. et al. 2002. Sonochemical method for the preparation of bismuth sulfide nanorods. *The Journal of Physical Chemistry B*, 106: 3848–3854.
Xia, B., Lenggoro, I. W., and Okuyama, K. 2001a. Novel route to nanoparticle synthesis by salt-assisted aerosol decomposition. *Advanced Materials*, 13: 1579–1582.
Xia, B., Lenggoro, I. W., and Okuyama, K. 2001b. Synthesis of CeO_2 nanoparticles by salt-assisted ultrasonic aerosol decomposition. *Journal of Materials Chemistry*, 11: 2925–2927.
Xie, Y., Zheng, X., Jiang, X. et al. 2002. Sonochemical synthesis and mechanistic study of copper selenides Cu_2-xSe, β-CuSe, and Cu_3Se_2. *Inorganic Chemistry*, 41: 387–392.
Xiong, H. M., Shchukin, D. G., Möhwald, H. et al. 2009. Sonochemical synthesis of highly luminescent zinc oxide nanoparticles doped with magnesium (II). *Angewandte Chemie International Edition*, 48: 2727–2731.
Yeung, S. A., Hobson, R., Biggs, S. et al. 1993. Formation of gold sols using ultrasound. *Journal of the Chemical Society, Chemical Communications*, 4: 378–379.
Zhang, J., Du, J., Han, B. et al. 2006. Sonochemical formation of single-crystalline gold nanobelts. *Angewandte Chemie International Edition*, 45: 1116–1119.
Zheng, T., Pang, J., Tan, G. et al. 2007. Surfactant templating effects on the encapsulation of iron oxide nanoparticles within silica microspheres. *Langmuir*, 23: 5143–5147.

Zhou, S. M., Feng, Y. S., and Zhang, L. D. 2003a. Sonochemical synthesis of large-scale single crystal CdS nanorods. *Materials Letters*, 57: 2936–2939.

Zhou, S. M., Feng, Y. S., and Zhang, L. D. 2003b. Sonochemical synthesis of large-scale single-crystal PbS nanorods. *Journal of Materials Research*, 18: 1188–1191.

Zhu, J., Aruna, S. T., Koltypin, Y. et al. 2000a. A novel method for the preparation of lead selenide: Pulse sonoelectrochemical synthesis of lead selenide nanoparticles. *Chemistry of Materials*, 12: 143–147.

Zhu, J., Koltypin, Y., and Gedanken, A. 2000b. General sonochemical method for the preparation of nanophasic selenides: Synthesis of ZnSe nanoparticles. *Chemistry of Materials*, 12: 73–78.

Zhu, J., Liu, S., Palchik, O. et al. 2000c. Shape-controlled synthesis of silver nanoparticles by pulse sonoelectrochemical methods. *Langmuir*, 16: 6396–6399.

Zhu, J. J. and Singh, V. 2005. Sonochemistry for the preparation of nanosized metal chalcogenide. In *Proc. SPIE 59290B-1*, Bellingham, WA.

Zhu, J. J., Wang, H., Xu, S. et al. 2002. Sonochemical method for the preparation of monodisperse spherical and rectangular lead selenide nanoparticles. *Langmuir*, 18: 3306–3310.

Zin, V., Pollet, B. G., and Dabala, M. 2009. Sonoelectrochemical (20 kHz) production of platinum nanoparticles from aqueous solution. *Electrochimica Acta*, 54: 7201–7206.

7 Ultrasound for Fruit and Vegetable Quality Evaluation

Amos Mizrach

CONTENTS

Introduction	129
Fundamentals of Ultrasound Technology	131
Utilization of Ultrasound Technology for Agricultural Purposes	131
Ultrasound Wave Propagation through Fresh Agricultural Tissue	132
Wave Modes and Velocities	133
Wave Attenuation and Amplitude	133
Directivity Distribution of Ultrasound Waves	134
Basic Ultrasound Hardware	138
Quality-Related Parameters Correlated with Ultrasound	138
Mechanical Parameters	139
Physicochemical Indices	139
Ultrasonic Measurement Techniques Developed for Fruits and Vegetables	140
Measurements of Tissue Segments	140
Measurements of Cut Half Fruit Specimens	141
Measurements of Whole Fruits	145
Portable Systems	147
Fixed-Load System Operating in the Time Domain	147
Fixed-Load System Operating in the Frequency Domain	148
Variable-Load System Operating in the Time Domain	148
Ultrasound Implementations and Quality Assessment of Selected Fruits and Vegetables	150
Avocado	150
Apple	154
Mango	154
Melon	156
Olive	156
Plum	157
Potato	157
Tomato	158
Conclusions	158
References	159

INTRODUCTION

With regard to produce, the term "quality" encompasses sensory properties, nutritive values, chemical constituents, mechanical properties, functional properties, and freedom from defects, each of which has been the subject of many studies (Abbott, 1999; Shewfelt and Bruckner, 2000). Increasing awareness of quality and enhanced perspicacity on the part of consumers are stimulating a strong drive for improved quality of fruit and vegetables, in both the fresh produce market and the food-processing industry. There is strong demand to evaluate fruit and vegetable quality during growth and maturation and in the course of the harvest period, storage, and shelf life, and this is an important issue for the growers as well as the distributors and consumers. Generally, people base their quality evaluation of fruits on combined inputs from several senses and, since human judgments are, in general, subjective, these evaluations are liable to be inconsistent and to lead to erroneous quality assessment of the fruit. Thus, there is an increasing need for better quality monitoring. Over the years, many methods and associated instrumentation, based on mimicking human sensory perceptions, have been developed to measure quality- and quality- related attributes; they have been accompanied by a vast array of instrumental sensors for real-time and nondestructive testing (Upchurch et al., 1994; Abbott, 1999). There have been numerous studies of technologies for nondestructive quality measurement of fruits and vegetables—visual, spectroscopic, acoustic, etc.—but Butz et al. (2005) surveyed the literature and found that, in the context of food, about 20% of nondestructive or noninvasive techniques used acoustic methods. This multitude of techniques encompasses an enormous range, from traditional human sensory evaluation by trained inspectors to sophisticated mechanical and electronic devices equipped with state-of-the-art sensors to measure and classify many different quality parameters of fruits and vegetables. The use of ultrasound obviously belongs among the acoustic methods, and Butz et al. (2005) included some examples from this research field in their survey. Ultrasonics is a rather recent addition to the acoustic technologies and methods used in agriculture, especially in fruit quality (Mizrach, 2000a,b). One of the major advantages of ultrasonic waves is their capability to penetrate the produce and nondestructively extract the material characteristic that might be correlated with quality indices of the fruits and vegetables (Mizrach et al., 1989; McClements et al., 1990).

The potential for application of ultrasound in the food industry has been recognized since the 1970s (Povey and Wilkinson, 1980), and developments in the technology have progressed rapidly during the years (Povey and McClements, 1988; Povey, 1998). However, development of ultrasound techniques as a means of evaluating food quality has not progressed as fast in the fresh fruit sector as in the food-processing industry. Lack of appropriate equipment that is sufficiently powerful to penetrate but, at the same time, sufficiently gentle to avoid damage to the sensitive tissues of fruits and vegetables has been an important deterrent (Porteous et al., 1981; Mizrach et al., 1989). However, some advances in equipment design, and availability of new instruments and sensors, mainly designed for industrial use with new composite materials, have facilitated progress and have stimulated more studies and development of ultrasonic methods and techniques to assess produce quality in the fresh fruit and vegetable markets (Mizrach et al., 1989). Since then, ultrasonic techniques have been investigated for use in sensory analysis of various quality parameters of agricultural produce and their influence on shelf life, durability, and quality. Various devices and measuring techniques based on ultrasonic waves have been developed for nondestructively monitoring some physicochemical, biochemical, and mechanical changes that occur in fruit tissues during the various stages of their pre- and postharvest existence. These stages include growth and maturation (Self et al., 1994; Chivers et al., 1995; Mizrach et al., 1999a,b; Gaete-Garreton et al., 2005), storage under various conditions (Flitsanov et al., 2000; Mizrach et al., 2000; Verlinden et al., 2004), and shelf life (Mizrach and Flitsanov, 1999; Mizrach, 2000a; Johnston et al., 2002).

In this chapter, quality evaluation of fresh fruits and vegetables and determination of their physicochemical, biochemical, and mechanical properties by means of ultrasound technology are addressed; therefore, it encompasses both aspects: Published data on quality-related properties of fruits and vegetables, as they affect acoustical parameters, are summarized, and a comprehensive survey of the latest developments and advances in ultrasonic technology for monitoring the quality of fruits and vegetables during growth, maturation, harvest, storage, and shelf life, is presented. This chapter describes various ultrasound techniques and their fundamental elements; various ultrasonic measurement methods, requirements, and instrument performance; and the links between some ultrasound parameters and several quality-related indices. It presents the progress to date toward realization of such a technology for examination of fruit tissues and whole fruits, in conjunction with quality indices, physical properties, harvesting time, and changes that occur during storage and shelf life. Finally, some ultrasound implementations applied to quality assessment of selected fruits and vegetables are presented.

FUNDAMENTALS OF ULTRASOUND TECHNOLOGY

Ultrasound technology has been known for many years; its main applications have been in medical diagnostics and industrial processes and inspections. At high frequencies and low power, it can be used as an analytical and diagnostic tool; at very high power, it can assist processing. Throughout its range of applications, ultrasound is generated in the same way: a ceramic crystal, within a device known as a transducer, is excited by a short electrical pulse that typically comprises several sine wave cycles and whose shape does not change in the course of each measurement in a fruit sample; the piezoelectric effect is used to convert this electrical energy into a mechanical (acoustic) wave that is propagated as a short ultrasonic pulse with the fundamental frequency of the transducer. This energy is emitted into the material or body under analysis and is propagated through it (Krautkramer and Krautkramer, 1990). The ultrasound signal that emerges from the test specimen is sensed by a piezoelectric element that acts as a receiver, converting any ultrasonic energy impinging on it back into electrical energy. When the system operates in "pulse-echo" mode, the same piezoelectric element acts alternately as transmitter and receiver; when a "through-transmission" mode is used, a second piezoelectric element acts as the receiver. Ultrasonic energy will propagate through a material until the sound wave encounters an impedance change, which indicates changes in the material density and/or the velocity of the sound wave (Kuttruff, 1991); such changes within the material may be associated with changes in the nature of the tissue or the presence of a void or reflecting body. Some of the sound energy is reflected, and the amount reflected depends on the impedance change and/or the size of the reflector. If there are no internal reflectors, the wave will continue until it reaches the far side of the test object or until the energy is totally attenuated. The energy attenuation of the ultrasound beam and the propagation speed of the wave depend on the nature of the material and its structure (Kuttruff, 1991). Most physical or chemical changes in the material cause changes in the attenuation and velocity of the propagated waves. In solid and liquid industrial materials, as well as in most biological tissues such as those of the human body, ultrasound energy is easily propagated, which facilitates diagnostic or detection procedures (Walls, 1969). However, unlike the relatively easy diagnosis or analysis offered by these materials, analysis of ultrasonic waves presents extreme difficulties when examining fruit and vegetable tissues.

UTILIZATION OF ULTRASOUND TECHNOLOGY FOR AGRICULTURAL PURPOSES

Despite the wide and successful use of ultrasound applications in industry and medicine, very little attention has been paid to the application of ultrasonic techniques to the solid media that constitute fresh agricultural products, such as fruit and vegetable tissues (Mizrach et al., 1989). For many years, it had proved impossible to apply ultrasound measurement techniques to most fruits and vegetables because their tissues have acoustical properties very different from those of biological or industrial materials (Povey, 1998). The primary attenuation mechanism in most fruits and vegetables is scattering from voids and pores; thus, these are highly attenuating materials, a characteristic that complicates the interpretation of ultrasound data (Mizrach et al., 1989; Povey, 1998). In fact, the frequency domains that are used for industrial and medical applications (usually between 0.5 and 30 MHz), with their associated attenuation, have been difficult to apply to fruit and vegetable tissues (Porteous et al., 1981; Sarkar and Wolfe, 1983; Mizrach et al., 1989).

Several studies have involved attempts to penetrate fruit specimens with ultrasonic waves in order to detect tissue inhomogeneities, such as discontinuities, voids, pits (seeds or stones), foreign materials, bruises, rots, or internal breakdown (Upchurch et al., 1985, 1986, 1987; Mizrach et al., 1989, 1991; Hansen et al., 1992; Cheng and Haugh, 1994; Self et al., 1994; Nielsen et al., 1998; Inomata and Suzuki, 2001; Camarena and Martinez-Mora, 2006). Tests for characterizing tissue were suggested in many studies (Upchurch et al., 1985, 1986, 1987; Mizrach et al., 1989, 1991; Self et al., 1994; Nielsen et al., 1998). Early pioneer work, in the 1980s, that attempted to use this method and equipment to detect bruises in fresh apples revealed difficulties in achieving penetration of ultrasonic waves into fruit and vegetable specimens (Porteous et al., 1981; Upchurch et al., 1987). Other studies found that the high attenuation of the signal made it impossible to examine the interior of fruits satisfactorily (Sarkar and Wolfe, 1983; Watts and Russell, 1985). The frequency of the transmitted ultrasonic waves used in those studies was found to be a very important factor in fruit tissue penetration. Sarkar and Wolfe (1983) reported very high attenuation of the ultrasonic wave when they used regular frequencies of 0.5 and 1 MHz in tissues of potato, cantaloupe, and apple, and they suggested that a lower frequency (between 100 and 500 kHz) and higher power of acoustic radiation might be beneficial in nondestructive testing of fresh fruits and vegetables. Mizrach et al. (1989) applied lower frequencies and modified a high-power, low-frequency ultrasound device (a novel device at that time), originally designed for an industrial application, for measurements in strongly attenuating agricultural specimens. They used 50 and 100 kHz transducers in the through-transmission mode to overcome the high attenuation of several fruit tissues. This mode was recommended for use with low frequencies (<100 kHz), because the pulse-echo mode produced high levels of signal noise and ringing (Krautkramer and Krautkramer, 1990). By using this measurement method, Mizrach et al. (1989) successfully measured three major acoustic parameters: reflective loss; propagation velocities of ultrasonic waves passing through cylindrical specimens of potato, avocado, cucumber, carrot, pumpkin, melon, and apple tissues; and attenuation coefficients for potato, carrot, and avocado. Specific data and results obtained for selected fruits and vegetables are presented in the following section. Mizrach et al. (1989) recommended this method as appropriate for characterizing tissue segments of highly attenuating agricultural materials; it also facilitates acoustical measurements of whole fruits and received international patent in Israel, the United States, and France (Mizrach et al., 1994b). These recommendations have been implemented in subsequent studies and new methods and systems have been developed; their operation and performance are discussed in the following sections.

The rapid development of sensors, microprocessors, and methods of signal analysis has opened up new possibilities for application of this simple, nondestructive testing method. Nevertheless, the existing industrial equipment still required modification to enable its application to fruit tissues: it was necessary to determine suitable frequency and power settings for application of ultrasonic excitation to various agricultural products and to select the appropriate equipment and measurement techniques. The equipment then needed to be tested and evaluated for use with tissue specimens or whole fruits (Mizrach et al., 1989).

ULTRASOUND WAVE PROPAGATION THROUGH FRESH AGRICULTURAL TISSUE

Ultrasonic parameters, such as wave velocity, attenuation, and directional characteristics of wave propagation, are widely documented in the literature, for many engineering materials and biological tissues (Walls, 1969; Krautkramer and Krautkramer, 1990). The use of ultrasonic waves has primarily focused on two major wave attributes: velocity and attenuation. Each of these parameters is dominated by differences between various kinds of materials; variations in their specific properties; their structural forms; whether they are in solid, liquid, or gaseous form; and whether their behavior is elastic, viscoelastic, or plastic.

In the following sections, the major important ultrasound parameters used in determining the quality of fresh agricultural produce are surveyed. The modes and forms of the waves, their attenuation and amplitude, and the directivity distribution of ultrasound wave propagation in compound media of fruit and vegetable tissues are presented.

Ultrasound for Fruit and Vegetable Quality Evaluation

Wave Modes and Velocities

When a vibration source such as a mechanical oscillator acts on a body, it creates waves that move through the medium in the form of oscillations of the material particles. The propagation of ultrasonic waves can be considered in terms of transport of energy through the vibrating medium as the disturbance moves. When the oscillations are longitudinal with respect to the orientation of the body, the motion of the disturbance is in the same direction as the oscillations, and the wave is called a longitudinal sound wave; it is expressed in terms of compressions and dilations in the direction of travel and, therefore, is also called a pressure or compression wave. This kind of wave can propagate through solids, liquids, and gases; therefore, it is considered the most important of the waves and is used in many engineering applications (Krautkramer and Krautkramer, 1990). When the oscillations occur perpendicularly to the surface, transverse or shear waves are created in which the particles no longer oscillate in the direction of propagation but at a certain angle to it, i.e., transversely. When such shear waves are transmitted, the particles in the medium vibrate in the plane of the originating source vibrations, but transmission of their transverse oscillations will lag in time, depending on the distance from the plane of excitation. Shear or transverse waves can be used only in solid bodies because gases and liquids are practically incapable of transmitting these kinds of waves. Another kind of wave that can occur in solid bodies is the surface or Rayleigh wave; in this case, the deformation is not strictly sinusoidal and is to some extent already known from studies of water waves. As the depth below the surface increases, the amplitude of the oscillation decreases rapidly, and the particles are practically at rest (Krautkramer and Krautkramer, 1990).

The velocities of the various kinds of sound waves can be determined by measuring the transit time required for the pulse to traverse the sample between an oscillator source and a collector probe. The velocity, C_p, is calculated from the following expression:

$$C_p = \frac{l}{t} \tag{7.1}$$

where
  l is the distance
  t is the transit time between the oscillator and a collector probe

There are known links between the various kinds of sound waves and the physical constants of the material concerned; these are, in the case of elastic materials, the modulus of elasticity, E; the density, ρ; and Poisson's ratio, μ (a dimensionless number). The equations describing the relationships between those constants are well presented in standard text books (e.g., Krautkramer and Krautkramer, 1990). Fruits and vegetables are viscoelastic in nature, since they comprise both elastic and viscous components, and usually exhibit a complex non-monotonic relationship with their physicochemical properties (Mizrach et al., 1996). Therefore, they are considered to be of higher complexity, which necessitates the use of advanced techniques for velocity calculations (Moreno et al., 2002). However, in many studies, wave velocities were found to be not sensitive enough to changes that occurred in the fruit tissue and were, therefore, ignored in subsequent studies (Mizrach et al., 1999a,b, 2000; Mizrach, 2007). Povey (1998) found that, in some food systems, ultrasonic wave velocity was a less useful property than other major measurable ultrasonic wave properties, such as attenuation.

Wave Attenuation and Amplitude

Attenuation of ultrasonic waves is one of the most important wave properties that can be used in assessing material structure and characteristics. In a real material, the sound wave continually loses part of its energy because of micro-scattering and absorption: part of the sound wave is scattered by microscopic interfaces along the transit path of the wave; another part is absorbed in the material through direct conversion of sound energy into heat by internal friction. Both effects—micro-scattering and internal friction—are traditionally included in an exponential term that represents the attenuation losses of planar waves in the material (Krautkramer and Krautkramer, 1990). If we consider only the decay of the wave maxima, the attenuation coefficient, α, of a material is defined by the exponential expression

$$A_d = A_0 e^{(-\alpha d)} \quad (7.2)$$

or by the logarithmic scale expression

$$\ln \frac{A_d}{A_0} = -\alpha d \quad (7.3)$$

where
A_0 and A_d are the ultrasonic signal amplitudes at the beginning and the end, respectively, of the propagation path of the wave
d is the distance the wave has travelled in the material

These equations describe the amplitude attenuation of plane harmonic waves of an infinite medium. The attenuation coefficient can easily be obtained by plotting the ratio of A_d/A_0 on a logarithmic scale Y-axis against d on a linear X-axis (Equation 7.3). The attenuation coefficient (α) is the inclination of the straight line obtained (e.g., Figure 7.1).

Directivity Distribution of Ultrasound Waves

For describing the directivity distributions of ultrasound waves through a solid medium, it is possible to imagine the body to consist of individual particles kept in position by elastic or viscoelastic forces. A source of imposed vibration such as mechanical oscillator acting on the body generates longitudinal, transverse, or shear waves and surface or Rayleigh waves that move through the medium, carried by oscillations of material particles. When a source of imposed vibrations, such as an ultrasonic wave transducer, is applied to an agricultural body such as fresh fruit or vegetable tissue, all of the described wave forms can be developed.

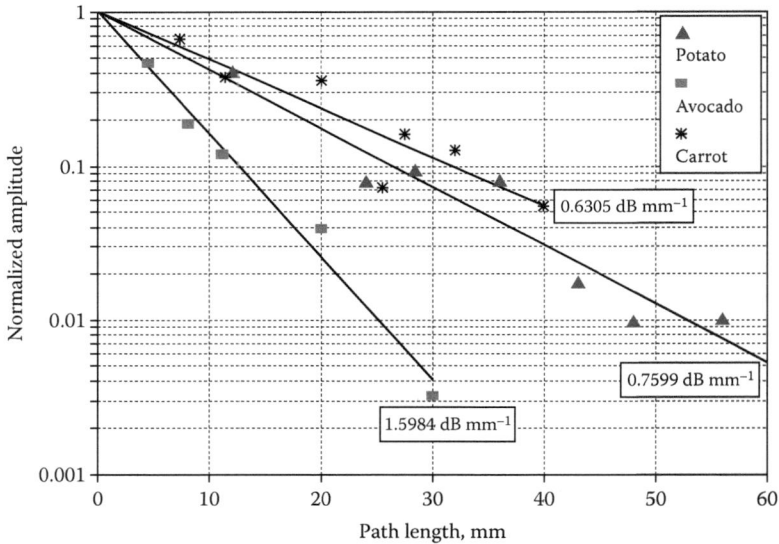

FIGURE 7.1 Example of attenuation coefficient calculation using Equation 7.3. (From Mizrach personal files.)

The directivity distribution of longitudinal waves in a half-space of fruit tissue was theoretically and practically investigated by Mizrach et al. (1992, 2009), who applied fluid-medium theory to the viscoelastic nature of fruit and vegetable tissue.

The ultrasonic field in a half-space material domain is commonly analyzed and illustrated by means of a directivity distribution function of the acoustic pressure caused by a circular piston in a fluid medium (e.g., Krautkramer and Krautkramer, 1990):

$$P(r,\theta) = P(r,0) \frac{2J_1(\beta \sin \theta)}{(\beta \sin \theta)} \tag{7.4}$$

where
$P(r, \theta)$ is the far-field acoustic pressure within the material
r and θ are the distance and angle of wave propagation in polar coordinates
$P(r, 0)$ is the pressure amplitude along the axis of symmetry ($\theta = 0$)
$J_1(\beta \sin \theta)$ is a Bessel function of the first kind
β is the directivity parameter of the material:

$$\beta = \pi \frac{D}{\lambda} = \pi \frac{Df}{C} \tag{7.5}$$

where
D is the diameter of the crystal
λ, f, and C are the sound wavelength, frequency of the crystal, and the sound velocity in the material, respectively

The length of the near-field zone, N, for cases in which the crystal diameter is much larger than the wavelength is $N = D^2/4\lambda$ (Krautkramer and Krautkramer, 1990). The far-field zone begins at an axial distance of about N from the crystal.

The acoustic pressure field of Equation 7.4 is characterized by a principal central beam and additional side lobes, according to the specific values of the directivity parameter, β. When $\beta \gg 1$, the divergence angle of the beam is small and the sound field has a classic pressure distribution of a long central beam and small side lobes. With decreasing β, the angle of divergence of the central beam increases up to 90° (when $\beta = 3.83$). For a very small source, namely, a point source ($\beta \ll 1$), the longitudinal pressure field is nearly spherical. Additional off-axis transverse waves, not included in Equation 7.4, may be developed in solid materials (Filipczynski, 1964; Krautkramer and Krautkramer, 1990). As the wavelength, β, depends on the material tested and the frequency of the transmitter, it may be possible to determine the appropriate transmitter for a desired pattern of directional wave propagation within the material.

An explicit expression for $P(r, 0)$ along the central axis according to Equation 7.4 following (Kinsler and Frey, 1972) is

$$P(r,0) = \frac{\rho_0 \pi D^2 f U_0}{4r} = \frac{\rho_0 \pi^2 D^2 f^2 A_0}{2r} \tag{7.6}$$

where
ρ_0 is the material density
U_0 and A_0 are the velocity and displacement amplitudes of the crystal, respectively

Equations 7.4 and 7.6 represent the geometrical spreading of the sound wave in an ideal, non-dissipative fluid. Thus, the intensity of the sound beam radiated from a probe would decrease with increasing distance from the source, unlike a plane wave in which there is no reduction of sound pressure along the propagation path. However, in a real material, the sound wave continually loses part of its energy because of micro-scattering and absorption: part of the sound wave is scattered by microscopic interfaces along the transit path of the wave; another part is absorbed in the material through direct conversion of sound energy into heat by internal friction. Both effects, micro-scattering and internal friction, are traditionally included in an exponential term that represents the attenuation of a planar wave in the material (Krautkramer and Krautkramer, 1990):

$$P_d = P_0 e^{(-\alpha d)} \tag{7.7}$$

where

P_0 and P_d, respectively, are the pressure of a planar sound wave at the start and end of a path element of length d

α is the "true" material sound attenuation coefficient for planar waves at a specific frequency

Equation 7.7 is sometimes used for evaluation of the "apparent attenuation" of non-planar waves as well, in order to quantify the total amplitude decrease in a certain direction.

In attempting to include the true wave attenuation of a material in the classical model of angular pressure distribution, Mizrach et al. (2009) assumed that the "true attenuation" losses of Equation 7.7 take place along the propagation path of each directional wave in the material:

$$P(r, \theta, \alpha) = P_0 e^{-\alpha(r-r_0)} \frac{2 r_0 j_1(\beta \sin \theta)}{r \beta \sin(\theta)} \tag{7.8}$$

in which P_0 is the far-field sound pressure amplitude at a reference distance $r_0 > 3N$ along the axis of symmetry. An explicit expression for P_0, according to Equations 7.6 and 7.7, was

$$P_0 = \frac{\rho_0 \pi^2 D^2 f^2 A_0}{2 r_0} e^{(-\alpha r_0)} \tag{7.9}$$

The directivity expression of the suggested model (Equation 7.8), which now includes an attenuation term of a real material, as well as other material and transmitter parameters (Equations 7.5 and 7.9), was used by Mizrach et al. (2009) in the analysis of sound pressure distribution on a half-space surface of fruit tissue. Three-dimensional presentations of the sound wave amplitude over the half-cut fruit surface in the Cartesian coordinate system were made by using graphical analysis software; an example of the acoustic field derived, e.g., for potato, is shown in Figure 7.2 (Mizrach et al., 2009).

The amplitude was presented as a percentage of the maximum measured amplitude for all observations; it decreased with increasing direct distance of the measurement points from the source and also laterally on each side of the axis of symmetry. The actual measurements were taken at 5 mm intervals, and the upper surface of the graph was derived from the measured points by means of an extrapolation program. Figure 7.2 illustrates the strong attenuation of the ultrasonic signal with increasing distance from the transducer over the cut surface plane and the directional characteristics of the acoustic wave over the cut half fruit field. The equipment and the method used by Mizrach et al. (2009) for preparing the attenuation field over a cut half surface is described in the following section and illustrated in Figure 7.3.

Ultrasound for Fruit and Vegetable Quality Evaluation

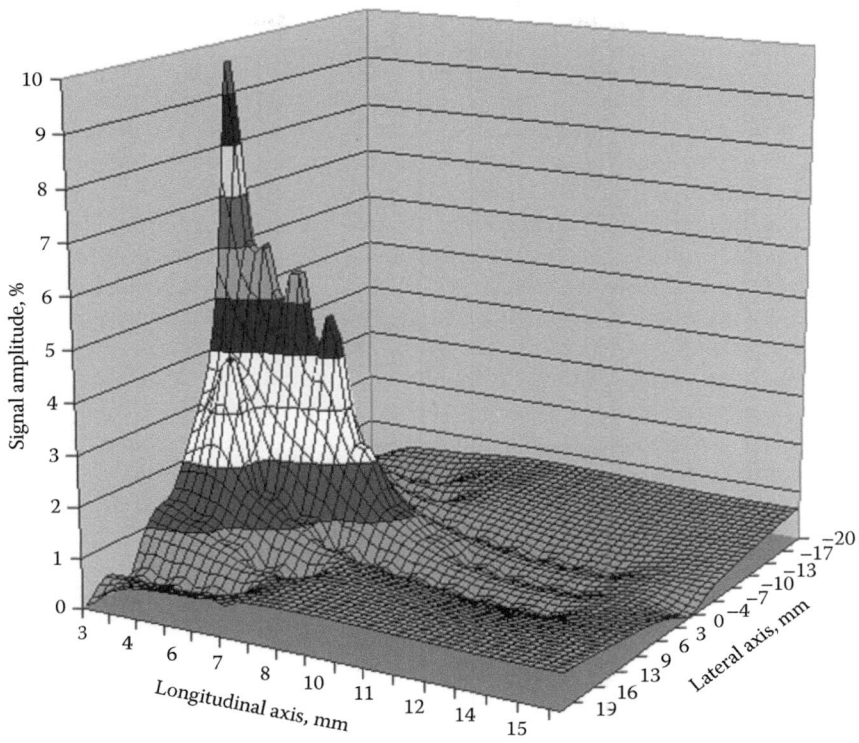

FIGURE 7.2 Ultrasonic three-dimensional field over the surface of a half-cut potato. (From Mizrach, A. et al., *Ultrasonics*, 49, 83, 2009. With permission.)

FIGURE 7.3 The experimental system for half-cut fruit testing: (a) transmitter, (b) tissue sample, (c) beam-focusing element, (d) receiver, (e) typical signal. (From Mizrach personal files.)

Basic Ultrasound Hardware

Ultrasound devices generally consist of four basic components: a wave generator (pulser/receiver), an emitter and receiver, a microprocessor (or computer) equipped with signal-processing software, and a display. Many different devices and arrangements can be used to generate, receive, and process ultrasound waves. Some of the components might be bundled into one unit, such as a wave generator together with a receiver (pulser/receiver). Some systems are equipped with a single transducer that alternately emits ultrasonic waves and receives echoes (pulse-echo mode); some with two transducers—a transmitter and a receiver—(through-transmission mode) and some with a single transducer that serves as both the transmitter and receiver. Some of the ultrasound devices might include advanced hardware and software for signal analysis and processing (Krautkramer and Krautkramer, 1990). The ultrasound devices and probes used in industry and medicine usually operate in frequency ranges of 0.5–30 MHz, at acoustic power levels ranging from a few watts to several kilowatts per unit beam area; both parameters are chosen according to the purpose or application of the device. The ultrasound components generally used in early studies of agricultural applications were adapted from medical and industrial applications, but later some components were modified to suit specific requirements. More recently, in several innovative studies, and for specific applications, dedicated systems were developed and installed.

Since this chapter aims to address quality assessments of fruits and vegetables by ultrasonic means, only equipment developed and used for this purpose is described here.

QUALITY-RELATED PARAMETERS CORRELATED WITH ULTRASOUND

Ultrasonic techniques used with food products form an entire field of applications and provide the user with a wide variety of information about the properties of materials being processed (Povey, 1998). In fruit and vegetable tissues, changes in their properties are part of the natural processes that occur during growth and maturation, and in the course of harvesting, storage, and shelf life. Various physiological and physicochemical changes occur during these processes, and each change is specifically determined by one or more factors, characteristic of the preharvest, harvest, and postharvest periods, respectively. The changes are expressed differently in each of the various periods and are mostly reflected in the quality of the final produce. Textural attributes of fruits and vegetables are among the main factors considered in quality assessment (Peacock et al., 1986) and are regularly used for determination of the stage of maturity of various kinds of fruits and vegetables (Abbott, 1999). Firmness is considered to be one of the main indices of maturity, and its changes during the ripening and softening process start on the tree and continue during harvesting, handling, and storage. Chemical contents and concentrations in fresh tissues are also important factors in determining the maturity of fruits and vegetables, but firmness is the factor most closely related to the stage of maturity (Peacock et al., 1986).

When acoustical measurements are used in conjunction with other physicochemical measurements, such as firmness, mealiness, dry weight percentage (DW), oil contents, total soluble solids (TSS) content, and acidity, a link between acoustical parameters and physicochemical indices enables the indirect assessment of the proper harvesting time, storage period, or shelf life (Mizrach et al., 1989; Abbott, 1999; Butz et al., 2005). The ultrasound technique has been adopted mainly for nondestructive, rapid, and accurate assessment of the changes involved, but has also been used to monitor additional natural and environmental factors that cause changes in quality-related parameters (Mizrach et al., 1989, 1999a, 2000).

The most common quality-related indices that were addressed in the various studies mentioned in this chapter and by Mizrach et al. (1997) and the indices that correlated with the changes in the mechanical and physicochemical parameters of the fruits in the course of the pre- and postharvest processes are described in the following.

MECHANICAL PARAMETERS

The most important mechanical property of fruits and vegetables that correlates with ultrasound characteristics is firmness. It is regarded as a very important parameter that reflects the changes in tissue texture during the course of growth, maturation, storage, and shelf life. Firmness is usually associated with ripeness, freshness, retention of good quality, and, therefore, with salability. It can be measured by compression or by puncturing with various probes, applied with a range of force or deformation levels, depending on the purpose of the measurement and how the quality attributes are defined (Abbott, 1999). Horticulturists tend to define firmness in terms of the maximum force applied, although sometimes the rupture or bioyield force is used (Abbott, 1999). Penetration tests are the most acceptable method for measuring firmness. Firmness of fruits is commonly measured with destructive penetration devices equipped with a puncture probe, or with a conical or curved head (manufactured, e.g., by Chatillon, New York). These techniques are generally destructive, not objective, and time-consuming; therefore demand arose for a rapid nondestructive technique. Ultrasound techniques were found to be among several that correlated well with firmness; therefore, ultrasound was suggested as the basis of a method that could fulfill these requirements during the course of the pre- and postharvest processes. In most cases described in this review, both destructive penetration measurements and ultrasound measurements of firmness were performed on the same fruit samples, in order to correlate results obtained with the two techniques; obviously, nondestructive ultrasound measurements were taken first. The nondestructive ultrasound measurements enabled correlations with firmness changes for a given fruit and even enabled the use of exactly the same measurement location during periods such as shelf life and storage (Mizrach, 2000a; Mizrach et al., 2000). In most cases, firmness measurements on whole fruits were taken in the radial direction at a point on the circumference of the largest cross section perpendicular to the blossom end–stem end axis (Mizrach et al., 1996). These correlations, for several selected fruits, are presented elsewhere in this chapter.

PHYSICOCHEMICAL INDICES

The most important quality-related physicochemical indexes that are usually measured in conjunction with ultrasound parameters are dry-weight (DW) content, oil content, total soluble solid (TSS) content, and acidity. Each kind of fruit has its own dominant physical parameters. In avocado fruits, oil and dry weight contents of the flesh are the basis of the most acceptable and reliable criteria of maturity, according to which the harvest date are determined (Lewis, 1978). The DW is usually determined according to Lee et al. (1983): samples taken from the fruit tissue for DW measurement are weighed and dried in an oven at 60°C for 3 days and are then reweighed. The oil content of the avocado fruit is determined by means of long-established refractive index–based techniques (Shannon, 1949; Harkness, 1954) that were described in detail by Lee (1981). Physicochemical indicators of maturity in mango flesh during the ripening process are decreasing acidity and increasing contents of sugars, soluble solids, and total solids, and these are associated with a concurrent reduction in titratable acidity (Fuchs et al., 1980). Acidity in mango is assessed by extraction procedures, described in detail by Fuchs et al. (1980): tissue samples are macerated with a commercial juice extractor, filtered and centrifuged at $10,000 \times g$ for 10 min. The supernatant juice is used for the determination of TSS and titratable acidity. The TSS is determined by a long-established technique that uses a refractometer (e.g., PR-1; Atago, Tokyo, Japan), an optical instrument used to determine the refractive index of a substance on its ocular surface and that yields a result expressed as degrees Brix or TSS (mg g^{-1}) (Horwitz, 1970). Titratable acidity is determined by titrating the supernatant juice with 0.1 N NaOH, and the result is expressed as the percentage of citric acid in the juice (Fuchs et al., 1980). In most cases described in this chapter, both destructive determination of physicochemical indices and ultrasound measurements were applied to the same sample of fruit for correlation purposes. Again, the nondestructive ultrasound measurements were done first; they enabled correlations with changes in the physicochemical indices for a given fruit, and even in exactly the same measurement location, during periods such as shelf life and storage (Mizrach, 2000a; Mizrach et al., 2000). These correlations, for several fruits, are given elsewhere in this chapter.

ULTRASONIC MEASUREMENT TECHNIQUES DEVELOPED FOR FRUITS AND VEGETABLES

Various ultrasound measurement methods and devices have been developed, modified, and used to measure the various quality-related attributes of fruit sand vegetables as they change during growth, maturation, storage, and shelf life. Various studies have been performed with existing or modified ultrasound systems in order to assess the response of biological materials to ultrasonic excitations (Mizrach, 2000a). The systems, originally intended for determination of physical properties of dense materials such as cement and composites, were usually modified from their industrial or medical configurations and were adapted for measurements of properties of the much more attenuated fruit and vegetable tissues (Mizrach et al., 1994b; Povey, 1998). Each system included the basic components for generating ultrasonic waves, such as emitter/receiver, transducers, and a microprocessor, as mentioned in the previous sections. The transducers were usually mounted in head structures designed to suit a specific measurement method, application to a tissue sample or a whole fruit, and the kind of fruit or vegetable examined.

Several categories of ultrasound devices were developed for each specific function, and according to the measurements needed: Single-touch systems were designed for measurements of tissue segments taken from fruits or vegetables; continuous-touch systems were developed specifically for nondestructive evaluation of tissue characteristics of cut-half fruit specimens and whole fruits or vegetables during maturation, storage, and shelf life; portable single-touch systems were developed, suitable for use in the field for rapid measurement of attenuation, in monitoring changing properties of fruits and vegetables during various pre- and postharvest processes. The various measurement methods have been used in numerous studies (e.g., Self et al., 1994; Mizrach et al., 1996, 1997, 1999a; Mizrach and Flitsanov, 1999; Kim et al., 2004; Bechar et al., 2005). Researchers used these systems for collecting the acoustical information emerging from tissue segments or whole fruit specimens, in order to extract the ultrasonic properties for determination of the quality of selected fruits and vegetables.

Tissue specimens in cylindrical or other shapes were studied to assess their acoustical parameters in conjunction with their physicochemical properties such as firmness, sugar content, and DW percentage (%DW) (Mizrach et al., 1989; De-Smedt, 2000; Gaete-Garreton et al., 2005), cut halves of fruits were used for studying ultrasonic wave paths within fruit tissues and for directional model development (Mizrach et al., 1992, 2009), and whole fruits were subjected to nondestructive determination of their physicochemical properties during growth, maturation, storage, and shelf life (Galili et al., 1993; Self et al., 1994; Nielsen et al., 1998; Mizrach, 1999a, 2000a,b).

MEASUREMENTS OF TISSUE SEGMENTS

Tissue specimens of fruits and vegetables are prepared, usually in a cylindrical shape, to assess their material characteristic and the acoustical parameters that basically are associated with the physicochemical properties of the tissue. The tissue measurements are performed with ultrasound systems typically available in industry and medicine, which were modified especially for measurement of tissue segments. Such measurements were, by definition, destructive tests. A typical tissue segment measurement device was equipped with two transducers, one acting as an emitter and the second as a receiver (in through-transmission mode). A single touch of either transducer, or its proximity to the tested object, activated the system. An electrical pulse, generated by an ultrasound device, excited the surface of the emitter which was in contact with the tested object, in which

Ultrasound for Fruit and Vegetable Quality Evaluation

FIGURE 7.4 Single-touch system for tissue segment measurements: (a) X–Y–Z fixation device, (b) receiver and energy concentrator, (c) ultrasound emitter/receiver, (d) transmitter and energy concentrator, (e) half-cut fruit, (f) gypsum casting, (g) adjustable-height jack, and (h) sample holder. (From Mizrach, A. et al., *Ultrasonics*, 49, 83, 2009. With permission.)

longitudinal mechanical waves were propagated. Echoes reflected from the far side of the object were detected by the same or a second transducer (in "pulse-echo" or "through-transmission" mode, respectively), and the resulting signals were transferred to the processor to calculate the object's acoustic properties.

An example of a tissue segment measurement device is shown in Figure 7.4 (Mizrach et al., 1994a).

The transducers have relatively large diameter excitation front surfaces to suit the relatively low ultrasonic frequency (~50 kHz). Beam-focusing elements were used to match the surface diameter of the wave emitter and receiver of each transducer to the desired area of contact with the fruit that was placed between the tips. The propagation velocity of the acoustic wave in the material was obtained by measuring the transit time required for the pulse to traverse the sample, by taking a specific reference point on each pulse function. The wave propagation velocity and the attenuation characteristics of the tissue segments were determined as mentioned above (Equations 7.1 and 7.2, respectively).

By using this method, Mizrach et al. (1989) measured the acoustic velocity and attenuation in potato, avocado, and carrot tissue specimens (Table 7.1). They placed a cylindrical specimen of fresh fruit tissue, 20 mm in diameter and 10, 15, or 20 mm in length, taken from an intact fruit, between a 50 kHz transmitter and a 50 kHz receiver that were coupled to the two opposite sides of the specimen. A polyamide rod, 20 mm in diameter and 90 mm long, was attached to the center of the transmitter, in order to insert the required delay time between the transmitted and the echo pulses, and to eliminate echoes of returning pulses that might pass through the air gap between the transducers. In a more advanced arrangement of this method, the polyamide rod was replaced with Plexiglas exponential energy concentrators that tapered exponentially from the 50 mm diameter of the probe to 6 mm diameter flat tips to match the 6 mm diameter tissue segments (Mizrach et al., 1994a). Results obtained with this system, for three selected fruits, were presented in the previous section and in Figure 7.1.

MEASUREMENTS OF CUT HALF FRUIT SPECIMENS

A study of the directional characteristics of ultrasonic waves across and in the vicinity of fruit surfaces used cut halves of fruits in order to overcome the experimental difficulties involved in direct measurement of pressure waves inside fruits (Mizrach et al., 1992, 2009). An experimental setup was designed and built to emit and receive an ultrasonic signal over a cut half fruit (Figure 7.4).

TABLE 7.1
Applications, Crops, Acoustic Parameters, and Aim of Measurements Using Ultrasound Technology in Pre- and Postharvest Processes

Crop	Aim	Ultrasonic Parameters			Reference
		Frequency (kHz)	Attenuation Coefficient (dB mm^{-1})[a]	Velocity Range (m s^{-1})	
Avocado (*Persea americana* Mill.) 'Ettinger', 'Fuerte'	Shelf life, firmness, DW, oil content	50	2.5–5.0 (increase with time)	200–400	Mizrach et al. (1989, 1992, 1994a,b, 1996, 1999a, 2000), Galili et al. (1993), Mizrach (2000a), Mizrach and Flitsanov, (1999), Self et al. (1994), Flitsanov et al. (2000)
Avocado (*Persea americana* Mill.) 'Ettinger'	Cold storage, shelf life, firmness, DW	50	2–4.3 (increase with storage time)	—	Flitsanov et al. (2000), Mizrach et al. (2000)
Avocado (*Persea americana* Mill.) 'Ettinger', 'Fuerte'	Ripeness	50	6–2.5 (decrease with increasing growth time)	350–200 (decrease with time)	Mizrach et al. (1999a), Self et al. (1994)
Avocado (*Persea americana* Mill.)	Firmness, ripeness	20.5	0.025–0.061[b] (increase with time)	—	Gaete-Garreton et al. (2005)
Apple (*Malus domestica*) 'Golden delicious', 'Jonagold', 'Cox'	Mealiness	100, 80	Energy[c] (V × s)	—	Bechar et al. (2005), De-Smedt (2000), Mizrach et al. (2003)
Apple (*Malus domestica*)	Damage detection, bruising	1000, 5000	1.3–2.6dB (reflection loss)	—	Upchurch et al. (1985, 1986, 1987)
Mango (*Mangifera indica* L.)	Maturity, firmness, TSS, acidity	50	4.7–2.16 (decrease with increasing time)	—	Mizrach (2000a), Mizrach et al. (1997, 1999b)
Tomato (*Lycopersicon esculentum* Mill) '870'	Firmness, TSS	50	3.9–2.7 (decrease with increasing time)	—	Mizrach (2007)
Tomato (*Lycopersicon esculentum* Mill) 'Tradiro'	Chilling injury	50	3–32[d] dB (increase with injury)	—	Verlinden et al. (2004)
Plume (*Prunus salicina*) 'Royal Z'	Firmness, TSS	50	5.2–1.0 (decrease with increasing time)	—	Mizrach (2004)

Product	Measured parameter	Frequency (kHz)	Attenuation / value	Other value	Reference
Melon (*Cucumis melo* L.) 'Galia'	Ripeness, TSS	50	2.6–7.1 (increase with ripeness)	61–90 (increase with ripeness)	Mizrach et al. (1991, 1994c)
Potato (*Solanum tuberosum*)	Hollow heart	100, 250	—	824	Watts and Russell (1985), Cheng and Haugh (1994), Ha et al. (1991), Mizrach et al. (1992)
Olive (*Olea europaea*)	Firmness, DW	50	9–19 (decrease with increasing time)	—	Mizrach et al. (2006b)
Potato (*Solanum tuberosum*)	Data collection	50	0.76 ($R^2 = 0.82$)	380 (std. = 6.6)	Mizrach et al. (1989)
Sweet potato (*Ipomea batatas* L.)	Weevil canals	5000, 7500	—	—	Hansen et al. (1992)
Carrot (*Daucus carota* L.)	Water loss, storage, data	37, 50	1.1[e], 0.63	500, 341	Nielsen et al. (1998), Mizrach et al. (1989)
Orange (*Citrus medica* L.) 'Valen.', 'Fort.', 'Navel', 'Salust.'	Turgidity and hydration of orange peel	200	1.8–3.7, 3.1–4.3, 3.3–3.7 at harvest, in room conditions, in cool conditions, respectively	130–240	Camarena and Martinez-Mora (2006)
Pear (*Pyrus pyrifolia* Nakai), apple (*Malus domestica*), peach (*Prunus persica* L.)	Firmness	500	Flight time, attenuation, peak frequency	—	Kim et al. (2004)

Source: Mizrach, A., *Postharv. Biol. Technol.*, 48, 315, 2008. With permission.

[a] Units of attenuation coefficient are dB mm^{-1} except if otherwise indicated.
[b] Surface absorption coefficient.
[c] Ultrasonic absorbed energy (V × s).
[d] Units in dB.
[e] Average value, LSD$_{0.95}$ = 0.52.

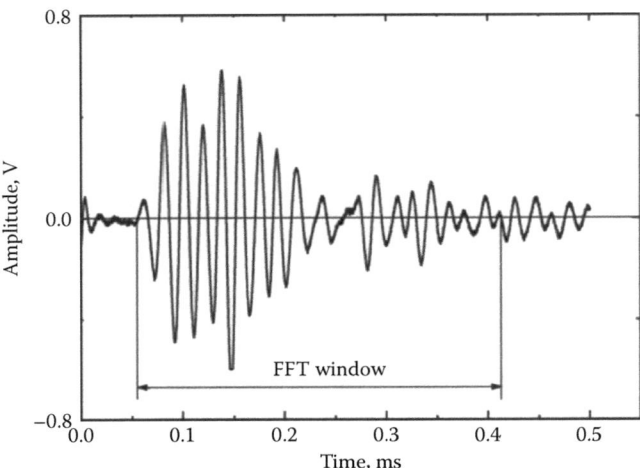

FIGURE 7.5 Typical ultrasonic wave signal. 50 kHz quartz transducer. (From Mizrach, A. et al., *Postharv. Biol. Technol.*, 16, 179, 1999b. With permission.)

The setup included an X–Y–Z linear fixation device (a, Figure 7.4), a receiver (b, Figure 7.4), and an emitter (d, Figure. 7.4). The original 55 mm diameter of the transducer crystals was tapered down to 6 mm by two Plexiglas ultrasonic energy exponential concentrators (after Makarov, 1964). The half-cut fruit (e, Figure 7.4) was placed in a gypsum casting (f, Figure 7.4) mounted on an adjustable-height jack (g, Figure 7.4). The sample holder (h, Figure 7.4) was equipped with an arm that held the fruit in its gypsum casting against the emitter, to impose the desired contact load. The emitter was attached to an adjustable mount that could move along the fruit–stem axis, to be clamped to match the desired fruit length. The emitter and its energy concentrator were horizontally attached to the side of the fruit, along its stem axis with ultrasonic gel. The receiver and its energy concentrator were mounted on the X–Y–Z linear fixation device, perpendicular to the plane of the fruit section, as shown in Figure 7.4. A low-frequency, high-power ultrasonic device (Pulser-receiver Model USL 33; Krautkramer, Germany) was used to excite the transmitted signal and an amplifier and receiver unit for signal reception. The time-domain waveform resembled a regular narrow-band, low-frequency ultrasonic pulse, with a central frequency of 50 kHz, similar to that described by Mizrach et al. (1999b) (Figure 7.5).

One of the analog output peaks in the time-domain waveform, and always the same one, was selected manually by means of the CRT monitor screen and was used as the reference amplitude throughout the measurement process. Mizrach et al. (1992, 2009) developed and used this technique for collecting the ultrasonic signal emerging from the X–Y plane of the tissue specimen and analyzed the data collected, by using two alternative coordinate systems: a polar (r–θ) coordinate system was used for a directional description of sound-wave amplitudes and a Cartesian (X–Y) system for three-dimensional presentation of the acoustic field. The r–θ measurements were taken along a semicircular arc of constant radius measured from the center of the transducer; the angular displacement steps were 5° (e.g., potato sample, Figure 7.6).

The X–Y sound-wave amplitude measurements were taken along (Y-axis) and across (X-axis) the fruit, where Y and X denote the longitudinal and lateral axes, respectively. The receiver was shifted along the lateral X-axis in 14 steps of 3 mm (7 steps on each side of the longitudinal center line) for several tissue samples. Measurements along the Y-axis directions were taken at 5 mm intervals, out to 33 mm from the transmitter head face. Mizrach et al. (1992, 2009) proposed a mathematical model for the directivity distribution of ultrasound waves over a cut half sample (see previous section); it simulated the ultrasonic wave distribution in strongly attenuating fruit tissue. The simulation model represented by Equation 7.8 enabled Mizrach et al. (1992, 2009) to qualitatively identify

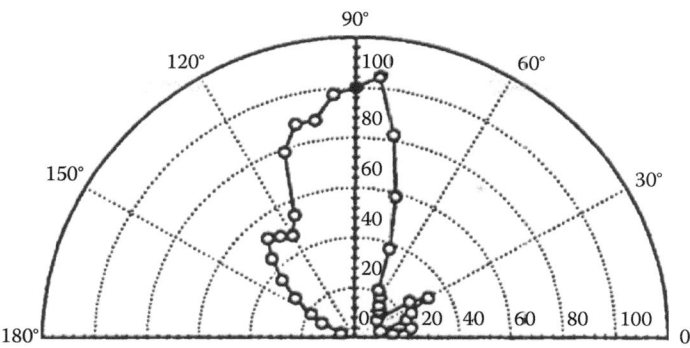

FIGURE 7.6 Ultrasonic two-dimensional field over the surface of a half-cut melon. (From Mizrach, A. et al., *Ultrasonics*, 49, 83, 2009. With permission.)

the combined effects of source characteristics and some fruit parameters on the sound propagation path and pressure distribution within the fruit. As β, the directivity parameter, and P_0, the input pressure amplitude, of the model depend on the material and emitter parameters (Equations 7.5 and 7.9), it is possible to select appropriate emitter dimensions and frequency to provide the required directional wave propagation within the fruit material. For example, increasing the probe diameter might increase both the input pressure amplitude and the penetration distance of the collimated beam for whole fruit penetration (side to side). A point-source probe could be more suitable for surface testing fruits in the vicinity of the transducers, because of the wide-angle wave propagation pattern of such a source. However, the input pressure amplitude depends on the size of the contact area of the transducer (Equation 7.9); therefore the point-source signal might be too weak for practical applications. In light of these considerations, a transducer with a chisel-type probe end was selected and used in further studies for local ultrasonic testing and evaluation of ultrasound wave paths within whole fruits (Mizrach et al., 1996, 1997, 1999a,b; Mizrach, 2000a). For practical considerations, this model indicated how to select the parameters of the ultrasonic transducer (oscillator dimensions, frequency, and amplitude) to control the magnitude and directivity of the ultrasonic waves in the fruit tissue.

MEASUREMENTS OF WHOLE FRUITS

Naturally, assessing quality-related properties of fruits or vegetables by means of ultrasound preferably would be performed nondestructively. However, in many kinds of fruits, the high attenuation of sound waves prevents their transmission through the whole fruit; therefore, ultrasonic tests need to be conducted locally, over a short distance on the peel. The assumption in this case is that the internal fruit tissue next to the peel can be regarded as representative of the flesh of the entire fruit. Several kinds of systems were developed specifically for nondestructive evaluation of tissue characteristics and of quality-related indices of whole fruits or vegetables during the course of maturation, storage, and shelf life. Early measurements of ultrasonic surface waves passing through the peel of a whole fruit were performed by Galili et al. (1993) and Mizrach et al. (1994a). In many studies that aimed to correlate physicochemical indices of various whole fruits and vegetables, emitter and receiver probes mounted on adjustable continuous-touch systems that facilitated wave-amplitude measurements over varied distances were used (Mizrach et al., 1994a, 2000; Mizrach, 2000a, 2004).

A typical setup of a system developed by Mizrach (2000a, 2004) and Mizrach et al. (1994a, 2000) to measure ultrasonic waves over peel of whole fruit consists of a high-power, low-frequency ultrasound emitter/receiver, two identical ultrasonic transducers, and a mounting structure that provides three-axis movement for the transducers, as shown in Figure 7.7.

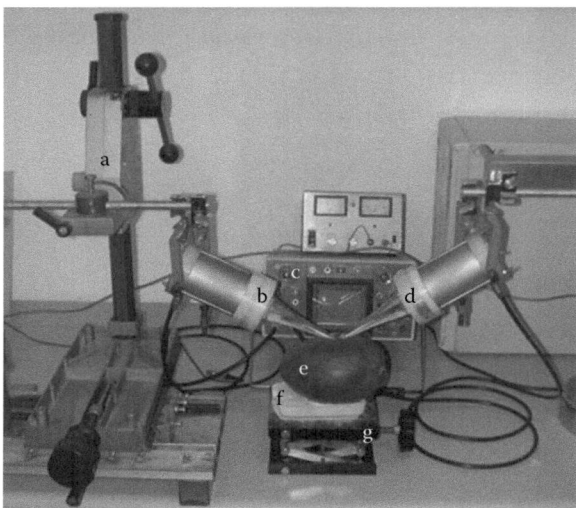

FIGURE 7.7 Diagram of the continuous-touch system for whole fruit: (a) X–Y–Z fixation device, (b) receiver and energy concentrator, (c) ultrasound emitter/receiver, (d) transmitter and energy concentrator, (e) whole fruit, (f) gypsum casting, and (g) adjustable-height jack. (From Mizrach personal files.)

The system used exponentially tapered beam-focusing elements with (0.5 × 6.0 mm) chisel profiles to match the 50 mm diameters of the wave emitter and receiver surfaces of the transducers to the desired fruit contact area (Mizrach et al., 1997). In the through-transmission mode, with one transducer acting as an emitter and the other as a receiver, the mounting structure enabled the transducers to move relative to one another, in order to adjust the gap between them and to apply a controlled contact force to the fruit peel. The angle between the major axes of the transducers was set to a constant 120°, and the gap between the transducer tips was varied. The probes could be moved to and fro, to vary the spacing between their tips, generally within a range of 5–18 mm. Emitted waves penetrated the peel and propagated through the adjacent tissue along the gap between the probe tips. The output pulse amplitude and the paths of the transmitted waves could be observed visually on a monitor screen and drawn on a graph. In parallel, a built-in peak detector and microprocessor-controlled serial interface captured the attenuated waves that crossed the predetermined gap and sent digitized data to an external microcomputer, which calculated the wave propagation velocities and the attenuation coefficient of the tissue for the signal passing through it in the vicinity of the external peel.

These kinds of systems were developed and patented by Mizrach et al. (1994b) and have been used during the last decade in numerous studies to enable the ultrasonic parameters to be chosen so as to correlate with physicochemical indices of various whole fruits and vegetables (Mizrach et al., 1994a, 2000; Mizrach, 2000a, 2004). The results obtained and the correlations found between quality-related indices and ultrasonic parameters for selected fruits and vegetables are presented in the following sections.

Another method for measuring the properties of whole avocado fruits, based on a surface absorption concept, was suggested by Gaete-Garreton et al. (2005); it was developed to assess the ripeness of whole avocado fruits; it is in common use in the classic acoustic evaluation of materials (Morse and Ingard, 1968) and was adapted by Gaete-Garreton et al. (2005) for their study. The setup included a cylindrical stainless steel chamber that functioned as a wave guide, with a wave generator installed at one end and the other end placed in contact with the fruit. The acoustic generator was a prestressed sandwich piezoelectric transducer coupled to a stepped horn made of titanium alloy. The device was equipped with an electronic system that kept the wave emission at a constant level and a needle probe to measure the field inside the chamber. The absorption coefficient of the material placed in the extreme end of the wave guide could be calculated by measuring the standing

Ultrasound for Fruit and Vegetable Quality Evaluation

wave ratio (Morse and Ingard, 1968). The authors validated the method and found their results comparable with those obtained by standard techniques such as use of the penetrometer. Also, they compared their proposed method with other ultrasonic systems (Mizrach et al., 1996) and suggested that further development of their proposed system could facilitate measurements with a single ultrasonic source and, possibly, the development of an online evaluation system for all kinds of fruits.

PORTABLE SYSTEMS

The single- and continuous-touch systems described above were built and used as bench-top devices for laboratory applications and for extended shelf-life measurements in commercial use. Portable devices were developed later for field applications, for monitoring pre- and postharvest processes by rapid measurements of attenuation. These systems include the basic hardware commonly found in most ultrasound devices but, since they were designed to measure distinctive types of fruit specimens, they included a special transducer mounting structure, and used different signal processing methods from those mentioned above, which were intended for bench-top systems. The following sections review the development of three different types of portable data collection systems that featured two different mounting methods and two modular signal processing routines (Mizrach et al., 1999b, 2003; Mizrach, 2000b).

Fixed-Load System Operating in the Time Domain

A portable single-touch, constant-load system was developed for field use (Mizrach, 2000b). It consisted of three 50 kHz ultrasonic transducers, each equipped with a round-tipped beam-focusing element. The transducers were mounted in a structure that allowed independent vertical movement of the ultrasonic probes, while maintaining fixed distances between their tips. One transducer functioned as an emitter and the other two as receivers. The angles between the axes of the three transducers were set to 120°. The distances between the tip of the emitter and those of the two receivers were set to provide two different gaps (e.g., 3 and 5 mm). A preset constant vertical load was applied to each of the three transducers; it was sufficient to ensure good dry contact between the tips and the fruit while minimizing possible damage to the peel. A built-in peak detector and a microprocessor-controlled interface collected the output pulse; the signal amplitude was processed by an internal microcomputer which calculated the attenuation coefficient of the fruit for the signal passing through it and presented it on a readout liquid crystal display (LCD). The attenuation was measured by the same procedure as that used in the continuous-touch system built for measuring properties of whole fruits, except that the pulse amplitude of the transmitted ultrasonic signal was measured for only two different spacings between the two probes. Unlike the continuous-touch method, there was no relative movement between the probes. The attenuation coefficient was calculated with the same equations that were used with the stationary continuous-touch systems.

A semicommercial version of the one-touch constant-load portable device ("Avo-Check," Figure 7.8) was developed for use by consumers in nondestructive prediction of the ripening duration of avocados, and thereby to encourage sales by facilitating inspection (Mizrach, 2000b).

The device weighed about 2.5 kg and was simple enough for the consumer to operate. It had a display screen that showed the buyer how many days to wait until a specific avocado "will be ripe and suitable for eating." This device was developed within the framework of a startup technology incubator (Avotec Ltd.), under a license from the Agricultural Research Organization (ARO), Israel, the patent assignee in Israel, USA, and France (Mizrach et al., 1994b). The results obtained showed that this model was able to predict the "ready-to-eat" softening stage (defined as a penetration force <6 N with a 6.25 mm diameter, 60° cone head) up to 12 days ahead with ±1 day accuracy, for avocado fruit cultivars 'Ettinger' and 'Fuerte'. This accuracy could not be achieved with other avocado varieties such as 'Haas', probably because their peel is not smooth (Mizrach, 2000a,b). A more advanced ultrasound device that might overcome the problems of contact between the probes and the nubbly skin that characterizes some fruits, e.g., 'Haas' avocado, was developed later (Mizrach et al., 2003).

FIGURE 7.8 Diagram of the one-touch system in the time domain, "Avocheck." (From Mizrach, A., *The XIV Memorial CIGR Word Congress 2000*, Paper No. R6210, Tokyo, Japan.)

Fixed-Load System Operating in the Frequency Domain

This system was based on an ultrasonic emitter/receiver device similar to the one that operated in the time domain, but it used only one emitter and one receiver.

The system was developed for nondestructive measurements of whole fruits and operated in the frequency domain (Matrix, 1989; Mizrach et al., 1999b). Two wide-band 100 kHz ultrasonic transducers, one acting as an emitter and the second as a receiver, were used to emit ultrasonic energy into the fruit and to measure the emerging frequency spectrum. The transducers were assembled together and mounted in a structure that maintained a constant gap (about 2 mm) and angle (120°) between their tips. The entire structure, with the two probes, was able to move only vertically and maintained a constant distance between the tips of the beam-focusing elements. The probes were installed beneath a tray on which the examined fruit was placed.

Each probe was held against the peel of the fruit with a controlled contact force, and a single touch of both probes on the peel activated the device to emit an ultrasonic pulse into the tissue. Although the fruit tissue affected the waveform of the pulse during its 2 mm transit along the peel, the information on tissue characteristics was obtained from the surrounding tissue volume. Generally, two or three measurements around each fruit could yield sufficient information to characterize the entire fruit (Mizrach et al., 1999b). When sensed by the receiver, the wave was transformed into the frequency domain. A frequency spectrum analysis was implemented in MATLAB® software (MATLAB, 1995); it included data processing of the stored signal and use of fast Fourier transform (FFT) procedures to extract the frequency amplitude spectra. A typical ultrasonic wave on which the FFT window is superimposed is shown in Figure 7.5. The FFT values obtained for each fruit were analyzed by means of the Spectra Matrix computer program (Matrix, 1989). Multilinear regression (MLR) and partial least squares (PLS) statistical analyses were used to develop models that related the FFT spectra to the physicochemical indices of each tested fruit (Mizrach et al., 1999b). This system was used to evaluate the physicochemical properties of mango fruits and enabled the establishment of the relationships between the nondestructive ultrasonic attenuation measurements and the major physicochemical quality indices of mango.

Variable-Load System Operating in the Time Domain

A more advanced, more accurate one-touch system, based on energy absorption techniques, was developed for nondestructive measurement of mealiness in whole apples (Figures 7.9 and 7.10) (Mizrach et al., 2003; Bechar et al., 2005).

Ultrasound for Fruit and Vegetable Quality Evaluation

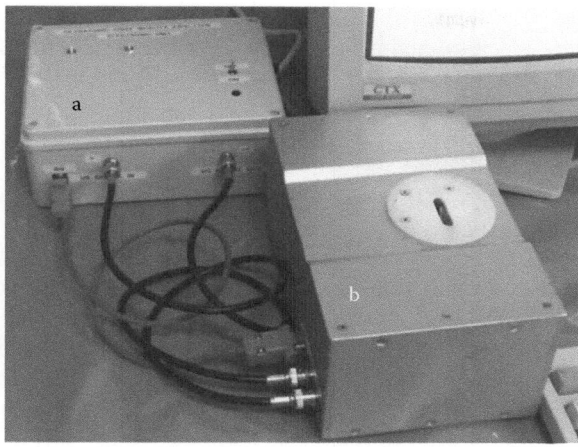

FIGURE 7.9 Diagram of the one-touch variable-load system in the time domain: (a) electronic unit and (b) mechanical unit. (From Bechar, A. et al., *Biosyst. Eng.*, 91, 329, 2005. With permission.)

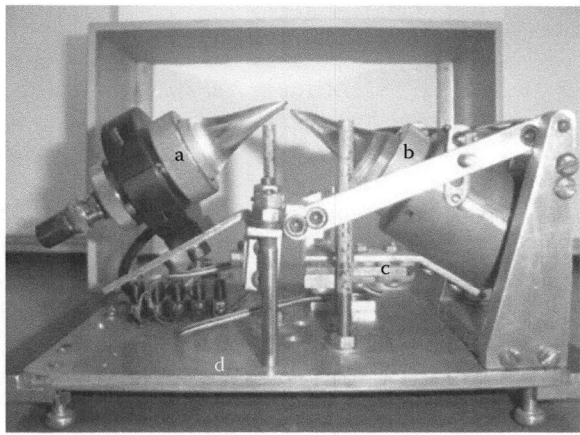

FIGURE 7.10 Transducer and sensor assembly of the mechanical unit in one-touch system: (a) transmitter and energy concentrator, (b) receiver and energy concentrator, (c) strain gage, and (d) base. (From Bechar, A. et al., *Biosyst. Eng.*, 91, 329, 2005. With permission.)

This system included mechanical, electronic, and microcomputer units. The mechanical part comprised a pair of 80 kHz ultrasonic transducers assembled together as emitter and receiver (a, b, respectively, Figure 7.10) and mounted with a constant gap (about 2 mm) and angle (120°) between their tips. Each transducer could move independently up and down while variable loads were applied to the fruit, and a fixed distance was maintained between the transducer tips. The transmitter was attached to a strain-gage element (c, Figure 7.10) that monitored the force applied to the fruit. The receiver was attached to a soft spring that allowed it to move easily in the vertical direction and to function as a sensitive signal pickup (d, Figure 7.10). The transmitted pulse passed through the apple tissue; part of its energy was absorbed, depending on the internal texture, and the rest of the energy, carried by the weakened emerging signal, was sensed by the receiver. The electronic part of the system comprised a high-power ultrasonic generator for activating and monitoring the transducers, a strain-gage controller, and a microprocessor for collecting and processing the ultrasonic signals and the strain gage readings. Processed data were sent to a data acquisition system for analysis. The operating principle of this system was based on findings of Mizrach et al. (2003), who found linear correlations between the force applied and the energy absorbed in the tissue; the gradients of these correlations depended on structural and physicochemical variations in the fruit tissue. The mealiness of two varieties of apples was successfully measured with this system (Mizrach et al., 2003; Bechar et al., 2005). Further details are presented in the following section.

ULTRASOUND IMPLEMENTATIONS AND QUALITY ASSESSMENT OF SELECTED FRUITS AND VEGETABLES

The ultrasound systems described in this chapter were adapted, modified, or developed for measuring quality-related indices in fruits and vegetables during their growth and postharvest processes. These systems offer the potential for rapid, nondestructive measurement of physicochemical changes in fruits and vegetables during growth, storage, and shelf life. Most systems were developed for a specific kind of fruit and were used to measure its physicochemical changes during one or more stages of growth and postharvest development and processing. During growth, an ultrasound system might obtain important information regarding the maturation processes and indicate the appropriate harvest time, i.e., when the fruit has reached its optimal maturity; it might also obtain information regarding quality-related physicochemical properties of the produce (Mizrach et al., 1999a). During storage and shelf life, ultrasound systems can sense changes in the physicochemical characteristics of fruit and vegetable tissues, such as firmness, softening, mealiness, and other quality indices, and thereby determine the appropriate marketing time or predict attainment of their "ready-to-eat" condition (Diederichs, 1996; Mizrach, 2000b; Mizrach et al., 2003; Gaete-Garreton et al., 2005). Ultrasonic systems were also used to measure physicochemical changes in fruits during long-term storage (Mizrach et al., 2000) or during storage and subsequent shelf life (Flitsanov et al., 2000).

The ultrasonic properties and quality parameters of many fruits and vegetables have been reported in the literature, and Table 7.1 presents those of a number of selected horticultural products. This section surveys the use of ultrasound technology in measurements, applications, and assessment of quality parameters of selected fruits and vegetables, whether as tissue segments or as whole-fruit specimens, during growth and postharvest processes. The measurement systems developed for collecting ultrasound parameters and determining quality indices of these products are also discussed.

Avocado

Avocado fruits were evaluated in practice by using ultrasound systems during growth, maturation, storage, marketing, and shelf life. Applications and the results they yielded in each pre- and postharvest stage are reviewed below.

By measuring cylindrical segments of avocado tissues, as described above, Mizrach et al. (1989) found that acoustic velocity and ultrasonic attenuation in the fruit ranged from 200 to 400 m s^{-1} and from 2 to 6 dB mm^{-1}, respectively, depending on cultivar and aim, i.e., what the measurements were intended to determine (Table 7.1). Later, Mizrach et al. (1992, 2009) evaluated the directional decay rate of the ultrasonic waves in avocado tissue. By measuring the directivity distribution of longitudinal waves in a fruit half-space along the X–Y principal axes, and applying Equation 7.7 to the measured data, they were able to measure the wave velocity and attenuation (Table 7.1), and the directional distribution of the decay rates of the apparent attenuation coefficients along the transverse and the longitudinal axes.

Mizrach et al. (1996) monitored avocado fruits in their laboratory and derived correlations between the attenuation of the ultrasonic waves, on the one hand, and the firmness and DW percentage of the fruits, on the other hand. Even though the continuous-touch method was capable of wave velocity measurements, it was not used for these measurements because Mizrach et al. (1996) showed that the changes in ultrasound wave velocity within the fruit were relatively small, and not significantly correlated with its physicochemical changes; they therefore ignored velocities in most of their subsequent studies. However, in more recent studies, the attenuation of the ultrasonic signal was found to change dramatically during storage, shelf life, postharvest softening, and ripening of avocado (Mizrach et al., 1997, 1999a,b; Mizrach and Flitsanov, 1999; Mizrach, 2000a).

The physicochemical changes in whole avocado during growth and maturation and determination of the appropriate harvest time were studied with continuous-touch ultrasonic systems by Mizrach et al. (1999a), who derived correlations between the changes in attenuation of the ultrasonic signal and those in DW content during the fruiting season. Attenuation of 50 kHz ultrasonic waves was measured during this preharvest stage, and changes in the physicochemical and chemical parameters of the fruits were correlated with the changes in ultrasonic attenuation. Because of their physicochemical nature, the oil and DW percentages of avocado fruits increase during growth and do not change after harvest (Degani et al., 1986). DW is an acceptable and convenient indicator for evaluation of oil content in avocado, when physicochemical tests are performed. A nonlinear regression procedure was used to relate variations in ultrasound attenuation and DW to growth time: a simple curve was fitted to the experimental results and used to determine the constants for the equations, and a parabolic function was fitted to the mean values to define the mathematical and statistical models relating the time variation of DW with that of attenuation, and to correlate them with one another. A monotonic decrease in the attenuation of the ultrasonic signal with time suggested that parabolic and linear curves be selected for 'Ettinger' and 'Fuerte', respectively (Mizrach et al., 1999a). Because the DW and attenuation were both time dependent, a direct relationship was drawn between these two parameters: an exponential expression was selected as the curve of "best fit" between ultrasonic attenuation and the DW percentage of 'Ettinger' fruits during growth (Mizrach et al., 1999a). The ultrasonic wave attenuation diminished with increasing DW content during fruit growth, but not at a constant rate: it fell sharply during the first 3 months of growth and then moderately as the fruit approached maturity during the last 2 months of growth. According to the attenuation–DW curve, Mizrach et al. (1999a) showed that the ultrasonic attenuation approached a constant asymptotic value (Figure 7.11).

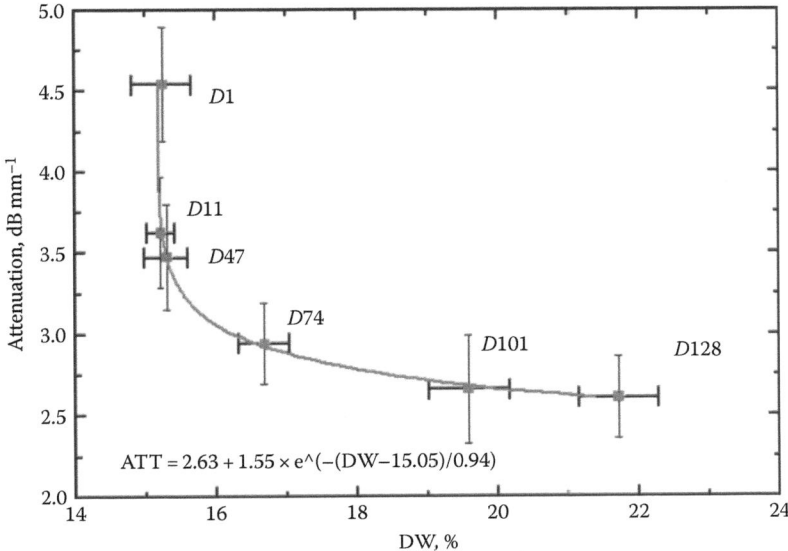

FIGURE 7.11 Attenuation versus dry weight (DW) content during growth, for avocado fruit cv. 'Ettinger'. Vertical and horizontal lines represent confidence intervals for attenuation and DW, respectively (confidence limit = 95%). (From Mizrach, A. et al., *Sci. Hort.*, 80, 173, 1999a. With permission.)

Entry into the asymptotic part of the curve indicates that the DW content is approaching the minimum, which is regarded as indicating maturity, and that the fruit has reached the condition appropriate for harvesting. Mizrach et al. (1999a) showed that it was possible to determine the maturity of avocado during growth, which suggests that the DW percentage in avocado could be evaluated by ultrasonic attenuation measurement during growth and maturation, and that this could be a basis for harvest-time determination.

After harvest, avocadoes are stored, and then put on shelves for marketing. During these stages, acoustical, mechanical, and physicochemical changes take place, and ultrasonic systems were used to monitor these changes. Relationships between acoustical, mechanical, and physicochemical parameters were proposed for avocado fruits, cvs. 'Ettinger' and 'Fuerte' (Mizrach et al., 1997, 1999a,b). The researchers used continuous-touch ultrasonic systems to assess the changes in the quality indices—firmness, DW, and oil content—during storage and shelf life. They correlated the ultrasound wave attenuation with these quality indices. The results of measurement of the ultrasonic parameters of avocado fruits showed the changes in the ultrasonic characteristics to be strongly related to storage time. The velocity of the ultrasonic wave in avocado tissue showed a complex non-monotonic relationship with the tissue's physicochemical properties and was found to be not sufficiently sensitive to changes in these properties for use as an indicator. The attenuation of the ultrasonic signal in avocado fruits was found to decrease as the fruit grew (Mizrach et al., 1999a) and to increase during the postharvest softening and ripening process, until the fruit became very soft (Mizrach et al., 1989, 1996; Mizrach and Flitsanov, 1999). The reversal of the direction of the attenuation changes in the course of time probably resulted from the simultaneous activity of several different physicochemical processes that have not yet been investigated.

Typical results of measurements of changes of velocity, attenuation, and firmness in whole avocadoes during shelf life are presented in Figure 7.12.

Nonlinear regression procedures were used to correlate variations in ultrasound wave attenuation and velocity with storage time. A simple curve-fitting program was used to express the experimental results as functions of time, to determine the curve shapes, and to formulate the equations and the constants that describe these curves; quadratic equations were chosen for attenuation versus time and cubic equations for velocity versus time in avocado (Mizrach et al., 1989, 1996; Mizrach and Flitsanov, 1999).

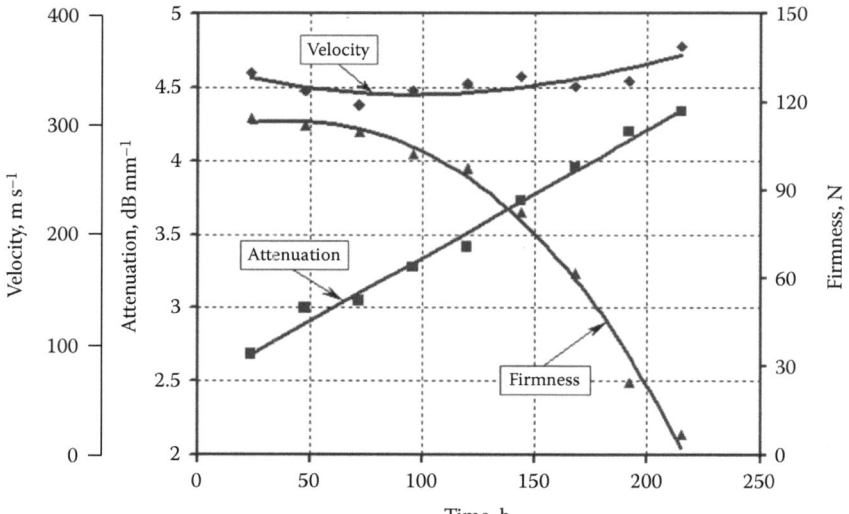

FIGURE 7.12 The mean values of firmness, wave attenuation, and velocity, as functions of storage time, and the suggested model curves. (From Mizrach, A. and Flitsanov, U., *J. Food Eng.*, 40, 139, 1999. With permission.)

The measured attenuation of ultrasonic waves in avocado fruits varied because of the changes in quality indices of the fruit. However, wide scattering in measured attenuation values was obtained in picked fruits of the same batch after harvesting. This could be caused by lack of homogeneity among the fruits because of differing maturities at picking time and differing locations on the tree and in the orchard. Mizrach et al. (1996) developed a practical statistical treatment to solve this problem. The DW and oil contents do not change after harvest; therefore, firmness remained the only major quality parameter that really changed. In order to analyze the results with respect to the same scale of maturity, it was hypothesized that firmness at the time the fruit becomes soft is a preferred indication of the fruit's biological age. At this time point, the scatter in firmness measurements was found to be minimal; therefore, the timescale of measurements was normalized to the day on which maximum ripeness was indicated. For analysis, the attenuation and velocity data were "time shifted" so that all measurements could be compared with the values measured on the day of maximum ripeness. This process extended the timescale of the results, but it reduced the scattering in attenuation measurements and enabled better prediction models, which have been used since then.

During low-temperature storage of avocado fruits, the attenuation initially decreased, and the lower the storage temperature, the less the measured attenuation, but it increased during the 4 weeks of storage. The differences in attenuation changes between fruits stored at different temperatures were found to be quite significant (Mizrach et al., 2000). Diminution of firmness during storage is a natural physicochemical process in avocado, but low-temperature storage slows the softening process; the firmness still decreases, but the lower the storage temperature, the slower the softening (Mizrach et al., 2000). Since the firmness and attenuation were both found to be time dependent, a direct relationship between these two parameters was defined. The ultrasonic wave attenuation increased with diminishing firmness at all storage temperatures (Figure 7.13).

The gradients for fruits stored at 6°C, 8°C, and 20°C were very close to one another, but differed from those for 2°C and 4°C, which were also very close to one another (Figure 7.13). This suggests that monitoring of fruit firmness by means of attenuation measurements should take account of differences between fruits stored at temperatures above and below 6°C.

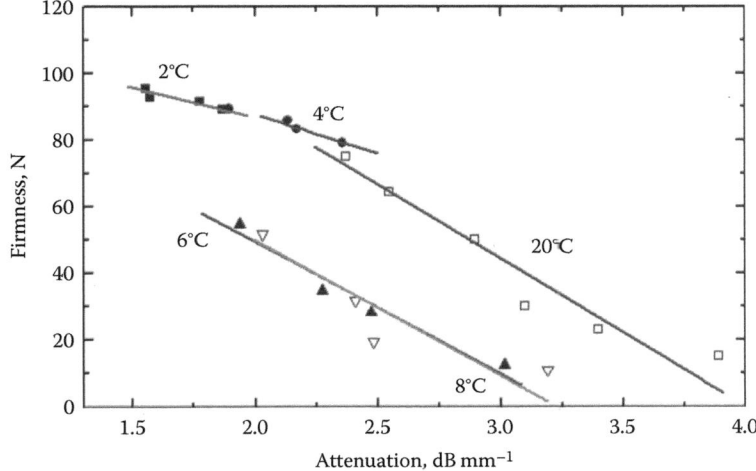

FIGURE 7.13 Attenuation versus firmness and linear regressions at several storage temperatures, for avocado fruit cv. 'Ettinger'. ■, 2°C; ●, 4°C; ▲, 6°C; ▽, 8°C; □, 20°C (control). (From Mizrach, A. et al., *Comput. Electron. Agric.*, 26, 199, 2000. With permission.)

Apple

In apple fruits, texture is seen to be a primary quality attribute, together with flavor and appearance. Crispness, firmness, hardness, juiciness, and mealiness are the most generally recognized texture attributes in apples; of these, mealiness impairs the quality and reduces market acceptability and price. Retailers have been found perfectly able to distinguish a mealy product from a fresh one, describing it correctly with terms such as non-juicy, soft, etc. (De-Smedt, 2000); nevertheless, there was a need for a reliable method, supported by appropriate sensors, for nondestructive evaluation and classification of apples in terms of their mealiness. The demand for high quality necessitates a reliable, rapid, nondestructive, noninvasive technique for measuring some of the texture attributes of the fruit, especially mealiness, that develop as it matures and that are indicative of its quality. Furthermore, such a technique must be usable by an untrained person. A method based on a patented development that enables the fruit quality attributes to be assessed by measuring the changes in ultrasonic sound waves passing through the peel and flesh (Mizrach et al., 1994b) was modified to facilitate the measurement of the acoustic parameters of whole apples with a single-touch, variable-load system operating in the time domain (Figures 7.9 and 7.10). This system, based on ultrasonic energy absorption, measured the ultrasonic attenuation in tissues of three mealiness levels—fresh, ripe, and overripe—in 'Jonagold' and 'Cox' apples (Mizrach et al., 2003; Bechar et al., 2005). The apples were also examined for differences in energy absorption between the red and the green sides of the fruits. The results showed that the mean calculated value of mealiness for the green side of 'Jonagold' apples increased with increasing mealiness and that there were significant differences between the fresh and overripe groups. The mean calculated values of mealiness for the red side of fresh fruits were found to be significantly lower than those for overripe ones. However, in 'Cox' apples, no significant differences were found between mealiness levels on the green and the red side of the fruit. Mizrach et al. (2003) concluded that the results obtained make it is possible to distinguish among the three mealiness levels in 'Jonagold' but not in 'Cox' apples; they suggested that this phenomenon be investigated further.

In a follow-on study, Bechar et al. (2005) analyzed the ultrasound waves that had been measured by Mizrach et al. (2003), in parallel with the determination of the mealiness level of the fruit by destructive measurements under confined compression. The comparison between the results of the ultrasound measurements and of the confined-compression tests showed good matching between attenuation and the fresh and the overripe mealiness levels for 'Cox' apples, but not for 'Jonagold' apples; the differences were found to be minor and statistically insignificant.

Mango

Mango fruits were monitored with ultrasound systems throughout the postharvest stages of shelf life and marketing. The physicochemical changes in whole mango fruits were studied with continuous-touch ultrasonic systems working in the time domain (Mizrach et al., 1997, 1999a), and with a single-touch, permanent-load system operating in the frequency domain (Mizrach et al., 1999b). Mizrach et al. (1997) measured the attenuation of the ultrasonic signal of mango fruits during 10 days of shelf life at room temperature and correlated the results with fruit firmness, sugar content, and acidity. They found an increase in attenuation from 2.7 dB mm^{-1} on the first day to 4.16 dB mm^{-1} at the end of the test. The trends were numerically and graphically analyzed by means of statistical and curve-fitting procedures, and a quadratic expression was found to be a good fit ($R^2 = 0.99$) to the changes in attenuation.

Changes in the physiological and chemical parameters of the fruits were correlated with the changes in ultrasonic attenuation. The fruits were subjected to ultrasonic nondestructive tests as well as to destructive penetration measurements of firmness, and to physicochemical tests for tissue sugar content and acidity. The data set was analyzed statistically, and calibration equations were developed to determine the relationships between the ultrasonic parameters and the firmness and

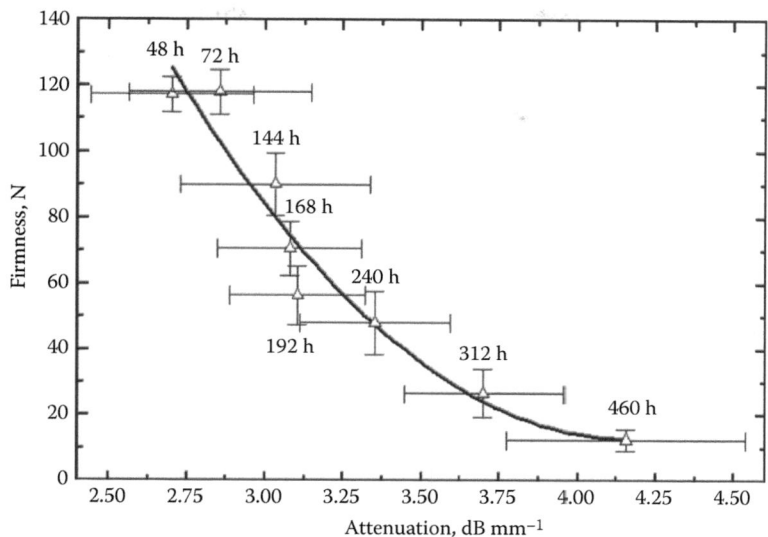

FIGURE 7.14 Parabolic expressions for the averaged firmness values of mango fruits (cv. 'Tommy Atkins') versus NDT ultrasonic attenuation. (From Mizrach, A. et al., *Trans. Am. Soc. Agric. Eng.*, 40(4), 1107, 1997.)

physicochemical indices. Since the firmness and attenuation were both time dependent, a direct relationship between them was defined. A parabolic expression was selected as the curve of "best fit" to describe this relationship for mango (Mizrach et al., 1997) (Figure 7.14).

The variations in ultrasound attenuation and in the chemical changes (sugar content and acidity) during the storage time were related to one another by means of a nonlinear regression procedure (Figure 7.15).

A simple curve was fitted to the experimental results and was used to determine the constants for its equations. A third-degree polynomial expression was selected to describe the direct relationship between attenuation and sugar content, and a parabolic expression was used for that between

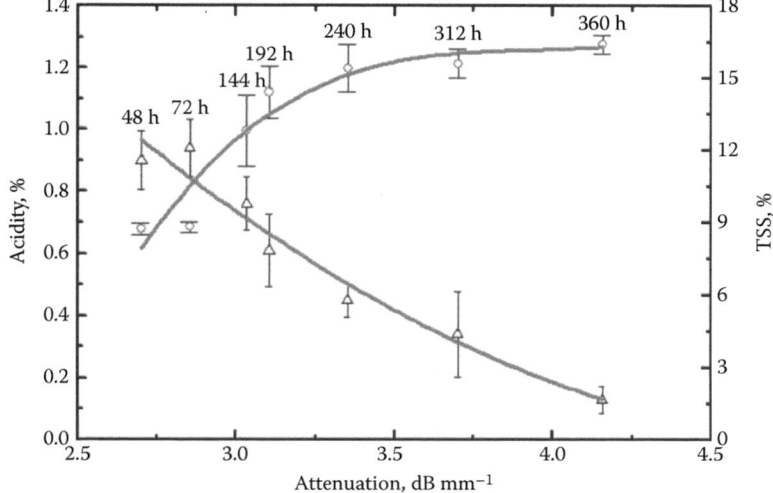

FIGURE 7.15 Polynomial expression of the averaged sugar contents (○) and acidity (△) of mango fruits (cv. 'Tommy Atkins') versus attenuation. (From Mizrach, A. et al., *Trans. Am. Soc. Agric. Eng.*, 40(4), 1107, 1997.)

attenuation and acidity. Both of these physiological parameters reflect internal changes in the mango fruit during ripening. The equations enable us to determine the sugar content and acidity, and to monitor the softening process in a batch of mango fruits in a packing house directly, by nondestructively measuring their ultrasonic attenuation (Mizrach et al., 1997). The relationships between the ultrasonic attenuation and the physiological parameters suggested by Mizrach et al. (1997) enable nondestructive determination of the physicochemical properties of fruits, estimation of their maturity, precise determination of the appropriate harvest time, and nondestructive assessment of their firmness.

Melon

A study of ultrasonic determination of the internal physicochemical parameters of ripe autumn-grown and winter-grown melons (cv. 'Galia') was carried out by Mizrach et al. (1991, 1994c). Acoustical, mechanical, and quality parameters of melon sectors sampled from several depths and the relationships among these parameters were measured and analyzed with single-touch ultrasonic systems. The authors found that the modulus of elasticity and the tangent modulus of the sample tissues decreased drastically, from 644 to 209 kPa, and the attenuation of a transmitted ultrasound pulse decreased from 3.71 to 1.1 dB mm^{-1}, as the sampling depth increased from 10 to 30 mm. They also found that the firmness and sugar content increased strongly with sampling depth. When Mizrach et al. (1991, 1994c) correlated these results, they found a strong dependence between attenuation measurements in the tissue sectors, on the one hand, and the physicochemical parameters, firmness, and sugar contents, on the other hand, and they concluded that this strong dependence on depth indicated a potential for using the attenuation coefficient for determination of internal fruit quality (Mizrach et al., 1991, 1994c). In a different phase of the study, Mizrach et al. (1994c) determined the ripeness status of melons from their peel color, as it changed from green to yellow. By applying the continuous-touch ultrasound system, they measured the surface wave velocity and attenuation in a whole melon fruit (Mizrach et al., 1994c); they found that the amplitude of the wave transmitted through the peel increased as the color changed from green to yellow, and that there was a good correlation between the peel color, i.e., the ripeness, and the acoustic attenuation. They concluded that nondestructive techniques might be used to determine the ripeness stage of other varieties of melons, in which the color change is not so conspicuous.

Olive

In olives destined for table use and for oil extraction, the most important quality parameter is the amount of oil in the mature fruit; this enables the determination of their optimum harvest time (Mizrach et al., 2006a). The growth process in the olive fruit extends from fruit set through green or black maturation and might last for 5–7 months, depending mainly on the variety. During this period, changes in fruit weight, oil content, and color are observed. The most important change occurring in the fruit involves the factor considered most significant with respect to quality—the accumulation of oil in its cells. In many fruit tree species, changes in visible color are used to determine the optimal harvest time (Mizrach et al., 1996), but in olive these changes do not provide a clear-cut criterion for harvest-time determination, and a delay in harvesting might cause fruit loss and reduction in quality, without a significant gain in oil yield. A continuous-touch ultrasonic system, based on measurement of acoustic wave attenuation in the tissue, was applied to whole olive fruits by means of ultrasonic probes in contact with the peel. The ultrasonic wave attenuation was measured and correlated with the oil content of the fruits (Mizrach et al., 2006b), and it was found that the ultrasonic attenuation in the oil cultivars 'Souri' and 'Barnea' decreased from 19 to 9 dB mm^{-1} as the oil content increased from 19% to 27% (Figure 7.16).

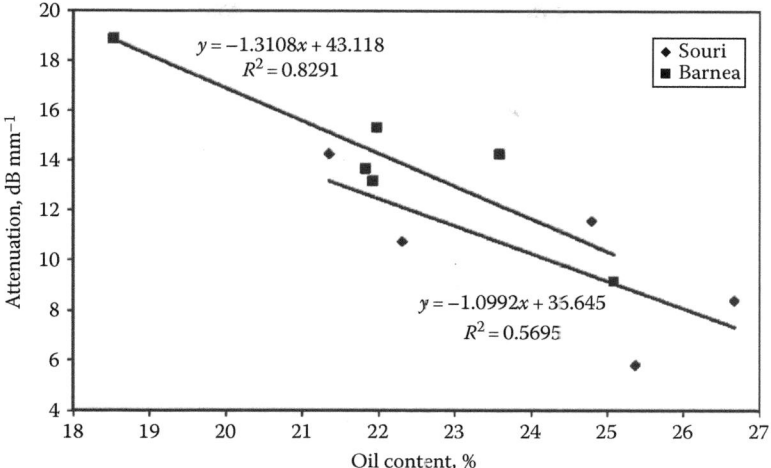

FIGURE 7.16 Average values and trend lines for attenuation versus oil content in olives (cvs. 'Souri' and 'Barnea'). (From Mizrach, A. et al., Maturity measurements of olive fruits using acoustic and compression methods, in *CIGR XVI Word Congress*, 2006b, Bonn, Germany.)

This suggested that the ultrasonic method could be used for nondestructive quality determination of the oil olives 'Souri' and 'Barnea' by estimation of the oil content in their fruits and therefore for the determination of their optimum harvest time.

PLUM

A nondestructive ultrasonic method was used to determine the maturity and sugar content of plum fruits (cv. 'Royal Z') during storage for up to 151 h (Mizrach, 2004); it involved a continuous-touch ultrasonic system based on measurement of acoustic wave attenuation in the fruit tissue by means of ultrasonic probes in contact with the peel. The differences in the acoustic signals transmitted through the tissue of fruits at different stages of maturity were measured and analyzed at intervals throughout the storage period. The fruits were also subjected to destructive penetration measurements of firmness and sugar content, and the relationships between the results of the ultrasonic attenuation measurements and those of the destructive measurements were determined during the course of shelf life. The measured attenuation and the firmness were both found to decrease in the course of shelf life, with extreme ranges of 5.41–0.54 dB mm^{-1}, and 1.8–0.25 lb, respectively. Good correlation ($R^2 = 0.72$) between the attenuation and the firmness was observed from 78 h after shelf-life entry until the end of the softening process. Mizrach (2004) suggested that this ultrasonic method might be used as a nondestructive technique for monitoring the firmness of plums at a series of time points during storage.

POTATO

The acoustic properties of potatoes were measured by the pulse transmission method, with longitudinal ultrasonic waves in a frequency range of 50 kHz–1 MHz, in a study that aimed to distinguish between good and defective potatoes (Ha et al., 1991). The average ultrasonic velocity was found to be 824 m s^{-1} at 100 kHz and the attenuation coefficient, as measured with several transducers at several frequencies, was found to be approximately proportional to $1.4 \times f$, where f is the frequency of the transducer. The attenuation in defective potatoes was found to be much higher than that in sound ones; therefore, the authors suggested that this was a practicable way of detecting defective potatoes (Ha et al., 1991; Hansen et al., 1992).

Hansen et al. (1992) attempted to use radiography- and ultrasound-based methods to detect weevil activities in weevil-infested sweet potatoes. They reported that no weevil life stage in the tuberous roots was clearly identified by radiographic methods, but uninfested sweet potatoes were clearly distinguishable from infested ones that contained feeding tunnels. They concluded that radiography could assist in the development of quarantine treatments; however, ultrasound could not penetrate the root surface and did not produce an image (Hansen et al., 1992; Cheng and Haugh, 1994). The lack of penetration ability could be attributed to the use of frequencies of 5 and 7.5 MHz, which are very strongly attenuated in potato tissue.

Lower-frequency probes were used in other studies: Cheng and Haugh (1994) used ultrasonic waves at 250 kHz in a through-transmission system applied to whole potato tubers to investigate the presence of hollow heart. They found that the waveform of ultrasonic signals transmitted through a hollow-heart potato differed from that of those through a sound potato. Furthermore, the defective tubers could be separated from the sound ones according to the amount of ultrasonic power transmitted through them (Cheng and Haugh, 1994; Abbott, 1999).

Tomato

Verlinden et al. (2004) used a continuous-touch ultrasonic technique to evaluate chilling injury in tomatoes, but found that the contact pressure between the probes and the tomato flesh caused difficulties. The contact force that had to be applied to obtain results caused destructive penetration into the tomatoes; therefore, Verlinden et al. (2004) suggested that a nondestructive ultrasonic technique be developed to make this method applicable.

A nondestructive ultrasonic method was used to monitor the physicochemical changes in firmness and sugar content in greenhouse tomatoes (cv. '870') during their shelf life (Mizrach, 2007). The acoustic wave attenuation in the fruit tissue was measured with a continuous-touch ultrasonic system that used ultrasonic probes in contact with the fruit peel, and the fruits were then also subjected to destructive penetration measurements of firmness. The results were analyzed statistically to determine the changes in nondestructive ultrasonic attenuation and the destructively measured firmness during 8 days of shelf life. The measured attenuation and firmness were found to decrease significantly in the course of shelf life: extreme ranges were 5.57 down to 2.05 dB mm^{-1}, and 8.9 down to 2.2 N, respectively. Since the attenuation and firmness were both measured in the same time frame, it was possible to directly examine the relationship between attenuation and firmness: attenuation was found to be linearly related to fruit firmness during 8 days following entry into storage, which suggests that this ultrasonic method might be used for nondestructive firmness monitoring of tomatoes during their shelf life (Mizrach, 2007).

CONCLUSIONS

This chapter presents the concepts, technologies, developments, modifications, and applications associated with the use of ultrasonic techniques for fruit and vegetable quality evaluation during pre- and postharvest processes. It surveys various ultrasound measurement methods and how they have been adapted for measuring physicochemical changes and quality indices of various tissues, specimens, and whole fresh fruits during the course of growth, maturation, harvest, storage, shelf life, and consumption. This chapter confirms that, in light of two decades of attempts to apply ultrasound technology to fresh agricultural produce, the technique is now feasible. There has been considerable progress since the early studies, which were hampered by limited knowledge of the responses of fruit and vegetable tissues to ultrasonic waves, the lack of suitable equipment or components, and the inappropriate frequency ranges or lack of power of the available transducers. New lines of equipment have been developed, and awareness has grown of the advantages to be gained through the use of ultrasonic technology for quality evaluation of fresh fruits and vegetables. Many reports, cited in this review, describe difficulties and limitations in applying the ultrasound technique for quality evaluation during the various biological processes, and, in fact, most of the ultrasound techniques still remain efficient research tools that are not yet applicable to quality-related measurements. This suggests that the technology is not yet ripe for commercial use and that there is yet much to be done in order to make it a widely used sorting tool. The wide range of practical uses, and the success of ultrasound applications in medicine and industry are encouraging, since they indicate that there is good scope for future successful applications of this technology in agriculture also. The knowledge we now have and the nondestructive nature of this technique also help to boost its attractiveness for applications in monitoring the quality of fruits and vegetables. Thus, continued development of equipment and techniques will ensure increasing implementation and uptake of this technology to meet the expanding requirements of the fresh and processed agricultural produce sectors. Commercial application of these techniques will be beneficial to growers as well as to consumers and distributors. Scientific advances will increase the availability of better-quality products; the income of fruit and vegetables growers will match the quality of their products; and distributors will be able to comply with consumer demands for uniformly high-quality products. The public demand for better-quality agricultural produce will increase in the future and will stimulate studies that lead to increased availability of sophisticated techniques, sensors, and user-friendly noninvasive devices for measuring quality indices; some of these will certainly apply ultrasound technologies.

REFERENCES

Abbott, J. A. 1999. Quality measurement of fruits and vegetables. *Postharvest Biology and Technology*, 15, 207–225.

Bechar, A., Mizrach, A., Barreiro, P., and Landahl, S. 2005. Determination of mealiness in apples using ultrasonic measurements. *Biosystems Engineering*, 91, 329–334.

Butz, P., Hofmann, C., and Tauscher, B. 2005. Recent developments in noninvasive techniques for fresh fruit and vegetable internal quality analysis. *Journal of Food Science*, 70, R131–R141.

Camarena, F. and Martinez-Mora, J. A. 2006. Potential of ultrasound to evaluate turgidity and hydration of the orange peel. *Journal of Food Engineering*, 75, 503–507.

Cheng, Y. and Haugh, C. G. 1994. Detecting hollow heart in potatoes using ultrasound. *Transactions of the American Society of Agricultural Engineering*, 37, 217–222.

Chivers, R. C., Russell, H., and Anson, L. W. 1995. Ultrasonic studies of preserved peaches. *Ultrasonics*, 33, 75–77.

Degani, C., Bechor, V., Albazri, R., and Blumenfeld, A. 1986. Dry weight content as an index for determination of maturity of avocado fruits. *Alon Hanotea*, 40, 1017–1022.

De-Smedt, V. 2000. Measurement and modeling of mealiness in apples. Catholic University of Leuven, Belgium.

Diederichs, R. 1996. UT in Israel. *NDTnet—The e–Journal of Nondestructive Testing*, 1, [cited December 10, 2006]. Available from http://www.ndt.net/article/map/il_map/il.htm

Filipczynski, L. 1964. Measurement of longitudinal and transverse waves radiated by a compressional source into elastic semispace. *Proceedings of Vibration Problems*, 5, 89–93.

Flitsanov, U., Mizrach, A., Liberzon, A., Akerman, M., and Zauberman, G. 2000. Measurement of avocado softening at various temperatures using ultrasound. *Postharvest Biology and Technology*, 20, 279–286.

Fuchs, Y., Pesis, E., and Zauberman, G. 1980. Changes in amylase activity, starch and sugars contents in mango pulp. *Scientia Horticulturae*, 13, 155–160.

Gaete-Garreton, L., Vargas-Hernandez, Y., Leon-Vidal, C., and Pettorino-Besnier, A. 2005. A novel noninvasive ultrasonic method to assess avocado ripening. *Journal of Food Science*, 70, E187–E191.

Galili, N., Mizrach, A., and Rosenhouse, G. 1993. Ultrasonic testing of whole fruit for nondestructive quality evaluation. ASAE Paper No. 936026.

Ha, K., Kanai, H., Chubachi, N., and Kamimura, K. 1991. A basic study on nondestructive evaluation of potatoes using ultrasound. *Japanese Journal of Applied Physics Part 1–Regular Papers Short Notes and Review Papers*, 30, 80–82.

Hansen, J. D., Emerson, C. L., and Signorotti, D. A. 1992. Visual detection of sweet-potato weevil by noninvasive methods. *Florida Entomologist*, 75, 369–375.

Harkness, R. W. 1954. Chemical and physical tests of avocado maturity. *Proceedings of the Florida State Horticultural Society*, 67, 248–250.

Horwitz, W. 1970. *Official Methods of Analysis*. The Association of Official Analytical Chemists, Washington, DC.
Inomata, Y. and Suzuki, K. 2001. Non-destructive measurement of watercore in Japanese pear (Pyrus pyrifolia Nakai) 'Hosui'. *Jarq—Japan Agricultural Research Quarterly*, 35, 125–129.
Johnston, J. W., Hewett, E. W., and Hertog, M. 2002. Postharvest softening of apple (*Malus domestica*) fruit: A review. *New Zealand Journal of Crop and Horticultural Science*, 30, 145–160.
Kim, K. B., Jung, H. M., Kim, M. S., and Kim, G. S. 2004. Evaluation of fruit firmness by ultrasonic measurement. *Advances in Nondestructive Evaluation*, Pt 1–3. Trans Tech Publications Ltd., Zurich-Uetikon,
Kinsler, L. E. and Frey, A. R. 1972. *Fundamentals of Acoustics*. John Wiley, New York.
Krautkramer, J. and Krautkramer, H. 1990. *Ultrasonic Testing of Materials*. Springer-Verlag, Heidelberg, Germany.
Kuttruff, H. 1991. *Ultrasonics: Fundamentals and Applications*. Elsevier, New York.
Lee, S. K. 1981. Method for percent oil analysis of avocado fruit. *Yearbook (California Avocado Society)*, 65, 133–41.
Lee, S. K., Young, R. E., Schiffman, P. M., and Coggins, C. W. J. 1983. Maturity studies of avocado fruit based on picking dates and dry weight. *Journal of the American Society for Horticultural Science*, 108, 390–394.
Lewis, C. E. 1978. The maturity of avocados—A general review. *Journal of the Science of Food and Agriculture*, 29, 857–866.
Makarov, L. O. 1964. Method of design of rod-type exponential ultrasonic concentrators. In Nozdreva, V. F. (Ed.) *Ultrasound in Industrial Processing and Control*. Academic Press, New York.
MATLAB. 1995. The Mathworks, Inc., South Natick, MA.
Matrix, S. 1989. Spectra matrix. In *Spectra–Matrix*, 1.8 edn. LT Industries, Inc., Rockville, MD.
McClements, D. J., Povey, M. J. W., Jury, M., and Betsanis, E. 1990. Ultrasonic characterization of a food emulsion. *Ultrasonics*, 28, 266–72.
Mizrach, A. 2000a. Determination of avocado and mango fruit properties by ultrasonic technique. *Ultrasonics*, 38, 717–722.
Mizrach, A. 2000b. Portable ultrasonic nondestructive fruit quality analyzer. In *The XIV Memorial CIGR Word Congress 2000*. Paper No. R6210, Tokyo, Japan.
Mizrach, A. 2004. Assessing plum fruit quality attributes with an ultrasonic method. *Food Research International*, 37, 627–631.
Mizrach, A. 2007. Nondestructive monitoring of tomato during shelf life storage utilizing ultrasonic method. *Postharvest Biology and Technology*, 46, 271–274.
Mizrach, A. 2008. Ultrasonic technology for quality evaluation of fresh fruit and vegetables in pre- and postharvest processes. *Postharvest Biology and Technology*, 48, 315–330.
Mizrach, A., Bechar, A., Grinshpon, Y., Hofman, A., Egozi, H., and Rosenfeld, L. 2003. Ultrasonic classification of mealiness in apples. *Transactions of the American Society of Agricultural Engineering*, 46, 397–400.
Mizrach, A. and Flitsanov, U. 1999. Nondestructive ultrasonic determination of avocado softening process. *Journal of Food Engineering*, 40, 139–144.
Mizrach, A., Flitsanov, U., Akerman, M., and Zauberman, G. 2000. Monitoring avocado softening in low-temperature storage using ultrasonic measurements. *Computers and Electronics in Agriculture*, 26, 199–207.
Mizrach, A., Flitsanov, U., El-Batsri, R., and Degani, C. 1999a. Determination of avocado maturity by ultrasonic attenuation measurements. *Scientia Horticulturae*, 80, 173–180.
Mizrach, A., Flitsanov, U., and Fuchs, Y. 1997. An ultrasonic nondestructive method for measuring maturity of mango fruit. *Transactions of the American Society of Agricultural Engineering*, 40, 1107–1111.
Mizrach, A., Flitsanov, U., Schmilovitch, Z., and Fuchs, Y. 1999b. Determination of mango physiological indices by mechanical wave analysis. *Postharvest Biology and Technology*, 16, 179–186.
Mizrach, A., Galili, N., Ganmor, S., Flitsanov, U., and Prigozin, I. 1996. Models of ultrasonic parameters to assess avocado properties and shelf life. *Journal of Agricultural Engineering Research*, 65, 261–267.
Mizrach, A., Galili, N., and Rosenhouse, G. 1989. Determination of fruit and vegetable properties by ultrasonic excitation. *Transactions of the American Society of Agricultural Engineering*, 32, 2053–2058.
Mizrach, A., Galili, N., and Rosenhouse, G. 1992. Half-cut fruit response to ultrasonic excitation. ASAE Paper No. 923017. American Society of Agricultural Engineers, St. Joseph, MI.
Mizrach, A., Galili, N., and Rosenhouse, G. 1994a. Determining quality of fresh products by ultrasonic excitation. *Food Technology*, 48, 68–71.

Mizrach, A., Galili, N., and Rosenhouse, G. 1994b. A method and a system for non-destructive determination of quality parameters in fresh produce. Israel Patent No. 109406, USA Patent No. 5589209, French Patent No. 9504869.

Mizrach, A., Galili, N., and Rosenhouse, G. 2009. 3-D Model of sound pressure field in a meridinal section plane of fruit. *Ultrasonics*, 49, 83–88.

Mizrach, A., Galili, N., Rosenhouse, G., and Teitel, D. C. 1991. Acoustical, mechanical, and quality parameters of winter-grown melon tissue. *Transactions of the American Society of Agricultural Engineers*, 34, 2135–2138.

Mizrach, A., Galili, N., Teitel, D. C., and Rosenhouse, G. 1994c. Ultrasonic evaluation of some ripening parameters of autumn and winter-grown galia melons. *Scientia Horticulturae*, 56, 291–297.

Mizrach, A., Schmilovitch, Z., and Avidan, B. 2006a. Maturity determination in olive fruit using acoustic and firmness measurements. In *Olivebioteq—The 2nd International Seminar*, Sicily, Italy.

Mizrach, A., Schmilovitch, Z., and Avidan, B. 2006b. Maturity measurements of olive fruits using acoustic and compression methods. In *CIGR XVI Word Congress*, Bonn, Germany.

Moreno, E., García, F., Castillo, M., Sotomayor, A., and Fuentes, M. 2002. Ultrasonic velocity measurement in viscoelastic material using the wavelet transform. In: Lee, H. (Ed.). *Acoustical Imaging*, Springer, Berlin, Germany.

Morse, P. M. and Ingard, K. U. 1968. *Theoretical Acoustics*. McGraw-Hill Book Company, New York.

Nielsen, M., Martens, H. J., and Kaack, K. 1998. Low frequency ultrasonics for texture measurements in carrots (Daucus carota L.) in relation to water loss and storage. *Postharvest Biology and Technology*, 14, 297–308.

Peacock, B. C., Murray, C., Kosiyachinda, S., Kosittrakul, M., and Tansiriyakul, S. 1986. Influence of harvest maturity of mangoes on storage potential and ripe fruit quality. *Association of South East Asian Nations Food Journal*, 2, 99–109.

Porteous, R. L., Muir, A. Y., and Wastie, R. L. 1981. The identification of diseases and defects in potato tubers from measurements of spectral reflectance. *Journal of Agricultural Engineering Research*, 26, 151–160.

Povey, M. J. W. 1998. Ultrasonics of food. *Contemporary Physics*, 39, 467–478.

Povey, M. J. W. and Mcclements, D. J. 1988. Ultrasonics in food engineering. Part I: Introduction and experimental methods. *Journal of Food Engineering*, 8, 217–245.

Povey, M. J. W. and Wilkinson, J. M. 1980. Application of ultrasonic pulse-echo techniques to egg albumin quality testing: A preliminary report. *British Poultry Science*, 21, 489–495.

Sarkar, N. and Wolfe, R. R. 1983. Potential of ultrasonic measurements in food quality evaluation. *Transactions of the American Society of Agricultural Engineering*, 26, 624–629.

Self, G. K., Ordozgoiti, E., Povey, M. J. W., and Wainwright, H. 1994. Ultrasonic evaluation of ripening avocado flesh. *Postharvest Biology and Technology*, 4, 111.

Shannon, A. F. 1949. Refractive index and other extraction methods for oil in avocados. *Bulletin of the California Department of Agriculture*, 38, 127–132.

Shewfelt, R. L. and Bruckner, B. 2000. *Fruit and Vegetables Quality: An Integrated View*. Technomic Publication, Lancaster, PA.

Upchurch, B. L., Affeldt, H. A., Aneshansley, D. J., Birth, G. S., Cavalieri, R. P., Chen, P., Miller, W. M., Sarig, Y., Schmilovitch, Z., Throop, J. A., and Tollner, B. W. 1994. *Detection of Internal Disorders*. ASAE publications, St. Joseph, MI.

Upchurch, B. L., Furgason, E. S., and Miles, G. E. 1985. Spectral analysis of acoustical signal for damage detection. ASAE Paper No. 85-Wt4.

Upchurch, B. L., Furgason, E. S., and Miles, G. E. 1986. Ultrasonic measurements for damage detection on agricultural products. *IEEE Transactions on Ultrasonics Ferroelectrics and Frequency Control*, 33, 101.

Upchurch, B. L., Miles, G. E., Stroshine, R. L., Furgason, E. S., and Emerson, F. H. 1987. Ultrasonic measurement for detecting apple bruises. *Transactions of the American Society of Agricultural Engineering*, 30, 803–809.

Verlinden, B. E., De Smedt, V., and Nicolai, B. M. 2004. Evaluation of ultrasonic wave propagation to measure chilling injury in tomatoes. *Postharvest Biology and Technology*, 32, 109–113.

Walls, P. N. T. 1969. *Physical Principles of Ultrasonic Diagnosis*. Academic Press, New York.

Watts, K. C. and Russell, L. T. 1985. A review of techniques for detecting hollow heart in potatoes. *Canadian Agricultural Engineering*, 27, 85–90.

8 Ultrasound in Food Technology

Taner Baysal and Aslihan Demirdoven

CONTENTS

Introduction 163
Ultrasound in Food Technology 164
Food-Processing Applications of Ultrasound 164
    Microbial and Enzyme Inactivation 165
    Extraction 168
        Extraction of Bioactive Compounds 170
        Oil and Protein Extraction 172
    Other Applications 173
Ultrasonic Quality Measurements and Control of Food Products 174
Conclusions 175
References 176

INTRODUCTION

Foods begin to lose their quality the moment they are harvested, through changes resulting from enzymatic, microbiological, chemical, or physical reactions. Food preservation prevents these deteriorative reactions, ensuring its safety and extending food's shelf life. Microorganisms and enzymes are the main agents responsible for food spoilage and therefore the targets of preservation techniques. Currently, heat is the main food preservation technique to act via inactivation, which is used substantially in the food industry (Raso and Barbosa-Canovas, 2003). Although thermal preservation methods provide safer foods, there is a loss of food quality associated with this processing method. Thermal treatment can cause undesirable alterations of sensory attributes, i.e., color, smell, flavor, texture, and nutritional (vitamins, proteins) qualities. Consumers now demand minimally processed fresh-like food with high-quality sensory and nutritional attributes. For this reason, targeted nonthermal food processing and preservation methods are gaining importance (Demirdoven and Baysal, 2009). Hence, the main objective of the nonthermal food preservation methods is to minimize the degradation of food quality through limiting heat damage of the food. The development of food-processing technology has been influenced by numerous factors; among them, consumer demands have undoubtedly oriented the new trends in the manufacturing, preservation, and control of food (Senorans et al., 2003).

One important trend in food processing is the development and optimization of novel food preservation processes. In particular are those used to obtain minimally processed foods (through hurdle technologies and/or with natural preservatives) as well as those based on emerging physical techniques (high hydrostatic pressure, pulsed electric fields, ultrasound, etc.), processes able to produce more nutritive, fresher, less processed, and safer foods (Mermelstein, 1999, 2000). Ultrasound is probably the most versatile and simplest method for the production of extracts and the disruption of cells; in addition, this method is efficient, safe, and reliable. Ultrasound is the energy generated by sound waves of 20,000 or more vibrations per second that are able to travel through gas, liquid, and solid materials. The use of ultrasound within the food industry has been a subject of researches for many years (Mason, 1990; Earnshaw, 1995; Mason et al., 1996; Earnshaw, 1998; Piyasena et al., 2003; Zenker et al., 2003).

The objectives of this chapter are to describe some of the most important food-processing applications of ultrasound, to review ultrasonic quality and control measurements of food products, and to outline possible applications of the technique in the food industry.

ULTRASOUND IN FOOD TECHNOLOGY

Ultrasound is defined as sound waves with frequencies above the threshold for human hearing (>16 kHz) and in its most basic definition refers to pressure waves with a frequency of 20 kHz or more (Butz and Tauscher, 2002). Generally, ultrasound equipment uses frequencies from 20 kHz to 10 MHz. And there are two distinct types of applications of ultrasound in the food industry: high and low intensity (Bjorno, 1991).

High-intensity ultrasound (power ultrasound) is used to physically alter the properties of a material through which it propagates. It utilizes relatively high power levels (>1 W cm^{-2}) and low frequencies (<0.1 MHz). There has been a growing interest lately in the use of high-intensity ultrasound as a preservation method (McClements, 1995a), and such treatment causes physical and chemical alterations to the food being processed (Villamiel et al., 1999).

Applications of power ultrasound have been tested for microbial and enzyme inactivation, biocomponent separation, emulsification, interface heat and mass transfer enhancement, cutting, crystallization enhancement, and extraction of bioactive component(s) in foods and plants. Due to new developments in ultrasound generation techniques, as well as increased understanding of cavitation phenomena, interest has increased in recent years to examine the use of power ultrasound as an alternative food-processing and preservation tool (Feng et al., 2008).

Low-intensity ultrasound is used to provide information about the physical properties of food materials. The power levels used are lower than those used in high-intensity applications (<0.1 W cm^{-2}) and the frequencies higher (0.1–100 MHz). Low-intensity ultrasound provides information about physicochemical properties, while high-intensity ultrasound is used to alter, either physically or chemically, the properties of foods, e.g., to generate emulsions, disrupt cells, promote chemical reactions, inhibit enzymes, tenderize meat, and modify crystallization processes (McClements, 1995a,b).

The effectiveness of ultrasound as a food-processing tool has been proven in the laboratory, and there are a number of examples of scale-up. In most cases, the frequency used has been that which is available commercially, i.e., 20 or 40 kHz, and this has proved quite satisfactory. In such cases, the variable parameters are temperature, treatment time, and acoustic power (Mason et al., 2005). The effects of ultrasound in liquid media depend on many variables, such as the characteristics of the treatment medium (viscosity, surface tension, vapor pressure, nature and concentration of the dissolved gas, and presence of solid particles), treatment parameters (pressure and temperature), ultrasound generator performance (frequency, power input), size, and geometry of the treatment vessel (Berlan and Mason, 1992). Ultrasound is known to disrupt biological structures and when applied with sufficient intensity has the potential to cause cell death (Harvey and Loomis, 1929; Hughes and Nyborg, 1962; Williams et al., 1970; Malo et al., 2005).

FOOD-PROCESSING APPLICATIONS OF ULTRASOUND

Ultrasound processes are used in food manufacturing for peeling, disintegration of cells, extracting (extract intracellular components or obtain cell-free bacterial enzyme), activation (acceleration) of an enzyme reaction in liquid foods, acceleration of a microbial fermentation, mixing, homogenizing, dispersion of a dry powder in a liquid, emulsifying of oil/fat in a liquid stream, spraying, degassing, inspection, e.g., in the beverage industry, deactivation of enzymes, microbial inactivation (preservation), crystallization of fats and sugars, foam breaking, meat tenderization, cleaning and surface decontamination, effluent treatment, humidifying and fogging, stimulation of living cells, and enhanced oxidation (Vollmer et al., 1998; Betts et al., 1999; Demirdoven and Baysal, 2009).

MICROBIAL AND ENZYME INACTIVATION

Ultrasound technology is being studied alone or combined with other preservation processes for inactivation of microorganisms (Table 8.1) and enzymes (Table 8.2) in food processing (Villamiel and de Jong, 2000a,b; Senorans et al., 2003). Critical processing factors are assumed to be the amplitude of the ultrasonic waves, the exposure/contact time with the microorganisms, the type of microorganism, the volume of food to be processed, the composition of the food, and the temperature of treatment (Anonymous, 2000; Senorans et al., 2003). Hence, different authors have attempted to use it in combination with other antimicrobial methods to increase its effect in microbial and enzyme inactivation (Hoover, 1997). Because of the complexity and sometimes protective nature of the food, the singular use of ultrasound as a preservation method is impracticable. Although ultrasound technology has a wide range of current and future applications in the food industry, including inactivation of microorganisms and enzymes, presently, most developments for food applications are nonmicrobial. There are not many data on inactivation of food microorganisms and enzymes

TABLE 8.1
Effects of Ultrasound on Microorganisms

Microorganism	Ultrasound Process	Remarks	References
L. moncytogenes	20 kHz (ambient temperature)	D value 4.3 min	Pagan et al. (1999a)
	20 kHz (ambient temperature) and 200 kPa pressure	D value 1.5 min	Pagan et al. (1999b)
Salmonella spp.	160 kHz, 100 W, 10 min	4 log reduction	Lee et al. (1989)
Salmonella Typhimurium	32–40 kHz and chlorine (0.5 ppm)	Synergetic effect on antimicrobial activity	Seymour et al. (2002)
Salmonella Senftenberg	117 μm, 60°C	½ log reduction after thermosonication	Manas et al. (2000a)
	117 μm, 60°C, 200 kPa pressure	3 log reduction after monothermosonication	
Saccharomyces cerevisiae	1–3 MHz	95.5% removal after 11.5 min	Limaye and Coakley (1998)
E. coli		95.5% removal after 4.5 min	
E. coli	70 kHz and antibiotic (gentamicin sulfate)	97% reduction after 2 h	Johnson et al. (1998)
Yersina enteroolitica	30°C, 21–150 μm, 200 kPa pressure	D value decreases from 4 to 0.37 min	Raso et al. (1998b)
	30°C, 21–150 μm, 600 kPa pressure	D value decreases from 1.52 to 0.2 min	
P aeruginosa	24 kHz, 2–30 min	Reduction increased by increasing the treatment time 68%–72%	Scherba et al. (1991)
B. subtilis		Reduction increased by increasing the treatment time 52%–76%	
S. aureus		Reduction increased by increasing the treatment time 42%–43%	

TABLE 8.2
Effects of Ultrasound on Enzymes

Enzyme	Product	Ultrasound Process	Remarks	References
Pectinmethylesterase	Tomato	Thermosonication (62.5°C)	D value 45 min for thermal treatment 0.85 min for ultrasound	Lopez et al. (1998)
Pectinmethylesterase	Tomato juice	Thermosonication (50°C–72°C)	Decreasing of D value to 0.3 min	Raviyan et al. (2005)
Pectinmethylesterase	Orange juice	Manothermosonication (72°C, 20 kHz, 200 kPa)	Inactivation rate was increased by a factor of 400	Vercet et al. (1999)
Polyphenoloxidase	Quava juice	35 kHz, 30 min	Increasing of PPO activity after ultrasound	Cheng et al. (2007)
Polyphenoloxidase	Model buffer system	Manothermosonication	Linear decrease in log D value	Lopez et al. (1994)
Peroxidase	Watercress	Thermosonication	90% POD inactivation at 90°C; a thermal treatment time of 70 s is necessary compared to 5 s for thermosonication	Cruz et al. (2006)
Lipoxygenase	Soybean	20 kHz	75%–85% inactivation of LOX	Thakur and Nelson (1997)

by ultrasound (Butz and Tauscher, 2002). Research activities must center on the combination of ultrasound with other preservation processes (e.g., heat and mild pressure) that appear to have the greatest potential for industrial applications.

Enzyme inactivation caused by ultrasound has been attributed to different mechanisms. It is generally agreed that sonication depolymerizes macromolecules (Lopez et al., 1998). The shear stress generated by stable cavitation is considered important, which can cause degradation of high-molecular-weight polymers even without the presence of bubble collapse. Beneficial combinations include thermosonication (heat and ultrasound), manosonication (pressure and ultrasound), and manothermosonication (pressure, heat, and ultrasound) (Raso et al., 1998b).

The lethal effect of ultrasound on some microorganisms was demonstrated first by Harvey and Loomis (1929); thus, ultrasound has been proposed as a means of sterilization of liquid foods (Jacobs and Thornley, 1954; Pagan et al., 1999a,b), and its use has been continually suggested for disinfection and food preservation (Paci, 1953; Jacobs and Thornley, 1954; Boucher, 1980; Gaboriaud, 1984). However, this technology has not been adopted, probably due to the long treatment needed for substantial microbial inactivation (Raso and Barbosa-Canovas, 2003).

The application of ultrasound and heat has been termed thermosonication. Heat combined with ultrasound (thermosonication) is considered to reduce process temperatures and processing times, for pasteurization or sterilization processes that achieve the same lethality values as with conventional processes (Mason et al., 1996; Villamiel, 1999) The heat resistance of *B. cereus*, *Bacillus licheniformis*, *B. stearothermophilus*, and thermoduric *Streptococci* decreased following ultrasonication treatment at 20 kHz (Burgos et al., 1972; Ordonez et al., 1984; Sanz et al., 1985; Garcia-Graells et al., 1998; Betts et al., 1999). The effect of the combined treatment of ultrasound and heat in a continuous process on microbial destruction was demonstrated by the comparison of the integrated time–temperature intensity (F value) of each treatment (Zenker et al., 2001). Reduction of the temperature and/or processing time should result in improved food quality (Piyasena et al., 2003; Zenker et al., 2003). Ultrasound applicability was predicted for the support of conventional thermal treatments, based on the possible synergy between low-frequency ultrasound and heat for bacterial inactivation (Ordenez et al., 1984; Zenker et al., 2003; Piyasena et al., 2003). And this combination markedly increases the lethality of heat treatments and consequent reductions in time and/or temperature of heat processes (Sala et al., 1995; Zenker et al., 2001). Moreover, thermosonication treatments have been reported to lower maximum processing temperatures by 25%–50%. After treatment, changes in color and ascorbic acid were minimal (Zenker et al., 2001).

In a milk thermosonication test at 60°C, the decimal reduction (D) value of *Listeria monocytogenes* is 0.3 min, which represents a sevenfold increase in inactivation rate compared to thermal treatment only at the same temperature (Earnshaw et al., 1995). However, an upper temperature limit exists for each microorganism beyond which the application of ultrasound to a food system does not introduce additional inactivation (Ugarte et al., 2007).

As ultrasound was initially discarded for food preservation because of its weak lethal effect, combined application of ultrasound with an external hydrostatic pressure of up to 600 kPa (manosonication) increases substantially the lethality of the treatment. It has been found that manosonication treatments sensitize spores of *Bacillus subtilis* to lysozyme. Therefore, it has been suggested that ultrasonic waves could damage the external layers of the spore, facilitating its rehydration and consequently reducing its extreme heat resistance. In contrast to the clear mechanisms of inactivation proposed for ultrasound, a much more complicated picture emerges for high hydrostatic pressure inactivation (Raso et al., 1998).

For example, an increment of hydrostatic pressure from 0 to 600 kPa at constant amplitude (150 μm) decreased the decimal reduction time (D value) of *Yersinia enterocolitica* eight times. The increased lethality of ultrasound under higher static pressure was more remarkable within a given pressure range (0–300 kPa). At constant hydrostatic pressure, microbial inactivation depended on the amplitude of the ultrasonic waves. At 200 kPa, the D value of *Y. enterocolitica* decreased 11 times when the amplitude increased from 21 to 150 μm. An exponential relationship was observed between the lethality of ultrasound and the amplitude. The influence of hydrostatic pressure and the amplitude on the lethality of monosonication treatments in different Gram-negative and Gram-positive bacteria was the same (Pagan et al., 1999a).

The concept of combination treatment has been further explored by introducing elevated static pressure in an ultrasound treatment chamber in a process called manothermosonication (Raso et al., 1998). The application of manothermosonication results in increased microbial and enzyme inactivation. Therefore, the same inactivation level is achieved over a shorter treatment period or at lower temperature. The lethality of ultrasound under pressure treatments is almost not modified by an increase in temperature unless lethal temperatures are reached (manothermosonication), in which case an additive lethal effect is generally attained, although in some cases the total lethal effect has been found to be synergistic (Pagan et al., 1999b; Alvarez et al., 2003). Manothermosonication has proved to be an efficient tool to inactivate microorganisms, especially in those conditions in which their thermotolerance is higher (Burgos et al., 1972, Burgos, 1999; Alvarez, 1998; Pagan et al., 1999b; Manas et al., 2000). While in most vegetative cells the lethal effect of manothermosonication was additive, on *Enterococcus faeciu*, *Bacillus subtilis*, *Bacillus coagulans*, *Bacillus cereus*, *Bacillus sterothermophilus*, *Saccharomyces cerevisiae*, and *Aeromonas hydrophila*, a synergistic effect was observed (Pagan et al., 1999; Raso et al., 1998). For example, the D value of tomato pectin methylesterase (PME) at 62.5°C was reduced 53-fold, from 45 min in thermal treatments to 0.85 min by manothermosonication (Lopez et al., 1998). This effect was even bigger when heat and ultrasound were applied simultaneously (Ordonez et al., 1987; Garcia et al., 1989).

Manothermosonication has also been used to deactivate peroxidase (Lopez and Burgos, 1995b; Gennaro et al., 1999), lipoxygenase (Lopez and Burgos, 1995a), lipase and protease (Vercet et al., 2002, 1997), and tomato or orange PME (Kuldiloke, 2002; Vercet et al., 2002, 1999) all with an increased inactivation. In tomato PME and polygalacturonase (PG) inactivation tests, a 52.9- and 26.3-fold increase in the inactivation rate for thermoresistant PGI and PGII, compared to thermal treatment alone, has been reported (Lopez et al., 1998). Consequently, this combination could be advantageous; due to the minimization of heat-induced damage in product quality (Lopez et al., 1994). Several mechanisms have been suggested to explain the synergistic effect of manothermosonication on enzyme inactivation. Propagation of ultrasonic waves in a liquid medium generates bubbles (cavities) that grow up to a critical size and then collapse (cavitational collapse) (Suslick, 1998).

Ultrasound effects are mainly related to the cavitation phenomenon. As a result of intense cavitation, water molecules can be broken, generating highly reactive free radicals that can react with and modify certain molecules. Mechanical stress, generated by shock waves derived from bubble implosion or from microstreaming derived from bubble's size oscillations, is also able to break large macromolecules or particles. This cavitational collapse creates strong shear stresses, extremely high pressures and temperatures in the so-called hot spots and water sonolysis, which produces free radicals (Berlan and Mason, 1996). The combination of these phenomena can promote enzyme denaturation, with the relative effect depending on the structure of the protein.

Manothermosonication effects have been mainly studied on enzymes and microorganisms (Lopez et al., 1998; Vercet et al., 2001a,b), but it is also possible to modify and improve textural and functional properties of tomato juice and milk proteins (Lopez and Burgos, 1995a; Vercet et al., 2002). Manothermosonication has also been proposed as an alternative to heat treatments in the processing of liquid eggs (Gimeno et al., 2006). As manothermosonication is suitable for treatment of liquid foods, two of the potential products to which manothermosonication could be applied are fruit juices and milk. Manothermosonication is an efficient tool to inactivate enzymes from psychrotrophic bacteria (Vercet et al., 2002) which are responsible for some quality problems of milk and some dairy products (Sorhaug and Stepaniak, 1997) and to inactivate thermoresistant PME in orange juice (Vercet et al., 2002) and pectic enzymes from tomato paste (Vercet et al., 1999).

Hereby, microbial inactivation tests are usually conducted at 20 kHz, a frequency at the low end of the power ultrasound frequency spectrum. It has been found that spores, and Gram-positive and coccal cells, are more resistant to ultrasound treatment than vegetative, Gram-negative, and rod-shaped bacteria. Cell-injury studies have demonstrated that thermosonication causes extensive physical damage on a cell envelope in the form of wrinkles, ruptures, and perforations (Ugarte et al., 2006). In sonication tests assisted with elevated external pressure, the survival Gram-positive and Gram-negative cells recovered in media with sodium chloride added are virtually identical to those recovered in a nonselective medium (Pagan et al., 1999b). The absence of sub-lethally injured cells in current studies has been attributed to irreversible physical damage to the outer membrane (Manas and Pagan, 2005). More ultrasound inactivation data are needed for vegetative cells, yeast, mold, spores, viruses, and food toxins. In addition, most microbial inactivation tests have been conducted in liquid media. Although surface decontamination of solid objects with airborne ultrasound has been proven effective in the inactivation of a virus, more studies are needed before one can draw general conclusions (Feng et al., 2008).

Ultrasonic inactivation of food enzymes mainly focuses on those endogenous enzymes more resistant to a thermal treatment than foodborne pathogens. Therefore, inactivation tests have been conducted to reduce enzyme activity in citrus, tomato, and dairy products. Inactivation of enzymes at sublethal temperatures has not proved very effective (Raviyan et al., 2005). Most experiments have been conducted at temperatures elevated high enough to cause microbial inactivation. The most effective inactivation is achieved with pressure-assisted sonication treatments (Lopez et al., 1998).

Extraction

The application of ultrasonic extraction (Table 8.3) in food processing is extremely interesting as it enables an increase in both the extraction yield and rate, leading to a significant reduction in the extraction time and a higher throughput (Mason et al., 2005; Dolatowski et al., 2007). It is possible to apply ultrasonic extraction to enhance the aqueous extraction and also in cases where organic solvents can be replaced with generally-recognized-as-safe (GRAS) solvents, which may

TABLE 8.3
Ultrasonic Extraction

Product	Ultrasound Process and Solvent	Remarks	References
Soy protein	20 kHz, 3 W, water and alkali (sodium hydroxide)	53% and 23% yield increase over ultrasonic bath conditions	Moultan and Wang (1982)
Soy isoflavones	24 kHz and solvent	Up to 15% increase in extraction efficiency	Rostagno et al. (2003)
Amino acid, polyphenol, and caffeine from green tea	40 kHz and water	Increased yield at 65°C, compared with 85°C	Xia et al. (2006)
Phenolic compounds, antioxidants, and anthocyanins from grape seeds	40 kHz, 250 W, 55°C–60°C, and ethanol	Enhance the yield of bioactive compounds	Ghafoor et al. (2009)
Oleuropein from olive leaves	20 kHz, 450 W, 40°C and ethanol	Faster and more effective than conventional method; 25 min–24 h for 100% yield	Japon-Lujan et al. (2006)
Antimicrobials from ginger, fingerroot, and turmeric	20 kHz, 6.8 W cm^{-2}, 5 min and hegzane/iso-propanol	Reduction of processing time and costs of spice essential oils with antimicrobial activity	Thongson et al. (2004)
Almond oil	20 kHz and supercritical carbon dioxide	30% increased yield and reduction of extraction time	Riera et al. (2004)
Peanut oil	400 kHz and hegzane	Yield increased by ultrasound application	Thomson and Sutherland (1955)
Carnoxic acid from rosemary	20 and 40 kHz; butanone and ethyl acetate	Reduction of extraction time	Albu et al. (2004)

provide economical, environmental, as well as health and safety benefits (Vilkhu et al., 2008). As previously mentioned, the mild processing temperature in ultrasonic extraction may also lead to an enhanced extraction of thermolabile food bioactives (Soria and Villamiel, 2010). Several papers have been published dealing with ultrasonically assisted extraction of different food material. One of the first citations concerning ultrasonic extraction (1952) was related to hop extraction in an aqueous medium, and ultrasonic extraction was compared with boiling extraction process. It was shown that during ultrasonic extraction it was possible to save 30%–40% of hops in beer production (Vinatoru, 2001). It can be avowable that high-frequency ultrasound did not increase the yield extraction significantly. In the case of low-frequency ultrasound, degradation becomes more important, especially when alcoloids are being extracted. This effect could be employed as a tool to help in the extraction of medical compounds by using low frequencies to assist in the degradation of toxic alkaloids during the process (Vinatoru, 2001).

In extraction process, ultrasound can be used for (1) the extraction of phenolic compounds from vacuolar structures by disrupting plant tissue; (2) the extraction of betacyanin (red pigments, e.g., from beets) and betaxanthin (yellow pigments); (3) the extraction of lipids and proteins from plant seeds, such as soybean (e.g., flour); (4) the improvement of oil extraction from oil seeds; (5) cell membrane permeabilization of fruits, such as grapes, plums, and mango; (6) processing of fruit juices (e.g., orange, grapefruit, mango, grape, and plum), purees, sauces (e.g., tomato, asparagus, bell pepper, and mushroom), dairy products; and (7) improving the stability of dispersions, such as orange juice, i.e., reduce settling (Knorr et al., 2004).

The key issues and observations relating to ultrasound extraction have been identified as follows: (a) the nature of the tissue being extracted and the location of the components to be extracted with respect to tissue structures, (b) pretreatment of the tissue prior to extraction, (c) the nature of the component being extracted, (d) the effects of ultrasonics primarily involve superficial tissue disruption, (e) increasing surface mass transfer (Balachandran et al., 2006; Jian-Bing et al., 2006), (f) intraparticle diffusion, (g) loading of the extraction chamber with substrate, (h) increased yield of extracted components, and (i) increased rate of extraction, particularly early in the extraction cycle enabling major reduction in extraction time and higher processing throughput (Moulton and Wang, 1982; Caili et al., 2006; Vilkhu et al., 2008).

Extraction of Bioactive Compounds

The extraction of bioactive compounds from plants or seeds has classically been based on the suitable combination of solvent, heat, and/or agitation (Patist and Bates, 2008). All the mechanical effects involved in ultrasound can accelerate the eddy and internal diffusion, giving rise to an increased mass transfer (Jian-Bing et al., 2006). In addition, they allow a greater penetration of solvent into the sample matrix (Rostagno et al., 2003). If the substrate is dry, then ultrasound may be used to facilitate swelling and hydration and cause an enlargement of the pores of the cell wall (Vinatoru, 2001).This can be significantly improved by the use of high-powered ultrasound, as the energy generated from collapsing cavitational bubbles provides greater penetration of the solvent into the cellular material and improves mass transfer to and from interfaces (Vinatoru, 2001; Knorr, 2003; Zhang et al., 2003; Li et al., 2004; Vilkhu et al., 2008). Ultrasonic extraction was as effective as any other high-temperature long-time extraction process because it could greatly decrease the extraction time. The efficiency of ultrasonic extraction could be explained by the fact that sonication simultaneously enhanced the hydration and fragmentation process while facilitating the mass transfer of solutes to the extraction solvent (Soria and Villamiel, 2010).

At higher ultrasonic intensities (W cm^{-2}), extraction processes can be further improved with the disruption of cell walls and the release of cellular materials (Patist and Bates, 2008). High-intensity ultrasound extraction is used as an inexpensive, reproducible, simple, and efficient alternative method of industrial relevance to improve the extraction process of food bioactives. Additional benefits result from the disruption of the biological cell walls during the ultrasonically induced cavitation to facilitate the release of contents (Dolatowski et al., 2007). Furthermore, mild operating conditions usually employed in ultrasonic-assisted extraction show no significant changes in the structural/molecular properties and functionality of most bioactives, this aspect being of paramount importance in the case of heat-sensitive food components of bioactive compounds (Soria and Villamiel, 2010).

The effect of ultrasound conditions (frequency and duration of ultrasonication) on the extraction efficiency of isoflavones and trans-resveratrol from peanuts has been recently studied by Chukwumah et al. (2009). The results showed that sonication at 80 kHz facilitates the extraction of biochanin A and trans-resveratrol, whereas sonication at 25 kHz was effective in the extraction of daidzein and genistein, multifrequency being more efficient than single frequency. The higher amount of analytes extracted by dual-frequency radiation could be explained by the increased cavitation bubble collision which caused further reduction in particle size and promoted leaching (Soria and Villamiel, 2010).

Thongson et al. (2004) used ultrasound to obtain extracts of ginger, fingerroot, and turmeric. The application of this technique reduced the time of extraction to 5 min as compared with the 24 h conventional extraction. However, a slightly reduced antimicrobial activity of the extracts against *Listeria* was observed, whereas that against *Salmonella* was maintained (Soria and Villamiel, 2010).

Zhao et al. (2006) applied high-intensity ultrasound in the extraction of a variety of biologically active compounds including carotenoids, and they found that one of the studied carotenoids, (all-E)-astaxanthin, was degraded to unidentified colorless compounds, the degradation being higher when both the treatment time and the ultrasonic power increased.

Vilkhu et al. (2008) have recently revised the main applications and opportunities for ultrasound-assisted extraction in the food industry. Nowadays, developments in ultrasonic equipment are such that it is feasible to consider commercial opportunities based on industrial-scale ultrasonic-aided extraction of bioactives from plant and animal materials, with worthwhile economics gains (Vinatoru, 2001; Hielscher, 2010; Soria and Villamiel, 2010). In addition, significantly enhanced contents of tea polyphenols, amino acid, and caffeine in tea infusions were recovered with ultrasound-assisted extraction when compared with conventional extraction. The sensory quality of tea infusion with ultrasound-assisted extraction was better than that of tea infusion with conventional extraction (Xia et al., 2006). Addition of sucrose or heating at temperatures up to 80°C had little effect on pigment stability. However, pigment stability and color were greatly improved by the addition of citric acid (Cai et al., 2003).

The effect of ultrasound on anthocyanins was studied for strawberry juice by Tiwari et al. (2008). They reported a slight increase (1%–2%) in the pelargonidin-3-glucoside content of the juice at lower amplitude levels and treatment times which may be due to the extraction of bound anthocyanins from the suspended pulp. Similarly, weak ultrasonic irradiation is reported to promote an increase in the amount of phenolic compounds found in red wine (Masuzawa et al., 2000). Literature indicates that ultrasound processing enhances the extraction of phenolic and other bioactive compounds from grape must or wine (Cocito et al., 1995). Ultrasound-assisted extraction of bioactive compounds and anthocyanins was recently reviewed by Vilkhu et al. (2008). The application of ultrasound-assisted extraction improves the extraction yield of bioactive compounds by between 6% and 35% (Vilkhu et al., 2008) compared to conventional processing. Zhao et al. (2006) reported degradation of (all-E)-astaxanthin into unidentified colorless molecule(s) during extraction with sonication at an increased power level and treatment time. Accordingly in a study by Tiwari et al. (2008), the anthocyanin content of the juice was found to degrade when higher amplitude levels were employed; however, the maximum observed degradation was <5% (Twari et al., 2009).

Enzymes, such as pectinases, cellulases, and hemicellulases are widely used in juice processing in order to degrade cell walls and improve the juice extraction. The disruption of the cell wall matrix also releases components, such as phenolic compounds into the juice. Ultrasound improves the extraction process and therefore can lead to an increase in the phenolic compound, alkaloids, and juice yield, commonly left in the press cake (Hielscher, 2010). The beneficial effects of ultrasonic treatment on the liberation of phenolic compounds and anthocyanins from grape and berry matrix, in particular from bilberries (*Vaccinium myrtillus*) and black currants (*Ribes nigrum*) into juice, were investigated after thawing, mashing, and enzyme incubation. The disruption of the cell walls by enzymatic treatment (Pectinex BE-3L for bilberries and Biopectinase CCM for black currents) was improved when combined with ultrasound enhances (Hielscher, 2010). Ultrasound treatment increases the concentration of phenolic compounds of bilberry juice by more than 15%. The influence of ultrasound was more significant with black currants, which are more challenging berries in juice processing than bilberries due to their high content of pectin and different cell wall architecture. The concentration of phenolic compounds in the juice increased 15%–25% by using ultrasound treatment after enzyme incubation (Okkila et al., 2004). The extraction variables, particularly extraction temperature and time, strongly influenced the ultrasonic extraction of compounds (Ghafoor et al., 2009).

Ultrasound has also been successfully applied combined with other alternative methods such as supercritical-CO_2 extraction for fractionation of isoflavones from soybeans (Rostagno et al., 2003) and gingerols from ginger (Balachandran et al., 2006), with improved rates and final yields. Since cavitational events in a supercritical fluid seem impossible due to the absence of liquid/gas phase boundaries, several other mechanisms, such as acoustic streaming and the presence of gas pockets in the solid causing cavitational collapse, are proposed (Patist and Bates, 2008). The ultrasonic enhancement of the supercritical extraction could be attributed to the disruption of the cell structures and an increase in the accessibility of the solvent to the internal particle structure, which enhances the intra-particle diffusivity. Furthermore, by reducing the substrate particle size, significant improvements in both the extraction efficiency and the time reduction could be achieved (Soria and Villamiel, 2010).

Therefore, ultrasound has a potential benefit in the extraction and isolation of novel potentially bioactive components, e.g., from non-utilized by-product streams formed in current processes. Ultrasound can also help to intensify the effects of enzyme treatment, and by this reduce the amount of enzyme needed or increase the yield of extractable relevant compounds (Hielscher, 2010).

Oil and Protein Extraction

Ultrasound is often used to improve the extraction of proteins and oils from plant seeds, such as soybeans (e.g., flour or defatted soybeans) or other oil seeds. In this case, the destruction of the cell walls facilitates the pressing (cold or hot) and thereby reduces the residual oil or fat in the pressing cake (Hielscher, 2010). Thus ultrasound may reduce the dependence on a solvent and enable the use of alternative solvents which may provide more attractive (a) economics, (b) environmental, and (c) health and safety benefits (Vilkhu et al., 2008).

Classical oil extraction technologies are based on the use of an appropriate solvent to remove lipophilic compounds from the interior of plant tissues. The choice of a suitable solvent in combination with sufficient mechanical agitation influences mass transport processes and subsequently efficiency of the extraction (Li et al., 2004). The most widely used solvent to extract edible oils from plant sources is hexane. Hexane is available at low cost and is efficient in terms of oil and solvent recovery (Mustakas, 1980; Serrato, 1981). More recently, the use of alternative solvents such as alcohols (isopropanol or ethanol) and supercritical carbon dioxide has increased due to environmental, health, and safety concerns (Dunnuck, 1991). Alternative solvents are often less efficient due to a decreased molecular affinity between solvent and solute, and costs for solvent and process equipment can be higher (Karnofsky, 1981; Baker and Sullivan, 1983; Freidrich and Pryde, 1984). For this reason, the application of ultrasound is practicable for extraction.

Ultrasound extraction has been recognized for application in the edible oil industry to improve efficiency and reduce extraction time (Babaei et al., 2006). This potential was based on ultrasound extraction increases in oil from soybeans and carvone. The ultrasonically induced cavitation was shown to increase the permeability of the plant tissues. Microfractures and disruption of cell walls in soybean flakes provided more evidence for the mechanical effects of ultrasound, thus facilitating the release of their contents, in contrast to conventional maceration or extraction. Importance was given to the effect of solvent surface tension on cavitation intensity and vapor pressure (Haizhou et al., 2004).

The benefit of using ultrasonic pretreatment before extracting oil from the seeds of *Jatropha curcas* L., almond, and apricot seeds by aqueous enzymatic oil extraction (AEOE) process was evaluated by Shah et al. (2005) and Sharma and Gupta (2006). Ultrasonic pretreatment of the almond and apricot seeds before aqueous oil extraction and aqueous enzymatic oil extraction provided significantly higher yield with reduction in extraction time. Since therefore, implementation of ultrasonic pretreatment reduced oil extraction time that may improve through put in commercial oil production process (Vilkhu et al., 2008).

In recent years, it has been shown that combination of ultrasound and supercritical CO_2 on extraction could be used to significantly improve extraction yield of amaranth oil from seeds (Bruni et al., 2002), tea seed oil (Rajaei et al., 2005); gingerols from ginger (Balachandran et al., 2006); almond oil (Riera et al., 2004); operating parameters such as temperature, pressure, and CO_2 flow for Adlay seed (*Coix lachrymal-jobi* L. var. Adlay) oil; and coixenolide from adlay seed (Ai-jun et al., 2006).

The influence of continuous ultrasonic extraction to the yield of dispersed protein was demonstrated by Moulton et al. (1982). The sonication increased the recovery of dispersed protein progressively as the flake/solvent ratio changed from 1:10 to 1:30. It was showed that ultrasound is capable to peptize soy protein at almost any commercial throughput and that the sonication energy required was the lowest, when thicker slurries were used. The continuous high-intensity application extracted 54% and 23% more protein for aqueous and alkali extraction respectively, compared with the batch extraction using comparable processing times and volumes. During the trials, it was estimated that the continuous process used 70% less energy than the batch system to extract the same amount of protein, and sonication efficiency improved with the greater load of thicker slurry, up to 1:10 (flake to solvent) ratio (Moulton et al., 1982; Hielscher, 2010).

The proposed benefits of ultrasound extraction for the food industry include (a) overall enhancement of extraction yield or rate, (b) enhancement of aqueous extraction processes where solvents cannot be used (juice concentrate processing), (c) providing the opportunity to use alternative (GRAS) solvents by improvement of their extraction performance, (d) enable sourcing/substitution of cheaper raw product sources (variety) while maintaining bioactive levels, and (e) enhancing extraction of heat-sensitive components under conditions, which would otherwise have low or unacceptable yields (Vilkhu et al., 2008).

OTHER APPLICATIONS

Ultrasonic homogenizing is very efficient for the reduction of soft and hard particles. The major advantage of ultrasonic homogenizers is the low number of wetted and moving parts. This reduces frictional wear and cleaning time. There are only two wetted parts: the sonotrode and the flow cell, and both have simple geometries and no small or hidden orifices. Another advantage is the exact control over the operational parameters influencing the cavitation (Hielscher, 2010). Ultrasonic homogenization can be used for fruit juices, mayonnaise and tomato ketchup, milk and yogurt processing (Wu et al., 2000).

Ultrasound produces a series of effects (microagitation, creation of microscopic channels, and water cavitation) which facilitate moisture removal from food (Mulet et al., 2003). The synergic effect of ultrasound and temperature in convective drying assisted by high-power ultrasound improves the rate of the process and allows dehydration to be carried out at milder temperatures (Garcia et al., 2007), thus preserving the bioactivity of heat-sensitive food constituents and giving rise to dehydrated food of premium quality (Soria and Villamiel, 2010). But no commercial scale installation has yet been developed for this application.

In addition, ultrasound can assist crystallization by controlling the nucleation and growth rate of crystals in frozen food (Luque de Castro and Priego-Capote, 2007). It also reduces textural softening and the release of cellular liquid on thawing (Zheng and Sun, 2006), this being of capital importance for the consumer's acceptance of meat, fruit, and vegetable products as well as for a better preservation of their nutrients and/or bioactives. Application of power ultrasound can also benefit ice cream manufacture by reducing crystal size, preventing incrustation on freezing surface, etc. The ability of power ultrasound in performing these functions is affected by a variety of parameters, such as the duration, intensity, or frequency of ultrasonic waves, etc. And more fundamental research is still needed to establish their relationships with the acoustic efficiency (Zheng and Sun, 2006).

Ultrasound technology also represents a clean and commercial alternative to conventional methods for defoaming, fermentation systems, carbonated beverages, and other food processes, where foaming adversely affects the product yield and quality (Villamiel et al., 2000; Gallego-Juarez et al., 2007). The breaking and destruction of foams by ultrasound-based defoamers is assumed to be a combination, among others, of the following effects: partial vacuum on the foam bubble surface produced by high acoustic pressure, resonance of the foam bubbles which create interstitial friction causing bubble coalescence, acoustic streaming, and cavitation (Mason et al., 2005). In this respect of manufacture, advances in ultrasonic defoamer systems have overcome the initial limitation of those based on aerodynamic acoustic sources regarding energy consumption and difficulties in sterilization (Gallego et al., 2002). The increased production throughput, the reduction or elimination of antifoam chemicals, and the reduction of wastage in bottling production lines are additional advantages of the use of ultrasound as an efficient additional step in food processing (Soria and Villamiel, 2010).

The use of ultrasound for separation of food components/constitutes has been investigated in recent years. In starch–protein separation experiments, sonication can recover 97.3%–99.5% of the total starch from degermed corn flour (67.5% total starch) and hominy feed (46.4% total starch)—two low-value dry-milling by-products. The quality of the resulting starch is comparable to regular commercial corn starch (Zhang et al., 2005). When applying ultrasonication to the fine fiber stream in a quick germ/quick fiber process, a process in which no SO_2 is added during steeping to enhance starch separation, starch yield (66.93%–68.72%) was almost as high as that of a traditional wet milling operation (68.92%). The quality of starch from the ultrasound treatment was comparable to or better than conventional wet milling starch, as evidenced by lower protein content, comparable color, and similar pasting properties measured with a rapid visco-analyzer (RVA) (Zhang et al., 2005b; Feng et al., 2008). Wet milling process of corn power ultrasound was used to remove corn pericarp prior to steeping, which resulted in a reduction in steeping time and improved the isolated starch gelatinization and pasting properties (Liu, 2002; Yang et al., 2002).

Power ultrasound treatment was used to extend the shelf life of roasted peanuts by removing oils on peanut kernel surfaces. Ten minute sonication removed, as can be discerned from microscopic examination, all the surface oil and increased the shelf life by up to 17% (Yang et al., 2005). In an ultrasound-assisted tomato peeling test, with a 2% lye solution, peel loss was reduced by 3%, compared to peeling under commercial conditions, where a lye solution of 10% was used (Lee and Feng, 2004).

ULTRASONIC QUALITY MEASUREMENTS AND CONTROL OF FOOD PRODUCTS

There are many situations in the food industry where ultrasound can be usefully applied. One of the earliest applications of this technology might have been level measurement. In many cases, there is a need to know the level within a tank, either liquid or particulate, or even the existence of product (Povey, 1998; Mulet et al., 2002). Ultrasound-based measurement methods have demonstrated potential in various food-processing applications, including concentration gauging, flow measurement, level detection (Ridgway et al., 1998), food shelf-life monitoring (Kulmyrzaev et al., 2000), food property assessment (Benedito et al., 2000; Mizrach, 2000; Zhao et al., 2003), and foreign body detection (Chivers et al., 1995). There is growing pressure within the food industry to improve the measurement of food quality, particularly within packaged food products (Morris et al., 2004). This inspection may include looking for time-dependent physical changes, nonuniformity of the product, or contamination.

Food control measurements consist mainly of determining changes linked to different compositions or textures and also detecting and analyzing signal reflections. One of the measurements is mostly used to characterize products, whereas the second is related with physical discontinuities. Velocity (time of flight) is the main parameter considered for these measurements, although attenuation and spectrum analysis is also considered in some applications (Mulet et al., 2002).

Ultrasound has the ability to differentiate between both the propagation velocity within various media and the differences in acoustic impedance between different regions within a given volume. Thus, using the usual contact or immersion techniques, ultrasound can be used to measure the moisture content of food products (Steele, 1974), oil properties at various temperatures (Chanamai et al., 1998), and the liquid level (Hull et al., 1995). In the case of palm oil, physiochemical changes such as crystallization are important in determining the condition of the finished products. The control of the crystallization process can be complicated, and it is difficult to monitor the process nondestructively (Marshall et al., 2000), and reprocessing is sometimes required. In some cases, the product can be rendered unusable. It is thus important to have a cost-effective and reliable system. Contact ultrasonic approaches have been reported in which the crystallization behavior of palm oil has been measured (Hodate et al., 1997). The study was performed using ultrasonic velocity measurements under both cooling and heating processes. The rates of crystallization of palm oil were monitored by the changes in the ultrasonic velocity values, which increase with increasing amount of fat crystals in the palm oil phase (Hodate et al., 1997; Tat et al., 2006).

There is also growing pressure within the food industry for the routine monitoring of changes in the quality of complex viscoelastic products after processing, with a view to extending shelf life. An example is milk-based products, where the quality changes according to temperature and environment, especially so when storing these products for a long period of time. Papadakis (1976) described how ultrasonic velocity and attenuation are very valuable properties when studying the physical properties of matter. Coagulating milk, being a liquid to semisolid system, is especially suitable for this type of ultrasonic measurement (Gunasekaran and Ay, 1994).

Other applications of attenuation measurements also related to changes in texture are found in maturity assessment for fruits like avocados, melons, and mangoes (Mizrach et al., 1996, 1997). In the meat industry, there are also many applications of ultrasound, many of them linked to composition, ranging from live cattle to processed meat. In live cattle, evaluation of subcutaneous fat can be performed, and also an assessment of carcasses can be carried out (Mulet et al., 1999). Composition of chicken, cod, and pork meat has also been addressed by measuring ultrasonic velocity at different temperatures (Cross and Belk, 1994; Ghaedian et al., 1997; Ghaedian, 1998; Mulet et al., 2002).

Ice content can be measured by a variety of methods, differential scanning calorimetry being the more precise, although it is destructive for the sample and impossible to be used for online monitoring. Other methods like nuclear magnetic resonance can also be employed, but it is technically complex and expensive for its use in food storage and processing applications. Different authors have performed ultrasound experiments to study the propagation of ultrasonic waves through partially frozen foods and the relationship between ultrasonic properties and ice content (Miles, 1974; Lee et al., 2004; Sigfusson et al., 2004; Gülseren and Coupland, 2007; Carcione et al., 2007). Miles's (1974) method requires the previous knowledge of the value of the speed of sound through the sample at 0°C and its water fraction.

Lee et al. (2004) measured the ultrasonic velocity and attenuation in partially frozen orange juice over a wide temperature range. Recently, Carcione et al. (2007) successfully used a poroelastic model to describe the propagation of ultrasonic waves through orange juice, which is subjected to a freezing process. They used the Kelvin's model to obtain the amount of unfrozen water in the juice as a function of temperature and the Biot's poroelastic theory (Biot, 1962) to calculate the ultrasonic properties of orange juice as a function of temperature, below the eutectic point (Aparicio et al., 2008).

CONCLUSIONS

Ultrasound is one of the simplest and most versatile methods for cellular disruption and for food extract production. This technology works best when used in conjunction with heat and pressure, but it can be used alone for fruit juices, sauces, purees, and dairy products. Ultrasound treatment has been found to be more effective when combined with other processes such as mano-and thermosonication, pressure, and/or heat. Furthermore ultrasonic extraction disrupts plant tissue phenolic compounds from their vacuolar structures, can be utilized to extract proteins, and oils from food materials. Ultrasound can also be used in emulsification, dispersing, homogenizing, and crystallization processes. Ultrasound has not yet been demonstrated to achieve major beneficial effects that warrant serious consideration for processing. In addition; ultrasonic measurements are quick, allowing one to monitor processes with sudden changes, which is an interesting feature for process control. Thus, successful applications of ultrasonics are not only reliable but also accurate for both processes and product control applications. Now research should concentrate on the microbial and enzyme inactivation mechanism and inactivation kinetics and identify the most resistant pathogenic microorganism and examine the effect of ultrasound on quality attributes of food. Moreover, considering the economic feasibility of an ultrasonic processing preservation of ultrasound technology should be required. The use of ultrasound in the food industry is widening, and this wide range of applications suggests that ultrasonics may enjoy an even larger number of future applications for the food technology and the food industry.

REFERENCES

Ai-Jun, H., Shuna, Z., Hanhua, L., Tai-Qiu, Q., and Guohua, C. 2006. Ultrasound assisted supercritical fluid extraction of oil and coixenolide from adlay seed. *Ultrasonics Sonochemistry*, 14: 219–224.
Albu, S., Joyce, E., Paniwnyk, L., Lorimer, P., and Mason, J. 2004. Potential for the use of ultrasound in the extraction of antioxidants from *Rosmarinus officinalis* for the food and pharmaceutical industry. *Ultrasonics Sonochemistry*, 11: 261–265.
Alvarez, I., Manas, P., Sala, F.J., and Condon, S. 2003. Inactivation of *Salmonella enteritidis* by ultrasonic waves under pressure at different water activities. *Applied Environmental Microbiology*, 69: 668–672.
Anonymous. 2000. Kinetics of microbial inactivation for alternative food processing technologies. Executive Summary, U.S. Food and Drug Administration. Center for Food Safety and Applied Nutrition, http://www.cfsan.fda.gov (March 09, 2009).
Aparicio, C., Otero, L., Guignon, B., Molina-Garcia, A.D., and Sanz, P.D. 2008. Ice content and temperature determination from ultrasonic measurements in partially frozen foods. *Journal of Food Engineering*, 88: 272–279.
Babaei, R., Jabbari, A., and Yamini, Y. 2006. Solid–liquid extraction of fatty acids of some variety of Iranian rice in closed vessel in the absence and presence of ultrasonic waves. *Asian Journal of Chemistry*, 18: 57–64.
Baker, E. and Sullivan, D. 1983. Development of a pilot-plant process for extraction of soy flakes with aqueous isopropyl alcohol. *Journal of the American Oil Chemist' Society*, 60(7): 1271–1276.
Balachandran, S., Kentish, E., Mawson, R., and Ashokkumar, M. 2006. Ultrasonic enhancement of the supercritical extraction from ginger. *Ultrasonics Sonochemistry*, 13: 471–479.
Benedito, J., Carcel, J.A., Sanjuan, N., and Mulet, A. 2000. Use of ultrasound to assess cheddar cheese characteristics. *Ultrasonics*, 38: 727–730.
Berlan, J. and Mason, T.J. 1992. Sonochemistry: From research laboratories to industrial plants. *Ultrasonics*, 30: 203–211.
Berlan, J. and Mason, T.J. 1996. Dosimetry for power ultrasound and sonochemistry. *Advances in Sonochemistry*, 4: 1–73.
Betts, G.D., Williams, A., and Oakley, R.M. 1999. Ultrasonic standing waves, inactivation of foodborne microorganisms using power ultrasound. In *Encyclopedia of Food Microbiology*, Robinson, R.K., Batt, C.A., and Patel, P.D. (Eds.). Academic Press, New York, pp. 2202–2208.
Biot, M.A. 1962. Mechanics of deformation and acoustic propagation in porous media. *Journal of Applied Physics*, 33(4): 1482–1498.
Bjorno, L. 1991. *Ultrasonics International 91 Conference Proceedings*. Butterworth-Heinemann, Oxford, U.K., p. 23.
Boucher, R.M.G. 1980. Process for ultrasonic pasteurization. U.S. Patent 4,211,744.
Bruni, R., Guerrini, A., Scalia, S., Romagnoli, C., and Sacchetti, G. 2002. Rapid techniques for the extraction of vitamin E isomers from *Amaranthus caudatus* seeds: Ultrasonic and supercritical fluid extraction. *Phytochemical Analysis*, 13: 257–261.
Burgos, J. 1999. Manothermosonication. In *Encyclopedia of Food Microbiology*, Robinson, R.K., Batt, C.A., and Patel, P.D. (Eds.). Academic Press, New York, pp. 1462–1469.
Burgos, J., Ordonez, J.A., and Sala, F.J. 1972. Effect of ultrasonic waves on the heat resistance of *Bacillus cereus* and *Bacillus licheniformis* spores. *Applied Microbiology*, 24: 497–498.
Butz, P. and Tauscher, B. 2002. Emerging technologies: Chemical aspects. *Food Research International*, 35(2/3): 279–284.
Cai, J., Liu, X., Li, Z., and An, C. 2003. Study on extraction technology of strawberry pigments and its physicochemical properties. *Food and Fermentation Industries*, 29: 69–73.
Caili, F., Haijun, T., Quanhong, L., Tongyi, C., and Wenjuan, D. 2006. Ultrasound-assisted extraction of xyloglucan from apple pomace. *Ultrasonics Sonochemistry*, 13: 511–516.
Carcione, J.M., Campanella, O.H., and Santos, J.E. 2007. A poroelastic model for wave propagation in partially frozen orange juice. *Journal of Food Engineering*, 80(1): 11–17.
Chanamai, R., Coupland, J.N., and Mcclements, D.J. 1998. Effect of temperature on the ultrasonic properties of oil-in-water emulsions. *Colloids and Surfaces A: Physiochemical and Engineering Aspects*, 139: 241–250.
Cheng, L.H., Soh, C.Y., Liew, S.C., and Teh, F.F. 2007. Effects of sonication and carbonation on guava juice quality. *Food Chemistry*, 104: 1396–1401.
Chivers, R.C., Russel, H., and Anson, L.W. 1995. Ultrasonic studies of preserved peaches. *Ultrasonics*, 33: 75–77.

Chukwumah, Y.C., Walker, L.T., Verghese, M., and Ogutu, S. 2009. Effect of frequency and duration of ultrasonication on the extraction efficiency of selected isoflavones and trans-resveratrol from peanuts (*Arachis hypogaea*). *Ultrasonics Sonochemistry*, 16: 293–299.

Cocito, C., Gaetano, G., and Delfini, C. 1995. Rapid extraction of aroma compounds in must and wine by means of ultrasound. *Food Chemistry*, 52: 311–320.

Cross, H.R. and Belk, K.E. 1994. Objective measurements of carcass and meat quality. *Meat Science*, 36: 191–202.

Cruz, R.M.S., Vieira, M.C., and Silva, C.L.M. 2006. Effect of heat and thermosonication treatments on peroxidase inactivation kinetics in watercress (*Nasturtium officinale*). *Journal of Food Engineering*, 72(1): 8–15.

Demirdoven, A. and Baysal, T. 2009. The use of ultrasound and combined technologies in food preservation. *Food Reviews International*, 25: 1–11.

Dolatowski, Z.J., Stadnik, J., and Stasiak, D. 2007. Applications of ultrasound in food technology. *Acta Scientiarum Polonorun, Technologia Alimentaria*, 6: 89–99.

Dunnuck, J. 1991. NTP technical report on the toxicity studies of *n*-hexane in mice. *Toxicity Report Series*, 2: 1–32.

Earnshaw, R.G. 1998. Ultrasound: A new opportunity for food preservation. In *Ultrasound in Food Processing*, Povey, M.J.W. and Mason, T.J. (Eds.). Blackie Academic & Professional, London, U.K., pp. 183–192.

Earnshaw, R.G., Appleyard, J., and Hurst, R.M. 1995. Understanding physical inactivation process: Combined preservation opportunities using heat, ultrasound and pressure. *International Journal of Food Microbiology*, 28: 197–219.

Feng, H., Yang, W., and Hielscher, T. 2008. Power ultrasound. *Food Science and Technology International*, 14: 433.

Freidrich, J.P. and Pryde, E.H. 1984. Supercritical CO_2 extraction of lipid bearing materials and characterization of the products. *Journal of the American Oil Chemist' Society*, 61(2): 223–228.

Gaboriaud, P.L.F. 1984. Sterilisation de liquides par ultrasons. French Patent 2 575 641 A1.

Gallego, J.A., Rodriguez, G., Acosta, V.M., Andre, S.E., Blanco, A., and Montoya, F. 2002. Procedimiento Y Sistema Ultraso'Nico De Desespumacio'N Mediante Emisores Con Placa Vibrante Escalonada. Spanish Paten 200202113.

Gallego-Juarez, J.A., Riera, E., De La Fuente, S., Rodriguez-Corral, G., Acosta-Aparicio, V.M., and Blanco, A. 2007. Application of high-power ultrasound for dehydration of vegetables: Processes and devices. *Drying Technology*, 25: 1893–1901.

Garcia, M.L., Burgos, J., Sanz, B., and Ordonez, J.A. 1989. Effect of heat and ultrasonic waves on the survival of two strains of *Bacillus subtilis*. *Journal of Applied Bacteriology*, 67: 619–628.

Garcia-Graells, C., Hauben, E.J.A., and Michiels, C.W. 1998. High-pressure inactivation and sublethal injury of pressure-resistant *Escherichia coli* mutants in fruit juices. *Applied Environmental Microbiology*, 64: 1566–1568.

Garcia Perez, J.V., Rosello, C., Carcel, J.A., De La Fuente, S., and Mulet, A. 2007. Effect of air temperature on convective drying assisted by high power ultrasound. *Defect and Diffusion Forum*, 258–260: 563–574.

Gennaro, D.L., Cavella, S., Romano, R., and Masi, P. 1999. The use of ultrasound in food technology I: Inactivation of peroxidase by thermosonication. *Journal of Food Engineering*, 39: 401–407.

Ghaedian, R., Coupland, J.N., Decker, E.A., and Mcclements, D.J. 1998. Ultrasonic determination of fish composition. *Journal of Food Engineering*, 35: 323–337.

Ghaedian, R., Decker, E.A., and Mcclements, D.J. 1997. Use of ultrasound to determine cod fillet composition. *Journal of Food Science*, 62: 500–504.

Ghafoor, K., Choi, Y.H., Jeon, J.Y., and Jo, I.H. 2009. Optimization of ultrasound-assisted extraction of phenolic compounds, antioxidants, and anthocyanins from grape (*Vitis vinifera*) seeds. *Journal of Agricultural and Food Chemistry*, 57: 4988–4994.

Gimeno, A.C.S., Vercet, A., and Buesa, P.L. 2006. Studies of ovalbumin gelation in the presence of carrageenans and after manothermosonication treatments. *Innovative Food Science and Emerging Technologies*, 7: 270–274.

Gülseren, I. and Coupland, J.N. 2007. Ultrasonic velocity measurements in frozen model food solutions. *Journal of Food Engineering*, 79(3): 1071–1078.

Gunasekaran, S. and Ay, C. 1994. Evaluating milk coagulation with ultrasonics. *Food Technology*, 48: 74–78.

Haizhou, L., Pordesimo, L., and Weiss, J. 2004. High intensity ultrasound-assisted extraction of oil from soybeans. *Food Research International*, 37: 731–738.

Harvey, E. and Loomis, A. 1929. The destruction of luminous bacteria by high frequency sound waves. *Journal of Bacteriology*, 17: 314–318. (From Rosa and Barbosa-Canovas, 2003).

Hielscher. 2010. Ultrasound in the food industry. http://www.hielscher.com/ultrasonics/food_01.htm (May 15, 2010).

Hodate, Y., Ueno, Y., Yano, J., Katsuragi, T., Tezuka, Y., Tagawa, T. et al. 1997. Ultrasonic velocity measurement of crystallization rates of palm oil in oil–water emulsions. *Colloids and Surfaces A: Physicochemical and Engineering Aspects*, 128: 217–224.

Hoover, D.G. 1997. Minimally processed fruits and vegetables: Reducing microbial load by nonthermal physical treatments. *Food Technology*, 51: 66–71.

Hughes, D.E. and Nyborg, W.L. 1962. Cell disruption by ultrasound. *Science*, 138: 108–144.

Hull, J.B., Muumbo, A.M., and Whalley, R. 1995. Controlling waste in food processing using ultrasound level monitoring technology. *Strategies for Monitoring, Control and Management Waste*, 1995: 35–48.

Jacobs, S.E. and Thornley, M.J. 1954. The lethal action of ultrasonic waves on bacteria suspended in milk and other liquids. *Journal of Applied Bacteriology*, 17: 38–56 (From Pagan et al., 1999).

Japon-Lujan, R., Luque-Rodriguez, J.M., and Luque De Castro, M.D. 2006. Dynamic ultrasound-assisted extraction of oleuropein and related biophenols from olive leaves. *Journal of Chromatography A*, 1108: 76–82.

Jian-Bing, J., Xiang-Hong, L., Mei-Qiang, C., and Zhi-Chao, X. 2006. Improvement of leaching process of geniposide with ultrasound. *Ultrasonics Sonochemistry*, 13: 455–462.

Johnson, L.L., Peterson, R.V., and Pitt, W.G. 1998. Treatment of bacterial biofilms on polymeric biomaterials using antibiotics and ultrasound. *Journal of Biomaterials Science. Polymer Ed.* 9(11): 1177–1185.

Karnofsky, G. 1981. Ethanol and isopropanol as solvents for full-fat cottonseed extraction. *Oil Mill Gazette*, 85(10): 34–36.

Kim, S.M. and Zayas, J.F. 1989. Processing parameter of chymosin extraction by ultrasound. *International Journal of Food Science*, 54: 700.

Knorr, D. 2003. Impact of non-thermal processing on plant metabolites. *Journal of Food Engineering*, 56: 131–134.

Knorr, D., Zenker, M., Heinz, V., and Lee, D. 2004. Applications and potential of ultrasonics in food processing. *Trends in Food Science & Technology*, 15: 261–266.

Kuldiloke, J. 2002. Effect of ultrasound, temperature and pressure treatments on enzyme activity and quality indicators of fruit and vegetable juices. Doctoral dissertation, Technical University of Berlin, Berlin, Germany.

Kulmyrzaev, A., Cancelliere, C., and Mcclements, D.J. 2000. Characterization of aerated foods using ultrasonic reflectance spectroscopy. *Journal of Food Engineering*, 46: 235–241.

Lee, J.W. and Feng, H. 2004. Tomato peeling with ultrasound. Unpublished Work. University of Illinois.

Lee, B.H., Kermasha, S., and Baker, B.E. 1989. Thermal, ultrasonic and ultraviolet inactivation of salmonella in thin films of aqueous media and chocolate. *Food Microbiology*, 6: 143–152.

Lee, S., Pyrak-Nolte, L.J., Cornillon, P., and Campanella, O. 2004. Characterisation of frozen orange juice by ultrasound and wavelet analysis. *Journal of the Science of Food and Agriculture*, 84(5): 405–410.

Li, H., Pordesimo, L., and Weiss, J. 2004. High intensity ultrasound-assisted extraction of oil from soybeans. *Food Research International*, 37: 731–738.

Limaye, M.S. and Coakley, W.T. 1998. Clarification of small volume microbial suspensions in an ultrasonic standing wave. *Journal of Applied Microbiology*, 84(6): 1035–1042.

Liu, Z. 2002. Ultrasound enhanced corn pericarp separation process. MS thesis. Department of Food Science, University of Arkansas, Fayetteville, AR.

Lopez, P. and Burgos, J. 1995a. Lipoxygenase inactivation by manothermosonication: Effects of sonication physical parameters, pH, KCl, sugars, glycerol, and enzyme concentration. *Journal of Agricultural and Food Chemistry*, 43: 620–625.

Lopez, P. and Burgos, J. 1995b. Peroxidase stability and reactivation after heat treatment and manothermosonication. *Journal of Food Science*, 60: 451–455.

Lopez, P., Sala, F.J., De La Fuente, J.L., Condon, S., Raso, J., and Burgos, J. 1994. Inactivation of peroxidase, lipoxygenase and polypohenol oxidase by manothermosonication. *Journal of Agricultural and Food Chemistry*, 42: 252–256.

Lopez, P., Vercet, A., Sanchez, A.C., and Burgos, J. 1998. Inactivation of tomato pectin enzymes by manothermosonication. *Zeitschrift Fur Lebensmittel–Untersuchung Und–Forschung A*, 207: 249–252.

Luque De Castro, M.D. and Priego-Capote, F. 2007. Ultrasound assisted crystallization (sonocrystallization). *Ultrasonics Sonochemistry*, 14: 717–724.

Malo, A.L., Palou, E., Fernandez, M.J., Alzamora, S.M., and Guerrero, S. 2005. Multifactorial fungal inactivation combining thermosonication and antimicrobials. *Journal of Food Engineering*, 67: 87–93.

Manas, P. and Pagan, R. 2005. Microbial inactivation by new technologies of food preservation. *Journal of Applied Microbiology*, 98: 1387–1399.

Manas, P., Pagan, R., Raso, J., Sala, F.J., and Condon, S. 2000. Inactivation of *S. typhimurium*, *S. enteritidis* and *S. senftenberg* by ultrasonic waves under pressure. *Journal of Food Protection*, 63: 451–456.

Marshall, T., Tebbutt, J.S., and Challis, R.E. 2000. Monitoring the crystallization from solution of a reactive dye by ultrasound. *Measurement Science and Technology*, 11: 509–517.

Mason, T.J. 1990. Chemistry with ultrasound. *Critical Reports on Applied Chemistry*, 28: 1–25.

Mason, T.J., Paniwnyk, L., and Lorimer, J.P. 1996. The uses of ultrasound in food technology. *Ultrasonics Sonochemistry*, 3: 253–260.

Mason, T., Riera, E., Vercet, A., and Lopez-Buesa, P. 2005. Applications of ultrasound. In *Emerging Technologies for Food Processing*, Sun, Da-Wen (Ed.). Elsevier Academic Press, London, U.K., Chap. 13, pp. 323–351.

Masuzawa, N., Ohdaira, E., and Ide, M. 2000. Effects of ultrasonic irradiation on phenolic compounds in wine. *Japanese Journal of Applied Physics*, 39: 2978–2979.

McClements, D.J. 1995a. Advances in the application of ultrasound in food analysis and processing. *Trends in Food Science and Technology*, 6: 293–299.

McClements, D.J. 1995b. Ultrasonic characterization of foods. In *Characterization of Food: Emerging Methods*, Gaonkar, A.G. (Ed.). Elsevier Science B.V., New York, pp. 93–116.

Mermelstein, N.H. 1999. IFT annual meeting and food expo. Annual meeting papers address minimal processing. *Food Technology*, 53: 118–122.

Mermelstein, N.H. 2000. Annual meeting papers address nonthermal processing methods. *Food Technology*, 54: 184–192.

Miles, C.A. 1974. The ice content of frozen meat and its measurements using ultrasonic waves. In *Meat Freezing: Why and How*, Cutting, C.L. (Ed.). AFRC Meat Research Institute, Langford, Bristol, U.K., pp. 151–157.

Mizrach, A. 2000. Determination of avocado and mango fruit properties by ultrasonic technique. *Ultrasonics*, 38: 717–722.

Mizrach, A., Flitsanov, U., and Fuchs, Y. 1997. An ultrasonic non-destructive method for measuring maturity of mango fruit. *Transactions of the American Society of Agricultural Engineering*, 40: 1107–1111.

Mizrach, A., Galili, N., Gan-Mor, S., Flitsanov, U., and Prigozin, I. 1996. Models of ultrasonic parameters to assess avocado properties and shelflife. *Journal of Agricultural Engineering Research*, 65: 261–267.

Morris, C., Brody, A.L., and Wicker, L. 2007. Non-thermal food processing/preservation technologies: A review with packaging implications. *Packaging Technology and Science*, 20(4): 275–286.

Moulton, J. and Wang, C. 1982. A pilot plant study of continuous ultrasonic extraction of soybean protein. *Journal of Food Science*, 47: 1127–1129.

Mulet, A., Benedito, J., Bon, J., and Sanjuan, N. 1999. Low intensity ultrasonics in food technology. *Food Science and Technology International*, 5(4): 285–297.

Mulet, A., Benedito, J., Golás, Y., and Cárcel, J.A. 2002. Noninvasive ultrasonic measurements in the food industry. *Food Reviews International*, 18: 123–133.

Mulet, A., Carcel, J.A., Sanjuan, N., and Bon, J. 2003. New food drying technologies-use of ultrasound. *Food Science and Technology International*, 9: 215–221.

Mustakas, G.C. 1980. Recovery of oil from soybeans. In *Handbook of Soy Oil Processing and Utilization*, Erickson, D.R. (Ed.). American Soybean Association and American Oil Chemists' Society, St. Louis, MN, pp. 123–125.

Okkila, M., Mustranta, A., Buchert, J., and Poutanen, K. 2004. Combining power ultrasound with enzymes in berry juice processing. In *Second International Conference on Biocatalysis of Food and Drinks*, September 19–22, 2004, Stuttgart, Germany.

Ordonez, J.A., Aguilera, M.A., Garcia, M.L., and Sanz, B. 1987. Effect of combined ultrasonic and heat treatment (thermoultrasonication) on the survival of a strain of *Staphylococcus aureus*. *Journal of Dairy Science*, 54: 61–67.

Ordonez, J.A., Sanz, B., Hernandez, P.E., and Lopez-Lorenzo, P. 1984. A note on the effect of combined ultrasonic and heat treatments on the survival of thermoduric *Streptococci*. *Journal of Applied Bacteriology*, 56: 175–177.

Paci, C. 1953. L'emploi des ultra-sons pour l'assainissement du lait. *Le Lait*, 33: 610–615 (From Rosa And Barbosa-Canovas, 2003).

Pagan, R., Manas, P., Palop, A., and Sala, F.J. 1999b. Resistance of heat-shocked cells of listeria monocytogenes to manosonication and to manothermosonication. *Letters in Applied Microbiology*, 28: 71–75.

Pagan, R., Manas, P., Raso, J., and Condon, S. 1999a. Bacterial resistance to ultrasonic waves under pressure at no lethal (manosonication) and lethal (manothermosonication) temperatures. *Applied and Environmental Microbiology*, 65: 297–300.

Papadakis, E.P. 1976. *Ultrasonic Velocity and Attenuation: Measurement Methods with Scientific and Industrial Application*. Academic Press, New York.

Patist, A. and Bates, D. 2008. Ultrasonic innovations in the food industry: From the laboratory to commercial production. *Innovative Food Science and Emerging Technologies*, 9: 147–154.

Piyasena, P., Mohareb, E., and Mckellar, R.C. 2003. Inactivation of microbes using ultrasound: A review. *International Journal of Food Microbiology*, 87: 207–216.

Povey, M.J.W. 1998. Rapid determination of food material properties. In *Ultrasound in Food Processing*, Povey, M.J.W. and Mason, T.J. (Eds.). Thomson Science, New York, pp. 30–65.

Rajaei, A., Barzegar, M., and Yamini, Y. 2005. Supercritical fluid extraction of tea seed oil and its comparison with solvent extraction. *European Food Research and Technology*, 220: 401–405.

Raso, J. and Barbosa-Canovas, G.V. 2003. Nonthermal preservation of foods using combined processing techniques. *Critical Reviews in Food Science and Nutrition*, 43(3): 265–285.

Raso, J., Pagan, R., Condon, S., and Sala, F.J. 1998a. Influence of treatment and pressure on the lethality of ultrasound. *Applied and Environmental Microbiology*, 64(2): 465–471.

Raso, J., Palop, A., Pagan, R., and Condon, S. 1998b. Inactivation of *Bacillus subtilis* spores by combining ultrasonic waves under pressure and mild heat treatment. *Journal of Applied Microbiology*, 85: 849–854.

Raviyan, P., Zhang, Z., and Feng, H. 2005. Ultrasonication for food enzyme inactivation: Effect of cavitation intensity and temperature on inactivation. *Journal of Food Engineering*, 70(2): 189–196.

Ridgway, J., Henthorn, K.S., and Hull. J.B. 1998. Controlling overfilling in food processing. In *Ultrasound in Food Processing*, Povey, M.J.W. and Mason, T. (Eds.). Blackie Academic & Professional, London, U.K., pp. 1–6.

Riera, E., Golás, Y., Blanco, A., Gallego, A., Blasco, M., and Mulet, A. 2004. Mass transfer enhancement in supercritical fluids extraction by means of power ultrasound. *Ultrasonics Sonochemistry*, 11: 241–244.

Rostagno, A., Palma, M., and Barroso, C. 2003. Ultrasound-assisted extraction of soy isoflavones. *Journal of Chromatography A*, 1012: 119–128.

Sala, F.J., Burgos, J., Condon, S., Lopez, P., and Raso, J. 1995. Effect of heat and ultrasound on microorganisms and enzymes. In *New Methods of Food Preservation*, Gould, G.W. (Ed.). Blackie Academic & Professional Publisher, London, U.K., pp. 177–203.

Sanz, B., Palacios, P., Lopez, P., and Ordonez, J.A. 1985. Effect of ultrasonic waves on the heat resistance of *Bacillus stearothermophilus* spores. In *Fundamental and Applied Aspects of Bacterial Spores*, Dring, G.J., Ellar, D.J. and Gould, G.W. (Eds.). Academic Press, London, U.K., pp. 251–259.

Scherba, G., Weigel, R.M., and O'Brien, W.D. 1991. Quantitative assessment of the germicidal efficacy of ultrasonic energy. *Applied and Environmental Microbiology*, 57(7): 2079–2084.

Senorans, F.J., Ibanez, E., and Cifuentes, A. 2003. New trends in food processing. *Critical Reviews in Food Science and Nutrition*, 43(5): 507–526.

Serrato, A.G. 1981. Extraction of oil from soybeans. *Journal of the American Oil Chemist' Society*, 3: 157–159.

Seymour, I.J., Burfoot, D., Smith, R.L., Cox, L.A., and Lockwook, A. 2002. Ultrasound decontamination of minimally processed fruits and vegetables. *International Journal of Food Science & Technology*, 37: 547–557.

Shah, S., Sharma, A., and Gupta, N. 2005. Extraction of oil from *Jatropha curcas* (L.) seed kernels by combination of ultrasonication and aqueous enzymatic oil extraction. *Bioresource Technology*, 96: 121–123.

Sharma, A. and Gupta, N. 2006. Ultrasonic pre-irradiation effect upon aqueous enzymatic oil extraction from almond and apricot seeds. *Ultrasonic Sonochemistry*, 13: 529–534.

Sigfusson, H., Ziegler, G.R., and Coupland, J.N. 2004. Ultrasonic monitoring of food freezing. *Journal of Food Engineering*, 62(3): 263–269.

Sorhaug, T. and Stepaniak, L. 1997. Psychrotrophs and their enzymes in milk and dairy products: Quality aspects. *Trends Food Science and Technology*, 8: 35–41.

Soria, A.C. and Villamiel, M. 2010. Effect of ultrasound on the technological properties and bioactivity of food: A review. *Trends in Food Science & Technology*, 21(7): 323–331. doi:10.1016/J.Tifs.2010.04.003.

Steele, D.J. 1974. Ultrasonics to measure the moisture content of food products. *British Journal Nondestructive Testing*, 16: 169–173.

Suslick, K.S. 1998. Homogeneous sonochemistry. In *Ultrasound: Its Chemical, Physical and Biological Effects*, Suslick, K.S. (Ed.). VCH, New York, pp. 123–163.

Tat, H.G., Parakash, P., and Hutchins, D.A. 2006. Non-contact ultrasonic quality measurements of food products. *Journal of Food Engineering*, 77: 239–247.

Thakur, B.R. and Nelson, P.E. 1997. Inactivation of lipoxygenase in whole soy flour suspension by ultrasonic cavitation. *Nahrung*, 41(5): 299–301.

Thomson, D. and Sutherland, D.G. 1955. Ultrasonic insonation effect on liquid–solid extraction. *Industrial and Engineering Chemistry*, 47: 955–1165.

Thongson, C., Davidson, P.M., Mahakarnchanakul, W., and Weiss, J. 2004. Antimicrobial activity of ultrasound-assisted solvent-extracted spices. *Letters in Applied Microbiology*, 39: 401–406.

Tiwari, B.K., O'Donnell, C.P., and Cullen, P.J. 2009. Effect of non thermal processing technologies on the anthocyanin content of fruit juices. *Trends in Food Science & Technology*, 20: 137–145.

Tiwari, B.K., O'Donnell, C.P., Patras, A., and Cullen, P.J. 2008. Anthocyanin and ascorbic acid degradation in sonicated strawberry juice. *Journal of Agricultural and Food Chemistry*, 56(21): 10071–10077.

Ugarte, E., Feng, H., and Martin, E.S. 2007. Inactivation of Shigella and *Listeria monocytogenes* with power ultrasound at sub-lethal and lethal temperatures. *Journal of Food Science*, 72(4): 103–107.

Ugarte, E., Feng, H., Martin, E.S., and Cadwallader, K.R. 2006. Inactivation of *Escherichia coli* with power ultrasound in apple cider. *Journal of Food Science*, 71(2): 102–108.

Vercet, A., Burgos, J., Crelier, S., and Lopez, B.P. 2001. Inactivation of proteases and lipase by ultrasound. *Innovative Food Science and Emerging Technologies*, 2: 139–150.

Vercet, A., Burgos, J., and Lopez, B.P. 2001. Manothermosonication of foods and food-resembling systems: Effect on nutrient content and nonenzymatic browning. *Journal of Agriculture and Food Chemistry*, 49: 483–489.

Vercet, A., Burgos, J., and Lopez, B.P. 2002. Manothermosonication of heat resistant lipase and protease from pseudomonas fluorescens: Effect of pH and sonication parameters. *Journal of Dairy Research*, 69: 243–254.

Vercet, A., Lopez, P., and Burgos, J. 1997. Inactivation of heat-resistant lipase and protease from pseudomonas fluorescens by manothermosonication. *Journal of Dairy Science*, 80: 29–36.

Vercet, A., Lopez, P., and Burgos, J. 1999. Inactivation of heat-resistant pectinmethylesterase from orange by manothermosonication. *Journal of Agriculture and Food Chemistry*, 47: 432–437.

Vercet, A., Oria, R., Marquina, P., Crelie, S., and Lopez Buesa, P. 2002. Rheological properties of yoghurt made with milk submitted to manothermosonication. *Journal of Agricultural and Food Chemistry*, 50: 6165–6171.

Vercet, A., Sanchez-Gimeno, C., Burgos, J., and Lopez Buesa, P. 2002. The effects of manothermosonication on tomato pectic enzymes and tomato paste rheological properties. *Journal of Food Engineering*, 53: 273–278.

Vilkhu, K., Mawson, R., Simons, L., and Bates, D. 2008. Applications and opportunities for ultrasound assisted extraction in the food industry—A review. *Innovative Food Science and Emerging Technologies*, 9: 161–169.

Villamiel, M. and De Jong, P. 2000a. Inactivation of *Pseudomonas fluorescens* and *Streptococcus thermophilus* in Trypticase® Soy Broth and total bacteria in milk by continuous-flow ultrasonic treatment and conventional heating. *Journal of Food Engineering*, 45: 171–179.

Villamiel, M. and De Jong, P. 2000b. Influence of high-intensity ultrasound and heat treatment in continuous flow on fat, proteins, and native enzymes of milk. *Journal of Agricultural and Food Chemistry*, 48: 472–478.

Villamiel, M., Van Hamersveld, E.H., and De Jong, P. 1999. Review: Effects of ultrasound processing on the quality of dairy products. *Milchwissenschaft*, 54: 69–73.

Villamiel, M., Verdurmen, R., and De Jong, P. 2000. Degassing of milk by high-intensity ultrasound. *Milchwissenschaft*, 55: 123–125.

Vinatoru, M. 2001. An overview of the ultrasonically assisted extraction of bioactive principles from herbs. *Ultrasonic Sonochemistry*, 8: 303–313.

Vollmer, A.C., Everbach, E.C., Halpern, M., and Kwakye, S. 1998. Bacterial stress responses to 1-megahertz pulsed ultrasound in the presence of microbubbles. *Applied Environmental Microbiology*, 64(10): 3927–3931.

Williams, A.R., Stafford, D.A., Callely, A.G., and Hughes, D.E. 1970. Ultrasonic dispersal of activated sludge flocks. *Journal of Applied Bacteriology*, 33: 656–663.

Wu, H., Hulbert, G.J., and Mount, J.R. 2000. Effects of ultrasound on milk homogenization and fermentation with yogurt starter. *Innovative Food Science & Emerging Technologies*, 1: 211–218.

Xia, T., Shi, S., and Wan, X. 2006. Impact of ultrasonic-assisted extraction on the chemical and sensory quality of tea infusion. *Journal of Food Engineering*, 74: 557–560.

Yang, W., Siebenmorgen, T.J., and Liu, Z. 2002. Rapid debranning of corn with power ultrasound. In *Proceedings of the 2002 AACC Annual Conference*, Montreal, Quebec, Canada, October 13–17.

Yang, W., Wambura, P., and Williams, L. 2005. Extending the capability of power ultrasound to cereal and oilseed processing for food and non-food applications. In *Proceedings IFT Annual Meeting*, New Orleans, LA, July 16–20, 2005.

Zenker, M., Heinz, V., and Knorr, D. 2001. Combined application of ultrasound and temperature for energy-saving and mild preservation of liquid food. In *Conference Proceedings of the Third European Congress of Chemical Engineering*, Nuremberg, Germany, June 26–28, 2001.

Zenker, M., Heinz, V., and Knorr, D. 2003. Application of ultrasound-assisted thermal processing for preservation and quality retention of liquid foods. *Journal of Food Protection*, 66: 1642–1649.

Zhang, Z., Niu, Y., Eckhoff, S.R., and Feng, H. 2005. Sonication enhanced starch separation in a milling process and its effect on the resulting starch. *Starch*, 57: 240–245.

Zhang, R., Xu, Y., and Shi, Y. 2003. The extracting technology of flavonoids compounds. *Food and Machinery*, 1: 21–22.

Zhao, B., Basir, O.A., and Mitt, G.S. 2003. Detection of metal, glass and plastic pieces in bottled beverages using ultrasound. *Food Research International*, 36: 513–521.

Zhao, L., Zhao, G., Chen, F., Wang, Z., Wu, J., and Hu, X. 2006. Different effects of microwave and ultrasound on the stability of (all-E)-astaxanthin. *Journal of Agricultural and Food Chemistry*, 54: 8346–8351.

Zheng, L. and Sun, D.W. 2006. Innovative applications of power ultrasound during food freezing processes: A review. *Trends in Food Science and Technology*, 17: 16–23.

9 Use of Ultrasound in Coordination and Organometallic Chemistry

Boris Ildusovich Kharisov, Oxana Vasilievna Kharissova, and Ubaldo Ortiz-Méndez

CONTENTS

Introduction 183
Complexes with Direct C–M Bonds (σ- and π-Organometallic Compounds) 184
    N-Containing Metal Complexes 189
    N-Macrocyclic Complexes 189
    O-Containing Metal Complexes 191
    S-Containing Metal Complexes 195
    Te-Containing Metal Complexes 196
    Halogen-Containing Metal Complexes 197
    N,O-Containing Metal Complexes 197
        Metalated Peptide Complexes 201
    S,O-Containing Metal Complexes 201
    N-,O-,S-Containing Metal Complexes 202
Coordination Polymers 202
Catalytic Applications of Ultrasonically Obtained Complexes and Composites 203
Other Applications 204
Concluding Remarks 205
Acknowledgments 205
References 205

INTRODUCTION

During the last few decades, various techniques, such as microwaves, laser, UV-, γ- and x-rays, electron and ion beams, cryosynthesis, flame (combustion) or electrochemical synthesis, arc discharge, and so on, have been successfully applied, to synthesize almost all types of metal complexes. In many cases, the conditions for synthesis have been milder and yield higher when compared with the conventional preparation methods. Ultrasound treatment of the reaction system belongs to the above-mentioned series of supportive techniques, which helps to obtain coordination and organometallic compounds by easy operation, much faster, and with considerably higher yields, sometimes changing the reaction route. Additional to classic metal complexes, σ- and π-organometallics can be obtained via ultrasonic irradiation of precursors, particularly in those systems using elemental metals.

Starting from the beginning of this century, a series of monographs (Mason and Lorimer, 2002; Capelo-Martínez, 2009), book chapters or sections in them (Suslick, 2001; Carruthers and Coldham, 2004; Lalena et al., 2008; Suslick and Skrabalak, 2008; Basset et al., 2009), and reviews (Suslick, 1995; Cains et al., 1998; Cravotto and Cintas, 2006; Suslick and Flannigan, 2008), dedicated to various aspects of sonochemistry for organic and inorganic synthesis, have been published. The ultrasonic horns and cells, used for laboratory purposes, are easily accessible and are described by SONITEK (2009) and Garnovskii and Kharisov (2003). This chapter highlights the recent advances in the application of ultrasound for synthesis, destruction, and transformation of a wide variety of metal complexes.

COMPLEXES WITH DIRECT C–M BONDS (σ- AND π-ORGANOMETALLIC COMPOUNDS)

Ultrasonically obtained or modified metallo-organic compounds are represented by a series of complexes containing a wide variety of organic ligands, in particular those containing single or multiple metal-carbon bonds. Zero-valent metals, their salts or complexes have been used as precursors in these reactions or have been the final reaction products (metals were frequently obtained in a variety of nanoforms) (Kharissova and Kharisov, 2008). The synthesis of nanostructured metals (Fe, Co, Ni) and metal alloys (Fe–Co, Pt–Pd, M50 steel) from the corresponding organometallic precursors by ultrasonic irradiation, among other methods (thermal decomposition, chemical vapor deposition (CVD), laser pyrolysis, and reduction), has been reviewed by Gonsalves et al. (2000). On the contrary, use of ultrasonic irradiation for forming organometallics and organic compounds, using zero-valent metals, was generalized in a series of publications (Ameta et al., 2001; Bian et al., 2002; Cintas, 2004) to specifically obtain the famous *Grignard reagents*, one of the most classic and well-studied σ-organometallic compounds, which continue to be attractive precursors in organic and organometallic synthesis in the twenty-first century also. During the last decade, a series of reports have been published on the use of ultrasonic irradiation, to obtain the Grignard reagents starting from elemental magnesium and their further interaction with organic compounds.

Thus, 1,4-bis(dimethylsilyl) benzene was formed at 5°C with 62% yield by the reaction of chlorodimethylsilane and di-Grignard reagent, which was prepared from 1,4-dibromobenzene and Mg (1:3) at 45°C, under ultrasonic irradiation, in tetrahydrofuran (THF) for 25 min (Shi and Chen, 2007a,b; Chen et al., 2007). The reaction of dibutyltin oxide with excessive C_{1-8} alkyl Grignard reagent under heating and assistance of the ultrasonic wave, removing the residual Grignard reagent with excess acid, sequentially extracting with organic nonpolar solvent and mixture of C_{1-6} alcohols and water led to obtaining dibutyldialkyltin with a low boiling point (Shen et al., 2008). Several target compounds [α,α-diphenyl-N-[(un)substituted benzyl]-2-pyrrolidinemethanol derivatives] were synthesized by a reaction of an intermediate L-proline-derived ester with a Grignard reagent under ultrasound radiation and a water- and oxygen-free nitrogen atmosphere (Li and Xue, 2008). A preliminary, agriculturally biological activity test indicated that some of them had good insecticidal activities. Synthesis methods of some organic compounds such as dimethylbenzyl carbinol, 4-methyl-3-decene-5-ol, 1-octene-3-ol, and 8-allyl-8-hydroxytricyclo-[5,2,1,02,6]-decane using the Grignard reaction were summarized using the supersonic wave as a revulsant (Yuan and Zhen, 2006). This method had characteristics such as short reaction time, few energetic resources, simple operation conditions, high yield, and so on, compared to the conventional methods. A facile reaction of aryl Grignard reagents with 4-alkylacetophenones under ultrasonic irradiation, within two minutes, with a conversion of 15%–95%, gave 1,1-diarylethene and tertiary alcohols with selectivity over 60%–96% (Chen et al., 2003).

In addition to the ultrasonic treatment of elemental magnesium, its *alloys* were also used, for instance the magnesium calcium alloy $MgCa_{30}$, whose reaction with chlorobenzene in THF, in the presence of ultrasound, yielded phenylmagnesium chloride (Kunz et al., 1999). Other elemental metals were also used as precursors for various organometallics, for example, zinc powder, which was applied for obtaining a series of homopropargyl alcohols from the reaction mixture of zinc powder, 1,2-diiodoethane, 3-bromo-1-propyne, and aldehyde or ketone in anhydrous THF, under ultrasound treatment (Lee et al., 2004).

A series of divinylzinc complexes, one of which represents the only structurally characterized zinc(II) π-complex, including vinylzinc reagents, $Zn[C(Me)=CH_2]_2$ (**9.1**) and $Zn[C(H)=CMe_2]_2$ (**9.2**), were ultrasonically synthesized (Scheme 9.1) and isolated as crystalline solids in 66% and 72%

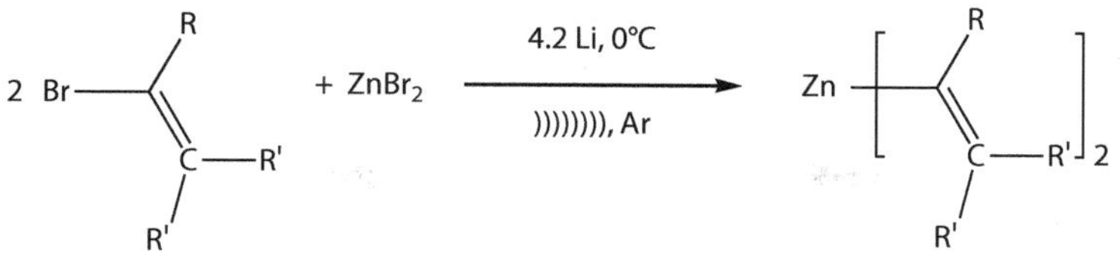

$$2\ Br-\underset{R'}{\underset{|}{C}}\hspace{-2pt}=\hspace{-2pt}\underset{R'}{\overset{R}{C}} + ZnBr_2 \xrightarrow[\text{))))))))), Ar}]{4.2\ Li,\ 0°C} Zn\left[\underset{R'}{\underset{|}{C}}\hspace{-2pt}=\hspace{-2pt}\underset{R'}{\overset{R}{C}}\right]_2$$

9.1, R = Me, R' = H, 66% yield
9.2, R = H, R' = Me, 72% yield

SCHEME 9.1

yields, respectively (Wooten, 2006). They possessed an infinite polymeric architecture in the solid state via a series of zinc-π (in **9.1**) and zinc-σ-bonded (in **9.2**) bridging interactions. The addition of chelating ligands to these divinylzinc compounds allowed isolation of the monomeric adducts (bipy)Zn[C(Me)=CH$_2$]$_2$ (**9.1** · bipy), (tmeda)Zn[C(Me)=CH$_2$]$_2$ (**9.1** · tmeda), (bipy)Zn[C(H)=CMe$_2$]$_2$ (**9.2** · bipy), and (tmeda)Zn[C(H)=CMe$_2$]$_2$ (**9.3** · tmeda). Furthermore, ultrasound was applied to prepare a (Wang et al., 2007) metal carborane-targeting drug delivery system by dissolving monometallic/bimetallic center (Fe, Ru, Co, and/or Rh) and carbon metal carborane carborane in an organic solvent and dispersing surface functional monodisperse Fe$_3$O$_4$ magnetic nanoparticles in an aqueous solution under ultrasonic treatment, to form a magnetic nanoparticle saturated suspending liquid, with further mixing of solutions, adjusting of pH, stirring, and ultrasonic emulsifying for 10–15 min, to obtain the final magnetic nanomaterial. The targeting drug delivery system could move to the tumor site under an external magnetic field, effectively realizing Boron neutron capture therapy.

The ultrasound-mediated reaction of the Fischer *carbene* complex [4-MeC$_6$H$_4$C(OMe):Cr(CO)$_5$] with PhC≡CPh in Bu$_2$O for 20 min followed by Ce(IV) oxidation, afforded 2,3-diphenyl-6-methyl-1,4-naphthoquinone in 69% yield (Harrity et al., 1993). A related compound, 2,3-diphenyl-1,4-naphthoquinone, was similarly prepared (65% yield) from [PhC(MeO):Cr(CO)$_5$], PhC≡CPh, and SiO$_2$ in hexane or Et$_2$O by adsorbing the reagents onto silica and heating. The reaction of chromium alkyl(alkoxy)carbenes with propargylic alcohols was investigated under thermal and ultrasound conditions (Caldwell et al., 1999). Both sets of conditions provided rapid access to alkyl-substituted β-lactone products.

Metal *carbonyls* as a hearth of organometallic chemistry are widely used in organic/organometallic reactions, especially under the ultrasonic treatment of the reaction system, sometimes yielding unexpected products or composites. Thus, a unique zigzag ···—Bi—Fe—··· chain was found to form the basis of the structure of [BuBiFe(CO)$_4$]$_∞$, which was the result of the cleavage of Bi—Fe bonds, upon the ultrasonication of the cyclic, dimeric product [{BuBiFe(CO)$_4$}$_2$], formed from the reaction of [Et$_4$N]$_3$[Bi{Fe(CO)$_4$}$_4$] with BuBr followed by the acidification with HOAc (Shieh et al., 2002). Moreover, metal carbonyls or products of their transformations, under ultrasonic irradiation, are frequently used as efficient catalysts, among other applications. Thus, Mo and Co oxide were precipitated under ultrasonication treatment from Mo(CO)$_6$ and Co(CO)$_3$NO dissolved in decalin (Landau et al., 2001). The final high-loading Co-Mo/Al-MCM-41 catalyst was 1.7 times more active in the hydrodesulfurization of dibenzothiophene. Large-diameter fullerenes with thin walls were prepared in a solid-state reactor, at 600°C–750°C, in the presence of an iron catalyst derived from iron carbonyls, and by applying ultrasound in one of the final steps of the process (Sheng and Wang, 2006). Carbon nanotube composites with nanometals or metal oxides uniformly distributed on the surface of carbon nanotubes were obtained by dissolving organometallic compounds with organic solvents, treating carbon nanotubes with nitric acid and ultrasonic wave, stirring, and standing to obtain carbon nanotubes with carboxyl, and/or carbonyl, and/or hydroxyl on their surface (Yu et al., 2005). The products could be used in sensors, nanoelectronic devices, superhigh magnetic recording multimedia, lithium batteries, and solar batteries based on the nanomaterial and the catalysts.

Phenylethyne cobalt hexacarbonyl complex (**9.3**) reacted with 2,5-dihydrofuran in hexane to give 37% oxabicyclooctenone (**9.4**) (Billington et al., 1988) (Scheme 9.2). Under similar conditions,

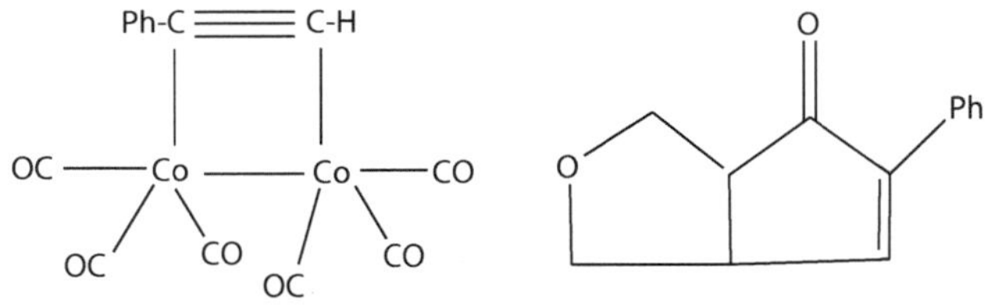

```
        Ph-C≡≡≡C-H
           |    |
   OC ── Co ──── Co ── CO
       / \    / \
      OC  CO OC  CO
          OC
         9.3
```

```
       O
       ‖
   O       Ph
    \    /
     [furanone structure]
         9.4
```

SCHEME 9.2

but in the presence of Bu_3PO, 2,5-dihydrofuran and (**9.3**) gave 69% of (**9.4**). It was proved that ultrasonic irradiation allowed this *Khand* reaction to be conducted rapidly at low temperatures; this effect of the combination of ultrasound and phosphine and phosphine oxides on the reaction was studied in detail. Other interesting results of ultrasound treatment of the systems containing carbonyls and phosphines are known, in particular the production of simple binary inorganic compounds, for instance, the interaction of $Fe(CO)_5$ and triethylphosphine, which was found to produce solid amorphous iron phosphide FeP (Sweet and Casadonte, 2001). Among processes where metal carbonyls have been used in combination with ultrasonic treatment, we also noted the preparation of hollow carbon nanocages (Wang and Sheng, 2006; Wang and Teng, 2009), manufacture of metallic nanoparticle/C nanofiber structures (Motoyama et al., 2006), and hardening of metallic surfaces (Dumbolov et al., 2006). Applications of metallic amorphous particles produced from metallic carbonyl, as well as uses of metallic nanomaterials obtained by ultrasonic reduction, were reviewed by Mizukoshi et al. (2001).

Other reported σ-bonded, metal carbon, organometallics are represented by several examples; some of them are on a silicium-organic basis. Thus, the ultrasound-mediated reaction of InBr with the THF adduct of $LiC(SiMe_3)_3$ afforded the deep violet tetraindium compound, $In_4[C(SiMe_3)_3]_4$ (a tetrahedral cluster of four monovalent In atoms) (Uhl et al., 2002). A yellow byproduct, containing a chain of three In atoms connected by In–In single bonds, was isolated and identified as $[In_3Br_3\{C(SiMe_3)_3\}_3]^-[Li(THF)_3]^+$ (**9.5**). $[Ba((\mu-CH_2SiMe_3)_2ZnCH_2SiMe_3)_2]$ underwent transmetalation with Ba in heptane under the application of ultrasound to give $Ba_4[Me_3SiCHZn(CH(SiMe_3)Zn(CH_2SiMe_3)_2)_2]_2$ (Westerhausen et al., 2001) (Scheme 9.3). The central $Ba_4Zn_2C_6$ core of the product was found to be regarded as a distorted double cube with a common Ba_2C_2 face. Ultrasonic treatment of $(Me_2PhSi)_2C=CH_2$ with Li in THF yields the Li complex $[\{(Me_2PhSi)_2C(CH_2)\}Li(THF)_n]_2$, which reacts in situ with one equivalent of KOBut in Et_2O to give the K salt $[\{(Me_2PhSi)_2C=CH_2\}K(THF)]_2$ (Izod et al., 2009). Similarly, ultrasonic treatment of $(Me_3Si)\{Ph_2P(BH_3)\}C=CH_2$ with Li in THF yields the Li complex $[[\{Ph_2P(BH_3)\}(Me_3Si)C(CH_2)]Li(THF)_3]_2 \cdot 2THF$.

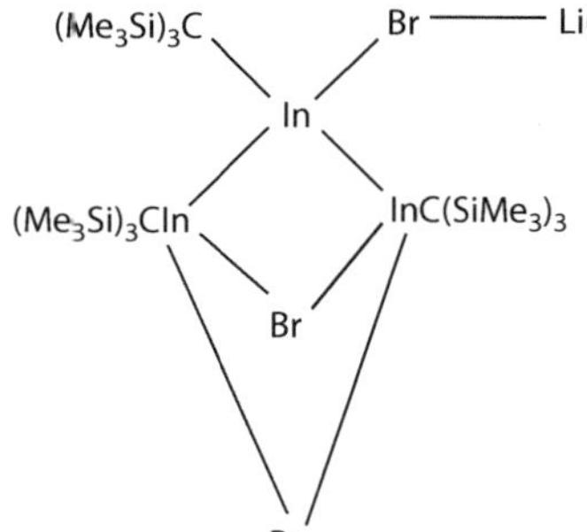

```
   (Me3Si)3C           Br ── Li(THF)3
            \        /
             In
            / \
   (Me3Si)3ClIn    InC(SiMe3)3
            \    /
             Br
             |
             Br
            9.5
```

SCHEME 9.3

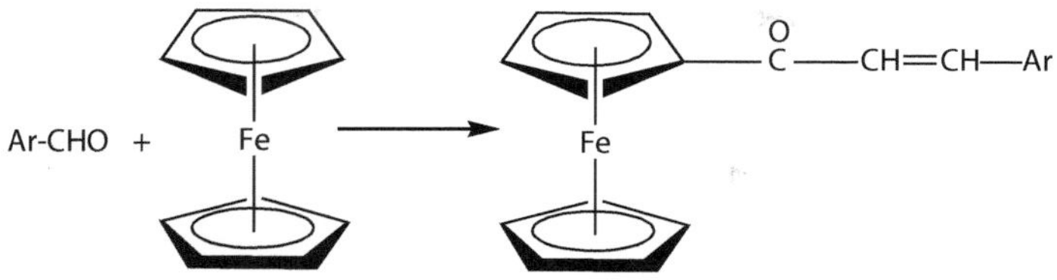

Ar–CHO + [ferrocene] → [ferrocene]–C(=O)–CH=CH–Ar

9.6

SCHEME 9.4

In case of carbonyls also, the ultrasound was extensively applied for the purpose of synthesis in the organometallic chemistry of *cyclopentadienyls*, particularly in their classic representative *ferrocene* and its derivatives. Thus, a facile synthesis of ferrocenylenone, accelerated by ultrasonic irradiation, as shown by Scheme 9.4, was reported by (Ji et al., 2003). It was found that ultrasonic irradiation was very simple and convenient for the synthesis of ferrocenylenones (**9.6**) at room temperature (r.t.), by using an ultrasonic cleaner with a frequency of 40 kHz and a nominal power of 100 W, with a higher yield, milder reaction conditions, and easier workup as compared to the traditional procedure. Moreover, the authors compared the microwave and ultrasonic routes, where yields were in a favor of ultrasound: 53%–63% and 83%–92% for different compounds, respectively. Further Michael addition of ferrocenylenones with aliphatic amines under ultrasound irradiation, in the absence of a solvent or catalyst at r.t., afforded 1-ferrocenyl-3-amino carbonyl compounds rapidly and in high yields, which is also efficient in the aza-Michael reaction of other α,β-unsaturated carbonyl compounds, such as, chalcone, carboxylic ester, and so on (Yang et al., 2005). As an example, 1-ferrocenyl-3-(4-chlorophenyl)prop-2-en-1-one reacted with piperidine when using ultrasound irradiation for 0.5 h, giving 1-ferrocenyl-3-(4-chlorophenyl)-3-(piperidin-1-yl)propan-1-one with 98% yield.

Ferrocene has been used as a precursor for a series of nanomaterials and composites on a carbon basis (mainly carbon nanotubes), under ultrasonic treatment. Thus, carbon-coated ferroferric oxide nanomaterial comprising of nano ferroferric oxide as the core and carbon as the coating layer, with the carbon-coating layer containing amino and hydroxy groups, was obtained by a series of steps from trichlorophenol and ferrocene as precursors and the application of ultrasound (Weng et al., 2009). BN/carbon nanotube composite particles in submicrometer-sizes were prepared with the help of the spray pyrolysis method, using an ultrasonic nebulizer, BN nanoparticles, ferrocene (as a catalyst), and ethanol (as a solvent and carbon source). These were sprayed into a tubular furnace (fixed temperature of 800°C) under an Ar flow of 1 L min^{-1} (Nandiyanto et al., 2009). This preparation method could be broadly applied in the fabrication of various composite materials, with the growth of CNT on the particle surface, using a simple, fast, and continuous process. Also, unsupported carbon nanotubes were obtained from ferrocene by an environment-friendly process, dissolving ferrocene in xylene at a ratio of 0.002–0.04 g mL^{-1} under ultrasonic treatment, feeding it into a quartz weighing bottle, and allowing a hydrothermal reaction at 600°C–800°C for 12–18 min to give the final product (Wu and Yuan, 2008). The same precursors were found to produce high-purity, SWCNTs by ultrasonication of a xylene-ferrocene mixture, but this time at r.t. and atmospheric pressure (Figure 9.1) (Srinivasan, 2005). High-purity vertically aligned films of multiwall carbon nanotubes were synthesized via an aerosol-assisted CVD method, using a solution of ferrocene in *m*-xylene under ultrasonication (Barreiro et al., 2006). Carbon fibers were synthesized by heating acetylene for 2 h at 750°C in the presence of Ni powder-coated graphite, and exposing the reaction system to ultrasound waves at 25 kHz and 260 dB, to give coil-shaped carbon fibers with a diameter of 1–1.5 μm and a coil layer thickness of 4–4.5 mm in 70% (on acetylene) yield (Motojima and Hishikawa, 2002).

Ferrocene and its derivatives have been used not only for the fabrication of carbon and/or iron nanomaterials, but they have also been obtained as nanomaterials (Wu et al., 2007) or nanoforms, for instance ferrocene nanocrystals (~40 nm), are successfully prepared by the ultrasonic-solvent-substitution method (Chen et al., 2007) or by dumbbell-like nanosuperstructures of

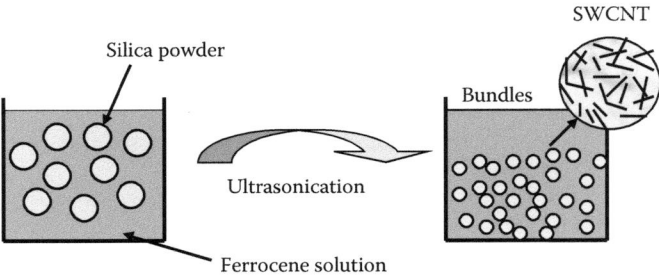

FIGURE 9.1 Schematic of the sonochemical route to SWCNTs. Silica powder is immersed in the solution (ferrocene–xylene mixture). Ultrasonication is carried out for 20 min at r.t. and atmospheric pressure. Sonication produces high-purity SWCNTs on the surface of the silica powder. (From Srinivasan, C., *Curr. Sci.*, 88(1), 12, 2005. With permission.)

(3-carboxy-1-acyl-propyl)-ferrocene being synthesized by the ultrasonic-pH controlling-reprecipitation method (Zhu et al., 2008). The latter products were composed of nanorods; their length was 6–12 μm, the diameters of the two polar coronas and the middle part (waist) were, respectively, 2–8 and 0.5–2 μm. The diameter of the nanorods that was the basic unit of the superstructures was 0.3 μm.

Among other reports on Cp derivatives, we noted the reaction of 30% aqueous KOH with thallium(I) sulfate in H_2O followed by treatment with freshly cracked cyclopentadiene under sonication, which gave an almost quantitative yield of cyclopentadienylthallium (Federman et al., 1997). Additionally, an efficient Cu- and ligand-free, catalyzed Sonogashira cross-coupling reaction of aryl iodides with ferrocenylacetylene by $PdCl_2$ under ultrasound irradiation at r.t., under balloon pressure of Ar, was developed by Fu et al. (2008).

Electrochemical studies carried out under ultrasonic irradiation include the electrochemical reduction of noble metal electrodes in the presence of a redox ionic liquid, [FcEMIM][TFSI], investigated by cyclic voltammetry (Ghilane et al., 2008). The reduced electrode exhibited the presence of ferrocene even after contact with air, after the ultrasound, and after physical polishing, highlighting the large stability of these organometallic phases formed in this media. A method for increasing the electrical capacity of the positive electrode of an organic electrochemical supercapacitor (comprising of ultrasonically dispersing cyclopentadienyl transition metal complexes, such as, ferrocene, nickelocene, cobaltocene, etc.) as an active additive in a weight amount of 1%–10% of the positive electrode material at a power of 100–600 W, for 1–3 h, during manufacture of a positive electrode (with vacuum drying and immersing the positive electrode in organic electrolyte solution for 6–12 h) was patented by Li and Gao (2009). The cyclopentadienyl transition metal complex provided redox pseudocapacitance, which reduced ohmic internal resistance and improved energy density by 5%–50%.

A few reports on the use of ultrasound in reactions of *arenes* and *dienes* are known. Thus, the nucleophilic aromatic substitution under ultrasound irradiation of a 1,4-$Cl_2C_6H_4$Fe η^6–complex with various secondary amines was reported (Raouafi et al., 2009). As a result, the reaction time at moderate temperatures was considerably shortened to 15 min compared to non-sonicated reaction conditions at r.t. (several days) or at solvent refluxing temperature (12–48 h). The reaction of [(arene)$RuCl_2$]$_2$ complexes with simple amines, using ultrasound, gave monomeric addition compounds (arene)$RuCl_2NR_3$ (Bates et al., 1990). Me_3SiN_3 was a particularly suitable source of the azide ligand, generating the structurally characterized dimer [(arene)Ru(μ–N_3)Cl]$_2$, undergoing dimer cleavage and halide loss reactions. The homoleptic CuI-arene complex [Cu(1,2-$F_2C_6H_4$)$_2$]$^+$[Al{OC(CF_3)$_3$}$_4$]$^-$, the CuI-methylene chloride complex [Cu(CH_2Cl_2)Al{OCMe(CF_3)$_2$}$_4$], and the donor-free dimer [CuAl{OCH(CF_3)$_2$}$_4$]$_2$ were synthesized in quantitative yields by sonicating Li[Al(ORF)$_4$] (RF=C(CF_3)$_3$, CMe(CF_3)$_2$, or CH(CF_3)$_2$), AgF, and a threefold excess of CuI in 1,2-$F_2C_6H_4$ or CH_2Cl_2. The synthesized substances were good starting materials for further CuI chemistry (Santiso-Quinones et al., 2009). The emulsion ethylene polymerization method with an adjustable molecular weight of polyethylene using bis(cycloocta-1,5-diene)nickel and Et 4,4,4-trifluoro-2-(tri-Ph phosphine ylide) acetoacetate as precursors was developed by Wei and Zhang (2009). Emulsion polyethylene, obtained by a series of steps including ultrasonic treatment, had a molecular weight of 1,165–11,600. Aldol condensation of 4-ferrocenyl-3-buten-2-one with aromatic aldehydes afforded 1-aryl-5-ferrocenyl-penta-1,4-dien-3-one in 83%–97% yields under ultrasound irradiation or in a solvent-free condition at r.t. (Zhou et al., 2004). This method had the advantages of simplicity, mild reaction conditions, good yields, and low costs.

N-Containing Metal Complexes

Ultrasonically obtained metal complexes with "pure" nitrogen-containing non-macrocyclic ligands are represented by a few examples of *amines* and *azines*. Thus, the effect of ultrasound treatment on the nature of the interaction and phase composition of compounds based on CdI_2 and PbI_2, with pyridine (PY), piperidine (PP) and $PhNH_2$ (AN) yielding the intercalation compounds $PbI_2(PP)_{8.3\pm0.2}$, $CdI_2(py)_{4.3\pm0.2}$, and $CdI_2(py)_{6.0\pm0.2}$, as well as, $PbI_2(AN)$, was studied (Konopleva et al., 2000). Nanoparticles of a new Bi(III) supramolecular compound, $[Bi_2(\mu-4,4'-bipy)Cl_{10}] \cdot 2(4,4'-Hbipy) \cdot (4,4'-H_2bipy) \cdot 2H_2O$, were synthesized by a sonochemical method (Soltanzadeh and Morsali, 2009). Calcination of its nanoparticles at 400°C under air yielded nanosized particles of α-Bi_2O_3.

N-Macrocyclic Complexes

The application of ultrasound for the preparation of *porphyrin* and closely related *phthalocyanine* metal complexes has been described in a series of reports, as having synthesis as its goal, destruction of these π-conjugated aromatic macrocycles, as well as preparation of materials possessing catalytic properties or suitable for medical purposes. Thus, the electrosynthesis of a nickel metalloporphyrin by using 5,10,15,20-tetraquis(*p*–hydroxyphenyl)porphyrin (TPPOH) as binding agent was carried out in a nondivided cell, with a nickel sacrificial anode, and at controlled potential, in order to favor the chemical formation of a metal complex by the reaction between the porphyrin dianion radical and Ni(II) ions, both of which were electro-generated during the process (Aguilera et al., 2009). The effect of the use of an ultrasonic wave of 20kHz in the metalloporphyrin electrosynthesis process was studied, and it showed that the application of the ultrasonic wave at the amplitude of the study favored the yield of the reaction and did not lead to changes in the electrochemical mechanism. In case of a nonmetallic porphyrin, the ultrasonic method (a combination of the reprecipitation and sonication method) was applied for the preparation of porphyrin nanoparticles with narrow-sized distribution and good dispersibility (Motlagh et al., 2008). These nanoparticles were stable in the solution, without precipitation, for at least 30 days. On the contrary, in certain conditions, ultrasonic waves could *destroy* the macromolecule structure. For example, nickel and vanadium porphyrins were subjected to oxidation with aqueous $H_2O_2/CHCl_3$ in order to release water-soluble metals and to decompose the porphyrin structure. It was shown that cavitation by ultrasound was quite effective for porphyrin degradation and metal release, and it occurred through consecutive reactions with an oxidation reaction intermediate (Tu and Yen, 1999).

The oxygen-reducing *catalyst*, on the basis of the cobalt porphyrin complex, obtained ultrasonically from the porphyrin monomer (tetra–Me,Ph-porphyrin or tetramethoxy-Ph-porphyrin) and cobalt acetate as precursors, can be used in proton-exchange membrane fuel cells (Xie et al., 2006). Ruthenium(II) porphyrin-catalyzed amidation of aromatic heterocycles (**9.8**) with iminoiodanes under mild conditions (CH_2Cl_2, molecular sieves, ultrasound, 40°C) was achieved (Scheme 9.5) in moderate to good yields (up to 84%) and conversions to form (**9.9**) (up to 99%) (He et al., 2004). As an example, ultrasound treatment of furan (**9.10**) (1 equivalent) with 10mol% [Ru^{II}(TTP)(CO)] (**9.7**) and PhI=NTs (1.5 equivalents), in CH_2Cl_2 containing 4 Å molecular sieves, at 40°C, gave N,N-ditosylamido-2-furan (**9.11**) in 73% isolated yield. Another catalytic use was reported for alkenes, which transformed them to their corresponding epoxides in high yields and with high selectivity by sodium periodate, under ultrasonic irradiation, in the presence of catalytic amounts of manganese porphyrin supported on ion-exchange resin (Mirkhani et al., 2000).

Porphyrins under ultrasonic irradiation have been used for *medical applications*. Thus, the interaction between hematoporphyrin zinc (HP-Zn) and bovine serum albumin (BSA) and damage of

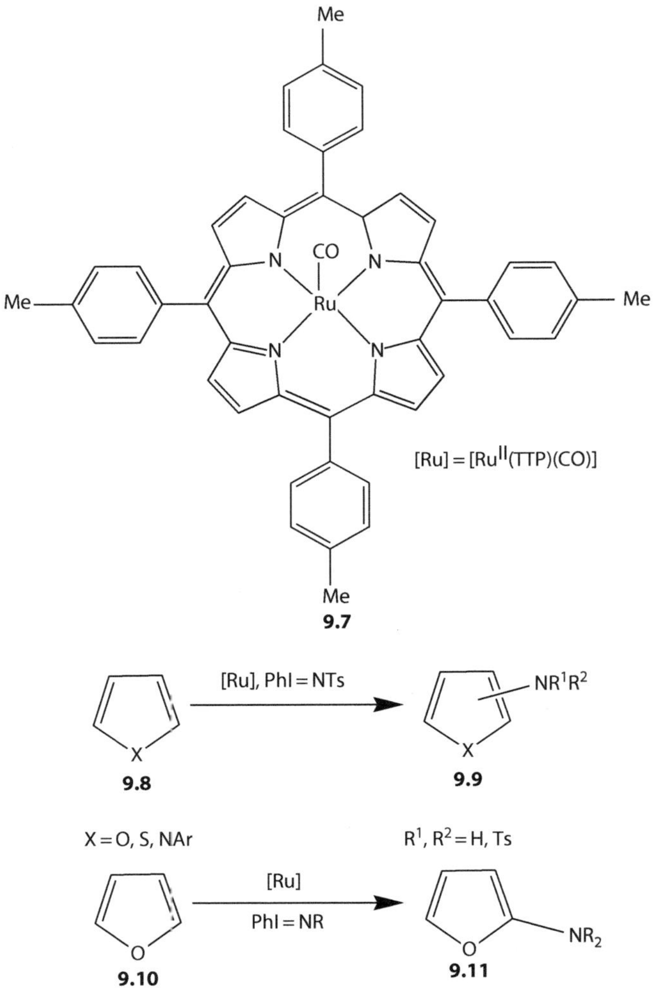

[Ru] = [Ru^{II}(TTP)(CO)]

9.7

9.8 → 9.9 via [Ru], PhI=NTs; substituent NR^{1}R^2

X = O, S, NAr

R^1, R^2 = H, Ts

9.10 → 9.11 via [Ru], PhI=NR; substituent NR$_2$

SCHEME 9.5

BSA in the presence of HP-Zn under ultrasonic irradiation were studied for driving sonodynamic treatment to a clinical application (Liu et al., 2009). The results showed that at 37°C, the binding site number, the apparent binding constant, and the binding distance were 1.0379, 6.2661 × 10^4, and 3.53 nm, respectively. Under different conditions, the damage degree of BSA rose with the increase in ultrasonic irradiation time, HP-Zn concentration, and pH value of the solution. Other relative metal porphyrin applications were related to sonodynamic therapy (synergistic effect of ultrasound and drugs on cells) (Miyoshi et al., 2003).

A series of reports on the systematic synthesis of non-substituted metal phthalocyanines (**9.13**) (PcM), starting from phthalonitrile (**9.12**) and elemental metals (Scheme 9.6) in different forms (in particular *Rieke* metals), obtained by the ultrasonic activation of the reaction system, were recently published (Kharisov et al., 2006a, 2007a). The metals, whose phthalocyaninates were the most common industrial phthalocyanine products (Cu, Ni, Mg, and Zn), were used as nonactivated powders, wires, or sheets, or as activated forms ("dry" pyrophoric *Raney* nickel or *Rieke* metals), reacting with phthalonitrile solutions in nonaqueous solvents (mainly low-weight alcohols and THF) under ultrasonic activation. Furthermore, magnesium was also used as an active magnesium–anthracene complex (source of a "soluble" metal) (Kharisov et al., 2005a). The mechanism of ultrasonically assisted destruction of metal aggregates, leading to the formation of preferential reaction sites and

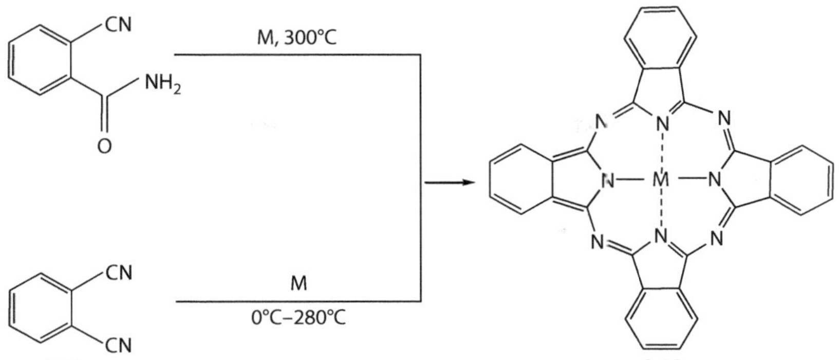

SCHEME 9.6

smaller metal particles, reacting further with phthalonitrile and forming the final metal phthalocyaninate, was offered (Figure 9.2) (Kharisov et al., 2004, 2006b). Copper and nickel were also used as metallic aggregates, supported in alumina (Figure 9.3) (Kharisov et al., 2007b). Comparing these methods, it was established that *Rieke* metals showed more activity when compared with supported metals at relatively low temperatures (0°C–50°C). Under ultrasonic treatment, metal phthalocyaninate yields were close to quantitative. Similar experiments were carried out using zeolites as a matrix for phthalonitrile cyclotetramerization (Kharisov et al., 2005b; De la Rosa et al., 2007), yielding the usefulness of ultrasonic treatment for the reaction mixture (Scheme 9.6).

The same researchers also applied the direct electrochemical synthesis, described earlier, for the preparation of porphyrins, to obtain phthalocyanines in a temperature range 0°C–120°C (Kharisov et al., 1999a,b, 2000), depending on the metal, solvent, and precursors. It was shown that in some reactions the use of ultrasound was a necessary procedure, as the electrode surfaces during the electrolysis process had became covered with the product layer. As a consequence, an unstable condition, due to increase of voltage in the cell, occurred. Use of ultrasound allowed it to partially or completely resolve this problem, eliminating the formed metal phthalocyaninate from the electrode surfaces and thus stabilizing the electrolysis.

Co-phthalocyanine/Fe nanocomposite particles were obtained by using the composite in situ method, with the mixture of carbonyl iron and solution of Co(II)-phthalocyanine (Co-Pc) ultrasonic dispersing in DMF (Gong et al., 2001). It was found that the Co-Pc-Fe nanocomposite particles were completely covered with Co-Pc and had the structure of Chinese gooseberry. Among other reports on ultrasonic application in phthalocyanine synthesis and product modification, we noted the production of CuPc nanofilms (Xue et al., 2009), surface modification of CuPc pigments (Bulychev et al., 2008), formation of LiPc nanotubes (its mechanism is shown in Figure 9.4) (Sostaric, 2006), as well as decolorization due to degradation of Ni(II) and vanadyl 2,9,16,23-tetraphenoxy-29H,31H-phthalocyanine in the ultrasonic field (Banks et al., 2004). The results for porphyrins were similar to the ones mentioned earlier. In a related research (Tu et al., 2002), with use of copper and nickel phthalocyanines in water, and organic solvents at r.t. and atmospheric pressure, in the presence of a catalytic amount of oxidant, under 20 kHz frequency, and at a power level of 37–59 W cm^{-2}, the pigments could be destroyed (92%–95%) within 50 min. Preliminary kinetic studies showed the process was of a Langmuir–Hinshelwood type. The reaction mechanism was similar to that of metalloporphyrins and the intermediate was found to be an unstable ion radical, which could release metal.

O-CONTAINING METAL COMPLEXES

In a difference of ligands with unique N-donor atoms, ultrasonically prepared O-containing metal complexes are very common. Thus, hydrolysis of *alcoholates*, such as, the adduct/(solvent-complex) of iron butylate with THF, Fe(OBut)$_2$(THF)$_2$, followed by ultrasonic and thermal treatment,

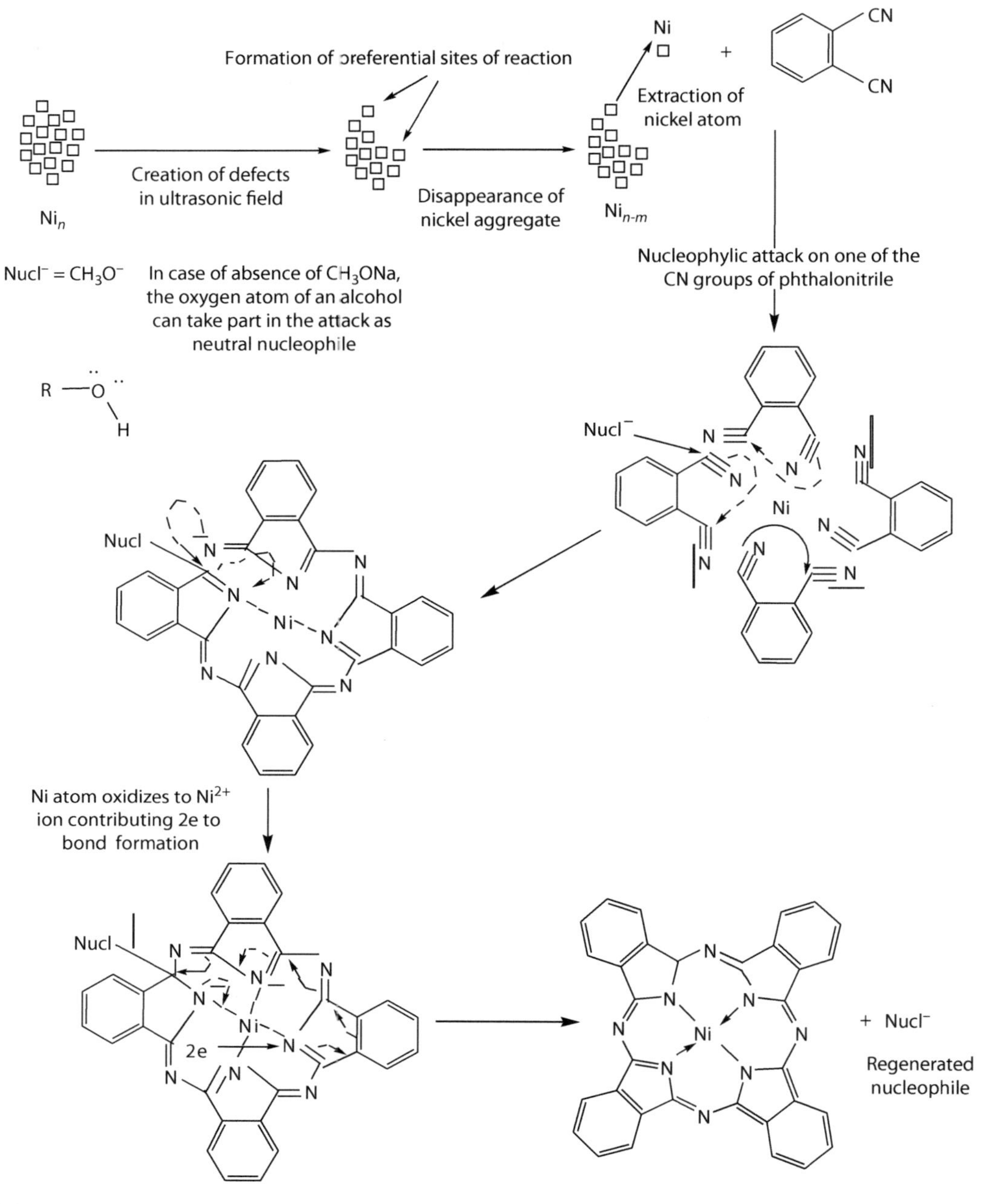

FIGURE 9.2 Proposed mechanism of PcM formation from *Rieke* metals.

yielded either nanoparticles of γ-Fe_2O_3 (maghemite, 9±2 nm), which are superparamagnetic and form unique needle-like assemblies of nanoparticle arrays, or Fe_3O_4 (19±2 nm) forming plate-like aggregates ~10 μm thick (Biddlecombe et al., 2001). Additionally, ($Fe[NC(C_6H_4)C(NSiMe_3)_2]_2Cl$, $Fe_2[O_2Si(C_6H_5)_2]_3$, and $[Fe(OBut)_3Na(THF)]_2$), can be also used as precursors (Gun'ko et al., 2001). Taking the example of *ketone complexes*, a difference between the application of ultrasound and a standard reflux in the synthesis of metal complexes, leading to distinct products, can be observed. Thus, when dpk was allowed to react with $[Mn(CO)_5Br]$ in dry Et_2O, under ultrasonic conditions, *fac*-$[Mn(CO)_3(dpk)Br]$ was isolated in good yield and when the same reaction was carried out under

FIGURE 9.3 Atoms of copper (a) and nickel (c) in the alumina structure before the treatment in the phthalonitrile solution and phthalocyanine, formed on the matrix of these agglomerates (b, Cu; d, Ni) in an ultrasonic field.

reflux conditions in toluene, in presence of water, traces of *fac*-[Mn(CO)$_3$(dpkO,OH)] were isolated (Bakir et al., 2003), which was different from *fac*-Mn(CO)$_3$(L–L)Br formed in the case of using di-2-pyridyl ketone hydrazones (see also *hydrazones* a little later in the text) of the type (C$_5$H$_4$N)$_2$C=N–NH–R=L–L (R = aryl) instead of dpk (Bakir et al., 2005).

β–*Diketonates*, traditionally well studied and possessing a lot of applications, are also much better represented in relation to ultrasound, among all other reported oxygen-containing complexes. Ultrasound in diketonate coordination chemistry has been used more for destruction of diketonates (in order to produce oxide- or carbide-based materials) than for their synthesis, although a few preparative examples have also been reported. Thus, a simple and convenient technique for the complex formation of a wide variety of transition metals/alkaline earth metal salts with 1,3-diketones (**9.14**) under sonication (Scheme 9.7) resulting in β-diketonates (**9.15**) has been offered by Nandurkar et al. (2008). This method shows a significant rate enhancement for metal complex formation in the presence of ultrasound, as compared to their silent counterpart, thereby providing higher yields. In case of using elemental metals, their interaction with β-diketones and other ligands in (or without) nonaqueous solutions (so-called direct synthesis of metal complexes from zero-valent metals and organic ligands) under ultrasonic treatment, was generalized by Garnovskii and Kharisov (1999).

Terbium acetylacetonate composite nanoparticles, prepared under vigorous ultrasonic irradiation, being water soluble, stable, and having extremely narrow emission bands and high internal quantum efficiencies, were used as fluorescence probes in the detection of enoxacin, based on the fluorescence enhancement of nanoparticles through fluorescence resonance energy transfer (Karim and Lee, 2008). Among other useful applications of metal acetylacetonates, we noted the preparation of the surfactant

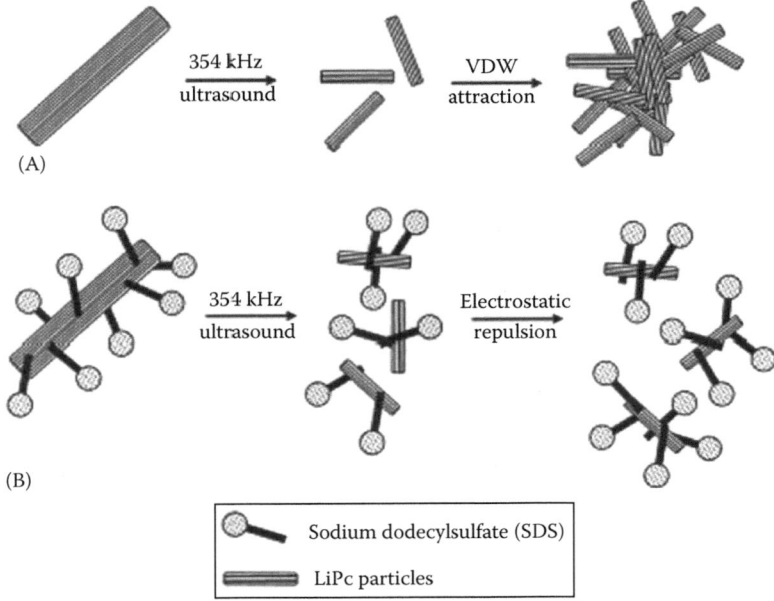

FIGURE 9.4 Depiction of the effect of ultrasound on LiPc particles in (A) the presence and (B) absence of SDS during sonolysis. In the absence of SDS, freshly formed smaller particles tend to agglomerate to form larger-sized particles. When sonolysis is conducted in the presence of SDS, the freshly formed smaller particles are stabilized through electrostatic repulsion. (With permission from Sostaric, J.Z., Pandian, R.P., Weavers, L.K., and Kuppusamy, P., Formation of lithium phthalocyanine nanotubes by size reduction using low- and high-frequency ultrasound, *Chem. Mater.*, 18(7), 4183–4189, 2006. Copyright 2006 American Chemical Society.)

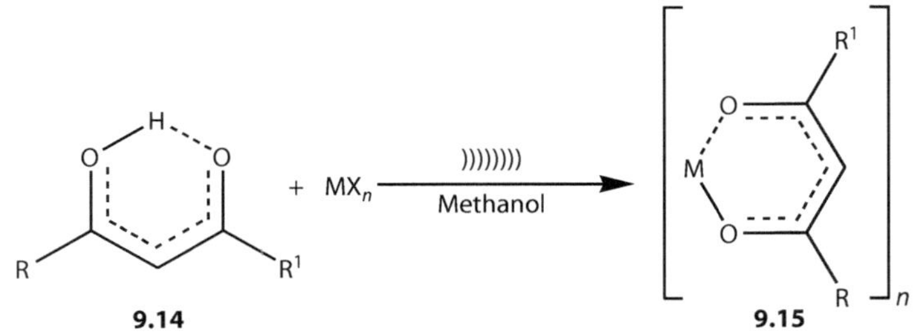

SCHEME 9.7

clad Mo-acetylacetonate (acac) catalyst (Zou, 2007) and anti-inflammatory agent Tolmetin (Ouyang et al., 2008). The last compound was synthesized from N-methylpyrrole and p-toluene formamide in the presence of phosphorus oxychloride via benzoylation, to form 1-methyl-2-(4-methylbenzoyl)-1H-pyrrole, which then reacted with triethylmethane tricarboxylate in an organic solvent, under ultrasonic treatment (25–80 kHz, 100–800 W), in the presence of transition metal complexes, for example $Mn(OAc)_3$, $Mn(acac)_3$, $Mn(OAc)_2$, $Co(OAc)_3$, and $Cu(OAc)_2$. The method has advantages of having a high yield (33.9%), having a cost-effective and easily available starting material, being a simple and safe process, with good product quality, and suitable for industrial manufacture.

β-Diketonates are convenient materials for fabrication of oxide thin films, powders, and other different micro- and nanoforms, as these precursors can be ultrasonically destroyed in combination with heating techniques. Thus, yttria thin films were deposited on silicon substrates using the ultrasonic spray pyrolysis technique, with the thermal decomposition of yttrium acetylacetonate $Y(acac)_3$

(Alarcon-Flores et al., 2008). Ultrasonic irradiation with the frequency of 20–22 kHz and absorbed acoustic power of about 0.4 W mL^{-1} caused degradation of An(IV) *tetrakis*-β-diketonates and AnL$_4$, where An(IV) was Th(IV), Np(IV), and Pu(IV), and HL was hexafluoroacetylacetone and dibenzoylmethane, in hexadecane solutions, in the presence of argon, resulting in a mixture of actinide carbides and partial degradation products (PDP) of initial metal β-diketonates (Nikitenko et al., 2000; 2004). The authors proposed that metal carbides were formed within the cavitating bubbles as a result of the high-temperature process, with participation of actinide(IV) β-diketonates and solvent vapors, meanwhile the PDP formation was attributed to the thermolysis of the complexes in a liquid reaction zone surrounding the cavitating bubble. The formation of MgO films by ultrasonic spray pyrolysis from magnesium β-diketonate was observed by Stryckmans et al. (1996). In addition to metal β-diketonates, carboxylate salts, metal alkoxides, or α-hydroxycarboxylate salts, such as metal citrates, tartrates, malates, lactates, or glycolates, could be subjected to ultrasonic wave energy (40–2000 kHz), to be decomposed by a series of steps, patented long ago by David (1986). This route was applied to produce metastable tetragonal ZrO$_2$ powder containing Y, Ca, Sr, Yb, Dy, or Ce, whose acetylacetonates were used as stabilizing agents in the formation of metastable ZrO$_2$ powders, as also Ni$_{0.7}$Zn$_{0.3}$Fe$_2$O$_4$ and MFe$_2$O$_4$ (M = Mn, Fe, Co, Ni, Cu, Zn, Cd, Mg, Ba, Sr). Titanium (isopropoxide + acetylacetonate) was found to be suitable as the starting salt for the preparation of Pb(Zr, Ti)O$_3$ powders by ultrasonic spray pyrolysis, in the viewpoint of the phase and morphology of the powders (Kim et al., 1995). The ultrasonic method for preparing shell-core fluorescent nanocomposite of europium-(trifluoroacetylacetone)$_3$-1,10-phenanthroline and silicon dioxide ((Eu(tfacac)$_3$phen/SiO$_2$) with SiO$_2$ as the shell) was offered by Guo and Zhao (2009). The obtained nanocomposite had high stability under UV irradiation, high fluorescence intensity, and UV absorbability.

Catecholates are represented by a tubular complex (C$_2$H$_9$N$_2$)$_2$(C$_2$H$_{10}$N$_2$)$_{0.5}$[MoO$_2$(OC$_6$H$_4$O)$_2$] (OC$_6$H$_4$O = catecholate, cations NH$_2$CH$_2$CH$_2$NH$_2$ are protonated) with a tube-like framework and two types of isomers (λ/δ configuration) for chiral anions [Mo^{V}O$_2$(OC$_6$H$_4$O)$_2$]$_3$, crystallizing in the tetragonal system, space group *P*4(2)/*n* (Wang et al., 2008), whose further sonication leads to the transformation from this bulk tubular complex to the helical nanostructure, a new morphology of inorganic–organic hybrid materials on the nanoscale level. Both left- and right-handed nanohelices have been detected as products. Additionally, the sonication of silica gel in basic solutions of diols cause a rapid breakdown of the inorganic polymer and the formation of monomeric Si complexes, for example *tris*–catecholate complex Na$_2$[(C$_6$H$_4$O$_2$)$_3$Si] forms in good yield at r.t. (Lickiss and Lucas, 1996; Li et al., 2008b). Its germanium analog Li$_2$[(C$_6$H$_4$O$_2$)$_3$Ge] may form similarly from GeO$_2$, but TiO$_2$, on the contrary, remains unaffected by the ultrasound.

A few examples of ultrasonically synthesized *carboxylates* have been reported. Thus, a 3-D metal-organic framework with 3-D channels, that is, Cu$_3$(BTC)$_2$, was synthesized by the reaction of cupric acetate and H$_3$BTC in a mixed solution of DMF/EtOH/H$_2$O (3:1:2, vol./vol.) under ultrasonic irradiation at r.t. for short reaction times (5–60 min), in high yields (62.6%–85.1%), (Li et al., 2008e). Compared to the traditional synthetic techniques, such as, solvent diffusion technique, and hydrothermal and solvothermal methods, the ultrasonic route was found to be highly efficient and environmentally friendly for the construction of porous MOFs. Nanocrystals of a fluorescent microporous MOF of the analogous compound Zn$_3$(BTC)$_2 \cdot$ 12H$_2$O were prepared similarly (Qiu et al., 2008).

S-Containing Metal Complexes

The ligands containing sulfur are represented by a considerably lesser number of examples than N and O ligands. Thus, the ultrasound-assisted reaction of the lithiation of *thiophene* gave 2-lithiothiophene, which on treatment with 2-formylthiophene gave 71% 2-(2′-thienylhydroxymethyl)thiophene (Garrigues et al., 2000). The splitting of the quasidegenerate electronic states in the dinuclear bis[(1,3-dithiole-2-thione-4,5-dithiolato)-di-(carbonyl)-cyclopentadienyliron(II)] complex with bridging, S–S-coupled, dimerized sulfur-rich *dithiolate* ligands, [Fe(C$_5$H$_5$)(CO)$_2$(C$_3$S$_5$–C$_3$S$_5$)Fe(C$_5$H$_5$)(CO)$_2$] (**9.16**) (Scheme 9.8), for whose synthesis the use of ultrasound irradiation was earlier (Matsubayashi et al.,

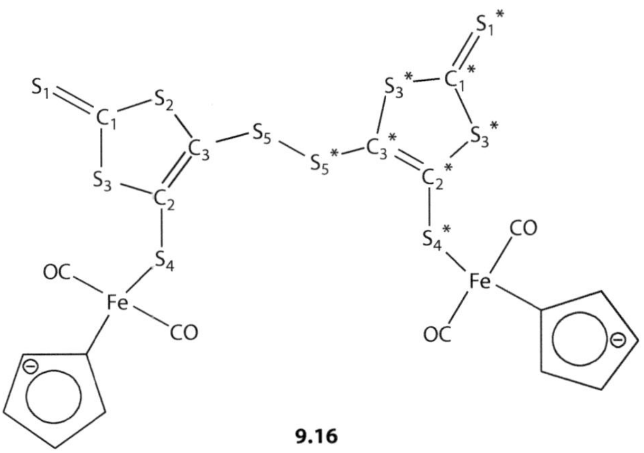

9.16

SCHEME 9.8

2002) reported, was found by the means of Mössbauer spectroscopy and by the measurement of the temperature dependence of magnetic susceptibility (2–300 K) (Vitushkina et al., 2009). An unusual effect of the splitting of electron states of the binuclear low symmetry complex, caused by the notable electron conductivity of the Fe−S_3−C_2−S_2−S'_2−C'_2−S'_3−Fe′ chain, was found in this compound.

Te-Containing Metal Complexes

Elemental tellurium was found to react with μ-alkylidyne complexes (**9.17**) [CpMFe(μ-CR)(CO)$_5$] (M = W, Mo; R = C$_6$H$_3$Me$_2$-2,6) under ultrasonic activation (Scheme 9.9) to provide the telluroaroyl [CpMoFe(μ-TeCR)(CO)$_5$] (**9.18**) or μ-telluride clusters (**9.19** and **9.20**) (R = 2,6-Me$_2$C$_6$H$_3$) (Hulkes et al., 2004).

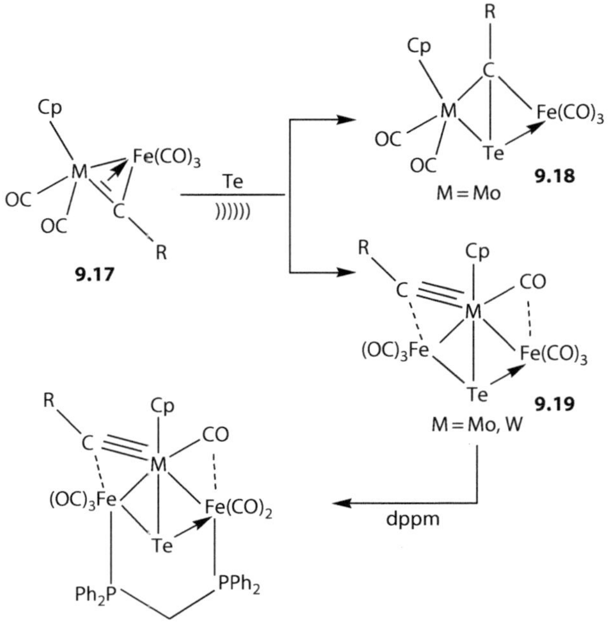

SCHEME 9.9

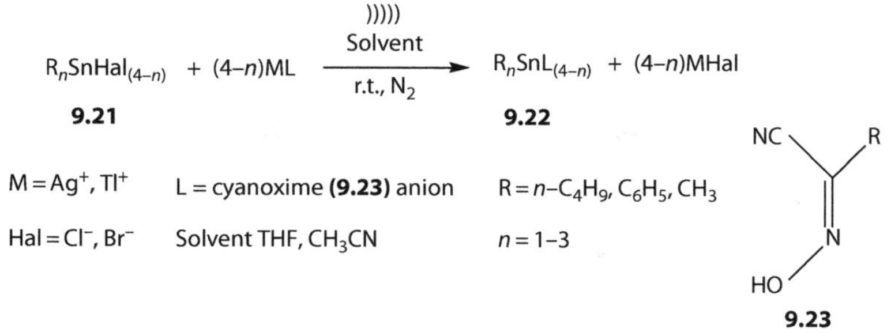

$R_nSnHal_{(4-n)}$ + (4−n)ML $\xrightarrow[\text{r.t., N}_2]{\text{)))))}\atop\text{Solvent}}$ $R_nSnL_{(4-n)}$ + (4−n)MHal

9.21   **9.22**

M = Ag⁺, Tl⁺   L = cyanoxime (**9.23**) anion   R = n-C_4H_9, C_6H_5, CH_3

Hal = Cl⁻, Br⁻   Solvent THF, CH_3CN n = 1–3

9.23

SCHEME 9.10

HALOGEN-CONTAINING METAL COMPLEXES

Halogen-containing complexes, prepared with ultrasonic assistance, are also rare. Nineteen organotin(IV) complexes (**9.22**), with nine different cyanoxime ligands (**9.23**), were anaerobically prepared by the heterogeneous metathesis reaction (Scheme 9.10) between the respective organotin(IV) halides (**9.21**) (Cl, Br) and ML (M = Ag, Tl; L = cyanoximate anion), using ultrasound in the MeCN at r.t. (Gerasimchuk et al., 2007). The crystal structures of the complexes revealed the formation of two types of Sn(IV) cyanoximates: mononuclear five-coordinated compounds of $R_{4-n}SnL_n$ composition (R = Me, Et, Bu, Ph; n = 1, 2; L = cyanoximate anion), and the tetranuclear $R_8Sn_4(OH)_2O_2L_2$ species (R = Bu, Ph). The latter complex contained a planar $[Sn_4(OH)_2O_2]^{2-}$ core, consisting of three adjacent rhombs with bridging oxo and hydroxo groups. The two dibutyltin(IV) cyanoximates showed cytotoxicity similar and greater, against several types of cancer, to that of *cis*-platin.

N,O-CONTAINING METAL COMPLEXES

Among a variety of reported titled complexes, *Schiff base* complexes should be mentioned first. Schiff bases, as a hearth of coordination chemistry, have attracted a lot of attention of the researchers, from the point of view of applying various techniques for obtaining their metal complexes, in a particular ultrasound. Thus, transition metal (M = Zn, Fe, and Co) Schiff base ($C_4N_3H_{13}$) complexes ([[2,2′-[iminobis(2,1-ethanediylnitrilomethylidyne)]bis[phenolato]](2−)−N,N′,N″,O,O′]M), supported on SiO_2, were prepared with 3-chloropropyltrimethoxysilane as the coupling agents, using the ultrasonic technique, under mild conditions and showed different catalytic activities, (a) toward CO_2 coupling with propylene oxide (Zhang et al., 2008) and (b) selective oxidation of styrene to benzaldehyde, with H_2O_2 as the oxidant (Zhang et al., 2006). These transition metal Schiff base complexes showed a higher catalytic activity than $SiO_2/C_4N_3H_{13}$ in the first process. For the second series of reactions, both high styrene conversion (>90%) and benzaldehyde selectivity (93.5%) were obtained. Additionally, the selectivities were found to be influenced not only by the reaction time but also by the types of transition metal in the catalysts. A macroheterocyclic Schiff base dinuclear zinc complex Zn(TTA)−Zn(L) (L is a Schiff base of 1,3-bis(*o*-aminophenoxy)propane and TTA) was also synthesized under ultrasonic catalysis (Li et al., 2004). It was found that one Zn bond to 1,5-diaza-8,12-dioxa-4-trifluoromethyl-2-(2-thenoyl)-6,7,13,14-dibenzocyclotetradeca-1,4-diene ligand with dsp^3 hybrid orbitals and another Zn bond to 2-thenoyltrifluoroacetone with sp^2 hybrid orbitals helped to obtain a tree-like dinuclear complex. Ultrasonic velocity was studied in solutions of furfurylidene-4-aminoacetanilide, 3-nitrobenzylidene-2-amino-4-chlorophenol, and 4-chlorobenzylidene-2-amino-4-chlorophenol, and their Co, Ni, and Cu antimicrobially active complexes in methanol at 303.15 K (Mishra et al., 2002).

    N-containing derivatives of *carboxylic acids*, generally having a wide variety of coordination modes with metal ions (Figure 9.5, formulae **9.24** through **9.37**), are also well represented

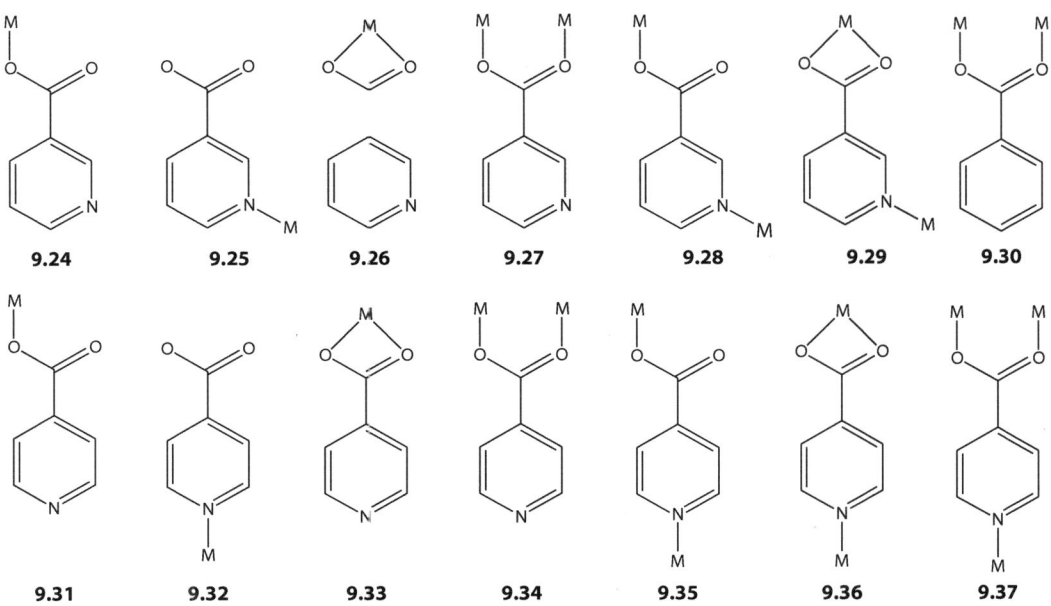

FIGURE 9.5 Various coordination modes of nicotinic (upper) and isonicotinic (bottom) acid with metal ions (M). (With permission from Chen, W. and Fukuzumi, S., Ligand-dependent ultrasonic-assistant self-assemblies and photophysical properties of lanthanide nicotinic/isonicotinic complexes, *Inorg. Chem.*, 48(8), 3800–3807, 2009. Copyright 2009 American Chemical Society.)

in respect of the application of ultrasound for their synthesis, although not by such a high number of examples as β-diketonates. Thus, two structural series, including two isomorphous homodinuclear complexes $Ln_2(H_2O)_4(C_6NO_2H_4)_6$ (Ln = Tb, Er) and four isostructural 1-D chain-like assemblies $[Ln(H_2O)_4(C_6NO_2H_4)_2]_n \cdot nCl$ (Ln = Sm, Eu, Tb, Dy) containing eight-coordinated lanthanide ions and bridging isonicotinic acid ligands, were rationally prepared through a facile ultrasonic synthesis (Chen et al., 2009). The 1-D polycationic chains and the isolated chloride anions are interconnected via hydrogen bonds and π–π interactions to form a three-dimensional supramolecular network.

N-containing carboxylates also served as precursors of metal nanoparticles. For instance, nanoscale Cu particles, together with Cu_2O, were prepared by sonochemical reduction of copper(II) hydrazine carboxylate $\{Cu(N_2H_3COO)_2 \cdot 2H_2O\}$ complex, in an aqueous medium under an Ar atmosphere for 23 h (Dhas et al., 1998). The probable reaction steps and explanation for the sonochemical reduction process are shown by Equations 9.1 through 9.5. Formation of Cu–Cu_2O mixture was explained by the partial oxidation of Cu by in situ generated H_2O_2 under the sonochemical conditions (Equations 9.4 and 9.5). However, in the presence of Ar and H_2 atmosphere, the formation of H_2O_2 can be arrested due to the scavenging of OH• radicals by hydrogen, thereby resulting in pure copper nanoparticles. The obtained Cu nanoparticles were catalytically active toward an *Ullmann* reaction (Scheme 9.11, 200°C, 5 h), that is, the condensation of aryl halides (**9.38**) to biphenyls (**9.39**), to an extent of 80%–90% conversion (particle size 50–70 nm), when compared to the thermal (particle size 200–250 nm, 79% yield) or commercial (particle size 500–600 nm, 43% yield) source of the copper catalyst.

$$H_2O))))) H^\bullet + OH^\bullet \tag{9.1}$$

$$Cu^{2+} + 2H^\bullet \rightarrow Cu^0 + 2H^+ \tag{9.2}$$

$$nCu^0 \rightarrow (Cu^0)_{n \text{ (aggregates)}} \tag{9.3}$$

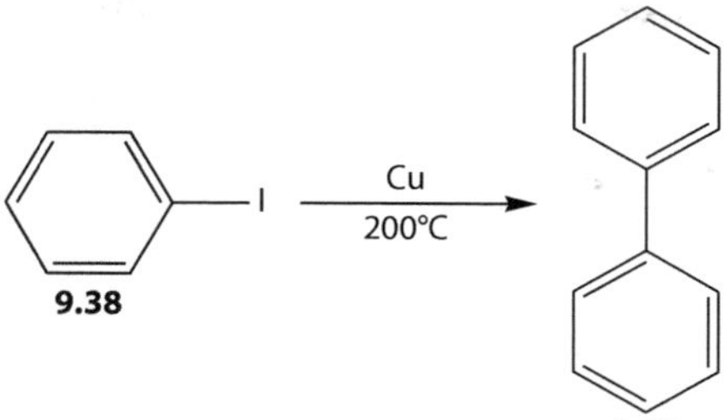

9.38

9.39

SCHEME 9.11

$$2H/2OH^{\bullet} \rightarrow H_2/H_2O_2/H_2O \quad (9.4)$$

$$Cu^0 + H_2O_2 \rightarrow Cu_2O + H_2O \quad (9.5)$$

In case of N-containing derivatives of *crown-ethers*, eight complexes of rare earth nitrates with lactam-type open-chain crown ethers (**9.40** through **9.42**) were synthesized by the ultrasonic method (Ni et al., 2001) (Scheme 9.12).

Among the O-containing *azine* derivatives, we noted that the bis(1-(2-pyridylazo)-2-naphtholato) zinc {$Zn(PAN)_2$} complex (with a rodlike morphology and a diameter of ~20–70 nm and a length of ~100–300 nm) was successfully synthesized via a facile sonochemical method (Pan et al., 2007a).

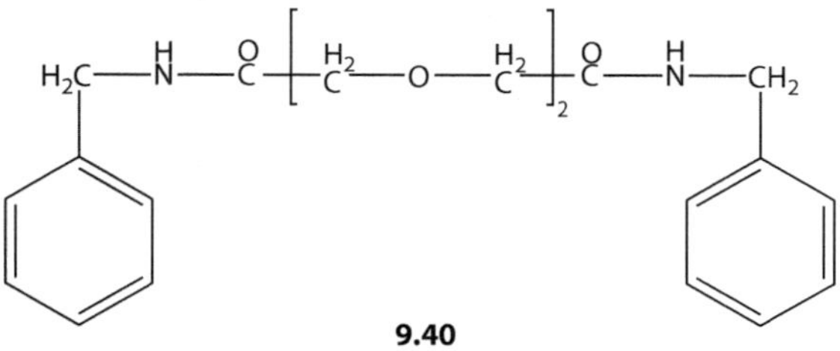

9.40

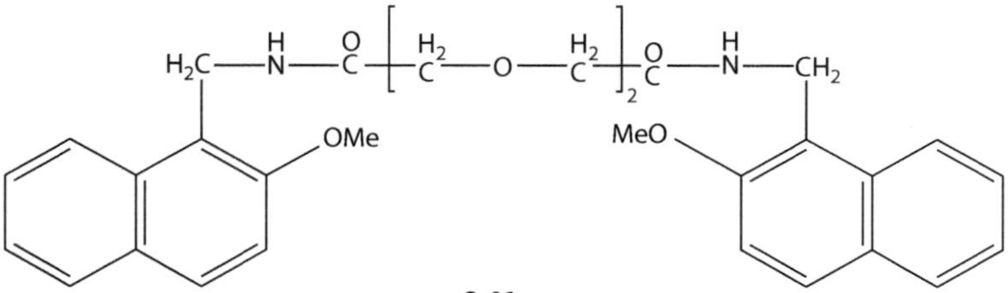

9.41

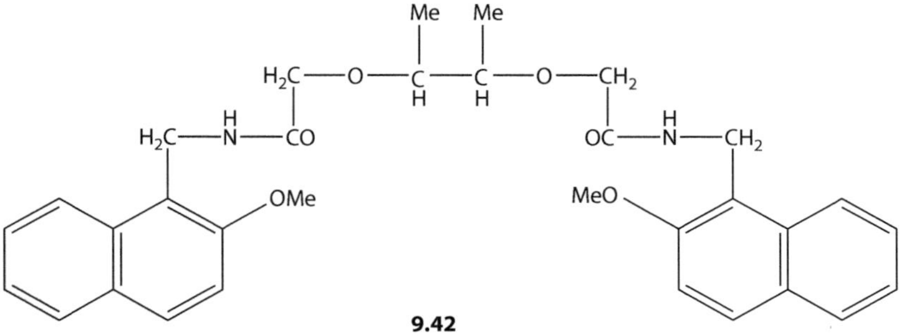

9.42

SCHEME 9.12

It was found that exciton coupling among neighbor Zn(PAN)$_2$ complex monomers in the nanorods produced resonance-enhanced light scattering. Additionally, the reaction between ZnCl$_2$ and dpkbh in MeCN, under ultrasonic or reflux conditions, gave a good yield of [ZnCl$_2$(η^3–N,N,O–dpkbh)]. Its solid-state IR spectra revealed the coordination of dpkbh, the presence of the amide proton, and the binding of the O atom of dpkbh. Furthermore, the optical measurements showed reversible interconversion between two forms, which could be due to complex–substrate interactions (Bakir et al., 2008). New materials of the *amine*-functionalized mesoporous silica (NH$_2$–MS) and ferrocene-functionalized mesoporous silica (Fc–CONH–MS) were obtained by post-synthesis grafting, where the peptide bond of the amine group (–NH$_2$) of mesoporous silica was linked with the carboxylic acid group (–COOH) at both ends of the ferrocene derivatives (Kwon and Lee, 2008). It was noted that the ferrocene attached to the amino-functionalized mesoporous silica pore outlet was cleavaged by ultrasound irradiation, which opened the closed-pore outlets, suggesting a possible application as a controlled release drug carrier. An example of *8-hydroxyquinoline* complexes, highly luminescent zinc(II)-bis(8-hydroxyquinoline) (Znq$_2$) complex nanorods (with a diameter of about 200–450 nm and a length of about 1–3 μm, Figure 9.6) were synthesized via a sonochemical route from a microemulsion containing zinc acetate and 8-hydroxyquinoline (Pan et al., 2007). A proposed mechanism (Figure 9.7) for the formation of Znq$_2$ nanorods included a possible collision and fusion of initial Znq$_2$ nuclei by ultrasound to form nanorods. The Znq$_2$ nanorods were found to be good protein probes for easy and highly sensitive detection.

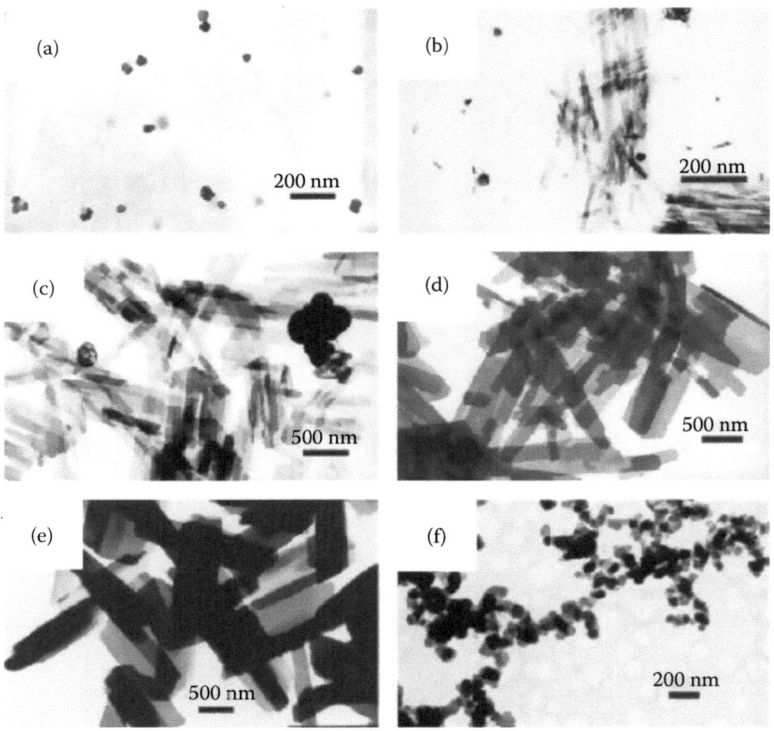

FIGURE 9.6 TEM images of the Znq$_2$ products obtained at different reaction conditions after ultrasound treatment for (a) 1, (b) 3, (c) 6, (d) 12, and (e) 40 min and under electromagnetic stirring instead of ultrasound treatment for (f) 45 min. (With permission from Pan, H.-C., Liang, F.-P., Mao, C.-J., Zhu, J.-J., and Chen, H.-Y., Highly luminescent zinc(II)-bis(8-hydroxyquinoline) complex nanorods: Sonochemical synthesis, characterizations, and protein sensing *J. Phys. Chem. B*, 111(20), 5767–5772, 2007. Copyright 2007 American Chemical Society.)

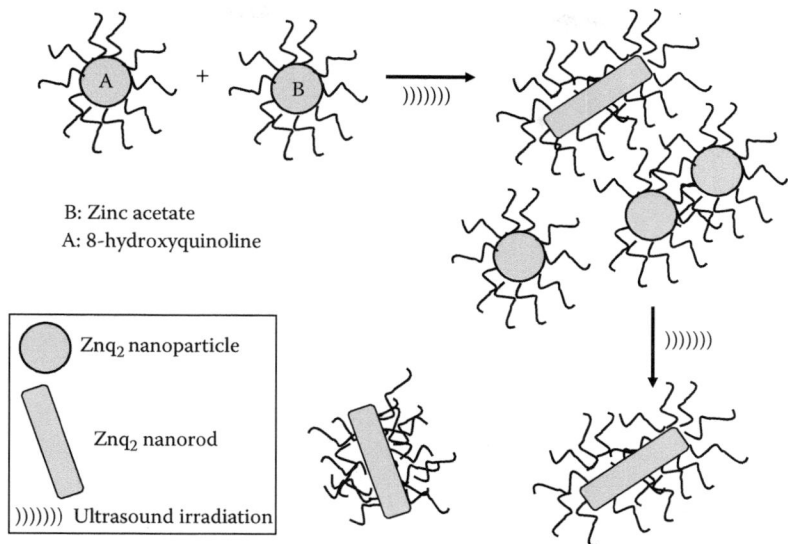

FIGURE 9.7 Schematic illustration of the formation mechanism of Znq$_2$ nanorods. (With permission from Pan, H.-C., Liang, F.-P., Mao, C.-J., Zhu, J,-J., and Chen, H.-Y., Highly luminescent zinc(II)-bis(8-hydroxyquinoline) complex nanorods: Sonochemical synthesis, characterizations, and protein sensing, *J. Phys. Chem. B*, 111(20), 5767–5772, 2007. Copyright 2007 American Chemical Society.)

Metalated Peptide Complexes

The first case of a reversible, remotely controlled, and rapid sol–gel transition by H-bonding aggregates was recently observed (Figure 9.8) on the example of palladium *o*-metalated N-dipeptidyl benzaldimine complexes [LXPd–1,2–C$_6$H$_4$CH:N–κN–(CH$_2$)$_n$OCOCH$_2$CH$_2$CH(NHFmoc)CONHCH(CONHBu)CH$_2$CH$_2$COO(CH$_2$)$_n$N:CHC$_6$H$_4$–1,2–PdXL] (L=PPh$_3$; X = Cl, n = 2; X = NCS, n = 2; X = Cl, n = 5), Fmoc[NHCHYCONH]$_m$Bu (Y = CH$_2$CH$_2$CO$_2$CH$_2$CH$_2$N:CH–1,2–C$_6$H$_4$PdLCl; m = 1–4) undergoing ultrasound-induced gelation, explained by the authors as extensive H-bond formation (Isozaki et al., 2007). By adjusting the sonication time, the gelation rates and heat-resistant properties of the aggregates could be controlled.

S,O-Containing Metal Complexes

S,O-containing metal complexes are represented by *sulfoxide* adducts. Thus, measurements of propagation speed at a frequency of 1 MHz and absorption coefficient of ultrasonic waves within

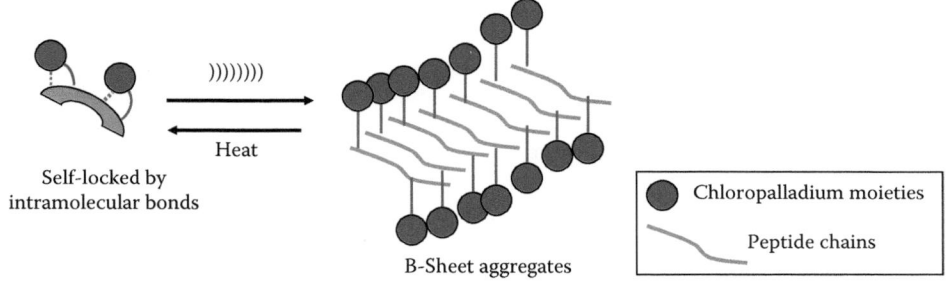

FIGURE 9.8 Ultrasound-assisted gelation. (Isozaki, K., Takaya, H., and Naota, T.: Ultrasound-induced gelation of organic fluids with metalated peptides. *Angew. Chem. Int. Ed.* 2007. 46(16). 2855–2857, S2855/1–S2855/31. Copyright Wiley-VCH Verlag GmbH & Co. KGaA. With permission.)

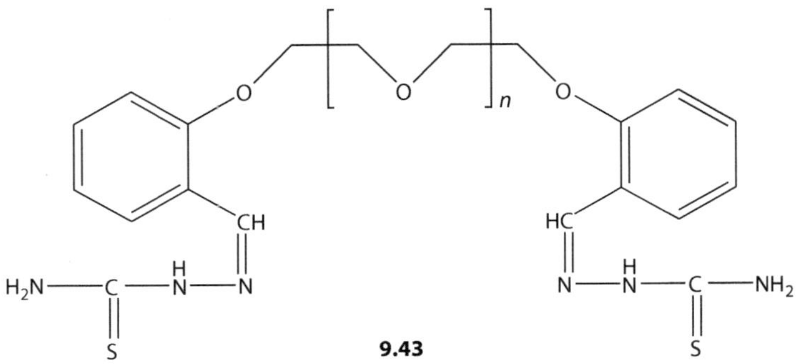

SCHEME 9.13

the range of 10–100 MHz in aqueous solutions of DMSO and $ZnCl_2$ showed the formation of solvatomers composed of $[(DMSO)_2Zn(H_2O)_2]^{2+}$ and $[(DMSO)Zn(H_2O)_3]^{2+}$ (Miecznik, 1993). The relaxation process discovered in the system DMSO–H_2O–$ZnCl_2$ was assigned to the formation and disintegration of solvatomers composed of the last complex.

N-,O-,S-CONTAINING METAL COMPLEXES

These ligands are represented by a unique example of *hydrazono podands* (**9.43**) L (n = 0, 1, 2), which reacted with $CuSO_4$ in EtOH under ultrasound treatment, to give $CuL(SO_4)$ (Fedorova and Ovchinnikova, 1995) (Scheme 9.13).

COORDINATION POLYMERS

The ultrasound is known as one of the most efficient techniques to produce shear forces in a solution, which, acting on polymers (Paulusse and Sijbesma, 2006), cause the chains to be stretched and broken. Selectivity of the mechanochemical chain scission in mixed palladium(II) and platinum(II) coordination polymers was studied by Paulusse and Sijbesma (2008). Ultrasonic scission on palladium(II)-phosphine-based coordination polymers was found to be highly reversible, and therefore, occurred exclusively at the palladium–phosphorus bond (Paulusse et al., 2007). The ultrasound-induced scission (with confirmed mechanochemical origin) of silver carbene coordination complexes with poly-THF-functionalized N-heterocyclic carbene ligands (**9.44**), with complete conversion to product (**9.45**), within 10 min (Scheme 9.14), when the polymers had a molecular weight of 6.7 kDa, was reported (Karthikeyan et al., 2008). The mechanochemical process at r.t. was found to be much faster than thermal scission at 60°C (30% conversion in 18 h).

A new gelation mechanism was proposed based on a readily available coordination polymer, $\{Zn(bibp)_2(OSO_2CF_3)_2\}_n$, in which the ultrasound dramatically changed the morphology of the material from sheet-like microparticles into nanofibers, which further got entangled with each other to form a three-dimensional fibrillar network, thereby resulting in the immobilization of organic fluids (Figure 9.9) (Zhang et al., 2009). A supramolecular metal-organic framework constructed by 2D infinite coordination polymers, $[Zn(BDC)(H_2O)]_n$ (yield 43.4%–53.2%), was synthesized by the reaction of zinc acetate with H_2BDC in DMF under ultrasonic irradiation at r.t. Samples with different morphologies, that is, nanobelts, nanosheets, and microcrystals, were obtained under ultrasound irradiation for different reaction times (Li et al., 2008a). Cadmium bis(6-mercaptopurinate) was heated at 200°C for 4 h to form an anhydrous coordination polymer of cadmium 6-mercaptopurinate, purified by ultrasonic dispersion and ultracentrifugation (Olea et al., 2005). Furthermore, a series of poly(carbosilazane-$CuCl_2$) metallopolymers was prepared by the reaction of varying amounts (5%–30%) of anhydrous $CuCl_2$ with polycarbosilazane $[-(CH_3)_2SiNH(CH_2)_2NH-]_n$, in THF, under continuous sonication (Arafa et al., 2005).

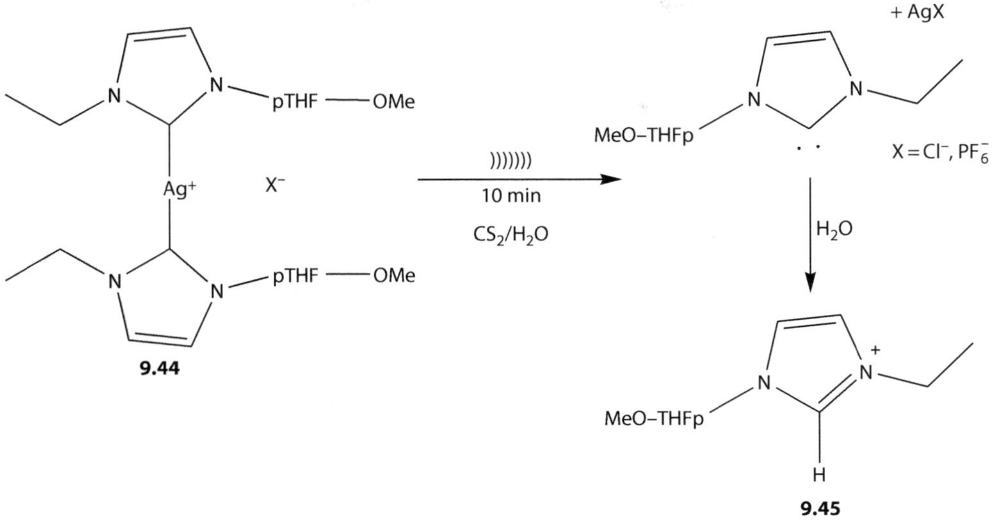

Use of Ultrasound in Coordination and Organometallic Chemistry

SCHEME 9.14

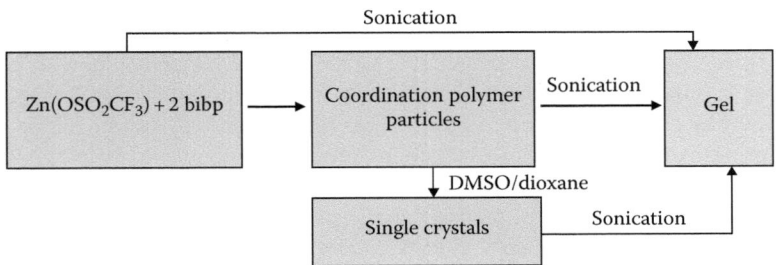

FIGURE 9.9 Morphological transformations of $\{Zn(bibp)_2(OSO_2CF_3)_2\}_n$. (With permission from Zhang, S., Yang, S., Lan, J., Tang, Y., Xue, Y., and You, J., Ultrasound-induced switching of sheetlike coordination polymer microparticles to nanofibers capable of gelating solvents, *J. Am. Chem. Soc.*, 131(5), 1689–1691. Copyright 2009 American Chemical Society.)

CATALYTIC APPLICATIONS OF ULTRASONICALLY OBTAINED COMPLEXES AND COMPOSITES

In addition to the above-mentioned catalytic uses of porphyrins and carbonyls, other coordination and organometallic compounds are widely used as catalysts under ultrasonic irradiation; also, the catalysts can be ultrasonically prepared and successfully used further without an ultrasound. All methodologies, based on the applications of metal complexes under ultrasound treatment in the organic synthesis have reported several advantages, such as, excellent yields, simple procedure, short reaction times, and milder conditions (Werner, 2005). Thus, a simple, efficient, and green procedure was offered for the synthesis of 2,4,5-triaryl imidazoles, for example, (**9.46**) (R^1 = H, Me, MeO, *i*-Pr, Cl, F, CN, NO_2, and so on; R^2 = H, MeO, NO_2, F; R^3 = H, MeO, F), catalyzed by zirconium(IV) acetylacetonate using ultrasonic irradiation (Khosropour et al., 2008). Different palladium complexes were found to be suitable catalysts for the preparation of biaryls (**9.47**) by using the *Suzuki* cross-coupling (Scheme 9.15), under ultrasonic irradiation, in the absence of high-yielding phosphine ligands (Silva et al., 2007). The catalyst was recycled up to three times with good to moderate activity. Titanium trichloride in EtOH, being reduced by Al to the corresponding low-valent titanium complexes, was found to be able to reduce some aromatic aldehydes and ketones

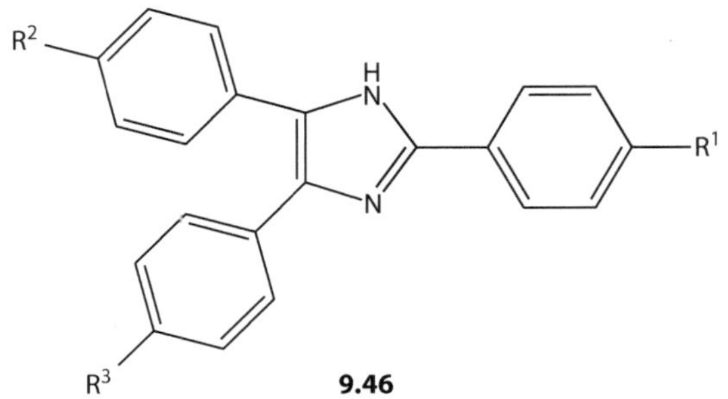

SCHEME 9.15

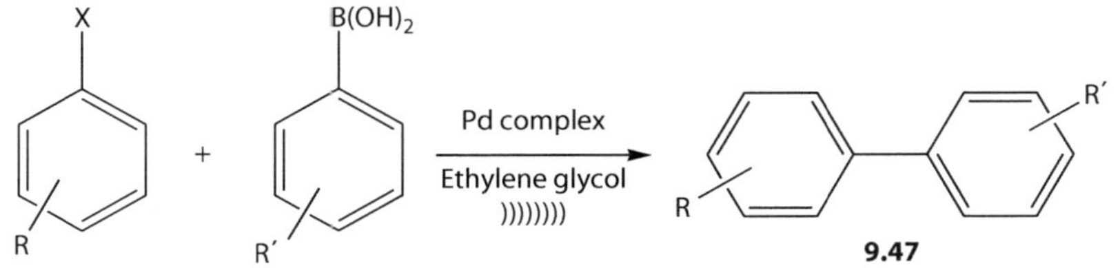

to the corresponding pinacols in 40%–82% yields within 30–90 min at r.t. under ultrasound irradiation (Li et al., 2004).

Several inorganic catalysts were prepared through the intermediate formation of meal complexes under ultrasonic irradiation. Thus, a catalytically high active nanocrystal Pt/C catalyst, possessing good selectivity, was obtained, starting with $PtCl_4$ as a precursor via its interaction with a mixed solvent of ketone (ethanone or acetone) and alcohol (methanol, propylene glycol, polyvinyl alcohol, or ethylene glycol) (at 1:1), and ultrasonic treatment at 20°C–55°C for 20–90 min, to obtain an intermediate platinum complex, by immersing and adsorption on the carrier carbon at 20°C–50°C for 2–12 h, and a further reduction step (Lei et al., 2008). The product had a series of advantages in comparison with similar products obtained by classic methods, in particular controllable particle sizes and dispersion, with good repeatability of the catalyst batches. Furthermore, highly active MoS_2 (particle size in the range of 25–50 nm) catalysts were obtained by the sonochemical preparation, from $(NH_4)_6Mo_7O_2 \cdot 44H_2O$, CH_3COSH, and polyethyleneglycol (Uzcanga, 2007). Using ligand-free palladium(II) acetate $[Pd(OAc)_2]$ in the range of 0.01–0.1 mol% or palladium-on-carbon (Pd/C) 10% in the range of 1.0–2.0 mol%, most aryl iodides and bromides gave high yields in Heck couplings under conventional heating (120°C) in 18 h (Palmisano, 2007). However, microwave irradiation alone or, better still, combined with high-intensity ultrasound, strongly promoted the reaction, generally decreasing reaction times to one hour.

OTHER APPLICATIONS

In addition to the applications of ultrasound in relation to the above-mentioned metal complexes throughout the text, this technique has been successfully applied for obtaining nanoparticles (particularly of elemental metals, see earlier in the text), which can be synthesized from metal-organic or organometallic precursors, either by classical thermal decomposition, ultrasound activation, photolysis, and hydrogenation, or hydrolysis reactions (Amiens and Chaudret, 2007). Thus, nanostructured $CoFe_2O_4$ particles (<5 nm) (bulk $CoFe_2O_4$ is a ferrimagnetic material with a partially inverse spinel structure, with the formula $(Co_xFe_{1-x})[Co_{1-x}Fe_{1+x}]$) were prepared by a sonochemical approach, first by preparation of the amorphous precursor powders (decomposition of solutions of volatile organic precursors, $Fe(CO)_5$ and $Co(NO)(CO)_3$, in decalin at 273 K, under an O_2 pressure of 100–150 kPa and ultrasonic treatment), followed by a heating treatment at relatively very low temperatures (Shafi et al., 1998). Certain radiochemical functionalities can be reached by using ultrasonic

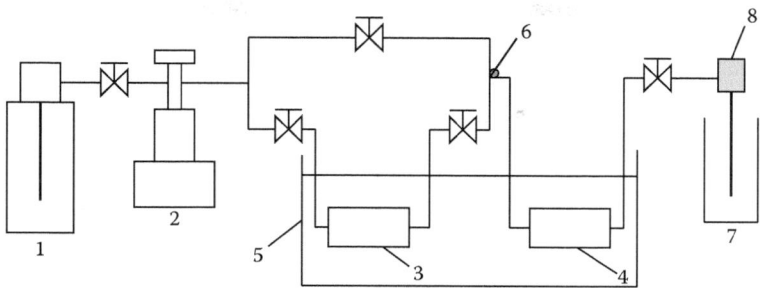

FIGURE 9.10 Schematic diagram of the experimental system for UO_2 dissolution in SF-CO_2: 1, CO_2 cylinder; 2, syringe pump; 3, ligand cell; 4, sample cell; 5, ultrasound device with a water bath; 6, T-shaped joint; 7, collection vial; 8, heater for the poly(ether ether ketone) (PEEK) restrictor.

irradiation. For instance, the application of the ultrasound significantly enhanced the dissolution of UO_2 powders placed on small glass beads in supercritical fluid CO_2, by using a CO_2-soluble tri-*n*-butylphosphate (TBP)/HNO_3/H_2O complexant as an extractant (Figure 9.10), probably contributing to the transfer of locally concentrated $UO_2(NO_3)_2 \cdot 2TBP$ from the surface of glass beads into the supercritical fluid CO_2 (Youichi Enokida et al., 2002).

CONCLUDING REMARKS

It was observed that the number of publications on the use of ultrasound, to synthesize coordination and organometallic compounds, is actually considerably lesser in comparison with hundreds of reports on the ultrasonic application in organic chemistry or nanotechnology. Ultrasound has been mainly applied to obtain metal complexes or composites, to transform them to other compounds, or to reduce them to elemental metals. Furthermore, a series of catalysts using ultrasonically prepared metal complexes has been reported. Metal complexes are widely used as catalysts under simultaneous ultrasonic treatment. The main advantages of ultrasonic application in organometallic and coordination chemistry are frequent higher yields in comparison with classic techniques, milder conditions (generally lower temperature), and a considerably shortened reaction time.

ACKNOWLEDGMENTS

The authors are grateful to Conacyt-Mexico y Paicyt/UANL (Monterrey, Mexico, project 2009–2010) for the financial support; to Professors Chien M. Wai, Jingsong You, Periannan Kuppusamy, Jun-Jie Zhu, Hikaru Takaya, Takeshi Naota, and C. Srinivasan; to the editorials Wiley; to the American Chemical Society for permission to reproduce images from their publications; and finally to Professor Alexander D. Garnovskii (who died in December 2010) for the critical analysis of the manuscript.

REFERENCES

Aguilera, E.N., Blanco, L.M., Huerta, A.M., and Obregon, L.A. 2009. Electrochemical metallization with Ni(II) from 5,10,15,20-tetraquis(*p*-hydroxiphenyl)porfirin. *Portugaliae Electrochimica Acta*, 27(3): 317–328.

Alarcon-Flores, G., Aguilar-Frutis, M., Garcia-Hipolito, M., Guzman-Mendoza, J., Canseco, M.A., and Falcony, C. 2008. Optical and structural characteristics of Y_2O_3 thin films synthesized from yttrium acetylacetonate. *Journal of Materials Science*, 43(10): 3582–3588.

Ameta, S.C., Punjabi, P.B., Swarnkar, H., Chhabra, N., and Jain, M. 2001. Sonochemistry: Past, present and future. *Journal of Indian Chemical Society*, 78(10–12): 627–633.

Amiens, C. and Chaudret, B. 2007. Organometallic synthesis of nanoparticles. *Modern Physics Letters*, *B* 21(18): 1133–1141.

Arafa, I.M., El-Ghanem, H.M., Hallak, A.B., and Jawad, S.A. 2005. Dielectric spectroscopy of polycarbosilazane-based $CuCl_2$ metallopolymers. *International Journal of Polymeric Materials*, 54(9): 857–870.

Bakir, M., Green, O., and Mulder, W.H. 2008. Synthesis, characterization and molecular sensing behavior of $[ZnCl_2(\eta^3$–N,N,O–dpkbh)] (dpkbh = di-2-pyridyl ketone benzoyl hydrazone). *Journal of Molecular Structure*, 873(1–3): 17–28.

Bakir, M., Gyles, C., and Green, O. 2005. Manganese carbonyl compounds of di-2-pyridyl ketone and its hydrazone derivatives. In *Abstracts of Papers, 230th ACS National Meeting*, Washington, DC, August 28–September 1, 2005, INOR-500.

Bakir, M., Hassan, I., and Green, O. 2003. Manganese carbonyl compounds of *N,N*-bidentate di-2-pyridylketone (dpk) and N,O,N-tridentate hydroxybis(2-pyridyl)methanolato (dpkO,OH). The structure of fac-$[Mn(CO)_3(dpkO,OH)]$. *Journal of Molecular Structure*, 657(1–3): 75–83.

Banks, C.E., Wylie, A.H., and Compton, R.G. 2004. Ultrasonically induced phthalocyanine degradation: Decoloration vs. metal release. *Ultrasonics Sonochemistry*, 11(5): 327–331.

Barreiro, A., Selbmann, D., Pichler, T., Biedermann, K., Gemming, T., Ruemmeli, M.H., Schwalke, U., and Buechner, B. 2006. On the effects of solution and reaction parameters for the aerosol-assisted CVD growth of long carbon nanotubes. *Applied. Physics A: Materials Science & Processing*, 82(4): 719–725.

Basset, J.-M., Psaro, R., Roberto, D., and Ugo, R. (Eds.). 2009. *Modern Surface Organometallic Chemistry*. Wiley-VCH, Weinheim, Germany, 725pp.

Bates, R.S., Begley, M.J., and Wright, A.H. 1990. Synthesis and reactions of arene-ruthenium complexes containing nitrogen-donor ligands; crystal structure of (*p*-cymene)$_2$Ru$_2$Cl$_2$(N$_3$)$_2$. *Polyhedron*, 9(8): 1113–1118.

Bian, Y.-J., Li, J.-T., and Li, T.-S. 2002. Applications of ultrasound in organic synthesis using metal reagents. *Youji Huaxue*, 22(4): 227–232.

Biddlecombe, G.B., Gun'ko, Y.K., Kelly, J.M., Pillai, S.C., Coey, J.M.D., Venkatesan, M., and Douvalis, A.P. 2001. Preparation of magnetic nanoparticles and their assemblies using a new Fe(II) alkoxide precursor. *Journal of Materials Chemistry*, 11(12): 2937–2939.

Billington, D.C., Helps, I.M., Pauson, P.L., Thomson, W., and Willison, D. 1988. The effect of ultrasound and of phosphine and phosphine oxides on the Khand reaction. *Journal of Organometallic Chemistry*, 354(2): 233–242.

Bulychev, N.A., Kisterev, E.V., Arutunov, I.A., and Zubov, V.P. 2008. Ultrasonic treatment assisted surface modification of inorganic and organic pigments in aqueous dispersions. *Journal of Balkan Tribological Association*, 14(1): 30–39.

Cains, P.W., Martin, P.D., and Price, C.J. 1998. The use of ultrasound in industrial chemical synthesis and crystallization. 1. Applications to synthetic chemistry. *Organic Process Research and Development*, 2(1): 34–48.

Caldwell, J.J., Harrity, J.P.A., Heron, N.M., Kerr, W.J., McKendry, S., and Middlemiss, D. 1999. Chromium-mediated β-lactone synthesis using ultrasonication. *Tetrahedron Letters*, 40(17): 3481–3484.

Caldwell, J.J., Kerr, W.J., and McKendry, S. 1999. Functionalized β-lactones from chromium alkyl(alkoxy) carbene complexes. *Tetrahedron Letters*, 40(17): 3485–3486.

Capelo-Martínez, J.-L. (Ed.). 2009. *Ultrasound in Chemistry: Analytical Applications*. Wiley-VCH, Weinheim, Germany, 171pp.

Carruthers, W. and Coldham, I. 2004. *Modern Methods of Organic Synthesis*, 4th edn. Cambridge University Press, England, U.K., 506pp.

Chen, X.-B., An, Z.-W., Li, J., Du, W.-S., and Gao, A.-A. 2003. Reactions of aryl Grignard reagents with 4-alkylacetophenone under ultrasonic irradiation. *Yingyong Huaxue*, 20(1): 100–102.

Chen, X., Cui, Y., Yin, G., and Jia, Z. 2007. Preparation and thermal property of poly(tetramethyl-*p*-silphenylene-dimethyl) siloxane. *Huagong Jinzhan*, 26(9): 1333–1337.

Chen, W. and Fukuzumi, S. 2009. Ligand-dependent ultrasonic-assistant self-assemblies and photophysical properties of lanthanide nicotinic/isonicotinic complexes. *Inorganic Chemistry*, 48(8): 3800–3807.

Chen, P., Wu, Q.-S., and Ding, Y.-P. 2007. Preparation of ferrocene nanocrystals by the ultrasonic-solvent-substitution method and their electrochemical properties. *Small*, 3(4): 644–649.

Cintas, P. 2004. Transition metals in radiation-induced reactions for organic synthesis: Applications of ultrasound. In Beller, M. and Bolm, C. (Eds.), *Transition Metals for Organic Synthesis*, 2nd edn., Vol. 2. Wiley-VCH, Weinheim, Germany, pp. 583–596.

Cravotto, G. and Cintas, P. 2006. Power ultrasound in organic synthesis: Moving cavitational chemistry from academia to innovative and large-scale applications. *Chemical Society Reviews*, 35(2): 180–196.

David, L.D. 1986. Microcrystalline metal oxides. U.S. Patent 4588575.
De La Rosa, J.R., Kharisov, B.I., Medina, A.M., Ortiz Méndez, U., and Ibarra Arvizu, A.K. 2007. Natural zeolites used as surface reaction for phthalocyanine synthesis. *Materials and Manufacturing Processes*, 22(3): 314–317.
Dhas, N.A., Raj, C.P., and Gedanken, A. 1998. Synthesis, characterization, and properties of metallic copper nanoparticles. *Chemistry of Materials*, 10(5): 1446–1452.
Dumbolov, D.U., Vasil'ev, V.V., Slezka, V.F., and Varnakov, V.V. 2006. Method of hardening surfaces of metal articles. Patent RU 2287020, 8 pp.
Federman Neto, A., Borges, A.D.L., Miller, J., and Darin, V.A. 1997. Improvements in the preparation of cyclopentadienylthallium and methylcyclopentadienylthallium and in their use in organometallic chemistry. *Synthesis and Reactivity in Inorganic, Metal-Organic, and Nano-Metal Chemistry*, 27(9): 1299–1314.
Fedorova, O.V. and Ovchinnikova, I.G. 1995. Sonochemical synthesis of cupric sulfate coordination compounds with hydrazono podands. *Zhurnal Obshchei Khimistry*, 65(4): 655–656.
Fu, N., Zhang, Y., Yang, D., Chen, B., and Wu, X. 2008. A rapid synthesis of 1-ferrocenyl-2-arylacetylenes under ultrasound irradiation. *Catalysis Communications*, 9(6): 976–979.
Garnovskii, A.D. and Kharisov, B.I. (Eds.). 1999. *Direct Synthesis of Coordination and Organometallic Compounds*. Elsevier Science, Lausanne, Switzerland, 254 pp.
Garnovskii, A.D. and Kharisov, B.I. (Eds.). 2003. *Synthetic Coordination & Organometallic Chemistry*. Marcel Dekker, New York, 513 pp.
Garrigues, B., Sanchez, M., Diallo, O., and Chemat, F. 2000. Ultrasound assisted in-situ synthesis of alkyl lithium compounds. *Journal of Nature*, 12(1): 25–28.
Gerasimchuk, N., Maher, T., Durham, P., Domasevitch, K.V., Wilking, J., and Mokhir, A. 2007. Tin(IV) cyanoximates: Synthesis, characterization, and cytotoxicity. *Inorganic Chemistry*, 46(18): 7268–7284.
Ghilane, J., Fontaine, O., Martin, P., Lacroix, J.-C., and Randriamahazaka, H. 2008. Formation of negative oxidation states of platinum and gold in redox ionic liquid: Electrochemical evidence. *Electrochemistry Communications*, 10(8): 1205–1209.
Gong, R.-Z., Guan, J.-G., Qu, W.-L., and Yuan, R.-Z. 2001. Model of in situ composite of cobalt(II)-phthalocyanine and carbonyl iron for organic/inorganic nanocomposites. *Journal of Wuhan University of Technology, Materials Science Edition*, 16(1): 1–4.
Gonsalves, K.E., Li, H., Perez, R., Santiago, P., and Jose-Yacaman, M. 2000. Synthesis of nanostructured metals and metal alloys from organometallics. *Coordination Chemistry Reviews*, 206–207: 607–630.
Gun'ko, Y.K., Pillai, S.C., and McInerney, D. 2001. Magnetic nanoparticles and nanoparticle assemblies from metallorganic precursors. *Journal of Materials Science: Materials in Electronics*, 12(4–6): 299–302.
Guo, G. and Zhao, L. 2009. Method for preparing shell-core fluorescent nanocomposite of europium-(trifluoroacetyl acetone)$_3$-1,10-phenanthroline and silicon dioxide. Patent CN 101560384, 11 pp.
Harrity, J.P.A., Kerr, W.J., and Middlemiss, D. 1993. Promotion of the chromium carbene Doetz annulation reaction under sonochemical and dry state adsorption conditions. *Tetrahedron*, 49(25): 5565–5576.
He, L., Chan, P.W.H., Tsui, W.-M., Yu, W.-Y., and Che, C.-M. 2004. Ruthenium(II) porphyrin-catalyzed amidation of aromatic heterocycles. *Organic Letters*, 6(14): 2405–2408.
Hulkes, A.J, Hill, A.F., Nasir, B.A., White, A.J.P., and Williams, D.J. 2004. Reactions of μ-alkylidyne complexes with tellurium. Telluroacyl versus μ-telluride formation. *Organometallics*, 23(4): 679–686.
Isozaki, K., Takaya, H., and Naota, T. 2007. Ultrasound-induced gelation of organic fluids with metalated peptides. *Angewandte Chemie International Edition*, 46(16): 2855–2857, S2855/1–S2855/31.
Izod, K., Bowman, L.J., Wills, C., Clegg, W., and Harrington, R.W. 2009. Synthesis and structural characterisation of alkali metal complexes of heteroatom-stabilized 1,4- and 1,6-dicarbanions. *Dalton Transactions*, (17): 3340–3347.
Ji, S.J., Shen, Z.L., and Wang, S.Y. 2003. Aldol condensation of acetylferrocene under ultrasound. *Chinese Chemical Letters*, 14(7): 663–666.
Karim, M.M. and Lee, S.H. 2008. Determination of enoxacin using Tb composite nanoparticles sensitized luminescence method. *Journal of Fluorescence*, 18(5): 827–833.
Karthikeyan, S., Potisek, S.L., Piermattei, A., and Sijbesma, R.P. 2008. Highly efficient mechanochemical scission of silver-carbene coordination polymers. *Journal of American Chemical Society*, 130(45): 14968–14969.
Kharisov, B.I., Blanco, L.M., and García-Luna, A. 1999b. Direct electrochemical synthesis of metal complexes. lanthanide phthalocyanines: Optimization of the synthesis. *Revista de la Sociedad Quimica de Mexico*, 43(2): 50–53.
Kharisov, B.I., Blanco, L.M., Torres-Martínez, L.M., and García-Luna, A. 1999a. Electrosynthesis of metal phthalocyanines: Influence of solvent. *Industrial and Engineering Chemistry Research*, 38(8): 2880–2887.

Kharisov, B.I., Cantú Coronado, C.E., Coronado Cerda, K.P., Ortiz Méndez, U., Jacobo Guzmán, J.A., and Ramírez Patlán, L.A. 2004. Use of elemental metals in different grade of activation for phthalocyanine preparation. *Inorganic Chemical Communications*, 7(12): 1269–1272.

Kharisov, B.I., Garnovskii, A.D., Kharissova, O.V., Ortiz Mendez, U., and Tsivadze, A. Yu. 2007a. Direct electrochemical synthesis of metal complexes of phthalocyanines and azomethines as model compounds: Advantages and problems of this method versus traditional synthetic techniques. *Journal of Coordination Chemistry*, 60(13): 1435–1455.

Kharisov, B.I., Garza-Rodríguez, L.A., Leija Gutiérrez, H.M., Ortiz Méndez, U., García Caballero, R., and Tsivadze, A.Yu. 2005a. Preparation of non-substituted metal phthalocyanines at low temperature using activated Rieke zinc and magnesium. *Synthesis and Reactivity in Inorganic, Metal-Organic, and Nano-Metal Chemistry*, 35(10): 755–760.

Kharisov, B.I., Medina, A.M., Rivera de la Rosa, J., and Ortiz Méndez, U. 2005b. Use of zeolites for phthalocyanine synthesis at low temperature. *Journal of Chemical Research*, 2005(6): 404–406.

Kharisov, B.I., Méndez-Rojas, M.A., and Ganich, E.A. 2000. Traditional and electrochemical methods of preparation of phthalocyanines. Influence of solvent. *Koordinatsionnaya Khimiya*, 26(5): 301–310.

Kharisov, B.I., Ortiz Méndez, U., Garza-Rodríguez, L.A., Leija Gutiérrez, H.M., Medina Medina, A., and Berdonosov, S.S. 2006b. Use of various activated forms of elemental nickel and copper for the synthesis of phthalocyanine at low temperature. *Journal of Coordination Chemistry*, 59(15): 1657–1666.

Kharisov, B.I., Ortiz Méndez, U., and Rivera de la Rosa, J. 2006a. Low-temperature synthesis of metal phthalocyaninates. *Russian Journal Coordination Chemistry*, 32(9): 643–658.

Kharisov, B.I., Rivera de la Rosa, J., Kharissova, O.V., Almaraz Garza, J.L., Almaguer Rodríguez, J.R., Puente, L.I., Ortiz Méndez, U., and Ibarra Arvizu, A.K. 2007b. Use of elemental copper and nickel, supported in alumina, for preparation of non-substituted metal phthalocyaninates at low temperature. *Journal of Coordination Chemistry*, 60(3): 355–364.

Kharissova, O.V. and Kharisov, B.I. 2008. Synthetic techniques and applications of activated nanostructured metals: Highlights up to 2008. *Recent Patents on Nanotechnology*, 2(2): 103–119.

Khosropour, A.R. 2008. Ultrasound-promoted greener synthesis of 2,4,5-trisubstituted imidazoles catalyzed by $Zr(acac)_4$ under ambient conditions. *Ultrasonics Sonochemistry*, 15(5): 659–664.

Kim, H.-B., Lee, J.-H., and Park, S.Ja. 1995. Influences of the starting salts on the powder characteristics of the $Pb(Zr, Ti)O_3$ powders prepared by ultrasonic spray pyrolysis. *Han'guk Chaelyo Hakhoechi*, 5(8): 905–912.

Konopleva, K.G., Venskovskii, N.U., Tupoleva, A.L., and Babushkina, T.A. 2000. Containerless ultrasound-assisted synthesis of cadmium and lead iodide intercalation compounds. *Zhurnal Neorganicheskoi Khimii*, 45(8): 1283–1287.

Kunz, U., Hoffmann, U., Rosenplaenter, A., and Horst, C. 1999. Use of ultrasonics for solid/liquid reaction. Patent DE 19808239, 20 pp.

Kwon, E.J. and Lee, T.G. 2008. Surface-modified mesoporous silica with ferrocene derivatives and its ultrasound-triggered functionality. *Applied Surface Science*, 254(15): 4732–4737.

Lalena, J.N., Cleary, D.A., Carpenter, E., and Dean, N.F. 2008. *Inorganic Materials Synthesis and Fabrication*. Wiley-Interscience, New York, 303pp.

Landau, M.V., Vradman, L., Herskowitz, M., Koltypin, Y., and Gedanken, A. 2001. Ultrasonically controlled deposition-precipitation of cobalt and molybdenum oxide in the preparation of hydrodesulfurization catalysts. *Journal of Catalysis*, 201(1): 22–36.

Lee, A.S.-Y., Chu, S.-F., Chang, Y.-T., and Wang, S.-H. 2004. Synthesis of homopropargyl alcohols via sonochemical Barbier-type reaction. *Tetrahedron Letters*, 45(7): 1551–1553.

Lei, D., Liu, H., Yang, J., and Feng, Z. 2008. Method for preparation of high active nano crystal Pt/C catalyst. Patent CN 101269324, 10 pp.

Li, J. and Gao, F. 2009. Method for increasing electric capacity of positive electrode of organic electrochemical supercapacitor using cyclopentadienyl transition metal complex. Patent CN 101546652, 9 pp.

Li, J.-T., Lin, Z.-P., Qi, N., and Li, T.-S. 2004. Pinacol coupling of aromatic aldehydes and ketones using $TiCl_3$–Al–EtOH under ultrasound irradiation. *Synthetic Communications*, 34(23): 4339–4348.

Li, L.-X., Liu, J.-D., Wu, A.-B., Qin, S.-X., Li, Q., Deng, L.-Z., and Huang, X.-L. 2004. Synthesis under ultrasonic catalysis and structural characterization of dinuclear Zn(II)-macroheterocyclic complex. *Youji Huaxue*, 24(3): 297–299.

Li, Z.-Q., Qiu, L.-G., Wang, W., Xu, T., Wu, Y., and Jiang, X. 2008a. Fabrication of nanosheets of a fluorescent metal-organic framework $[Zn(BDC)(H_2O)]_n$ (BDC = 1,4-benzenedicarboxylate): Ultrasonic synthesis and sensing of ethylamine. *Inorganic Chemistry Communications*, 11(11): 1375–1377.

Li, Z.-Q., Qiu, L.-G., Xu, T., Wu, Y., Wang, W., Wu, Z.-Y., and Jiang, X. 2008b (volume date 2009). Ultrasonic synthesis of the microporous metal-organic framework $Cu_3(BTC)_2$ at ambient temperature and pressure: An efficient and environmentally friendly method. *Materials Letters*, 63(1): 78–80.

Li, X.-F. and Xue, S.-J. 2008. Synthesis of α,α-diphenyl-1-(phenylmethyl)-2-pyrrolidinemethanol derivatives and determination of their activity as pesticides. *Youji Huaxue*, 28(6): 1079–1082.

Lickiss, P.D. and Lucas, R. 1996. Ultrasonic activation of SiO_2 and GeO_2 in basic solutions of diols. *Polyhedron*, 15(12): 1975–1979.

Liu, B., Zhang, Y.-Y., Wang, J., Wang, X., Liu, L., Guo, Y., and Zhang, L.-Q. 2009. Study on damage of BSA in the presence of HP-Zn complex under ultrasonic irradiation by spectrophotomerty. *Bohai Daxue Xuebao, Ziran Kexueban*, 30(1): 5–12.

Mason, T.J. and Lorimer, J.P. 2002. *Applied Sonochemistry: Uses of Power Ultrasound in Chemistry and Processing*. Wiley-VCH, Weinheim, Germany, 314 pp.

Matsubayashi, G.-E., Ryowa, T., Tamura, H., Nakano, M., and Arakawa, R. 2002. Preparation and properties of dinuclear bis[dicarbonyl(cyclopentadienyl)]diiron(II) complexes with S–S coupled, dimerized sulfur-rich dithiolate ligands. *Journal of Organometallic Chemistry*, 645(1–2): 94–100.

Miecznik, P. 1993. Investigations of complexing in aqueous solutions of dimethyl sulfoxide and zinc chloride by an ultrasonic spectroscopy method. *Acustica*, 78(1): 36–45.

Mirkhani, V., Tangestaninejad, S., Moghadam, M., and Yadollahi, B. 2000. Efficient and selective epoxidation of alkenes by supported manganese porphyrin under ultrasonic irradiation. *Journal of Chemical Research, Synopses*, 2000(11): 515–517.

Mishra, A.P. and Gautam, S.K. 2002. Acoustical studies of some novel antimicrobially active 3d-metal-Schiff base complexes. *Journal of Indian Chemical Society*, 79(9): 725–728.

Miyoshi, N., Sostaric, J.Z., and Riesz, P. 2003. Correlation between sonochemistry of surfactant solutions and human leukemia cell killing by ultrasound and porphyrins. *Free Radical Biology & Medicine*, 34(6): 710–719.

Mizukoshi, K., Maeda, Y., and Nagata, Y. 2001. Concerted amplification control in ultrasonic special reaction field and preparation of functional micro particles. *Funsai*, 44: 27–33.

Motlagh, M.M., Rahimi, R., and Kachousangi, M.J. 2008. Ultrasonic method for the preparation of organic nanoparticles of porphyrin. In Seijas, J.A. and Vazquez, T.M.P. (Eds.), *International Electronic Conference on Synthetic Organic Chemistry, 12th*, Nov. 1–30, 2008, MOTL/1–MOTL/7.

Motojima, S. and Hishikawa, Y. 2002. Manufacture of carbon fibers by gas-phase growth in high yield by simultaneously pyrolyzing hydrocarbons or carbon monoxide in the presence of metal catalysts and exposing the reaction mixtures to ultrasound waves. Patent JP 2002088590, 7 pp.

Motoyama, Y., Nagashima, H., Takasaki, M., Yoon, S.H., and Mochida, I. 2006. Manufacturing method of metallic nanoparticle/C nano fiber structure. Patent JP 2006281201, 15 pp.

Nandiyanto, A.B.D., Kaihatsu, Y., Iskandar, F., and Okuyama, K. 2009. Rapid synthesis of a BN/CNT composite particle via spray routes using ferrocene/ethanol as a catalyst/carbon source. *Materials Letters*, 63(21): 1847–1850.

Nandurkar, N.S., Patil, D.S., and Bhanage, B.M. 2008. Ultrasound assisted synthesis of metal-1,3-diketonates. *Inorganic Chemistry Communications*, 11(7): 733–736.

Ni, J., Geng, J.-L., Qi, W.-B., Wang, Z.-L., Luo, Q.-H., Lu, G.-Y., and Feng, R. 2001. Ultrasonic synthesis of rare earth complexes with lactam type open chain crown ethers. *Wuji Huaxue Xuebao*, 17(5): 713–717.

Nikitenko, S.I., Moisy, P., Blanc, P., and Madic, C. 2004. Sonolysis of actinide(IV) β-diketonates in alkanes. *Comptes Rendus Chimie*, 7(12): 1191–1199.

Nikitenko, S.I., Moisy, Ph., Tcharushnikova, I.A., Blanc, P., and Madic, C. 2000. Volatile metal β-diketonates—New precursors for the sonochemical synthesis of nanosized materials—Sonolysis of thorium(IV) β-diketonates. *Ultrasonics Sonochemistry*, 7(4): 177–182.

Olea, D., Alexandre, S.S., Amo-Ochoa, P., Guijarro, A., de Jesus, F., Soler, J.M., de Pablo, P.J., Zamora, F., and Gomez-Herrero, J. 2005. From coordination polymer macrocrystals to nanometric individual chains. *Advanced Materials*, 17(14): 1761–1765.

Ouyang, J., Yu, L., Shen, H., Yu, J., and Zhang, Y. 2008. Method for synthesis of nonsteroid anti-inflammatory agent Tolmetin. CN 101177411, 10 pp.

Palmisano, G., Bonrath, W., Boffa, L., Garella, D., Barge, A., and Cravotto, G. 2007. Heck reactions with very low ligandless catalyst loads accelerated by microwaves or simultaneous microwaves/ultrasound irradiation. *Advanced Synthesis and Catalysis*, 349(14 + 15): 2338–2344.

Pan, H.-C., Liang, F.-P., Mao, C.-J., and Zhu, J.-J. 2007a. Sonochemical synthesis and resonance light scattering effect of Zn(II)bis(1-(2-pyridylazo)-2-naphthol)nanorods. *Nanotechnology*, 18(19): 195606/1–195606/6.

Pan, H.-C., Liang, F.-P., Mao, C.-J., Zhu, J.-J., and Chen, H.-Y. 2007b. Highly luminescent zinc(II)-bis(8-hydroxyquinoline) complex nanorods: Sonochemical synthesis, characterizations, and protein sensing. *Journal of Physical Chemistry B*, 111(20): 5767–5772.

Paulusse, J.M.J., Huijbers, J.P.J., and Sijbesma, R.P. 2007. Probing selectivity of ultrasound induced chain scission in reversible coordination polymers. *Polymer Preprints (American Chemical Society, Division of Polymer Chemistry)*, 48(2): 639–640.

Paulusse, J.M.J. and Sijbesma, R.P. 2006. Ultrasound in polymer chemistry: Revival of an established technique. *Journal of Polymer Science, Part A: Polymer Chemistry*, 44(19): 5445–5453.

Paulusse, J.M.J. and Sijbesma, R.P. 2008. Selectivity of mechanochemical chain scission in mixed palladium(II) and platinum(II) coordination polymers. *Chemistry Communications*, (37): 4416–4418.

Qiu, L.-G., Li, Z.-Q., Wu, Y., Wang, W., Xu, T., and Jiang, X. 2008. Facile synthesis of nanocrystals of a microporous metal-organic framework by an ultrasonic method and selective sensing of organoamines. *Chemical Communications*, (31): 3642–3644.

Raouafi, N., Belhadj, N., Boujlel, K., Ourari, A., Amatore, C., Maisonhaute, E., and Schoellhorn, B. 2009. Ultrasound-promoted aromatic nucleophilic substitution of dichlorobenzene iron(II) complexes. *Tetrahedron Letters*, 50(15): 1720–1722.

Santiso-Quinones, G., Higelin, A., Schaefer, J., Brueckner, R., Knapp, C., and Krossing, I. 2009. Cu[Al(ORF)$_4$] starting materials and their application in the preparation of [Cu(S$_n$)]$^+$ (n = 12, 8) complexes. *Chemistry—A European Journal*, 15(27): 6663–6677, S6663/1–S6663/110.

Shafi, K.V.P.M., Gedanken, A., Prozorov, R., and Balogh, J. 1998. Sonochemical preparation and size-dependent properties of nanostructured CoFe$_2$O$_4$ particles. *Chemistry of Materials*, 10(11): 3445–3450.

Shen, Q., Zhou, L., Zhou, R., and Wu, R. 2008. Method for accurately detecting content of dibutyltin oxide. Patent CN 101241111, 5 pp.

Sheng, Z. and Wang, J. 2006. Thin-walled large-diameter fullerene synthesis in solid-state reactor with iron carbonyl-based growth catalysts. Patent CN 1810634, 9 pp.

Shi, M. and Chen, X. 2007a. Preparation of 1,4-bis(dimethylsilyl)benzene under ultrasonic irradiation. *Fujian Fenxi Ceshi*, 16(3): 70–73.

Shi, M.-H. and Chen, X.-J. 2007b. Preparation of 1,4-bis(dimethylsilyl)benzene under ultrasonic irradiation. *Huagong Jishu Yu Kaifa*, 36(9): 8–10, 34.

Shieh, M., Liou, Y., Hsu, M.-H., Chen, R.-T., Yeh, S.-J., Peng, S.-M., and Lee, G.-H. 2002. A unique bismuth-iron chain polymer containing the…–Bi–Fe–… link: Formation and structure of [nBuBiFe(CO)$_4$]$_\infty$. *Angewandte Chemie International Edition*, 41(13): 2384–2386.

Silva, A. da C., de Souza, A.L.F., and Antunes, O.A.C. 2007. Phosphine-free Suzuki cross-coupling reactions under ultrasound. *Journal of Organometallic Chemistry*, 692(14): 3104–3107.

Soltanzadeh, N. and Morsali, A. 2009. Syntheses and characterization nano-structured bismuth(III) oxide from a new nano-sized bismuth(III) supramolecular compound. *Polyhedron*, 28(7): 1343–1347.

SONITEK: Ultrasonic horns. http://www.usedultrasonicwelders.com/Horn%20Brochure.pdf (accessed December 1, 2009).

Sostaric, J.Z., Pandian, R.P., Weavers, L.K., and Kuppusamy, P. 2006. Formation of lithium phthalocyanine nanotubes by size reduction using low- and high-frequency ultrasound. *Chemistry of Materials*, 18(17): 4183–4189.

Srinivasan, C. 2005. A 'sound' method for synthesis of single-walled carbon nanotubes under ambient conditions. *Current Science*, 88(1): 12–13.

Stryckmans, O., Segato, T., and Duvigneaud, P.H. 1996. Formation of MgO films by ultrasonic spray pyrolysis from β-diketonate. *Thin Solid Films*, 283(1–2): 17–25.

Suslick, K.S. 1995. Applications of ultrasound to materials chemistry. *MRS Bulletin*. April: 29–34.

Suslick, K.S. 2001. Sonochemistry and sonoluminescence. In *Encyclopedia of Physical Science and Technology*, 3rd edn., Vol. 17. Academic Press, San Diego, CA, 363–376.

Suslick, K.S. and Flannigan, D.J. 2008. Sonoluminescence. *Annual Reviews of Physical Chemistry*, 59: 659–683.

Suslick, K.S. and Skrabalak, S.E. 2008. Sonocatalysis. In Ertl, G., Knözinger, H., Schüth, F., and Weitkamp, J. (Eds.), *Handbook of Heterogeneous Catalysis*, Vol. 4. Wiley-VCH, Weinheim, Germany, p. 2006-17.

Sweet, J.D. and Casadonte Jr., D.J. 2001. Sonochemical synthesis of iron phosphide. *Ultrasonics Sonochemistry*, 8(2): 97–101.

Tu, S.P., Kim, D., and Yen, T.F. 2002. Decolorization and destruction of metallophthalocyanines in aqueous medium by ultrasound: A feasibility study. *Journal of Environmental Engineering and Science*, 1(3): 237–246.

Tu, S.-P. and Yen, T.F. 1999. Degradation and metal recovery of metalloporphyrins by ultrasonic oxidation. *Preprints—American Chemical Society, Division of Petroleum Chemistry*, 44(2): 209–212.

Uhl, W., Schmock, F., and Geiseler, G. 2002. [In$_3$Br$_3${C(SiMe$_3$)$_3$}$_3$][Li(THF)$_3$] with a chain of three indium atoms—A remarkable by-product of the synthesis of the tetraindium cluster In$_4$[C(SiMe$_3$)$_2$]$_4$. *Zeitschrift für Anorganische und Allgemeine Chemie*, 628(9–10): 1963–1966.

Uzcanga, I., Vrinat, M., Bezverkhyy, I., Afanasiev, P., and Scott, C.E. 2007. Ultrasound assisted synthesis, in aqueous media, of nanometric MoS$_2$, and its promotion with Co. *Preprints—American Chemical Society, Division of Petroleum Chemistry*, 52(2): 27–28.

Vitushkina, S.V., Ziolkovskiy, D.V., Starodub, V.A., Presniakov, I.A., Sobolev, A.V., Kajnakov, M., Radvakova, A., and Feher, A. 2009. *Research Letters in Inorganic Chemistry*, Article ID 243296. Hindawi Publishing Corporation, New York, 4 pp.

Wang, T., Lu, X.M., Han, L., Pang, X.L., and Ye, C.H. 2008. Helical nanostructure of tubular metal-organic complex synthesized by sonochemical process. *Science in China, Series B: Chemistry*, 51(10): 971–975.

Wang, J. and Sheng, Z. 2006. Hollow nano-cage carbon nanoparticles prepared by reaction of acetylene in presence of iron carbonyl catalysts. Patent CN 1810635, 9 pp.

Wang, J. and Teng, S. 2009. Method for preparation of hollow carbon nanocages by heat treatment with iodine. Patent CN 101544365, 7 pp.

Wang, X., Wu, C., Song, M., Li, J., Zhang, R., Wu, D., Yan, H., Gu, Z., and Chen, B. 2007. Method for preparing organic metal carborane targeting drug delivery system. Patent CN 1970087, 12 pp.

Wei, L. and Zhang, Y. 2009. Emulsion ethylene polymerization method with adjustable molecular weight of polyethylene. Patent CN 101445572, 7 pp.

Weng, J., Lai, L., Wang, X., and Zhang, Q. 2009. Hydroxy- and amino-functionalized sea urchins shaped carbon coated ferroferric oxide nanomaterial and preparation method thereof. Patent CN 101486491, 10 pp.

Werner, B. 2005. Ultrasound supported catalysis. *Ultrasonics Sonochemistry*, 12(1–2): 103–106.

Westerhausen, M., Gueckel, C., and Mayer, P. 2001. Synthesis and structure of a dimeric alkyldibariumtriszincate with a tetraanionic triszincate ligand and a unique central Ba$_4$Zn$_2$C$_6$ moiety. *Angewandte Chemie International Edition*, 40(14): 2666–2268.

Wooten, A., Carroll, P.J., Maestri, A.G., and Walsh, P.J. 2006. Unprecedented alkene complex of zinc(II): Structures and bonding of divinylzinc complexes. *Journal of American Chemical Society*, 128(14): 4624–4631.

Wu, Q., Chen, P., Liu, W., Wang, X., and Ding, Y. 2007. Method for preparing nanomaterials of ferrocene and its derivatives. Patent CN 1900097, 8 pp.

Wu, Q. and Yuan, P. 2008. Method for preparing carbon nanotubes from ferrocene. Patent CN 101244817, 8pp.

Xie, X., Ma, Z., and Ren, Q. 2006. Method for preparing carbon-supported cobalt porphyrin complex oxygen reducing catalyst. Patent CN 1824385, 8 pp.

Xue, M., Su, J., Ma, N., Sheng, Q., Zhang, Q., and Liu, Y. 2009. Production of copper phthalocyanine nanofilms. Patent CN 101372757, 9 pp.

Yang, J.-M., Ji, S.-J., Gu, D.-G., Shen, Z.-L., and Wang, S.-Y. 2005. Ultrasound-irradiated Michael addition of amines to ferrocenylenones under solvent-free and catalyst-free conditions at room temperature. *Journal of Organometallic Chemistry*, 690(12): 2989–2995.

Youichi, E., Samir, A.E.-F., and Chien, M.W. 2002. Ultrasound-enhanced dissolution of UO$_2$ in supercritical CO$_2$ containing a CO$_2$-philic complexant of tri-*n*-butylphosphate and nitric acid. *Industrial and Engineering Chemistry Research*, 41(9): 2282–2286.

Yu, Y., Ma, L., Huang, W., and Li, J. 2005. Manufacture of carbon nanotube composites with nano metals or metal oxides uniformly distributed on the surface. Patent CN 1569623, 11 pp.

Yuan, J. and Zhen, H. 2006. Synthesis of aromatics by Grignard reaction with supersonic wave as revulsant. *Jingxi Huagong Zhongjianti*, 36(4): 37–39.

Zhang, X.-H., Guan, L.-X., Li, J.-P., Wei, W., Zhao, N., Dong, M.-X., and Sun, Y.-H. 2006. Preparation of grafted transition metal Schiff base complex and their catalytic activity. *Huaxue Xuebao*, 64(24): 2479–2485.

Zhang, S., Yang, S., Lan, J., Tang, Y., Xue, Y., and You, J. 2009. Ultrasound-induced switching of sheetlike coordination polymer microparticles to nanofibers capable of gelating solvents. *Journal of American Chemical Society*, 131(5): 1689–1691.

Zhang, X.-H., Yang, X.-H., Xu, B., Han, W.-R., Zhang, Y.-X., and Zhang, X.-C. 2008. Preparation of grafted materials and their catalytic activity. *Hebei Shifan Daxue Xuebao, Ziran Kexueban*, 32(6): 784–787.

Zhou, W.-J., Ji, S.-J., and Shen, Z.-L. 2004. Facile synthesis of 1-aryl-5-ferrocenyl-penta-1,4-dien-3-ones under ultrasound irradiation or solvent-free condition. *Hecheng Huaxue*, 12(5): 421–424.

Zhu, T., Wu, Q., Chen, P., and Ding, Y. 2008 (volume date 2009). A novel waist-regulable dumbbell-like nanosuperstructure of (3-carboxy-1-acyl-propyl)-ferrocene. *Journal of Organometallic Chemistry*, 694(1): 21–26.

Zou, C. 2007. Preparation and application of surfactant clad Mo-acetylacetonate catalyst. Patent CN 101007286, 7 pp.

10 Ultrasound in Synthetic Applications and Organic Chemistry

Murlidhar S. Shingare and Bapurao B. Shingate

CONTENTS

Introduction	214
C–C Bond Formation Reactions	214
C–N Bond Formation Reactions	221
C–O Bond Formation Reactions	224
Esterification	225
Etherification	225
C-Alkylation/Acylation	226
N-Alkylation	227
S-Acylation	227
O-Acylation	227
N-Acylation	228
Sulfonation	228
Multicomponent Reactions	228
Name Reactions	233
Knoevenagel Condensation	233
Biginelli Reaction	233
Ullmann Reaction	235
Curtius Rearrangement	235
Cyclocondensation Reactions	236
Ring Opening Reactions	244
Oxidation	245
Epoxidation	247
Reduction	247
Nucleophilic Substitution Reactions	250
Oxime Deprotection	251
Preparations of Ionic Liquids	251
Miscellaneous Reactions	252
Conclusions	253
Abbreviations	253
References	254

INTRODUCTION

Ultrasound can be used in organic reactions since it provides specific activation based on a physical phenomenon called as acoustic cavitation. The chemical consequences of high-intensity ultrasound do not arise from an interaction of acoustic waves and matter at a molecular or atomic level. Instead, in liquids irradiated with high-intensity ultrasound, acoustic cavitation, that is, the formation, growth, and collapse of bubbles, provides the primary mechanism for sonochemical effects. In cavitation, bubble collapse produces intense local heating, high pressures, and very short lifetimes; these transient, localized hot spots drive high-energy chemical reactions. It is well established fact that these hot spots have temperatures of above $5000 \pm °C$, pressures at around 1000 atm, and heating and cooling rates above 1010 K/s. Thus, cavitation serves as a means of concentrating the diffuse energy of sound into a unique set of conditions to produce unusual materials from dissolved and generally volatile precursors.

As far as solids or solid–gas systems are concerned, chemical reactions are not generally seen in the ultrasonic irradiation. In addition, the interfacial region around cavitation bubbles has very large temperature, pressure, and possibly electric field gradients. Liquid motion in this vicinity also generates very large shear and strain gradients; these are caused by the very rapid streaming of solvent molecules around the cavitation bubble, as well as the intense shockwaves emanated on collapse. There are various applications of ultrasound in organic synthesis:

1. Homogeneous ultrasound chemistry
   a. Aqueous medium
   b. Nonaqueous media
2. Heterogeneous ultrasound chemistry
   a. Phase transfer catalysis
   b. Reactions with metals
   c. Heterogeneous catalysis
3. Enzyme reactions preferred in ultrasound

Some of the new advances in organic synthesis using ultrasound are discussed in this chapter.

C–C BOND FORMATION REACTIONS

Carbon–carbon bond-forming reactions are at the heart of synthetic organic chemistry; they allow for constructing simple feedstock chemicals as well as complex pharmaceuticals. Among them are reactions that directly couple two different carbon centers to form a new compound. Carbon–carbon bond formation is the essence of organic synthesis and provides the foundation for generating more complicated organic compounds from the simpler ones. In recent years, there has been increased recognition that organic reactions can precede well using ultrasound irradiation.

Javed et al. (1995) have shown that ultrasonic irradiations provide efficient promotion of reaction between 1,3-dienes and 1,4-diones in a Diels–Alder cycloaddition to afford bicyclo[4.4.0]-fused ring system in high yields (Scheme 10.1).

Guilet et al. (1998) have disclosed the influence of ultrasound power on the C-alkylation of phenyl acetonitrile by ethyl bromide under solid–liquid phase transfer catalysis in the presence of KOH and n-tetrabutyl ammonium hydrogen sulfate (TBAHS) (Scheme 10.2).

Liu et al. (2008) developed simple, novel, and efficient synthetic method for the synthesis of 3-indolylbenzoquinones using a catalytic amount of molecular iodine under ultrasound irradiation at room temperature (Scheme 10.3).

Li et al. (2007a) have described an efficient and convenient method for the preparation of pinacols from some aromatic aldehydes and ketones by using $TiCl_4$ in ethyl acetate under ultrasound irradiation (Scheme 10.4).

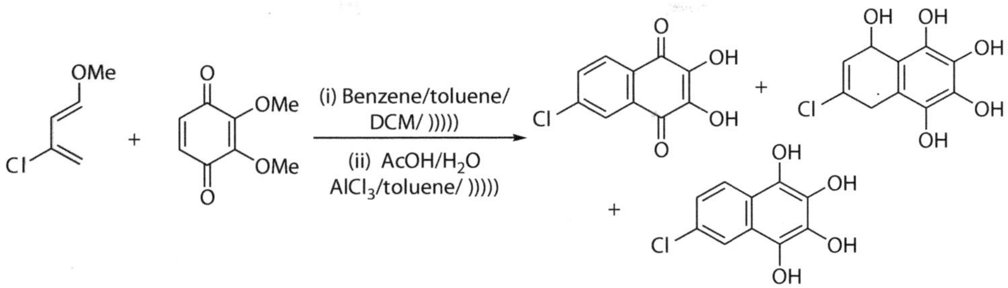

SCHEME 10.1 (From *Ultrason. Sonochem.*, 2, Javed, T., Mason, T.J., Phull, S.S., Baker, N.R., and Robertson, A., Influence of ultrasound on the Diels-Alder cyclization reaction: Synthesis of some hydroquinone derivatives and lonapalene, an anti-psoriatic agent, 3–4, Copyright (1995), with permission from Elsevier.)

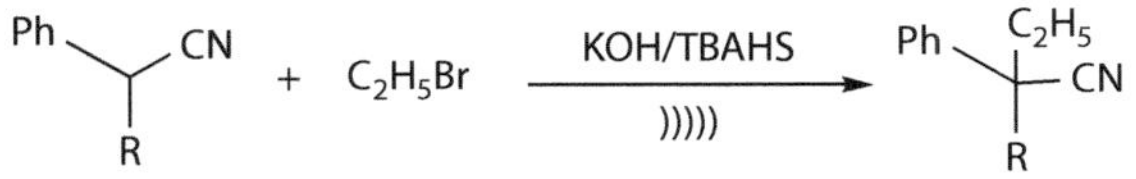

SCHEME 10.2 (From *Ultrason. Sonochem.*, 5, Guilet, R., Berlan, J., Louisnard, O., and Schwartzentruber, J., Influence of ultrasound power on the alkylation of phenyl acetonitrile under solid-liquid phase transfer catalysis condition, 21–25, Copyright (1998), with permission from Elsevier.)

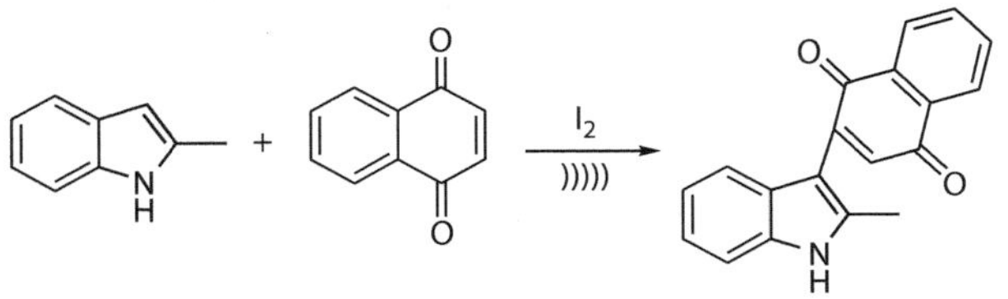

SCHEME 10.3 (From Liu, B. et al., *Synth. Commun.*, 38, 1279, 2008. With permission.)

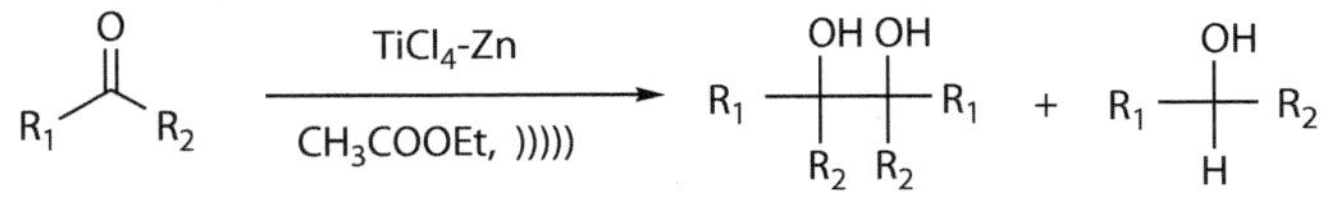

SCHEME 10.4 (From Li, J.T. et al., *Ind. J. Chem.*, 46B, 1303, 2007.)

Li et al. (2008) described under ultrasonication the regioselective alkylation at the 3-position of indole through conjugate addition-type reaction with 1,5 diaryl-1,4-pentadien-3-ones (Scheme 10.5).

Zhang et al. (2008) proved that ultrasound can facilitate the heterogeneous reaction of Suzuki coupling of phenylboronic acid with aryl halides in the presence of TBAB (Scheme 10.6).

Li et al. (2006a) have found an efficient and practical procedure for preparation of 2-(1*H*-indol-3-yl)(aryl)methyl malonitriles from some aryl methylenemalonitriles and indole under ultrasound irradiation (Scheme 10.7).

Jin et al. (2008) carried out the synthesis of chalcoide-like compounds catalyzed by $KF-Al_2O_3$ under ultrasound (Scheme 10.8).

Wang et al. (2008a) have introduced Meldrum's acid as an efficient organocatalyst in water for synthesis of bis (indo-3-yl) methane derivatives at ambient temperature under ultrasound irradiation (Scheme 10.9).

Fu et al. (2008) have developed a general synthesis that might prove a synthetically useful method for preparation of 1-ferrocenyl-2-arylacetylenes under ultrasound (Scheme 10.10).

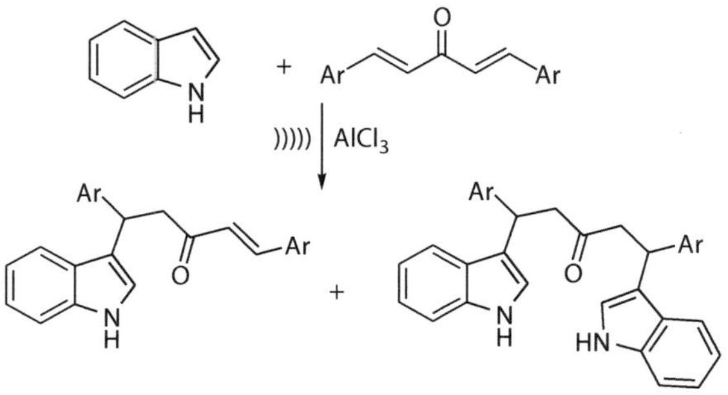

SCHEME 10.5 (From Li, J.T. et al., *Ind. J. Chem.*, 47B, 283, 2008.)

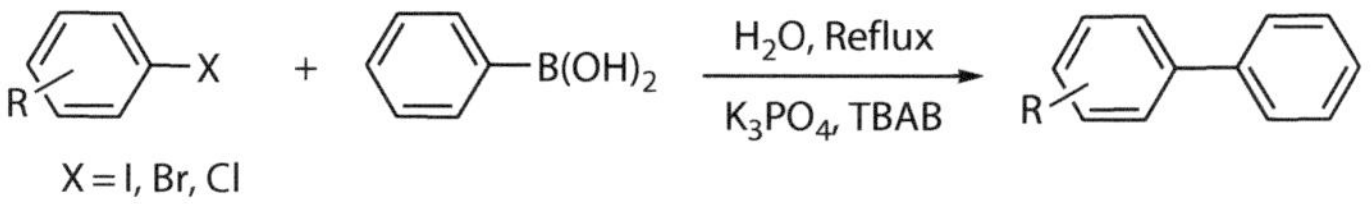

SCHEME 10.6 (From *Ultrason. Sonochem.*, 15, Zhang, J., Yang, F., Ren, G., Mak, T.C.W., Song, M., and Wu, Y., Ultrasonic irradiation accelerated cyclopalladated ferrocenylimines catalyzed Suzuki reaction in neat water, 115–118, Copyright (2008), with permission from Elsevier.)

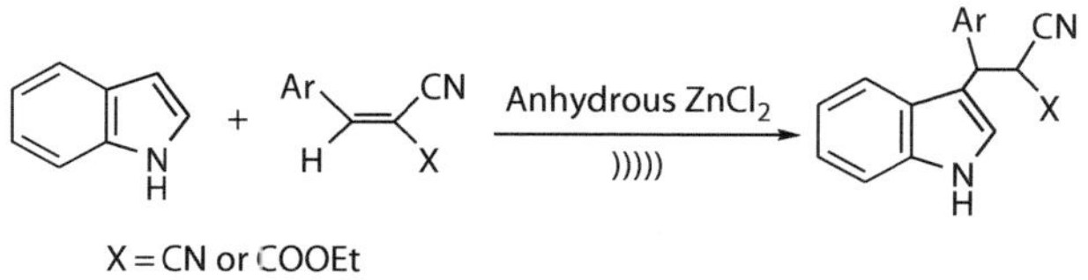

SCHEME 10.7 (From *Ultrason. Sonochem.*, 13, Li, J.T., Dai, H.G., Xu, W.Z. and Li, T.S., An efficient and practical synthesis of bis(indolyl)methanes catalyzed by aminosulfonic acid under ultrasound, 24–27, Copyright (2006), with permission from Elsevier.)

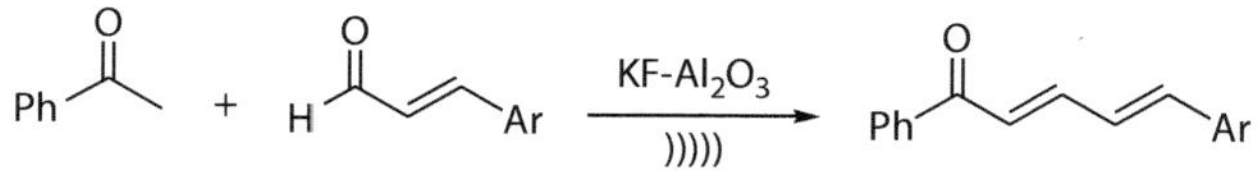

SCHEME 10.8 (From *Ultrason. Sonochem.*, 15, Jin, H., Xiang, L., Wen, F., Tao, K., Liu, Q., and Hou, T., Improved synthesis of chalconoid-like compounds under ultrasound irradiation, 681–683, Copyright (2008), with permission from Elsevier.)

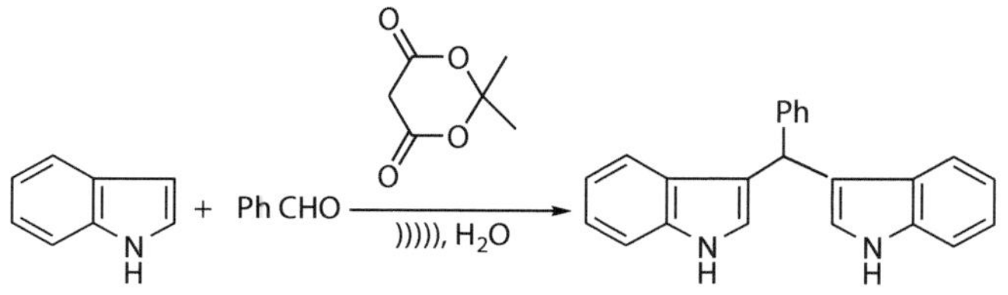

SCHEME 10.9 (From Wang, S.Y. et al., *Chin. J. Chem.*, 26, 22, 2008. With permission.)

Palmisano et al. (2007) showed that the Heck reactions can conveniently be carried out under ultrasound irradiation (Scheme 10.11).

Li et al. (2006a) described that amino sulfonic acid was found to be a cheap, novel, convenient, and efficient catalyst for the synthesis of BIM from indole and carbonyl compounds under ultrasound (Scheme 10.12).

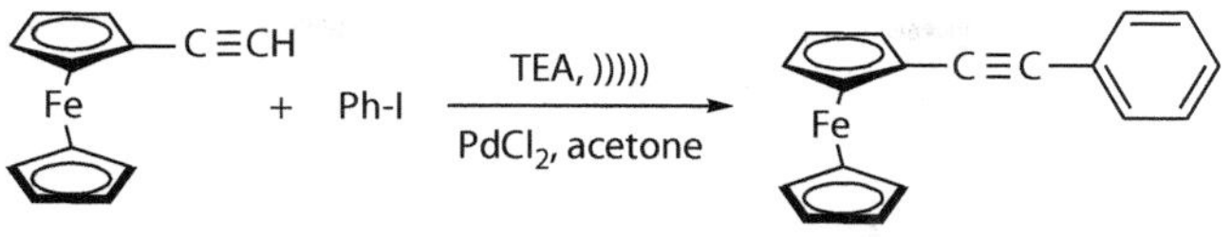

SCHEME 10.10 (From *Catal. Commun.*, 9, Fu, N., Zhang, Y., Yang, D., Chen, B., and Wu, X., A rapid synthesis of 1-ferrocenyl-2-arylacetylenes under ultrasound irradiation, 976–979, Copyright (2008), with permission from Elsevier.)

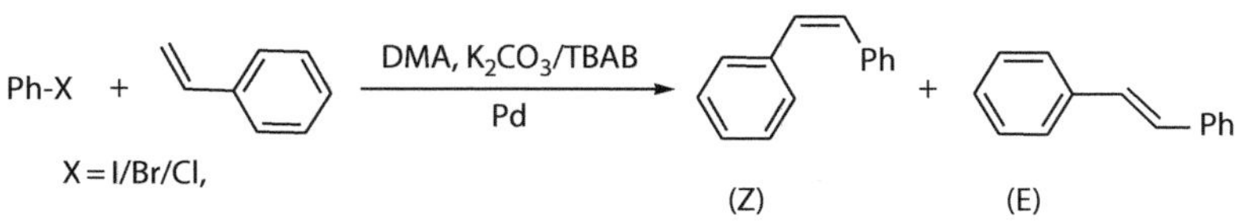

SCHEME 10.11 (Palmisano, G., Bonrath, W., Boffa, L., Garell, D., Barge, A., and Carvotto, G: Heck reactions with very low ligandless catalyst loads accelerated by microwaves or simultaneous microwaves/ultrasound irradiation. *Adv. Synth. Catal.* 349. 2338–2344. Copyright Wiley-VCH Verlag GmbH & Co. KGaA. With permission.)

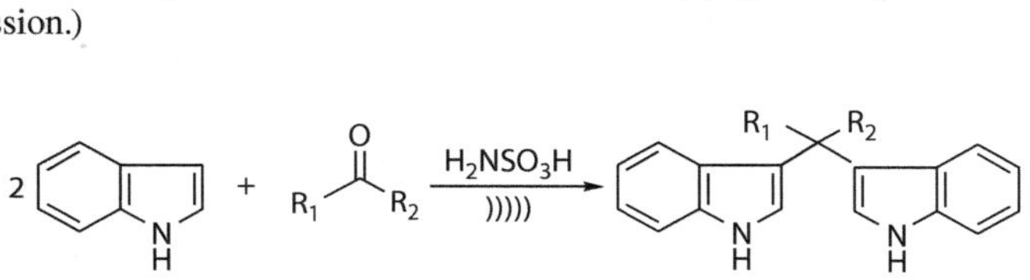

SCHEME 10.12 (From *Ultrason. Sonochem.*, 13, Li, J.T., Dai, H.G., Xu, W.Z., and Li, T.S., An efficient and practical synthesis of bis(indolyl)methanes catalyzed by aminosulfonic acid under ultrasound, 24–27, Copyright (2006), with permission from Elsevier.)

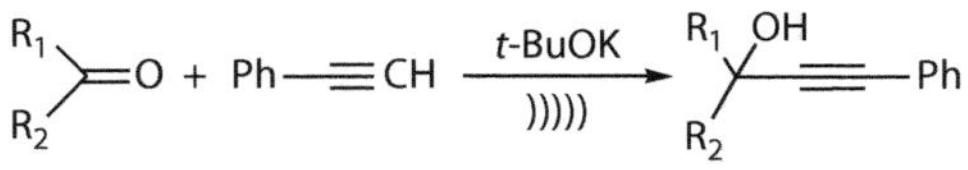

SCHEME 10.13 (From *Ultrason. Sonochem.*, 12, Ji, S.J., Shen, Z.L., Gu, D.G., and Hauang, X.Y., Ultrasound-promoted alkylation of ethylbenzene to ketones under solvent-free condition, 161–163, Copyright (2005), with permission from Elsevier.)

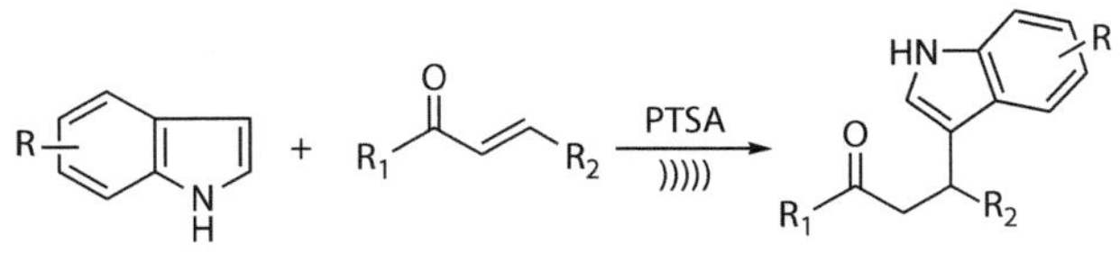

SCHEME 10.14 (From *Ultrason. Sonochem.*, 12, Ji, S.J., and Wang, S.Y., An expeditious synthesis of β-indolylketones catalyzed by p-toluenesulfonic acid (PTSA) using ultrasonic irradiation, 339–343, Copyright (2005), with permission from Elsevier.)

Ji and Wang (2005) have developed the preparation of propargylic alcohols catalyzed by potassium *tert*-butoxide under ultrasonication (Scheme 10.13).

Ji et al. (2005) found that PTSA as cheap, novel, convenient, and efficient catalyst for the synthesis of β-indolylketone derivatives from indoles and α–β unsaturated carbonyl ketones in the presence of ultrasound irradiation (Scheme 10.14).

Wang et al. (2005) have found an efficient and convenient method for pinacol coupling reaction of some aromatic aldehydes by using aqueous Vanadium (II) solution under ultrasound irradiation (Scheme 10.15).

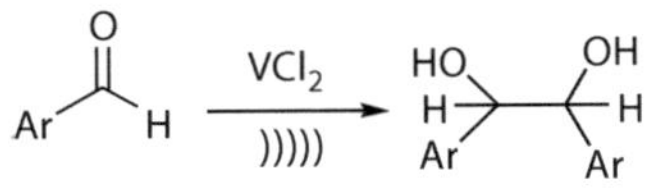

SCHEME 10.15 (From Wang, S.X. et al., *Synth. Commun.*, 35, 2387, 2005. With permission.)

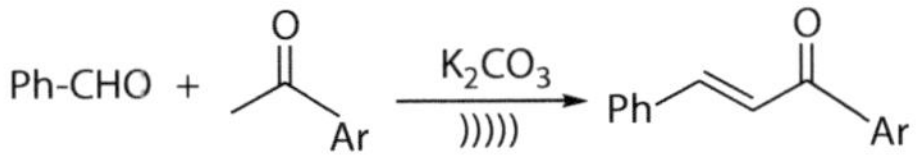

SCHEME 10.16 (From *Ultrason. Sonochem.*, 12, Wei, W., Qunrong, W., Liqin, D., Aiqing, Z., and Duoyuan, W., Synthesis of dinitrochalcones by using ultrasonic irradiation in the presence of potassium carbonate, 411–414, Copyright (2005), with permission from Elsevier.)

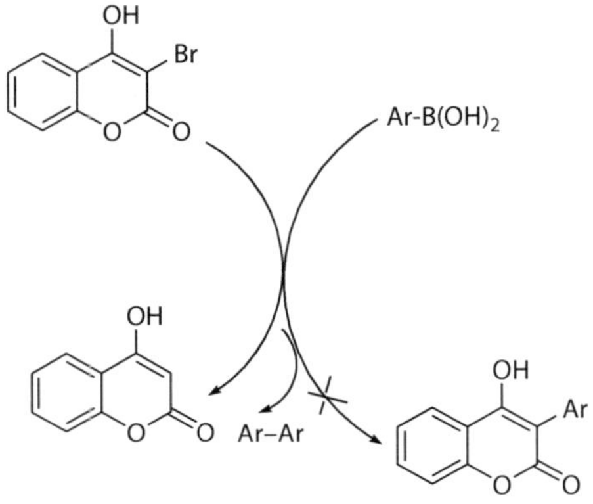

SCHEME 10.17 (From *Ultrason. Sonochem.*, 12, Cravotto, G., Palmisano, G., Tollari, S., Nano, G.M., and Penoni, A., The Suzuki homocoupling reaction under high-intensity ultrasound, 91–94, Copyright (2005), with permission from Elsevier.)

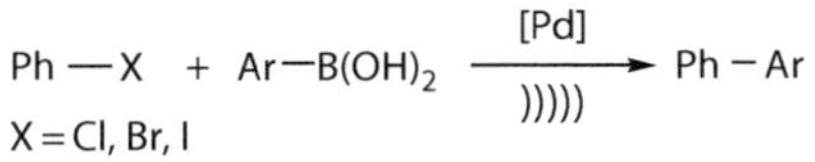

SCHEME 10.18 (From *Ultrason. Sonochem.*, 12, Polackova, V., Hutka, M., and Toma, S., Ultrasound effect on Suzuki reactions. 1. Synthesis of unsymmetrical biaryls, 99–102, Copyright (2005), with permission from Elsevier.)

Wei et al. (2005) described the synthesis of chalcones through the condensation between benzaldehyde, and acetophenone was performed in milder conditions by using ultrasound irradiation (Scheme 10.16).

Cravotto et al. (2005) described the Suzuki–Miyaura cross-coupling arylboronic acids with coumarino bromides in aq. media under ultrasound (Scheme 10.17).

Polackova et al. (2005) stated that ultrasound can facilitate the heterogeneous reaction of iodoarenes with different arylboronic acids, catalyzed by Pd/C and KF as base (Scheme 10.18).

Synthesis of bis indolyl alkanes can be carried out in excellent yields in the presence of silica-supported preyssler nanoparticles reported by Heravi et al. (2009a) and 1-hexene sulfonic acid sodium salt reported by Joshi et al. (2010) under ultrasound irradiation technique (Scheme 10.19).

Li et al. (2003a) have reported the Michael reaction of chalcones with various active methylene compounds such as diethyl malonate, nitromethane, cyclohexanone, ethyl acetoacetate, and acetyl acetone catalyzed by KF/basic alumina within shorter reaction times under ultrasonic irradiation (Scheme 10.20).

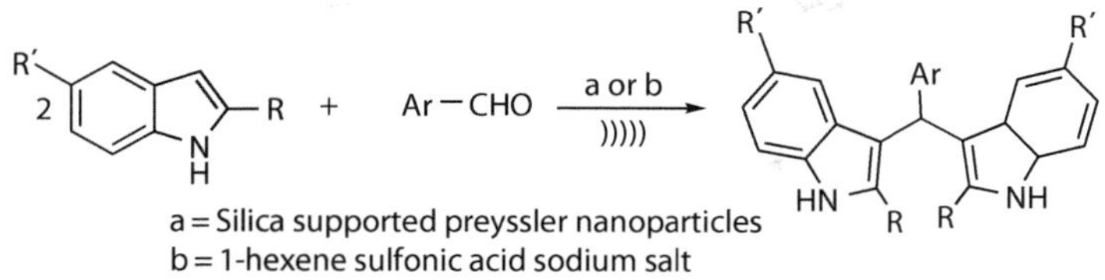

SCHEME 10.19 (From *Ultrason. Sonochem.*, 16, Heravi, M.M., Sadjadi, S., Sadjadi, S., Oskooie, H.A., and Bamoharram, F.F., A convenient synthesis of bis(indolyl)alkanes under ultra sonic irradiation using silica-supported Preyssler nano particles, 718–720, Copyright (2009), with permission from Elsevier; From *Ultrason. Sonochem.*, 17, Joshi, R.S., Mandhane, P.G., Diwakar, S.D., and Gill, C.H., Ultrasound assisted green synthesis of bis(indol-3-yl)methanes catalyzed by 1-hexenesuphonic acid sodium salt, 298–300, Copyright (2010), with permission from Elsevier.)

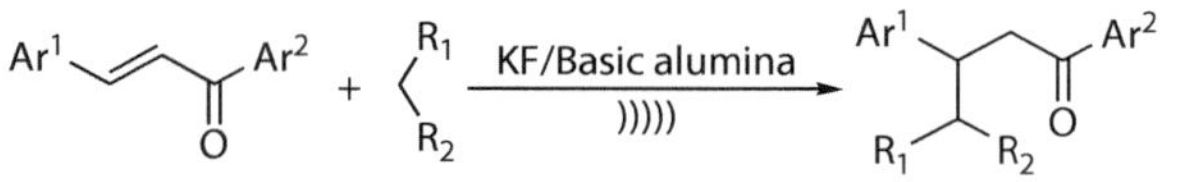

SCHEME 10.20 (From *Ultrason. Sonochem.*, 10, Li, J.T., Chen, G.F., Xu, W.Z., and Li, T.S., The Michael reaction catalyzed by KF/basic alumina under ultrasound irradiation, 115–118, Copyright (2003), with permission from Elsevier.)

SCHEME 10.21 (From *Ultrason. Sonochem.*, 16, Mamaghani, M. and Dastmard, S., An efficient ultrasound-promoted synthesis of the Baylis-Hillman adducts catalyzed by imidazole and L-proline, 445–447, Copyright (2009), with permission from Elsevier.)

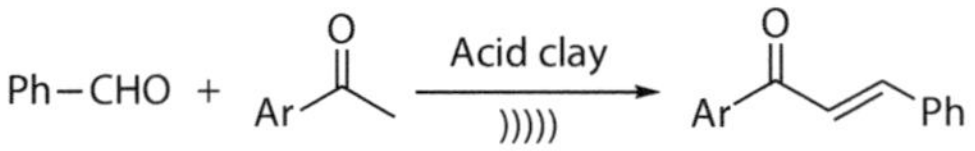

SCHEME 10.22 (From *Ultrason. Sonochem.*, 17, Chtourou, M., Abdelhédi, R., Frikha, M.H., and Trabelsi, M., Solvent free synthesis of 1,3-diaryl-2-propenones catalyzed by commercial acid-clays under ultrasound irradiation, 246–249, Copyright (2010), with permission from Elsevier.)

Convenient protocol for the synthesis of Baylis–Hillman adducts in the presence of imidazole and L-proline under ultrasound waves is reported by Mamaghani and Dastmard (2009) (Scheme 10.21).

Synthesis of *trans*-chalcones catalyzed by commercial acid clay (montmorillonite-KSF) under ultrasound irradiation was reported by Chtourou et al. (2010) (Scheme 10.22).

Li et al. (2010a) described ultrasound promoted synthesis of 3-aryl-3-hydroxy-2-1*H* indol-3-yl-1-phenyl-1-propanone via the cleavage of epoxides with indole in K10-ZnCl$_2$ (Scheme 10.23).

Claisen–Schmidt condensation of furfural with cycloalkanones or acetophenones to furnish α, α′-bis (substituted furfurylidine) cycloalkanones and chalcones has been shown by Li et al. (1999a) (Scheme 10.24).

Bian et al. (2009) have carried out the allylation reactions of aromatic aldehydes with allyl bromide using Sb-H$_2$O-KF-CH$_3$OH system under ultrasound irradiation (Scheme 10.25).

Lin et al. (2003) developed ultrasound irradiation technique has also been proved to be useful in the generation of dichlorocarbene from carbon tetrachloride and magnesium (Scheme 10.26).

Wei et al. (2004) have shown that ultrasonic irradiation effectively promotes the Diels–Alder reaction of substituted furans with reactive dienophiles (Scheme 10.27).

Gholap et al. (2005) have exploited the applications of ultrasound waves for the copper and ligand free Sonogashira reaction catalyzed by Pd(0) nanoparticles under ultrasonic irradiation (Scheme 10.28).

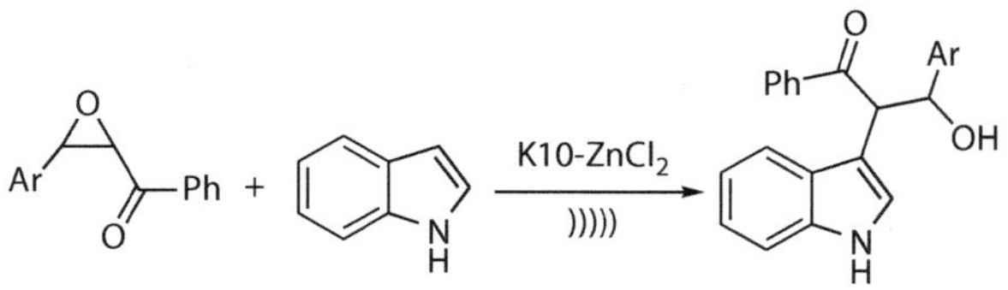

SCHEME 10.23 (From *Ultrason. Sonochem.*, 17, Li, J.T., Sun, M.X., and Yin, Y., Ultrasound promoted efficient method for the cleavage of 3-aryl-2,3-epoxyl-1-phenyl-1-propanone with indole, 359–362, Copyright (2010), with permission from Elsevier.)

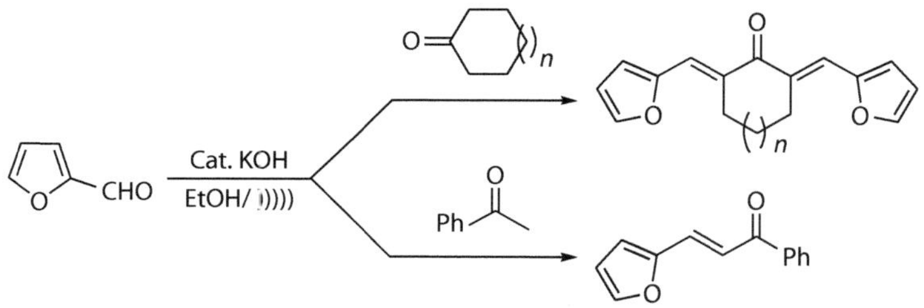

SCHEME 10.24 (From Li, J.T. et al., *Synth. Commun.*, 29, 965, 1999. With permission.)

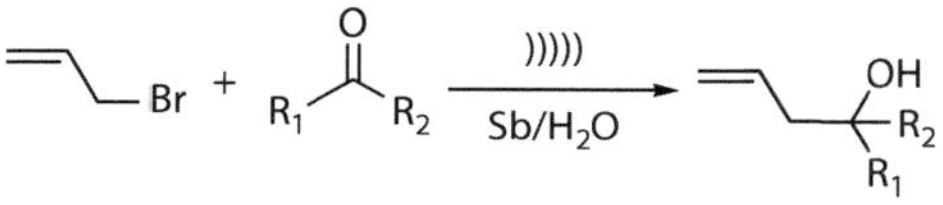

SCHEME 10.25 (From Bian, Y.J. et al., *Synth. Commun.*, 39, 2370, 2009. With permission.)

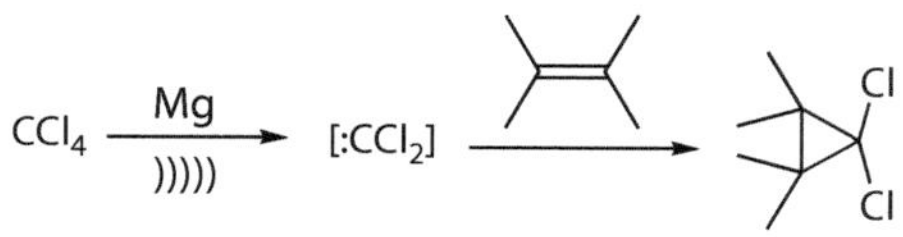

SCHEME 10.26 (From Lin, H. et al., *Molecules*, 8, 608, 2003.)

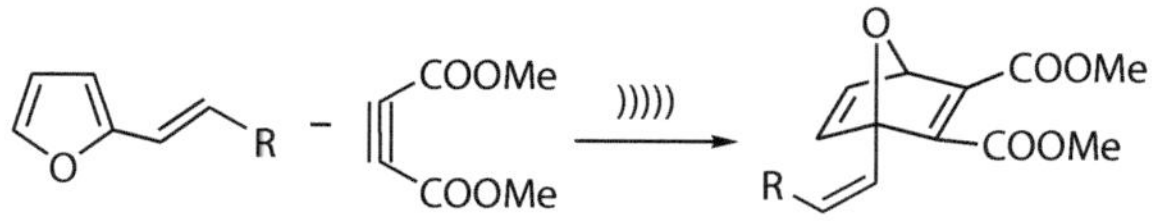

SCHEME 10.27 (With permission from Wei, K., Gao, H.T., and Li, W.D.Z., Facile synthesis of oxabicyclic alkenes by ultrasonication promoted Diels–Alder cycloaddition of furano dienes, *J. Organ. Chem.*, 69, 5763–5765, 2004.)

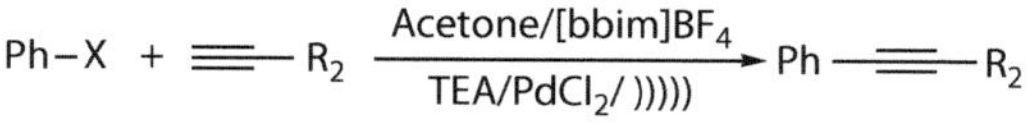

SCHEME 10.28 (With permission from Gholap, A.R., Venkatesan, K., Pasricha, R., Daniel, T., Lahoti, R.J. and Srinivasan, K.V., Copper and ligand-free Sonogashira reaction catalyzed Pd(0) nanoparticles at ambient conditions under ultrasound irradiation, *J. Organ. Chem.*, 70, 4869–4872, 2005.)

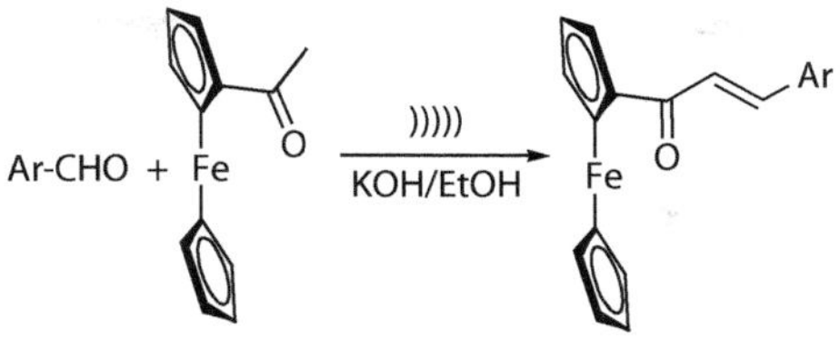

SCHEME 10.29 (From Ji, S.J. et al., *Chin. Chem. Lett.*, 14, 663, 2003.)

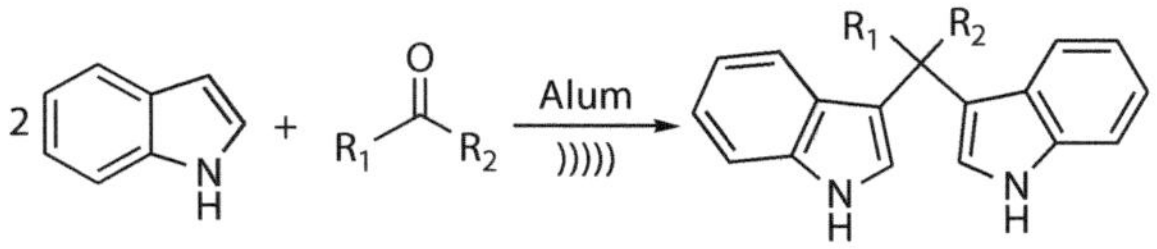

SCHEME 10.30 (From Sonar, S.S. et al., *Bull. Korean Chem. Soc.*, 30, 825, 2009. With permission.)

Ji et al. (2003) have established a practical condensation protocol for the synthesis of ferrocenyl under ultrasound irradiation (Scheme 10.29).

Sonar et al. (2009) described alum as an easily available, inexpensive, efficient, and safe catalyst for the synthesis of bis(indolyl)-methane derivatives from various aryl aldehydes by ultrasound irradiation (Scheme 10.30).

C–N BOND FORMATION REACTIONS

Carbon–nitrogen bond formation is the essence of organic synthesis and provides the foundation for generating more complex organic compounds from the simpler ones. In the latest decade, there has been increased recognition that the synthesized organic compounds have great interest from the viewpoint of their pharmaceutical as well as medicinal significance due to their biodynamic properties. Li et al. (2010b) have synthesized 5-aryl-1,3-diphenyl pyrazoles in the presence of ultrasound irradiation using hydrochloric acid as catalyst under essentially ecofriendly reaction conditions (Scheme 10.31).

Ultrasound irradiations have also found applications in the glycouril synthesis in the presence of potassium hydroxide reported by Li et al. (2010c) (Scheme 10.32).

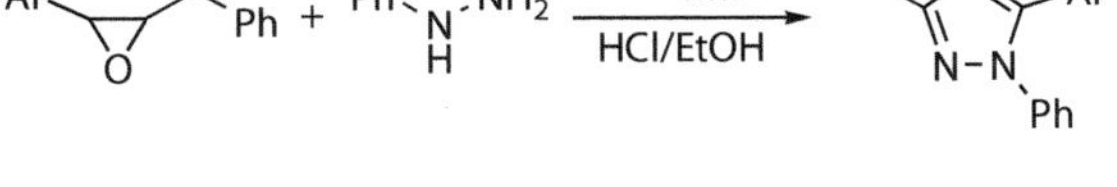

SCHEME 10.31 (From *Ultrason. Sonochem.*, 17, Li, J.T., Yin, Y., Li, L., and Sun, M. X., A convenient and efficient protocol for the synthesis of 5-aryl-1,3-diphenylpyrazole catalyzed by hydrochloric acid under ultrasound irradiation, 11–13, Copyright (2010), with permission from Elsevier.)

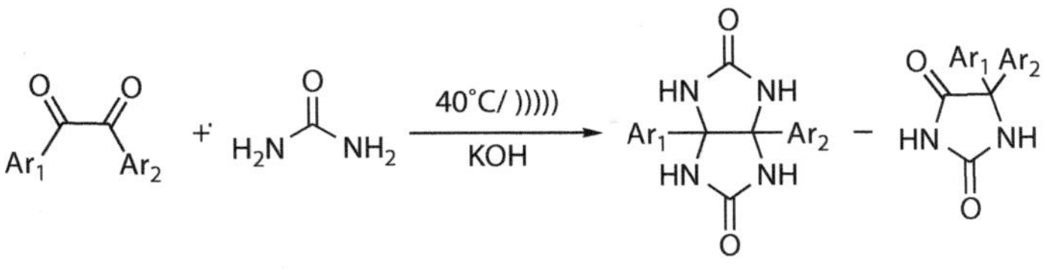

SCHEME 10.32 (From *Ultrason. Sonochem.*, 17, Li, J.T., Liu, X.R., and Sun, M.X., Synthesis of glycouril catalyzed by potassium hydroxide under ultrasound irradiation, 55–57, Copyright (2010), with permission from Elsevier.)

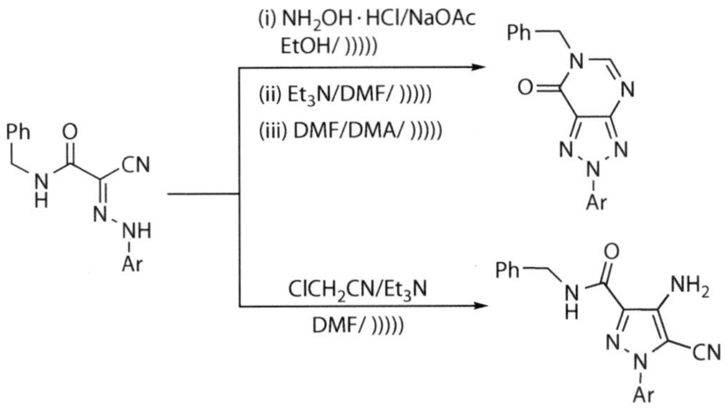

SCHEME 10.33 (From *Ultrason. Sonochem.*, 16, Al-Zaydi, K.M., A simplified green chemistry approaches to synthesis of 2-substituted 1,2,3-triazoles and 4-amino-5-cyanopyrazole derivatives conventional heating versus microwave and ultrasound as ecofriendly energy sources, 805–809, Copyright (2009), with permission from Elsevier.)

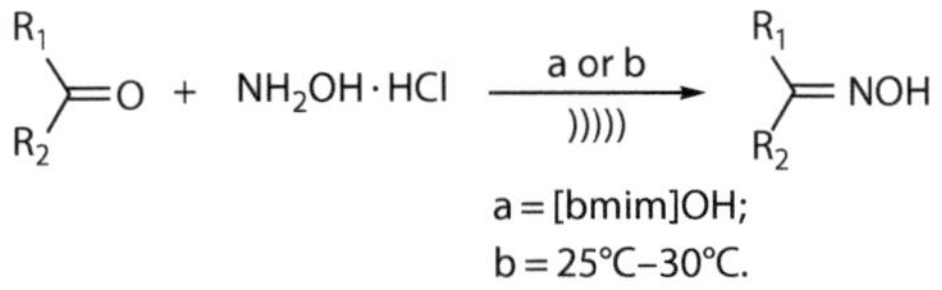

SCHEME 10.34 (From *Ultrason. Sonochem.*, 16, Zang, H., Wang, M., Cheng, B.W., and Song, J., Ultrasound-promoted synthesis of oximes catalyzed by a basic ionic liquid [bmIm]OH, 301–303, Copyright (2009), with permission from Elsevier.)

Al Zaydi (2009) has synthesized a novel substituted 2-aryl-1,2,3-triazoles and 4-amino pyrazoles from aryl hydrazono nitriles by applying ultrasound technique (Scheme 10.33).

Influence of ultrasound irradiations have made remarkable improvement in the efficient transformation of carbonyl compounds into its corresponding oximes (Scheme 10.34) reported by Zang (2009) and Li (2006b).

Kamal et al. (2004) provided sonochemical nitration of phenols with $ZnCl_2/HNO_3$ in high regioselectivity within shorter reaction times (Scheme 10.35).

Meciarova et al. (2003) described the application of ultrasound in the nucleophilic aromatic substitution reactions on haloarenes with a variety of amines (Scheme 10.36).

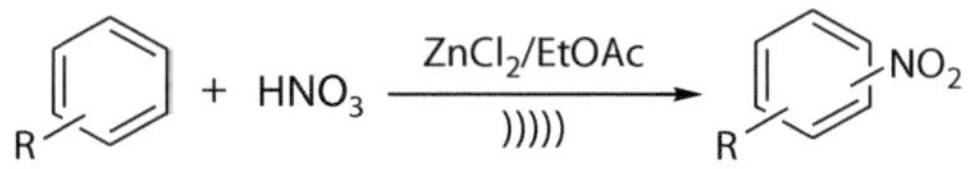

SCHEME 10.35 (From *Ultrason. Sonochem.*, 11, Kamal, A., Ashwini Kumar, B., Arifuddin, M., and Patric, M., An efficient and facile nitration of phenols with nitric acid/zinc chloride under ultrasonic conditions, 455–457, Copyright (2004), with permission from Elsevier.)

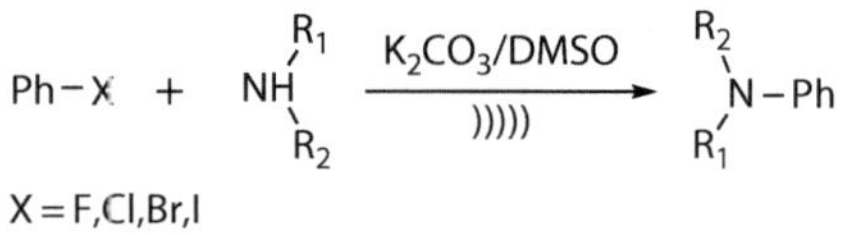

SCHEME 10.36 (From *Ultrason. Sonochem.*, 10, Meciarova, M., Toma, S., and Magdolen, P., Ultrasound effect on the aromatic nucleophilic substitution reactions on some haloarenes, 265–270, Copyright (2003), with permission from Elsevier.)

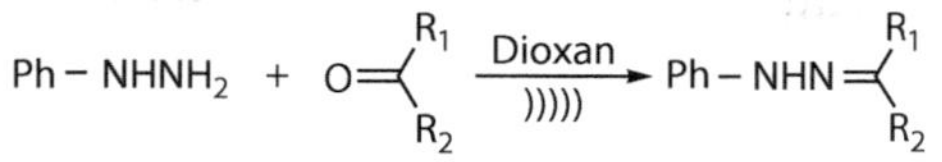

$$Ph-NHNH_2 + O{=}{<}^{R_1}_{R_2} \xrightarrow[\text{)))))}]{\text{Dioxan}} Ph-NHN{=}{<}^{R_1}_{R_2}$$

SCHEME 10.37 (From *Ultrason. Sonochem.*, 10, Jarikote, D.V., Deshmukh, R.R., Rajagopal, R., Lahoti, R.J., Daniel, T., and Srinivasan, K.V., Ultrasound promoted facile synthesis of arylhydrazones at ambient conditions, 45–48, Copyright (2003), with permission from Elsevier.)

$$Ar-OH \xrightarrow[\text{Ethyl ammonium nitrate/)))))}]{\text{Ferric nitrate/Clayfan}} Ar-NO_2$$

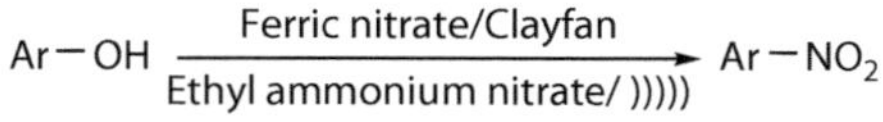

SCHEME 10.38 (From *Ultrason. Sonochem.*, 10, Rajagopal, R. and Srinivasan, K.V., Ultrasound promoted para-selective nitration of phenols in ionic liquid, 41–43, Copyright (2003), with permission from Elsevier.)

$$R\text{-COOH} \xrightarrow[\text{AcCN/)))))}]{\text{ClCO}_2\text{Et/K}_2\text{CO}_3} [R\text{-CO-O-CO-OEt}] \xrightarrow[\text{AcCN/)))))}]{R'NH_2} R\text{-CO-NHR}'$$

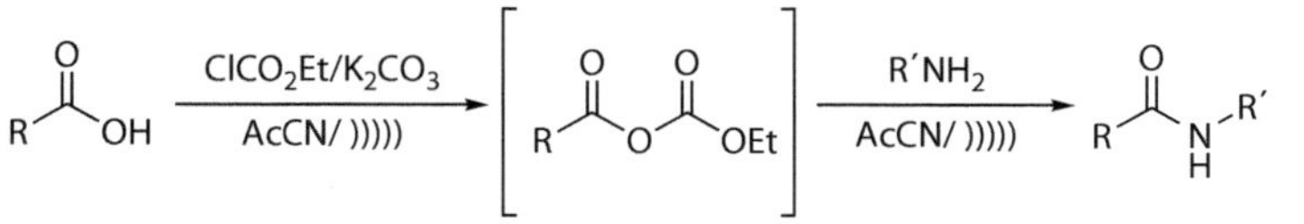

SCHEME 10.39 (From *Ultrason. Sonochem.*, 16, Srivastava, R.M., Neves Filho, R.A.W., da Silva, C.A., and Bortoluzzi, A.J., First ultrasound-mediated one-pot synthesis of N-substituted amides, 737–742, Copyright (2009), with permission from Elsevier.)

Synthesis of aryl hydrazones using phenyl hydrazines and carbonyl compounds has been successfully carried out in the presence of ultrasound by Jarikote et al. (2003) (Scheme 10.37).

Rajgopal and Srinivasan (2003) have found the application of sonochemistry in the para nitration of phenols using ferric nitrate and clayfan in ethyl ammonium nitrate (Scheme 10.38).

Srivastva et al. (2009) have developed a speedy and clean protocol for the synthesis of secondary amides using ultrasound waves (Scheme 10.39).

Three-component reaction of substituted amine, fluoroaldehyde, and dialkyl phosphite to afford novel α-amino phosphonates has been carried out by Song et al. (2006) in the presence of ultrasound irradiations and $BF_3 \cdot Et_2O$ (Scheme 10.40).

Leite et al. (2008) have discovered mild and rapid procedure for ultrasound-accelerated synthesis of aryl hydrazones from aryl aldehyde, ketone, and hydrazides in aqueous medium (Scheme 10.41).

Zhang et al. (2005) have synthesized 2-cyanoacrylates containing pyridinyl moiety under ultrasound irradiations (Scheme 10.42).

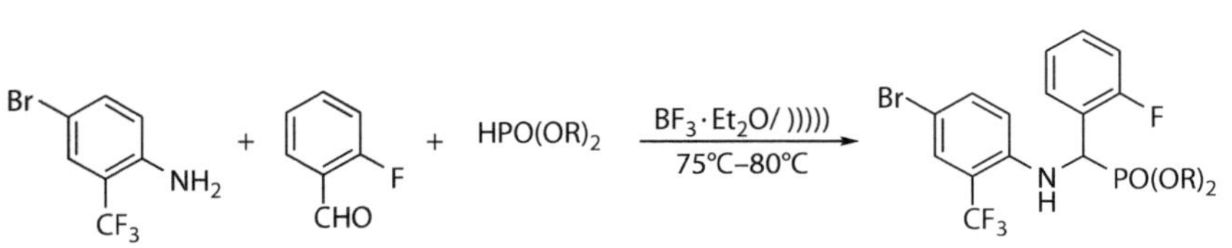

SCHEME 10.40 (From *Ultrason. Sonochem.*, 13, Song, B.A., Zang, G.P., Yang, S., Hu, D.Y., and Jin, L.H., Synthesis of N-(4-bromo-2-trifluoromethylphenyl)-1-(2-fluorophenyl)-O,O-dialkyl-α-aminophosphonates under ultrasonic irradiation, 139–142, Copyright (2006), with permission from Elsevier.)

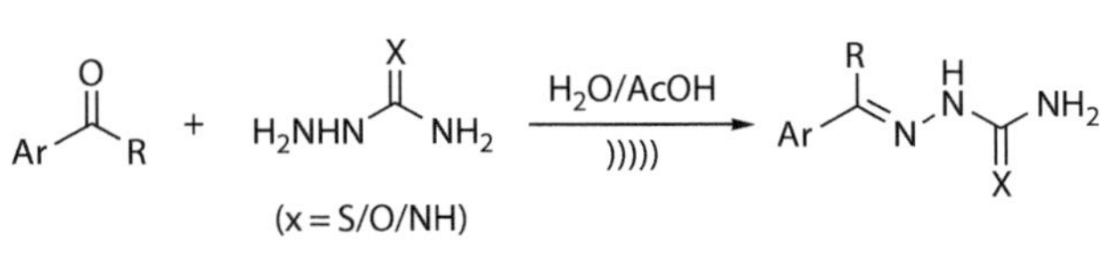

SCHEME 10.41 (From *Tetrahedron Lett.*, 49, Leite, A.C.L., Moreira, D.R. de M., Coelho, L.C.D., Menezes, F.D. de., and Brondani, D.J., Synthesis of aryl-hydrazones via ultrasound irradiation in aqueous medium, 1538–1541, Copyright (2008), with permission from Elsevier.)

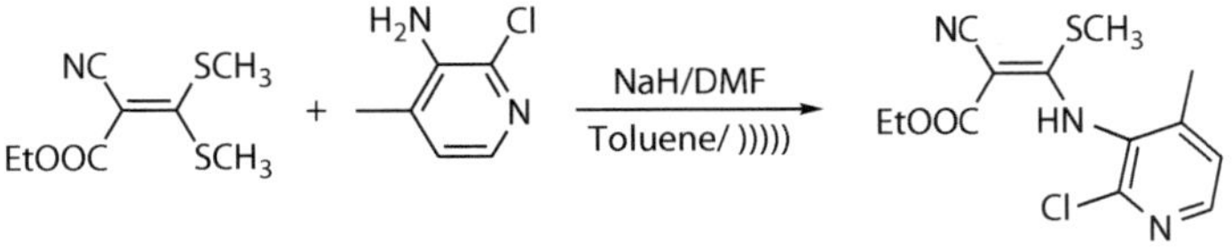

SCHEME 10.42 (From Zhang, H. et al., *J. Heterocycl. Chem.*, 42, 1211, 2005. With permission.)

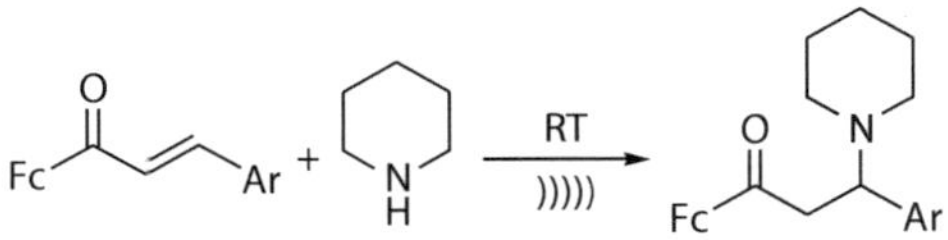

SCHEME 10.43 (From *J. Organometall. Chem.*, 690, Yang, J.M., Ji, S.J., Gu, D.G., Shen, Z.L., and Wang, S.Y., Ultrasound-irradiated Michael addition of amines to ferrocenylenones under solvent-free and catalyst-free conditions at room temperature, 2989–2995, Copyright (2005), with permission from Elsevier.)

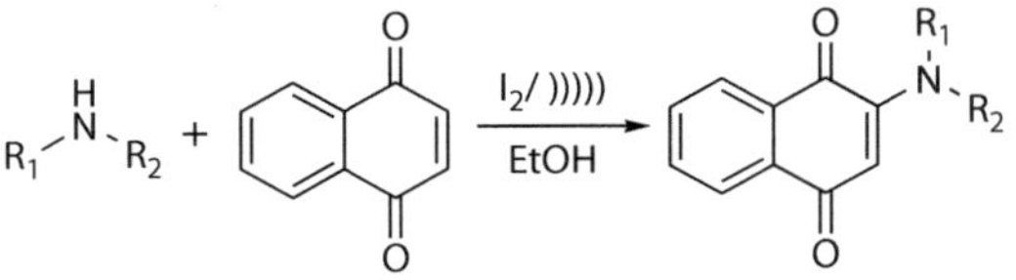

SCHEME 10.44 (From Liu, B. and Ji, S.J., *Synth. Commun.*, 38, 1201. With permission.)

Yang et al. (2005) have reported a facile Michael addition of ferrocenyl enones with aliphatic amines under ultrasound irradiation in the absence of solvent and catalyst to afford 1-ferrocenyl-2-amino carbonyl compounds (Scheme 10.43).

Liu and Ji (2008) have developed iodine promoted protocol for the synthesis of 2-amino-1,4-naphthoquinolines using ultrasound irradiations at room temperature (Scheme 10.44).

C—O BOND FORMATION REACTIONS

Mitsunobu coupling reaction of sterically hindered phenols and alcohols by the use of high reaction concentrations in combination with sonication has been reported by Lepore and He (2003) (Scheme 10.45).

A new one-pot method for the preparation of alkynyl sulfonate esters from terminal alkynes using metal-assisted, ultrasound-enhanced nucleophilic acetylinic displacement was discovered by Tuncay et al. (1999) (Scheme 10.46).

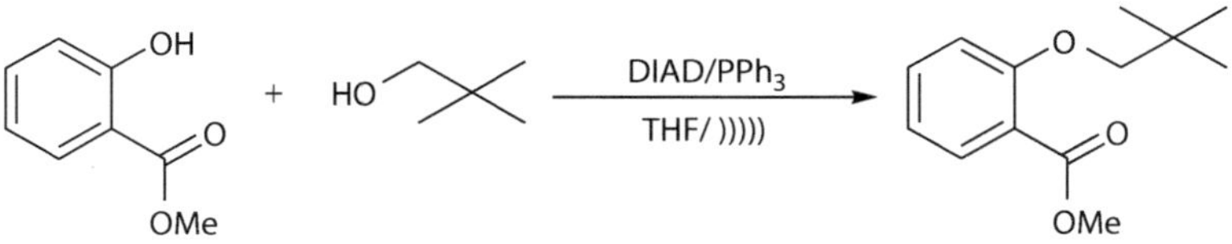

SCHEME 10.45 (From Lepore, S.D. and He, Y., *J. Organ. Chem.*, 68, 8261, 2003. With permission.)

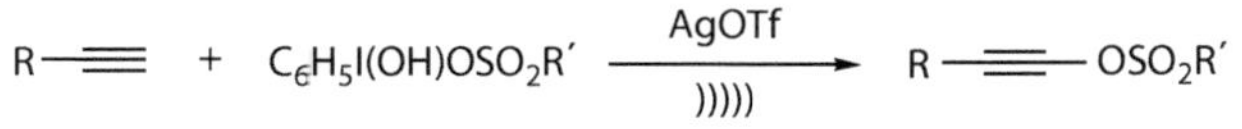

SCHEME 10.46 (From *Tetrahedron Lett.*, 40, Tuncay, A., Anaclerio, B.M., Zolodz, M., and Suslick, K.S., New one-pot method for the synthesis of alkynyl sulfonate esters using ultrasound, 599–602, Copyright (1999), with permission from Elsevier.)

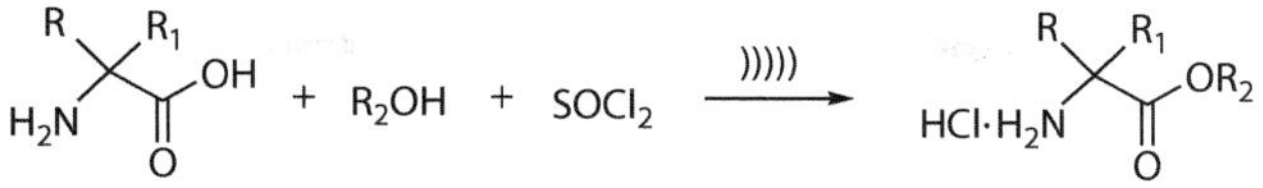

$$\underset{H_2N}{\overset{R}{\underset{|}{C}}}\underset{\parallel}{\overset{R_1}{\underset{O}{C}}}OH + R_2OH + SOCl_2 \xrightarrow{)))))} \underset{HCl\cdot H_2N}{\overset{R}{\underset{|}{C}}}\underset{\parallel}{\overset{R_1}{\underset{O}{C}}}OR_2$$

SCHEME 10.47 (From Kantharaju and Babu, V.V.S., *Ind. J. Chem.*, 45B, 1942, 2006.)

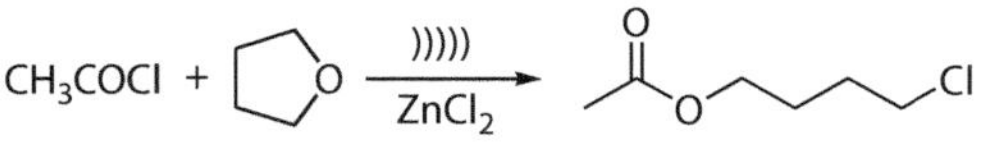

$$CH_3COCl + \underset{O}{\bigcirc} \xrightarrow[ZnCl_2]{)))))} \underset{O}{\overset{O}{\parallel}}\! \diagdown\!\!O\!\!\diagdown\!\!\diagdown\!\!Cl$$

SCHEME 10.48 (From *Ultrason. Sonochem.*, 13, Pasha, M.A. and Myint, Y.Y., Ultrasound assisted synthesis of δ-chloroesters from tetrahydrofuran and acyl chlorides in the presence of catalytic zinc dust, 175–179, Copyright (2006), with permission from Elsevier.)

$$Ph\underset{\parallel}{\overset{O}{C}}OH + R\text{-}OH \underset{}{\overset{)))))}{\rightleftharpoons}} Ph\underset{\parallel}{\overset{O}{C}}OR$$

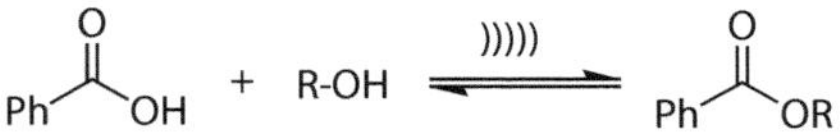

SCHEME 10.49 (With kind permission from Springer Science + Business Media: *Catalysis Letters*, Nitric acid-oxidized carbon for the preparation of esters under ultrasonic activation, 87, 2003, 143–47, Nevskaia, D.M. and Martin-Aranda, R.M.)

ESTERIFICATION

Ultrasound technique has been proved in various protections and deprotection methods of amino acids. Efficient protocol for esterification of wide range of amino acids under ultrasound has been developed by Kantharaju and Babu (2006) (Scheme 10.47).

Pasha and Myint (2006) have shown the application of ultrasound irradiation technique in transformation of acyl chlorides into its corresponding δ-chloroesters (Scheme 10.48).

This protocol describes the use of nitric acid–oxidized carbon for the esterification of benzoic and phenyl acetic acids with alcohols under ultrasound irradiation by Nevskaia Martin-Aranda (2003) (Scheme 10.49).

ETHERIFICATION

Rama and Pasha (2005) described that regioselective synthesis of β-iodoethers can be carried out effectively in the presence of ultrasound irradiations from olefin, alcohol, and iodine to afford Markovnikov addition product (Scheme 10.50).

Peng and Song (2002) have reported the synthesis of ethers by greener pathways employing ultrasound technique along with microwave irradiations (Scheme 10.51).

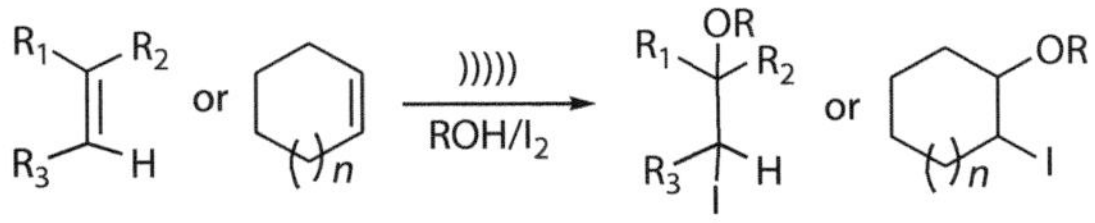

SCHEME 10.50 (From *Ultrason. Sonochem.*, 12, Rama, K., Pasha, M.A., Ultrasound promoted regioselective synthesis of β-iodoethers from olefin-I2-alcohol, 437–440, Copyright (2005), with permission from Elsevier.)

$$Ar\text{-}OH + R\text{-}Cl \xrightarrow[SMUI]{NaOH/H_2O} Ar\text{-}O\text{-}R$$

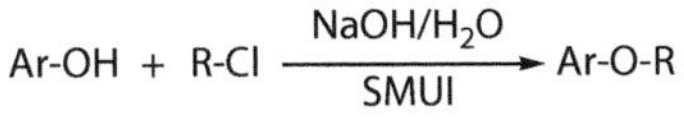

SCHEME 10.51 (From Peng, Y. and Song, G., Combined microwave and ultrasound assisted Williamson ether synthesis in the absence of phase-transfer catalysts, *Green Chem.*, 4, 349–351, 2002, by permission of The Royal Society of Chemistry.)

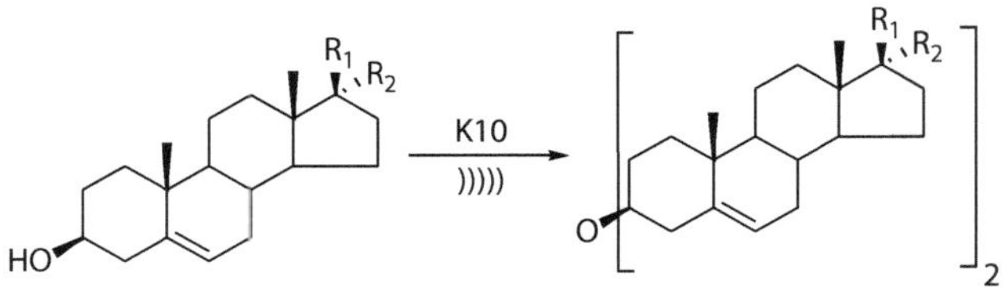

SCHEME 10.52 (From *Ultrason. Sonochem.*, 5, Li, J.T., Li, T.S., Li, L.J., and Yang, Z.Q., Ultrasound-promoted preparation of disteryl ethers catalyzed by montmorillonite K 10, 83–85, Copyright (1998), with permission from Elsevier.)

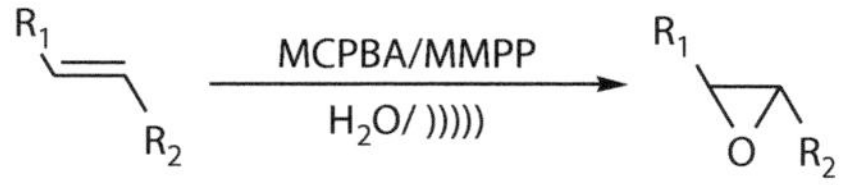

SCHEME 10.53 (From *Ultrason. Sonochem.*, 2, Marcel, S.F., Lie, K.J., and Lam, C.K., Ultrasound-assisted epoxidation reaction of long-chain unsaturated fatty esters, s11–s14, Copyright (1995), with permission from Elsevier.)

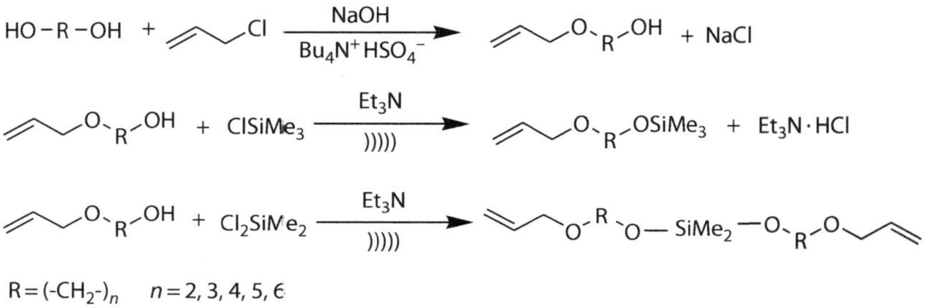

SCHEME 10.54 (From *Ultrason. Sonochem.*, 11, Urbala, M. and Antoszczyszyn, M., The synthesis of allyl ether functionalized siloxane monomers under ultrasonic irradiation at ambient conditions, 409–414, Copyright (2004), with permission from Elsevier.)

Li et al. (1998) have shown that ultrasound irradiation can accelerate the formation of disteryl ethers in the presence of montmorillonite K10 (Scheme 10.52).

Ultrasound technique has been applied for the epoxidation of unsaturated fatty esters using MCPBA in water by Marcel et al. (1995) (Scheme 10.53).

Urbala and Antoszczyszyn (2004) reported the synthesis of allyl ether–functionalized siloxane monomers under ultrasonic irradiation at ambient condition (Scheme 10.54).

C-ALKYLATION/ACYLATION

Hofmann et al. (2003) have shown the effect of ultrasound waves on reaction rate in the presence of phase transfer catalyst for the C-alkylation of benzyl cyanide (Scheme 10.55).

Yadav and Rahuman (2003) have used solid acids with the ultrasound technique for the acylation of 2-methoxy naphthalene in efficient ways (Scheme 10.56).

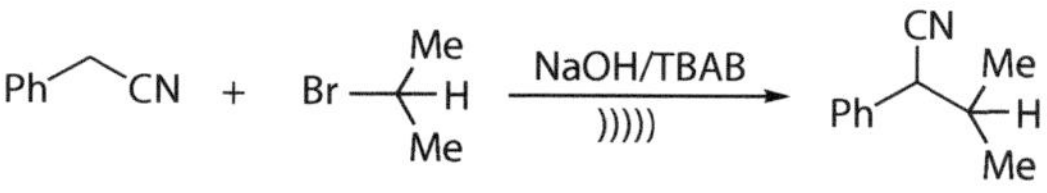

SCHEME 10.55 (From *Ultrason. Sonochem.*, 10, Hofmann, J., Freier, U., and Wecks, M., Ultrasound promoted C-alkylation of benzyl cyanide—Effect of reactor and ultrasound parameters, 271–275, Copyright (2003), with permission from Elsevier.)

SCHEME 10.56 (From *Ultrason. Sonochem.*, 10, Yadav, G.D. and Mujeebur Rahuman, M.S.M., Synergism of ultrasound and solid acids in intensification of Friedel–Crafts acylation of 2-methoxynaphthalene with acetic anhydride, 135–138, Copyright (2003), with permission from Elsevier.)

N-ALKYLATION

Yim et al. (1997) developed the N-alkylation of pyrrole by alkylating reagents using potassium superoxide as base in the presence of 18-crown-6 under ultrasound irradiation (Scheme 10.57).

S-ACYLATION

Duarte et al. (2010) have developed the method using ultrasound promotion for the synthesis of some thioesters from 2-mercaptobezoxa (thia) zoles. The method used as replacement for conventional thermal synthetic methodology (Scheme 10.58).

O-ACYLATION

To enhance reaction rates and yields or selectivity of reactions, ultrasound irradiation was used by Gholap et al. (2003) during the reaction of alcohols with acetic anhydride to corresponding esters using a room-temperature ionic liquid as the medium as well as a promoter (Scheme 10.59).

SCHEME 10.57 (From *Ultrason. Sonochem.*, 4, Yim, E.S., Park, M.K., and Han, B.H., Ultrasound promoted N-alkylation of pyrrole using potassium superoxide as base in crown ether, 95–98, Copyright (1997), with permission from Elsevier.)

SCHEME 10.58 (From *Ultrason. Sonochem.*, 17, Duarte, A., Cunico, W., Pereira, C.M.P., Flores, A.F.C., and Freitag, R.A., Ultrasound promoted synthesis of thioesters from 2-mercaptobenzoxa(thia)zoles, 281–283, Copyright (2010), with permission from Elsevier.)

SCHEME 10.59 (From Gholap, A.R., Venkatesan, K., Daniel, T., Lahoti, R.J., and Srinivasan, K.V., Ultrasound promoted acetylation of alcohols in room temperature ionic liquid under ambient conditions, *Green Chem.*, 5, 693–696, 2003, by permission of The Royal Society of Chemistry.)

N-ACYLATION

Anuradha and Ravindranath (1997) performed acylation of unprotected amino acids using ultrasound irradiation. N-carboxy anhydrides of amino acids reacted with other amino acids, dipeptides, and tripeptides (Scheme 10.60).

SULFONATION

Qureshi et al. (2008) described an efficient protocol for regioselective sulfonation of aromatic compounds under solvent-free conditions using ultrasound (Scheme 10.61).

MULTICOMPONENT REACTIONS

Multicomponent reactions (MCRs) have emerged as an important tool for building diverse and complex organic molecules through carbon–carbon and carbon–heteroatom bond formation taking place in tandem manner. These MCRs are of increasing importance in organic and medicinal chemistry, because the strategies of MCRs offer significant advantages over conventional linear-type syntheses. MCRs leading to interesting heterocyclic scaffolds are particularly useful for the creation of diverse chemical libraries of drug-like molecules for biological screening. Designing of multicomponent reactions in water is another attractive area in chemistry, because water is a cheap, safe, and environmentally benign solvent.

Zhang et al. (2009) have found that $Ga(OTf)_3$ can efficiently catalyze the three-component Mannich reaction of aryl aldehydes, aryl amines, and ketones in water under ultrasonic irradiation to afford β-carbonyl compounds (Scheme 10.62).

Tei et al. (2009) exploited the Ugi four-component reaction in a single-step bifunctional ditopic chelator using DOTA monoamide (DOTAMA) derivatives as amino acid components (Scheme 10.63).

Recently, Niralwad et al. (2010a) have reported the synthesis of α-aminophosphonates by the coupling of aldehydes/ketones, an amine, and triethyl phosphite using 1-hexanesulfonic acid

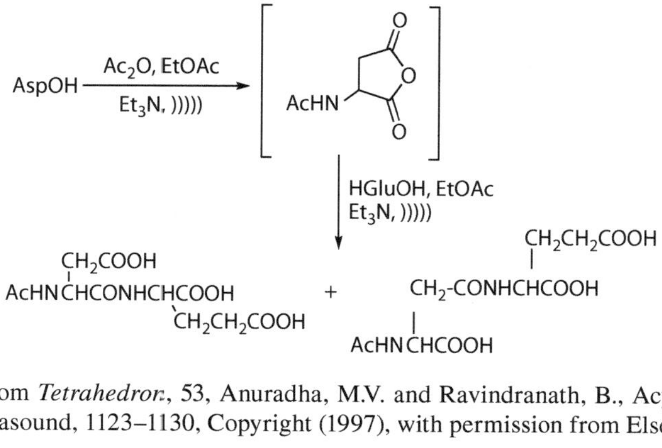

SCHEME 10.60 (From *Tetrahedron*, 53, Anuradha, M.V. and Ravindranath, B., Acylation of unprotected amino acids using ultrasound, 1123–1130, Copyright (1997), with permission from Elsevier.)

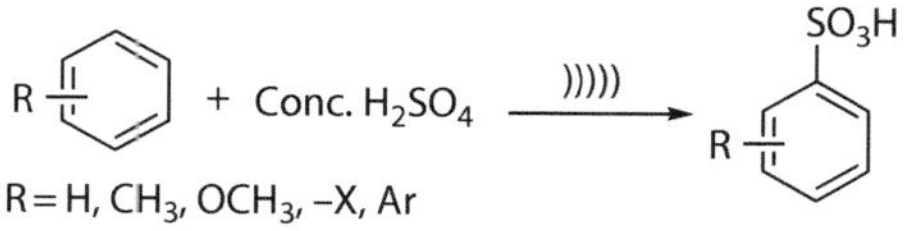

SCHEME 10.61 (From *Ultrason. Sonochem.*, 16, Qureshi, Z.S., Deshmukh, K.M., Jagtap, S.R., Nandurkar, N.S., and Bhanage, B.M., Ultrasound assisted regioselective sulfonation of aromatic compounds with sulfuric acid, 308–311, Copyright (2008), with permission from Elsevier.)

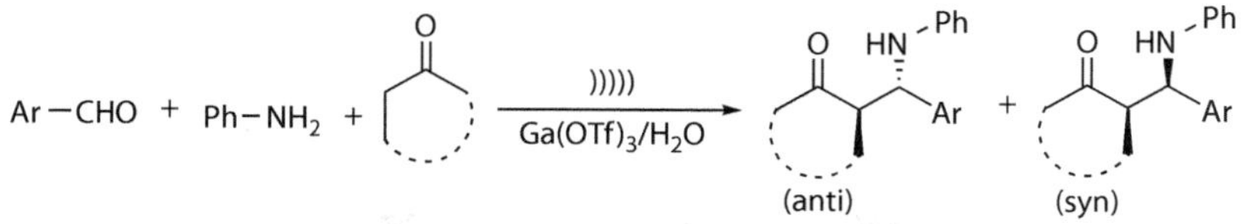

SCHEME 10.62 (From Zhang, G. et al., *Chin. J. Chem.*, 27, 1967, 2009.)

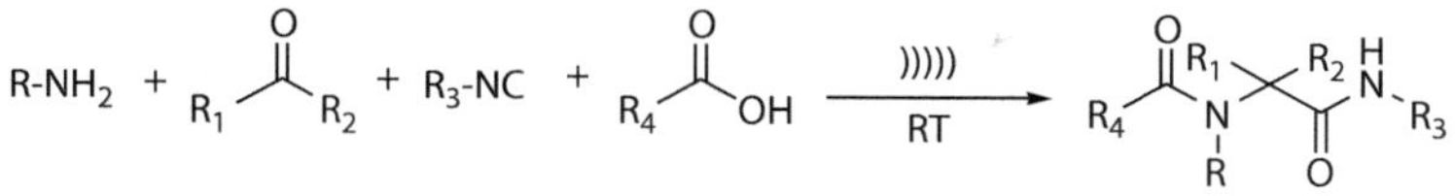

SCHEME 10.63 (From Tei, L. et al., *Organ. Biol. Chem.*, 7, 4406, 2009. With permission.)

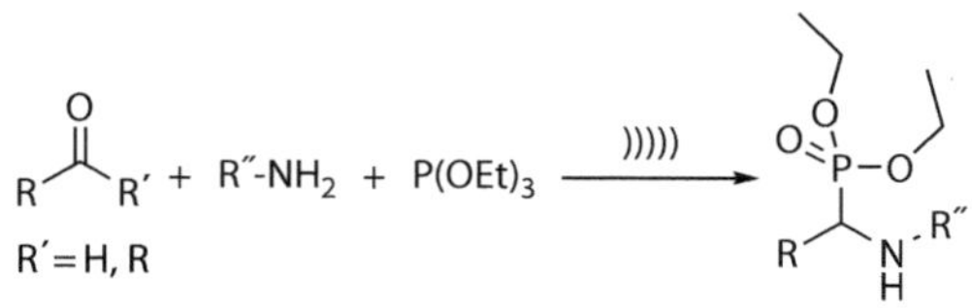

SCHEME 10.64 (From Sonar, S.S. et al., *Phosphorous Sulf. Silicon Relat. Elem*, 185, 65, 2010; *Ultrason. Sonochem.*, 17, Niralwad, K.S., Shingate, B.B., and Shingare, M.S., Solvent-free sonochemical preparation of α-aminophosphonates catalyzed by 1-hexanesulphonic acid sodium salt, 760–763, Copyright (2010), with permission from Elsevier.)

sodium salt under ultrasound irradiation at ambient temperature in good to excellent yield under solvent-free condition and Sonar et al. (2010) reported that *p*-TSA is an efficient catalyst for the synthesis of novel oxazepine α-aminophosphonates by the reaction of quino[2,3-b][1,5]benzoxazepines (4a-j) with triethyl phosphite using an ultrasonic approach. (Scheme 10.64).

One-pot three-component synthesis of spiro[indoline-3,4'-pyrazolo(3,4-b)pyridine]-2,6'(1*H*)-diones in water under ultrasonic irradiation gives excellent yield as reported by Bazgir et al. (2010) (Scheme 10.65).

The ultrasound-assisted three-component synthesis of 3-(5-amino-1*H*-pyrazol-4-yl)-3-(2-hydroxy-4,4-dimethyl-6-oxocyclohex-1-enyl)indolin-2-ones in aqueous media by using readily available catalyst, gives excellent yields with easy work up procedure as reported by Khorrami et al. (2010) (Scheme 10.66).

Nikpassand et al. (2010) studied ultrasound-promoted regioselective synthesis of fused polycyclic 4-aryl-3-methyl-4,7-dihydro-1*H*-pyrazole[3,4-b]pyridines in ethanol gives shorter reaction time with excellent yields (Scheme 10.67).

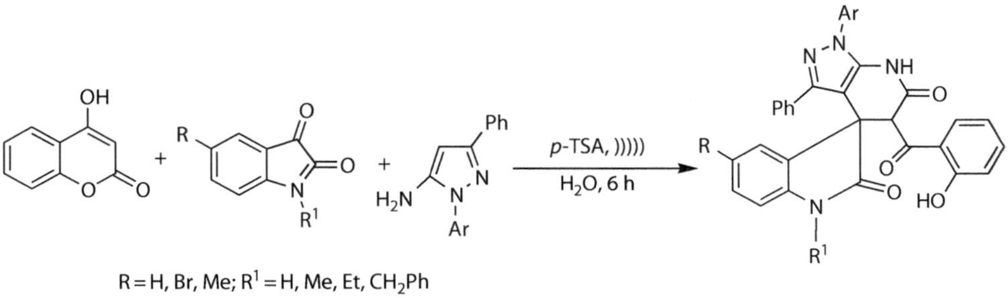

R = H, Br, Me; R¹ = H, Me, Et, CH₂Ph

SCHEME 10.65 (From *Ultrason. Sonochem.*, 17, Bazgir, A., Ahadi, S., Ghahremanzadeh, R., Khavasi, H., and Mirzaei, P., Ultrasound-assisted one-pot, three-component synthesis of spiro[indoline-3,4'-pyrazolo[3,4-b]pyridine]-2,6'(1'H)-diones in water, 447–452, Copyright (2010), with permission from Elsevier.)

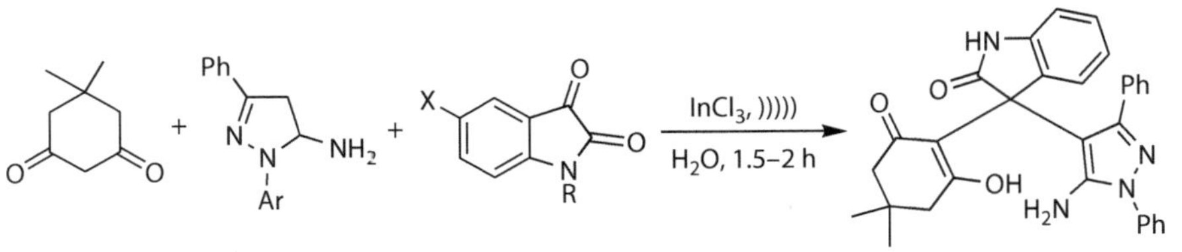

SCHEME 10.66 (From *Ultrason. Sonochem.*, 17, Khorrami, A.R., Faraji, F., and Bazgir, A., Ultrasound-assisted three-component synthesis of 3-(5-amino-1H-pyrazol-4-yl)-3-(2-hydroxy-4,4-dimethyl-6-oxocyclohex-1-enyl)indolin-2-ones in water, 587–591, Copyright (2010), with permission from Elsevier.)

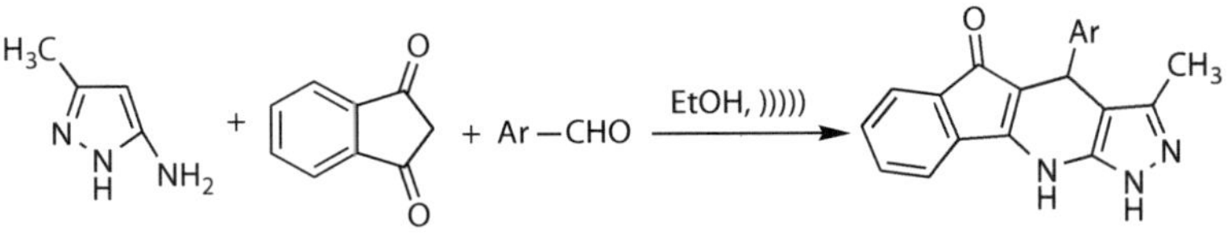

SCHEME 10.67 (From *Ultrason. Sonochem.*, 17, Nikpassand, M., Mamaghani, M., Shirini, F., and Tabatabaein, K., A convenient ultrasound-promoted regioselective synthesis of fused polycyclic 4-aryl-3-methyl-4,7-dihydro-1H-pyrazolo[3,4-b]pyridines, 301–305, Copyright (2010), with permission from Elsevier.)

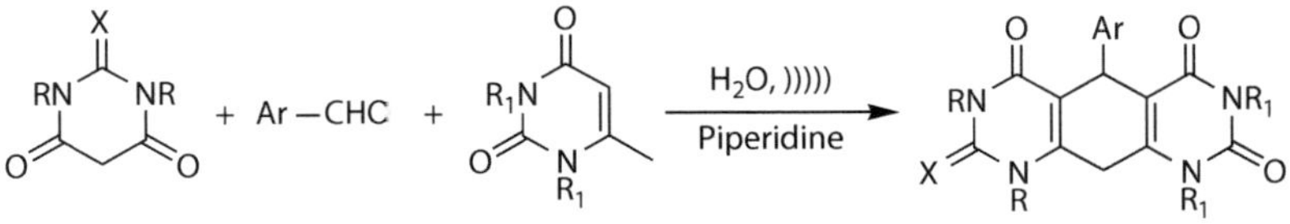

SCHEME 10.68 (From *Ultrason. Sonochem.*, 17, Mosslememin, M.H. and Nateghi, M.R., Rapid and efficient synthesis of fused heterocyclic pyrimidines under ultrasonic irradiation, 162–167, Copyright (2010), with permission from Elsevier.)

Mosselemin and Nateghi (2010) studied the synthesis of fused heterocyclic pyrimidine in the presence of piperidine in water gives good yield in very short time period and easy work-up procedure under ultrasound irradiation (Scheme 10.68).

Nabid et al. (2010) reported that one-pot synthesis of 1*H*-Pyrazolo[1,2-b]phthalazine-5,10-dionesunder ultrasound irradiation in the presence of triethyl amine in ethanol gives higher yields (Scheme 10.69).

One-pot three-component Mannich-type reaction was reported by Zeng and Shao (2009) using sulfamic acid catalyst under ultrasound irradiation at room temperature (Scheme 10.70).

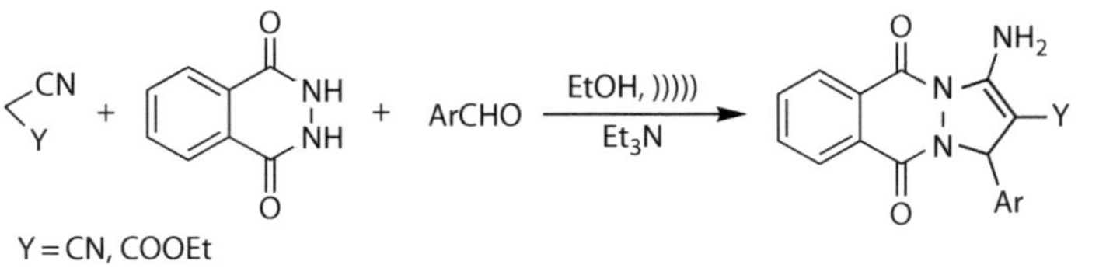

SCHEME 10.69 (From *Ultrason. Sonochem.*, 17, Nabid, M.R., Rezaei, S.J.T., Ghahremanzadeh, R., and Bazgir, A., Ultrasound-assisted one-pot, three-component synthesis of 1H-pyrazolo[1,2-b]phthalazine-5,10-diones, 159–161, Copyright (2010), with permission from Elsevier.)

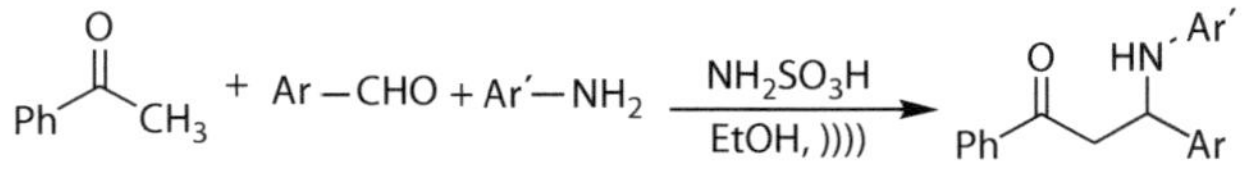

SCHEME 10.70 (From *Ultrason. Sonochem.*, 16, Zeng, H. and Shao, H.L.W., One-pot three-component Mannich-type reactions using sulfamic acid catalyst under ultrasound irradiation, 758–762, Copyright (2009), with permission from Elsevier.)

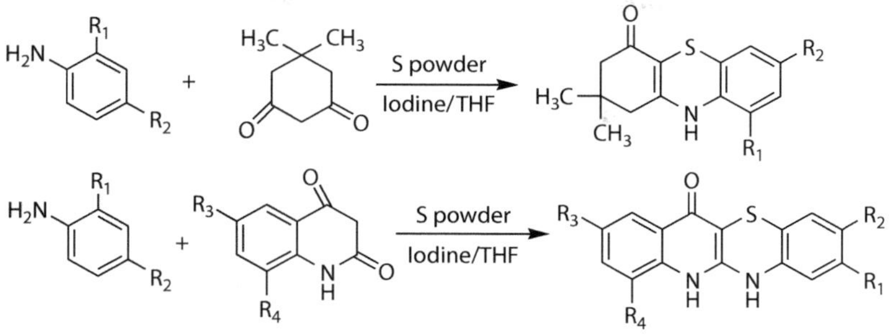

SCHEME 10.71 (From Dabholkar, V.V. and Ansari, F.Y., *Ind. J. Chem.*, 47B, 1759, 2008.)

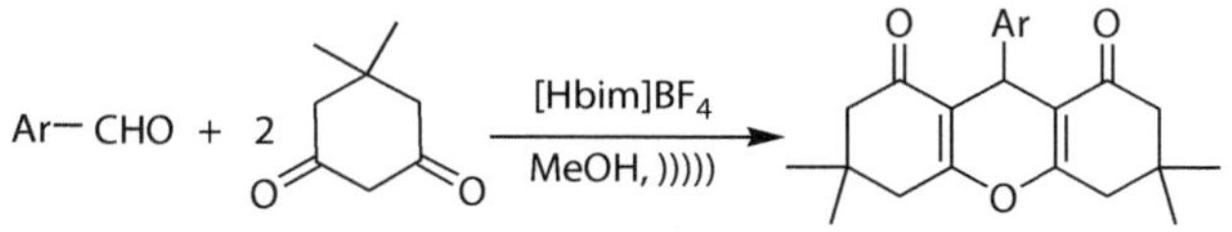

SCHEME 10.72 (From *Ultrason. Sonochem.*, 15, Venkatesan, K., Pujari, S.S., Lahoti, K.V., and Srinivasan, K.V., An efficient synthesis of 1,8-dioxo-octahydro-xanthene derivatives promoted by a room temperature ionic liquid at ambient conditions under ultrasound irradiation, 548–553, Copyright (2008), with permission from Elsevier.)

Dabholkar and Ansari (2008) have synthesized thiazines by using sulfur powder and iodine as a catalyst in THF under ultrasonication (Scheme 10.71).

Venkatesan et al. (2008) reported the synthesis of 1,8-dioxo-octahydro-xanthene in the presence of ionic liquid [(Hbim)BF$_4$] at the ambient condition at room temperature under ultrasound irradiation in excellent yield (Scheme 10.72).

Xia and Lu (2007) reported one-pot synthesis of α-amino phosphonates under solvent-free and catalyst conditions under ultrasound irradiation in good to excellent yields (Scheme 10.73).

The synthesis of some biologically active 1,2,5,6-tetrahydropyrimidines under ultrasound irradiation in high yields described by Muravyova et al. (2007) (Scheme 10.74).

Patil et al. (2007) reported one-pot three component synthesis of 1-amidoalkyl-2-naphthols in the presence of sulfamic acid using ultrasound at ambient condition in excellent yield with short time period (Scheme 10.75).

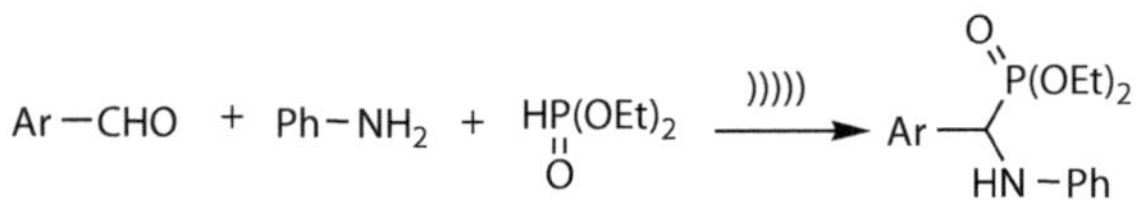

SCHEME 10.73 (From *Ultrason. Sonochem.*, 14, Xia, M. and Lu, Y.D., Ultrasound-assisted one-pot approach to α-amino phosphonates under solvent-free and catalyst-free conditions, 235–240, Copyright (2007), with permission from Elsevier.)

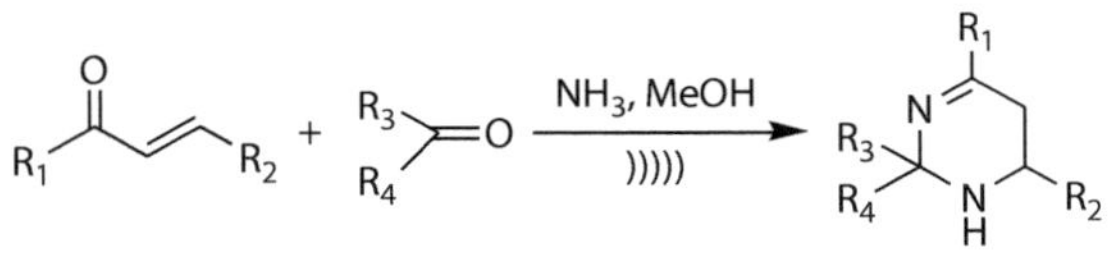

SCHEME 10.74 (With permission from Muravyova, E.A., Desenko, S.M., Musatov, V.I., Knyazeva, I.V., Shishkina, S.V., Shishkin, O.V. and Chebanov, V.A., Ultrasonic-promoted three-component synthesis of some biologically active 1,2,5,6-tetrahydropyrimidines, *J. Combinat. Chem.*, 9, 797–803, 2007. Copyright 2007 American Chemical Society.)

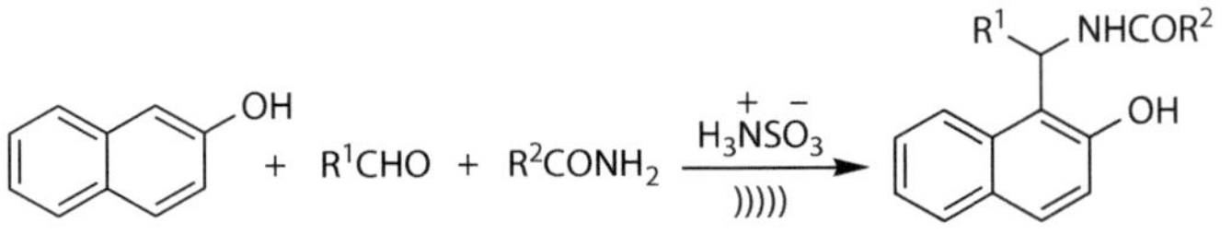

SCHEME 10.75 (From *Ultrason. Sonochem.*, 14, Patil, S.B., Singh, P.R., Surpur, M.P., and Samant, S.D., Ultrasound-promoted synthesis of 1-amidoalkyl-2-naphthols via a three-component condensation of 2-naphthol, ureas/amides, and aldehydes, catalyzed by sulfamic acid under ambient conditions, 515–518, Copyright (2007), with permission from Elsevier.)

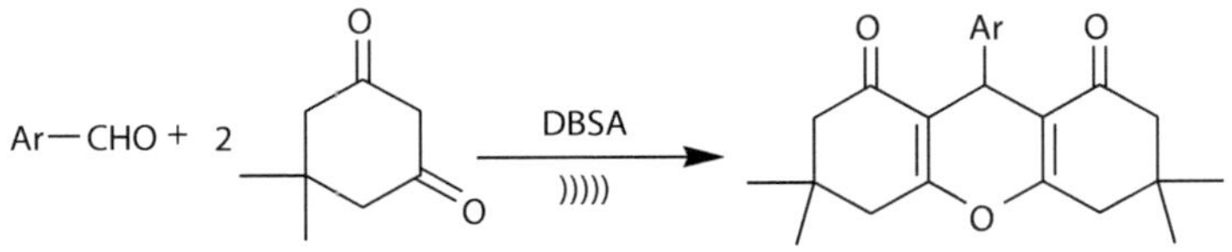

SCHEME 10.76 (From *Ultrason. Sonochem.*, 13, Jin, T.S., Zhang, J.S., Wang, A.Q., and Li, T.S., Ultrasound-assisted synthesis of 1,8-dioxo-octahydroxanthene derivatives catalyzed by p-dodecylbenzenesulfonic acid in aqueous media, 220–224, Copyright (2006), with permission from Elsevier.)

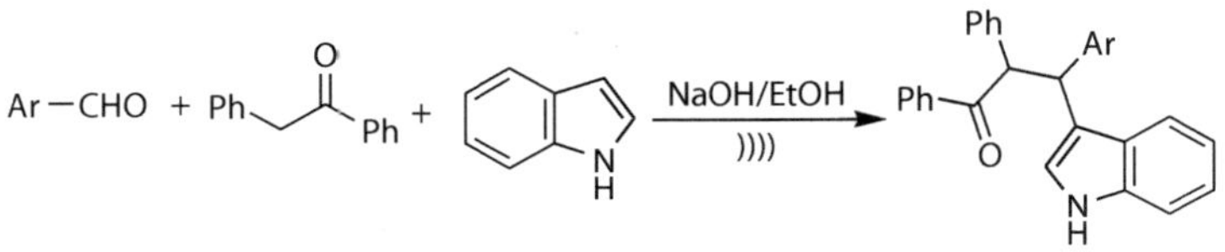

SCHEME 10.77 (From *Tetrahedron*, 61, Shen, Z.L., Ji, S.J., Wang, S.Y., and Zeng, X.F., A novel base-promoted synthesis of β-indolylketones via a three-component condensation under ultrasonic irradiation, 10552–10558, Copyright (2005), with permission from Elsevier.)

Jin et al. (2006) described ultrasound-assisted synthesis of 1.8-dioxo-octahydroxanthene derivatives catalyzed by *p*-dodecylbenzene sulfamic acid in aqueous media in excellent yields with a simple work-up procedure (Scheme 10.76).

Shen et al. (2005) have reported the novel-base-promoted synthesis of β-indolyl ketone via three-component condensation under ultrasound irradiation in good to excellent yields (Scheme 10.77).

General and practical route to 2-amino-2-chromenes in water in the presence of cetyltrimethyl ammonium bromide (CTABr) as a catalyst is described under ultrasound-irradiation by Jin et al. (2004) (Scheme 10.78).

Shinde et al. (2010) developed an expedient and clean protocol for the synthesis of 2-amino-3,5-dicarbonitrile-6-thio-pyridines. The use of ultrasound irradiations has decreased the reaction time (Scheme 10.79).

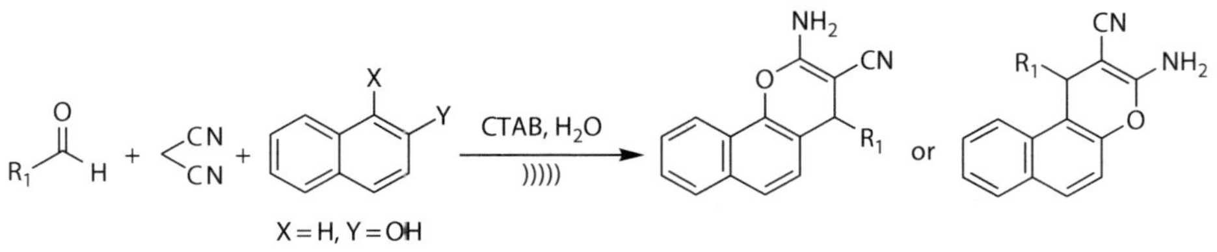

SCHEME 10.78 (From *Ultrason. Sonochem.*, 11, Jin, T.S., Xiao, J.C., Wang, S.J., and Li, T.S., Ultrasound-assisted synthesis of 2-amino-2-chromenes with cetyltrimethylammonium bromide in aqueous media, 393–397, Copyright (2004), with permission from Elsevier.)

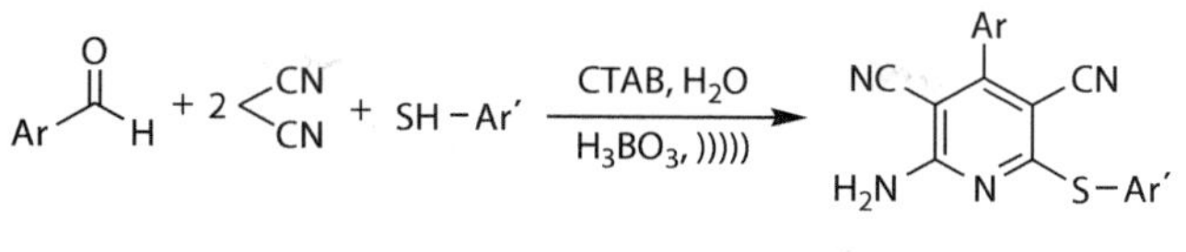

$$\text{Ar-C(O)-H} + 2 \, \text{CH}_2(\text{CN})_2 + \text{SH-Ar'} \xrightarrow[\text{H}_3\text{BO}_3, \,)))))]{\text{CTAB, H}_2\text{O}} \text{2-amino-3,5-dicarbonitrile-6-thio-pyridine}$$

SCHEME 10.79 (From *Tetrahedron Lett.*, 51, Shinde, P.V., Sonar, S.S., Shingate, B.B., and Shingare, M.S., Boric acid catalyzed convenient synthesis of 2-amino-3,5-dicarbonitrile-6-thio-pyridines in aqueous media, 1309–1312, Copyright (2010), with permission from Elsevier.)

NAME REACTIONS

KNOEVENAGEL CONDENSATION

A practical condensation procedure for the preparation of ethyl α-cyano cinnamates employing ultrasound irradiations was developed by Li et al. (1999b) (Scheme 10.80).

Kakade et al. (2008) developed a method that is environmentally benign, eco-friendly, and a cleaner methodology for rapid Knoevenagel condensation of 2-chloroquinoline-3-carbaldehyde with ethyl cyanoacetate under ultrasonic irradiation in presence of 1, 8-Diazabicyclo-undec-7-ene (DBU) catalyst (Scheme 10.81).

Shindalkar et al. (2005a) described condensation reaction of 4-oxo-(4H)-1-benzopyran-3-carbaldehyde with 3-methyl-1-phenylpyrazolin-5-(4H)-one under ultrasonic irradiation at room temperature (Scheme 10.82).

BIGINELLI REACTION

This reaction was developed by Pietro Biginelli in 1891. The acid-catalyzed, three-component reaction between an aldehyde, a β-ketoester, and urea constitutes a rapid and facile synthesis of dihydropyrimidones (DHPMs) (Scheme 10.83), which are interesting compounds and are widely used in the pharmaceutical industry as calcium channel blockers, antihypertensive agents, and alpha-1-a-antagonists.

DHPMs, named Biginelli compounds, are known to exhibit a wide range of biological activities such as antiviral, antitumor, and antibacterial. Moreover, several marine alkaloids containing the DHPMS core unit have shown interesting biological properties.

Aryl-dihydropyrimidines first synthesized by Hantzsch and Eisner et al. have played an important role in medicinal chemistry as vasodilators and antihypertensive anents. Various aryl

$$\text{RCHO} + \text{NCCH}_2\text{CO}_2\text{Et} \xrightarrow{)))))} \text{RHC} = \text{C(CN)CO}_2\text{Et}$$

SCHEME 10.80 (From *Ultrason. Sonochem.*, 6, Li, J.T., Li, T.S., Li, L.J., and Cheng, X., Synthesis of ethyl α-cyanocinnamates under ultrasound irradiation, 199–201, Copyright (1999), with permission from Elsevier.)

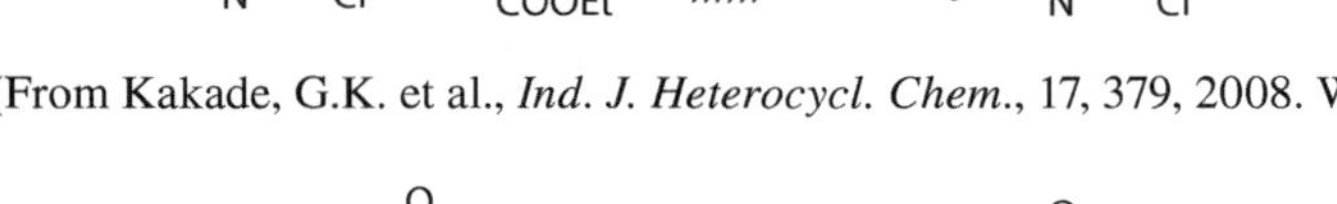

SCHEME 10.81 (From Kakade, G.K. et al., *Ind. J. Heterocycl. Chem.*, 17, 379, 2008. With permission.)

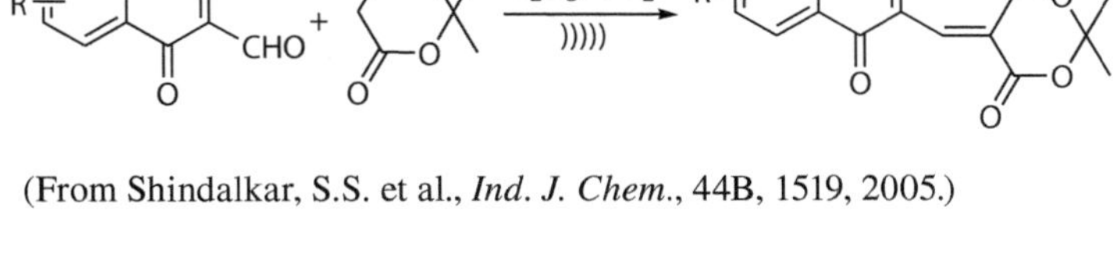

SCHEME 10.82 (From Shindalkar, S.S. et al., *Ind. J. Chem.*, 44B, 1519, 2005.)

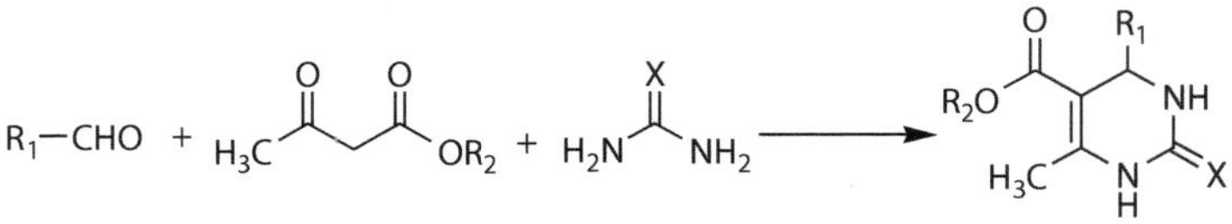

SCHEME 10.83

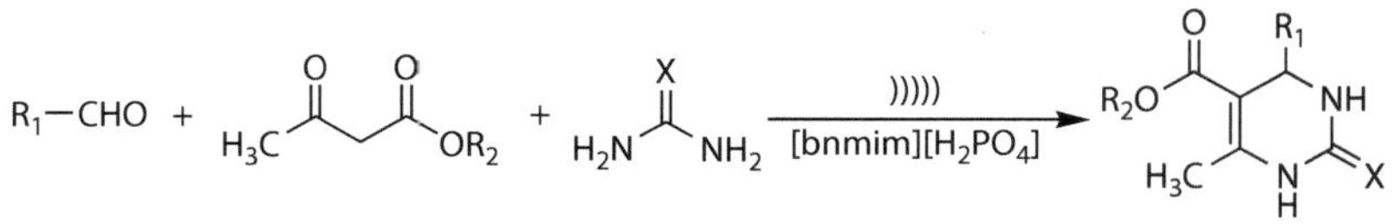

SCHEME 10.84 (From Shelke, K.F et al., *Organ. Chem. Ind. J.*, 4, 277, 2008. Copyright © Trade Science Inc. With permission; Gholap, A.R., Venkatesan, K., Danial, T., Lahoti, R.J., and Srinivasan, K.V., *Green Chem.*, 6, 147–150, 2004, by permission of The Royal Society of Chemistry.)

dihydropyrimidines like Nifedipine and SKF 24260 have been found to be highly effective calcium antagonist and vasodilators.

Shelke et al. (2008) and Gholap et al. (2004) described the Biginelli reaction in the presence of ionic liquid under ultrasound irradiation (Scheme 10.84).

Niralwad et al. (2010b) have reported the synthesis of octahydroquinazolinone derivatives through condensation of dimedone, aldehyde, and urea/thiourea using ionic liquid under ultrasonication (Scheme 10.85).

Wang et al. (2008b) reported the Biginelli reaction using aldehyde, ethyl acetoacetate, and ammonium acetate without using catalyst and solvent under ultrasound irradiation (Scheme 10.86).

Li et al. (2003b) also reported the synthesis of dihydropyrimidinones catalyzed by NH_2SO_3H under ultrasound irradiation with 85%–98% yield within short time span, that is, 25–60 min (Scheme 10.87).

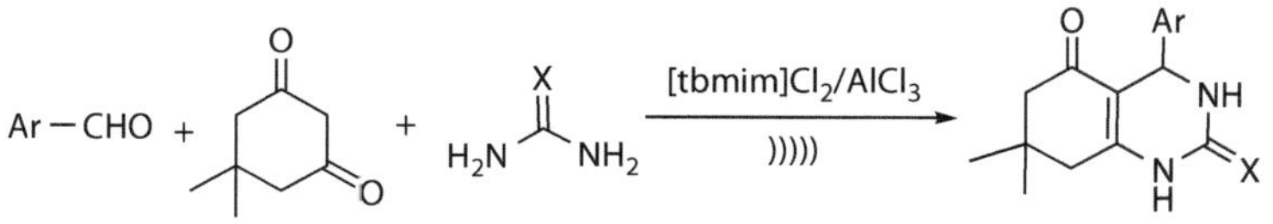

SCHEME 10.85 (From Niralwad, K.S., Shingate, B.B., and Shingare, M.S., *J. Chin. Chem. Soc.*, 57, 89–92, 2010. With permission.)

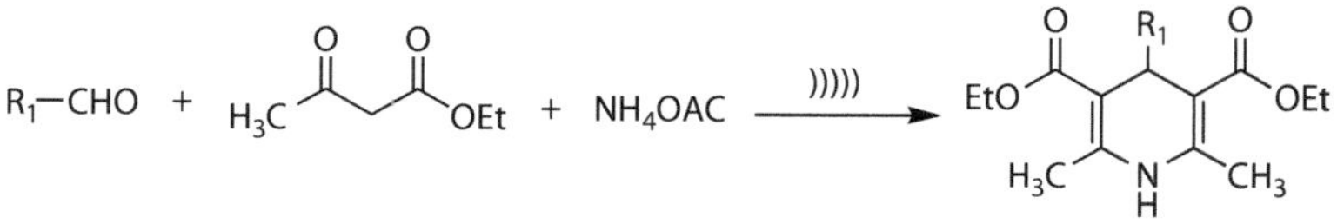

SCHEME 10.86 (From *Ultrason. Sonochem.*, 15, Wang, X.H., Li, Z.Y., Zhang, J.C., and Li, J.T., The solvent-free synthesis of 1,4-dihydropyridines under ultrasound irradiation without catalyst, 677–680, Copyright (2008), with permission from Elsevier.)

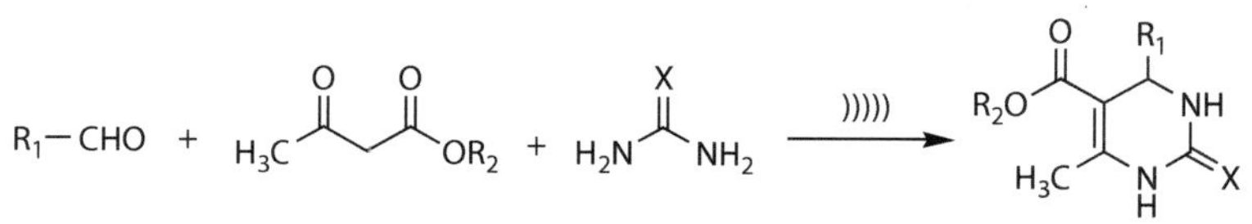

SCHEME 10.87 (From *Ultrason. Sonochem.*, 10, Li, J.T., Han, J.F., Yang, J.H., and Li, T.S., An efficient synthesis of 3,4-dihydropyrimidin-2-ones catalyzed by NH_2SO_3H under ultrasound irradiation. 119–122, Copyright (2003c), with permission from Elsevier.)

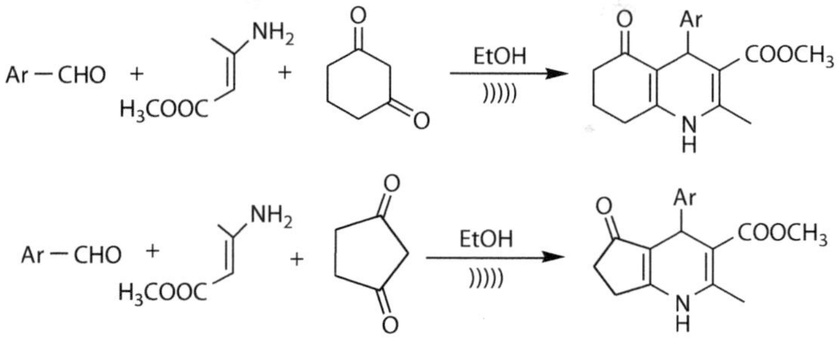

SCHEME 10.88 (From Thirumalai, D. et al., *Ind. J. Chem.*, 45B, 335–338, 2006. With permission.)

Thirumalai et al. (2006) reported synthesis of arylhexahydro-quinoline and arylcyclopentanopyridine derivatives carried out by the condensation of cyclic 1,3-diones with aldehydes and β-amino-crotonate using conventional and nonconventional methods (Scheme 10.88).

ULLMANN REACTION

Pellon and Docampoz (2007) reported a mild method for Ullmann reaction of 2-chlorobenzoic acids and aminothiazoles or aminobenzothiazoles under ultrasonic irradiation. Here, the authors synthesized 5H-[1,3]thiazolo[2,3-b]quinazolin-5-one and 12H-[1,3] benzothiazolo[2,3-b] quinazolin-12-one using copper (Scheme 10.89).

CURTIUS REARRANGEMENT

Sureshbabu et al. (2008) reported efficient synthesis of o-succinimidyl-(ter-butoxycarbonylamino) methyl carbamates derived from alpha-amino acids using N-methylmorphine (NMM) by ultrasound application to the synthesis of ureidopeptides (Scheme 10.90).

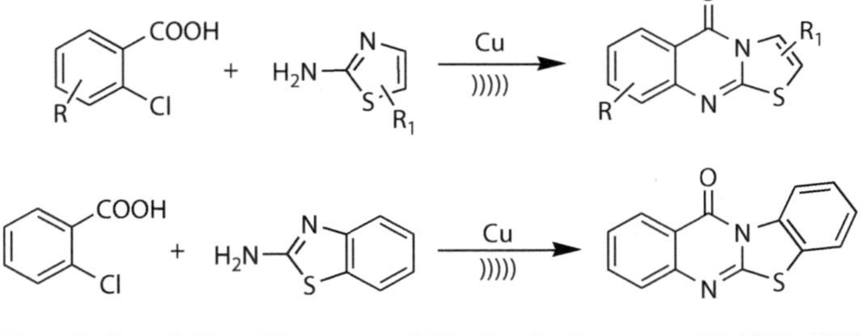

SCHEME 10.89 (From Pellon, R.F. and Docampo, M.L. *Synth. Commun.*, 37, 1853, 2007. With permission.)

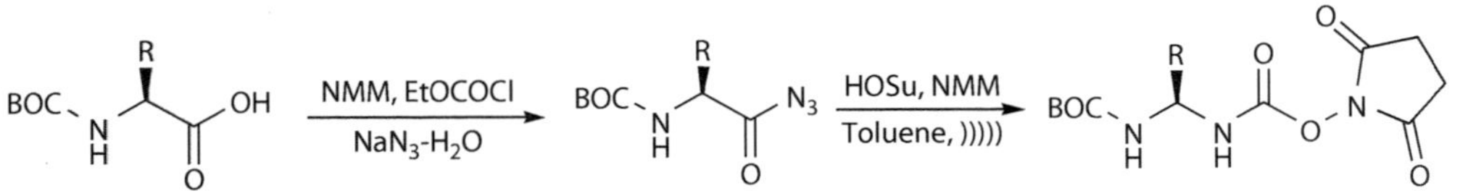

SCHEME 10.90 (From Sureshbabu, V.V. et al., *Synth. Commun.*, 38, 2168, 2008. With permission.)

CYCLOCONDENSATION REACTIONS

A convenient ultrasound-mediated synthesis of substituted pyrazolones under solvent-free conditions has been studied by Mojtahedi et al. (2008) by cyclocondensation of phenyl hydrazine with various β-keto esters (Scheme 10.91).

A synthesis of 2-aryl-4H-1-benzopyran-4-ones under phase transfer catalyst in one-pot protocol under ultrasound by Pathak et al. (2008) has been reported (Scheme 10.92).

Khosropour et al. (2008) have synthesized the 2,4,5-trisubstituted imidazole derivatives in the presence of Zr(acac)$_4$ under ambient conditions at room temperature with excellent yields (Scheme 10.93).

Convenient procedure for the synthesis of 1,3,5-triaryl-2-pyrazolines in sodium acetate-acetic acid aqueous solution at room temperature under ultrasound irradiation has been developed by Li et al. (2007b) (Scheme 10.94).

Flash synthesis of 4H-pyrano [2,3-c]pyrazoles in aqueous media is performed under ultrasound irradiation by Peng et al. (2006) (Scheme 10.95).

Du et al. (2006) demonstrated the synthesis of 3-carboxycoumarines in aqueous media under ultrasound irradiation avoiding addition of catalyst (Scheme 10.96).

The comparative study of thiadiazoles, triazoles, and oxadiazoles using conventional and ultrasound irradiations has been carried out by Narwade et al. (2006). That ultrasound-promoted synthesis is more superior than conventional method in the context of reaction rate and product yield is also described (Scheme 10.97).

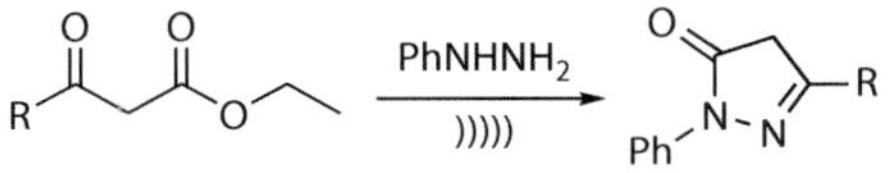

SCHEME 10.91 (From *Ultrason. Sonochem.*, 15, Mojtahedi, M.M., Javadpour, M., and Abaee, M.S. Convenient ultrasound mediated synthesis of substituted pyrazolones under solvent-free conditions, 828–832, Copyright (2008), with permission from Elsevier.)

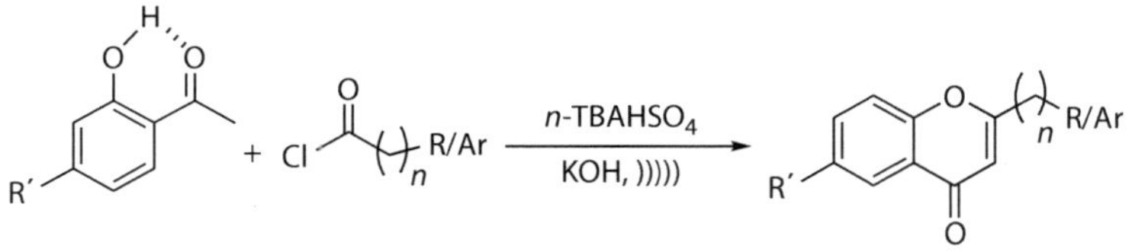

SCHEME 10.92 (From *J. Heterocycl. Chem.*, 45, Pathak, V.N., Gupta, R., and Varshney, B., A one-pot synthesis of 2-aryl-4H-1-benzopyran-4-ones under coupled microwave phase transfer catalysis (PTC) and ultrasonic irradiation PTC, 589–592, Copyright (2008), with permission from Elsevier.)

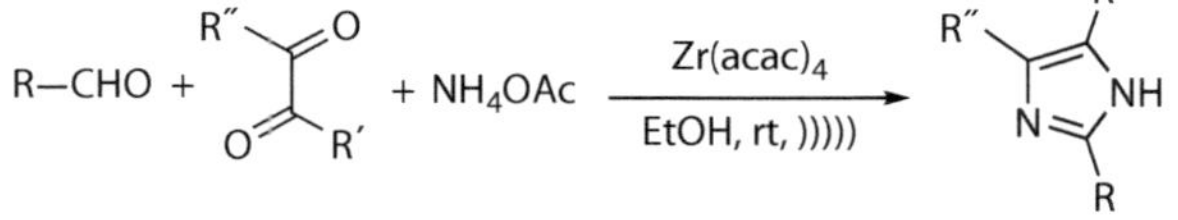

SCHEME 10.93 (From Khosropour, A.R., *Ultrason. Sonochem.*, 15, Ultrasound-promoted greener synthesis of 2,4,5-trisubstituted imidazoles catalyzed by Zr(acac)4 under ambient conditions, 659–664, Copyright (2008), with permission from Elsevier.)

SCHEME 10.94 (From Li, J.T. et al., *Beilstein J. Organ. Chem.*, 3, 13, 2007. With permission.)

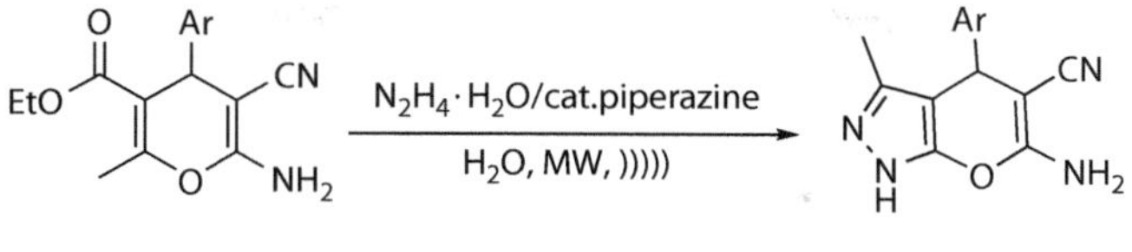

SCHEME 10.95 (From Peng, Y., Song, G., and Dou, R., Surface cleaning under combined microwave and ultrasound irradiation: flash synthesis of 4H-pyrano[2,3-c]pyrazoles in aqueous media, *Green Chem.*, 8: 573–575, 2006, by permission of The Royal Society of Chemistry.)

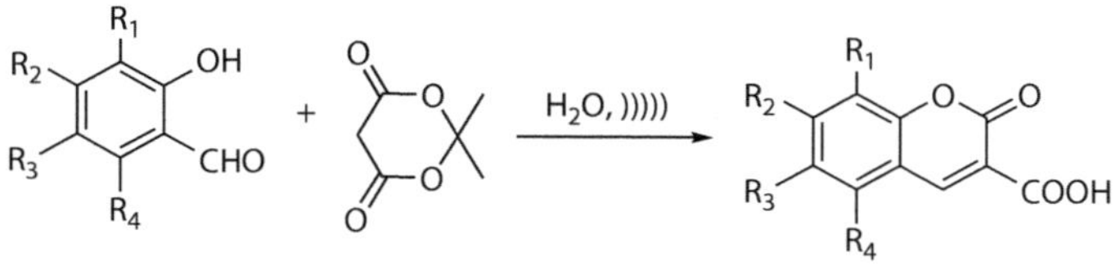

SCHEME 10.96 (From Du, J.L. et al., *E-J. Chem.*, 3, 1, 2006.)

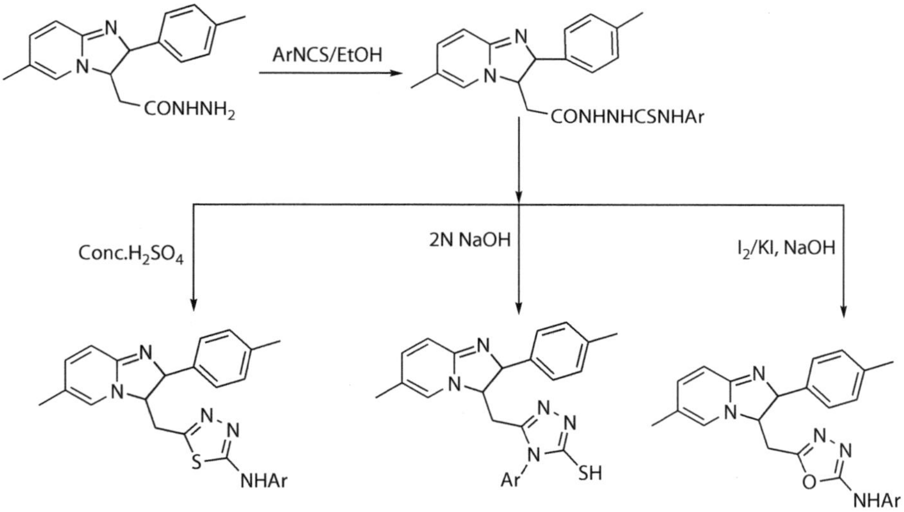

SCHEME 10.97 (From Narwade, S.K. et al., *Ind. J. Chem.*, 45B, 2776, 2006. With permission.)

Zhou et al. (2006) has described an ultrasound-assisted synthesis of ferrocenyl substituted 3-cyanopyridine derivatives via the condensation of ferrocenyl substituted chalcones with malanonitrile in sodium alkoxide solution (Scheme 10.98).

Cyclization of 1,1,1-trihalo-4-alkoxy-3-alken-2-ones with hydroxylamine and anilines gives 5-hydroxy-5-trihalo-4,5-dihydroisoxazoles and β-enamino trihalomethylketones, respectively, using water as a solvent under ultrasound irradiation has been reported by Martins et al. (2006) (Scheme 10.99).

Li et al.(2005) described the cyclocondensation via double Michael addition of 1,5-diaryl-1,4-pentadien-3-one with various active methylene compounds such as dimethyl malonate, diethyl malonate, methyl cyanoacetate, and ethyl cyanoacetate catalyzed by KF/basic alumina under ultrasound irradiation (Scheme 10.100).

Shen and Fuchigami (2004) described that the synthesis of oxindole and 3-oxo-tetrahydroisoquinoline derivatives was achieved under the influence of electrolytic condition. The desired cyclization was accelerated effectively under ultrasound irradiation (Scheme 10.101).

Ultrasound-promoted synthesis of β-lactams in the presence of polymer-supported reagent was achieved by Donati et al. (2004) (Scheme 10.102).

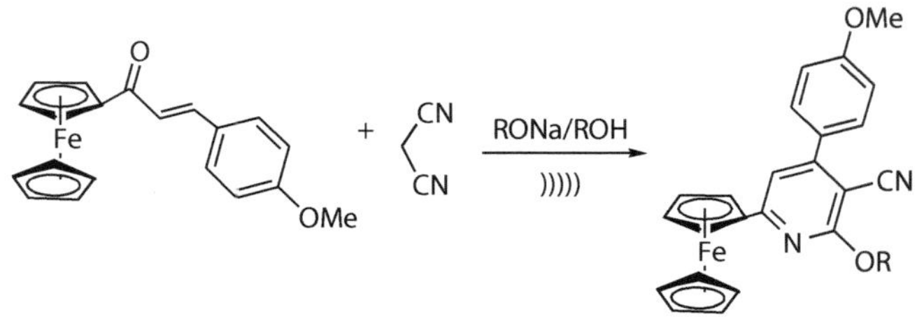

SCHEME 10.98 (From *J. Organometall. Chem.*, 691, Zhou, W.J., Ji, S.J., and Shen, Z.L., An efficient synthesis of ferrocenyl substituted 3-cyanopyridine derivatives under ultrasound irradiation, 1356–1360, Copyright (2006), with permission from Elsevier.)

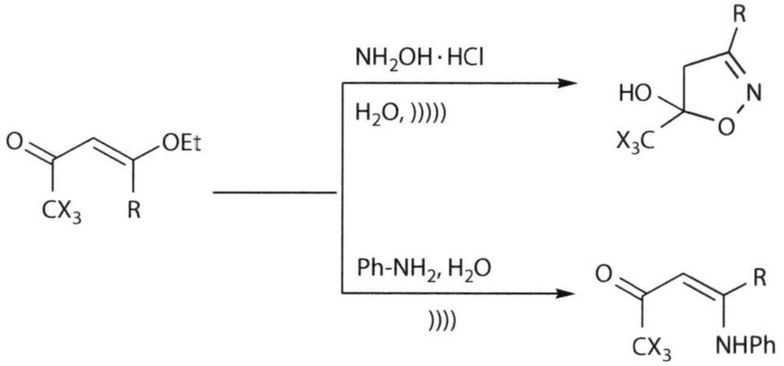

SCHEME 10.99 (From *Ultrason. Sonochem.*, 13, Martins, M.A.P., Pereira, C.M.P., Cunico, W., Moura, S., Rosa, F.A., Peres, R.L., Machado, P., Zanatta, N., and Bonacorso, H.G. Ultrasound promoted synthesis of 5-hydroxy-5-trihalomethyl-4,5-dihydroisoxazoles and β-enamino trihalomethyl ketones in water, 364–370, Copyright (2006), with permission from Elsevier.)

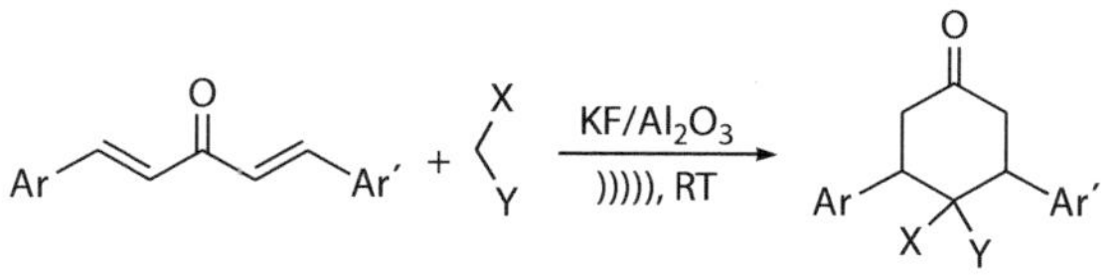

SCHEME 10.100 (From *Ultrason. Sonochem.*, 12, Li, J.T., Zu, W.Z., Chen, G.F., and Li, T.S., Synthesis of 1,1-disubstituted-2,6-diarylcyclohexane-4-ones catalyzed by KF/basic Al_2O_3 under ultrasound, 473–476, Copyright (2005), with permission from Elsevier.)

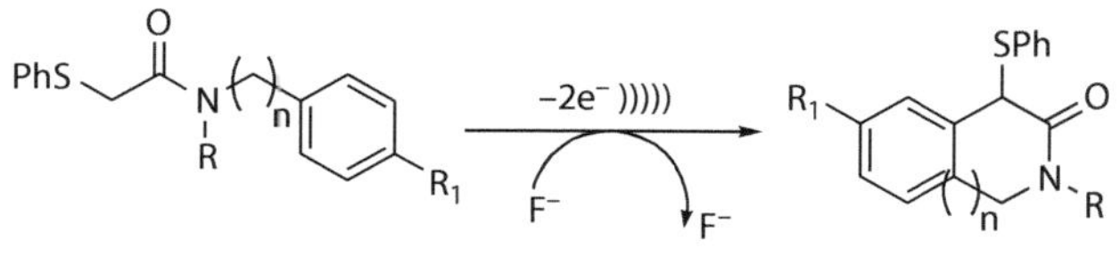

SCHEME 10.101 (From Shen, Y. and Fuchigami, M., *Organ. Lett.*, 6, 2441, 2004. With permission.)

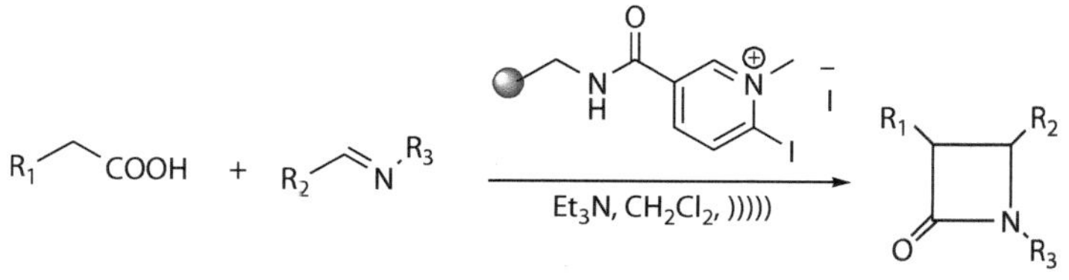

SCHEME 10.102 (From Donati, D. et al., *J. Organ. Chem.*, 69, 9316, 2004. With permission.)

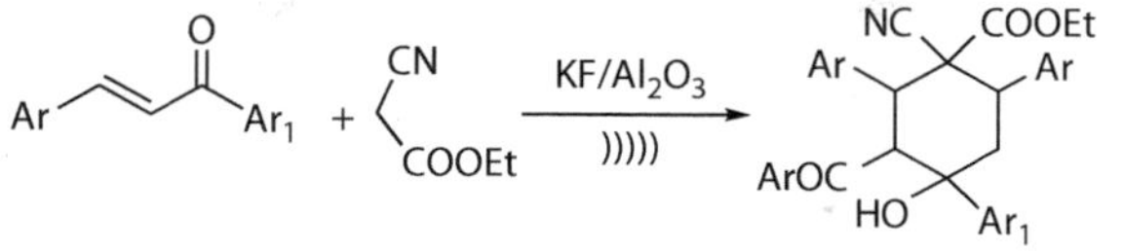

SCHEME 10.103 (From *Ultrason. Sonochem.*, 10, Li, J.T., Chen, G.F., Yang, W.Z., and Li, J.S., Ultrasound promoted synthesis of 2-aroyl-1,3,5-triaryl-4-carboethoxy-4-cyanocyclohexanols, 123–126, Copyright (2003), with permission from Elsevier.)

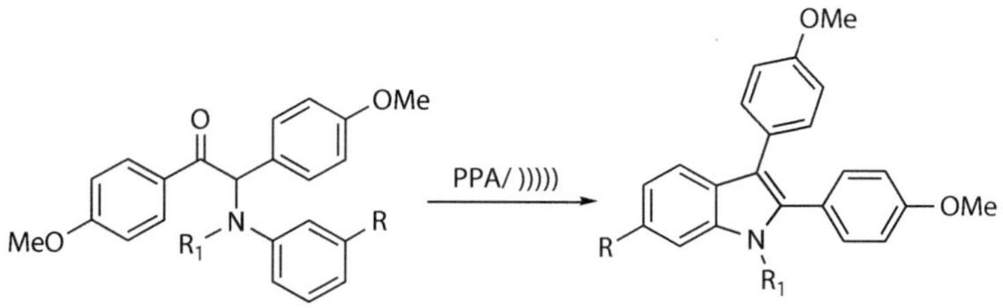

SCHEME 10.104 (From Koulocheri, S.D. and Haroutounian, S.A., *Eur. J. Organ. Chem.*, 1723, 2001. With permission.)

Ultrasound-assisted synthesis of 2-aryl-1,3,5-triaryl-4-carbethoxy-4-cyanocyclohexanols has been carried out from chalcones with ethyl cyanoacetate by Li et al. (2003a) (Scheme 10.103).

Synthesis of 2,3-bis(4-hydroxyphenyl)indoles, promoted under ultrasound irradiation was achieved by Koulocheri and Haroutounian (2001) (Scheme 10.104).

Synthesis of pyrazoles and pyrazolinones in the presence of K-10 under ultrasonic irradiation for 5 h has been achieved by Valduga et al. (1998) (Scheme 10.105).

An efficient and convenient ultrasound-promoted procedure for the synthesis of 5,5-disubstituted hyndantoins has been carried out by Li et al. (1996) (Scheme 10.106).

Mosslemin and Nateghi (2010) developed a simple and efficient and green protocol for the synthesis of pyrimidine derivative by one-pot and three-component reaction under ultrasound irradiation (Scheme 10.107).

Li et al. (2010f) described an efficient one-pot synthesis of some 3-aza-6,10-diaryl-2-oxa-spiro[4.5] decane-1,4,8-trione from 1,5-diaryl-1,4-pentadien-3-one by using dimethyl malonate, 1,5-diaryl-1,4-pentadien-3-one, sodium hydroxide, methanol under ultrasonic irradiation (Scheme 10.108).

Ni et al. (2010) described improved synthesis of the diethyl 2,6-dimethyl-4-aryl-4H-pyran-3,5-dicarbonylates(1) from aryl aldehyde and 1,3-diketone catalyzed by $ZnCl_2$ under ultrasonic

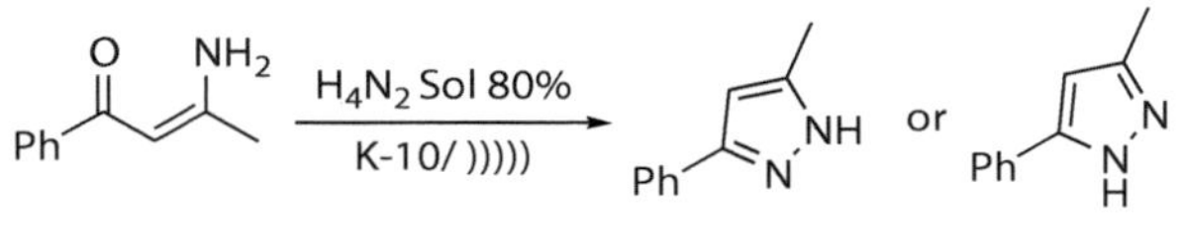

SCHEME 10.105 (From Valduga, C.J. et al., *J. Heterocycl. Chem.*, 35, 189, 1998.)

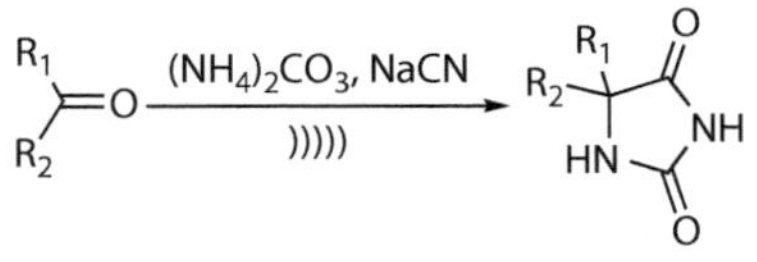

SCHEME 10.106 (From *Ultrason. Sonochem*, 3, Li, J., Li, L., Li, T., Li, H., and Liu, J., An efficient and convenient procedure for the synthesis of 5,5-disubstituted hydantoins under ultrasound, 141–143, Copyright (1996), with permission from Elsevier.)

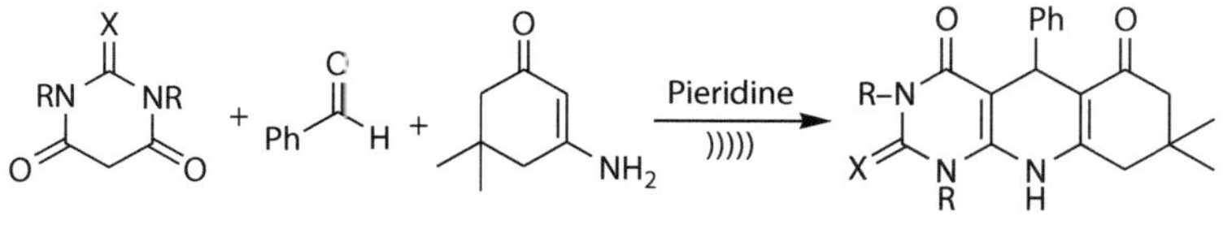

SCHEME 10.107 (From *Ultrason. Sonochem.*, 17, Mosslememin, M.H. and Nateghi, M.R., Rapid and efficient synthesis of fused heterocyclic pyrimidines under ultrasonic irradiation, 162–167, Copyright (2010), with permission from Elsevier.)

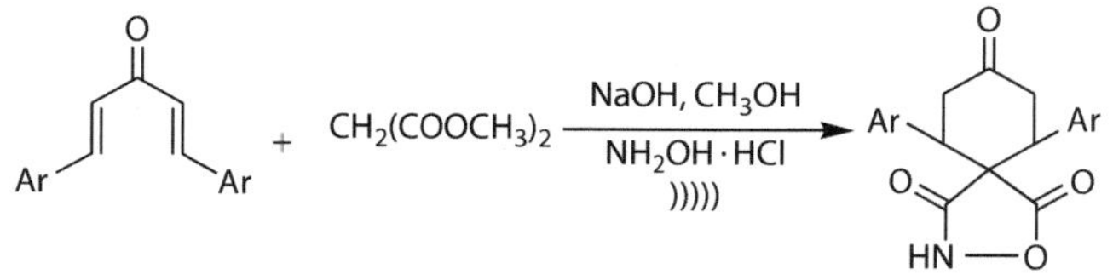

SCHEME 10.108 (From *Ultrason. Sonochem.*, 17, Li, J.T., Zhai, X.L., and Chan, G.F., Ultrasound promoted one-pot synthesis of 3-aza-6,10-diaryl-2-oxa-spiro[4.5]decane-1,4,8-trione, 356–358, Copyright (2010), with permission from Elsevier.)

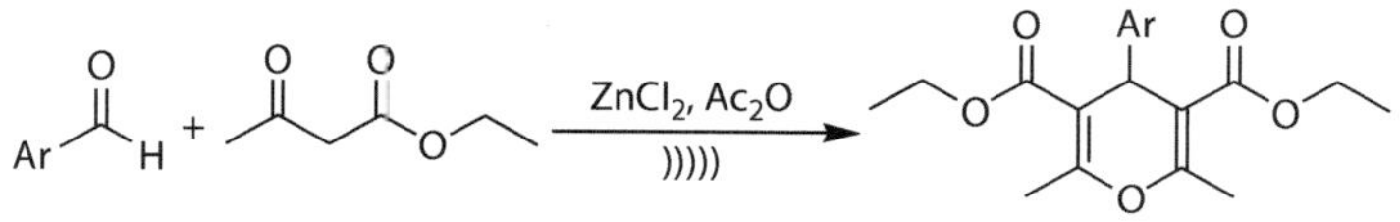

SCHEME 10.109 (From *Ultrason. Sonochem.*, 17, Ni, C.L., Song, X.H., Yan, H., Song, X.Q., and Zhong, R. Improved synthesis of diethyl-2,6-dimethyl-4-aryl-4H-pyra-3,5-dicarboxylate under ultrasound irradiation, 367–369, Copyright (2010), with permission from Elsevier.)

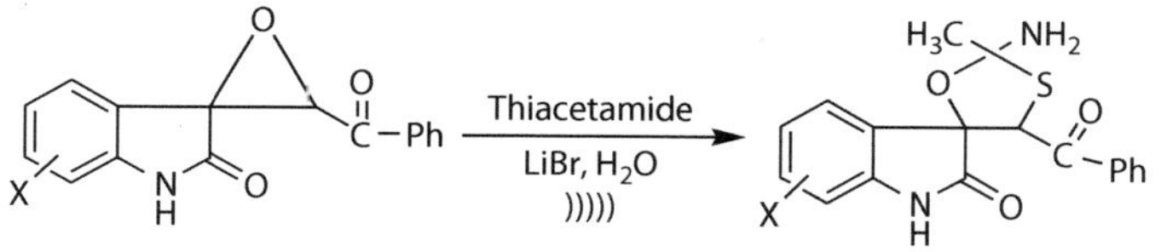

SCHEME 10.110 (From Dandia, A., Sing, R., and Bhaskaran, S., *Ultrason. Sonochem.*, 17, 399, 2010. With permission.)

irradiation, the effect of changes in the ultrasonic power, temperature, and reaction time are discussed (Scheme 10.109).

Dandia et al. (2010) developed an effective methodology for synthesis of Spiro [indole-3,5′-[1,3] oxathiolanes by using Spiro [indole-1,3-oxiranes] with thioacetamide in presence of LiBr under ultrasonic irradiation and water as the reaction medium (Scheme 10.110).

Dabholkar and Wadkar (2009) explained synthesis and characterization of fused and Spiro heterocycles under ultrasonic irradiation (Scheme 10.111).

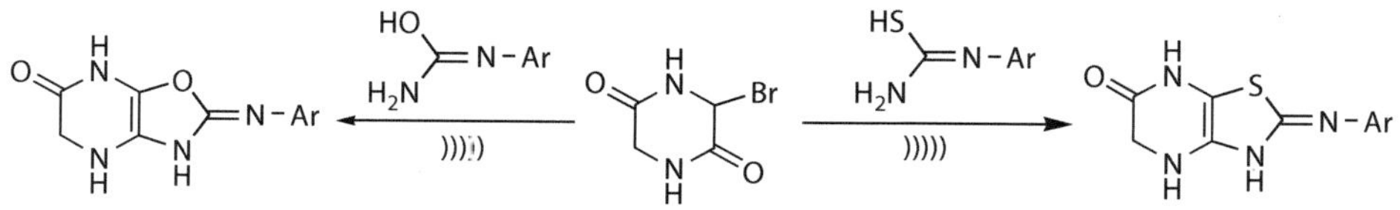

SCHEME 10.111 (From Dabholakar, V.V. and Wadkar, M.M., *Ind. J. Chem.*, 48B, 1027, 2009.)

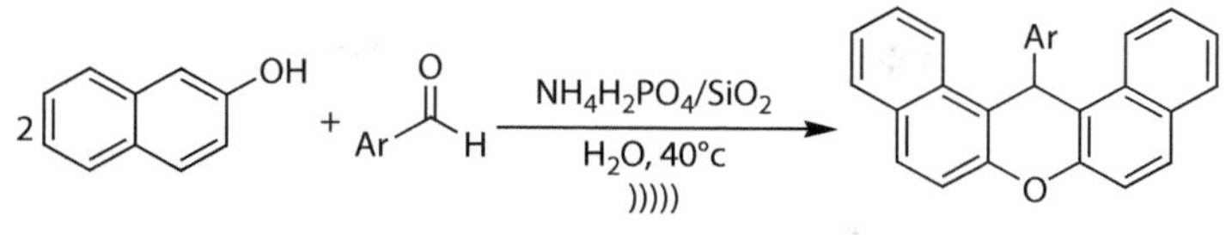

SCHEME 10.112 (Reprinted from *Ultrason. Sonochem.*,16, Mahdavinia, G.H., Rostamizadeh, S., Amani, A.M., and Emdadi, Z., Ultrasound-promoted greener synthesis of aryl-14-H-dibenzo[a,j]xanthenes catalyzed by $NH_4H_2PO_4/SiO_2$ in water, 7–10, Copyright (2009), with permission from Elsevier.)

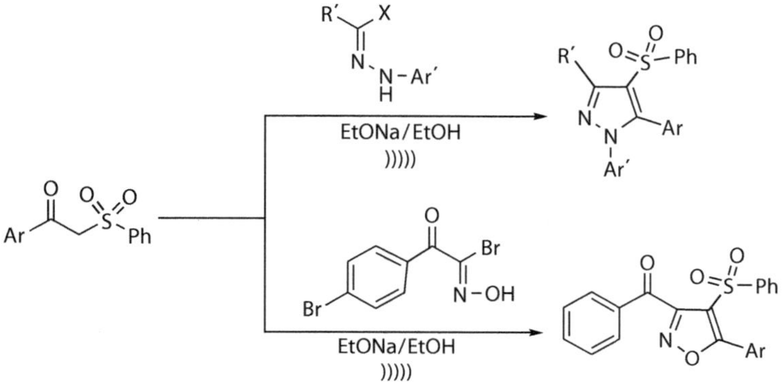

SCHEME 10.113 (From *Ultrason. Sonochem.*, 16, El-Rahman, N.M.A. and Saleh, T.S., Ultrasound promoted synthesis of substituted pyrazoles and isoxazoles containing sulphone moiety, 237–242, Copyright (2009), with permission from Elsevier.)

Mahadavinia et al. (2009) explains ammonium dihydrogen phosphate adsorbed on silica gel catalyzed highly efficient one-pot, green protocol for the synthesis of aryl-14-H-dibenzo[aj] xanthenes by the condensation of an aldehyde and β-naphthol under ultrasound irradiation (Scheme 10.112).

El-Rahman and Saleh (2009) developed new methodology for the synthesis of substituted pyrazoles and isoxazoles containing sulfone moiety under ultrasonic irradiation (Scheme 10.113).

El-Rahman et al. (2009) carried out synthesis of novel, substituted 1,3,4,-thiadiazole and bi(1,3,4,-thiadiazole) derivatives under ultrasonic irradiation (Scheme 10.114).

Flores et al. (2009) developed a new methodology and synthesized several 1-thiocarbamoyl-3,5-diaryl-4,5-dihydro-1H-pyrazole by using chalcone and thiosemicarbazide with ethanol and KOH under ultrasonic irradiation (Scheme 10.115).

Puri et al. (2009) described copper perchlorate–catalyzed, highly efficient, one-pot, green protocol for synthesis of 2H-chromen-2-ones by reaction of substituted phenols and β-keto esters under ultrasound irradiation (Scheme 10.116).

Mantu et al. (2009) introduced facile method for preparation of N-substituted-pyridazinone under ultrasonic irradiation. It was noticed that substituents from 3-(6)-position of pyridazone heterocycle have a substantial influence concerning reactivity, while the influence of those one from 1-(2)-position seems to be of minor importance (Scheme 10.117).

Heravi et al. (2009b) developed a new methodology for synthesis of 4(3H)-quinazolinones by using silica-supported preyssler nanoparticles under ultrasonic irradiation (Scheme 10.118).

Khosropour and Noei (2009) described green procedure for synthesis of 2,4-diarylthiazoles under ambient temperature in [bmim]Bf_4 under ultrasound irradiation (Scheme 10.119).

Heravi (2009) developed a methodology for the synthesis of quinolines by employing ionic liquid [Hbim] [Bf_4] in reaction of o-aminoaryl ketone with α-methylene ketone under ultrasonic irradiation (Scheme 10.120).

Shelke et al. (2009a,b) developed a methodology for efficient and convenient route to the construction of 2,4,5-triarylimidazole using CAN as catalyst under ultrasonic irradiation and also

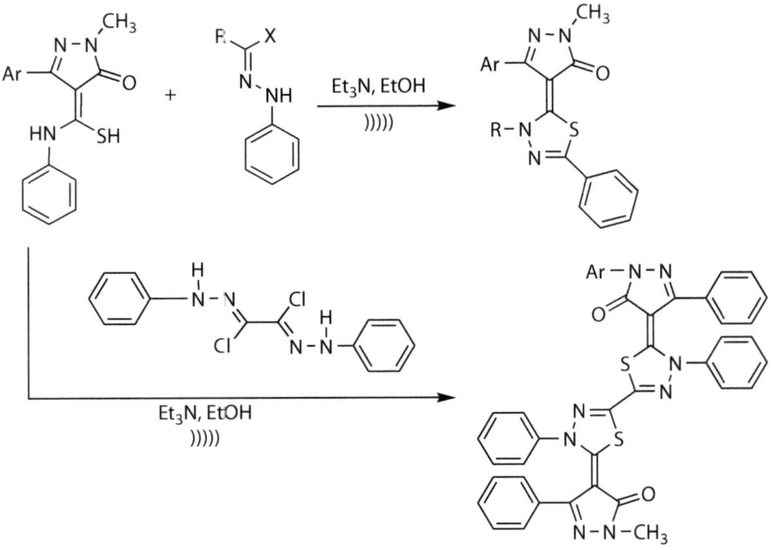

SCHEME 10.114 (Reprinted from *Ultrason. Sonochem.*, 16, El-Rehaman, N.M.A., Saleh, T.S., and Mady, M.F., Ultrasound assisted synthesis of some new 1,3,4-thiadiazole and bi(1,3,4-thiadiazole)derivatives incorporating pyrazolone moiety, 70–74, Copyright (2009), with permission from Elsevier.)

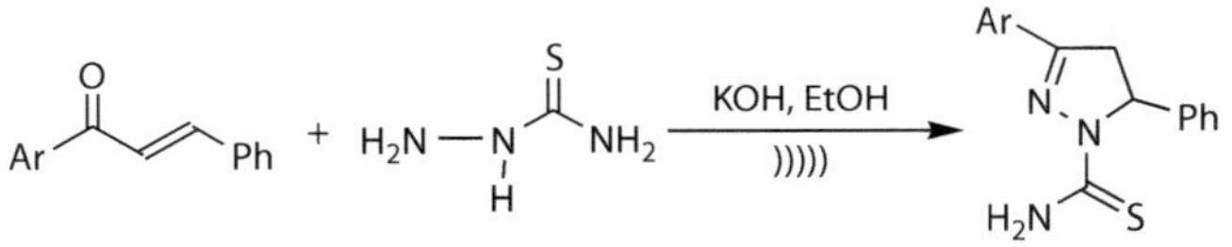

SCHEME 10.115 (From *Ultrason. Sonochem.*, 16, Pizzuti, L., Piovesan, L.A., Flores, A.F.C., Quina, F.H., and Pereira, C.M.P., Environmentally friendly sonocatalysis promoted preparation of 1-thiocarbamoyl-3,5-diaryl-4,5-dihydro-1H-pyrazoles, 728–731, Copyright (2009), with permission from Elsevier.)

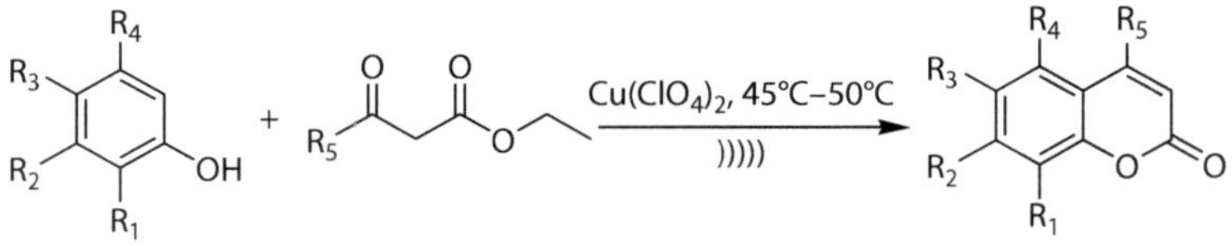

SCHEME 10.116 (From *Ultrason. Sonochem.*, 16, Puri, S., Kaur, B., Parmar, A., and Kumar, H., Ultrasound promoted greener synthesis of 2H-chromene-2-ones catalyzed by copper perchlorate in solventless media, 705–707, Copyright (2009), with permission from Elsevier.)

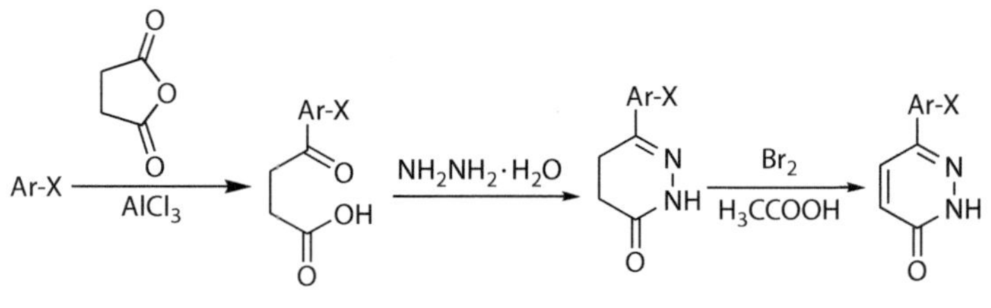

SCHEME 10.117 (From *Ultrason. Sonochem.*, 16, Mantu, D., Maldoveanu, C., Nicolesu, A., Deleanu, C., and Mangalagiu, I.I., A facile synthesis of pyridazinone derivatives under ultrasonic irradiation, 425–454, Copyright (2009), with permission from Elsevier.)

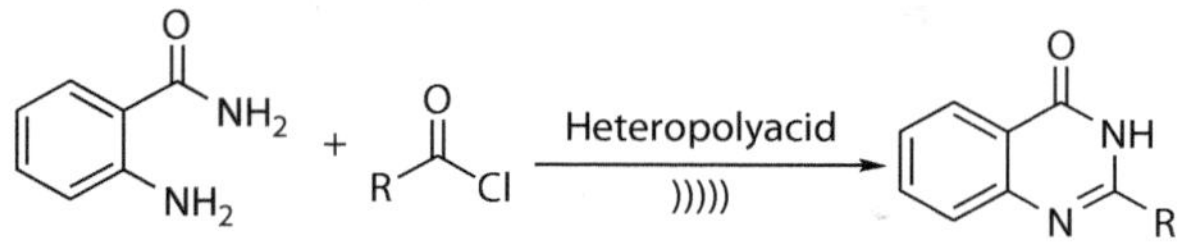

SCHEME 10.118 (From *Ultrason. Sonochem.*, 16, Heravi, M.M., Sadjadi, S., Sadjadi, S., Oskooie, H.A., and Bamoharram, F.F., Rapid and efficient synthesis of 4(3H)-quanazolinones under ultrasonic irradiation using silica-supported preyssler nano particles, 708–710, Copyright (2009), with permission from Elsevier.)

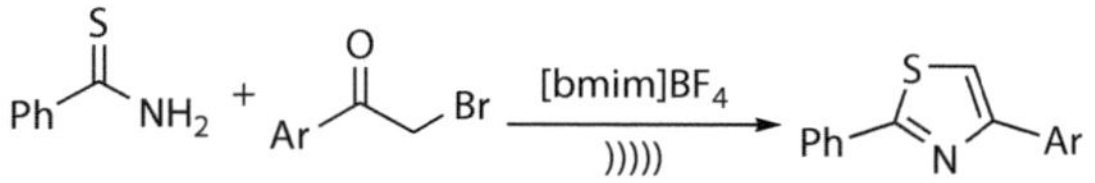

SCHEME 10.119 (From *Ultrason. Sonochem.*, 16, Noei, J. and Khosropour, A.R., Ultrasound-promoted a green protocol for the synthesis of 2,4-diarylthiazoles under ambient temperature in [bmim]BF$_4$, 711–717, Copyright (2009), with permission from Elsevier.)

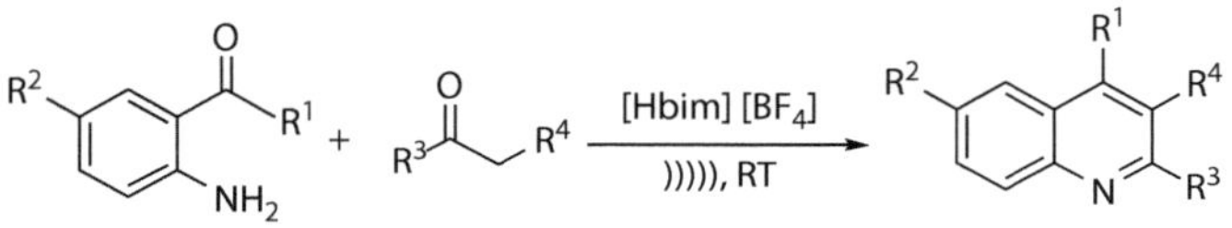

SCHEME 10.120 (Reprinted from *Ultrason. Sonochem.*, 16, Heravi, M.R.P., An efficient synthesis of quinlines derivatives promoted by a room temperature ionic liquid at ambient conditions under ultrasound irradiation via the tandem addition/annulation reaction of o-aminoaryl ketones with α-methylene ketones, 361–366, Copyright (2009c), with permission from Elsevier.)

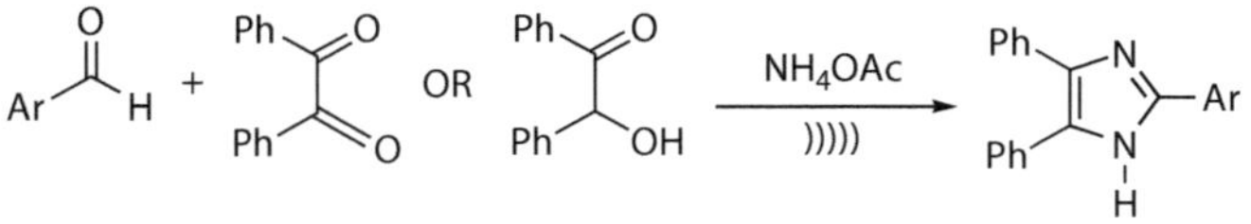

SCHEME 10.121 (From Shelke, K.F. et al., Ultrasound-assisted one-pot synthesis of 2,4,5-triarylimidazole derivatives catalyzed by ceric(IV)ammonium nitrate in aqueous media, *Chin. Chem. Lett.*, 20, 283, 2009. With permission; Shelke, K.F. et al., *Bull. Korean Chem. Soc.*, 30, 1057, 2009. With permission.)

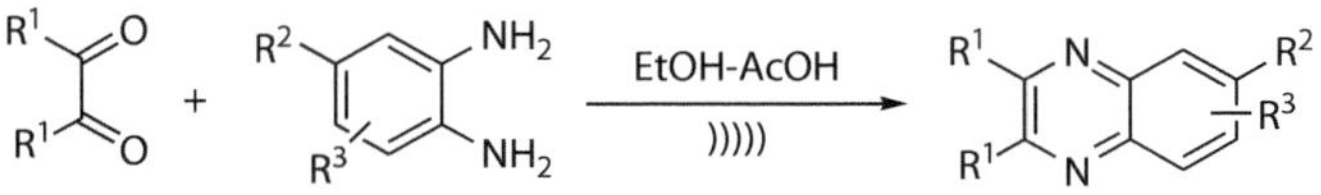

SCHEME 10.122 (From Gao, W.X. et al., *J. Braz. Chem. Soc.*, 9, 1674, 2009. With permission.)

developed an ultrasound-assisted, efficient, and convenient method for the one-pot, three-component synthesis of 2,4,5-triarylimidazole derivatives using cheap and readily available boric acid as a catalyst (Scheme 10.121).

Gao et al. (2009) explained synthesis of quinoxaline derivatives from the reaction of 1,2-diketone and different aryl diamines without catalyst under ultrasonic irradiation (Scheme 10.122).

Tu et al. (2008) developed an efficient procedure for synthesis of furo [3′,4′: 5,6]-pyrido[2,3-d] pyrimidine and [2′,1′: 5,6] pyrido[2,3-d]-pyrimidine derivatives under ultrasonic irradiation (Scheme 10.123).

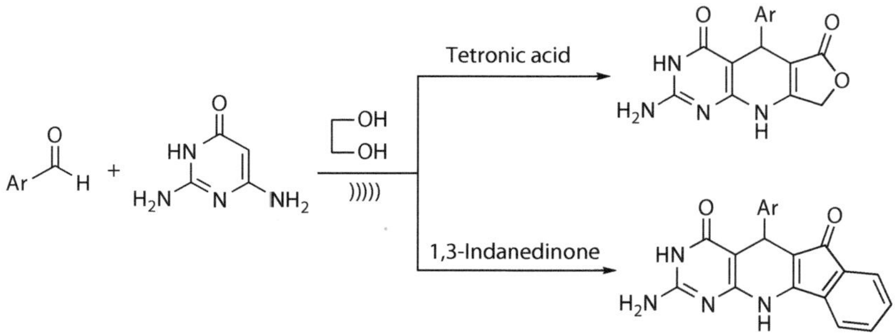

SCHEME 10.123 (From *Ultrason. Sonochem.*, 15, Tu, S., Cao, L., Zhang, Y., Shao, Q., Zhou, D., and Li, C., An efficient synthesis of pyrido[2,3-d]pyrimidine derivatives and related compounds under ultrasound irradiation without catalyst, 217–221, Copyright (2008), with permission from Elsevier.)

RING OPENING REACTIONS

Dalvi et al. (2006) synthesized 1-(2-hydroxy-phenyl)-3-piperidin-1-yl-propenone by the treatment of 3-formylchromones with piperidine in dry ethanol under the influence of ultrasonic irradiation (Scheme 10.124).

Kamal and Arifuddin (2005) have developed an efficient protocol for cleavage of epoxide with aromatic amines in the presence of $FeCl_3$ promoted by ultrasound irradiation (Scheme 10.125).

Xu et al. (1997) described the tellurium mediated nucleophilic epoxide ring opening under ultrasound irradiation under solvent-free condition (Scheme 10.126).

Chou et al. (1991) explained the reductive C–S bond cleavage reactions by using ultrasonically dispersed potassium (UDP) in premeasured amount of water in THF (Scheme 10.127).

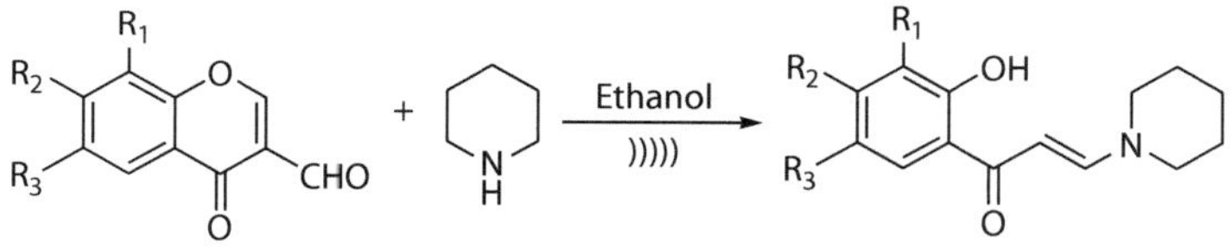

SCHEME 10.124 (From Dalvi, N.R. et al., *Synth. Commun.*, 37, 1421, 2006. With permission.)

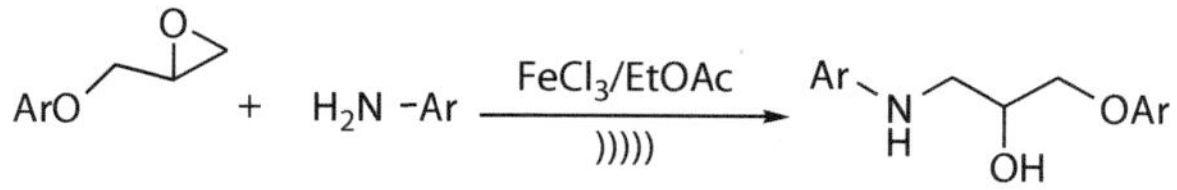

SCHEME 10.125 (Reprinted from *Ultrason. Sonochem.*, 12, Kamal, A., Adil, S.F., and Arifuddin, M., Ultrasonic activated efficient method for the cleavage of epoxides with aromatic amines, 429–431, Copyright (2005), with permission from Elsevier.)

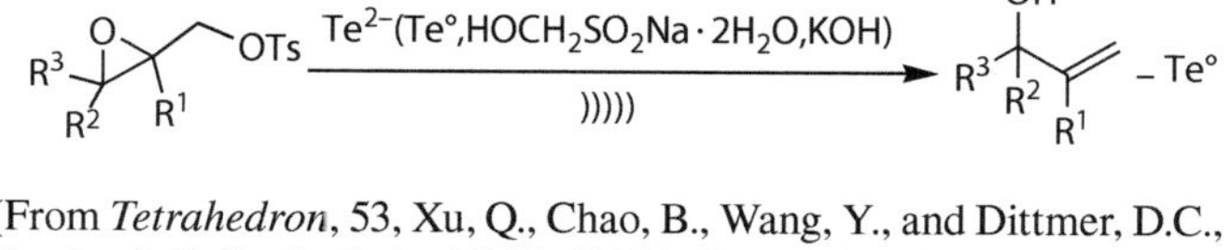

SCHEME 10.126 (From *Tetrahedron*, 53, Xu, Q., Chao, B., Wang, Y., and Dittmer, D.C., Tellurium in the "no-solvent" organic synthesis of allylic alcohols, 12131–12146, Copyright (1997), with permission from Elsevier.)

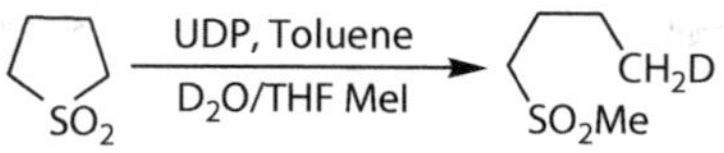

SCHEME 10.127 (Reprinted from *Tetrahedron Lett.*, 32, Chou, T., Hung, S.H., Peng, M.L., and Lee, S.J., Ultrasonically dispersed potassium in organic synthesis. Water-acceleration in reductive C–S bond cleavage reactions, 29, 3551–3554, Copyright (1991), with permission from Elsevier.)

OXIDATION

Yang et al. (1997) described oxidation of various substituted carbonyl compounds to their corresponding carboxylic acid derivative in aqueous media under ultrasonic irradiation in the presence of sodium hypochlorite (Scheme 10.128).

Mamarian et al. (2010) explains the dehydrogenation of 5-acetyl-3,4-dihydropyrimidin-2(*1H*)-ones by using peroxysulfate in aqueous acetonitrile under ultrasonic irradiation at 70°C in 10–40 min (Scheme 10.129).

Joshi et al. (2006) have developed a one-pot, two-step dehydrogenation and oxidation of arylpropanes with excess DDQ in dioxane containing a few drops of acetic acid gave (E) cinnamaldehyde under ultrasound sonication (Scheme 10.130).

Kumar et al. (2007) developed a DDQ-catalyzed benzylic acetoxylation protocol wherein the application of ultrasound imparted exquisite control of the oxidation process (Scheme 10.131).

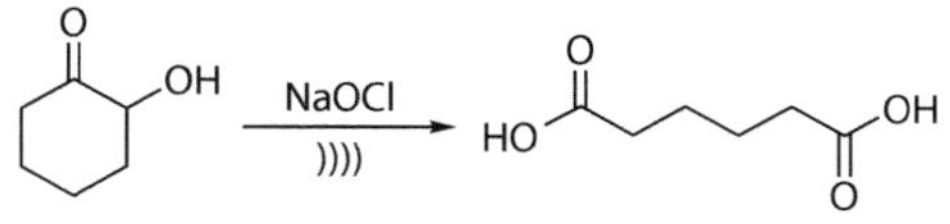

SCHEME 10.128 (From Yang, D.T.C. et al., *Synth. Commun.*, 9, 1601, 1997. With permission.)

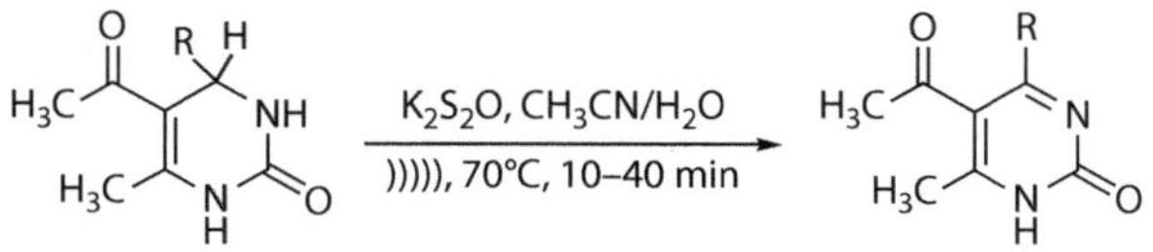

SCHEME 10.129 (From *Ultrason. Sonochem.*, 17, Mamarian, H.R., Farhadi, A., and Sabzyan, H., Ultrasound assisted dehydrogenation of 5-acetyle-3,4-dihydropyrimidin-2(1H)-ones, 579–586, Copyright (2010), with permission from Elsevier.)

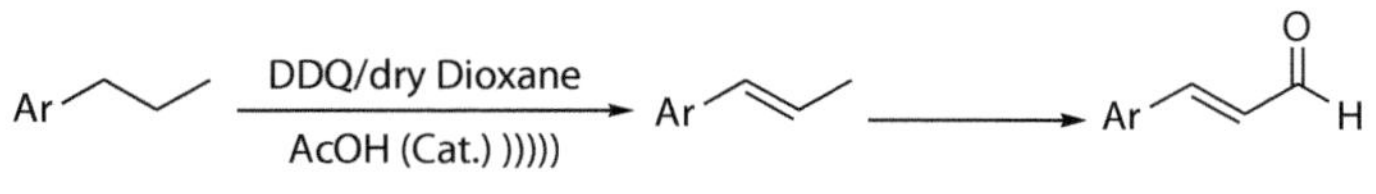

SCHEME 10.130 (From *Tetrahedron*, 62, Joshi, B.P., Sharma, A., and Sinha, A.K., Efficient one-pot, two-step synthesis of (E)-cinnmaldehydes by dehydrogenation-oxidation of arylpropanes using DDQ under ultrasonic irradiation, 2590–2593, Copyright (2006), with permission from Elsevier.)

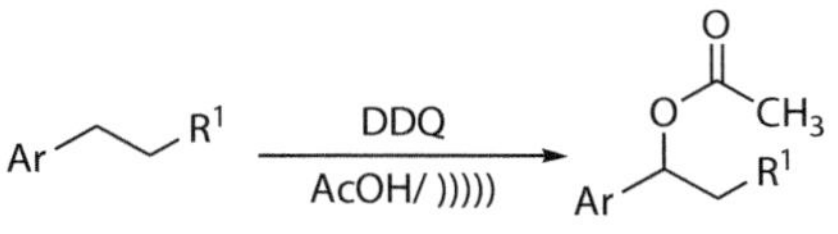

SCHEME 10.131 (From *Tetrahedron*, 63, Kumar, V., Sharma. A., Sharma, M., Sharma, U., and Sinha, A.K., DDQ catalyzed benzylic acetoxylation of arylalkanes: A case of exquisitely controlled oxidation under sonochemical activation, 9718–9723, Copyright (2007), with permission from Elsevier.)

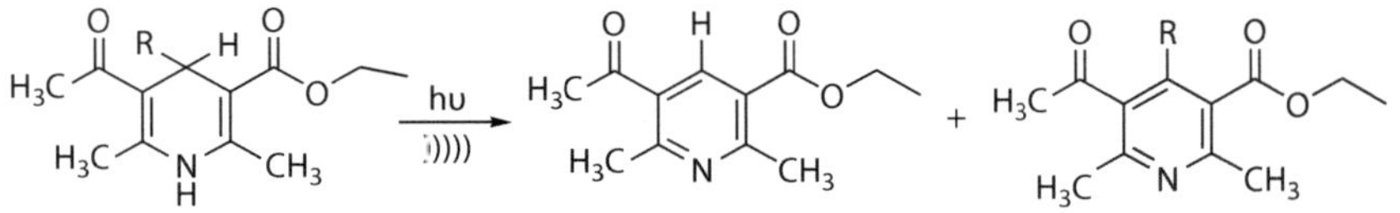

SCHEME 10.132 (From *Ultrason. Sonochem.*, 15, Memarian, H.R. and Senejani, M.A., Ultrasound-assisted photochemical oxidation of unsymmetrically substituted 1,4-dihydropyridines, 110–114, Copyright (2008), with permission from Elsevier.)

Memarian and Senejani (2008) described that ultrasound can seriously affect photo-oxidation of unsymmetrical 1,4-dihydropyridines predominantly by the perfect homogenization of the reactants and excited states in the solution (Scheme 10.132).

Luu et al. (2008) introduced well-defined material potassium permanganate absorbed on copper (II) sulfate pentahydrate (PP/4-CSP) as a highly efficient oxidant toward the oxidation of alcohols in solvent-free condition under ultrasonic irradiation (Scheme 10.133).

Mahamuni et al. (2006) studied the role of ultrasound in the synthesis of benzaldehyde from benzyl alcohol using H_2O_2 in the presence of ultrasound irradiation (Scheme 10.134).

Polackova et al. (1996) explained the cannizzaro reaction of *p*-chlorobenzaldehyde in the presence of phase transfer catalyst under ultrasonication (Scheme 10.135).

Soudagar and Samant (1995) investigated the heterogenous oxidation of unsubstituted arylalkanes using aqueous potassium permangnate under ultrasonic irradiation (Scheme 10.136).

Ruano et al. (2008) developed a new method to obtain disulfides from thiols by using air along with base-like Et_3N and DMF as solvent under ultrasound irradiation (Scheme 10.137).

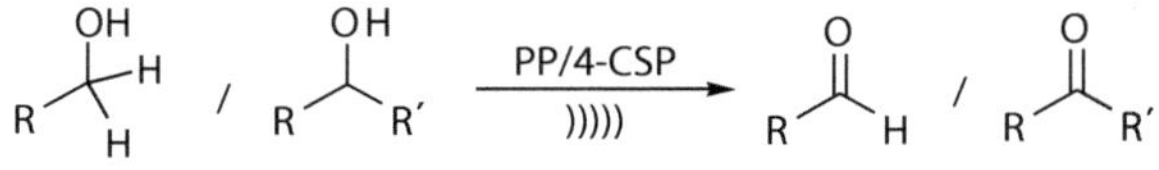

SCHEME 10.133 (From Luu, T.X.T. et al., *Synth. Commun.*, 38, 2011, 2008. With permission.)

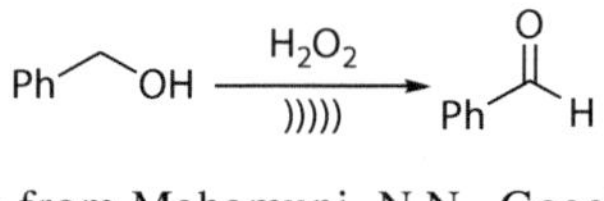

SCHEME 10.134 (With permission from Mahamuni, N.N., Gogate P.R., and Pandit, A.B., Kinetics of solvent-free C-alkylation of phenylacetonitrile using ultrasonic irradiation. *Ind. Eng. Chem. Res.*, 45, 98–108. Copyright (2006) American Chemical Society.)

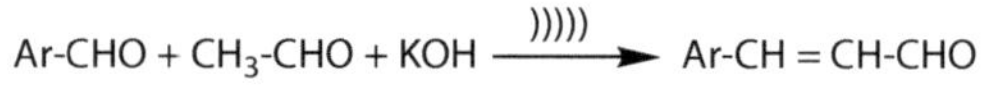

Ar-CHO + CH_3-CHO + KOH $\xrightarrow{))))}$ Ar-CH=CH-CHO

SCHEME 10.135 (From *Ultrason. Sonochem.*, 3, Polackova, V., Tomova, V., Elecko, P., and Toma, S., Ultrasound-promoted Cannizzaro reaction under phase-transfer conditions, 15–17, Copyright (1996), with permission from Elsevier.)

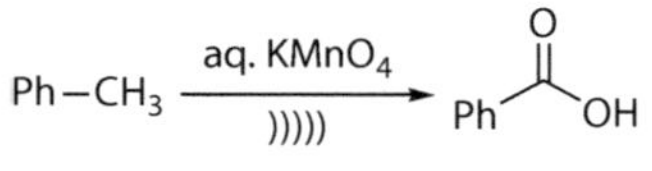

SCHEME 10.136 (From *Ultrason. Sonochem.*, 2, Soudagar, S.R. and Samant, S.D., Investigation of ultrasound catalyzed oxidation of arylalkanes using aqueous potassium permanganate, s15–s18, Copyright (1995), with permission from Elsevier.)

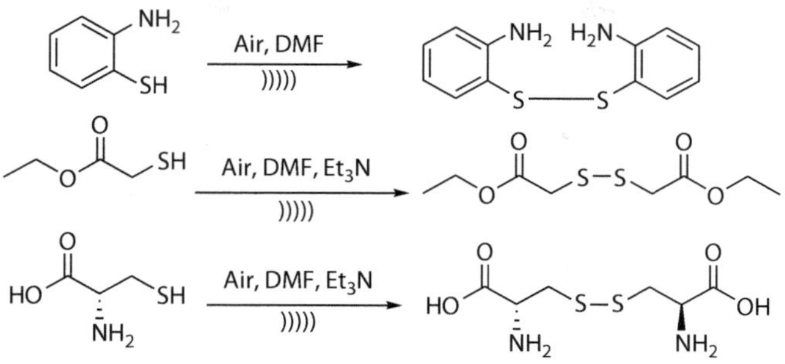

SCHEME 10.137 (From Ruano, J.L., Parra, A., and Aleman, J., Efficient synthesis of disulfides by air oxidation of thiols under sonication, *Green Chem.*, 10, 706–711, 2008. By permission of The Royal Society of Chemistry.)

EPOXIDATION

Li et al. (2010d) found an efficient and practical protocol for the synthesis of some 2,3-epoxyl-1,3-diaryl-1-propanone directly from benzaldehyde and acetophenones in the presence of aqueous potassium hydroxide under ultrasound irradiation (Scheme 10.138).

Jin et al. (2009) carried out an efficient epoxidation of chalcones with urea hydrogen peroxide (UHP) by ultrasound irradiation (Scheme 10.139).

REDUCTION

Reduction is one of the frequently used reactions in organic synthesis and a vast variety of reducing agents have been introduced for this achievement. However, modifying the reducing power of reducing agent by the chemical applications of ultrasound, "Sonochemistry" have become an exciting field of research during the past two decades. Compared with traditional methods the procedure is more convenient. A large number of organic reactions, for example, reduction and cyclocondensation reactions can be carried out in higher yield, shorter reaction time, and milder reaction condition under ultrasound irradiation than the classical methods.

Li et al. (2009) studied the one-pot synthesis of benzylacetamide from oxime in the presence of zinc dust in anhydrous acetic acid/acetic anhydride at 35°C–40°C under ultrasound irradiation (Scheme 10.140).

An aromatic nitro compound having different functionalities like halogen, carbonyl, nitrile, and ester has been subjected for selective reduction of nitro group into its corresponding amine group in the presence of iron or $SnCl_2 \cdot 2H_2O$ under ultrasonic irradiation by Gamble et al. (2007) (Scheme 10.141).

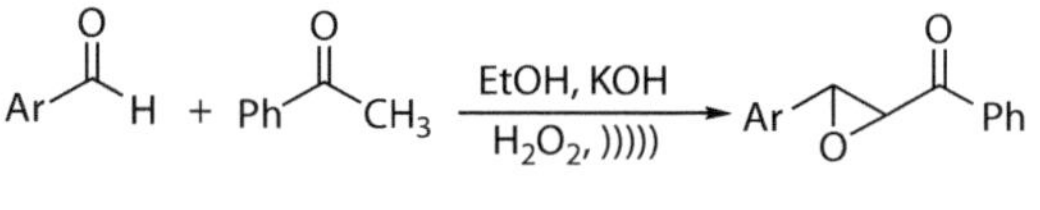

SCHEME 10.138 (From *Ultrason. Sonochem.*, 17, Li, J.T., Yin, Y. and Sun, M.X., An efficient one-pot synthesis of 2,3-epoxyl-1,3-diaryl-1-propanone directly from acetophenones and aromatic aldehydes under ultrasound irradiation, 363–366, Copyright (2010), with permission from Elsevier.)

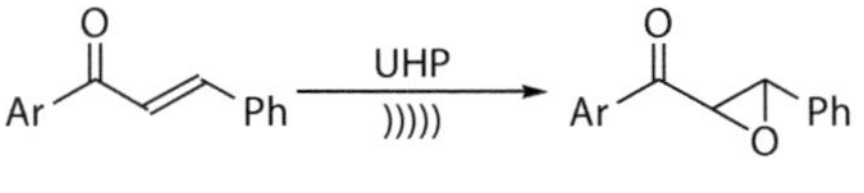

SCHEME 10.139 (From *Ultrason. Sonochem.*, 16, Jin, H., Zhao, H., Zhao, F., Li, S., Liu, W., Zhou, G., Tao, K., and Hou, T., Efficient epoxidation of chalcones with urea-hydrogen peroxide under ultrasound irradiation, 304–307, Copyright (2009), with permission from Elsevier.)

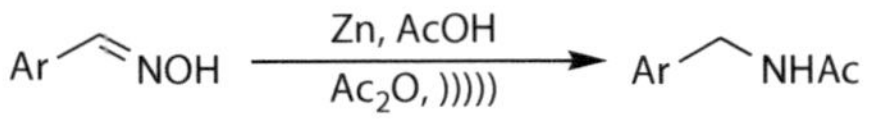

$$\text{Ar-CH=NOH} \xrightarrow[\text{Ac}_2\text{O,)))))}]{\text{Zn, AcOH}} \text{Ar-CH}_2\text{-NHAc}$$

SCHEME 10.140 (From *Ultrason. Sonochem.*, 16, Li, J.T., Meng, X.T., and Zhai, X.L., One-pot synthesis of benzylacetamide from oxime under ultrasound irradiation, 590–592, Copyright (2009), with permission from Elsevier.)

$$\text{MeO-C}_6\text{H}_4\text{-NO}_2 \xrightarrow[\text{)))))}]{\text{Fe/SnCl}_2 \cdot 2\text{H}_2\text{O}} \text{MeO-C}_6\text{H}_4\text{-NH}_2$$

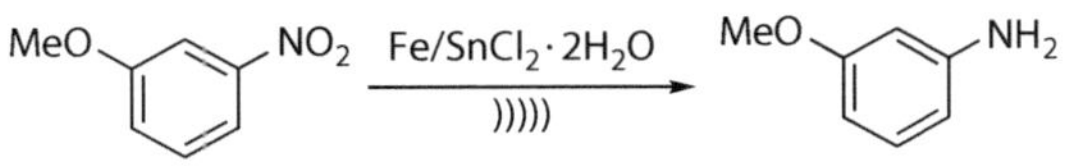

SCHEME 10.141 (From Gamble, A.B. et al., *Synth. Commun.*, 37, 2777, 2007. With permission.)

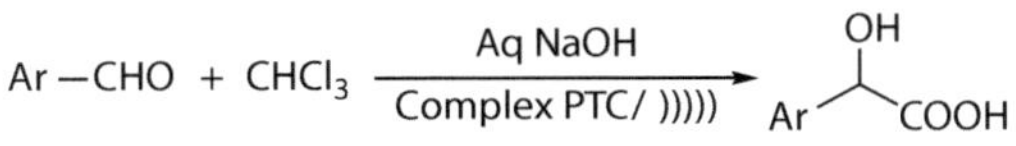

$$\text{Ar-CHO} + \text{CHCl}_3 \xrightarrow[\text{Complex PTC/)))))}]{\text{Aq NaOH}} \text{Ar-CH(OH)-COOH}$$

SCHEME 10.142 (From *Ultrason. Sonochem.*, 15, Xu, H. and Chen, Y., An efficient and practical synthesis of mandelic acid by combination of complex phase transfer catalyst and ultrasonic irradiation, 930–932, Copyright (2008), with permission from Elsevier.)

$$\underset{\text{N-O}}{\overset{R'}{\text{isoxazole-CO}_2\text{Me}}} \xrightarrow[\text{Aq MeOH,)))))}]{\text{Zn, CuI, RI}} \underset{\text{N-O}}{\overset{R'}{\text{isoxazoline-CO}_2\text{Me, R}}}$$

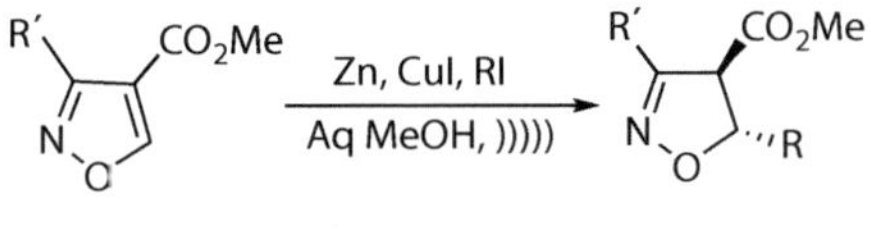

SCHEME 10.143 (With permission from Lee, C. et al., *J. Org. Chem.*, 3221–3231. Copyright 2006 American Chemical Society.)

Xu and Chen (2008) have converted benzaldehyde into mandelic acid using chloroform in the presence of TEBA/PEG-800 as a complex PTC at 60°C in alkaline medium (Scheme 10.142).

Ultrasound-promoted synthesis of substituted methyl trans 3-[2,4,6-trimethyl phenyl]-isoxazolines in the presence of Zn dust/CuI at 5°C has been carried out by Lee et al. (2006) (Scheme 10.143).

Disselkamp et al. (2005) have performed the ultrasound-assisted hydrogenation of cinnamaldehyde using Pd black and Raney Ni as a catalyst at 298 K (Scheme 10.144).

Peng et al. (2005) reported the reduction of benzophenone under ultrasound irradiation in the presence of Zn/EtOH in alkaline medium at room temperature (Scheme 10.145).

Reduction of aryl nitro compounds to azoarenes/aryl amines by Al/NaOH in methanol under ultrasound irradiation has been described by Pasha and Jayashankara (2005) (Scheme 10.146).

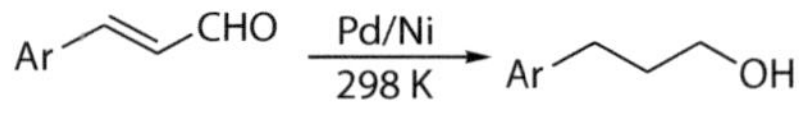

$$\text{Ar-CH=CH-CHO} \xrightarrow[\text{298 K}]{\text{Pd/Ni}} \text{Ar-CH}_2\text{-CH}_2\text{-CH}_2\text{OH}$$

SCHEME 10.144 (From *Ultrason. Sonochem.*, 12, Disselkamp, R.S., Hart T.R., Williams, A.M., White, J.F., and Peden, C.H.F., Ultrasound-assisted hydrogenation of cinnamaldehyde, 319–324, Copyright (2005), with permission from Elsevier.)

$$\text{Ph-CO-Ph} \xrightarrow[\text{)))))}]{\text{Zn, NaOH, EtOH}} \text{Ph-CH(OH)-Ph}$$

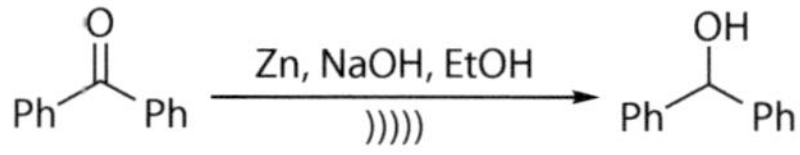

SCHEME 10.145 (From *Ultrason. Sonochem.*, 12, Peng, Y., Zhong, W., and Song, G., Efficient and mild room temperature reduction of benzophenones under ultrasound irradiation, 169–172, Copyright (2005), with permission from Elsevier.)

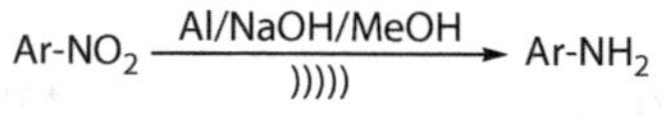

$$\text{Ar-NO}_2 \xrightarrow[\text{)))))}]{\text{Al/NaOH/MeOH}} \text{Ar-NH}_2$$

SCHEME 10.146 (From *Ultrason. Sonochem.*, 12, Pasha, M.A. and Jayashankara, V.P., Reduction of arylnitro compounds to azoarenes and/or arylamines by Al/NaOH in methanol under ultrasonic conditions, 433–435, Copyright (2005), with permission from Elsevier.)

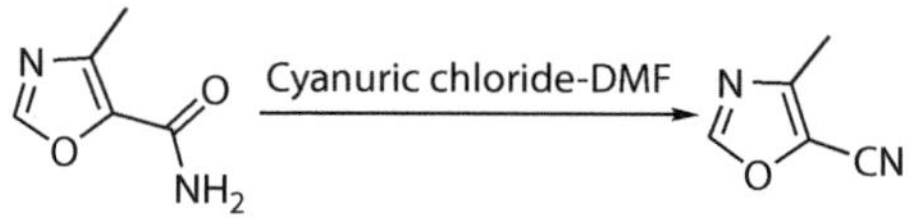

SCHEME 10.147 (From *Ultrason. Sonochem.*, 11, Bonrath, W., Chemical reactions under "non-classical conditions", microwaves and ultrasound in the synthesis of vitamins, 1–4, Copyright (2004), with permission from Elsevier.)

Bonrath (2004) reported the dehydration of 4-methyl-oxazole-5-carboxylic acid amide into respective 4-methyloxazole-5-carbonitrile in the presence of cyanuric chloride-DMF under ultrasound energy (Scheme 10.147).

Wang et al. (2000) described ultrasonically dispersed potassium (UDP) shows the effective catalytical performance for the conversion of azoxy-arene into azoarene at room temperature within shorter reaction time (Scheme 10.148).

Balazsik et al. (1999) studied the hydrogenation of trifluoromethyl ketone over Pt catalyst using heterogeneous as well as homogeneous asymmetric reaction conditions (Scheme 10.149).

Ultrasound-assisted hydrogenation of α, β-unsaturated ketones in the presence of raney Ni as a catalyst is a valuable strategy in which chemo selective C=C hydrogenation is *effectively carried out in cyclic as well as* non-cyclic α, β-unsaturated ketones by Wang et al. (1999) (Scheme 10.150).

Tsuzuki et al. (1995) proposed the synthesis of deuteriated aliphatic amides from the reduction of α,β-unsaturated amides under ultrasound irradiation in the presence of Cu–Al alloy in deuteriated reagents (Scheme 10.151).

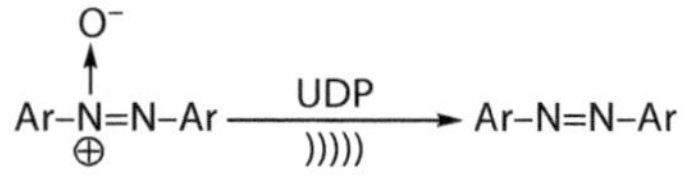

SCHEME 10.148 (From Wang X. et al., *Synth. Commun.*, 30, 2253, 2000. With permission.)

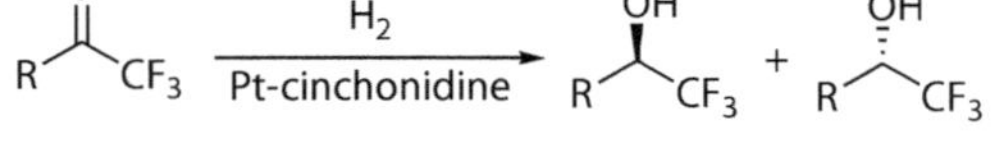

SCHEME 10.149 (From *Ultrason. Sonochem.*, 5, Balazsik, K., Torok, B., Felfoldi, K., and Bartok, M., Homogeneous and heterogeneous asymmetric reactions: Part 11: Sonochemical enantioselective hydrogenation of trifluoromethyl ketones over platinum catalysts, 149–155, Copyright (1999), with permission from Elsevier.)

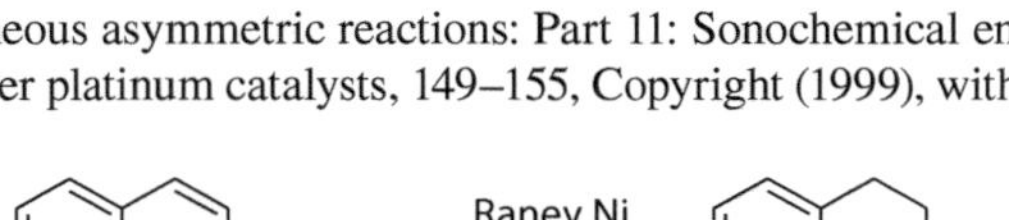

SCHEME 10.150 (From Wang. H. et al., *Synth. Commun.*, 29, 129, 1999. With permission.)

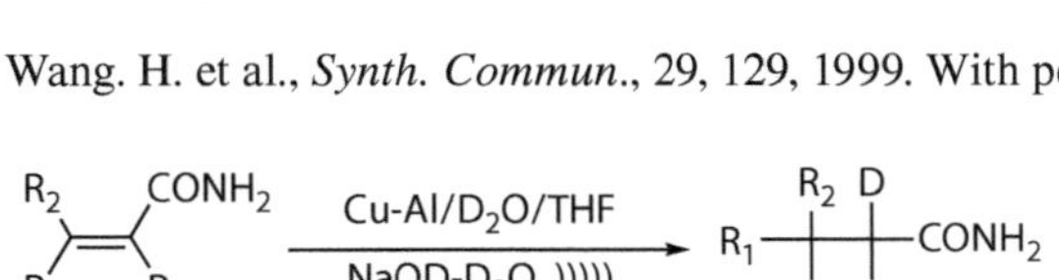

SCHEME 10.151 (From Tsuzuki, H. et al., *J. Lab. Comput. Radiopharmacol.*, 38, 385, 1995.)

SCHEME 10.152 (From Marchand, A.P. and Reddy, G.M., *Synthesis*, 198, 1991. With permission. Copyright Thieme Publishing.)

The selective reduction of α, β-unsaturated γ-dicarbonyl compounds using Zn/AcOH under ultrasonication has been performed by Marchand and Reddy (1991) (Scheme 10.152).

NUCLEOPHILIC SUBSTITUTION REACTIONS

Luzzio and Chen (2008) performed the use of ultrasound irradiation for the conversion of *p*-methoxybenzyl alcohol to PMB-Cl (4-methoxy benzyl chloride) using hydrochloric acid followed by the nucleophilic substitution of phenols to the corresponding ethers (Scheme 10.153).

Deng et al. (2006) has been demonstrated that sonication can be an excellent energy source for accelerating a wide variety of organic reactions used for complex carbohydrate synthesis (Scheme 10.154).

Palacios and Comdom (2003) has been described an improved synthesis of salicylic acid from 2-chloro benzoic acid using ultrasonic irradiation in copper-pyridine-H_2O system (Scheme 10.155).

Lennox et al. (2001) developed a method for the enantiospecific synthesis of 7-azabicyclo[2.2.1] heptano [2,3-c] pyridines from D-glutamic acid: the cyclization of the corresponding iodopyridicyl praline methyl ester, obtained via ultrasound-facilitated chloro-iodo exchange (Scheme 10.156).

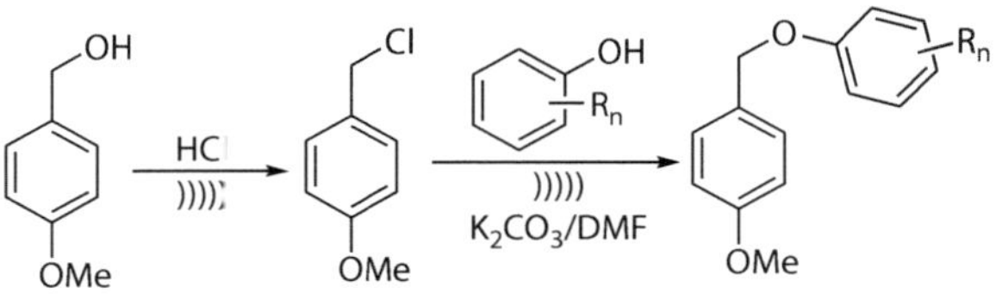

SCHEME 10.153 (With permission from Luzzio, F.A. and Chen, J., Efficient preparation and processing of the 4-methoxybenzyl (PMB) group for phenolic protection using ultrasound, *J. Organ. Chem.*, 73, 5621–5624, 2008. Copyright 2008 American Chemical Society.)

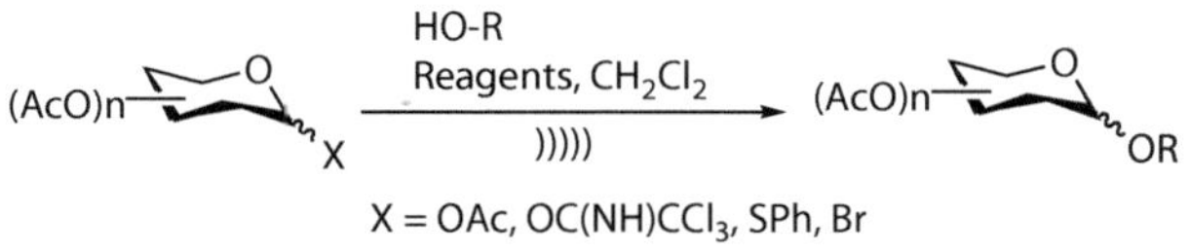

SCHEME 10.154 (From Deng, S. et al., *J. Organ. Chem.*, 71, 5179, 2006. With permission.)

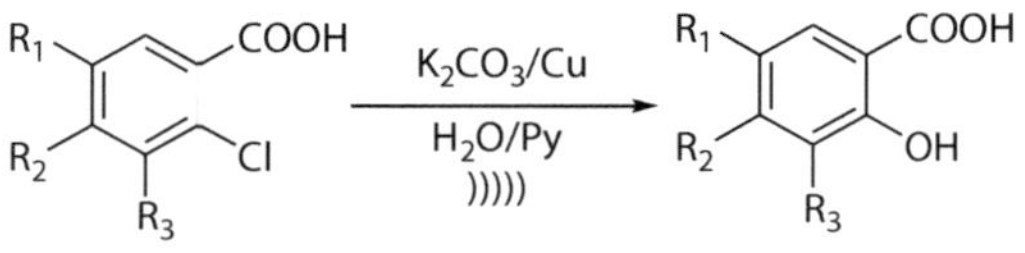

SCHEME 10.155 (From Palacios, M.L.D. and Comdom, R.F.P., *Synth. Commun.*, 33, 1783, 2003. With permission.)

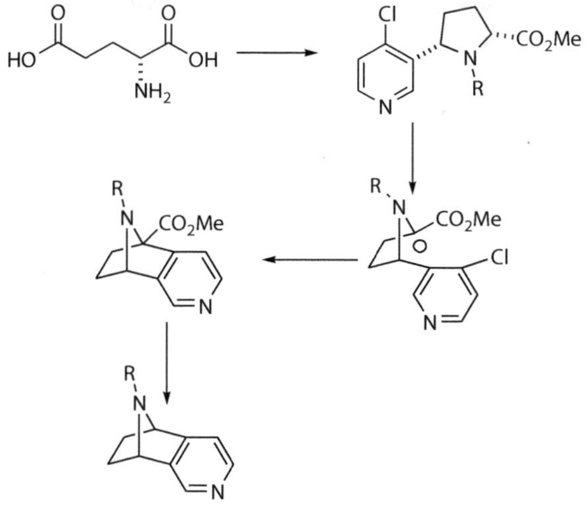

SCHEME 10.156 (With permission from Lennox, J.R., Turner, S.C., and Rapoport, H., Enantiospecific synthesis of annulated nicotine analogues from D-glutamic acid. 7-Azabicyclo[2.2.1]heptano[2.3-c]pyridines, *J. Organ. Chem.*, 66, 7078–7083, 2001. Copyright 2001 American Chemical Society.)

OXIME DEPROTECTION

Li et al. (2010e) carried out the deprotection of oximes to corresponding carbonyl compounds in silica sulfuric/surfactant/paraformaldehyde under the influence of ultrasound irradiation in aqueous medium (Scheme 10.157).

Shaabani et al. (2007) used $NaBrO_3$/ion exchange resin (IER) for the cleavage of oximes to carbonyl compounds promoted by ultrasound irradiation (Scheme 10.158).

PREPARATIONS OF IONIC LIQUIDS

Ionic liquid is receiving renewed attention as green solvents. Therefore, preparation of ionic liquids is important nowadays. Leveque et al. (2002) reported an improved preparation of 1-butyl-3-methyl imidazolium salts (BMIX) as an ionic liquid (Scheme 10.159).

Zhao et al. (2010) reported an efficient ultrasound-assisted synthesis of imidazolium and pyridinium salts based on Zincke reaction. In this reaction, tertiary nitrogen nucleophiles such as pyridines and imidazoles can be alkylated with primary amine by simply using their ammonium form Zincke salt (Scheme 10.160).

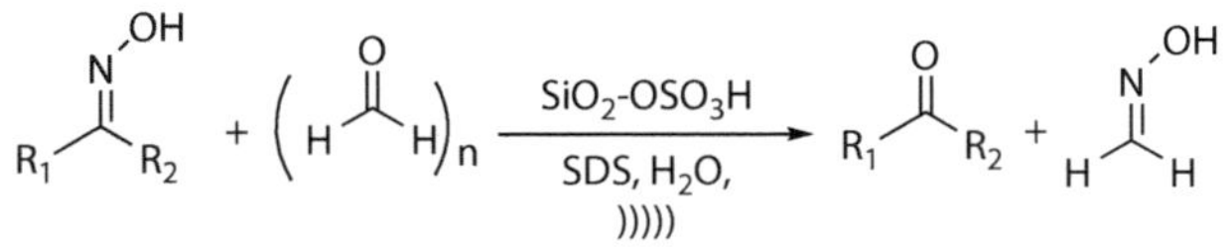

SCHEME 10.157 (From *Ultrason. Sonochem.*, 17, Li, J.T., Menq, X.T., Bai, B., and Sun, M.X., An efficient deprotection of oximes to carbonyls catalyzed by silica sulfuric acid in water under ultrasound irradiation, 14–16, Copyright (2010), with permission from Elsevier.)

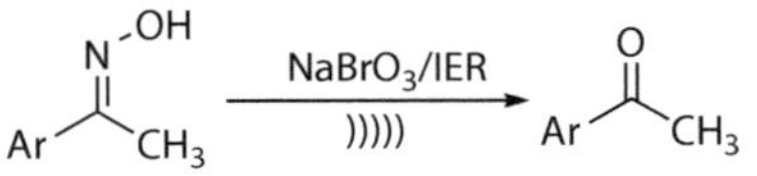

SCHEME 10.158 (From Shaabani, A. et al., *Synth. Commun.*, 37, 4035, 2007. With permission.)

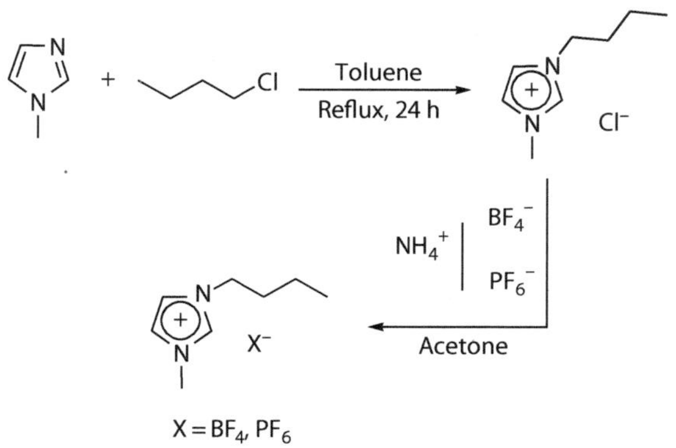

SCHEME 10.159 (From Leveque, J.M., Luche, J.L., Petrier, C., Roux, R., and Bonrath, W., An improved preparation of ionic liquids by ultrasound, *Green Chem.*, 4, 357–360, 2002, by permission of The Royal Society of Chemistry.)

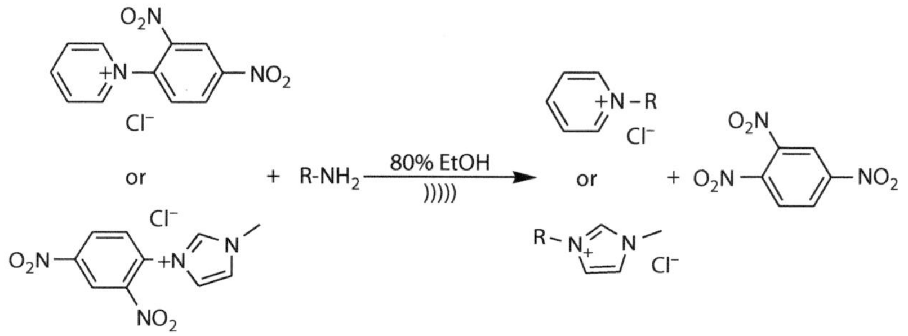

SCHEME 10.160 (From *Ultrason. Sonochem.*, 17, Zhao, S., Xu, X., Zheng, L., and Liu, H., An efficient ultrasonic-assisted synthesis of imidazolium and pyridinium salts based on the Zincke reaction, 685–689, Copyright (2010), with permission from Elsevier.)

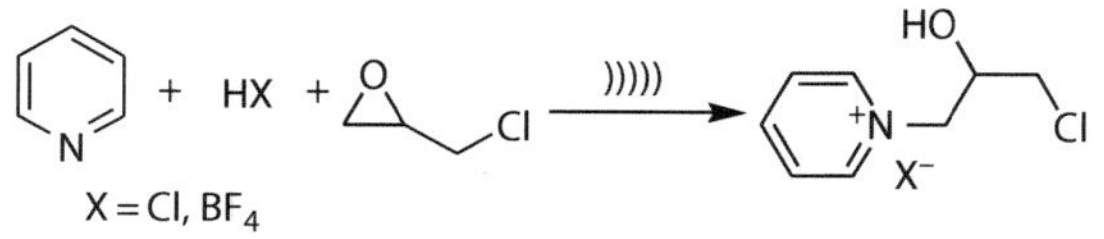

SCHEME 10.161 (From *Ultrason. Sonochem.*, 15, Zhao, S., Zhao, E., Shen, P., Zhao, M., and Sun, J., An atom-efficient and practical synthesis of new pyridinium ionic liquids and application in Morita–Baylis–Hillman reaction, 955–959, Copyright (2008), with permission from Elsevier.)

Zhao et al. (2008) have developed a simple synthetic method to prepare new ionic liquids containing hydroxyl group using ultrasonic bath (Scheme 10.161).

MISCELLANEOUS REACTIONS

Langle et al. (2003) have developed simple and efficient procedure for the synthesis of organogermanium compounds and styrenes with para substitution (Scheme 10.162).

Bremner and Mitchell (1999) have investigated the silylation of several bromothiophene using chlorotrimethylsilane in the presence of ultrasound irradiation (Scheme 10.163).

Sadaphal et al. (2009) developed a green, efficient, cost-effective, and solvent-free method for the synthesis of α-hydroxy phosphonates using ultrasound irradiation (Scheme 10.164).

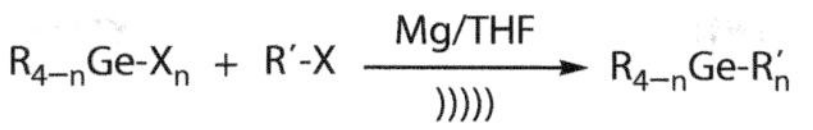

Ultrasound in Synthetic Applications and Organic Chemistry

$$R_{4-n}Ge\text{-}X_n + R'\text{-}X \xrightarrow[\text{)))))}]{\text{Mg/THF}} R_{4-n}Ge\text{-}R'_n$$

SCHEME 10.162 (From *J. Organometall. Chem.*, 671, Langle, S., David-Quillot, F., Balland, A., Abarbri, M., and Duchene, A., General access to para-substituted styrenes, 113–119, Copyright (2003), with permission from Elsevier.)

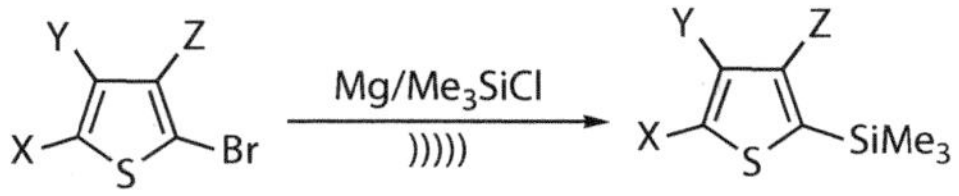

SCHEME 10.163 (From *Ultrason. Sonochem.*, 6, Bremner, D.V. and Mitchell, S.R., Sonochemical reaction of bromothiophenes with chlorotrimethylsilane, 171–173, Copyright (1999), with permission from Elsevier.)

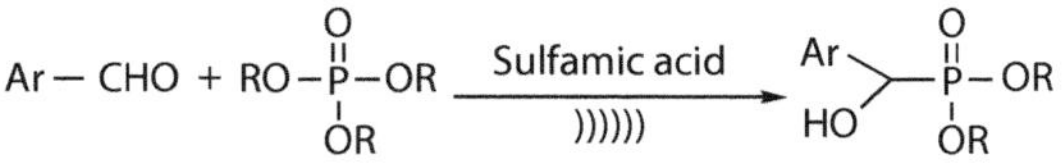

SCHEME 10.164 (From Sadaphal, S.A. et al., *J. Korean Chem. Soc.*, 53, 536, 2009. With permission.)

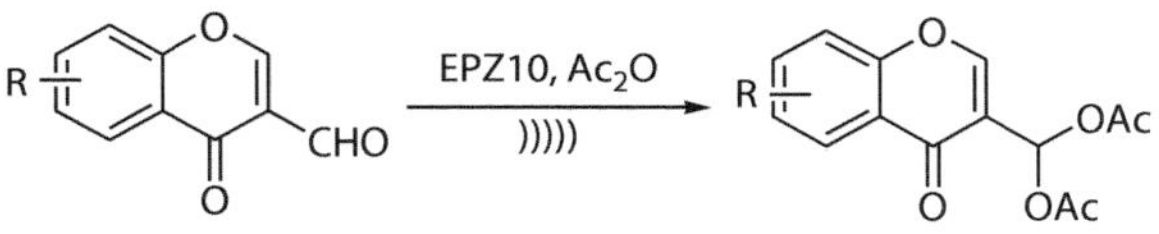

SCHEME 10.165 (From Shindalkar, S.S. et al., *Ind. J. Chem.*, 44B, 1519, 2005. With permission.)

Shindalkar et al. (2005b) developed a environmentally benign, eco-friendly, safe, and cleaner method for the preparation of acylals from 4-Oxo-(4*H*)-1-benzopyran-3-carbaldehyde using reusable Envirocat EPZ10^R catalyst under ultrasonic irradiation (Scheme 10.165).

CONCLUSIONS

While studying the role of ultrasound in synthetic organic chemistry, crucial mechanistic performances have been observed. The energy provided through cavitation by ultrasound waves to organic molecules activates different functionalities in the reacting molecules. Other reagents such as catalysts and solvents and also temperature effect facilitate further steps that are responsible to convert reactants into desired products. The features of ultrasound-promoted organic reactions such as selectivity, ease of experimental manipulation, and enhanced reaction rate are connected with aspects of sustainability. By considering the energy strength of ultrasound, it has been noticed that unexpected bond breaking is not favored to produce side products controlling wastage of reacting materials. Ultrasonic techniques when compared with conventional methods like extraction, crystallization, evaporation, sonication, and sonolysis transdermal drug delivery methods appear to be more effective. Therefore, nowadays, ultrasound sonochemistry attracts immense interest of chemists exhibiting versatile contribution during reaction processes.

ABBREVIATIONS

CTAB	cetyl trimethyl ammonium bromide
DBSA	*para*-dodecylbenzene sulfonic acid
DCM	dichloromethane
DDQ	2,3-dichloro-5,6-dicyano-1,4-benzoquinone
DMF	dimethyl formamide

DMSO   dimethyl sulfoxide
MCPBA   meta chloro perbenzoic acid
MMP   magnesium mono-peroxy phthalate
PPA   polyphosphoric acid
PTSA   para toluene sulfonic acid
RT   room temperature
TBAB n-tetra butyl ammonium bromide
TBAHS n-tetra butyl ammonium hydrogen sulfate
TEA   triethylamine
UDP   ultrasonically dispersed potassium
UHP   urea hydrogen peroxide

REFERENCES

Al-Zaydi, K.M. 2009. A simplified green chemistry approaches to synthesis of 2-substituted 1,2,3-triazoles and 4-amino-5-cyanopyrazole derivatives conventional heating versus microwave and ultrasound as eco-friendly energy sources. *Ultrasonics Sonochemistry*, 16: 805–809.

Anuradha, M.V. and Ravindranath, B. 1997. Acylation of unprotected amino acids using ultrasound. *Tetrahedron*, 53: 1123–1130.

Balazsik, K., Torok, B., Felfoldi, K., and Bartok, M. 1999. Homogeneous and heterogeneous asymmetric reactions: Part 11: Sonochemical enantioselective hydrogenation of trifluoromethyl ketones over platinum catalysts. *Ultrasonics Sonochemistry*, 5: 149–155.

Bazgir, A., Ahadi, S., Ghahremanzadeh, R., Khavasi, H., and Mirzaei, P. 2010. Ultrasound-assisted one-pot, three-component synthesis of spiro[indoline-3,4'-pyrazolo[3,4-b]pyridine]-2,6'(1'H)-diones in water. *Ultrasonics Sonochemistry*, 17: 447–452.

Bian, Y.J., Zhao, H.M., and Yu, X.G. 2009. Allylation reactions of aromatic aldehydes with antimony in aqueous media under ultrasonic irradiation. *Synthetic Communications*, 39: 2370–2377.

Bonrath, W. 2004. Chemical reactions under "non-classical conditions," microwaves and ultrasound in the synthesis of vitamins. *Ultrasonics Sonochemistry*, 11: 1–4.

Bremner, D.V. and Mitchell, S.R. 1999. Sonochemical reaction of bromothiophenes with chlorotrimethylsilane. *Ultrasonics Sonochemistry*, 6: 171–173.

Chou, T., Hung, S.H., Peng, M.L., and Lee, S.J. 1991. Ultrasonically dispersed potassium in organic synthesis. Water-acceleration in reductive C–S bond cleavage reactions. *Tetrahedron Letters*, 32: 3551–3554.

Chtourou, M., Abdelhédi, R., Frikha, M.H., and Trabelsi, M. 2010. Solvent free synthesis of 1,3-diaryl-2-propenones catalyzed by commercial acid-clays under ultrasound irradiation. *Ultrasonics Sonochemistry*, 17: 246–249.

Cravotto, G., Palmisano, G., Tollari, S., Nano, G.M., and Penoni, A. 2005. The Suzuki homocoupling reaction under high-intensity ultrasound. *Ultrasonics Sonochemistry*, 12: 91–94.

Dabholkar, V.V. and Ansari, F.Y. 2008. Synthesis of thiazines using an unusual means-sonication. *Indian Journal Chemistry*, 47B: 1759–1761.

Dabholkar, V.V. and Wadkar, M.M. 2009. Synthesis and characterization of fused and spiro heterocycles by ultrasonic methods. *Indian Journal of Chemistry*, 48B: 1027–1030.

Dalvi, N.R., Shelke, S.N., Karale, B.K., and Gill, C.H. 2006. Synthesis of 1-(2-hydroxy-phenyl)-3-piperidin-1-yl-propenone by ultrasonic irradiation. *Synthetic Communications*, 37: 1421–1424.

Dandia, A., Sing, R., and Bhaskaran, S. 2010. Ultrasound promoted greener synthesis of spiro[indole-3,5'-(1,3)]oxathiolanes in water. *Ultrasonics Sonochemistry*, 17: 399–402.

Deng, S., Gangadharmath, U., and Chang, C.W.T. 2006. Sonochemistry: A powerful way of enhancing the efficiency of carbohydrate synthesis. *Journal of Organic Chemistry*, 71: 5179–5185.

Disselkamp, R.S., Hart T.R., Williams, A.M., White, J.F., and Peden, C.H.F. 2005. Ultrasound-assisted hydrogenation of cinnamaldehyde. *Ultrasonics Sonochemistry*, 12: 319–324.

Donati, D., Morelli, C., Porcheddu, A., and Taddei, M. 2004. A new polymer-supported reagent for the synthesis of β-lactams in solution. *Journal of Organic Chemistry*, 69: 9316–9318.

Du, J.L., Li, L.J., and Zhang, D. 2006. Ultrasound promoted synthesis of 3-carboxycoumarins in aqueous media. *E-Journal of Chemistry*, 3: 1–4.

Duarte, A., Cunico, W., Pereira, C.M.P., Flores, A.F.C., and Freitag, R.A. 2010. Ultrasound promoted synthesis of thioesters from 2-mercaptobenzoxa(thia)zoles. *Ultrasonics Sonochemistry*, 17: 281–283.

El-Rahman, N.M.A. and Saleh, T.S. 2009. Ultrasound promoted synthesis of substituted pyrazoles and isoxazoles containing sulphone moiety. *Ultrasonics Sonochemistry*, 16: 237–242.

El-Rahman, N.M.A., Saleh, T.S., and Mady, M.F. 2009. Ultrasound assisted synthesis of some new 1,3,4-thiadiazole and bi(1,3,4-thiadiazole)derivatives incorporating pyrazolone moiety. *Ultrasonics Sonochemistry*, 16: 70–74.

Flores, A.F.C., Pizzuti, L., Piovesan, L.A., Quina, F.H., and Pereira, C.M.P. 2009. Environmentally friendly sonocatalysis promoted preparation of 1-thiocarbamoyl-3,5-diaryl-4,5-dihydro-1H-pyrazoles. *Ultrasonics Sonochemistry*, 16: 728–731.

Fu, N., Zhang, Y., Yang, D., Chen, B., and Wu, X. 2008. A rapid synthesis of 1-ferrocenyl-2-arylacetylenes under ultrasound irradiation. *Catalysis Communications*, 9: 976–979.

Gamble, A.B., Garner, J., Gordon, C.P., O'Conner, S.M.J., and Keller, P.A. 2007. Aryl nitro reduction with iron powder or stannous chloride under ultrasonic irradiation. *Synthetic Communications*, 37: 2777–2786.

Gao, W.X., Jin, H.L., Chen, J.X., Chen, F., Ding, J.C., and Wu, H.Y. 2009. An efficient catalyst-free protocol for the synthesis of quinoxaline derivatives under ultrasound irradiation. *Journal of Brazilian Chemical Society*, 9: 1674–1679.

Gholap, A.R., Venkatesan, K., Daniel, T., Lahoti, R.J., and Srinivasan, K.V. 2003. Ultrasound promoted acetylation of alcohols in room temperature ionic liquid under ambient conditions. *Green Chemistry*, 5: 693–696.

Gholap, A.R., Venkatesan, K., Danial, T., Lahoti, R.J., and Srinivasan, K.V. 2004. Ionic liquid promoted novel and efficient one pot synthesis of 3,4-dihydropyrimidin-2-(1H)-ones at ambient temperature under ultrasound irradiation. *Green Chemistry*, 6: 147–150.

Gholap, A.R., Venkatesan, K., Pasricha, R., Daniel, T., Lahoti, R.J., and Srinivasan, K.V. 2005. Copper and ligand-free Sonogashira reaction catalyzed Pd(0) nanoparticles at ambient conditions under ultrasound irradiation. *Journal of Organic Chemistry*, 70: 4869–4872.

Guilet, R., Berlan, J., Louisnard, O., and Schwartzentruber, J. 1998. Influence of ultrasound power on the alkylation of phenyl acetonitrile under solid-liquid phase transfer catalysis condition. *Ultrasonics Sonochemistry*, 5: 21–25.

Heravi, M.R.P. 2009. An efficient synthesis of quinlines derivatives promoted by a room temperature ionic liquid at ambient conditions under ultrasound irradiation via the tandem addition/annulation reaction of o-aminoaryl ketones with α-methylene ketones. *Ultrasonics Sonochemistry*,16: 361–366.

Heravi, M.M., Sadjadi, S., Sadjadi, S., Oskooie, H.A., and Bamoharram, F.F. 2009a. A convenient synthesis of bis(indolyl)alkanes under ultra sonic irradiation using silica-supported Preyssler nano particles. *Ultrasonics Sonochemistry*, 16: 718–720.

Heravi, M.M., Sadjadi, S., Sadjadi, S., Oskooie, H.A., and Bamoharram, F.F. 2009b. Rapid and efficient synthesis of 4(3H)-quanazolinones under ultrasonic irradiation using silica-supported preyssler nano particles. *Ultrasonics Sonochemistry*, 16: 708–710.

Hofmann, J., Freier, U., and Wecks, M. 2003. Ultrasound promoted C-alkylation of benzyl cyanide-effect of reactor and ultrasound parameters. *Ultrasonics Sonochemistry*, 10: 271–275.

Jarikote, D.V., Deshmukh, R.R., Rajagopal, R., Lahoti, R.J., Daniel, T., and Srinivasan, K.V. 2003. Ultrasound promoted facile synthesis of arylhydrazones at ambient conditions. *Ultrasonics Sonochemistry*, 10: 45–48.

Javed, T., Mason, T.J., Phull, S.S., Baker, N.R., and Robertson, A. 1995. Influence of ultrasound on the Diels-Alder cyclization reaction: synthesis of some hydroquinone derivatives and lonapalene, an anti-psoriatic agent. *Ultrasonics Sonochemistry*, 2: 3–4.

Ji, S.J., Shen, Z.L., Gu, D.G., and Hauang, X.Y. 2005. Ultrasound-promoted alkylation of ethylbenzene to ketones under solvent-free condition. *Ultrasonics Sonochemistry*, 12: 161–163.

Ji, S.J., Shen, Z.L., and Wang, S.Y. 2003. Aldol condensation of acetylferrocene under ultrasound. *Chinese Chemical Letters*, 14: 663–666.

Ji, S.J. and Wang, S.Y. 2005. An expeditious synthesis of β-indolylketones catalyzed by p-toluenesulfonic acid (PTSA) using ultrasonic irradiation. *Ultrasonics Sonochemistry*, 12: 339–343.

Jin, H., Xiang, L., Wen, F., Tao, K., Liu, Q., and Hou, T. 2008. Improved synthesis of chalconoid-like compounds under ultrasound irradiation. *Ultrasonics Sonochemistry*, 15: 681–683.

Jin, T.S., Xiao, J.C., Wang, S.J., and Li, T.S. 2004. Ultrasound-assisted synthesis of 2-amino-2-chromenes with cetyltrimethylammonium bromide in aqueous media. *Ultrasonics Sonochemistry*, 11: 393–397.

Jin, T.S., Zhang, J.S., Wang, A.Q., and Li, T.S. 2006. Ultrasound-assisted synthesis of 1,8-dioxo-octahydroxanthene derivatives catalyzed by p-dodecylbenzenesulfonic acid in aqueous media. *Ultrasonics Sonochemistry*, 13: 220–224.

Jin, H., Zhao, H., Zhao, F., Li, S., Liu, W., Zhou, G., Tao, K., and Hou, T. 2009. Efficient epoxidation of chalcones with urea–hydrogen peroxide under ultrasound irradiation. *Ultrasonics Sonochemistry*, 16: 304–307.

Joshi, R.S., Mandhane, P.G., Diwakar, S.D., and Gill, C.H. 2010. Ultrasound assisted green synthesis of bis(indol-3-yl)methanes catalyzed by 1-hexenesuphonic acid sodium salt. *Ultrasonics Sonochemistry*, 17: 298–300.

Joshi, B.P., Sharma, A., and Sinha, A.K. 2006. Efficient one-pot, two-step synthesis of (E)-cinnmaldehydes by dehydrogenation-oxidation of arylpropanes using DDQ under ultrasonic irradiation. *Tetrahedron*, 62: 2590–2593.

Kakade, G.K., Madje, B.R., Pokalwar, R.U., Ware, M.N., and Shingare, M.S. 2008. DBU catalyzed Knoevenagel condensation under ultrasonic irradiation. *Indian Journal of Heterocyclic Chemistry*, 17: 379–380.

Kamal, A., Adil, S.F., and Arifuddin, M. 2005. Ultrasonic activated efficient method for the cleavage of epoxides with aromatic amines. *Ultrasonics Sonochemistry*, 12: 429–431.

Kamal, A., Ashwini Kumar, B., Arifuddin, M., and Patric, M. 2004. An efficient and facile nitration of phenols with nitric acid/zinc chloride under ultrasonic conditions. *Ultrasonics Sonochemistry*, 11: 455–457.

Kantharaju and Babu, V.V.S. 2006. Ultrasound accelerated synthesis of proteinogenic and α,α'-dialkylamino acid ester salts. *Indian Journal of Chemistry*, 45B: 1942–1944.

Khorrami, A.R., Faraji, F., and Bazgir, A. 2010. Ultrasound-assisted three-component synthesis of 3-(5-amino-1H-pyrazol-4-yl)-3-(2-hydroxy-4,4-dimethyl-6-oxocyclohex-1-enyl)indolin-2-ones in water. *Ultrasonics Sonochemistry*, 17, 587–591.

Khosropour, A.R. 2008. Ultrasound-promoted greener synthesis of 2,4,5-trisubstituted imidazoles catalyzed by Zr(acac)$_4$ under ambient conditions. *Ultrasonics Sonochemistry*, 15: 659–664.

Khosropour, A.R. and Noei, J. 2009. Ultrasound-promoted a green protocol for the synthesis of 2,4-diarylthiazoles under ambient temperature in [bmim]BF$_4$. *Ultrasonics Sonochemistry*, 16: 711–717.

Koulocheri, S.D. and Haroutounian, S.A. 2001. Ultrasound promoted synthesis of 2,3-bis(4-hydroxyphenyl) indole derivatives as inherently ligands for the estrogen receptor. *European Journal of Organic Chemistry*, 1723–1729.

Kumar, V., Sharma. A., Sharma, M., Sharma, U., and Sinha, A.K. 2007. DDQ catalyzed benzylic acetoxylation of arylalkanes: A case of exquisitely controlled oxidation under sonochemical activation. *Tetrahedron*, 63: 9718–9723.

Langle, S., David-Quillot, F., Balland, A., Abarbri, M., and Duchene, A. 2003. General access to para-substituted styrenes. *Journal of Organometallic Chemistry*, 671: 113–119.

Lee, C.K.Y., Herlt, A.J., Simpson, G.W., Willis A.C., and Easton, C.J. 2006. 4-alkoxycarbonyl- and aminocarbonyl-substituted isoxazoles as masked acrylates in the asymmetric synthesis of Δ^2-isoxazolines. *Journal of Organic Chemistry*, 71: 3221–3231.

Leite, A.C.L., Moreira, D.R. de M., Coelho, L.C.D., Menezes, F.D. de., and Brondani, D.J. 2008. Synthesis of aryl-hydrazones via ultrasound irradiation in aqueous medium. *Tetrahedron Letters*, 49: 1538–1541.

Lennox, J.R., Turner, S.C., and Rapoport, H. 2001. Enantiospecific synthesis of annulated nicotine analogues from D-glutamic acid. 7-Azabicyclo[2.2.1]heptano[2.3-c]pyridines. *Journal of Organic Chemistry*, 66: 7078–7083.

Lepore, S.D. and He, Y. 2003. Use of sonication for the coupling of sterically hindered substrates in the phenolic Mitsunobu reaction. *Journal of Organic Chemistry*, 68: 8261–8263.

Leveque, J.M., Luche, J.L., Petrier, C., Roux, R., and Bonrath, W. 2002. An improved preparation of ionic liquids by ultrasound. *Green Chemistry*, 4: 357–3560.

Li, J.T., Chen, G.F., Wang, J.X., and Li, T.S. 1999a. Ultrasound promoted synthesis of α,α'-bis(substituted furfurylidene) cycloalkanones and chalcones. *Synthetic Communications*, 29: 965–971.

Li, J.T., Chen, G.F., Xu, W.Z., and Li, T.S. 2003a. The Michael reaction catalyzed by KF/basic alumina under ultrasound irradiation. *Ultrasonics Sonochemistry*, 10: 115–118.

Li, J.T., Chen, G.F., Yang, W.Z., and Li, J.S. 2003b. Ultrasound promoted synthesis of 2-aroyl-1,3,5-triaryl-4-carboethoxy-4-cyanocyclohexanols. *Ultrasonics Sonochemistry*, 10: 123–126.

Li, J.T., Dai, H.G., Xu, W.Z., and Li, T.S. 2006a. An efficient and practical synthesis of bis(indolyl)methanes catalyzed by aminosulfonic acid under ultrasound. *Ultrasonics Sonochemistry*, 13: 24–27.

Li, J.T., Han, J.F., Yang, J.H., and Li, T.S. 2003c. An efficient synthesis of 3,4-dihydropyrimidin-2-ones catalyzed by NH$_2$SO$_3$H under ultrasound irradiation. *Ultrasonics Sonochemistry*, 10: 119–122.

Li, J.T., Li, X.L., and Li, T.S. 2006b. Synthesis of oximes under ultrasound irradiation. *Ultrasonics Sonochemistry*, 13: 200–202.

Li, J.T., Li, T.S., Li, L.J., and Cheng, X. 1999b. Synthesis of ethyl α-cyanocinnamates under ultrasound irradiation. *Ultrasonics Sonochemistry*, 6: 199–201.

Li, J., Li, L., Li, T., Li, H., and Liu, J. 1996. An efficient and convenient procedure for the synthesis of 5,5-disubstituted hydantoins under ultrasound. *Ultrasonics Sonochemistry*, 3: 141–143.

Li, J.T., Li, T.S., Li, L.J., and Yang, Z.Q. 1998. Ultrasound-promoted preparation of disteryl ethers catalyzed by montmorillonite K 10. *Ultrasonics Sonochemistry*, 5: 83–85.

Li, J.T., Lin, Z.P., and Liu, C.T. 2008. Study on the conjugate addition of indole with 1,5-diaryl-1,4-pentadien-3-ones catalyzed by $AlCl_3$ under ultrasonication. *Indian Journal of Chemistry*, 47B: 283–290.

Li, J.T., Liu, X.R., and Sun, M.X. 2010a. Synthesis of glycoluril catalyzed by potassium hydroxide under ultrasound irradiation. *Ultrasonics Sonochemistry*, 17: 55–57.

Li, J.T., Meng, X.T., and Zhai, X.L. 2009. One-pot synthesis of benzylacetamide from oxime under ultrasound irradiation. *Ultrasonics Sonochemistry*, 16: 590–592.

Li, J.T., Menq, X.T., Bai, B., and Sun, M.X. 2010b. An efficient deprotection of oximes to carbonyls catalyzed by silica sulfuric acid in water under ultrasound irradiation. *Ultrasonics Sonochemistry*, 17: 14–16.

Li, J.T., Sun, X.L., Lin, Z.P., Chen, Y.X., and Li, T.S. 2007a. Pinacol coupling of aromatic aldehydes and ketones mediated by $TiCl_4$-Zn in ethyl acetate under ultrasound. *Indian Journal of Chemistry*, 46B: 1303–1307.

Li, J.T., Sun, M.X., and Yin, Y. 2010c. Ultrasound promoted efficient method for the cleavage of 3-aryl-2,3-epoxyl-1-phenyl-1-propanone with indole. *Ultrasonics Sonochemistry*, 17: 359–362.

Li, J.T., Yin, Y., Li, L., and Sun, M.X. 2010e. A convenient and efficient protocol for the synthesis of 5-aryl-1,3-diphenylpyrazole catalyzed by hydrochloric acid under ultrasound irradiation. *Ultrasonics Sonochemistry*, 17: 11–13.

Li, J.T., Yin, Y., and Sun, M.X. 2010d. An efficient one-pot synthesis of 2,3-epoxyl-1,3-diaryl-1-propanone directly from acetophenones and aromatic aldehydes under ultrasound irradiation. *Ultrasonics Sonochemistry*, 17: 363–366.

Li, J.T., Zhai, X.L., and Chan, G.F. 2010f. Ultrasound promoted one-pot synthesis of 3-aza-6,10-diaryl-2-oxa-spiro[4.5]decane-1,4,8-trione. *Ultrasonics Sonochemistry*, 17: 356–358.

Li, J.T., Zhang, X.H., and Lin, Z.P. 2007b. An improved synthesis of 1,3,5-triaryl-2-pyrazolines in acetic acid aqueous solution under ultrasound irradiation. *Beilstein Journal of Organic Chemistry*, 3: 13–16.

Li, J.T., Zu, W.Z., Chen, G.F., and Li, T.S. 2005. Synthesis of 1,1-disubstituted-2,6-diarylcyclohexane-4-ones catalyzed by KF/basic Al_2O_3 under ultrasound. *Ultrasonics Sonochemistry*, 12: 473–476.

Lin, H., Yang, M., Huang, P., and Cao, W. 2003. A facile procedure for the generation of dichlorocarbene from the reaction of carbon tetrachloride and magnesium using ultrasonic irradiation. *Molecules*, 8: 608–613.

Liu, B. and Ji, S.J. 2008. Facile synthesis of 2-amino-1,4-naphthoquinones catalyzed by molecular iodine under ultrasonic irradiation. *Synthetic Communications*, 38: 1201–1211.

Liu, B., Ji, S.J., Su, X.M., and Wang, S.Y. 2008. Novel synthesis of 3-indolylquinones catalyzed by molecular iodine under ultrasonic irradiation. *Synthetic Communications*, 38: 1279–1290.

Luu, T.X.T., Ctristensen, P., Duus, F., and Le, T.N. 2008. Microwave- and ultrasound-accelerated green oxidation of alcohols by potassium permanganate absorbed on copper(II) sulfate pentahydrate. *Synthetic Communications*, 38: 2011–2024.

Luzzio, F.A. and Chen, J. 2008. Efficient preparation and processing of the 4-methoxybenzyl (PMB) group for phenolic protection using ultrasound. *Journal of Organic Chemistry*, 73: 5621–5624.

Mahadavinia, G.H., Rostamizadeh, S., Amani, A.M., and Emdadi, Z. 2009. Ultrasound-promoted greener synthesis of aryl-14-H-dibenzo[a,j]xanthenes catalyzed by $NH_4H_2PO_4/SiO_2$ in water. *Ultrasonics Sonochemistry*, 16: 7–10.

Mahamuni, N.N., Gogate P.R., and Pandit, A.B. 2006. Kinetics of solvent-free C-alkylation of phenylacetonitrile using ultrasonic irradiation. *Industrial and Engineering Chemistry Research*, 45: 98–108.

Mamaghani, M. and Dastmard, S. 2009. An efficient ultrasound-promoted synthesis of the Baylis-Hillman adducts catalyzed by imidazole and L-proline. *Ultrasonics Sonochemistry*, 16: 445–447.

Mamarian, H.R., Farhadi, A., and Sabzyan, H. 2010. Ultrasound-assisted dehydrogenation of 5-acetyl-3,4-dihydropyrimidin-2(1H)-ones. *Ultrasonics Sonochemistry*, 17: 579–586.

Mantu, D., Maldoveanu, C., Nicolesu, A., Deleanu, C., and Mangalagiu, I.I. 2009. A facile synthesis of pyridazinone derivatives under ultrasonic irradiation. *Ultrasonics Sonochemistry.*, 16: 425–454.

Marcel, S.F., Lie, K.J., and Lam, C.K. 1995. Ultrasound-assisted epoxidation reaction of long-chain unsaturated fatty esters. *Ultrasonics Sonochemistry*, 2: s11–s14.

Marchand, A.P. and Reddy, G.M. 1991. Mild and highly selective ultrasound-promoted zinc/acetic acid reduction of C:C bonds in α,β-unsaturated γ-dicarbonyl compounds. *Synthesis*, 198–200.

Martins, M.A.P., Pereira, C.M.P., Cunico, W., Moura, S., Rosa, F.A., Peres, R.L., Machado, P., Zanatta, N., and Bonacorso, H.G. 2006. Ultrasound promoted synthesis of 5-hydroxy-5-trihalomethyl-4,5-dihydroisoxazoles and β-enamino trihalomethyl ketones in water. *Ultrasonics Sonochemistry*, 13: 364–370.

Meciarova, M., Toma, S., and Magdolen, P. 2003. Ultrasound effect on the aromatic nucleophilic substitution reactions on some haloarenes. *Ultrasonics Sonochemistry*, 10: 265–270.

Memarian, H.R. and Senejani, M.A. 2008. Ultrasound-assisted photochemical oxidation of unsymmetrically substituted 1,4-dihydropyridines. *Ultrasonics Sonochemistry*, 15: 110–114.

Mojtahedi, M.M., Javadpour, M., and Abaee, M.S. 2008. Convenient ultrasound mediated synthesis of substituted pyrazolones under solvent-free conditions. *Ultrasonics Sonochemistry*, 15: 828–832.

Mosslememin, M.H. and Nateghi, M.R. 2010. Rapid and efficient synthesis of fused heterocyclic pyrimidines under ultrasonic irradiation. *Ultrasonics Sonochemistry*, 17: 162–167.

Muravyova, E.A., Desenko, S.M., Musatov, V.I., Knyazeva, I.V., Shishkina, S.V., Shishkin, O.V., and Chebanov, V.A. 2007. Ultrasonic-promoted three-component synthesis of some biologically active 1,2,5,6-tetrahydropyrimidines. *Journal of Combinatorial Chemistry*, 9: 797–803.

Nabid, M.R., Rezaei, S.J.T., Ghahremanzadeh, R., and Bazgir, A. 2010. Ultrasound-assisted one-pot, three-component synthesis of 1H-pyrazolo[1,2-b]phthalazine-5,10-diones. *Ultrasonics Sonochemistry*, 17: 159–161.

Narwade, S.K., Halnor, V.B., Dalvi, N.R., Gill, C.H., and Karale, B.K. 2006. Conventional and ultrasound mediated synthesis of some thiadiazoles, triazoles and oxadiazoles. *Indian Journal of Chemistry*, 45B: 2776–2780.

Nevskaia, D.M. and Martin-Aranda, R.M. 2003. Nitric acid-oxidized carbon for the preparation of esters under ultrasonic activation. *Catalysis Letters*, 87: 143–147.

Ni, C.L., Song, X.H., Yan, H., Song, X.Q., and Zhong, R. 2010. Improved synthesis of diethyl2,6-dimethyl-4-aryl-4H-pyra-3,5-dicarboxylate under ultrasound irradiation. *Ultrasonics Sonochemistry*, 17: 367–369.

Nikpassand, M., Mamaghani, M., Shirini, F., and Tabatabaein, K. 2010. A convenient ultrasound-promoted regioselective synthesis of fused polycyclic 4-aryl-3-methyl-4,7-dihydro-1H-pyrazolo[3,4-b]pyridines. *Ultrasonics Sonochemistry*, 17: 301–105.

Niralwad, K.S., Shingate, B.B., and Shingare, M.S. 2010a. Solvent-free Sonochemical Preparation of α-aminophosphonates catalyzed by 1-hexanesulphonic acid sodium salt. *Ultrasonics Sonochemistry*, 17: 760–763.

Niralwad, K.S., Shingate, B.B., and Shingare, M.S. 2010b. Ultrasound-assisted one-pot synthesis of octahydroquinazolinone derivatives catalyzed by acidic ionic liquid [tbmim]Cl_2. *Journal of the Chinese Chemical Society*, 57: 89–92.

Palacios, M.L.D. and Comdom, R.F.P. 2003. Synthesis of salicylic acid derivatives in presence of ultrasonic irradiation using water as solvent. *Synthetic Communications*, 33: 1783–1787.

Palmisano, G., Bonrath, W., Boffa, L., Garell, D., Barge, A., and Carvotto, G. 2007. Heck reactions with very low ligandless catalyst loads accelerated by microwaves or simultaneous microwaves/ultrasound irradiation. *Advances in Synthetic Catalysis*, 349: 2338–2344.

Pasha, M.A. and Jayashankara, V.P. 2005. Reduction of arylnitro compounds to azoarenes and/or arylamines by Al/NaOH in methanol under ultrasonic conditions. *Ultrasonics Sonochemistry*, 12: 433–4335.

Pasha, M.A. and Myint, Y.Y. 2006. Ultrasound assisted synthesis of δ-chloroesters from tetrahydrofuran and acyl chlorides in the presence of catalytic zinc dust. *Ultrasonics Sonochemistry*, 13: 175–179.

Pathak, V.N., Gupta, R., and Varshney, B. 2008. A one-pot synthesis of 2-aryl-4H-1-benzopyran-4-ones under coupled microwave phase transfer catalysis (PTC) and ultrasonic irradiation PTC. *Journal of Heterocyclic Chemistry*, 45: 589–592.

Patil, S.B., Singh, P.R., Surpur, M.P., and Samant, S.D. 2007. Ultrasound-promoted synthesis of 1-amidoalkyl-2-naphthols via a three-component condensation of 2-naphthol, ureas/amides, and aldehydes, catalyzed by sulfamic acid under ambient conditions. *Ultrasonics Sonochemistry*, 14: 515–518.

Pellon, R.F. and Docampo, M.L. 2007. Mild method for Ullmann reaction of 2-chlorobenzoic acids and aminothiazoles or aminobenzothiazoles under ultrasonic irradiation. *Synthetic Communications*, 37: 1853–1864.

Peng, Y. and Song, G. 2002. Combined microwave and ultrasound assisted Williamson ether synthesis in the absence of phase-transfer catalysts. *Green Chemistry*, 4: 349–351.

Peng, Y., Song, G., and Dou, R. 2006. Surface cleaning under combined microwave and ultrasound irradiation: flash synthesis of 4H-pyrano[2,3-c]pyrazoles in aqueous media. *Green Chemistry*, 8: 573–575.

Peng, Y., Zhong, W., and Song, G. 2005. Efficient and mild room temperature reduction of benzophenones under ultrasound irradiation. *Ultrasonics Sonochemistry*, 12: 169–172.

Polackova, V., Hutka, M., and Toma, S. 2005. Ultrasound effect on Suzuki reactions. 1. Synthesis of unsymmetrical biaryls. *Ultrasonics Sonochemistry*, 12: 99–102.

Polackova, V., Tomova, V., Elecko, P., and Toma, S. 1996. Ultrasound-promoted Cannizzaro reaction under phase-transfer conditions. *Ultrasonics Sonochemistry*, 3: 15–17.

Puri, S., Kaur, B., Parmar, A., and Kumar, H. 2009. Ultrasound promoted greener synthesis of 2H-chromene-2-ones catalyzed by copper perchlorate in solventless media. *Ultrasonics Sonochemistry*, 16: 705–707.

Qureshi, Z.S., Deshmukh, K.M., Jagtap, S.R., Nandurkar, N.S., and Bhanage, B.M. 2008. Ultrasound assisted regioselective sulfonation of aromatic compounds with sulfuric acid. *Ultrasonics Sonochemistry*, 16: 308–311.

Rajagopal, R. and Srinivasan, K.V. 2003. Ultrasound promoted para-selective nitration of phenols in ionic liquid. *Ultrasonics Sonochemistry*, 10: 41–43.

Rama, K. and Pasha, M.A. 2005. Ultrasound promoted regioselective synthesis of β-iodoethers from olefin-I2-alcohol. *Ultrasonics Sonochemistry*, 12: 437–440.

Ruano, J.L., Parra, A., and Aleman, J. 2008. Efficient synthesis of disulfides by air oxidation of thiols under sonication. *Green Chemistry*, 10: 706–711.

Sadaphal, S.A., Sonar, S.S., Pokalwar, R.U., Shitole, N.V., and Shingare, M.S. 2009. Sulphamic acid: an efficient catalyst for the synthesis of α-hydroxy phosphonates using ultrasound irradiation. *Journal of Korean Chemistry Society*, 53: 536–541.

Shaabani, A., Rahmati, A., and Naderi, S. 2007. Ultrasound-Promoted Rapid Oxidative Cleavage of Oximes with NaBrO$_3$/Ion Exchange Resin. *Synthetic Communications*, 37: 4035–4042.

Shelke, K.F., Madje, B.R., Sadaphal, S.A., Shitole, N.V., and Shingare, M.S. 2008. A facile and efficient one-pot synthesis of dihydropyrimidinones in ionic liquid under solvent-free conditions. *Organic Chemistry: An Indian Journal*, 4: 277–280.

Shelke, K.F., Sapkal, S.B., and Shingare, M.S. 2009a. Ultrasound-assisted one-pot synthesis of 2,4,5-triarylimidazole derivatives catalyzed by ceric(IV)ammonium nitrate in aqueous media. *Chinese Chemical Letters*, 20: 283–287.

Shelke, K.F., Sapkal, S.B., Sonar, S.S., Madje, B.R., Shingate, B.B., and Shingare, M.S. 2009b. An efficient synthesis of 2,4,5-triaryl-1H-imidazole derivatives catalyzed by boric acid in aqueous media under ultrasound-irradiation. *Bulletin of the Korean Chemical Society*, 30: 1057–1060.

Shen, Y. and Fuchigami, M. 2004. Electroorganic synthesis using a fluoride ion mediator under ultrasound irradiation: Synthesis of oxindole and 3-oxotetrahydroisoquinoline derivatives. *Organic Letters*, 6: 2441–2444.

Shen, Z.L., Ji, S.J., Wang, S.Y., and Zeng, X.F. 2005. A novel base-promoted synthesis of β-indolylketones via a three-component condensation under ultrasonic irradiation. *Tetrahedron*, 61: 10552–10558.

Shindalkar, S.S., Madje, B.R., and Shingare, M.S. 2005a. A simple procedure for the preparation of acylals from 4-oxo-4H-1-benzopyran-3-carbaldehyde using Envirocat EPZ10R catalyst under ultrasonic irradiation. *Indian Journal of Heterocyclic Chemistry*, 15: 81–82.

Shindalkar, S.S., Madje, B.R., and Shingare, M.S. 2005b. Ultrasonically accelerated Knoevenagel condensation reaction at room temperature in distilled water. *Indian Journal of Chemistry*, 44B: 1519–1521.

Shinde, P.V., Sonar, S.S., Shingate, B.B., and Shingare, M.S. 2010. Boric acid catalyzed convenient synthesis of 2-amino-3,5-dicarbonitrile-6-thio-pyridines in aqueous media. *Tetrahedron Letters*, 51: 1309–1312.

Sonar, S.S., Sadaphal, S.A., Kategaonkar, A.H., Pokalwar, R.U., Shingate, B.B., and Shingare, M.S. 2009. Alum catalyzed simple and efficient synthesis of Bis(indolyl)methanes by ultrasound approach. *Bulletin of the Korean Chemical Society*, 30: 825–828.

Sonar, S.S., Sadaphal, S.A., Labade, V.B., Shingate, B.B., and Shingare, M.S. 2010. An efficient synthesis and antibacterial screening of novel oxazepine α-aminophosphonates by ultrasound approach. *Phosphorus Sulfur Silicon and Related Elements*, 185: 65–73.

Song, B.A., Zang, G.P., Yang, S., Hu, D.Y., and Jin, L.H. 2006. Synthesis of N-(4-bromo-2-trifluoromethylphenyl)-1-(2-fluorophenyl)-O,O-dialkyl-α-aminophosphonates under ultrasonic irradiation. *Ultrasonics Sonochemistry*, 13: 139–142.

Soudagar, S.R. and Samant, S.D. 1995. Investigation of ultrasound catalyzed oxidation of arylalkanes using aqueous potassium permanganate. *Ultrasonics Sonochemistry*, 2: s15–s18.

Srivastava, R.M., Neves Filho, R.A.W., da Silva, C.A., and Bortoluzzi, A.J. 2009. First ultrasound-mediated one-pot synthesis of N-substituted amides. *Ultrasonics Sonochemistry*, 16: 737–742.

Sureshbabu, V.V., Sudarshan, N.S., and Kantharaju 2008. Efficient synthesis of O-succinimidyl-(tert-butoxycarbonylamino)methyl carbamates derived from α-amino acids accelerated by ultrasound: application to the synthesis of ureidodipeptides. *Synthetic Communications*, 38: 2168–2184.

Tei, L., Gugliotta, G., Avedano, S., Giovenzana, G.B., and Botta, M. 2009. Application of the Ugi four-component reaction to the synthesis of ditopic bifunctional chelating agents. *Organic and Biological Chemistry*, 7: 4406–4414.

Thirumalai, D., Murugan, P., and Ramkrishna, V.T. 2006. Synthesis of 4-aryl-5oxo-1-*H*, 4*H*-5,6,7,8-tetrahydroquinoline and 4-aryl-5-oxo-1*H*-4,5,6,7-tetrahydrocyclopenteno[*b*]pyridine derivatives by ultrasound irradiation and by conventional methods. *Indian Journal of Chemistry*, 45B: 335–338.

Tsuzuki, H., Harada, T., Mukumoto, M., Mataka, S., Tsikinoki, T., Kakinami, T., Nagano, Y., and Tashiro, M. 1995. Ultrasound-assisted reduction of cyanides to deuteriated aliphatic amines. *Journal of Laboratory and Computational Radiopharmacology*, 38: 385–393.

Tu, S., Cao, L., Zhang, Y., Shao, Q., Zhou, D., and Li, C. 2008. An efficient synthesis of pyrido[2,3-d]pyrimidine derivatives and related compounds under ultrasound irradiation without catalyst. *Ultrasonics Sonochemistry*, 15: 217–221.

Tuncay, A., Anaclerio, B.M., and Zolodz, M. 1999. New one-pot method for the synthesis of alkynyl sulfonate esters using ultrasound. *Tetrahedron Letters*, 40: 599–602.

Urbala, M. and Antoszczyszyn, M. 2004. The synthesis of allyl ether functionalized siloxane monomers under ultrasonic irradiation at ambient conditions. *Ultrasonics Sonochemistry*, 11: 409–414.

Valduga, C.J., Braibante, H.S., and Braibante, E.F. 1998. Reactivity of *p*-phenyl substituted β-enamino compounds using K-10/ultrasound. Synthesis of pyrazoles and pyrazolinones. *Journal of Heterocyclic Chemistry*, 35: 189–190.

Venkatesan, K., Pujari, S.S., Lahoti, K.V., and Srinivasan, K.V. 2008. An efficient synthesis of 1,8-dioxo-octahydro-xanthene derivatives promoted by a room temperature ionic liquid at ambient conditions under ultrasound irradiation. *Ultrasonics Sonochemistry*, 15: 548–553.

Wang, S.Y., Ji, S.J., and Ming, S.X. 2008a. A Meldrum's acid catalyzed synthesis of bis(indolyl)methanes in water under ultrasonic condition. *Chinese Journal of Chemistry*, 26: 22–24.

Wang, X.H., Li, Z.Y., Zhang, J.C., and Li, J.T. 2008b. The solvent-free synthesis of 1,4-dihydropyridines under ultrasound irradiation without catalyst. *Ultrasonics Sonochemistry*, 15: 677–680.

Wang. H., Lian, H., Chen, J., Pan, Y., and Shi, Y. 1999. Ultrasonic accelerated hydrogenation of α,β-unsaturated ketones with Raney nickel catalyst. *Synthetic Communications*, 29: 129–134.

Wang, S.X., Wang, K., and Li, J.T. 2005. Pinacol coupling reaction of aromatic aldehydes mediated by aqueous Vanadium(II) solution under ultrasound irradiation. *Synthetic Communications*, 35: 2387–2394.

Wang X., Xu, M., Chen, J., Pan, Y., and Shi Y. 2000. A new procedure to azoarene from azoxyarene by ultrasonic dispersed potassium. *Synthetic Communications*, 30: 2253–2257.

Wei, K., Gao, H.T., and Li, W.D.Z. 2004. Facile synthesis of oxabicyclic alkenes by ultrasonication promoted Diels-Alder cycloaddition of furano dienes. *Journal of Organic Chemistry*, 69: 5763–5765.

Wei, W., Qunrong, W., Liqin, D., Aiqing, Z., and Duoyuan, W. 2005. Synthesis of dinitrochalcones by using ultrasonic irradiation in the presence of potassium carbonate. *Ultrasonics Sonochemistry*, 12: 411–414.

Xia, M. and Lu, Y.D. 2007. Ultrasound-assisted one-pot approach to α-amino phosphonates under solvent-free and catalyst-free conditions. *Ultrasonics Sonochemistry*, 14: 235–240.

Xu, H. and Chen, Y. 2008. An efficient and practical synthesis of mandelic acid by combination of complex phase transfer catalyst and ultrasonic irradiation. *Ultrasonics Sonochemistry*, 15: 930–932.

Xu, Q., Chao, B., Wang, Y., and Dittmer, D.C. 1997. Tellurium in the "no-solvent" organic synthesis of allylic alcohols. *Tetrahedron*, 53: 12131–12146.

Yadav, G.D. and Mujeebur Rahuman, M.S.M. 2003. Synergism of ultrasound and solid acids in intensification of Friedel-Crafts acylation of 2-methoxynaphthalene with acetic anhydride. *Ultrasonics Sonochemistry*, 10: 135–138.

Yang, J.M., Ji, S.J., Gu, D.G., Shen, Z.L., and Wang, S.Y. 2005. Ultrasound-irradiated Michael addition of amines to ferrocenylenones under solvent-free and catalyst-free conditions at room temperature. *Journal of Organometallic Chemistry*, 690: 2989–2995.

Yang, D.T.C., Zhang, C.J., Fu, P.P., and Kabalka, G.W. 1997. Oxidation of α-substituted carbonyl compounds to carboxylic acids in aqueous media using ultrasound. *Synthetic Communications*, 9: 1601–1605.

Yim, E.S., Park, M.K., and Han, B.H. 1997. Ultrasound promoted N-alkylation of pyrrole using potassium superoxide as base in crown ether. *Ultrasonics Sonochemistry*, 4: 95–98.

Zang, H., Wang, M., Cheng, B.W., and Song, J. 2009. Ultrasound-promoted synthesis of oximes catalyzed by a basic ionic liquid [bmIm]OH. *Ultrasonics Sonochemistry*, 16: 301–303.

Zeng, H. and Shao, H.L.W. 2009. One-pot three-component Mannich-type reactions using Sulfamic acid catalyst under ultrasound irradiation. *Ultrasonics Sonochemistry*, 16: 758–762.

Zhang, G., Huang, Z., and Zou, J. 2009. Three-component Mannich Reaction in Water Promoted by Ultrasound Irradiation. *Chinese Journal of Chemistry*, 27: 1967–1974.

Zhang, H., Song, B., Zhong, H., Yang, S., Jin, L., Hu, D., and He, W. 2005. Synthesis of 2-cyanoacrylates containing pyridinyl moiety under ultrasound irradiation. *Journal of Heterocyclic Chemistry*, 42: 1211–1214.

Zhang, J., Yang, F., Ren, G., Mak, T.C.W., Song, M., and Wu, Y. 2008. Ultrasonic irradiation accelerated cyclopalladated ferrocenylimines catalyzed Suzuki reaction in neat water. *Ultrasonics Sonochemistry* 15: 115–118.

Zhao, S., Xu, X., Zheng, L., and Liu, H. 2010. An efficient ultrasonic-assisted synthesis of imidazolium and pyridinium salts based on the Zincke reaction. *Ultrasonics Sonochemistry*, 17: 685–689.

Zhao, S., Zhao, E., Shen, P., Zhao, M., and Sun, J. 2008. An atom-efficient and practical synthesis of new pyridinium ionic liquids and application in Morita-Baylis-Hillman reaction. *Ultrasonics Sonochemistry*, 15: 955–959.

Zhou, W.J., Ji, S.J., and Shen, Z.L. 2006. An efficient synthesis of ferrocenyl substituted 3-cyanopyridine derivatives under ultrasound irradiation. *Journal of Organometallic Chemistry*, 691: 1356–1360.

11 Ultrasound in Synthetic Applications and Organic Chemistry

Rodrigo Cella

CONTENTS

Introduction 263
Historical Background 264
Cavitation: Origin and Theories 265
Condensation Reactions 266
Michael Additions 269
Mannich Reactions 270
Cross-Coupling Reactions 272
Heterocycles Synthesis 274
Miscellaneous 276
Conclusions 277
References 277

INTRODUCTION

The use of ultrasound to promote chemical reactions is called sonochemistry. The effects of ultrasound observed during organic reactions are due to cavitation, a physical process that creates, enlarges, and implodes gaseous and vaporous cavities in an irradiated liquid. Cavitation induces very high local temperatures and pressures inside the bubbles (cavities), leading to turbulent flow of the liquid and enhanced mass transfer.

One of the most fascinating areas of sonochemistry in organic chemistry is sonochemical switching. In some cases, application of ultrasound may completely change the distribution of products or even cause the formation of different substances. The first example in which ultrasound induced a divergent pathway relative to thermal conditions was reported by Ando et al. more than 30 years ago (Ando et al.,1984). Herein, we will also discuss some cases of sonochemical switching.

Recently, the sustainability of chemical reactions has gained relevance in scientific and political discussions. In this context, sonochemistry is discussed as a complementary technique for promoting chemical reactions. These (often called "green") techniques can help to reduce the amount of undesired hazardous chemicals and solvents, reduce energy consumption, and increase the selectivity toward the given product(s).

Ultrasound has been utilized to accelerate a wide number of synthetically useful organic reactions (Suslick, 1990; Fillion and Luche, 1998; Cravotto and Cintas, 2006). In addition to the field of organic chemistry, sonochemistry has also been used in the preparation of micro and nanomaterials, that is, protein microspheres (Peters, 1996; Gedanken, 2004, 2008). Ultrasound also has many therapeutic and diagnostic applications, that is, medical ultrasonography and teeth cleaning;

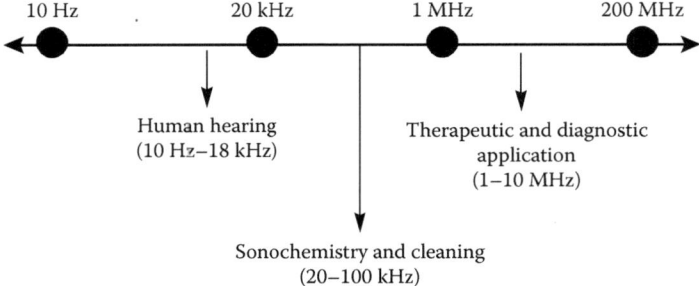

FIGURE 11.1 Ultrasound range diagram.

however, a higher frequency (1–10 MHz) is used in these cases than in sonochemistry (20–100 kHz) (Figure 11.1).

The availability of many works on sonochemistry clearly indicates the impact of ultrasound on organic synthesis in the last past 30 years. In this chapter, we will mostly discuss the use of ultrasound in the concepts of green chemistry, and where possible, it will be compared with cases where the conventional conditions (thermal and stirring) are employed.

HISTORICAL BACKGROUND

Most modern ultrasonic devices rely on transducers (energy converters), which are composed of piezoelectric materials. The basis for present-day generation of US devices was established around 1880, with the discovery of the piezoelectric effect by the brothers Pierre and Jacques Curie. Piezoelectric materials respond to the application of an electrical potential across opposite faces with a small change in dimension. If the potential is alternated at high frequencies, the crystal converts the electrical energy to mechanical vibration energy; at sufficiently high alternating potential,

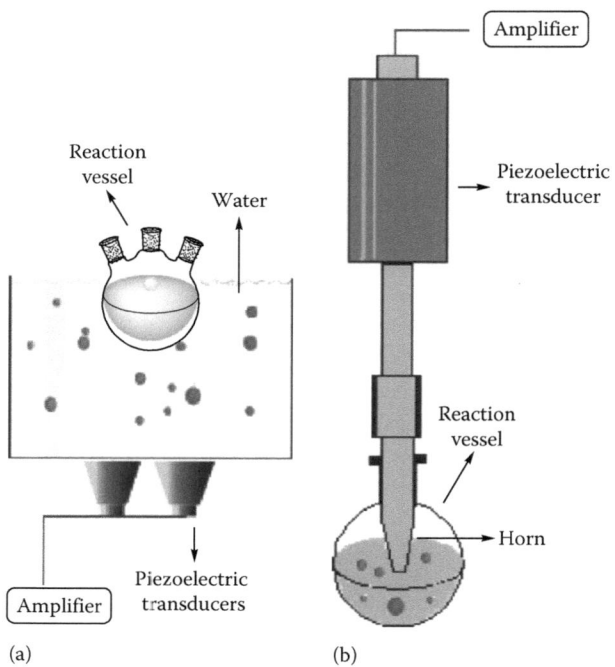

FIGURE 11.2 (a) Ultrasonic cleaning bath. (b) Ultrasonic probe.

high frequency sound (ultrasound) is generated. However, cavitation as a phenomenon was first identified and reported in 1895 by John Thornycroft and Sidney Barnaby (Suslick, 1990). During field tests of high-speed torpedo boats, they observed that the formation and collapse of large bubbles caused erosion of the ship's propeller.

In 1927, Richards and Loomis (1927) noticed the first chemical effects of US. With some exceptions, the field was quite forgotten for nearly 60 years. However, in the 1980s, sonochemistry was reborn and began to be widely used in many different areas. The reason for this growth was the availability of inexpensive and appropriate laboratory equipment, such as ultrasonic cleaning baths (low intensity) or ultrasonic probes (high intensity) (Figure 11.2).

CAVITATION: ORIGIN AND THEORIES

Since US (waves of compression and expansion) is generated by a piezoelectric ceramic in a probe or cleaning bath, it will pass through a liquid, with the expansion cycles exerting negative pressure on the liquid. If this applied negative pressure is strong enough to break down the intermolecular van der Waals force of the liquid, small cavities or gas-filled microbubbles are formed. Cavitation is considered as a nucleated process, meaning that these micrometer-scale bubbles will be formed at preexisting weak points in the liquid, such as gas-filled crevices in suspended particulate matter or transient microbubbles from prior cavitation events. Most liquids are sufficiently contaminated by small particles such that cavitation can be readily initiated at moderate negative pressures.

As microbubbles are formed, they absorb energy from US waves and grow. However, it will reach a stage where it can no longer absorb energy as efficiently. Without the energy input, the cavity can no longer sustain itself and implodes. It is this implosion of the cavity that creates an unusual environment for chemical reactions (Cravotto and Cintas, 2006).

There are a few factors that can affect the efficiency of bubble collapse, such as (Sehgal and Wang, 1981): (1) vapor pressure; (2) temperature; (3) thermal conductivity; (4) surface tension and viscosity; (5) the US frequency; and (6) acoustic intensity.

Since the wavelength of US between successive compression waves measures approximately from 10 to 10^{-3} cm, it does not directly interact with molecules to induce chemical change. Basically, two theories have been proposed to explain the effect of cavitation on chemical reactions: the "hot spot" (Suslick et al., 1986; Flint and Suslick, 1991) and electrical microdischarge theories (Margulis, 1990). Because the latter is not well established, it will not be discussed here; however, it cannot be entirely ruled out due to the complex nature of cavitation.

The "hot spot" theory relies on bubble collapse in the liquid to produce enormous amounts of energy from conversion of the kinetic energy of liquid motion into heating of the bubble contents. Compression of the bubbles during cavitation is more rapid than thermal transport, resulting in the generation of short-lived localized hot spots. Experimental results have shown that these bubbles have temperatures around 5000 K, pressures of approximately 1000 atm, and heating and cooling rates above 10^{10} K s^{-1}. Three classes of sonochemical reactions exist:

1. *Homogeneous sonochemistry*: homogeneous systems that proceed via radical or radical-ion intermediates. This implies that sonication is able to affect reactions proceeding through radicals and, furthermore, that it is unlikely to affect ionic reactions. In the case of volatile molecules, the bubbles (or cavities) are believed to act as a microreactor; as the volatile molecules enter the microbubbles and the high temperature and pressure produced during cavitation break their chemical bonds, short-lived chemical species are returned to the bulk liquid at room temperature, thus reacting with other species. Compounds of low volatility, which are unlikely to enter bubbles and thus be directly exposed to these extreme conditions, still experience a high energy environment resulting from the pressure changes associated with the propagation of the acoustic wave or with bubble collapse (shock waves); alternatively, they can react with radical species generated by sonolysis of the solvent.

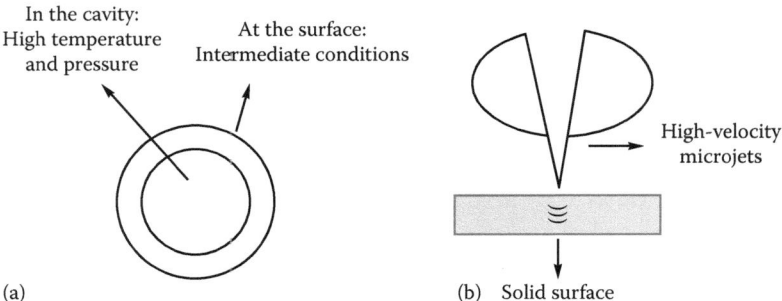

FIGURE 11.3 (a) Cavitation bubble in a homogeneous system. (b) Cavitation bubble in a heterogeneous system.

2. Heterogeneous sonochemistry (liquid–liquid or solid–liquid systems): heterogeneous systems that proceed via ionic intermediates. Here, the reaction is influenced primarily through the mechanical effects of cavitation, such as surface cleaning, particle size reduction, and improved mass transfer. When cavitation occurs in a liquid near a solid surface, the dynamics of cavity collapse change dramatically. In homogeneous systems, the cavity remains spherical during collapse because its surroundings are uniform (Figure 11.3a). Close to a solid boundary, cavity collapse is very asymmetric and generates high-speed jets of liquid (with velocities of approximately $111\,\text{m s}^{-1}$; Figure 11.3b). These jets hit the surface with tremendous force. This process can cause harsh damage at the point of impact and produce newly exposed highly reactive surfaces.
3. Sonocatalysis (overlap homogeneous and heterogeneous sonochemistry): heterogeneous reactions that include a radical and ionic mechanism. Radical reactions will be chemically enhanced by sonication, but the general mechanical effect described above may very well still apply. If radical and ionic mechanisms lead to different products, US should favor the radical pathway, potentially leading to a change in the nature of the reaction products.

CONDENSATION REACTIONS

The Knoevenagel condensation is a classic general method for the preparation of important intermediates. This condensation can be carried out using organic base, Lewis base, or acid as the catalyst and employing conventional heating method. However, there are some disadvantages due to long reaction time, difficult workup, low yield, or environmental concerns. Recent results showed that under ultrasound the Knoevenagel condensation could be carried out in milder conditions, shorter reaction time, and higher yield.

The pyridine-catalyzed Knoevenagel condensation (Li et al., 1999b) of ethyl cyanoacetate **11.1** with a variety of aromatic aldehydes **11.2** under ultrasonic conditions afforded ethyl α-cyanocinnamates **11.3** in good to excellent yields (Table 11.1). The same reaction under conventional techniques showed lower yields in most of the cases as well as harsher conditions (Popp, 1960; Cabello et al., 1984).

The reaction between **11.1** and **11.2** can also be catalyzed by KF supported by Al_2O_3 to afford the ethyl α-cyanocinnamates **11.3** (Li et al., 2002) in excellent yields (Table 11.1). The reaction was carried out in ethanol and the KF–Al_2O_3 could be recycled without significant loss in the yields.

Malononitrile **11.4** can condensate with aromatic aldehydes **11.2** under ultrasound conditions in the absence of any catalyst to form arylmethylenemalononitrile **11.5** (Li et al., 2003a), Scheme 11.1. The reaction was tolerant to electron withdrawing and donating groups attached to the aldehydes **11.2** and compounds **11.5** were obtained in excellent yields at room temperature using ethanol or methanol as solvent.

TABLE 11.1
Knoevenagel Condensation under Ultrasound Waves

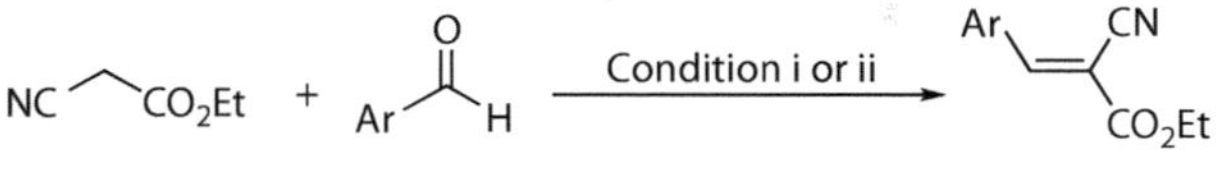

NC–CH₂–CO₂Et (**11.1**) + Ar–CHO (**11.2**) →[Condition i or ii] Ar–CH=C(CN)(CO₂Et) (**11.3**)

Condition i: Pyridine (12 mol%),))), 20°C–40°C, 2–3 h.
Condition ii: KF–Al$_2$O$_3$,))), 20°C–40°C, 35–180 min.

Entry	Ar	Condition i Yield (%)[a]	Condition ii Yield (%)[a]
1	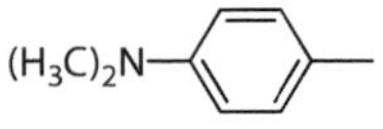 4-(H$_3$C)$_2$N–C$_6$H$_4$–	95 (87)	97 (87)
2	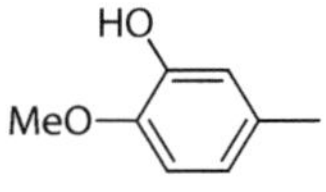 3-HO-4-MeO–C$_6$H$_3$–	95 (92)	97 (86)
3	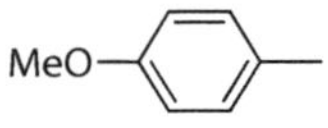 4-MeO–C$_6$H$_4$–	93 (52)	98 (86)
4	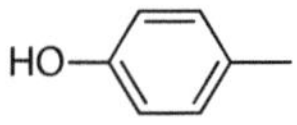 4-HO–C$_6$H$_4$–	94 (58)	98 (80)
5	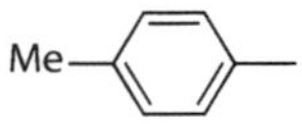 4-Me–C$_6$H$_4$–	96 (62)	99 (78)
6	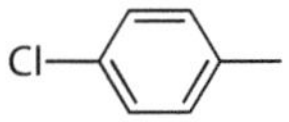 4-Cl–C$_6$H$_4$–	86 (47)	99 (92)
7	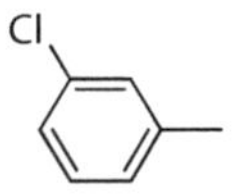 3-Cl–C$_6$H$_4$–	91 (80)	97 (89)

[a] Yields in parenthesis are under no ultrasound conditions.

Chalcones are important in the preparation, as well as the central core for a variety of several biological compounds; one of the more popular ways to prepare chalcones is the Claisen–Schmidt condensation. Claisen–Schmidt condensation is the condensation of aromatic aldehydes with ketone using reagents such as NaOH, KOH, or Ba(OH)$_2$ as the catalyst, by conventional method. However, there are always some problems due to long reaction time or difficulty in workup.

In 1987, Fuentes et al. have reported a sonochemical synthesis of chalcones **11.7** catalyzed by an activated Ba(OH)$_2$ catalyst (Table 11.2). The sonochemical process took place at room temperature and with a lower catalyst loading and reaction time than the thermal process. KOH also has been used as catalyst in the condensation between **11.2** and **11.6** under ultrasound (Li et al., 1999a, 2002). While under ultrasound the reaction is carried out in 25 min, the reaction takes 5 h under stirring

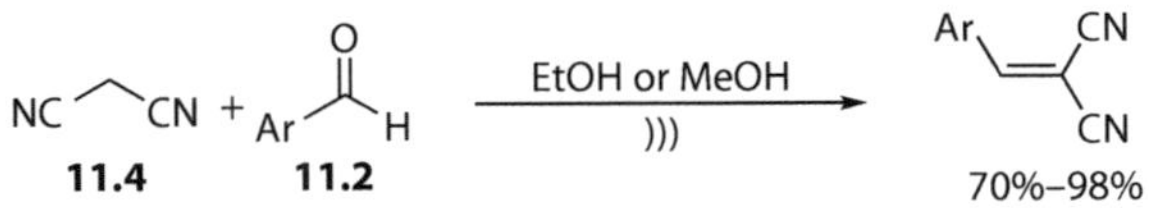

NC–CH$_2$–CN (**11.4**) + Ar–CHO (**11.2**) →[EtOH or MeOH,)))] Ar–CH=C(CN)$_2$ (**11.5**) 70%–98%

SCHEME 11.1

TABLE 11.2
Synthesis of Chalcones 7 under Ultrasound Conditions

$R^1COCH_2R^2$ (**11.6**) + $ArCHO$ (**11.2**) $\xrightarrow{\text{Condition A-D}}$ Ar-CH=C(R²)-CO-R¹ (**11.7**)

Condition A: Ba(OH)$_2$ (10 mol%), EtOH))).
Condition B: NaOH (16 mol%), EtOH, 20°C–40°C, 4–50 min,))).
Condition C: KF–Al$_2$O$_3$, MeOH, 25°C–40°C, 5–240 min,))).
Condition D: KSF, r.t., 30–240 min,))).

Entry	Product	Condition A Yield (%)	Condition B Yield (%)	Condition C Yield (%)	Condition D Yield (%)
1	Ph-CH=CH-CO-Ph	36	80	86	79
2	p-MeO-Ph-CH=CH-CO-Ph	47	79	84	95
3	p-Me-Ph-CH=CH-CO-Ph	36	93	85	nd
4	p-O$_2$N-Ph-CH=CH-CO-Ph	52	91	97	88
5	p-Cl-Ph-CH=CH-CO-Ph	80	91	90	86
6	furyl-CH=CH-CO-Ph-OMe-p	nd	90	nd	nd
7	2,6-bis(furylmethylene)cyclohexanone	nd	85	nd	nd

conditions (Table 11.2) and the chalcones **11.7** were obtained in slightly better yields. A reusable catalyst KF–Al$_2$O$_3$ is also able to catalyze the Claisen–Schmidt condensation between **11.2** and **11.6** (Li et al., 2002). Recently, an acid-montmorillonite (KSF) was used as the catalyst for a solvent-free Claisen–Schmidt reaction (Cella et al., 2010). In both cases, the compounds **11.7** were obtained in excellent yields (Table 11.2).

Condensation of nitromethane **11.8** and aromatic aldehydes **11.2** can be ultrasound-mediated to synthesize nitroalkenes **11.9** (McNulty et al., 1998). The nitroaldol reaction was catalyzed by NH$_4$OAc–HOAc system and it worked well either with electron-withdrawing as with electron-donating groups attached to the ring of **11.2** (Scheme 11.2).

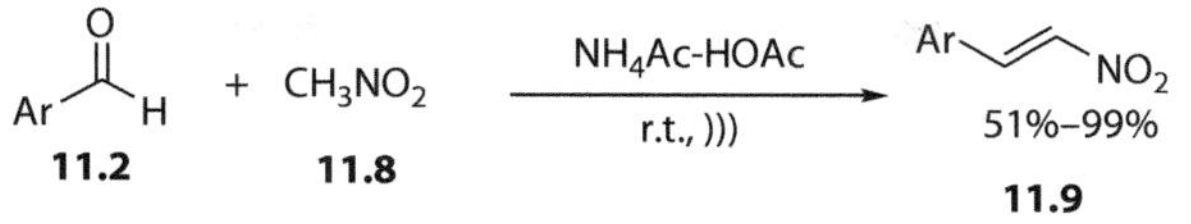

Ultrasound in Synthetic Applications and Organic Chemistry

$$\text{Ar-CHO} + \text{CH}_3\text{NO}_2 \xrightarrow[\text{r.t.,)))}]{\text{NH}_4\text{Ac-HOAc}} \text{Ar-CH=CH-NO}_2$$

11.2 + **11.8** → **11.9** (51%–99%)

SCHEME 11.2

MICHAEL ADDITIONS

Michael addition is the nucleophilic addition of a carbanion or another nucleophile to an a, b-unsaturated carbonyl compound. It belongs to the larger class of conjugate additions. It is an important atom-economical method for diastereoselective and enantioselective C–C bond formation. Bases or Lewis acids under homogeneous conditions classically catalyze these reactions. However, there were some disadvantages due to long reaction time, difficult workup, low yield, or environmental concerns. As in the condensation reactions, the ultrasound was used for active methylene compounds, in addition to chalcones, to give good results.

The potassium hydroxide–catalyzed Michael addition of ethyl acetoacetate **11.10** to chalcone **11.11** under ultrasound (Li et al., 2003a) despite the milder conditions gives different products that react under thermal conditions (Garcia-Raso et al., 1982) (Scheme 11.3). Under ultrasound waves, a cyclic product **11.12** is obtained in good yields, whereas under thermal conditions, an acyclic compound **11.13** is obtained. The cyclohexenone **11.12** probably is obtained from the intermediate **11.14**, which was in several cases isolated just by shorter time of irradiation.

The acyclic product **11.16** can also be formed under ultrasound condition (Li et al., 2003a). However, basic $KF-Al_2O_3$ is used as a catalyst (Scheme 11.4) once again conventional Michael reactions gave adducts in lower yields and needed much longer reaction time (Watanabe et al., 1982).

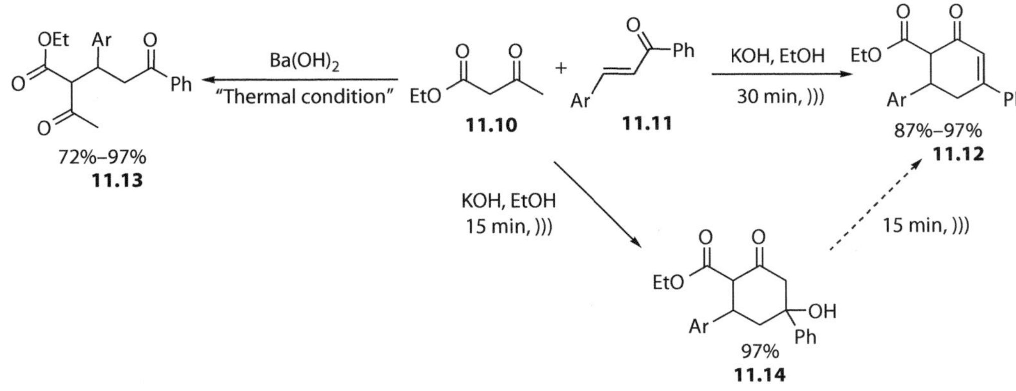

SCHEME 11.3

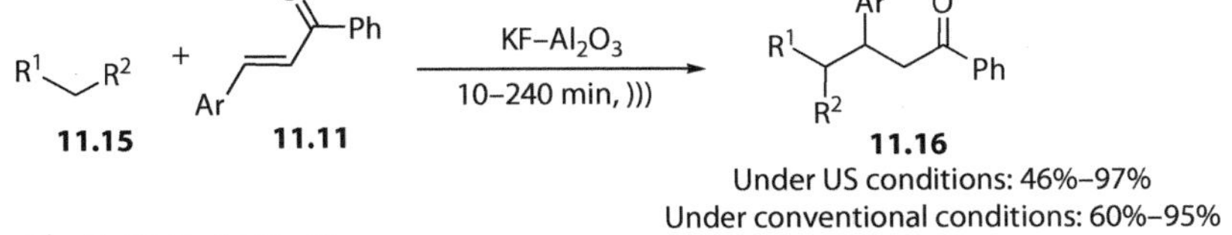

$R^1 = H, CO_2Et, COMe, Ph$
$R^2 = CO_2Et, NO_2, COMe$
$Ar = Ph, p\text{-Cl-Ph}, p\text{-Me-Ph}, p\text{-MeO-Ph}, p\text{-NO}_2\text{-Ph}, ...$

Under US conditions: 46%–97%
Under conventional conditions: 60%–95%

SCHEME 11.4

MANNICH REACTIONS

The Mannich reaction is a one-pot three-component organic reaction, which consists of an amino alkylation of an acidic proton placed next to a carbonyl functional group with aldehyde and ammonia or any primary or secondary amine. The final product is a β-amino-carbonyl compound also known as a Mannich base. Reactions between aldimines and α-methylene carbonyls are also considered as Mannich reactions because these imines form between amines and aldehydes.

Sulfamic acid has been used as an efficient, inexpensive, and recyclable green catalyst for the ultrasound-assisted Mannich reaction of acetophenone **11.17** with aldehydes **11.2** and amines **11.18** (Zeng et al., 2009). The Mannich bases **11.19** were obtained in excellent yields and when compared with the conventional conditions despite the better yields the reaction under ultrasound irradiation was much shorter (Scheme 11.5). This ultrasound protocol has advantages of high yield, mild condition, no environmental pollution, and simple workup procedures.

The ultrasound-promoted three-component Mannich reaction catalyzed by $Ga(OTf)_3$ (Zhang et al., 2009) of cycloketones **11.20** with aromatic aldehydes **11.2** and aromatic amines **11.18** in water gives the corresponding β-amino cycloketones in good-to-excellent yields and good *anti* selectivities (Scheme 11.6).

The preferred *anti*-adduct formation may be explained by the relative stability of transition states of A, B, C, and D. As seen in Figure 11.4, A and D leading to the *anti*-adduct are more stable than B and C leading to the *syn*-adduct.

Excellent yields and selectivity in Mannich reaction were obtained by the use of (*S*) -proline as a catalyst (Kantam et al., 2006). Ultrasonic irradiation was applied to three-component Mannich reactions between hydroxyacetone **11.22**, aromatic aldehydes **11.2**, and *p*-anisidine **11.23** (Scheme 11.7). In most cases, they delivered the corresponding Mannich products **11.24** with very good diastereo- and enantioselectivities in good yields.

The proline-catalyzed Mannich reaction has also been realized between hydroxyacetone **11.22** and preformed imine **11.25** (Scheme 11.8). As in the three-component case, a pronounced

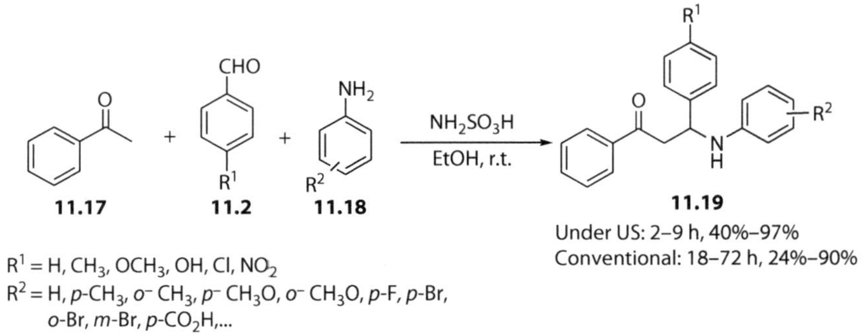
CHO
NH2
NH2SO3H
EtOH, r.t.
R1
R2
11.17
11.2
11.18
11.19
Under US: 2–9 h, 40%–97%
Conventional: 18–72 h, 24%–90%
R1 = H, CH3, OCH3, OH, Cl, NO2
R2 = H, p-CH3, o- CH3, p- CH3O, o- CH3O, p-F, p-Br,
o-Br, m-Br, p-CO2H,...

SCHEME 11.5

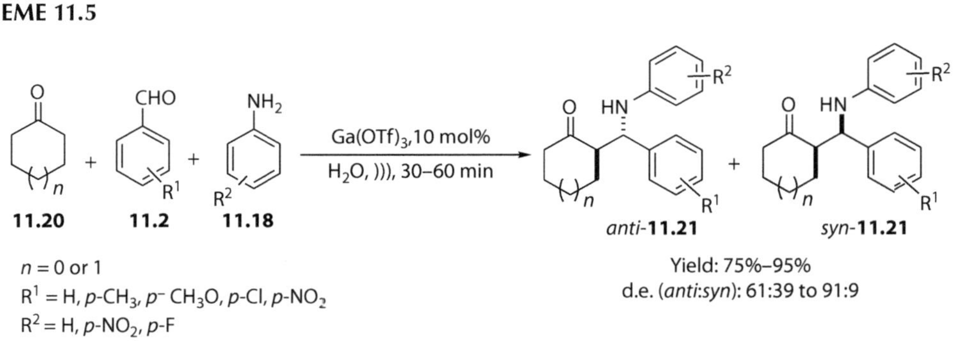
CHO
NH2
Ga(OTf)3,10 mol%
H2O,))), 30–60 min
11.20
11.2
11.18
anti-11.21
syn-11.21
n = 0 or 1
R1 = H, p-CH3, p- CH3O, p-Cl, p-NO2
R2 = H, p-NO2, p-F
Yield: 75%–95%
d.e. (anti:syn): 61:39 to 91:9

SCHEME 11.6

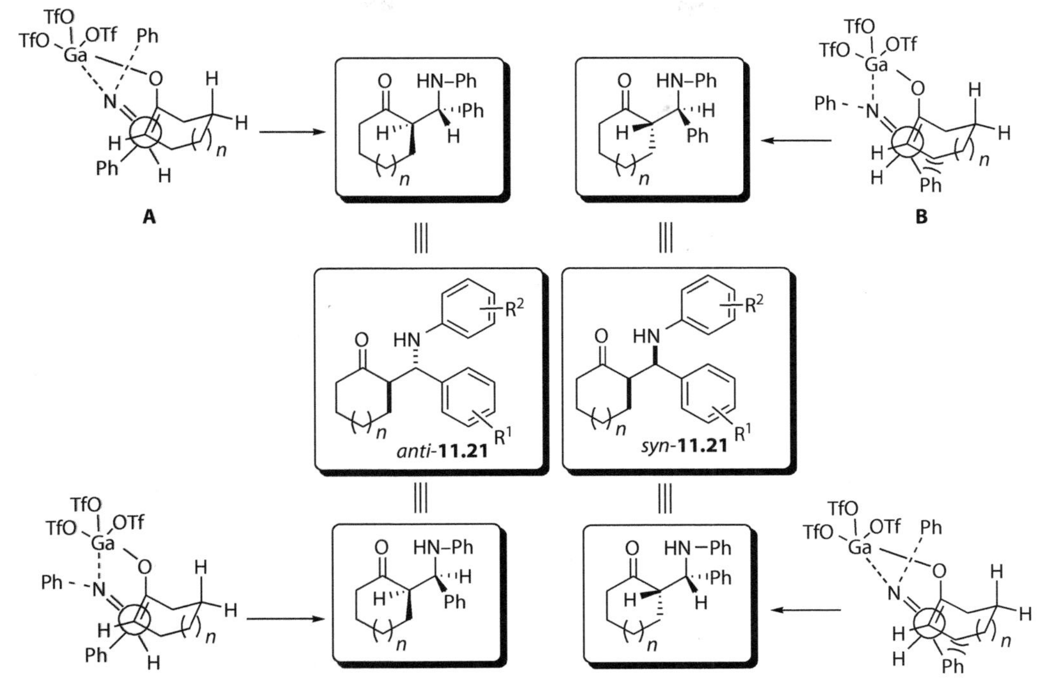

FIGURE 11.4 Transition state for Ga(OTf)$_3$-catalyzed Mannich reaction.

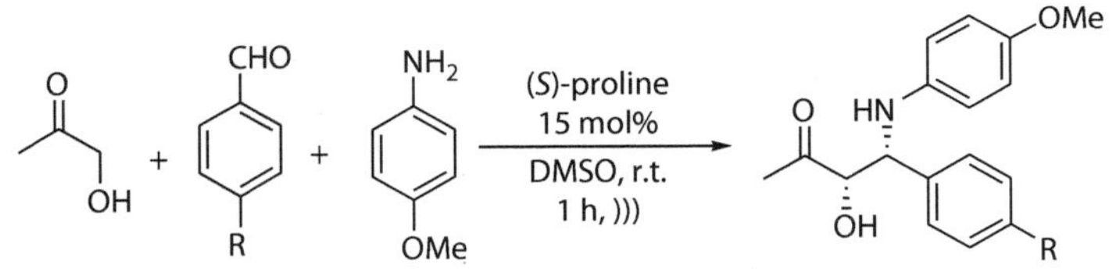

SCHEME 11.7

11.22 + 11.2 + 11.23 → (S)-proline 15 mol%, DMSO, r.t., 1 h,))) → 11.24

Yield: 85%–98%
d.e. (syn:anti): 75:25 to 96:4
ee: 66%–99%

SCHEME 11.8

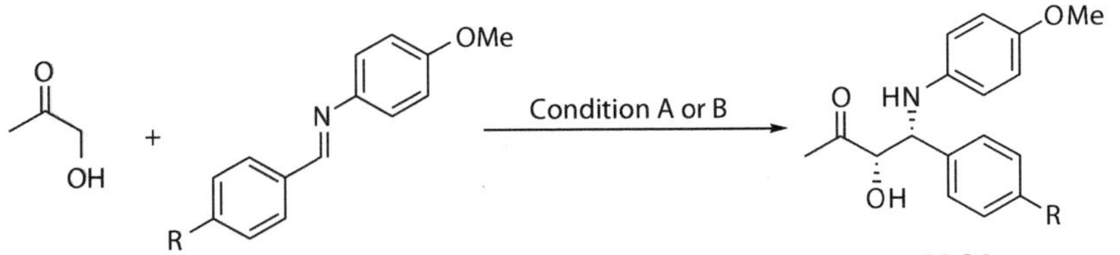

11.22 + 11.25 → Condition A or B → 11.24

Condition A: (S)-proline (15 mol%), DMSO, r.t.,1 h,)))
Yield: 90%–98%
d.e. (syn:anti): 75:25 to 96:4
e.e. 81%–>99%

Condition B: (S)-proline (15 mol%), DMSO, Δ, 1 h
Yield: 85%–93%
d.e. (syn:anti): 75:25 to 90:10
e.e. 65%–93%

acceleration was observed as well as the better yields and selectivity than the same reaction under conventional condition.

CROSS-COUPLING REACTIONS

In the past several decades, transition metal catalyzed cross-coupling reactions have been well developed and widely applied in organic synthesis, which provides useful methods to construct complicated scaffolds. For example, Suzuki–Miyaura coupling, Stille coupling, Kumada coupling, Hiyama coupling, and Negishi coupling, have been well studied as powerful methods in the toolbox of organic chemists. It is well known that traditional cross-coupling involves two kinds of fully functionalized starting materials, including organic halides and organometallic reagents. Many of them require prolonged heating and may be plagued by the formation of side or unwanted products. The current drive toward cleaner chemistry and chemical engineering has spurred a search for more selective and energy-saving protocols, prompting a reconsideration of some metal-catalyzed processes that were once regarded as ideal syntheses. In this context, ultrasound has been shown to considerably reduce reaction times, increase product yields, and enhance product purity by reducing or even eliminating side reactions.

Heck reaction (Deshmukh et al., 2001) between iodobenzenes **11.26** and activated alkenes **11.27** has been performed at room temperature in the presence of 2 mol% of Pd(OAc)$_2$ with considerably enhanced reaction rates by the combined use of ultrasonic irradiation and ionic liquids as solvent (Scheme 11.9). The same reaction was also performed in the presence of heterogeneous Pd/C (Ambulgekar et al., 2005). After reaction completion, solubilized Pd was redeposited onto the support by using sodium formate as reducing agent. The reaction did not take place in the absence of US, which shows that sonication had a key role in accelerating it.

Pd/C is air stable and recyclable; however, under conventional conditions it usually requires high temperatures and long reaction times. It is known that heterogeneous reactions, which are slow due to poor mass transfer, are accelerated by sonication because of cavitation effects. Ultrasound irradiation promotes some palladium leaches out into the medium and the reactions are mainly catalyzed by this dissolved fraction, which can subsequently be returned to the support by thermal or chemical redeposition, so that the catalyst is recyclable without loss of activity.

A regioselective Heck reaction of 2,4-diiodoalkylbenzenes **11.29** (Scheme 11.10) catalyzed *in situ* by palladium nanoparticles in an aqueous medium under US at room temperature has been reported

Condition i: Pd(OAc)$_2$ (2 mol%), NaOAc, [bbim]$^+$Br$^-$/[bbim]$^+$BF$_4^-$, r.t.,))) (73–87%).

Condition ii: 10% Pd/C, Et$_3$N, NMP, r.t.,))) (70%).

R = H, MeO, Cl

EWG = H, CO$_2$Me, CO$_2$Et

SCHEME 11.9

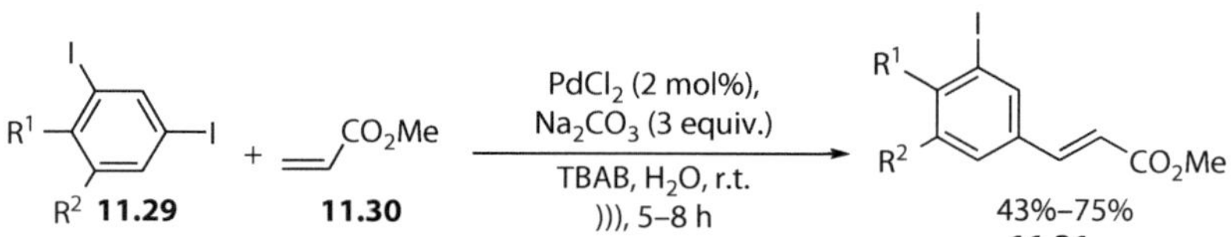

PdCl$_2$ (2 mol%), Na$_2$CO$_3$ (3 equiv.)

TBAB, H$_2$O, r.t.

))), 5–8 h

43%–75%

11.31

R^1 = MeO, MOMO, NO2, MeOCNH-

R^2 = H or I

SCHEME 11.10

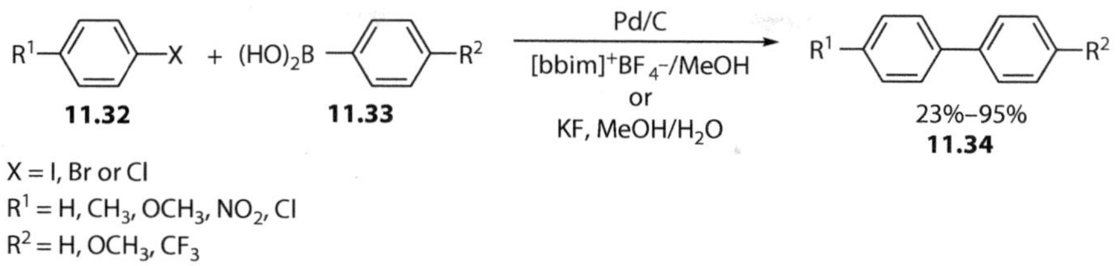

R^1—⟨⟩—X + (HO)$_2$B—⟨⟩—R^2 →[Pd/C, [bbim]$^+$BF$_4^-$/MeOH or KF, MeOH/H$_2$O] R^1—⟨⟩—⟨⟩—R^2

11.32 **11.33**   23%–95%
                                                               11.34

X = I, Br or Cl
R^1 = H, CH$_3$, OCH$_3$, NO$_2$, Cl
R^2 = H, OCH$_3$, CF$_3$

SCHEME 11.11

by Wang and Zhou (2006). Palladium nanoparticles were prepared *in situ* from PdCl$_2$ and it could be reused for multiple reactions. The regioselectivity was *para-* over *ortho-*substitution.

A sonochemical Pd/C-catalyzed Suzuki reaction (Rajagopal et al., 2002) was carried out in an ionic liquid in the absence of any other additive (Scheme 11.11). Cross-coupling reactions of halobenzenes **11.32** including chlorobenzenes with phenylboronic acid **11.33** have been achieved at room temperature under these conditions. Toma et al. (2005) used ultrasound to promote the heterogenous reaction of haloarenes **11.32** with different aryl boronic acids **11.33**, catalyzed by Pd/C and KF in methanol–water (Scheme 11.11). These reactions were complete in about 1 h, while a 4 h heating was necessary to achieve comparable results under reflux.

Aqueous Suzuki–Miyaura reaction was mediated by ultrasound irradiation using very low loading of palladium nanoparticles stabilized in polyvinylpyrrolidone (PVP) as a catalyst system (de Souza et al., 2008). Ultrasonic irradiation can dramatically accelerate the heterogenous Suzuki coupling (Zhang et al., 2008) of phenylboronic acid **11.33** with aryl halides **11.32** in neat water, in the presence of TBAB and ligand-free cyclopalladated ferrocenylimines **11.35** and **11.36** (Figure 11.5). Even aryl chlorides were able to cross-couple with boronic acids.

The ultrasound-assisted cross-coupling reaction between organo tellurides **11.37** (used as an alternative to the traditional organyl halides) and potassium aryl trifluoroborate salts **11.38** catalyzed by palladium(0) tetrakis(triphenylphosphine) has been widely studied by Stefani et al. (2006, 2008). All the reactions required the presence of a silver salt, a wide range of coupled products **11.39** were obtained in intermediate to excellent yields and all the reactions were highly stereo- and chemoselective, preferring to react in the telluride moiety instead of halide when both were present in the reactant (Scheme 11.12).

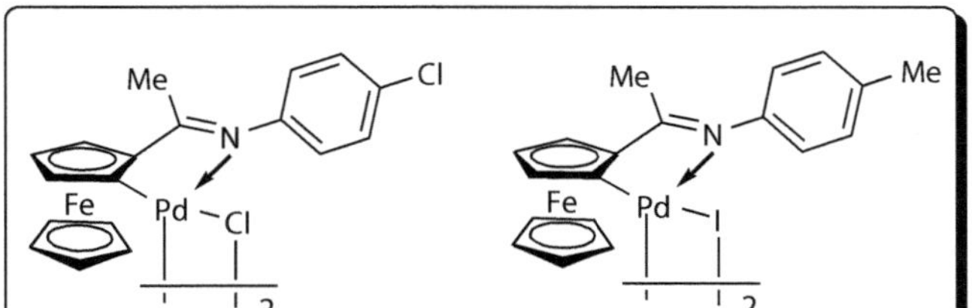

FIGURE 11.5 Cyclopalladated ferrocenylimines.

R^1—TeBu-*n* + KF$_3$B—R^2 →[Pd[P(Ph$_3$)]$_4$ (8–20 mol%), Ag$_2$O or AgOAc, MeOH, Et$_3$N,)))] R^1– R^2

11.37   **11.38**   48%–91%
                                                                  11.39

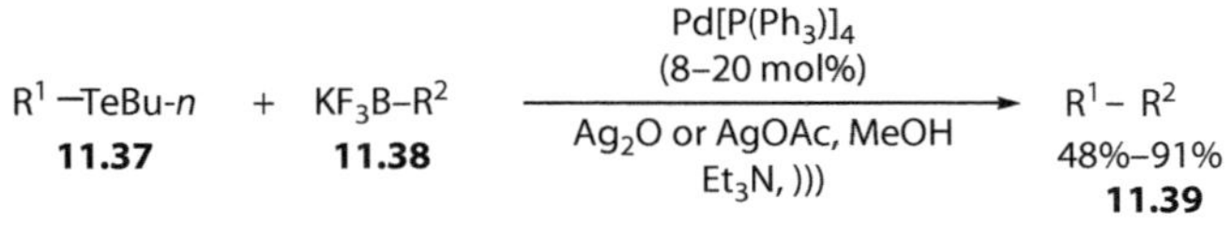

SCHEME 11.12

HETEROCYCLES SYNTHESIS

The importance of heterocycles in many fields of science (including organic, bioorganic, agricultural, industrial, pharmaceutical and medicinal chemistry, as well as material science) can hardly be overemphasized, and justifies a long lasting effort to work out new synthetic protocols for their production. A particularly attractive approach is based on ultrasound-promoted heterocyclization reactions of suitably functionalized substrates, which can allow the regioselective synthesis of highly functionalized heterocycles using readily available starting materials under mild and selective conditions (Cella and Stefani, 2009).

Ultrasound irradiation has been used in the synthesis of 1H-benzotriazoles **11.41** from the cyclization of o-phenylenediamine **11.40** with sodium nitrite in acetic acid (Guzen et al., 2007). Products **11.41** were obtained in good-to-excellent yields and then subjected to an acylation reaction under US irradiation to obtain 1-acylbenzotriazoles **11.42** (Scheme 11.13).

The 1,4-disubstituted 1,2,3-triazoles **11.45** have been successfully synthesized from the classical Huisgen 1,3-dipolar cycloaddition between sodium azide, terminal alkynes **11.43**, and alkyl/aryl halides **11.44** (Sheedhar and Reddy, 2007). The reactions were catalyzed by 10 mol% of CuI in an aqueous medium under US irradiation at room temperature and products **11.45** were obtained in good-to-excellent yields and high regioselectivity (Scheme 11.14).

Pyrazole rings **11.48** or **11.49** were prepared from α-oxo thioxoester **11.46** or from α-oxoketene O,N-acetals **11.47**, respectively, and hydrazine derivatives using Montmorillonite K-10 as a solid support under US irradiation. Regiospecific pyrazoles were obtained in low to moderate yields (Scheme 11.15).

Pyrazole rings can also be functionalized using ultrasound irradiation as energy source (Stefani et al., 2005). 3,5-Dimethyl pyrazoles **11.50** suffered the halogenations at position 4 when

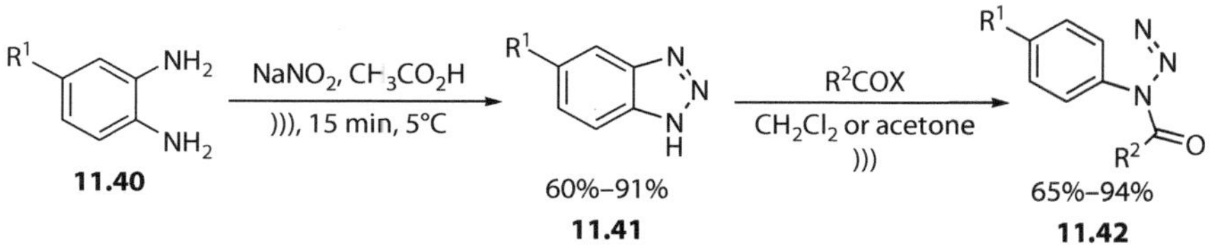

SCHEME 11.13

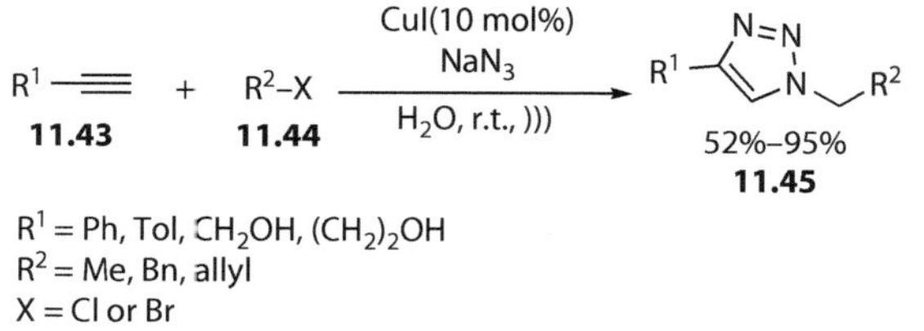

SCHEME 11.14

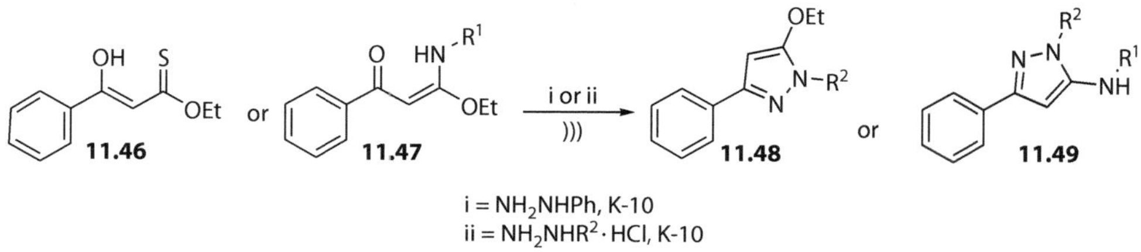

SCHEME 11.15

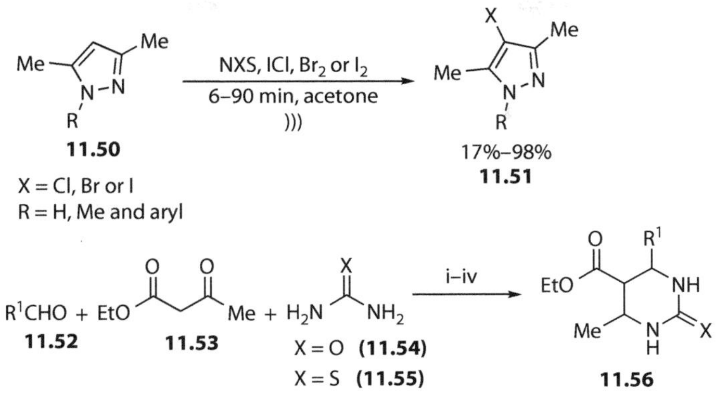

SCHEME 11.16

irradiated in the presence of *N*-halosuccinimides, ICl, or molecular I_2 and Br_2 (Scheme 11.16). Halogenated products **11.51** were obtained in good-to-excellent yields in shorter time and easier workup compared with the same reaction using traditional conditions.

The Biginelli reaction is a multiple-component chemical reaction that creates 3,4-dihydropyrimidin-2(1*H*)-ones from ethyl acetoacetates, an aldehyde, and urea. It is named after the Italian chemist Pietro Biginelli. The reaction can be catalyzed by Brønsted acids and/or by Lewis acids. Dihydropyrimidinones, the products of the Biginelli reaction, are widely used in the pharmaceutical industry as calcium channel blockers, antihypertensive agents, etc. Recent studies showed that under ultrasound, the Biginelli reaction could be carried out in milder conditions, shorter reaction time, higher yield, and easier workup.

Zhidovinova et al. (2003) showed that the classical Biginelli reaction (EtOH and HCl) is accelerated by a factor of 40 times or more as a result of US irradiation. The three-component reaction among aldehydes **11.52**, ethyl acetoacetate **11.53**, and urea **11.54**, or thiourea **11.55** in the presence of a catalytic amount of HCl was completed within 2–5 min at room temperature, and dihydropyrimidinones **11.56** were obtained in excellent yields (Scheme 11.17). A solvent-free Biginelli reaction has been described (Singh et al., 2006), it was catalyzed by HCl (1 mol%) or trifluoroacetic acid (5 mol%) and completed within 15–45 min in reactions involving urea **11.54** and 60–90 min in reactions involving thiourea **11.55** (Scheme 11.17).

Biginelli reactions can also be performed in the absence of any catalyst (Gholap et al., 2004). The reaction was carried out in 1-*n*-butylimidazolium tetrafluoroborate ([Hbim]BF_4), a nonvolatile ionic liquid, under ultrasound irradiation in a very short reaction time (Scheme 11.17). Ionic liquid could be recovered and reused in the same reaction at least three times without a decrease in yield. It has been postulated that the ionic liquid plays an important role in this multicomponent reaction, acting as an inherent Brønsted acid.

Biginelli adducts **11.58** have been produced by utilization of inexpensive ammonium chloride as a mediator of the reaction under US irradiation (Stefani et al., 2006). The Biginelli reaction was carried out in methanol and irradiated for 3–5 h in a cleaning bath (Scheme 11.18). The antioxidant

i = EtOH, HCl (drops),))), 2–5 min (90%–95%)	ii = Solvent-free, HCl (1 mol%),))), 15–45 min (72%–97%)
iii = Solvent-free, TFA (5 mol%),))), 45–90 min (67%–92%)	iv = [Hbim]BF_4,))), 40–90 min (83%–97%)

SCHEME 11.17

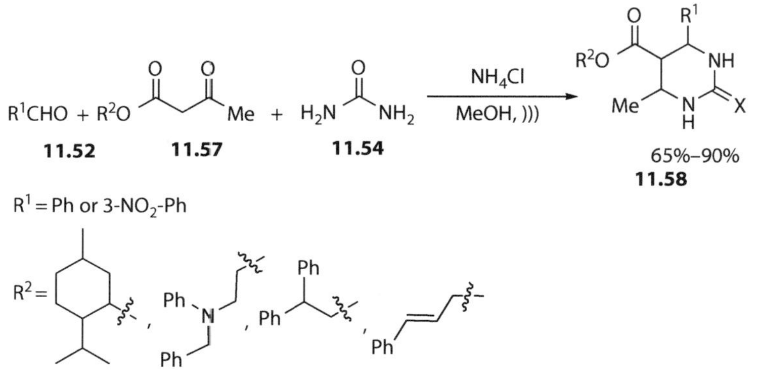

SCHEME 11.18

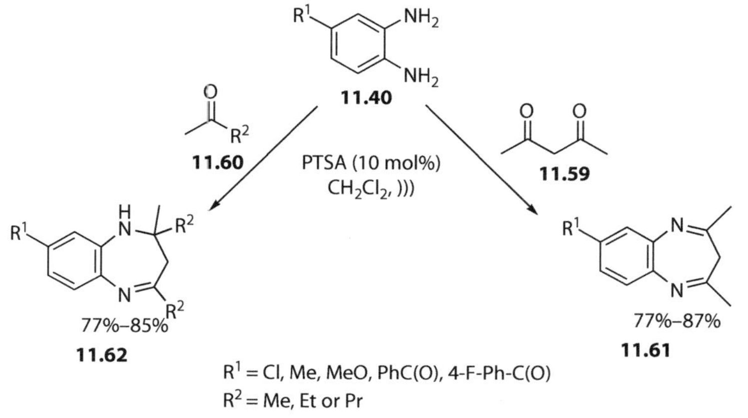

R^1 = Ph or 3-NO$_2$-Ph

SCHEME 11.19

activity of **11.58** was evaluated, and some of these compounds exhibited strong activity against lipid peroxidation induced by Fe and EDTA.

An ultrasound-mediated condensation of *o*-phenylenediamine derivative **11.40** with 2,4-pentadione **11.59** or ketones **11.60** has been used in the synthesis of 1,5-benzodiazepinic rings **11.61** or **11.62**, respectively (Guzen et al., 2006). The reaction was catalyzed by 10 mol% of *p*-toluenesulfonic acid (PTSA), and the products were obtained in good yields (Scheme 11.19), containing either electron-withdrawing or electron-donating groups attached to the diamine **11.40**.

MISCELLANEOUS

An ultrasound-mediated reaction of arylacetylenes **11.63** and metallic lithium with a variety of electrophiles provides an efficient, mild, practical, and inexpensive route to obtain functionalized arylacetylenes **11.64** (Stefani et al., 2005). This methodology avoids the use of strong bases (e.g., n-BuLi, NH$_2$Na or Grignard's reagents) and compounds **11.64** were obtained in low-to-excellent yields (Scheme 11.20); however, for reasons not yet understood, this reaction does not work for aliphatic alkynes.

Ultrasound irradiation has been used in the preparation of a series of imines **11.66** (Guzen et al., 2007). The ultrasound-promoted reaction of aldehydes **11.2** and primary amines **11.65** catalyzed by silica and products **11.66** were obtained in high yields even in large scale synthesis (Scheme 11.21). The reuse of silica was evaluated and it could be reused through the fourth time with slight decrease in the yield.

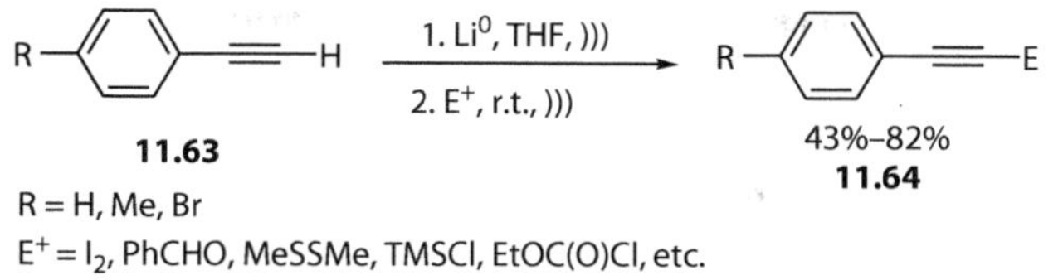

Scheme 11.20

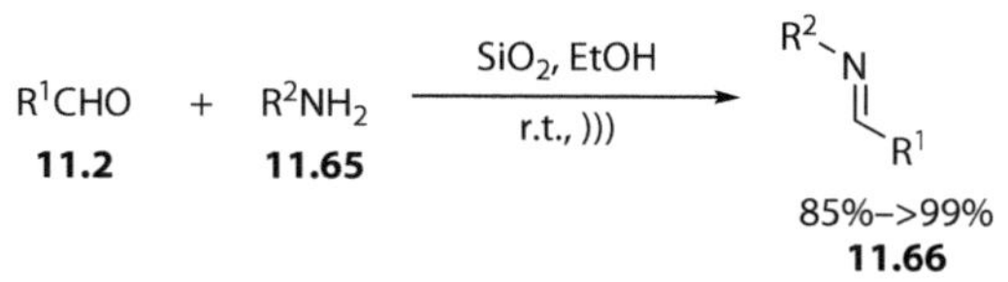

R—C₆H₄—C≡C—H → (1. Li⁰, THF,))) ; 2. E⁺, r.t.,))) → R—C₆H₄—C≡C—E

11.63 → **11.64** (43%–82%)

R = H, Me, Br
E⁺ = I$_2$, PhCHO, MeSSMe, TMSCl, EtOC(O)Cl, etc.

SCHEME 11.20

Scheme 11.21

R^{1}CHO + R^{2}NH$_2$ → (SiO$_2$, EtOH, r.t.,))) → R^2—N=CH—R^1

11.2 + **11.65** → **11.66** (85%–>99%)

SCHEME 11.21

CONCLUSIONS

To show all the ultrasound uses in organic synthesis, we would need an entire book; however, we expect that here we were able to resume the main uses of this amazing technique nowadays.

Synthetic organic reactions performed under nontraditional conditions are gaining popularity, primarily to circumvent growing environmental concerns (*Green Chemistry*). The features of US-assisted organic transformations, namely, the selectivity, ease of experimental manipulation, and enhanced reaction rates, were highlighted. The use of this nontraditional tool aids in overcoming many of the difficulties associated with conventional reactions, and offers both process-related and environmental advantages in organic synthesis.

REFERENCES

Ambulgekar, G.V., Bhanage, B.M., and Samant, S.D. 2005. Low temperature recyclable catalyst for Heck reactions using ultrasound. *Tetrahedron Letters*, 46:2483–2485.

Ando, T., Sumi, S., Kawate, T., Ichihara, J., and Hanafusa, T. 1984. Sonochemical switching of reaction pathways in solid-liquid two-phase reactions. *Journal of Chemical Society and Chemical Communications*, 439–440.

Cabello, J.A., Campelo, J.M., Garcia, A., Luna, D., and Marinas, J.M. 1984. Knoevenagel condensation in the heterogeneous phase using AlPO$_4$-Al$_2$O$_3$ as a new catalyst. *Journal of Organic Chemistry*, 49:5195–5197.

Cella, R., Orfão, A.T.G., and Stefani, H.A. 2006. Palladium-catalyzed cross-coupling of vinylic tellurides and potassium vinyltrifluoroborate salt: Synthesis of 1,3-dienes. *Tetrahedron Letters*, 47:5075–5078.

Cella, R. and Stefani, H.A. 2006. Ultrasound-assisted synthesis of Z- and E-stylbenes by Suzuki cross-coupling reactions of organotellurides with potassium organotrifluoroborate salts. *Tetrahedron*, 62:5656–5662.

Cella, R. and Stefani, H.A. 2009. Ultrasound in heterocycles chemistry. *Tetrahedron*, 65:2619–2641.

Chtourou, M., Abdelhédi, R., Frikha, M.H., and Trabelsi, M. 2010. Solvent free synthesis of 1,3-diaryl-2-propenones catalyzed by commercial acid-clays under ultrasound irradiation. *Ultrasonics Sonochemistry*, 17:246–249.

Cravotto, G. and Cintas, P. 2006. Power ultrasound in organic synthesis: Moving cavitational chemistry from academia to innovative and large-scale applications. *Chemical Society Reviews*, 35:180–196.

Deshmukh, R.R., Rajagopal, R., and Srinivasan, K.V. 2001. Ultrasound promoted C–C bond formation: Heck reaction at ambient conditions in room temperature ionic liquids. *Chemical Communications*, 1544–1545.

De Souza, A.L.F., da Silva, L.C., Oliveira, B.L., and Antunes, O.A.C. 2008. Microwave- and ultrasound-assisted Suzuki-Miyaura cross-coupling reaction catalyzed by Pd/PVP. *Tetrahedron Letters*, 49:3895–3898.

Fillion, H. and Luche, J.L. 1998. *Synthetic Organic Sonochemistry*. Plenum Press: New York.

Flint, E.B. and Suslick, K.S. 1991. The temperature of cavitation. *Science*, 253:1397–1399.

Fuentes, A., Marinas, J.M., and Sinisterra, J.V. 1987. Catalyzed synthesis of chalcones under interfacial solid-liquid conditions with ultrasound. *Tetrahedron Letters*, 28:4541–4544.

Garcia-Raso, A., Garcia-Raso, J., Campaner, B., Mestres, R., and Sinisterra, J.V. 1982. An improved procedure for the Michael reaction of chalcones. *Synthesis*, 12:1037–1041.

Gedanken, A. 2004. Using sonochemistry for the fabrication of nanomaterials. *Ultrasonics Sonochemistry*, 11:47–55.

Gedanken, A. 2008. Preparation and properties of proteinaceous microspheres made sonochemically. *Chemistry - A European Journal*, 14:3840–3853.

Gholap, A.R., Venkatesan, K., Daniel, T., Lahoti, R.J., and Srinivasan, K.V. 2004. Ionic liquid promoted novel and efficient one pot synthesis of 3,4-dihydropyrimidin-2 –(1*H*) -ones at ambient temperature under ultrasound irradiation. *Green Chemistry*, 6:147–150.

Guadagnin, R.C., Suganuma, C.A., Singh, F.V., Vieira, A.S., Cella, R., and Stefani, H.A. 2008. Chemoselective cross-coupling Suzuki-Miyaura reaction of (Z)-(2-chlorovinyl)tellurides and potassium aryltrifluoroborate salts. *Tetrahedron Letters*, 49:4713–4716.

Guzen, K.P., Cella, R., and Stefani, H.A. 2006. Ultrasound enhanced synthesis of 1,5-benzodiazepinic heterocyclic rings. *Tetrahedron Letters*, 47:8133–8136.

Guzen, K.P., Guarezemini, A.S., Órfão, A.T.G., Cella, R., Pereira, C.M.P., and Stefani, H.A. 2007. Eco-friendly synthesis of imines by ultrasound irradiation. *Tetrahedron Letters*, 48:1845–1848.

Kantam, M.L., Rajasekhar, C.V., Gopikrishna, G., Reddy, K.R., and Choudary, B.M. 2006. Proline catalyzed three-component and self-asymmetric Mannich reactions promoted by ultrasonic conditions. *Tetrahedron Letters*, 47:5965–5967.

Li, J.T., Chen, G.F., Wang, J.X., and Li, T.S. 1999a. Ultrasound promoted synthesis of α,α'-bis(substituted furfurylidene) cycloalkanones and chalcones. *Synthetic Communications*, 29:965–971.

Li, J.T., Chen, G.F., Xu, W.Z., and Li, T.S. 2003a. The Michael reaction catalyzed by KF/basic alumina under ultrasound irradiation. *Ultrasonics Sonochemistry*, 10:115–118.

Li, J.T., Cui, Y., Chen, G.F., Cheng, Z.L., and Li, T.S. 2003b. Michael addition catalyzed by potassium hydroxide under ultrasound. *Synthetic Communications*, 33:353–359.

Li, J.T., Li, T.S., Li, L.J., and Cheng, X. 1999b. Synthesis of ethyl α-cyanocinnamates under ultrasound irradiation. *Ultrasonic Sonochemistry*, 6:199–201.

Li, J.T., Yang, W.Z., Wang, S.X., Li, S.H., and Li, T.S. 2002. Improved synthesis of chalcones under ultrasound irradiation. *Ultrasonics Sonochemistry*, 9:237–239.

Margulis, M.A. 1990. *Advances in Sonochemistry*. JAI Press: London, U.K.

McNulty, J., Steere, J.A., and Wolf, S. 1998. The ultrasound promoted Knoevenagel condensation of aromatic aldehydes. *Tetrahedron Letters*, 39:8013–8016.

Pereira, C.M.P., Stefani, H.A., Guzen, K.P., and Orfão, A.T.G. 2007. Improved synthesis of benzotriazoles and 1-acylbenzotriazoles by ultrasound irradiation. *Letters in Organic Chemistry*, 4:43–46.

Peters, D. 1996. Ultrasound in materials chemistry. *Journal of Material Chemistry*, 6:1605–1618.

Popp, F.D. 1960. Synthesis of 3-hydroxypyridines. I. Condensation of aromatic aldehydes with ethyl cyanoacetate. *Journal of Organic Chemistry*, 25:646–647.

Rajagopal, R., Jarikote, D.V., and Srinivasan, K.V. 2002. Ultrasound promoted Suzuki cross-coupling reactions in ionic liquid at ambient conditions. *Chemical Communications*, 616–617.

Richards, W.T. and Loomis, A.L. 1927. The chemical effects of high frequency sound waves: I A preliminary survey. *Journal of America Chemical Society*, 49:3086.

Sehgal, C.M. and Wang, S.Y. 1981. Threshold intensities and kinetics of sonoreaction of thymine in aqueous solutions at low ultrasonic intensities. *Journal of the America Chemical Society*, 103:6606–6611.

Sheedhar, B. and Reddy, P.S. 2007. Sonochemical synthesis of 1,4-disubstituted 1,2,3-triazoles in aqueous medium. *Synthetic Communications*, 37:805–812.

Singh, K., Singh, S., and Kaur, P. 2006. Efficacious preparation of Biginelli compounds. A comparative study of different reaction techniques. *Letters in Organic Chemistry*, 3:201–203.

Stefani, H.A., Cella, R., Dörr, F.A., Pereira, C.M.P., Gomes, F.P., and Zeni, G. 2005a. Ultrasound-assisted synthesis of functionalized arylacetylenes. *Tetrahedron Letters*, 46:2001–2003.

Stefani, H.A., Oliveira, C.B., Almeida, R.B., Pereira, C.M.P., Braga, R.C., Cella, R., Borges, V.C., Savegnago, L., and Nogueira, C.W. 2006. Dihydropyrimidin -(2H)-ones obtained by ultrasound irradiation: A new class of potential antioxidant agents. *European Journal of Medicinal Chemistry*, 41:513–518.

Stefani, H.A., Pereira, C.M.P., Almeida, R.B., Braga, R.C., Guzen, K.P., and Cella, R. 2005. A mild and efficient method for halogenation of 3,5-dimethyl pyrazoles by ultrasound irradiation using *N*-halosuccinimides. *Tetrahedron Letters*, 46:6833–6837.

Suslick, K.S. 1990. Sonochemistry. *Science*, 247:1439–1445.

Suslick, K.S., Hammerton, D.A., and Cline, Jr., R.E. 1986. Sonochemical hot spot. *Journal of American Chemical Society*, 108:5641–5642.

Wang, S.X., Li, J.T., Yang, W.Z., and Li, T.S. 2002. Synthesis of ethyl α-cyanocinnamates catalyzed by KF-Al$_2$O$_3$ under ultrasound irradiation. *Ultrasonic Sonochemistry*, 9:159–161.

Watanabe, K.I., Miyazu, K.I., and Irie, K. 1982. Michael additions catalyzed by metal (II) complexes. *Bulletin of the Chemical Society of Japan*, 55:3212–3215.

Zeng, H., Li, H., and Shao, H. 2009. One-pot three-component Mannich-type reaction using sulfamic acid catalyst under ultrasound irradiation. *Ultrasonics Sonochemistry*, 16:758–762.

Zhang, G., Huang, Z., and Zou, J. 2009. Ga(OTf)$_3$-catalyzed three-component Mannich reaction in water promoted by ultrasound irradiation. *Chinese Journal of Chemistry*, 27:1967–1974.

Zhang, J., Yang, F., Ren, G., Mak, T.C.W., Song, M., and Wu, Y. 2008. Ultrasonic irradiation accelerated cyclopalladated ferrocenylimines catalyzed Suzuki reaction in neat water. *Ultrasonic Sonochemistry*, 15:115–118.

Zhang, Z., Zha, Z., Gan, C., Pan, C., Zhou, Y., Wang, Z., and Zhou, M.M. 2006. Catalysis and regioselectivity of the aqueous Heck reaction by Pd(0) nanoparticles under ultrasonic irradiation. *Journal of Organic Chemistry*, 71:4339–4342.

Zhidovinova, M.S., Fedorova, O.V., Rusinov, G.L., and Ovchinnikova, I.G. 2003. Sonochemical synthesis of Biginelli compounds. *Russian Chemical Bulletin*, 52:2527–2528.

12 Ultrasound Applications in Synthetic Organic Chemistry

Mohammad Majid Mojtahedi and Mohammad Saeed Abaee

CONTENTS

Introduction	281
Cycloaddition Reactions	282
Multicomponent Reactions	285
Nucleophilic Addition Reactions	287
Condensation Reactions	289
Organometallic Chemistry	291
Heterocycles	296
Three-Membered Rings	296
Five-Membered Rings	297
Six-Membered Rings	299
Spiro Heterocycles	301
Oxidation and Reduction Reactions	303
Protection and Deprotection Procedures	305
Carbon–Heteroatom (C–X) Bond Formation Reactions	307
Conclusions	308
References	309

INTRODUCTION

Scientific discussions and political mandates in recent decades have enforced both academic and industrial chemical societies to move toward designing and employing more environmentally friendly techniques and methodologies in order to minimize the quantities of chemical disposals and energy consumptions. In this regard, ultrasonic irradiation has emerged as one of the most successful means to significantly help develop the so-called "green" chemistry by enhancing the reactivity and selectivity of the reactions and lowering the energy uses at the same time (Mason and Cintas, 2002; Mason and Lorimer, 2002; Cravotto and Cintas, 2007; Bruckmann et al., 2008). The interaction between the matter and the ultrasonic waves and the subsequent chemical changes is called sonochemistry and is attributed to cavitation, a physical process caused during ultrasonic irradiation of liquids by creation, enlargement, and collapse of bubbles (Bremner, 1990). The cavitation would result in induction of extremely high local pressure and temperature, enhancing mass transfer and mechanical effects in the reaction mixture. On many occasions, irradiation leads to more effective mixing of the reaction phases, and, therefore, increased rates and yields are observed because of better mechanical effects of the ultrasound (US) waves (Flint and Suslick, 1991; Suslick and Kemper, 1993; Brennen, 1995). In other cases, ultrasonic energy causes the formation of new reactive intermediates and species that are not usually formed in regular thermal reactions. This is called "true sonochemistry" (Cravotto and Cintas, 2006) or "sonochemical switching" (Cintas and Luche, 1999), where changes in the reaction mechanism occur and thus the distribution of the products and the reaction selectivity are altered.

In 1927, Richards and Loomis reported the first chemical reaction induced by US irradiation (Rechards and Loomis, 1927). But it took nearly 60 years until sonochemistry could start finding applications in different areas of chemistry. Advancements in technology grew the availability of laboratory size equipment with lower expenses and easier operational procedures. This led to extensive application of US in various fields of chemistry, including organic synthesis, and consequently there was an overwhelming increase in the number of publications. Several excellent reviews were published in the last decade (Einhorn et al., 1989; Mason, 1997; Luche, 1998; Margulis, 2004; Li et al., 2005d) summarizing the effects of US in organic synthesis. However, to the best of our knowledge, no comprehensive review has been carried out at least in the past decade from merely a synthetic point of view. In this chapter, we have tried to summarize densely as many investigations as possible performed on various synthetic transformations using US energy. Special attention is paid to those studies that were carried out in the last two decades or were not covered in previous reviews. For this reason, all examples of advantageous applications of the ultrasonic techniques in organic reactions are included, regardless of their true sonochemical, thermal, or mechanical natures. The second to sixth sections mainly cover the carbon–carbon bond formation reactions. The rest of the chapter deals with heterocyclic chemistry, oxidation/reduction reactions, protection/deprotection procedures, and the formation of carbon–heteroatom bonds.

CYCLOADDITION REACTIONS

The very first example of US-promoted Diels–Alder reaction was reported by Lee and Snyder (1989), where they could synthesize a group of naturally occurring *o*-quinonic derivatives that are found in some traditional Chinese medicine plants. While under thermal conditions poor yields were obtained, sonication of the reaction mixtures enhanced both yields and the regioselectivity, similar to those observed for the same process conducted at high pressure. As a result, the reaction led to 65% formation of adducts favoring the natural regioisomer **12.3** in a ratio of 3.5:1 over its opposite regioisomer (Scheme 12.1).

One of the pioneering investigations on the effects of sonochemistry on Diels–Alder reactions was carried out by Caulier and Reisse (1996). Their study showed remarkable improvement in both kinetic and stereoselectivity of the cycloaddition reaction between cyclopentadiene and methyl vinyl ketone in halogenated solvents. Although a Lewis acid–catalyzed pathway is possible via in situ formation of $TiCl_4$ or $TiBr_4$ (formed from the reaction of halogenated solvents with the titanium horn of the US transducer), experiments excluded intervention of such pathway or also a radical mechanism. In contrast, the use of small quantities of gaseous HCl under silent conditions led to similar kinetic and stereochemical enhancements, supporting possible involvement of a Bronsted acid catalytic pathway caused via sonolytic formation of hydrogen halides from the halogenated solvents. The same reaction was also investigated in an emulsion of supercritical carbon dioxide and water (Timko et al., 2006). The sonication of the mixture led to increased rate and improved *endo* stereoselectivity (up to 16:1). This study suggested that water acts as a solvent and a catalyst, carbon dioxide performs as a carrier for water-insoluble reactants, and US increases the mass transfer and mixes the two phases.

Chiral derivatives of tetrahydronaphthalene structure were successfully synthesized using US irradiation. In a one-pot process (Scheme 12.2), *o*-quinodimethane diene **12.4** was generated in situ

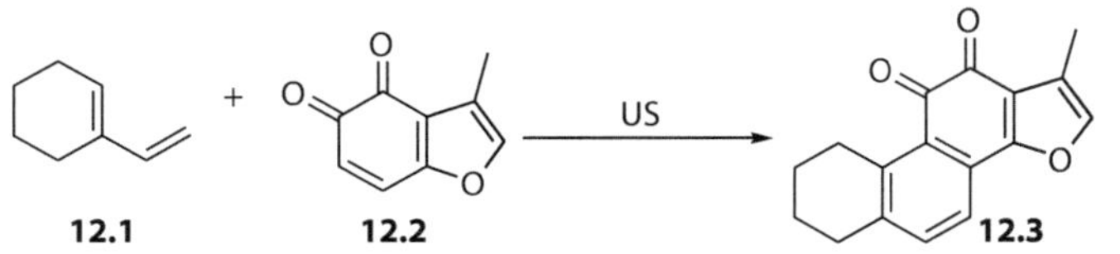

SCHEME 12.1

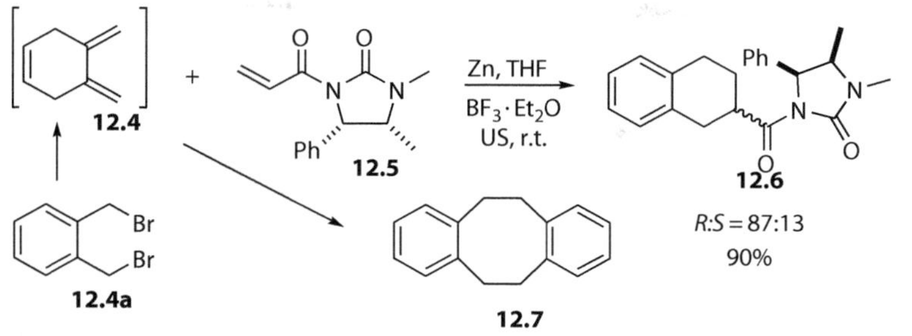

SCHEME 12.2

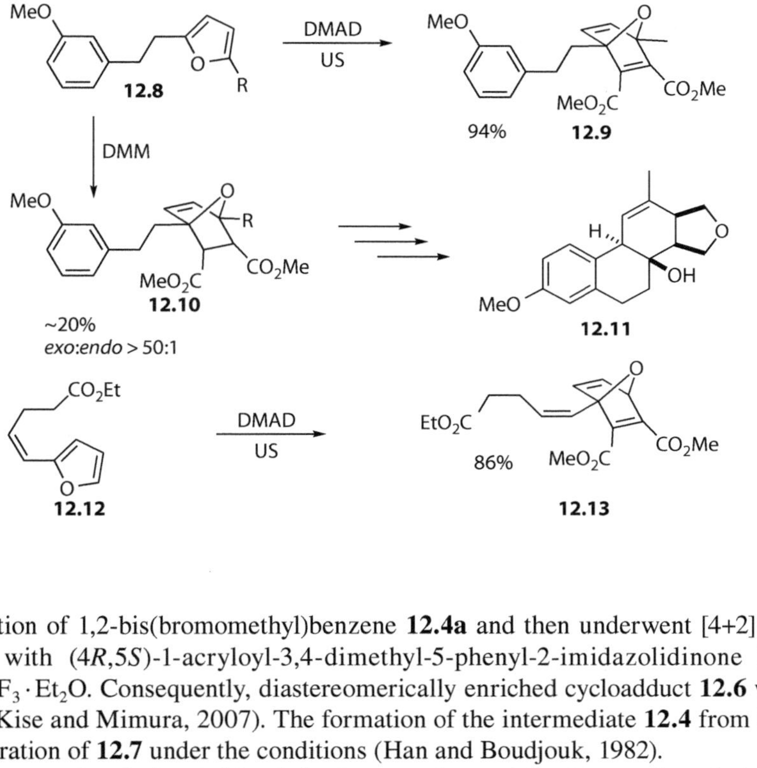

SCHEME 12.3

via zinc reduction of 1,2-bis(bromomethyl)benzene **12.4a** and then underwent [4+2] Diels–Alder cycloaddition with (4R,5S)-1-acryloyl-3,4-dimethyl-5-phenyl-2-imidazolidinone **12.5** in the presence of BF$_3$·Et$_2$O. Consequently, diastereomerically enriched cycloadduct **12.6** was obtained in 90% yield (Kise and Mimura, 2007). The formation of the intermediate **12.4** from **12.4a** is supported by separation of **12.7** under the conditions (Han and Boudjouk, 1982).

A significant sonochemical switching behavior was observed for furano diene derivatives, which are normally very sluggish or unreactive toward cycloaddition conditions. Under US irradiation, a neat mixture of **12.8** reacted with DMAD at room temperature to solely furnish **12.9** in high yield (Scheme 12.3). Less-reactive dienophile DMM also reacted with the same diene to produce lower yields of **12.10**, a precursor to access multicyclic structures of type **12.11**, but in excellent diastereoselectivity. Even for exocyclic triene **12.12**, the furao diene portion reacted selectively under the conditions to produce **12.13**. This was a remarkable result since such systems are known to remain unreactive, undergo undesired side reactions, or participate in cycloaddition processes by their exocyclic diene portion (Wei et al., 2004).

In another sonochemical DA investigation (Scheme 12.4), furan derivatives acted as dienophile when subjected to react with o-quinones (Avalos et al., 2000). The oxidation of methyl vanillate **12.14** with diacetoxyiodobenzene (DAIB) under US irradiation led to the formation of the corresponding o-quinone monoketal **12.15**, which further reacted in situ with furan derivatives to give bridged products in moderate yields. Reactions proceeded with high diastereoselectivity. Interestingly, for unsymmetrically substituted furanes, out of four possible regioisomers, only the formation of product **12.16** is observed.

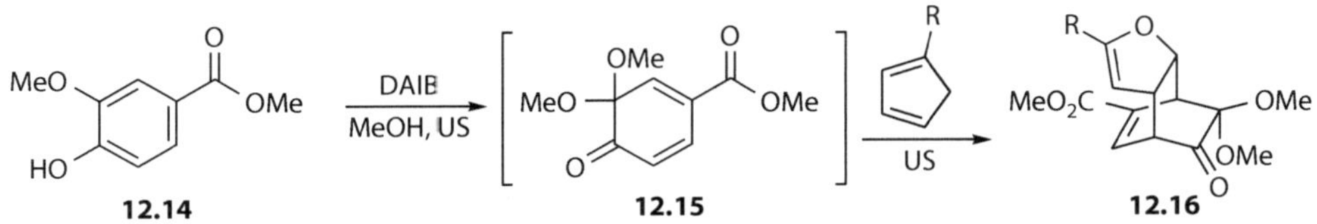

SCHEME 12.4

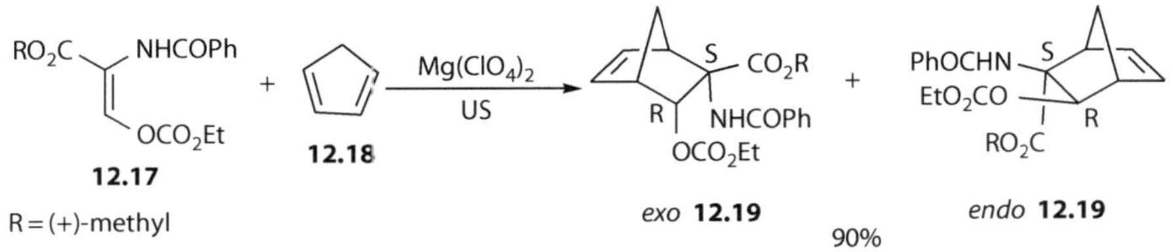

SCHEME 12.5

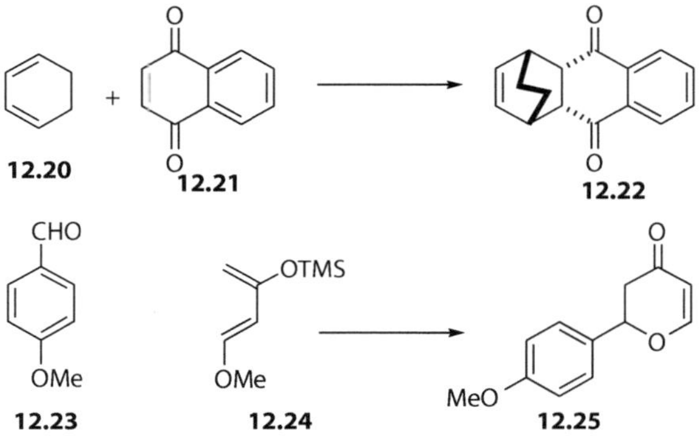

SCHEME 12.6

In search for designing synthetic strategies to access constrained functionalized carbocyclic amino acids, Abbiati et al. studied the Diels–Alder reaction between cyclopentadiene and different dienophiles (Abbiati et al., 2001) under various conditions. When US irradiation was employed, addition of enantiomerically pure **12.17** to cyclopentadiene **12.18** led to the formation of all possible four diastereomers with overall 82% formation of *exo* **12.19** and *endo* **12.19** as the major stereomers in a 3.5:1 ratio (Scheme 12.5).

Alfaro and McClusky conducted a mechanistic survey to study the effects of US on both homo and hetero Diels–Alder reactions (Alfaro and McClusky, 2001). While both reactions were enhanced by US irradiation, cavitation appeared to be responsible for cycloaddition of cyclohexadiene **12.20** to napthoquinone **12.21** since higher US power led to increased reaction rates. However, for the zinc chloride—catalyzed hetero–Diels–Alder reaction of *p*-anisaldehyde **12.23** with Danishefsky's diene **12.24**, the small rate enhancement was attributed to the increased mechanical mixing (Scheme 12.6).

In comparison to thermal conditions, a dramatic reactivity enhancement was observed for US-induced [4+2] cycloaddition of allenic trichloromethyl sulfoxides with cyclopentadiene (Raj et al., 1998). When a mixture of cyclopentadiene and **12.26** was irradiated for 1 h at 0°C, *endo* **12.27** was obtained as the major DA stereomer, while the same mixture gave only 8% of **12.27** after 2 h treatment at 0°C. In contrast, the 3-substituted substrates did not react under the same conditions and underwent isomerization to dienes of type **12.28** (Scheme 12.7).

Yinghuai et al. dehalogenated polyhalides with zinc or magnesium metals in ionic liquid medium to produce the carbene moieties in situ. The carbenes then reacted with fullerene to give its various cyclopropyl fused derivatives, for example, **12.29**, in the presence of US irradiation. Repeat the process

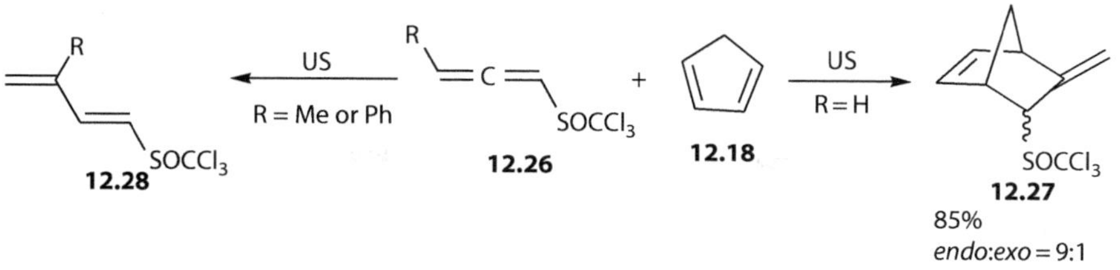

SCHEME 12.7

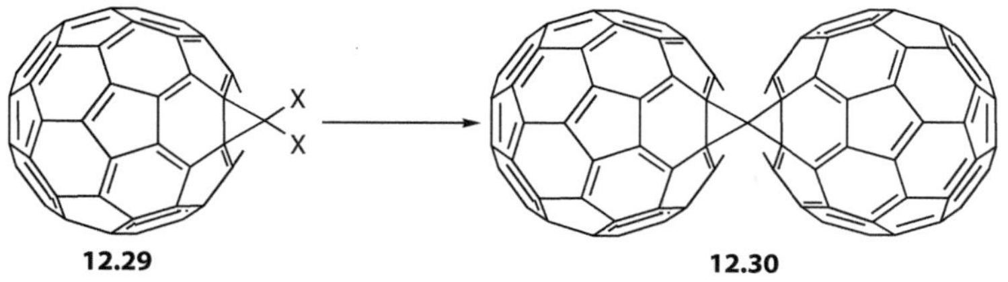

SCHEME 12.8

for **12.29** furnished dimmers of C_{60} **12.30** (Scheme 12.8). Although formation of several isomeric dimmers is possible, only a single isomer was detected in the reaction mixture (Yinghuai et al., 2003).

MULTICOMPONENT REACTIONS

Today, multicomponent reactions (MCRs) include a vast variety of synthetically useful organic procedures. MCRs facilitate one-pot combination of three or more reactants and allow the access to complex target products and libraries of desired molecules in one step. Similar to many other reaction types, US irradiation is used to enhance the scope of MCRs. The Mannich and Baylis–Hillman (BH) reactions are discussed in this section in more detail. Some other MCRs, which are of more special topics, would be mentioned in the sixth and seventh sections. Mannich reaction is a three-component aminoalkylation reaction that provides direct access to β-amino carbonyl compounds as key substructural units in the skeletone of many natural products. In addition, β-amino carbonyls are versatile intermediates in synthetic organic chemistry. Chronologically, Peng et al. (2005a) used US for the first time to boost the Mannich reaction between acetophenone derivatives and a formaldehyde equivalent with various amines in aqueous media and compared the results with those of the same reactions conducted under conventional or microwave (MW) heating conditions (Peng et al., 2005a). However, they only observed a slight improvement in the rate and yield of US-mediated reactions. Mannich reactions of various ketones with aromatic aldehydes and amines are also investigated under ultrasonic conditions (Zeng et al., 2009; Zhang et al., 2009). A rapid and asymmetric three-component procedure was reported by Choudary and colleagues, where products could be obtained in high yields with good stereoselectivities (Kantam et al., 2006). Interestingly, US-mediated self-Mannich reaction of propanal **12.31** with *p*-anisidine **12.32** followed by $NaBH_4$ reduction gave 80% of the amino alcohol **12.33** with 91% ee within 1 h (Scheme 12.9).

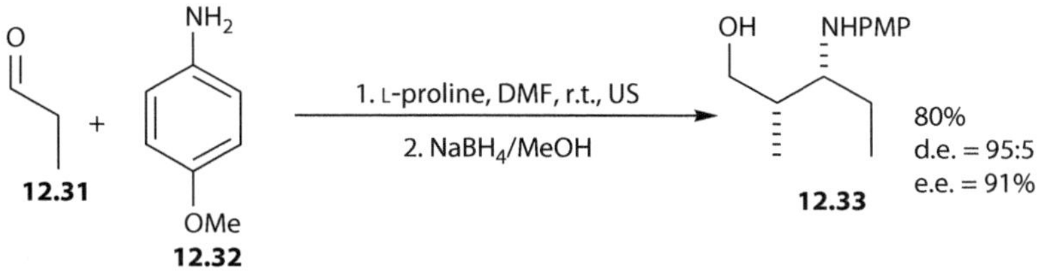

SCHEME 12.9

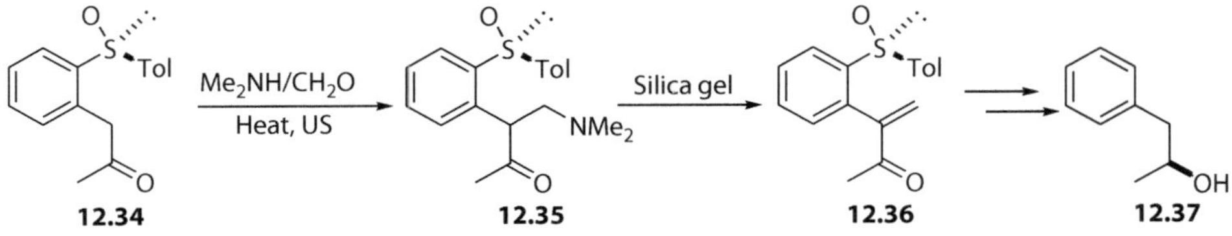

SCHEME 12.10

Elsewhere, highly enantioselective synthesis of allylic alcohol **12.37** was furnished by the preparation of the required starting α-methylene ketone **12.36**. The intermediate **12.36** itself was obtained via US-mediated Mannich reaction of ketone **12.34** with formaldehyde and dimethylamine followed by a silica gel–assisted elimination of dimethylamine from **12.35** (Ruano et al., 2009) (Scheme 12.10).

Among the well-known procedures for carbon–carbon bond formations, the BH reaction has been one of the most frequently reported reactions in recent years. The BH process is able to produce densely functionalized products from simple substrates via a one-pot procedure. Extensive investigations including US-mediated procedures are carried out to expand the scope of the BH reaction. Almeida and Coelho obtained moderate yields of products for the addition of various aromatic aldehydes to methyl acrylate (Almeida and Coelho, 1998). Similarly, Mamaghani and Dastmard developed an efficient synthetic route to BH adducts derived from the reaction of ethyl vinyl ketone and aromatic aldehydes using US irradiation and L-proline at room temperature (Mamaghani and Dastmard, 2009).

Coelho et al. further showed that under US conditions DABCO is a very effective catalyst for the BH reactions of both aliphatic and aromatic aldehydes (Coelho et al., 2002). As a result, under the conditions, moderate to good yields were observed for the reactions of several α,β-unsaturated reactants with a variety of aldehydes, many of which did not react at all in the absence of irradiation or required long reaction times to produce low quantities of their respective BH adducts. In addition, a significant rate and yield enhancement was observed for many of the US-mediated reactions. Later, Coelho employed chiral α-amino aldehydes to obtain chiral skeletons via stereoselective BH reactions under US irradiation. A variety of N-Boc-protected substrates provided their corresponding chiral BH adducts with no or negligible racemization within relatively short time periods. One of the best results was obtained for the reaction of N-Boc-L-phenylalaninal **12.38** with methyl acrylate **12.39** to produce 75% of the *anti*-isomer of **12.40** as the major product of the reaction with ee of >99% (Coelho et al., 2006) (Scheme 12.11).

A strong synergic effect along with a significant increase in the rate and yield was observed when BH reaction of **12.41** and methyl acrylate **12.39** was irradiated in an ionic liquid medium. In the absence of irradiation, only 7% of **12.42** was obtained at 0°C within 30 min, while with US and in ionic liquid the same reaction led to 92% of **12.42** in a ratio of approximately 2:1 in favor of the *anti*-adduct (Porto et al., 2009) (Scheme 12.12).

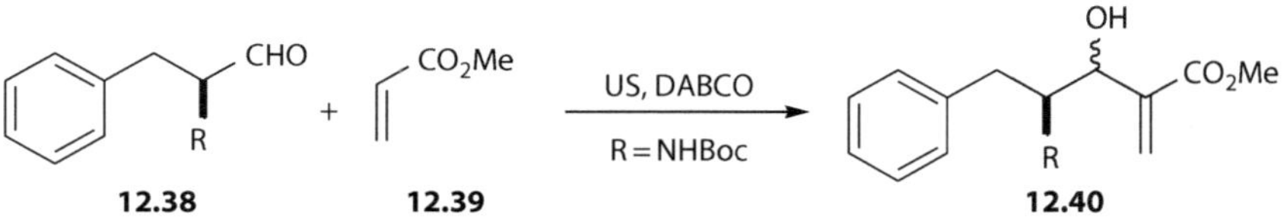

SCHEME 12.11

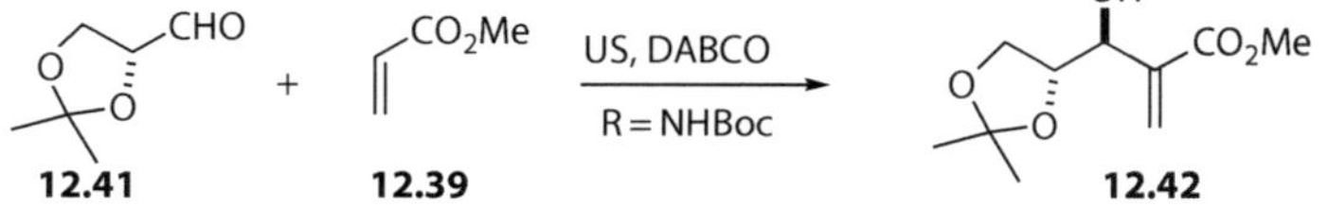

SCHEME 12.12

NUCLEOPHILIC ADDITION REACTIONS

Nucleophilic addition reactions constitute an important group of transformation that allow the interconversion of single and multiple bonds into a vast array of important functional groups. This involves the addition of nucleophiles to carbon–heteroatom and carbon–carbon multiple bonds and small rings such as epoxides. In this section, the Michael reaction and ring-opening of epoxides are discussed. The Michael reaction usually involves the addition of a nucleophile to an unsaturated carbon–carbon multiple bonds conjugated with a carbonyl group and is one of the most useful methods for the formation of carbon–carbon bonds. Like many other key synthetic transformations, US irradiation is used in recent years to accelerate and develop this reaction as well.

Sonochemical zinc-mediated conjugate addition of halides to α,β-unsaturated ketones was developed by Luche et al. offering many advantages in comparison with other related methods (Luche et al., 1983; Luche and Allavena, 1988). Sarandeses and coworkers developed the diastereoselective variant of the reaction by conducting a zinc–copper conjugate addition of alkyl iodides to chiral α,β-unsaturated substrates in aqueous media (as shown for conversion of **12.43** to **12.44**) (Scheme 12.13) and postulated that the stereoselectivity is gained either during the radical nucleophilic addition or when the protonation of the enolate occurs (Suárez et al., 2002).

Michael reaction of chalcones **12.45** with diethyl malonate, nitromethane, and ethyl acetoacetate resulted in the formation of high yields of products **12.46** under US irradiation in relatively short times (Li et al., 2003a,b). Double addition of ethyl cyanoacetate to **12.45** followed by intramolecular aldol cyclization gave derivatives of cyclohexanol **12.47** in high yields (Li et al., 2003c) (Scheme 12.14). Others reported similar sonochemistry for the formation of **12.48** from the addition of indole to chalcones using silica sulfuric acid (Li et al., 2006a), ceric ammonium nitrate (Ji and Wang, 2003), *p*-toluenesulfonic acid (Ji and Wang, 2005), or $KHSO_4$ (Zeng et al., 2007). Ji also performed sonochemical addition of indole to adducts derived from the Knoevenagel condensation of various aldehydes with malononitrile or ethyl cyanoacetate in the presence of anhydrous zinc chloride (Li and Lin, 2008). The same group was successful as well in US cyclization of 1,5-diaryl-1,4-pentadien-3-ones with malonate derivatives to prepare disubstituted-diarylcyclohexane-4-ones in high yields and in different media (Li et al., 2005d; Li et al., 2008).

Conjugate addition of heteroatom nucleophiles to α,β-unsaturated moieties are also studied. A facile solvent-free aza-Michael reaction of various amines was carried out by Duan et al. using

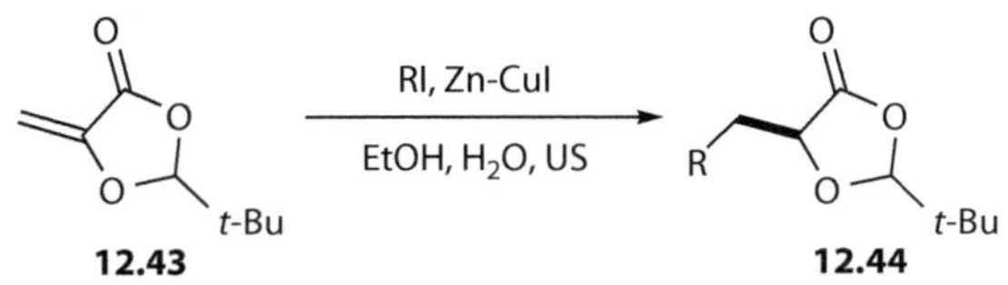

SCHEME 12.13

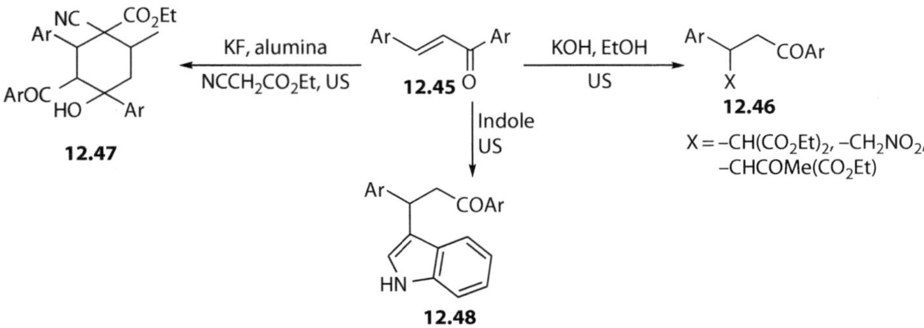

SCHEME 12.14

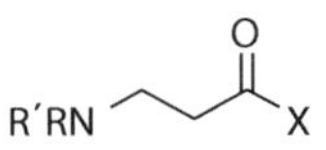
12.49

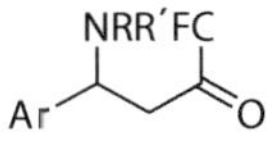
12.50

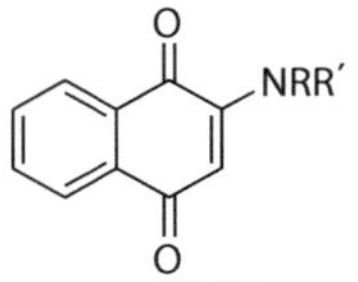
12.51

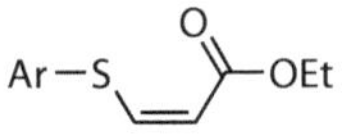
12.52

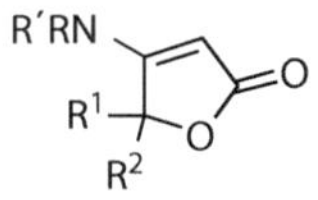
12.53

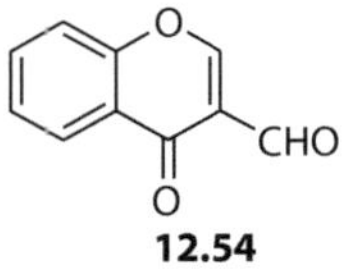
12.54

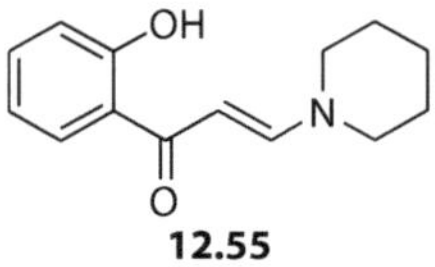
12.55

SCHEME 12.15

cerium ammonium nitrate under US irradiation (Duan et al., 2006), where both aromatic and aliphatic amines added efficiently to ethyl acrylate and acrylonitrile to form **12.49** (X=CO_2Et or CN). Additive-free Michael addition of various amines to ferrocenylenones at room temperature rapidly afforded the corresponding ferrocenyl-3-amino carbonyl **12.50** in high yields (Yang et al., 2005) (Scheme 12.15). Similar sonochemistry was developed by Liu and Ji for molecular iodine–catalyzed conjugate addition of amines to 1,4-naphthoquinone to obtain moderate to high yields of 2-amino-1,4-naphthoquinones **12.51** (Liu et al., 2008a,b). Aqueous conditioned Michael addition of aromatic thiols to 4-hydroxy-2-alkynoates proceeded at room temperature under US irradiation to get **12.52**. When amines were employed instead of thiols, the conjugate Michael addition was followed by an in situ lactonization reaction leading to 4-amino-furan-2-one derivatives **12.53** (Arcadi et al., 2009). Finally, reactions of piperidine with 3-formylchromene **12.54** gave products of type **12.55**, presumably through an ultrasonic addition–elimination mechanism (Dalvi et al., 2007).

Ring-opening of epoxides by various nucleophiles are among key methods that provide access to important synthetic and biologically active moieties such as β-aminoalcohols, 1,2-diols, and 2-hydroxy sulfides. Many modern methods including ultrasonic techniques are developed in recent years to enhance the synthetic scope of epoxides ring cleavages (Mojtahedi et al., 2007a; Abaee et al., 2008a; Dalpozzo et al., 2009; Chimni et al., 2010). Kamal et al. developed a procedure where simple epoxides were opened with aromatic amines in the presence of $FeCl_3$ and US (Kamal et al., 2005). An improved alternate of the method was later introduced for stereoselective ring-opening of a variety of epoxides with both aromatic and aliphatic amines under aqueous conditions and in the presence of no catalyst or additive to form β-aminoalcohols **12.56**. In comparison with other methods, the main advantages of this procedure were the chemoselectivity of reactions of different amines competing for a particular epoxide and lack of requirement for pH adjustment, which is usually required in procedures conducted under aqueous conditions (Abaee et al., 2008b).

The US-assisted ring-opening of epoxides was used to synthesize new ionic liquids containing pyridinium cation (**12.57**) by the reaction of pyridine with acid and 3-chloro-propylene oxide at room temperature (Zhao et al., 2008). Another successful sonicated (with or without simultaneous US irradiation) cleavage of epoxides by nucleophiles was performed by Palmisano et al. (2007a) in an aqueous medium. Consequently, a series of epoxides rapidly reacted with azide ion and 1-(3-chlorophenyl)piperazine to form their respective substituted alcohols **12.58** in good yields without any intervention of water as a nucleophile (Palmisano et al., 2007a) (Scheme 12.16).

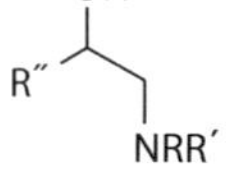
12.56

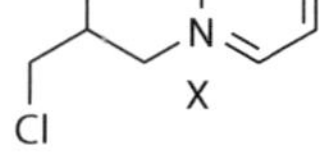
12.57

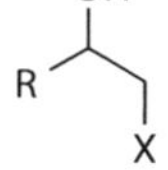
12.58

X = N_3 or
1-(3-chlorophenyl)piperazine

SCHEME 12.16

Besides nitrogen, similar reactions of other nucleophiles were also investigated. Liu and coworkers studied the ring-opening of epoxides with oxygen nucleophiles. Reactions with primary, secondary, and tertiary alcohols in the presence of amberlyst-15 led to regioselective ring-opening of epoxides and formed various β-alkoxy alcohols under US irradiation (Liu et al., 2008c). Similar reactions with hydrogen peroxide were catalyzed by antimony trichloride/SiO_2 to result in the formation of their corresponding β-hydroperoxy alcohols in high yields (Liu et al., 2008d). Finally, room-temperature addition of indole, as a carbon nucleophile, to epoxides led to efficient synthesis of 3-hydroxy-2-indolyl propanone derivatives using montmorillonite-K10 and $ZnCl_2$ (Li et al., 2010a).

CONDENSATION REACTIONS

In a condensation reaction, two or more molecules or functional groups of one molecule react together to form one single product and simultaneously liberate another molecule. In most cases, the lost molecule is water and the condensation reaction is therefore a dehydration process. Condensation reactions have important applications in the synthesis of many polymers, silicates, and polyphosphates. In addition, several biochemical transformations such as synthesis of polypeptides, polyketides, and terpenes take place via condensations and also a large number of transformations in synthetic organic chemistry are based on condensation reactions. Available reports on US-mediated condensation reactions mainly include aldol condensation, Knoevenagel reaction, and synthesis of bis(indolyl)methanes. In addition, there are a few reports on benzoin condensation as well (Hagu et al., 2007; Estager et al., 2007a; Tuulmets et al., 2007).

Aldol condensation of carbonyl compounds with aldehydes and ketones is one of the most useful carbon–carbon double bond formation reactions and has been extensively employed for the synthesis of α,β-unsaturated compounds such as chalcones (Abaee et al., 2009) and bisarylmethylideneketones (Abaee et al., 2005, 2006, 2007a,b,c, 2008c) with their many synthetic and biological applications. Many efforts including US irradiation have been devoted in recent years to widen the synthetic scope of aldol condensation reactions. Several groups investigated the effect of US activation on the synthesis of *trans*-chalcones **12.59** from aryl methyl ketones and aromatic aldehydes in the presence of either montmorillonites (Chtourou et al., 2010), activated carbons (Calvino et al., 2006), or KF/alumina (Li et al., 2002). Similarly, Xin et al. adopted the aldol condensation for the synthesis of various 1,5-diarylpenta-2,4-dien-2-ones **12.60** from conjugated aldehydes and acetophenone derivatives in the presence of activated barium hydroxide and US irradiation (Xin et al., 2009) (Scheme 12.17).

Sonochemically activated synthesis of bisarylmethylidenes of ketones **12.61** is also carried out in the presence of K_2CO_3 (Ding et al., 2007) and KF/Al_2O_3 (Li et al., 2003). The reaction mechanism and the factors influencing the products were also discussed. At room temperature, acetylferrocene was condensed with various aromatic aldehydes to form high yields of the respective products **12.62** under US conditions (Li et al., 2003b). Synthesis of **12.63** was accomplished via a sequential aldol condensation-decarboxylation process using the reaction of 2-(4-nitrophenyl)acetic acid with chromene-4-one derivatives (Dalvi et al., 2005).

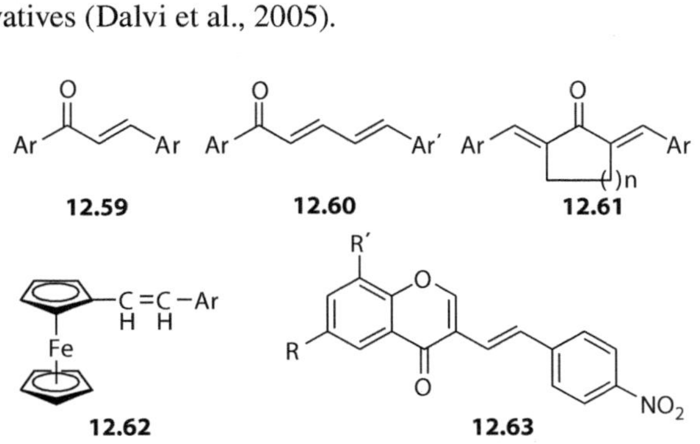

SCHEME 12.17

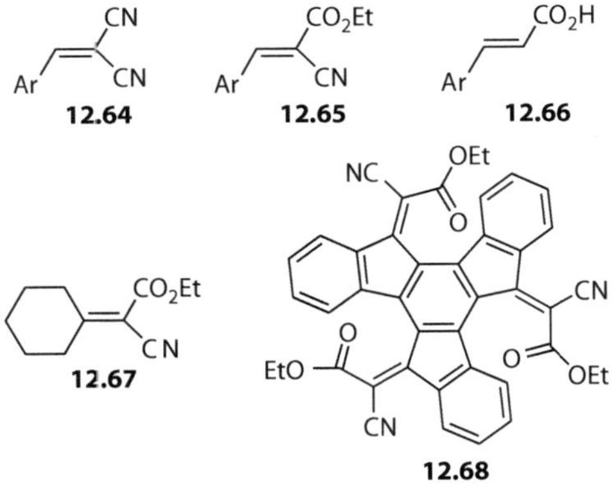

SCHEME 12.18

Knoevenagel condensation is another important reaction for the formation of C=C bonds and is widely employed for the synthesis of arylmethylenemalononitriles, derivatives of cinnamic acid, and other polysubstituted olefins, which are useful intermediates in the preparation of many target compounds important from biological, industrial, or synthetic points of view. The reaction is traditionally between active methylene compounds and various aldehydes and ketones using thermal conditions and bases. Consequently, the use of green and environmentally friendly methods for more efficient Knoevenagel condensations is of interest. In anhydrous ethanol, Knoevenagel condensation of malononitrile with aromatic aldehydes was used for the preparation of arylmethylenemalononitriles **12.64** in high yields within a few minutes under US irradiation using recoverable montmorillonite-K10-supported $ZnCl_2$ catalyst (Li et al., 2004). Similar reactions were reported by the same group using KF/Al_2O_3 for the synthesis of **12.64** (Li et al., 2005b) and **12.65** (Wang et al., 2002), the latter was prepared by the reactions of aldehydes with ethyl cyanoacetate (Scheme 12.18).

Simultaneous application of MW and US in aqueous medium led to solvent-free formation of good yields of 3-aryl acrylic acids **12.66** from aromatic aldehydes and malonic acid in the presence of K_2CO_3 and piperidine (Peng and Song, 2003). Synthesis of ethyl 2-cyano-2-cycloalkylideneacetates such as **12.67** was furnished by sonication of cyclic ketones with ethyl cyanoacetate in $NH_4OAC/AcOH$ (Li et al., 2001). Multicyclic structure **12.68**, an n-type organic semiconductor, was synthesized via Knoevenagel condensation of truxenone and ethyl cyanoacetate in the presence of $TiCl_4$ and N-methyl morpholine. When the reaction was conducted under US, the yield improved significantly (Zhang et al., 2006a).

Indole derivatives including bisindolylalkanes are important due to possessing a variety of biological and pharmaceutical properties. Despite numerous related procedures available in the literature, the facile synthesis of these compounds is still of great interest in synthetic organic chemistry. Condensation of two indole molecules with an aldehyde or a ketone is the main pathway to produce bis(indol-3-yl)methanes. Derivatives of bis(indol-3-yl)methane **12.69** (R=R′=H) were synthesized in good-to-excellent yield by sonication of aqueous mixtures of indole with various aliphatic and aromatic aldehydes under catalysis of 1-hexenesulfonic acid sodium salt at ambient conditions (Joshi et al., 2010). Again **12.69** (R=H) were prepared by the reaction of indole with ketones and aromatic aldehydes using either alum ($KAl(SO_4)_2 \cdot 12H_2O$) (Sonar et al., 2009) or H_2NSO_3H (Li et al., 2006b). In another work, indole and methylindole reacted with various aldehydes and ketones to afford **12.69** (R=H and R=Me) in high yields using bismuth salts (Scheme 12.19). Under the conditions, selective condensation of indoles with aldehydes in the presence of ketones was observed (Mohammadpoor-Baltork et al., 2006). Unsymmetrical bisindols **12.70** were synthesized under sonic waves via the reactions of indole with (1H-indol-3-yl)(alkyl)methanol catalyzed by ceric ammonium nitrate (Zeng et al., 2005). Ceric ammonium nitrate was also employed to catalyze the reaction of isatin with various indoles to afford derivatives of 3,3-di(indolyl)indolin-2-ones **12.71** under US conditions (Wang and Ji, 2006).

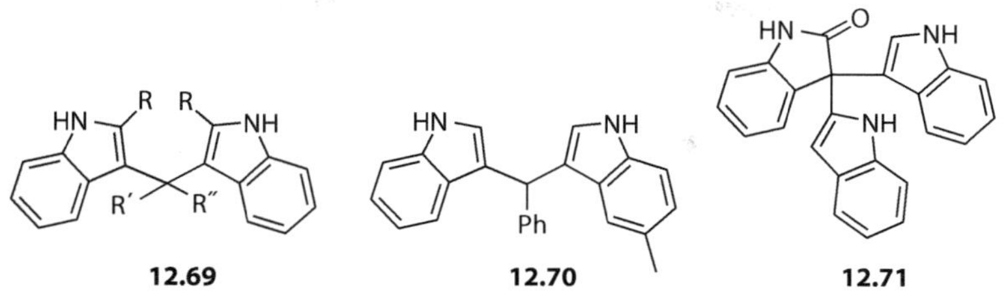

12.69 **12.70** **12.71**

SCHEME 12.19

ORGANOMETALLIC CHEMISTRY

Organometallic species have found practical applications as useful intermediates and catalysts in today's synthetic organic manipulations, especially those involved in carbon–carbon bond formation reactions, petroleum processing, and production of polymers. One of the classic organometallic transformations is the Reformatsky reaction, a zinc-mediated addition of α-haloesters to aldehydes or ketones to form β-hydroxyesters. The low reactivity of zinc and also control of the subsequent exothermic addition of the organozinc intermediate to the carbonyl group are the main limitations of the Reformatsky reaction. Thus, there have been numerous efforts through the years to overcome these limitations. In 1982, Han and Boudjouk had a report on the use of low-intensity US, which caused a significant increase in the rate and yields of Reformatsky reaction, although the use of distilled dioxane and activation of the zinc dust were still required (Han and Boudjouk, 1982). When high-intensity US was employed, no zinc activation was necessary, and reagent-grade dioxane was used without distillation to rapidly obtain high yields of β-hydroxyesters or its lactone equivalent from the reaction of α-bromoester, zinc dust, and catalytic amounts of iodine. When DL-ethyl α-bromopropionate was used as the α-bromoester component, no improvement in the diastereoselectivity of the reaction was observed as compared with the thermal reaction (Ross and Bartsch, 2003). This approach was also used by the same investigators to obtain β-aminoesters or their respective lactams from similar reactions of imines (Ross et al., 2004) and also to synthesize **12.76** from the reactions of α-bromoesters with a ketodibenzo-16-crown-5 (Ross and Bartsch, 2001) (Scheme 12.20).

Several modified versions of the reaction are also studied under US. An alternative procedure promoted by in situ–produced indium was developed for the addition of ethyl bromoacetate **12.77** to aldehydes or ketones affording high yields of β-hydroxyesters. Under the conditions, substrates like 2- or 3-hydroxybenzaldehyde bearing acidic hydrogens also reacted efficiently. The organoindium intermediates were prepared in situ and did not react with themselves while no use of promoters like iodine was required (Bang et al., 2002). Reformatsky-type reactions are also investigated for the synthesis of β-ketoesters of type **12.78** or δ-hydroxy-β-oxoesters of type **12.79** from the reaction of **12.77** with appropriate reactants in the presence of zinc (Narkunan and Uang, 1998) or indium (Young et al., 2005), respectively (Scheme 12.21).

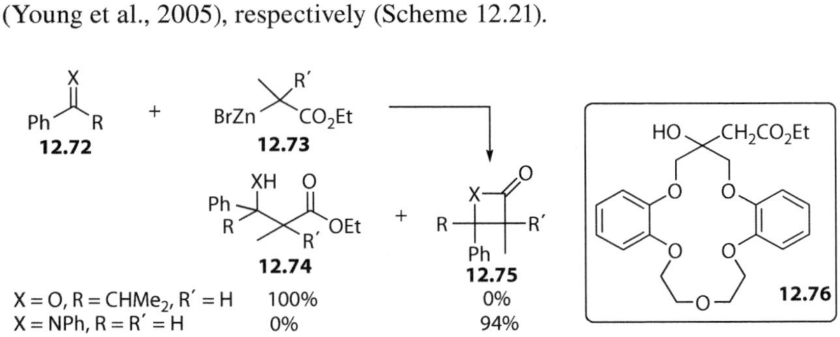

12.72 **12.73** **12.74** **12.75** **12.76**

X = O, R = CHMe$_2$, R′ = H   100%   0%
X = NPh, R = R′ = H   0%   94%

SCHEME 12.20

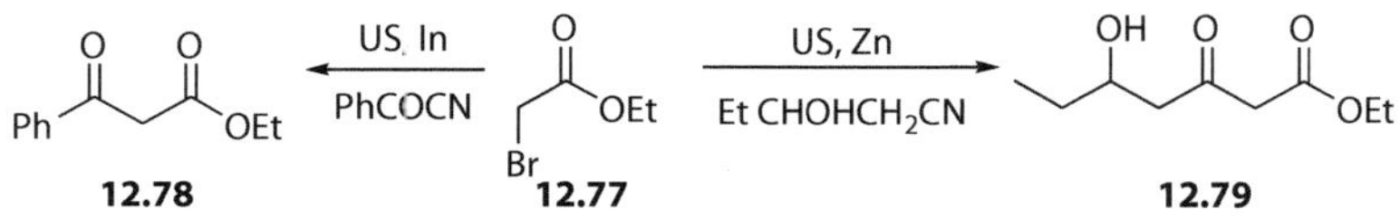

SCHEME 12.21

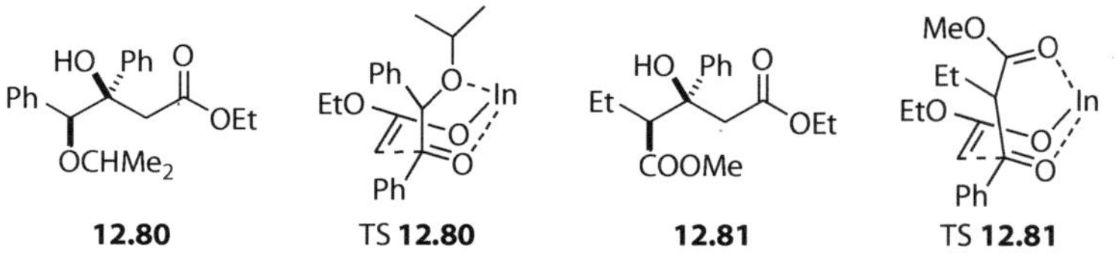

SCHEME 12.22

A remarkable *syn* selective indium-mediated procedure was developed by Babu and coworkers for the reaction of α-alkoxy ketones with β-keto esters and ethyl 2-bromoacetate to afford **80** and **81**, respectively, under sonication (Scheme 12.22). The observed outstanding diastereoselections were attributed to the presence of chelation-controlled transition states, which induce strong reactivity and selectivity in the reactions (Babu et al., 2005).

Another palladium-catalyzed coupling process is the Suzuki reaction between derivatives of phenylboronic acid and organohalides. This coupling is an extremely versatile carbon–carbon bond formation reaction and is very powerful in the synthesis of biaryls and engineering materials. In comparison with other organometallic reagents like Grignard and organozinc compounds, the boronic acid reagent and the reactants of the Suzuki coupling are more stable to the reaction conditions and a simple workup would yield the products. Many studies are carried out and the reaction is still under extensive investigation to develop more environmentally compatible Suzuki protocols. First example on US-promoted Suzuki reaction was communicated in 2002 (Rajagopal et al., 2002). Table 12.1 summarizes several studies carried out in the presence of ultrasonic irradiation for this coupling. Entry **1** shows the reaction conducted in ethylene glycol under phosphine-free conditions to obtain biaryls in high yields using a recyclable palladium catalyst (Silva et al., 2007). Ferrocenylimine organopalladium catalyzed Suzuki reaction of phenylboronic acid with

TABLE 12.1
US-Mediated Suzuki Reactions

Entry	Boronic Acid	Aryl Halide	Conditions	Yield (%)
1	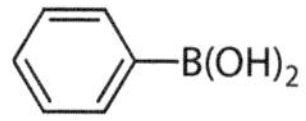 Ph–B(OH)$_2$	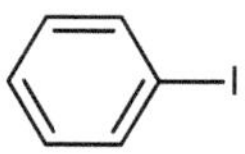 Ph–I	Pd$_2$(dba)$_3$, TBAB, K$_2$CO$_3$	94
2	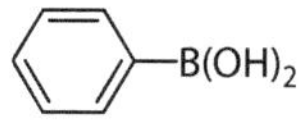 Ph–B(OH)$_2$	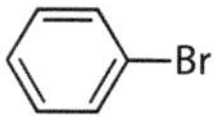 Ph–Br	K$_3$PO$_4$, TBAB	73
3	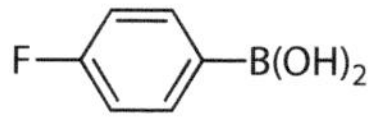 F–C$_6$H$_4$–B(OH)$_2$	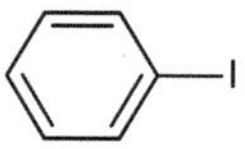 Ph–I	Pd/PVP, K$_2$CO$_3$	94
4	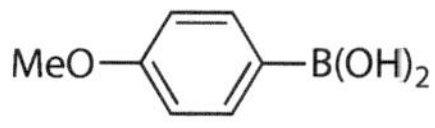 MeO–C$_6$H$_4$–B(OH)$_2$	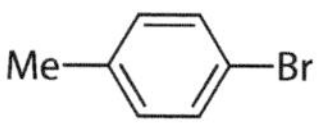 Me–C$_6$H$_4$–Br	PdCl$_2$, TEBA, K$_2$CO$_3$	82
5	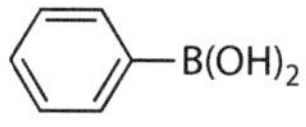 Ph–B(OH)$_2$	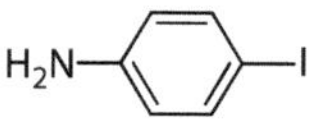 H$_2$N–C$_6$H$_4$–I	Pd(OAc)$_2$, TEBA, Cs$_2$CO$_3$	90
6	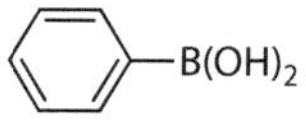 Ph–B(OH)$_2$	none	Pd/C, KF, oxidant	87

arylhalides mediated by water proceded efficiently with both electron-withdrawing and electron-releasing groups (entry **2**) (Zhang et al., 2008).

In another study, comparable results were obtained for MW, US, and thermal reactions. In sonicated reactions, the medium was recoverable with no significant loss of activity (entry **3**) (de Souza et al., 2008). Heterogeneous Pd/KF-catalyzed addition of arylboronic acids to iodoarenes in methanol/water medium took only 1 h irradiation to produce high yields of the expected products (entry **4**) (Poláčková et al., 2005). An aqueous medium and Pd/C were again employed for both homo- and cross-coupling reactions of boronic acids with aryl halides under high-intensity US and phosphine-free conditions. In the case of electron-deficient aryl chlorides, the reaction required the use of palladium(II) acetate as the catalyst (entry **5**) (Cravotto et al., 2005a). In an interesting experiment, self-coupling of boronic acids was observed in the presence of an oxidant (molecular oxygen or 3-bromo-4-hydroxycoumarin) to solely afford symmetric biaryls under heterogeneous Pd/C catalysis and no use of phosphine ligand. An oxidative pathway through which the boronic acid dimerizes and the oxidant reduces to 4-hydroxycoumarin was suggested for this homocoupling process (entry **6**) (Cravotto et al., 2005b).

Alternative approaches to classic palladium-catalyzed Suzuki reaction involve the use of organotellurium or hypervalent iodine compounds. A Suzuki-type coupling reaction was reported in water using hypervalent iodine and sodium tetraphenylborate in the absence of a catalyst or a base (Shi, 2007). Stefani et al. practiced the use of organotelluriums in the Suzuki-Miyaura reaction (Zeni et al., 1999) and subsequently developed the US-assisted version of the reaction in the presence of Et_3N, $Pd(PPh_3)_4$, and silver salts, as presented in Table 12.2. They accomplished stereodefined synthesis of 1,3-dienes by using organotrifluoroborates via the coupling of vinylic tellurides and potassium β-styryl trifluoroborate salt. The procedure led to the synthesis of various 1,3-dienes in short time periods and under mild conditions (entry **1**) (Cella et al., 2006). Similarly, cross-coupling reaction of potassium aryltrifluoroborate with butyl(2-chloro-2-phenylvinyl)tellane derivatives resulted in efficient synthesis of a variety of functionalized vinylic chloride skletons (entry **2**) (Guadagnin et al., 2008). The extension of the methodology to the synthesis of functionalized 1,3-enynes (entry **3**) (Singh et al., 2008) and stilbens (entry **4**) (Cella and Stefani,

TABLE 12.2
US-Mediated Suzuki Reactions of Organotellurides

Entry	Organotellurium	Organoborane	Product	Yield (%)
1	Br-C₆H₄-CH=CH-TeBu	Ph-CH=CH-BF₃K	Br-C₆H₄-CH=CH-CH=CH-Ph	82
2	BuTe-C(Cl)=CH-Ph	MeO-C₆H₄-BF₃K	MeO-C₆H₄-C(Cl)=CH-Ph	80
3	BuTe-C(Ph)=CH₂	Ph—≡—BF₃K	Ph-C≡C-C(Ph)=CH₂	79
4	BuTe-CH=CH-Ph	KF₃B–Ph	Ph-CH=CH-Ph	82
5	F₃C-C₆H₄-TeBu	none	F₃C-C₆H₄-C₆H₄-CF₃	82

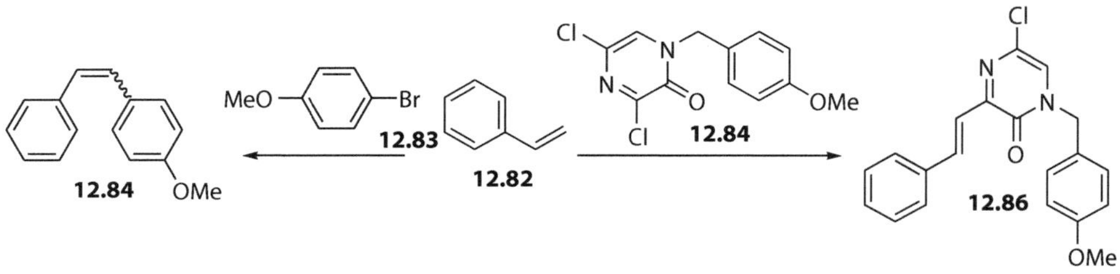

SCHEME 12.23

2006) is also reported by the same group. A Suzuki-type homocoupling of various aryltellurides was conducted under the catalysis of Pd(PPh$_3$)$_4$ to produce symmetrical biaryls in good yields (entry **5**) (Singh and Stefani, 2008).

Another important orgamometal-mediated C–C bond formation is the Heck reaction, a palladium coupling involving the addition of alkenes or alkynes to aryl or vinyl halides in polar media. Srinivasan and group revealed the first report on US-promoted Heck reaction (Deshmukh et al., 2001). They completed the coupling at ambient temperature in an ionic liquid solvent within 1.5–3 h using Pd(OAc)$_2$. This was a significant improvement to the conventional versions of the reaction, which usually require long-time treatment of the reactants at higher temperatures. The simultaneous irradiation of MW and US significantly enhanced the Heck coupling of several aryl halides with styrene in the presence of Pd(OAc)$_2$ or Pd/C. For example, the reaction with 4-bromoanisole gave 96% of **12.84** in a ratio of 1:13 in favor of the *trans* isomer within 1.5 h (Palmisano et al., 2007b). A similar procedure was used for the synthesis of **12.86** in 80% yield after 20 min irradiation (Garella et al., 2010) (Scheme 12.23). Comparative studies suggested that a synergic phenomenon comes into effect when MW and US are used together, enhancing the outcome of the reactions.

Sonochemical Heck reactions to obtain acrylate derivatives were also studied (Ambulgekar et al., 2005). At room temperature and in the presence of in situ–formed nanoparticles, various alkenes including acrylates underwent Heck reaction with iodobenzene derivatives to furnish high yields of the corresponding products. In addition, excellent regiochemistry was observed for Heck reactions of polyiodobenzenes giving *para*- as opposed to the *ortho*-substitution (Zhang et al., 2006b). Elsewhere and in the framework of an extensive program to develop recyclable linking systems for solid supported reactions, Enders et al. conducted a sonolytic Heck reaction of halogenated resin **12.87** with *tert*-butyl acrylate to furnish high yields of product **12.88** after treating **12.87** with Pd(OAc)$_2$, PPh$_3$, and NEt$_3$ in DMF followed by a cleavage process (Brase et al., 1998) (Scheme 12.24).

The palladium-catalyzed coupling of aryl halides and terminal acetylenes is usually conducted in the presence of a copper cocatalyst and an amine. The process is named after Sonogashira and offers an efficient method for the synthesis of substituted acetylenes. A limitation to the reaction is the occasional formation of side products caused by the additives of the reaction. To overcome this limitation, copper-, phosphine-, and amine-free versions of the Sonogashira reaction have been developed in recent years to gain reactions with much better performance and selectivities. In this regard, the application of US has been very useful to soften the conditions so that the reaction would occur at room temperature. Srinivasan et al. designed a set of experiments where the Sonogashira reaction proceeded at ambient conditions in an ionic liquid solvent. Excellent yields of a variety of

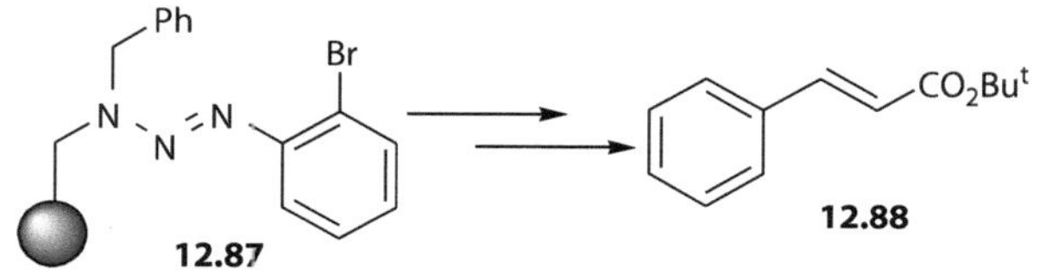

SCHEME 12.24

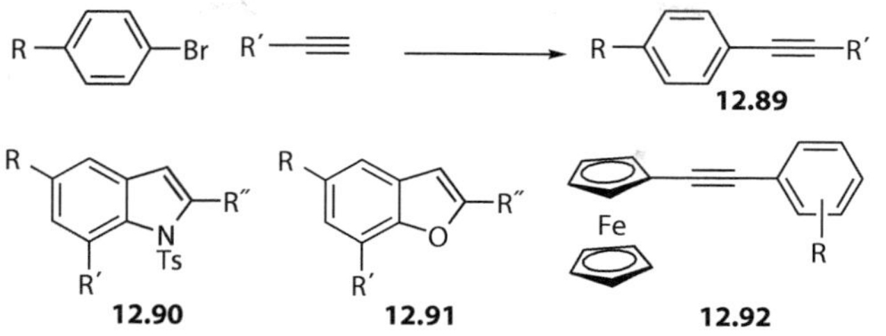

SCHEME 12.25

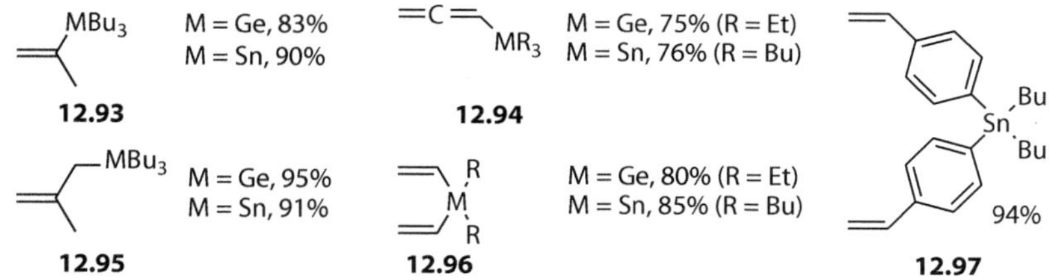

SCHEME 12.26

1,2-diarylacetylenes **12.89** were obtained in short time and with very good chemoselectivity using no copper cocatalyst or phosphine ligand (Scheme 12.25). Experiments showed that Pd nanoparticles are formed in situ under US irradiation to catalyze the reaction (Gholap et al., 2005).

A significant rate enhancement was observed for a sonochemical one-pot process consisting of a palladium acetate–catalyzed Sonogashira coupling of iodobenzenes with acetylene derivatives followed by the cyclization of the intermediate to yield 2-substituted indoles **12.90** at room temperature. The conditions were tolerable for both electron-withdrawing and electron-releasing groups and no copper, ligand, or amine was used in the process (Palimkar et al., 2006). A similar approach was used for the synthesis of benzofuran derivatives **12.91** (Palimkar et al., 2008) (Scheme 12.26). The copper- and ligand-free approach was also employed for room-temperature Sonogashira cross-coupling of aryl iodides with ferrocenylacetylene to produce **12.92** in the presence of $PdCl_2$ (Fu et al., 2008).

There are a few other reports on US-promoted reactions of other organometals (Sanchez et al., 2001). Langle et al. (2003) developed two efficient one-pot procedures for the synthesis of various organogermanium (Langle et al., 2003) and organotin (Lamandé-Langle et al., 2009) compounds. The reactions took place via a magnesium metal–mediated Barbier reaction of organohalides with tin or germanium halides producing moderate to high yields of the products. Some of these products (**12.93–12.97**) are shown in Scheme 12.26. Under these conditions, transfer of more than one group was also feasible.

The reaction between alkyl halides and carbonyl compounds is another carbon–carbon bond formation reaction and is mediated with the use of various metals. Several investigations are carried out in recent years to conduct such reactions under environmentally safer conditions. Table 12.3 shows representative examples on the application of US irradiation in metal-mediated alkylation of carbonyl compounds. Bian et al. (2010) allylated aromatic aldehydes and ketones in water using $SnCl_2$ at room temperature. Their study showed that reactions proceed with more efficiency when mixtures are irradiated (entry **1**) (Bian et al., 2010). The same group also studied the reactions of allyl bromide with aromatic aldehydes in $Sb/H_2O/KF/MeOH$ (entry **2**) (Bian et al., 2009), Li/THF (entry **3**) (Bian et al., 2007), or $Zn/NH_4Cl/H_2O/THF$ (entry **4**) (Bian et al., 2006a) systems to obtain allyic alcohols in high yields. Similar procedures are developed as well using propargyl bromide (Lee et al., 2004) and trichlorofluorocarbon (Barkakaty et al., 2006) to furnish allenyl alcohols (entry **5**) and chlorofluoro carbinols (entry **6**), respectively.

TABLE 12.3
Various Metal-Mediated Alkylation of Carbonyl Compounds under US

Entry	RX	Carbonyl Compound	Product	Conditions	Yield (%)
1	Allyl-Br	PhCHO	Ph-CH(OH)-CH₂-CH=CH₂	SnCl₂, H₂O, 5 h	86
2	Allyl-Br	Furfural	2-furyl-CH(OH)-CH₂-CH=CH₂	Sb, H₂O, KF, MeOH, 2.5 h	97
3	Allyl-Br	Ph₂C=O	Ph₂C(OH)-CH₂-CH=CH₂	Li, THF, 40 min	70
4	Allyl-Br	PhCOCH₃	Ph-C(CH₃)(OH)-CH₂-CH=CH₂	Zn, NH₄Cl, H₂O, THF, 10 min	99
5	HC≡C-CH₂Br	pentyl-CHO	CH₃(CH₂)₄-CH(OH)-CH₂-C≡CH	Zn, ICH₂CH₂I, THF, 2.5 h	96
6	CFCl₃	PhCHO	Cl₂FC-CH(OH)-Ph	Al, SnCl₂, DMF, 8 h	85

HETEROCYCLES

Undoubtedly, heterocycles constitute key groups of organic compounds with many important applications in various fields of science, engineering, and industry. Consequently, continuing efforts have been devoted during the past years to the study and exploration of various features of heterocycles. Accordingly, the effect of US on the synthesis of heterocycles is also investigated extensively in recent decades. In 2009, Cella and Stefani (2009) had a comprehensive review on various US-promoted heterocyclization reactions. Thus, this section aims to mainly review the articles that published afterward or were not covered by Cella and Stefani (2009).

THREE-MEMBERED RINGS

There are several investigations on US-promoted synthesis of epoxide rings involving the epoxidation of simple alkenes with different reagents. The reaction of cyclohexene with iodine in aqueous dioxane and in the presence of Cu(OAc)$_2$.H$_2$O resulted in formation of the respective iodohydrine, which further cyclized to cyclohexene oxide after sonication in the presence of Na$_2$CO$_3$ (Fernandes et al., 2007). Mirkhani et al. succeeded to rapidly and efficiently convert a variety of alkenes to their respective epoxides using nanoparticle supported polyoxometalates and hydrogene peroxide under US irradiation (Salavati et al., 2010a). The same group also reported similar conversions by employing other oxidants and ultrasonic conditions (Mirkhani et al., 2000; Tangestaninejad et al., 2006, 2008; Salavati et al., 2010b). In situ generation of peroxycarboximidic acids by sonication of mixture of hydrogen peroxide with different nitriles was reported to be responsible for the epoxidation of cyclohexene (Braghiroli et al., 2006). In another work, industrial synthesis of cyclohexene oxide from cyclohexene was reported at room temperature using molecular oxygen, isobutyraldehyde, and US irradiation (Zhang et al., 2007a).

Efficient epoxidation of chalcones with urea–hydrogen peroxide was carried out under US irradiation in relatively short reaction times (Jin et al., 2009). An improved version of the same process was offered by Li et al. where products **12.100** were obtained by room-temperature sonication of

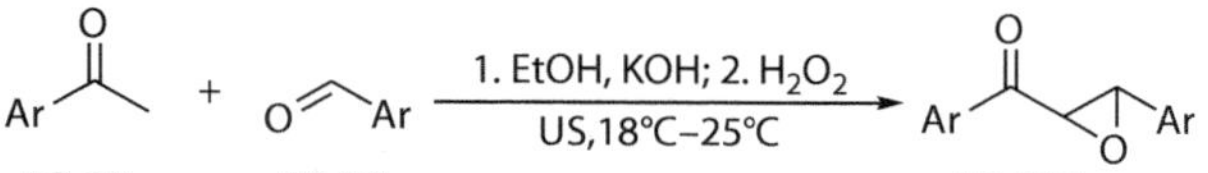

12.98 12.99 12.100

SCHEME 12.27

acetophenone derivatives **12.98** and aromatic aldehydes **12.99** in the presence of H_2O_2 (Scheme 12.27). In comparison with related multistep reactions, the procedure reached to completion in shorter time periods without isolation of any intermediate or the use of toxic solvents (Li et al., 2010b).

Five-Membered Rings

Among five-membered heterocyclic rings, imidazoline derivatives are of particular importance due to having outstanding biological and pharmacological activities and also being used as catalysts and synthetic intermediates. Entezari investigated the synthesis of 1,2,5-trisubstituted imidazole **12.102** in the presence of US irradiation (Entezari and Asghari, 2008) (Scheme 12.28). While the yield of the process reached to 90% within half an hour sonication, it took 72 h to produce 70% of the same product in the absence of US. Alternatively, 2,4,5-trisubstituted derivative of the same skeleton (**12.103**) was synthesized under US conditions by two different methods. Zang et al. (2010a) irradiated mixtures of benzil with aromatic aldehydes in an ionic liquid medium to obtain the expected products in 70%–96% yield within 45–90 min, while irradiation of similar mixtures in water led to relatively higher yields in slightly shorter times (Shelke et al., 2009). An alternative sonochemical synthetic pathway was also demonstrated by condensation of aromatic aldehydes with *o*-phenylenediamine using a heteropoly acid (Fazaeli and Aliyan, 2009).

Sonochemical synthesis of 2-imidazoline derivatives has also been investigated. Martins and his colleagues irradiated aqueous mixtures of aromatic aldehydes and ethylenediamine in the presence of NBS to obtain high yields of **12.106** within 12–18 min (Sant' Anna et al., 2009) (Scheme 12.29). An improved high scalable version of the reaction was introduced by Mirkhani et al. where the same condensation was performed in the presence of catalytic amounts of sulfur giving various derivatives of **12.106** in shorter time periods (Mirkhani et al., 2006).

Several related reports are also found on the synthesis of various pyrazole derivatives, a group of heterocycles found as the key substructure in many biologically active compounds and pharmaceuticals. Three recent investigations were reported on ultrasonic synthesis of dihydro-1*H*-pyrazole derivatives (Pizzuti et al., 2009, 2010; Nabid et al., 2010). A variety of 1,3,5-triarylpyrazole **12.107**

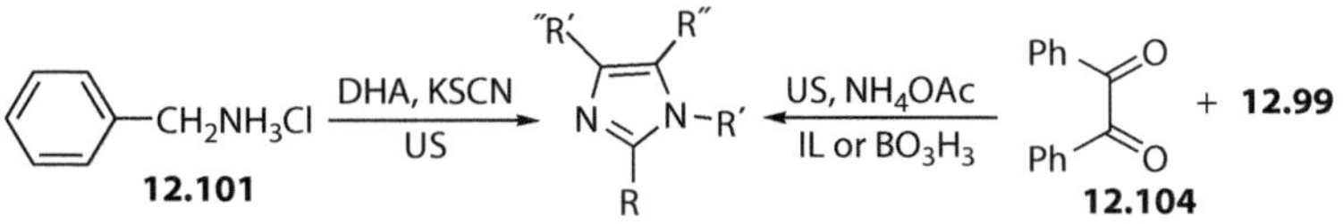

12.101 12.104

12.102: R = SH, R′ = CH_2Ph, R″ = CH_2OH, R‴ = H
12.103: R = Ar, R′ = H, R″ = R‴ = Ph

SCHEME 12.28

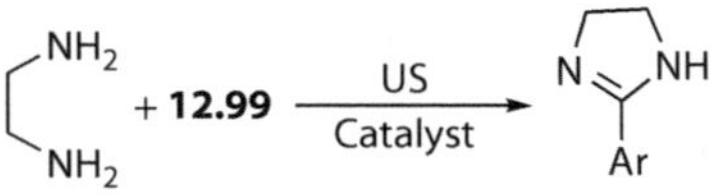

12.105 12.106

SCHEME 12.29

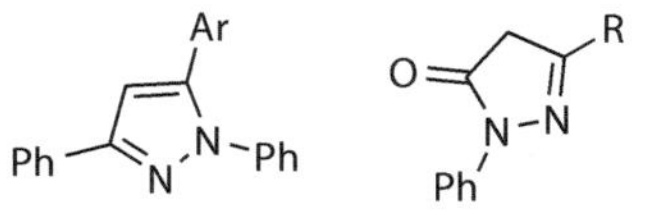

12.107 12.108

SCHEME 12.30

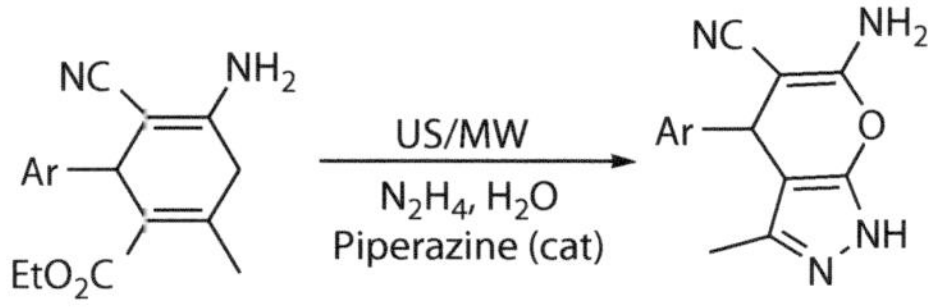

12.109 12.110

SCHEME 12.31

derivatives were synthesized by Li et al. via the reaction of appropriate 3-aryl-2,3-epoxy-1-phenyl-1-propanones with phenylhydrazine. Irradiation of the mixtures at room temperature gave the products **12.107** in 69%–99% yields (Li et al., 2010c) (Scheme 12.30). An alternative solvent-free method for room-temperature synthesis of 3-alkyl-1*H*-pyrazolones **12.108** was carried out via the condensation of hydrazine derivatives with various β-keto esters under ultrasonic conditions without using any extra additives (Mojtahedi et al., 2008a).

A dramatic rate enhancement was observed for the synthesis of pyranopyrazoles **12.110** under combined use of MW and US in heterogeneous conditions (Scheme 12.31). The dramatic kinetic acceleration was attributed to the effect of US irradiation through continuous cleaning of the surface of the substrate particles resulting in significant increase in mass and heat transfer between the phases (Peng et al., 2006).

A multistep synthesis of 1,2,3-triazoles was reported by Al-Zaydi (2009a) using the coupling of cyanoacetamides with aromatic diazonium salts followed by the reaction of the product of the former step with hydroxylamine and a cyclization process (Al-Zaydi, 2009a). An interesting one-pot palladium-catalyzed coupling was developed by Chen and coworkers, where the US-promoted combination of an acid chloride **12.112** with a terminal acetylenes **12.111** efficiently prepared the required intermediate to in situ react with sodium azide and give the 1,2,3-triazole products **12.113** (Li et al., 2009a) (Scheme 12.32).

Sonochemical synthesis of oxazoline derivatives was investigated by Khistiaev et al. As a result, novel 4-fluoro-3-oxazolines **12.116** were synthesized by three-component combination of diarylmethanimines **12.114**, CF_2Br_2, and trifluoroacetophenones **12.115** in the presence of lead metal (Scheme 12.33). The reaction presumably proceeds via two sequential steps: the formation of an

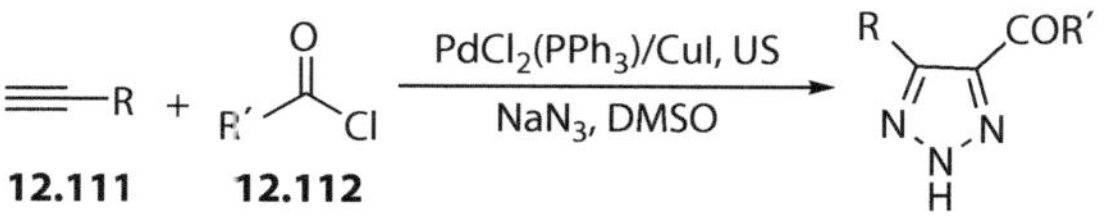

12.111 12.112

12.113

SCHEME 12.32

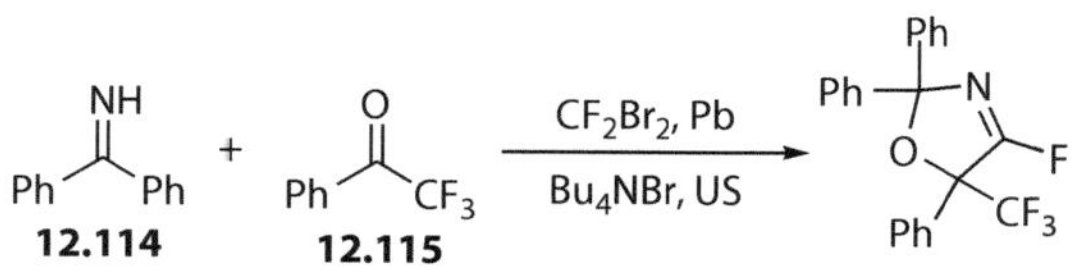

12.114 12.115

12.116

SCHEME 12.33

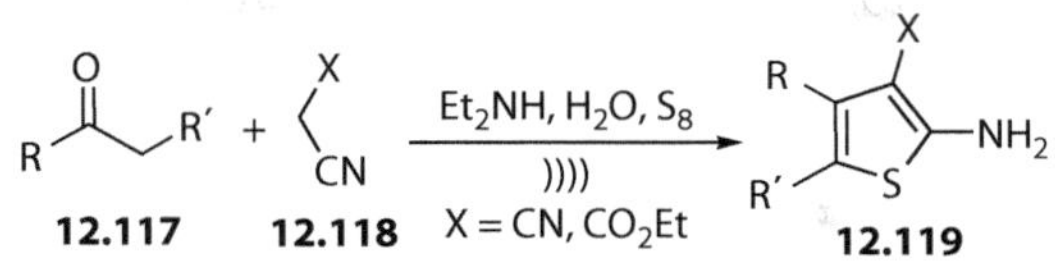

SCHEME 12.34

intermediate ylide and a subsequent 1,3-dipolar cycloaddition of the ylide with the acetophenone moiety (Khistiaev et al., 2008).

A convenient ultrasonic method was offered for the synthesis of 2-aminothiophene derivatives through acceleration of the Gewald reaction. While the attempts for the synthesis of thiophene derivatives via Gewald reaction was reported to fail under US irradiation (Šibor and Pazdera, 1996), the use of diethylamine in aqueous conditions caused a rapid method for high-yield synthesis of **12.119** via the reaction of **12.117** with **12.118** and elemental sulfur (Scheme 12.34). Due to the polarity of the medium, products precipitated spontaneously and were easily separated from the reaction mixtures by filtration (Mojtahedi et al., 2010).

Six-Membered Rings

Ultrasonic synthesis of a variety of six-membered heterocyclic systems is studied (Jie et al., 2001; Tei et al., 2009; Mandhane et al., 2010). 1,4-Dihydropyridines (DHPs) constitute one of the most frequently occurring subunits in the structure of many natural and synthetic heterocycles and have received great deals of attention due to exhibiting a wide range of pharmaceutical and biological activities. A well-established strategy for the synthesis of DHPs involves the one-pot Hantzsch condensation of aldehydes and ethyl acetoacetate with ammonia equivalents. To overcome the inherent disadvantages associated with classical Hantzsch-type reactions such as long reaction time, harsh conditions, and low yields of products, sonochemical procedures are shown in recent years as one of the most efficient alternative methods for the synthesis of DHPs. In this regard, Shaabani et al. reported rapid Hantzsch condensation of aromatic aldehydes with acetoacetate derivatives and ammonium acetate to produce DHPs **12.120** in high yields at room temperature in a recoverable ionic liquid medium under US irradiation (Shaabani et al., 2006) (Scheme 12.35). An improved solvent-free version of the same reaction was later reported by Wang et al. (2008a) to furnish high yields of **12.120** in shorter time periods. Similar symmetrical structures were synthesized using three-component Hantzsch-type reactions, where barbituric acid or dimedone (Muscia et al., 2009; Mosslemin et al., 2010) condensed with amines and aromatic aldehydes under US conditions to produce high yields of **12.121** or **12.122**, respectively (Scheme 12.35).

The one-pot strategy was also adopted for the preparation of tetrahydrobenzo-acridin-one **12.123** by treatment of dimedone, 1-naphthylamine, and aromatic aldehydes in ethanol (Zang et al., 2010b).

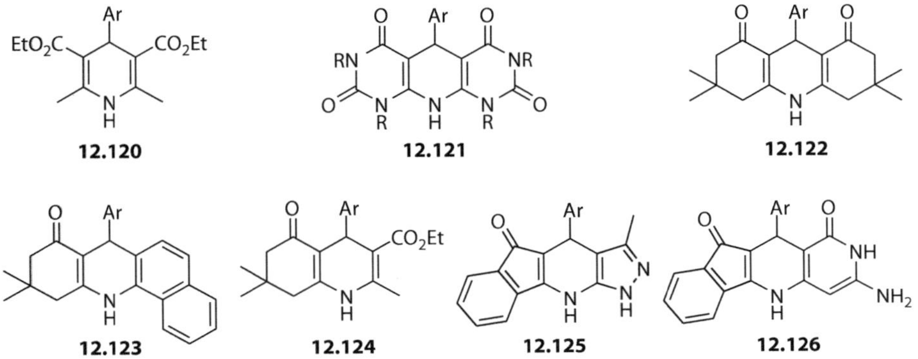

SCHEME 12.35

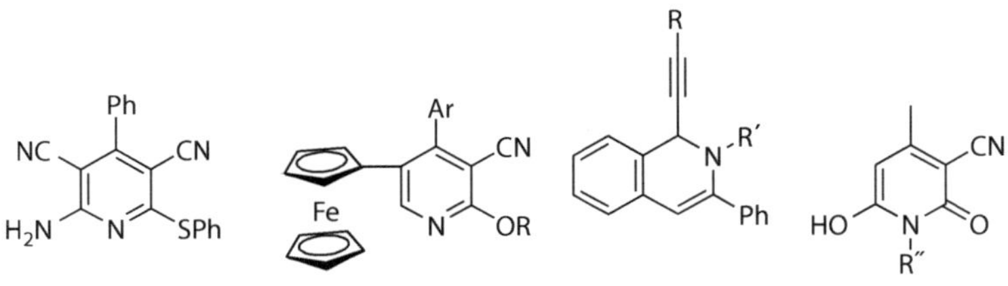

12.127 **12.128** **12.129** **12.130**

SCHEME 12.36

A strong acceleration was observed when a mixture of dimedone, aldehydes, ethyl acetoacetate, and ammonium acetate was irradiated in aqueous micelles to form **12.124** in high yields (Kumar and Maurya, 2008). A similar route was also reported using the same reactants; while thermal synthesis of **12.124** required 3h reflux in ethanol, sonication of the same mixture at ambient temperature gave the product within 2–5 min (Ahluwalia et al., 1997). Similar chemistry was chosen for the synthesis of **12.125** by a combination of 5-amino-3-methyl-1H-pyrazole, 2H-indene-1,3-dione, and an aldehyde in EtOH (Nikpassand et al., 2010). Replacement of the starting pyrazole with diamino-pyrimidine-4-one gave **12.126** (Tu et al., 2008) (Scheme 12.36).

Ultrasonic syntheses of pyridine and its 1,2-dihydro moieties are also investigated. A one-pot condensation of aldehydes, malononitrile, and thiophenol was reported by Shinde et al. (2010) The reaction was catalyzed by boric acid in aqueous medium to afford pyridines of type **12.127** in short reaction times and high yields. Condensation of ferrocenyl chalcones with malononitrile in sodium alkoxide produced 2-alkoxy-4-aryl-6-ferrocenyl-3-cyanopyridines **12.128** (Zhou et al., 2006). A three-component reaction of 2-alkynylbenzaldehydes with aromatic anilines and phenylacetylenes in water and in the presence of a combined Lewis acidic-surfactant catalyst led to the synthesis of a variety of alkynyl substituted dihydroisoquinolines **12.129** (Ye et al., 2008). Also, Al-Zaydi (2009b) and Al-Zaydi et al., (2009) reported ultrasonic synthesis of novel 2-oxo-1,2-dihydropyridines **12.130** by treatment of ethyl cyanoacetate and amines followed by the reaction of the product with ethyl acetoacetate (Scheme 12.36).

Xanthene and 4H-pyran derivatives contain an oxygen atom in their central six-membered rings and both are known as key heterocyclic moieties in medicinal and synthetic organic chemistry and numerous methods are available for their syntheses. In this regard, the use of US-mediated methodologies has provided the opportunity to prepare these heterocycles with several advantages such as mild reaction conditions, high yields, and simple experimental procedures. The use of silica supported $NH_4H_2PO_4$ as a recyclable heterogeneous catalyst led to facile synthesis of various aryl-dibenzo-xanthenes **12.131** from the condensation of various phenols with aromatic aldehydes in water and under US irradiation (Mahdavinia et al., 2009) (Scheme 12.37). 4H-Pyran derivatives of **12.132** were prepared by the reaction of aryl aldehyde and ethyl acetoacetate under $ZnCl_2$ catalysis and sonication in acetic anhydride. Variation in the US power, temperature, and the reaction time showed that the optimized conditions are obtained at 50°C in half an hour irradiation with sonication power of 100 W (Ni et al., 2010). Synthesis of several heterocycles containing the 4-hydroxy-coumarin core such as **12.133** was carried out in aqueous media (Scheme 12.37). Comparison of irradiated reactions with those performed by conventional methods showed the enhancing effect of sonication in the reactions (Cravotto et al., 2003a).

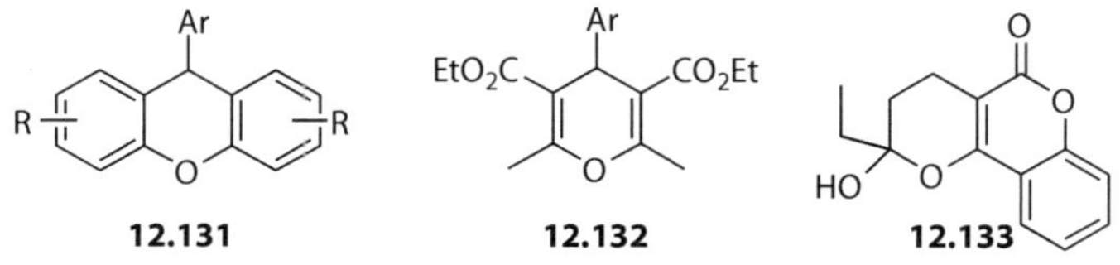

12.131 **12.132** **12.133**

SCHEME 12.37

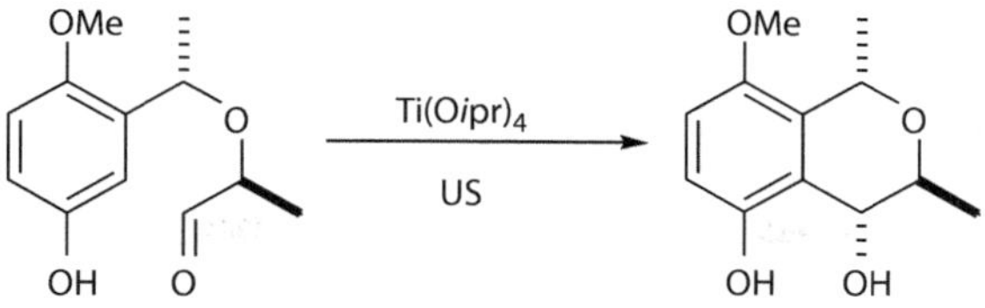

SCHEME 12.38

Asymmetric synthesis of 2-benzopyrans was reported by Giles and Joll via intramolecular diastereoselective annulation of titanium phenolates of phenolic aldehydes. For example, irradiation of the mixture of phenolic aldehyde **12.134** with titanium tetraisopropoxide diastereoselectively afforded high yield of alcohol **12.135** (Scheme 12.38). The mild conditions of the method prevented racemization of the α stereogenic center of both the starting and final materials. In addition, use of US was shown to be crucial for the reaction to complete (Giles et al., 1999).

SPIRO HETEROCYCLES

Molecules with spiro heterocycles in their structures are among very important organic compounds because many of them possess vast biological and pharmaceutical activities. Thus, the development of efficient methods to access such compounds has always been of great interest to organic chemists. Conceptually, sonochemical procedures are adopted as one of the versatile means to develop green and environmentally safe methods for the synthesis of spiro heterocycles. Several recent reports are available in the literature on the synthesis of nitrogen containing spiro rings (Bazgir et al., 2010; Li et al., 2010e). For example, synthesis of spiro-pyrrolo-pyrrolidine derivatives was reported by moldoveanu et al. (2009). Suresh Babu and Raghunathan (2007) approached a multicomponent [3+2] cycloaddition strategy by using isatin **12.136**, oxoindolino-ylidene acetophenone **12.137**, and L-proline **12.138** to produce similar spiro-pyrrolopyrrolidine structures (**12.139**) (Scheme 12.39). The reactions proceeded regioselectively in high yields under sonication in acetonitrile and in the presence of silica.

Cycloaddition strategy was also adopted by Jadidi et al. for the synthesis of novel pyrrolizidines **12.141** under both ultrasonic and thermal conditions by combination of acenaphtenequinone **12.140**, L-proline, and appropriate dipolarophiles (Scheme 12.40). Improvements in rates and experimental conditions were observed when sonication was used. The diastereoselectivity and regioselectivity of the reactions were confirmed by NMR and x-ray crystallographic analyses (Jadidi et al., 2008).

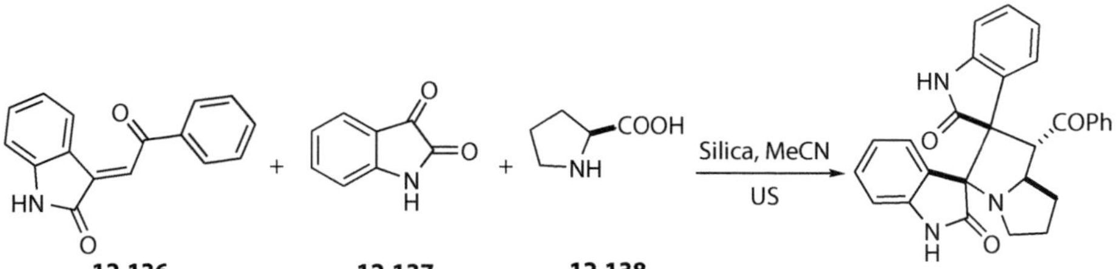

SCHEME 12.39

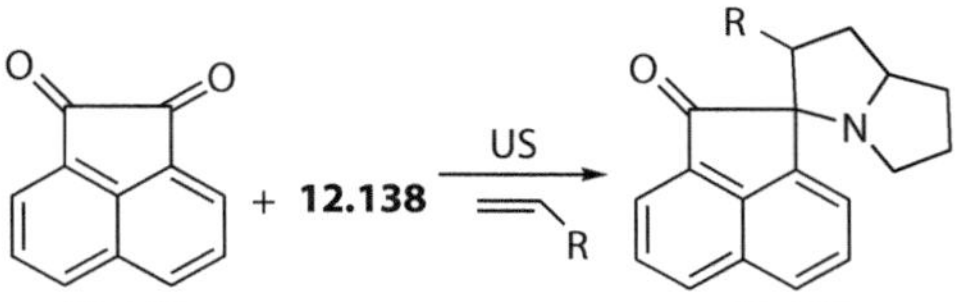

SCHEME 12.40

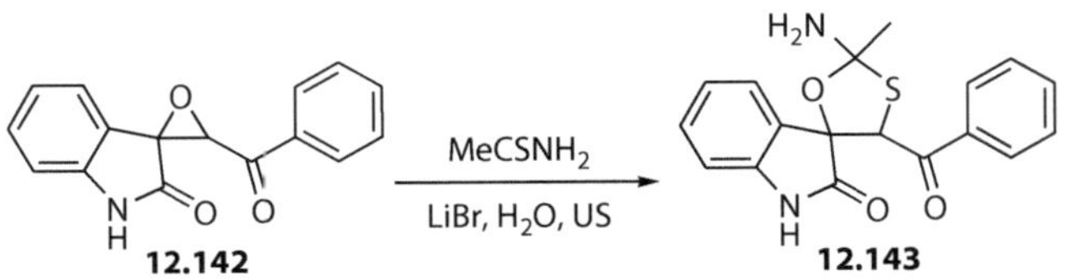

SCHEME 12.41

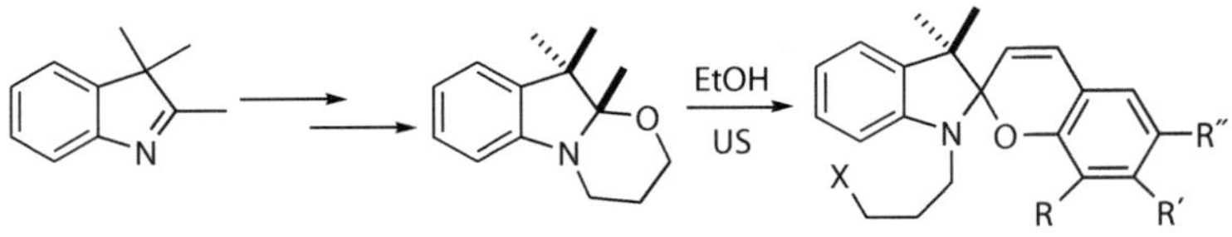

SCHEME 12.42

A ring expansion of spiroepoxides was used for the synthesis of potential biologically active spiro[indole-3,5′-[1,3]oxathiolan]-2-ones under US irradiation (Dandia et al., 2010). In the presence of LiBr and under sonication, aqueous mixture of **12.142** and thioacetamide gave 84% of **12.143** within 7 min, while the same reactions took longer times under stirring or even MW conditions (Scheme 12.41). A plausible mechanism was suggested in which the nucleophilic attack of bromide ion on the less hindered side of the epoxide followed by the reaction of the alkoxide intermediate with acetamide led to the formation of the product.

Beyer and Wagenknecht could significantly improve the synthesis of spiropyrans with the use of sonochemistry. They prepared a wide range of products with iodo, hydroxyl, ethinyl, or azido substituents, which are precursors for the synthesis of molecular switches and dyads (Beyer and Wagenknecht, 2010). Two series of spiropyrans **12.146** with hydroxyl or iodo groups (X = OH or I) were synthesized. While in the key step of conversion of **12.145** into **12.146**, conventional methods gave side products and caused the formation of low yields of the desired compounds, US radiation gave moderate to high yields of **12.146** in shorter time periods (Scheme 12.42). The observed rate enhancement was attributed to more effective mixing of the reagents

Spiro systems with two six-membered rings are also investigated. A variety of pyrazolopyridopyrimidines were obtained by a multicomponent process involving the combination of azomethines, barbituric acid derivatives, and various aromatic aldehydes to synthesize spiro products of type **12.147** under various conditions (Scheme 12.43). Moderate to good yields of products were obtained when the reactions were conducted under US irradiation (Muravyova et al., 2009). Attempts for multicomponent reactions of piperidin-4-one with malononitrile and carbonyl compounds under various conditions to access spiro systems did not furnish outstanding results. Under conventional conditions, Knoevenagel condensation of piperidin-4-one with malononitrile followed by a dimerization process was observed to give **12.148**, while with US irradiation the room-temperature condensation of piperidin-4-one and malononitrile with quinolinone derivatives produced the respective spiro products regioselectively, as exemplified by **12.149** (Dandia et al., 2007) (Scheme 12.43).

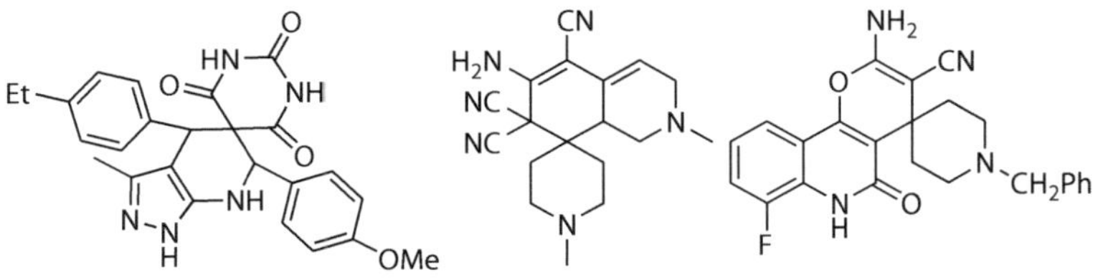

SCHEME 12.43

OXIDATION AND REDUCTION REACTIONS

This is an obvious fact that oxidation and reduction processes are involved in a substantial number of various reactions and transformations in organic chemistry and thus this issue is of prime importance to synthetic chemists. One of the fundamental reactions related to this topic is the oxidation of alcohols since it can provide the access to a broad range of carbonyl compounds. Numerous procedures for oxidation of alcohols are available in the literature. However, the use of stoichiometric amounts of traditionally used metallic oxidants leads to the production of environmentally harmful disposals and therefore the development of new procedures employing safer reagents and energy sources would be of great demand. There are a few reports in the literature on sonochemical oxidation of hydrobenzoin **12.150** to different products (Scheme 12.44). Li et al. (2009b) employed ammonium chlorochromate (ACC)/silica gel system for the oxidation of benzoins **12.151** and cleavage of **12.150** to the corresponding benzils **12.152** and aldehydes (Li et al., 2009c), respectively. A similar procedure was developed by the same group for the oxidation of **12.150** to **12.152** (Li and Sun, 2006) (Scheme 12.44).

Several ultrasonic procedures are reported for the oxidation of hydroxyl moieties into the corresponding carbonyl groups (Olah et al., 1989; Mills and Holland, 1997). A solvent-free oxidation of alcohols was reported to obtain the corresponding carbonyl derivatives. Under the conditions, methanol, secondary alcohols, and hydroquinone were oxidized very efficiently to the corresponding ketones at ambient temperature by $KMnO_4/CuSO_4 \cdot 5H_2O$ system (Luu et al., 2008). The photocatalytic oxidation of 2-propanol to acetone and ethanol to acetaldehyde was studied under US using TiO_2 suspension in aqueous media. A significant reactivity enhancement was observed for the oxidation of 2-propanol to acetone. Irradiation also caused agglomeration of the TiO_2 photocatalyst. The observed enhancement was attributed to US activation of the photocatalyst and improved mass transfer rate (Kado et al., 2001). Permanganate/$CuSO_4 \cdot 5H_2O$ oxidation of benzyl alcohol derivatives proceeded with significantly higher yields of the corresponding carbonyl compounds under sonochemical conditions, when compared to silent experiments. Reactions took place at room temperature in shorter time periods and were applicable to the oxidation of alkylarenes as well (Mečiarova et al., 2000).

In another work, the effect of US on RuI_3-mediated degradation of phenol was studied. Irradiation showed a significant role in the reaction by increasing the catalyst surface area via fragmentation of the catalyst particles, preventing agglomeration of the particles, and increasing the accessibility of phenol and the oxidant to the catalyst active sites. The increase in the efficiency of the irradiated process comparing to the silent reaction was attributed to production of active oxidants like •OH, •HO_2, and •I_2, and better catalytic performance of the catalyst. A free radical mechanism was proposed for the US oxidation reaction and was experimentally supported by the use of radical scavengers (Rokhina et al., 2009).

Oxidation of benzylic carbons to carbonyl group is also investigated under sonochemical conditions such as the report by Cum et al. for the conversion of indane to inden-1-one (Cum et al., 1988). Luzzio and Moore could prepare and study the applications of chromyl chloride in oxidation reactions by US mediation. Under the situation, oxidation of arylalkanes to the corresponding arylalkylketones completed faster than it did in the silent experiment. Similarly, oxidation of *trans*-2-octene gave 3-chloro-2-octanone and 2-chloro-3-octanone. Acceleration of the reactions with US irradiation was attributed to efficient sonochemical homogeneousity and local heating caused by cavitations (Luzzio and Moore, 1993). Other sonochemical oxidations include the Baeyer–Villiger oxidation of cyclohexanone to ε-caprolactone (Zhang et al., 2006c), air oxidation of thiols to disulfides (Ruano et al., 2008),

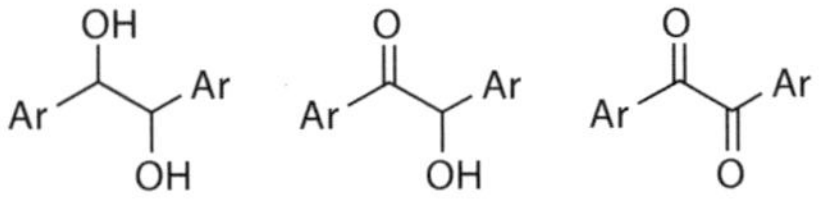

SCHEME 12.44

synthesis of sulfoxides from sulfides (Mahamuni et al., 2006), sonooxidation of 4-piperidinone skeleton (Pétrier et al., 1992), oxidative cleavage of alkenes by RuO_4 (Rup et al., 2010), oxidative coupling of naphthyl derivatives (Bhor et al., 2008), and dehydrogenation reactions (Costa et al., 1999; Memarian and Farhadi, 2008; Memarian and Abdoli-Senejani, 2008;).

Although the reduction of alkenes is very well documented in the literature, there is high interest to find new procedures and experimental conditions for more efficient and selective hydrogenation reactions. One of the useful means for this purpose is the application of US irradiation in the hydrogenation of alkenes. Carcenac et al. conducted room-temperature synthesis of several 1,4-disubstituted cyclohexanes via hydrogenation of sterically hindered and electron poor perfluoroalkyl alkenes using hydrogen at ambient pressure. A dramatic increase in the yield and rate of the reactions was observed when the process was carried out under sonication (Carcenac et al., 2005). Although hydrogenation of fluorinated acrylic acids or their benzyl esters did not proceed efficiently under conventional conditions, sonication of the same mixtures in the presence of Pd/C in methanol improved the reaction significantly (Kitazume et al., 1989). Sonochemical reduction of C=C bonds and desulfurization of benzothiophene was also performed using formic acid as the source of hydrogen at ambient temperature and pressure (Grobas et al., 2007).

Selective hydrogenation of carbonyl compounds to make their respective alcohols, in the presence of competing functional groups, has been a challenge in synthetic organic chemistry for many years and has received considerable attention due to its numerous applications in medicinal and industrial chemistry. As a result, growing numbers of diverse methods are developed for the reduction of various carbonyl compounds. Several reports on the application of US in reduction of carbonyls exist in the literature. Hydrogenation of cinnamaldehyde to obtain cinnamyl alcohol was performed with excellent selectivity at atmospheric pressure using a Ru catalyst under sonication (Li et al., 2006d). $Rh(PPh_3)_3Cl$-catalyzed hydrosilylation of alkyl substituted cyclohexanones was carried out with high yields and selectivities under sonochemical conditions (Felföldi et al., 2000). Similar procedures were also developed for room-temperature reduction of benzophenones derivatives to benzhydrols (Peng et al., 2005b), enantioselective hydrogenation of α-ketoesters to hydroxy derivatives (Török et al., 2000), reductive deoxygenation of ketones and aldehydes to olefins (Nayak and Banerji, 1991), and conversion of carboxylates to ketones by Barbier reaction (Aurell et al., 1995).

Nitroaromatics are industrially produced in large quantities and are employed as the precursors for the synthesis of a variety of other useful organic products. Depending on the conditions, various nitrogen-containing compounds such as amines, hydroxylamines, and hydrazines could be obtained from the reduction of nitroaromatics and therefore, the control of the selectivity during their manipulation is a great challenge. By the use of hydrazine and in the presence of Raney nickel, high-intensity US could reduce bulky nitroaromatic compounds to their respective amines in a much faster fashion than the silent conditions can cause (Heropoulos et al., 2005). Similarly, zinc/NH_4Cl reduction of nitroaromatics rapidly afforded the respective arylhydroxylamines (Ung et al., 2005). Alternatively, zinc and ammonium chloride were used under high-intensity US for chemoselective reduction of nitroarenes to azo and azoxy compounds (Cravotto et al., 2006).

Coupling of aldehydes and ketones to their respective pinacol equivalents is reported in several occasions. These include the use of various reductive systems based on magnesium (Li et al., 2002; Li et al., 2005a; Wang et al., 2005a), zinc (Yang et al., 2004), and titanium salts in combination with different metals (Li et al., 2005c; Lin et al., 2006), aluminum (Bian et al., 2002), neodymium (Bian et al., 2006c), lanthanum (Bian et al., 2006b), and vanadium (Bian et al., 2006c). An amino pinacolization version of the reaction was performed by dimerization of imines to 1,2-diamines using US irradiation and lithium metal in ethereal lithium perchlorate solution (Mojtahedi et al., 2001). Pinacolization of benzophenone was also studied under photochemical/US irradiation conditions. It was observed that by simultaneous use of sonication and UV irradiation, the rates and yields of the reaction increase. This was attributed to better photoconversion of benzophenone caused by sonolytic decomposition of the light-absorbing transient species and formation of the triplet state quenching as a result of a better collisional deactivation process (Gaplovsky et al., 2000).

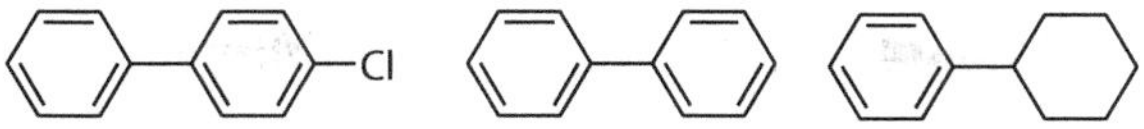

SCHEME 12.45   12.153   12.154   12.155

There are some other scattered US-mediated reductions of organic compounds in the literature like the reductive deacetoxylation of Passerini adducts with zinc/NH_4Cl in methanol to afford the respective β-keto amides (Neo et al., 2005) and reductive deoxygenation of alkyl halides to alkyl hydroperoxides (Nakamura et al., 1995). Also, conversion of chlorinated biphenyls has been studied under US conditions using Raney Ni–Al alloy in alkali hydroxides and carbonate aqueous solutions. For example, formation of mixture of **12.154** and **12.155** from dechlorination of **12.153** was selectively directed to solely obtain **12.154** by controlling the amount of NaOH in the reaction mixture (Scheme 12.45). Under silent conditions, the dechlorination did not reach to completion even after long reaction times (Liu et al., 2009).

PROTECTION AND DEPROTECTION PROCEDURES

Protection and deprotection of organic functional groups play essential roles in accomplishing multistep syntheses (Wuts and Green, 2007). Depending on the simplicity of the process, ease of operation and workup, yield of the desired product and its stability to the reaction conditions, and the overall expenses of the process, suitable conditions will be chosen. On these grounds, diverse arrays of different methods are developed during the years for protection/deprotection of various functional groups. Accordingly, several US-mediated functional group protection strategies are offered in the literature. This section covers the related reports for the masking of hydroxyl groups, amines, and various carbonyl functions.

Protection of hydroxyl functionalities is of fundamental importance in both organic synthesis and analytical chemistry. There is at least one-step protection of alcohols during many of multistep syntheses and organic transformations. One of the most common approaches for the protection of alcohols is the conversion of OH groups into their corresponding silyl ethers (Mojtahedi et al., 2006a,b, 2008b; Kadam and Kim, 2010; Weickgenannt et al., 2010). Gholap et al. (2003) used US irradiation to rapidly obtain excellent yields of various alkyl acetates via the reactions of the corresponding alcohols with acetic anhydride at ambient conditions in the absence of any additive using an ionic liquid medium. After isolation of the products, the ionic liquid was recycled and reused efficiently. Singh et al. (2006) used catalytic amounts of ionic liquids for monotetrahydropyranylation of diols and alcohols under US irradiation. Comparison of the results with those obtained in the absence of irradiation showed the efficiency of the method, while the ionic liquid was also recycled.

Another ultrasonic method was introduced for efficient TMS protection of various types of alcohols and phenols by HMDS at room temperature without the use of any solvent or additive. Under the conditions, competitive protections resulted in good-to-excellent chemoselectivity in favor of sterically less hindered alcohols. In addition, phenols could also be exclusively protected in the presence of aromatic amines (Mojtahedi et al., 2007b). The synthesis of **12.157** and its subsequent reaction with phenols were conveniently conducted under power US to obtain protected phenolic ether **12.158** within 15 min (Scheme 12.46). In the case of multisubstituted phenols including sensitive phenolic aldehydes, reactions showed much more efficiency than silent procedures. In addition, no use of undesired halogenated reagents was involved and the operational and workup procedures were not cumbersome (Luzzio and Chen, 2008).

N-tert-Butyloxycarbonylation (*N*-Boc) is perhaps the most commonly used group to protect the amine functions due the stability and the ease of protection and deprotection associated with this group. Two different procedures are offered for US-mediated installation of the *N*-Boc onto the

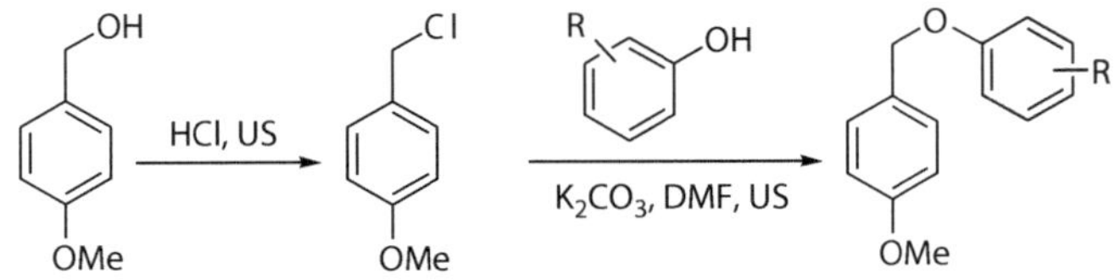

SCHEME 12.46

amine functional moieties. First, conventional reactions of primary and secondary amines with di-*tert*-butyl dicarbonate and $NaHCO_3$ in alcoholic solvents took 24 h to complete, while irradiation of the same mixtures gave complete formation of the protected products within a few minutes (Einhorn et al., 1991). In the second work, the chemoselective *N*-Boc protection of various aliphatic, aromatic, acyclic, and heterocyclic amines was performed with excellent yields in the presence of catalytic amounts of sulfamic acid. Under the conditions, the use of sterically hindered and electron-poor amines in the reaction was also successful (Upadhyaya et al., 2007).

Carbonyl moieties are protected in different forms such as oximes, arylhydrazones, dithianes, imines, and enamines. The protection practice is important from the synthetic point of application as well since the masked products in many cases are the precursors for the preparation of many other synthetically important compounds. Oximes are traditionally prepared via treatment of alcoholic solutions of aldehydes or ketones with hydroxylamine hydrochloride and pyridine under refluxing conditions. However, toxicity and flammability risks associated with the use of pyridine impose limitations on such methods and therefore the development of safer methods would be of high demand. Accordingly, sonochemistry has served as a useful mean to access milder methods for the synthesis of oximes from carbonyl compounds and vice versa (Li et al., 2006c). Condensation of aldehydes and ketones with hydroxylamine hydrochloride was efficiently promoted under the influence of basic [bmim]OH ionic liquid. The efficiency of the method was illustrated by obtaining much lower efficiency for the same reaction under silent conditions (Zang et al., 2009). A deprotection method was also developed under US conditions to convert oximes to their corresponding carbonyl compounds using a silica sulfuric acid/surfactant/paraformaldehyde system in water (Li et al., 2010d).

Protection of aldehydes/ketones as arylhydrazone moieties is also studied under US irradiation. Leite at al. converted various aromatic aldehydes to their corresponding hydrazides using acidic aqueous conditions. Reactions proceeded at room temperature and produced high yields of various arylhydrazones (Leite et al., 2008). Excellent yields of arylhydrazones of various aromatic and aliphatic carbonyls were also obtained at ambient temperature by Jarikote et al. (2003) without using any additive. The efficiency of the process was shown to be strongly dependent on the amount of molecular oxygen present in the atmosphere of the reaction.

Synthesis of various imines **12.159**, mainly from aromatic aldehydes and primary amines, was facilitated by US irradiation in the presence of several solid catalysts (Scheme 12.47). Under the conditions, silica showed the best performance to give good yields of the products. The reaction was applicable to high-scale synthesis of imines as well (Guzen et al., 2007). Various amines were also used by Brandt et al. for solvent-free synthesis of β-enamino esters **12.160** from the corresponding 1,3-dicarbonyl compounds in the presence of acetic acid and US irradiation (Brandt et al., 2004). In another work, an US-mediated procedure was developed by Duarte et al. for the protection of aromatic acid chlorides as thioesters **12.161** starting from benzoyl chlorides and 2-mercaptobenzoxa(thia) zoles (Duarte et al., 2010) (Scheme 12.47). A rapid, practical, and base-free sonochemical protection procedure was reported by Sureshbabu et al. (2008a) by the synthesis of N^α-protected thiopeptide esters **12.162** from their corresponding peptide esters using P_2S_5 (Sureshbabu et al., 2008a). Montmorillonite-K10 supported deprotection of 1,3-dithianes and 1,3-dithiolanes **12.163** gave high yields of the corresponding aldehydes and ketones using $Cu(NO_3)_2 \cdot 2.5H_2O$ and US waves at room temperature (Oksdath-Mansilla and Peñéñory, 2007) (Scheme 12.47).

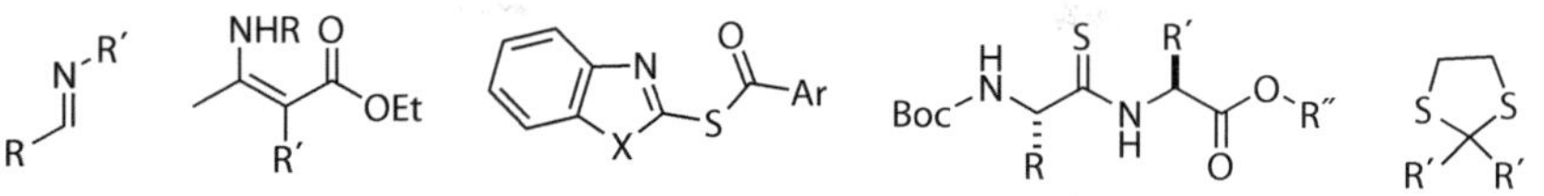

SCHEME 12.47

12.159 **12.160** **12.161** **12.162** **12.163**

CARBON–HETEROATOM (C–X) BOND FORMATION REACTIONS

The presence of heteroatoms such as nitrogen, oxygen, silicone, halides, and sulfur in the structure of organic molecules is found to be responsible for diverse functions of pharmaceuticals, proteins, polymers, carbohydrates, and other naturally occurring and synthetic compounds (Hartwig, 2008). This section focuses on summarizing the US-mediated procedures for the construction of carbon–heteroatom bond formation reactions.

The carbon–oxygen bond formation reaction has major applications in the synthesis of ethers, O-alkylation reactions, and protection of functional groups such as enols, alcohols, and oximes. The Williamson synthesis, Ullman-type reaction, and Mitsunobo reaction are the main pathways usually employed for the preparation of ethers. In 1992, an Ullmann-type procedure was reported to give moderate yields of diaryl ethers **12.164** under US irradiation (Smith and Jones, 1992) (Scheme 12.48). Almost a decade later, Peng and Song (2002) took the advantage of the use of combined MW and US irradiation for the Williamson synthesis of a variety of **12.164** and benzyl aryl ethers **12.165** in an aqueous sodium hydroxide medium (Scheme 12.48). The process took place in the absence of the additives, the use of which has been necessary in many other related procedures. A Mitsunobu procedure was used for monoalkylation of dihydroxycoumarins under sonochemical conditions to selectively produce 6- or 7-O-protected dihydroxycoumarins **12.166** in good yields (Cravotto et al., 2003b). Other related procedures include selective mono O-alkylation of calyx[6]arenas (Semwal et al., 2002), synthesis of oxime ethers (Li et al., 2009d), O-alkylation of N-hydroxyphthalimide (Wang et al., 2008b), conversion of 2-chlorobenzoic acids to their respective salicylic acid derivatives (Docampo Palacios and Pellón Comdom, 2003a), and alkylation of furoin (Zhang et al., 2007b).

Ultrasonic carbon–nitrogen bond formation reactions have wide applications in alkylation of amines (Li et al., 2000) and synthesis of N-alkylated heterocycles (Calvino-Casilda et al., 2004, 2008; Zhao et al., 2010). The strategy specially finds importance in N-alkylation of imidazole rings in the synthesis of a diverse array of N-alkyl imidazolium–based ionic liquids (Lévêque et al., 2002; López-Pestaña et al., 2002; Namboodiri and Varma, 2002; Durán-Valle et al., 2004; Costarrosa et al., 2006; Estager et al., 2007b; Cravotto et al., 2007, 2008; Calvino-Casilda et al., 2008; Ferrera-Escudero et al., 2010), a green and reusable group of solvents with many uses in synthetic organic chemistry. Various other approaches are used in combination with US techniques to convert carbon–hydrogen (Liu and Ji, 2008a), carbon–oxygen (Peng and Song, 2001), and carbon–halogen bonds to carbon–nitrogen bonds. In this regard, nucleophilic addition of amines to carbon–halogen bonds is employed in several cases (Balan et al., 2009). The substitution of chlorine atom in dichlorobenzene iron η^6-complex **12.167** with various secondary amines produced aromatic amines **12.170** and **12.171** (Scheme 12.49). In comparison with thermal conditions, reactions proceeded considerably faster under sonication. By the variation of the solvent and the nucleophilicity of amines, the desired mono- or di-substituted products were synthesized (Raouafi et al., 2009).

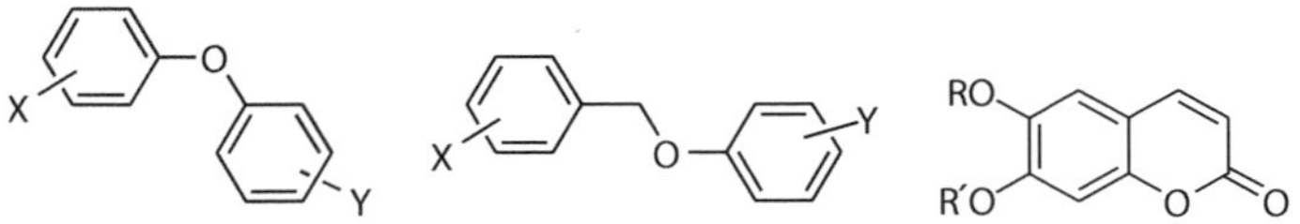

12.164 **12.165** **12.166**

SCHEME 12.48

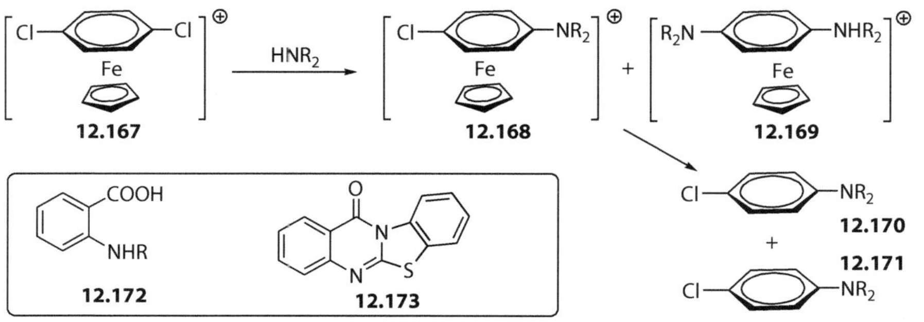

SCHEME 12.49

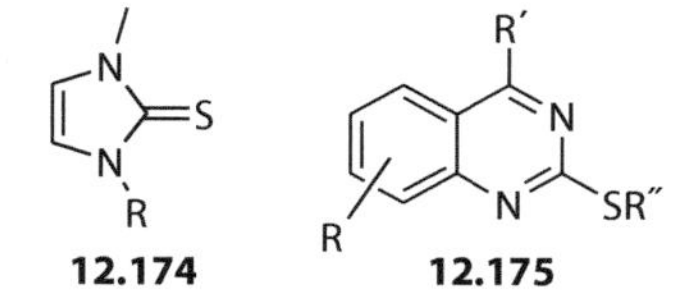

SCHEME 12.50

Pellon et al. used US irradiation for the synthesis of various 2-alkylamino derivatives of benzoic acid **12.172** (R = alkyl) from the corresponding 2-chloro substrates via the Ullmann condensation (Scheme 12.49). Under the conditions and in aqueous mixtures, high yields of the products were achieved within short time periods (Pellón et al., 2005). Similarly, the 2-arylamino derivatives of **12.172** (R = Ar) were obtained using the same procedure (Docampo Palacios and Pellón Comdom, 2003b). The application of the methodology was also used by the same group for one-pot synthesis of thiazoloquinazolinone derivatives **12.173** (Pellón et al., 2007) (Scheme 12.49).

Other C-heteroatom formation reactions include the synthesis of carbon–sulfur and carbon–halogen bonds. 1,3-Dialkylimidazole-2-thiones **12.174** were prepared via cathodic reduction of 1,3-dialkylimidazolium ionic liquids followed by the reaction of the carbine intermediates with sulfur under US irradiation (Scheme 12.50). Reaction proceeded cleanly with no use of any other additive while formation of side-products was not observed (Feroci et al., 2009). A variety of heteroaromatic thiols **12.175** were selectively S-alkylated via the reaction with alkyl bromides and iodides under sonochemical irradiation (Deligeorgiev et al., 2010).

Sadeghi et al. reported rapid sonochemical fluorination of methine and methylene groups neighbored by a nitro function and a heterocyclic ring to obtain the respective mono- or di-fluorinated products. Reactions took place using 1-chloromethyl-4-fluoro-1,4-diazoniabicyclo[2,2,2]octane bis-tetrafluoroborate (Selectfluor) in the presence of DBU (Sadeghi et al., 2006a,b). Again, by using Selectfluor and a room-temperature desilylation–fluorination process, alkenyl fluorides were selectively synthesized from alkenyltrimethylsilanes using US. The same reaction proceeded with much lower efficiency in the absence of irradiation (Ranjbar-Karimi, 2010). Other sonolytic halogenations include the halogenation of 3,5-dimethyl Pyrazoles (Stefani et al., 2005) and halogenation of alcohol derivatives with *tert*-butyl halides (Ranu and Jana, 2005).

CONCLUSIONS

It appears that sonochemistry has a long way to go yet, although it has more than eight decades of history. Organic chemists have enjoyed the convenience associated with the use of sonochemistry to overcome the synthetic limitations in many cases, enhance the reactivity of various transformations, and gain improved selectivities from the reactions, which otherwise proceed with much lower efficiency. However, the real power of the US lies on its potential to switch the reactions to obtain surprising products and unexpected selectivities through "true sonochemistry." Thus, it seems that this is "physical organic sonochemistry" that can help synthetic chemists design successful strategies to overcome current barriers in the synthesis of molecular targets. The use of sonochemistry in combination with other green chemistry techniques (aqueous media, ionic liquids, organocatalysts, etc.) needs more explorations and could lead to new horizons of surprising results.

In this chapter, it was tried to comprehensively cover the US-mediated studies in synthetic organic chemistry. As appeared in the earlier sections, the majority of the papers published in the last two decades, especially those that were not included in the previous reviews, could fall into the selected categories of the reactions. As far as we could investigate using search engines, not too many recent papers on synthetic applications of US are left outside the selected categories. Perhaps a few scattered papers on some reactions such as alkylation (Hofmann et al., 2003; Ji et al., 2005), dealkylation (Katohgi et al., 2000; Katohgi and Togo, 2001), desulfonylation (Addie et al., 2000), acylation (Mehrabi, 2008), dehydration (Aquino et al., 2005), rearrangement (Sureshbabu et al., 2008b), aldol (Cravotto et al., 2003c; Ji-Tai et al., 2004), and Wittig reactions (Riccaboni et al., 2010) could be mentioned as well. The extension of the sonochemistry to other synthetic transformations remains to be explored.

REFERENCES

Abaee, M.S., Hamidi, V., and Mojtahedi, M.M. 2008b. Ultrasound promoted aminolysis of epoxides in aqueous media: A rapid procedure with no pH adjustment for additive-free synthesis of α-aminoalcohols. *Ultrasonics Sonochemistry*, 15: 823–827.

Abaee, M.S., Mojtahedi, M.M., Abbasi, H., and Fatemi, E.R. 2008a. Additive-free thiolysis of epoxides in water: A green and efficient regioselective pathway to β-hydroxy sulfides. *Synthetic Communications*, 38: 282–289.

Abaee, M.S., Mojtahedi, M.M., Forghani, S., Ghandchi, N.M., Forouzani, M., Sharifi, R., and Chaharnazm, B. 2009. A green, inexpensive and efficient organocatalyzed procedure for aqueous aldol condensations. *Journal of the Brazilian Chemical Society*, 20: 1895–1900.

Abaee, M.S., Mojtahedi, M.M., Hamidi, V., Mesbah, A.W., and Massa, W. 2008c. The first synthesis of bis(arylmethylidene)dioxan-5-ones: Potential scaffolds to access vicinal tricarbonyl derivatives. *Synthesis*, 2122–2126.

Abaee, M.S., Mojtahedi, M.M., Sharifi, R., Zahedi, M.M., Abbasi, H., and Tabar-Heidar, K. 2006. Facile synthesis of bis(arylmethylidene)cycloalkanones mediated by lithium perchlorate under solvent-free conditions. *Journal of the Iranian Chemical Society*, 3: 293–296.

Abaee, M.S., Mojtahedi, M.M., Sharifi, R., and Zahedi, M.M. 2007a. A highly efficient method for solvent-free synthesis of bis(arylmethylidene)piperidinones. *Journal of Heterocyclic Chemistry*, 44: 1497–1499.

Abaee, M.S., Mojtahedi, M.M., and Zahedi, M.M. 2005. An efficient and improved method for the synthesis of bis(arylmethylidene)thiopyranones. *Synlett*, 2317–2320.

Abaee, M.S., Mojtahedi, M.M., Zahedi, M.M., Sharifi, R., and Khavasi, H. 2007b. Efficient synthesis of novel 3-substituted thiopyran-4-ones. *Synthesis*, 39: 3339–3344.

Abaee, M.S., Mojtahedi, M.M., Zahedi, M.M., and Sharifi, R. 2007c. A highly efficient method for solvent-free synthesis of bisarylmethylidenes of pyranones and thiopyranones. *Heteroatom Chemistry*, 18: 44–49.

Abbiati, G., Clerici, F., Gelmi, M.L., Gambini, A., and Pilati, T. 2001. Asymmetric synthesis of 2-amino-3-hydroxynorbornene-2-carboxylic acid derivatives. *Journal of Organic Chemistry*, 66: 6299–6304.

Addie, M.S. and Taylor, R.J.K. 2000. New routes to 5-substituted oxazoles, *Journal of the Chemical Society, Perkin Transactions 1*, 31: 527–531.

Ahluwalia, V.K., Goyal, B., and Das, U. 1997. One-pot syntheses of 5-oxo-1,4,5,6,7,8-hexahydroquinolines and pyrimido[4,5-b]quinolines using microwave irradiation and ultrasound. *Journal of Chemical Research*, 28: 266.

Alfaro, R. and McClusky, J.V. 2001. Ultrasonic irradiation and the homo– and hetero–Diels–Alder reaction. *Synthetic Communications*, 31: 2513–2522.

Almeida, W.P. and Coelho, F. 1998. Piperonal as electrophile in the Baylis–Hillman reaction. A synthesis of hydroxy-β-piperonyl-γ-butyrolactone derivative. *Tetrahedron Letters*, 39: 8609–8612.

Al-Zaydi, K.M. 2009a. A simplified Green Chem. approaches to synthesis of 2-substituted 1,2,3-triazoles and 4-amino-5-cyanopyrazole derivatives conventional heating versus microwave and ultrasound as eco-friendly energy sources. *Ultrasonics Sonochemistry*, 16: 805–809.

Al-Zaydi, K.M. 2009b. Microwave and ultrasound promoted synthesis of substituted new arylhydrazono pyridinones. *Arabian Journal of Chemistry*, 2: 55–58.

Al-Zaydi, K.M., Borik, R.M., and Elnagdi, M.H. 2009. Studies with arylhydrazonopyridinones: Synthesis of new arylhydrazono thieno[3,4-c]pyridinones as novel D2T2 dye class; classical verse green methodologies. *Ultrasonics Sonochemistry*, 16: 660–668.

Ambulgekar, G.V., Bhanage, B.M., and Samant, S.D. 2005. Temperature recyclable catalyst for Heck reactions using ultrasound. *Tetrahedron Letters*, 46: 2483–2485.

Aquino, F., Bonrath, W., Paz Schmidt, R.A., and Schiefer, G. 2005. Dehydration reaction of hydroxenin monoacetate in carbon tetrachloride and an aliphatic alcohol under ultrasound irradiation. *Ultrasonics Sonochemistry*, 12: 107–114.

Arcadi, A., Alfonsi, M., and Marinelli, F. 2009. Facile reaction of thiols and amines with alkyl 4-hydroxy-2-alkynoates in water under neutral conditions and ultrasound irradiation. *Tetrahedron Letters*, 50: 2060–2064.

Aurell, M.J., Danhui, Y., Einhorn, J., Einhorn, C., and Luche, J.L. 1995. A direct access to ketones from lithium carboxylates via the sonochemical Barbier reaction. *Synlett*, 459–460.

Avalos, M., Babiano, R., Bravo, J.L., Cabello, N., Cintas, P., Hursthouse, M.B., Jiménez, J.L., Light, M.E., and Palacios, J.C. 2000. Sonochemical cycloadditions of *o*-quinones. The search for a cation radical pathway. *Tetrahedron Letters*, 41: 4101–4105.

Babu, S.A., Yasuda, M., Shibata, I., and Baba, A. 2005. In- or In(I)-employed tailoring of the stereogenic centers in the Reformatsky-type reactions of simple ketones, α-alkoxy ketones, and β-keto esters. *Journal of Organic Chemistry*, 70: 10408–10419.

Balan, A.M., Florea, O., Moldoveanu, C., Zbancioc, G., Iurea, D., and Mangalagiu, I.I. 2009. Diazinium salts with dihydroxyacetophenone skeleton: Syntheses and antimicrobial activity. *European Journal of Medicinal Chemistry*, 44: 2275–2279.

Bang, K., Lee, K., Park, Y.K., and Lee, P.H. 2002. Sonochemical Reformatsky reaction using indium. *Bulletin of the Korean Chemical Society*, 23: 1272–1276.

Barkakaty, B., Takaguchi, Y., and Tsuboi, S. 2006. Addition of $CFCl_3$ to aromatic aldehydes under ultrasonic irradiation. *Synthesis*, 959–962.

Bazgir, A., Ahadi, S., Ghahremanzadeh, R., Khavasi, H.R., and Mirzaei, P. 2010. Ultrasound-assisted one-pot, three-component synthesis of spiro[indoline-3,4′-pyrazolo[3,4-b]pyridine]-2,6′ (1′H)-diones in water. *Ultrasonics Sonochemistry*, 17: 447–452.

Beyer, C. and Wagenknecht, H.A. 2010. Synthesis of spiropyrans as building blocks for molecular switches and dyads. *Journal of Organic Chemistry*, 75: 2752–2755.

Bhor, M.D., Nandurkar, N.S., Bhanushali, M.J., and Bhanage, B.M. 2008. Ultrasound promoted selective synthesis of 1,1′-binaphthyls catalyzed by Fe impregnated pillared montmorillonite K10 in presence of TBHP as an oxidant. *Ultrasonics Sonochemistry*, 15: 195–202.

Bian, Y.J., Fan, C.R., Hu, X.H., and Li, J.T. 2006a. Zinc-mediated allylation reactions of aldehydes and ketones in aqueous media under ultrasonic irradiation. *Indian Journal of Chemistry—Section B Organic and Medicinal Chemistry*, 45: 1587–1590.

Bian, Y.J., Liu, S.M., Li, J.T., and Li, T.S. 2002. Pinacol coupling of aromatic aldehydes using aluminium under ultrasound irradiation. *Synthetic Communications*, 32: 1169–1173.

Bian, Y.J., Wang, H.L., Wu, B., and Li, J.T. 2006. Studies on the neodymium induced pinacol coupling of aromatic aldehydes and ketones in aqueous media. *Chinese Chemical Letters*, 17: 501–594.

Bian, Y.J., Xue, W.L., and Yu, X.G. 2010. The allylation reactions of aromatic aldehydes and ketones with tin dichloride in water. *Ultrasonics Sonochemistry*, 17: 58–60.

Bian, Y.J., Yu, X.G., Peng, H.W., and Li, J.T. 2006b. Studies on the lanthanum-induced pinacol coupling of aromatic aldehydes and ketones in aqueous media. *Synthetic Communications*, 36: 2513–2518.

Bian, Y.J., Zhang, J.Y., and Li, J.T. 2007. Allylation reactions of aromatic aldehydes and ketones with lithium in THF under ultrasonic irradiation. *Journal of Chemical Research*, 162–163.

Bian, Y.J., Zhao, H.M., and Yu, X.G. 2009. Allylation reactions of aromatic aldehydes with antimony in aqueous media under ultrasonic irradiation. *Synthetic Communications*, 39: 2370–2377.

Braghiroli, F.L., Barboza, J.C.S., and Serra, A.A. 2006. Sonochemical epoxidation of cyclohexene in R-CN/H_2O_2 system. *Ultrasonics Sonochemistry*, 13: 443–445.

Brandt, C.A., Da Silva, A.C.M.P., Pancote, C.G., and Brito, C.L. 2004. Efficient synthetic method for β-enamino esters using ultrasound. *Synthesis*, 1557–1559.

Brase, S., Enders, D., Kobberling, J., and Avemaria, F. 1998. A surprising solid-phase effect: Development of a recyclable "traceless" linker system for reactions on solid support. *Angewandte Chemie-International Edition*, 37: 3413–3415.

Bremner, D. 1990. In *Advances in Sonochemistry*, ed. T.J. Mason. London, U.K.: JAI Press.

Brennen, C.E. 1995. Cavitation and bubble dynamics. Oxford, U.K.: Oxford University Press.

Bruckmann, A., Krebs, A., and Bolm, C. 2008. Organocatalytic reactions: Effects of ball milling, microwave and ultrasound irradiation. *Green Chemistry*, 10: 1131–1141.

Calvino, V., Picallo, M., López-Peinado, A.J., Martín-Aranda, R.M., and Durán-Valle, C.J. 2006. Ultrasound accelerated Claisen–Schmidt condensation: A green route to chalcones. *Applied Surface Science*, 252: 6071–6074.

Calvino-Casilda, V., López-Peinado, A.J., Martín-Aranda, R.M., Ferrera-Escudero, S., and Durán-Valle, C.J. 2004. Ultrasound-promoted N-propargylation of imidazole by alkaline-doped carbons. *Carbon*, 42: 1363–1366.

Calvino-Casilda, V., Martín-Aranda, R.M., López-Peinado, A.J., Bejblová, M., and Čejka, J. 2008. Sonocatalysis and zeolites: An efficient route to prepare N-alkylimidazoles. Kinetic aspects. *Applied Catalysis A: General*, 338: 130–135.

Carcenac, Y., Tordeux, M., Wakselman, C., and Diter, P. 2005. Convenient synthesis of fluorinated alkanes and cycloalkanes by hydrogenation of perfluoroalkylalkenes under ultrasound irradiation. *Journal of Fluorine Chemistry*, 126: 1347–1355.

Caulier, T.P. and Reisse, J. 1996. On sonochemical effects on the Diels–Alder reaction. *Journal of Organic Chemistry*, 61: 2547–2548.

Cella, R., Orfão, A.T.G., and Stefani, H.A. 2006. Palladium-catalyzed cross-coupling of vinylic tellurides and potassium vinyltrifluoroborate salt: Synthesis of 1,3-dienes. *Tetrahedron Letters*, 47: 5075–5078.

Cella, R. and Stefani, H.A. 2006. Ultrasound-assisted synthesis of Z and E stilbenes by Suzuki cross-coupling reactions of organotellurides with potassium organotrifluoroborate salts. *Tetrahedron*, 62: 5656–5662.

Cella, R. and Stefani, H.A. 2009. Ultrasound in heterocycles chemistry. *Tetrahedron*, 65: 2619–2641.

Chimni, S.S., Bala, N., Dixit, V.A., and Bharatam, P.V. 2010. Thiourea catalyzed aminolysis of epoxides under solvent free conditions. Electronic control of regioselective ring opening. *Tetrahedron*, 66: 3042–3049.

Chtourou, M., Abdelhédi, R., Frikha, M.H., and Trabelsi, M. 2010. Solvent free synthesis of 1,3-diaryl-2-propenones catalyzed by commercial acid-clays under ultrasound irradiation. *Ultrasonics Sonochemistry*, 17: 246–249.

Cintas, P. and Luche, J.L. 1999. The sonochemical approach. *Green Chemistry*, 1999, 115–125.

Coelho, F., Almeida, W.P., Veronese, D., Mateus, C.R., Silva Lopes, E.C., Rossi, R.C., Silveira, G.P.C., and Pavam, C.H. 2002. Ultrasound in Baylis–Hillman reactions with aliphatic and aromatic aldehydes: Scope and limitations. *Tetrahedron*, 58: 7437–7447.

Coelho, F., Diaz, G., Abella, C.A.M., and Almeida, W.P. 2006. The Baylis–Hillman reaction with chiral α-amino aldehydes under racemization-free conditions. *Synlett*, 435–439.

Costa, S.C.P., Moreno, M.J.S.M., Sae Melo, M.L., and Campos Neves, A.S. 1999. Ultrasound assisted remote functionalization of non-activated carbon atoms: Efficient in situ formation of tetrahydrofurans by sonolysis of bromohydrins with (diacetoxyiodo)benzene/I_2. *Tetrahedron Letters*, 40: 8711–8714.

Costarrosa, L., Calvino-Casilda, V., Ferrera-Escudero, S., Durán-Valle, C.J., and Martín-Aranda, R.M. 2006. Alkylation of imidazole under ultrasound irradiation over alkaline carbons. *Applied Surface Science*, 252: 6089–6092.

Cravotto, G., Beggiato, M., Penoni, A., Palmisano, G., Tollari, S., Lévêque, J.M., and Bonrath, W. 2005a. High-intensity ultrasound and microwave, alone or combined, promote Pd/C-catalyzed aryl-aryl couplings. *Tetrahedron Letters*, 46: 2267–2271.

Cravotto, G., Boffa, L., Bia, M., Bonrath, W., Curini, M., and Heropoulos, G.A. 2006. An easy access to aromatic azo compounds under ultrasound/microwave irradiation. *Synlett*, 2605–2608.

Cravotto, G., Boffa, L., Lévêque, J.M., Estager, J., Draye, M., and Bonrath, W. 2007. A speedy one-pot synthesis of second-generation ionic liquids under ultrasound and/or microwave irradiation. *Australian Journal of Chemistry*, 60: 946–950.

Cravotto, G., Chimichi, S., Robaldo, B., and Boccalini, M. 2003b. Monoalkylation of dihydroxycoumarins via Mitsunobu dehydroalkylation under high intensity ultrasound. The synthesis of ferujol. *Tetrahedron Letters*, 44: 8383–8386.

Cravotto, G. and Cintas, P. 2006. Power ultrasound in organic synthesis: Moving cavitational chemistry from academia to innovative and large-scale applications. *Chemical Society Reviews*, 35: 180–196.

Cravotto, G. and Cintas, P. 2007. The combined use of microwaves and ultrasound: Improved tools in process chemistry and organic synthesis. *Chemistry European Journal*, 13: 1902–1909.

Cravotto, G., Demetri, A., Nano, G.M., Palmisano, G., Penoni, A., and Tagliapietra, S. 2003c. The aldol reaction under high-intensity ultrasound: A novel approach to an old reaction. *European Journal of Organic Chemistry*, 4438–4444.

Cravotto, G., Gaudino, E.C., Boffa, L., Lévêque, J.M., Estager, J., and Bonrath, W. 2008. Preparation of second generation ionic liquids by efficient solvent-free alkylation of N-heterocycles with chloroalkanes. *Molecules*, 13: 149–156.

Cravotto, G., Nano, G.M., Palmisano, G., and Tagliapietra, S. 2003a. The reactivity of 4-hydroxycoumarin under heterogeneous high-intensity sonochemical conditions. *Synthesis*, 1286–1291.

Cravotto, G., Palmisano, G., Tollari, S., Nano, G.M., and Penoni, A. 2005b. The Suzuki homocoupling reaction under high-intensity ultrasound. *Ultrasonics Sonochemistry*, 12: 91–94.

Cum, G., Gallo, R., Spadaro, A., and Galli, G. 1988. Effect of static pressure on the ultrasonic activation of chemical reactions. Selective oxidation at benzylic carbon in the liquid phase. *Journal of the Chemical Society, Perkin Transactions 2*, 375–383.

Dalpozzo, R., Nardi, M., Oliverio, M., Paonessa, R., and Procopio, A. 2009. Erbium(III) triflate is a highly efficient catalyst for the synthesis of β-alkoxy alcohols, 1,2-diols and β-hydroxy sulfides by ring opening of epoxides. *Synthesis*, 3433–3438.

Dalvi, N.R., Karale, B.K., and Gill, C.H. 2005. Synthesis of 3-[2-(4-nitrophenyl)vinyl]-chromon-4-one by ultrasonic irradiation. *Indian Journal of Chemistry—Section B Organic and Medicinal Chemistry*, 44: 1522–1523.

Dalvi, N.R., Shelke, S.N., Karale, B.K., and Gill, H.C. 2007. Synthesis of 1-(2-hydroxy-phenyl)-3-piperidin-1-yl-propenone by ultrasonic irradiation. *Synthetic Communications*, 37: 1421–1424.

Dandia, A., Gautam, S., and Jain, A.K. 2007. An efficient synthesis of fluorine-containing substituted spiro[piperidine-4,4′-pyrano[3,2-c]quinoline]-3′-carbonitrile by nonconventional methods. *Journal of Fluorine Chemistry*, 128: 1454–1460.

Dandia, A., Singh, R., and Bhaskaran, S. 2010. Ultrasound promoted greener synthesis of spiro[indole-3,5′-[1,3] oxathiolanes in water. *Ultrasonics Sonochemistry*, 17: 399–402.

Deligeorgiev, T., Kaloyanova, S., Lesev, N., and Vaquero, J.J. 2010. An easy and fast ultrasonic selective S-alkylation of hetaryl thiols at room temperature. *Ultrasonics Sonochemistry*, 17: 783–788.

Deshmukh, R.R., Rajagopal, R., and Srinivasan, K.V. 2001. Ultrasound promoted C–C bond formation: Heck reaction at ambient conditions in room temperature ionic liquids. *Chemical Communications*, 1544–1545.

de Souza, A.L.F., da Silva, L.C., Oliveira, B.L., and Antunes, O.A.C. 2008. Microwave- and ultrasound-assisted Suzuki-Miyaura cross-coupling reactions catalyzed by Pd/PVP. *Tetrahedron Letters*, 49: 3895–3898.

Ding, L., Wang, W., and Zhang, A. 2007. Synthesis of 1,5-dinitroaryl-1,4-pentadien-3-ones under ultrasound irradiation. *Ultrasonics Sonochemistry*, 14: 563–567.

Docampo Palacios, M.L. and Pellón Comdom, R.F. 2003a. Synthesis of salicylic acid derivatives in presence of ultrasonic irradiation using water as solvent. *Synthetic Communications*, 33: 1783–1787.

Docampo Palacios, M.L. and Pellón Comdom, R.F. 2003b. Synthesis of N-phenylanthranilic acid derivatives using water as solvent in the presence of ultrasound irradiation. *Synthetic Communications*, 33: 1771–1775.

Duan, Z., Xuan, X., Li, T., Yang, C., and Wu, Y. 2006. Cerium(IV) ammonium nitrate (CAN) catalyzed aza-Michael addition of amines to α,β-unsaturated electrophiles. *Tetrahedron Letters*, 47: 5433–5436.

Duarte, A., Cunico, W., Pereira, C.M.P., Flores, A.F.C., Freitag, R.A., and Siqueira, G.M. 2010. Ultrasound promoted synthesis of thioesters from 2-mercaptobenzoxa(thia)zoles. *Ultrasonics Sonochemistry*, 17: 281–283.

Durán-Valle, C.J., Ferrera-Escudero, S., Calvino-Casilda, V., Díaz-Terán, J., and Martín-Aranda, R.M. 2004. The effect of ultrasound on the catalytic activity of alkaline carbons: Preparation of N-alkyl imidazoles. *Applied Surface Science*, 238: 97–100.

Einhorn, C., Einhorn, J., and Luche, J.L. 1989. Sonochemistry—The use of ultrasonic waves in synthetic organic chemistry, *Synthesis*, 787–813.

Einhorn, J., Einhorn, C., and Luche, J.L. 1991. A mild and efficient sonochemical *tert*-butoxycarbonylation of amines from their salts. *Synlett*, 37–38.

Entezari, M.H. and Asghari, A. 2008. Ultrasound improves the synthesis of 5-hydroxymethyl-2-mercapto-1-benzylimidazole as a base compound of some pharmaceutical products. *European Journal of Medicinal Chemistry*, 43: 2835–2839.

Estager, J., Lévêque, J.M., Cravotto, G., Boffa, L., Bonrath, W., and Draye, M. 2007b. One-pot and solventless synthesis of ionic liquids under ultrasonic irradiation. *Synlett*, 2065–2068.

Estager, J., Lévêque, J.M., Turgis, R., and Draye, M. 2007a. Neat benzoin condensation in recyclable room-temperature ionic liquids under ultrasonic activation. *Tetrahedron Letters*, 48: 755–759.

Fazaeli, R. and Aliyan, H. 2009. Heterogeneous catalyst for efficient and green synthesis of 2-arylbenzothiazoles and 2-arylbenzimidazoles. *Applied Catalysis A: General*, 353: 74–79.

Felföldi, K., Szőri, K., Török, B., and Bartók, M. 2000. Sonochemical hydrosilylation of 2-substituted cyclohexanones in the presence of Wilkinson complex. *Ultrasonics Sonochemistry*, 7: 15–17.

Fernandes, V.S., Barboza, J.C.S., and Serra, A.A. 2007. Sonochemical formation of iodohydrin and epoxide from cyclohexene. *Synthetic Communications*, 37: 1433–1436.

Feroci, M., Orsini, M., and Inesi, A. 2009. An efficient combined electrochemical and ultrasound assisted synthesis of imidazole-2-thiones. *Advanced Synthesis and Catalysis*, 351: 2067–2070.

Ferrera-Escudero, S., Perozo-Rondón, E., Calvino-Casilda, V., Casal, B., Martín-Aranda, R.M., López-Peinado, A.J., and Durán-Valle, C.J. 2010. The effect of ultrasound on the N-alkylation of imidazole over alkaline carbons: Kinetic aspects. *Applied Catalysis A: General*, 378: 26–32.

Flint, E.B. and Suslick, K.S. 1991. The temperature of cavitation. *Science*, 253: 1397–1399.

Fu, N., Zhang, Y., Yang, D., Chen, B., and Wu, X. 2008. A rapid synthesis of 1-ferrocenyl-2-arylacetylenes under ultrasound irradiation. *Catalysis Communications*, 9: 976–979.

Gaplovsky, A., Gaplovsky, M., Toma, S., and Luche, J.L. 2000. Ultrasound effects on the photopinacolization of benzophenone. *Journal of Organic Chemistry*, 65: 8444–8447.

Garella, D., Tagliapietra, S., Mehta, V.P., Van Der Eycken, E., and Cravotto, G. 2010. Straightforward functionalization of 3,5-dichloro-2-pyrazinones under simultaneous microwave and ultrasound irradiation. *Synthesis*, 136–140.

Gholap, A.R., Venkatesan, K., Daniel, T., Lahoti, R.J., and, Srinivasan, K.V. 2003. Ultrasound promoted acetylation of alcohols in room temperature ionic liquid under ambient conditions. *Green Chemistry*, 5: 693–996.

Gholap, A.R., Venkatesan, K., Pasricha, R., Daniel, T., Lahoti, R.J., and Srinivasan, K.V. 2005. Copper- and ligand-free Sonogashira reaction catalyzed by Pd(0) nanoparticles at ambient conditions under ultrasound irradiation. *Journal of Organic Chemistry*, 70: 4869–4872.

Giles, R.G.F. and Joll, C.A. 1999. The asymmetric synthesis of 2-benzopyrans and their quinones through intramolecular diastereoselective ring-closure of titanium phenolates of phenolic aldehydes. *Journal of the Chemical Society, Perkin Transactions 1*, 3039–3048.

Grobas, J., Bolivar, C., and Scott, C.E. 2007. Hydrodesulfurization of benzothiophene and hydrogenation of cyclohexene, biphenyl, and quinoline, assisted by ultrasound, using formic acid as hydrogen precursor. *Energy and Fuels*, 21: 19–22.

Guadagnin, R.C., Suganuma, C.A., Singh, F.V., Vieira, A.S., Cella, R., and Stefani, H.A. 2008. Chemoselective cross-coupling Suzuki-Miyaura reaction of (Z)-(2-chlorovinyl)tellurides and potassium aryltrifluoroborate salts. *Tetrahedron Letters*, 49: 4713–4716.

Guzen, K.P., Guarezemini, A.S., Órfão, A.T.G., Cella, R., Pereira, C.M.P., and Stefani, H.A. 2007. Eco-friendly synthesis of imines by ultrasound irradiation. *Tetrahedron Letters*, 48: 1845–1848.

Hagu, H., Salmar, S., and Tuulmets, A. 2007. Impact of ultrasound on hydrophobic interactions in solutions: Ultrasonic retardation of benzoin condensation. *Ultrasonics Sonochemistry*, 14: 445–449.

Han, B.H. and Boudjouk, P. 1982. Ultrasound-promoted reaction of zinc with α,α'-dibromo-o-xylene. Evidence for facile generation of o-xylylene. *Journal of Organic Chemistry*, 47: 751–752.

Hartwig, J.F. 2008. Carbon-heteroatom bond formation catalysed by organometallic complexes. *Nature*, 455: 314–322.

Heropoulos, G.A., Georgakopoulos, S., and Steele, B.R. 2005. High intensity ultrasound-assisted reduction of sterically demanding nitroaromatics. *Tetrahedron Letters*, 46: 2469–2473.

Hofmann, J., Freier, U., and Wecks, M. 2003. Ultrasound promoted C-alkylation of benzyl cyanide—Effect of reactor and ultrasound parameters. *Ultrasonics Sonochemistry*, 10: 271–275.

Jadidi, K., Gharemanzadeh, R., Mehrdad, M., Darabi, H.R., Khavasi, H.R., and Asgari, D. 2008. A facile synthesis of novel pyrrolizidines under classical and ultrasonic conditions. *Ultrasonics Sonochemistry*, 15: 124–128.

Jarikote, D.V., Deshmukh, R.R., Rajagopal, R., Lahoti, R.J., Daniel, T., and Srinivasan, K.V. 2003. Ultrasound promoted facile synthesis of arylhydrazones at ambient conditions. *Ultrasonics Sonochemistry*, 10: 45–48.

Ji, S.J., Shen, Z.L., Gu, D.G., and Huang, X.Y. 2005. Ultrasound-promoted alkynylation of ethynylbenzene to ketones under solvent-free condition. *Ultrasonics Sonochemistry*, 12: 161–163.

Ji, S.J., Shen, Z.L., and Wang, S.Y. 2003b. Aldol condensation of acetylferrocene under ultrasound. *Chinese Chemical Letters*, 14: 663–666.

Ji, S.J. and Wang, S.Y. 2003. Ultrasound-accelerated Michael Addition of Indole to α,β-Unsaturated ketones catalyzed by ceric ammonium nitrate (CAN). *Synlett*, 2074–2076.

Ji, S.J. and Wang, S.Y. 2005. An expeditious synthesis of β-indolylketones catalyzed by p-toluenesulfonic acid (PTSA) using ultrasonic irradiation. *Ultrasonics Sonochemistry*, 12: 339–343.

Jie, M., Lau, M.M.L., and Kalluri, P. 2001. Cyclodehydration reactions of methyl 9,10-;10,12-; and 9,12-dioxostearates with 1,2-diaminoethane under ultrasonic irradiation *Lipids*, 36: 201–204.

Jin, H., Zhao, H., Zhao, F., Li, S., Liu, W., Zhou, G., Tao, K., and Hou, T. 2009. Efficient epoxidation of chalcones with urea-hydrogen peroxide under ultrasound irradiation. *Ultrasonics Sonochemistry*, 16: 304–307.

Ji-Tai, L.P., Xu, W.Z., Chao, X., and Li, T.S. 2004. Aldol reaction under high-intensity ultrasound: A novel approach to an old reaction. *European Journal of Organic Chemistry*, 4438–4444.

Joshi, R.S., Mandhane, P.G., Diwakar, S.D., and Gill, C.H. 2010. Ultrasound assisted green synthesis of bis(indol-3-yl)methanes catalyzed by 1-hexenesulphonic acid sodium salt. *Ultrasonics Sonochemistry*, 17: 298–300.

Kadam, S.T. and Kim, S.S. 2010. Catalyst-free silylation of alcohols and phenols by promoting HMDS in CH_3NO_2 as solvent. *Green Chemistry*, 12: 94–98.

Kado, Y., Atobe, M., and Nonaka, T. 2001. Ultrasonic effects on electroorganic processes—Part 20. Photocatalytic oxidation of aliphatic alcohols in aqueous suspension of TiO_2 powder. *Ultrasonics Sonochemistry*, 8: 69–74.

Kamal, A., Adil, S.F., and Arifuddin, M. 2005. Ultrasonic activated efficient method for the cleavage of epoxides with aromatic amines. *Ultrasonics Sonochemistry*, 12: 429–431.

Kantam, M.L., Rajasekhar, C.V., Gopikrishna, G., Rajender Reddy, K., and Choudary, B.M. 2006. Proline catalyzed two-component, three-component and self-asymmetric Mannich reactions promoted by ultrasonic conditions. *Tetrahedron Letters*, 47: 5965–5967.

Katohgi, M. and Togo, H. 2001. Oxidatively sonochemical dealkylation of various N-alkylsulfonamides to free sulfonamides and aldehydes. *Tetrahedron*, 57: 7481–7486.

Katohgi, M., Yokoyama, M., and Togo, H. 2000. Novel sonochemical dealkylation of N-alkylsulfonamides in the presence of (diacetoxyiodo)benzene and iodine. *Synlett*, 1055–1057.

Khistiaev, K.A., Novikov, M.S., Khlebnikov, A.F., and Magull, J. 2008. gem-Difluorosubstituted NH-azomethine ylides in the synthesis of 4-fluorooxazolines via the three-component reaction of imines, trifluoroacetophenones and CF_2Br_2. *Tetrahedron Letters*, 49: 1237–1240.

Kitazume, T., Ohnogi, T., Miyauchi, H., Yamazaki, T., and Watanabe, S. 1989. Ultrasound-promoted synthesis of α-difluoromethylated carboxylic acids. *Journal of Organic Chemistry*, 54: 5630–5632.

Kise, N. and Mimura, R. 2007. Diastereoselective cycloaddition of chiral 1-acryloyl-2-imidazolidinone and o-quinodimethane generated by reduction of 1,2-bis(bromomethyl)benzene with zinc. *Tetrahedron Asymmetry*, 18: 988–993.

Kumar, A. and Maurya, R.A. 2008. Efficient synthesis of Hantzsch esters and polyhydroquinoline derivatives in aqueous micelles. *Synlett*, 883–885.

Lamandé-Langle, S., Abarbri, M., Thibonnet, J., and Duchêne, A. 2009. A novel mode of access to polyfunctional organotin compounds and their reactivity in Stille cross-coupling reaction. *Journal of Organometallic Chemistry*, 694: 2368–2374.

Langle, S., David-Quillot, F., Balland, A., Abarbri, M., and Duchêne, A. 2003. General access to para-substituted styrenes. *Journal of Organometallic Chemistry*, 671: 113–119.

Lee, A.S.Y., Chu, S.F., Chang, Y.T., and Wang, S.H. 2004. Synthesis of homopropargyl alcohols via sonochemical Barbier-type reaction. *Tetrahedron Letters*, 45: 1551–1553.

Lee, J. and Snyder, J.K. 1989. Ultrasound-promoted Diels–Alder reactions: Syntheses of tanshinone IIA, nortanshinone, and (±)-tanshindiol B. *Journal of the American Chemical Society*, 111: 1522–1524.

Leite, A.C.L., Moreira, D.R.d.M., Coelho, L.C.D., de Menezes, F.D., and Brondani, D.J. 2008. Synthesis of aryl-hydrazones via ultrasound irradiation in aqueous medium. *Tetrahedron Letters*, 49: 1538–1541.

Lévêque, J.M., Luche, J.L., Pétrier, C., Roux, R., and Bonrath, W. 2002. An improved preparation of ionic liquids by ultrasound. *Green Chemistry*, 4: 357–360.

Li, H., Ma, C.J., and Li, H.X. 2006d. Ultrasound-assisted cinnamaldehyde hydrogenation to cinnamyl alcohol at atmospheric pressure over Ru-B amorphous catalyst. *Chinese Journal of Chemistry*, 24: 613–619.

Li, J.T., Bian, Y.J., Zang, H.J., and Li, T.S. 2002. Pinacol coupling of aromatic aldehydes and ketones using magnesium in aqueous ammonium chloride under ultrasound. *Synthetic Communications*, 32: 547–551.

Li, J.T., Chen, G.F., Wang, S.X., He, L., and Li, T.S. 2005b. Synthesis of arylmethylenemalononitriles catalyzed by $KF–Al_2O_3$ under ultrasound. *Australian Journal of Chemistry*, 58: 231–233.

Li, J.T., Chen, G.F., Xu, W.Z., and Li, T.S. 2003b. The Michael reaction catalyzed by KF/basic alumina under ultrasound irradiation. *Ultrasonics Sonochemistry*, 10: 115–118.

Li, J.T., Chen, G.F., Yang, W.Z., and Li, T.S. 2003c. Ultrasound promoted synthesis of 2-aroyl-1,3,5-triaryl-4-carbethoxy-4-cyanocyclohexanols. *Ultrasonics Sonochemistry*, 10: 123–126.

Li, J.T., Chen, Y.X., and Li, T.S. 2005a. Ultrasound-promoted magnesium-ammonium bromide-mediated pinacol coupling of aromatic aldehydes and ketones. *Synthetic Communications*, 35: 2831–2837.

Li, J.T., Cui, Y., Chen, G.F., Cheng, Z.L., and Li, T.S. 2003a. Michael addition catalyzed by potassium hydroxide under ultrasound. *Synthetic Communications*, 33: 353–359.

Li, J.T., Dai, H.G., Xu, W.Z., and Li, T.S. 2006a. Michael addition of indole to α,β-unsaturated ketones catalysed by silica sulfuric acid under ultrasonic irradiation. *Journal of Chemical Research*, 37: 41–42.

Li, J.T., Dai, H.G., Xu, W.Z., and Li, T.S. 2006b. An efficient and practical synthesis of bis(indolyl)methanes catalyzed by aminosulfonic acid under ultrasound. *Ultrasonics Sonochemistry*, 13: 24–27.

Li, J.T., Li, X.L., and Li, T.S. 2006c. Synthesis of oximes under ultrasound irradiation. *Ultrasonics Sonochemistry*, 13: 200–202.

Li, J., Li, X., Liu, X., and Ma, J. 2009d. Synthesis of O-benzyl oximes by combination of phase transfer catalysis and ultrasound irradiation. *Frontiers of Chemistry in China*, 4: 58–62.

Li, J.T. and Lin, Z.P. 2008. An efficient and practical synthesis of 2-((1H-indol-3-yl)(aryl)methyl)malononitriles under ultrasound irradiation. *Ultrasonics Sonochemistry*, 15: 265–268.

Li, J.T., Liu, X.R., and Liu, X.F. 2009c. Oxidative cleavage of hydrobenzoin by ACC/silica gel under ultrasound irradiation. *Ultrasonics Sonochemistry*, 16: 4–6.

Li, J.T., Liu, X.R., and Wang, W.F. 2009b. An efficient oxidation of benzoins to benzils by ACC/silica gel under ultrasound irradiation. *Ultrasonics Sonochemistry*, 16: 331–333.

Li, J.T., Lin, Z.P., and Li, T.S. 2005c. Pinacol coupling of aromatic aldehydes and ketones using $TiCl_3$-Mg under ultrasound irradiation. *Ultrasonics Sonochemistry*, 12: 349–352.

Li, J.T., Meng, X.T., Bai, B., and Sun, M.X. 2010d. An efficient deprotection of oximes to carbonyls catalyzed by silica sulfuric acid in water under ultrasound irradiation. *Ultrasonics Sonochemistry*, 17: 14–16.

Li, X., Santos, J., and Bu, X.R. 2000. Phase transfer catalysis and ultrasound in tandem alkylation of azo dyes for bifunctional molecules. *Tetrahedron Letters*, 41: 4057–4059.

Li, J.T. and Sun, X.L. 2006. An efficient synthesis of benzils from hydrobenzoins by CrO_3-NH_4 clunder ultrasound irradiation. *Letters in Organic Chemistry*, 3: 842–844.

Li, J.T., Sun, M.X., and Yin, Y. 2010a. Ultrasound promoted efficient method for the cleavage of 3-aryl-2,3-epoxyl-1-phenyl-1-propanone with indole. *Ultrasonics Sonochemistry*, 17: 359–362.

Li, J.H., Wang, D., Zhang, Y.Q., Li, J.T., and Chen, B.H. 2009a. Facile one-pot synthesis of 4,5-disubstituted 1 2,3-(NH)-triazoles through Sonogashira coupling/1,3-dipolar cycloaddition of acid chlorides, terminal acetylenes, and sodium azide. *Organic Letters*, 11: 3024–3027.

Li, J.T., Wang, S.X., Chen, G.F., and Li, T.S. 2005d. Some applications of ultrasound irradiation in organic synthesis. *Current Organic Synthesis*, 2: 415–436.

Li, J.T., Xu, W.Z., Chen, G.F., and Li, T.S. 2005d. Synthesis of 1,1-disubstituted-2,6-diarylcyclohexane-4-ones catalyzed by KF/basic Al_2O_3 under ultrasound. *Ultrasonics Sonochemistry*, 12: 473–476.

Li, J.T., Xing, C.Y., and Li, T.S. 2004. An efficient and environmentally friendly method for synthesis of arylmethylenemalononitrile catalyzed by Montmorillonite K10-$ZnCl_2$ under ultrasound irradiation. *Journal of Chemical Technology and Biotechnology*, 79: 1275–1278.

Li, J.T., Yang, W.Z., Chen, G.F., and Li, T.S. 2003. A facile synthesis of α,α'-bis(substituted benzylidene) cycloalkanones catalyzed by KF/Al_2O_3 under ultrasound irradiation. *Synthetic Communications*, 33: 2619–2625.

Li, J.T., Yang, W.Z., Wang, S.X., Li, S.H., and Li, T.S. 2002. Improved synthesis of chalcones under ultrasound irradiation. *Ultrasonics Sonochemistry*, 9: 237–239.

Li, J.T., Yin, Y., and Sun, M.X. 2010b. An efficient one-pot synthesis of 2,3-epoxyl-1,3-diaryl-1-propanone directly from acetophenones and aromatic aldehydes under ultrasound irradiation. *Ultrasonics Sonochemistry*, 17: 363–366.

Li, J.T., Yin, Y., Li, L., and Sun, M.X. 2010c. A convenient and efficient protocol for the synthesis of 5-aryl-1,3-diphenylpyrazole catalyzed by hydrochloric acid under ultrasound irradiation. *Ultrasonics Sonochemistry*, 17: 11–13.

Li, J.T., Zang, H.J., Meng, L.H., Li, L.J., Yin, Y.H., and Li, T.S. 2001. Synthesis of ethyl alkylidene α-cyanoacetates under ultrasound irradiation. *Ultrasonics Sonochemistry*, 8: 93–95.

Li, J.T., Zhai, X.L., and Chen, G.F. 2010e. Ultrasound promoted one-pot synthesis of 3-aza-6,10-diaryl-2-oxaspiro[4.5]decane-1,4,8-trione. *Ultrasonics Sonochemistry*, 17: 356–358.

Li, J.T., Zhai, X.L., Lin, Z.P., and Zhang, X.H. 2008. An efficient synthesis of 7,11-diarylspiro[5.5] undecane-1,9-dione by the Michael condensation under ultrasound irradiation in aqueous and organic two phase in the presence of phase-transfer catalyst. *Letters in Organic Chemistry*, 5: 579–582.

Lin, Z.P., Li, J.T., and Li, T.S. 2006. Pinacol coupling of aromatic aldehydes using $TiCl_3$-Al-H_2O under ultrasound irradiation. *Letters in Organic Chemistry*, 3: 278–281.

Liu, B. and Ji, S.J. 2008a. Facile synthesis of 2-amino-1,4-naphthoquinones catalyzed by molecular iodine under ultrasonic irradiation. *Synthetic Communications*, 38: 1201–1211.

Liu, B., Ji, S.J., Su, X.M., and Wang, S.Y. 2008b. Novel synthesis of 3-indolylquinones catalyzed by molecular iodine under ultrasonic irradiation. *Synthetic Communications*, 38: 1279–1290.

Liu, Y.H., Liu, Q.S., and Zhang, Z.H. 2008c. Amberlyst-15 as a new and reusable catalyst for regioselective ring-opening reactions of epoxides to β-alkoxy alcohols. *Journal of Molecular Catalysis A: Chemical*, 296: 42–46.

Liu, G.B., Tashiro, M., and Thiemann, T. 2009. A facile method for the dechlorination of mono- and dichlorobiphenyls using Raney Ni-Al alloy in dilute aqueous solutions of alkali hydroxides or alkali metal carbonates. *Tetrahedron*, 65: 2497–2505.

Liu, Y.H., Zhang, Z.H., and Li, T.S. 2008d. Efficient conversion of epoxides into β-hydroperoxy alcohols catalyzed by antimony trichloride/SiO_2. *Synthesis*, 3314–3318.

López-Pestaña, J.M., Ávila-Rey, M.J., and Martín-Aranda, R.M. 2002. Ultrasound-promoted N-alkylation of imidazole. Catalysis by solid-base, alkali-metal doped carbons. *Green Chemistry*, 4: 628–630.

Luche, J.L. 1998. *Synthetic Organic Sonochemistry*. New York: Plenum Press.

Luche, J.L. and Allavena, C. 1988. Ultrasound in organic synthesis. Optimisation of the conjugate additions to α,β-unsaturated carbonyl compounds in aqueous media *Tetrahedron Letters*, 29: 5369–5372.

Luche, J.L., Petrier, C., Lansard, J.P., and Greene, A.E. 1983. Ultrasound in organic synthesis. 4. A simplified preparation of diarylzinc reagents and their conjugate addition to α-enones. *Journal of Organic Chemistry*, 48: 3837–3839.

Luu, T.X.T., Christensen, P., Duus, F., and Le, T.N. 2008. Microwave- and ultrasound-accelerated green oxidation of alcohols by potassium permanganate absorbed on copper(II) sulfate pentahydrate. *Synthetic Communications*, 38: 2011–2024.

Luzzio, F.A. and Chen, J. 2008. Efficient preparation and processing of the 4-methoxybenzyl (PMB) group for phenolic protection using ultrasound. *Journal of Organic Chemistry*, 73: 5621–5624.

Luzzio, F.A. and Moore, W.J. 1993. Ultrasound-mediated preparation and applications of chromyl chloride. *Journal of Organic Chemistry*, 58: 512–515.

Mahamuni, N.N., Gogate, P.R., and Pandit, A.B. 2006. Ultrasound-accelerated green and selective oxidation of sulfides to sulfoxides. *Industrial and Engineering Chemistry Research*, 45: 8829–8836.

Mahdavinia, G.H., Rostamizadeh, S., Amani, A.M., and Emdadi, Z. 2009. Ultrasound–promoted greener synthesis of aryl-14-*H*-dibenzo[a,j]xanthenes catalyzed by $NH_4H_2PO_4/SiO_2$ in water. *Ultrasonics Sonochemistry*, 16, 7–10.

Mamaghani, M. and Dastmard, S. 2009. An efficient ultrasound-promoted synthesis of the Baylis–Hillman adducts catalyzed by imidazole and L-proline. *Ultrasonics Sonochemistry*, 16: 445–447.

Mandhane, P.G., Joshi, R.S., Nagargoje, D.R., and Gill, C.H. 2010. An efficient synthesis of 3,4-dihydropyrimidin-2(1*H*)-ones catalyzed by thiamine hydrochloride in water under ultrasound irradiation. *Tetrahedron Letters*, 51: 3138–3140.

Margulis, M.A. 2004. Sonochemistry as a new promising area of high energy chemistry. *High Energy Chemistry*, 38: 135–142.

Mason, T.J. 1997. Ultrasound in synthetic organic chemistry. *Chemical Society Reviews*, 26: 443–451.

Mason, T.J. and Cintas, P. 2002. In *Handbook of Green Chemistry and Technology*, ed. J. Clark, and D. Macquarrie. Oxford, U.K.: Blackwell Science Ltd.

Mason, T.J. and Lorimer, J.P. 2002. *Applied Sonochemistry: The Uses of Power Ultrasound in Chemistry and Processing*. Weinheim, Germany: Wiley-VCH.

Mečiarova, M., Toma, Š., and Heribanová, A. 2000. Ultrasound assisted heterogeneous permanganate oxidations. *Tetrahedron*, 56: 8561–8566.

Mehrabi, H. 2008. Synthesis of α-oximinoketones under ultrasound irradiation. *Ultrasonics Sonochemistry*, 15: 279–282.

Memarian, H.R. and Abdoli-Senejani, M. 2008. Ultrasound-assisted photochemical oxidation of unsymmetrically substituted 1,4-dihydropyridines. *Ultrasonics Sonochemistry*, 15: 110–114.

Memarian, H.R. and Farhadi, A. 2008. Sono-thermal oxidation of dihydropyrimidinones *Ultrasonics Sonochemistry*, 15: 1015–1018.

Mills, A. and Holland, C. 1997. Investigation into the nature of the oxoruthenate species used to mediate the oxidation of an organic substrate by hypochlorite in a biphasic system. *Journal of Chemical Research*, 368–369.

Mirkhani, V., Moghadam, M., Tangestaninejad, S., and Kargar, H. 2006. Rapid and efficient synthesis of 2-imidazolines and bis-imidazolines under ultrasonic irradiation. *Tetrahedron Letters*, 47: 2129–2132.

Mirkhani, V., Tangestaninejad, S., Moghadam, M., and Yadollahi, B. 2000. Efficient and selective epoxidation of alkenes by supported manganese porphyrin under ultrasonic irradiation. *Journal of Chemical Research*, 515–517.

Mohammadpoor-Baltork, I., Memarian, H.R., Khosropour, A.R., and Nikoofar, K. 2006. $BiOClO_4 \cdot xH_2O$ and $Bi(OTf)_3$ as efficient and environmentally benign catalysts for synthesis of Bis(indolyl)methanes in solution and under ultrasound irradiation. *Letters in Organic Chemistry*, 3: 768–772.

Mojtahedi, M.M., Abaee, M.S., and Eghtedari, M. 2008b. Superparamagnetic iron oxide as an efficient and recoverable catalyst for rapid and selective trimethylsilyl protection of hydroxyl groups. *Applied Organometallic Chemistry*, 22: 529–532.

Mojtahedi, M.M., Abaee, M.S., and Hamidi, V. 2007a. Efficient solvent-free aminolysis of epoxides and oxetanes under $MgBr_2 \cdot OEt_2$ catalysis. *Catalysis Communications*, 8: 1671–1674.

Mojtahedi, M.M., Abaee, M.S., Mahmoodi, P., and Adib, M. 2010. Convenient synthesis of 2-aminothiophene derivatives by acceleration of Gewald reaction under ultrasonic aqueous conditions. *Synthetic Communications*, 40: 2067–2074.

Mojtahedi, M.M., Abbasi, H., and Abaee, M.S. 2006a. $MgBr_2 \cdot OEt_2$ mediated protection of alcohols with hexamethyldisilazane: An efficient catalytic route for the preparation of silyl ethers under solvent-free conditions. *Journal of Molecular Catalysis A: Chemical*, 250: 6–8.

Mojtahedi, M.M., Abbasi, H., and Abaee, M.S. 2006b. A novel efficient method for the silylation of alcohols using hexamethyldisilazane in an ionic liquid. *Phosphorus Sulfur and Silicon and the Related Elements*, 181: 1541–1544.

Mojtahedi, M.M., Javadpour, M., and Abaee, M.S. 2008a. Convenient ultrasound mediated synthesis of substituted pyrazolones under solvent-free conditions. *Ultrasonics Sonochemistry*, 15: 828–832.

Mojtahedi, M.M., Saeed Abaee, M., Hamidi, V., and Zolfaghari, A. 2007b. Ultrasound promoted protection of alcohols: An efficient solvent-free pathway for the preparation of silyl ethers in the presence of no additive. *Ultrasonics Sonochemistry*, 14: 596–598.

Mojtahedi, M.M., Saidi, M.R., Shirzi, J.S., and Bolourtchian, M. 2001. Ultrasound accelerated reductive coupling of imine or iminium ion generated in 5m lithium perchlorate solution by lithium metal. *Synthetic Communications*, 31: 3587–3592.

Moldoveanu, C.C., Jones, P.G., and Mangalagiu, I.I. 2009. Spiroheterocyclic compounds: Old stories with new outcomes. *Tetrahedron Letters*, 50: 7205–7208.

Mosslemin, M.H. and Nateghi, M.R. 2010. Rapid and efficient synthesis of fused heterocyclic pyrimidines under ultrasonic irradiation. *Ultrasonics Sonochemistry*, 17: 162–167.

Muravyova, E.A., Shishkina, S.V., Musatov, V.I., Knyazeva, I.V., Shishkin, O.V., Desenko, S.M., and Chebanov, V.A. 2009. Chemoselectivity of multicomponent condensations of barbituric acids, 5-aminopyrazoles, and aldehydes. *Synthesis*, 1375–1385.

Muscia, G.C., Buldain, G.Y., and Asís, S.E. 2009. Only acridine derivative from Hantzsch-type one-pot three-component reactions. *Monatshefte für Chemie*, 140: 1529–1532.

Nabid, M.R., Rezaei, S.J.T., Ghahremanzadeh, R., and Bazgir, A. 2010. Ultrasound-assisted one-pot, three-component synthesis of 1*H*-pyrazolo[1,2-b]phthalazine-5,10-diones. *Ultrasonics Sonochemistry*, 17: 159–161.

Nakamura, E., Sato, K., and Imanishi, Y. 1995. Sonochemical synthesis of alkylhydroperoxides by aerobic reductive oxygenation of alkyl-halides. *Synlett*, 525–5226.

Namboodiri, V.V. and Varma, R.S. 2002. Solvent-free sonochemical preparation of ionic liquids. *Organic Letters*, 4: 3161–3163.

Narkunan, K. and Uang, B.J. 1998. Synthesis of δ-hydroxy-β-oxo esters using sonochemical Blaise reaction. *Synthesis*, 1713–1714.

Nayak, S.K. and Banerji, A. 1991. Stereocontrolled reductive deoxygenation using low-valent titanium: Effects of ultrasound waves and solvents. *Journal of Organic Chemistry*, 56: 1940–1942.

Neo, A.G., Delgado, J., Polo, C., Marcaccini, S., and Marcos, C.F. 2005. A new synthesis of β-keto amides by reduction of Passerini adducts. *Tetrahedron Letters*, 46: 23–26.

Ni, C.L., Song, X.H., Yan, H., Song, X.Q., and Zhong, R.G. 2010. Improved synthesis of diethyl 2,6-dimethyl-4-aryl-4*H*-pyran-3,5-dicarboxylate under ultrasound irradiation. *Ultrasonics Sonochemistry*, 17: 367–369.

Nikpassand, M., Mamaghani, M., Shirini, F., and Tabatabaeian, K. 2010. A convenient ultrasound-promoted regioselective synthesis of fused polycyclic 4-aryl-3-methyl-4,7-dihydro-1*H*-pyrazolo[3,4-b]pyridines. *Ultrasonics Sonochemistry*, 17: 301–305.

Oksdath-Mansilla, G. and Peñéñory, A.B. 2007. Simple and efficient deprotection of 1,3-dithianes and 1,3-dithiolanes by copper(II) salts under solvent-free conditions. *Tetrahedron Letters*, 48: 6150–6154.

Olah, G.A., Wu, A.H., and Farooq, O. 1989. Synthetic methods and reactions. Ultrasound-assisted preparation of di-*tert*-butyl-,di-1,1′ adamantyl- and (1-adamantyl) -*tert*-butylketenes. *Synthesis*, 1989: 566–567.

Palimkar, S.S., Harish Kumar, P., Lahoti, R.J., and Srinivasan, K.V. 2006. Ligand-, copper-, and amine-free one-pot synthesis of 2-substituted indoles via Sonogashira coupling 5-endo-dig cyclization. *Tetrahedron*, 62: 5109–5115.

Palmikar, S.S., More, V.S., and Srinivasan, K.V. 2008. Ultrasound promoted copper-, ligand- and amine-free synthesis of benzo[b]furans/nitro benzo[b]furans via Sonogashira coupling-5-endo-dig-cyclization. *Ultrasonics Sonochemistry*, 15: 853–862.

Palmisano, G., Bonrath, W., Boffa, L., Garella, D., Barge, A., and Cravotto, G. 2007b. Heck reactions with very low ligandless catalyst loads accelerated by microwaves or simultaneous microwaves/ultrasound irradiation. *Advanced Synthesis and Catalysis*, 349: 2338.

Palmisano, G., Tagliapietra, S., Barge, A., Binello, A., Boffa, L., and Cravotto, G. 2007a. Efficient regioselective opening of epoxides by nucleophiles in water under simultaneous ultrasound/microwave irradiation. *Synlett*, 2041–2044.

Pellón, R.F., Docampo, M.L., and Fascio, M.L. 2007. Mild method for Ullmann reaction of 2-chlorobenzoic acids and aminothiazoles or aminobenzothiazoles under ultrasonic irradiation. *Synthetic Communications*, 37: 1853–1864.

Pellón, R.F., Estévez-Braun, A., Docampo, M.L., Martín, A., and Ravelo, A.G. 2005. Use of ultrasound in the synthesis of 2- (alkylamino)benzoic acids in water. *Synlett*, 1606–1608.

Peng, Y., Dou, R., Song, G., and Jiang, J. 2005a. Dramatically accelerated synthesis of β-aminoketones via aqueous Mannich reaction under combined microwave and ultrasound irradiation. *Synlett*, 2245–2247.

Peng, Y. and Song, G. 2001. Simultaneous microwave and ultrasound irradiation: A rapid synthesis of hydrazides. *Green Chemistry*, 3: 302–304.

Peng, Y. and Song, G. 2002. Combined microwave and ultrasound assisted Williamson ether synthesis in the absence of phase-transfer catalysts. *Green Chemistry*, 4: 349–351.

Peng, Y. and Song, G. 2003. Combined microwave and ultrasound accelerated Knoevenagel-Doebner reaction in aqueous media: A green route to 3-aryl acrylic acids. *Green Chemistry*, 5: 704–706.

Peng, Y., Song, G., and Dou, R. 2006. Surface cleaning under combined microwave and ultrasound irradiation: Flash synthesis of 4*H*-pyrano[2,3-c]pyrazoles in aqueous media. *Green Chemistry*, 8: 573–575.

Peng, Y., Zhong, W., and Song, G. 2005b. Efficient and mild room temperature reduction of benzophenones under ultrasound irradiation. *Ultrasonics Sonochemistry*, 12: 169–172.

Pétrier, C., Jeunet, A., Luche, J.L., and Reverdy, G. 1992. Unexpected frequency effects on the rate of oxidative processes induced by ultrasound. *Journal of the American Chemical Society*, 114: 3148–3150.

Pizzuti, L., Martins, P.L.G., Ribeiro, B.A., Quina, F.H., Pinto, E., Flores, A.F.C., Venzke, D., and Pereira, C.M.P. 2010. Efficient sonochemical synthesis of novel 3,5-diaryl-4,5-dihydro-1*H*-pyrazole-1-carboximidamides. *Ultrasonics Sonochemistry*, 17: 34–37.

Pizzuti, L., Piovesan, L.A., Flores, A.F.C., Quina, F.H., and Pereira, C.M.P. 2009. Environmentally friendly sonocatalysis promoted preparation of 1-thiocarbamoyl-3,5-diaryl-4,5-dihydro-1*H*-pyrazoles. *Ultrasonics Sonochemistry*, 16: 728–731.

Poláčková, V., Hut'Ka, M., and Toma, Š. 2005. Ultrasound effect on Suzuki reactions. Synthesis of unsymmetrical biaryls. *Ultrasonics Sonochemistry*, 12: 99–102.

Porto, R.S., Amarante, G.W., Cavallaro, M., and Poppi, R.J. 2009. Improved catalysis of Morita–Baylis–Hillman reaction. The strong synergic effect using both an imidazolic ionic liquid and a temperature. *Tetrahedron Letters*, 50: 1184–1187.

Raj, C.P., Dhas, N.A., Cherkinski, M., Gedanken, A., and Braverman, S. 1998. Sonochemical synthesis of norbornane derivatives using allene cyclopentadiene Diels–Alder cycloaddition. *Tetrahedron Letters*, 39: 5413–5416.

Rajagopal, R., Jarikote, D.V., and Srinivasan, K.V. 2002. Ultrasound promoted Suzuki cross-coupling reactions in ionic liquid at ambient conditions. *Chemical Communications*, 616–617.

Ranjbar-Karimi, R. 2010. Acceleration of alkenyltrimethylsilane fluorination under mild conditions using ultrasound. *Ultrasonics Sonochemistry*, 17: 768–769.

Ranu, B.C. and Jana, R. 2005. Direct halogenation of alcohols and their derivatives with *tert*-butyl halides in the ionic liquid [pmIm]Br under sonication conditions—A novel, efficient and green methodology. *European Journal of Organic Chemistry*, 2005: 755–758.

Raouafi, N., Belhadj, N., Boujlel, K., Ourari, A., Amatore, C., Maisonhaute, E., and Schollhőrn, B. 2009. Ultrasound-promoted aromatic nucleophilic substitution of dichlorobenzene iron(II) complexes. *Tetrahedron Letters*, 50: 1720–1722.

Rechards, W.T. and Loomis, A.L. 1927. *Journal of the American Chemical Society*, 49: 3086–4000.

Riccaboni, M., La Porta, E., Martorana, A., and Attanasio, R. 2010. Effect of phase transfer chemistry, segmented fluid flow, and sonication on the synthesis of cinnamic esters. *Tetrahedron*, 66: 4032–4039.

Rokhina, E.V., Lahtinen, M., Nolte, M.C.M., and Virkutyte, J. 2009. The influence of ultrasound on the RuI_3-catalyzed oxidation of phenol: Catalyst study and experimental design. *Applied Catalysis B: Environmental*, 87: 162–170.

Ross, N.A. and Bartsch, R.A. 2001. Sonochemical Reformatsky reactions of β-bromoesters with sym- (keto) dibenzo 16-crown-5. *Journal of Heterocyclic Chemistry*, 38: 1255–1258.

Ross, N.A. and Bartsch, R.A. 2003. High-intensity ultrasound-promoted Reformatsky reactions. *Journal of Organic Chemistry*, 68: 360–366.

Ross, N.A., MacGregor, R.R., and Bartsch, R.A. 2004. Synthesis of β-lactams and β-aminoesters via high intensity ultrasound-promoted Reformatsky reactions. *Tetrahedron*, 60: 2035–2041.

Ruano, J.L.G., Fernández-Ibáñez, M.Á., Fernández-Salas, J.A., Maestro, M.C., Márquez-López, P., and Rodríguez-Fernández, M.M. 2009. Remote stereocontrol mediated by a sulfinyl group: Synthesis of allylic alcohols via chemoselective and diastereoselective reduction of γ-methylene δ-ketosulfoxides. *Journal of Organic Chemistry*, 74: 1200–1207.

Ruano, J.L.G., Parra, A., and Alemán, J. 2008. Efficient synthesis of disulfides by air oxidation of thiols under sonication. *Green Chemistry*, 10: 706–711.

Rup, S., Sindt, M., and Oget, N. 2010. Catalytic oxidative cleavage of olefins by RuO_4 organic solvent-free under ultrasonic irradiation. *Tetrahedron Letters*, 51: 3123–3126.

Sadeghi, M.M., Loghmani-Khouzani, H., Ranjbar-Karimi, R., Butler, P., and Golding, B.T. 2006a. Sonochemical fluorination of heterocyclic nitro compounds with Selectfluor. *Tetrahedron Letters*, 47: 4519–4522.

Sadeghi, M.M., Loghmani-Khouzani, H., Ranjbar-Karimi, R., and Golding, B.T. 2006b. Sonochemical fluorination of heterocyclic nitro compounds with Selectfluor™ (F-TEDA-BF_4). *Tetrahedron Letters*, 47: 2455–2457.

Salavati, H., Tangestaninejad, S., Moghadam, M., Mirkhani, V., and Mohammadpoor-Baltork, I. 2010a. Sonocatalytic epoxidation of alkenes by vanadium-containing polyphosphomolybdate immobilized on multi-wall carbon nanotubes. *Ultrasonics Sonochemistry*, 17: 453–459.

Salavati, H., Tangestaninejad, S., Moghadam, M., Mirkhani, V., and Mohammadpoor-Baltork, I. 2010b. Sonocatalytic oxidation of olefins catalyzed by heteropolyanion-montmorillonite nanocomposite. *Ultrasonics Sonochemistry*, 17: 145–152.

Sanchez, M., Diallo, O., Oussaid, A., Oussaid, B., and Garrigues, B. 2001. In-situ syntheses of alkyllithium compounds under ultrasonic irradiation. *Phosphorus Sulfur and Silicon and the Related Elements*, 173: 235–242.

Sant' Anna, G.d.S., Machado, P., Sauzem, P.D., Rosa, F.A., Rubin, M.A., Ferreira, J., Bonacorso, H.G., Zanatta, N., and Martins, M.A.P. 2009. Ultrasound promoted synthesis of 2-imidazolines in water: A greener approach toward monoamine oxidase inhibitors. *Bioorganic and Medicinal Chemistry Letters*, 19: 546–549.

Semwal, A., Bhattacharya, A., and Nayak, S.K. 2002. Ultrasound mediated selective monoalkylation of 4-*tert*-butylcalix[6]arene at the lower rim. *Tetrahedron*, 58: 5287–5290.

Shaabani, A., Rezayan, A.H., Rahmati, A., and Sharifi, M. 2006. Ultrasound-accelerated synthesis of 1,4-dihydropyridines in an ionic liquid. *Monatshefte für Chemie*, 137: 77–81.

Shelke, K.F., Sapkal, S.B., Sonar, S.S., Madje, B.R., Shingate, B.B., and Shingare, M.S. 2009. An efficient synthesis of 2,4,5-triaryl-1*H*-imidazole derivatives catalyzed by boric acid in aqueous media under ultrasound-irradiation. *Bulletin of the Korean Chemical Society*, 30: 1057–1060.

Shi, Q. 2007. Rapid catalyst-free carbon-carbon bonds coupling reaction under ambient conditions and ultrasonic irradiation. *Journal of Chemical Research*, 617–618.

Shinde, P.V., Sonar, S.S., Shingate, B.B., and Shingare, M.S. 2010. Boric acid catalyzed convenient synthesis of 2-amino-3,5-dicarbonitrile-6-thio-pyridines in aqueous media. *Tetrahedron Letters*, 51: 1309–1312.

Šibor, J. and Pazdera, P. 1996. Synthesis of some new five-membered heterocycles containing selenium and tellurium. *Molecules*, 1: 157–162.

Silva, A.d.C., de Souza, A.L.F., and Antunes, O.A.C. 2007. Phosphine-free Suzuki cross-coupling reactions under ultrasound. *Journal of Organometallic Chemistry*, 692: 3104–3107.

Singh, J., Gupta, N., Kad, G.L., and Kaur, J. 2006. Efficient role of ionic liquid (bmim)HSO_4 as novel catalyst for monotetrahydropyranylation of diols and tetrahydropyranylation of alcohols. *Synthetic Communications*, 36: 2893–2900.

Singh, F.V. and Stefani, H.A. 2008. Ultrasound-assisted synthesis of symmetrical biaryls by palladium-catalyzed homocoupling of aryl *n*-butyl tellurides. *Synlett*, 3221–3225.

Singh, F.V., Weber, M., Guadagnin, R.C., and Stefani, H.A. 2008. Ultrasound-assisted synthesis of functionalized 1,3-enynes by palladium-catalyzed cross-coupling reaction of α-styrylbutyltelluride with alkynyltrifluoroborate salts. *Synlett*, 1889–1893.

Smith, K. and Jones, D. 1992. A superior synthesis of diaryl ethers by the use of ultrasound in the Ullmann reaction. *Journal of the Chemical Society, Perkin Transactions 1*, 23: 407–408.

Sonar, S.S., Sadaphal, S.A., Kategaonkar, A.H., Pokalwar, R.U., Shingate, B.B., and Shingare, M.S. 2009. Alum catalyzed simple and efficient synthesis of Bis(indolyl)methanes by ultrasound approach. *Bulletin of the Korean Chemical Society*, 30: 825–828.

Stefani, H.A., Pereira, C.M.P., Almeida, R.B., Braga, R.C., Guzen, K.P., and Cella, R. 2005. A mild and efficient method for halogenation of 3,5-dimethyl pyrazoles by ultrasound irradiation using N-halosuccinimides. *Tetrahedron Letters*, 46: 6833–6837.

Suárez, R.M., Sestelo, J.P., and Sarandeses, L.A. 2002. Diastereoselective ultrasonically induced zinc-copper conjugate addition to chiral α,β-unsaturated carbonyl systems in aqueous media. *Synlett*, 1435–1438.

Sureshbabu, V.V., Nagendra, G., and Venkataramanarao, R. 2008a. Ultrasound accelerated conversion of N$^\alpha$-urethane protected peptide esters to their thiopeptides using P_2S_5. *Ultrasonics Sonochemistry*, 15: 927–929.

Suresh Babu, A.R. and Raghunathan, R. 2007. Ultrasonic assisted-silica mediated [3+2] cycloaddition of azomethine ylides-a facile multicomponent one-pot synthesis of novel dispiroheterocycles. *Tetrahedron Letters*, 48: 6809–6813.

Sureshbabu, V.V., Sudarshan, N.S., and Kantharaju. 2008b. Efficient synthesis of O-succinimidyl- (*tert*-butoxycarbonylamino)methyl carbamates derived from α-amino acids accelerated by ultrasound: Application to the synthesis of ureidodipeptides. *Synthetic Communications*, 38: 2168–2184.

Suslick, K.S.K. and Kemper, A. 1993. The effect of fluorocarbon gases on sonoluminescence: A failure of the electrical hypothesis. *Ultrasonics*, 31: 463–465.

Tangestaninejad, S., Mirkhani, V., Moghadam, M., Mohammadpoor-Baltork, I., Shams, E., and, Salavati, H. 2008. Hydrocarbon oxidation catalyzed by vanadium polyoxometalate supported on mesoporous MCM-41 under ultrasonic irradiation. *Ultrasonics Sonochemistry*, 15: 438–447.

Tangestaninejad, S., Moghadam, M., Mirkhani, V., and Kargar, H. 2006. Efficient and selective hydrocarbon oxidation with sodium periodate under ultrasonic irradiation catalyzed by polystyrene-bound Mn (TPyP). *Ultrasonics Sonochemistry*, 13: 32–36.

Tei, L., Gugliotta, G., Avedano, S., Giovenzana, G.B., and Botta, M. 2009. Application of the Ugi four-component reaction to the synthesis of ditopic bifunctional chelating agents. *Organic and Biomolecular Chemistry*, 7: 4406–4414.

Timko, M.T., Allen, A.J., Danheiser, R.L., Steinfeld, J.A., Smith, K.A., and Tester, J.W. 2006. Improved conversion and selectivity of a Diels–Alder cycloaddition by use of emulsions of carbon dioxide and water. *Industrial and Engineering Chemistry Research*, 45: 1594–1503.

Török, B., Balázsik, K., Török, M., Szöllösi, G., and Bartók, M. 2000. Enantioselective hydrogenation of α-ketoesters over platinum catalysts. *Ultrasonics Sonochemistry*, 7: 151–155.

Tu, S., Cao, L., Zhang, Y., Shao, Q., Zhou, D., and Li, C. 2008. An efficient synthesis of pyrido[2,3-d]pyrimidine derivatives and related compounds under ultrasound irradiation without catalyst. *Ultrasonics Sonochemistry*, 15: 217–221.

Tuulmets, A., Hagu, H., Salmar, S., Cravotto, G., and Järv, J. 2007. Ultrasonic evidence of hydrophobic interactions. Effect of ultrasound on benzoin condensation and some other reactions in aqueous ethanol. *Journal of Physical Chemistry B*, 111: 3133–3138.

Ung, S., Falguières, A., Guy, A., and Ferroud, C. 2005. Ultrasonically activated reduction of substituted nitrobenzenes to corresponding N-arylhydroxylamines. *Tetrahedron Letters*, 46: 5913–5917.

Upadhyaya, D.J., Barge, A., Stefania. R., and Cravotto, G. 2007. Efficient, solventless N-Boc protection of amines carried out at room temperature using sulfamic acid as recyclable catalyst. *Tetrahedron Letters*, 48: 8318–8322.

Wang, S.Y. and Ji, S.J. 2006. Facile synthesis of 3,3-di(heteroaryl)indolin-2-one derivatives catalyzed by ceric ammonium nitrate (CAN) under ultrasound irradiation. *Tetrahedron*, 62: 1527–1535.

Wang, S.X., Li, X.W., and Li, J.T. 2008b. Synthesis of N-alkoxyphthalimides under ultrasound irradiation. *Ultrasonics Sonochemistry*, 15: 33–36.

Wang, J.S., Li, J.T., Lin, Z.P., and Li, T.S. 2005a. Magnesium-induced pinacol coupling of aromatic aldehydes and ketones under ultrasound irradiation. *Synthetic Communications*, 35: 1419–1424.

Wang, S.X., Li, J.T., Yang, W.Z., and Li, T.S. 2002. Synthesis of ethyl α-cyanocinnamates catalyzed by $KF-Al_2O_3$ under ultrasound irradiation. *Ultrasonics Sonochemistry*, 9: 159–161.

Wang, S.X., Li, Z.Y., Zhang, J.C., and Li, J.T. 2008a. The solvent-free synthesis of 1,4-dihydropyridines under ultrasound irradiation without catalyst. *Ultrasonics Sonochemistry*, 15: 677–680.

Wang, S.X., Wang, K., and Li, J.T. 2005b. Pinacol coupling reaction of aromatic aldehydes mediated by aqueous vanadium(II) solution under ultrasound irradiation. *Synthetic Communications*, 35: 2387–2394.

Wei, K., Gao, H.T., and Li, W.D.Z. 2004. Facile synthesis of oxabicyclic alkenes by ultrasonication-promoted Diels–Alder cycloaddition of furano Dienes. *Journal of Organic Chemistry*, 69: 5763–5765.

Weickgenannt, A., Mewald, M., Muesmann, T.W.T., and Oestreich, M. 2010. Catalytic asymmetric Si-O coupling of simple achiral silanes and chiral donor–functionalized alcohols. *Angewandte Chemie–International Edition*, 49: 2223–2226.

Wuts, P.G.M. and Greene, T.W. 2007. Greene's protective groups in organic chemistry. Hoboken: John Wiley & Sons.

Xin, Y., Zang, Z.H., and Chen, F.L. 2009. Ultrasound-promoted synthesis of 1,5-diarylpenta-2,4-dien-1-ones catalyzed by activated barium hydroxide. *Synthetic Communications*, 39: 4062–4068.

Yang, J.M., Ji, S.J., Gu, D.G., Shen, Z.L., and Wang, S.Y. 2005. Ultrasound-irradiated Michael addition of amines to ferrocenylenones under solvent-free and catalyst-free conditions at room temperature. *Journal of Organometallic Chemistry*, 690: 2989–2995.

Yang, J.H., Li, J.T., Zhao, J.L., and Li, T.S. 2004. Pinacol coupling reaction of aromatic aldehydes mediated by Zn in acid aqueous media under ultrasound irradiation. *Synthetic Communications*, 34: 993–1000.

Ye, Y., Ding, Q., and Wu, J. 2008. Three-component reaction of 2-alkynylbenzaldehyde, amine, and nucleophile using Lewis acid-surfactant combined catalyst in water. *Tetrahedron*, 64: 1378–1382.

Yinghuai, Z., Bahnmueller, S., Chibun, C., Carpenter, K., Hosmane, N.S., and Maguire, J.A. 2003. An effective system to synthesize methanofullerenes: Substrate-ionic liquid-ultrasonic irradiation. *Tetrahedron Letters*, 44: 5473–5476.

Young, S.P., Jung, H.H., Yoo, B., Kyung, I.C., Joong, H.K., Cheol, M.Y., and Yoo, B.W. 2005. Facile synthesis of β-ketoesters by indium-mediated reaction of acyl cyanides with ethyl bromoacetate under ultrasonication. *Bulletin of the Korean Chemical Society*, 26: 878–879.

Zang, H., Su, Q., Mo, Y., Cheng, B.W., and Jun, S. 2010a. Ionic liquid EMIMOAc under ultrasonic irradiation towards the first synthesis of trisubstituted imidazoles. *Ultrasonics Sonochemistry*, 17: 749–751.

Zang, H., Wang, M., Cheng, B.W., and Song, J. 2009. Ultrasound-promoted synthesis of oximes catalyzed by a basic ionic liquid [bmIm]OH. *Ultrasonics Sonochemistry*, 16: 301–303.

Zang, H., Zhang, Y., Zang, Y., and Cheng, B.W. 2010b. An efficient ultrasound-promoted method for the one-pot synthesis of 7,10,11,12-tetrahydrobenzo[c]acridin-8(9H) -one derivatives. *Ultrasonics Sonochemistry*, 17: 495–499.

Zeng, X.F., Ji, S.J., and Shen, S.S. 2007. Conjugate addition of indoles to α,β-unsaturated ketones (chalcones) catalyzed by $KHSO_4$ under ultrasonic conditions. *Chinese Journal of Chemistry*, 25: 1777–1780.

Zeng, X.F., Ji, S.J., and Wang, S.Y. 2005. Novel method for synthesis of unsymmetrical bis(indolyl)alkanes catalyzed by ceric ammonium nitrate (CAN) under ultrasound irradiation. *Tetrahedron*, 61: 10235–1041.

Zeng, H., Li, H., and Shao, H. 2009. One-pot three-component Mannich-type reactions using Sulfamic acid catalyst under ultrasound irradiation. *Ultrasonics Sonochemistry*, 16: 758–762.

Zeni, G., Chieffi, A., Cunha, R., Zukerman-Schpector, J., Stefani, H.A., and Comasseto, J.V. 1999. Addition reaction of *p*-methoxyphenyltellurium trichloride to 3-hydroxy alkynes. *Organometallics*, 18: 803–806.

Zhang, X.R., Chao, W., Chuai, Y.T., Ma, Y., Hao, R., Zou, D.C., Wei, Y.G., and Wang, Y. 2006a. A new n-type organic semiconductor synthesized by Knoevenagel condensation of truxenone and ethyl cyanoacetate. *Organic Letters*, 8: 2563–2566.

Zhang, G., Huang, Z., and Zou, J. 2009. $Ga(OTf)_3$-catalyzed three-component Mannich reaction in water promoted by ultrasound irradiation. *Chinese Journal of Chemistry*, 27: 1967–1974.

Zhang, F., Sun, J., Gao, D., Li, Y., Zhang, Y., Zhao, T., and Chen, X. 2007b. An efficient and convenient procedure for the synthesis of 2-alkyl-2-alkoxy-1,2-di(furan-2-yl)ethanone under ultrasound in the presence of solid-liquid phase transfer catalysis conditions. *Ultrasonics Sonochemistry*, 14: 493–496.

Zhang, P., Yang, M., and Lü, X. 2007a. Epoxidation of cyclohexene with oxygen in ultrasound airlift loop reactor. *Chinese Journal of Chemical Engineering*, 15: 196–199.

Zhang, P., Yang, M., Lu, X., Han, P., and Wang, Y. 2006c. Baeyer-Villiger oxidation of cyclohexanone to ε-caprolactone in airlift sonochemical reactor. *Ultrasonics*, 44: e393–e395.

Zhang, J., Yang, F., Ren, G., Mak, T.C.W., Song, M., and Wu, Y. 2008. Ultrasonic irradiation accelerated cyclopalladated ferrocenylimines catalyzed Suzuki reaction in neat water. *Ultrasonics Sonochemistry*, 15: 115–118.

Zhang, Z., Zha, Z., Gan, C., Pan, C., Zhou, Y., Wang, Z., and Zhou, M.M. 2006b. Catalysis and regioselectivity of the aqueous Heck reaction by Pd(0) nanoparticles under ultrasonic irradiation. *Journal of Organic Chemistry*, 71: 4339–4342.

Zhao, S.H., Xu, X.M., Zheng, L., and Liu, H. 2010. *Ultrasonics Sonochemistry*, 17: 685–689.

Zhao, S., Zhao, E., Shen, P., Zhao, M., and Sun, J. 2008. An atom-efficient and practical synthesis of new pyridinium ionic liquids and application in Morita–Baylis–Hillman reaction. *Ultrasonics Sonochemistry*, 15: 955–959.

Zhou, W.J., Ji, S.J., and Shen, Z.L. 2006. An efficient synthesis of ferrocenyl substituted 3-cyanopyridine derivatives under ultrasound irradiation. *Journal of Organometallic Chemistry*, 691: 1356–1360.

13 Ultrasound-Assisted Anaerobic Digestion of Sludge

Ackmez Mudhoo and Sanjay K. Sharma

CONTENTS

Introduction 323
    Sludge Management and Anaerobic Digestion 323
    Innovative Pretreatment of Sludge 325
    Ultrasound Technology and Anaerobic Digestion 325
    Sludge Minimization 326
Principles of Anaerobic Digestion 326
    Anaerobic Digestion Biotechnology 326
    Merits and Demerits of Anaerobic Biotechnology 327
Ultrasound and Sonication in Anaerobic Digestion 328
    Principles of Ultrasound: Cavitation 328
    Merits and Demerits of Ultrasound Technology 329
    Application of Ultrasound in Sludge Anaerobic Digestion 329
    Sludge Disintegration and Ultrasound Control Parameters 330
    Effects of Sonication and Parameter Monitoring 332
        Ultrasound and Sludge Dewaterability 332
        Ultrasound and Enhanced Biogas Production 336
Research Avenues and Concluding Remarks 337
Acknowledgment 337
References 337

INTRODUCTION

SLUDGE MANAGEMENT AND ANAEROBIC DIGESTION

Sludge treatment has long become one of the most challenging problems in wastewater treatment plants (Zhang et al., 2007). As a result of the wide application and utilization of the waste activated sludge process, excess sludge presents a serious disposal problem (Neyens and Baeyens, 2003; Hao et al., 2007). The management of excess activated sludge also imposes great economic costs on the operation and maintenance of wastewater treatment plants and hence represents in itself significant technical challenges (Li et al., 2008) as a results of environmental, economic, social, and legal factors (Chu et al., 2009). Many efforts have been devoted to reduce the excess sludge burden (Naddeo et al., 2009) by treatments such as digestion and dewatering. Some sludge treatment technologies include pretreatment and sludge minimization, anaerobic digestion, aerobic digestion, alkaline stabilization, composting, dewatering, drying, and innovative technologies (Fitzmorris et al., 2009). Table 13.1 lists a selected few innovative sludge pretreatment and management research that have been undertaken.

It has been known for many years that a thermal pretreatment usually gives an improvement in the dewaterability of sludges (Neyens and Baeyens, 2003; Eskicioglu et al., 2006; Wilson

TABLE 13.1
Recent Innovative Sludge Management and/or Conditioning Techniques

Innovative Pretreatment Conditions	Main Observations	Reference
Batch anaerobic digesters were used to stabilize microwave-irradiated waste activated sludge	Waste activated sludge, microwaved to 96°C, produced the greatest improvement in cumulative biogas production with $15 \pm 0.5\%$ and $20 \pm 0.3\%$ increases over controls after 19 days of digestion at low and high waste activated sludge concentrations Dewaterability of microwaved sludge was enhanced after anaerobic digestion	Eskicioglu et al. (2007)
Gamma irradiated sludge for its suitability as a soil amendment in agriculture	Growth parameters and yield of carrot was not significantly different from controls	Rathod et al. (2008)
Bench-scale ozonation of waste activated sludge	For an ozone contact time of 12 days: total chemical oxygen demand removal 91.1%	
Microwave pretreatment for enhanced anaerobiosis of secondary sludge	The soluble chemical oxygen demand concentration increased up to 22% as microwave irradiation time increased, which indicated the sludge particles disintegrated	Park et al. (2004)
Ultrasonic irradiation	Biogas production linked to the soluble part of sludge increased with ultrasonic power	Bougrier et al. (2005)

and Novak, 2009) and bioavailability and biodegradability characteristics of the organic fraction (Borges and Chernicharo, 2009). The optimum treatment conditions to obtain an enhanced dewaterability and digestibility of sludge have been relatively widely studied and tested. The main commercial hydrolysis processes developed to this end are the Cambi, Porteous, and Zimpro processes.

Two kinds of sludges are produced tremendously everyday in biological wastewater treatment plants worldwide (Mao et al., 2004). The primary sludge generally comes from the settling of easily settleable solids and the secondary sludge comes from the biomass after biological treatment has taken place in the specific biological treatment system. Secondary sludge is particularly troublesome to stabilize, and notorious for its difficulty in dewatering and digestion. Anaerobic digestion has now become a commonly applied biological process for stabilization of sewage sludges (Aitken et al., 2005; Arnaiz et al., 2006). The process is more beneficial among several sludge stabilization methods as it is capable to produce a net energy gain (Mao et al., 2004; Bohn et al., 2007; Lu et al., 2008) in the form of methane gas, leading to cost-effectiveness (Mao et al., 2004). Due to the rate limiting step of hydrolysis, however, anaerobic digestion is a very slow process (Vavilin et al., 2008; Zhao et al., 2009) and large fermenters are required to bring together the necessary chemical, biological, and physical conditions that are conducive to an optimum biochemistry in the reactor tanks. It hence becomes important to equip treatment plants with sufficiently large digesters or alternatively incorporate technological aids to overcome the inherent enzymatic limitations. Moreover, sludge disintegration has recently gained renewed and heightened attention in the context of using renewable energy sources as it might be a way to improve anaerobic digestion for a better conversion of biomass to biogas (Ward et al., 2008). Biogas is a clean environment friendly fuel (Harasimowicz et al., 2007; Schievano et al., 2008). Raw biogas contains about 55%–65% methane, 30%–45% carbon dioxide, traces of hydrogen sulfide, carbon monoxide (Fantozzi and Buratti, 2009), and fractions of water vapor. Pure methane has a calorific value of $38,074.4\,kJ\,m^{-3}$ at $15.5°C$ and 1 atm; the calorific value of biogas varies from $20,064$ to $28,842\,kJ\,m^{-3}$ (Harasimowicz et al., 2007).

INNOVATIVE PRETREATMENT OF SLUDGE

The biodegradability of waste sludge can be improved by using thermal energy (Bougrier et al., 2008), enzymes and bacteria (Li et al., 2009b), ozonation (Dytczak et al., 2007; Zhang et al., 2009), acidification (Liu et al., 2009), alkaline addition (López Torres and Espinosa Lloréns, 2008), high pressure homogenization (Kidak et al., 2009), mechanical disintegration, and ultrasound (Chu et al., 2001) pretreatments. Some investigations have discussed the combined treatment of alkaline addition and ultrasound. Among these processes of physical pretreatments, ultrasonication is viewed as an environmentally and economically sound pretreatment (Mao and Show, 2007; Show et al., 2007) that exhibits the benefit of not being hazardous to the environment and hence being "green" (Cintas and Luche, 1999; Chu et al., 2001; Nikolopoulos et al., 2006). In recent yesteryears, relevant interest has been devoted to activated sludge disintegration and solubilization techniques in order to cope with the biological limitations related to particulate degradation (Braguglia et al., 2006). Mechanical disintegration with ultrasound irradiation can efficiently transform insoluble organics into a soluble form (Nasseri et al., 2006). This solubilized organic matter is released from the cells to the bulk phase and hence speeds up the rate that determines the hydrolysis step of the digestion process (Braguglia et al., 2006). Hence, the radiation technology, which also encompasses ultrasound irradiation, may be regarded to be a promising alternative for its high efficiency in pathogen inactivation, organic pollutants oxidation, odor nuisance elimination, and some other characteristics enhancement, which will facilitate the downstream process of sludge treatment and disposal (Wang and Wang, 2007).

ULTRASOUND TECHNOLOGY AND ANAEROBIC DIGESTION

It has been recognized for many years that ultrasound power has great potential in a wide variety of processes in the chemical and allied industries (Mason, 2000). Some of these processes have been known for many years and continue to prosper as major commercial applications like plastic welding and cleaning (Shoh, 1975). Sonochemistry could be successfully combined with biotechnology with the aim of enhancing the efficiency of bioprocesses (Zabaneh and Bar, 1991; Aliyu and Hepher, 2000; Xie et al., 2007), including biofuel production (Khanal et al., 2007b), bioprocess monitoring, enzyme biocatalysts (Chisti, 2003; Lee et al., 2008), biosensors, and biosludge treatment (Rokhina et al., 2009).

Activated sludge processes, which are fundamentally anaerobic processes, are key technologies in wastewater treatment. These biological processes produce huge amounts of waste activated sludge and other biosolids (Naddeo et al., 2009). With regard to the application of sonication (i.e., irradiation with ultrasound) to anaerobic digestion processes, high-power ultrasound is a relatively new and innovative approach to disintegrate bacterial cells (Tiehm et al., 2001; Mason et al., 2003; Braguglia et al., 2006). The sonication of sewage sludge can be used as a pretreatment to anaerobic digestion. The anaerobic digestion process can potentially be made more efficient through the breakup and solubilization of solid sludge particles (Nickel and Neis, 2007; Pérez-Elvira et al., 2009). Biological cell lysis is known to be the rate-limiting step of anaerobic biosolids degradation (Benabdallah El-Hadj et al., 2007; Naddeo et al., 2009), but shear forces generated by low-frequency ultrasound assist in the disintegration of the bacterial cells in the sewage sludge (Tiehm et al., 2001; Geciova et al., 2002; Palmowski et al., 2006). Thus, the quantity of dissolved organic substrate is increased (Nickel and Neis, 2007) and the degradation rate, chemical oxygen demand solubilization (Naddeo et al., 2009), and the biodegradability of organic biosolids mass are improved as a result of subsequent enhancements of the acidogenesis, acetogenesis, and methanogenesis reactions. Several fundamental pilot studies with ultrasound application to anaerobic digestion have shown a significantly accelerated biosolids degradation with less digested sludge being produced and an increased biogas production being attained (Forster et al., 2000; Wang et al., 2005) under controlled sonolysis conditions of ultrasound density, sonication time, and specific energy (Naddeo et al., 2009).

SLUDGE MINIMIZATION

Sludge volume minimization technologies have been available for several decades. However, recent developments have brought some sludge minimization technologies to the forefront of applied research and engineering. All of the technologies utilize one or more of the three basic approaches to minimize the amount of waste activated sludge produced by an activated sludge process. These are cell lysis (Ward et al., 1999), cyclic oxic environments, and long solid retention time. Sludge minimization refers, in principle, to the optimum reduction of the mass of the sludge or biosolids produced at a wastewater treatment plant. The sludge minimization technologies that have emerged perform their main solids reduction mechanisms within the activated sludge process, prior to sludge stabilization and conversion to biosolid.

Early attempts at sludge minimization focused on long solids-retention times within the activated sludge process, and the reduced sludge production was seen as a benefit of extended aeration plants. Ultrasonic cell lysis was first developed through laboratory-scale research in the 1960s (Glauert, 1962) and was initially uneconomical due to limitations of the ultrasound equipment available at that time. Advances in ultrasound technology in the last decade have now enabled wider commercial application of the technology for wastewater applications for sludge minimization in the activated sludge process or in anaerobic digestion (Comninellis et al., 2008). The Cannibal™ process has shown recent success, and is marketed by Siemens USFilter. IDI has developed a competing process known as Biolysis® "O." Both of these technologies emphasize the cyclic alternation between aerobic, anoxic, and anaerobic environments. The MicroSludge™ homogenization process is another recent development for sludge minimization, relying on chemical pretreatment and mechanical shear forces to lyse bacterial cells.

PRINCIPLES OF ANAEROBIC DIGESTION

ANAEROBIC DIGESTION BIOTECHNOLOGY

Anaerobic processes are defined as biological processes in which organic matter is metabolized in an environment free of dissolved oxygen or its precursors (Khanal, 2008). The anaerobic process is classified as either anaerobic fermentation (Valdez-Vazquez et al., 2005; Ren et al., 2006) or anaerobic respiration (Rhoads et al., 2005), depending on the type of electron acceptors (Khanal, 2008).

In an anaerobic fermentation, organic matter is catabolized in the absence of an external electron acceptor by facultative anaerobes through internally balanced oxidation–reduction reactions under dark conditions (Khanal, 2008; Vatsala et al., 2008). The product generated during the process accepts the electrons released during the breakdown of organic matter. Thus, organic matter acts as both electron donor and acceptor. During the fermentation reactions, the substrate is only partially oxidized, and therefore, only a small amount of the energy stored in the substrate is conserved (Khanal, 2008). The major portion of the adenosine triphosphate (ATP) or energy is generated by substrate-level phosphorylation (Atlante et al., 2005; Sgarbi et al., 2009; Lemire et al., 2009). Anaerobic respiration, on the other hand, requires external electron acceptors for the disposal of electrons released during the degradation of organic matter. The electron acceptors in this case could be CO_2, SO_4^{2-}, or NO_3^-. Both substrate-level phosphorylation and oxidative phosphorylation generate energy (or ATP) (Khanal, 2008). The energy released under such a condition is much greater than anaerobic fermentation (Skoog et al., 2007). Skoog et al. (2007) have reported that at in situ geochemical conditions, where large numbers of heterotrophic microorganisms inhabit hydrothermal systems, for aldose being reacted upon by these microbial populations, fermentation yields 220–420 kJ mol^{-1} of energy while anaerobic respiration releases 500–2400 kJ mol^{-1}.

The anaerobic digestion process is characterized by a series of biochemical transformations brought on by different consortia of bacteria (Fantozzi and Buratti, 2009). The anaerobic digestion of organic matter basically follows the following stages: hydrolysis, acidogenesis, acetogenesis, and methanogenesis (Appels et al., 2008; Vavilin et al., 2008; Fountoulakis et al., 2008; Fantozzi and Buratti, 2009). Despite the successive steps, hydrolysis is generally considered as rate limiting (Appels et al., 2008) and the rate of hydrolysis depends on the pH, temperature, composition, and concentration of intermediate compounds (Fantozzi and Buratti, 2009). The hydrolysis step degrades both insoluble organic material and high molecular weight compounds such as lipids, polysaccharides, proteins, and nucleic acids, into soluble organic substances (e.g., amino acids and fatty acids) (Appels et al., 2008) by extracellular hydrolytic enzymes produced by hydrolytic bacteria and then dissolved into solution. The components formed during hydrolysis are further split during acidogenesis—the second step. Volatile fatty acids, alcohols (Fantozzi and Buratti, 2009) are produced by acidogenic bacteria (Bengtsson et al., 2008) along with ammonia, carbon dioxide, hydrogen sulfide, and other by-products (Göblös et al., 2008). This phase is accompanied by decrease of pH due to production of acids and protonic acidification. If the reactor is overloaded, low pH value may inhibit the process (Chen et al., 2008). The main species identified as responsible for the biological hydrogen production during the acidogenesis of the carbohydrates are *Enterobacter*, *Bacillus*, and *Clostridium* (Hawkes et al., 2002; Kotay and Das, 2007; Davila-Vazquez et al., 2008; Cai et al., 2009). It shows that, from a strict theoretical standpoint, fermentation pathways that produce acetate and butyrate are those that are mainly responsible for hydrogen production (Vavilin et al., 1995; Hawkes et al., 2002; Cheong and Hansen, 2007; Aceves-Lara et al., 2008). On the other hand, the pathways that produce ethanol, lactate, and propionate are unable to produce hydrogen, because they also consume their biochemical intermediates, like nicotinamide adenine dinucleotide (NADH) (Aceves-Lara et al., 2006). During the acidogenesis process, several operating conditions must be optimized to increase the hydrogen production. Among these conditions, the hydraulic retention time pH in the reactor and agitation of the liquid phase are considered as very important parameters. The pH is a key factor because a pH < 5 can induce the solvatogenesis (Sauer et al., 1995; Hawkes et al., 2002) or the bacterial sporulation (Sauer et al., 1995).

The third stage in anaerobic digestion is acetogenesis, where the higher organic acids and alcohols produced by acidogenesis (Dogan et al., 2009; Shida et al., 2009) are further digested by acetogens to mainly produce acetic acid as well as CO_2 and H_2. This conversion is controlled, to a large extent, by the partial pressure of H_2 in the mixture (Appels et al., 2008). The final stage of methanogenesis produces methane (Tatsuzawa et al., 2006) by two groups of methanogenic bacteria (Narihiro and Sekiguchi, 2007): the first group splits acetate into methane and carbon dioxide and the second group uses hydrogen as electron donor and carbon dioxide as acceptor to produce methane. The bacteria involved in the methanogenesis stage are sensitive to low as well as to high pH, which must be kept within a range of 6.5–8. Depending on the microorganism species, three components are used to produce methane: (i) mixture of CO_2 and H_2, (ii) acetic acid (CH_3COOH), and (iii) methanol (CH_3OH). Within the anaerobic environment, various important parameters affect the rates of the different steps of the digestion process (Appels et al., 2008). These are pH and alkalinity, temperature, solids and hydraulic retention times (Khanal, 2008), biomass yield, substrate utilization rate (Noike et al., 1985), microbiology (Tang et al., 2004), reactor configuration (Young and Yang, 1989), start-up time, and volatile acids/alkalinity ratio (Ince et al., 1995; Yacob et al., 2006).

Merits and Demerits of Anaerobic Biotechnology

Anaerobic biotechnology is becoming widely popular due to its potential to produce renewable biofuels and value-added products from low-value feedstock such as waste streams (Khanal, 2008). In addition, it provides an opportunity for the removal of pollutants from liquid and solid wastes more economically than the aerobic processes (Boyd et al., 1983; Govind et al., 1991; Field et al., 1995; O'Neill et al., 2000; Marttinen et al., 2003; Khanal, 2008). The merits and demerits of anaerobic digestion are outlined as follows.

Recovery of bioenergy and biofuels: Biomethane production, biohydrogen production, butanol production, biodiesel production from biogas, and electricity generation using microbial fuel cell

Recovery of value-added products: Recovery of acetic acid and production of nisin and lactic acid

Waste treatment: Less energy requirement, less sludge generation, less nutrients (N and P) requirement, higher volumetric organic loading rate, and ability to reduce concentrations of refractory organics.

Although the anaerobic process has many inherent benefits, it is not a panacea for the treatment of all types of wastewaters and sludges (Khanal, 2008). Some of the limitations of anaerobic treatment system are long start-up time, long recovery time, specific nutrients, and trace metal requirements, more susceptible to changes in environmental conditions, treatment of high-sulfate wastewater, and constant meticulous operational attention.

ULTRASOUND AND SONICATION IN ANAEROBIC DIGESTION

PRINCIPLES OF ULTRASOUND: CAVITATION

Using ultrasonic technology for sludge treatment is a relatively new application in the biosolids treatment. The development of ultrasound technology is based on the principle of using high-intensity ultrasound, at a frequency of 20 kHz (Hogan et al., 2004). The origin of the power of ultrasound in a liquid is primarily cavitation (Barber, 2005) arising from a pressure wave leading to rise to local temperature and a pressure above 500 bar. Ultrasound frequencies range from 20 kHz to 10 MHz. Ultrasound of high acoustic intensities causes cavitation in water bodies, if the energy applied exceeds the binding energy of the molecular attractive forces (Neis et al., 2000). During sound oscillation the local pressure in the aqueous phase falls below the evaporating pressure, resulting in the explosive formation of microscopic bubbles (Neis et al., 2000). These bubbles oscillate in the sound field over several oscillation periods and grow by a process termed rectified diffusion (Neis et al., 2000). These oscillations occur in an extremely small interval of time (microseconds) releasing large magnitudes of energy simultaneously at millions of such locations in water with contaminants (Barber, 2005). The subsequent implosion of the gas and vapor filled bubbles leads to high mechanical shear forces, which are apt to disintegrate bacterial cell material. Thus, ultrasonic treatment is a suitable method to disintegrate sewage sludge and to overcome the slow biological sludge hydrolysis. The physical cavitations caused by the ultrasonic probe disintegrate cellular material in sludge within short period of time.

The effect of sonication on filamentous organisms has also been reported by Wünsch et al. (2002). In the control sample analyzed, a large number of filaments were present causing a threadlike structure of the sludge. Even a low energy input of 0.05 kWh kg^{-1} could cause damage to the filamentous structure, and produce a large number of broken filament segments. The subsequent reduction of the filament length was reported to be sufficient to improve the settling behavior (Wünsch et al., 2002). An increase of the energy input of about 0.2 kWh kg^{-1} also caused a further cut of the thread segments and improved the settling behavior. Another mechanism that occurs when sludge is sonicated is acoustic streaming (Marmottant et al., 2006; Kuznetsova and Coakley, 2007; Mahulkar et al., 2009). Acoustic streaming has been studied since 1831 and occurs at the solid/liquid (sludge) interface when the solid interface experiences harmonic vibrations. The main benefit of streaming in sludge processing is mixing, which facilitates the uniform distribution of ultrasound energy (Kumar et al., 2007) within the sludge mass, convection of the liquid, and distribution of any heating that occurs.

Similar cell lysis step can take up to 8 days in a conventional mesophilic anaerobic digester. The mechanism of action, which consists of cavitation and sonochemical reactions, of ultrasonic radiation is such that it acts as a potential tool for enhanced biodegradation and sonodegradation of waste (Gupta et al., 2006), and also recovery of resources from treated wastes. In ultrasonic waste treatment technology, the frequency of ultrasound, and treatment time are significant factors in determining optimal reaction conditions (Wu et al., 2008). The optimum frequency is substrate specific and low frequency is suitable for sewage sludge treatment (Gupta et al., 2006). Another desirable advantage of ultrasonic technology is that it does not require chemicals or extreme environmental conditions such as pressure or temperature. Ultrasonic technology is used to increase solids hydrolysis by lysing waste activated sludge organisms, resulting in enhanced anaerobic digestion (Barber, 2005). Other applications of this technology include control of filamentous growth in wastewater treatment plants biological nutrient and sludge reductions, and increase in digester gas production.

Merits and Demerits of Ultrasound Technology

Ultrasound disintegration is essentially a physical process and, therefore, it neither generates secondary toxic compounds nor contributes additional chemical compounds. In addition to physical sludge disintegration, many toxic and recalcitrant organic pollutants, such as aromatic compounds, chlorinated aliphatic compounds, surfactants, and organic dyes are also broken down into simpler forms. This is due to generation of the highly oxidative reactive radicals—hydroxyl, hydrogen, and hydroperoxyl and hydrogen peroxide during ultrasound pretreatment, which lead to the oxidative breakdown of these recalcitrant compounds. Some other merits of ultrasound pretreatment are as follows:

- Compact design and easy retrofit within existing systems
- Low cost and efficient operation compared to several other pretreatments
- Production of an in situ carbon source for denitrification plants
- Complete process automation
- Potential to control filamentous bulking and foaming in the digester (Neis et al., 2000)
- Better digester stability
- Improved volatile solids destruction and biogas production
- Better sludge dewaterability
- Improved biosolids quality (biosolids with low residual biodegradable organics and low pathogen counts)

Ultrasound pretreatment also faces some challenges. One of the major issues is the high capital and operating costs of ultrasound units. The cost may go down as the technology becomes mature but that may take some years more. Similarly, long-term performance data of full-scale ultrasound systems are still limited. These collectively discourage design and waste engineers to recommend ultrasound systems for full-scale application.

Application of Ultrasound in Sludge Anaerobic Digestion

During recent decades, the anaerobic digestion process has been extensively studied and various methods for process enhancement have been explored (Clark and Nujjoo, 2007), but have not proved to be economically competitive despite their technical merits. For wastewater applications, it has been shown that ultrasound is most beneficial when applied on biological secondary solids (sludge), where rapid hydrolysis can be induced (Neis et al., 2000). Most of the work on ultrasound in wastewater applications has been done in Europe; in North America, California has been at the forefront of ultrasound development, with demonstrations conducted by Orange County Sanitation District and the Los Angeles County Sanitation District (Neis et al., 2000).

Ultrasound technology is also being implemented in Australia and Singapore. The Sonix™ technology has been evaluated around the globe and demonstration trials have been undertaken, resulting in a number of full-scale installations and has proven that the use of ultrasound can effectively enhance anaerobic digestion.

The use of low-frequency (10–60 kHz) ultrasound for enhancement of various biotechnological processes, include sludge pretreatment and conditioning for anaerobic digestion, has received increased attention as a rapid and reagentless (and hence green) method (Rokhina et al., 2009). Ultrasound-assisted anaerobic digestion is a promising alternative in the analysis of solid sludge and organic matter samples, when either simple dissolution or direct analysis is not applicable (Priego-Capote and Luque de Castro, 2007). However, the field of application of ultrasonic sample digestion is still small in comparison with classical digestion alternatives and, particularly, with microwave-assisted digestion (Priego-Capote and Luque de Castro, 2007). The close control at low temperatures of ultrasound applications allows the implementation of ultrasonic-assisted steps in biochemical analyses, and in this connection, ultrasonic enzymatic digestion and assistance of ultrasound for cell disruption (Mahulkar et al., 2009) are the key areas of application of ultrasound in anaerobic digestion (Priego-Capote and Luque de Castro, 2007).

Biosludge, which contains large quantities of water, biomass, and extracellular polymeric substances, is difficult to be dewatered as a compactable sludge (Yin et al., 2004). A combination of ultrasound or other method could agglomerate the sludge, improve the activities of biomass, enhance anaerobic process, and decrease over 10% final water content of sludge. The mechanisms of ultrasonic influence on sludge are not very clear, but the application of ultrasound to industrial process is relatively easy and possible. The application of ultrasonic pretreatment in anaerobic digestion of sludge focuses on the effects on the physical (floc size, filterability, settleability, bound water content, and surface charge), chemical (chemical oxygen demand, biochemical oxygen demand, and concentrations of divalent cations in supernatant), and biological (survival ratios of heterotrophic bacteria and of total coliform) characteristics of a waste activated sludge.

SLUDGE DISINTEGRATION AND ULTRASOUND CONTROL PARAMETERS

The main goal of sludge disintegration is to rupture the cell wall and to facilitate the release of intracellular matter in the aqueous phase. This accelerates the subsequent degradation and reduces the retention time needed during digestion. Sludge disintegration also disrupts other organic particles to low molecular weight compounds, and it is also important to appreciate that sludge disintegration is not just limited to sludge digestion. Sludge disintegration may also provide on-site soluble substrate to wastewater treatment plants that employ biological nutrient removal. Disintegration of cellular structures is most significant at low frequencies, because the bubble radius is inversely proportional to the frequency and large bubbles mean strong shear forces. Therefore, the ultrasound frequency of 20 kHz is considered the most appropriate.

The efficiency of the ultrasonic disintegration is governed by several factors. These factors can be broadly classified into three categories: sludge (solid) characteristics, sonication conditions, and design of ultrasonic components (Khanal et al., 2007a). The sludge characteristics such as type of sludge (primary solids, waste activated sludge, or animal manure), total solids content, and particle size have significant effects on ultrasonic disintegration. The definitions of some parameters important in qualifying the nature of the ultrasonic treatment are mainly

Ultrasound intensity, I (W m^{-2}), which is the energy flux per unit of emitting area. The cavitation threshold value for water is approximately 40 W cm^{-2}
Treatment time, t (s)
Power, P (W), which is the energy consumed during sonication per unit of time
Specific ultrasound power, PV (W m^{-3}), which is the power input per unit of sonicated volume

The sonication conditions comprising sonication time, intensity and density, temperature, pH, oscillation frequency, amplitude, and power input are some of the important parameters that affect the ultrasonic disintegration (Khanal et al., 2007a). Shen et al. (2007) have observed that ultrasound pretreatment can advance the quantity of chemical oxygen demand in sludge supernatant fluid, which increases with ultrasound intensity and sonication time. The degree of ultrasound disintegration was also found to increase with the specific energy input. When the specific energy input is 10,000 kJ kg^{-1} of total dry solids, the degree of ultrasonic sludge disintegration reached 40%. Akin et al. (2006) examined the effectiveness of ultrasound pretreatment on waste activated sludge disintegration at different specific energy inputs, ultrasonic densities, and total solids contents. Akin et al. (2006) found that the cut diameter (d_{50}) for waste activated sludge with a 2% total solids content declined nearly 6.5-fold at an ultrasonic density of 0.67 W mL^{-1}, while for higher total solids contents of 4% and 6%, higher densities of 1.03 and 0.86 W mL^{-1}, respectively, were needed to achieve the same degree of particle size reduction. Also, the efficacy of ultrasonic disintegration measured as soluble chemical oxygen demand release was primarily found to be governed by ultrasonic density, whereas ultrasonic density did not show a significant effect on the protein release at all total solids levels. Akin et al. (2006) also observed that sludge disintegration efficiency declined significantly at higher total solids content. Thus, there is most seemingly a limiting total solids concentration that could be effectively disintegrated by ultrasound, and this is governed by the capability of the ultrasonic unit in producing the desired and required cavitation effects. Based on turbidity and settling velocity measurements, Feng et al. (2009) deduced that the energy used for sonication strongly influences the physical–chemical characteristics of sludge, and 1000 kJ kg^{-1} total solids is recommended as an optimal specific energy input for improving sludge settling.

Moving from the established fact that ultrasonic treatment can disintegrate sludge, enhance microbial activity, and improve sludge dewaterability at different energy inputs, Li et al. (2009a) investigated the interrelationship among these three phenomena during ultrasonic treatment synchronously. An experimental model was also established to describe the process of ultrasonic sludge disintegration. Their analysis and results showed that the changes of sludge microbial activity and dewaterability were dependent on the extent of sludge disintegration during ultrasonic treatment. When sludge disintegration degree was lower than 20%, sludge flocs were disintegrated into microfloc aggregates and the microbial activity increased over 20%. However, when the sludge disintegration degree was over 40%, most cells were destroyed at different degrees, and the sludge activity decreased drastically. It was only at a sludge disintegration degree of 2%–5% that sludge dewaterability was improved with the conditioning of FeCl$_3$. Li et al. (2009a) equally found that sonication with low density and long durations was more efficient than sonication with high density and short duration with the same energy input for sludge disintegration.

Earlier, Show et al. (2007) examined the correlation of sonication operating condition, sludge property, formation, and behavior of cavitation bubbles in sludge disruption under low-frequency ultrasound sonication. The influence of sonication time, sonication density, type of sludge, and solids content on the disruption was evaluated. The most vigorous particle disruption was achieved in the initial period of sonication, which subsided subsequently. While sonication density exhibited the most significant role in cavitation bubble formation and behavior, Show et al. (2007) observed that particle disruption could be optimized for energy input by sonicating at higher density and shorter time. Based on theoretical consideration, Show et al. (2007) deduced that within an optimum sludge solids content ranging between 2.3% and 3.2%, superior particle disruption could be accomplished within a minute for secondary sludge sonicated at a density of 0.52 W mL^{-1}. Mao et al. (2004) had also noted a greater decrease in particle size and increase in soluble organics of sludge during the ultrasound treatment of primary and secondary sludges. They had consequently deduced that secondary sludge has a more remarkable improvement after sonication over primary sludge. With respects to the extent of disintegration and energy consumption, higher sonication density performed more effectively in terms of specific energy, and it was also established that there is evidently an optimal solids concentration range for both the sludges for optimum sonication.

EFFECTS OF SONICATION AND PARAMETER MONITORING

The main aim of ultrasound pretreatment being to destroy the cell wall of microbes and to release the intracellular materials to the aqueous phase, the quantitative evaluation of the effects of the pretreatment (Khanal et al., 2007a) becomes much valuable in providing useful data for the design and process optimization of an ultrasonic system. All the more, the quantitative evaluation and assessment of the ultrasound pretreatment performance becomes critically important to judge how efficiently the sludge is being disintegrated. To this end, different parameters have been employed to evaluate sludge disintegration efficiency. They can be collectively classified into physical (change in particle size distribution and microscopic examination), chemical (increase in soluble chemical oxygen demand concentration and ammonia concentration, release of protein), and biological (oxygen uptake rate and heterotrophic bacterial count) disintegration (Khanal et al., 2007a). Tan and Guodong (2010) have reported that *Pseudomonas* sp., *Comamonas* sp., and *Diaphorobacter* sp. had been identified as being able to utilize carbazole as a carbon source, survive in an anaerobic and ultrahigh temperature environment, and thereafter even become the dominant bacterial taxa during the with-ultrasound stage in a study on the status of, and changes in, the bacterial communities at two acclimation stages (with and without ultrasound) in a small 70°C ultrasound-enhanced anaerobic reactor for treating carbazole-containing wastewater using polymerase chain reaction combined with denaturing gradient gel electrophoresis (PCR-DGGE) and real-time PCR techniques. Tan and Guodong (2010) deduced that the total bacterial density in the with-ultrasonic stages was 10 times higher than in the without-ultrasonic treatment.

Particle size analysis, microscopic image, turbidity, and sludge dewaterability are some of the techniques adopted to assess the effectiveness of ultrasonic disintegration. Physical evaluation, especially particle size distribution and microscopic image analysis, has been widely employed for simplicity as qualitative measures of sludge disintegration. Chemical evaluation is far more quantitative for measuring sludge disintegration than physical (Khanal et al., 2007a). It primarily measures the solubilization of the waste activated sludge in the aqueous phase. The biological evaluation includes heterotrophic plate counts and specific oxygen uptake rate. Since waste activated sludge consists of heterotrophic bacteria, the measure of their survival during ultrasonic treatment could also provide representative data on the efficacy of ultrasonic disintegration (Khanal et al., 2007a). Table 13.2 presents a summary of parameters that have been monitored and the techniques used to assess the effectiveness of ultrasonic irradiation.

The sections to follow highlight some (latest) research findings of studies conducted to analyze the effects of ultrasound pretreatment on sludge properties and anaerobic digestion of sludge. Table 13.3 summarizes some other studies that report colateral improvements observed in the anaerobic digestion of sonicated sludge.

Ultrasound and Sludge Dewaterability

Yin et al. (2006) have reported in their essay, the influences of low frequency ultrasound (20 kHz) on the dewaterability and anaerobic digestion behaviors of activated sewage sludge, obtained from Yangzi Water Treatment Plant, Yangzi Petrochemical Corporation. Yin et al. (2006) found that ultrasound pretreatment had enhanced the filtration progress and decreased the moisture content of the sludge from 99% to 80%. After 2–4 min treatment of ultrasound under intensity of 400 W m^{-2}, the bound water of sludge decreased from 16.7 g g^{-1} (dry basis) to above 2.0 g g^{-1} (dry basis). Yin et al. (2006) also observed that the ultrasound pretreatment could enhance digestion and reduce digestion

TABLE 13.2
Parameters Assessed and Techniques for Assessment of Effectiveness of Ultrasonic Irradiation

Parameter(s) Assessed and Techniques	Reference
Fourier transform infrared (FTIR) spectra of centrifugation pellets	Laurent et al. (2009)
Potentiometric titration coupled with proton surface complexation modeling	Laurent et al. (2009)
Floc size and settleability	Laurent et al. (2009)
Monitoring changes in pyridine concentration, pH, dissolved oxygen, and chemical oxygen demand	Sistla (2005)
Calorimetry or acoustical measurements; image analysis particle counting was used to measure the size distribution of particles	Gibson et al. (2009)
Electrothermal atomic absorption spectrometry	Kazi et al. (2009)
Monitoring of amount of liberated iodine and the number of DNA double-strand breaks	Kondo and Kano (1988)
pH values, sucrose concentrations, and sound intensities	Sakakibara et al. (1996)
High-performance liquid chromatography–electrospray–mass spectrometry	Destaillats et al. (2000)
Saturating gas, initial pollutant concentration, ultrasonic power density, the category, and consumption of catalyst	Ning et al. (2005)
Oxygen uptake rate, proteases activity, and dehydrogenases activity of sludge	Zeng et al. (2006)

time (to the same resolution ratio, such as 49%, the digestion time of sludge with ultrasound pretreatment was 7 days less than that without ultrasound). Na et al. (2007) have equally investigated the dewaterability and physiochemical properties of digested sludge after treatment with ultrasonic energy for the purpose of reducing sludge. Their study involved laboratory experimentation under varying test conditions of treatment time, volume of sludge, and ultrasonic energy. The results of the experiments of Na et al. (2007) showed that particle size (dp_{50}, dp_{10}, and U) of the ultrasonically treated sludge had decreased due to the separation of sludge flocs. Capillary suction times had also decreased significantly, while turbidity, volatile dissolved solids/volatile solids, and soluble chemical oxygen demand/total chemical oxygen demand ratios had increased after ultrasonic treatment. From these results, Na et al. (2007) found that the ultrasonic treatment specified by the supplied energy can, in addition to improving the dewaterability, also reduce the volume and mass and change the chemical properties of sludge. More recently, Braguglia et al. (2009) have studied and compared the dewaterability parameters of untreated and "presonicated" sludge during semicontinuous anaerobic digestion by particle charge density and sludge filterability measurements. Braguglia et al. (2009) noted that in all the tests, despite the higher specific charge density and soluble chemical oxygen demand values of the sonicated feed, the digested sludge at steady state presented, in both reactors, statistically comparable values, independently of the pretreatment. Braguglia et al. (2009) explained that the biological hydrolysis of the untreated sludge causes a large release of dispersed charged fines whereas the digestion of the sonicated sludge is characterized by a significant removal of fines and colloids already present in the pretreated feed. Noting that the ultrasound pretreatment did not improve the dewaterability of the digested sludge significantly as compared to the unsonicated one, optimization of both disintegration degree and hydraulic residence time have then been proposed as being necessary to improve the dewaterability of the digested sonicated sludge. Lately, Feng et al. (2009) have investigated the potential benefits of ultrasound-conditioned sludge dewatering treatments with specific energy dosages from 0 to 35,000 kJ kg^{-1} total solids. Their results indicated that the application of low specific energy dosages of less than 4400 kJ kg^{-1} total solids slightly enhanced sludge dewaterability, but larger specific energy dosages of more than 4400 kJ kg^{-1} total solids significantly deteriorated sludge dewaterability. The optimal specific energy to give maximal and satisfactory dewaterability characteristics was found to be 800 kJ kg^{-1} total solids, which generated sludge with particle size distribution of 80–90 μm diameter.

TABLE 13.3
Improvements Observed in Sludge Properties and Anaerobic Digestion Performance of Sonicated Sludge

Sonicated Substrate(s)	Sonication Conditions	Improvements Observed	Reference
Mixed anaerobic sludge	Effect of low-frequency ultrasound pretreatment to inoculum on performance of microbial fuel cell was evaluated	Maximum power density during polarization in a microbial fuel cell inoculated with ultrasonication pretreatment to the sludge for 5 min (40 kHz, 120 W) was 2.5 times higher than that obtained without any pretreatment to the inoculum sludge. Substrate removal was higher in the microbial fuel cell with ultrasonicated inoculum, than inoculum without any pretreatment and combined pretreated with ultrasonication and heating	More and Ghangrekar (2010)
Anaerobic sludge	Low-intensity ultrasound irradiation on anaerobic sludge activity dehydrogenate activity and the content of coenzyme F420 monitored	Biological activity was enhanced dramatically under optimal conditions (35 kHz, 0.2 W cm^{-2} for 10 min). Chemical oxygen demand removal efficiency was increased and was 30% lower than that of the control (without exposure)	Xie et al. (2009)
Sonicated activated sludge	Anaerobic digestions were compared in reactors	Anaerobic biodegradability was improved for a sonication treatment of 108,000 kJ kg TS^{-1} due to the increase of the instantaneous specific soluble chemical oxygen demand uptake rate. Sonication led to an increase of biogas production due to the increase of available soluble chemical oxygen demand	Salsabil et al. (2009)
Waste activated sludge	Alkaline and ultrasonic sludge disintegration were studied	For waste activated sludge samples with combined pretreatment, the released chemical oxygen demand levels were higher than those with ultrasonic or alkaline pretreatment alone. Using combined NaOH and ultrasonic pretreatment with optimal parameters, the degradation efficiency of organic matter was increased from 38.0% to 50.7%	Jin et al. (2009)
Sewage sludge	Effect of low power ultrasonic radiation on anaerobic biodegradability was studied	Optimal parameters were found to be an exposure time of 15 min, ultrasonic intensity of 0.35 W cm^{-2}, and ultrasonic power density of 0.25 W mL^{-1}. Under optimal conditions, anaerobic biodegradability of sewage sludge was increased by 67.6%	Liu et al. (2009)

Substrate	Method	Results	Reference
Wastewater sludge (different wastewater sludge solids concentrations, ultrasonication intensities, and exposure times of pretreatment were investigated)	Ultrasonic waves at frequency of 20 kHz using fully automated lab-scale ultrasonication equipment were employed	Optimal conditions of ultrasonic pretreatment were 0.75 W cm^{-2} ultrasonication intensity, 60 min and 23 g L^{-1} total solids concentration	Pham et al. (2009)
Excess sludge from the wastewater treatment system	Conication and cryptic growth were studied in subsequent biological reactors	Flowability of ultrasonicated sludge in terms of viscosity showed exponential behavior at different total solids concentrations, and pseudoplastic and thixotropic behavior similar to raw sludge Amount of excess sludge was reduced by 58.8%	Li et al. (2008)
Extracellular proteins, polysaccharides and five types of hydrolytic enzymes (protease, α-amylase, α-glucosidase, alkaline-phosphatase and acid-phosphatase) from sludge flocs	Sonication pretreatment	Optimal sludge sonication ratio was 1/7, the ultrasonic intensity was 1.6 W mL^{-1}, the irradiation time was 15 min and the ultrasonic frequency was 25 kHz Optimum ultrasonic pretreatment conditions: 10 min and density of 3 kW L^{-1} at the frequency of 20 kHz At optimum ultrasonic pretreatment, sludge reduction for total suspended solids in aerobic digestion was 42.7% in which the part of 11.8% was removed by the ultrasonic pretreatment, compared with 20.9% for control	Yu et al. (2008)
Concentrated activated sludge	Concentrated sludge was sonicated in an extra chamber for short period and then returned to the activated sludge system	Results showed that the bioactivity of the activated sludge, expressed as oxygen utilization rate was enhanced by ultrasonic irradiation Optimal sonication conditions were sound frequency of 25 kHz, power density of 0.2 W mL^{-1} and duration of 30 s under which the sludge oxygen utilization rate increased by 28%, the biomass growth rate increased by 12.5%, and the wastewater chemical oxygen demand and total nitrogen removal efficiency increased by 5%–6%	Zhang et al. (2008)
Excess sludge from the sequential batch reactor system	Partial sludge was disintegrated into dissolved substrates by ultrasound in an external sono-tank and was then returned to the SBR for biodegradation	Most effective conditions for sludge reduction were: sludge sonication ratio of 3/14, ultrasound intensity of 120 kW kg^{-1} DS, and sonication duration of 15 min Amount of excess sludge was reduced by 91.1% to 7.8 mg L^{-1} d^{-1}	Zhang et al. (2007)
Biosolids	Low energy system	A mean 17% increase in daily biogas production, a mean increase in additional total solids destruction of 6.2% as well as a mean increase of 5.9% of volatile solids destruction More stable biosolids produced with a 10%–24% reduction in organo-sulfur odor potential	

Ultrasound and Enhanced Biogas Production

Quarmby et al. (1999) have compared the performance of anaerobic digestion when fed with unsonicated sludge and sludges sonicated at two different intensities in a series of batch flasks and three 100 L anaerobic digesters. The results from the batch tests clearly indicated the positive effect on anaerobic digestion through an enhanced gas production increasing by 15% and volatile fatty acid production. Although a little difference was observed between the volumes of gas produced in the digesters, there was an increase in methane production, volatile solids reduction, and soluble chemical oxygen demand by up to 6%, 5.5%, and 15%, respectively, when a comparison was made between sonicated and unsonicated sludges. Later, Forster et al. (2000) treated thickened waste-activated sludge by ultrasound and demonstrated, based on the residual turbidity and the release of soluble carbohydrate, the optimum dose was $1.5–3.0\,kJ\,g^{-1}$ of total solids. About 500 mL digesters, which were fed daily, were used to compare the gas yields obtained from sonicated and unsonicated waste activated sludges. Forster et al. (2000) found that, with a 10 day hydraulic retention time, sonication increased the biogas yield by 15%. Both digested sludge and waste activated sludges, which had been thickened to more than $15\,g\,L^{-1}$ total solids, had a distinct yield stress suggesting in the end that sonication could be used as a pretreatment to enhance thermophilic digestion and as a posttreatment to improve the pumping characteristics of sludge. McDermott et al. (2001) have assessed the effectiveness of ultrasonication as a pretreatment method for the psychrophilic anaerobic treatment of aquaculture effluents in 4 L solids digesters. A 10% enhancement in the removal of chemical oxygen demand by anaerobic digestion and a concurrent increase in total biogas production from 0.29 to $0.45\,L\,day^{-1}$ with a corresponding 10% increase in methane concentration were the most noteworthy results. Lafitte-Trouqué and Forster (2002) have examined the effect of ultrasound as pretreatment for the anaerobic digestion of waste activated sludge at both mesophilic and thermophilic temperatures. The sonication time was 90 s using a Soniprep 150 (MSE Scientific Instruments), which had been operated at 23 kHz and had been adjusted to give an output of 47 W. The digesters were operated in a semicontinuous mode, being fed with fresh sludge every 24 h at hydraulic retention times of 8, 10, and 12 days. It was found that the thermophilic digestion performed better than mesophilic digestion in terms of biogas production, volatile solids reductions, and specific methane. Bohdziewicz et al. (2005) determined the influence of ultrasonic field on biodegradation of refractory compounds in leachate and on enhancement of treatment efficiency during anaerobic digestion process. It was found that in the case of leachate ultrasonication for 300 s and at the amplitude of 14 m, the chemical oxygen demand removal efficiency was by 7% higher compared with that in fermentation of nonconditioned wastewater. An increase in a biogas production was also observed and the specific methane yield was by 22% higher compared with that of nonconditioned leachate. Benabdallah El-Hadj et al. (2007) also focused their study on the effect of ultrasonic pretreatment on raw sewage sludge before being fed to mesophilic and thermophilic anaerobic digestion. In concert with previous results, it was found that the use of pretreated sludge improved significantly the chemical oxygen demand removal efficiency and biogas production in lab-scale anaerobic digesters when compared with the performance without pretreatment, especially under mesophilic conditions. Bougrier et al. (2005) have also observed that ultrasound led to an increase in biogas production. Kim and Lee (2005) applied ultrasound to enhance the activity of anaerobic granules and found that the specific methanogenic activity had increased by 26%–84% (St. Louis plant) and 163%–220% (Newark plant) under the conditions of 50, 100, and 150 W for 5 min at a frequency of 40 kHz. Braguglia et al. (2006) carried out experiments with bench scale anaerobic reactors fed with either untreated or disintegrated excess sludge, added with a biomass inoculum taken from a full scale anaerobic digester. Beneficial results had been recorded for biogas production with a maximum gain of 25% at 0.5 feed/inoculum ratio. In their investigations on the influence of ultrasonication on hydrolysis, acidogenesis, and methanogenesis in the anaerobic decomposition of sludge, Mao and Show (2007) found that sonicated sludge exhibited prehydrolysis and preacidogenesis effects in the anaerobic decomposition process. They equally noted that digesters fed with sonicated sludge demonstrated enhanced methanogenesis over the control unit. The overall quantitative results from Mao and Show (2007) unanimously suggested that ultrasonication could enhance anaerobic decomposition of sludge (Grönroos et al., 2005; Ding et al., 2006), resulting in an accelerated bioconversion, improved organics degradation, improved biogas production, and an increased methane content.

RESEARCH AVENUES AND CONCLUDING REMARKS

Ultrasonic irradiation application for sludge pretreatment for enhanced anaerobic digestion is an emerging area of research. Although a good number of publications are now available in the literature on this burgeoning research, there are still a number of issues that need to be researched further on. There are inconsistencies in many of the previously published papers but a well-designed standardization of methodologies and experimental protocols may surely make comparisons between findings of different researchers easier, consistent, interpretable, and ultimately scalable. Additionally, there is also a much justified need to perform a thorough study of the effects of different total solids content, sludge characteristics, operating temperature, pH, ultrasonic density, specific energy input, and ultrasonic intensity among others for elucidating the individual mechanistic aspects of each parameter on the rates of ultrasound-assisted sludge disintegration, and thereafter also attempt to model these disintegration phenomena and eventually determine the relative significance of each of these factors through a robust multiparameter sensitivity analysis. Such an intricate, highly demanding (in terms of computational power), but robust and reliable sensitivity analysis may be designed based on the procedures described earlier by Parker (1997). The expected results of the sensitivity analysis shall depict those physical, chemical, and/or biological process parameters whose effects are most significant in controlling the rate(s) of ultrasound-assisted anaerobic digestion of sludge. These important findings shall most hopefully eventually assist in better reactor design and optimization of ultrasonic pretreatment units at pilot and commercial scales.

Sonication, hence, shows great potential in sludge and other related biosolid residuals pretreatment, and its application in sludge disintegration may significantly improve the overall biodegradability of biological sludge during anaerobic digestion. Nevertheless, thorough cost-benefit analyses of ultrasonic-integrated systems are also a must for minimizing and justifying the economics of the process in full-scale applications.

ACKNOWLEDGMENT

The authors are grateful to all the researchers whose valuable research findings and discussions reported in their respective publication(s) listed below have been of significance in synthesizing this chapter. The authors are also much thankful to Dr. Samir K. Khanal (Department of Molecular Biosciences and Bioengineering, University of Hawaii at Mānoa) for providing valuable data and discussions points and the anonymous reviewers whose suggestions and criticisms have helped in improving this chapter.

REFERENCES

Aceves-Lara, C.A., Latrille, E., Buffière, P., Bernet, N., and Steyer, J.P. 2008. Experimental determination by principal component analysis of a reaction pathway of biohydrogen production by anaerobic fermentation. *Chemical Engineering and Processing: Process Intensification*, 47:1968–1975.

Aceves-Lara, C.A., Latrille, E., Conte, T., Bernet, N., Buffière, P., and Steyer, J.P. 2006. Optimization of hydrogen production in anaerobic digestion processes. *16th World Hydrogen Energy Conference, (WHEC)*, Lyon, France.

Aitken, M.D., Walters, G.W., Crunk, P.L., Willis, J.L., Farrell, J.B., Schafer, P.L., Arnett, C., and Turner, B.G. 2005. Laboratory evaluation of thermophilic-anaerobic digestion to produce Class A biosolids. 1. Stabilization performance of a continuous-flow reactor at low residence time. *Water Environment Research*, 77:3019–3027.

Akin, B., Khanal, S.K., Sung, S., Grewell, D., and Van Leeuwen, J. 2006. Ultrasound pre-treatment of waste activated sludge. *Water Science and Technology*, 6:35–42.

Aliyu, M. and Hepher, M.J. 2000. Effects of ultrasound energy on degradation of cellulose material. *Ultrasonics Sonochemistry*, 7:265–268.

Appels, L., Baeyens, J., Degrève, J., and Dewil, R. 2008. Principles and potential of the anaerobic digestion of waste-activated sludge. *Progress in Energy and Combustion Science*, 34:755–781.

Arnaiz, C., Gutierrez, J.C., and Lebrato, J. 2006. Biomass stabilization in the anaerobic digestion of wastewater sludges. *Bioresource Technology*, 97:1179–1184.

Atlante, A., Giannattasio, S., Bobba, A., Gagliardi, S., Petragallo, V., Calissano, P., Marra, E., and Passarella, S. 2005. An increase in the ATP levels occurs in cerebellar granule cells en route to apoptosis in which ATP derives from both oxidative phosphorylation and anaerobic glycolysis. *Biochimica et Biophysica Acta (BBA)—Bioenergetics*, 1708:50–62.

Barber, W.P. 2005. The effects of ultrasound on sludge digestion. *Journal of the Chartered Institution of Water and Environmental Management*, 19:2–7.

Benabdallah El-Hadj, T., Dosta, J., Márquez-Serrano, R., and Mata-Álvarez, J. 2007. Effect of ultrasound pre-treatment in mesophilic and thermophilic anaerobic digestion with emphasis on naphthalene and pyrene removal. *Water Research*, 41:87–94.

Bengtsson, S., Hallquist, J., Werker, A., and Welander, T. 2008. Acidogenic fermentation of industrial wastewaters: Effects of chemostat retention time and pH on volatile fatty acids production. *Biochemical Engineering Journal*, 40:492–499.

Bohdziewicz, J., Kwarciak, A., and Neczaj, E. 2005. Influence of ultrasound field on landfill leachate treatment by means of anaerobic process. *Environment Protection Engineering*, 31:61–71.

Bohn, I., Björnsson, L., and Mattiasson, B. 2007. The energy balance in farm scale anaerobic digestion of crop residues at 11–37°C. *Process Biochemistry*, 42:57–64.

Borges, E.S.M. and Chernicharo, C.A.L. 2009. Effect of thermal treatment of anaerobic sludge on the bioavailability and biodegradability characteristics of the organic fraction. *Brazilian Journal of Chemical Engineering*, 26:469–480.

Bougrier, C., Carrère, H., and Delgenès, J.P. 2005. Solubilisation of waste-activated sludge by ultrasonic treatment. *Chemical Engineering Journal*, 106:163–169.

Bougrier, C., Delgenès, J.P., and Carrère, H. 2008. Effects of thermal treatments on five different waste activated sludge samples solubilisation, physical properties and anaerobic digestion. *Chemical Engineering Journal*, 139:236–244.

Boyd, S.A., Shelton, D.R., Berry, D., and Tiedje, J.M. 1983. Anaerobic biodegradation of phenolic compounds in digested sludge. *Applied and Environmental Microbiology*, 46:50–54.

Braguglia, C.M., Gianico, A., and Mininni, G. 2009. Effect of ultrasound on particle surface charge and filterability during sludge anaerobic digestion. *Water Science and Technology*, 60:2025–2033.

Braguglia, C.M., Mininni, G., Tomei, M.C., and Rolle, E. 2006. Effect of feed/inoculum ratio on anaerobic digestion of sonicated sludge. *Water Science and Technology*, 54:77–84.

Cai, J.L., Wang, G.C., Li, Y.C., Zhu, D.L., and Pan, G.H. 2009. Enrichment and hydrogen production by marine anaerobic hydrogen-producing microflora. *Chinese Science Bulletin*, 54:2656–2661.

Chen, Y., Cheng, J.J., and Creamer, K.S. 2008. Inhibition of anaerobic digestion process: A review. *Bioresource Technology*, 99:4044–4064.

Cheong, D.Y. and Hansen, C.L. 2007. Feasibility of hydrogen production in thermophilic mixed fermentation by natural anaerobes. *Bioresource Technology*, 98:2229–2239.

Chisti, Y. 2003. Ultrasound—The power of a silent gong. *Biotechnology Advances*, 21:1.

Chu, C.P., Chang, B.V., Liao, G.S., Jean, S.D., and Lee, D.J. 2001. Observations on changes in ultrasonically treated waste-activated sludge. *Water Research*, 35:1038–1046.

Chu, L., Yan, S., Xing, X.H., Sun, X., and Jurcik, B. 2009. Progress and perspectives of sludge ozonation as a powerful pretreatment method for minimization of excess sludge production. *Water Research*, 43:1811–1822.

Cintas, P. and Luche, J.L. 1999. Green chemistry: The sonochemical approach. *Green Chemistry*, 1:115–125.

Clark, P.B. and Nujjoo, I. 2007. Ultrasonic sludge pretreatment for enhanced sludge digestion. *SO: Water and Environment Journal*, 14:66–71.

Comninellis, C., Kapalka, A., Malato, S., Parsons, S.A., Poulios, I., and Mantzavinos, D. 2008. Advanced oxidation processes for water treatment: Advances and trends for R&D. *Journal of Chemical Technology & Biotechnology*, 83:769–776.

Davila-Vazquez, G., Arriaga, S., Alatriste-Mondragón, F., de León-Rodríguez, A., Rosales-Colunga, L.M., and Razo-Flores, E. 2008. Fermentative biohydrogen production: Trends and perspectives. *Reviews in Environmental Science and Biotechnology*, 7:27–45.

Destaillats, H., Hung, H.M., and Hoffmann, M.R. 2000. Degradation of alkylphenol ethoxylate surfactants in water with ultrasonic irradiation. *Environmental Science and Technology*, 34:311–317.

Ding, W., Li, D., Zeng, X., and Long, T. 2006. Enhancing excess sludge aerobic digestion with low intensity ultrasound. *Journal of Central South University of Technology*, 13:408–411.

Dogan, E., Dunaev, T., Erguder, T.H., and Demirer, G.N. 2009. Performance of leaching bed reactor converting the organic fraction of municipal solid waste to organic acids and alcohols. *Chemosphere*, 74:797–803.

Dytczak, M.A., Londry, K.L., Siegrist, H., and Oleszkiewicz, J.A. 2007. Ozonation reduces sludge production and improves denitrification. *Water Research*, 41:543–550.

Eskicioglu, C., Kennedy, K.J., and Droste, R.L. 2006. Characterization of soluble organic matter of waste activated sludge before and after thermal pretreatment. *Water Research*, 40:3725–3736.

Eskicioglu, C., Kennedy, K.J., and Droste, R.L. 2007. Enhancement of batch waste activated sludge digestion by microwave pretreatment. *Water Environment Research*, 79:2304–2317.

Fantozzi, F. and Buratti, C. 2009. Biogas production from different substrates in an experimental continuously stirred tank reactor anaerobic digester. *Bioresource Technology*, 100:5783–5789.

Feng, X., Deng, J., Lei, H., Bai, T., Fan, Q., and Li, Z. 2009. Dewaterability of waste activated sludge with ultrasound conditioning. *Bioresource Technology*, 100:1074–1081.

Field, J.A., Stams, A.J.M., Kato, M., and Schraa, G. 1995. Enhanced biodegradation of aromatic pollutants in cocultures of anaerobic and aerobic bacterial consortia. *Antonie van Leeuwenhoek*, 67:47–77.

Fitzmorris, K.B., Sarmiento, F., and O'Callaghan, P. 2009. Biosolids and sludge management. *Water Environment Research*, 1376–1393.

Forster, C.F., Chacin, E., and Fernandez, N. 2000. The use of ultrasound to enhance the thermophilic digestion of waste activated sludge. *Environmental Technology*, 21:357–362.

Fountoulakis, M.S., Stamatelatou, K., and Lyberatos, G. 2008. The effect of pharmaceuticals on the kinetics of methanogenesis and acetogenesis. *Bioresource Technology*, 99:7083–7090.

Geciova, J., Bury, D., and Jelen, P. 2002. Methods for disruption of microbial cells for potential use in the dairy industry—A review. *International Dairy Journal*, 12:541–553.

Gibson, J.H., Hon, H., Farnood, R., Droppo, I.G., and Seto, P. 2009. Effects of ultrasound on suspended particles in municipal wastewater. *Water Research*, 43:2251–2259.

Glauert, A.M. 1962. The fine structure of bacteria. *British Medical Bulletin*, 18:245–250.

Göblös, Sz., Portörő, P., Bordás, D., Kálmán, M., and Kiss, I. 2008. Comparison of the effectivities of two-phase and single-phase anaerobic sequencing batch reactors during dairy wastewater treatment. *Renewable Energy*, 33:960–965.

Govind, R., Flaherty, P.A., and Dobbs, R.A. 1991. Fate and effects of semivolatile organic pollutants during anaerobic digestion of sludge. *Water Research*, 25:547–555.

Grönroos, A., Kyllönen, H., Korpijärvi, K., Pirkonen, P., Paavola, T., Jokela, J., and Rintala, J. 2005. Ultrasound assisted method to increase soluble chemical oxygen demand (SCOD) of sewage sludge for digestion. *Ultrasonics Sonochemistry*, 12:115–120.

Gupta, S.K., Behari, J., and Kesari, K.Kr. 2006. Low frequencies ultrasonic treatment of sludge. *Asian Journal of Water, Environment and Pollution*, 3:101–105.

Hao, X.-D., Zhang, L.-P., and Li, L. 2007. Global overview of excess sludge treatment and disposal methods. *China Water & Wastewater*, 23:1–5.

Harasimowicz, M., Orluk, P., Zakrzewska-Trznadel, G., and Chmielewski, A.G. 2007. Application of polyimide membranes for biogas purification and enrichment. *Journal of Hazardous Materials*, 144:698–702.

Hawkes, F.R., Dinsdale, R., Hawkes, D.L., and Hussy, I. 2002. Sustainable fermentative hydrogen production: Challenges for process optimization. *International Journal of Hydrogen Energy*, 27:1339–1347.

Hogan, F., Mormede, S., Clark, P., and Crane, M. 2004. Ultrasonic sludge treatment for enhanced anaerobic digestion. *Water Science and Technology*, 50:25–32.

Ince, O., Anderson, G.K., and Kasapgil, B. 1995. Control of organic loading rate using the specific methanogenic activity test during start-up of an anaerobic digestion system. *Water Research*, 29:349–355.

Jin, Y., Li, H., Mahar, R.B., Wang, Z., and Nie, Y. 2009. Combined alkaline and ultrasonic pretreatment of sludge before aerobic digestion. *Journal of Environmental Sciences*, 21:279–284.

Kazi, T.G., Jamali, M.K., Arain, M.B., Afridi, H.I., Jalbani, N., Sarfraz, R.A., and Ansari, R. 2009. Evaluation of an ultrasonic acid digestion procedure for total heavy metals determination in environmental and biological samples. *Journal of Hazardous Materials*, 161:1391–1398.

Khanal, S.K. 2008. Overview of anaerobic biotechnology. In *Anaerobic Biotechnology for Bioenergy Production: Principles and Applications*, pp. 1–27, Chap. 1, John Wiley & Sons and Blackwell Publishing, New York.

Khanal, S.K., Grewell, D., Sung, S., and Van Leeuwen, J. 2007a. Ultrasound applications in wastewater sludge pretreatment: A review. *Critical Reviews in Environmental Science and Technology*, 37:277–313.

Khanal, S.K., Montalbo, M., van Leeuwen, J., Srinivasan, G., and Grewell, D. 2007b. Ultrasonic enhanced liquefaction and saccharification of corn for bio-fuel production. American Society of Agricultural and Biological Engineers (Abstract available at http://asae.frymulti.com//abstract.asp?aid=23391&t=1. Accessed on November 17, 2009), 2007b ASAE Annual Meeting 072710.

Kidak, R., Wilhelm, A.-M., and Delmas, H. 2009. Effect of process parameters on the energy requirement in ultrasonical treatment of waste sludge. *Chemical Engineering and Processing: Process Intensification*, 48:1346–1352.

Kim, Y. and Lee, J. 2005. Effect of ultrasound on methanogenic activity of anaerobic granules. *Japanese Journal of Applied Physics*, 44:8259–8261.

Kondo, T. and Kano, E. 1988. Effect of free radicals induced by ultrasonic cavitation on cell killing. *International Journal of Radiation Biology*, 54:475–486.

Kotay, S.M. and Das, D. 2007. Microbial hydrogen production with *Bacillus coagulans* IIT-BT S1 isolated from anaerobic sewage sludge. *Bioresource Technology*, 98:1183–1190.

Kumar, A., Gogate, P.R., and Pandit, A.B. 2007. Mapping the efficacy of new designs for large scale sonochemical reactors. *Ultrasonics Sonochemistry*, 14:538–544.

Kuznetsova, L.A. and Coakley, W.T. 2007. Applications of ultrasound streaming and radiation force in biosensors. *Biosensors and Bioelectronics*, 22:1567–1577.

Lafitte-Trouqué, S. and Forster, C.F. 2002. The use of ultrasound and γ-irradiation as pre-treatments for the anaerobic digestion of waste activated sludge at mesophilic and thermophilic temperatures. *Bioresource Technology*, 84:113–118.

Laurent, J., Casellas, M., Pons, M.N., and Dagot, C. 2009. Flocs surface functionality assessment of sonicated activated sludge in relation with physico-chemical properties. *Ultrasonics Sonochemistry*, 16:488–494.

Lee, S.H., Nguyen, H.M., Koo, Y.M., and Ha, S.H. 2008. Ultrasound-enhanced lipase activity in the synthesis of sugar ester using ionic liquids. *Process Biochemistry*, 43:1009–10012.

Lemire, J., Mailloux, R., Puiseux-Dao, S., and Appanna, V.D. 2009. Aluminum-induced defective mitochondrial metabolism perturbs cytoskeletal dynamics in human astrocytoma cells. *Journal of Neuroscience Research*, 87:1474–1483.

Li, H., Jin, Y., Mahar, R.B., Wang, Z., and Nie, Y. 2009a. Effects of ultrasonic disintegration on sludge microbial activity and dewaterability. *Journal of Hazardous Materials*, 161:1421–1426.

Li, X., Ma, H., Wang, Q., Matsumoto, S., Maeda, T., and Ogawa, H.I. 2009b. Isolation, identification of sludge-lysing strain and its utilization in thermophilic aerobic digestion for waste activated sludge. *Bioresource Technology*, 100:2475–2481.

Li, W., Zhang, G., Zhang, P., and Liu, H. 2008. Waste activated sludge reduction using sonication and cryptic growth. *International Journal of Biotechnology*, 10:64–72.

Liu, C., Xiao, B., Dauta, A., Peng, G., Liu, S., and Hu, Z. 2009. Effect of low power ultrasonic radiation on anaerobic biodegradability of sewage sludge. *Waste Management*, 28:2614–2622.

López Torres, M. and Espinosa Llorérs, M.C. 2008. Effect of alkaline pretreatment on anaerobic digestion of solid wastes. *Waste Management*, 28:2229–2234.

Lu, J., Gavala, H.N., Skiadas, I.V., Mladenovska, Z., and Ahring, B.K. 2008. Improving anaerobic sewage sludge digestion by implementation of a hyper-thermophilic prehydrolysis step. *Journal of Environmental Management*, 88:881–889.

Mahulkar, A.V., Riedel, C., Gogate, P.R., Neis, U., and Pandit, A.B. 2009. Effect of dissolved gas on efficacy of sonochemical reactors for microbial cell disruption: Experimental and numerical analysis. *Ultrasonics Sonochemistry*, 16:635–643.

Mao, T., Hong, S.-Y., Show, K.-Y., Tay, J.-H., and Lee, D.-J. 2004. A comparison of ultrasound treatment on primary and secondary sludges. *Water Science and Technology*, 50:91–97.

Mao, T. and Show, K.-Y. 2007. Influence of ultrasonication on anaerobic bioconversion of sludge. *Water Environment Research*, 79:436–441.

Marmottant, P., Versluis, M., de Jong, N., Hilgenfeldt, S., and Lohse, D. 2006. High-speed imaging of an ultrasound-driven bubble in contact with a wall: "Narcissus" effect and resolved acoustic streaming. *Experiments in Fluids*, 41:147–153.

Marttinen, S.K., Kettunen, R.H., Sormunen, K.M., and Rintala, J.A. 2003. Removal of bis(2-ethylhexyl) phthalate at a sewage treatment plant. *Water Research*, 37:1385–1393.

Mason, T.J. 2000. Large scale sonochemical processing: Aspiration and actuality. *Ultrasonics Sonochemistry*, 7:145–149.

Mason, T.J., Joyce, E.E., Phull, S.S., and Lorimer, J.P. 2003. Potential uses of ultrasound in the biological decontamination of water. *Ultrasonics Sonochemistry*, 10:319–323.

McDermott, B.L., Chalmers, A.D., and Goodwin, J.A.S. 2001. Ultrasonication as a pre-treatment method for the enhancement of the psychrophilic anaerobic digestion of aquaculture effluents. *Environmental Technology*, 22:823–830.

Mines Jr., R.O., Lackey, L.W., and Tribble, D. 2008. Bench-scale ozonation of waste activated sludge. *Proceeding of World Environmental and Water Resources Congress 2008*, Ahupua'A, pp. 1–8.

More, T.T. and Ghangrekar, M.M. 2010. Improving performance of microbial fuel cell with ultrasonication pretreatment of mixed anaerobic inoculum sludge. *Bioresource Technology*, 101:562–567.

Mottet, A., Steyer, J.P., Déléris, S., Vedrenne, F., Chauzy, J., and Carrère, H. 2009. Kinetics of thermophilic batch anaerobic digestion of thermal hydrolysed waste activated sludge. *Biochemical Engineering Journal*, 46:169–175.

Na, S., Kim, Y.U., and Khim, J. 2007. Physiochemical properties of digested sewage sludge with ultrasonic treatment. *Ultrasonics Sonochemistry*, 14:281–285.

Naddeo, V., Belgiorno, V., Landi, M., Zarra, M., and Napoli, R.M.A. 2009. Effect of sonolysis on waste activated sludge solubilisation and anaerobic biodegradability. *Desalination*, 249:762–767.

Narihiro, T. and Sekiguchi, Y. 2007. Microbial communities in anaerobic digestion processes for waste and wastewater treatment: A microbiological update. *Current Opinion in Biotechnology*, 18:273–278.

Nasseri, S., Vaezi, F., Mahvi, A.H., Nabizadeh, R., and Haddadi, S. 2006. Determination of the ultrasonic effectiveness in advanced wastewater treatment. *Iranian Journal of Environmental Health Science & Engineering*, 3:109–116.

Neis, U., Nickel, K., and Tiehm, A. 2000. Enhancement of anaerobic sludge digestion by ultrasonic disintegration. *Water Science and Technology*, 42:73–80.

Neyens, E. and Baeyens, J. 2003. A review of thermal sludge pre-treatment processes to improve dewaterability. *Journal of Hazardous Materials*, 98:51–67.

Nickel, K. and Neis, U. 2007. Ultrasonic disintegration of biosolids for improved biodegradation. *Ultrasonics Sonochemistry*, 14:450–455.

Nikolopoulos, A.N., Igglessi-Markopoulou, O., and Papayannakos, N. 2006. Ultrasound assisted catalytic wet peroxide oxidation of phenol: Kinetics and intraparticle diffusion effects. *Ultrasonics Sonochemistry*, 13:92–97.

Ning, P., Bart, H.J., Jiang, Y., de Haan, A., and Tien, C. 2005. Treatment of organic pollutants in coke plant wastewater by the method of ultrasonic irradiation, catalytic oxidation and activated sludge. *Separation and Purification Technology*, 41:133–139.

Noike, T., Endo, G., Chang, J.-E., Yaguchi, J.I., and Matsumoto, J.-I. 1985. Characteristics of carbohydrate degradation and the rate-limiting step in anaerobic digestion. *Biotechnology and Bioengineering*, 27:1482–1489.

O'Neill, C., Lopez, A., Esteves, S., Hawkes, F.R., Hawkes, D.L., and Wilcox, S. 2000. Azo-dye degradation in an anaerobic-aerobic treatment system operating on simulated textile effluent. *Applied Microbiology and Biotechnology*, 53:249–254.

Palmowski, L., Simons, L., and Brooks, R. 2006. Ultrasonic treatment to improve anaerobic digestibility of dairy waste streams. *Water Science and Technology*, 53:281–288.

Park, B., Ahn, J.H., Kim, J., and Hwang, S. 2004. Use of microwave pretreatment for enhanced anaerobiosis of secondary sludge. *Water Science and Technology*, 50:17–23.

Parker, W.J. 1997. A multi-parameter sensitivity analysis of a model describing the fate of volatile organic compounds in trickling filters. *Journal of the Air & Waste Management Association*, 47:871–880.

Pérez-Elvira, S., Fdz-Polanco, M., Plaza, F.I., Garralón, G., and Fdz-Polanco, F. 2009. Ultrasound pre-treatment for anaerobic digestion improvement. *Water Science and Technology*, 60:1525–1532.

Pham, T.T.H., Brar, S.K., Tyagi, R.D., and Surampalli, R.Y. 2009. Ultrasonication of wastewater sludge—Consequences on biodegradability and flowability. *Journal of Hazardous Materials*, 163:891–898.

Priego-Capote, F. and Luque de Castro, M.D. 2007. Ultrasound-assisted digestion: A useful alternative in sample preparation. *Journal of Biochemical and Biophysical Methods*, 70:299–310.

Quarmby, J., Scott, J.R., Mason, A.K., Davies, G., and Parsons, S.A. 1999. The application of ultrasound as a pre-treatment for anaerobic digestion. *Environmental Technology*, 20:1155–1161.

Rathod, P.H., Patel, J.C., Shah, M.R., and Jhala, A.J. 2008. Evaluation of gamma irradiation for bio-solid waste management. *International Journal of Environment and Waste Management*, 2:37–48.

Ren, N., Li, N., Li, B., Wang, Y., and Liu, S. 2006. Biohydrogen production from molasses by anaerobic fermentation with a pilot-scale bioreactor system. *International Journal of Hydrogen Energy*, 31:2147–2157.

Ren, N., Wang, B., and Huang, J.C. 1997. Ethanol-type fermentation from carbohydrate in high rate acidogenic reactor. *Biotechnology and Bioengineering*, 54:428–433.

Rhoads, A., Beyenal, H., and Lewandowski, Z. 2005. Microbial fuel cell using anaerobic respiration as an anodic reaction and biomineralized manganese as a cathodic reactant. *Environmental Science and Technology*, 39:4666–4671.

Rokhina, E.V., Lens, P., and Virkutyte, J. 2009. Low-frequency ultrasound in biotechnology: State of the art. *Trends in Biotechnology*, 27:298–306.

Sakakibara, M., Wang, D., Takahashi, R., Takahashi, K., and Mori, S. 1996. Influence of ultrasound irradiation on hydrolysis of sucrose catalyzed by invertase. *Enzyme and Microbial Technology*, 18:444–448.

Salsabil, M.R., Prorot, A., Casellas, M., and Dagot, C. 2009. Pre-treatment of activated sludge: Effect of sonication on aerobic and anaerobic digestibility. *Chemical Engineering Journal*, 148:327–335.

Sauer, U., Santangelo, J.D., Treuner, A., Buchholz, M., and Dürre, P. 1995. Sigma factor and sporulation genes in Clostridium. *FEMS Microbiology Reviews*, 17:331–340.

Schievano, A., Pognani, M., D'Imporzano, G., and Adani, F. 2008. Predicting anaerobic biogasification potential of ingestates and digestates of a full-scale biogas plant using chemical and biological parameters. *Bioresource Technology*, 99:8112–8117.

Sgarbi, G., Casalena, G.A., Baracca, A., Lenaz, G., DiMauro, S., and Solaini, G. 2009. Human NARP mitochondrial mutation metabolism corrected with α-ketoglutarate/aspartate a potential new therapy. *Archives of Neurology*, 66:951–957.

Shen, J., Yin, X., Gu, H., and Lü, X. 2007. Studies of ultrasound disintegration of residual sludge and its energy consumption in water treatment of petrochemical plant. *Frontiers of Chemical Engineering in China*, 1:395–398.

Shida, G.M., Barros, A.R., Marques dos Reis, C., Cavalcante de Amorim, E.L., Damianovic, M.H.R.Z., and Silva, E.L. 2009. Long-term stability of hydrogen and organic acids production in an anaerobic fluidized-bed reactor using heat treated anaerobic sludge inoculum. *International Journal of Hydrogen Energy*, 34:3679–3688.

Shoh, A. 1975. Industrial applications of ultrasound—A review. I. High-power ultrasound. *IEEE Transactions on Sonics and Ultrasonics*, SU–22:60–71.

Show, K.Y., Mao, T., and Lee, D.J. 2007. Optimisation of sludge disruption by sonication. *Water Research*, 41:4741–4747.

Sistla, S. 2005. Degradation of pyridine by ultrasound: A common refractory pollutant in wastewater effluents. *Asian Journal of Water, Environment and Pollution*, 2:89–93.

Skoog, A., Vlahos, P., Rogers, K.L., and Amend, J.P. 2007. Concentrations, distributions, and energy yields of dissolved neutral aldoses in a shallow hydrothermal vent system of Vulcano, Italy. *Organic Geochemistry*, 38:1416–1430.

Tan, Y. and Guodong, J. 2010. Bacterial community structure and dominant bacteria in activated sludge from a 70°C ultrasound-enhanced anaerobic reactor for treating carbazole-containing wastewater. *Bioresource Technology*, 101:174–180.

Tang, Y., Shigematsu, T., Ikbal Morimura, S., and Kida, K. 2004. The effects of micro-aeration on the phylogenetic diversity of microorganisms in a thermophilic anaerobic municipal solid-waste digester. *Water Research*, 38:2537–2550.

Tatsuzawa, T., Hao, L., Ayame, S., Shimomura, T., Kataoka, N., and Miya, A. 2006. Population dynamics of anaerobic microbial consortia in thermophilic methanogenic sludge treating paper-containing solid waste. *Water Science and Technology*, 54:113–119.

Tiehm, A., Nickel, K., Zellhorn, M., and Neis, U. 2001. Ultrasonic waste activated sludge disintegration for improving anaerobic stabilization. *Water Research*, 35:2003–2009.

Valdez-Vazquez, I., Sparling, R., Risbey, D., Rinderknecht-Seijas, N., and Poggi-Varaldo, H.M. 2005. Hydrogen generation via anaerobic fermentation of paper mill wastes. *Bioresource Technology*, 96:1907–1913.

Vatsala, T.M., Mohan Raj, S., and Manimaran, A. 2008. A pilot-scale study of biohydrogen production from distillery effluent using defined bacterial co-culture. *International Journal of Hydrogen Energy*, 33:5404–5415.

Vavilin, V.A., Fernandez, B., Palatsi, J., and Flotats, X. 2008. Hydrolysis kinetics in anaerobic degradation of particulate organic material: An overview. *Waste Management*, 28:939–951.

Vavilin, V.A., Rytow, S.V., and Lokshina, L.Y. 1995. Modelling hydrogen partial pressure change as a result of competition between the butyric and propionic groups of acidogenic bacteria. *Bioresource Technology*, 54:171–177.

Wang, J. and Wang, J. 2007. Application of radiation technology to sewage sludge processing: A review. *Journal of Hazardous Materials*, 143:2–7.

Wang, F., Wang, Y., and Ji, M. 2005. Mechanisms and kinetics models for ultrasonic waste activated sludge disintegration. *Journal of Hazardous Materials*, 123:145–150.

Ward, A.J., Hobbs, P.J., Holliman, P.J., and Jones, D.L. 2008. Optimisation of the anaerobic digestion of agricultural resources. *Bioresource Technology*, 99:7928–7940.

Ward, M., Wu, J., and Chiu, J.F. 1999. Ultrasound-induced cell lysis and sonoporation enhanced by contrast agents. *The Journal of the Acoustical Society of America*. 105:2951–2957.

Wilson, C.A. and Novak, J.T. 2009. Hydrolysis of macromolecular components of primary and secondary wastewater sludge by thermal hydrolytic pretreatment. *Water Research*, 43:4489–4498.

Wu, T., Zivanovic, S., Hayes, D.G., and Weiss, J. 2008. Efficient reduction of chitosan molecular weight by high-intensity ultrasound: Underlying mechanism and effect of process parameters. *Journal of Agricultural and Food Chemistry*, 56:5112–5119.

Wünsch, B., Heine, W., and Neis, U. 2002. Combating bulking sludge with ultrasound. In TU Hamburg-Harburg Reports on Sanitary Engineering, Neis, U. (ed): *Ultrasound in Environmental Engineering II*, 35:201–212.

Xie, B., Liu, H., and Yan, Y. 2009. Improvement of the activity of anaerobic sludge by low-intensity ultrasound. *Journal of Environmental Management*, 90:260–264.

Xie, R., Xing, Y., Ghani, Y.A., Ooi, K., and Ng, S. 2007. Full-scale demonstration of an ultrasonic disintegration technology in enhancing anaerobic digestion of mixed primary and thickened secondary sewage sludge. *Journal of Environmental Engineering and Science*, 6:533–541.

Yacob, S., Shirai, Y., Ali Hassan, M., Wakisaka, M., and Subash, S. 2006. Start-up operation of semi-commercial closed anaerobic digester for palm oil mill effluent treatment. *Process Biochemistry*, 41:962–964.

Yin, X., Han, P., Lu, X., and Wang, Y. 2004. A review on the dewaterability of bio-sludge and ultrasound pre-treatment. *Ultrasonics Sonochemistry*, 11:337–348.

Yin, X., Lu, X., Han, P., and Wang, Y. 2006. Ultrasonic treatment on activated sewage sludge from petro-plant for reduction. *Ultrasonics*, 44:397–399.

Young, J.C. and Yang, B.S. 1989. Design considerations for full-scale anaerobic filters. *Research Journal of the Water Pollution Control Federation*, 61:1576–1587.

Yu, G.H., He, P.J., Shao, L.M., and Zhu, Y.S. 2008. Extracellular proteins, polysaccharides and enzymes impact on sludge aerobic digestion after ultrasonic pretreatment. *Water Research*, 42:1925–1934.

Zabaneh, M. and Bar, R. 1991. Ultrasound-enhanced bioprocess. II: Dehydrogenation of hydrocortisone by *Arthrobacter simplex*. *Biotechnology and Bioengineering*, 37:998–1003.

Zeng, X.L., Long, T.R., Ding, W.C., Xu, L., and Zou, L. 2006. Improvement of biological activity of aerobic sludge by low energy ultrasonic irradiation. *China Water & Wastewater*, 22:88–91.

Zhang, G., Yang, J., Liu, H., and Zhang, J. 2009. Sludge ozonation: Disintegration, supernatant changes and mechanisms. *Bioresource Technology*, 100:1505–1509.

Zhang, G., Zhang, P., Gao, J., and Chen, Y. 2008. Using acoustic cavitation to improve the bio-activity of activated sludge. *Bioresource Technology*, 99:1497–1502.

Zhang, G., Zhang, P., Yang, J., and Chen, Y. 2007. Ultrasonic reduction of excess sludge from the activated sludge system. *Journal of Hazardous Materials*, 145:515–519.

Zhao, B.H., Yue, Z.B., Ni, B.J., Mu, Y., Yu, H.Q., and Harada, H. 2009. Modeling anaerobic digestion of aquatic plants by rumen cultures: Cattail as an example. *Water Research*, 43:2047–2055.

14 Ultrasound Application in Analyses of Organic Pollutants in Environment

Senar Ozcan, Ali Tor, and Mehmet Emin Aydin

CONTENTS

Introduction 345
Experimental Methodology 348
    Reagents and Solvents 348
    Chromatographic Analysis 348
    Cleanup Procedure 349
    Liquid–Liquid Extraction, Solid Phase Extraction, Shake-Flask Extraction,
    and Soxhlet Extraction 349
    Ultrasonic Solvent Extraction (USE) 350
    Real-Water, Soil, and Air Samples 351
Results and Discussions 351
    Water Analysis 351
    Soil Analysis 356
    Air Sample Analysis 362
Conclusions 367
References 368

INTRODUCTION

Polychlorinated biphenyls (PCBs) are hazardous substances due to their persistence, hydrophobic character, and toxic properties (ATSDR, 2000). Although they have been banned in the industrialized countries for years and in some instances for decades, PCBs are still routinely found throughout the world and continue to cause many ecotoxicological problems (Rezaei et al., 2008). Because of their persistence and hydrophobicity, PCBs accumulate in soils where they are likely to be retained for many years. Consequently, soils are an important reservoir for these compounds (Chekol et al., 2004). PCBs may enter the atmosphere from transformers, incinerators, landfills, and sludge drying beds (Murphy et al., 1985; Hermanson and Hites, 1989; Hsu et al., 2003). Additionally, these compounds can cause various human health problems, such as neurotoxicity, dermatological, and pulmonary diseases (ATSDR, 2000; Orlinskii et al., 2001). Therefore, the analysis of PCBs in environmental samples should be continued by developing analytical methods.

    Polycyclic aromatic hydrocarbons (PAHs) are by-products of the incomplete combustion of organic matters. PAHs are ubiquitous environmental contaminants that originate from different emission sources, mainly associated with human activities, such as the incomplete combustion of fossil fuels, industrial processes, waste incineration plants, oil refining, or the use of motor vehicles. In addition, an important natural source is the biomass burning that occurs in forest fires (Kolb et al., 1995; Cecinato et al., 1997; Junker et al., 2000; Oros and Simoneit, 2001). Most PAHs in soil come from air via wet and dry deposition and they accumulate in soils due to their hydrophobicity. Atmospheric fate, transport, and dry and wet deposition of these compounds are influenced considerably by the distribution of substances between gaseous and particles phases (Falconer and Bidleman, 1994; Cotham and Bidleman, 1995; Hoff et al., 1996). The determination of PAHs in environmental samples is an important topic because of their toxicity to humans and deteriorative effects on soil organisms and plants (ATSDR, 1995).

Trace analysis of PCBs and PAHs in water is usually performed by gas chromatography (GC) combined with a previous extraction or a preconcentration step including traditional liquid–liquid extraction (LLE) (Tor et al., 2003; Filipkowska et al., 2005; Zaater et al., 2005), solid phase extraction (SPE) (Aydin et al., 2004; Westbom et al., 2004; Filipkowska et al., 2005; Werres et al., 2009), solid phase microextraction (SPME) (King et al., 2004; Lambropoulou et al., 2006), and the more recently developed dispersive liquid–liquid microextraction (DLLME), single-drop microextraction (SDME), and hollow-fiber microextraction (HF-LPME) (Rezaei et al., 2008; Sarafraz-Yazdi and Amiri, 2010).

LLE and SPE are the oldest procedures for the extraction of PCBs and PAHs from aqueous matrices. LLE is probably the most widely used method for the extraction of PCBs and PAHs from aqueous samples (Tor et al., 2003; Aydin et al., 2004). However, LLE needs relatively large volumes of organic solvents and samples and is a time-consuming as well as a labor-intensive method. The LLE method has some complications such as the formation of stable emulsions. SPE has been used as an alternative method to LLE for the extraction of PCBs and PAHs from water samples because it uses less solvent and is less time-consuming than LLE. Nevertheless, SPE demands a large volume of organic solvents and samples. However, SPE is a relatively expensive method. In this method, analytes may be adsorbed, and complex matrices can cause settling in of cartridges (Lambropoulou et al., 2006). LLE and SPE methods complicate and cause difficultly in automation. Using large amounts of organic solvents can cause environmental pollution and health hazards for laboratory personnel and extra operational costs for waste treatment (Sarafraz-Yazdi and Amiri, 2010).

Therefore, in order to overcome disadvantages of these methods, an efficient, fast, easy, economical, and comparable sample preparation method such as SPME (Lambropoulou et al., 2006) and different modes of liquid–liquid microextraction (LLME), termed as liquid phase microextraction (LPME) or solvent microextraction (SME), for example, SDME (Psillakis et al., 2003a; Tor et al., 2006; Tor, 2006), hollow-fiber liquid phase microextraction (LPME) (Ho et al., 2002; Psillakis et al., 2003b; Rasmussen and Pedersen-Bjergaard, 2004), headspace LPME (Zhao et al., 2004; Vidal et al., 2005), dynamic LPME (Wu et al., 2005), and DLLME (Berijani et al., 2006; Rezaee et al., 2006) have been developed in recent years (He and Lee, 1997). Among these methods, SPME is based on the partitioning of analytes between sample matrixes and the polymer-coated fiber. While SPME has some important advantages such as rapid, simple, and solvent free, the main disadvantages of SPME method are relatively high price and fragile coating layer of fiber. Fiber also can degrade with time and the partial loss of stationary phase can cause coelution with the analytes. In addition, sample carryover has been frequently reported for SPME method (Psillakis et al., 2003b).

LLME is based on the distribution of the analytes between a microvolume of organic solvent and the aqueous solution (Jeannot and Cantwell, 1996, 1997; He and Lee, 1997). These alternative techniques such as SDME, LPME, and DLLME have advantages, such as short extraction time, small volumes of solvent and water requirement, rapid, easy, and low cost. Compared to the SPME, SDME has many advantages including no sample carryover, wide selection of available solvents, simplicity and ease of use, short preconcentration time, requiring no conditioning (as is the case with the fiber in the SPME), no need for instrument modification, etc. Nevertheless, these techniques also have some drawbacks. For example, SDME method has difficulty in automation, instability of droplet, and relative low precisions (Xu et al., 2007). In comparison to the traditional LLE and SPE, LPME procedure has many advantages including wide selection of available solvents, low cost, simplicity and ease of use, minimal solvent use, short preconcentration time, and possibility of automation. Furthermore, compared to the SPME, LPME also has other advantages, such as no sample carryover, requiring no conditioning, and no need for instrument modification. (Khajeh et al., 2006). Nevertheless, some drawbacks, such as instability of droplet and relative low precision, were reported for LPME procedure (Xu et al., 2007). DLLME is based on the formation of tiny droplets of the extractant in the sample solution using water-immiscible organic solvent (extractant) dissolved in a water-miscible organic dispersive solvent. The advantages of DLLME could be given as rapid, simple, short extraction time, low cost, high recovery of analytes. However, general drawbacks of this method are difficulty in automation and it requires the use of dispersive solvent, which usually decreases the partition coefficient of analytes into the extraction solvent (Rezaee et al., 2006; Pena-Pereira et al., 2009).

The analysis of trace levels of organic pollutants in complex matrices such as soil, sediment usually requires several steps. An extraction step is followed by a cleanup of the extract prior to the chromatographic analysis. Extraction is a critical sample-preparation step for the analysis of PCBs and PAHs in soil samples, because these hydrophobic compounds are strongly sorbed to the soil material. Various extraction procedures including soxhlet (Wobst et al., 1999; Fatoki and Awofolu, 2003; Bakan and Ariman, 2004), shake flask (Kolb et al., 1995; Pozo et al., 2001; Nawab et al., 2003), sonication (Babic et al., 1998; Castro et al., 2001; Banjoo and Nelson, 2005; Gonçalves and Alpendurada, 2005; Tor et al., 2006), microwave-assisted extraction (MAE) (Camel, 2000; Ericsson and Colmsjo, 2000; Pino et al., 2000; Jayaraman et al., 2001), supercritical fluid extraction (SFE) (Reindl and Hofler, 1994; Barnabas et al., 1995; Koinecke et al., 1997; Benner, 1998; Morselli et al., 1999), and pressured liquid extraction (Richter, 2000; Lundstedt et al., 2000; Ramos et al., 2000) can be used for the extraction of target compounds from soil. Moreover, determination of PCBs and PAHs in soil can be carried out by using German standard method (DFG S-19 multimethod) (DFG, 1987) and ISO 10382 (ISO, 2002). The preference of each technique mainly depends on the efficiency, recovery, reproducibility, minimal solvent use, simplicity, and ease of use.

Soxhlet extraction is considered to be the standard method used for the extraction of PCBs and PAHs from soils (Barco-Bonilla et al., 2009). The soxhlet and shake-flask extractions are time-consuming and require large volume of organic solvents. Therefore, in order to reduce the extraction time, amount of solvent required, as well as sample amount, new extraction procedures, that is, SFE (Librando et al., 1994), MAE (Li et al., 2003), and pressurized liquid extraction (PLE) (Ramos et al., 2000; Barco-Bonilla et al., 2009) have been developed as alternative techniques. More recent procedures, that is, SFE, MAE, and accelerated solvent extraction (ASE), gave shorter extraction time and reduced the solvent consumption because these extraction procedures are working at high temperatures above the boiling point of the solvent. Except for SFE, reconcentration and cleanup steps have to be performed for MAE and ASE procedures. On the other hand, time and cost needed for both SFE and ASE are quite high (Berset et al., 1999).

PCB and PAH partitions between the gas and particulate phase are based on their concentrations, vapor pressures, the ambient air temperature and the concentration of particulate matter present in the air. PAHs in the atmosphere are generally associated with airborne particles. In air, smaller molecular weight (less than three rings) PAHs are mostly found in the gaseous phase while those of higher molecular weight (more than five rings) PAHs are generally associated with the particle phase (Venkataraman et al., 1994). Because of the persistence, moderate vapor pressure, and lipophilic features of the PCBs, the same association between these pollutants and airborne particles is expected. PCBs of higher molecular weight (containing five or more chlorine atoms) are deposited easily on surfaces of plants, soil, and waters because they have lower vapor pressures than PCBs with lower degree of chlorination (three or four chlorine atoms) (Nielsen et al., 1996; Van der Hoff and Van Zoonen, 1999; ATSDR, 2000). Analysis procedures of PCBs and PAHs in air samples are given by EPA method TO-13A and method TO-4A, respectively. There are various studies reporting the determination of the PAHs and PCBs in gas phase and airborne particles (Vasconcellos et al., 2003; Yeo et al., 2003; Tasdemir et al., 2004a,b; Ozcan and Aydin, 2009). In these studies, soxhlet extraction was used for isolating target compounds from air samples.

Ultrasonication is being used more and more in analytical chemistry, enabling different steps in the analytical process, particularly in sample preparation, such as the extraction of organic and inorganic compounds from different matrices (Mierzwa et al., 1997; Ashley et al., 2001; Aydin et al., 2006; Ozcan et al., 2009a). Ultrasonic radiation is a powerful means for acceleration of various steps in analytical procedure for both solid and liquid samples (Priego-López and Luque de Castro, 2003; Aydin et al., 2006; Tor et al., 2006). This type of energy is of great help in the pretreatment of samples as it facilitates and accelerates operations such as the extraction of organic and inorganic compounds. In ultrasound-assisted LLE, it facilitates the emulsification phenomenon and accelerates the mass-transfer process between two immiscible phases. This leads to an increment in the extraction efficiency of the procedure in a minimum time (Luque de Castro and Priego-Capote, 2006, 2007). The most widely accepted mechanism for ultrasound-assisted emulsification is based on the cavitation effect. The implosion bubbles generated by the cavitation phenomenon produce intensive shockwaves in the surrounding liquid and high velocity liquid jets. Such microjets can cause droplet disruption in the vicinity of collapsing bubbles and, thus, improve emulsification by generating smaller droplet size of the dispersed phase, right after disruption (Luque de Castro and Priego-Capote, 2006). Submicron-sized droplet results in significant enlargement of the contact surface between both immiscible liquids improving the mass transfer between the phases.

Additionally, ultrasonication offers several advantages that make it an ideal method for pretreating a large number of samples. These advantages include high extraction efficiency, lower equipment costs, ease of operation, lower extraction temperatures, etc. Therefore, in this chapter, the application of ultrasonic extraction procedures for residue analysis of PCBs and PAHs in water, soil, and air samples was described. The applicability of the ultrasonic extraction was evaluated by comparison with traditional extraction methods (LLE and SPE for water samples; shake-flask, soxhlet extraction, and large-scale ultrasonic extraction for soil samples; and soxhlet extraction for air samples).

EXPERIMENTAL METHODOLOGY

REAGENTS AND SOLVENTS

All chemicals used were of analytical grade. PCBs mixed standard including PCB 28, 52, 101, 138, 153, and 180 and the EPA 16 PAHs mixed standard including naphthalene (NAP), acenaphthylene (ACY), acenaphthene (ACE), fluorene (FLO), phenanthrene (PHE), anthracene (ANT), fluoranthene (FLA), pyrene (PYR), benzo[a]anthracene (BaA), chrysene (CHR), benzo[b]fluoranthene (BbF), benzo[k]fluoranthene (BkF), benzo[a]pyrene (BaP), indeno[1,2,3-cd]pyrene (IcdP), dibenzo[a,h]anthracene (DahA), benzo[g,h,i]perylene (BghiP) were from Accustandard Co. (New Haven, CT). Solvents of residue grade purity including acetone, dichloromethane, chloroform, n-hexane, methanol, ethylacetate, diethylether, and petroleum ether (40°C–60°C) were obtained from Merck Co. (Darmstadt, Germany). Sodium chloride and sodium sulfate were also from Merck Co. Octadecyl (C_{18}) SPE cartridges were obtained from J and T Baker (Deventer, Holland). Alumina 90 active, neutral, [(0.063–0.200 mm), (70–230 mesh ASTM)] and silica gel 60 (0.063–0.200 mm) were also from Merck Co. Standard stock solution of target compounds (1 mg L^{-1} of each PCB, 1 g mL^{-1} of each PAH) was prepared in methanol. All solutions were stored in the dark at 4°C. Working solutions were prepared by dilution of standard stock solution with distilled water.

CHROMATOGRAPHIC ANALYSIS

The determination of PCBs and PAHs was carried out by GC, (Agilent 6890N, Agilent Technologies, Palo Alto, CA) equipped with mass-selective (MS) detector, (Agilent 5973, Agilent Technologies, Foster City, CA). The features and operating conditions of GC–MS system were as follows: GC, equipped with programmed temperature vaporizing (PTV) injector, DB-5 MS 5% phenylmethyl siloxane fused silica capillary column (30 m length, 0.25 mm i.d., and 0.25 μm film thickness) and helium (purity 99.999%) was used as carrier gas. PTV program was as follow: 80°C, 12°C s^{-1} to 350°C, and hold at 350°C for 2 min. Injections were performed by an Agilent 7683 B Series automatic injector (Agilent Technologies, Palo Alto, CA). The temperature of the ion source and MS transfer line were maintained at 170°C and 280°C, respectively.

The temperature programs for the analysis of PCBs and PAHs by GC–MS were as follows: PCBs: initial temperature 70°C for 2 min, 25°C min^{-1} to 150°C, 3°C min^{-1} to 200°C, 8°C min^{-1} to 280°C hold for 10 min (run time: 41.87 min); PAHs: initial temperature 60°C for 4 min, 15°C min^{-1} to 160°C, 3°C min^{-1} to 300°C, hold at 300°C for 10 min (run time: 67.33 min). MS detector was operated in selected ion monitoring (SIM) mode and the used masses (m/z) of PCBs and PAHs were selected according to Aydin et al. (2006) and Ozcan et al. (2010).

Cleanup Procedure

The activation and deactivation of the column sorbent material, aluminum oxide and silica gel-60, were performed as follows. The aluminum oxide and silica gel were activated at 210°C for 4 h. It was allowed to cool down in a desiccator and then deactivation and homogenization were carried out by adding certain amounts of deionized water (2% or 5%) and shaking the sorbents in a horizontal shaker at 210 rpm for 2 h.

The preparation of the traditional cleanup column filled with 10 g of deactivated column sorbent material was described in a previous paper (Tor et al., 2006). The cleanup column, length of 30 cm and 1 cm of internal diameter, was prepared according to slurry packing technique (Jaouen-Madoulet et al., 2000). The extract, reduced in volume to 1 mL, was transferred quantitatively onto the top of the column. A volume of 70 mL n-hexane was used to elute PCBs while the elution of PAHs was carried out with 60 mL of n-hexane/ethylacetate (1/1, v/v) and concentrated to exactly 1 mL, using a rotary evaporator (Buchi B-160 Vocabox, Switzerland) and nitrogen stream, prior to GC–MS analysis.

The microscale cleanup column consisted of a pasteur pipette, length of 10 and 0.5 cm internal diameter, fitted at its base with a plug of glass wool. A detailed description of the preparation of a miniaturized cleanup column filled with 0.5 g of deactivated column sorbent material was also given in a previous paper (Aydin et al., 2006). The volume of the extract from the ultrasonic extraction was reduced and was transferred quantitatively onto the top of the column. The target compounds were eluted from the pipette under gravity (flow rate of approximately two drop s^{-1}) with 5 mL of n-hexane for PCBs and 5 mL of n-hexane/ethyl acetate (1/1, v/v) for PAHs. The eluate was concentrated to 250 μL prior to GC–MS analysis.

Liquid–Liquid Extraction, Solid Phase Extraction, Shake-Flask Extraction, and Soxhlet Extraction

The LLE procedure was adopted from US EPA Method 3510C (US EPA, 1996a). Two hundred milliliter water sample was placed in a 250 mL separatory funnel. The extraction was carried out three times with 20 mL of dichloromethane. The extracts were combined and dried with anhydrous sodium sulfate. The resulting extract was concentrated to exactly 1 mL using rotary evaporator (Buchi B-160 Vocabox, Flawil one, Switzerland) and gentle nitrogen stream. Then, GC–MS analysis was performed as described in "Chromatographic analysis" section.

SPE procedure was carried out as described by Aydin et al. (2004). Octadecyl (C_{18}) SPE cartridge was used for the extraction of PCBs and PAHs from water sample. The cartridge was consecutively washed with 10 mL of methanol and 8 mL of n-hexane/ethyl acetate (5/3, v/v). Then, it was conditioned with 10 mL of methanol and 2 × 5 mL of distilled water. A 200 mL water sample was passed through the cartridge in vacuum. After the cartridge was dried for 10 min by maintaining vacuum, elution of PCBs and PAHs from the cartridge was carried out with 10 mL of n-hexane/ethyl acetate

(7/3, v/v). The extract was dried with sodium sulfate and concentrated to exactly 1 mL by rotary evaporator and under gentle nitrogen stream. Then, GC–MS analysis was carried out as described in "Chromatographic analysis" section.

For shake-flask extraction, a 10 g soil sample was suspended in 50 mL of extraction solvent (petroleum ether/acetone mixture (1/1, v/v) for PCBs and ethylacetate for PAHs) and shaken on a horizontal shaker for 12 h. Then, the extract was filtered and concentrated to exactly 1 mL by using rotary evaporator and nitrogen stream, respectively (Aydin et al., 2006; Tor et al., 2006). For soxhlet extraction, a 10 g soil sample was put into the extraction thimble and extracted with 150 mL extraction solvent for 18 h petroleum ether/acetone mixture (1/1, v/v) for PCBs and ethylacetate for PAHs (Tor et al., 2006). The extract was reduced to exactly 1 mL using a rotary evaporator and under a gentle stream of nitrogen. The concentrated extract was transferred onto the traditional cleanup column filled with 10 g of aluminum oxide (5% deactivated) and elution was performed as described in *Cleanup Procedure* section. GC–MS analysis were performed as described in "Chromatographic analysis" section.

For soxhlet extraction of filter and polyurethane foam (PUF) plug, filter and PUF plug was put into the extraction thimble and extracted with 150 mL of a solvent mixture of diethylether:n-hexane (1/9, v/v) for 16 h (US EPA, 1996b). The extract was reduced to exactly 1 mL using a rotary evaporator and under a gentle stream of nitrogen. After each extraction procedure, the concentrated extract was transferred onto the traditional cleanup column filled with 10 g of silica gel (2% deactivated) and the extracts were cleaned up and GC–MS analysis were performed as described in *Cleanup procedure* and *Chromatographic Analysis* sections, respectively.

ULTRASONIC SOLVENT EXTRACTION (USE)

A 10 mL water sample was placed in a 10 mL glass centrifuge tube. As an extraction solvent, chloroform (100–200 µL) was added into the water sample and mixed. The resulting mixture was immersed into an ultrasonic bath (frequency 35 kHz, 320 W, Super RK 510, Sonorex, Bandelin, Germany) for 15 min at 25°C. During the sonication, the solution became turbid due to the dispersion of fine chloroform droplets into the aqueous bulk. The emulsification phenomenon favored the mass-transfer process of PCBs and PAHs from the aqueous bulk to the organic phase. The emulsion was centrifuged at 4000 rpm for 5 min to disrupt the emulsions and separate the solvent from the aqueous phase. After centrifugation, extraction solvent was removed from the bottom of the tube by using a 250 µL Hamilton syringe (Hamilton Bonaduz AG, Switzerland) and transferred into the microvial. Then, GC–MS analysis was performed as described in *Chromatography Analysis* section.

For analyses of PCBs compounds in soil, 1.5 g sample was sonicated three times for 5 min with 2 mL of mixture of acetone-petroleum ether (1/1, v/v) in an ultrasonic bath. For analyses of PAHs compounds, a 0.5 g soil sample was sonicated three times for 5 min with 5 mL of ethyl acetate in an ultrasonic bath. The extracts were combined and were filtered by using Whatman filter paper. The filtrates were reduced to 1 mL with a rotary evaporator (Buchi B-160 Vacobox, Switzerland) and adjusted to exactly 250 µL by using a gentle nitrogen stream. The concentrated extract was transferred onto the microscale cleanup column and elution was performed as described in *Cleanup Procedure* section.

For PCBs and PAHs, extractions from filters with the ultrasonic extraction method were carried out once with 75 mL extraction solvent n-hexane/petroleum ether (1/1, v/v) and 45 min of sonication. PUF plugs were sonicated for 45 min with 300 mL diethylether/n-hexane (1/9, v/v) of extraction solvent in an ultrasonic bath. The extracts were reduced with a rotary evaporator and under a gentle nitrogen stream. The extracts were cleaned up with traditional cleanup column filled with 10 g of silica gel (2% deactivated) and the amount of extracted target compounds were determined by GC–MS as described in *Cleanup Procedure* section.

REAL-WATER, SOIL, AND AIR SAMPLES

The efficiency of the ultrasonic extraction procedure was also compared with traditional LLE and SPE procedures on the real water samples including tap water, well water, domestic, and industrial wastewater samples. Tap water was obtained from the laboratory and well water came from deep-groundwater in Konya (Turkey). Domestic and industrial wastewater samples were taken from the sewage system in residential area and industrial zone in Konya (Turkey), respectively. All samples were collected free of air bubbles in glass containers and they were stored in the dark at 4°C. Tap and well water samples were analyzed without previous treatment or filtration. The domestic and industrial wastewater samples were filtered through a membrane filter with 0.45 μm pore size before the extraction procedures.

Real soil samples were also obtained from the Department of Soil, Agricultural Faculty of Selcuk University (Konya, Turkey). The textures of the soil samples were as follows: Sample A, sand: 49.2%, silt: 32.3%, clay: 18.5%, organic matter: 1.90%, pH (0.01 M $CaCl_2$): 6.5, and maximum water capacity: 19.6%. Sample B, sand: 42.2%, silt: 31.5%, clay: 24.5, organic matter: 1.80%, pH (0.01 M $CaCl_2$): 7.2, and maximum water capacity: 20.4%. The soil samples were dried and sieved to <2 mm and stored at 4°C until analysis.

The modified high volume air sampler of Model GPS-11 (Thermo Andersen Inc.) was used to collect the air samples. Particles were collected by passing air through the GF/A type glass microfiber filter with a diameter of 90 mm and pore size of 1.6 μm. Pollutants in gas phase were collected in PUF plug with length of 5 cm, diameter of 6.5 cm, and density of 0.0225 g cm^{-3}. Samples with an average sampling duration of 24 h were taken on the campus area of Selcuk University (Konya, Turkey). The mean flow rate and sampling volume were about 0.25 m^3 min^{-1} and 360 m^3, respectively.

RESULTS AND DISCUSSIONS

WATER ANALYSIS

The recovery experiments were carried out for the determination of the USE efficiency of selected PCBs and PAHs in water samples. After the choice of the most suitable solvent and extraction time, several other parameters including solvent volume, centrifugation time, and ionic strength of the water sample were optimized. The efficiency of USE procedure was compared with LLE and SPE methods on the real water samples. At the beginning of the experiments, the extraction efficiency of dichloromethane, 1,2-dichlorobenzene, 1,2,4-tichlorobenzene, chloroform, and bromoform was determined. Ten milliliter aliquots of distilled water including PCB were extracted by using 100 μL of each solvent in ultrasonic bath for 5 min. The choice of extraction solvent is critical for developing an efficient USE procedure since physicochemical properties of the solvent govern the emulsification phenomenon, and consequently, the extraction efficiency. Moreover, the extraction solvent should have good affinity for target compounds and it should have excellent gas chromatographic behavior. Emulsification was observed in all cases with the exception of dichloromethane. Dichloromethane was completely dissolved in the aqueous solution. The results revealed that chloroform was of the highest extraction efficiency among the examined solvents. In the second set of experiments, the optimum extraction time was determined. This optimization experiment was carried out by using chloroform, which gave the highest recovery for the studied PCBs. In order to determine optimum extraction time, 10 mL aliquots of fortified distilled water with no ionic strength adjustment were extracted by using 100 μL of chloroform for 1, 5, 10, 15, and 25 min. Time plays an important role in the emulsification and mass-transfer phenomena. Both phenomena affect the extraction efficiency of the analytes. The extraction time interval was defined, as the time elapsed between addition of chloroform and the end of the sonication stage. It was observed that extraction efficiency increased with increase in the extraction time up to 10 min. Then, the extraction efficiency remained constant. Therefore, 10 min was chosen as an extraction time in further experiments.

TABLE 14.1
Design Matrix for Factorial Design and Average Recoveries of PCBs for the Effect of Parameters on the Ultrasonic Solvent Extraction Method (Ozcan et al., 2009b)

Number	Codified Variables			No Codified Variables			Average Recovery (%)
	X_1 (μL)	X_2 (min)	X_3 (%)	X_1 (μL)	X_2 (min)	X_3 (%)	
1–9	−	−	−	100	5	0	72
2–10	+	−	−	200	5	0	94
3–11	−	+	−	100	10	0	73
4–12	+	+	−	200	10	0	94
5–13	−	−	+	100	5	10	61
6–14	+	−	+	200	5	10	82
7–15	−	+	+	100	10	10	62
8–16	+	+	+	200	10	10	83

Note: X_1, extraction solvent volume; X_2, centrifugation time; X_3, ionic strength of the sample.

After the choice of chloroform and 10 min as the optimum extraction solvent and extraction time, respectively, several other factors affecting the ultrasonic extraction procedure, such as extraction solvent volume, centrifugation time, and ionic strength of the sample were optimized by using a 2^3 factorial experimental design. The lower and higher level for each factor was designated as "−" and "+" signs, respectively (Table 14.1). A full 2^3 design would have required eight experiments, which were duplicated in order to calculate the residual error. The experiments were performed in a randomized order to avoid any systematic error. After processing the data by analysis of variance (ANOVA) using Tool Pak in Microsoft Excel, the ANOVA tables were constructed to test the significance of the effect of each factor on the extraction efficiency. At significance level of 5%, the factor with F-value over critical F-value (5.318) has a significant effect on the extraction efficiency.

After each extraction, the emulsion was centrifuged for 5 min at 4000 rpm. Then, extraction solvent was removed from the bottom of the tube by using a 250 μL syringe and transferred into the micro vial. Then, GC–MS analysis was performed as described earlier.

The solvent volume was a significant factor with positive effect on the extraction of some PCBs. The main effect of the ultrasound in LLE is that the fragmentation of one of the phases to form emulsions with submicron droplet size that enormously extend the contact surface between both liquids (Abismail et al., 1999). Therefore, it is expected that increasing the volume of chloroform from 100 to 200 μL increases the number of submicron droplet. Hence, a higher mass transfer or extraction efficiency is obtained. Centrifugation time was not a significant factor for studied PCBs. Namely, for present study, 5 min of centrifugation was adequate to break down the emulsion, hence the phase separation. The ionic strength of the sample was also a significant factor with negative effect on the extraction of all studied PCBs. As it is well-known, the ionic strength affects the partitioning coefficients of analytes between an aqueous and organic phase. On the other hand, as the ionic strength of the medium increases, the viscosity and density of the solution increase. This causes a diminishing in the efficiency of the mass-transfer process and consequently, the extraction efficiency of the procedure (Fontana et al., 2009). Additionally, the ultrasound waves can be absorbed and dispersed in a viscous medium as calorific energy; thus, the cavitation process could be withdrawn, reducing the emulsification phenomenon (Mason and Lorimer, 2002). In this study, an increase in the ionic strength of the sample from 0% to 10% decreased. Additionally, interactions between the solvent volume and both the centrifugation time and the ionic strength were significant with positive effect on the extraction of PCBs. However, interaction between the centrifugation time and ionic strength was significant with negative effect.

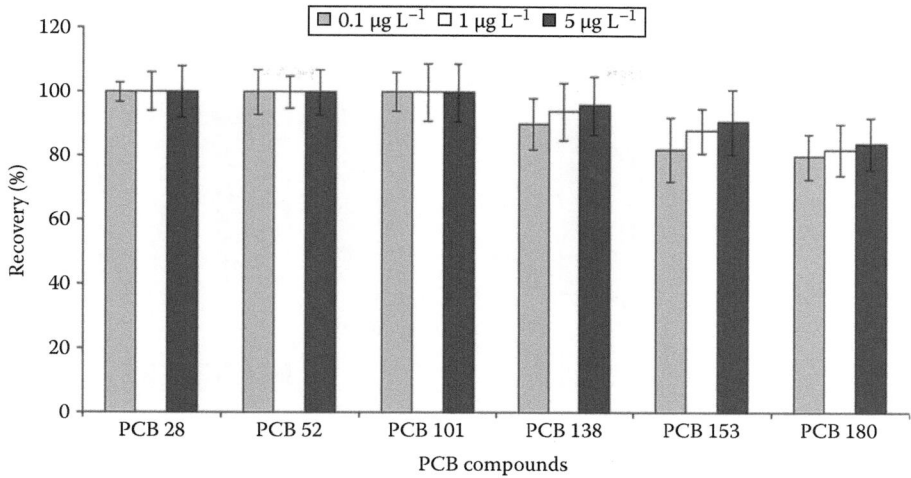

FIGURE 14.1 Recoveries of PCBs in spiked distilled water with three fortification levels using USE method [$n = 8$]. (Extraction conditions; extraction solvent: Chloroform, extraction time: 10 min, sample volume: 10 mL, extraction solvent volume: 200 µL, extraction time: 10 min, centrifugation time: 5 min, ionic strength: 0%, ambient temperature: 25°C.)

According to the results, the optimum conditions for ultrasonic extraction procedure of PCBs from water were as follows: chloroform as an extractant; solvent volume, 200 µL; extraction time, 10 min with no addition of sodium chloride at 25°C and centrifugation time, 5 min.

The results of recoveries for the fortified distilled water with three different fortification levels are given in Figure 14.1. According to fortification level 1 (0.1 µg L^{-1}), recoveries ranged from 80(±7)% to 100(±7)%. Comparable recoveries were also obtained from fortification levels 2 (1 µg L^{-1}) and 3 (5 µg L^{-1}) (Figure 14.1). When statistical evaluation was carried out between recoveries of PCBs from fortification level 1 and level 2, no significant differences ($p > 0.05$) were observed. Additionally, no significant differences were observed when the same statistical evaluations were carried out between fortification levels 1–3 and 2–3. This indicated that developed USE method was of considerable efficiency in order to extract PCBs from water samples.

The validation of the USE procedure was carried out using both fortified water and wastewater samples. In addition, the efficiency of the method was also compared with traditional LLE and SPE techniques on the fortified real water samples. The recoveries are given in Figure 14.2, which indicate that the recoveries of studied PCBs are higher than 78% with RSD in the range of 5%–10%. Analyses of real water samples showed that sample matrices had no adverse effect on the efficiency of the USE procedure. When recoveries of PCBs were gauged against absolute limits of 70% and 130% (US EPA, 1995), it was seen that the method gave satisfactory results. The efficiency of the USE was also compared with those involving traditional LLE and SPE method on the same fortified real samples. As seen in Figure 14.2, the method gave comparable results with traditional LLE and SPE methods. However, it should be emphasized that the USE is not a time-consuming procedure and it is not necessary for a reconcentration step prior to the GC analysis. Furthermore, it needs much lower volumes of solvent than the traditional LLE and SPE techniques.

After the choice of the most suitable solvent and solvent volume, several other parameters (extraction time, centrifugation time, and ionic strength of the water sample) for PAHs in waters were optimized. The highest recoveries of PAHs were obtained with chloroform in examined solvents including 1,2-dichlorobenzene, bromoform. A lower surface tension of chloroform (at 20°C, bromoform: 41.5 mN m^{-1}, 1,2-dichlorobenzene: 37 mN m^{-1}, chloroform: 27.5 mN m^{-1}) would enable a higher cavitation under ultrasound irradiation, and hence, a higher efficiency in emulsion formation. Therefore, chloroform was selected as the extraction solvent for further experiments. To increase

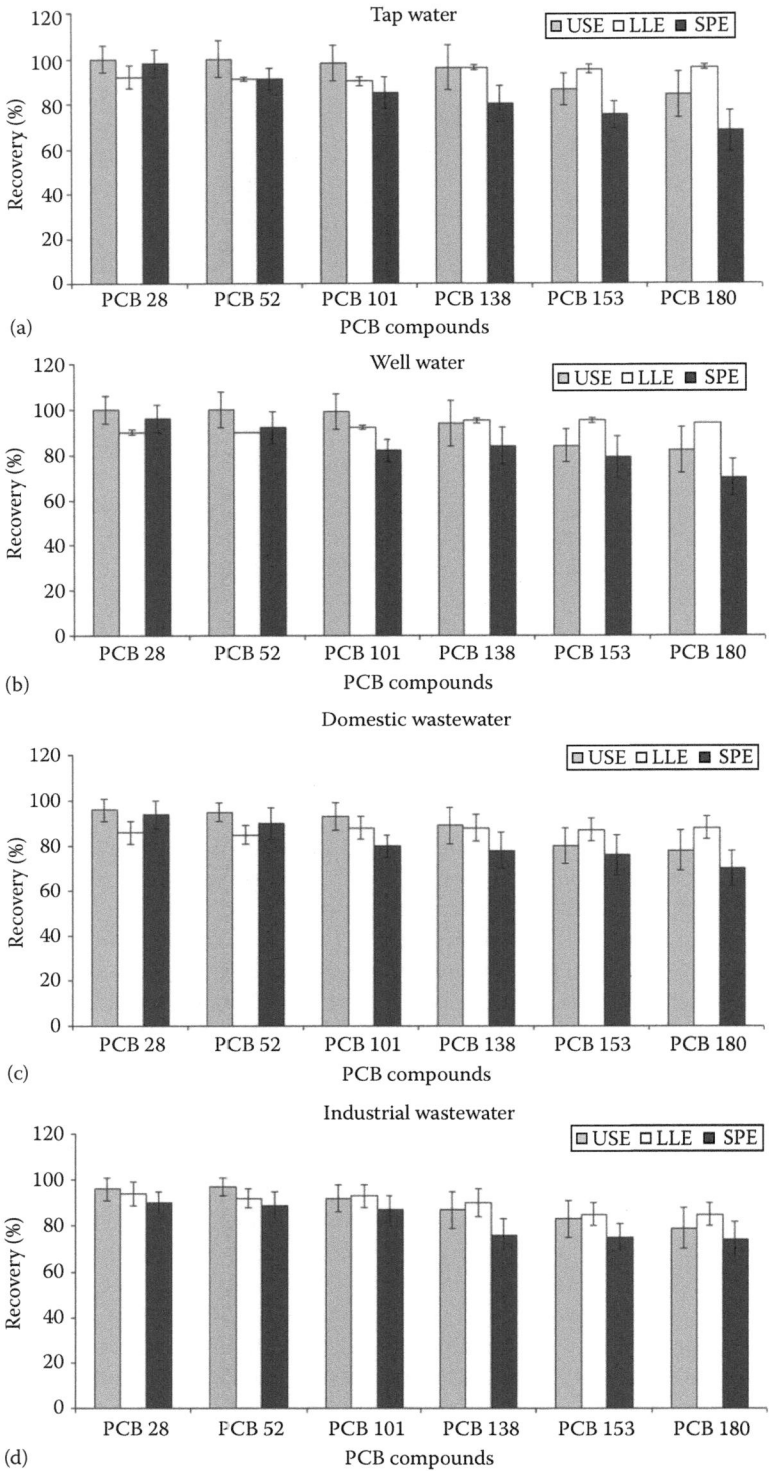

FIGURE 14.2 Comparison of extraction efficiency of the USE method with LLE and SPE for PCBs in fortified real water samples (a) tap water, (b) well water, (c) domestic wastewater (d) industrial wastewater (fortification concentration for each compound: 1 µg L^{-1}), [n = 4].

TABLE 14.2
Design Matrix for Factorial Design and Average Recoveries of PAHs for the Effect of Parameters on the Ultrasonic Solvent Extraction Method (Ozcan et al., 2010)

Number	Codified Variables			No Codified Variables			Average Recovery (%)
	X_1 (min)	X_2 (min)	X_3 (%)	X_1 (min)	X_2 (min)	X_3 (%)	
1–9	−	−	−	5	5	0	78
2–10	+	−	−	15	5	0	99
3–11	−	+	−	5	10	0	84
4–12	+	+	−	15	10	0	97
5–13	−	−	+	5	5	10	40
6–14	+	−	+	15	5	10	47
7–15	−	+	+	5	10	10	43
8–16	+	+	+	15	10	10	51

Note: X_1, extraction solvent volume; X_2, centrifugation time; X_3, ionic strength of the sample.

the sensitivity of the USE method, different volumes of chloroform were examined and increasing the chloroform volume from 100 to 300 µL resulted in a decrease of detector response for all PAHs. This observation might be attributed to dilution effect. After selection of the most suitable extraction solvent (chloroform) and its volume (100 µL), the other factors affecting the efficiency of the USE were optimized by using a 2^3 factorial experimental design. The lower and higher level for each factor was designated as "−" and "+" signs, respectively (Table 14.2). After processing the data by ANOVA, the ANOVA tables were constructed to test the significance of the effect of each factor on the extraction efficiency. At the significance level of 5%, the factor with F-value over critical F-value (5.318) has a significant effect on the extraction efficiency.

The extraction time was a significant factor with positive effect on the extraction of all PAHs. An increase in the extraction time from 5 to 15 min improved the recoveries of all PAHs. The centrifugation time was not a significant factor for all PAHs. Namely, 5 min of centrifugation was adequate to break down the emulsion; hence, the phase separation was achieved. The effect of the ionic strength on the extraction efficiency was evaluated by increasing NaCl concentration of sample from 0% to 10%. The ionic strength of the sample was a significant factor with negative effect on the extraction of all PAHs and decreased the extraction recovery of all PAHs.

Additionally, interaction between the extraction and centrifugation times was significant with positive effect on the efficiency of USE method. However, interactions between the extraction time and ionic strength of the sample and centrifugation time and ionic strength of the sample were significant with negative effect. As a result, the optimum conditions for USE of PAHs from water were as follows: chloroform as an extraction solvent, solvent volume: 100 µL; extraction time: 15 min at 25°C with no addition of NaCl and centrifugation time: 5 min.

The results of PAHs recoveries for the fortified distilled water with three different fortification levels are given in Figure 14.3. The recoveries of PAHs from distilled water fortified with 0.5 µg L^{-1} of each compound ranged from 92% ± 7% to 98% ± 6%. Comparable results were obtained at fortification levels of 2 µg L^{-1} (recoveries between 94% ± 5% and 102% ± 8%) and of 5 µg L^{-1} (recoveries between 94% ± 8% and 105% ± 6%). The recoveries at three different fortification levels were not significantly different ($p > 0.05$) indicating the high efficiency of USE method for extraction of PAHs from water.

Real water samples are expected to represent very complex matrices. Therefore, in order to study possible matrix effects, USE method was applied to fortified real water samples, including tap water and well water as well as domestic and industrial wastewater samples. The efficiency of USE method was also compared with those involving LLE and SPE on the fortified real water samples.

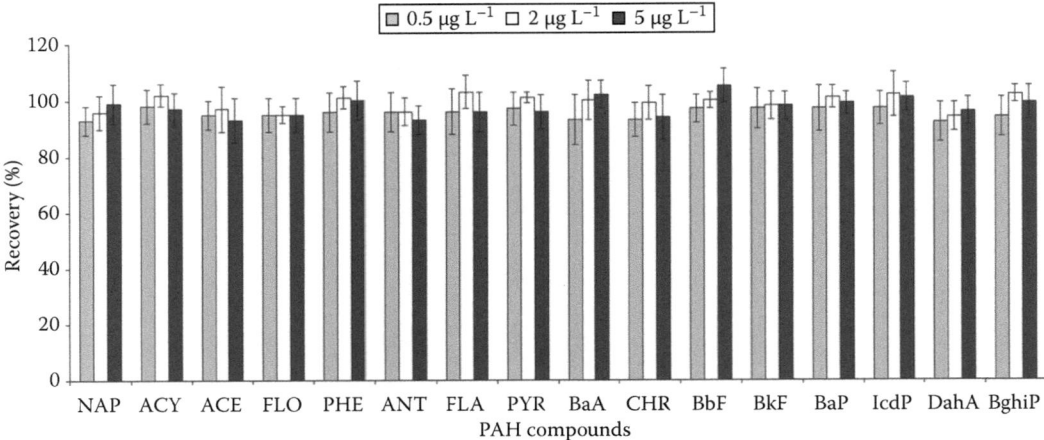

FIGURE 14.3 Recoveries of PAHs from fortified distilled water with three fortification levels using USE method [$n = 8$]. (Extraction conditions; extraction solvent: chloroform, sample volume: 10 mL, extraction solvent volume: 100 µL, extraction time: 15 min, centrifugation time: 5 min (4000 rpm), without addition of NaCl into the sample, temperature: 25°C.)

The USE procedure showed comparable results especially with LLE for unfortified real water samples. As seen in Figure 14.4, the recoveries of 16 PAHs were in the range of 88%–101% with RSD below 9% LLE showed comparable recoveries (81%–108% with RSD below 7%). However, the recoveries obtained from SPE were in the range of 61%–87% with RSD below 9%. When all recoveries of 16 PAHs from the fortified real water samples were gauged against absolute limits of 70% and 130% (USEPA, 1995), it was seen that developed USE method gave satisfactory results. The results also showed that efficiency of the developed method was higher than that of SPE method. Moreover, the developed method showed comparable efficiency with LLE (Figure 14.4). It should also be emphasized that USE is not a time-consuming procedure. Furthermore, it needs much lower volumes of extraction solvent than LLE and SPE, and it is not necessary to concentrate the sample for GC analysis.

Soil Analysis

The recovery experiments were carried out for optimization of an USE of selected PCBs and PAHs from soil samples. The factors affecting the performance of USE (i.e., amount of sample, volume of extraction solvent, and the number of extraction step) were optimized by using a 2^3 factorial experimental design. The sample amount considered according to the level indicated in the respective trial (Table 14.3) was extracted by means of ultrasound. The number of extraction step and the amount of extraction solvent were also optimized. The experimental design matrix is constituted as shown in Table 14.3. The lower and higher level for each factor was designated as "−" and "+" signs, respectively (Table 14.3). After processing the data by ANOVA, the ANOVA tables were constructed to test the significance of the effect of each factor on the extraction efficiency. At significance level of 5%, the factor with F-value over critical F-value (5.318) has a significant effect on the extraction efficiency.

The applicability of the USE was tested by a comparison with conventional soxhlet and shake-flask extraction of real soil samples with spiked PCBs and PAHs.

Samples (0.5–1.5 g amount of soil spiked with PCBs) were placed into a 10 mL capacity of vial. Each extraction step was performed in an ultrasonic bath for 5 min. After each extraction, extract was reduced in volume to 300 µL by a gentle stream of nitrogen. Cleanup of the extract was carried out with microscale cleanup column, as described earlier. The eluents were reduced to 300 µL and

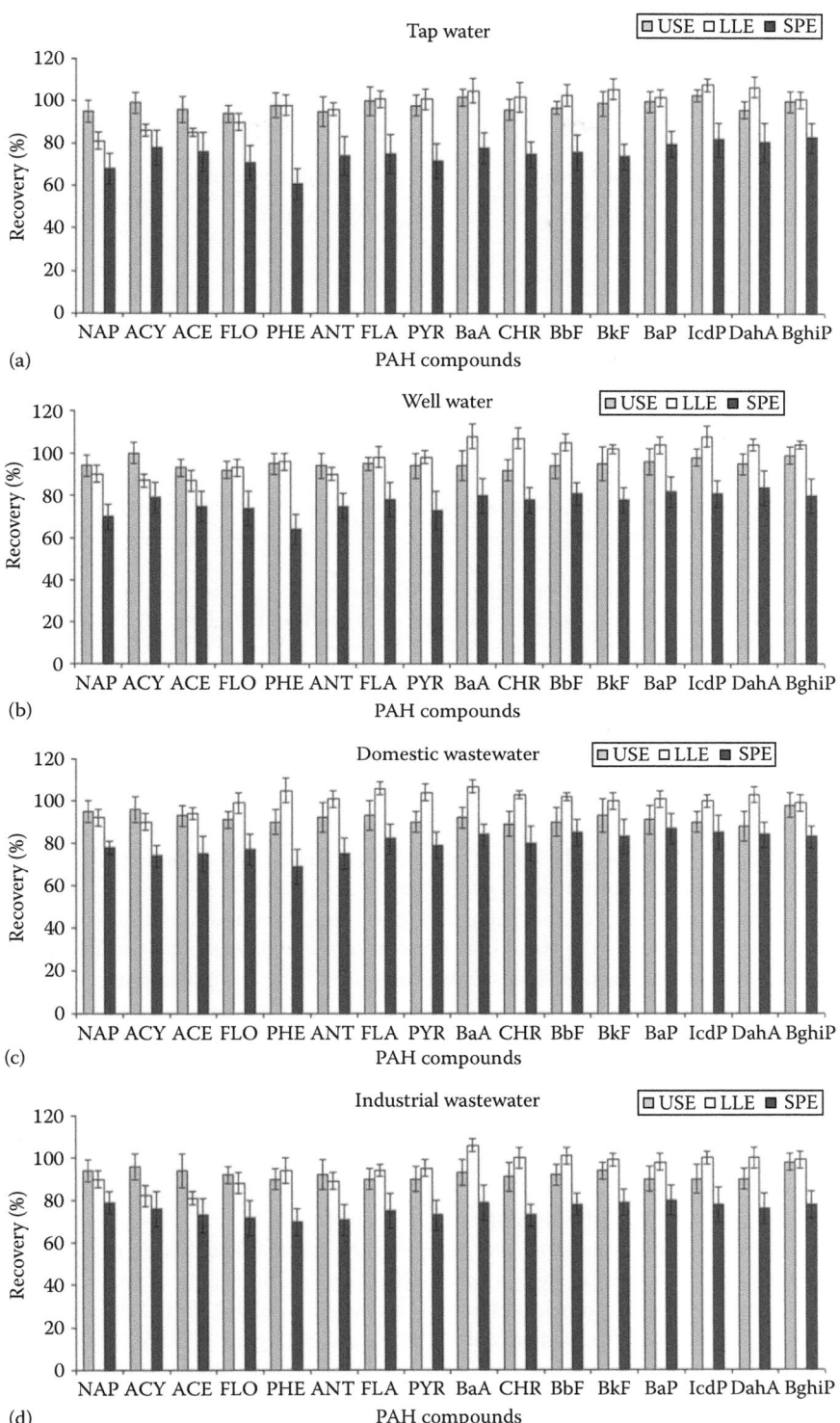

FIGURE 14.4 Comparison of extraction efficiency of the USE method with LLE and SPE for PAHs in fortified real water samples (a) tap water, (b) well water, (c) domestic wastewater, (d) industrial wastewater (fortification concentration for each compound: 2 μg L^{-1}), [$n = 4$].

TABLE 14.3
Design Matrix for Factorial Design and Average Recoveries of PCBs for the Effect of Parameters on the Ultrasonic Solvent Extraction Method (Aydin et al., 2006)

Number	Codified Variables			No Codified Variables			Average Recovery (%)
	X_1 (min)	X_2 (min)	X_3 (%)	X_1 (min)	X_2 (min)	X_3 (%)	
1–9	−	−	−	0.5	2	1	50
2–10	+	−	−	1.5	2	1	63
3–11	−	+	−	0.5	4	1	53
4–12	+	+	−	1.5	4	1	76
5–13	−	−	+	0.5	2	3	68
6–14	+	−	+	1.5	2	3	96
7–15	−	+	+	0.5	4	3	90
8–16	+	+	+	1.5	4	3	90

Note: X_1, soil sample amount; X_2, extraction solvent volume; X_3, number of extraction step.

were analyzed by GC–MS. In the optimization of ultrasonic extraction, a mixture of petroleum ether and acetone (1/1, v/v) was used for sonication. In soil and sludges, PCBs is adsorbed on or in soil aggregates. Especially in aged materials, it is important to get access to the PCBs adsorbed inside the aggregates. Therefore, it was advised that acetone in combination with petroleum ether should be used for the extraction of PCBs from these kinds of samples (ISO, 2002; Banjoo and Nelson, 2005). Acetone, in combination with some mechanical forces and petroleum ether, will disintegrate the aggregates and improve the extraction. Similar result was also reported by Tor et al. (2006) in which it is indicated that higher efficiencies were obtained with solvent mixture of acetone and petroleum ether (1/1, v/v) as compared to n-hexane, ethyl acetate, and acetone for extracting OCPs in soil samples. Therefore, the mixture of petroleum ether and acetone (1/1, v/v) was used as extraction solvent in further optimization experiments.

It was observed that the significant factors were the sample amount and number of the extraction step for all PCB compounds in soil. However, solvent volume was not significant. Additionally, interactions between sample amount and solvent volume, sample amount and number of the extraction step, and solvent volume and number of the extraction step were significant and they were affected by positive sign. It can be seen that sample amount was affected by a positive sign for all PCBs; therefore, 1.5 g amount of sample is better than 0.5 g; solvent volume had also positive sign, but it had no significant effect. Therefore, 2 mL of acetone/petroleum ether mixture is better than 4 mL of the same solvent mixture. Last, number of the extraction step had positive sign for all PCBs. According to the results, the optimum conditions for USE of PCBs from soil were chosen as follows: sample amount: 1.5 g; solvent volume: 2 mL (mixture of acetone/petroleum ether, 1/1, v/v) and number of extraction step: 3, with 5 min of sonication time.

The optimum extraction procedure was examined by using three different fortification levels (levels 1, 40 μg kg^{-1}; level 2, 80 μg kg^{-1}; level 3, 120 μg kg^{-1}). The results of recoveries are given in Figure 14.5. According to fortification level 1, recoveries ranged from 91 (±7)% to 96 (±7)%. Comparable recoveries were also obtained from fortification levels 2 and 3 (Figure 14.5). When statistical evaluation was carried out between quantities of PCBs extracted from fortification levels 1 and 2, no significant differences ($p > 0.05$) were observed. Moreover, the same statistical evaluations were carried out between fortification levels 1–3 and 2–3, no significant differences were observed. This indicated that optimized USE was of considerable efficiency in order to extract PCBs from soil sample.

The applicability of the USE method to the real soil samples was investigated by comparing with the soxhlet and shake-flask extraction method. The analyses for two different soil samples were

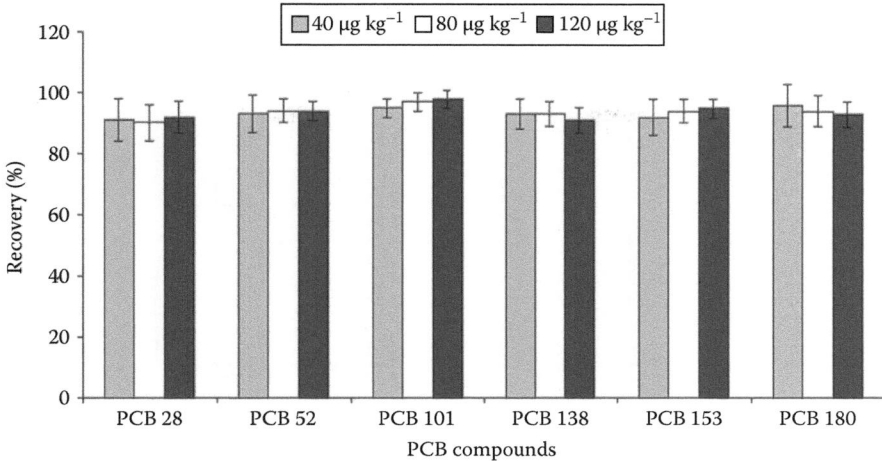

FIGURE 14.5 Recoveries of PCBs from spiked soil with three fortification levels using USE method [$n = 5$]. (Extraction conditions; extraction solvent: acetone/petroleum ether (1/1, v/v), soil sample amount: 1.5 g, extraction solvent volume: 2 mL, number of extraction step: 3, extraction time: 5 min, temperature: 25°C.)

carried out. Soil sample A and B were spiked with PCBs (spike level for each compound: 80 µg kg^{-1}) and analysis was performed by using USE, soxhlet, and shake-flask extraction. As seen in Figure 14.6, USE gave comparable results with soxhlet and shake-flask extraction methods. Statistical evaluation indicated no significant differences ($p > 0.05$) between the quantities of the PCBs extracted by USE, soxhlet, shake-flask extraction. When recoveries of PCBs for the ultrasonic extraction were gauged against absolute limits of 70% and 130% (USEPA, 1995), it was seen that USE gave satisfactory results.

Soxhlet and shake-flask extractions have been the traditional methods used for the extraction of PCBs from soils (ISO, 2002). The main disadvantages of these methods are that there are needs for more volume of solvent, long time for extraction, reconcentration, and cleanup steps (Bøwadt et al., 1995; Hartonen et al., 1997; Schantz et al., 1998). Ultrasonication allows an intensive contact between soil particles and solvent and it reduces the extraction time. Therefore, ultrasonic extraction can be used to extract PCBs from soil as an alternative to common soxhlet and shake-flask extractions (USEPA, 1996; Sporring et al., 2005). Soxtec extraction, based on soxhlet system, is a two-step extraction procedure, involving a boiling and rinsing step, which drastically reduces the total time of extraction. However, reconcentration and cleanup steps are also required for both ultrasonic and soxtec extraction techniques (Pastor et al., 1997; Popp et al., 1997; Nilsson et al., 2001). Apart from these methods, three more recent techniques from literature, including SFE (Bøwadt and Hawthorne, 1995), MAE (Eskilsson and Björklund, 2000), and ASE (Björklund et al., 2000) were also compared for the extraction of PCBs from soil. The main key to shorter extraction times and reduced solvent consumption with these techniques is the possibility of working at elevated temperatures above the boiling point of the solvent. Thereby, the extraction process is facilitated due to increased analyte desorption and diffusion from the solid matrix (Sporring et al., 2005). SFE and ASE techniques need much lower volumes of organic solvents than other extraction techniques. SFE showed to be particularly attractive as no more than 1.5 mL was necessary in order to elute the analytes from the reversed phase trap directly into a GC vial. Except for SFE technique, reconcentration and cleanup steps have to be performed for MAE and ASE techniques (Berset et al., 1999). Obviously, the contamination risk for those extraction techniques, which require reconcentration and cleanup steps, is higher than that of SFE. On the other hand, method development, time, and costs for SFE are quite high as well as for ASE (Berset et al., 1999; Sporring

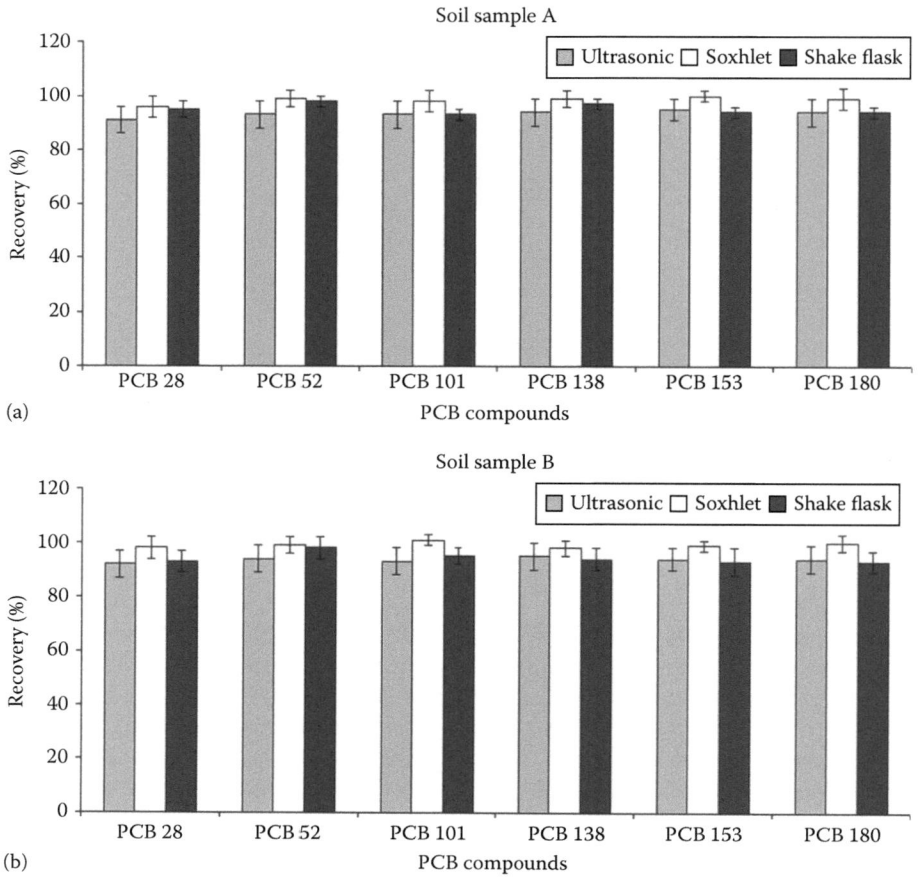

FIGURE 14.6 Comparison of extraction efficiency of the ultrasonic method with soxhlet and shake-flask for PCBs in fortified real soil samples (a) soil sample A, (b) soil sample B (fortification concentration for each compound: 80 μg kg^{-1}), [$n = 5$].

et al., 2005). Compared to the conventional soxhlet and shake-flask extraction techniques, the USE in this study has many advantages including minimal solvent use, short extraction and preconcentration time, low cost, simplicity, and ease of use. In addition, this method is cheaper and easier than MAE, ASE, and SFE techniques.

Recovery experiments were carried out for the determination of the efficiency of the USE of PAHs from soil. At the beginning of the experiments, the extraction efficiency of different solvents was compared. For that, a 0.5 g fortified with PAHs soil sample was sonicated for 5 min with 5 mL of each investigated extraction solvent or solvent mixture in an ultrasonic bath. The extracts were filtered by using filter paper and the filtrates were reduced to 1 mL with a rotary evaporator and adjusted to exactly 250 μL by using a gentle nitrogen stream. The concentrated extract was transferred onto the microscale cleanup column and elution was performed, as described earlier. The amounts of extracted PAHs were determined by GC–MS and the recoveries (%) were calculated. The other parameters affecting the extraction efficiency of the USE procedure (i.e., amount of sample, volume of extraction solvent, and number of extraction steps) were optimized by using a 2^3 factorial experimental design. The experimental design matrix is constituted, as shown in Table 14.4. The lower and higher level for each factor was designated as "−" and "+" signs, respectively (Table 14.4). After each extraction, the determination of PAHs in the extracts was determined by GC–MS, as described earlier.

TABLE 14.4
Design Matrix for Factorial Design and Average Recoveries of PAHs for the Effect of Parameters on the Ultrasonic Solvent Extraction Method (Ozcan et al., 2009a)

Number	Codified Variables			No Codified Variables			Average Recovery (%)
	X_1 (min)	X_2 (min)	X_3 (%)	X_1 (min)	X_2 (min)	X_3 (%)	
1–9	−	−	−	0.5	2	1	25
2–10	+	−	−	1.5	2	1	17
3–11	−	+	−	0.5	5	1	76
4–12	+	+	−	1.5	5	1	45
5–13	−	−	+	0.5	2	3	58
6–14	+	−	+	1.5	2	3	27
7–15	−	+	+	0.5	5	3	96
8–16	+	+	+	1.5	5	3	56

Note: X_1, soil sample amount; X_2, extraction solvent volume; X_3, number of extraction step.

The extraction solvent has a significant effect on the efficiency of extraction of the target compounds (Luque de Castro and Garcia-Ayuso, 1998). To achieve better extraction efficiencies of PAHs from the soil sample, we first examined different extraction solvents with a wide polarity range like *n*-hexane, ethyl acetate, acetone, a mixture of *n*-hexane and acetone (1/1 v/v), a mixture of *n*-hexane and ethyl acetate (1/1 v/v), and a mixture of petroleum ether and acetone (1/1 v/v). The recovery results indicated that the ethyl acetate gave the highest recoveries followed by acetone, a mixture of *n*-hexane and acetone (1/1 v/v), a mixture of *n*-hexane and ethyl acetate (1/1 v/v), a mixture of petroleum ether and acetone (1/1 v/v), and *n*-hexane (volume of each solvent or solvent mixture: 5 mL) in USE. Therefore, ethyl acetate was used as the extraction solvent in further optimization experiments.

After the choice of ethyl acetate as the optimum extraction solvent, and using 5 min sonication time, several other factors affecting the efficiency of the extraction procedure, such as sample amount, solvent volume, and number of extraction times, were optimized. For all PAHs compounds, the significant parameters were sample amount, solvent volume, and number of extraction times. Additionally, interactions between the sample amount and solvent volume and between the sample amount and number of extraction times were found to be significant. Lastly, interaction between the solvent volume and number of extraction times was also significant. Regarding the direction of the effects, sample amount was of a negative sign. It is expected that a high sample amount may require longer sonication times for the extraction of all PAHs from soil. In other words, a fixed sonication time (5 min) was insufficient for extraction of 1.5 g of sample. Another reason may be that 2 mL of extraction solvent is not adequate for complete extraction of PAHs from 1.5 g of soil sample. Hence, 0.5 g of sample is better than 1.5 g for the extraction of PAHs with 2 mL of solvent and 5 min of sonication time. Solvent volume had a positive sign; hence, 5 mL is better than 2 mL for the extraction. In soil and sludges, pollutants are adsorbed on or in soil aggregates. Especially in aged materials, it is important to get access to the pollutants adsorbed inside the aggregates. In combination with ultrasonication, when the volume of extraction solvent is increased, disintegration of the aggregates, and, hence, extraction efficiency will be improved (Banjoo and Nelson, 2005). Namely, in comparison to 2 mL, 5 mL of extraction solvent is better for disintegration of the soil aggregates and extraction of PAHs from soil. The number of extractions also had a positive sign for all studied PAHs. Increasing the number of extraction steps also increased the extraction efficiency. Thus, recoveries obtained from three extractions are higher than those from a single-step extraction.

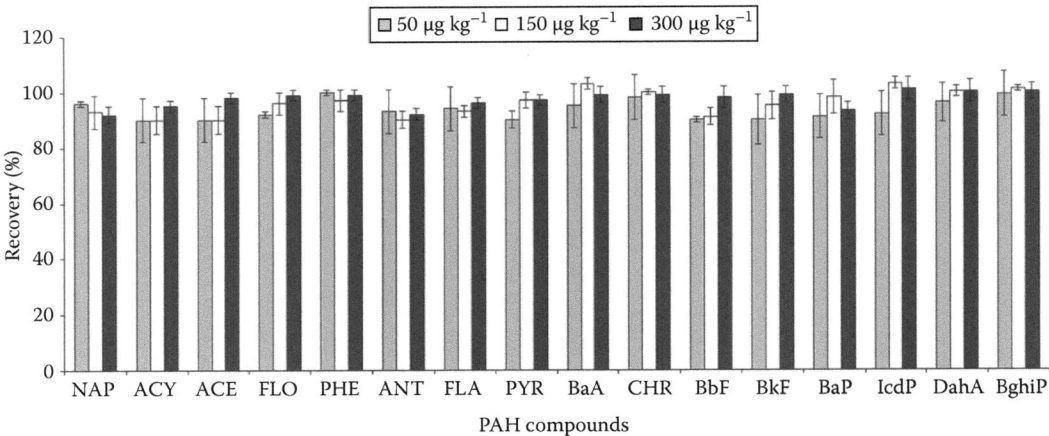

FIGURE 14.7 Recoveries of PAHs from spiked soil with three fortification levels using USE method [$n = 5$]. (Extraction conditions; extraction solvent: ethyl acetate, soil sample amount: 0.5 g, extraction solvent volume: 5 mL, number of extraction step: 3, extraction time: 5 min, temperature: 25°C.)

The optimum conditions for miniaturized USE of PAHs from soil were determined as follows: sample amount: 0.5 g; solvent volume: 5 mL with ethyl acetate and number of extraction steps: 3, with 5 min of sonication time.

The USE procedure was examined on a soil with spiked three different fortification levels (level 1: 50 µg kg^{-1}, level 2: 150 µg kg^{-1} and level 3: 300 µg kg^{-1}).

The recovery results are given in Figure 14.7. According to fortification level 1, recoveries ranged from 90% to 100%, with RSDs in the range of <1%–15%. Comparable recoveries were also obtained from fortification level 2 and 3 (see Figure 14.7). When statistical evaluation was carried out between the quantities of PAHs extracted from fortification level 1 and level 2, no significant differences ($p > 0.05$) were observed. Moreover, when the same statistical evaluations were carried out between fortification levels 1 and 3 as well as 2 and 3, no significant differences were observed. Furthermore, when recoveries of PAHs for the USE were gauged against absolute limits of 70% and 130% (US EPA, 1995), it can be seen that USE gave satisfactory results.

The USE was compared with traditional soxhlet and shake-flask extractions by means of different real soil samples with spiked PAHs analysis. In our comparison study, because ethyl acetate gave the maximum extraction efficiency for PAHs, it was also chosen as the extraction solvent for soxhlet and shake-flask extraction procedures. Figure 14.8 shows the results of PAH recoveries for the real soil sample. As seen in Figure 14.8, the proposed USE gave comparable results to the other extraction procedures. A statistical evaluation indicated no significant differences ($p > 0.05$) between the quantities of PAH extracted by USE and other extraction procedures. As a result, the ultrasonic extraction is superior to the other examined soxhlet and shake-flask extraction techniques in terms of sample requirement, solvent consumption, and extraction time. Furthermore, the method is cheaper and easier than other extraction techniques reported in the literature, such as SFE (Librando et al., 2004), MAE (Li et al., 2003), and matrix solid phase dispersion (Pena et al., 2007; Sánchez-Brunete et al., 2007).

AIR SAMPLE ANALYSIS

There are many works that report extracting different organic pollutants from airborne particles using soxhlet extraction (Vasconcellos et al., 2003; Cindoruk and Tasdemir, 2007; Tasdemir and Esen, 2007). However, the main disadvantage of soxhlet extraction is that it takes long time

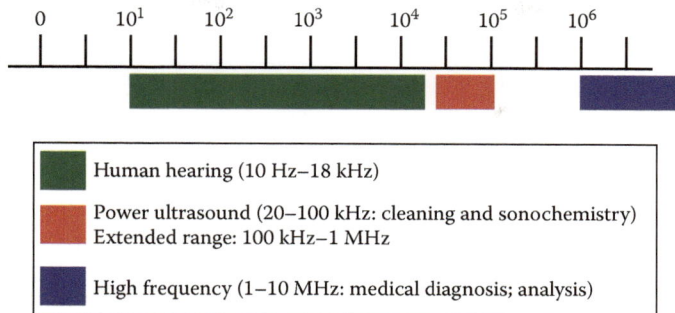

FIGURE 2.1 Sound frequencies (scale in Hz).

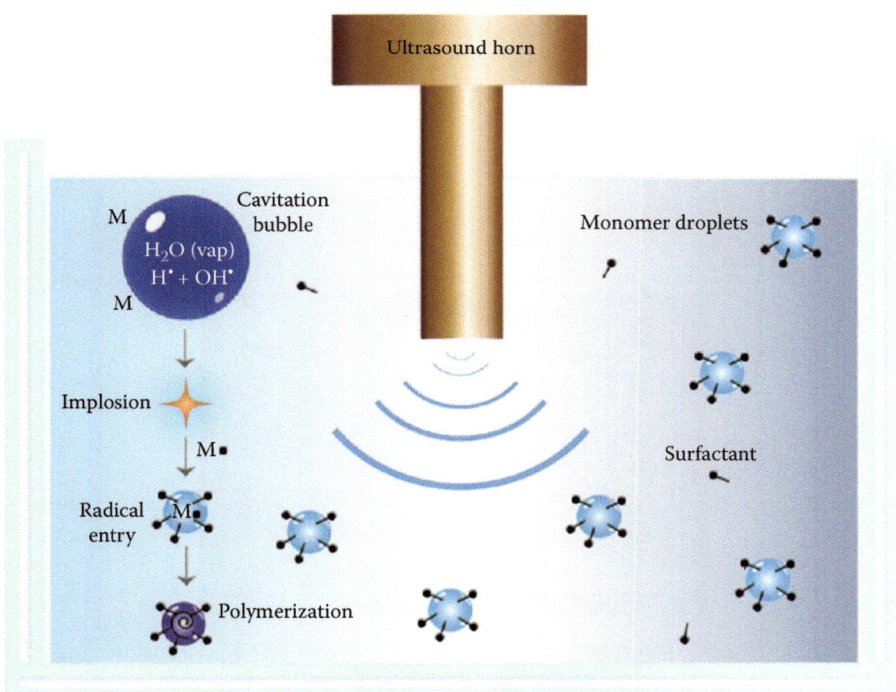

FIGURE 19.4 Diagram of the proposed sonochemical miniemulsion polymerization pathway. The ultrasound from the horn tip produces cavitation bubbles that upon collapse generate the conditions that lead to primary radical formation and emulsification of the monomer. Monomeric radicals are mainly formed at the surface of the cavitation bubbles and subsequently enter into monomer droplets producing latex particles.

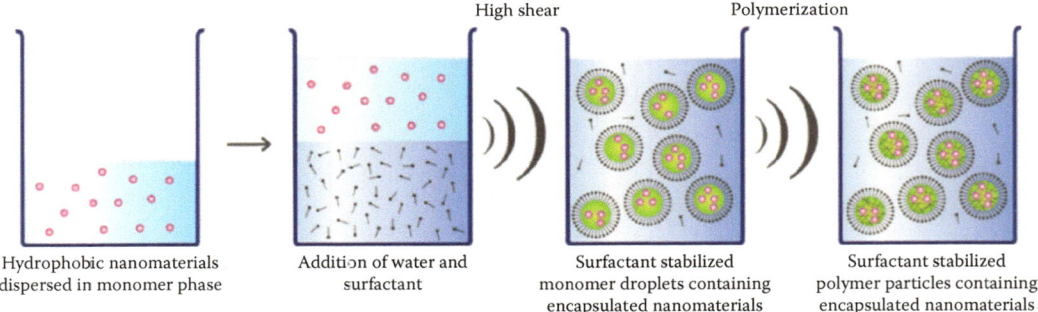

FIGURE 19.7 Schematic diagram of the procedure for the encapsulation of hydrophobic nanomaterials and the 1:1 copy of monomer droplets to latex particles by the sonochemically driven miniemulsion polymerization pathway.

FIGURE 23.3 Laser visualization at 20 kHz. (From *Ultrason. Sonochem.*, 11, Viennet, R., Ligier, V., Hihn, J.-Y., Bereiziat, D., Nika, P., and Doche, M.-L., Visualisation and electrochemical determination of the actives zones in an ultrasonic reactor using 20 and 500 kHz frequencies, 125–129, Copyright (2004), with permission from Elsevier.)

FIGURE 23.4 Horizontal cutting close to the transducer. (From *Ultrason. Sonochem.*, 11, Viennet, R., Ligier, V., Hihn, J.-Y., Bereiziat, D., Nika, P., and Doche, M.-L., Visualisation and electrochemical determination of the actives zones in an ultrasonic reactor using 20 and 500 kHz frequencies, 125–129, Copyright (2004), with permission from Elsevier.)

FIGURE 23.5 Vertical light sheet in the entire reactor volume. (From *Ultrason. Sonochem.*, 11, Viennet, R., Ligier, V., Hihn, J.-Y., Bereiziat, D., Nika, P., and Doche, M.-L., Visualisation and electrochemical determination of the actives zones in an ultrasonic reactor using 20 and 500 kHz frequencies, 125–129, Copyright (2004), with permission from Elsevier.)

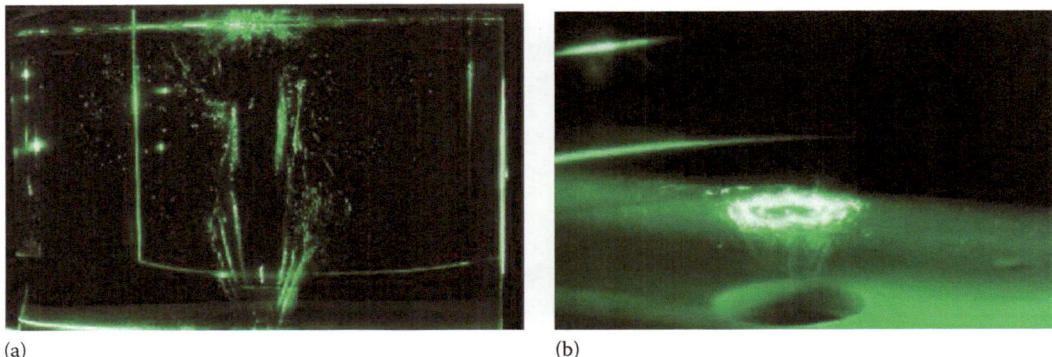

FIGURE 23.6 (a) Visualization in the entire reactor volume; (b) radial distribution close to the transducer. (From *Ultrason. Sonochem.*, 11, Viennet, R., Ligier, V., Hihn, J.-Y., Bereiziat, D., Nika, P., and Doche, M.-L., Visualisation and electrochemical determination of the actives zones in an ultrasonic reactor using 20 and 500 kHz frequencies, 125–129, Copyright (2004), with permission from Elsevier.)

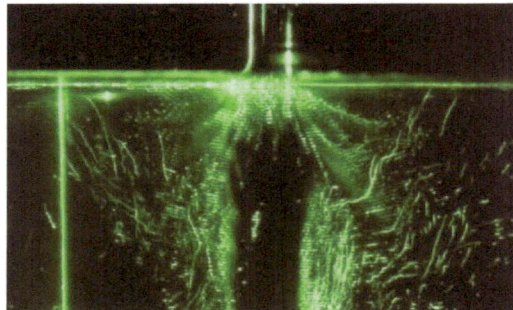

FIGURE 23.7 Details close to the water/air interface at high frequency. (From *Ultrason. Sonochem.*, 11, Viennet, R., Ligier, V., Hihn, J.-Y., Bereiziat, D., Nika, P., and Doche, M.-L., Visualisation and electrochemical determination of the actives zones in an ultrasonic reactor using 20 and 500 kHz frequencies, 125–129, Copyright (2004), with permission from Elsevier.)

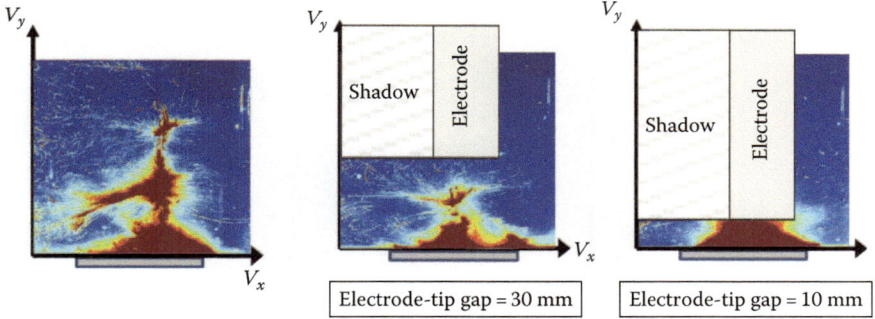

FIGURE 23.8 Laser visualization of ultrasonic actives zones in electrode presence.

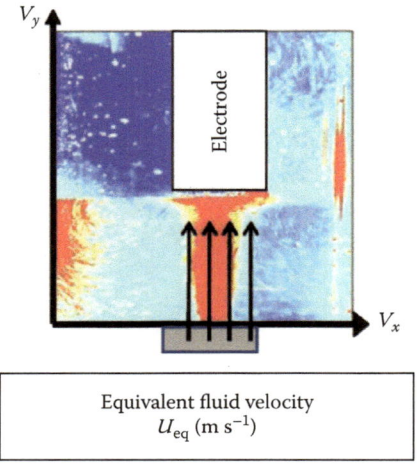

FIGURE 23.18 Equivalent fluid velocity—coordinate system.

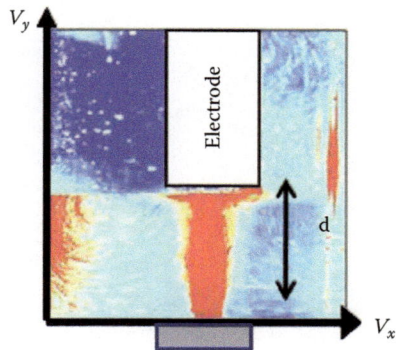

FIGURE 23.19 Determination of the "equivalent flow" in the zone close to the transducer.

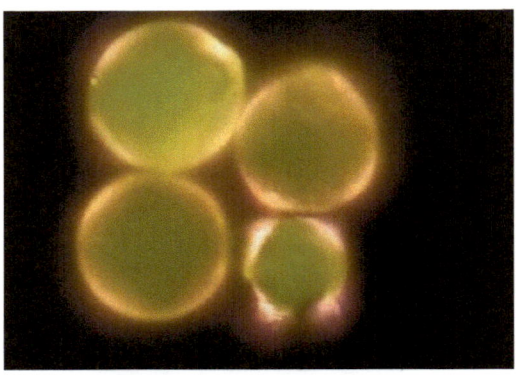

FIGURE 23.21 Fluorescent tracers. (From *Ultrason. Sonochem.*, 16, Mandroyan, A., Doche, M.-L., Hihn, J.-Y., Viennet, R., Bailly, Y., and Simonin, L., Modification of the ultrasound induced activity by the presence of an electrode in a sono-reactor working at two low frequencies (20 and 40 kHz). Part II: Mapping flow velocities by particle image velocimetry (PIV), 97–104, Copyright (2009b), with permission from Elsevier.)

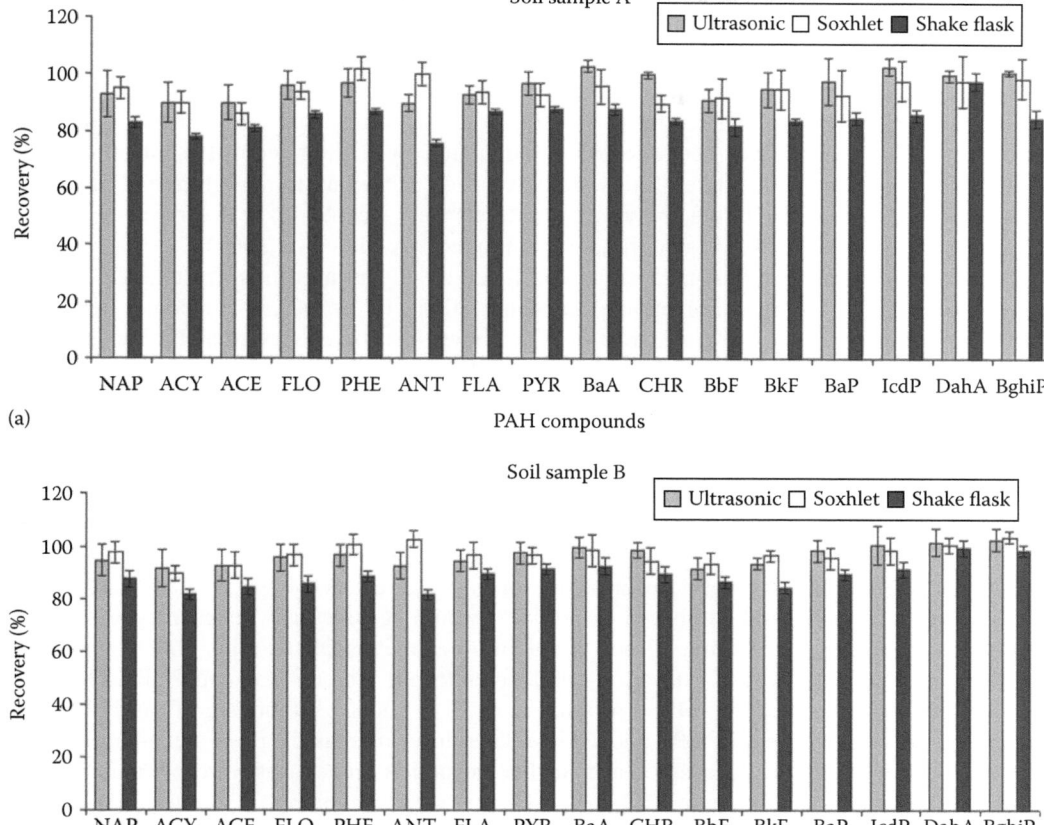

FIGURE 14.8 Comparison of extraction efficiency of the ultrasonic method with soxhlet and shake-flask for PAHs in fortified real soil samples (a) soil sample A, (b) soil sample B (fortification concentration for each compound: 150 μg kg^{-1}), [$n = 5$].

(Hartonen et al., 1997). Ultrasonication allows an intensive contact between sample and solvent and it reduces the extraction time. Therefore, ultrasonic extraction can be used to extract target compounds from samples as an alternative to traditional soxhlet extraction (Sporring et al., 2005). Therefore, USE for the determination of PCBs and PAHs from airborne particles on glass microfiber filter and from gas phase in PUF plugs were carried out.

For extraction of PCBs and PAHs from air samples, the recovery efficiency of the USE method (1999a,b) has been evaluated by the analysis of precleaned glass microfiber filter and PUF plug samples spiked with known amounts of PCBs and PAHs (McConnell et al., 1998; Alegria et al., 2006; Mandalakis and Stephanou, 2007).

At the beginning of this study, the extraction efficiency of n-hexane:diethyl ether (9/1, v/v), dichloromethane:petroleum ether (1/4, v/v), and n-hexane:petroleum ether (1/1, v/v) were compared. For that, spiked glass microfiber filters were prepared by adding a standard mixture of PCBs and PAHs. And then the glass microfiber filters were sonicated for 45 min with 100 mL of solvents in an ultrasonic bath. The extracts were reduced to 2 mL with a rotary evaporator and adjusted to exactly 1 mL under a gentle nitrogen stream. The concentrated extract was transferred onto the cleanup column and elution was performed, as described earlier. The amount of extracted target compounds were determined by GC–MS and the recoveries were calculated. In the second set of experiments, the optimum volume of solvent, optimum sonication time, and optimum repetition of extractions were determined. In order to determine optimum volume of extraction solvent, spiked glass microfiber filters were sonicated for 45 min with 25, 50, 75, and 100 mL of extraction solvent. In order to determine the optimum sonication time, spiked glass microfiber filters were sonicated with 75 mL of extraction solvent for 15, 30, 45, 60 min. The extraction of spiked microfiber filters with 25 mL of extraction solvent for 15 min was repeated three times. The cumulative recoveries were calculated by adding recoveries of compounds determined in each extract. Spiked filters were extracted according to the optimized ultrasonic extraction conditions. Prior to the analysis, the extracts were cleaned up as described earlier and recoveries were determined for each group of compounds (Aydin et al., 2007).

In the optimization of ultrasonic extraction, different solvent mixtures, including n-hexane: diethyl ether (9/1, v/v), dichloromethane:petroleum ether (1/4, v/v), and n-hexane: petroleum ether (1/1, v/v), were compared for extraction efficiency. Ultrasonic extraction efficiencies of the solvents were checked by calculating the recoveries. When statistical evaluation was carried out between extraction efficiencies of n-hexane:diethyl ether (9/1, v/v), dichloromethane:petroleum ether (1/4, v/v), and n-hexane:petroleum ether (1/1, v/v) for PCBs, no significant differences ($p > 0.05$) were observed. The mixture of n-hexane: petroleum ether (1/1, v/v) gave the highest recoveries for PAHs among the solvent mixtures used. The mixture of n-hexane:petroleum ether (1/1, v/v) was used in further experiments for the extraction of all target compounds. The aim of the optimization procedure was to improve the extraction efficiency with minimum solvent and time consumption. Therefore, a careful optimization of the other extraction parameters (i.e., solvent volume, extraction time, and number of extractions) would be necessary in order to get satisfactory results. Generally, increasing of the volume of n-hexane:petroleum ether (1/1, v/v) from 25 to 75 mL resulted in increased extraction efficiencies for all target PCBs and PAHs. However, increasing of the volume of the solvent mixture from 75 to 100 mL did not significantly improve the extraction efficiencies of PCBs and PAHs. Therefore, increasing the n-hexane:petroleum ether (1/1, v/v) volume was interrupted and further extraction procedures were carried out using 75 mL of n-hexane:petroleum ether (1/1, v/v). The next step was the optimization of the sonication time. The sonication time was changed between 15 and 60 min. Especially for PAHs, it was found that the recoveries increased up to 45 min and there were few fluctuations in the recoveries for increasing time from 45 to 60 min. Additionally, 45 min was enough for the ultrasonic extraction of PCBs. Thus, the optimum sonication time for extraction was selected as 45 min. If the extraction solvent volume is divided into two or three aliquots and extraction is carried out separately with these solvents twice or three times, the extraction efficiency increases (Skoog et al., 1996). For PCBs and PAHs compounds extracted from filters with the ultrasonic extraction method using an optimum solvent volume of 75 mL, optimum extraction time was determined as 45 min. A volume of 75 mL extraction solvent was divided into three aliquots and filters were extracted using 25 mL of solvent, 15 min sonication time three times sequentially. Extracted solvents were combined and reduced to 1 mL for GC analysis. For PCB compounds, comparison of extractions were performed using a solvent volume of 75 mL and 45 min of sonication was carried out three times, with a solvent volume of 25 mL and 15 min of sonication. Statistical evaluation was performed on the efficiencies of these two extraction procedures, no significant differences were determined ($p > 0.05$). Statistical evaluation of both extraction procedures for the recoveries of PAHs showed no significant differences ($p > 0.05$). Therefore, 75 mL of extraction solvent n-hexane/petroleum ether (1/1, v/v) was used in the extraction of filters and the extraction time was 45 min.

For USE of PCBs and PAHs from PUF plug, the extraction efficiency of n-hexane:diethyl ether (9/1, v/v), dichloromethane:petroleum ether (1/4, v/v), and n-hexane:petroleum ether (1/1, v/v) were compared. Spiked PUF plugs were sonicated for 45 min with 300 mL of solvents in an ultrasonic bath. The extracts were reduced to 2 mL with a rotary evaporator and adjusted to exactly 1 mL under a gentle nitrogen stream. The concentrated extract was transferred onto the cleanup column and elution was performed, as described earlier. The amount of extracted target compounds were determined by GC–MS and the recoveries were calculated. The mixture of n-hexane:diethyl ether (9/1, v/v) gave the highest recoveries for PCBs and PAHs among the solvent mixtures used.

For the ultrasonic extraction of PUF plugs, 300 mL of extraction solvent is needed. However, the time needed for the ultrasonic extraction is shorter than those of other methods. Therefore, optimization experiments for PUF plugs were conducted by employing the USE method.

Investigations were carried out in order to determine the matrix effects of samples on PCBs, and PAHs extraction, fractionation, and cleanup steps from coextracted interfering nontarget compounds, prior to the GC analysis. A volume of 360 m^3 air sample was passed through the glass microfiber filter and PUF plug at 0.25 m^3 min^{-1} flow rate in 24 h by using the modified high volume air sampler. After, the filter and PUF plug were extracted ultrasonic and soxhlet extraction procedure, the volumes were adjusted to exactly 1 mL and standard mixture of PCBs and PAHs were spiked in the extract. Then, the extract was transferred onto the top of the column and fractionations were carried out consecutively using 70 mL n-hexane and 60 mL n-hexane:ethyl acetate (1/1, v:v). Then, the volumes of the elutions were reduced to 1 mL and GC–MS was performed. The recovery from filters values for the target compounds were

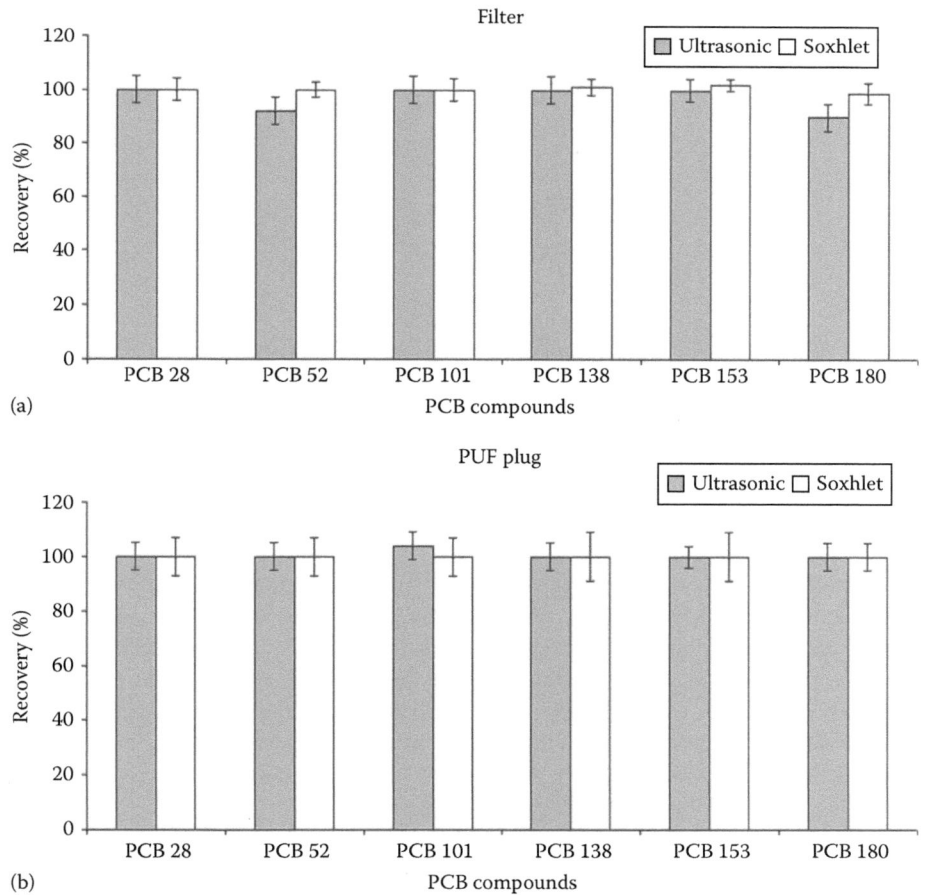

FIGURE 14.9 Comparison of extraction efficiency of the ultrasonic method with soxhlet for PCBs in fortified filter (a) and PUF plug (b) (fortification concentration for each compound: 0.2 ng m^{-3}), [$n = 5$].

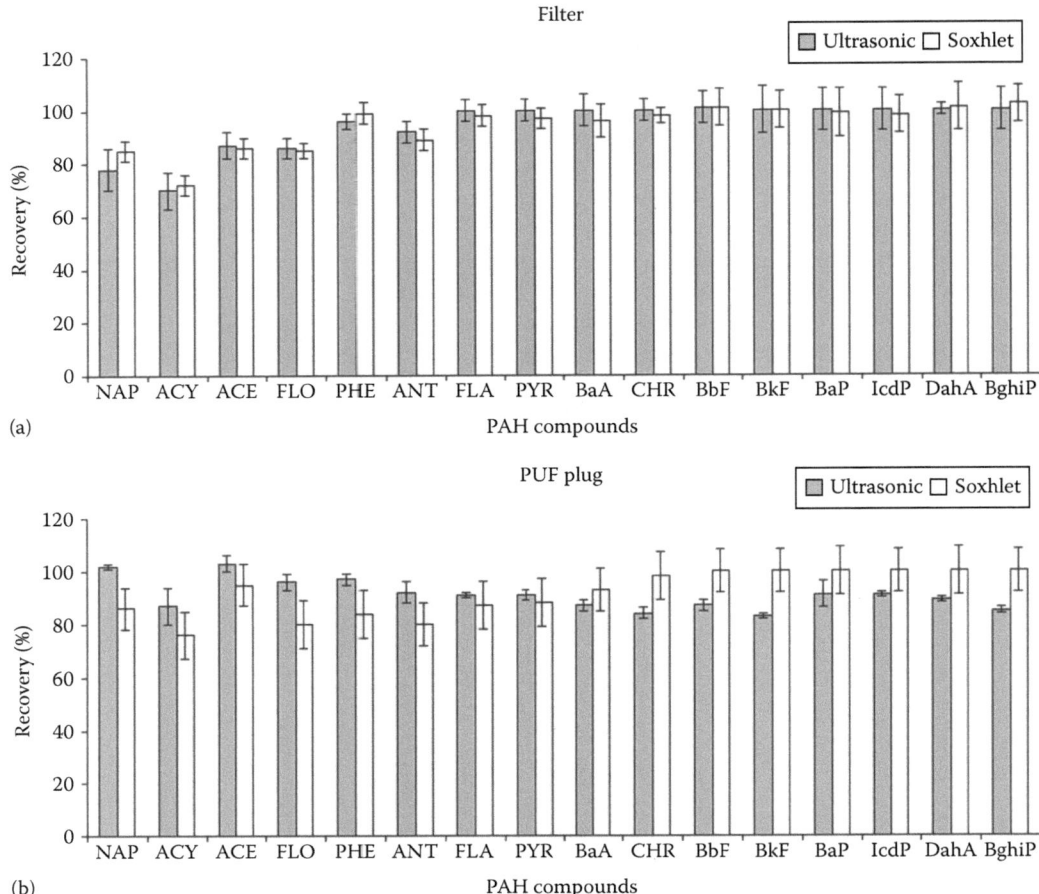

FIGURE 14.10 Comparison of extraction efficiency of the ultrasonic method with soxhlet for PAHs in fortified filter (a) and PUF plug (b) (fortification concentration for each compound: 2 ng m^{-3}), [$n = 5$].

obtained between 92% ± 5% and 100% ± 3% with RSD < ±5% for PCBs, between 78% ± 8% and 101% ± 6% with RSD < ±9% for PAHs. Recoveries for all two groups of compounds were in acceptable levels in comparison to limits given by the US EPA, which are between 60% and 120% for multiple compounds (US EPA, 1995).

The USE method was compared with the soxhlet extraction method, which is used for the extraction of PCB and PAH compounds from filters and PUF plugs. As seen in Figures 14.9 and 14.10, USE gave comparable results to the soxhlet extraction procedure. A statistical evaluation indicated no significant differences ($p > 0.05$) between the quantities of PCBs and PAH extracted by USE and other extraction procedure. The most important disadvantages of the soxhlet extraction method are the long extraction time requirement and higher solvent consumption. However, using the ultrasonic extraction method reduces the hazardous solvent consumption, analysis time, and expenses in the analysis of trace levels organic pollutants from filters.

A comparison of general parameters of the extraction methods of PCBs and PAHs compounds from water, soil, and air samples is shown in Table 14.5, which indicates that the ultrasound-assisted extraction method needs less extraction time, sample amount as well as extraction solvent, consequently lower cost, compared to other extraction procedures, including LLE, SPE, soxhlet, and shake-flask extractions.

TABLE 14.5
Comparison of General Parameters for the Different Extraction Techniques Used for Determination of PCBs and PAHs in Water, Soil, and Air

	Water		
General Parameters	Ultrasonic Solvent Extraction	Liquid–Liquid Extraction	Solid Phase Extraction
Sample volume (mL)	10	200	200
Extraction time (h)	0.08	1	4
Solvent volume (mL)	0.1–0.2	60	38
Reconcentration step	No	Yes	Yes
Need for cleanup	No	Yes	No
Cost	Very low	Medium	High

	Soil		
General Parameters	Ultrasonic Solvent Extraction	Soxhlet Extraction	Shake Flask Extraction
Sample amount (g)	0.5–1.5	10	10
Extraction time (h)	0.25	16–18	12
Solvent volume (mL)	15	150	50
Reconcentration step	Yes	Yes	Yes
Need for clean up	Yes	Yes	Yes
Cost	Very low	Medium	Medium

	Air			
	Filter		PUF Plug	
General Parameters	Ultrasonic Solvent Extraction	Soxhlet Extraction	Ultrasonic Solvent Extraction	Soxhlet Extraction
Extraction time (h)	0.75	16	0.75	16
Solvent volume (mL)	75	150	300	150
Reconcentration step	Yes	Yes	Yes	Yes
Need for clean up	Yes	Yes	Yes	Yes
Cost	Low	Medium	Medium	Medium

CONCLUSIONS

This chapter has outlined the successful development and application of USE procedure for the determination of PCBs and PAHs in water, soil, and air samples by using GC–MS. Analyses of real samples showed that sample matrices had no adverse effect on the efficiency of USE procedure. As a consequence, the USE method is precise, reproducible, and rapid and easy for the analyses of PCBs and PAHs in water, soil, and air samples. It also requires only small volumes of extraction solvent and sample materials. In addition, the USE method has been demonstrated to be viable, rapid, and easy to use for the qualitative and quantitative analysis of PCBs and PAHs in different water, soil, and air samples. Additionally, the USE method uses less solvent than traditional approaches (i.e., liquid–liquid, solid phase, shake-flask, soxhlet extractions), reducing the costs associated with solvent purchase and waste disposal. The USE method will reduce laboratory expenses without substantial new equipment and without compromising accuracy and precision. Furthermore, USE method is cheaper and easier than LLE, SPE, SPME, MAE, ASE, and SFE techniques and it can be concluded that most commercial laboratories can efficiently use the proposed method for the extraction of PCBs and PAHs from water, soil, and air.

REFERENCES

Abismail, B., Canselier, J.P., Wilhelm, A.M., Delmas, H., and Gourdon, C. 1999. Emulsification by ultrasound: Drop size distribution and stability. *Ultrasonics Sonochemistry*, 6: 75–84.
Alegria, H., Bidleman, T.F., and Figueroa, M.S. 2006. Organochlorine pesticides in the ambient air of Chiapas, Mexico. *Environmental Pollution*, 140: 484–491.
Ashley, K., Andrews, R.N., Cavazos, L., and Demange, M. 2001. Ultrasonic extraction as a sample preparation technique for elemental analysis by atomic spectrometry. *Journal of Analytical and Atomic Spectrometry*, 16: 1147–1153.
ATSDR (Agency for Toxic Substances and Disease Registry, Division of Toxicology/Toxicology Information Branch). 1995. *Toxicological Profile for Polycyclic Aromatic Hydrocarbons (PAHs)*, Atlanta, GA, pp. 11–111.
ATSDR (Agency for Toxic Substances and Disease Registry, Division of Toxicology/Toxicology Information Branch). 2000. *Toxicological Profile for Polychlorinated Biphenyls (PCBs)*, Atlanta, GA, pp. 477–507.
Aydin, M.E., Ozcan, S., and Tor, A. 2007. Ultrasonic solvent extraction of persistent organic pollutants from airborne particles. *Clean: Soil, Air, Water*, 35: 660–668.
Aydin, M.E., Tor, A., and Ozcan, S. 2006. Determination of selected polychlorinated biphenyls in soil by miniaturized ultrasonic solvent extraction and gas chromatography-mass selective detection. *Analytical Chimica Acta*, 577: 232–237.
Aydin, M.E., Wichmann, H., and Bahadir, M. 2004. Priority organic pollutants in fresh and wastewaters of Konya-Turkey. *Fresenius Environmental Bulletin*, 13: 118–123.
Babic, S., Petrovic, M., and Kastelan-Macan, M. 1998. Ultrasonic solvent extraction of pesticides from soil. *Journal of Chromatography A*, 823: 3–9.
Bakan, G. and Ariman, S. 2004. Persistent organochlorine residues in sediments along the coast of mid-Black Sea region of Turkey. *Marine Pollution Bulletin*, 48: 1031–1039.
Banjoo, D.R. and Nelson, P.K. 2005. Improved ultrasonic extraction procedure for the determination of polycyclic aromatic hydrocarbons in sediments, *Journal of Chromatography A*, 1066: 9–18.
Barco-Bonilla, N., Martinez Vidal, J.L., Garrido Frenich, A., and Romero-Gonzalez, R. 2009. Comparison of ultrasonic and pressurized liquid extraction for the analysis of polycyclic aromatic compounds in soil samples by gas chromatography coupled to tandem mass spectrometry. *Talanta*, 78: 156–164.
Barnabas, I.J., Dean, J.R., Tomlinson, W.R., and Owen, S.P. 1995. Experimental design approach for the extraction of polycyclic aromatic hydrocarbons from soil using supercritical carbon dioxide. *Analytical Chemistry*, 67: 2064–2069.
Benner, B.A. 1998. Summarizing the effectiveness of supercritical fluid extraction of polycyclic aromatic hydrocarbons from natural matrix environmental samples. *Analytical Chemistry*, 70: 4594–4601.
Berijani, S., Assadi, Y., Anbia, M., Milani Hosseini, M.R., and Aghaee, E. 2006. Dispersive liquid-liquid microextraction combined with gas chromatography-flame photometric detection, very simple, rapid and sensitive method for the determination of organophosphorus pesticides in water. *Journal of Chromatography A*, 1123: 1–9.
Berset, J.D., Ejem, M., Holzer, R., and Lischer, P. 1999. Comparison of different drying, extraction and detection techniques for the determination of priority polycyclic aromatic hydrocarbons in background contaminated soil samples. *Analytical Chimica Acta*, 383: 263–275.
Björklund, E., Nilsson, T., and Bøwadt, S. 2000. Pressurised liquid extraction of persistent organic pollutants in environmental analysis. *Trends in Analytical Chemistry*, 19: 434–445.
Bøwadt, S. and Hawthorne, S. 1995. Supercritical fluid extraction in environmental analysis. *Journal of Chromatography A*, 703: 549–571.
Bøwadt, S., Johansson, B., Wunderli, S., Zennegg, M., de Alencastro, L.F., and Grandjean, D. 1995. Independent comparison of soxhlet and supercritical fluid extraction for the determination of PCBs in an industrial soil. *Analytical Chemistry*, 67: 2424–2430.
Camel, V. 2000. Microwave-assisted solvent extraction of environmental samples. *Trends in Analytical Chemistry*, 19: 229–248.
Castro, J., Sanchez-Brunete, C., and Tadeo, J.L. 2001. Multiresidue analysis of insecticides in soil by gas chromatography with electron-capture detection and confirmation by gas chromatography-mass spectrometry. *Journal of Chromatography A*, 918: 371–380.
Cecinato, A., Ciccioli, P., Brancaleoni, E., Brachetti, A., and Vasconcellos, P.C. 1997. PAH as Candidate Markers for Biomass Burning in the Amazonia Forest Area. *Annals Chimica*, 87: 555–569.
Chekol, T., Vough, L.R., and Chaney, R.L. 2004. Phytoremediation of polychlorinated biphenyl-contaminated soils: The rhizosphere effect. *Environmental International*, 30: 799–804.

Cindoruk, S. and Tasdemir, Y. 2007. Characterization of gas/particle concentrations and partitioning of polychlorinated biphenyls (PCBs) measured in an urban site of Turkey. *Environmental Pollution*, 148: 325–333.

Cotham, W.E. and Bidleman, T. 1995. Polycyclic aromatic hydrocarbons and polychlorinated biphenyls in air at an urban and a rural site near lake Michigan. *Environmental Science Technology*, 29: 2782–2789.

DFG (Deutsche Forschungsgemeinschaft) Pesticide Commission 1987. *Manual of Pesticide Residue Analysis*, VCH, Weinheim, Germany.

Ericsson, M. and Colmsjo, A. 2000. Dynamic microwave-assisted extraction. *Journal of Chromatography A*, 877: 141–151.

Eskilsson, C. and Björklund, E. 2000. Analytical-scale microwave-assisted extraction. *Journal of Chromatography A*, 902: 227–250.

Falconer, R.L. and Bidleman, T.F. 1994. Vapour pressures and predicted particle/gas distributions of polychlorinated biphenyl congeners as functions of temperature and orthochlorine substitution. *Atmospheric Environment*, 28: 547–554.

Fatoki, O.S. and Awofolu, R.O. 2003. Methods for selective determination of persistent organochlorine pesticide residues in water and sediments by capillary gas chromatography and electron-capture detection. *Journal of Chromatography A*, 983: 225–236.

Filipkowska, A., Lubecki, L., and Kowalewsk, G., 2005. Polycyclic aromatic hydrocarbon analysis in different matrices of the marine environment. *Analytical Chimica Acta*, 547: 243–254.

Fontana, A.R., Wuilloud, R.G., Martínez, L.D., and Altamirano, J.C. 2009. Simple approach based on ultrasound-assisted emulsification-microextraction for determination of polibrominated flame retardants in water samples by gas chromatography-mass spectrometry. *Journal of Chromatography A*, 1216: 147–153.

Gonçalves, C. and Alpendurada, M.F. 2005. Assessment of pesticide contamination in soil samples from an intensive horticulture area, using ultrasonic extraction and gas chromatography-mass spectrometry. *Talanta*, 65: 1179–1189.

Hartonen, K., Bøwadt, S., Hawthorne, S.B., and Riekkola, M.L. 1997. Supercritical fluid extraction with solid-phase trapping of chlorinated and brominated pollutants from sediment samples, *Journal of Chromatography A*, 774: 229–242.

He, Y. and Lee, H.K. 1997. Liquid-phase microextraction in a single drop of organic solvent by using a conventional microsyringe. *Analytical Chemistry*, 69: 4634–4640.

Hermanson, M.H. and Hites, R.A. 1989. Long-term measurements of atmospheric polychlorinated biphenyls in the vicinity of superfund dumps, *Environmental Science Technology*, 23: 1253–1258.

Ho, T.S., Pedersen-Bjergaard, S., and Rasmussen, K.E. 2002. Liquid-phase microextraction of protein-bound drugs under non-equilibrium conditions. *Analyst*, 127: 608–613.

Hoff, R.M., Strachan, W.M.J., Sweet, C.W. et al. 1996. Atmospheric deposition of toxic chemicals to the great lakes: A review of data through 1994. *Atmospheric Environmental*, 30: 3505–3527.

Hsu, Y.K., Holsen, T.M., and Hopke, P.K. 2003. Locating and quantifying PCB sources in Chicago: Receptor modeling and field sampling. *Environmental Science Technology*, 37: 681–690.

ISO 10382, International Organization for Standardization 2002, Geneva, Switzerland.

Jaouen-Madoulet, A., Abarnou, A., Le Guellec, A.M., Loizeau, V., and Leboulenger, F. 2000. Validation of an analytical procedure for polychlorinated biphenyls, coplanar polychlorinated biphenyls and polycyclic aromatic hydrocarbons in environmental samples. *Journal of Chromatography A*, 886: 153–173.

Jayaraman, S., Pruell, R.J., and McKinney, R. 2001. Extraction of organic contaminants from marine sediments and tissues using microwave energy. *Chemosphere*, 44: 181–191.

Jeannot, M.A. and Cantwell, F.F. 1996. Solvent microextraction into a single drop. *Analytical Chemistry*, 68: 2236–2240.

Jeannot, M.A. and Cantwell, F.F. 1997. Mass transfer characteristics of solvent extraction into a single drop at the tip of a syringe needle. *Analytical Chemistry*, 69: 235–239.

Junker, M., Kasper, M., Roosli, M. et al. 2000. Airborne particle number profile, particle mass distribution and particle-bound PAH concentrations within the city environment of Basel: An assessment as part of the BRISKA Project. *Atmospheric Environment*, 34: 3171–3181.

Khajeh, M., Yamini, Y., and Hassan, J. 2006. Trace analysis of chlorobenzenes in water samples using headspace solvent microextraction and gas chromatography/electron capture detection. *Talanta*, 69: 1088–1094.

King, A.J., Readman, J.W., and Zhou, J.L. 2004. Determination of polycyclic aromatic hydrocarbons in water by solid-phase microextraction-gas chromatography-mass spectrometry, *Analytical Chimica Acta*, 523: 259–267.

Koinecke, A., Kreuzig, R., and Bahadir, M. 1997. Effects of modifiers, adsorbents and eluents in supercritical fluid extraction of selected pesticides in soil. *Journal of Chromatography A*, 786: 155–161.

Kolb, M., Böhm, H.B., and Bahadir, M. 1995. Analytical multimethod for the determination of low volatile organic pollutants in sediments and sewage sludges. *Fresenius Journal Analytical Chemistry*, 351: 286–296.

Lambropoulou, D.A., Konstantinou, I.K., and Albanis, T.A. 2006. Sample pretreatment method for the determination of polychlorinated biphenyls in bird livers using ultrasonic extraction followed by headspace solid-phase microextraction and gas chromatography-mass spectrometry. *Journal of Chromatography A*, 1124: 97–105.

Li, K., Landriault, M., Fingas, M., and Llompart, M. 2003. Accelerated solvent extraction (ASE) of environmental organic compounds in soils using a modified supercritical fluid extractor. *Journal of Hazardous Materials*, 102: 93–104.

Librando, V., Hutzinger, O., Tringali, G., and Aresta, M. 2004. Supercritical fluid extraction of polycyclic aromatic hydrocarbons from marine sediments and soil samples. *Chemosphere*, 54: 1189–1197.

Lundstedt, S., Van Bavel, B., Haglund, P., Tysklind, M., and Oberg, L. 2000. Pressurised liquid extraction of polycyclic aromatic hydrocarbons from contaminated soils. *Journal of Chromatography A*, 883: 151–162.

Luque de Castro, M.D. and Garcia-Ayuso, L.E. 1998. Soxhlet extraction of solid materials: An outdated technique with a promising innovative future. *Analytical Chimica Acta*, 369: 1–10.

Luque de Castro, M.D. and Priego-Capote, F. 2006. *Analytical Applications of Ultrasound*, Elsevier, Amsterdam, the Netherlands.

Luque de Castro, M.D. and Priego-Capote, F. 2007. Ultrasound-assisted preparation of liquid samples. *Talanta*, 72: 321–334.

Mandalakis, M. and Stephanou, E.G. 2007. Atmospheric concentration characteristic and gas-particle partitioning of PCBs in rural area of eastern Germany. *Environmental Pollution*, 147: 211–221.

Mason, T.J. and Lorimer, J.P. 2002. *Applied Sonochemistry: Uses of Power Ultrasound in Chemistry and Processing*, Wiley VCH Verlag GmbH, Weinheim, Germany.

McConnell, L.L., Bidleman, T.F., Cotham, W.E., and Walla, M.D. 1998. Air concentrations of organochlorine insecticides and polychlorinated biphenyls over Green Bay, WI, and the four lower great lakes. *Environmental Pollution*, 101: 391–399.

Mierzwa, J., Sun, Y.C., and Yang, M.H. 1997. Determination of Co and Ni in soils and river sediments by electrothermal atomic absorption spectrometry with slurry sampling. *Analytical Chimica Acta*, 355: 277–282.

Morselli, L., Setti, L., Iannuccilli, A., Maly, S., Dinelli, G., and Quattroni, G. 1999. Supercritical fluid extraction for the determination of petroleum hydrocarbons in soil. *Journal of Chromatography A*, 845: 357–363.

Murphy, T., Formanski, L.J., Brownawell, B.B., and Meyer, J.A. 1985. PCB emissions to the atmosphere in the great lakes region, municipal landfills and incinerators. *Environmental Science Technology*, 19: 942–946.

Nawab, A., Aleem, A., and Malik, A. 2003. Determination of organochlorine pesticide in agriculture soil with special reference to γ-HCH degradation by Pseudomonas strains. *Bioresource Technology*, 88: 41–46.

Nielsen, T., Jorgensen, H.E., Larsen, J.C., and Poulsen, M. 1996. City air pollution of polycyclic aromatic hydrocarbons and other mutagens: Occurrence, sources and health effects. *Science of the Total Environment*, 189/190: 41–49.

Nilsson, M., Waldeback, M., Liljegren, G., Kylin, H., and Markides, K. 2001. Pressurized-fluid extraction (PFE) of chlorinated paraffins from the biodegradable fraction of source-separated household waste. *Fresenius Journal of Analytical Chemistry*, 370: 913–918.

Orlinskii, D., Priputina, I., Popova, A. et al. 2001. Influence of environmental contamination with PCBs on human health. *Environmental Geochemical Health*, 23: 317–332.

Oros, D.R. and Simoneit, B.R.T. 2001. Identification and emission factors of molecular tracers in organic aerosols from biomass burning, Part 2, Deciduous trees. *Applied Geochemistry*, 16: 1545–1565.

Ozcan, S. and Aydin, M.E. 2009. Polycyclic aromatic hydrocarbons, polychlorinated biphenyls and organochlorine pesticides in urban air of Konya, Turkey. *Atmospheric Research*, 93: 715–722.

Ozcan, S., Aydin, M.E., and Tor, A. 2008. Chromatographic separation and analytic procedure for priority organic pollutants in urban air. *Clean: Soil, Air, Water*, 36: 969–977.

Ozcan, S., Tor, A., and Aydin, M.E. 2009a. Determination of polycyclic aromatic hydrocarbons in soil by miniaturized ultrasonic extraction and gas chromatography-mass selective detection. *Clean: Soil, Air, Water*, 37: 811–817.

Ozcan, S., Tor, A., and Aydin, M.E. 2009b. Determination of selected polychlorinated biphenyls in water samples by ultrasound-assisted emulsification-microextraction and gas chromatography-mass-selective detection. *Analytica Chimica Acta*, 647: 182–188.

Ozcan, S., Tor, A., and Aydin, M.E. 2010. Determination of polycyclic aromatic hydrocarbons in waters by ultrasound-assisted emulsification-microextraction and gas chromatography-mass spectrometry. *Analytical Chimica Acta*, 65: 193–199.

Pastor, A., Vazquez, E., Ciscar, R., and De la Guardia, M. 1997. Efficiency of the microwave-assisted extraction of hydrocarbons and pesticides from sediments. *Analytical Chimica Acta*, 344: 241–249.

Pena, M.T., Casais, M.C., Mejuto, M.C., and Cela, R. 2007. Optimization of the matrix solid-phase dispersion sample preparation procedure for analysis of polycyclic aromatic hydrocarbons in soils: Comparison with microwave-assisted extraction. *Journal of Chromatography A*, 1165: 32–38.

Pena-Pereira, F., Lavilla, I., and Bendicho, C. 2009. Miniaturized preconcentration methods based on liquid-liquid extraction and their application in inorganic ultratrace analysis and speciation: A review. *Spectrochimica Acta B*, 64: 1–15.

Pino, V., Ayala, J.H., Afonso, A.M., and Gonzalez, V. 2000. Determination of polycyclic aromatic hydrocarbons in marine sediments by high-performance liquid chromatography after microwave-assisted extraction with micellar media. *Journal of Chromatography A*, 869: 515–522.

Popp, P., Kiel, P., Möder, M., Paschke, A., and Thuss, U. 1997. Application of accelerated solvent extraction followed by gas chromatography, high-performance liquid chromatography and gas chromatography; mass spectrometry for the determination of polycyclic aromatic hydrocarbons, chlorinated pesticides and polychlorinated dibenzo-p-dioxins and dibenzofurans in solid wastes. *Journal of Chromatography A*, 774: 203–211.

Pozo, O., Pitarch, E., Sancho, J.V., and Hernandez, F. 2001. Determination of the herbicide 4-chloro-2-methylphenoxyacetic acid and its main metabolite, 4-chloro-2-methylphenol in water and soil by liquid chromatography-electrospray tandem mass spectrometry. *Journal of Chromatography A*, 923: 75–85.

Priego-López, E. and Luque de Castro, M.D. 2003. Ultrasound-assisted derivatization of phenolic compounds in spiked water samples before pervaporation, gas chromatographic separation, and flame ionization detection. *Chromatographia*, 57: 513–518.

Psillakis, E. and Kalogerakis, N. 2003a. Developments in liquid-phase microextraction. *Trends in Analytical Chemistry*, 22: 565–574.

Psillakis, E. and Kalogerakis, N. 2003b. Hollow-fibre liquid-phase microextraction of phthalate esters from water. *Journal of Chromatography A*, 999: 145–153.

Ramos, L., Vreuls, J.J., and Brinkman, U.A.T. 2000. Miniaturized pressurized liquid extraction of polycyclic aromatic hydrocarbons from soil and sediment with subsequent large-volume injection-gas chromatography. *Journal of Chromatography A*, 891: 275–286.

Rasmussen, K.E. and Pedersen-Bjergaard, S. 2004. Developments in hollow fibre-based, liquid-phase microextraction. *Trends in Analytical Chemistry*, 23: 1–10.

Reindl, S. and Hofler, F. 1994. Optimization of the parameters in supercritical fluid extraction of polynuclear aromatic hydrocarbons from soil samples. *Analytical Chemistry*, 66: 1808–1816.

Rezaee, M., Assadi, Y., Milani Hosseini, M.R. Aghaee, E., Ahmadi, F., and Berijani, S. 2006. Determination of organic compounds in water using dispersive liquid-liquid microextraction. *Journal of Chromatography A*, 1116: 1–9.

Rezaei, F., Bidari, A., Birjandi, A.P., Hosseini, M.R.M., and Assadi, Y. 2008. Development of a dispersive liquid-liquid microextraction method for the determination of polychlorinated biphenyls in water. *Journal of Hazardous Materials*, 158: 621–627.

Richter, B.E. 2000. Extraction of hydrocarbon contamination from soils using accelerated solvent extraction. *Journal of Chromatography A*, 874: 217–224.

Sánchez-Brunete, C., Miguel, E., and Tadeo, J.L. 2007. Analysis of 27 polycyclic aromatic hydrocarbons by matrix solid-phase dispersion and isotope dilution gas chromatography-mass spectrometry in sewage sludge from the Spanish area of Madrid. *Journal of Chromatography A*, 1148: 219–227.

Sarafraz-Yazdi, A. and Amiri, A. 2010. Liquid-phase microextraction, *Trends in Analytical Chemistry*, 29: 1–14.

Schantz, M., Bøwadt, S., Brenner Jr., B., Wise, S., and Hawthorne, S. 1998. Comparison of supercritical fluid extraction and Soxhlet extraction for the determination of polychlorinated biphenyls in environmental matrix standard reference materials. *Journal of Chromatography A*, 816: 213–220.

Skoog, D.A., West, D.M., and Holler, F.J. 1996. *Fundamentals of Analytical Chemistry*, 7th edn., Saunders College Publication, Fort Worth, TX.

Sporring, S., Bøwadt, S., Swensmark, B., and Bjorklund, E. 2005. Comprehensive comparison of classic soxhlet extraction with soxtec extraction, ultrasonication extraction, supercritical fluid extraction, microwave assisted extraction and accelerated solvent extraction for the determination of polychlorinated biphenyls in soil. *Journal of Chromatography A*, 1090: 1–9.

Tasdemir, Y. and Esen, F. 2007. Dry deposition fluxes and deposition velocities of PAHs at an urban site in Turkey. *Atmospheric Environment*, 41: 1288–1301.

Tasdemir, Y., Odabasi, M., Vardar, N., Sofuoglu, A., Murphy, T.J., and Holsen, T.M. 2004a. Dry deposition fluxes and velocities of polychlorinated biphenyls (PCBs) associated with particles. *Atmospheric Environment*, 38: 2447–2456.

Tasdemir, Y., Vardar, N., Odabasi, M., and Holsen, T.M. 2004b. Concentrations and gas/particle partitioning of PCBs in Chicago. *Environmental Pollution*, 131: 35–44.

Tor, A. 2006. Determination of chlorobenzenes in water by drop-based liquid-phase microextraction and gas chromatography-electron capture detection. *Journal of Chromatography A*, 1125: 129–132.

Tor, A. and Aydin, M.E. 2006. Application of liquid-phase microextraction to the analysis of trihalomethanes in water. *Analytical Chimica Acta*, 575: 138–143.

Tor, A., Aydin, M.E., and Ozcan, S. 2006. Ultrasonic solvent extraction of organochlorine pesticides from soil. *Analytical Chimica Acta*, 559: 173–180.

Tor, A., Cengeloglu, Y., Aydin, M.E., Ersoz, M., Wichmann, H., and Bahadir, M. 2003. Polychlorinated biphenyls (PCB) and polycyclic aromatic hydrocarbons (PAH) in wastewater samples from the sewage system of Konya-Turkey. *Fresenius Environmental Bulletin*, 12: 732–735.

US Environmental Protection Agency (USEPA). 1995. QA/QC Guidance for sampling and analysis of sediments, water, and tissues for dredged material analysis.

US Environmental Protection Agency (USEPA). 1996a. Method 3510C. Separatory funnel liquid-liquid extraction (Revision 3-December, 1996), SW846 CH 4.2.1.

US Environmental Protection Agency (USEPA). 1996b. Method 3540 C: Soxhlet extraction. Test Methods SW-846.

US Environmental Protection Agency (USEPA). 1999a. *Compendium of Methods for the Determination of Toxic Organic Compounds in Ambient Air*, 2nd edn., Compendium method TO 4A, Cincinnati, OH.

US Environmental Protection Agency (USEPA). 1999b. *Compendium of Methods for the Determination of Toxic Organic Compounds in Ambient Air*, 2nd edn., Compendium method TO 13A, Cincinnati, OH.

Van der Hoff, G.R. and Van Zoonen, P. 1999. Trace analysis of pesticides by gas chromatography. *Journal of Chromatography A*, 843: 301–322.

Vasconcellos, P.C., Zacarias, D., Pires, M.A.F., Pool, C.S., and Carvalho, L.R.F. 2003. Measurements of polycyclic aromatic hydrocarbons in airborne particles from the metropolitan area of Sao Paulo city, Brazil, *Atmospheric Environment*, 37: 3009–3018.

Venkataraman, C., Lyons, J.M., and Friedlander, S. 1994. Size distributions of aromatic hydrocarbons and elemental carbon. 1. Sampling, measurement methods and source characterization. *Environmental Science and Technology*, 28: 555–562.

Vidal, L., Canals, A., Kalogerakis, N., and Psillakis, E. 2005. Headspace single-drop microextraction for the analysis of chlorobenzenes in water samples. *Journal of Chromatography A*, 1089: 25–30.

Werres, F., Balsaa, P., and Schmidt, T.C. 2009. Total concentration analysis of polycyclic aromatic hydrocarbons in aqueous samples with high suspended particulate matter content. *Journal of Chromatography A*, 1216: 2235–2240.

Westbom, R., Thorneby, L., Zorita, S., Mathiasson, L., and Bjorklund, E. 2004. Development of a solid-phase extraction method for the determination of polychlorinated biphenyls in water. *Journal of Chromatography A*, 1033: 1–8.

Wobst, M., Wichmann, H., and Bahadir, M. 1999. Surface contamination with PASH, PAH and PCDD/F after fire accidents in private residences. *Chemosphere*, 38: 1685–1691.

Wu, J.M., Ee, K.H., and Lee, H.K. 2005. Automated dynamic liquid-liquid-liquid microextraction followed by high-performance liquid chromatography-ultraviolet detection for the determination of phenoxy acid herbicides in environmental waters. *Journal of Chromatography A*, 1082: 121–127.

Xu, L., Basheer, C., and Lee, H.K. 2007. Developments in single-drop microextraction. *Journal of Chromatography A*, 1152: 184–192.

Yeo, H.G., Choi, M., Chun, M.Y., and Sunwoo, Y. 2003. Concentration distribution of polychlorinated biphenyls and organochlorine pesticides and their relationship with temperature in rural air of Korea. *Atmospheric Environment*, 37: 3831–3839.

Zaater, M., Tahboub, Y., and Qasrawy, S. 2005. Monitoring of polychlorinated biphenyls in surface water using liquid extraction, GC/MS, and GC/ECD. *Analytical Letters*, 38: 2231–2245.

Zhao, R., Lao, W., and Xu, X. 2004. Headspace liquid-phase microextraction of trihalomethanes in drinking water and their gas chromatographic determination. *Talanta*, 62: 751–756.

15 Applications of Ultrasound in Water and Wastewater Treatment

Dong Chen

CONTENTS

Fundamentals of Ultrasound 373
Ultrasonic Factors 376
    Power of Ultrasound 376
    Frequency of Ultrasound 376
    Pulsed or Continued Sonication 377
Property of Contaminants 377
    Volatility 377
    Hydrophobicity 378
Ultrasonic Degradation of Anthropogenic Contaminants 379
Degradation of Natural Organic Matter 382
    NOM Property Changes through Sonication 382
        TOC Reduction of NOM 385
        Decrease in UV/Vis Absorbance 385
        Decrease in Hydrophobicity of NOM 386
        ^{13}C NMR Changes to NOM 386
        Decrease in Molecular Weight of NOM 386
        Increase in Total Acidity of NOM 389
        Implications of Sonochemical Degradations of NOM 390
Ultrasound and Disinfection 390
Control Membrane Fouling in Filtration Processes 392
    Membrane Filtration and Membrane Fouling 392
    Mechanism of Ultrasonic Control of Membrane Fouling 393
    Integrity of Membranes under Ultrasonic Irradiation 395
    Ultrasonic Factors for Membrane Fouling Control 396
    Effects of Solution Chemistry on Ultrasonic Control of Membrane Fouling 397
Acknowledgments 398
References 399

FUNDAMENTALS OF ULTRASOUND

The application of ultrasonic technology has been receiving wide attention in water and wastewater treatment and environmental remediation areas, such as degradation of recalcitrant organic pollutants in aqueous phase (Hiskia et al., 2001; Song et al., 2006; Yang et al., 2006; David, 2009; Neppolian et al., 2009; Vecitis et al., 2010; Cheng et al., 2010), decontamination of sediments (Lu and Weavers, 2002; He et al., 2005), assistance of membrane filtration for membrane cleaning and fouling control (Chai et al., 1998; Kobayashi et al., 1999, 2003; Chen et al., 2006a,b,c), and disinfection (Hua and Thompson, 2000; Arrojo et al., 2008).

Ultrasound is a longitudinal wave with a frequency typically between 16 kHz and 500 MHz (Ensminger, 1973; Thompson and Doraiswamy, 1999). When ultrasound is introduced into liquid (e.g., water), it creates oscillating regions of positive and negative pressure. Correspondingly, the liquid molecules experience periodic compression and expansion cycles. When the pressure amplitude exceeds the tensile strength of liquid during the rarefaction of ultrasonic waves, cavitational bubbles are formed. Cavitational bubbles collapse during the compression cycle of ultrasonic wave. Localized hot spots are formed, which reach temperatures and pressures around 5000 K and 500 atm, respectively (Suslick, 1990; Flint and Suslick, 1991), depending on factors such as ultrasonic power, frequency, hydrostatic pressure, temperature, solvent property, and dissolved gas. According to the temperature profile, there are three zones associated with a cavitational bubble (Figure 15.1). (1) Thermolytic center, which is the core of the bubble with localized hot temperature (~5000 K) and high pressure (~500 atm) during final collapse of cavitation. The high temperature results in thermolysis of volatile chemical compounds and water vapor (Flint and Suslick, 1991), producing radical species including •OH and •H radical. The reactions between volatile chemical compounds and •OH radicals in gaseous phase also happen in this region. (2) Interfacial region between the cavitational bubble and bulk liquid. In this region, the thickness of the liquid is estimated to be about 200 nm from the bubble surface to the bulk and a lifetime is less than 2 μs (Suslick, 1990, 1989; Flint and Suslick, 1991). The temperature is about 2000 K at the final cavitational collapse (Riesz et al., 1985). There are vast gradients of temperature and pressure. Thermolysis and oxidation by •OH radicals of hydrophobic, nonvolatile, and hydrophilic compounds occur in the region. In this interfacial region, hydrophobic compounds are more concentrated than the bulk solution, and hydrophilic compounds have the same concentration as the bulk solution. (3) The bulk region with ambient temperature and pressure. The self combination of •OH radicals produces hydrogen peroxide in the solution (Reaction 15.1) along with a small amount of •OH radicals that react with hydrophilic, including ionic compounds in bulk region. The combination rate constant of •OH radicals measured in water under ambient conditions is 5.5×10^9 L mol^{-1} s^{-1} (Buxton et al., 1988).

$$2 \cdot OH \rightarrow H_2O_2 \tag{15.1}$$

•OH radicals can oxidize broad organic pollutants in water, similar to advanced oxidation processes (AOPs). Since •OH radicals are produced as a result of cavitational collapse, they spatially concentrate

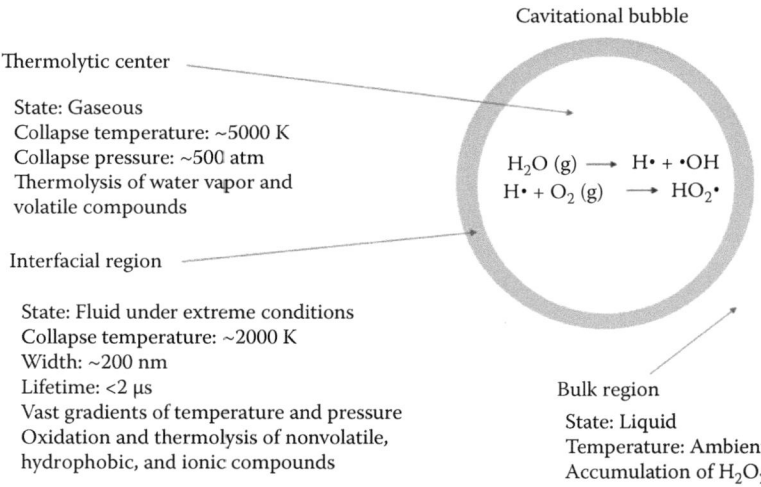

FIGURE 15.1 Diagram of a cavitational bubble.

inside the bubble and in the region of bubble–bulk interface during collapse. Consequently, volatile and hydrophobic compounds or parts are exposed to more •OH radicals in addition to thermolysis than hydrophilic ones and thus subject to more rapid degradation, because they tend to accumulate in these two regions.

Fenton's reagent is often used to enhance the degradation initiated by •OH radical. As shown in reaction (15.1), formation of hydrogen peroxide decreases the concentration of •OH radical, which has stronger oxidative potential than hydrogen peroxide. However, Fenton's reaction uses ferrous ion (Fe^{2+}) that catalytically produces •OH radicals from hydrogen peroxide (Walling and Kato, 1971; Walling, 1975) (reactions (15.2) through (15.5)). As a result, faster degradations have been observed in sonolysis when combined with Fenton's reagent (Beckett and Hua 2003; Liang et al. 2007a).

$$Fe^{2+} + H_2O_2 \rightarrow Fe^{3+} + \bullet OH + OH^- \tag{15.2}$$

$$Fe^{3+} + H_2O_2 \rightarrow Fe-OOH^{2+} + H^+ \tag{15.3}$$

$$Fe-OOH^{2+} \rightarrow Fe^{2+} + HOO\bullet \tag{15.4}$$

$$Fe^{3+} + HOO\bullet \rightarrow Fe^{2+} + O_2 + H^+ \tag{15.5}$$

Fenton's reagent requires low pH to keep Fe^{2+} and Fe^{3+} soluble. Liang et al. (2007b) reported that 200 kHz sonolysis with Fenton's reagent was able to completely decompose 4-chlorophenol in 2 min at pH 3 when compared with 1 h at pH 5.6. At pH 5.6, the concentration of iron ions in the solution was very low.

Besides the chemical effects described above, ultrasound also has significant physical effects, that is, sonophysical effects. More specifically, vibration and acoustic streaming produced by ultrasonic waves along with microstreaming, microstreamers, microjets, and shock waves produced by cavitational bubbles result in turbulent fluid movement and a great microscale velocity gradient in the vicinity of cavitational bubbles (Leighton, 1994; Roy, 1999; Chen et al., 2006a,b). Acoustic streaming is the result of liquid medium absorbing the acoustic energy of sound waves (Leighton, 1994). Consequently, liquid medium flows along the propagation direction of the sound waves. Microstreaming is the result of oscillations of an acoustically driven bubble in the sound field, that is, the pulsation of bubble wall with the alternate compression and expansion movements under the impact of acoustic waves (Roy, 1999). Microstreamers are caused by cavitation bubbles travel within the liquid to nodes or antinodes driven by *Bjerknes forces*, in which bubbles travel in ribbon like structures along tortuous paths (Luther et al., 2001). Microjets are formed when asymmetric collapse of cavitational bubbles occurs near a surface. The observed jet speed was over 100 m s^{-1} toward the surface and caused pittings and erosions on the solid surface (Vogel et al., 1989; Suslick, 1990; Roy, 1999). Shock waves are exhibited as an abrupt pressure of huge amplitude (in GPa), which is produced within nano seconds following the violent collapse of a cavitational bubble (Pecha and Gompf, 2000). Shock waves can break particles and macromolecules from the final collapse of cavitational bubbles (Basedow and Ebert, 1977; Hickenboth et al., 2007). For macromolecules, this is referred to as shear degradation of ultrasound.

The fluid movement enhances the physical mass-transfer processes between the solid–bulk and gas–bulk interfaces. As a result, sonophysical effects of ultrasound can facilitate mixing, break down particles and macromolecules, desorption, extraction, and cleaning processes. During sonication, the effects of ultrasound are combinations of both sonochemical and sonophysical effects. The mechanisms of ultrasound make it unique when compared with other AOPs. The advantages of ultrasound include potential chemical-free and simultaneous oxidation, thermolysis, shear degradation, and enhanced mass-transfer processes together. The yield of sonication depends on both ultrasonic factors and properties of contaminants.

ULTRASONIC FACTORS

POWER OF ULTRASOUND

The power intensity of ultrasound is the power delivered to the liquid divided by the surface area of the ultrasonic transducer. The relationship between the ultrasonic power intensity and the acoustic pressure may be expressed as (Mason and Lorimer, 1988),

$$I = \frac{P_0^2}{2\rho C} \tag{15.6}$$

where
I is the power intensity of a sound wave
P_0 is the acoustic pressure
ρ is the density of the liquid
C is the sound speed in the liquid

Consequently, higher acoustic pressure (amplitude of vibration), greater amounts of cavitational events, and more violent cavitational collapse happen at elevated power intensity of ultrasound. However, optimum power intensity has been observed corresponding to the highest reaction rate (Gutierrez and Henglein, 1990; Hatanaka et al., 2002). Beyond that point, a further increase in power causes a decline of the reaction rate. This phenomenon may be explained by bubble shielding effect. When the power intensity is high enough, a dense cloud of cavitational bubbles accumulate around the ultrasonic transducer. The cavitation bubbles attenuate sound waves due to both scattering and absorption and thus impede the propagation of sound waves, especially at the resonant size (Pace et al., 1997; Roy, 1999). The scattering and absorption result in a decrease of sound wave intensity when compared with fewer bubbles. Consequently, the sound wave intensity decreases more rapidly with distance from the source at a very high power intensity when compared with the optimized power intensity. Consistently, van Iersel et al. (2007) observed improved oxidation of potassium iodide when bubble shielding lessened.

FREQUENCY OF ULTRASOUND

The frequency of ultrasound directly affects the generation, oscillation, the resonant size, and final collapse of cavitational bubbles in terms of both quantity (the amounts of collapse) and the quality (the violence of collapse). Generally, the cavitational threshold increases with increasing ultrasonic frequency (Mason and Lorimer, 1988). In other words, a higher acoustic pressure is required to overcome the tensile strength of liquid molecules to produce cavitation at a higher ultrasonic frequency.

Low-frequency ultrasound has less bubble events, bigger resonant bubble size, and more violent collapse than high-frequency ultrasound (Leighton, 1994; Crum, 1995; Adewuyi, 2001; Beckett and Hua, 2001). Generally low-frequency ultrasound has stronger sonophysical effects than high-frequency ultrasound. When ultrasonic frequency rises, more cavitational bubbles increase both the production of •OH radicals and diffusion of gases and volatile compounds into the bubble (Hung and Hoffman, 1999). Consequently, more rapid destruction of organic pollutants increases with the ultrasonic frequency until about 358 kHz (Beckett and Hua, 2001). However, when ultrasonic frequency continues to rise, the cavitational effect is reduced because either (i) the rarefaction cycle of the sound wave produces a negative pressure, which is insufficient in its duration and/or intensity to initiate cavitation or let cavitation grow bigger; or (ii) the compression cycle occurs faster than the time required for the micro cavitational bubbles to collapse (Thompson and Doraiswamy, 1999).

PULSED OR CONTINUED SONICATION

Alternative to continued mode of sonication, pulsed mode may be adopted for destruction of organic contaminants, cleaning of water filtration membranes, and other applications. Pulsed sonication not only saves energy, but also affects the bubble oscillation and growth dynamics (Chen et al., 2006b). During sonication, some bubbles grow by rectified diffusion to sizes greater than their resonance size. Such bubbles are ineffective at producing cavitational effects (Hill et al., 1969; Leighton, 1994), and cause scattering and absorption of ultrasonic waves (i.e., bubble shielding) (Roy, 1999). Therefore, in continued ultrasound, some bubbles are ineffective (Hill et al., 1969) and some ultrasonic energy is wasted. The acoustic energy is dissipated as heat and vibration. However, during pulse intervals of pulsed sonication, bubble sizes are reduced back to below resonance size by dissolution, and some bubbles coalesce and consequently float to the water surface (Clarke and Hill, 1970; Leighton, 1994; Roy, 1999). As a result, bubble size and bubble quantity may decrease during the pulse intervals and thus reduce the shielding effect of sound waves by bubbles. Suitable pulse times and pulse intervals may reduce bubble shielding as well as make use of surviving bubbles from previous pulse cycles to generate effective cavitational collapses for degradation of contaminants or cleaning processes. In addition, pulse intervals provide extra time for hydrophobic contaminants to transfer from bulk solution to adsorb on the bubble–bulk interface of the survived bubbles (Yang et al., 2006). This process may facilitate the subsequent degradations led by cavitational collapse during the on-cycle of sonication. Moreover, "silent" reactions might occur during the pulsed intervals, which are initiated by •OH radicals (Neppolian et al., 2009). These effects bring extra sonochemical yields when compared with continued sonication.

In literature, yields of sonochemical reactions such as sonoluminescence (Leighton et al., 1989), oxidation of As (III) to As (V) (Neppolian et al., 2009), degradation of nonvolatile surfactants (Yang et al., 2006), generation of free iodine from KI solution (Clarke and Hill, 1970; Casadonte et al., 2005), oxidizing acid orange (Casadonte et al., 2005), and DNA degradation (Clarke and Hill, 1970) increased with pulsed sonication, due to more effective use of ultrasonic energy when compared with continued sonication. In the study of Yang et al. (2006), longer pulse intervals enabled more adsorption of less-diffusivity dodecylbenzenesulfonate (DBS) surfactant, which was more surface active on the bubble–bulk interfaces than 4-octylbenzene sulfonate, resulting in a more significant degradation of DBS than shorter pulse intervals. Casadonte et al. (2005) used power-modulated pulsed ultrasound, in which the peak power of pulsed ultrasound was increased to match the net acoustic input power of continued ultrasound per unit time. The results showed that pulsed ultrasound had a degradation rate increase by a factor of three as compared with continued irradiation. In ultrasound assisted membrane filtrations, sonication is used to prevent and clean water filtration membranes from being clogged by particles and macromolecules. Chen et al. (2006b) found that a short pulse interval (1.0 s on/0.1 s off) had the relative membrane permeate flux improvement of 73% ± 4%, which was similar to continued sonication (75% ± 7%) plus roughly 9% of energy savings.

PROPERTY OF CONTAMINANTS

Besides ultrasonic factors, the properties of contaminants such as volatility and hydrophobicity also greatly affect the degradation rate of sonication.

VOLATILITY

Volatile compounds more readily escape from the bulk solution and enter the gaseous phase of cavitational bubbles through rectified diffusion processes. Consequently they are subject to thermolysis and oxidation by •OH radicals in gaseous phase during the final collapse of cavitation. Henry's law constant H (atm L mol^{-1}) is used to quantify the volatility of compounds. Henry's law constant H is the ratio of the partial pressure P_i (atm) of a compound i exerts in gaseous phase over its aqueous phase concentration C_i (mol L^{-1}) at equilibrium status.

$$H = \frac{P_i}{C_i} \tag{15.7}$$

Henry's law is valid for dilute solutions. Greater Henry's law constant means higher volatility of a compound. Sonolysis degradation rate increases with higher values of Henry's law constant of different compounds (Weavers, 2001). However, besides the intrinsic volatility of a compound, increasing temperature will increase Henry's law constant and the volatility of a compound. Different from the case of intrinsic volatility, a higher solution temperature brings more gaseous compounds including solvent vapour in the cavitational bubble, which may quench the final collapse and decrease the magnitude of collapsing temperature and pressure. Therefore, the degradation rate might decrease with elevated temperatures (Thompson and Doraiswamy, 1999).

Thermolysis is a unique mechanism of ultrasound for destruction of contaminants as compared with other AOPs. It is especially suitable to treat dilute solutions. As a result, a few volatile and/or hydrophobic contaminants, which are oxidatively stable, can be thermolyzed by sonolysis. Examples of these contaminants include carbon tetrachloride (Francony and Petrier, 1996; Hung and Hoffmann, 1999; Lee and Oh, 2010) and fluorinated surfactants (Moriwaki et al., 2005; Schröder and Meesters, 2005; Vecitis et al., 2010; Cheng et al., 2010).

HYDROPHOBICITY

Sonication produces a gaseous phase (cavitational bubbles) within the bulk liquid. As a result, nonpolar neutral or hydrophobic organics in the solution will tend to accumulate on the bubble–bulk interface. Hydrophobic compounds may be destructed by thermolysis and •OH radicals in the interfacial region (Figure 15.1). The degree of hydrophobicity is commonly determined by K_{ow}, octanol–water partition coefficient, which is the ratio of the concentration of a compound in octanol phase C_o (mg L^{-1} or μg L^{-1}) over its concentration in aqueous phase C_w (mg L^{-1} or μg L^{-1}) at equilibrium status.

$$K_{ow} = \frac{C_o}{C_w} \tag{15.8}$$

Chemicals with a high value of K_{ow} tend to be hydrophobic and accumulate on bubble–water interface. In contrast, chemicals with a low value of K_{ow} tend to be hydrophilic and remain in the bulk aqueous phase. Consequently, more rapid degradation of a compound of a greater K_{ow} value has been observed (Weavers, 2001; Vecitis et al., 2010).

In addition to the intrinsic hydrophobicity of a compound, solution conditions may affect the degree of hydrophobicity as well. When pH is much lower than the pKa value of a compound, the compound becomes unionized or neutral. In other words, the compound is more hydrophobic than its ionized form at higher pH in water. Consistently, Jiang et al. (2002) observed that the ultrasonic degradation rate of 4-NP decreased with increasing pH, because the neutral hydrophobic species are more easily diffused to and accumulated at the interface of bubble–water in comparison with their corresponding ionic forms.

Besides the pH effect, ionic strength also affects the hydrophobicity of an ionic compound in water. High ionic strength causes charge screening effect on aqueous ionic species. Based on Debye and Hückel theory, Equation 15.8 of octanol–water partition coefficient can be rewritten to include the activity coefficient.

$$K_{ow} = \frac{C_o}{\gamma C_w} \tag{15.9}$$

$$\log \gamma = -0.5 Z^2 \frac{\sqrt{\mu}}{1+\sqrt{\mu}} \qquad (15.10)$$

$$\mu = \frac{1}{2} \sum_i C_i Z_i^2 \qquad (15.11)$$

where
  γ is the activity coefficient of the compound in water
  Z is the charge of the compound for which the activity coefficient is being determined
  μ is the ionic strength of the solution
  C_i is the molar concentration of ionic species i in the solution
  Z_i is the charge of ionic specie i in the solution

According to Equation 15.10, activity coefficient γ decreases with higher ionic strength μ. As a consequence, octanol–water partition coefficient K_{ow} increases with higher ionic strength. It means that the compound becomes more hydrophobic at higher ionic strength. This effect is only valid for ionic compounds as indicated in Equation 15.10. Seymour and Gupta (1997) reported salt-induced hydrophobicity of compounds including p-ethylphenol and phenol. Sodium chloride increased the hydrophobicity of these contaminants reflected by elevated partitioning coefficient in diethyl ether-aqueous phases. As a result, several orders of enhancement in sonolysis degradation rate were observed. However, adding salt (NaCl) brings extra chloride ions to the solution. The scavenging effect of •OH radicals by chloride ions might be a concern.

ULTRASONIC DEGRADATION OF ANTHROPOGENIC CONTAMINANTS

Decomposition and removal of anthropogenic hazardous contaminants from surface waters, groundwaters, sediments, and soils are very important in environmental remediation. The power of ultrasound has been employed to degrade numerous environmental pollutants (Zhang and Hua, 2000), and this unique mechanism of ultrasound-mediated pollutant degradation combines simultaneous oxidation, thermolysis, shear degradation of shock waves, microjets pitting, and enhanced mass transfer and mixing together. As a result, sonication is a very attractive and interesting technique in environmental remediation, especially for decontamination of recalcitrant and hazardous compounds. There are over 100 anthropogenic contaminants that have been studied by sonolytic degradation (Thompson and Doraiswamy, 1999; Adewuyi, 2005, 2001; Liang et al., 2007a; Belgiorno et al., 2007; Chowdhury and Viraraghavan, 2009; Pham et al., 2009). Due to limited space, only selected studies on typical and important contaminants are summarized here.

Perfluorinated chemicals such as perfluorooctane sulfonate (PFOS) and perfluorooctanoate (PFOA) are globally distributed, bioaccumulative, metabolically and photochemically inert, and oxidatively recalcitrant (Key et al., 1998; Moriwaki et al., 2005; Vecitis et al., 2010). In the research conducted by Moriwaki et al. (2005), 200 kHz sonication cleaved the perfluorocarbon chains. The half-life of PFOS and PFOA degradations was 43 and 22 min, respectively under the atmosphere of argon. Because PFOS and PFOA molecules have both hydrophobic (perfluoroalkyl group) and hydrophilic group (acid group), they behave like an anionic surfactant: the hydrophobic group migrate and accumulate in the bubble–bulk interfacial region and are subject to pyrolysis (or thermolysis) and •OH radical attack. Since PFOS and PFOA are nonvolatile, the pyrolysis in the gaseous phase of cavitation should be ruled out. The results of Fenton experiments suggested that the compounds were not decomposed by •OH radicals either. Consequently, it was concluded that most of the PFOS and PFOA molecules were pyrolyzed at the bubble–bulk interfacial region, where the temperature was still enough high for pyrolysis reactions. Consistently, Vecitis et al. (2010) indicated that pyrolytic cleavage of the C–S bond of PFOS in the bubble–bulk interfacial region was the initial degradation step. The complicated organic matrix of aqueous film-forming foams only had a minor effect on the sonochemical degradation of PFOS, even though the total organic concentration of the matrix was 50 times the PFOS concentration. This result suggested the superior surfactant properties of fluorochemicals, which were highly competitive in adsorption to the bubble–bulk interfacial region. Besides the influence of organic matrix, Cheng et al. (2010) observed a decrease in sonochemical degradation rates of PFOA and PFOS in the presence of inorganic anions such as HCO_3^- and SO_4^{2-} in groundwater. It was hypothesized that inorganic ions partitioned to and interacted with the bubble–water interface. Initial solution pH enhanced the degradation rates of PFOA and PFOS at 3, but had negligible effects over the pH range of 4–11.

Similar to perfluorinated chemicals, carbon tetrachloride is oxidatively recalcitrant and unreactive toward the hydroxyl radical with half-life longer than 330 years (Cox et al., 1976). However, sonolytic degradation of CCl_4 in aqueous solution has been shown effective (Hua and Hoffmann, 1996; Francony and Petrier, 1996; Hung and Hoffmann, 1999; Lee and Oh, 2010). Because of its high vapor pressure (113.83 mmHg) and great hydrophobicity (log K_{ow} = 2.83), CCl_4 appears to undergo pyrolysis in the gas-phase interior of the cavitational bubbles as well as in the interfacial region (Hua and Hoffmann 1996). Using 20 kHz and 112.5 W cm^{-2} ultrasound, the observed first-order degradation rate constant in an Ar-saturated solution was 3.3×10^{-3} s^{-1} when the initial CCl_4 concentration was 1.95×10^{-4} mol L^{-1}. Sonication byproducts included hexachloroethane, tetrachloroethylene, chloride ion, and hypochlorous acid. Hung and Hoffmann (1999) investigated ultrasonic degradation of CCl_4 at six different frequencies ranging from 20 to 1078 kHz. The rate of degradation increased with increasing ultrasonic frequency with the optimal degradation rate at 500 kHz. Hexachloroethane was found as the primary intermediate in the degradation of CCl_4. Consistently, Francony and Petrier (1996) reported that a faster degradation rate of CCl_4 occurred at 500 kHz than 20 kHz. Sonication led to almost complete mineralization of CCl_4 for a relatively short irradiation time. Radical-trap addition did not change the reaction rate, since thermolysis is the mechanism of sonolytic degradation. Similarly, Lee and Oh (2010) confirmed that the addition of t-BuOH, a hydroxyl radical scavenger did not affect the degradation rate of carbon tetrachloride.

Ultrasound can also be used to degrade environmental emerging contaminants, such as pharmaceutical and personal care products, especially endocrine disruptor compounds. The conventional microbiological processes used in municipal wastewater treatment plants are not designed or capable to remove these contaminants. Suri et al. (2007) investigated sonolytic destruction of estrogen hormones in aqueous solution, including 17α-estradiol, 17β-estradiol, estrone, estriol, equilin, 17α-dihydroequilin, 17α-ethinyl estradiol, and norgestrel. The results showed that 20 kHz sonolysis destructed 80%–90% of individual estrogens at an initial concentration of 10 μg L^{-1} within 40–60 min of reaction. The first-order degradation rate constant of individual estrogen increased with higher power intensity. However, the energy efficiency of the reactor was higher at lower power density. As a result, the choice of reactor and ultrasonic power were important to achieve optimized kinetics and energy efficiency.

Isariebel et al. (2009) investigated sonolysis of pharmaceutical compounds of levodopa and paracetamol in aqueous solutions. Levodopa is the most frequently prescribed drug for the treatment of Parkinson disease (Lara et al., 2006) and paracetamol is a widely used non-steroidal anti-inflammatory recalcitrant drug found in water bodies (Bedner and MacCrehan, 2006). Experiments were performed at 574, 860, and 1134 kHz of ultrasound with initial concentrations of 25, 50, 100, and 150 mg L^{-1} of both compounds. Sonochemical degradations of both compounds followed pseudo first-order reaction kinetics. Contaminants and COD degradations were found to decrease with increasing the initial solute concentration and decreasing power. The best result was obtained at 574 kHz. Using 1-butanol as •OH radical scavenger and H_2O_2 as promoter revealed that •OH radical attack was the principal degradation mechanism. During the reactions, some intermediates were found recalcitrant and long lived to sonolysis.

Torres et al. (2008) studied sonolytic degradation of bisphenol A (BPA), an endocrine disruptor largely used in polycarbonate plastics and has been detected in surface waters (Belfroid et al., 2002). The effect of saturating gas (oxygen, argon, and air), BPA initial concentration (0.15–460 µmol L^{-1}), ultrasonic frequency (300–800 kHz), and power (20–80 W) were evaluated. With an initial concentration of 118 µmol L^{-1}, BPA was readily eliminated by sonolysis in about 90 min at 300 kHz, 80 W, and with oxygen as the saturating gas. However, even after long ultrasonic irradiation time (9 h), more than 50% of chemical oxygen demand (COD) and 80% of total organic carbon (TOC) remained in the solution. Analyses of intermediates using HPLC–MS identified several intermediate byproducts: Monohydroxylated bisphenol A, 4-isopropenylphenol, quinone of monohydroxylated bisphenol A, dihydroxylated bisphenol A, quinone of dihydroxylated bisphenol A, monohydroxylated-4-isopropenylphenol, and 4-hydroxyacetophenone. The presence of these hydroxylated aromatic structures showed that the main ultrasonic BPA degradation pathway was related to the reaction of BPA with •OH radical. After 2 h, these early products were converted into biodegradable aliphatic acids. In their earlier study (Torres et al., 2007), 300 kHz and 80 W sonolysis was compared with Fenton's reagent (100 µmol L^{-1} ferrous sulfate and continuous H$_2$O$_2$ addition) at pH 3 to degrade BPA. Identical BPA elimination rate and primary intermediates were observed for both processes. It was suggested that the main chemical pathways involved reactions with •OH radicals. COD and TOC analyses showed that the Fenton's process was slightly more efficient than ultrasonic treatment for the removal of BPA byproducts in deionized water. However, in natural water (pH 7.6, main ions concentration: Ca^{2+} = 486 mg L^{-1}, Na$^+$ = 9.1 mg L^{-1}, Cl$^-$ = 10 mg L^{-1}, SO$_4^{2-}$ = 1187 mg L^{-1}, and HCO$_3^-$ = 402 mg L^{-1}), inhibition of the Fenton process was evidenced; while the ultrasonic process was not hampered.

Consistently, using 20 kHz ultrasound, Guo and Feng (2009) indicated that •OH radical induced oxidation was identified as the major destruction pathway during sonolysis of BPA. Inoue et al. (2008) investigated different power intensities on sonolysis of BPA. At 404 kHz, 0.5 mmol L^{-1} BPA was completely degraded after 10, 3, and 2 h of sonication at the power intensity of 3.5, 9.0, and 12.9 kW m^{-2}, respectively. The intermediates such as 3-hydyroxybisphenol A, formaldehyde, and organic acids were detected. At pH 3, the addition of ferrous sulfate (FeSO$_4$) did not increase BPA degradation rate. Instead, more TOC reduction was observed with increasing of ferrous sulfate concentration up to 4.0 mmol L^{-1} at 404 kHz and 9.0 kW m^{-2}.

Methyl tert-butyl ether (MTBE) is of special concern because of its wide distribution in the environment. MTBE is manufactured in a large quantity annually as a popular additive as a fuel oxygenate (up to 15% by volume of gasoline) (Johnson et al., 2000). It is a suspected carcinogen and has been found in various environmental media, including troposphere, surface and groundwaters, and storm water (Cooper et al., 2009). Since it is volatile, MTBE is suitable for sonolytic degradation in aqueous phase. Kang et al. (1999) investigated sonolysis of MTBE with or without ozone under different ultrasonic frequencies and applied powers. In the frequency range of 205–1078 kHz, the higher overall reaction rates were observed at 358 and 618 kHz and then at 205 and 1078 kHz. The observed pseudo first-order rate constant for MTBE degradation increased with higher power density up to 250 W L^{-1}. The reaction rate constant also increased with increasing ozone dosage from 0 to 0.19 mmol L^{-1}. In the presence of natural organic matter (NOM) (Fluka AG) up to 4.2 mg L^{-1}, negligible effect of NOM on the MTBE ([MTBE]$_0$ = 0.05 mmol L^{-1}) decomposition rate was observed. The explanation was that the major reaction site for MTBE was in the vapor phase of cavitational bubbles, where thermolysis and reaction with •OH radicals took place. As a result, highly volatile compounds such as MTBE and CCl$_4$ were more rapidly decomposed during sonolysis than were semivolatile or nonvolatile compounds, like NOM.

In the study by Neppolian et al. (2002), the degradation kinetics and intermediate byproducts of MTBE degradation were investigated with 20 kHz ultrasound. The observed pseudo first-order rate constant decreased from 1.25×10^{-4} to 5.32×10^{-5} s^{-1} as the concentration of MTBE increased from 2.84×10^{-2} to 2.84×10^{-1} mmol L^{-1}. The rate of degradation of MTBE increased with increasing power density of ultrasound and also with the rise in system temperature. In the presence of an oxidizing agent, potassium persulfate, the sonolytic degradation rate of MTBE was accelerated substantially. *Tert*-butyl formate and acetone were found to be the major intermediates of the degradation of MTBE. It was also found that the ultrasound coupled with Fenton reagent (Fe^{2+}/H_2O_2) effectively degraded more than 95% of MTBE (2.84×10^{-2} mmol L^{-1}) along with its intermediate products in 5 h, which was much more efficient than sonolysis alone. More recently, Selli et al. (2005) investigated the degradation kinetics of MTBE in water with an O_2/Ar 80:20 atmosphere by sonolysis at 20 kHz. MTBE concentration decreased following a first-order kinetic parameters.

Another important group of anthropogenic contaminants detected in natural water bodies is pesticides and herbicides. Collings and Gwan (2010) performed a study on sonochemical degradation of DDT, chlordane, atrazine, 2, 4, 5-T, and endosulfan in sand–water slurries. Destruction rates of about 70% for 10 min of 20 kHz sonication at 150 W were obtained for DDT, chlordane, atrazine, and endosulfan, while about 50% for 2, 4, 5-T at 50 wt.% slurries. For a lower slurry concentrations, that is, 20 g of sand in 100 mL of water, there was little to no change in the destruction of atrazine but a big improvement in the destruction of 2, 4, 5-T. The absence of breakdown products suggested that •OH radical reactions were not prevailing. Instead, the dominant mechanism of destruction of the contaminants in slurry was thermal pyrolysis. Yao et al. (2010) investigated sonolytic degradation of parathion, an example of typical organophosphorus pesticides. The results indicated that the degradation followed a pseudo first-order kinetic. The degradation rate decreased with increasing initial concentration and decreasing power. The optimal frequency for parathion degradation was 600 kHz in the range of 200–800 kHz. The bubble–bulk interfacial regions were the effective reaction sites for sonochemical degradation of parathion. The reaction could be described as a gas/liquid heterogeneous reaction, which obeyed a kinetic model based on Langmuir–Hinshelwood model. •OH radical reactions predominated in the sonochemical degradation of parathion. It was indicated that the N_2 in air took part in the parathion degradation through the formation of •NO_2 under ultrasonic irradiation. Parathion was decomposed into paraoxon and 4-nitrophenol in the first step via two different pathways, respectively, which was in agreement with the theoretical molecular orbital calculations. Liu et al. (2008a) studied combination of O_3 and 40 kHz sonolysis to degrade dimethoate, another type of organophosphorous pesticides. The system imposed a synergistic effect combining sonochemical merit with high O_3 transfer rate. Under the optimal operation conditions, that is, O_3 flow rate was 0.41 m^3 h^{-1}, ultrasonic intensity was 4.64 W cm^{-2}, pH value was 10.0, reaction temperature was 25°C, and the initial concentration of dimethoate was 20 mg L^{-1}, the degradation rate of dimethoate increased to 90.8% in 4 h.

DEGRADATION OF NATURAL ORGANIC MATTER

NOM Property Changes through Sonication

When nature waters are involved, the presence of NOM may significantly influence the effectiveness of ultrasound in reactions. NOM is so significant that it deserves a separate discussion. NOM is ubiquitous and present in natural waters as a result of microbial and geochemical degradation of plant and animal debris. The typical NOM concentrations in surface waters, groundwaters, soils, and sediments are listed in Table 15.1. NOM is described as refractory, dark-colored, and heterogeneous natural organic compound (Stevenson, 1994). It is a significant global carbon pool (Cole et al., 2007). The chemical structure of NOM is ill-defined (Aiken and Malcolm, 1987), varies depending on the source, but typically includes aromatic, aliphatic, carboxylic, and phenolic functional groups (Stevenson, 1994). In drinking water treatment, the presence of NOM may result in color, bad tastes, and toxic or carcinogenic disinfection byproducts (Liu et al., 2008b). In addition, NOM can undergo a variety of reactions in natural and engineered systems and often interferes with treatment processes, including particle coagulation, bioavailability, as a nutrient for microorganisms, toxicity of trace metals, mobility of contaminants by binding organic and inorganic contaminants, and scavenging reactive species in water (Chin et al., 1994; Edwards et al., 1996; Meier et al., 1999; Perdue and Ritchie, 2005; Kitis et al., 2002).

TABLE 15.1
Typical NOM Concentrations in Environmental Media

Media	NOM Concentration (as Organic Carbon)	Reference
Surface waters	4.5–7.7 mg L^{-1}	Dobbs et al. (1972)
Groundwaters	0.1–15 mg L^{-1} (median 0.7 mg L^{-1}, average 1.2 mg L^{-1})	Leenheer et al. (1974)
Soils and sediments	1–100 (mg g^{-1})	Mayer (1994)

The chemical nature or structure of NOM depends on its source materials and biogeochemical processes that take place at the site of their formation (Aiken and Cotsaris, 1995). Based on solubility under acidic or alkaline conditions, NOM can be fractioned into: (i) humin, the insoluble fraction of NOM at all pH levels; (ii) humic acid, the fraction soluble under alkaline conditions but not acidic conditions (pH ≤ 2); (iii) fulvic acid, the fraction soluble under all pH conditions (Stevenson, 1994). For scientific research, there are several types of model NOM commonly used, such as Suwannee river humic acid, Aldrich humic acid, and Pahokee peat humic acid. The purification steps of aquatic Suwannee river humic acid include filtration, acidification to precipitate humic acid out of fulvic acid, and isolation with XAD-8 resin (Aiken, 1985). Aldrich humic acid is extracted from coal by base extraction (Bob and Walker, 2000). Pahokee peat humic acid is by base extraction from Pahokee peat, which is a typical agricultural peat soil of the Florida everglades (IHSS website). Generally, Aldrich and peat humic acid are more aromatic and have a higher molecular weight than humic materials isolated from natural waters (Hatcher et al., 1980; Chin et al., 1994).

The properties of humic substances are important in complexation with other environmental components. Higher hydrophobicity, higher aromaticity, and greater molecular weight of humic substances result in stronger affinity to nonpolar neutral or hydrophobic organics (Chin et al., 1997). Higher acidity of humic substances, on the other hand, facilitates increased complexation with metallic cations and other positively charged species (Bowles et al., 1994; Chin et al., 1997). Important characteristics of humic substances include concentration of organic carbon (measured by TOC or dissolved organic carbon [DOC]), color (measured at a visible wavelength, e.g., Color$_{465}$, the UV–Vis absorbance at 465 nm), hydrophobicity (measured by K_{ow} or specific UV absorbance [SUVA] at 254 nm), aromaticity (measured by SUVA at 280 nm or ^{13}C NMR), molecular weight (directly measured by high pressure size exclusion chromatography [HPSEC] or indirectly reflected by E_4/E_6 ratio), and acidity (measured by potentiometric titration) (Thurman, 1985; Chen et al., 2004). Of these parameters, color relates to chromophores in NOM, including conjugated double bonds, aromatic rings, and phenolic functional groups. These groups serve as color centers in humic substances (Schnitzer and Khan, 1972). SUVA (m^{-1} L [mg C] $^{-1}$) expressed by the ratio of UV absorbance at 254 or 280 nm over DOC concentration, increases with increasing hydrophobicity and aromaticity of NOM, because π–π* electron transitions occur at these wavelengths (Chin et al., 1994; Croue et al., 1999; Westerhoff et al., 1999). E_4/E_6 is the ratio of UV–Vis absorbance at 465 nm over at 665 nm (i.e., Color$_{465}$/Color$_{665}$). E_4/E_6 ratio positively correlates to the oxygen content ($r = 0.82$) and inversely correlates to the reduced viscosity (molecular size or molecular weight) ($r = -0.95$) of NOM (Chen et al., 1977). At pH close to neutral, Chen et al. (1977) determined that E_4/E_6 ratio was in the range of 5.44–5.7 for humic acids and 8.50–8.88 for fulvic acids; while Kukkonen (1992) found E_4/E_6 ratio was about 3.8–5.8 for humic acids and 7.6–11.5 for fulvic acid.

The possible mechanisms of sonochemical degradation of NOM are: (i) •OH radical attack; (ii) thermolysis in the bubble–bulk interfacial region; (iii) dynamic shearing of shock waves. NOM can be hydrophobic or contains both hydrophilic (carboxylic and phenolic groups) and hydrophobic

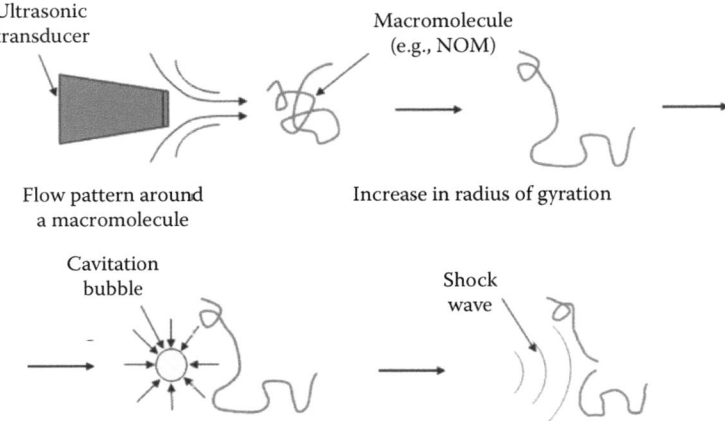

FIGURE 15.2 Proposed mechanism of shearing degradation of NOM macromolecules by shock waves.

functional groups (nonionized groups). As a consequence, hydrophobic NOM or part is more subject to thermolysis and •OH radical attack in the bubble–water interfacial region. Original NOM is nonvolatile macromolecular organic. Therefore it is unlikely to be thermolyzed in the vapor phase of cavitational bubbles. However, when NOM macromolecule is degraded to smaller molecules through sonication, they may volatilize into the gaseous phase of cavitation bubbles, and consequently are thermolyzed during the collapse of cavitational bubbles (Chen et al. 2004). Figure 15.2 illustrates the mechanism of mechanical shearing degradation by shock waves (Basedow and Ebert, 1977; Price, 1990; Taghizadeh and Asadpour, 2009). In the diagram, first, the flow pattern generated by ultrasonic wave unfolds and increases the radius of gyration of a macromolecule, like NOM. Subsequently, the shock waves produced by the final collapse of cavitation bubbles may break the covalent bond of the macromolecule and reduce the molecular weight. Cleavage of a covalent bond can occur in two ways: homolytically, resulting in one electron from the bond going to each fragment to produce radical species; or heterolytically, with both electrons associating with one fragment, leading to formation of an ion pair (Price, 1990). Both of these possibilities have been observed during polymer degradation by ultrasound (Melville and Murray, 1950; Henglein, 1955; Thomas and Vries, 1959; Price, 1990). In addition to shear degradation, macromolecules may be broken down by sonochemically produced •OH radicals as well (Nagata et al., 1996; Chemat et al., 2001; Chen et al., 2004).

Since sonochemically generated •OH radical is important for oxidation, the measurement of H_2O_2 formation was used to indirectly reflect the quantities of •OH radical produced during sonication (Hua and Hoffman, 1997; Frim and Weavers, 2003; Chen et al., 2004). •OH radical can self-combination to form hydrogen peroxide (reaction (15.1)). Using frequency at 20 or 354 kHz and power density of 120 or 450 W L^{-1}, Figure 15.3 shows the concentration of H_2O_2 formed with sonication in the absence of NOM. The highest concentration of H_2O_2 (0.4 mmol L^{-1} after 4 h) was produced at high frequency (354 kHz) and high power density (450 W L^{-1}), followed by high frequency and low power density (120 W L^{-1}), and the lowest concentration of H_2O_2 was formed at low frequency (20 kHz) even with a high power density (450 W L^{-1}). Higher •OH radical production has been observed around 354 kHz in other studies (Beckett and Hua, 2001; Frim and Weavers, 2003). However, mechanical shearing, a type of sonophysical effects is not reflected by Figure 15.3. As discussed before, mechanical shearing may reduce the molecular weight of macromolecules. Cavitational bubbles at 20 kHz ultrasound are thought to undergo more violent collapse and thus have stronger mechanical shearing effects than at 354 kHz (Crum, 1995).

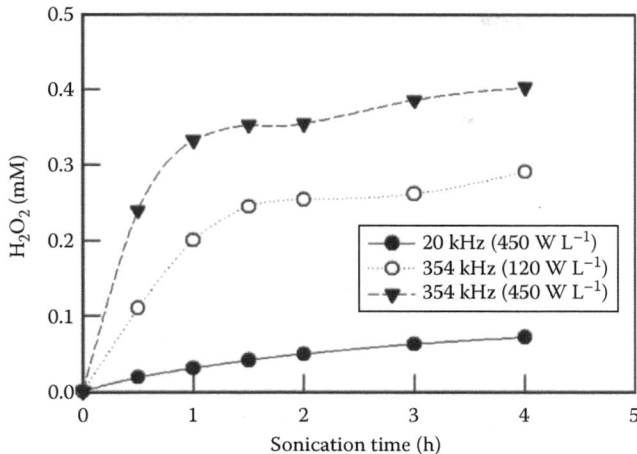

FIGURE 15.3 The formation of H_2O_2 with sonication under different ultrasonic frequencies and power densities in the absence of NOM. (With kind permission from Springer Science+Business Media: *Research on Chemical Intermediates*, 30, 2004, 735–753, Chen, D., He, Z., Weavers, L.K., Chin, Y.P., Walker, H.W., and Hatcher, P.G., Copyright 2004 Springer.)

TOC Reduction of NOM

Since NOM is refractory, mineralization is difficult unless enough •OH radicals are produced through cavitational collapse. As a result, no or slight degradation of TOC or DOC was observed by sonolysis alone without optimizing ultrasonic frequency, increasing ultrasonic power, adding oxidants, or removing •OH radicals scavengers (Olson and Barbier, 1994; Chen et al., 2004; Kreller et al., 2005).

In the study of Chen et al. (2004), no significant TOC reductions were observed for either Aldrich or Pahokee peat NOM through 4 h of sonication except at 354 kHz with the higher power density (450 W L^{-1}), in which more •OH radicals were produced (Figure 15.3). At this condition, TOC decreased from 22.5 mg L^{-1} to 15.0 and 18.2 mg L^{-1} for Aldrich and Pahokee peat NOM, respectively. Kreller et al. (2005) compared DOC degradation between 640 kHz sonolysis and γ-radiolysis of ^{60}Co. With sparging of oxygen, only minor decreases in DOC were observed after sonolysis of 10 h. Kim et al. (2007) indicated that more significant TOC removal of NOM by 20 kHz sonication occurred with an increase in hydrogen peroxide concentration from 1 to 10 mmol L^{-1}. In the study of Olson and Barbier (1994), unless ozone was added, no decrease in TOC was observed after sonolysis of 10 mg TOC L^{-1} fulvic acid solution with 20 kHz ultrasound. In addition, removal of •OH radical scavenger, bicarbonate through acidification of groundwater enhanced TOC reduction.

Decrease in UV/Vis Absorbance

Even when no significant TOC or DOC reduction is observed after sonolysis, it does not necessarily mean there is no molecular structural change of NOM as a result of sonication. UV/Vis absorbance can be used to examine the molecular structural changes to NOM, such as chromophores and aromaticity as described before.

Chen et al. (2004) showed that $Color_{465}$ of Aldrich and Pahokee peat NOM decreased with sonication at both 20 and 354 kHz. Again, a more significant decrease of $Color_{465}$ with sonication time occurred at 354 kHz with the higher power density (450 W L^{-1}). These results indicated that sonication caused the destruction of chromophores in NOM, such as conjugated double bonds, aromatic rings, and phenolic functional groups (Schnitzer and Khan, 1972).

Consistently, Naffrechoux et al. (2003) observed a decline of UV absorbance at 254 nm after sonolysis of 3 h with 500 kHz and 25 W calorimetric power. The small difference in hydrogen peroxide concentration between pure water and 20 mg L^{-1} Aldrich humic acid solution suggested limited reactivity between humic acid and •OH radicals. Moreover, a pseudo first-order decrease of fluorescence was observed as a result of sonolysis of NOM. Similarly, Taylor et al. (1999) showed a pseudo first-order decay of fluorescence intensity of polycyclic aromatic hydrocarbons (PAHs) through 20 kHz sonolysis. Kreller et al. (2005) indicated decreases in absorbance of NOM at 200, 280, and 400 nm wavelength, respectively. The greatest decrease in absorbance occurred at 400 nm, followed by 280 and 200 nm. With 1 mg min^{-1} ozone dosage, Olson and Barbier (1994) reported a decrease in UV absorbance of NOM at 230 nm with 20 kHz sonication. Higher ultrasonic power caused more significant decrease in UV absorbance. Qi et al. (2004) found the absorbance of NOM at 230 and 254 nm first increased with sonication, then decreased after 60 min of sonication. Similar trend was observed with E_4/E_6 ratio. However, it is unclear what the reason was leading to the initial increase of UV absorbance or E_4/E_6 ratio.

Decrease in Hydrophobicity of NOM

SUVA value directly correlates to the hydrophobicity of NOM. Experimental results showed that SUVA of Aldrich and Pahokee peat NOM decreased with sonication time at both 20 kHz and 354 kHz (Chen et al., 2004). For example, SUVA at 254 nm decreased from 6.65 to 5.99 and from 5.56 to 4.42 m^{-1} L (mg C)$^{-1}$ for Aldrich and Pahokee peat NOM, respectively, through 4 h of sonication at 20 kHz. Again, a more significant decrease in SUVA at 254 nm or at 280 nm was observed at 354 kHz with the higher power density (450 W L^{-1}). The decrease in SUVA at 254 and 280 nm suggested a decrease in the hydrophobicity and the aromaticity of NOM through sonication.

Similarly, Kreller et al. (2005) examined the intensity of log K_{ow} at 1.65 and 1.35, respectively, with different sonication time of NOM. The log K_{ow} value of 1.65 lies on the hydrophobic tail of the distribution; while the log K_{ow} of 1.35 lies approximately on the center of the main distribution peak. Sonolysis caused a steady decrease in the intensity at log K_{ow} = 1.65, while the intensity at log K_{ow} = 1.35 increased to a maximum during the first 30 min before decline. It was suggested that sonolysis converted highly hydrophobic fractions of NOM to medium hydrophobic fractions.

^{13}C NMR Changes to NOM

^{13}C NMR spectra have been used to directly determine changes in molecular structure of NOM as a result of sonication (Chen et al., 2004). The ^{13}C NMR spectra were integrated according to the following regions: 0–45 ppm, paraffinic carbons; 45–60 ppm, methoxyl; 60–90 ppm, carbohydrate carbons; 90–112 ppm, carbohydrate and proton-substituted aromatic carbons; 112–140 ppm, carbon-substituted aromatic carbons; 140–160 ppm, oxygen-substituted aromatic carbons; 160–190 ppm, carboxyl and aliphatic amide carbons; 190–220 ppm, aldehyde and ketone carbons (Knicker and Lüdemann, 1995; Chefetz et al., 2000; Dria et al., 2002). Total aromaticity was calculated by expressing aromatic C as a percentage of the aliphatic plus aromatic C (Hatcher et al., 1981; Dria et al., 2002).

As shown in Figure 15.4, generally the aromatic peak area (determined from the spectra between 112 and 160 ppm) of Aldrich and Pahokee peat NOM decreased and the aliphatic peak area (determined from the spectra between 0 and 90 ppm) increased through sonication, mostly at 354 kHz. As indicated in Table 15.2, sonication generally caused a decrease in the percentage of aromatic carbon, and an increase in the percentage of aliphatic carbon of both Aldrich and Pahokee peat NOM. More significant changes to the ^{13}C NMR spectra happened at 354 kHz than 20 kHz sonication. With the decrease in aromaticity, the hydrophobicity of NOM was expected to decline with sonication. Therefore, ^{13}C NMR spectral analysis was in agreement with the SUVA results that showed a decrease in hydrophobicity and aromaticity of NOM through sonication.

Decrease in Molecular Weight of NOM

In addition to the molecular structural changes to NOM, the molecular weight is expected to decrease through sonication as well. Figure 15.5 shows the spectra of HPSEC of NOM before and after sonication (Chen et al., 2004). Consistent with the results of UV absorbance, the decline of the peak area suggested that sonication destructed the chromophores of both Aldrich and Pahokee peat NOM

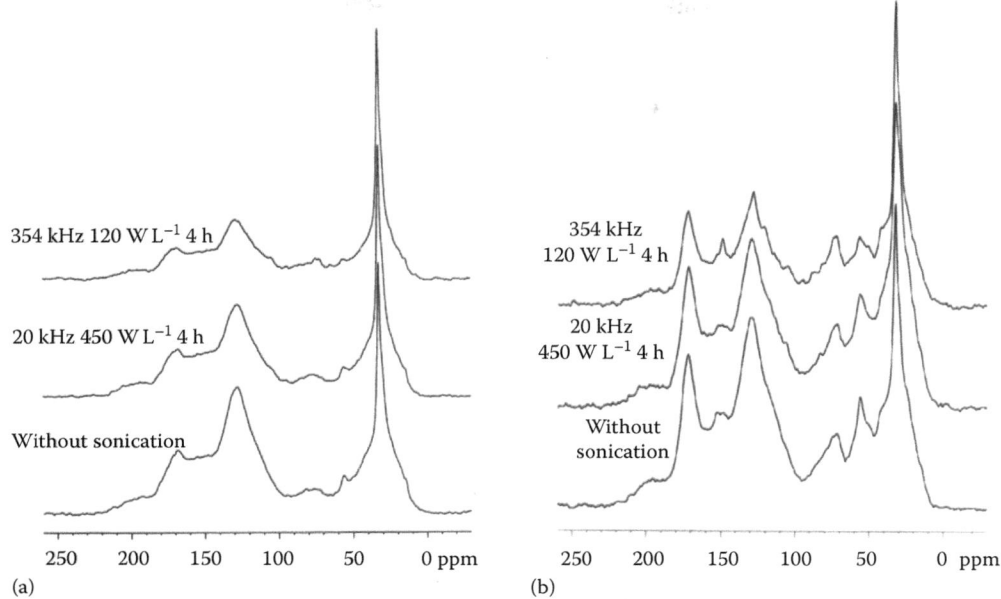

FIGURE 15.4 ^{13}C NMR spectra of NOM before and after 4 h of sonication. Y-axis represents relatively intensity. (a) Aldrich NOM. (b) Pahokee peat NOM. (With kind permission from Springer Science+Business Media: *Research on Chemical Intermediates*, 30, 2004, 735–753, Chen, D., He, Z., Weavers, L.K., Chin, Y.P., Walker, H.W., and Hatcher, P.G., Copyright 2004 Springer.)

molecules centered around 3000 Da. Sonication at 354 kHz resulted in a greater decline of the peak area than 20 kHz sonication, especially at the higher power intensity. Comparing sonication at 20 kHz 450 W L^{-1} (spectrum (2)) with 354 kHz 120 W L^{-1} (spectrum (3)), a broadening of the peak of spectrum (3) and a shifting of the peak of spectrum (2) to longer retention times occurred for both Aldrich and Pahokee peat NOM. This result suggested that 20 kHz ultrasound preferentially degraded large molecules (larger than 6400 Da based on a retention time of 8.5 min in Figure 15.5. The time of 8.5 min is the cross point between the spectra of 20 kHz 450 W L^{-1} and 354 kHz 120 W L^{-1}.).

The weight-averaged molecular weights of NOM before and after 4 h of sonication are shown in Figure 15.6. For both NOM, the most apparent decrease in molecular weight occurred at high frequency (354 kHz) with high power (450 W L^{-1}), followed by low frequency (20 kHz). No significant decrease in molecular weight was found at high frequency (354 kHz) with low power (120 W L^{-1}). Obviously, •OH radical (or H_2O_2) concentration alone could not explain that more decrease in molecular weight occurred at 20 kHz than 354 kHz with 120 W L^{-1}, although less •OH radicals were produced at 20 kHz.

As mentioned before, in addition to •OH radical oxidation, dynamic shearing caused by shock waves may also play a role in breaking covalent bonds of macromolecules (Basedow and Ebert, 1977; Price et al., 2002). Cavitational bubbles at 20 kHz sonication are thought to undergo more violent collapse and thus have stronger mechanical shearing effects than at 354 kHz (Crum, 1995). The effect of dynamic shearing is not reflected by •OH radical measurement as shown in Figure 15.3. Dynamic shearing is more effective in the degradation of polymers with higher molecular weight than lower molecular weight (Price, 1990). This phenomenon is observed because when the molecule is small enough, the physical dimension of polymer is too small to be impacted by shock waves (Chen et al., 2004). Therefore dynamic shearing of cavitational bubbles likely contributed to the breakdown of Aldrich and Pahokee peat NOM macromolecules, especially at 20 kHz, although H_2O_2 produced at 354 kHz with 120 W L^{-1} was more than four times as much as 20 kHz. The most significant decrease in NOM molecular weight happened at high frequency (354 kHz) with high power (450 W L^{-1}), because this

TABLE 15.2
Comparison of the Aromaticity of NOM as a Result of Sonication

	Before Sonication			20 kHz 450 W L^{-1} after 4 h			354 kHz 120 W L^{-1} after 4 h		
	Aliphatic Carbon (%)	Aromatic Carbon (%)	Aromaticity (%)	Aliphatic Carbon (%)	Aromatic Carbon (%)	Aromaticity (%)	Aliphatic Carbon (%)	Aromatic Carbon (%)	Aromaticity (%)
Aldrich NOM	40.5	43.4	51.7	44.1	40.3	47.8	49.6	35.8	41.9
Pahokee peat NOM	51.0	33.2	39.4	50.5	34.1	40.3	53.5	31.6	37.1

Source: With kind permission from Springer Science+Business Media: *Research on Chemical Intermediates*, 30, 2004, 735–753, Chen, D., He, Z., Weavers, L.K., Chin, Y.P., Walker, H.W., and Hatcher, P.G., Copyright 2004 Springer.

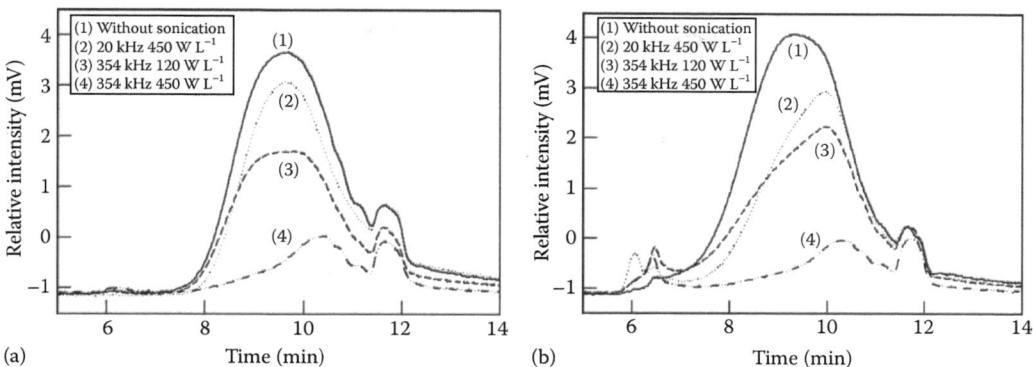

FIGURE 15.5 HPSEC spectra of NOM before and after 4 h of sonication. (a) Aldrich NOM. (b) Pahokee peat NOM. (With kind permission from Springer Science+Business Media: *Research on Chemical Intermediates*, 30, 2004, 735–753, Chen, D., He, Z., Weavers, L.K., Chin, Y.P., Walker, H.W., and Hatcher, P.G., Copyright 2004 Springer.)

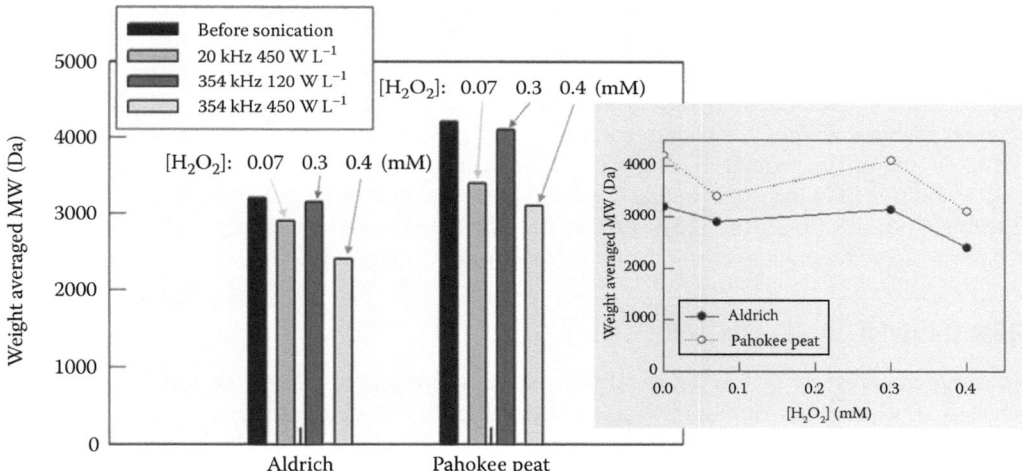

FIGURE 15.6 The relation between molecular weight of NOM and sonochemically produced hydrogen peroxide concentration. (Produced from Chen, D. et al., *Res. Chem. Intermed*, 30, 735, 2004.)

condition had the strongest combination of both •OH radical attack and shearing degradation; although its H_2O_2 concentration was only 0.1 mmol L^{-1} higher than 120 W L^{-1}, the mechanical shearing is stronger at 450 W L^{-1} than 120 W L^{-1} at the same 354 kHz.

In other studies, Nagata et al. (1996) indicated a decrease in molecular weight from 4,800 to 3,400, from 200,000 to 100,000, and from 10,500 to 6,400 Da of three types of humic acid, respectively, after 60 min sonication at 200 kHz with a power of 200 W. Kreller et al. (2005) found that sonolysis decreased the mass of the high molecular weight NOM fraction with a substantial gain of mass in the intermediate to low molecular weight fraction. Naffrechoux et al. (2003) showed the initial weight-averaged molecular weight of Aldrich humic acid was 13,000 Da. Elevated UV absorbance peaks at 3,200, 5,500, and 7,250 Da were observed after 3 h of sonication at 200 kHz and 25 W. However, there was no significant decrease in molecular weight of NOM through sonolysis after 3 h.

Increase in Total Acidity of NOM

Acidity is a fundamental characteristic of NOM. The acidity of NOM is primarily due to the presence of carboxylic (strong acid) and phenolic (weak acid) functional groups (Bowles et al., 1994).

In the study conducted by Chen et al. (2004), the total acidity of Aldrich NOM increased from 12.7 to 61.5 meg (g TOC) $^{-1}$ after 4 h of sonication at 354 kHz and 450 W L^{-1}. Meanwhile, the total acidity of Pahokee peat NOM increased from 10.6 to 16.3 meg (g TOC)$^{-1}$ under the same sonication condition. More apparent increase in the total acidity occurred at high frequency (354 kHz) with the higher power density (450 W L^{-1}) for both Aldrich and Pahokee peat NOM, while insignificant change was observed at 20 kHz ultrasound. Kreller et al. (2005) measured the total organic acidity of NOM in the pH range 4–9. Sonolysis at 640 kHz increased the total organic acidity of NOM from 8.6 ± 1.2 to 10.8 ± 0.4 equiv (kg C) $^{-1}$. The analyte proton buffering capacity of ultrasonically treated NOM increased significantly compared with the control sample.

IMPLICATIONS OF SONOCHEMICAL DEGRADATIONS OF NOM

A decrease in the aromaticity, the hydrophobicity, and the molecular weight, and an increase in acidity of NOM were observed through sonication. The property changes of NOM through sonication may significantly affect the physical–chemical and microbiological processes of NOM and the complexation between NOM and environmental components (Chen et al., 2004). More specifically, sonication is expected to (i) weaken hydrophobic interactions between NOM and hydrophobic organic compounds, since NOM becomes more hydrophilic through sonication; (ii) increase the bioavailability and biodegradability of NOM, as decrease in the aromaticity, the hydrophobicity, and the molecular weight of NOM have been observed as a result of sonication; and (iii) enhance the electrostatic interactions between NOM and environmental components, because sonication increases the total acidity (the sum of carboxylic and phenolic functional groups) of NOM.

The degree of the property changes to NOM depends on the frequency of ultrasound, power density, addition of oxidants, as well as sonication time. In practice, if property changes to NOM are favorable, optimized frequency (e.g., around 354 kHz), higher power density, addition of oxidants such as H_2O_2 or O_3, and longer sonication times would be preferred.

ULTRASOUND AND DISINFECTION

Ultrasonic disinfection is also chemical-free and less sensitive to particulate and UV-absorbance materials in water. Microstreaming and cavitational collapse produced shock wave, localized hot temperature, and hydroxyl radicals may disrupt and damage the cell membrane and consequently inactivate microorganisms in water (Scherba et al., 1991; Hua and Thompson, 2000; Oyane et al., 2005). In addition, sonication may break the agglomeration of microorganism clusters and flocs in solution/effluents, thus they are more susceptible to other disinfectants (Hua and Thompson, 2000; Blume and Neis, 2004).

Microorganisms are mostly hydrophobic. They may act as nuclei to induce cavitation in ultrasonic field. This effect may enhance disinfection efficiency. Overall, the inactivation rate of microorganisms depends on duration of sonication, ultrasonic power level, frequency, dissolved gas, and the properties of microorganisms, including the size and the shape of the cell, stage of development, and species (Thacker, 1973; Blume and Neis, 2004; Gogate, 2007). Blume and Neis (2004) reported that gram-positive streptococci seem less vulnerable to ultrasound exposure than thinner-walled gram-negative bacteria like the entire group of coliforms. Previous study has shown that sonication alone can inactivate microorganisms (Broekman et al., 2010), although sonication is usually combined with other disinfection techniques such as UV, chlorine, and ozone to obtain synergistic effects. Hulsmans et al. (2010) found that the specific energy, treatment time of water with ultrasound, and number of passages through ultrasonic reactors (Telsonic and Bandelin) are crucial influential parameters of ultrasonic disinfection of contaminated water in a pilot scale water disinfection system.

Many studies indicated that the inactivation of microorganisms followed pseudo first-order kinetics. More microorganisms were inactivated with longer sonication time. Oyane et al. (2005)

found that 26.6 kHz 30 W ultrasound inactivated 97% of *Cryptosporidium parvum* oocysts (initial concentration of 2260 oocysts mL^{-1}) in 5.2 min with an inactivation rate of 33 mL min^{-1}. The reduction of oocysts was due to complete disruption of the oocyst wall by ultrasonic cavitation. After the ultrasonic irradiation, the cell wall of the oocysts burst and the nuclei protruded. As a result, it was assumed that the oocysts were disrupted by shock waves rather than by •OH radicals, because the shape of oocysts changed. Ince and Belen (2001) showed that adding granular activated carbon improved the pseudo first-order inactivation rate of *E. coli*, which was due to increased cavitational nuclei. Neis and Blume (2003) found inactivation of *E. coli* and *Streptococci* followed first-order kinetics with 20 kHz 400 W L^{-1} ultrasound. The first-order reaction rate was 0.11 and 0.03 min^{-1} for *E. coli* and *Streptococci*, respectively.

Power level of ultrasound directly affects the rate and extent of disinfection, because normally more cavitational events, more violent cavitational collapse, and a greater amount of •OH radicals were generated at a higher ultrasonic power level. Huang and Myoda (2007) reported that the inactivation rate coefficient of *Cryptosporidium* was 0.035, 0.096, 0.098, and 0.100 min^{-1} with 20 kHz ultrasonic at energy intensity of 83, 248, 413, and 496 W cm^{-2} (or 2.8, 8.2, 14.0, and 16.5 W mL^{-1}), respectively. It seems the inactivation rate coefficient was not linearly proportional to the power intensity of ultrasound. In the study of Oyane et al. (2005), sonication at 52 W inactivated 72.5% of the *Cryptosporidium parvum* oocysts after 60 s; while at 126 W, 94.9% were inactivated. Hua and Thompson (2000) tested the inactivation of *E. coli* with power intensity from 4.6 to 74 W cm^{-2}. The inactivation rates increased with the power intensity of ultrasound. After 60 min of sonication, the log-based pseudo first-order inactivation rate was 0.319, 0.132, and 0.0299 min^{-1} corresponding to the power intensity of 74.1, 18.5, and 4.56 W cm^{-2} or power density of 0.26, 0.28, and 0.47 W mL^{-1}, respectively. Scherba et al. (1991) conducted a study on ultrasonic inactivation of typical microorganisms in common-use water facilities including bacteria of *E. coli*, *Staphylococcus aureus*, *Bacillus subtilis*, and *Pseudomonas aeruginosa*, fungus of *Trichophyton mentagrophytes*, and viruses of feline herpesvirus type 1 and feline calicirvirus. At 26 kHz and 39°C ± 0.3°C, there was a significant effect of ultrasonic intensity for fungus and all bacteria except *E. coli*. Thacker (1973) investigated the effects of ultrasonic power on inactivation of yeast cells of haploid and diploid. With ultrasonic intensities of 1, 2, and 4 W cm^{-2} and frequency at 1 MHz, fewer yeast cells survived with higher intensity. It was also observed that the dividing or the diploid cells were more susceptible to ultrasonic disinfection than non-dividing or haploid cells. Comparing the inactivation of cells in saline or yeast extract-peptone-dextrose medium (an effective •OH radicals scavenger), no big difference was found. Therefore, it was suggested that the inactivation of yeast cells was due to cavitationally produced mechanical stresses. In Blume and Neis's study (2004), 20 s of 20 kHz sonication at 30 W L^{-1} degraded the mean diameter of bioparticles from 70 to 11 µm, which facilitated the subsequent UV disinfection process.

The frequency of ultrasound also plays an important role in the effectiveness of disinfection. Hua and Thompson (2000) examined the frequencies at 205, 358, 618, and 1017 kHz on ultrasonic inactivation of *E. coli*. It was found that 205 kHz had the maximum inactivation rate coefficient of 0.078 min^{-1}, which was greater than 358 kHz (0.064 min^{-1}) and more than twice as 1017 kHz (0.030 min^{-1}); although the formation rate of hydrogen peroxide at 358 kHz (4.7 ± 0.71 µmol L^{-1} min^{-1}) was higher than 205 kHz (3.7 ± 0.33 µmol L^{-1} min^{-1}). Since the formation rate of hydrogen peroxide indirectly reflects the formation rate of •OH radicals, it was suggested that •OH radical and hydrogen peroxide was not the sole mechanism of disinfection. Microstreaming and cavitational collapse generated shock waves and localized heat might also contribute to the inactivation of bacteria, because 205 kHz ultrasound is supposed to produce more violent collapse and stronger sonophysical effects than 358 kHz (Crum, 1995). This sonophysical effect is not reflected by the formation rate of hydrogen peroxide. Nakanishi et al. (2001) investigated disinfection of *Cryptosporidium* oocysts at ultrasonic frequencies of 28, 45, and 100 kHz. Results showed that 28 kHz was the most effective in inactivating oocysts. After sonication of 10 min at 28 kHz, 10% of the total oocysts disappeared.

The oocyst cell wall ruptured and the nuclei burst from 97% of the remaining oocysts. The infectivity at 28 kHz was less than 1% of the base line, which was much less than 40% and 100% infectivity at frequencies of 45 and 100 kHz, respectively. Thacker (1973) examined the effect of ultrasonic frequency on inactivation of yeast cells at 20 kHz and 1 MHz, respectively. Greater rates of killing occurred at 20 kHz than 1 MHz at a specific power intensity. Consistently, Mason et al. (2003) reported that 27 kHz ultrasound had a better efficiency in inactivating *B. subtilis* in 20 L (approximately 73% of after 60 min) than higher ultrasonic frequencies.

CONTROL MEMBRANE FOULING IN FILTRATION PROCESSES

MEMBRANE FILTRATION AND MEMBRANE FOULING

Membrane filtration plays a critical role in advanced water and wastewater treatment processes, because of the technology's high removal capacity, ability to meet stringent treatment goals and small footprint. With membranes having high removal thresholds, the common contaminants in water such as dissolved salts, organic matters, viruses, bacteria, and particles can be removed in a single step. Consequently, high purity fresh water can be produced.

Despite the big prominence of the membrane technology, one of the main barriers to its greater applications is membrane fouling. Generally, membrane fouling is caused by the accumulation of water impurities (i.e., membrane foulants), such as colloidal particles, organic matters, microorganisms, and limiting salts on the membrane surface and/or within the membrane pores. Consequently, the membrane gets clogged over filtration time. Membrane fouling causes an increase in membrane resistance resulting in a significant decline of the permeate flux with filtration time. For example, the membrane permeate may decrease to 50% after several hours of process time (Chang et al., 2002). Membrane fouling affects both the quantity (permeate flux) and the quality (solute concentration) of the product water (Zhu and Elimelech, 1997). The characteristics and location of membrane foulants may play an important role in determining the extent and reversibility of permeate flux decline (Wiesner and Aptel, 1996). Irreversible fouling may require replacement of the membrane, and thus shortens membrane life. Detrimental effects of fouling on membrane performance ultimately cause an increase in operation, maintenance, and capital cost due to membrane replacement.

Ultrasonic technique has significant advantages over conventional cleaning methods for membrane fouling control. The conventional membrane cleaning includes increasing crossflow rate, backwashing, bubble scouring, and chemical cleaning of the membranes (Gutman, 1987; Li et al., 2009). However, these methods have their own problems. Increasing crossflow rate, back washing, and bubble scouring are not very effective, but require high energy cost. It is also very difficult to remove the strongly adhered foulants on the membranes, such as organic matters. In addition, chemical cleaning may damage the membrane and cause secondary pollution (Li et al., 2002). It also requires special storage, handling, and disposal of hazardous chemicals. For membrane cleaning, the filtration process must be shutdown, which reduces clean water production and increases the complexity of the membrane process (Chen et al., 2006a,b,c). The advantages of ultrasound include effectiveness, chemical free, and simultaneous membrane cleaning and fouling prevention during the filtration process. As a result, no downtime of filtration is necessary.

Results indicated that with the assistance of ultrasound, the clean water permeate flux of the membrane can be maintained throughout the duration of filtration (Chen et al., 2006a). It means ultrasound could completely eliminate membrane fouling. The effectiveness of ultrasound-assisted membrane filtration has been reported by several researchers (Tarleton and Wakeman, 1992; Kokugan et al., 1995; Matsumoto et al., 1996; Chai et al., 1998; Kobayashi et al., 1999, 2003; Masselin et al., 2001; Li et al., 2002; Lamminen et al., 2004, 2006; Muthukumaran et al., 2004, 2005; Chen et al., 2006a,b,c). All of these studies demonstrated that ultrasound effectively controlled membrane fouling and enhanced permeate flux for both polymeric and ceramic membranes.

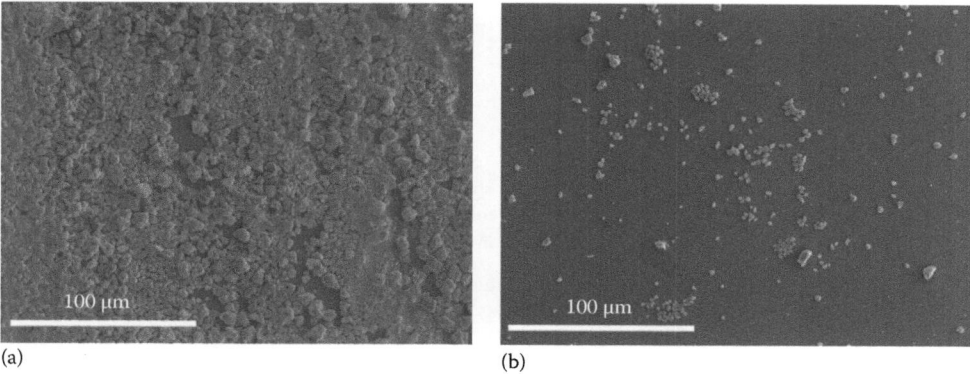

FIGURE 15.7 SEM images of the membrane surface after 240 min filtration with 8 mg L^{-1} NOM and 0.3 g L^{-1} silica particles at pH 9.2. (a) Without ultrasound. (b) With ultrasound. (Reprinted from *J. Membr. Sci.*, 276, Chen, D., Weavers, L.K., Walker, H.W., and Lenhart, J.J., Ultrasonic control of ceramic membrane fouling by natural organic matter and silica particles, 135–144. Copyright (2006), with permission from Elsevier.)

Figure 15.7 shows SEM images of NOM and silica particles fouled γ-Al$_2$O$_3$ ultrafiltration membranes (0.02 μm pore size) with or without ultrasound during filtration (Chen et al. 2006c). In Figure 15.7a, without ultrasound the membrane surface was largely covered by a foulant layer composed of NOM and silica particles. However, during filtration with ultrasound as shown in Figure 15.7b, the foulant layer at the membrane surface was significantly removed, revealing the membrane surface clearly.

MECHANISM OF ULTRASONIC CONTROL OF MEMBRANE FOULING

Ultrasound generates acoustic streaming and cavitation bubbles in water. Cavitation bubbles produce microstreaming, microstreamers, microjets, and shock waves (Leighton, 1994). When the membrane is outside the cavitational region, acoustic streaming and cavitationally generated turbulence transformed from microstreaming, microstreamers, microjets, and shock waves produce shear forces near the membrane surface that dislodge foulants from the membrane and/or prevent the deposition of foulants that causes membrane fouling (Chen et al., 2006a). Although the effective distance is in the range of microns for a single event of microjet, shock wave, and microstreaming, there are numerous cavitational bubbles collapsing at the same time during sonication. Consequently, the turbulence generated by the transformation of microjets, shock waves, and microstreaming is considerable (Chen et al., 2006a).

When the membrane is inside the cavitational region, all of these mechanisms directly contribute to membrane cleaning. Figure 15.8 illustrates the cleaning mechanism of ultrasound for membrane fouling control. However, microjets and shock waves are so energetic that they may damage the membranes by forming surface pittings, erosions, and cracks (Chen et al., 2006b). A major difference between the fluid movement within or outside the cavitational region is the energy density, which is extremely high within the cavitational region where the high velocity fluid movement occurs at the micron-scale (Crum, 1988). When the membrane is inside the cavitational region, the membrane and the foulants such as particles, macromolecules, and microorganisms may act as nuclei to induce ultrasonic cavitation, because cavitation bubbles preferentially form at the gas–liquid and solid–liquid interfaces where discontinuity in free energy occur (Collings and Gwan, 2010). This effect may promote ultrasonic cleaning under the premise that the integrity of the membrane is maintained.

Membrane fouling is initiated by deposition of foulants (macromolecules, colloids and particles, scales, and microorganisms) on the membrane surface or pores. As illustrated in Figure 15.9, the

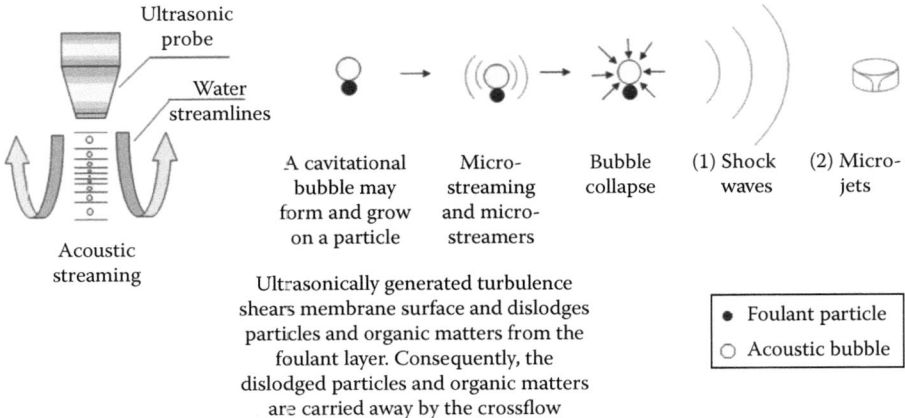

FIGURE 15.8 Proposed mechanism of ultrasonic cleaning of water filtration membranes.

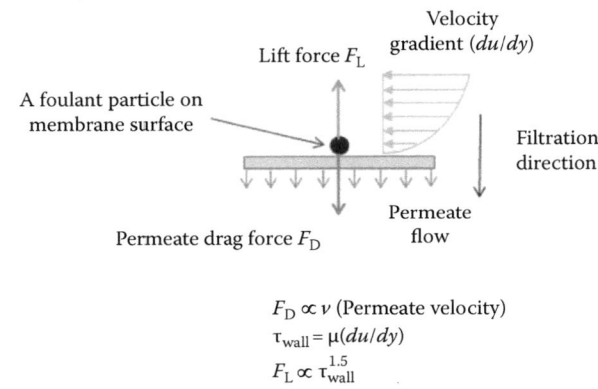

FIGURE 15.9 Force balance analysis of a foulant particle on the membrane surface. When the lift force generated by ultrasonically produced turbulence and/or crossflow is greater than the permeate drag force, the foulant particle will be dislodged from the membrane surface (Other forces such as electrostatic and hydrophobic interactions among the foulant–foulant and foulant–membrane were not considered in this case.)

deposition is driven by the permeate drag force (F_D) caused by membrane permeate flux (Altmann and Ripperger, 1997).

$$F_D = 3\pi\eta a_p v \tag{15.12}$$

where
a_p is the diameter of the foulant particle
η is the kinematic fluid viscosity
v is the velocity of the permeate flux

Ultrasonically generated turbulence and/or crossflow cause velocity gradient (du/dy) at the membrane surface. The velocity gradient is expected to increase as the membrane is located closer to the ultrasonic source. The shearing stress on the membrane surface (τ_{wall}) equals the dynamic fluid viscosity (μ) of the solution times the local velocity gradient (du/dy).

$$\tau_{wall} = \mu\left(\frac{du}{dy}\right) \qquad (15.13)$$

The lift force (F_L) acting on a foulant particle on the membrane surface is proportional to the shearing stress ($\tau_{wall}^{1.5}$) (Altmann and Ripperger, 1997).

$$F_L = 0.761 \frac{\tau_{wall}^{1.5} a_p^{3} \rho^{0.5}}{\eta} \qquad (15.14)$$

where ρ is the density of the fluid. When the lift force (F_L) is greater than the permeate drag force (F_D), the foulant particle will be dislodged from the membrane surface or pores. This effect will also prevent future deposition of foulants that causes membrane fouling. Consequently, the membrane can maintain a clean condition during filtration.

INTEGRITY OF MEMBRANES UNDER ULTRASONIC IRRADIATION

The integrity of membranes is of critical importance to maintain the membrane's capacity to reject contaminants during filtration. However, ultrasonic irradiation may potentially damage the membranes through the mechanisms of (Crum, 1988; Philipp and Lauterborn, 1998): (i) shock waves, which may create micro-fissures or pittings on the membranes; (ii) microjets, which may cause pittings and surface erosion; (iii) oxidation of organic membrane materials by •OH radicals. Both microjets and shock waves have extremely high energy density, but their direct working distance is in the range of microns (Suslick, 1990; Philipp and Lauterborn, 1998; Roy, 1999; Pecha and Gompf, 2000). Therefore, when the membrane is located outside the ultrasonic cavitation region, the membrane surface is unlikely to be directly impacted by microjets and/or shock waves generated from the final collapse of cavitational bubbles. As a result, the ideal location of a filtration membrane is just outside but close to the region of cavitation; where the membrane may still receive great turbulence generated from ultrasound while the integrity is maintained (Chen et al., 2006b).

There is a discrepancy in the literature regarding the integrity of membranes following exposure to ultrasound. Masselin et al. (2001) observed damages to polyethersulfone membranes by ultrasound, while other researchers (Chai et al., 1998; Kobayashi et al., 1999, 2003; Li et al., 2002; Muthukumaran et al., 2004, 2005; Chen et al., 2006a,c) showed that the integrity of membranes was maintained throughout sonication. Unfortunately, no enough information of the cavitational region in these studies was given. Sonochemically induced chemiluminescence (SCL) is a useful method to determine the spatial distribution of the cavitational region in the membrane cell (Chen et al., 2006a,b). When the membrane is inside the cavitational region, it is prone to damage. As shown in Figure 15.10, membrane damages by surface pitting were observed when the membrane was inside the ultrasonic cavitational region. The pittings revealed the underlying structure of a γ-Al_2O_3 membrane of $0.02\,\mu m$ pore size (Chen et al., 2006b). The existence of both microjets and shock waves has been established during cavitation collapse, but their relative importance is a matter of debate (Suslick, 1990). According to Crum's report (1988) of erosion of metals by cavitational collapse, the pitting on metal surfaces is more likely caused by microjets. Consistently, Juang and Lin (2004) found slight damages to the structure of a hydrophilic regenerated cellulose membrane (YM10 from Amicon) when the distance between the transducer tip and the membrane was 10 mm with more than 80 W power. However, when

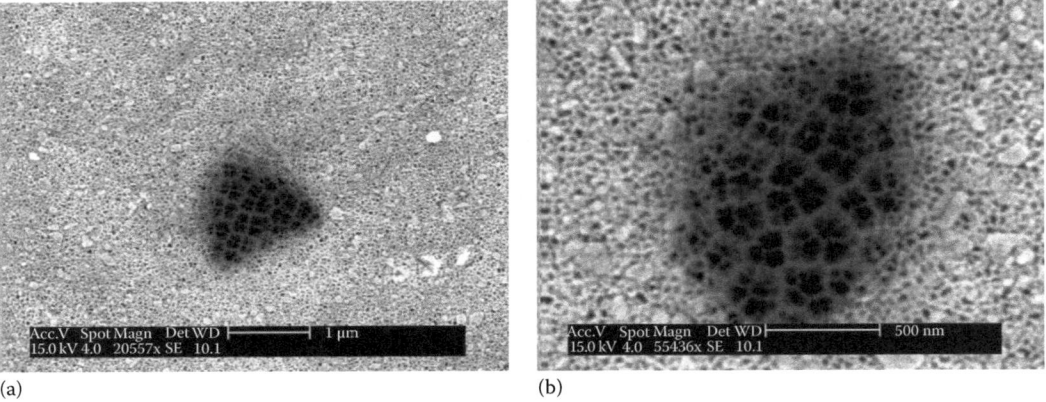

FIGURE 15.10 Examples of pittings on a γ-Al_2O_3 ultrafiltration membrane surface caused by microjets and/or shock waves after 5 min of sonication at 20 kHz and power intensity of 3.8 ± 0.1 W cm^{-2}. The filtration pressure was 5 psi. In both (a) and (b), the distance between the ultrasonic probe and the membrane was 1.3 cm. The extent of the cavitational region was 1.5 cm from the probe based on the photo of SCL. So the membrane was within the ultrasonic cavitational region.

the distance increased to greater than 20 mm, the membrane was not damaged even with a power over 200 W. In addition to the location of the membranes with respect to the cavitational region, the membrane material also affects the possibility and the degree of cavitational damage. Masselin et al. (2001) reported that polyethersulfone was broken by ultrasonic irradiation, while PVDF and polyacrylonitrile membranes seemed unaffected.

ULTRASONIC FACTORS FOR MEMBRANE FOULING CONTROL

Ultrasonic frequency, power level, and the distance between the ultrasonic transducer and the membrane are important factors affecting membrane cleaning and fouling control. First, sonication did not affect the intrinsic permeability of the membranes when the integrity of the membranes was unaffected (Kobayashi et al., 1999; Lamminen et al., 2004). In the study of Chen et al. (2006a), a slight increase in the clean water permeate flux of γ-Al_2O_3 membrane was observed due to the elevated temperature caused by sonication, which reduced the viscosity of water solution.

As mentioned before, more significant mechanical effects occur at low-frequency ultrasound. As a result, a better cleaning effect at low-frequency ultrasound has been observed. Kobayashi et al. (2003) tested ultrasonic frequency at 28, 45, and 100 kHz with 23 W cm^{-2} output power to assist ultrafiltration and microfiltration of peptone and milk aqueous solutions. The ultrafiltration membrane was made of polysulfone and the microfiltration membrane was made of cellulose. Membrane operations were performed by crossflow filtration with 60 kPa operating pressure in the ultrasonic field. More effective ultrasonic cleaning was found at 28 kHz.

Increasing ultrasonic power normally increases acoustic streaming, microstreaming, the violence of cavitational collapse, and the extent of cavitational region. As a consequence, stronger turbulence is produced by ultrasound and improved cleaning effects have been observed (Kobayashi et al., 1999; Lamminen et al., 2006). Kobayashi et al. (1999) tested ultrasonic power intensity in the range of 2.5–3.3 W cm^{-2} on a permeate flux of dextran solutions through polyacrylonitrile ultrafiltration membranes. The extent of the enhancement of permeate flux was found improved with increasing ultrasonic intensity. In addition, the irradiation direction of ultrasound relative to the membrane was also an important factor. In the study of Lamminen et al. (2006), elevated ultrasonic power improved the flux of the latex particle-fouled PVDF membrane. However, when the ultrasonic power reached the highest level of 12.2 W, some damages to the polyvinylidene fluoride membrane were observed. At lower applied powers, no damage to the membrane was detected. Visualization of the cavitational region by SCL indicated that greater extent of the cavitational region happened at higher power level of ultrasound.

EFFECTS OF SOLUTION CHEMISTRY ON ULTRASONIC CONTROL OF MEMBRANE FOULING

Solution chemistry affects the affinity among foulant–foulant and foulant–membrane. Apparently, a better ultrasonic cleaning effect is expected when the affinity among them is weak. The affinity among foulant–foulant and foulant–membrane includes specific chemical bonds, electrostatic and hydrophobic interactions, and van der Waals force. For example, NOM is a major membrane foulant during filtration of natural waters (Nystrom et al., 1996; Zhang et al., 2003; Kim et al., 2010). NOM may block membrane pores, form a gel layer, or bind particles together to form a low permeability NOM/particle cake layer on the membrane surface (Zhang et al., 2003). Generally, more severe membrane fouling by NOM occurs at low pH, at high ionic strength, in the presence of divalent cations, and hydrophobic membranes (Jucker and Clark, 1994; Hong and Elimelech, 1997; Braghetta et al., 1998; Yuan and Zydney, 1999; Amy and Cho, 1999; Schäfer et al., 2000; Jones and O'Melia, 2000; Kim et al., 2010). Chen et al. (2006c) investigated ultrasonic control of γ-Al_2O_3 membrane fouling by 8 mg L^{-1} (measured by TOC) purified Aldrich NOM and 0.3 g L^{-1} silica particles (mean diameter of 1.56 μm) in a crossflow filtration system. The results indicated that ultrasound significantly reduced membrane fouling caused by NOM and silica particles. However, the solution chemistry significantly affected ultrasonic cleaning. At high pH, electrostatic repulsions among NOM, silica particles, and the membrane caused the membrane foulants to be more readily removed by ultrasound than did low pH. Consistently, Fourier transform infrared (FTIR) spectra in Figure 15.11 show that ultrasound reduced the magnitude of the characteristic NOM vibrations at both pH 9.2 and 4.0.

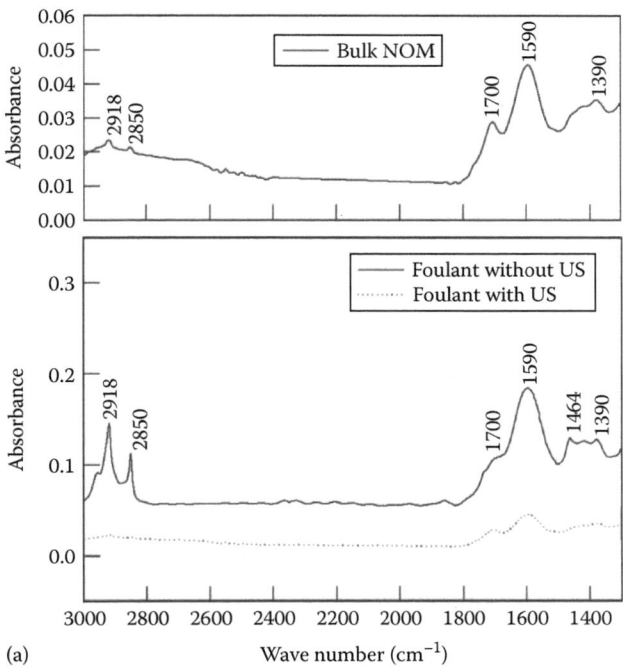

FIGURE 15.11 (a) FTIR spectra of the bulk NOM at pH 4.0 and NOM foulant on the membrane after 240 min filtration with 8 mg L^{-1} NOM at pH 4.0 with and without ultrasound (US).

(*continued*)

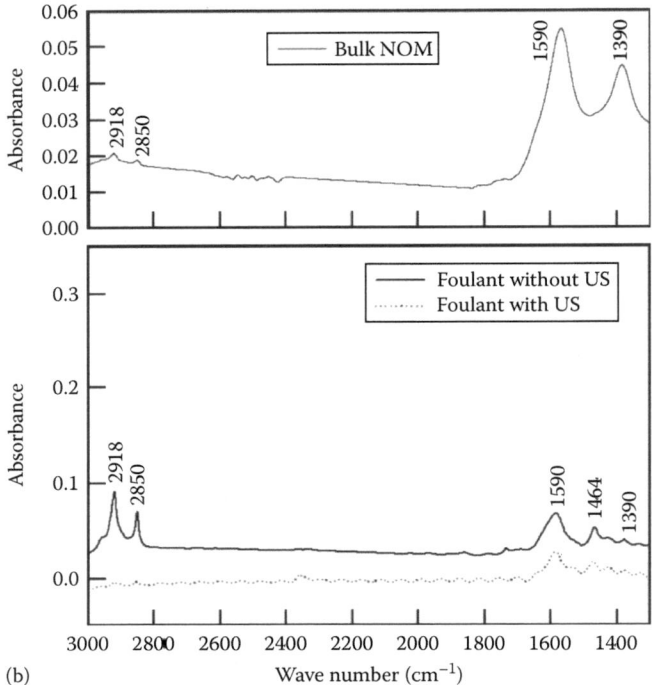

FIGURE 15.11 (continued) (b) FTIR spectra of the bulk NOM at pH 9.2 and NOM foulant on the membrane after 240 min filtration with 8 mg L^{-1} NOM at pH 9.2 with and without ultrasound (US). No silica particle was used in either (a) or (b). (Reprinted from *J. Membr. Sci.*, 276, Chen, D., Weavers, L.K., Walker, H.W., and Lenhart, J.J., Ultrasonic control of ceramic membrane fouling by natural organic matter and silica particles, 135–144. Copyright (2006), with permission from Elsevier.)

However, at pH 9.2, fewer carboxylate (carboxylate asymmetric and ring C=C stretching from 1550 to 1630 cm^{-1}, and carboxylate symmetric stretching from 1380 to 1420 cm^{-1}) and aliphatic functional groups (aliphatic C–H stretching from 2800 to 3000 cm^{-1}, and C–H deformation of aliphatic CH$_2$ or CH$_3$ groups at 1464 cm^{-1}) were retained on the membrane than at pH 4.0. Ultrasound at pH 9.2 cleaned the membrane to the baseline more close to a virgin membrane. In addition, at high ionic strength, charge screening among NOM macromolecules, silica particles, and the membrane caused weaker foulant–foulant and foulant–membrane repulsions. Consequently, the cleaning effectiveness of ultrasound decreased. Because Ca^{2+} causes charge neutralization and bridging among NOM, silica particles, and the membrane (Hong and Elimelech, 1997), addition of 1 mM Ca^{2+} deteriorated membrane fouling and reduced ultrasonic cleaning effect. In Figure 15.12, FTIR spectra indicate that Ca^{2+} increased the retention of carboxylate and the aliphatic functional groups on the membrane regardless the use of ultrasound. Moreover, the results also showed that ultrasound restored the NOM rejection rate of the clean membrane and did not damage the membrane.

ACKNOWLEDGMENTS

This work was supported by new faculty starting fund and summer faculty research fund from the Office of Research and External Support, Indiana University–Purdue University, Fort Wayne, Indiana. The author would greatly appreciate Linda weavers and Harold Walker for their valuable advice and support during his Ph.D., study at The Ohio State University.

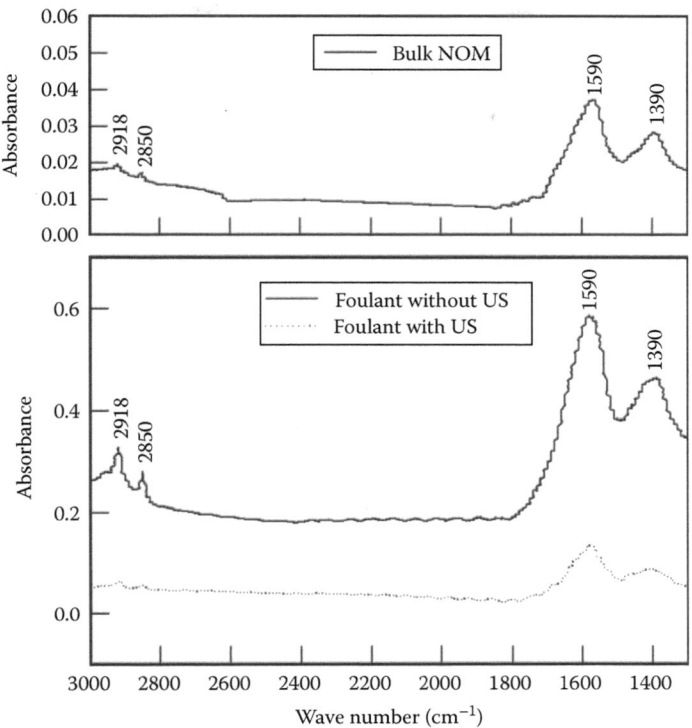

FIGURE 15.12 FTIR spectra of the bulk NOM at pH 9.2 in the presence of 1 mmol L^{-1} CaCl$_2$ and NOM foulant on the membrane after 240 min filtration with 8 mg L^{-1} NOM at pH 9.2 in the presence of 1 mmol L^{-1} CaCl$_2$ with and without ultrasound (US). No silica particle was used. (Reprinted from *J. Membr. Sci.*, 276, Chen, D., Weavers, L.K., Walker, H.W., and Lenhart, J.J., Ultrasonic control of ceramic membrane fouling by natural organic matter and silica particles, 135–144. Copyright (2006), with permission from Elsevier.)

REFERENCES

Adewuyi, Y.G. 2001. Sonochemistry: Environmental science and engineering applications. *Industrial & Engineering Chemistry Research*, 40:4681–4715.

Adewuyi, Y.G. 2005. Sonochemistry in environmental remediation. Combinative and hybrid sonophotochemical oxidation processes for the treatment of pollutants in water. *Environmental Science and Technology*, 39:3409–3420.

Aiken, G.R. 1985. Isolation and concentration techniques for aquatic humic substances. In *Humic Substances in Soil, Sediment and Water: Geochemistry and Isolation*, Aiken, G.R., McKnight, D.M., Wershaw, R.L., and MacCarthy, P. (Eds.). New York: Wiley-Interscience.

Aiken, G. and Cotsaris, E. 1995. Soil and hydrology: Their effect on NOM. *Journal of American Water Works Association*, 87:36–45.

Aiken, G.R. and Malcolm, R.L. 1987. Molecular weight of aquatic fulvic acids by vapor pressure osmometry. *Geochimica et Cosmochimica Acta*, 51:2177–2184.

Altmann, J. and Ripperger, S. 1997. Particle deposition and layer formation at the crossflow microfiltration. *Journal of Membrane Science*, 124:119–128.

Amy, G. and Cho, J. 1999. Interactions between natural organic matter (NOM) and membranes: Rejection and fouling. *Water Science and Technology*, 40:131–139.

Arrojo, S., Benito, Y., and Tarifa, A.M. 2008. A parametrical study of disinfection with hydrodynamic cavitation. *Ultrasonics Sonochemistry*, 15:903–908.

Basedow, A.M. and Ebert, K.H. 1977. Ultrasonic degradation of polymers in solution. *Advances in Polymer Science*, 22:83–148.

Beckett, M.A. and Hua, I. 2001. Impact of ultrasonic frequency on aqueous sonoluminescence and sonochemistry. *The Journal of Physical Chemistry A*, 105:3796–3802.

Beckett, M.A. and Hua, I. 2003. Enhanced sonochemical decomposition of 1,4-dioxane by ferrous iron. *Water Research*, 37:2372–2376.

Bedner, M. and MacCrehan, W.A. 2006. Transformation of acetaminophen by chlorination produces the toxicants 1,4-benzoquinone and n-acetyl-p-benzoquinone imine. *Environmental Science and Technology*, 40:516–522.

Belfroid, A., Velzen, M., van der Horst, B., and Vethaak, D. 2002. Occurrence of bisphenol A in surface water and uptake in fish: Evaluation of field measurements. *Chemosphere*, 49:97–103.

Belgiorno, V., Rizzo, L., Fatta, D., Rocca, C.D., Lofrano, G., Nikolaou, A., Naddeo, V., and Meric, S. 2007. Review on endocrine disrupting-emerging compounds in urban wastewater: Occurrence and removal by photocatalysis and ultrasonic irradiation for wastewater reuse. *Desalination*, 215:166–176.

Blume, T. and Neis, U. 2004. Improved wastewater disinfection by ultrasonic pretreatment. *Ultrasonics Sonochemistry*, 11:333–336.

Bob, M.M. and Walker, H.W. 2000. Effect of natural organic coatings on the polymer-induced coagulation of colloidal particles. *Colloids and Surfaces A: Physicochemical and Engineering Aspects*, 177:215–222.

Bowles, E.C., Antweiler, R.C., and MacCarthy, P. 1994. Acid–base titration and hydrolysis of fulvic acid from the Suwannee river. In *Humic Substances in the Suwannee River, Georgia: Interactions, Properties, and Proposed Structures*, Averett, R.C., Leenheer, J.A., McKnight, D.M., and Thorn, K.A. (Eds.). Washington, DC: U.S. Geological Survey, pp. 115–127.

Braghetta, A., DiGiano, F.A., and Ball, W.P. 1998. NOM accumulation at NF membrane surface: Impact of chemistry and shear. *Journal of Environmental Engineering*, 124:1087–1098.

Broekman, S., Pohlmanna, O., Beardwooda, E.S., and de Meulenaer, E.C. 2010. Ultrasonic treatment for microbiological control of water systems. *Ultrasonics Sonochemistry*, 17:1041–1048.

Buxton, G.V., Greenstock, C.L., Helman, W.P., and Ross, A.B. 1988. Critical review of rate constants for reactions of hydrated electrons, hydrogen atoms and hydroxyl radicals (OH/O$^-$) in aqueous solution. *Journal of Physical and Chemical Reference Data*, 17:513–886.

Casadonte Jr., D.J., Flores, M., and Petrier, C. 2005. Enhancing sonochemical activity in aqueous media using power-modulated pulsed ultrasound: An initial study. *Ultrasonics Sonochemistry*, 12:147–152.

Chai, X., Kobayashi, T., and Fujii, N. 1998. Ultrasound effect on cross-flow filtration of polyacrylonitrile ultra-filtration membranes. *Journal of Membrane Science*, 148:129–135.

Chang, I.S., Clech, P.L., Jefferson, B., and Judd, S. 2002. Membrane fouling in membrane bioreactors for wastewater treatment. *Journal of Environmental Engineering*, 128:1018–1029.

Chefetz, B., Deshmukh, A.P., and Hatcher, P.G. 2000. Pyrene sorption by natural organic matter. *Environmental Science & Technology*, 34:2925–2930.

Chemat, F., Teunissen, P.G.M., Chemat, S., and Bartels, P.V. 2001. Sono-oxidation treatment of humic substances in drinking water. *Ultrasonics Sonochemistry*, 8:247–250.

Chen, D., He, Z., Weavers, L.K., Chin, Y.P., Walker, H.W., and Hatcher, P.G. 2004. Sonochemical reactions of dissolved organic matter. *Research on Chemical Intermediates*, 30:735–753.

Chen, D., Weavers, L.K., and Walker, H.W. 2006a. Ultrasonic control of ceramic membrane fouling: Effect of particle characteristics. *Water Research*, 40:840–850.

Chen, D., Weavers, L.K., and Walker, H.W. 2006b. Ultrasonic control of ceramic membrane fouling by particles: Effect of ultrasonic factors. *Ultrasonics Sonochemistry*, 13(5):379–387.

Chen, D., Weavers, L.K., Walker, H.W., and Lenhart, J.J. 2006c. Ultrasonic control of ceramic membrane fouling by natural organic matter and silica particles. *Journal of Membrane Science*, 276:135–144.

Chen, Y., Senesi, N., and Schnitzer, M. 1977. Information provided on humic substances by E_4/E_6 ratios. *Soil Science Society of America Journal*, 41:352–358.

Cheng, J., Vecitis, C.D., Park, H., Mader, B.T., and Hoffmann, M.R. 2010. Sonochemical degradation of perfluorooctane sulfonate (PFOS) and perfluorooctanoate (PFOA) in groundwater: Kinetic effects of matrix inorganics. *Environmental Science & Technology*, 44:445–450.

Chin, Y.P., Aiken, G.R., and Danielsen, K.M. 1997. Binding of pyrene to aquatic and commercial humic substances: The role of molecular weight and aromaticity. *Environmental Science & Technology*, 31:1630–1635.

Chin, Y.P., Aiken, G.R., and O'Loughlin, E. 1994. Molecular weight, polydispersivity, and spectroscopic properties of aquatic humic substances. *Environmental Science & Technology*, 28:1853–1858.

Chowdhury, P. and Viraraghavan, T. 2009. Sonochemical degradation of chlorinated organic compounds, phenolic compounds and organic dyes—A review. *Science of the Total Environment*, 407:2474–2492.

Clarke, P.R. and Hill, C.R. 1970. Physical and chemical aspects of ultrasonic disruption of cells. *The Journal of the Acoustical Society of America*, 47:649–653.

Cole, J.J., Prairie, Y.T., Caraco, N.F., McDowell, W.H., Tranvik, L.J., Striegl, R.G., Duarte, C.M., Kortelainen, P., Downing, J.A., Middelburg, J.J., and Melack, J. 2007. Plumbing the global carbon cycle: Integrating inland waters into the terrestrial carbon budget. *Ecosystems*, 10:171–184.

Collings, A.F. and Gwan, P.B. 2010. Ultrasonic destruction of pesticide contaminants in slurries. *Ultrasonics Sonochemistry*, 17:1–3.

Cooper, W.J., Cramer, C.J., Martin, N.H., Mezyk, S.P., O'Shea, K.E., and Sonntag, C.V. 2009. Free radical mechanisms for the treatment of methyl tert-butyl ether (MTBE) via advanced oxidation/reductive processes in aqueous solutions. *Chemical Reviews*, 109:1302–1345.

Cox, R.A., Derwent, R.G., Eggleton, A.E.J., and Lovelock, J.E. 1976. Photochemical oxidation of halocarbons in the troposphere. *Atmospheric Environment*, 10:305–308.

Croue, J.P., Violleau, D., Bodaire, C., and Legube, B. 1999. Removal of hydrophobic and hydrophilic constituents by anion exchange resin. *Water Science and Technology*, 40(9):207–214.

Crum, L.A. 1988. Cavitation microjets as a contributory mechanism for renal calculi disintegration in ESWL. *The Journal of Urology*, 140(6):1587–1590.

Crum, L.A. 1995. Comments on the evolving field of sonochemistry by a cavitation physicist. *Ultrasonics Sonochemistry*, 2:147–152.

David, B. 2009. Sonochemical degradation of PAH in aqueous solution. Part I: Monocomponent PAH solution. *Ultrasonics Sonochemistry*, 16:260–265.

Dobbs, R.A., Wise, R.H., and Dean, R.B. 1972. The use of ultra-violet absorbance for monitoring the total organic carbon content of water and wastewater. *Water Research*, 6:1173–1180.

Dria, K.J., Sachleben, J.R., and Hatcher, P.G. 2002. Solid-state carbon-13 nuclear magnetic resonance of humic acids at high magnetic field strengths. *Journal of Environmental Quality*, 31:393–401.

Edwards, M., Benjamin, M.M., and Ryan, J.N. 1996. Role of organic acidity in sorption of natural organic matter (NOM) to oxide surfaces. *Colloids and Surfaces A: Physicochemical and Engineering Aspects*, 107:297–307.

Ensminger, D. 1973. *Ultrasonics: The Low-and High-Intensity Applications*. New York: Marcel Dekker.

Flint, E.B. and Suslick, K.S. 1991. The temperature of cavitation. *Science*, 253:1397–1399.

Francony, A. and Petrier, C. 1996. Sonochemical degradation of carbon tetrachloride in aqueous solution at two frequencies: 20 kHz and 500 kHz. *Ultrasonics Sonochemistry*, 3:S77–S82.

Frim, J.A. and Weavers, L.K. 2003. Sonochemical destruction of free and metal-binding ethylenediaminetetraacetic acid. *Water Research*, 37(13):3155–3163.

Gogate, P.R. 2007. Application of cavitational reactors for water disinfection: Current status and path forward. *Journal of Environmental Management*, 85:801–815.

Guo, Z. and Feng, R. 2009. Ultrasonic irradiation-induced degradation of low-concentration bisphenol A in aqueous solution. *Journal of Hazardous Materials*, 163:855–860.

Gutierrez, M. and Henglein, A. 1990. Chemical action of pulsed ultrasound: Observation of an unprecedented intensity effect. *The Journal of Physical Chemistry*, 94:3625–3628.

Gutman, R.G. 1987. *Membrane Filtration: The Technology of Pressure-Driven Crossflow Processes*. Bristol, England: Adam Hilger, pp. 84–129.

Hatanaka, S., Yasuib, K., Kozukab, T., Tuziutib, T., and Mitome, H. 2002. Influence of bubble clustering on multibubble sonoluminescence. *Ultrasonics*, 40:655–660.

Hatcher, P.G., Rowan, R., and Mattingly, M.A. 1980. ^1H and ^{13}C NMR of marine humic acids. *Organic Geochemistry*, 2:77–85.

Hatcher, P.G., Schnitzer, M., Dennis, L.W., and Maciel, G.E. 1981. Aromaticity of humic substances in soils. *Soil Science Society of America Journal*, 45:1089–1094.

He, Z., Traina, S.J., Bigham, J.M., and Weavers, L.K. 2005. Sonolytic desorption of mercury from aluminum oxide. *Environmental Science & Technology*, 39:1037–1044.

Henglein, A. 1955. The reaction of α,α-diphenyl-β-pikryl-hydrazyl with macroradicals formed in the ultrasonic degradation of polymethylmethacrylate (in German). *Macromolecular Chemistry and Physics*, 15:188–210.

Hickenboth, C.R., Moore, J.S., White, S.R., Sottos, N.R., Baudry, J., and Wilson, S.R. 2007. Biasing reaction pathways with mechanical force. *Nature*, 446:423–427.

Hill, C.R., Clarke, P.R., Crowe, M.R., and Hannick, J.W. 1969. Biophysical effects of cavitation in a 1 MHz ultrasonic beam. In *Ultrasonics for Industry, 7th and 8th Conference Papers*. Crawford, A.H. (Ed.). Surrey, England: Iliffe Science and Technology, pp. 26–30.

Hiskia, A., Ecke, M., Troupis, A., Kokorakis, A., Hennig, H., and Papaconstantinou, E. 2001. Sonolytic, photolytic, and photocatalytic decomposition of atrazine in the presence of polyoxometalates. *Environmental Science & Technology*, 35:2358–2364.

Hong, S. and Elimelech, M. 1997. Chemical and physical aspects of natural organic matter (NOM) fouling of nanofiltration membranes. *Journal of Membrane Science*, 132:159–181.

Hua, I. and Hoffmann, M.R. 1996. Kinetics and mechanism of the sonolytic degradation of CCl_4: Intermediates and byproducts. *Environmental Science & Technology*, 30:864–871.

Hua, I. and Hoffmann, M.R. 1997. Optimization of ultrasonic irradiation as an advanced oxidation technology. *Environmental Science & Technology*, 31:2237–2243.

Hua, I. and Thompson, J.E. 2000. Inactivation of *Escherichia coli* by sonication at discrete ultrasonic frequencies. *Water Research*, 34:3888–3893.

Huang, C.P. and Myoda, S.P. 2007. Sonochemical treatment of wastewater effluent for the removal of pathogenic protozoa exemplified by *Cryptosporidium*. *Practice Periodical of Hazardous, Toxic, and Radioactive Waste Management*, 11:114–122.

Hulsmans, A., Joris, K., Lambert, N. et al. 2010. Evaluation of process parameters of ultrasonic treatment of bacterial suspensions in a pilot scale water disinfection system. *Ultrasonics Sonochemistry*, 17:1004–1009.

Hung, H.M. and Hoffman, M.R. 1999. Kinetics and mechanism of the sonolytic degradation of chlorinated hydrocarbons: Frequency effects. *The Journal of Physical Chemistry A*, 103:2734–2739.

IHSS website: http://www.ihss.gatech.edu/

Ince, N.H. and Belen, R. 2001. Aqueous phase disinfection with power ultrasound: Process kinetics and effect of solid catalysts. *Environmental Science & Technology*, 35:1885–1888.

Inoue, M., Masuda, Y., Okada, F., Sakurai, A., Takahashi, I., and Sakakibara, M. 2008. Degradation of bisphenol A using sonochemical reactions. *Water Research*, 42:1379–1386.

Isariebel, Q.P., Carine, J.L., Ulises-Javier, J.H., Anne-Marie, W., and Henri, D. 2009. Sonolysis of levodopa and paracetamol in aqueous solutions. *Ultrasonics Sonochemistry*, 16:610–616.

Jiang, Y., Pétrier, C., and Waite, T.D. 2002. Effect of pH on the ultrasonic degradation of ionic aromatic compounds in aqueous solution. *Ultrasonics Sonochemistry*, 9:163–168.

Johnson, R., Pankow, J., Bender, D., Price, C., and Zogorski, J. 2000. MTBE: To what extent will past releases contaminate community water supply wells? *Environmental Science & Technology News* May 1st, 2A–9A.

Jones, K.L. and O'Melia, C.R. 2000. Protein and humic acid adsorption onto hydrophilic membrane surfaces: Effects of pH and ionic strength. *Journal of Membrane Science*, 165:31–46.

Juang, R.S. and Lin, K.H. 2004. Flux recovery in the ultrafiltration of suspended solutions with ultrasound. *Journal of Membrane Science*, 243:115–124.

Jucker, C. and Clark, M.M. 1994. Adsorption of aquatic humic substances on hydrophobic ultrafiltration membranes. *Journal of Membrane Science*, 97:37–52.

Kang, J.W., Hung, H.M., Lin, A., and Hoffmann, M.R. 1999. The sonolytic destruction of methyl tertiary butyl ether (MTBE) by ultrasonic irradiation: Effects of ozone and frequency. *Environmental Science & Technology*, 33:3199–3205.

Key, B.D., Howell, R.D., and Criddle, C.S. 1998. Defluorination of organofluorine sulfur compounds by *Pseudomonas sp.* strain D2. *Environmental Science & Technology*, 32:2283–2287.

Kim, J., Cai, Z., and Benjamin, M.M. 2010. NOM fouling mechanisms in a hybrid adsorption/membrane system. *Journal of Membrane Science*, 349:35–43.

Kim, I., Hong, S., Hwang, I., Kwon, D., Kwon, J., and Huang, C.P. 2007. TOC and THMFP reduction by ultrasonic irradiation in wastewater effluent. *Desalination*, 202:9–15.

Kitis, M., Karanfil, T., Wigton, A., and Kilduff, J.E. 2002. Probing reactivity of dissolved organic matter for disinfection by-product formation using XAD-8 resin adsorption and ultrafiltration fractionation. *Water Research*, 36:3834–3848.

Knicker, H. and Lüdemann, H.D. 1995. N-15 and C-13 CPMAS and solution NMR studies of N-15 enriched plant material during 600 days of microbial degradation. *Organic Geochemistry*, 23:329–341.

Kobayashi, T., Chai, X., and Fujii, N. 1999. Ultrasound enhanced cross-flow membrane filtration. *Separation and Purification Technology*, 17:31–40.

Kobayashi, T., Kobayashi, T., Hosaka, Y., and Fujii, N. 2003. Ultrasound-enhanced membrane-cleaning processes applied water treatments: Influence of sonic frequency on filtration treatments. *Ultrasonics*, 41:185–190.

Kokugan, T. et al. 1995. Ultrasonic effect on ultrafiltration properties of ceramic membrane. *Membrane*, 20:213–223.

Kreller, D.I., Turner, B.F., Dejanovic, K.N., and Maurice, P.A. 2005. Comparison of the effects of sonolysis and γ-radiolysis on dissolved organic matter. *Environmental Science & Technology*, 39:9732–9737.

Kukkonen, J. 1992. Effects of lignin and chlorolignin in pulp mill effluents on the binding and bioavailability of hydrophobic organic pollutants. *Water Research*, 26:1523–1532.

Lamminen, M.O., Walker, H.W., and Weavers, L.K. 2004. Mechanisms and factors influencing the ultrasonic cleaning of particle-fouled ceramic membranes. *Journal of Membrane Science*, 237:213–223.

Lamminen, M.O., Walker, H.W., and Weavers, L.K. 2006. Cleaning of particle-fouled membranes during cross-flow filtration using an embedded ultrasonic transducer system. *Journal of Membrane Science*, 283:225–232.

Lara, G., Cuadrado-Gamarra, J.I., de Pedro-Cuesta, J., Esteban, E.M., Giménez-Roldán, S., and Luis-González, S. 2006. Epidemiological assessment of levodopa use in Cuba: 1993–1998. *Pharmacoepidemiology and Drug Safety*, 15:521–526.

Lee, M. and Oh, J. 2010. Sonolysis of trichloroethylene and carbon tetrachloride in aqueous solution. *Ultrasonics Sonochemistry*, 17:207–212.

Leenheer, J.A., Malcolm, R.L., McKinley, P.W., and Eccles, L.A. 1974. Occurrence of dissolved organic carbon in selected ground-water samples in the United States. *Journal of Research of the US Geological Survey*, 2:361–369.

Leighton, T. 1994. *The Acoustic Bubble*. London, U.K.: Academic Press.

Leighton, T.G., Pickworth, M.J.W., Walton, A.J., and Dendy, P.P. 1989. The pulse enhancement of unstable cavitation by mechanisms of bubble migration. *Proceedings of the Institute of Acoustics*, 11:461–469.

Li, S., Heijman, S.G.J., Verberk, J.Q.J.C., Verliefde, A.R.D., Kemperman, A.J.B., van Dijk, J.C., and Amy, G. 2009. Impact of backwash water composition on ultrafiltration fouling control. *Journal of Membrane Science*, 344:17–25.

Li, J., Sanderson, R.D., and Jacobs, E.P. 2002. Ultrasonic cleaning of nylon microfiltration membranes fouled by Kraft paper mill effluent. *Journal of Membrane Science*, 205:247–257.

Liang, J., Komarov, S., Hayashi, N., and Kasai, E. 2007a. Recent trends in the decomposition of chlorinated aromatic hydrocarbons by ultrasound irradiation and Fenton's reagent. *Journal of Material Cycles and Waste Management*, 9:47–55.

Liang, J., Komarov, S., Hayashi, N., and Kasai, E. 2007b. Improvement in sonochemical degradation of 4-chlorophenol by combined use of Fenton-like reagents. *Ultrasonics Sonochemistry*, 14:201–207.

Liu, Y.N., Jin, D., Lu, X.P., and Han, P.F. 2008a. Study on degradation of dimethoate solution in ultrasonic airlift loop reactor. *Ultrasonics Sonochemistry*, 15:755–760.

Liu, S., Lim, M., Fabris, R., Chow, C., Drikas, M., and Amal, R. 2008b. TiO_2 photocatalysis of natural organic matter in surface water: Impact on trihalomethane and haloacetic acid formation potential. *Environmental Science & Technology*, 42:6218–6223.

Lu, Y. and Weavers, L.K. 2002. Sonochemical desorption and destruction of 4-chlorobiphenyl from synthetic sediments. *Environmental Science & Technology*, 36:232–237.

Luther, S., Mettin, R., Koch, P., and Lauterborn, W. 2001. Observation of acoustic cavitation bubbles at 2250 frames per second. *Ultrasonics Sonochemistry*, 8:159–162.

Mason, T.J., Joyce, E., Phull, S.S., and Lorimer, J.P. 2003. Potential uses of ultrasound in the biological decontamination of water. *Ultrasonics Sonochemistry*, 10:319–323.

Mason, T.J. and Lorimer, J.P. 1988. *Sonochemistry: Theory, Applications and Uses of Ultrasound in Chemistry*. Chichester, U.K.: Ellis Horwood.

Masselin, I., Chasseray, X., Durand-Bourlier, L., Laine, J.M., Syzaret, P.Y., and Lemordant, D. 2001. Effect of sonication on polymeric membranes. *Journal of Membrane Science*, 181:213–220.

Matsumoto, Y., Miwa, T., Nakao, S., and Kimura, S. 1996. Improvement of membrane permeation performance by ultrasonic microfiltration. *Journal of Chemical Engineering of Japan*, 29:561–567.

Mayer, L.M. 1994. Relationships between mineral surfaces and organic carbon concentrations in soils and sediments. *Chemical Geology*, 114:347–363.

Meier, M., Dejanovic, K.N., Maurice, P.A., Chin, Y.P., and Aiken, G.R. 1999. Fractionation of aquatic natural organic matter upon sorption to goethite and kaolinite. *Chemical Geology*, 157:275–284.

Melville, H.W. and Murray, A.J.R. 1950. The ultrasonic degradation of polymers. *Transactions of the Faraday Society*, 46:996–1009.

Moriwaki, H., Takagi, Y., Tanaka, M., Tsuruho, K., Okitsu, K., and Maeda, Y. 2005. Sonochemical decomposition of perfluorooctane sulfonate and perfluorooctanoic acid. *Environmental Science & Technology*, 39:3388–3392.

Muthukumaran, S., Kentish, S., Lalchandani, S., Ashokkumar, M., Mawson, R., Stevens, G.W., and Grieser, F. 2005. The optimization of ultrasonic cleaning procedures for dairy fouled ultrafiltration membranes. *Ultrasonics Sonochemistry*, 12:29–35.

Muthukumaran, S., Yang, K., Seuren, A., Kentish, S., Ashokkumar, M., Stevens, G.W., and Grieser, F. 2004. The use of ultrasonic cleaning for ultrafiltration membranes in the dairy industry. *Separation and Purification Technology*, 39:99–107.

Naffrechoux, E., Combet, E., Fanget, B., and Petrier, C. 2003. Reduction of chloroform formation potential of humic acid by sonolysis and ultraviolet irradiation. *Water Research*, 37:1948–1952.

Nagata, Y., Hirai, K., Bandow, H., and Maeda, Y. 1996. Decomposition of hydroxybenzoic and humic acids in water by ultrasonic irradiation. *Environmental Science & Technology*, 30:1133–1138.

Nakanishi, M., Mukai, S., Kimata, I., Iseki, M., and Maeda, Y. 2001. Inactivation of *Cryptosporidium parvum* oocysts in drinking water by high-intensity ultrasonic waves. *Proceedings of the American Water Works Association*, 2001 Annual Conference, Washington, DC, June 17–21.

Neis, U. and Blume, T. 2003. Ultrasonic disinfection of wastewater effluents for high-quality reuse. *Water Science and Technology: Water Supply*, 3:261–267.

Neppolian, B., Doronila, A., Grieser, F., and Ashokkumar, M. 2009. Simple and efficient sonochemical method for the oxidation of arsenic (III) to arsenic (V). *Environmental Science & Technology*, 43:6793–6798.

Neppolian, B., Jung, H., Choi, H., Lee, J.H., and Kang, J.W. 2002. Sonolytic degradation of methyl *tert*-butyl ether: The role of coupled Fenton process and persulphate ion. *Water Research*, 36:4699–4708.

Nystrom, M., Ruohomaki, K., and Kaipia, L. 1996. Humic acid as a fouling agent in filtration. *Desalination*, 106:79–87.

Olson, T.M. and Barbier, P.F. 1994. Oxidation kinetics of natural organic matter by sonolysis and ozone. *Water Research*, 28:1383–1391.

Oyane, I., Furuta, M., Stavarache, C.E., Hashiba, K., Mukai, S., Nakanishi, M., Kimata, I., and Maeda, Y. 2005. Inactivation of Cryptosporidium parvum by ultrasonic irradiation. *Environmental Science & Technology*, 39:7294–7298.

Pace, N.G., Cowley, A., and Campbell, A.M. 1997. Short pulse acoustic excitation of microbubbles. *The Journal of the Acoustical Society of America*, 102(3):1474–1479.

Pecha, R. and Gompf, B. 2000. Microimplosions: Cavitation collapse and shock wave emission on a nanosecond time scale. *Physical Review Letters*, 84:1328–1330.

Perdue, E.M. and Ritchie, J.D. 2005. Dissolved organic matter in fresh waters. In *Surface and Ground Water, Weathering, Erosion and Soils*, J.I. Drever (Ed.), vol. 5. Oxford, U.K.: Elsevier, pp. 273–318.

Pham, T.D., Shrestha, R.A., Virkutyte, J., and Sillanpaa, M. 2009. Recent studies in environmental applications of ultrasound. *Canadian Journal of Civil Engineering*, 36:1849–1858.

Philipp, A. and Lauterborn, W. 1998. Cavitation erosion by single laser-produced bubbles. *Journal of Fluid Mechanics*, 361:75–116.

Price, G.J. 1990. The use of ultrasound for the controlled degradation of polymer solutions. *Advances in Sonochemistry*, 1:231–287.

Price, G.J., Lenz, E.J., and Ansell, C.W.G. 2002. The effect of high intensity ultrasound on the ring opening polymerisation of cyclic lactones. *European Polymer Journal*, 38:1753–1760.

Qi, B.C., Aldrich, C., and Lorenzen, L. 2004. Effect of ultrasonication on the humic acids extracted from lignocellulose substrate decomposed by anaerobic digestion. *Chemical Engineering Journal*, 98:153–163.

Riesz, P., Berdahl, D., and Christman, C.L. 1985. Free radical generation by ultrasound in aqueous and non-aqueous solutions. *Environmental Health Perspectives*, 64:233–252.

Roy, R.A. 1999. Cavitation sonophysics. In *Sonochemistry and Sonoluminescence*, Crum, L.A., Mason, T.J., Reisse, J.L., and Suslick, K.S. (Ed.). Dordrecht, the Netherlands: Kluwer Academic Publishers, pp. 25–38.

Schäfer, A.I., Schwicker, U., Fischer, M.M., Fane, A.G., and Waite, T.D. 2000. Microfiltration of colloids and natural organic matter. *Journal of Membrane Science*, 171:151–172.

Scherba, G., Weigel, R.M., and O'Brien Jr., W.D. 1991. Quantitative assessment of the germicidal efficacy of ultrasonic energy. *Applied and Environmental Microbiology*, 57:2079–2084.

Schnitzer, M. and Khan, S.U. 1972. *Humic Substances in the Environment*. New York: Marcel Dekker, p. 327.

Schröder, H.F. and Meesters, R.J. 2005. Stability of fluorinated surfactants in advanced oxidation processes—A follow up of degradation products using flow injection-mass spectrometry, liquid chromatography-mass spectrometry and liquid chromatography-multiple stage mass spectrometry. *Journal of Chromatography A*, 1082(1):110–119.

Selli, E., Bianchi, C.L., Pirola, C., and Bertelli, M. 2005. Degradation of methyl tert-butyl ether in water: Effects of the combined use of sonolysis and photocatalysis. *Ultrasonics Sonochemistry*, 12:395–400.

Seymour, J.D. and Gupta, R.B. 1997. Oxidation of aqueous pollutants using ultrasound: Salt-induced enhancement. *Industrial & Engineering Chemistry Research*, 36:3453–3457.

Song, W., de la Cruz, A.A., Rein, K., and O'Shea, K.E. 2006. Ultrasonically induced degradation of microcystin-LR and -RR: Identification of products, effect of pH, formation and destruction of peroxides. *Environmental Science & Technology*, 40:3941–3946.

Stevenson, F.J. 1994. *Humus Chemistry: Genesis, Composition, Reactions*, 2nd edn. New York: John Wiley & Sons.

Suri, R.P.S., Nayak, M., Devaiah, U., and Helmig, E. 2007. Ultrasound assisted destruction of estrogen hormones in aqueous solution: Effect of power density, power intensity and reactor configuration. *Journal of Hazardous Materials*, 146:472–478.

Suslick, K.S. 1989. The chemical effects of ultrasound. *Scientific American*, 260:80–86.

Suslick, K.S. 1990. Sonochemistry. *Science*, 247:1439–1445.

Taghizadeh, M.T. and Asadpour, T. 2009. Effect of molecular weight on the ultrasonic degradation of poly(vinylpyrrolidone). *Ultrasonics Sonochemistry*, 16:280–286.

Tarleton, E.S. and Wakeman, R.J. 1992. Electro-acoustic crossflow microfiltration. *Filtration and Separation*, 29:425–432.

Taylor Jr., E., Cook, B.B., and Tarr, M.A. 1999. Dissolved organic matter inhibition of sonochemical degradation of aqueous polycyclic aromatic hydrocarbons. *Ultrasonics Sonochemistry*, 6:175–183.

Thacker, J. 1973. An approach to the mechanism of killing of cells in suspension by ultrasound. *Biochimica et Biophysica Acta*, 304:240–248.

Thomas, J.R. and deVries, L. 1959. Sonically induced heterolytic cleavage of polymethylsiloxane. *The Journal of Physical Chemistry*, 63:253–256.

Thompson, L.H. and Doraiswamy, L.K. 1999. Sonochemistry: Science and engineering. *Industrial & Engineering Chemistry Research*, 38:1215–1249.

Thurman, E.M. 1985. Humic substances in groundwater. In *Humic Substances in Soil, Sediment, and Water: Geochemistry, Isolation, and Characterization*, Aiken, G.R., McKnight, D.M., Wershaw, R.L., and MacCarthy, P. (Ed.). New York: John Wiley & Sons, pp. 87–103.

Torres, R.A., Abdelmalek, F., Combet, E., Petrier, C., and Pulgarin, C. 2007. A comparative study of ultrasonic cavitation and Fenton's reagent for bisphenol A degradation in deionised and natural waters. *Journal of Hazardous Materials*, 146:546–551.

Torres, R.A., Petrier, C., Combet, E., Carrier, M., and Pulgarin, C. 2008. Ultrasonic cavitation applied to the treatment of bisphenol A. Effect of sonochemical parameters and analysis of BPA by-products. *Ultrasonics Sonochemistry*, 15:605–611.

van Iersel, M.M., van den Manacker, J.P.A.J., Benes, N.E., and Keurentjes, J.T.F. 2007. Pressure-induced reduction of shielding for improving sonochemical activity. *The Journal of Physical Chemistry B*, 111:3081–3084.

Vecitis, C.D., Wang, Y., Cheng, J., Park, H., Mader, B.T., and Hoffmann, M.R. 2010. Sonochemical degradation of perfluorooctane sulfonate in aqueous film-forming foams. *Environmental Science & Technology*, 44:432–438.

Vogel, A., Lauterborn, W., and Timm, R. 1989. Optical and acoustic investigations of the dynamics of laser-produced cavitation bubbles near a solid boundary. *Journal of Fluid Mechanics*, 206:299–338.

Walling, C. 1975. Fenton's reagent revisited. *Accounts of Chemical Research*, 8:125–131.

Walling, C. and Kato, S. 1971. The oxidation of alcohols by Fenton's reagent: The effect of copper ion. *Journal of the American Chemical Society*, 93:4275–4281.

Weavers, L.K. 2001. Sonolytic ozonation for the remediation of hazardous pollutants. In *Advances in Sonochemistry*, Mason, T.J. (Ed.), vol. 6. Stamford, CT: JAI Press Ltd., pp. 111–139.

Westerhoff, P., Aiken, G., Amy, G., and Debroux, J. 1999. Relationships between the structure of natural organic matter and its reactivity towards molecular ozone and hydroxyl radicals. *Water Research*, 33(10):2265–2276.

Wiesner, M.R. and Aptel, P. 1996. Mass transport and permeate flux and fouling in pressure-driven processes. In *Water Treatment Membrane Processes*, Mallevialle, J., Odendaal, P.E., and Wiesner, M.R. (Ed.). New York: McGraw-Hill, pp. 4.1–4.30.

Yang, L., Rathman, J.F., and Weavers, L.K. 2006. Sonochemical degradation of alkylbenzene sulfonate surfactants in aqueous mixtures. *The Journal of Physical Chemistry B*, 110(37):18385–18391.

Yao, J.J., Gao, N.Y., Li, C., Li, L., and Xu, B. 2010. Mechanism and kinetics of parathion degradation under ultrasonic irradiation. *Journal of Hazardous Materials*, 75:138–145.

Yuan, W. and Zydney, A.L. 1999. Effects of solution environment on humic acid fouling during microfiltration. *Desalination*, 122:63–76.

Zhang, G. and Hua, I. 2000. Ultrasonic degradation of trichloroacetonitrile, chloropicrin and bromobenzene: Design factors and matrix effects. *Advances in Environmental Research*, 4:219–224.

Zhang, M., Li, C., Benjamin, M.M., and Chang, Y. 2003. Fouling and natural organic matter removal in adsorbent/membrane systems for drinking water treatment. *Environmental Science & Technology*, 37(8):1663–1669.

Zhu, X. and Elimelech, M. 1997. Colloidal fouling of reverse osmosis membranes: Measurements and fouling mechanism. *Environmental Science & Technology*, 31:3654–3662.

16 Ultrasound and Sonochemistry in the Treatment of Contaminated Soils by Persistent Organic Pollutants

Reena Amatya Shrestha, Ackmez Mudhoo,
Thuy-Duong Pham, and Mika Sillanpää

CONTENTS

Introduction 407
Ultrasound and Sonochemistry 409
Effects of Ultrasound in the Treatment of Organic-Contaminated Soils 411
    Effect of Ultrasound on Desorption of Organic Contaminants 411
    Effect of Ultrasound in Destroying Organic Contaminants 412
Ultrasonication as Assistant Process in Organic-Contaminated Soil Remediation 412
    Ultrasonically Enhanced Soil-Flushing 412
    Ultrasonically Assisted Advanced Oxidative Soil Remediation 413
    Ultrasonically Enhanced Electrokinetic Remediation 414
    Ultrasonically Enhanced Activated Carbon Amendment 415
    Ultrasonically Enhanced the Surfactant-Aided Soil-Washing 415
Conclusions 416
References 416

INTRODUCTION

Soil contamination concerns a serious environmental problem all over the world because of its significance as a threat to health through the food system and groundwater. Contaminants could originate from the careless human activities and the accidental or deliberate spills or discharge from industrial, agricultural, urban, and maritime sources. Contaminants accumulate in soil and sediment receptors. Since soil is the medium that produces most of the food required for most living creatures, soil and sediment contamination is a major environmental issue because of its potential toxic effects on biological resources and eventually on human health. As a key component of environmental chemical cycles, soil contamination often contributes to water and air pollution. Any hazardous substance present in a soil matrix represents a threat to public health and ground water (Pamukcu and Huang, 2001). There are major types of pollutants found in soil and sediments:

Nutrients, including phosphorous and nitrogen compounds such as ammonia.

Bulk organics, a class of hydrocarbons that includes oil and grease.

Halogenated hydrocarbons, compounds of "dirty dozen" are in this category.

Polycyclic aromatic hydrocarbons (PAHs), a group of organic chemicals that includes several petroleum products and byproducts.

Metals, including, for example, iron, manganese, lead, cadmium, zinc, and mercury, and metalloids such as arsenic and selenium.

Persistent organic pollutants (POPs), a group of chemicals that are more resistant to decomposition are taken into account here as the most important pollutants. This group covers large terrific groups such as dirty dozen and PAHs. In 1995, the Governing Council of the United Nations Environment Programme (UNEP) decided to take global action on POPs, as those compounds can persist in the environment, bio-accumulate through the food web, and pose a risk of causing adverse effects to human health and the environment. Twelve POPs, such as DDT, hexachlorobenzene, aldrin, and various PCBs, were placed in the dirty dozen. Most of them are highly hydrophobic in nature. Therefore, decontamination of soils through the removal of POPs becomes more important. Site conditions, contaminant types, contaminant sources, and the potential impacts of the possible remedial measure determine the choice of a remediation strategy and technology. No single technology is appropriate for all contaminant types and various site-specific conditions (Khan et al., 2004). For organic contamination particularly, a variety of site remediation technologies are available. These methods are generally separated into two main groups: biological treatment and physicochemical treatment.

Biological technologies include the following:

Biodegradation: Using microorganisms to break down organic contaminants (notably light hydrocarbons). Through the digestion process, bacteria transform the contaminants into water and carbon dioxide.

Bioventing: Combination of soil venting and biological treatment. Air circulating through the soil stimulates biodegradation. The resulting molecules produced by microorganisms are extracted along with vapors from the soil.

Phytoremediation: Using living plants to take up contaminants into their leaves or roots. The plants are then pulled up and incinerated.

Physicochemical technologies include the following:

Soil vapor extraction: Extraction wells pull volatile contaminants out of the ground. The extracted vapors are condensed to liquids, adsorbed onto active charcoal, or incinerated.

In situ washing: Water and surfactants injected into the soil dissolve organic contaminants, which can then be pumped to the surface and separated in a settling tank.

Thermal desorption: Soil contaminated by organic products (even chlorinated) is roasted at a temperature of less than 500°C. This vaporizes the contaminants without destroying the soil.

Incineration: Soil is excavated and then heated to very high temperatures to destroy all organic molecules.

In other terms, these technologies can be categorized as *ex situ* and *in situ* treatments. The main advantage of *ex situ* treatments such as thermal desorption and incineration is that they generally require shorter time periods in comparison to *in situ* ones, and there is more certainty about the uniformity of treatment because of the ability to screen, to homogenize, and to continuously mix the soils. However, since *ex situ* treatments involve soil excavation, they can be costly. In addition, those remedial options may cause habitat alteration, requiring large-scale material handling and long-term management. Therefore, *in situ* stabilization methods that do not involve soil relocation or capping are more attractive. *In situ* treatments allow soil to be treated without being excavated and transported, resulting in a significant cost saving. However, the conventional *in situ* treatments are usually very site specific. Bioremediation is limited by a number of technical difficulties such as acclimation of microorganisms while few contaminants can effectively be removed by soil washing. Moreover, *in situ* technologies often work best on homogenous, permeable soils but are difficult to apply on low-permeable soils. Accounting for all of these obstacles, there is a necessity to develop new alternatives for *in situ* soil clean-up (Pamukcu and Huang, 2001).

Electrokinetic (EK) treatment has emerged as a potential technique for *in situ* soil decontamination (Pamukcu and Huang, 2001). Electrokinetic remediation technique is based on the application of low-level direct current, which is used to solubilize and mobilize contaminants via electro-migration, electro-osmotic, and electrophoresis phenomena. Electrokinetic treatment is especially unique because of the ability to work in low-permeable soils as well as high permeability soils and the applicability to a broad range of organic and inorganic contaminants. However, electrokinetic remediation is mostly used for metal removal as it is more efficient with charged and soluble contaminants than non-charged and low-soluble organic pollutants (Virkutyte et al., 2002).

Ultrasonic irradiation can overcome the above problem in organic decontamination of soil. Its application into contaminated soils can increase desorption, mobilization of contaminants, and porosity and permeability of soil through the development of cavitation (Chung and Kamon, 2005). Moreover, ultrasonic waves can promote the formation of free strong oxidative radicals that involve the oxidation of contaminants (Flores et al., 2007; Mason, 2007a), and the high local temperature and pressure forming during ultrasonic cavitation can destroy the contaminants through pyrolysis processes (Adewuyi, 2001). The use of ultrasound offers several advantages such as lack of dangerous breakdown products; low energy demand and technology can be made quite compact, transportable, allowing on-site treatment (Collings et al., 2006).

ULTRASOUND AND SONOCHEMISTRY

Ultrasound refers to inaudible sound waves with frequencies in the range of 16 kHz to 500 MHz, greater than the upper limit of human hearing. It can be transmitted through any elastic medium including water, gas-saturated water, and slurry. The use of ultrasound has been recognized for many years in various fields (Mason, 2007b) (Figure 16.1).

In terms of frequency, ultrasound can be categorized into two main strands: (1) high frequency (2–10 MHz)—low power diagnostic ultrasound, involving medical imaging, nondestructive testing, and (2) low-to-medium frequency (20–1000 kHz) frequency—high-power ultrasound, involving other applications in industry, nanotechnology, ultrasonic therapy, and sonochemistry.

Sonochemistry is the chemistry that deals with sonic waves on chemical reactions. Effect of sonochemistry on the chemical system generates from acoustic cavitations. Cavitation is the formation, growth, and the implosion of bubbles in a liquid by ultrasound. Like any sound wave, ultrasound is propagated via a series of compression and rarefaction waves induced in the molecules of the medium through which it passes. Compression cycles push molecules together, while expansion cycles pull them apart. At sufficiently high power the rarefaction cycle may exceed the attractive forces of the molecules of the liquid and cavitation bubbles will form. Cavitation bubble collapse is a remarkable phenomenon induced throughout the liquid. Cavitational collapse produces temperature as high

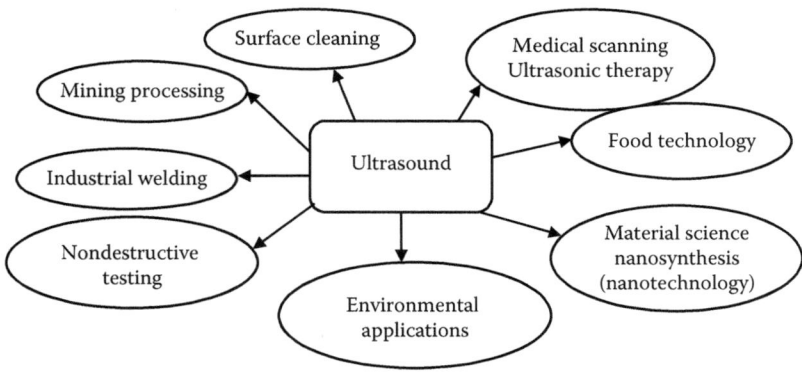

FIGURE 16.1 Diverse applications of ultrasound.

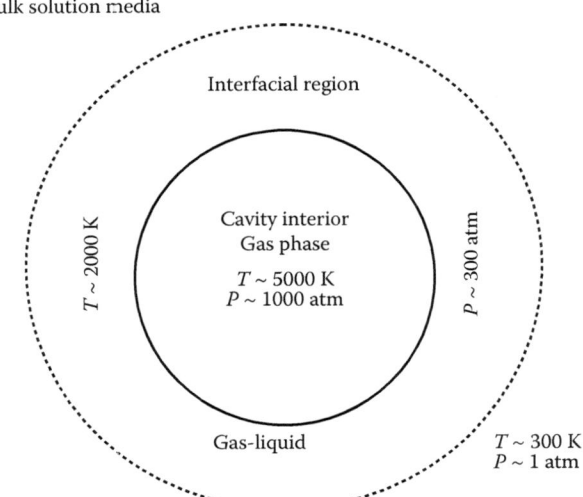

FIGURE 16.2 Three reaction zones in the cavitation process.

as those on surface of the sun (~5000 K) and pressure as great as those at the bottom of the ocean (~1000 atm) with very short lifetimes, implying the existence of extremely high heating and cooling rates (>10^9 K s^{-1}). It has been shown that transient supercritical water is obtained during the collapse of cavitation bubbles generated sonolytically (Hoffmann et al., 1996). Acoustic cavitation provides a unique interaction of energy and matter, and ultrasonic irradiation of liquids causes high energy chemical reactions to occur (Suslick and Crum, 1997). Acoustic cavitation is mechanical effect through a process whereas sonochemistry is chemical reaction that is initiated by high-intensity ultrasound.

According to Adewuyi (2001), so far four theories have been proposed to explain the sonochemical events: "hot-spot" theory, "electrical" theory, "plasma discharge" theory, and "supercritical" theory. These have led to several modes of reactivity being proposed: pyrolytic decomposition, hydroxyl radical oxidation, plasma chemistry, and supercritical water oxidation. Generally, most studies in environmental sonochemistry have adopted the "hot-spot" concepts to explain experimental results rather than the other theories. In the hot-spot model (Adewuyi, 2001), three regions are postulated (Figure 16.2): (1) a hot gaseous nucleus, (2) an interfacial region, and (3) bulk solution at ambient temperature. Reactions involving free radicals can occur within the collapsing bubble, at the interface of the bubble, and in the surrounding liquid. Within the center of the bubble, high temperatures and pressures generated during cavitation provide the activation energy required for bond breakage and dissociation of solvents and other vapors or gases, leading to the formation of free radicals or excited species. The radicals generated either react with each other to form new molecules and radicals or diffuse into the bulk liquid to serve as oxidants.

The second reaction site is the liquid shell immediately surrounding the imploding cavity, which has been estimated to heat up to approximately 2000 K during cavity implosion. In this solvent layer surrounding the hot bubble, both combustion and free-radical reactions (involving •OH derived from the decomposition of H_2O) occur. Reactions here are comparable to pyrolysis reactions. Pyrolysis in the interfacial region is predominant at high solute concentrations, while at low solute concentrations, free-radical reactions are likely to predominate. It has been shown that the majority of degradation takes place in the bubble–bulk interface region.

In the bulk liquid, no primary sonochemical activity takes place although subsequent reactions with ultrasonically generated intermediates may occur. A small number of free radicals produced in the cavities or at the interface may move into the bulk–liquid phase and react with the substrate present there in secondary reactions to form new products. Depending on their physical properties and concentrations, molecules present in the medium will be burned in close to the bubble (pyrolysis) or will undergo radical reactions.

There are basically three types of sonochemical processes: homogeneous sonochemistry of liquid, heterogeneous sonochemistry of solid–liquid system, and some type that represents a mix of both homogeneous and heterogeneous sonochemistry. Example of homogeneous sonochemistry is isomerization of 1,2-dichloroethene induced by the reversible addition of a bromine radical by ultrasonic irradiation (Caulier et al., 1995). It is totally different when there is cavitation in solid–liquid system. Two mechanisms, micro jet impact and shockwave damage, take place. When cavitation bubbles are created at the boundary of the liquid and solid, strong deformation can occur. This activity helps to create shockwaves and forms cavity collapse in the liquid. Both microjet and shockwaves are responsible for localized erosion. The examples of this type sonochemistry are in ultrasonic cleaning, sonocation effects on heterogeneous reactions.

These chemical effects (sonochemistry) explained above are utilized in most of the ultrasonic applications in environmental remediation, especially in organic decontamination.

EFFECTS OF ULTRASOUND IN THE TREATMENT OF ORGANIC-CONTAMINATED SOILS

Although ultrasonic applications in environmental areas are still in lab-scale and developing stage, they are growing rapidly, attracting more interest, because of the many advantages they offer: environmental friendly (no toxic chemicals are used or produced), low energy demands, and compact and transportable method that can be used on-site. Environmental remediation by ultrasonication involves mostly with organic pollutant destruction, through thermal decomposition (pyrolysis) and the formation of oxidative species like hydroxyl radical that enhance the mineralization of pollutants. Moreover, in soil treatment, ultrasonic waves increase the porosity of the soil and percolation rate thus accelerating the desorption and facilitating the removal of entrapped contaminants. On the other hand, ultrasound applied in environmental analysis also provides benefits such as shorter time, simplified procedure, and higher purity of the final product.

Effect of Ultrasound on Desorption of Organic Contaminants

Similar to the application of ultrasonic leaching for metal removal, ultrasound has been known for promoting organics desorption from soils and sediments (Kim and Wang, 2003; Pee, 2008). According to Feng and Aldrich (2000), the likely mechanism of ultrasonic desorption can be explained by considering the different effects of ultrasound on heterogeneous media. First, the high temperatures in localized hot spots enhance the breaking of physical bonds between the adsorbate (contaminants) and the adsorbent surface. Second, acoustic cavitation produces high-speed microjets and high-pressure shockwaves that impinge on the surface and erode the adsorbate (Suslick et al., 1987; Stephanis et al., 1997). Finally, ultrasound produces acoustic vortex microstreaming within the pores of the solid particles, as well as the solid–liquid interface. This phenomenon arises by the increase in momentum brought about as the liquid absorbs energy from the propagating sound waves, even in the absence of cavitation (Ley and Low, 1989). These effects may possibly be the cause of enhanced desorption rates (Feng and Aldrich, 2000).

In their study, Feng and Aldrich (2000) investigated the influence of factors such as slurry concentration, ultrasonic power intensity, duration of irradiation, particle size, diesel content, slurry pH, salinity, surfactant dosage during the remediation of stimulated soil contaminated with diesel fuel in the presence of ultrasound. Ultrasonic treatment performed more effective than high-speed mechanical agitation at the same energy input. However, prolonged ultrasonic irradiation could not increase the efficiency of diesel removal, possibly due to equilibrium between the desorption and re-adsorption of these hydrocarbons from and onto the particle surfaces. A multistage sonochemical treatment process for the remediation of sand contaminated with diesel was proposed as better results were obtained from this approach than from the single-stage treatment process.

EFFECT OF ULTRASOUND IN DESTROYING ORGANIC CONTAMINANTS

Ultrasonication not only assists the desorption of the contaminants from the soil, but also promotes the formation of the strong oxidant, ·OH radical (Flores et al., 2007). Ultrasonic energy can destroy the contaminants through oxidation by free radicals and pyrolysis processes, not only transport the contaminants from one place to other place as in conventional soil washing.

Collings et al. (2006) have developed high-power ultrasound to destroy persistent organic pollutants (POPs) in soils and sediments. They have worked successfully on major contaminants, atrazine, simazine, total petroleum hydrocarbons, DDT, lindane, endosulfan, 2,4,5-T, tetrachloronaphthalene, and TBT. The range of contaminants they have studied is sufficiently broad to suggest that high-power ultrasound will be effective for most adsorbed large molecules. The results indicate several advantages of high-power ultrasonic technology compared with conventional methods. These include high destruction rates, the lack of dangerous breakdown products and low energy demands leading to low cost. Moreover, the technology can be made quite compact and transportable, allowing on-site treatment.

The feasibility of ultrasonication on treatment of different kinds of highly contaminated soils (synthetic clay, natural farm clay and kaolin) (Shrestha et al., 2009) was investigated by using two target persistent organic pollutants (POPs); hexachlorobenzene (HCB) and phenanthrene (PHE). Experimental results showed that ultrasonication has a potential to reduce the high concentrations of these POPs. The treatment of soil by ultrasonication requires some amount of water for sonochemistry effects to perform. The reasonable moisture ratio of the slurry could be from 2:1 to 3:1 water and soil, the higher the better, particularly kaolin needed more amount of water than other clays to perform well. The removal efficiency increased but not very much after long ultrasonication time. Ultrasonication did not affect the pH values of slurries. The heating and irritated noise problems of ultrasonication should be considered carefully in larger scale applications. The removal rates of POPs in soils vary with soil type, power, and frequency of the ultrasound applied.

ULTRASONICATION AS ASSISTANT PROCESS IN ORGANIC-CONTAMINATED SOIL REMEDIATION

In most of the cases, ultrasound is used as a supplemental method to enhance the soil remediation process.

ULTRASONICALLY ENHANCED SOIL-FLUSHING

Soil contamination caused by underground petroleum and other organic pollutant leaks poses a strong environmental threat. A considerable amount of these pollutants can be held in voids in the soil in the form of residual saturation and can lead to long-term contamination of ground water through the action of rain water, if not removed in time (Feng et al., 2001). Soil washing is a promising *ex situ* method for the treatment of soil contaminated with oil and other organic pollutants. The soil is typically in intimate contact with washwater in a mechanical scrubber, which promotes transport of contaminants from the soil phase to the liquid phase. These contaminants can be floated and skimmed off in a subsequent flotation process (Feng et al., 2001). The principal value of a standard washing process with water is the production of a clean stream of sand or soil from which the contaminants have been removed.

An ultrasonically enhanced soil-flushing method for *in situ* remediation of the ground contaminated by non-aqueous phase liquid (NAPL) hydrocarbons was investigated by Kim and Wang (2003). Crisco vegetable oil was chosen as the model compound. The soil-flushing tests were conducted in two conditions—without ultrasound and with 20 kHz ultrasonic waves. Experimental results indicated that ultrasonication can enhance oil removal considerably. The degree of enhancement depends on factors such as ultrasonic power, water washing flow rate and soil type. Increasing ultrasonic power will increase pollutant extraction only up to the level where cavitation occurs. The effectiveness of ultrasonication decreases with flushing rate, but eventually becomes constant under higher flow rates (Kim and Wang, 2003).

Mason et al. (2004) had reported some laboratory research on ultrasonic soil washing of organic contaminants like pesticide DDT, PCB, and PAH. Initial concentrations of DDT (250 ppm), PCB (250 ppm), and PAH (400 ppm) in sand (200 g contaminated fine sand in 200 g water) were removed ultrasonically (20 kHz, 170 W) by 70% after 10 25, and 3 min, respectively. The potential for the scale-up of this soil washing using acoustic energy was also reported there. Two basic mechanisms for acoustically enhanced soil washing that have been suggested are abrasion of surface cleaning and leaching out of more deeply entrenched material. According to Mason and coworkers, factors that contribute toward improvement in efficiency by the influence of ultrasound include the following: (i) the high-speed microjets formed during asymmetric cavitation bubble collapse in the vicinity of the solid surface enhance transport rates and also increased surface area through surface pitting; (ii) particles fragmentation through collisions increase surface area; (iii) diffusion is enhanced by the ultrasonic capillary effect.

Shrestha et al. (2009) showed the feasibility of ultrasound on treatment of contaminated soils (synthetic clay, natural farm clay, and kaolin) by using two target persistent organic pollutants (POPs): hexachlorobenzene (HCB) and phenanthrene (PHE). The soils were highly contaminated in 500 mg kg^{-1}. The reasonable moisture ratio of the slurry could be in range of 2:1–3:1. The great advantage of this process was no change in pH values of soils. Kim et al. (2007) investigated the effect of ultrasound on diesel removal from soils by conducting lab-scale soil-flushing experiments for various conditions involving ultrasonic power, particle size, and diesel concentration, using specially designed and fabricated equipment. Their test results indicated that the rate of contaminant extraction had increased significantly with increasing ultrasonic power and that the degree of enhancement varied with test conditions. Additionally, Kim et al. (2007) performed the physical imaging of the specimens during tests and the images showed the disintegration of soil grains and oil drops resulting in a variation in contaminant removal efficiency.

ULTRASONICALLY ASSISTED ADVANCED OXIDATIVE SOIL REMEDIATION

The contamination of soil by means of bio-recalcitrant organic compounds, is becoming a matter of concern for scientific community and public opinion. The performance of traditional processes for the treatment of contaminated sites such as Air Sparging, Pump-and-Treat, Bioventing, and Soil Vapor Extraction, are limited by resistance to mass transport, which makes them effective only during the first phase of the treatment, and gradually less effective when the remediation goals are approached (tailing phenomena). Besides, rebound phenomena may occur after the site closure, requiring further remediation efforts. Hence, it becomes necessary to develop alternative processes and to assess their performance on those contamination cases of national concern. Among these, advanced oxidation processes (AOPs) could represent a potential solution to be applied for remediating contamination by bio-recalcitrant organic compounds. Their operative principle is based upon the idea of generating a pool of highly-oxidative species (radicals and non-). The AOPs differ only by the way in which this pool is generated. Once formed, these species are capable of effectively reacting with most of the common pollutants such as hydrocarbons, chlorinated solvents, polycyclic aromatic hydrocarbons, and polychlorobiphenyls until their complete oxidation to carbon dioxide and water, or at worst their transformation to more bio-degradable products. Moreover, some AOPs are able to effectively tackle sorbed compounds, since oxidative radicals can desorb these compounds from the soil surface, thus allowing their oxidation in aqueous phase. In view of these considerations, AOPs and especially sono-oxidative processes for environmentally remediation are acquiring more attention and their development, optimization, and application is currently growing up (Belgiorno et al., 2007).

Recently, Virkutyte et al. (2010) have tested the sono-Fenton-like process to degrade naphthalene in spiked soil utilizing mineral iron as a catalyst to generate radical species as well as to evaluate the efficiency of the proposed method and to optimize the treatment conditions. The sono-oxidation was performed with naphthalene contaminated soil (200, 400, and 800 mg kg^{-1} dry weight) to mimic industrially contaminated soil conditions in the presence of naturally occurring mineral iron, various ultrasound irradiations (100, 200, and 400 W), and hydrogen peroxide concentrations (100, 200, and 600 mg L^{-1}). Control experiments showed that, in the absence of hydrogen peroxide, naphthalene degradation up to 35% was achieved, which suggested that mineral iron was able to catalyze the production of hydroxyl radicals when ultrasound irradiation was used as an oxidizing agent. It was concluded that mineral iron was able to catalyze the degradation of naphthalene in the presence of ultrasound (up to 78% at 100 W and 97% at 400 W) and at various concentrations of hydrogen peroxide. A more critical analysis of the experiments, indicate that the Fenton-like oxidation of naphthalene in the presence of ultrasound has a potential to be used for practical purposes. Nevertheless, the upscaling of the lab-scale set up to a larger scale applicable for practical use warrants further research regarding the use of industrially contaminated soil with multiple organic contaminants. Moreover, improvement in the reactor design is much required to prolong the lifetime of the sonotrode, which tends to wear off due to ultrasound waves that propagate back to the sonotrode from the walls and the bottom of the reactor. A new process for remediation of soil contaminated with organic compounds (toluene and xylenes) has been proposed by Flores et al. (2007). The innovation combined the advanced oxidation method using Fenton-type catalyst, with the application of ultrasonic energy (47 kHz, 147 W, 10 min duration time for 20 g soil in 40 g aqueous solution). Experimental results showed that application of ultrasound not only assists desorption of the contaminants from the soil, but also promotes the formation of hydroxyl radicals, which are the main oxidant agent involved in the decontamination process. The global efficiency of the process was noticeably enhanced when applying ultrasonic energy, due to a synergistic effect in conjunction with the hydrogen peroxide concentration and Fenton catalyst (Flores et al., 2007). Despite the current difficulties, ultrasound has a bright future in on-site soil remediation field and may become one of the most feasible options and environmentally sound techniques.

On a more innovative tone of sono-oxidative remediation research, Dai et al. (2011) experimentally probed the characteristics of the use of Fe^{2+} or Cu^{2+} ions in the ultrasound-assisted oxidation desulfurization (UAODS) of diesel fuels. Dai et al. (2011) observed that the UAODS of diesel fuels fitted the pseudo-first-order kinetics and apparently in the UAODS of diesel fuels the apparent reaction rate constants could be greatly enhanced by addition of metal ions and/or using ultrasound. More interestingly, Dai et al. (2011) brought forward that the combination of ultrasound and the metal ions could also reduce the apparent activation energy rapidly and favor the sono-oxidative remediation processes in order of the apparent reaction rate constants in UAODS of diesel fuels as follows: US-Fe^{2+}-H_2O_2 system > US-Cu^{2+}-H_2O_2 system > US-H_2O_2 system > H_2O_2 system.

Ultrasonically Enhanced Electrokinetic Remediation

To enhance the transport of contaminant complexes and/or organic contaminants, the electrokinetic process has the potential to remove pollutants, such as PAHs from the soil by improving flow and soil–solution–contaminant interaction in limited permeability soils (Reddy and Saichek, 2004). Previous studies showed that electrokinetic technique was applied to remove mainly heavy metals and the ultrasonic technique was applied to remove mainly organic substances in contaminated soil. Thus, combination of the two techniques can predictably be helpful. Chung and Kamon (2005) have studied electrokinetic and ultrasonic remediation technologies for the removal of heavy metal and polycyclic aromatic hydrocarbon (PAH) in contaminated soils. The study emphasized the coupled effects of electrokinetic and ultrasonic techniques on migration as well as clean-up of contaminants in soils. Natural clay was used as a test specimen; Pb and phenanthrene were used as contaminants. Pb is a positive charged ionic contaminant; on the other hand, phenanthrene is a neutrally charged nonionic contaminant. The ultrasonic processor had a maximum power output of 200 W with a frequency of excitation equal to 30 kHz. Chung et al. (2006) carried out laboratory experiments involving electrokinetic remediation, ultrasonic remediation, and combined electrokinetic and ultrasonic remediation tests for the analysis of transportation and removal mechanism of sand soil spiked with 500 mg L^{-1} ethylene glycol. The test specimen was then subjected to ultrasonic waves at 30 kHz frequency from ultrasonic test setup and to electric power at 1.0 V cm^{-1} from electrokinetic test setup. Chung et al. (2006) demonstrated that the ultrasonic technique was the most effective in enhancing the removal efficiency of ethylene glycol from contaminated soil.

When ultrasonic energy was applied into contaminated soil, the viscosity of fluid phase decreased and flow rate increased, the molecular movement increased, sorbed contaminants mobilized, the cavitation developed and porosity and permeability increased. The removal efficiency of contaminant was higher for combined electrokinetic-ultrasonic test than for simple electrokinetic test alone. Therefore, the introduction of enhancement technique like ultrasonic process into electrokinetic process could be effective for increasing of contaminant removal rate from the contaminated soil.

Tests were also conducted using ultrasound alone, ultrasound as an enhancement for electrokinetic test and electrokinetic test alone to compare the removal performance of the three persistent organic pollutants, hexachlorobenzene, phenanthrene, fluoranthene (Pham et al., 2009a,b,c), and chrysene from low-permeable kaolin in reactors and pans with and without iron anodes (Shrestha et al., 2010). Results from experiments show that combined electrokinetic and ultrasonic treatment did prove positive coupling effect in PAHs removal than each single process alone, though the level of enhancement was not significant. Results indicated that the removal was more effective with lower concentrations of organic pollutants. The average removal was better in pan experiment with EKUS with iron anode (Shrestha et al., 2010). This might be due to increase in electroconductivity by iron ions. The assistance of ultrasound in electrokinetic remediation can help reduce POPs from clayey soil by improving the mobility of hydrophobic organic compounds and degrading these contaminants through pyrolysis and oxidation. Ultrasonication also sustains higher current and increases electroosmotic flow in combined EK-US test than in EK test alone.

ULTRASONICALLY ENHANCED ACTIVATED CARBON AMENDMENT

Addition of carbon particles to sediment provides strong sorption sites for the hydrophobic organic contaminants and reduces these freely dissolved compounds' concentrations. Thus, a powdered activated carbon (PAC) amendment assisted with sonication was used to reduce the bioaccessibility of polycyclic aromatic hydrocarbons in three creosote contaminated sediments (Pee, 2008). The study revealed that sonochemically induced switching of phenanthrene and pyrene from sediment to PAC was more effective than mechanical mixing in decreasing the bioavailability of these PAHs. The enhancement effect performed in sediment treated with sonication was explained to be attributed to the facilitation of desorption of PAHs through localized turbulent liquid movement, micro jets formation, and particles fragmentation.

ULTRASONICALLY ENHANCED THE SURFACTANT-AIDED SOIL-WASHING

The use of ultrasound as an enhancement mechanism in the surfactant-aided soil-washing process was examined by conducting desorption tests of soils contaminated with naphthalene or diesel–oil (Na et al., 2007). The experiments were conducted to elucidate the effect of ultrasound on the mass transfer from soil to the aqueous phase using naphthalene-contaminated soil. In addition, the use of ultrasound for the diesel–oil-contaminated soil was investigated under a range of conditions of surfactant concentration, sonication power, duration, soil/liquid ratio, particle size, and initial diesel-oil concentration. The ultrasound used in the soil-washing process significantly enhanced the mass transfer rate from the solid phase to the aqueous phase. The removal efficiency of diesel-oil from the soil phase generally increased with longer sonication time, higher power intensity, and large particle size.

CONCLUSIONS

Soil contamination in general and particularly by persistent organic pollutants (POPs) is a critical issue because of their long life span and toxicity, which can be a threat to public health, food system, and groundwater. Among many soil treatment technologies, electrokinetics has emerged as a potential technique for *in situ* soil remediation and is especially unique because of the ability to work in low-permeable soil. On the other hand, as a young, new and rapidly growing science, the applications of ultrasound in environmental technology hold a promising future. Compared to conventional methods, ultrasonication can bring several benefits such as environmentally friendly treatment (no toxic chemical are used or produced), low cost, and compact, allowing on-site treatment. Ultrasonic energy applied into contaminated soils can increase desorption and mobilization of contaminants and porosity and permeability of soil through developing of cavitation. Removal of nonpolar contaminants like most organic compounds are transported primarily by electroosmosis in electrokinetic remediation, thus the process is effective only if the contaminants are soluble in pore fluid. Thus, enhancement is needed to improve mobility of these hydrophobic compounds, which tend to adsorb strongly to the soil, particularly low-permeable soil. The coupling effect of combination of the two techniques, electrokinetics and ultrasonication, in persistent organic pollutant removal from contaminated low-permeable soil (with kaolin as a model medium) can be feasible treatment of highly contaminated soil by persistent organic pollutants. The laboratory experiments must consider various conditions (moisture, frequency, power, duration time, and initial concentration) to examine the effects of these parameters on the treatment process. These conditions play vital role in treatment process. Experimental results showed that ultrasonication has a potential to remove POPs, although the removal efficiencies were not high with short duration time. The study also suggested intermittent ultrasonication over longer time as an effective means to increase the removal efficiencies.

Then, experiments were conducted to compare the performances among electrokinetic process alone and electrokinetic processes combined with surfactant addition and mainly with ultrasonication, in open pans and in designed cylinders (with filter cloth separating central part and electrolyte parts). Combined electrokinetic and ultrasonic treatment did prove positive coupling effect compared to each single process alone, though the level of enhancement is not significant. The assistance of ultrasound in electrokinetic remediation can help reduce POPs from clayey soil by improving the mobility of hydrophobic organic compounds and degrading these contaminants through pyrolysis and oxidation. Ultrasonication also sustains higher current and increases electroosmotic flow. Initial contaminant concentration is an essential input parameter that can determine the removal effectiveness despite different treatment processes.

REFERENCES

Adewuyi, Y.G. 2001. Reviews—Sonochemistry: Environmental science and engineering applications. *Industrial and Engineering Chemistry Research*, 40: 4681–4715.

Belgiorno, V., Rizzo, L., Fatta, D. et al. 2007. Review on endocrine disrupting-emerging compounds in urban wastewater: Occurrence and removal by photocatalysis and ultrasonic irradiation for wastewater reuse. *Desalination*, 215: 166–176.

Caulier, T.P., Maeck, M., and Reisse, J. 1995. Homogeneous Sonochemistry: A study of the induced isomerization of 1,2-dichloroethene under ultrasonic irradiation. *Journal of Organic Chemistry*, 60: 272–273.

Chung, H.I., Chun, B.S., and Lee, Y.J. 2006. The combined electrokinetic and ultrasonic remediation of sand contaminated with heavy metal and organic substance. *KSCE Journal of Civil Engineering*, 10: 325–331.

Chung, H.I. and Kamon, M. 2005. Ultrasonically enhanced electrokinetic remediation for removal of Pb and phenanthrene in contaminated soils. *Engineering Geology*, 77: 233–242.

Collings, A.F., Farmer, A.D., Gwan, P.B., Pintos, A.P.S., and Leo, C.J. 2006. Processing contaminated soils and sediments by high power ultrasound. *Minerals Engineering*, 19: 450–453.

Dai, Y., Zhao, D., and Qi, Y. 2011. Sono-desulfurization oxidation reactivities of FCC diesel fuel in metal ion/H_2O_2 systems. *Ultrasonics Sonochemistry*, 18(1): 264–268.

Feng, D. and Aldrich, C. 2000. Sonochemical treatment of simulated soil contaminated with diesel. *Advances in Environmental Research*, 4: 103–112.

Feng, D., Lorenzen, L., Aldrich, C., and Maré, P.W. 2001. Ex situ diesel contaminated soil washing with mechanical methods. *Minerals Engineering,* 14: 1093–1100.

Flores, R., Blass, G., and Dominguez, V. 2007. Soil remediation by an advanced oxidative method assisted with ultrasonic energy. *Journal of Hazardous Materials*, 140: 399–402.

Hoffmann, M.R., Hua, I., and Höchemer, R. 1996. Application of ultrasonic irradiation for the degradation of chemical contaminants in water. *Ultrasonics Sonochemistry*, 3: 163–172.

Khan, F.I., Husain, T., and Hejazi, R. 2004. An overview and analysis of site remediation technologies. *Journal of Environmental Management*, 71: 95–122.

Kim, Y., Park, J.H., Kim, S.M., and Khim, J. 2007. Ultrasonically enhanced diesel removal from soil. *Japanese Journal of Applied Physics*, 46: 4912–4914.

Kim, Y.U. and Wang, M.C. 2003. Effect of ultrasound on oil removal from soils. *Ultrasonics*, 41: 539–542.

Ley, S.V. and Low, C.M.R. 1989. *Ultrasound in Synthesis*, Chap. 2. Berlin, Germany: Springer-Verlag.

Mason, T.J. 2007a. Review—Developments in ultrasound—Non-medical. *Progress in Biophysics and Molecular Biology*, 93: 166–175.

Mason, T.J. 2007b. Sonochemistry and the environment—Providing a "green" link between chemistry, physics and engineering. *Ultrasonics Sonochemistry*, 14: 476–483.

Mason, T.J., Collings, A., and Sumel, A. 2004. Sonic and ultrasonic removal of chemical contaminants from soil in the laboratory and on a large scale. *Ultrasonics Sonochemistry*, 11: 205–210.

Na, S., Park, Y., Hwang, A. et al. 2007. Effect of ultrasound on surfactant-aided soil washing. *Japanese Journal of Applied Physics*, 46: 4775–4778.

Pamukcu, S. and Huang, C.P. 2001. *In-Situ Remediation of Contaminated Soils by Electrokinetic Processes in Hazardous and Radioactive Waste Treatment Technologies Handbook*, Chang Ho Oh, Ed. Boca Raton, FL: CRC Press LLC.

Pee, G.Y. 2008. Sonochemical remediation of freshwater sediments contaminated with polycyclic aromatic hydrocarbons. PhD dissertation, The Ohio State University.

Pham, T.D., Shrestha, R.A., Virkutyte, J., and Sillanpää, M. 2009a. Combined ultrasonication and electrokinetic remediation for persistent organic removal from contaminated kaolin. *Electrochimica Acta*, 54: 1403–1407.

Pham, T.D., Shrestha, R.A., and Sillanpää, M. 2009b. Electrokinetic and ultrasonic treatment of kaoline contaminated by POPs. *Separation Science and Technology*, 44: 2410–2420.

Pham, T.D., Shrestha, R.A., Virkutyte, J., and Sillanpää, M. 2009c. Recent studies in environmental applications of ultrasound. *Journal of Environmental Engineering and Science*, 36: 1849–1858.

Reddy, K.R. and Saichek, R.E. 2004. Enhanced kinetics removal of phenanthrene from clay soil by periodic electric potential application. *Journal of Environmental Science and Health, Part A*, 39: 1189–1212.

Shrestha, R.A., Pham, T.D., and Sillanpää, M. 2009. Effect of ultrasound on removal of persistent organic pollutants (POPs) from different types of soils. *Journal of Hazardous Materials*, 170: 871–875.

Shrestha, R.A., Pham, T.D., and Sillanpää, M. 2010. Electro ultrasonic remediation of polycyclic aromatic hydrocarbons from contaminated soil. *Journal of Applied Electrochemistry*, 40: 1407–1413.

Stephanis, C.G., Hariris, J.G., and Mourmouras, D.E. 1997. Process (mechanism) of erosion of soluble brittle materials caused by cavitation. *Ultrasonics Sonochemistry*, 4: 269–271.

Suslick, K.S., Casadonte, D.J., Green, M.L.H., and Thompson, M.E. 1987. Effects of high intensity ultrasound on inorganic solids. *Ultrasonics*, 25: 56–61.

Suslick, K.S. and Crum, L.A. 1997. *Sonochemistry and Sonoluminescence in Encyclopedia of Acoustics*, Vol. 1, pp. 271–282. New York: Wiley-Interscience.

Virkutyte, J., Sillanpää, M., and Latostenmaa, P. 2002. Electrokinetic soil remediation—Critical overview. *Science of the Total Environment*, 289: 97–121.

Virkutyte, J., Vičkačkaite, V., and Padarauskas, A. 2010. Sono-oxidation of soils: Degradation of naphthalene by sono-Fenton-like process. *Journal of Soils and Sediments*, 10: 526–536.

17 Role of Heterogeneous Catalysis in the Sonocatalytic Degradation of Organic Pollutants in Wastewater

Juan A. Melero, Fernando Martínez, Raul Molina, and Yolanda Segura

CONTENTS

Introduction 419
Role of Solid Particles in Sonochemical Degradation of Organic Pollutants in Aqueous Solution 420
Improved Heterogeneous Sonocatalytic Systems by Addition of Oxidants 429
    Sono-Fenton Degradation Processes 429
    US/TiO$_2$ Systems with Hydrogen Peroxide 435
    Sono-Enzyme Peroxide Degradation Systems 436
Hybrid Methods: Integration of Sonocatalysis with Photoassisted Processes
for Wastewater Treatment 436
Conclusions and Future Outlook 441
References 442

INTRODUCTION

The application of ultrasound (US) irradiation for the degradation of organic pollutants in water has been broadly described in literature in the past. The thermal decomposition of pollutants by direct pyrolysis and/or oxidation by means of the reactive radicals coming from water and oxygen dissociation in the presence of US irradiation have been proposed in literature as the main degradation mechanisms. Consequently, hydrophobic pollutants with high vapor pressure are decomposed mainly by pyrolytic degradation, whereas hydrophilic pollutants with low vapor pressure are decomposed by hydroxyl radical oxidation. However, the rate of pollutant degradation of ultrasonic irradiation is rather low to be applied in practice, especially for highly hydrophilic compounds. Hence, one strategy to enhance the degradation efficiency of organic pollutants in water is to combine US irradiation with the presence of a solid catalyst. The presence of solid particles provides additional nuclei for the cavitation phenomena, increasing the number of cavitation events that result in the enhancement of the degradation performance activity. Likewise, in a biphasic solid–liquid medium, irradiated by power US, the solid particle size is reduced leading to an increase of surface area with the subsequent increase rate of phase mixing and mass transfer. Nevertheless, a high amount of solids might lead to the scattering of the sound waves with the decrease in the transferred energy to the reaction medium.

In those cases, where the degradation is controlled by free radicals within the liquid phase, the presence of different oxidants such as hydrogen peroxide and ozone has shown to amplify ultrasonic action. Particularly, the presence of iron-based catalysts in a US/H_2O_2 system combining acoustic cavitation with Fenton-like reactions has allowed the increase of hydroxyl radical generation and, hence, higher rates of pollutant removal (so-called sono-Fenton processes). In these cases, the enhancement of pollutant removals has been attributed to the increase of pollutant adsorption over the solid surface, the continuous cleaning and activation of catalyst surface, as well as the enhanced rate of mass transport, resulting from the turbulent effects of the cavitation. Likewise, the reduction of ferric ions (Fe^{3+}) to ferrous (Fe^{2+}) in the Fenton system and the splitting of H_2O_2 molecules into hydroxyl radicals have been proven to occur under ultrasonic irradiation. Both facts enhance the formation of radical active species. Other authors have also described the integration of sono-Fenton processes with ultraviolet (UV) irradiation (so-called sonophoto-Fenton processes). Sonophotocatalytic systems have also been discussed in literature. The main drawbacks of photocatalysis are related to the low efficiency of photocatalysts in continuous operation due to the blocking of UV-activated sites as well as severe mass transfer limitations. The above-mentioned turbulence induced by the cavitation phenomena could avoid those problems. In these systems, by combining several advanced oxidation processes, some synergism effects have been reported. In some cases, rates of degradation in the combined systems are higher as compared to individual operations.

In this chapter, we will analyze and discuss the role of solid particles in US systems that can use different oxidants, and also coupling systems based on the combination of UV and US irradiations for the enhancement of pollutant degradation.

ROLE OF SOLID PARTICLES IN SONOCHEMICAL DEGRADATION OF ORGANIC POLLUTANTS IN AQUEOUS SOLUTION

The presence of particulate matter can play a critical role in different ways for the sonochemical oxidation of pollutants in water solution. Thus, the presence of dispersed particles in the liquid solution during sonication provides additional nucleation sites for cavitation events over its surface, enhancing the number of microbubbles in the solution. Moreover, this fact has been proven to be highly influenced by the roughness of the particles (Figure 17.1). Particles can also act as a wall for the bubbles transmission, producing an asymmetric collapse of the cavitation bubbles and leading to the generation of a large number of tiny bubbles (Figure 17.1). Both phenomena produce an increase of microcavities that enhance the yield of the sonochemical oxidation. The asymmetric collapse of microbubbles over the solid surface also offers the additional degradation of adsorbed pollutants by the energy released during the in situ implosion of cavities. Keck et al. (2002) reported that the presence of quartz particles (3–8 µm) during sonolysis of pure water at low frequencies (<250 kHz) led to higher hydrogen peroxide production than in the absence of particles. Additionally, the elimination rates of different aromatic compounds (2-chlorobenzoic acid, salicylic acid and p-toluene sulfonic acid) were up to double in the presence of quartz particles. These benefits of the solid particles have also been proven by Tuziuti et al. (2005) with the addition of alumina (Al_2O_3) solid particles in a bath-type reactor. In this work, it was remarked that particles of Al_2O_3 with lower than 10 µm do not necessarily act as a wall to cause asymmetric collapse of the bubbles. A plausible explanation could be that these small particles are in motion with the surrounded bubbles, minimizing the number of interactions among them.

On the other hand, some components of the solid particles can have a catalytic activity for the decomposition of the hydrogen peroxide generated under sonicated conditions. For instance, this is the case of solid particles containing iron or other metallic species (Ge et al., 2003; Dai et al., 2006) that provide additional hydroxyl radicals from the hydrogen peroxide decomposition for the oxidation of pollutants (Figure 17.1).

Several authors have investigated thoroughly the influence of different solids ion the sonochemical degradation of organic compounds. A representative selection is summarized in Table 17.1.

Particularly, TiO_2 has been widely employed in different works with the main purpose of exploring future combinations between ultrasonic and UV-visible irradiation (Pandit et al., 2001).

Role of Heterogeneous Catalysis in the Sonocatalytic Degradation of Organic Pollutants

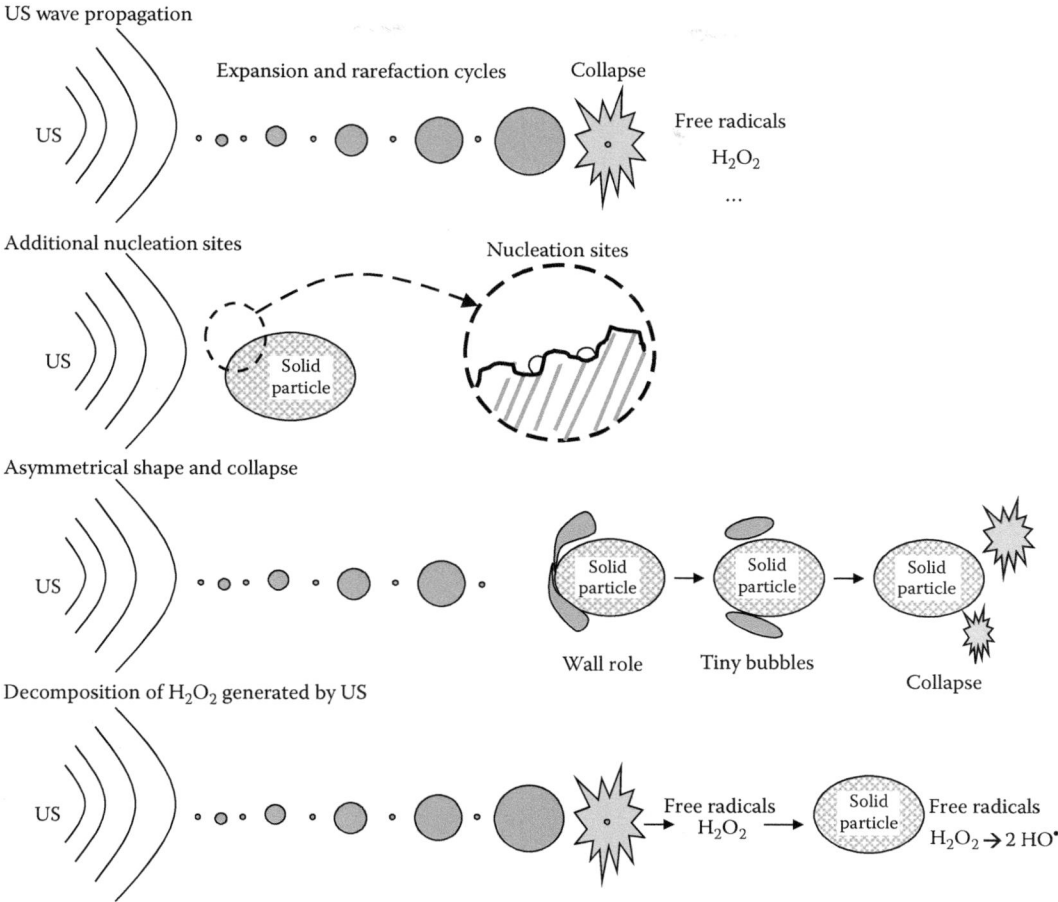

FIGURE 17.1 Interaction of solid particles and US waves or cavitation events.

Nevertheless, TiO_2 has also been used in ultrasonic-assisted systems in the absence of UV irradiation. Several authors have proposed that the efficiency of the US/TiO_2 system is not only due to the generation of nucleation sites for cavitation bubbles by the presence of TiO_2 particles, but also as the result of the sonoluminescence effect. Sonoluminescence is the phenomenon of light emission from the collapse of gas bubbles driven by a US field in a liquid (Walton et al., 1984; Gaitan et al., 1992). Thus, ultrasonic irradiation of liquids can result in the formation of intense UV light with wavelengths below 375 nm, in the same range than that necessary to activate TiO_2 photocatalyst. Activation of TiO_2 produces pairs of electron/hole by transition of an electron from the valence band into the conduction band, and hydroxyl radicals by dissociation of H_2O molecules (Ogi et al., 2002).

The US/TiO_2 system has been applied for the degradation of dye pollutants generated in textile industries. Several works demonstrated that rutile TiO_2 particles exhibit better catalytic properties than anatase TiO_2 in the sonochemical degradation of methyl orange and congo red dyes (Wang et al., 2005, 2006, 2007). The US/TiO_2 system has also been successfully used for the treatment of different organic pollutant in aqueous solution, such as phenol and 2,4-dinitrophenol (Ogi et al., 2002), 1,4-dioxan (Nakajima et al., 2007) or 2,4,6-trichlorophenol (Pandit et al., 2001). In the same way that UV/TiO_2 system, a major drawback of this process is that the sono-generated electrons and holes recombine easily, limiting the activity of the system. Doping TiO_2 with transition metal ions can reduce the band-gap energy for the TiO_2, maximizing the efficiency of hydroxyl radicals generation. Wang et al. (2009) studied the activity of Co-doped and Cr-doped mixed crystal TiO_2,

TABLE 17.1
Sonodegradation of Pollutants in Presence of Solid Particles

Reference	Solid Catalyst	Pollutant	Experimental Conditions	Highlight of Work
TiO_2				
Pandit et al. (2001)	TiO_2 (0.1–0.5 g L^{-1})	2, 4, 6 Trichlorophenol (100 mg L^{-1})	Ultrasonic horn probe 22.7 kHz, 600 W, pulse mode Room temperature	Overall disappearance is a combination of adsorption, desorption, and oxidation. Oxidation in the bulk solution. Up to 50% degradation of the pollutant
Dadjour et al. (2005)	TiO_2 (300–2000 g L^{-1}) Al_2O_3 (300 g L^{-1})	*Escherichia Coli* (10^3–10^6 colony formation units, CFU mL g L^{-1})	Ultrasonic bath 39 kHz, 200 W $T = 20°C$ Atmospheric pressure	TiO_2 induces the formation of reactive oxygen species in the solution. The enhancement of activity is proportional to the TiO_2 concentration. Disinfection of *E. coli*. $N/N_0 = 0.016$–0.032 (ca. 98% reduction of viable cells)
Wang et al. (2005)	TiO_2 (0.5–0.75 g L^{-1})	Methyl orange dye (10–20 mg L^{-1})	Ultrasonic horn-probe 40–60 kHz, 50 W pH = 3.0–5.0 $T = 40°C$	Degradation of methyl orange of ca. 90% for an initial methyl orange concentration of 10 mg L^{-1}, rutile TiO_2 of 0.5 g L^{-1}, pH = 3.0 and 150 min of reaction
Dadjour et al. (2006)	TiO_2 (200–1000 g L^{-1})	*Legionella* (4000–6000 CFU mL^{-1})	Ultrasonic bath 36 kHz, 300 W $T = 20°C$	Reduction of the concentrations of viable cells of ca. 94% in the presence of 0.2 g m L^{-1} TiO_2 after a 30 min of treatment period, while only an 18% reduction was observed in the absence of TiO_2. Hydroxyl radicals have a primary role in the disinfection process
Wang et al. (2006)	TiO_2 (0.5–0.75 g L^{-1})	Methyl orange dye (10–20 mg L^{-1})	Ultrasonic horn-probe 40–60 kHz, 50 W pH = 3.0–5.0 $T = 40°C$	The sonocatalytic activity of reused nanometer anatase TiO_2 catalyst was reduced by about 10% every time as compared with the fresh nanometer anatase TiO_2 powder. Degradation of methyl orange dye of ca. 100% after 120 min
Wang et al. (2007)	TiO_2 (0.5–1.5 g L^{-1})	Congo red dye (0–25 mg L^{-1})	Ultrasonic horn-probe 20–80 kHz, 50 W pH = 3.0–5.0 $T = 50°C$	Degradation of ca. 80% for an initial concentration of 25.0 mg L^{-1} congo red, 500 mg L^{-1} nanometer rutile TiO_2 powder and pH = 5.0

Reference	Catalyst	Pollutant	Conditions	Results
Nakajima et al. (2007)	TiO_2 (200 g L^{-1})	1,4-dioxane aqueous solution (50 mg L^{-1})	Ultrasonic horn-probe 20 kHz, 20 W, pulse mode $T = 23°C$	Thermally excited holes are generated in the TiO_2 surface by ultrasonic irradiation. Degradation efficiency of 1,4-dioxane were increased by the addition of reduced TiO_2 powder from 50% to 70%
Wang et al. (2009)	Co-doped and Cr-doped mixed crystal TiO_2 powders (0.2–1.25 g L^{-1})	Azo fuchsine solution (10 mg L^{-1})	Ultrasonic bath 40 kHz, 50 W pH = 10 $T = 20°C$	The sonocatalytic activity of Cr-doped mixed crystal TiO_2 powder was higher than that of Co-doped and undoped mixed crystal TiO_2 powder. Degradation grades ranging from 10% to 75%
Wang et al. (2010a)	CeO_2/TiO_2 (1 g L^{-1}) SnO_2/TiO_2 (1 g L^{-1}) ZrO_2/TiO_2 (1 g L^{-1})	Acid Red B (10 mg L^{-1})	Ultrasonic bath 40 kHz, 50 W pH = 7.0 $T = 25°C$	Degradation rate: $CeO_2/TiO_2 > SnO_2/TiO_2 > TiO_2 > ZrO_2/TiO_2 > SnO_2 > CeO_2 > ZrO_2$. Respective degradation grades: 91.3%, 67.4%, 65.3%, 41.7%, 28.3%, 26.7%, and 23.3%. The degradation ratio is only 16.7% under onefold ultrasonic irradiation
Carbon and activated carbon				
Ince et al. (2001)	Activated carbon Zinc Ceramic granules (0.12 g L^{-1}, each one)	*Escherichia coli* (1200–2000 CFU mL^{-1})	Ultrasonic horn probe 20 kHz, 180 W	Ultrasonic inactivation of bacteria: AC > ceramic > zinc. Time for killing 50% bacteria: 1.21, 2.90, and 3.41 min, respectively. The effect of solid catalysts is more significant at high concentrations of bacteria
Bernardo et al. (2006)	Activated carbon (1 g L^{-1})	Saccharin (100 mg L^{-1})	Ultrasonic cup horn 500 kHz, 15 W $T = 25°C$ Ar and O_2/N_2 (20/80 vol %)	Saccharin removal after 180 min of ultrasonication under Ar and O_2/N_2 atmospheres: 38% and 26%, respectively. Saccharin removal by AC adsorption without US pretreatment: 40%. Saccharin removal by AC adsorption with ultrasonic pretreatment: 75%. The pretreatment of sonication under O_2/N_2 produces the increase in the amount of saccharin adsorbed on AC
Li et al. (2007)	Exfoliated graphite (1 g L^{-1})	Azo dye direct scarlet 4BS (100 mg L^{-1})	Ultrasonic bath 28 kHz, 500 W $T = 30°C–50°C$ pH = 2–7	Color removal of 94% with a concentration of exfoliated graphite of 0.6 g L^{-1} (300 mL g^{-1} of exfoliated volume), pH = 2.0 and 50°C

(continued)

TABLE 17.1 (continued)
Sonodegradation of Pollutants in Presence of Solid Particles

Reference	Solid Catalyst	Pollutant	Experimental Conditions	Highlight of Work
Zero valence iron/zero valance copper/other metals				
Papadaki et al. (2004)	Pt, Pd and Ru supported on Al_2O_3 (5% w/w, 0.05 g L^{-1}) CuO·ZnO supported on Al_2O_3 (0.05 mg L^{-1})	Sodium dodecylbenzene sulfonate, SDBS (1000 mg L^{-1})	Ultrasonic horn probe 20 kHz, 125 W	CuO·ZnO supported on alumina catalyst enhances both SDBS fragmentation and total oxidation rates as well as hydrogen peroxide formation
Dai et al. (2006)	ZVI (Fe^0), 2 g L^{-1} Cu powder, 2 g L^{-1}	Pentachlorophenol, 10 mg L^{-1}	Ultrasonic bath 40 kHz, 600 W pH = 3.5	US increases the adsorption process by increasing the surface area of iron and copper particles. Degradation of pentachlorophenol in the US/Cu^0 system: ca. 60%. Degradation of pentachlorophenol in the US/Fe^0 system: ca. 85%
Liu et al. 2007	Zero valent iron/granular activated carbon (Fe^0/GAC) $[Fe^0]_0 = 120$ g L^{-1} $[GAC]_0 = 23$ g L^{-1}	Azo dye Acid Orange 7 (1000 mg L^{-1})	Ultrasonic bath 40 kHz, 100 W $T = 22°C$	Optimum Fe^0/GAC ratio: 1:1 (V/V). TOC degradation and color removal values ca. 55% and 80%, respectively. A decrease of the initial pH from 12.0 to 4.0 led to the increase of degradation efficiency
Lin et al. (2008)	Cast iron (Fe^0), 10–50 g L^{-1}	Azo dye CI Acid Red 14 (50–250 mg L^{-1})	Ultrasonic bath 59 kHz pH = 2.0–8.0 $T = 25°C$	A TOC removal of ca. 43% was observed after 20 min. The first-order rate constant of AR14 degradation by cast iron was 7.50×10^{-2} min^{-1} while that by US-cast iron was 2.58×10^{-1} min^{-1}. The decolorization efficiency decreased with the increasing pH or initial dye concentration
Zhou et al. (2008)	Zero valent iron, 5–50 g L^{-1}	2,4-dichlorophenol (50 and 100 mg L^{-1})	Ultrasonic horn probe 20 kHz $[EDTA]_0 = 0.28$–1.28 mmol L^{-1} pH = 6.5 $T = 10°C$–$50°C$ Air bubbled = 1 L min^{-1}	EDTA break down O–O bond of oxygen producing H_2O_2 in an Fe/EDTA system. TOC degradation: 81% TOC. DCP degradation: ca. 100%. EDTA elimination: 89%

Reference	Catalyst	Pollutant	Conditions	Results
Zhou et al. (2009a)	ZVI (Fe^0), 25 g L^{-1}	Azo dye Reactive Black 5 (RB5), 100 mg L^{-1}	Ultrasonic horn probe 20 kHz [EDTA]$_0$ = 0.4 mmol L^{-1} pH = 2–5 T = 10°C–50°C Air bubbled = 1 L min^{-1}	The removal rates of RB5, EDTA, Total organic carbon, and chemical oxygen demand were 100%, 96.5%, 68.6%, and 92.2%, respectively, after 3 h. At pH of the system below 2, mineralization of the organics in the wastewater was almost inhibited completely. Aeration is necessary to cover the oxygen demand for H$_2$O$_2$ generation
Zhou et al. (2009b)	ZVI (Fe^0), 25 g L^{-1}	4-Chlorophenol (100 mg L^{-1})	Ultrasonic horn probe 20 kHz [EDTA]$_0$ = 0.32 mmol L^{-1} pH = 2–5 T = 10°C–50°C Air bubbled = 1.0 L min^{-1}	4-Chlorophenol was completely dechlorinated and some low molecule organic acids were identified as the main final products after 1 h. TOC degradation: ca. 70%
Eren et al. (2010)	ZVC (Cu^0), 1–3 g L^{-1} TiO$_2$, 0.1–1 g L^{-1}	CI Direct Yellow 9 (DY9, 20 mg L^{-1}) and CI Reactive Red 141 (RR141, 50 mg L^{-1})	Ultrasonic horn probe 20 kHz, 180 W Transducer bottom plate 577, 861, and 1145 kHz, 120 W T = 20°C pH = 6.9 pH = 3.0 (with ZVC)	Sonodegradation process at low frequencies shows a higher efficiency than that obtained at high frequencies Degradation of DY9 with ZVC and TiO$_2$: ca. 55 and 70%, respectively Degradation of RR141 with ZVC and TiO$_2$: ca. 17 and 99%, respectively
Metallic oxides				
Ge et al. (2003)	MnO$_2$ (1 g L^{-1})	Azo dye acid red B, ARB (100 mg L^{-1})	Ultrasonic bath 50 kHz, 150 W T = 22°C pH = 1–10 Oxygen and argon atmosphere	Degradation grade higher than 90% at pH = 3. Degradation values close to 100% were achieved at pH = 1. Counter anions (SO$_4^{2-}$ and NO$_3^-$) greatly inhibit the decolorization process. Oxygen used as saturated gas is more favorable for mineralization of the dye
Ait et al. (2007a)	Fe@Fe$_2$O$_3$ core-shell nanowires (1 g L^{-1})	Rhodamine B (5 mg L^{-1})	Ultrasonic bath 25 kHz, 100 W pH = 2 Air bubbled = 0.1 m^3 h^{-1}	Decoloration close to 100% and TOC degradation: ca. 60% achieved in 60 min

(continued)

TABLE 17.1 (continued)
Sonodegradation of Pollutants in Presence of Solid Particles

Reference	Solid Catalyst	Pollutant	Experimental Conditions	Highlight of Work
Ait et al. (2007b)	Fe@Fe$_2$O$_3$ core-shell nanowires (1 g L^{-1})	Rhodamine B (5 mg L^{-1})	Ultrasonic bath 25 kHz, 100 W pH = Neutral Air bubbled = 0.1 m^3 h^{-1}	Addition of tiny Fe^{2+} (5 mg L^{-1}) enhances the degradation rate of RhB. Over 92% of RhB was degraded by the sono-Fenton process with Fe^{2+} and Fe@Fe$_2$O$_3$ for 60 min at neutral pH
Nakui et al. (2007)	Coal ash (0–15 × 10^{-3} g L^{-1})	Phenol (10 mg L^{-1})	Ultrasonic cup horn oscillator 200 kHz, 200 W T = 20°C	Maximum degradation of phenol (ca. 80%) for a coal ash concentration of 5–6 mg L^{-1}
Chen et al. (2010)	4A-zeolite supported α-Fe$_2$O$_3$ (0.5 g L^{-1})	Orange II (10–20 mg L^{-1})	Ultrasonic bath 40 kHz, 100 W pH = 2.7 and 6.8	Fe-4A composite showed a low iron ion dissolution level, high reactivity stability and a long lifetime under neutral condition in the sono-Fenton reaction. Degradation grade: 80%–90%
Nakui et al. (2009)	Coal ash (2–5 × 10^{-3} g L^{-1})	Hydrazine solution (3.5 mg L^{-1})	Ultrasonic cup horn oscillator 200 kHz, 200 W. pH = 2, 4 and 8. T = 20°C	Degradation of 100% after 60 minutes of reaction at pH = 4. Critical influence of pH, promoting changes in the coal ash surface, hydrazine protonation and enhancing adsorption of hydrazine on the coal ash
Others				
Tauber et al. (2005)	Laccase enzyme (5 × 10^{-9} kat L^{-1} laccase activity)	Acid Orange 5 (100 μmol L^{-1}) Acid Orange 52 (100 μmol L^{-1}) Direct Blue 71 (100 μmol L^{-1}) React. Black 5 (100 μmol L^{-1}) React. Orange 16 (100 μmol L^{-1}) React. Orange 107 (100 μmol L^{-1})	Ultrasonic horn probe 850 kHz, 90 W T = 40°C, pH = 4.5	US treatment can decolorize all tested dyes after 3 h at a high energy input, and prolonged sonication leads to nontoxic ionic species. Maximum decoloration 80%
Sagave et al. (2006)	Enzyme cellulose, blend of endoglucanase, cellobiohydrolase and β-glucosidase. Activity of 100 FPU g^{-1} of powder	Distillery wastewater (TOC and COD of 55,000 and 10,000 mg L^{-1}, respectively)	Ultrasonic bath 22 kHz, 120 W Combination with a biological aerobic oxidation pH = 3.8–3.9	The combined pretreatment technique (US + E + AO) yielded the highest COD reduction of 62.25% as compared to 34.9%, 36.61%, 42.26% COD reduction for the untreated, enzymatically pretreated (only E + AO), US pretreated (only US + AO), respectively, after 36 h of treatment

obtaining that the presence of M^{3+} (metallic ions) produces an additional consumption of electrons by reduction to M^{2+}, remaining as holes on the surface of the TiO_2. Thus, the decomposition of H_2O toward hydroxyl radicals is favored (Wang et al., 2010a). In the same way, authors tested different composites of CeO_2/TiO_2, SnO_2/TiO_2, and ZrO_2/TiO_2 as an alternative. These composite materials, based on different semiconductors, also restrain the recombination of the electron/hole pairs due to the different energy levels of conduction and valence bands for CeO_2, SnO_2, ZrO_2, and TiO_2. These differences promote the transition of the electrons generated by the ultrasonic irradiation from TiO_2 to the other semiconductors through the crystal frontier, resulting in the complete separation of electrons and holes, especially for the CeO_2/TiO_2 composite material.

Combination of US with activated carbon has also been studied in literature for the degradation of organic pollutants. US has been used as pretreatment on the removal of saccharin by activated carbon adsorption. The obtained results showed an increase in the AC adsorption of saccharin and the by-products, formed during ultrasonic treatment (Bernardo et al., 2006). A further step could be the combination of activated carbon and ultrasonic irradiation in a simultaneous system. In this way, exfoliated graphite, a type of porous carbon adsorption material has been used in combination with ultrasonic irradiation (ultrasonic bath, 28 kHz) for the decolorization of azo dye direct scarlet 4BS solutions. The higher efficiency of this process, as compared to silent adsorption, is attributed to the rupture of exfoliated graphite particles by sonication, with a decrease in the particle size and the consequent increase in the surface area available for the adsorption of the dye (Li et al., 2007).

Elemental iron zero valent iron (ZVI) has been used for the sonodegradation of organic pollutants as low-cost solid catalyst. It is known that the use of ZVI, under acidic conditions, promotes its oxidation to Fe^{2+} (Reaction 17.1). Additionally, Fe^0 in the presence of water can be oxidized by Reaction 17.2. The Fe^{2+} formed by Reactions 17.1 and 17.2 reacts rapidly with the sono-generated H_2O_2 molecules to produce additional hydroxyl radicals in the bulk medium (Reaction 17.3). Finally, the Fe^{2+} can be regenerated by the reaction between Fe^{3+} (usually in form of iron oxides) and Fe^0 (Reaction 17.4).

$$Fe^0 + 2H^+ \rightarrow Fe^{2+} + H_2 \tag{17.1}$$

$$Fe^0 + H_2O \rightarrow Fe^{2+} + \tfrac{1}{2}H_2 + HO^- \tag{17.2}$$

$$Fe^{2+} + H_2O_2 \rightarrow Fe^{3+} + HO^\bullet + HO^- \tag{17.3}$$

$$2Fe^{3+} + Fe^0 \rightarrow 3Fe^{2+} \tag{17.4}$$

In this process, the release of ferrous ions (Fe^{2+}) from the surface of the ZVI particle is identified as the rate-limited step, since the formation of a layer of iron oxide may inhibit the Fe^{2+} generation (Reaction 17.1). As it has been previously mentioned, the collapse of the cavitation bubbles near the solid surface causes microjets that hits the surface and produce asymmetrical shock waves. This phenomenon results in the well-known cleaning action of US, which eliminates the iron oxide layer, and consequently, it regenerates and reactivates the catalyst surface (Liu et al., 2007). In this way, Dai et al. (2006), found that the degradation of pentachlorophenol under ultrasonic irradiation could be enhanced by the presence of ZVI. The authors attributed this effect to the increase of the cavitation events induced by the solid particles, and the catalytic action of the Fe^{2+} ions released to the solution, which promotes the production of hydroxyl radicals. Different additives such as EDTA can form ligands with Fe^{2+} (Reaction 17.5). Fe(II)EDTA complex can be combined with an oxygen molecule initiating a series of reactions that activate the oxygen and finally generating hydrogen peroxide (Reactions 17.6 through 17.9, Zhou et al., 2009a). The effect of EDTA has been deeply studied in the degradation of model pollutants (i.e., 2,4-dichlorophenol) as well as wastewaters coming from textile industries (Zhou et al., 2008, 2009a,b).

$$Fe^{2+} + EDTA + H_2O \rightarrow [Fe^{II}(EDTA)(H_2O)]^{2-} \quad (17.5)$$

$$[Fe^{II}(EDTA)(H_2O)]^{2-} + O_2 \leftrightarrow [Fe^{II}(EDTA)(O_2)]^{2-} + H_2O \quad (17.6)$$

$$[Fe^{II}(EDTA)(O_2)]^{2-} \rightarrow [Fe^{III}(EDTA)(O^-_2)]^{2-} \quad (17.7)$$

$$Fe^{II}(EDTA)(O_2)]^{2-} + [Fe^{III}(EDTA)(O^-_2)]^{2-} \rightarrow [(EDTA)Fe^{III}(O_2^{-2})Fe^{III}(EDTA)]^{4-} + H_2O \quad (17.8)$$

$$[(EDTA)Fe^{III}(O_2^{-2})Fe^{III}(EDTA)]^{4-} \rightarrow (H^+, Fast) \rightarrow 2[Fe^{III}(EDTA)H_2O]^- + H_2O_2 \quad (17.9)$$

Another possible combination is the US/Fe⁰/carbon systems in which electrochemical batteries can be formed between the iron and the carbon. The Fe⁰ is oxidized to Fe²⁺ (anode, Reaction 17.10), whereas activated carbon acts as cathode (Reaction 17.11) by producing hydrogen from adsorbed protons (Liu et al., 2007). Again, the role of the US is the cleaning of reactive intermediates or products from the ZVI surface, reactivating the surface for subsequent reactions. A similar approach has been described by Lin et al. (2008) using a cast iron, an alloy of which the main elements are Fe and C. Alternative to ZVI, zero valent copper has also been used following a similar reaction mechanism mentioned for the preparation of ZVI (Eren et al., 2010).

$$\text{Anode (Fe): } Fe \rightarrow Fe^{2+} + 2e^- \quad (17.10)$$

$$\text{Cathode (C): } 2H^+ + 2e^- \rightarrow H_2 \quad (17.11)$$

As it has been previously mentioned, another benefit of the presence of certain solid particles during the sonication is the decomposition of the hydrogen peroxide generated by ultrasonic irradiation to form highly reactive and oxidizing hydroxyl radicals. Within the kind of solids that can promote this reaction, iron oxides as themselves or with support have been widely used for the degradation of different pollutants in aqueous solution (Ai et al., 2007a,b; Chen et al., 2010). Likewise, other metal oxides, such as MnO_2 (Ge et al., 2003) and $CuO \cdot ZnO$ supported on Al_2O_3 have also been used for their catalytic role in the decomposition of the sono-generated hydrogen peroxide. In some cases, $CuO \cdot ZnO$ supported on Al_2O_3 has been doped with noble metals such as Ru or Pd to enhance this effect (Papadaki et al., 2004). Another alternative is the coal ash generated as waste in thermal power stations, characterized as a mixture of different oxides, such as SiO_2, Al_2O_3, Fe_2O_3, or CaO (Nakui et al., 2007, 2009). A relevant conclusion obtained in these works is that exists an optimum in the amount of solid particles present in the aqueous medium. Higher amounts of solid particles produce negative effects in terms of H_2O_2 generation by cavitation events. In that case, large amounts of the coal ash induce scattering or adsorption of the ultrasonic wave, reducing the formation of the nucleation sites for cavitation bubbles.

Finally, it is remarkable that a combined treatment of USs and enzymes has also been proposed as oxidation process for depuration of mixtures of different dyes in aqueous solution and disinfection of distillery wastewater, obtaining promising results, as compared to the individual processes (Tauber et al., 2005; Sangave et al., 2006).

The above-mentioned works are related to the degradation of different organic pollutants (phenol, dyes, etc.) but several attempts have also been carried out for disinfection, using US/solid systems. US has been used in disinfection due to its capacity of producing cell lysis and other damaging effects, including functional and biochemical changes. Several authors reported that microbial cells were inactivated more effectively by the combination of US and TiO_2 (Dadjour et al., 2005, 2006), being the chemical oxidation by hydroxyl radicals—the main action responsible for damaging the membrane cells and finally causing cell death. Additionally, activated carbon together with other solid materials (zinc, ceramics) has been used in combination with US in disinfection, accelerating the overall reduction of *E. coli* with respect to the sonolytical disinfection (Ince et al., 2001).

IMPROVED HETEROGENEOUS SONOCATALYTIC SYSTEMS BY ADDITION OF OXIDANTS

Alternatively to the benefits of the presence of solid particles during the sonication, it should be noted the intrinsic activity of certain metals in the solid particles as catalyst for the generation of additional hydroxyl radicals from the autogenerated H_2O_2. For these solids with catalytic activity, the combination of heterogeneous sonocatalytic systems with different oxidants has been studied as novel advanced oxidation processes. In these cases, ultrasonic irradiation can improve the catalytic activity of the solid particles by increasing the surface area as a result of the particles fragmentation and by reducing mass transfer limitations. A brief review of the most recent works related to this issue is summarized in Table 17.2.

Although the H_2O_2 is the most relevant oxidant used in ultrasonic systems, other oxidants are also being used. For instance, a synergistic effect has been found for the combination of US and ozone, in a so-called sonozation process, which has been widely studied in literature. Theoretically, the combination of the sono-ozonolysis and heterogeneous catalysts should enhance the efficiency of the processes, by providing additional nucleation sites for the generation of cavitation bubbles in which thermal decomposition of the ozone could take place. Additionally, these catalysts could decompose the H_2O_2, formed during the sonication, generating hydroxyl radicals. The application of copper and iron craps as heterogeneous catalysts in combination with ozone at different ultrasonic frequencies irradiations for the treatment of phenolic aqueous solutions have been studied by Chand et al. (2009). However, it must be pointed out that better results were obtained by the substitution of ozone by hydrogen peroxide.

SONO-FENTON DEGRADATION PROCESSES

The addition of hydrogen peroxide to ultrasonic irradiation systems in the presence of iron-containing materials as Fenton-like processes can increase the hydroxyl radical production, enhancing the overall activity and degradation rate of the so-called sono-Fenton system. There are many examples in literature of heterogeneous sono-Fenton systems, using different iron oxides as catalysts (hematite and goethite). Neppolian et al. (2004) reported the combination of US/Goethite/H_2O_2 for the treatment of *p*-chlorobenzoic acid, obtaining a remarkable enhancement in the activity with respect to the goethite/H_2O_2 processes. Authors attributed this improvement to the continuous cleaning and chemical activation of the FeOOH surface by the physical and chemical effects of US. Thus, the turbulent effects of the cavitation promotes not only an increase in the surface particle area, but also the reduction of the goethite iron oxide to Fe^{2+} ions and dissolution in the reaction medium (Reaction 17.12), involving the subsequent homogeneous Fenton reaction (Reaction 17.13).

$$2FeOOH + US \rightarrow 2Fe^{2+} + O_2 + 2OH^- \quad (17.12)$$

$$Fe^{2+} + H_2O_2 \rightarrow Fe^{3+} + HO^\bullet + HO^- \quad (17.13)$$

The sono-Fenton processes using goethite have been widely reported for the treatment of wastewaters coming from textile industries, in which different dyes can be found as major pollutants

TABLE 17.2
Sonocatalytic Systems Combined with Hydrogen Peroxide

Reference	Solid Catalyst	Pollutant	Experimental Conditions	Highlight of Work
Sono-Fenton-like systems				
Nepppolian et al. (2004)	Goethite (0.25–4 g L^{-1})	p-chlorobenzoic acid (0.5 mg L^{-1})	Ultrasonic horn probe, 20 kHz [H_2O_2]$_0$ = 68–2680 mg L^{-1} pH = 3, 7 and 9; T = 20°C	Degradation of 65% under optimal conditions (pH = 3, hydrogen peroxide concentration of 20 mM, goethite loading corresponding to 1000 mg L^{-1})
Nikolopoulos et al. (2006)	Al-Fe pillard clay, FAZA (5 g L^{-1})	Phenol (50 mg L^{-1})	Ultrasonic cup horn and horn probe, 20 kHz [H_2O_2]$_0$ = 3400 mg L^{-1} (2 mL h^{-1}) pH = 3.5–4.0; T = 20°C	US enhances internal mass transfer phenomena, especially in the cup horn, where the effective diffusion coefficient was two orders of magnitude higher than in the horn probe. Degradation after 4 h in the ultrasonic horn probe: 100%. Degradation after 4 h in the ultrasonic cup horn: 80%
Molina et al. (2006)	Fe$_2$O$_3$/SBA-15 (0.2–1 g L^{-1})	Phenol (235 mg L^{-1})	Ultrasonic horn probe at 20 kHz [H_2O_2]$_0$ = 1190–4760 mg L^{-1} pH = 3.0; T = 22°C	An increase in the hydrogen peroxide concentration enhances the mineralization of phenol up to an optimum values corresponding to twice the stoichiometric amount of oxidant. 100% of phenol degradation after 60 minutes of reaction, and TOC mineralization of ca. 30% after 4 h. High stability of the supported iron species (less than 3% of iron leached out from the catalyst)
Liang et al. (2007)	Iron powder, basic oxygen furnace slag and mill scale (1 and 10 g L^{-1})	4-Chlorophenol (100 mg L^{-1})	Ultrasonic bottom plate Transducer, 200 kHz [H_2O_2]$_0$ = 10–100 mg L^{-1} pH = 3 and 5.6; T = 25°C	Total degradation of the 4-chlorophenol after 1 h in sonocatalytic oxidation with 1000 mg L^{-1} of iron or mill scale powder, 100 mg L^{-1} of H_2O_2 at pH = 3 and 5.6
Kim et al. (2007)	CuO, Cu/Al$_2$O$_3$ and CuO-ZnO/Al$_2$O$_3$ (1 g L^{-1})	4-Chlorophenol (200 mg L^{-1})	Ultrasonic horn probe, 20 kHz [H_2O_2]$_0$ = 1600 mg L^{-1} T = 25°C	High 4-CP and TOC removal obtained with supported catalysts, especially Cu/Al (ca. 100% and 40%, respectively). The good dispersion of the fragmented catalyst plays a critical role in the degradation efficiency

Reference	Pollutant	Catalyst	Conditions	Results
Muruganandham et al. (2007)	Direct Orange 39 (10 mg L^{-1})	Goethite (2.5–15 g L^{-1})	Ultrasonic bath, 35 kHz; $[H_2O_2]_0 = 5{,}000$–$15{,}000$ mg L^{-1}; pH = 3.0; $T = 25°C$	Optimum hydrogen peroxide concentration of 10 g L^{-1}. The goethite surface area, pore size, and pore volume by ultrasonic irradiation is increased in the first cycle and then begins to decrease until the fourth cycle. Goethite catalyst retains its activity up to the four successive runs. Iron dissolved <3 mg L^{-1}
Melero et al. (2008)	Phenol (235 mg L^{-1})	Fe_2O_3/SBA-15 Hematite Goethite (0.6 g L^{-1})	Ultrasonic horn probe, 20 kHz; $[H_2O_2]_0 = 2{,}380$ mg L^{-1}; pH = 3.0; $T = 22°C$	Hematite and goethite particles are agglomerated during the sonication, obtaining a decrease in the surface area. This effect is inhibited by supporting the iron species over the SBA-15. Activity of the iron containing mesoporous materials is several times higher than unsupported iron oxides
Namkung et al. (2008)	Phenol (36–360 mg L^{-1})	Zero valent iron bars (surface loading 72.6 cm^2 L^{-1})	Ultrasonic bath, 30 kHz; Ultrasonic cup horn, 20 kHz; $[H_2O_2]_0 =$ Solution 3%, flow rate 14.4–60 mL h^{-1}; pH = 2–2.5; $T = 20°C$	Ultrasonic effects were quite marked with the cup-horn but relatively small enhancements of TOC removal were observed with a bath type sonicator. TOC removals ranging from 30% to 40%
Bremner et al. (2009)	Phenol (60–470 mg L^{-1})	Fe_2O_3/SBA-15 (0.2–1 g L^{-1})	Ultrasonic bottom plate transducer, 300–1140 kHz; $[H_2O_2]_0 = 1190$–4760 mg L^{-1}; pH = 3.0; $T = 22°C$	Optimum frequency at 584 kHz. TOC degradation of ca. 40% after 4 h. The optimal hydrogen peroxide concentration is the stoichiometric amount in the presence of 600 mg L^{-1} of solid catalyst. Reusability of the catalyst up to 5 cycles without appreciable change in the catalytic activity and stability
Rokhina et al. (2009)	Phenol (100 mg L^{-1})	RuI_3 (0.5–2 g L^{-1})	Ultrasonic horn probe, 24 kHz; $[H_2O_2]_0 = 200$–1200 mg L^{-1}; pH = 5.5; $T = 25°C$	Maximum degradation grade: 55% after 4 h, using medium values of the catalyst and H_2O_2 concentrations. RuI_3 retains the catalytic activity up to three cycles of reusing for the degradation of phenol, but it is reduced after the fourth run
Song and Li (2009)	CI Direct Black 168 (100–1000 mg L^{-1})	Fly ash (0.2–2.4 g L^{-1})	Ultrasonic bath, 40 kHz; $[H_2O_2]_0 = 6.6$–133 mg L^{-1}; pH = 1.0–5.0; $T = 30°C$	Degradation grade of 99% achieved at initial concentration 100 mg L^{-1}, pH 3.0, dosage of fly ash of 2000 mg L^{-1} and 100 mg L^{-1} of H_2O_2

(continued)

TABLE 17.2 (continued)
Sonocatalytic Systems Combined with Hydrogen Peroxide

Reference	Solid Catalyst	Pollutant	Experimental Conditions	Highlight of Work
Zhang et al. (2009a)	Goethite (0.2–0.4 g L^{-1})	CI Acid Orange 7 (79.5 mg L^{-1})	Ultrasonic horn probe, 20 kHz; $[H_2O_2]_0 = 13$–265 mg L^{-1}; $T = 20°C$	The increase of hydrogen peroxide concentration leads to the increase of decolorization efficiency. Optimal initial pH of 3. TOC degradation after 90 min of reaction = ca. 50%
Zhang et al. (2009b)	Zero valent iron (0.1–0.25 g L^{-1})	CI Acid Orange 7 (200 mg L^{-1})	Ultrasonic horn probe, 20 kHz; $[H_2O_2]_0 = 170$–510 mg L^{-1}; pH = 2–5; $T = 25°C$	The decolorization rate increased with the increase of hydrogen peroxide concentration and power density, but decreased with the increase of initial pH value. Degradation grade of ca. 80% after 20 min. using 150 mg L^{-1} H_2O_2, 100 mg L^{-1} of ZVI and pH = 3.0
Rokhina et al. (2010)	RuI_3 (1 g L^{-1})	Phenol (100 mg L^{-1})	Ultrasonic horn probe, 24 kHz; $[H_2O_2]_0 = 600$ mg L^{-1}; Neutral pH; $T = 25°C$–70°C	US decreases the induction period and activation energy associated to the RuI_3 system, reducing half-life of the oxidation reaction. Phenol degradation = 80% after 8 h
Wang et al. (2010b)	Fe_3O_4 magnetic nanoparticles (0–1 g L^{-1})	Rhodamine B (9.5 mg L^{-1})	Ultrasonic horn probe, 20 kHz; $[H_2O_2]_0 = 200$–1360 mg L^{-1}; pH = 3–9; $T = 10°C$–50°C	Degradation of Rhodamine B at pH = 3 and 50°C: 95%. Increase in the degradation rate respect US or US/catalyst systems
US/TiO_2 systems				
Shimizu et al. (2007)	TiO_2 (400–2000 g L^{-1})	Methylene Blue (64–320 mg L^{-1})	Ultrasonic bath, 39 kHz; $[H_2O_2]_0 = 680$–3400 mg L^{-1}; pH = 3–12; $T = 20°C$	The dye degradation increases from 22% in absence of hydrogen peroxide up to 85% when a concentration of 1700 mg L^{-1} of H_2O_2 is added. The highest degradation ratio was observed at around pH 7
Abbasi et al. (2008)	Nano-TiO_2 (0.2 g L^{-1})	Basic Blue 41 dye (15–240 mg L^{-1})	Ultrasonic bath, 35 kHz; $[H_2O_2]_0 = 100$–1200 mg L^{-1}; pH = 4.5–8; $T = 25°C$	The decolorization efficiency decreased with the increasing of initial dye concentration. Decolorization efficiency of 89.5% in 180 min for initial dye concentration of 15 mg L^{-1}. Maximum decolorization with 1000 mg L^{-1} hydrogen peroxide concentration and initial pH of 8

Reference	Catalyst	Pollutant	Conditions	Remarks
Abdullah et al. (2010)	TiO_2 (1–3 g L^{-1})	Congo red, methyl orange, and methylene blue (10–50 mg L^{-1})	Ultrasonic horn probe, 20 kHz [H_2O_2]$_0$ = 0–600 mg L^{-1} pH = 2–7; T = 30°C	Optimal conditions: 1500 mg L^{-1} of catalyst loading and 450 ppm of H_2O_2 for a congo red removal efficiency of ca. 80% in 180 min. The reused catalyst retains up to 90% of the activity respect the first run
Sono-enzyme peroxide systems				
Entezari et al. (2003)	Horseradish Peroxidase (HRP) (0.8 unit mL^{-1})	Phenol, p-chlorophenol, p-bromophenol, p-iodophenol, p-methoxyphenol, p-cresol and p-nitrophenol (1.6 mmol L^{-1} each one)	Ultrasonic bottom place transducer, 423 kHz [H_2O_2]$_0$ = 2 mmol L^{-1} (68 mg L^{-1}) pH = 7; T = 28°C ± 1°C	Sono-enzyme oxidation was the best for the degradation of phenol and its halogenated substitutes. Enzyme treatment is a suitable method for p-methoxyphenol and p-cresol. USs as itself is sufficient for the destruction of p-nitrophenol
Entezari et al. (2004)	Horseradish peroxidase (HRP) (2.4 unit mL^{-1})	Phenol (2.4 mmol L^{-1}, 225 mg L^{-1})	Ultrasonic bottom plate transducer, 423 kHz [H_2O_2]$_0$ = 2 mmol L^{-1} (68 mg L^{-1}) pH = 7; T = 30°C ± 1°C	USs preserved the activity of peroxidase in the period of sonication. More than 95% of phenol was removed after about 30 min of sonication, in comparison to 70% in a silent experiment
Entezari et al. (2005)	Horseradish peroxidase (HRP) (0.8 unit mL^{-1})	Phenol, p-chlorophenol, p-cresol, p-nitrophenol and p-chlorophenol (1.6 mmol L^{-1}, and double mixtures of 0.8 mmol L^{-1} each)	Ultrasonic bottom plate transducer, 423 kHz [H_2O_2]$_0$ = 2 mmol L^{-1} (68 mg L^{-1}) pH = 7; T = 30°C ± 1°C	Most of the compounds were degraded outside the microcavities by the hydroxyl radical attacks. The rate of degradation in case of single component solutions is approximately the same for all individual compounds. In case of double component solutions the rate of reaction is governed by the magnitude of its bimolecular rate coefficient, which is higher for reactive compounds. Reactive in the sono-enzyme degradation: p-methylphenol > p-chlorophenol > Phenol > p-nitrophenol
Entezari et al. (2006)	Horseradish peroxidase (HRP) (0.078–0.168 unit mL^{-1})	2-Chlorophenol (0.1 mmol L^{-1}, 9.4 mg L^{-1})	Ultrasonic bottom plate transducer, 500 kHz [H_2O_2]$_0$ = 1 mmol L^{-1} (34 mg L^{-1}) pH = 7; T = 25°C–42°C	The higher rate induced by ultrasonic sonication is reduced by the HRP dosage Degradation of ca. 100% in approximately 30 min

(Muruganandham et al., 2007; Zhang et al., 2009a). These studies demonstrated the increase of surface area of the iron oxides after successive cycles, keeping its activity constant. The influence of the hydrogen peroxide concentration on the degradation process was also evaluated, indicating an increase of the decoloration efficiency until reaching an optimum value of hydrogen peroxide concentration. Higher oxidant concentrations had a negative effect in the degradation rate of the pollutants (Wang et al., 2010b).

The immobilization of the iron on inert support surfaces has also been widely studied in literature. Thus, mixed pillared clays (Al-Fe), FAZA, conformed by extrusion in extrudates, have been tested as solid catalyst, exhibiting beneficial properties in terms of high oxidation activity and low leachability of the catalyst (Nikolopoulos et al., 2006). Employing the FAZA clay, they observed that an induction period previous to the degradation was reduced due to the enhancement of the intraparticle diffusion by the ultrasonic irradiation. It is noteworthy that the calculated effective diffusion coefficients were increased for indirect sonication by means of an external cup horn, as compared to direct sonication, using an immersed horn probe. These results were explained taking into account that under direct sonication, the particles are forced to follow the quick motion of the water from zones of high energy (just under the probe) to zones of low energy. On the other hand, in the indirect sonication system, the solid particles are located within the ultrasonic regime for the total reaction time. Another example of supported iron-containing catalyst is the Fe_2O_3/SBA-15 material based on the immobilization of hematite and other iron oxides on a mesoporous SBA-15 silica support. This catalyst has exhibited high activity and stability in the oxidation of phenolic aqueous solutions by sono-Fenton processes in a wide range of frequencies (20–1140 kHz, Bremner et al., 2009). Moreover, it was demonstrated that there was an increase in the degradation rate up to an optimum concentration of hydrogen peroxide as well as relatively low oxidant consumption along the sonocatalytic process (Molina et al., 2006; Melero et al., 2008). On the contrary, high excess of the oxidant produced a negative effect on the phenol mineralization as a result of a major contribution of radical recombination on the catalyst surface, as it is proposed in Figure 17.2.

On the other hand, Melero et al. (2008) also demonstrated a different effect of the cavitation on the particle size for Fe_2O_3/SBA-15 and unsupported iron oxides (hematite and goethite). Nonsupported iron oxides exhibited a remarkable aggregation of particles, which was related to the attraction of magnetic iron phases produced as a result of the particle reduction of goethite under ultrasonic irradiation. In contrast, hematite supported on SBA-15 evidenced the expected fragmentation of particles after the sono-Fenton process. This different behavior influenced dramatically in the TOC degradation achieved with both catalysts.

A novel heterogeneous sono-Fenton-like system was proposed by the combination of US and the advanced Fenton process (AFP) (Namkung et al., 2008). In AFP, H_2O_2 is catalytically decomposed by Fe^{2+} (Reaction 17.3) coming from the corrosion of ZVI under acidic conditions (Reactions 17.1 and 17.2). US plays a critical role in the AFP system, cleaning the surface particles of iron oxides,

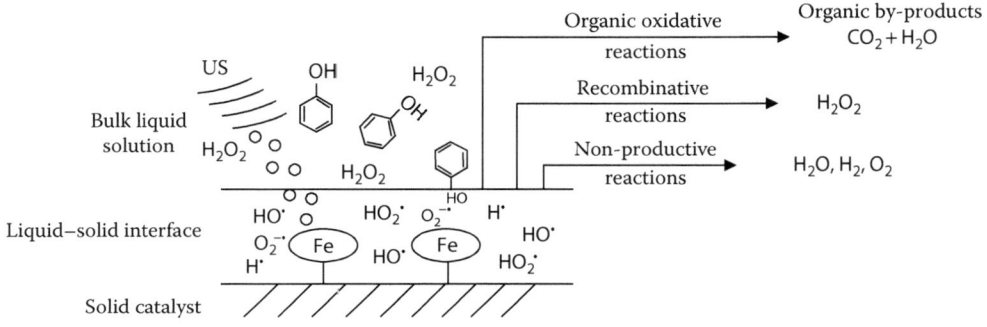

FIGURE 17.2 Simplified scheme of radical oxidation and recombination reactions in heterogeneous sono-Fenton systems.

providing more Fe⁰ surface available for Reactions 17.1 and 17.2, and enhancing the hydroxyl radicals production (Hardcastle et al., 2000). In this way, Liang et al. (2007) found a significant increase in the degradation of 4-chlorophenol by sonication of an AFP system with a bottom plate transducer. The US/AFP system has also been employed for the treatment of Acid Orange dye, obtaining a decrease in the decolorization grade with the increase in the initial pH (Zhang et al., 2009b). Additionally, a remarkable decrease in the efficiency of the processes was observed when an inert gas is dissolved in the reaction medium (i.e., nitrogen). Although dissolved gas affects aqueous sonochemical processes providing nucleation sites for cavitation events, in the case of nitrogen it might scavenge the free radicals from the bulk solution, inhibiting the oxidation of the CI Acid Orange 7 by the hydroxyl radicals. Namkung et al. (2008) treated phenolic aqueous solutions by means of continuous reactors with two indirect sonicator setups: a cup-horn and an ultrasonic bath. Results revealed a high influence of the initial concentration of phenol and hydrogen peroxide concentration on the TOC removal grade. However, the benefits of the ultrasonic bath were hardly significant with respect to the silent AFP oxidation.

Another attractive alternative has been the use of fly ash residues from a thermal power plant as source of iron-containing solids. These materials showed a remarkable decoloration of CI direct black 168 aqueous solutions by sono-Fenton systems (Song et al., 2009).

Besides iron oxides and iron-containing materials, other metallic oxides such as copper oxide as itself or supported over Al_2O_3 or ZnO/Al_2O_3 supports have demonstrated a high activity as solid-Fenton catalyst under ultrasonic irradiation (Kim et al., 2007). A remarkable efficiency of the copper solids in the degradation of p-chlorophenol was observed as compared to analogous catalytic runs in the absence of US. These results were attributed to the benefits of ultrasonic irradiation on the particles fragmentation, providing higher surface areas and the activation of the copper oxides in the production of hydroxyl radicals by Fenton-like reactions.

Finally, another typical solid sono-Fenton-like catalyst for the treatment of organic pollutants in aqueous solutions is those based on noble metals, outstanding the RuI_3 (Rokhina et al., 2009). The Fenton-like oxidation of phenol with RuI_3 exhibits an induction period similar to that obtained for the FAZA clay, which is also minimized by US. In this case, a decrease in the activation energy for the oxidation is reported as plausible reason, indicating that no changes were obtained in the reaction pathway, achieving a TOC degradation of ca. 55% after 5 h of reaction, higher than that obtained in the absence of US (ca. 35%, Rokhina et al., 2010).

US/TiO₂ SYSTEMS WITH HYDROGEN PEROXIDE

As it has been previously mentioned, ultrasonic irradiation involves generation of intense UV wavelength light, which can cause the transition of an electron from the valence band into the conduction band in the TiO_2 semiconductor, leaving a positive hole charge responsible of dissociation of water molecules to form hydroxyl free radicals. Addition of hydrogen peroxide to the US/TiO_2 system could promote further benefits to the degradation efficiency. Tuziuti et al. (2004) proposed that under ultrasonic irradiation, hydrogen peroxide can modify the TiO_2 surface promoting the formation of titanium peroxide (TiO_3) into the solution, which is active for oxidation reactions. Additionally, hydrogen peroxide can react with the pair electron/hole generated in the TiO_2 particles by sonoluminescence, producing hydroxyl radicals (Reaction 17.14).

$$TiO_2(e^-) + H_2O_2 \rightarrow HO^\bullet + HO^- \tag{17.14}$$

Consequently, the addition of hydrogen peroxide enhances the overall oxidation efficiency of the process. Shimizu et al. (2007) demonstrated that the degradation of methylene blue by a US/TiO_2 system occurs by hydroxyl radicals generated in the process, and the addition of H_2O_2 accelerated the degradation of the dye for the TiO_2 containing system up to an optimum value. Previously, Abbasi et al. (2008) had found a similar behavior in the sonochemical degradation of Basic Blue 41 dye in aqueous solution with nano-TiO_2 and H_2O_2. More recently, Abdullah et al. (2010) found that anatase TiO_2 with certain amount of rutile phase exhibited better sonocatalytic performance than pure anatase. It was attributed to the role of rutile to avoid hole-electron recombination. Authors also remarked that the physical effect of US for the disintegration of the TiO_2 particles might reduce the micro and mesoporosity leading to lower surface area. On the other hand, the above-mentioned reactivation of the sonocatalyst particles due to the cleaning effect of the ultrasonic streams enhances the efficiency of the degradation. The net effect of both is accounted for the activity of the US/TiO_2 system (Abdullah et al., 2010).

Sono-Enzyme Peroxide Degradation Systems

Several enzymes (oxidoreductases) have been studied for the removal of phenolic derivates from industrial wastewater and drinking water. Particularly, the application of the enzyme horseradish peroxidase (HRP) in combination with H_2O_2 and US has been proposed for the treatment of phenolic solutions, generating insoluble polymers as products. These polymers can be easily removed from the aqueous solution by filtration or sedimentation. Entezari et al. (2003, 2004) demonstrated that the sono-enzyme oxidation with hydrogen peroxide is a feasible treatment method for phenol and chlorophenol derivates. Additionally, the same authors found the following reaction rate order for different mixtures of monochloro-substituted phenolic compounds: p-methylphenol > p-chlorophenol > phenol > p-nitrophenol (Entezari et al., 2005). US irradiation allows cleaning the polymeric shell generated during the enzyme oxidation, reducing one of the major causes of inactivation of the HRP. Additionally, the HRP dosage is reduced in the sono-enzyme oxidation with respect to the silent one—a critical economic point for this kind of processes (Entezari et al., 2006).

HYBRID METHODS: INTEGRATION OF SONOCATALYSIS WITH PHOTOASSISTED PROCESSES FOR WASTEWATER TREATMENT

Most of the oxidation technologies for wastewater degradation is barely used individually for cost-effective operations, and this also includes sonocatalysis. The high operating costs of ultrasonic treatment processes are mostly due to electrical and very high capital costs, and also due to the use of large amount of energy for treating very small reaction volumes (Mahamuni et al., 2009).

As previously mentioned, in the case of reactions where the controlling mechanism is the free radical attack, sonocatalysis can be used together with other oxidation agents, such as ozone (Zhang et al., 2008; Zhaobing et al., 2009) or hydrogen peroxide (Lin et al., 1996), in order to enhance the rates of degradation due to additional generation of free radicals. The aim of it is to mineralize the organic compounds completely or only partially, converting the initial pollutants into less harmful compounds in order to treat them in a biological system. There are different combination techniques studied in literature for a variety of contaminants (Gogate, 2008). However, in most cases, sonocatalysis is applied in combination with UV light. Both processes are based on the generation of radicals and the subsequent attack of them on the contaminants.

Table 17.3 shows an overview of hybrid methods of the recent works undertaken in this field describing US in combination with UV light.

Although photocatalysis and sonolysis have been extensively employed individually for the degradation of several organic species in water, their combined use (i.e., sonophotocatalysis) has received noticeably less attention. This combination is especially attractive as US irradiation seems to overcome some drawbacks associated with photocatalysis such as the mass transfer limitations and fouling of the solid catalyst. During the acoustic irradiation the mass transfer resistance is dramatically reduced, and the increase of turbulence helps in the cleaning of the catalyst (Gogate, 2008).

TABLE 17.3
Hybrid Methods: Integration of Sonocatalysis with Photoassisted Processes for Wastewater Treatment

Reference	Catalyst	Integration System	Pollutant	Experimental Conditions	Highlight of Work
Sonophotocatalysis: US + photocatalysis (TiO_2)					
Shirgaonkar et al. (1998)	Anatase TiO_2 (0.1 g L^{-1})	Photocatalysis and Sonophotocatalysis	2,4,6 Trichlorophenol (100 mg L^{-1})	Sono: 22 kHz Photo: 15 W V: 1000 mL	Efficiency is influenced by the intensity of US, temperature, and type of US equipment. Independent of mode of UV transmission. Enhancement at lower sonication intensity and higher temperatures
Davydov et al. (2001)	TiO_2 (0.25 g L^{-1})	Photocatalysis and Sonophotocatalysis	Salicylic acid (250 mg L^{-1})	Sono: 20 kHz Photo: 7 UV lamps (4 W) Temperature: 30°C ± 2°C	Synergistic effect for the catalysts with smaller particle size. Nonenhancement for largest particle size. Degussa P25, the highest overall activity for degradation
Chen et al. (2002)	TiO_2 (0.25 g L^{-1})	UV and US, separately and simultaneously	Phenol and chlorophenols (1 mmol L^{-1})	Sono: 20 kHz, 65–75 W Photo: 450 W Temperature: 30°C ± 2°C V: 100 mL	Enhancement by reducing reaction volume and increasing average of ultrasonic powder. Solubility influences the performance of UV and US
Gogate et al. (2002)	TiO_2 (0.1–1 g L^{-1})	Simultaneous or sequentially	Formic acid (100–1000 mg L^{-1})	Sono: 20, 30, 50 kHz Photo: 8 W Temperature: 28°C ± 2°C V: 700 mL	Synergistic effect. Combined systems gives higher degradation compared to UV or US alone. Simultaneous mode, higher degradation than sequential combination
Kritikos et al. (2007)	TiO_2 (0.25 g L^{-1})	Photocatalysis and sonophotocatalysis	Diazo dye black 5 (60 mg L^{-1})	Sono: 80 kHz, 135 W Photo: 9 W Temperature: 25°C	Efficacy of photocatalysis increases by increasing of Ti loading, decreasing dye concentration and solution pH. The simultaneous application of UV and US resulted in increased decoloration compared to that achieved by US and UV separately
Sonophotocatalysis assisted with hydrogen peroxide: US + photocatalysis (TiO_2) + H_2O_2					
Mrowetz et al. (2003)	TiO_2 (P25) (0.1 g L^{-1})	Simultaneous sonophotocatalysis/H_2O_2	2-Chlorophenol (5×10^{-4} mol L^{-1}) Orange 8, Acid red 1 ($2–7 \times 10^{-5}$ mol L^{-1})	Sono: 20 kHz Photo: 15 W Temperature: 35°C ± 1°C V: 400 mL	Synergistic effect. USs increases surface area and thus, catalytic performance. Higher degradation by adding small amounts of Fe(III)

(continued)

TABLE 17.3 (continued)
Hybrid Methods: Integration of Sonocatalysis with Photoassisted Processes for Wastewater Treatment

Reference	Catalyst	Integration System	Pollutant	Experimental Conditions	Highlight of Work
Yano et al. (2005)	TiO_2 (0–1.6 g L^{-1})	Photocatalysis and Sonophotocatalysis/H_2O_2	Propyzamide (30 µmol L^{-1})	Sono: 200 W Photo: 100 W V: 150 mL	Photocatalytic system is accelerated when temperature and pH were increased. By the addition of H_2O_2 and USs: completely mineralization of propyzamide
Silva et al. (2007)	TiO_2 (P25) (0.75 g L^{-1})	Simultaneous Sonophotocatalytic/H_2O_2	Simulated olive mill wastewater (50 mg L^{-1} each)	Sono: 80 kHz, 120 W Photo: 9–250–400 W	Catalyst does not suffer composition or morphology changes during treatment. Although surface area increases Complete mineralization in 120 min
Sonophotocatalysis assisted with the Fenton's reagent: US + photocatalysis (TiO_2) + homogeneous Iron + H_2O_2					
Berberidou et al. (2007)	TiO_2 (P25) and Fe(II) (0.1–0.5 and 2.5 10^{-3}–5.0 10^{-3} g L^{-1}, respectively)	Individual and simultaneous Sonophoto-Fenton and sonophotocatalytic system	Malachite green	Sono: 80 kHz Photo: 9 W Temperature: 25°C–30°C V: 350 mL, pH: 5.5	For individual US: degradation increases by increasing US power and decreasing initial concentration Synergetic effect of sonophotocatalytic systems due to increased surface area of catalyst and higher formation of HO•
Solar sonophotocatalysis assisted with the Fenton's reagent: US + photo (solar) (TiO_2) + homogeneous iron + H_2O_2					
Mendez-Arriaga et al. (2009)	TiO_2, $FeSO_4$ (0.01 and 0.1 g L^{-1}, respectively)	Individual sonolysis and UV, sonophoto-Fenton Sonophotocatalysis	Ibuprofen (0.039 mmol L^{-1})	Sono: 300 kHz, 80 W Photo: solar simulated illumination T^a: 25°C, acid pH	The presence of US (to photocatalysis) improves de iron catalytic activity. The presence of both Fe and Ti catalysts promotes the highest DOC removal (90%)
Torres-Palma et al. (2010)	TiO_2 (P25) (0.01 g L^{-1}) $FeSO_4$ (5.6 10^{-3} g L^{-1})	Individual and combination of US, photo-Fenton and TiO_2 solar photocatalysis process	Bisphenol A (118 µmol L^{-1})	Sono: 300 kHz, 80 W Photo: solar lamp Temperature: 22°C ± 2°C V: 600 mL, acid pH	Synergistic effect due to combination of 3 processes. US offers the generation of H_2O_2 for the elimination of contaminants by photo-Fenton and photocatalytic activity of TiO_2
Heterogeneous sonophoto-Fenton: US + photo-Fenton (H_2O_2/Fe) heterogeneous iron catalyst					
Segura et al. (2009)	Fe_2O_3/SBA-15 (0.6 g L^{-1})	Individual sono- and photo-Fenton, sequential and simultaneous modes	Phenol (235 mg L^{-1})	Sono: 20 kHz Photo: 150 W T^a: 22°C ± 2°C V: 550 mL, acid pH	Sequentially integrated system enhances degradation, obtaining up to 90% of TOC (due to fragmentation of catalyst and surface area increases during first step of US)

The catalyst employed for almost all the integrated US with photocatalytic technologies is TiO_2. Although available in various crystalline forms, a commercial catalyst, Degussa P25, shows exceptional activity compared to other grades of TiO_2, and this is probably due to the morphology of its crystallites. Davydov et al. (2001) investigated the influence on the sonophotocatalytic destruction of salicylic acid on four commercial titania powders, exhibiting the highest overall activity for P25 TiO_2 catalyst. The combination of the action of ultrasonic and UV-assisted photocatalysis yielded higher activity for catalysts with smaller particle size, compared to largest particle size photocatalysts.

Although, the grade of enhancement depends on experimental and operating conditions, the synergetic effect when US is combined with photocatalysis (due to the benefits of US) is a well-known fact. Even more, the rates of degradation for the combined US and UV irradiations are at times higher in order of magnitude when compared to an individual operation. However, the influence of different parameters (such as reaction volume, US power, etc.) is not clear yet, although it is known to play an important role in the overall effect. There are other important factors that need to be considered while designing the optimum sonophotocatalytic process in order to achieve the maximum degradation with the highest energy efficiency. Most of the related works published in literature showed the importance of operating simultaneously rather than having sequential irradiation of US, followed by photocatalytic oxidation. However, it has also been published that sequential operation might also yield similar or higher rates of degradation for some specific cases such as equilibrium-controlled operations with very low rates of adsorption (Gogate et al., 2004).

Chen et al. (2002) have shown a synergistic effect to the application of US on the photocatalytic degradation of phenol and chlorophenols in the presence of TiO_2. They observed a remarkable enhancement by reducing the reaction volume or by increasing the average of ultrasonic power density. They also evaluated the influence of ionic strength and anions on sonophotocatalysis, suggesting that chloride anions (Cl^-) inhibit the action of sonophotocatalysis at low pH values, and that the solubility of the organic compounds has a significant influence on the performances of the photocatalytic and sonochemical reactions.

Ragaini et al. (2001) also evaluated the influence of the ultrasonic power density on the degradation of a phenolic aqueous solution, either separately or simultaneously, using different types of TiO_2 and different gas mixtures. Degussa P25 and systems using $Ar-O_2$ mixture showed the best catalytic performances with a noteworthy synergistic effect of the two degradation techniques for reaction volumes greater than 300 mL, concluding that excessive cavitation seems to have a detrimental effect.

Promising results of synergistic effects of the integrated US-UV processes have been obtained in the degradation of model pollutant, such as salicylic acid (Davydov et al., 2001), 2-chlorophenol (Ragaini et al., 2001), formic acid (Chen et al., 2002), methyl *tert*-butyl ether (Ragaini et al., 2005), and diazo dyes (Selli et al., 2005; Berberidou et al., 2007; Kritikos et al., 2007). As previously mentioned, the action of US toward UV is summarized by the increment of the catalyst surface area due to the deaggregation action of US, the improvement of mass transfer of organic compounds between the liquid phase and the catalysis surface, and the promotion of additional hydroxyl radicals due to the residual H_2O_2 generated. On the other hand, photocatalysis increases organic degradation due to the catalytic activity of TiO_2, not only by UV irradiation but also by the sonoluminescence effect of ultrasonic irradiation.

In simulated agro-industrial effluents containing 13 phenolic compounds typically found in olive mill wastewaters, Silva et al. (2007) found that combined process was considerably more effective that the respective individual treatments, and even more, process efficacy was further enhanced in the presence of H_2O_2 acting as hydroxyl radical source. They attributed the synergistic action to an increase in the production of hydroxyl radicals via water sonolysis and H_2O_2 cleavage as well as an increase in catalyst surface area. Further results related to synergistic effects of US and photocatalysis (TiO_2) systems enhanced by the addition of hydrogen peroxide were reported by different groups (Mrowetz et al., 2003; Yano et al., 2005). In this sense,

Gogate et al. (2002) also indicated the positive effect of H_2O_2 to increase the removal of different organic compounds due to the enhanced dissociation of H_2O_2 into hydroxyl radicals under the action of US and UV irradiation. They found better results with simultaneous operation as compared to sequential combinations, which confirmed the positive effects of US for increasing the production of free radicals and continuous cleaning of the catalyst particles.

In another work, Shirgaonkar et al. (1998) also studied the influence of different parameters on the sonophotochemical degradation of 2,4,6-trichlorophenol following two different setups. In one case, the UV source was placed parallel to the reactor, which used an ultrasonic horn probe as US device (Figure 17.3a), and in the other case the UV source was placed at the top of an ultrasonic bath (Figure 17.3b). The degradation was found to be dependent on the intensity of sonication, temperature of the reaction, and the type of ultrasonic equipment used, but was independent of the mode of UV transmission. It was observed that the degradation was similar in both cases indicating that there was no attenuation of any radiation by the glass material of the reactor.

Enhancement in the degradation rate was observed at higher temperatures of the solution and at lower sonication intensity. The latter effect was attributed to the higher deagglomeration of the catalyst at higher intensities, which reduces the overall transmission of the UV radiation. On the other hand, photolysis is known to be adversely affected with an increase in temperature. However, the observed enhancement in the degradation might be due to the red-shift of the bandgap edge of TiO_2 catalyst at higher temperatures.

In a further attempt to enhance the degradation extent of the contaminants in wastewater by sonophotocatalysis, Berberidou et al. (2007) have explored the use of adding homogenous Fe^{2+} to the sonophotocatalytic system, where the presence of both catalysts dissolved Fe^{2+} and heterogeneous TiO_2, promotes the highest dissolved organic carbon (DOC) removal as compared to the individual and simultaneous sonophotocatalytic processes in the degradation of a dye. Mrowetz et al. (2003) pointed out that small quantity of ferric ions can modify the TiO_2 surface, and, therefore, increase the adsorption of organic pollutants and yield a better catalytic activity. Moreover, Murakami et al. (2008) have suggested that Fe^{3+} ions on a TiO_2 surface can act as electron acceptor under UV irradiation.

Solar photocatalysis has gained considerable attention and several studies have reported the use of natural or simulated sunlight irradiation for water treatment. Méndez-Arriaga et al. (2009) have demonstrated the degradation of ibuprofen in distillate water using TiO_2 and/or Fe(II) in sonophotocatalytic systems. In the case of sonophoto-Fenton process the mineralization (60%) was attained mainly with photo-Fenton, although the presence of ultrasonic irradiation slightly

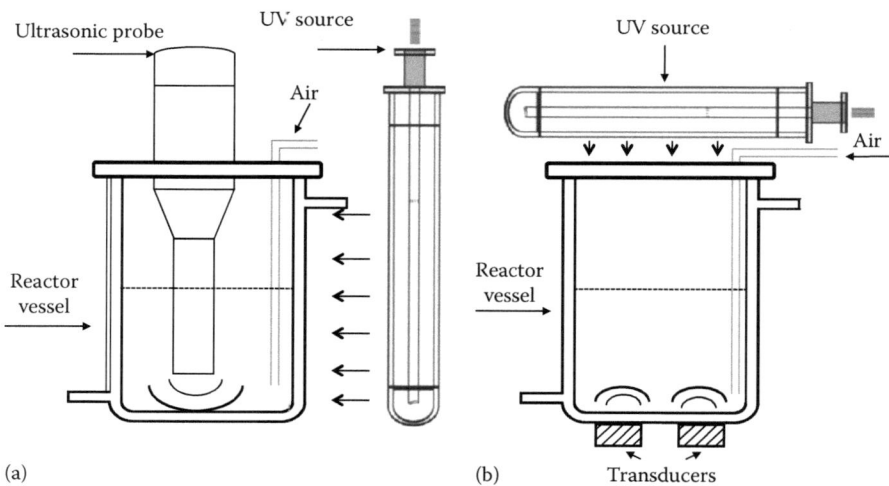

FIGURE 17.3 Setup schemes used for the combined processes of UV and US irradiations.

improves the iron catalytic activity. The presence of both, homogeneous (Fe^{2+}) and heterogeneous (TiO_2) catalysts, under illuminated and ultrasonic irradiation, promotes the highest DOC removal (ca. ≥90%). One further study has also been recently published by Torres-Palma et al. (2010) about the solar photocatalytic degradation of bisphenol in combination with US, using $FeSO_4$ and TiO_2 photocatalyst. They have evaluated the effect of different Fe^{2+} and TiO_2 concentrations, obtaining almost a complete removal of DOC (93%) after 4 h with the optimal catalysts loadings (using 10 and 5.6 mg L^{-1} of TiO_2 and Fe^{2+}, respectively) compared to 5%, 6%, and 22% of DOC removal when using individual processes alone (photocatalysis, US, and photo-Fenton, respectively). They observed a high synergy among the three systems. Water sonication can yield the H_2O_2 necessary for the photo-Fenton process. In other words, in this system, US seems to have the principal role of eliminating the initial substrate and providing hydrogen peroxide for the photocatalytic systems, while photo-Fenton and TiO_2 photocatalysis are mainly responsible for the transformation of the intermediates in CO_2 and H_2O.

The integration of US and UV–Vis with Fenton reagent, using heterogeneous iron containing catalysts and H_2O_2 as oxidant, has hardly been reported in literature. Most studies have stated the benefits of this hybrid system using homogeneous iron salts. In these cases, the synergism between processes—sonolysis, and photo-Fenton—is due to the efficiency of the photo-Fenton promoted by the reaction of H_2O_2 yielded by US.

To the best of our knowledge, the only work devoted to an heterogeneous sonophoto-Fenton process, is the work of Segura et al. (2009) that studied an integrated system combining US, with UV-vis and the Fenton reagent (photo-Fenton), using a heterogeneous Fe_2O_3 supported on SBA-15 material and phenol as the model pollutant. It appears that sequential sono-Fenton followed by photo-Fenton oxidation was more efficient compared to simultaneous and sequential photosono-Fenton, and it showed best cost-effective ratio for degradation of phenol under selected experimental conditions. The extent of degradation using this configuration can be possibly enhanced due to the benefits of US for the partial degradation of amenable aromatic compounds toward further oxidized by-products along with the fragmentation of catalyst to finer particles that increases their activity in the following photo-Fenton step. This hybrid process was able to get degradations up to 90% in terms of total organic carbon (TOC) mineralization under studied experimental conditions, compared to 15%, 20%, and 60% for individual sono-Fenton, photo-Fenton, and simultaneous combination, respectively.

CONCLUSIONS AND FUTURE OUTLOOK

In this chapter, the critical role of solid catalysts in ultrasonic systems has been outlined. Beside the promotion of additional nuclei sites for cavitation events, other significant effects can be pointed out, depending on the nature of the solid: exploitation of the sonoluminescence in the case of TiO_2, the continuous regeneration of the Fe^0 surface for the production of ferrous ions, and the catalyzed decomposition of sono-generated H_2O_2 molecules, among other effects. All of them enhance the overall results of the ultrasonic systems by increasing either the oxidation rate or the degradation grade of the organic pollutants. Additionally, a combination of heterogeneous US/catalyst systems with H_2O_2 promotes the oxidation of pollutants in different ways, although the final result is an increase in the hydroxyl radical production during the sonication process. In the case of the solid catalysts, a significant variable is the size of the particles, as well as the nature of the solid, whereas when hydrogen peroxide is added to the sonicated system, the initial concentration of the oxidant has a major influence on the overall reaction. It should be pointed out that a minimum concentration is necessary to obtain the above-mentioned benefits, although the hydrogen peroxide consumption in the process is usually far from 100%.

With regard to hybrid sonophoto systems, it should be noted here that all the studies related to hybrid systems are restricted to volumes <1 L, and, therefore, the scale-up of these reactors for actual wastewater treatment applications will be difficult, as the individual techniques (ultrasonic irradiation and photoassisted processes) have inherent limitations when considered for large-scale applications.

Overall, it can be said that combination of US with other oxidation techniques, such as photoassisted processes, leads to enhanced generation of the hydroxyl radicals, which eventually results in higher oxidation rates. The further addition of hydrogen peroxide or ozone to this hybrid system until an optimum value (additional source of free radicals) also increases the extent of destruction. Generally, the expected synergism between different hybrid methods discussed in this part of the work, is mainly due to the free radical attack mechanism. However, as it has been previously mentioned, the efficacy of the process and the extent of synergism depend not only on the enhancement in the production of free radicals but also the conditions and configuration leading to a better catalytic activity. The optimum value in terms of degradation strongly depends on operation conditions (frequency of irradiation as well as total power input by both US transducers and UV irradiation) and also the type of studied pollutant.

REFERENCES

Abbasi, M. and Asl, N.R. 2008. Sonochemical degradation of Basic Blue 41 dye assisted by nanoTiO$_2$ and H$_2$O$_2$. *Journal of Hazardous Materials*, 153: 942–947.

Abdullah, A.Z. and Ling, P.Y. 2010. Heat treatment effects on the characteristics and sonocatalytic performance of TiO$_2$ in the degradation of organic dyes in aqueous solution. *Journal of Hazardous Materials*, 173: 159–167.

Ai, Z.H., Lu, L., Li, J. et al. 2007a. Fe-Fe$_2$O$_3$ core-shell nanowires as iron reagent. 1. Efficient degradation of rhodamine B by a novel sono-Fenton process. *The Journal of Physical Chemistry C*, 111: 4087–4093.

Ai, Z.H., Lu, L., Li, J. et al. 2007b. Fe-Fe$_2$O$_3$ core-shell nanowires as the iron reagent. 2. An efficient and reusable sono-Fenton system working at neutral pH. *The Journal of Physical Chemistry C*, 111: 7430–7436.

Berberidou, C., Poulios, I., Xekoukoulotakis, N.P. et al. 2007. Sonolytic, photocatalytic and sonophotocatalytic degradation of malachite green in aqueous solutions. *Applied Catalysis B: Environmental*, 74: 63–72.

Bernardo, E.C., Fukuta, T., Fujita, T. et al. 2006. Enhancement of saccharin removal from aqueous solution by activated carbon adsorption with ultrasonic treatment. *Ultrasonics Sonochemistry*, 13: 13–18.

Bremner, D.H., Molina, R., Martinez, F. et al. 2009. Degradation of phenolic aqueous solutions by high frequency sono-Fenton systems (US-Fe$_2$O$_3$/SBA-15-H$_2$O$_2$). *Applied Catalysis B: Environmental*, 90: 380–388.

Chand, R., Ince, N.H., Gogate, P.R. et al. 2009. Phenol degradation using 20, 300 and 520 kHz ultrasonic reactors with hydrogen peroxide, ozone and zero valent metals. *Separation and Purification Technology*, 67: 103–109.

Chen, F., Li, Y., Cai, W. et al. 2010. Preparation and sono-Fenton performance of 4A-zeolite supported α-Fe$_2$O$_3$. *Journal of Hazardous Materials*, 177: 743–749.

Chen, Y. and Smirniotis, P. 2002. Enhancement of photocatalytic degradation of phenol and chlorophenols by ultrasounds. *Industrial and Engineering Chemistry Research*, 41: 5958–5965.

Dadjour, M.F., Ogino, C., Matsumura, S. et al. 2005. Kinetics of disinfection of Escherichia coli by catalytic ultrasonic irradiation with TiO$_2$. *Biochemical Engineering Journal*, 25: 243–248.

Dadjour, M.F., Ogino, C., Matsumura, S. et al. 2006. Disinfection of *Legionella pneumophila* by ultrasonic treatment with TiO$_2$. *Water Research*, 40: 1137–1142.

Dai, Y., Li, F., Ge, F. et al. 2006. Mechanism of the enhanced degradation of pentachlorophenol by ultrasound in the presence of elemental iron. *Journal of Hazardous Materials*, B137: 1424–1429.

Davydov, L., Reddy, E.P., France, P. et al. 2001. Sonophotocatalytic destruction of organic contaminants in aqueous systems on TiO$_2$ powders. *Applied Catalysis B: Environmental*, 32: 95–105.

Entezari, M.H., Mostafai, M., and Sarafraz-yazdi, A. 2006. A combination of ultrasound and a bio-catalyst: Removal of 2-chlorophenol from aqueous solution. *Ultrasonics Sonochemistry*, 13: 37–41.

Entezari, M.H. and Petrier, C. 2003. A combination of ultrasound and oxidative enzyme: Sono-biodegradation of substituted phenols. *Ultrasonics Sonochemistry*, 10: 241–246.

Entezari, M.H. and Petrier, C. 2004. A combination of ultrasound and oxidative enzyme: Sono-biodegradation of phenol. *Applied Catalysis B: Environmental*, 53: 257–263.

Entezari, M.H. and Petrier, C. 2005. A combination of ultrasound and oxidative enzyme: Sono-enzyme degradation of phenols in a mixture. *Ultrasonics Sonochemistry*, 12: 283–288.

Eren, Z. and Ince, N.H. 2010. Sonolytic and sonocatalytic degradation of azo dyes by low and high frequency ultrasound. *Journal of Hazardous Materials*, 177: 1019–1024.

Gaitan, D.F., Crum, L.A., Church, C.C. et al. 1992. Sonoluminescence and bubble dynamics for a single, stable, cavitation bubble. *The Journal of Acoustic Society of America*, 91: 3166–3183.

Ge, J. and Qu, J. 2003. Degradation of azo dye acid red B on manganese dioxide in the absence and presence of ultrasonic irradiation. *Journal of Hazardous Materials*, B100: 197–207.

Gogate, P. 2008. Treatment of wastewater streams containing phenolic compounds using hybrid techniques based on cavitation: A review of the current status and the way forward. *Ultrasonics Sonochemistry*, 15: 1–15.

Gogate, P.R., Mujumdar, S., and Pandit, A.B. 2002. A sonophotochemical reactor for the removal of formic acid from wastewater. *Industrial and Engineering Chemistry Research*, 41: 3370–3378.

Gogate, P.R. and Pandit, A.B. 2004. A review of imperative technologies for wastewater treatment II: Hybrid methods. *Advances in Environmental Research*, 8: 553–597.

Hardcastle, J.L., Ball, J.C., Hong, Q. et al. 2000. Sonoelectrochemical and sonochemical effects of cavitation: Correlation with interfacial cavitation induced by 20 kHz ultrasound. *Ultrasonics Sonochemistry*, 7: 7–14.

Ince, N.H. and Belen, R.A. 2001. Aqueous phase disinfection with power ultrasound: Process kinetics and effect of solid catalysts. *Environmental Science and Technology*, 35: 1885–1888.

Keck, A., Gilbert, E., and Köster, R. 2002. Influence of particles on sonochemical reactions in aqueous solutions. *Ultrasonics*, 40: 661–665.

Kim, J.K., Martinez, F., and Metcalfe, I.S. 2007. The beneficial role of use of ultrasound in heterogeneous Fenton-like system over supported copper catalysts for degradation of *p*-chlorophenol. *Catalysis Today*, 124: 224–231.

Kritikos, D.E., Xekoukoulotakis, N.P., Psillakis, E. et al. 2007. Photocatalytic degradation of reactive black 5 in aqueous solutions: Effect of operating conditions and coupling with ultrasound irradiation. *Water Research*, 41: 2236–2246.

Li, J.-T., Li, M., Li, J.-H. et al. 2007. Decolorization of azo dye direct scarlet 4BS solution using exfoliated graphite under ultrasonic irradiation. *Ultrasonics Sonochemistry*, 14: 241–245.

Liang, J., Komarov, S., Hayashi, N. et al. 2007. Improvement in sonochemical degradation of 4-chlorophenol by combined use of Fenton-like reagents. *Ultrasonics Sonochemistry*, 14: 201–207.

Lin, J., Chang, C., and Wu, J. 1996. Decomposition of 2-chlorophenol in aqueous solution by ultrasound/H_2O_2 process. *Water Science and Technology*, 33: 75–81.

Lin, J.-J., Zhao, X.-S., Liu, D. et al. 2008. The decoloration and mineralization of azo dye C.I. Acid Red 14 by sonochemical process: Rate improvement via Fenton's reactions. *Journal of Hazardous Materials*, 157: 541–546.

Liu, H., Li, G., Qua, J. et al. 2007. Degradation of azo dye Acid Orange 7 in water by Fe^0/granular activated carbon system in the presence of ultrasound. *Journal of Hazardous Materials*, 144: 180–186.

Mahamuni, N.N. and Adewuyi, Y.G. 2009. Advanced oxidation processes (AOPs) involving ultrasound for waste water treatment: A review with emphasis on cost estimation. *Ultrasonics Sonochemistry*, 17: 990–1003.

Melero, J.A., Martínez, F., and Molina, R. 2008. Effect of ultrasound on the properties of heterogeneous catalysts for sono-Fenton oxidation processes. *Journal of Advanced Oxidation Technologies*, 11: 75–83.

Méndez-Arriaga, F., Torres-Palma, F.R.A., Pétrier, C. et al. 2009. Mineralization enhancement of a recalcitrant pharmaceutical pollutant in water by advanced oxidation hybrid processes. *Water Research*, 43: 3984–3991.

Molina, R., Martínez, F., Melero, J.A. et al. 2006. Mineralization of phenol by a heterogeneous ultrasound/Fe-SBA-15/H_2O_2 process: Multivariate study by factorial design of experiments. *Applied Catalysis B: Environmental*, 66: 198–207.

Mrowetz, M., Pirola, C., and Selli, E. 2003. Degradation of organic water pollutants through sonophotocatalysis in presence of TiO_2. *Ultrasonics Sonochemistry*, 10: 247–254.

Murakami, N., Chiyoya, T., Tsubota, T. et al. 2008. Switching redox site of photocatalytic reaction on titanium (IV) oxide particles modified with transition-metal ion controlled by irradiation wavelength. *Applied Catalysis A: General*, 348: 148–152.

Muruganandham, M., Yang, J.-S., and Wu, J.J. 2007. Effect of ultrasonic irradiation on the catalytic activity and stability of goethite catalyst in the presence of H_2O_2 at acidic medium. *Industrial and Engineering Chemistry Research*, 46: 691–698.

Nakajima, A., Sasaki, H., Kameshima, Y. et al. 2007. Effect of TiO_2 powder addition on sonochemical destruction of 1,4-dioxane in aqueous systems. *Ultrasonics Sonochemistry*, 14: 197–200.

Nakui, H., Okitsu, K., Maeda, Y. et al. 2007. Effect of coal ash on sonochemical degradation of phenol in water. *Ultrasonics Sonochemistry*, 14: 191–196.

Nakui, H., Okitsu, K., Maeda, Y. et al. 2009. Sonochemical decomposition of hydrazine in water: Effects of coal ash and pH on the decomposition and adsorption behaviour. *Chemosphere*, 76: 716–720.

Namkung, K.C., Burgess, A.E., Bremner, D.H. et al. 2008. Advanced Fenton processing of aqueous phenol solutions: A continuous system study including sonication effect. *Ultrasonics Sonochemistry*, 15: 171–176.

Neppolian, B., Park, J.-P., and Choi, H. 2004. Effect of Fenton-like oxidation on enhanced oxidative degradation of para-chlorobenzoic acid by ultrasonic irradiation. *Ultrasonics Sonochemistry*, 11: 273–279.

Nikolopoulos, A.N., Igglessi-Markopoulou, O., and Papayannakos, N. 2006. Ultrasound assisted catalytic wet peroxide oxidation of phenol: Kinetics and intraparticle diffusion effects. *Ultrasonics Sonochemistry*, 13: 92–97.

Ogi, H., Hirao, M., and Shimoyama, M. 2002. Activation of TiO_2 photocatalyst by single-bubble sonoluminescence for water treatment. *Ultrasonics*, 40: 649–650.

Pandit, A., Gogate, P., and Mujumdar, S. 2001 Ultrasonic degradation of 2:4:6 trichlorophenol in presence of TiO_2 catalyst. *Ultrasonics Sonochemistry*, 8: 227–231.

Papadaki, M., Emery, R.J., Abu-Hassan, M.A. et al. 2004. Sonocatalytic oxidation processes for the removal of contaminants containing aromatic rings from aqueous effluents. *Separation and Purification Technology*, 34: 35–42.

Ragaini, V., Selli, E., Bianchi, C.L. et al. 2001. Sono-photocatalytic degradation of 2-chlorophenol in water: Kinetic and energetic comparison with other techniques. *Ultrasonics Sonochemistry*, 8: 251–258.

Ragaini, R., Selli, S., Bianchi, C.L. et al. 2005. Degradation of methyl tert-butyl ether in water: Effects of the combined use of sonolysis and photocatalysis. *Ultrasonics Sonochemistry*, 12: 395–400.

Rokhina, E.V., Lahtinen, M., Nolte, M.C.M. et al. 2009. The influence of ultrasound on the RuI_3-catalyzed oxidation of phenol: Catalyst study and experimental design. *Applied Catalysis B: Environmental*, 87: 162–170.

Rokhina, E.V., Repo, E., and Virkutyte, J. 2010. Comparative kinetic analysis of silent and ultrasound-assisted catalytic wet peroxide oxidation of phenol. *Ultrasonics Sonochemistry*, 17: 541–546.

Sangave, P.C. and Pandit, A.B. 2006. Ultrasound and enzyme assisted biodegradation of distillery wastewater. *Journal of Environmental Management*, 80: 36–46.

Segura, Y., Molina, R., Martínez, F. et al. 2009. Integrated heterogeneous sono-photo Fenton processes for the degradation of phenolic aqueous solutions. *Ultrasonics Sonochemistry*, 16: 417–424.

Selli, E., Bianchi, C.L., Pirola, C. et al. 2005. Degradation of methyl *tert*-butyl ether in water: Effects of the combined use of sonolysis and photocatalysis. *Ultrasonics Sonochemistry*, 12: 395–400.

Shimizu, N., Ogino, C., Dadjour, M.F. et al. 2007. Sonocatalytic degradation of methylene blue with TiO_2 pellets in water. *Ultrasonics Sonochemistry*, 14: 184–190.

Shirgaonkar, I.Z. and Pandit, A.B. 1998. Sonophotochemical destruction of aqueous solution of 2,4,6-trichlorophenol. *Ultrasonics Sonochemistry*, 5(2): 53–62.

Silva, A.M.T., Nouli, E., Carmo-Apolinario, A.C. et al. 2007. Sonophotocatalytic/H_2O_2 degradation of phenolic compounds in agro-industrial effluents. *Catalysis Today*, 124: 232–239.

Song, Y.-L. and Li, J.-T. 2009. Degradation of C.I. Direct Black 168 from aqueous solution by fly ash/H_2O_2 combining ultrasound. *Ultrasonics Sonochemistry*, 16: 440–444.

Tauber, M.M., Georg, M.G.M., and Rehorek, A. 2005. Degradation of azo dyes by laccase and ultrasound treatment. *Applied and Environmental Microbiology*, 71: 2600–2607.

Torres-Palma, R.A., Nieto, J.I., Combet, E. et al. 2010. An innovative ultrasound, Fe^{2+} and TiO_2 photoassisted process for bisphenol a mineralization. *Water Research*, 44: 2245–2252.

Tuziuti, T., Yasui, K., Iida, Y. et al. 2004. Effect of particle addition on sonochemical reaction. *Ultrasonics*, 42: 597–601.

Tuziuti, T., Yasui, K., Sivakumar, M. et al. 2005. Correlation between acoustic cavitation noise and yield enhancement of sonochemical reaction by particle addition. *Journal of Physical Chemistry A*, 109: 4869–4872.

Walton, A. and Reynolds, G. 1984. Sonoluminescence. *Advances in Physics*, 33: 595–660.

Wang, J., Guo, B., Zhang, X. et al. 2005. Sonocatalytic degradation of methyl orange in the presence of TiO_2 catalysts and catalytic activity comparison of rutile and anatase. *Ultrasonics Sonochemistry*, 12: 331–337.

Wang, J., Jiang, Y., Zhang, Z. et al. 2007. Investigation on the sonocatalytic degradation of congo red catalyzed by nanometer rutile TiO_2 powder and various influencing factors. *Desalination*, 216: 196–208.

Wang, J., Lv, Y., Zhang, Z. et al. 2009. Sonocatalytic degradation of azo fuchsine in the presence of the Co-doped and Cr-doped mixed crystal TiO_2 powders and comparison of their sonocatalytic activities. *Journal of Hazardous Materials*, 170: 398–404.

Wang, J., Lv, Y., Zhang, L. et al. 2010a. Sonocatalytic degradation of organic dyes and comparison of catalytic activities of CeO_2/TiO_2, SnO_2/TiO_2 and ZrO_2/TiO_2 composites under ultrasonic irradiation. *Ultrasonics Sonochemistry*, 17: 642–648.

Wang, J., Ma, T., Zhang, Z. et al. 2006. Investigation on the sonocatalytic degradation of methyl orange in the presence of nanometer anatase and rutile TiO_2 powders and comparison of their sonocatalytic activities. *Desalination*, 195: 294–305.

Wang, N., Zhu, L., Wang, M. et al. 2010b. Sono-enhanced degradation of dye pollutants with the use of H_2O_2 activated by Fe_3O_4 magnetic nanoparticles as peroxidase mimetic. *Ultrasonics Sonochemistry*, 17: 78–83.

Yano, J., Matsuura, J., Ohura, H. et al. 2005. Complete mineralization of propyzamide in aqueous solution containing TiO_2 particles and H_2O_2 by the simultaneous irradiation of light and ultrasonic waves. *Ultrasonics Sonochemistry*, 12: 197–203.

Zhang, H., Fu, H., and Zhang, D. 2009a. Degradation of C.I. Acid Orange 7 by ultrasound enhanced heterogeneous Fenton-like process. *Journal of Hazardous Materials*, 172: 654–660.

Zhang, H., Lv, Y., and Zhang, D. 2008. Degradation of C.I. Acid Orange by ultrasound enhanced ozonitation in a rectangular air-lift reactor. *Chemical Engineering Journal*, 138: 231–238.

Zhang, H., Zhang, J., Zhang, C. et al. 2009b. Degradation of C.I. Acid Orange 7 by the advanced Fenton process in combination with ultrasonic irradiation. *Ultrasonics Sonochemistry*, 16: 325–330.

Zhaobing, G. and Feng, R. 2009. Ultrasonic irradiation-induced degradation of low-concentration bisphenol A in aqueous solution. *Journal of Hazardous Materials*, 163: 855–860.

Zhou, T., Li, Y., Wong, F.-S. et al. 2008. Enhanced degradation of 2,4-dichlorophenol by ultrasound in a new Fenton like system (Fe/EDTA) at ambient circumstance. *Ultrasonics Sonochemistry*, 15: 782–790.

Zhou, T., Lima, T.-T., Lu, X. et al. 2009b. Simultaneous degradation of 4CP and EDTA in a heterogeneous Ultrasound/Fenton like system at ambient circumstance. *Separation and Purification Technology*, 68: 367–374.

Zhou, T., Lua, X., Wang, J. et al. 2009a. Rapid decolorization and mineralization of simulated textile wastewater in a heterogeneous Fenton like system with/without external energy. *Journal of Hazardous Materials*, 165: 193–199.

18 Degradation of Organic Pollutants Using Ultrasound

Kandasamy Thangavadivel, Mallavarapu Megharaj, Ackmez Mudhoo, and Ravi Naidu

CONTENTS

Introduction 447
Principles of Ultrasound-Aided Organic Pollutant Degradation 448
    Ultrasound 451
    Ultrasonic Frequency and Intensity 452
    Operating Pressure and Temperature 453
    Ultrasonic Reactor 453
Limits to Ultrasonic-Aided Organic Pollutant Degradation 454
Benefits of Ultrasound-Aided Degradation 455
Experimental Results 455
    Ultrasound-Assisted Dye Degradation 465
    Ultrasound-Assisted Pesticides, Insecticides, and Herbicides Degradation 467
    Ultrasound-Assisted Pharmaceuticals Degradation 468
    Ultrasound-Assisted Hormones Degradation 469
Suggested Sonication System for Organic Pollutant Degradation 469
Conclusion 470
References 470

INTRODUCTION

Most organic pollutants are hydrocarbon based and when these are halogenated they become more persistent in the environment and more hazardous to humans and other living organisms (Andrea et al., 2001). During degradation, hydrocarbon pollutants are broken down into simpler molecules such as short-chain organic acids or carbon dioxide and water and/or inorganic ions (Adewuyi, 2001). Today, organic pollutants are generally remediated using biological, chemical, physical, and physicochemical process or a combination of these (Andrea et al., 2001).

Biological treatment is currently the single largest organic remediation process in use worldwide. Municipal wastewater treatment plants in virtually all countries use aerobic and anaerobic processes to remediate municipal sewerage water. Bioremediation is also extensively used to remediate hydrocarbon-contaminated soils (Thangavadivel et al., 2009). The main advantages of the bioremediation process are its low cost and the fact that it does not require any chemicals other than nutrients. However, it is a slow process, do not withstand shock loading, and is not efficient at very high concentrations of organic pollutants or where these are in combination with other toxic cocontaminants. It also generates large volumes of sludge as a by-product, which requires disposal (Khanal et al., 2007). Because of the slowness of the process, it requires big reactors, which demand significant capital investment, as in the case of municipal wastewater plants. Also, microbes are unable to effectively mineralize halogenated hydrocarbons such as DDT [1,1,1-trichloro-2,2-bis(p-chloro phenyl) ethane] and perfluorooctanyl sulfonate (PFOS). These factors limit the application of this process for remediating industrial wastewater or soil contaminated by pesticides and other persistent organic contaminated compounds (Thangavadivel et al., 2009).

A combination of physical and chemical absorption and adsorption processes are widely used to remove the hydrocarbons from wastewater, especially in the pretreatment section of water treatment plants. Because of its excellent absorption capacity, activated carbon is mainly used for this purpose. During this physical and chemical absorption or adsorption process, organic waste is transferred from the liquid waste stream to the solid waste on the absorbent surface. This requires sufficient contact time with the carbon for effective removal of the organics from the liquid. The absorbent materials then need to go to land fill for safe disposal after being fully loaded with organic pollutant, which can become a long-term liability. This process is quite expensive, especially in terms of operating cost. At present, the only technically viable option for remediating halogenated hydrocarbons is incineration. This is not only expensive but can also generate very harmful by-products if incineration is not well controlled.

Oxidizing agent such as H_2O_2 are sometimes used to degrade organic pollutants, especially for small quantity of wastewater, as they oxidize the organics much faster than biological methods (Gogate and Pandit, 2004). But they also require special reactors and generate chemical sludge by reacting with other inorganic contaminants. Bulk chemicals are required for this remediation process to work efficiently. All these factors make the process less economic and pose risks in terms of operational safety.

Cheap zero valent iron products are widely used to remediate halogenated organic chemicals. For this process to be effective, the organic pollutant needs to be in close contact with the zero valent iron so that mineralization can take place. The zero valent iron surface need to be continuously corroded so that electrons are released. Due to mass transport limitation of pollutants to the zero valent iron surface and low corrosion rates of zero valent iron, this process is very slow in soil but quite fast in the liquid medium (Thangavadivel et al., 2009). Efficient mixing of the iron with the wastewater is critical for this process and it is influenced by reactor design. The level of iron content in the treated water is a possible further problem. After removing the halogens using zero valent iron, the remaining organic pollutants need to be treated by other methods. UV light is another methods being used to remediate organically contaminated water in the pretreatment section of water treatment plants. Here, the main issues are light penetration of the water and the contact time necessary for remediation to occur. The life of the quartz sleeves, UV lamp, the high power requirement, lamp disposal, and cooling requirements are other factors that contribute to a high cost of operation. Also, unspent OH radicals may attack upstream equipment, which is another significant factor (Gogate and Pandit, 2004).

Ultrasound (US) is one of the most advanced oxidation technologies. It uses acoustic cavitation to achieve physical as well as chemical effects within the solution, which help to degrade organic pollutants. It operates in normal atmospheric conditions, does not generate sludge, does not require chemicals, is easy to install and operate (Thompson and Doraiswamy, 1999). This section covers the ultrasonic-enhanced degradation of organic pollutants in water.

PRINCIPLES OF ULTRASOUND-AIDED ORGANIC POLLUTANT DEGRADATION

The hot-spot theory, electrical theory, and plasma discharge theory are three popular theories in sonochemistry (Chowdhury and Viraraghavan, 2009). Among these, hot-spot theory or cavitation theory is now widely accepted in sonochemistry. All laboratory experiments based on sonochemical degradation of organics are being interpreted based on this cavitation theory alone (Thompson and Doraiswamy, 1999).

US is applied to the liquid medium via a transducer. The mechanical vibration of the transducer is transmitted to the liquid and forms a pressure wave. This pressure wave moves in the medium as a sinusoidal wave. According to cavitation theory, during wave propagation, a repeating pattern of compression and rarefaction is generated in the medium. The rarefactions are

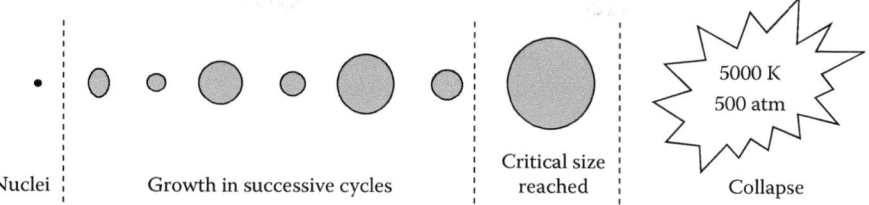

FIGURE 18.1 Bubble formation, growth, and collapse. (From Chowdhury, P. and Viraraghavan, T., *Sci. Total Environ.*, 407, 2474, 2009.)

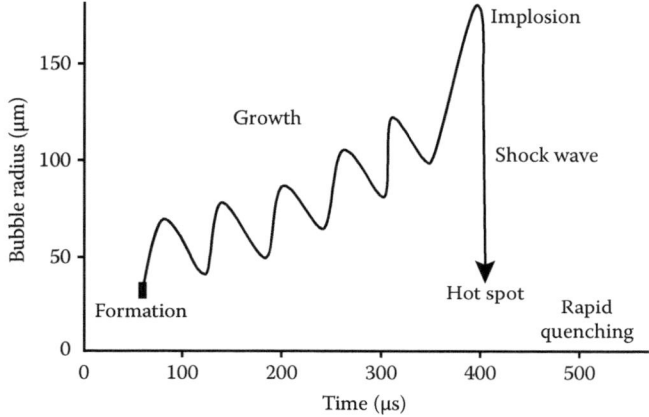

FIGURE 18.2 Graphical representation of bubble formation, growth, and implosion at 20 kHz ultrasound application in water. (Reprinted from *Sci. Total Environ.*, 407 Chowdhury, P. and Viraraghavan, T., Sonochemical degradation of chlorinated organic compounds, phenolic compounds and organic dyes—A review, 2474–2492, Copyright (2009), with permission from Elsevier.)

regions of low pressure (excessively large negative pressure) in which the liquid is literally "torn apart." As a result of these reduced pressures, microbubbles form in the rarefactions regions, with dissolved gases, fine particles, and other contaminants acting as a nuclei (Thompson and Doraiswamy, 1999). As it absorbs US wave energy, the bubble will grow during the low pressure cycle and shrink during the high pressure cycle (Figures 18.1 and 18.2). During the bubble growth time, water vapor, dissolved gases, and volatile organic matter enter the bubbles. As the bubble shrink, they also partially escape. Overall, more water vapor, dissolved gases, and organic vapor will enter the bubble during expansion than will escape during contraction because of the effect of bubble surface area (Figures 18.1 and 18.2). This process is called rectified diffusion (Chowdhury and Viraraghavan, 2009). When the bubble reaches its critical size (in μm) after few cycles (in few hundred μs), it will implode and temperature within the short-lived hot spot rises instantaneously to over 10,000°C, with the average temperature and pressure being 5000°C and 500 atm, respectively, for 20 kHz (Figure 18.1, Table 18.1).

Due to the very high temperatures, volatile organic matter is pyrolyzed and reformed as CO_2 and H_2O within the imploded bubble (Figure 18.3). The water vapor in the bubble becomes OH and H radicals. Other dissolved gases also produce radicals (Thompson and Doraiswamy, 1999). Generally, the pyrolysis is termed physical activity and the radical production contributing to degradation is called sonochemical activity. At the bubble surface and its surroundings, high temperatures and pressures create supercritical conditions in the water [e.g., about 20 nm from the bubble for 20 kHz US application in water (Table 18.1)]. This ionizes the surrounding water, including its organic pollutants, resulting in their decomposition. In the bulk liquid, the temperature remains at a level similar to room temperature because cavitation is an adiabatic process (Thompson and Doraiswamy, 1999). Due to

TABLE 18.1
Summary of Estimated Bubble Parameters (Average) under 20 kHz Sonication in Water at Room Temperature

Description	Value
Frequency	20 kHz
Wavelength in water	75 mm
Cycle time	50 μs
Minimum ultrasonic energy required for bubble formation	0.33 W cm^{-2}
Bubble collapse time	16.2 μm
Bubble resonance radius	178 μm
Bubble surface area	3.96×10^5 μm^2
Bubble volume	2.34×10^7 μm^3
Number of bubble per liters at 200 W/L	5.07×10^3
Bubble collapse temperature	5000°C
Bubble collapse pressure	500 atm
Bubble cooling rate	10^9 K s^{-1}
Reaction zone from bubble surface	200 nm
Reaction zone life time	<2 μs
Produced water jet velocity	100 m s^{-1}

Sources: Thompson, L.H. and Doraiswamy, L.K., *Ind. Eng. Chem. Res.*, 38, 1215, 1999; Adewuyi, Y.G., *Environ. Sci. Technol.*, 39, 3409, 2005; Chowdhury, P. and Viraraghavan, T., *Sci. Total Environ.*, 407, 2474, 2009.)

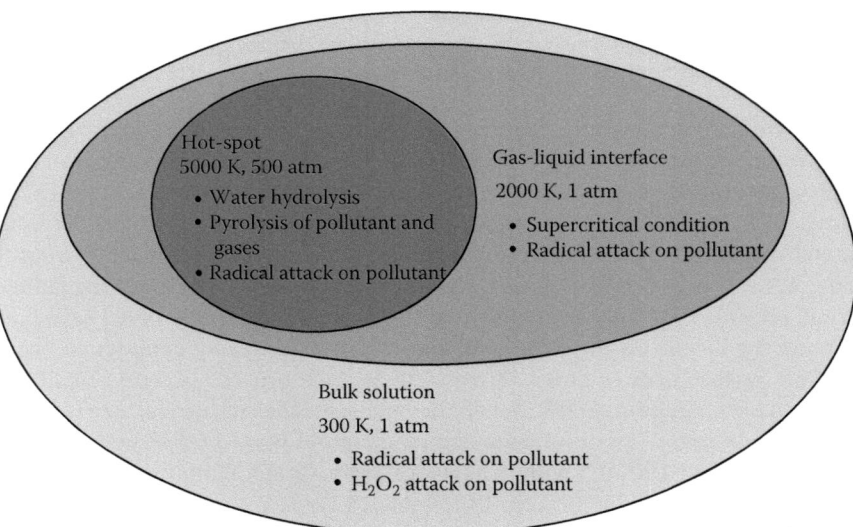

FIGURE 18.3 Reaction zone in cavitation process. (From Chowdhury, P. and Viraraghavan, T., *Sci. Total Environ.*, 407, 2474, 2009.)

this low temperature in the bulk liquid, the short-lived OH and H radicals remain in the cavity, recombine and form H_2O_2, which also degrades the pollutants in the bulk solution. The temperature is then lost to the bulk medium from the cavitation and helps to heat it, thereby reducing the intensity of cavitation. During cavity collapse, water jets will form with a speed of about 100 m s^{-1}. This contributes to the formation of an acoustic stream (Adewuyi, 2001).

The number of transient cavitation events per unit of time in a given volume of medium and the intensity of cavitation are the keys to achieving maximum remediation of the organic pollutant(s). Sonochemical activity takes place high in the frequency range of 200–600 kHz (Ashokkumar and Grieser, 2005).

When the intensity of US increases, the amplitude of the wave also increases. This helps the bubbles to absorb more ultrasonic energy and increases critical bubble sizes and, hence, the intensity of cavitation (Thompson and Doraiswamy, 1999). Intensity of cavitation is very important to obtain maximum pressure and temperature during bubble collapse and hence the optimal degradation rate.

For bubble formation, subsequent cavity collapse and energy release, unhindered nucleation is essential; dissolved gases or fine particles are used as nuclei (Kanthale et al., 2008). Thermal conductivity varies among the gases; however, gases with high thermal conductivity may heat up the solution and thereby reduce cavitation intensity. On the other hand, gases with a higher polytropic index will release more energy during cavity collapse (Adewuyi, 2001). This will lead to higher hot-spot temperatures and improved cavitation. Compared with a less soluble gas, the more soluble gases will tend to achieve better cavitation.

The Rayleigh-Plesset (RP) equation (Equation 18.1) is the mathematical equation that depicts the dynamics and mechanical aspects of the bubbles (Lin et al., 2002; Gehannin et al., 2009). The RP equation is normally solved to estimate the radius and pressure-pulse magnitude and history of bubble oscillation for a variety of conditions and is also used to analyze the sonochemical and sonoluminescence effects of cavitation. The solutions to the RP equation have been used to predict the maximum pressures and temperatures reached during the collapse of transient cavitation bubbles.

$$\frac{d\dot{R}}{dt} = \frac{1}{\rho_L}\left[P_{go}\left(\frac{R^{3\alpha}}{R^{3\alpha+1}}\right) + \frac{P_V}{R} - \frac{2\sigma}{R^2} - \frac{(P_b - P_a \sin \omega t)}{R}\right] - \frac{3\dot{R}}{2R} \quad (18.1)$$

The initial conditions are: at $t = 0$, $R = R_0$, and $\dot{R} = dR/dt = 0$, R is the instantaneous radius of the bubble or cavity, $\dot{R}$ is the bubble wall velocity, P_b is the atmospheric pressure (Ph), P_v is the vapor pressure in the bubble, P_{go} is the initial gas pressure inside the bubble, α ($=\kappa$) is the polytropic index of the saturated gas, which varies from γ (C_p/C_v) for adiabatic conditions to 1 for isothermal conditions.

The rate of OH radical formation in an ultrasonic process is estimated to be 19.8 μmol L^{-1} min^{-1} under argon and 14.7 μmol L^{-1} min^{-1} under air, while that of hydrogen formation is 20 μmol L^{-1} min^{-1} under argon in the sonolysis of pure water at 200 W and 200 kHz (Adewuyi, 2005). An increase in the ionic strengths of the electrolytes leads to higher degradation of water soluble organic compounds. However, ion content can help to heat up the solution easily, making additional cooling water necessary, which may adversely affect the cost-effectiveness of the process. The presence of fine inert particle will help to accelerate the degradation process by acting as a nuclei, but they may also attenuate the energy (Lifka et al., 2003). In general, the organic degradation process follows zero order at higher concentration and then become the first order with decrease in concentration.

ULTRASOUND

US energy is being supplied to the liquid pollutant in the form of sound waves. The absorbed US energy in the bubble is then released at the short-lived hot spot when the bubble implodes. The chemical and physical impacts of the US are produced at the hot spot and helps to degrade the pollutant. US also provides an excellent mass transport of the pollutant into the bubble by promoting turbulent mixing conditions. Henry's law constant and diffusion coefficient are the two major limiting parameters affecting the diffusion of organic compounds into the cavity, which determine the efficiency of pollutant degradation (Pétrier et al., 2007). In laboratory experiments, mmol L^{-1} level concentrations of organic pollutant are easily remediated in an average time of 30–90 min (Mahamuni and Adewuyi, 2009).

The ultrasonic process not only helps to degrade the pollutant within the cavity but also promotes faster degradation through millions of cavitation events occurring throughout the reactor at any particular point in time. It completely breaks down organic pollutants into simple molecules such as water and carbon dioxide and therefore produces no sludge. The formation of intermediates and their composition may vary, depending on treatment time and cavitation intensity. The most volatile organic compound undergoes degradation first, followed by the next most volatile and so on (Adewuyi, 2001).

Ultrasonic Frequency and Intensity

The environmental remediation industry generally uses ultrasonic frequencies ranging from 20 kHz to 2 MHz to remediate organic pollutants (Adewuyi, 2001). As the robustness of the ultrasonic transducer reduces with the increase in frequency, most industrial applications use the lowest frequency unit (Gallego-Juarez, 1989). These low frequency units are not only cheap but also available at different power ratings to suit various uses ranging from laboratory applications to large-scale industrial plants.

At low frequency, the physical effect of the US dominates, and is mainly used for cleaning and processing of food products (Vilkhu et al., 2008). However, the availability of high-intensity US makes it ideal for use in violent mixing, where extreme high temperatures and pressures are necessary to get the desired result. Because of violent cavitation of bigger cavities, the water jet produced is very strong and will agitate the solution vigorously. The number of cavitation events in low-frequency US are too less compared to high-frequency US. Since the bubbles are bigger under the low-frequency US compared to high-frequency US, it will have lot of water vapor into the cavities and it may not produce sufficient OH radicals similar to high-frequency US. The presence of high water vapor acts as a cushion and reduces the intensity of cavitation. Transient cavitation events are essential to produce faster remediation. However, this produces heat that results in a hotter solution, which in turn requires cooling, and this increases the treatment cost. To obtain the higher intensity, the process should be operated in the temperature range between 10°C and 20°C.

Inherently, low frequency units have very high intensity and as this rises, the cavitation will become far more violent. Too high an intensity may cause a decoupling effect between the vibrating surface and the liquid medium and, thus, lower the energy transmission to the liquid. It may also cause the bubbles to coalesce and, thus, reduce the cavitation intensity (Sunartio et al., 2007). These high-powered units are more suitable for remediation of very viscous liquids because the threshold of nucleation is much higher.

At a frequency range of 200–600 kHz, the chemical effect of cavitation will dominate, compared to physical effect (Thompson and Doraiswamy, 1999). Strong oxidizing agents and OH radicals will be produced in large quantities. This will accelerate the rate of pollutant degradation within the cavities, at the bubble surface and in the bulk medium. It has been reported that the sonochemical activities are high in the frequency range between 200 and 600 kHz (Thangavadivel et al., 2009). Although the bubble life and size are small compared with those generated by low-frequency US, nevertheless, the number of bubbles and homogeneity of their distribution throughout the medium is very high. Generally, the rise in temperature of the bulk solution is far less using high-frequency US than with low frequency. To minimize the nucleation threshold intensity, the operation should aim for a temperature range of 20°C–40°C.

In the MHz frequency range, generally, bubble lifetime and bubble size are too small for effective remediation to occur (Thangavadivel et al., 2009). Because of the extremely short cycle time, bubble vibration is too high and as a result, the intensity of cavitation is far less than that of low-frequency US. Smaller volumes, shorter durations, and greater vibration of bubble all limit the mass transport of the pollutant into a bubble compared to what occurs using lower-frequency US. High-frequency US does not produce much acoustic streaming. The solution temperature rise is minimal and the attenuation of energy very high (Thangavadivel et al., 2009).

Operating Pressure and Temperature

The temperature and pressure of the solution to be treated are important parameters in determining the intensity of the cavitation. When the intensity of cavitation is high, it will release high energy and, thus, degrade the pollutant more rapidly. Generally, if the temperature of the solution is high, then it forms the bubbles more readily because of the lower viscosity and surface tension of the liquid. At higher temperatures also, water is easily vaporized and enters the bubbles during their growth cycle. Therefore, for bubbles with a larger radius and longer cycle time, high fluid temperature is not very helpful because more water vapor occupies the cavities, where it serves as a cushion and thus reduces the cavitation effect. For smaller radius bubbles (high frequency), water vapor is not of much issue. Because of this, low-frequency US (20 kHz) demands a fluid temperature of about 10°C–20°C and 200–600 kHz US demands a temperature of 20°C–40°C, while the MHz range frequency requires 30°C–40°C.

Higher operating pressures hinder the nucleation and bubble growth process, but they help to accumulate more energy in the bubbles and so enhance the intensity of cavitation. Normally, high-frequency transducers are very fragile and they are being operated at atmospheric pressures (Gallego-Juarez, 1989). At higher operating pressures, the threshold of cavitation also increases, while the amount of water vapor entering the cavities declines. This results in lower OH radical production at high frequencies. However, at low frequencies it will help to increase the intensity of cavitation and also help to eliminate the cushioning effect of water vapor within the bubble.

Ultrasonic Reactor

Sonochemical reactor design plays a vital role in the degradation of pollution using US (Thompson and Doraiswamy, 1999). US is applied to the liquid medium via., ultrasonic transducer and, depending on the liquid medium's flow ability within the reactor, it can either use a batch system or flow cell can be employed. If the liquid is in close contact with the vibrating medium, then it is known as the direct application of US, whereas if the pollutant container is immersed in the liquid medium where US is being applied, it is called indirect sonication.

An ultrasonic horn produces a very high intensity US and can be easily applied to liquid in a sealed reactor vessel of varying shape. These reactors are mostly made of glass with water jacketing for cooling, equipped with ports for gas injection and withdrawal, sampling, and temperature and pressure measurement. An ultrasonic bath is also used in the laboratory, mainly for the indirect application of US to the pollutant. This produces low-intensity US compared with the standard probe unit. Generally, laboratory US reactors handle a volume range from 50 to 300 mL. Most batch reactors in laboratory and industry are standing wave reactors. Here, the liquid level in the standing wave reactor is very important for optimal intensity distribution (Little et al., 2007). The liquid level needs to be nodule plane of the ultrasonic wave, where sound wave pressure is zero so that there is minimal energy loss to the air. Apart from its size, the shape of the reactor is very important for achieving high intensity. A cone-shaped reactor provides the best intensity distribution among the standing wave type reactors.

To ensure the minimum nucleation threshold throughout the reactor, a sufficient number of transducers must be located throughout the reactor, based on size and shape of the reactor. To take full advantage of the intense physical effect of the process, flow cells are used. Since contact time is very short in the cell, a series of flow cells is used to ensure continuity of the remediation process (Gogate and Pandit, 2004).

Due to cavity collapse at the transducer and reactor surface, pitting occurs in both vessel and transducer. For this reason, reactors are, generally, made of glass or stainless steel, while ultrasonic probes are mostly made from aluminum and titanium. Since the greatest intensity of cavitation occurs near the vibrating area, it is better to have two or more transducers, to even out the intensity throughout the reactor and enhance the degradation rate. This also helps to minimize the pitting at the transducer surface. Overcoming wear and tear to the transducer material is the main challenge to overcome with respect to the operating cost of US. Most reactors are operated at atmospheric pressure and ambient temperature only, so they do not need to be particularly robust. Reactor temperature during sonication can be controlled automatically by adjusting the cooling water flow rate. Automation is also possible to protect the transducer from overheating and failing. Although many laboratory experimental results have been published on these issues, relatively few pilot scale results and very few industrial plant performances have so far been reported. During the scale-up of an US reactor, the intensity distribution, overall energy transfer efficiency, cooling water requirement, and method of US application are essential considerations. Modeling and simulation of these can be used to optimize reactor design. Energy transfer efficiency can be determined by (1) physical characterization using hydrophones, sensors, and power measurements; (2) calorimetry; and (3) chemical dosimeter. For optimizing reactor performance, hydrophones are ideal (Chowdhury and Viraraghavan, 2009).

LIMITS TO ULTRASONIC-AIDED ORGANIC POLLUTANT DEGRADATION

US has so far been used at mmol L^{-1} concentration ranges to remediate various organic pollutants in the laboratory (Adewuyi, 2001). However, if the pollutant concentration is very high, then the fluid viscosity also increases; creating a barrier to nucleation that must be overcome. Viscous bubbles may also hinder the transport of water vapor, dissolved gases and volatile organics into the bubble during rectified diffusion processes. This will in turn lessen OH radical production and reduce the degradation rate. It is, therefore, important to use plenty of energy to reach the nucleation threshold, and especially at high frequencies this can be a big challenge. Higher fluid viscosity also causes higher attenuation of energy as the viscous layers form on the transducer vibrating area, preventing energy transfer into the solution and heating the transducer/solution interface. Normally, organic pollutants occur in several combinations and under ultrasonication, the most volatile pollutants are degraded first, then the next most volatile and so on. This means it takes longer to completely degrade a solution containing a complex combination of pollutants. Bubbling a gas such as oxygen through the fluid can help to degrade the pollutants concurrently.

Overall, energy transfer efficiency is quite low in the ultrasonic process and is comparable to the cost of incineration. Electrical energy must be converted to mechanical vibration, and then to cavitation energy, and finally to pollution degradation via gross physical and chemical effects. It has been reported that the overall energy transfer efficiency of this process is below 10%.

Because of strong vibration of the liquid medium, degassing takes place at a very high rate, making continuous gas bubbling or a closed reactor with a gas saturated solution essential to ensure the nucleation process throughout the remediation process, especially for volatile organics remediation.

The US transducers are made of limited materials and these need to be chosen to be compatible with the pH of their operating conditions. Because of the pitting issue, transducer tips or cone will need to be changed frequently. Most standard units require special cooling for the reactor, which adds to the cost.

Some systems use a blend of chemicals and US to remediate organic pollutants. These chemicals not only contribute to the remediation rate but also help to heat up the solution and so affect the intensity of remediation (H_2O_2) (Adewuyi, 2001). For example, Fe is often added to remediate chlorinated hydrocarbon waste but, due to US-enhanced corrosion, this Fe dissolved in water becomes a pollutant. Depending on each individual circumstance, the system needs to be sized and fine tuned. Although the initial capital cost of using US is attractive, its operating cost can be high in terms of power consumption and the need to change transducer tips and horns. To be most effective, the ultrasonic medium should be liquid based. To obtain OH free radical, it needs to be water based. In the wastewater, organic degradation may be slow as the natural wastewater stream has a mixture of the pollutant and radical scavengers, which will limit the performance of the ultrasonic-enhanced remediation.

Flow cells are continuous treatment units that can supply very high-intensity US to the pollutant and so enhance and accelerate the degradation process (Thangavadivel et al., 2008). They are ideal for industrial applications and are being widely used in the food processing industry. After the flow cell, the upstream water needs to be cooled down and OH radicals removed to protect the piping and other equipment. US is used in combination with other oxidation technologies such as $UV/O_3/H_2O_2$/zero valent iron to enhance the degradation rate of pollutants by producing more OH free radicals (Adewuyi, 2001).

Although, several reports about laboratory-based US-assisted remediation are available, reports on field-scale remediation of contaminants using US are rare. Frequency, power transfer efficiency, applied intensity, treatment time, transducer position, reactor volume and shape, operating temperature, bubbling gas and flow rate, particle size and concentration, ionic strength, pollution composition, and concentration are all key parameters that need to be compared across the published reports.

It has been claimed that US can remediate almost any organic pollutant. However, because of its low energy transfer efficiency, the cost of treatment is high and this may explain why it is not more widely used. A clear research priority is to increase the overall energy transfer efficiency through innovative design and modeling of the transducer or reactor system, or else the OH radicals need to be produced at lower cost by other means. Only then will US treatment become a viable technology at the large scale.

BENEFITS OF ULTRASOUND-AIDED DEGRADATION

US is a proven technology for cleaning and food processing applications (Vilkhu et al., 2008). Recently, more research has been directed toward US application in organic pollutant remediation at a laboratory scale and the results are very promising. Degradation kinetics, reaction rate, concentration range, pollution composition, reactor configuration, tested frequency, temperature, pressure, intensity, bubbling gas, chemical used, conversion path way, treatment time, and achieved removal efficiency have been reported for many pollutants. Transducers, ultrasonic reactors, and accessories are available in the market. Therefore, more than ever, US is evolving as a promising green technology in environmental remediation field.

Most of the time, US does not require the use of any additional chemical and, therefore, is referred to as a chemical free technology or green technology (Mason, 2007). One of its major advantages is that it converts the pollutant into CO_2, H_2O, and other inorganic ions. Generally, this technology does not produce any harmful chemicals or sludge, which eliminates the need for costly sludge handling and disposal. This is a major benefit.

Most of the ultrasonic reactors are being operated in the atmospheric conditions and, hence, does not require costly process units. The technology is simple to install and operate, with all the spare parts readily available in the market. There is active research going on to further improve this technology.

Although US can be used as a standalone process, it can also be used in conjunction with other processes to improve efficiency and reduce costs. US can also be used in the pretreatment phase of a standard wastewater treatment plant, where it will not only help to accelerate the biological degradation process but also reduce sludge production.

EXPERIMENTAL RESULTS

Tables 18.2 and 18.3 consist of the degradation performance of certain selected organic pollutants under ultrasonic-enhanced degradation in batch treatment. It clearly shows that almost a large variety of organic pollutants can be remediated by using US at the laboratory scale under various

TABLE 18.2
Ultrasound-Assisted Remediation of Persistent Organic Pollutants

Contaminant	Concentration	Experimental Conditions	Degradation Efficiency/Other Result/Remarks	Reference
Phenol	75 and 115 mg L^{-1}	22 kHz horn, 240 W, 5.2% calorimetric efficient, 3.46 cm^2 radiating face, 95 mL without O$_3$ and 550 mL with O$_3$, pH 6.2, 31°C. 22 kHz bath, 120 W, 34.7% calorimetric efficient, 225 cm^2 radiating face, 3.15 L. NaCl and CCl$_4$ as an additive to the solution	1. Initial rate of phenol degradation in min^{-1}, for horn 0.0015, bath 0.0082, ozonation (35 mL s^{-1}) 0.1048, 2% NaCl with horn 0.0017, 8% NaCl with horn 0.0022, 76 mg L^{-1} CCl$_4$ with horn 0.001, 413 mg L^{-1} CCl$_4$ with horn 0.0047 2. Initial rate drop after 7–90 min treatment. The total treatment time is about 120 min additive enhanced the phenol degradation by forming intermediate product without additive, 15% removal efficiency attained in 60 min	Mahamuni and Pandit (2006)
Phenol	10 mg L^{-1}	200 kHz vibrator plate on bottom of water bath, 200 W, 33.19 cm^2 radiating face, 100 mL in a 50 mm diameter beaker kept in a water bath of 20°C by external cooling. The beaker kept in the nodal plane of the sound wave from vibrator. Coal ash also used to enhanced degradation rate	In 60 min, phenol degradation rate was stir > ash (0.5 wt%) > US > US + ash (0.5 wt%). The optimum amount of coal ash was 0.4–0.6 wt% with 53–106 μm in particle size. The surface roughness of the coal ash act as a nucleation site for bubble. This help to produce more OH radical and accelerate phenol degradation in 60 min, about 85% removal efficiency attained	Nakui et al. (2007)
Phenol	5 mmol L^{-1} = 470.55 mg L^{-1}	1. 20 kHz horn, 180 W, 60.8% calorimetric efficient, 1.13 cm^2 radiating face, 80 mL water jacketed reactor 2. 300 kHz flat transducer on bottom of reactor, 25 W, 56% calorimetric efficient, 150 mL reactor 3. 520 kHz flat transducer on bottom of reactor, 100 W, 84% calorimetric efficient, 1200 mL reactor	Phenol degradation efficiency 1. 300 kHz > 520 kHz > 20 kHz 2. Acidic > neutral > alkaline 3. 5 > 2 > 1 > 0.5 mM initial concentration 4. Air > argon gas flow 5 L min^{-1} resonating bubble size and nature of the bubble collapse were the reason for frequency effect treatment time was about 90 min max of 60% removal achieved	Kidak and Ince (2006)
Phenol	0.662–0.678 mmol L^{-1} = 63 mg L^{-1}	1. 20 kHz horn fitted on bottom of the reactor, 50 W calorimetric power, 12.57 cm^2 radiating face, 350 mL water jacketed reactor at 20°C 2. 500 kHz disk fitted on bottom of the reactor, 50 W calorimetric power, 12.57 cm^2 radiating face, 350 mL water jacketed reactor at 20°C 3. 35 kHz sonitube, 50 W calorimetric power, 3.143 cm^2 radiating face, 350 mL water jacketed reactor at 20°C	Phenol degradation efficiency 1. 35 kHz > 500 kHz > 20 kHz 2. At 35 kHz, the reaction rate is four times that of 500 kHz 3. By using 35 kHz US, 100% removal obtained in 80 min	Entezari et al. (2003)

Pollutant	Concentration	Reactor details	Observations	Reference
2-Chlorophenol	0.1 mmol L^{-1} = 12.9 mg L^{-1}	500 kHz disk fitted on bottom of the reactor, 4.8, 15.1, and 25.2 W calorimetric power, 12.57 cm^2 radiating face, 350 mL water jacketed	2-Chlorophenol degradation 1. Increases with temperature under US 2. Increases with temperature under US + enzyme 3. Increases with US power 4. Increases with US + enzyme dosage increment it took about 30–60 min to degrade 2-chlorophenol of 0.1 mmol L^{-1}	Entezari et al. (2004)
4-Chlorophenol	500 μmol L^{-1} = 12.8 mg L^{-1}	1. 200 kHz disk fitted on bottom of the reactor, 30 W, 12.57 cm^2 radiating face, 250 mL reactor at 20°C 2. 500 kHz disk fitted on bottom of the reactor, 30 W, 12.57 cm^2 radiating face, 250 mL reactor at 20°C 3. 800 kHz disk fitted on bottom of the reactor, 30 W, 12.57 cm^2 radiating face, 250 mL reactor at 20°C 4. 20 kHz horn, 30 W, 9.63 cm^2 radiating face, 250 mL reactor at 20°C	At 10°C, $r = 1.2$ μmol L^{-1} min^{-1}, 20 kHz At 20°C, $r = 1.05$ μmol L^{-1} min^{-1}, 20 kHz At 30°C, $r = 0.9$ μmol L^{-1} min^{-1}, 20 kHz At 40°C, $r = 0.8$ μmol L^{-1} min^{-1}, 20 kHz At 10°C, $r = 2.5$ μmol L^{-1} min^{-1}, 500 kHz At 20°C, $r = 2.8$ μmol L^{-1} min^{-1}, 500 kHz At 30°C, $r = 3$ μmol L^{-1} min^{-1}, 500 kHz At 40°C, $r =$ μmol L^{-1} min^{-1}, 500 kHz At 20°C, 20 kHz, $r = 1$ μmol L^{-1} min^{-1} At 20°C, 800 kHz, $r = 2$ μmol L^{-1} min^{-1} At 20°C, 500 kHz, $r = 3$ μmol L^{-1} min^{-1} At 20°C, 200 kHz, $r = 6$ μmol L^{-1} min^{-1} It takes about 88 min to reduce the 4-chlorophenol concentration to 250 μ μmol L^{-1} (from 500 μ μmol L^{-1})	Jiang et al. (2006)
4-Chlorophenol	100 mg L^{-1}	28, 50, 100, 200, and 600 kHz disk fitted on bottom of the 1 L, 50 W calorimetric power, 126.73, 126.73, 50.3, 50.3, and 50.3 cm^2 are the radiating face respectively, reactor jacketed at 20°C–30°C	Removal efficiency 1. 600 kHz > 200 kHz > 50 kHz > 28 kHz 2. 100 mg/L of H$_2$O$_2$ gives higher degradation rate 3. 1 g L^{-1} iron powder with smaller particle size give good performance 4. pH = 3 with 6 μmol L^{-1} particle size and 100 mg L^{-1} H$_2$O$_2$ gives best result	Liang et al. (2007)
4-Chlorophenol and chlorobenzene	500 and 405 μmol L^{-1} respectively = 64.2 and 45.58 mg L^{-1}, respectively	300 kHz flat transducer on bottom of 300 mL reactor, 47 W calorimetric power, 20°C	More volatile component goes to the bubble and degrade first. When O$_2$ added, both component degrade simultaneously it takes about 80–160 min to degrade the mixture	Petrier et al. (2007)

(continued)

TABLE 18.2 (continued)
Ultrasound-Assisted Remediation of Persistent Organic Pollutants

Contaminant	Concentration	Experimental Conditions	Degradation Efficiency/Other Result/Remarks	Reference
p-Chlorophenol	0.09 mmol L^{-1} = 11.57 mg L^{-1}	1. 850 kHz transducer fitted in the bottom of water bath where 100 mL beaker is kept 0.5 cm above transducer, 140 W, 17% calorimetric efficiency, 70.91 cm^2 vibrating area, 36°C–39°C water bath 2. 20 kHz cup-horn fitted on the bottom of water bath where 100 mL beaker is kept 0.5 cm above transducer, 475 W, 4% calorimetric efficiency, 38.5 cm^2 vibrating cup-horn, 36°C–39°C water bath. 3. 20 kHz probe is inverted into the 100 mL beaker, 475 W, 4% calorimetric efficiency, 1.77 cm^2 vibrating area 36°C–39°C water bath	1. Initial reduction rate, probe > beaker > cup-horn 2. Overall reduction rate, beaker > probe > cup-horn in 120 min, complete degraded in beaker system 3. Up to 20 mmol L^{-1} of H_2O_2, p-chlorophenol's degradation rate increases. Then up to 40 mmol L^{-1}, it is flat and then it is decreases. pH reduces with time from 6 to 3	Teo et al. (2001)
p-Chlorophenol	35 and 2010 mg L^{-1}	1.7 MHz, 35 W, 3.14 cm^2 vibrating area fitted on the bottom of the reactor, 25°C water jacketed reactor, air tight reactor	Pyrolysis is the dominant mechanism and no intermediate detected. 35 mg L^{-1} and 2010 mg L^{-1} p-chlorophenol's degradation is 20.1% and 2.64% respectively in 1 h	Hao et al. (2004)
2,3,5-Trichlorophenol (TCP)	156–188 mg L^{-1}	41 kHz and 3.2 MHz fitted on the bottom of 500 mL reactor, 40 W calorimetric power, 25 cm^2 vibrating area each, 20°C water jacketed. (one transducer at a reactor)—60 min operation 360 kHz fitted on the bottom of 500 mL reactor, 100 W calorimetric power, 25 cm^2 vibrating area, 20°C water jacketed—240 min operation	1. After 60 min of TCP degradation, frequency (% degradation), 360 kHz (40%) > 206 kHz (35%) > 618 kHz (27%) > 1068 kHz (24%) > 41 kHz (2%) > 3217 kHz (0%) 2. Operating pH = 4 3. At 180 min, 188 mg L^{-1} TCP's degradation completed by using 360 kHz US	Tiehm and Neis (2005)
Pentachlorophenol (PCP)	100 mg L^{-1}	500 Hz, 20 kW, 3 L, water jacketed to avoid variation of temperature below 3°C, 0.165 W mL^{-1}	1. Sonication helps to dissolve more O_3 than non sonication and stirring 2. Ozonation with sonication enhance the degradation rate of PCP 3. Audible frequency increases the rate constant than ultrasonication because audible frequency produces strong turbulence 90% PCP of 100 mg L^{-1} degrade in 2 min under 24 mg min^{-1} ozone feed, pH 9.4 and sonication	Zeng and McKinley (2006)

Pollutant	Concentration	Conditions	Observations	Reference
Pentachlorophenol (PCP)	37.5 μmol L^{-1} = 10 mg L^{-1}	40 kHz, 600 W ultrasonicator fixed at the bottom of the water bath where two reactor with water jacketed at 20°C kept with stirrer	2 g L^{-1} of Fe with US gives faster rate of PCP degradation. A combination of US and iron is highly effective in organic pollutant removal. OH radical is the main for degrading PCP about 60% removal efficiency obtained in 120 min	Dai et al. (2006)
Trichloroethylene (TCE)	50 mg L^{-1}	20 kHz probe, 600 W, 2.836 cm^2 vibrating surface, 1 L reactor in a water bath at 10°C (reactor temp is 24°C)	1. Both from experiments and mass transfer, less volatile compounds (Hv < 0.1) Henry's constant exerts a positive influence on the sonolytic degradation, but the effect is not continual. Eventually, for highly volatile compounds (Hv > 1), the influence of Henry's constant on degradation is negligible 2. Liquid phase diffusion coefficient is an important parameter in the degradation process 3. Sonolytic degradation rates and mass transfer rates of volatile solutes increases with increasing power density 4. 40% degradation achieved in 10 min	Ayyildiz et al. (2007)
1,1,1-Trichloroethane (TCA)	80 mg L^{-1}	20 kHz horn with 11.406 cm^2 radiating area, 200 W, 250 mL water jacketed reactor where water is flowing for 10 min to complete a reactor volume. Sample taken in every 10 min	Optimum conditions for US treatment of TCA are at the pressure of 0.212 MPa, temperature of 14.2°C and pH 10.9	Gaddam and Cheung (2001)
1,1,1-Trichloroethane (MC) Tri-chloroethylene (TCE) Tetrachloroethylene (PCE)	Each 70 μmol L^{-1} = 9 mg L^{-1}	100 kHz plate transducer fixed on the bottom of 1300 mL vessel, 7.07 cm^2 radiating area, 140 W, 44.3% calorimetric efficiency, water jacketed at 20°C, batch and continuous system	Rate of degradation 1. Batch experiment, MC > PCE > TCE 2. Continuous flow experiment, conversion depends on flow rate 70%–90% yield obtained. Used flow rate was 7–30 mL min^{-1} 3. 90% removed in 40 min in batch and continuous system	Yim et al. (2001)
Chlorobenzene (CB) trichloroethylene (TCE)	Each up to 3440 μmol L^{-1} = 400 mg L^{-1}	520 kHz, 9.4 W calorimetric power, 150 mL water jacketed reactor at 29°C, pH = 7	1. CB and TCE's rate constant decreases with increase in concentration 2. In mixture, the rate constant of each component is lower than single component of same concentration 3. Pyrolysis is the major mechanism and OH radical too contributing 4. It took about 120 min to degrade 25 μmol L^{-1} (2.8 mg L^{-1}) of CB	Dewulf et al. (2001)

(continued)

TABLE 18.2 (continued)
Ultrasound-Assisted Remediation of Persistent Organic Pollutants

Contaminant	Concentration	Experimental Conditions	Degradation Efficiency/Other Result/Remarks	Reference
Tetrachloride trichloroethylene (TCE) 1,2,3-Tri-chloropropane	$100\,mg\,L^{-1}$	20 kHz horn, 60 W, $1.3278\,cm^2$ radiating area, 350 mL water jacketed reactor	For TCE 1. When the temperature increases, rate constant reduces and half-life increases 2. At 10°C, half-life is 60 min 3. When the power intensity increases, the rate constant increases and half-life reduces 4. $Ar > air > N_2$ the main mechanism is pyrolysis in the bubble	Lim et al. (2007)
$CHCl_3$ CCl_4 $CHBrCl_2$ $CHBrCl$	$15.79\,\mu g\,L^{-1}$ $10.43\,\mu g\,L^{-1}$ $3.19\,\mu g\,L^{-1}$ $4.75\,\mu g\,L^{-1}$	20 kHz horn, 500 W, 55 W calorimetric power, $0.2828\,cm^2$ vibrating area, 50 mL water jacketed reactor at 25°C	Removal efficiency in 1 h is, component (removal efficiency), $CHCl_3$ (48.2%), CCl_4 (64.6%), $CHCl_2Br$ (58.3), and $CHClBr_2$ (54.6) degradation rate from a mixture is $CCl_4 > CHCl_2Br > CHClBr_2 > CHCl_3$	Guo et al. (2006)
Naphthalene (N) Phenanthrene (Phe) Anthracene (Ant) Pyrene (Pyr)	$227\,\mu g\,L^{-1}$ $97\,\mu g\,L^{-1}$ $110\,\mu g\,L^{-1}$ $118\,\mu g\,L^{-1}$	1. 20 kHz horn, 30 W calorimetric power, $4.91\,cm^2$ vibrating face, 150 mL water jacketed reactor at 20°C 2. 500 kHz flat transducer with $12.5\,cm^2$ of radiating surface fitted on the bottom of the reactor, 30 W calorimetric power, 150 mL water jacketed reactor at 20°C	Degradation rate 1. For all component, high-frequency US yield higher rate constant than low-frequency US 2. $N > Phe > Ant > Pyr$ 3. Pyrolysis is the dominant mechanism and OH radical is a minor contribution in degradation 4. In 40 min, above 90% degrade	David (2009)

Pollutant	Concentration	Experimental conditions	Observations	References
PFOS PFOA	$100\,\mu g\,L^{-1}$ $100\,\mu g\,L^{-1}$	1. 354 kHz transducer fixed on the bottom of 600 mL reactor, 72% calorimetric efficiency, 250 W L^{-1}, water jacketed at 10°C 2. 612 kHz transducer fixed on the bottom of 600 mL reactor, 72% calorimetric efficiency, 250 W L^{-1}, water jacketed at 10°C PFOS and PFOA were remediated in landfill waste as well as DI water	PFOS and PFOA degradation rate 1. PFOA > PFOS for both frequency and DI water as well as ground water 2. For both components, Milli Q > groundwater 3. PFOS degrade 63% from groundwater contaminant in 140 min	Cheng et al. (2008)
Ammonia	5%–25% (vol)	1. 1.7 MHz plate transducer fitted on the bottom of the 400 mL reactor, 9.5 W, 3.1428 cm^2 vibrating area, no temperature control (18°C–33°C) 2. 2.4 MHz plate transducer fitted on the bottom of the 400 mL reactor, 9.5 W, 3.1428 cm^2 vibrating area, no temperature control	1. Ammonia removal efficiency was higher at 2.4 MHz than 1.7 MHz for all concentration 2. Removal efficiency reduces with liquid level increment and also with increment in ammonia concentration 3. In 2 h sonication, the maximum ammonia removal efficiency obtained was 32% by using 2.4 MHz US with 5% ammonia concentration	Matouq and Al-Anber (2007)

DDT and PCB	$250\,mg\,L^{-1}$ and $700\,mg\,L^{-1}$, respectively	20 kHz horn, 170 W, 1.23 cm^2 vibrating face, 50% sand slurry of 400 mL 100 Hz, 75 kW, 50% sand slurry	1. 20 kHz
			  a. In 10 min, DDT dropped by 70% from $250\,mg\,L^{-1}$ of DDT from 50% sand slurry
			  b. In 60 min, PCB dropped by 90% from $250\,mg\,L^{-1}$ of PCB from 50% sand slurry
			  c. In 5 min, PAH dropped by 80% from $400\,mg\,L^{-1}$ of PAH from 50% soil slurry
			2. 100 Hz $700\,mg\,L^{-1}$ of PCB dropped to $<2\,mg\,L^{-1}$ in 105 min and the power was 0.2 kW L^{-1}
				Mason et al. (2004)

TABLE 18.3
Ultrasound-Assisted Degradation of Dyes Found in Industrial and Wastewater Treatment Plant (WWTP) Effluents

Pollutant(s)	Degradation Conditions	Degradation Performance	References
Basic Red 29 (BR29)	Ultrasonic degradation of dye in the presence of Co^{2+}–H_2O_2. 20 mg L^{-1} initial dye concentration. Ultrasonic bath at a frequency of 40 kHz	Best experimental conditions were as follows: 1000 mg L^{-1} (Co^{2+}) Act, 1000 mg L^{-1} H_2O_2, 40°C and original pH of 6.70. 30 min of sonication in the presence of Co^{2+}–H_2O_2, BR29 removal efficiency of practically 100% was achieved	Yavuz et al. (2009)
Rhodamine B (RhB)	Sono-enhanced degradation with H_2O_2 as a green oxidant and Fe_3O_4 magnetic nanoparticles (MNPs) as a peroxidase mimetic	Fe_3O_4 MNPs could catalyze the break of H_2O_2 to remove RhB in a wide pH range from 3.0 to 9.0 and its peroxidase-like activity was significantly enhanced by the US irradiation. At pH 5.0 and temperature 55°C, the US-assisted H_2O_2-Fe_3O_4 catalysis removed about 95% of RhB (0.02 mmol L^{-1}) in 15 min with an apparent rate constant of 0.15 min^{-1} for the degradation of RhB, being 6.5 and 37.6 folds of that in the simple catalytic H_2O_2-Fe_3O_4 system, and the simple ultrasonic US-H_2O_2 systems, respectively	Wang et al. (2010)
Congo Red (CR)	CR degradation by ultrasonic waves (50 kHz) was investigated at 25°C	After sonication for 60 min, the CR concentration gradually decreased from 100 to 27.7 mg L^{-1}. There is 68% chemical oxygen demand removal efficiency after US treatment	Srinivas and Chintalapati (2008)
Rhodamine B	The influence of bicarbonate and carbonate ions on sonolytic degradation of RhB in water	The results clearly demonstrated the significant intensification of sonolytic destruction of RhB in the presence of bicarbonate and carbonate, especially at lower dye concentrations. Carbonate radicals sonochemically generated are suitable for total removal of COD of sonicated RhB solutions. In the presence of bicarbonate, degradation rate reached a maximum at 3 g L^{-1} bicarbonate, and in RhB solutions containing carbonate, the oxidation rate gradually increased with increasing carbonate concentration up to 10 g L^{-1}	Merouani et al. (2010)

Pollutant	Description	Findings	Reference
Basic Blue 41 (BB41)	Sonolysis of BB41 in aqueous solution 35 kHz using ultrasonic power of 160 W (25°C, 180 min). TiO_2 nanoparticles were used as a catalyst	During the decolorization, all nitrogen atoms and aromatic groups of BB41 were converted to urea, nitrate, formic acid, acetic acid, and oxalic acid. The results showed that power US can be regarded as an appropriate tool for degradation of azo dyes to nontoxic end products	Abbasi and Asl (2008)
CI Acid Orange 7 (AO7)	Combination of 20 kHz US and ozone for the degradation of CI AO7	The decolorization rate increased with the increase of power density and gas flow rate, but decreased with the increasing initial dye concentration. Low frequency US-enhanced ozonation process for the decolorization of CI AO7 and that was mainly a direct reaction rather than radical reaction	Zhang et al. (2008)
AO7	Degradation of AO7 using zero valent iron/granular activated carbon (Fe^0/GAC) in the absence and presence of US	AO7 degradation efficiency by Fe^0/GAC was dramatically enhanced by US. US alone had a little effect on the degradation. The pseudo first-order rate constant of AO7 degradation by Fe^0/GAC was 8.74×10^{-3} min^{-1} while that by US-Fe^0/GAC was 3.91×10^{-2} min^{-1}. A significant synergetic effect was observed between US and Fe^0/GAC	Liu et al. (2007)
Reactive Black 5 (RB5)	Removal of RB5 from aqueous solutions by the sorption process in the presence and in the absence of US	RB5 could be removed by higher frequency apparatus (500 kHz) without sorbent in about 60 min sonication. The rate of removal was higher at the higher frequency than at the lower one	Entezari et al. (2008)
Linear alkylbenzene sulfonates (LAS)	Sonochemical reactor for the degradation of LAS from WWTP effluents investigated with LAS solution using MB active substances (MBAS) method. A frequency of 130 kHz, acoustic power value of 400 W, temperature of 18°C–20°C and pH value of 6.8–7 were employed. Initial LAS concentrations: 0.2 mg L^{-1}, 0.5 mg L^{-1}, 0.8 mg L^{-1}, and 1 mg L^{-1}	LAS degradation increased with increasing sonication time. As LAS concentration increased, the degradation rate decreased	Debghani et al. (2010)

experimentally imposed degradation conditions. Table 18.4 summarizes some studies where different catalysts and/or chemicals have been used to facilitate the sonolytic degradation of pollutants. Recent studies have indicated that the presence of titanium dioxide (TiO_2), known as a photocatalyst, accelerates the generation of hydroxyl radicals during sonication, and that the process is mediated through the induction of cavitation bubbles in irradiating solutions (Shimizu et al., 2010).

Numerous articles report in-depth study of the optimization of the degradation process. It has been estimated that the ultrasonic process should not exceed more than 50 W L^{-1}, in order to be economically viable at industrial scales (Mahamuni and Adewuyi, 2009). However, most of the laboratory processes consume more energy than this. Torres et al. (2008) observed that US was better able to eliminate bisphenol A (BPA) as compared to photocatalysis, which brings about a more efficient mineralization. Using the combined system of sonication and photocatalysis, an interesting synergistic effect, which depended on the TiO_2 catalyst loading, was observed for BPA mineralization.

TABLE 18.4
Catalysts Used in Ultrasound-Assisted Degradation of Pollutants

Pollutant	Catalyst(s)	Degradation Conditions	References
Bisphenol A	Fe^{2+} and TiO_2	AOP that combining sonolysis	Torres-Palma et al. (2010)
Nonylphenol ethoxylate surfactant	Au–TiO_2 nanoparticles photocatalysts	Catalytic activities of these nanomaterials were compared for the degradation of a polydisperse nonylphenol ethoxylate, Teric GN9 by photocatalysis and sonophotocatalysis under visible light/high-frequency US irradiation	Anandan and Ashokkumar (2009)
Methylene blue	TiO_2 nanotube	Sonophotocatalytic activity of TiO_2 nanotube array was evaluated 27 kHz	Yuan et al. (2009)
Malachite green	Heterogeneous (TiO_2) and homogeneous photocatalysis (photo-Fenton)	80 kHz of US irradiation was provided by a horn-type sonicator, while a 9 W lamp was used for UV-A irradiation	Berberidou et al. (2007)
2chloro-5methyl phenol	TiO_2 and H_2O_2; inorganic ions	Effect of inorganic ions on degradation rate of 2chloro-5methyl phenol were found to be in the order of $Cl^- > SO_4^{2-} > HPO_4^{2-} > HCO_3^-$	Nalini Vijaya Laxmi et al. (2010)
Methyl parathion (organophosphorus insecticide)	Micron-sized and nanosized rutile TiO_2	US of low power was used as an irradiation source to induce the catalytic activity of the rutile TiO_2 particles	Wang et al. (2007)
2-Chlorophenol	Horse radish peroxidase (HRP)	Enzyme treatment: utilization of HRP in presence of H_2O_2 to oxidize organic pollutant Sonication treatment: Ultrasonic waves irradiation alone for pollutant degradation Sono-enzyme degradation: combination of ultrasonic waves and HRP	Entezari et al. (2006)
Azo fuchsine dye	Fe-doped mixed crystal TiO_2 powder	Degradation process of dye and irradiation time, doping Fe^{3+} ion content, added amount of catalyst, and initial concentration of azo fuchsine solution, on the degradation were investigated by UV-vis spectra, ion chromatography, and HPLC	Wang et al. (2008a)

The best synergistic effect was found at a low catalyst loading of 0.05 g L^{-1} of titanium dioxide with 4 h of combined treatment causing 62% of dissolved organic carbon (DOC) to be eliminated being in contrast, 6% or 12% of DOC being removed by US alone or photocatalysis alone, respectively. Pétrier et al. (2007) examined the degradation of chlorobenzene and 4-chlorophenol in a dilute solution of a mixture of these compounds saturated with argon when subjected to sonication at 300 kHz. The two compounds exhibited sequential degradation with more volatile chlorobenzene entering the cavitation bubble and being destroyed first, while the 4-chlorophenol degradation occurred, subsequently, only when chlorobenzene had been completely destroyed. Pétrier et al. (2007) have also used these two compounds when sonicated in water saturated with oxygen. Under these conditions the two compounds were, however, degraded simultaneously. This observation was a remarkable one and Pétrier et al. (2007) proposed the following explanation, both of which are based on the formation of additional OH radical species. Pétrier et al. (2007) explained that there is a shell of supercritical water that surrounded the residual bubble (hot spot) on the point of collapse and the presence of oxygen could increase the production of OH radicals in this shell. Additionally, more OH radicals could also be derived from the combustion of chlorobenzene within the cavitation bubble itself, and by this mechanism the more volatile component (chlorobenzene) could induce the generation of more OH radicals, which then degraded the organic with lower volatility. The major corollary from the study of Pétrier et al. (2007) is the ability to produce conducive conditions for the simultaneous elimination of two organic compounds by the use of oxygen in the developing field of ultrasonic water and wastewater remediation.

Sivasankar and Moholkar (2008) stated that the fundamental physical phenomenon behind sonochemical degradation of pollutants is radial motion of cavitation bubbles. Their study (Sivasankar and Moholkar, 2008) implemented a dual approach to the problem by coupling results of their experiments under different conditions to a mathematical model that addressed the physics and chemistry of the cavitation bubbles. Their concurrent analysis of the experimental and simulation results revealed that overall degradation of the pollutant under analysis (nitroaromatics) achieved for a given combination of experimental conditions was a function of competing effects of the extent of radical production from the bubble, thickness of the liquid shell surrounding the bubble that gets heated up during transient collapse, the concentration of the pollutant in the interfacial region, and the extent of radical scavenging in the medium.

On a more innovative tone of research, Midathana and Moholkar (2009) have attempted to discern the physical mechanism of enhancement of adsorption of nitrobenzene, phenol, and p-nitrophenol with the application of US. Midathana and Moholkar (2009) observed that the correlation of their experimental and simulation results supported that the extent of adsorption in the presence of US showed an optimum with the intensity of convection generated in the medium by the cavitation bubbles. The microturbulence generated by cavitation bubbles would seemingly be responsible for a favorable contribution to the enhancement of adsorption. This was, therefore, attributed to the continuous nature of microturbulence with moderate liquid velocities. However, acoustic waves emitted by the cavitation bubbles apparently rendered an adverse effect on the process, and this was attributable to the discrete nature and high pressure amplitude of the waves, which created excessively high convection in the medium, causing desorption of the pollutant. Midathana and Moholkar (2009) concluded that the chemical nature of the pollutant, hence, also tends to influence the enhancement effect of US. As an important corollary to the study of Midathana and Moholkar (2009), for hydrophobic pollutants, ultrasonic enhancement is more pronounced than for hydrophilic pollutants under, otherwise, similar conditions.

Ultrasound-Assisted Dye Degradation

Dyes are an important class of pollutants, and can even be identified by the human eye. Large amounts of dyes are annually produced and applied in many different industries, including the textile, cosmetic, paper, leather, pharmaceutical, and food industries (Hai et al., 2007). There are more than 100,000 commercially available dyes with an estimated annual production of over 7×10^5 ton

(Hai et al., 2007), 15% of which is lost during the dyeing process. The presence of even trace concentrations of dyes in effluent is highly visible and undesirable (Hai et al., 2007). The release of colored wastewater in the ecosystem is a remarkable source of esthetic pollution, eutrophication, and perturbations in aquatic life. Dye effluent usually contains chemicals, including dye itself, that are toxic, carcinogenic, mutagenic, or teratogenic to various microbiological and fish species (Daneshvar et al., 2003). Concern arises, as many dyes are made from known carcinogens such as benzidine and other aromatic compounds (Hai et al., 2007). Also azo and nitro compounds have been reported to be reduced in sediments of aquatic bodies, consequently yielding potentially carcinogenic amines that spread in the ecosystem. The presence of dyes or their degradation products in water can also cause human health disorders such as nausea, hemorrhage, and ulceration of skin and mucous membranes, and can cause severe damage to the kidney, reproductive system, liver, brain, and central nervous system (Hai et al., 2007).

Disposal of dyes in precious water resources must be avoided, however; and for that, various treatment technologies are in use. Several industrial-scale decolorization systems are commercially available (Willmott et al., 1998). These include adsorption, filtration, precipitation, and activated sludge systems. All of these technologies work by concentrating the dyestuffs and transferring them to a solid phase that subsequently needs disposal. Among various methods, the sonolytic degradation of a variety of dyes is being studied and the results are promising. The decolorization of rhodamine B, methylene blue dye (basic dyes), and acid orange II, acid scarlet red 3R (acid dyes) solutions by cobalt activated persulfate (PS), and ultrasonication have been investigated by Gayathri et al. (2010). Gayathri et al. (2010) concluded that the decolorization efficiency were in the order of PS < PS + Co < PS + US < PS + US + Co for all the four dye solutions. Under the optimum condition, the decolorization obeyed the first-order kinetics and interestingly nearly 90%–97% of decolorization was achieved with chemical oxygen demand and total organic carbon removal of about 65%–73% and 53%–62%, respectively, achieved within an hour of sonication. Earlier, Minero et al. (2008) have observed that the sonochemical degradation rate of the charged substrates Acid Blue 40 (AB40) and methylene blue (MB) is enhanced by scavengers of hydroxyl radicals such as bicarbonate, carbonate, bromide, iodide, and (only in the case of AB40) nitrite. Minero et al. (2008) proposed that the degradation enhancement could occur if these radicals were sonochemically formed on the surface of the collapsing cavitation bubbles and underwent there radical–radical recombination at a lesser extent than •OH. In this way, the radicals would be more available than •OH for substrate degradation, both at the bubble surface and in the solution bulk, which could more than compensate for their lower intrinsic reactivity. The varied reactivity toward different substrates of the sonochemically formed radical species could then explain why nitrite inhibits MB degradation while enhancing that of AB40. The sonochemical formation of $Br_2^{•-}$, $I_2^{•-}$, and $•NO_2$ could give rise to halogenation and nitration in addition to oxidation processes. Wang et al. (2008b) studied the degradation of reactive brilliant red K-BP in aqueous solution by means of ultrasonic cavitation for a variety of operating conditions. Besides concluding that the degradation of reactive brilliant red K-BP in aqueous solution followed a pseudo first-order reaction kinetics and that the degradation rate was dependent on the initial concentration of reactive brilliant red K-BP, the temperature and acidity of the aqueous medium, results indicated convincingly that the sonochemical degradation rate of brilliant red K-BP in aqueous solution was substantially accelerated by Fe^{2+}, NaCl, or addition of Fenton reagent. Ghodbane and Hamdaoui (2009) studied the sonolytic degradation of an anthraquinonic dye, CI acid blue 25 (AB25), in aqueous phase using 1700 kHz US waves for an acoustic power of 14 W. The results were much conclusive and the methods analyzed clearly showed the production of oxidizing species during sonication and well-reflected the sonochemical effects of high-frequency ultrasonic irradiation. In addition, the combination of US with hydrogen peroxide also looked to be a promising option in increasing the generation of free radicals; and hence the concentration of hydrogen peroxide is most plausibly crucial in deciding the extent of enhancement obtained for the combined process involving sonication. Ghodbane and Hamdaoui (2009) concluded that the US/H_2O_2 and US/Fe(II) processes studied were efficient for the degradation of AB25 in aqueous solutions by high-frequency ultrasonic irradiation.

ULTRASOUND-ASSISTED PESTICIDES, INSECTICIDES, AND HERBICIDES DEGRADATION

Pesticide classes mostly detected in polluted natural and certain wastewaters involve herbicides used extensively in corn, cotton, and rice production, organophosphorus insecticides as well as the banned organochlorines insecticides due to their persistence in the aquatic environment (Konstantinou et al., 2006). The compounds most frequently detected atrazine, simazine, alachlor, metolachlor, and trifluralin of the herbicides, diazinon, parathion methyl of the insecticides and lindane, endosulfan, and aldrin of the organochlorine pesticides. Biological decomposition of pesticides is the most important and effective way to remove these compounds from the environment. The microorganisms have the ability to interact, both chemically and physically, with substances, leading to structural change or the complete degradation of the target molecule. However, the principal cause of pesticide persistence in soil is commonly the lack of favorable conditions for microbial degradation. To develop novel processes for the remediation of pesticides, the effects of US irradiation on the degradation of these molecules are being widely studied (Matouq et al., 2008).

Recently, Yao et al. (2010) found that the degradation rate of parathion decreased with increasing initial concentration and decreasing power for an optimal frequency for parathion degradation at 600 kHz. Yao et al. (2010) postulated that free radical reactions predominated in the sonochemical degradation of parathion and the reaction zones were predominately at the bubble interface and, to a much lesser extent, in bulk solution. The main pathways of parathion degradation by ultrasonic irradiation were also proposed and it was indicated that the N_2 in air took part in the parathion degradation through the formation of •NO_2 under ultrasonic irradiation, with the parathion decomposing into paraoxon and 4-nitrophenol in the first step via., two different pathways, respectively. Monocrotophos (MCP) is an organophosphate insecticide that has been found as a pollutant in aqueous environments—all the more, the sonolytic, photocatalytic, and sonophotocatalytic degradation of MCP in the presence of homogeneous (Fe^{3+}) and heterogeneous photocatalysts (TiO_2) were recently studied by Madhavan et al. (2010a). The photocatalytic degradation rate using TiO_2 was found to be lower than that of sonolysis alone due to the interference of phosphate ions formed as an intermediate product. On the other hand, a 15-fold enhancement in the degradation rate was found when photolysis was carried out in the presence of Fe^{3+} compared to the rate observed with photolysis alone. The combination of sonolysis and photocatalysis (using either TiO_2 or Fe^{3+}) showed a detrimental effect. Synergy indices of 0.62 and 0.87 were found for the sonophotocatalytic degradation of MCP in the presence of TiO_2 and Fe^{3+}, respectively. A TOC analysis revealed that the mineralization process was additive for both TiO_2 and Fe^{3+} sonophotocatalysis, and further analyses indicated that the sonication of MCP led to the formation of dimethyl phosphate, dimethylphosphonate, 3-hydroxy 2-buteneamide, and N-methyl 3-oxobutanamide as the intermediate products. Torres et al. (2009) have conducted experiments in natural and deionized conditions and found that alachlor degradation by US is practically unaffected by the presence of potential •OH radical scavengers: Bicarbonate, sulfate, chloride, and oxalic acid; but in both cases, alachlor was readily eliminated after 75 min of sonication.

Katsumata et al. (2010) performed the sonochemical photodegradation of fenitrothion, which is one of phosphorothiate insecticides, in the presence of Fe(III) and oxalate, and it was found that an initial fenitrothion concentration of 10 mg L^{-1} was completely degraded after 30 min of sonication at pH 6. The decrease of TOC as a result of mineralization of fenitrothion was observed during US/ferrioxalate/UV process. In addition, Katsumata et al. (2010) observed the formations of nitrite and sulfate ions as end-products of the degradation and concluded that the US/ferrioxalate/UV system could be a useful technology for the treatment of wastewater containing fenitrothion. Zhang et al. (2010) have treated apple juice (13°Brix) spiked with malathion and chlorpyrifos (2–3 mg L^{-1} of each compound) under different ultrasonic irradiations and found that the ultrasonic treatment was effective for the degradation of malathion and chlorpyrifos in apple juice, and the output power and treatment time significantly influenced the degradation of both pesticides with maximum degradations of 41.7% for malathion and 82.0% for chlorpyrifos after the ultrasonic treatment at 500 W for

120 min. Malaoxon and chlorpyrifos oxon were identified as the degradation products of malathion and chlorpyrifos, and the oxidation pathway through the hydroxyl radical attack on the P=S bond of pesticide molecules was proposed. Earlier, Bahena et al. (2008) have analyzed the degradation of alazine and gesaprim by the combined effects of photocatalysis with sonolysis (sonophotocatalysis), using a US source of 20 kHz. Bahena et al. (2008) observed that over 90% of the active component in the gesaprim was abated and those in alazine were completely degraded. Also, over 80% of chemical oxygen demand abatement was attained for both herbicides with sonophotocatalysis at 150 min of irradiation time.

ULTRASOUND-ASSISTED PHARMACEUTICALS DEGRADATION

Pharmaceutical are a diverse group of chemicals treated like potential environmental pollutants. Recently, pharmaceuticals have been detected in measurable amounts in surface and groundwater resources especially those receiving wastewater effluents. Sonolytic irradiation, an advanced oxidation process (AOP), has received increased attention lately as a possible remediation treatment for these pollutants.

Diclofenac (DF), as one of the most popular antiphlogistics, is produced in great quantities. Nowadays, this drug is ubiquitously present in the aquatic environment due to its resistance to biodegradation (Hartmann et al., 2008). Degradation by ultrasonic irradiation is a possibility to eliminate DF from water without the addition of chemicals. Hartmann et al. (2008) found that the degradation of DF by sonolysis of an aqueous solution at 617 kHz followed the first-order kinetics and that TiO_2 catalyst increased the rate of degradation. Within the first 30 min of irradiation, the relative concentration of DF decreased from 100% to 16%, and chlorinated anilines, phenols, and carboxylic acid derivatives the products of sonolysis. The sonolytic, photocatalytic, and sonophotocatalytic degradation of DF using three photocatalysts (TiO_2, ZnO, and Fe-ZnO) were studied recently by Madhavan et al. (2010b). The degradation of DF followed first-order like kinetics. It was observed that the sonophotocatalytic degradation using TiO_2 under UV-vis radiation showed a slight synergistic enhancement in the degradation of DF. Also, Naddeo et al. (2009) have evaluated the ultrasonic degradation of amoxicillin and carbamazepine in single solutions as well as in mixtures spiked in urban wastewater effluent at 25–100 W L^{-1}, initial pollutant concentrations of 2.5–10 mg L^{-1} and at pH 3–11. Naddeo et al. (2009) found that the US-induced amoxicillin and carbamazepine bioconversion was enhanced at increased applied power densities (25–100 W L^{-1}) and liquid bulk temperatures, acidic conditions, and in the presence of dissolved air or oxygen. Later, Naddeo et al. (2010) studied the 20 kHz US-induced degradation of DF conversion (2.5–80 mg L^{-1}) at 25–100 W L^{-1} and drew a similar conclusion as Naddeo et al. (2009). Both Naddeo et al. (2009, 2010) summed up their findings by indicating that the degradation rate for the pharmaceuticals under test increased with increasing substrate concentration in the range 2.5–5 mg L^{-1} but remained constant in the range 40–80 mg L^{-1}, indicating different kinetic regimes.

Ibuprofen (IBP), a widely used nonsteroidal antiinflammatory recalcitrant drug is found in water and wastewater streams. In their study on the degradation of IBP, Méndez-Arriaga et al. (2008) found that a 300 kHz US irradiation for 30 min increased the degradation of IBP from 30% to 98%. The initial rate of IBP degradation was in the range of 1.35 and 6.1 µmol L^{-1} min^{-1} for initial concentrations of 2–21 mg L^{-1} or 9.7–101 µmol L^{-1}, respectively. More interestingly, a complete removal of IBP was achieved under sonication but some DOC remained in solution showing that long-lived intermediates were still recalcitrant to the US irradiation. However, chemical and biological oxygen demands indicated that the sonication assisted degradation process completely oxidized IBP to biodegradable substances readily removable in a subsequent biological treatment process. Lately, Madhavan et al. (2010c) studied the sonolytic, photocatalytic, and sonophotocatalytic degradations of IBP in the presence of homogeneous (Fe^{3+}) and heterogeneous photocatalysts (TiO_2). Madhavan et al. (2010c) found that when compared with sonolysis and photocatalysis, a higher degradation rate was observed for sonophotocatalysis in the presence of TiO_2 or Fe^{3+} and also a slight synergistic enhancement was found with a synergy index of 1.3 and 1.6, respectively. Although TiO_2 sonophotocatalysis showed an additive process effect in the mineralization, a significant synergy effect was observed for the sonophotocatalysis in the presence of Fe^{3+}. Madhavan et al. (2010c) attributed this synergy to the formation of photoactive complexes between Fe^{3+} and IBP degradation products, such as carboxylic acids.

ULTRASOUND-ASSISTED HORMONES DEGRADATION

Estrogen compounds are being detected in significant concentrations in surface water, wastewater, soil, sediments, and groundwater (Stegeman and Zhou, 2008). These estrogenic compounds influence the growth and performance of the reproductive system. Several reports indicate that the major source of these contaminants to the ecosystem is the effluents from wastewater treatment plants. These contaminants are found at significant concentrations in the effluent and the water bodies into which they are discharged. Reports also indicate that these estrogens are found in trace level concentrations in drinking water. There are many reports documenting the adverse effects, such as feminization of fish, of estrogen hormones in the environment. One of the major sources of these compounds is from municipal wastewater effluents (Suri et al., 2007). The removal of estrogen hormones from water and wastewater is hence of importance due to their adverse effects toward ecosystems and potential risks to human health.

The biological processes at municipal wastewater treatment plants cannot completely remove these compounds (Stegeman and Zhou, 2008). Conventional treatment technologies are not designed to completely remove these pharmaceutically active chemicals (PhACs). Therefore, there is a need to develop new treatment technologies in addition to the existing technologies. In point of fact, US is an AOP that effectively destroys hormones. Suri et al. (2007) have used US to destroy estrogen compounds in water. Their study examined the effect of US power density and power intensity on the destruction of various estrogen compounds, which included 17α-estradiol, 17β-estradiol, estrone, estriol, equilin, 17α-dihydroequilin, 17α-ethinyl estradiol, and norgestrel in single component batch and flow through reactors using 0.6, 2, and 4 kW US sources. Suri et al. (2007) recorded net positive results since the sonolysis process produced 80%–90% destruction of individual estrogens at initial concentration of $10\,\mu g\,L^{-1}$ within 40–60 min of contact time. Fu et al. (2007) have also studied the US-induced destruction of 17α-estradiol, 17β-estradiol, ethinyl estradiol, estrone, equilin, gestodene, levonorgestrel, and norgestrel in aqueous solutions in a batch reactor using a $1.1\,W\,mL^{-1}$ sonication unit and in a continuous flow reactor using a $2.1\,W\,mL^{-1}$ sonication unit. The latter indicated that the reaction likely took place in the interfacial region, where supercritical environment was produced upon cavity implosion and in the bulk solution with radical species. A low solution pH and low solution temperature were more favorable for the destruction of the estrogens. Earlier, Chimchirian et al. (2005) had also reported that US power showed that degradation of estrogens is possible and effective, and about 65% removal of total estrogen concentration from wastewater may be observed in 120 min of reaction time using a 0.6 kW unit; and about 95% removal in 26 min using the 2 kW unit in clean water.

SUGGESTED SONICATION SYSTEM FOR ORGANIC POLLUTANT DEGRADATION

Most of the laboratory work on organic pollutant degradation is being carried out in batch reactor mode with reactor volumes ranging between 50 and 300 mL. Ultrasonic probes are commonly used at 20 kHz or other low frequency levels (Thompson and Doraiswamy, 1999). Since the probe directly applies US to the target pollutants, it is far more effective compared with other indirect application. However, high-frequency transducers are usually fixed at the bottom of the reactor. To prevent the transducer from pitting and to minimize the acoustic streaming effect, two or more transducers should be installed. Appropriate bubbling gas selection and flow rate is essential for effective remediation. For the MHz frequency range application, a 50 mL reactor is suitable because of very high attenuation of the energy.

During sonication, volatile organics may evaporate and escape into atmosphere making it advisable always to use a closed reactor for treating these contaminants. If the reaction is pH sensitive, it should be monitored because the CO_2 produced by the reaction will redissolve and reduce the pH of the solution.

For the continuous remediation application, flow cells are recommended, as most of the conversion of the contaminant may not occur at a single stage, for this reason recirculation or multiflow cells in series are recommended to ensure complete conversion. Since flow cells can be operated at high intensity, this may accelerate the remediation process although it will also add pumping costs. Under high-intensity operation, the tip of the sono rod will not last long under flow cell application and its replacement will also add to costs. Thangavadivel et al. (2009) demonstrated that high-frequency US, as an alternative to high-cost incineration, could assist in the remediation of DDT from sand and soil slurries. However, Thangavadivel et al. (2009) also brought forward that due to intensity limitations in currently available equipment and higher attenuation of energy, high-frequency US would be having a low volume coverage and would, thus, require circulation of the slurry past the sonotrode, multiple sonotrodes, larger sonotrode area, and, in addition, lower slurry densities may still be required.

CONCLUSION

US is currently emerging as a green remediation technology for dealing with a wide range of organic pollution. It is being researched in the laboratories worldwide and results appear very promising, with a deeper understanding of the US-enhanced remediation process including transducer, reactor, pollution, operating condition, degradation mechanism, and degradation rate being reported in the recent literature. However, there are still only a few large-scale plants in operation and more work needs to be done on the scale-up of the system. Since the energy transfer efficiency of this sonication system is low, reactor design is critical. As there are numerous laboratory scale studies on organic degradation, more focus now needs to be given to scale-up of this technology. Also, more research should be directed at enhancing treatment efficiency by using US in combination with other remediation processes especially biodegradation. This will lead to a reduction in the overall cost of the treatment.

REFERENCES

Abbasi, M. and Asl, N.R. 2008. Sonochemical degradation of Basic Blue 41 dye assisted by nanoTiO$_2$ and H$_2$O$_2$. *Journal of Hazardous Materials*, 153: 942–947.

Adewuyi, Y.G. 2001. Sonochemistry: Environmental science and engineering applications. *Industrial and Engineering Chemistry Research*, 40: 4681–4715.

Adewuyi, Y.G. 2005. Sonochemistry in environmental remediation. 1. Combinative and hybrid sonophotochemical oxidation processes for treatment of pollutants in water. *Environmental Science and Technology*, 39: 3409–3420.

Anandan, S. and Ashokkumar, M. 2009. Sonochemical synthesis of Au-TiO$_2$ nanoparticles for the sonophotocatalytic degradation of organic pollutants in aqueous environment. *Ultrasonics Sonochemistry*, 16: 316–320.

Andrea, L., Eduardo, G., and Stanislov, M. 2001. Overview of remediation technologies for persistent toxic substances. In *ICS-UNIDO Workshop on Contamination of Food and Agroproducts*, vol. 52, pp. 253–280.

Ashokkumar, M. and Grieser, F. 2005. A comparison between multibubble sonoluminescence intensity and the temperature within cavitation bubbles. *Journal of the American Chemical Society*, 127: 5326–5327.

Ayyildiz, O., Peters, R.W., and Anderson, P.R. 2007. Sonolytic degradation of halogenated organic compounds in ground water: Mass transfer effects. *Ultrasonics Sonochemistry*, 14: 163–172.

Bahena, C.L., Martínez, S.S., Guzmán, D.M., and del Refugio Trejo Hernández, M. 2008. Sonophotocatalytic degradation of alazine and gesaprim commercial herbicides in TiO_2 slurry. *Chemosphere*, 71: 982–989.

Berberidou, C., Poulios, I., Xekoukoulotakis, N.P., and Mantzavinos, D. 2007. Sonolytic, photocatalytic and sonophotocatalytic degradation of malachite green in aqueous solutions. *Applied Catalysis B: Environmental*, 74: 63–72.

Cheng, J., Vecitis, D., Park, H., Mader, T., and Hoffmann, R. 2008. Sonochemical degradation of perfluorooctane sulfonate (PFOS) and perfluorooctanoate (PFOA) in landfill groundwater: Environmental matrix effects. *Environmental Science and Technology*, 42: 8057–8063.

Chimchirian, R.F., Suri, R.P.S., Velicu, M., Constable, R., and Helmig, E. 2005. Sonolytic destruction of estrogen hormones in water using ultrasound irradiation. In *Proceedings of the Water Environment Federation, WEFTEC 2005*: Session 61 through Session 70, pp. 5171–5183.

Chowdhury, P. and Viraraghavan, T. 2009. Sonochemical degradation of chlorinated organic compounds, phenolic compounds and organic dyes—A review. *Science of the Total Environment*, 407: 2474–2492.

Dai, Y., Li, F., Ge, F., Zhu, F., Wu, L., and Yang, X. 2006. Mechanism of the enhanced degradation of pentachlorophenol by ultrasound in the presence of elemental iron. *Journal of Hazardous Materials*, 137: 1424–1429.

Daneshvar, N., Ashassi-Sorkhabi, H., and Tizpar, A. 2003. Decolorization of orange II by electrocoagulation method. *Separation and Purification Technology*, 31: 153–162.

David, B. 2009. Sonochemical degradation of PAH in aqueous solution. Part 1: Monocomponent PAH solution. *Ultrasonics Sonochemistry*, 16: 260–265.

Dehghani, M.H., Najafpoor, A.A., and Azam, K. 2010. Using sonochemical reactor for degradation of LAS from effluent of wastewater treatment plant. *Desalination*, 250: 82–86.

Dewulf, J., Van Langenhove, H., De Visscher, A., and Sabbe, S. 2001. Ultrasonic degradation of trichloroethylene and chlorobenzene at micromolar concentrations: Kinetics and modeling. *Ultrasonics Sonochemistry*, 8: 143–150.

Entezari, M.H., Mostafai, M., and Sarafraz-Yazdi, A. 2004. A combination of ultrasound and a bio-catalyst: Removal of 2-chlorophenol from aqueous solution. *Ultrasonics Sonochemistry*, 13: 137–141.

Entezari, M.H., Mostafai, M., and Sarafraz-Yazdi, A. 2006. A combination of ultrasound and a bio-catalyst: Removal of 2-chlorophenol from aqueous solution. *Ultrasonics Sonochemistry*, 13: 37–41.

Entezari, M.H., Petrier, C., and Devidal, P. 2003. Sonochemical degradation of phenol in water: A comparison of classical equipment with a new cylindrical reactor. *Ultrasonics Sonochemistry*, 10: 103–108.

Entezari, M.H., Sharif Al-Hoseini, Z., and Ashraf, N. 2008. Fast and efficient removal of Reactive Black 5 from aqueous solution by a combined method of ultrasound and sorption process. *Ultrasonics Sonochemistry*, 15: 433–437.

Fu, H., Suri, R.P.S., Chimchirian, R.F., Helmig, E., and Constable, R. 2007. Ultrasound-induced destruction of low levels of estrogen hormones in aqueous solutions. *Environmental Science and Technology*, 41: 5869–5874.

Gaddam, K. and Cheung, H.M. 2001. Effects of pressure, temperature, and pH on the sonochemical destruction of 1,1,1-trichloroethane in dilute aqueous solution. *Ultrasonics Sonochemistry*, 4: 103–109.

Gallego-Juarez, J.A. 1989. Piezoelectric ceramics and ultrasonic transducers. *Journal of Physics E: Scientific Instruments*, 22: 804–816.

Gayathri, P., Dorathi, R.P.J., and Palanivelu, K. 2010. Sonochemical degradation of textile dyes in aqueous solution using sulphate radicals activated by immobilized cobalt ions. *Ultrasonics Sonochemistry*, 17: 566–571.

Gehannin, J., Arghir, M., and Bonneau, O. 2009. Evaluation of Rayleigh–Plesset equation based cavitation models for squeeze film dampers. *Journal of Tribology*, 131: 4 p. doi:10.1115/1.3063819.

Ghodbane, H. and Hamdaoui, O. 2009. Degradation of Acid Blue 25 in aqueous media using 1700 kHz ultrasonic irradiation: Ultrasound/Fe(II) and ultrasound/H_2O_2 combinations. *Ultrasonics Sonochemistry*, 16: 593–598.

Gogate, P.R. and Pandit, A.B. 2004. A review of imperative technologies for wastewater treatment I: Oxidation technologies at ambient conditions. *Advances in Environmental Research*, 3: 501–551.

Guo, Z., Gu, C., Zheng, Z., Feng, R., Jiang, F., Gao, G., and Zheng, Y. 2006. Sonodegradation of halomethane mixtures in chlorinated drinking water. *Ultrasonics Sonochemistry*, 13: 487–492.

Hai, F.I., Yamamoto, K., and Fukushi, K. 2007. Hybrid treatment systems for dye wastewater. *Critical Reviews in Environmental Science and Technology*, 37: 315–377.

Hao, H., Chen, Y., Wu, M., Wang, H., Yin, Y., and Lu, Z. 2004. Sonochemistry of degrading *p*-chlorophenol in water by high frequency ultrasound. *Ultrasonics Sonochemistry*, 1: 43–46.

Hartmann, J., Bartels, P., Mau, U., Witter, M., Tümpling, W.V., Hofmann, J., and Nietzschmann, E. 2008. Degradation of the drug diclofenac in water by sonolysis in presence of catalysts. *Chemosphere*, 70: 453–461.

Jiang, Y., Petrier, C., and Waite, T.D. 2006. Sonolysis of 4-chlorophenol in aqueous solution: Effects of substrate concentration, aqueous temperature and ultrasonic frequency. *Ultrasonics Sonochemistry*, 13: 415–422.

Kanthale, P., Ashokkumar, M., and Grieser, F. 2008. Sonoluminescence, sonochemistry (H_2O_2 yield) and bubble dynamics: Frequency and power effects. *Ultrasonic Sonochemistry*, 15: 143–150.

Katsumata, H., Okada, T., Kaneco, S., Suzuki, T., and Ohta, K. 2010. Degradation of fenitrothion by ultrasound/ferrioxalate/UV system. *Ultrasonics Sonochemistry*, 17: 200–206.

Khanal, S.K., Grewell, D., Sung, S., and van Leeuwen, J. 2007. Ultrasound applications in wastewater sludge pretreatment: A review. *Critical reviews in Environmental Science and Technology*, 37: 277–313.

Kidak, R., and Ince, N.H. 2006. Effects of operating parameters on sonochemical decomposition of phenol. *Journal of Hazardous Materials*, 3: 1453–1457.

Konstantinou, I.K., Hela, D.G., and Albanis, T.A. 2006. The status of pesticide pollution in surface waters (rivers and lakes) of Greece. Part I. Review on occurrence and levels. *Environmental Pollution*, 141: 555–570.

Liang, J., Komarov, S., Hayashi, N., and Kasai, E. 2007. Improvement in sonochemical degradation of 4-chlorophenol by combined use of Fenton-like reagents. *Ultrasonics Sonochemistry*, 14: 201–207.

Lifka, J., Ondruschka, B., and Hofmann, J. 2003. The use of ultrasound for the degradation of pollutants in water: Aquasonolysis—A review. *Engineering and Life Sciences*, 3: 253–262.

Lim, M.H., Kim, S.H., Kim, Y.U., and Khim, J. 2007. Sonolysis of chlorinated compounds in aqueous solution. *Ultrasonics Sonochemistry*, 14: 93–98.

Lin, H., Storey, B.D., and Szeri, A.J. 2002. Inertially driven inhomogeneities in violently collapsing bubbles: The validity of the Rayleigh–Plesset equation. *Journal of Fluid Mechanics*, 452: 145–162.

Little, C., El-Sharif, M., and Hepher, M. 2007. The effect of solution level on calorific and dosimetric results in a 70 kHz tower type sonochemical reactor. *Ultrasonics Sonochemistry*, 14: 375–379.

Liu, H., Li, G., Qu, J., and Liu, H. 2007. Degradation of azo dye Acid Orange 7 in water by Fe^0/granular activated carbon system in the presence of ultrasound. *Journal of Hazardous Materials*, 144: 180–186.

Madhavan, J., Grieser, F., and Ashokkumar, M. 2010c. Combined advanced oxidation processes for the synergistic degradation of ibuprofen in aqueous environments. *Journal of Hazardous Materials*, 178: 202–208.

Madhavan, J., Kumar, P.S.S., Anandan, S., Grieser, F., and Ashokkumar, M. 2010a. Sonophotocatalytic degradation of monocrotophos using TiO_2 and Fe^{3+}. *Journal of Hazardous Materials*, 177: 944–949.

Madhavan, J., Kumar, P.S.S., Anandan, S., Zhou, M., Grieser, F., and Ashokkumar, M. 2010b. Ultrasound assisted photocatalytic degradation of diclofenac in an aqueous environment. *Chemosphere*, 80: 747–752.

Mahamuni, N. and Adewuyi, Y.G. 2009. Advanced oxidation processes (AOPs) involving ultrasound for waste water treatment: A review with emphasis on cost estimation. *Ultrasonics Sonochemistry*, 17: 990–1003.

Mahamuni, N. and Pandit, A. 2006. Effect of additives on ultrasonic degradation of phenol. *Ultrasonics Sonochemistry*, 13: 165–174.

Mason, T. 2007. Developments in ultrasound- non-medical—Review. *Progress in Biophysical and Molecular Biology*, 93: 166–175.

Mason, T.J., Collings, A., and Sumel, A. 2004. Sonic and ultrasonic removal of chemical contaminants from soil in the laboratory and on a large scale. *Ultrasonics Sonochemistry*, 11: 205–210.

Matouq, M.A. and Al-Anber, Z.A. 2007. The application of high frequency ultrasound waves to remove ammonia from simulated industrial wastewater. *Ultrasonics Sonochemistry*, 14: 393–397.

Matouq, M.A., Al-Anber, Z.A., Tagawa, T., et al. 2008. Degradation of dissolved diazinon pesticide in water using the high frequency of ultrasound wave. *Ultrasonics Sonochemistry*, 15: 869–874.

Méndez-Arriaga, F., Torres-Palma, R.A., Pétrier, C., Esplugas, S., Gimenez, J., and Pulgarin, C. 2008. Ultrasonic treatment of water contaminated with ibuprofen. *Water Research*, 42: 4243–4248.

Merouani, S., Hamdaoui, O., Saoudi, F., Chiha, M., and Pétrier, C. 2010. Influence of bicarbonate and carbonate ions on sonochemical degradation of Rhodamine B in aqueous phase. *Journal of Hazardous Materials*, 175: 593–599.

Midathana, V.R. and Moholkar, V.S. 2009. Mechanistic studies in ultrasound-assisted adsorption for removal of aromatic pollutants. *Industrial and Engineering Chemistry Research*, 48: 7368–7377.

Minero, C., Pellizzari, P., Maurino, V., Pelizzetti, E., and Vione, D. 2008. Enhancement of dye sonochemical degradation by some inorganic anions present in natural waters. *Applied Catalysis B: Environmental*, 77: 308–316.

Naddeo, V., Belgiorno, V., Kassinos, D., Mantzavinos, D., and Meric, S. 2010. Ultrasonic degradation, mineralization and detoxification of diclofenac in water: Optimization of operating parameters. *Ultrasonics Sonochemistry*, 17: 179–185.

Naddeo, V., Meriç, S., Kassinos, D., Belgiorno, V., and Guida, M. 2009. Fate of pharmaceuticals in contaminated urban wastewater effluent under ultrasonic irradiation. *Water Research*, 43: 4019–4027.

Nakui, H., Okitsu, K., Maeda, Y., and Nishimura, R. 2007. Effect of coal ash on sonochemical degradation of phenol in water, *Ultrasonics Sonochemistry*, 14: 191–196.

Nalini Vijaya Laxmi, O., Saritha, P., Rambabu, N., Himabindu, V., and Anjaneyulu, Y. 2010. Sonochemical degradation of 2chloro-5methyl phenol assisted by TiO_2 and H_2O_2. *Journal of Hazardous Materials*, 174: 151–155.

Pétrier, C., Combet, E., and Mason, T. 2007. Oxygen-induced concurrent ultrasonic degradation of volatile and non-volatile aromatic compounds. *Ultrasonics Sonochemistry*, 14: 117–121.

Shimizu, N., Ninomiya, K., Ogino, C., and Rahman, M.M. 2010. Potential uses of titanium dioxide in conjunction with ultrasound for improved disinfection. *Biochemical Engineering Journal*, 48: 416–423.

Sivasankar, T. and Moholkar, V.S. 2008. Physical features of sonochemical degradation of nitroaromatic pollutants. *Chemosphere*, 72: 1795–1806.

Srinivas, S. and Chintalapati, S. 2008. Sonochemical degradation of Congo Red. *International Journal of Environment and Waste Management*, 2: 309–319.

Stegemann, J.A. and Zhou, Q. 2008. Ultrasound assisted removal of estrogen hormones. *WIT Transactions on Ecology and the Environment*, 1: 13–19.

Sunartio, D., Ashokkumar, M., and Grieser, F. 2007. Study of the coalescence of acoustic bubbles as a function of frequency, power, and water-soluble additives. *Journal of the American Chemical Society*, 129: 6031–6036.

Suri, R.P.S., Nayak, M., Devaiah, U., and Helmig, E. 2007. Ultrasound assisted destruction of estrogen hormones in aqueous solution: Effect of power density, power intensity and reactor configuration. *Journal of Hazardous Materials*, 146: 472–478.

Teo, K.C., Xu, Y., and Yang, C. 2001. Sonochemical degradation for toxic halogenated organic compounds. *Ultrasonics Sonochemistry*, 8: 241–246.

Thangavadivel, K., Megharaj, M., Smart, R.St.C., Lesniewski, P.J., and Naidu, R. 2008. Influence of Sonochemical reactor shape on acoustic pressure distribution under applied 20 kHz ultrasound field: A modelling study. In *Australian Institute of Physics (AIP) 18th National Congress incorporating the 27th AINSE Plasma Science Conference*, November 30–December 5, 2008, The University of Adelaide, Adelaide, South Australia, p. 144.

Thangavadivel, K., Megharaj, M., Smart, R.St.C., Lesniewski, P.J., and Naidu, R. 2009. Application of high frequency ultrasound in the destruction of DDT in contaminated sand and water. *Journal of Hazardous Materials*, 168: 1380–1386.

Thompson, L.H. and Doraiswamy, L.K. 1999. Sonochemistry: Science and engineering. *Industrial and Engineering Chemistry Research*, 38: 1215–1249.

Tiehm, A. and Neis, U. 2005. Ultrasonic dehalogenation and toxicity reduction of trichlorophenol. *Ultrasonics Sonochemistry*, 1: 121–125.

Torres, R.A., Mosteo, R., Pétrier, C., and Pulgarin, C. 2009. Experimental design approach to the optimization of ultrasonic degradation of alachlor and enhancement of treated water biodegradability. *Ultrasonics Sonochemistry*, 16: 425–430.

Torres, R.A., Nieto, J.I., Combet, E., Pétrier, C., and Pulgarin, C. 2008. Influence of TiO_2 concentration on the synergistic effect between photocatalysis and high-frequency ultrasound for organic pollutant mineralization in water. *Applied Catalysis B: Environmental*, 80: 168–175.

Torres-Palma, R.A., Nieto, J.I., Combet, E., Pétrier, C., and Pulgarin, C. 2010. An innovative ultrasound, Fe^{2+} and TiO_2 photoassisted process for bisphenol a mineralization. *Water Research*, 44: 2245–2252.

Vilkhu, K., Mawson, R., Simons, L., and Bates, D. 2008. Applications and opportunities for ultrasound assisted extraction in the food industry—A review. *Innovative Food Science and Emerging Technologies*, 9: 161–169.

Wang, J., Sun, W., Zhang, Z. et al. 2007. Sonocatalytic degradation of methyl parathion in the presence of micron-sized and nano-sized rutile titanium dioxide catalysts and comparison of their sonocatalytic abilities. *Journal of Molecular Catalysis A: Chemical*, 272: 84–90.

Wang, J., Sun, W., Zhang, Z. et al. 2008a. Preparation of Fe-doped mixed crystal TiO_2 catalyst and investigation of its sonocatalytic activity during degradation of azo fuchsine under ultrasonic irradiation. *Journal of Colloid and Interface Science*, 320: 202–209.

Wang, X., Yao, Z., Wang, J., Guo, W., and Li, G. 2008b. Degradation of reactive brilliant red in aqueous solution by ultrasonic cavitation. *Ultrasonics Sonochemistry*, 15: 43–48.

Wang, N., Zhu, L., Wang, M., Wang, D., and Tang, H. 2010. Sono-enhanced degradation of dye pollutants with the use of H_2O_2 activated by Fe_3O_4 magnetic nanoparticles as peroxidase mimetic. *Ultrasonics Sonochemistry*, 17: 78–83.

Willmott, N., Guthrie, J., and Nelson, G. 1998. The biotechnology approach to colour removal from textile effluent. *Journal of the Society of Dyers and Colourists*, 114: 38–41.

Yao, J.J., Gao, N.Y., Li, C., Li, L., and Xu, B. 2010. Mechanism and kinetics of parathion degradation under ultrasonic irradiation. *Journal of Hazardous Materials*, 175: 138–145.

Yavuz, Y., Savaş Koparal, A., Artık, A., and Öğütveren, U.B. 2009. Degradation of C.I. Basic Red 29 solution by combined ultrasound and Co^{2+}–H_2O_2 system. *Desalination*, 249: 828–831.

Yim, B., Okuno, H., Nagata, Y., and Maeda, Y. 2001. Sonochemical degradation of chlorinated hydrocarbon using a batch and continuous flow system. *Journal of Hazardous Materials*, 81: 253–263.

Yuan, S., Yu, L., Shi, L., et al. 2009. Highly ordered TiO_2 nanotube array as recyclable catalyst for the sonophotocatalytic degradation of methylene blue. *Catalysis Communications*, 10: 1188–1191.

Zeng, L. and McKinley, J.W. 2006. Degradation of pentachlorophenol in aqueous solution by audible-frequency sonolytic ozonation. *Journal of Hazardous Materials*, 135: 218–225.

Zhang, H., Lv, Y., Liu, F., and Zhang, D. 2008. Degradation of C.I. Acid Orange 7 by ultrasound enhanced ozonation in a rectangular air-lift reactor. *Chemical Engineering Journal*, 138: 231–238.

Zhang, Y., Xiao, Z., Chen, F., et al. 2010. Degradation behavior and products of malathion and chlorpyrifos spiked in apple juice by ultrasonic treatment. *Ultrasonics Sonochemistry*, 17: 72–77.

19 Applications of Ultrasound to Polymer Synthesis

Boon Mian Teo, Franz Grieser, and Muthupandian Ashokkumar

CONTENTS

Introduction 475
Acoustic Cavitation and Sonochemistry 476
Free Radical Polymerization 477
    Emulsion Polymerization 479
        Compartmentalization (Gilbert, 1995) 480
        Pseudo-Bulk System (Gilbert, 1995) 480
        Compartmentalized Pseudo-Bulk System (Gilbert, 1995) 480
        Zero–One System (Gilbert, 1995) 480
    Miniemulsion Polymerization 480
Ultrasound and Polymer Chemistry 481
    Ultrasonic Depolymerization 481
    Ultrasonic Polymer Synthesis 482
    Sonochemical Polymerization in Homogenous Systems 483
    Sonochemical Polymerization in Heterogeneous Systems 483
    Initiation 486
    Entry 486
    Propagation 486
    Termination 486
    Sonochemical Synthesis of Polymer Nanocomposites 488
Sonochemical Preparation of Microspheres 494
Concluding Remarks 495
Acknowledgment 496
References 496

INTRODUCTION

As a relatively unexplored technique, the application of ultrasound to chemical systems shows great promise in promoting a wide variety of chemical processes such as the synthesis of proteinaceous microspheres (Zhou et al., 2010), nanoparticle synthesis (Ashokkumar and Grieser, 1999; Didenko and Suslick, 2005), and the degradation of a range of pollutants (Petrier et al., 1992). The wide applications of ultrasound in chemical processes have attracted intense attention in various fields of chemistry, materials science, and chemical engineering. It exploits the effect of acoustic cavitation (Leighton, 1994); microbubbles present in the solution grow and collapse when sound waves pass through a liquid. This results in the generation of radicals, excited state species, enhancement of reaction rates, and excellent mixing of multiphase systems.

This chapter briefly introduces the essential fundamental concepts involved in the field of sonochemistry and free radical polymerization, and in the later part, a detailed discussion on the effect and application of ultrasound to the synthesis of polymers is provided. It does not provide an exhaustive bibliography, rather it provides an overview of some important research in this area.

ACOUSTIC CAVITATION AND SONOCHEMISTRY

When ultrasound is passed through a liquid, it induces a range of chemical effects, collectively referred to as sonochemistry, as well as various physical effects. It is through a process known as acoustic cavitation (Figure 19.1), a term describing the growth and violent collapse of microscopic bubbles in liquids under the influence of a sound field. During cavitation, the collapse of microscopic bubbles produces intense local heating, high pressures, and short lifetimes. Such intense, localized hot spots are responsible for many high-energy chemical reactions.

These bubbles grow from nuclei over many acoustic cycles through a process referred to as rectified diffusion (Strasberg, 1961; Hsieh and Plesset, 1961), which is illustrated in Figure 19.2. This process is explained in terms of two contributing effects, namely, area and shell effects. Gas diffuses in and out of the bubble as the acoustic cycle alternates between the rarefaction and compression phases. During the expansion cycle, the average surface area of the bubble is larger than that during the compression cycle; hence, there will be a net inflow of gas into the bubble over several cycles. The differences in the gas concentration gradient in the fluid shell surrounding the bubble also contribute to this effect. Since the diffusion rate of gases into the bubble is proportional to the concentration gradient of dissolved gas, the net inflow of gas is essentially higher during the expansion of the bubble. Both the area and shell effects promote bubble growth.

When acoustic bubbles reach a critical size range during their growth, they undergo a large radial excursion to a maximum size and subsequently undergo violent collapse. There have been many theories proposed to explain the chemical effects stemming from the collapse of cavitation bubbles such as the hot spot theory proposed by Noltingk and Neppiras (1950) and the electrical discharge theory (Margulis, 1985, 1995). The electrical discharge theory states that an electrical charge forms on the surface of the cavitation bubbles as it begins to break into tiny microbubbles, thus forming a huge electric field gradient across the bubble. This theory, however, has been discounted by many researchers as a mechanism for sonoluminescence (SL) and sonochemistry, for example, Lepoint-Mullie et al. (1996). The most widely accepted theory is the hot spot theory and according to this

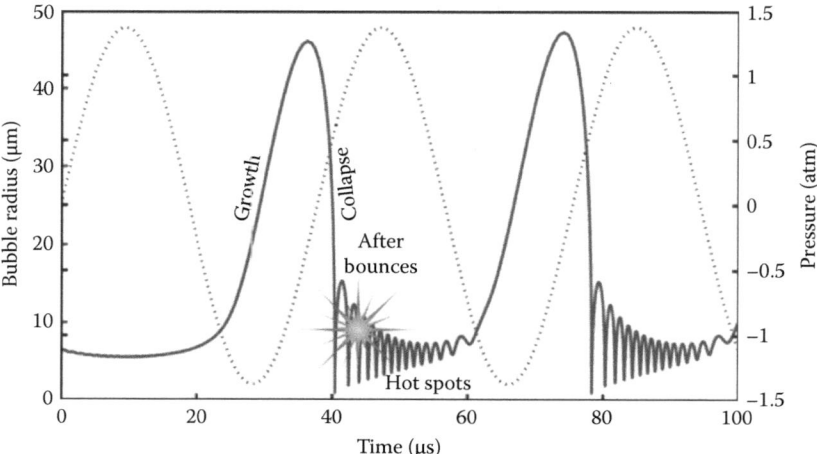

FIGURE 19.1 Acoustic pressure and bubble radius–time curve of an acoustically driven bubble. Rayleigh–Plesset equation for a bubble of initial radius 6.5 μm driven in a 15 W, 26.5 kHz field in an inert gas atmosphere and standard temperature and pressure.

Applications of Ultrasound to Polymer Synthesis

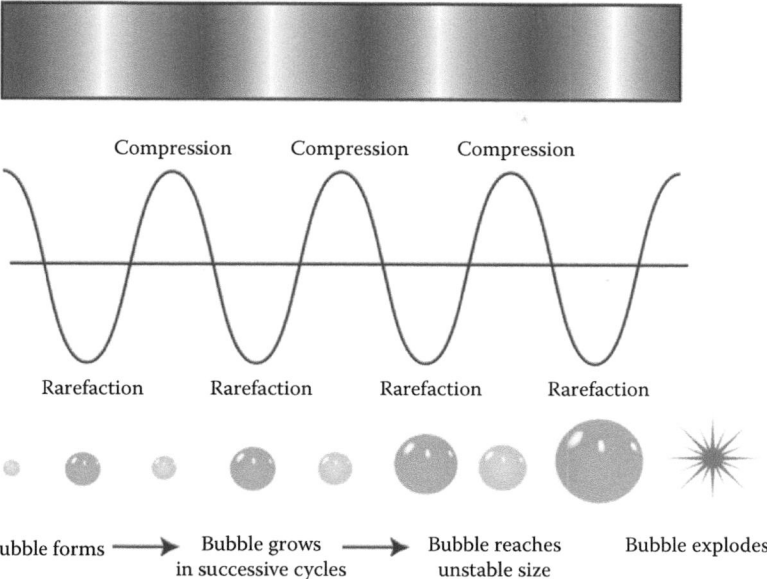

FIGURE 19.2 Schematic diagram of bubble growth and collapse in a liquid under an acoustic field resulting in the formation of localized hot spots.

theory, prior to the collapse, bubble growth proceeds under isothermal conditions. However, the system becomes almost adiabatic upon bubble collapse. As the bubble continues to collapse, the gas present within the bubble core is rapidly compressed; hence, temperatures of thousands of degrees and pressures more than hundreds of atmospheres can be generated. In addition to the extreme conditions experienced within the hot spots, a secondary region of a thin layer of liquid surrounding the collapsed bubble is also transiently heated, albeit to a lesser extent. This interface of approximately 200 nm in thickness (Suslick et al., 1986; Suslick, 1987) may be as hot as 2000 K. Due to the transient nature of the hot spot having an approximate life time in the nanosecond timescale, direct measurement of the hot spot temperature profile during collapse has not been achieved, although theoretical calculations of the profile have been made (Yasui et al., 2004).

Ultrasound-induced cavitation in a liquid gives rise to a number of physical effects, such as shock waves and microjets, which can affect the efficiencies of chemical reactions. Shock waves are created during bubble collapse, producing a great deal of turbulence in the solution. In heterogeneous systems containing solids, the highly energetic collapse of bubbles is no longer symmetrical as the solid surfaces prevent the even flow of fluid causing the collapse to be asymmetric, often having a "doughnut" like shape. This asymmetrical collapse results in jets of high-speed liquid directed at the solid surface at speeds of greater than 100 m s^{-1}. Shock waves and microjet-induced agitation can create emulsions in systems containing two immiscible liquids (Lauterborn and Vogel, 1984; Thompson and Doraiswamy, 1999) and are also responsible for the collision-induced fusing dispersed metal particles in sonicated solutions (Doktycz and Suslick, 1990).

FREE RADICAL POLYMERIZATION

The mechanism governing a free radical polymerization process involves three steps: Initiation, propagation, and termination (Flory, 1937; Rudin, 1999; Odian, 2004). As shown in Reactions 19.1 and 19.2, there are two steps involved in the initiation process. First, the dissociation of initiators to form free radicals and second, the reaction of the free radical with a monomer molecule:

$$I \xrightarrow{k_d} 2R_i \qquad (19.1)$$

$$R_i + M \xrightarrow{k_i} M_I \tag{19.2}$$

where
  R_i is the free radical formed
  k_d is the rate constant for initiator decomposition
  k_i is the rate constant for initiation process

The initiator can be decomposed into free radicals by means of heat, light, or redox reactions. The rate expression for initiation is given by Equation 19.3:

$$\frac{d[R]}{dt} = R_i = 2fk_d[I] \tag{19.3}$$

where
  R is the rate of initiation
  f is the initiator efficiency (Guan et al., 1993)

The initiator efficiency is defined as the fraction of free radicals that can react with the monomer. The value of f is usually less that 1 and typically falls in the range of 0.3–0.8. Once formed, free radicals can either recombine, terminate with another free radical, or initiate chain propagation (Capek, 1996). The initiation step plays a very important role in the overall rate of polymerization and in determining the ensuing molecular weight of the final polymer.

The constant addition of monomer molecules to growing polymer radicals in the propagation stage takes place very rapidly (Reactions 19.4 and 19.5):

$$R + M \xrightarrow{k_p} M_i \tag{19.4}$$

$$M_i + M \xrightarrow{k_p} M_{i+1} \tag{19.5}$$

The propagation rate constant, k_p, usually falls in the range of 10^2–10^4 mol^{-1} L^{-2} s^{-1} (Odian, 2004) and can be calculated using experimental conversion and time data in conjunction with the initiator concentration measurements. The relationship is shown below (Equation 19.6):

$$\frac{dx}{dt} = k_p[R][I-x] \tag{19.6}$$

where
  x is the mole fraction of the monomer
  k_p is the pseudo-propagation rate constant

As with the initiation stage, the propagation rate exhibits a significant decrease as the extent of monomer conversion increases.

The termination step involves the cessation of polymer chain growth and takes place either by combination or by disproportionation. Combination includes the reaction between two growing polymeric radicals (Reaction 19.7), whereas disproportionation includes the formation of two distinct polymeric molecules (Reaction 19.8):

$$M_i + M_j \xrightarrow{k_{tc}} P_{ij} \tag{19.7}$$

Applications of Ultrasound to Polymer Synthesis

$$M_i + M_j \xrightarrow{k_{td}} P_i + P_j \tag{19.8}$$

The rate constant for termination depends on various parameters such as the overall composition and the chain length of the propagating species. The rate constants for the combination reaction and the disproportionation reaction are often averaged to give a single termination rate constant. The termination rate also decreases as the conversion to polymer increases. In some cases, chain propagation continues as the termination rate is decreased and as a result, high molecular weight polymers are produced. This phenomenon is commonly referred to as the "gel effect" and is particularly evident in bulk polymerization reactions. In other cases, a polymer chain can also be prevented from growing by the removal of an atom from some substance present in the system to give a new radical, which may or may not start another new chain (Flory, 1937). This process is known as chain transfer and can be represented by Reactions 19.9 and 19.10.

$$P_i^\bullet + RX \rightarrow P_i R + X^\bullet \quad \text{Chain transfer reaction} \tag{19.9}$$

$$X^\bullet + M \rightarrow P_i^\bullet \quad \text{Generation of new kinetic chain} \tag{19.10}$$

Free radical polymerizations can be classified as either homogeneous or heterogeneous reactions. In homogeneous polymerization, all components, that is, the monomer chemical initiator and the polymer, are in the same phase throughout the reaction. In heterogeneous polymerization, at least one component is insoluble at some point during the reaction.

EMULSION POLYMERIZATION

The polymerization of direct emulsions is usually carried out with a mixture of insoluble monomer, water, surfactant, and initiator. The emulsion is formed by mechanical agitation of the mixture in the presence of the surfactant, also known as the emulsifier. The result of this is the formation of monomer droplets dispersed in an aqueous phase. Micelles are present when the surfactant concentration in the aqueous phase is higher than its critical micelle concentration (CMC). Emulsion polymerization, for the production of materials such as synthetic rubber, latex paints, adhesives, and coatings, has been widely used in many industries due to its attractive advantages over bulk polymerization (Gomes, 2005; Dar et al., 2005). For example, heat dissipated during the reaction can be easily controlled by heat transfer to the aqueous phase and a faster polymerization rate with high conversion to polymer can be achieved. The resulting polymers usually have high molecular weights that can be easily controlled by the addition of chain transfer agents. In addition, the viscosity of the solution during emulsion polymerization is close to that of water since water is the continuous phase. A fundamental understanding of emulsion polymerization is of great academic and commercial interest.

The theory of emulsion polymerization centers around the following equation (Odian, 2004):

$$R_p = k_p [M_p] N_p \frac{\bar{n}}{N_A} \tag{19.11}$$

where
  R_p is the rate of polymerization
  k_p is the propagation rate coefficient
  $[M_p]$ is the concentration of monomer in a particle
  $\bar{n}$ is the average number of radicals in a particle
  N_p is the number concentration of particles
  N_A is Avogadro's constant

According to this equation, the rate of polymerization, R_p depends on the following parameters. The value of $\bar{n}$ is critical in determining R_p. Through the mathematical analysis by Smith and Ewart, three scenarios can be considered in which the particle size and radical exit from the emulsion droplet rate may vary. These three situations are generated by considering a balance between N_{pi} (the number of particles containing i radicals) assuming that the number of particles remains constant (i.e., no nucleation occurs). In the first situation (Case 1), where the particles are small, monomers are relatively soluble in water and desorption of radicals from the particles is probable, $\bar{n}$ is very low and polymerization is slow. In the second situation (Case 2), radical exit is almost insignificant. When a radical enters a particle, polymerization occurs until the next radical comes along and both instantaneously terminate (zero–one kinetics). Under such conditions, $\bar{n} = 0.5$. In the last situation (Case 3), the particles are large such that more than two radicals can exist together in the particle without instantaneous termination. In this case, $\bar{n}$ can be greater than 1.

Compartmentalization (Gilbert, 1995)

Compartmentalization is the term used to describe the fact that in an emulsion polymerization, radicals are compartmentalized, that is, propagating radicals contained in one latex particle are physically separated from radicals contained in another particle. As a result of this property, bimolecular termination cannot occur between radicals in different particles.

Pseudo-Bulk System (Gilbert, 1995)

In this system, the number of radicals per particle is relatively high such that the emulsion polymerization resembles a bulk polymerization system. The average number of radicals is always greater than 0.5. Compartmentalization does not have an effect on the kinetics of polymerization in such a system and termination is always diffusion controlled.

Compartmentalized Pseudo-Bulk System (Gilbert, 1995)

This system, also known as the zero–one–two kinetics, arises when termination is not instantaneous and $\bar{n}$ is low. Zero–one–two kinetics are favored by high propagation rate coefficients when radicals grow rapidly and bimolecular termination of radicals is no longer instantaneous. As the name implies, the latex particles may contain zero, one, or two radicals.

Zero–One System (Gilbert, 1995)

The term zero–one typically describes the situation found in most emulsion polymerizations whereby all latex particles contain either zero or one active radical (i.e., two or more radicals cannot coexist in one particle). This is because the particles are very small and can result in instantaneous bimolecular termination. Therefore, the maximum value of the average number of radicals per latex particle, $\bar{n}$, is 0.5. In such systems, compartmentalization becomes important in the kinetic events of emulsion polymerization reactions.

MINIEMULSION POLYMERIZATION

A miniemulsion polymerization is performed by shearing a mixture containing two immiscible liquids, one of which being the organic monomer and a hydrophobe, and the other, an aqueous surfactant solution. Monomer droplets of size ranging from 50 to 500 nm are usually obtained. The formation of miniemulsion droplets relies on the combination of the shear force treatment, the amount and type of surfactant, and the hydrophobe. The surface area of a dispersion of even just a few weight percent of micro–nano droplets is very large, and at typical surfactant concentrations used, all the surfactant molecules are adsorbed at the droplet surfaces such that little free surfactant molecules can be found in the aqueous phase. In an ideal case of a miniemulsion polymerization process, the monomer droplet is the primary locus of the polymerization reaction and it is generally known as the 1:1 copy of the original monomer droplet to latex particle. Every monomer droplet

TABLE 19.1
Summary and Comparison of the Important Features of Macroemulsions, Miniemulsions and Microemulsions

Emulsion Type	Macroemulsion	Miniemulsion	Microemulsion
Droplet size range	>1 µm	50–500 nm	10–100 nm
Duration of stability	Seconds to hours	Hours to months	Indefinitely
Diffusional stabilization	Kinetic	Kinetic	Thermodynamic
Nucleation mechanism	Micellar, homogenous	Droplet	Droplet
Emulsifier concentration	Moderate	Moderate	High
Costabilizer type	None	Hexadecane, cetyl alcohol	Hexanol, pentanol
Homogenization method	None	Mechanical, ultrasonic	None
Particle size range	50–500 nm	50–500 nm	10–100 nm
N_p range (per L H_2O)	10^{16}–10^{19}	10^{16}–10^{19}	10^{18}–10^{21}

acts as an independent nanoreactor and polymerization inside the droplet occurs via a suspension polymerization type reaction. The system is called a direct miniemulsion if the dispersed phase is hydrophobic whereas an inverse miniemulsion is made up of a hydrophilic dispersed phase.

Ugelstad et al. (1973) first introduced the concept of miniemulsion polymerization in 1973 when they successfully synthesized submicron-sized styrene particles using an emulsifier mixture of sodium dodecyl sulfate and cetyl alcohol. The water-soluble initiator, potassium persulfate, was used as the chemical initiator for the polymerization reaction after the miniemulsion styrene droplets were formed. The sizes of the monomer droplets and the latex particles were observed to be similar and it was proposed that nucleation occurred primarily inside the droplets. It is necessary to understand the differences in the different types of emulsion polymerization processes. A comparison of the properties of these three systems is given in Table 19.1.

ULTRASOUND AND POLYMER CHEMISTRY

The effect of the application of ultrasound on polymers consists of three main fields, namely, the degradation of polymers (depolymerization), the synthesis of polymers, and the synthesis of polymer nanocomposites.

ULTRASONIC DEPOLYMERIZATION

The degradation of a polymer molecule is an irreversible process that decreases the polymer chain length by cleavage. It has been long known that irradiating a polymer solution with ultrasound reduces the viscosity of the solution. Brohult (1937) made the discovery that biological polymer solutions exposed to ultrasound could lead to the degradation of the polymer, and subsequently, Schmid and Rommel (1939) were the first to notice an irreversible reduction in the viscosity of a range of polymers in solutions due to the breakage of the covalent bonds in the polymer backbone. They observed a rapid initial depolymerization rate, which slowed down and stopped when a minimum molecular weight was approached. Their findings led them to postulate that depolymerization is a physical process that is independent of the nature of the polymer in question but instead depends on the polymer chain length in solution. There have been many attempts to explain the degradation of polymer under ultrasound irradiation. When high-intensity ultrasound is applied to a polymer in a solvent, sufficiently long polymer chains may be stretched amidst the solvent flow ensuing from the movement of fluid surrounding the collapsing cavitation bubbles and from the propagation of the resulting shock waves (Glyn et al., 1972; Van der Hoff and Gall, 1977; Madras et al., 2000). Contrary to other chemical or thermal decomposition reactions, ultrasonic depolymerization is a

TABLE 19.2
Parameters Influencing Polymer Degradation under Ultrasound Irradiation

Parameter	Effects of Depolymerization
Molecular mass of polymer	A larger initial molecular weight of the polymer increases the degree of depolymerization
	There exists a molecular mass limit below which no degradation will occur
Concentration of polymer	Decreasing the concentration of polymer in the solution leads to an increase in the degradation process
Nature of solvent	Higher vapor pressure of the solvent leads to less violent collapse of acoustic bubbles and a lesser extent of degradation
	Acoustic cavitation occurs more readily in solvents with low viscosity and surface tension
External temperature	Increasing the external temperature leads to an increase in vapor pressure leading to bubble collapse being less violent, decreasing the degradation rate
Acoustic intensity	Increasing the acoustic intensity increases the rate and extent of degradation

nonrandom process with cleavage occurring preferentially near the middle of the polymer chain. The strong velocity gradients coming from the fluid flow surrounding the bubble collapse are sufficient to cleave large polymer chains into smaller ones. This permits the removal of the large molecular weight polymers in a polydispersed system. Van der Hoff and Gall (1977) investigated the degradation of polystyrene in tetrahydrofuran and found a Gaussian distribution of the scission around the middle of the polymer chain. This center cleavage model is consistent with the stretching and breakage mechanism. Below a certain limiting molecular weight, ultrasound depolymerization does not take place and this results in initially broad molecular weight distributions becoming more narrow during irradiation (El' Piner, 1964). The existence of a limiting molecular weight beyond which no further degradation can occur was also established by other researchers (Price et al., 1992, 1996; Van der Hoff and Gall, 1977; Price, 1990). The depolymerizing effect of applied ultrasound is more pronounced for higher molecular weight polymers such that the degradation proceeds at a faster rate. The rate of degradation also depends on the irradiation time, the concentration of polymer in solution, the nature of polymer and solvent, and the ultrasonic parameters (power intensity and frequency). Table 19.2 outlines the range of parameters that influence polymer degradation under a sound field. In dilute solution, the polymer chains are not entangled and are free to move around the cavitation bubbles (Basedow and Ebert, 1975; Odell and Keller, 1986). Hence, it can be expected that in more dilute solutions, degradation of the polymer is more effective. In addition, degradation can be more efficient at higher power intensities due to the larger size and number of bubbles generating stronger shear forces. Thus, by suitable manipulation of the experimental conditions, some control over the degradation process can be achieved. Chun and Park (1994) applied ultrasound to anionic xanthan polyelectrolyte and nonionic schizophyllan rodlike biopolymers to obtain polymer fractions of different molecular weights and studied the ionic strength dependence of intrinsic viscosity as a function of molecular weight. Kim et al. (1998) investigated the turbulent drag reduction characteristics of sonicated polymer fractions of rodlike polysaccharide xanthan gum of different molecular weights (Kim et al., 1998; Choi et al., 2000; Sohn et al., 2001). There have been extensive studies in the area of ultrasonic degradation of polymers and these have been summarized in a review by Price (1990).

Ultrasonic Polymer Synthesis

Most synthetic polymers are prepared from monomers that contain a reactive double bond, which can undergo chain growth or addition reactions. The most common polymer synthesis method is via a free radical initiation process. As already mentioned, acoustic cavitation in liquids can produce highly reactive radicals. Thus, the application of ultrasound to solutions containing vinyl monomers constitutes an alternative route to polymer synthesis.

Applications of Ultrasound to Polymer Synthesis

The earliest work on the chemical effects of ultrasound on macromolecules biopolymers, egg albumin and plastein, was reported by Florsdorf and Chambers (1933). They found that the polymers in aqueous medium, when exposed to ultrasound, coagulated instantly at 30°C and that the hydrolysis of sucrose was accelerated at temperatures as low as 5°C comparable with the rate of the boiling point in the absence of polymerization.

It was only in the early 1980s that extensive work on the effects of ultrasound on polymer systems was conducted. Ultrasound-initiated polymer synthesis can be divided into two categories. First, the homogenous system commonly referred to as bulk polymerization allows for pure monomer to be polymerized. Second, in the heterogeneous system, which can be subdivided into three subcategories: precipitation, suspension, and emulsion systems, the polymer formed is insoluble in the reaction medium. Precipitation polymerization begins as homogeneous polymerization but is quickly transformed into heterogeneous polymerization. This occurs in the polymerization of monomers in bulk solution with the ensuing polymer products being insoluble in the reaction medium.

Sonochemical Polymerization in Homogenous Systems

The extensive work by Price and coworkers (Price et al., 1991, 1992; Price and Smith, 1993; Price, 1996) has illustrated some important features of using ultrasound as a means of polymerization. They found that during the sonochemical polymerization of vinyl monomers, high molecular weight polymers were formed in the earlier stages of polymerization but the average chain length decreased dramatically with prolonged sonication, due to the degradation of the polymer in the solution. For a methyl methacrylate (MMA) system, conversion of approximately 12% was achieved after 6 h of sonication and after this time, the viscosity of the medium increased, and cavitation in the solution was essentially nonexistent. As a consequence, there was no further radical formation, and further conversion of monomer to polymer could not occur. Radical trapping studies conducted to estimate the rate of the initiation process in the MMA system showed that the crucial step in polymer formation is the formation of radicals by the breakdown of solvent vapor in the cavitation bubbles. It is these primary radicals that initiate the sonochemical polymerization process. They reported that below an acoustic intensity of approximately $12\,W\,cm^{-2}$, there were no radicals generated. This is the minimum intensity required to generate cavitation in the systems studied.

Kruus et al. (Kruus and Patraboy, 1985; Kruus, 1991) conducted a detailed study of the mechanism of sonochemical polymerization of MMA and their findings showed that under certain reaction conditions, pyrolysis of the monomer could take place within the cavitation bubbles, which resulted in the formation of a large amount of insoluble chars in addition to linear polymers. However, if the monomer was purified and the system deoxygenated, soluble, high molecular weight polymers of MMA and styrene could be produced. They also reported that the rate of conversion of monomer to polymer exhibited an inverse relationship with the reaction temperature and that the polymerization reaction stopped when the ultrasound generator was switched off. This indicated that the formation of polymer was due to the radicals produced by acoustic cavitation and not due to a thermal initiation process.

Cass et al. (2010) reported the polymerization of water-soluble acrylic monomers by ultrasound generated to form hydrogels. Water-soluble additives such as glycerol, sorbitol, or glucose were essential for the formation of the hydrogels. They proposed that ultrasound may find application in the field of biomaterial synthesis since the use of chemical initiators are not required.

Sonochemical Polymerization in Heterogeneous Systems

An alternative method for free radical polymerizations is through a latex dispersion as an emulsion or suspension in an aqueous reaction medium. One of the major drawbacks of using chemical initiators and stabilizers in such systems is that they can alter the properties of the final polymer product (Bradley and Grieser, 2002). Such problems can be prevented by employing techniques that

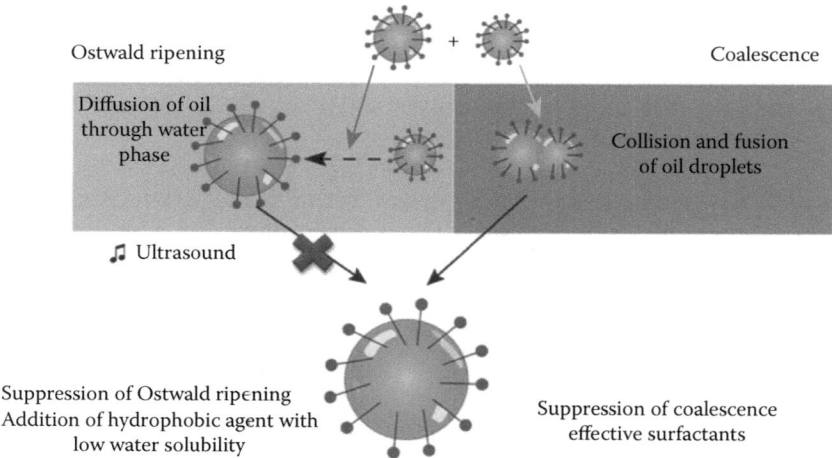

FIGURE 19.3 Schematic diagram of the growth of monomer droplets by Ostwald ripening and coalescence. Growth of the monomer droplets by diffusion of monomer through the water phase can be greatly reduced by the addition of a hydrophobe, or in the case where ultrasound is present, a hydrophobe is not required, and growth of the monomer droplets by collision and fusion of monomer droplets can be minimized by the addition of surfactants.

do not require chemical initiators and stabilizers. Using ultrasound as a means of emulsification, polymerization is an effective way of circumventing these drawbacks. Due to the high localized shear gradients generated, ultrasound allows for efficient mixing and dispersion of the emulsion mixture, minimizes Ostwald ripening (Figure 19.3), and maintains a small and narrow distribution of monomer droplet sizes (Eldik and Hubbard, 1996; Bradley and Grieser, 2002; Xia et al., 2002). In addition, sonochemical emulsion polymerization permits good control over the overall reaction, in that the polymerization reaction effectively stops once sonication ceases. At fast polymerization rates with high monomer to polymer conversion at ambient temperature, polymers of high molecular weights can be produced, compared with conventional methods.

The first report of applying ultrasound to this type of polymerization reaction dates back to the early 1950s. Ostroski and Stambaugh (1950) reported that when conventional emulsion polymerization was conducted under sonication, better dispersion of styrene was obtained, which in turn dramatically enhanced the rate of polymerization. They attributed the enhancement in the rate of styrene polymerization to a faster decomposition rate of the chemical initiator in the solution. This is a consequence of the faster and more efficient emulsification produced by ultrasound agitation. Since then, there have been a number of studies reporting on sonochemical emulsion polymerization.

Biggs and Grieser (1995) conducted mechanistic study on the synthesis of poly(styrene) latex particles by ultrasound irradiation at ambient temperature. From their experimental results, they drew the following conclusions: the rate of polymerization increased with increasing surfactant concentration, the average size of the latex particles was very small (approximately 50 nm), polymer molecular weights exceeded 10^6 g mol^{-1}, and there was a continuous formation of polymer particles. They concluded that their polymerization system bears similarities with the conventional microemulsion polymerization system but with a much lower surfactant concentration. They extended their work to explore the effects of varying the acoustic intensity on the polymerization rate. An apparent increase in polymerization rate as a function of sonication time was observed as the intensity was increased. It was also reported that increasing the intensity did not affect the ensuing particle size range of the latexes, which was found to be in the range of 40–50 nm. They suggested that the narrow particle size range and high conversion rates obtained were due to a continuous nucleation of monomer droplets that then scavenged the sonochemically produced free radicals throughout the polymerization process.

Applications of Ultrasound to Polymer Synthesis

In another study, Cooper et al. (1996) found that by using a horn-type sonicator, it was possible to produce polymer latex particles with particle sizes smaller than those generated in the conventional process at low levels of or even using no surfactants. They observed a faster polymerization rate for butyl acrylate (BA) than vinyl acetate and related this finding to the differences in the vapor pressures of the monomers. In contrast with the solution or bulk polymerization systems as mentioned earlier, high conversion of monomer to polymer (approximately 100%) was routinely achieved. Okudaira et al. (2003) reported a successful suspension sonopolymerization in the absence of surfactant and chemical initiator. The monomer styrene was added drop wise into water and sonicated at 40 kHz to form monomer droplets dispersed in the aqueous phase. Polymerization of the monomer in the droplets was initiated by ultrasonic irradiation at 200 kHz. Both light-scattering data and TEM images indicated monodispersed polystyrene beads with an average diameter of 50 nm. Surprisingly, they reported that no radical species were produced under low-frequency irradiation and it was exclusively used as a dispersant to generate monomer droplets in aqueous phase.

The mechanistic details suggested by Bradley and Grieser (2002) provide a better insight to sonochemical polymerization in miniemulsion systems. They reported that the polymerization process is initiated by the reaction of the primary radicals with free monomer molecules, leading to the creation of monomeric radicals in the emulsion system (Figure 19.4). Upon collapse of the micron-sized bubbles in an aqueous solution, H• and •OH radicals are produced (Reaction 19.12). Although the direct formation of MMA radicals is not crucial to ultrasonically initiated miniemulsion polymerization, MMA is volatile and has the capacity to evaporate into a cavitation bubble and be decomposed into hydrocarbon products. The H• and •OH radicals produced are intercepted

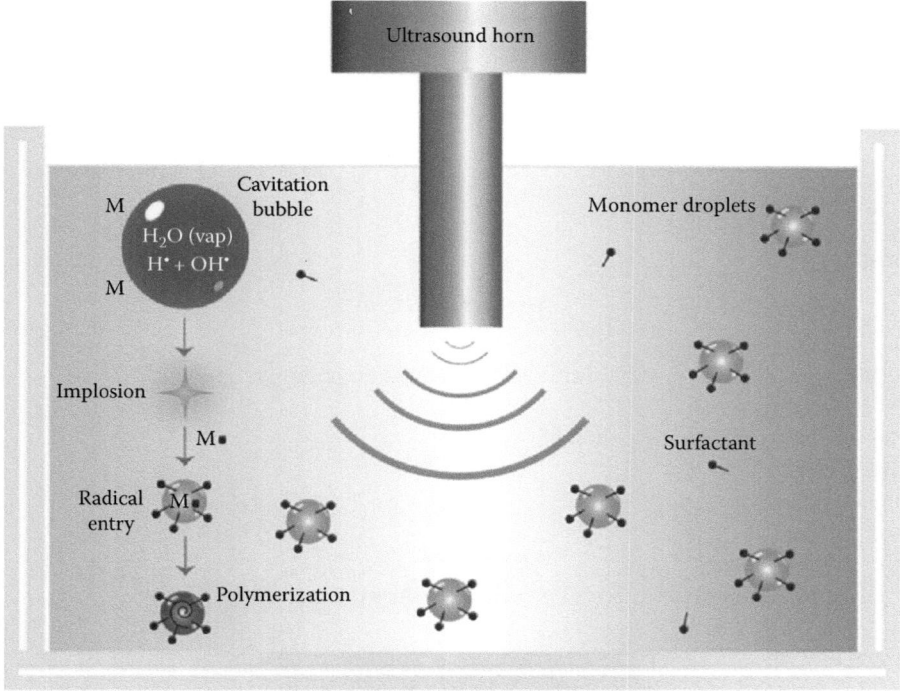

FIGURE 19.4 (See color insert.) Diagram of the proposed sonochemical miniemulsion polymerization pathway. The ultrasound from the horn tip produces cavitation bubbles that upon collapse generate the conditions that lead to primary radical formation and emulsification of the monomer. Monomeric radicals are mainly formed at the surface of the cavitation bubbles and subsequently enter into monomer droplets producing latex particles.

by the solutes at the bubble–solution interface before they reach the aqueous phase. These solutes, present in the emulsion mixture, are mainly monomer and surfactant molecules. The primary radicals can react with the monomer adsorbed at the bubble–solution interface forming monomeric radicals in the bulk solution (Reactions 19.13a and b). The monomeric radicals enter the monomer droplets and propagate the polymerization process (Reaction 19.14). Reactions 19.15 and 19.16 represent a propagating radical undergoing polymerization. Termination of the polymerization reaction occurs when the growing radical reacts with another growing radical by disproportionation (Reaction 19.17). It should be noted that hydrocarbon-based radicals produced from the thermal decomposition of the monomer within the hot bubble core may also react during the synthesis of the H$^\bullet$ and $^\bullet$OH radicals and contribute to the polymerization process.

INITIATION

$$H_2O \;)))\; OH^\bullet + H^\bullet \tag{19.12}$$

$$OH^\bullet + M_s \rightarrow HOM_s^\bullet \equiv \left(M_s^\bullet\right) \tag{19.13a}$$

$$H^\bullet + M_s \rightarrow HM_s^\bullet \equiv \left(M_s^\bullet\right) \tag{19.13b}$$

ENTRY

$$RM^\bullet(aq) \rightarrow RM^\bullet(d) \tag{19.14}$$

PROPAGATION

$$RM^\bullet(p) + M(p) \rightarrow P_2^\bullet \tag{19.15}$$

$$P_i^\bullet + M^\bullet \rightarrow P_{i+1}^\bullet \tag{19.16}$$

TERMINATION

$$P_i^\bullet + P_j^\bullet \rightarrow P_i + P_j \tag{19.17}$$

Notes: Mechanism for ultrasound-initiated miniemulsion polymerization
M represents the monomer
s is the surface of the cavitation bubble
R is the radical
d is the monomer droplet
p is the particle
P is the polymer chain
i and j denote the length of different growing polymer chains

Similarly, Chou and Stoffer (1999) carried out a comprehensive investigation of the emulsion sonopolymerization of MMA at ambient temperature using sodium dodecyl sulfate as the surfactant. The effects of cavitation type, nature, and source of the free radical for the initiation process and a range of experimental conditions were explored in depth. Contrary to what was found by Bradley and Grieser, they stated that the surfactant molecules were degraded and acted as free radicals for the polymerization process. The rate of polymerization was found to be enhanced when the acoustic intensity, argon flow rate, and surfactant concentration were increased. The molecular weight of the polymer increased with an increase in the monomer concentration to a certain level and then became independent of the monomer concentration. It was also found that the molecular weight of the polymer decreased with an increase in the surfactant concentration, which was ascribed to an increase in the amount of radicals generated resulting from an increase in the number of surfactant molecules acting as initiators. The assumption that surfactant molecules serve as radicals for the polymerization reaction is unlikely because Bradley and Grieser (2002) found that they could be successfully removed from the polymer samples after the reaction by dialysis. This indicates that the surfactant does not play a major role in the initiation process for sonochemical polymerization reactions. The evidence that Chou and Stoffer put forward is thus not strong enough to support their assertion that the source of radicals was from the degradation of the surfactant molecules. These radicals were identified using radical trapping experiments in aqueous sodium dodecyl sulfate solutions in the absence of monomer, clearly an unrepresentative system.

In a more recent work by Wang et al. (Wang et al., 2001a,b; Liao et al., 2001) on the sonochemical polymerization of MMA in an emulsion system with sodium dodecyl sulfate as the surfactant, they found that with increasing surfactant concentration, the conversion of monomer to polymer increased significantly, but when no surfactant was added, no polymer was formed. Therefore, they suggested that the surfactant plays an important role in the initiation process. In addition, they observed an increase in monomer to polymer conversion when the reaction temperature was increased and that increasing the N_2 sparging rate increased the conversion percentage. They were able to obtain a monomer conversion of 67% and polymer molecular weights in the order of several millions daltons under their experimental conditions. They have also investigated the ultrasonic emulsion polymerization of BA to study the factors that affect the induction period and rates of polymerization, and hence proposed a mechanism for sonochemical emulsion polymerization. They found that by increasing the N_2 sparging rate, temperature, surfactant concentration, and power intensity, and decreasing the monomer concentration, a decrease in the induction period and an increase in polymerization rate occurred. Under their experimental conditions, the conversion of BA to poly(butyl acrylate) (PBA) reached approximately 90% within 10 min of sonication. They suggested that the mechanism of ultrasound-initiated emulsion polymerization was analogous to conventional emulsion polymerization in that a high surfactant concentration gives rise to a larger number of micelles, leading to more polymerization loci and enhanced polymerization rates. Through analysis of the polymer product with nuclear magnetic resonance spectroscopy and Fourier transform infrared spectroscopy, their polymers were found to be slightly crosslinked and branched. Overall, the structure of the polymers obtained via sonochemistry was said to be different from those obtained through conventional methods.

In a later study by Teo et al. (2008b), they reported on the sonochemical miniemulsion polymerization of a family of methacrylate monomers. Their results suggest that the physicochemical properties (e.g., vapor pressure and water solubility) of the monomers play an important role in determining the rates of polymerization in ultrasonic initiated polymerization (Figure 19.5A). Their results also indicate that the polymerization reactions proceed via pseudo first-order kinetics, which supports the use of a zero–one model for polymerization such that radical entering a particle already containing a growing radical will lead to pseudo-instantaneous termination (Figure 19.5B). By performing particle sizing analysis using the dynamic light-scattering technique, they concluded that their results are consistent with a conventional miniemulsion polymerization mechanism in that there is a 1:1 copying of the monomer droplets to polymer particles.

Moholkar and coworkers (Morya et al., 2008) attempted to explain the sonochemical emulsion polymerization mechanism using the Keller-Miksis equation as a mathematical model for the radial motion of cavitation bubble. They concluded that the experimental parameters, the extent of radical production from cavitation bubbles, magnitude of microturbulence or shear velocity and shock waves produced by the bubbles, glass transition temperature of the polymer, and the population density of the polymer particles, all contributed to the final average size and size distribution of the polymer particles.

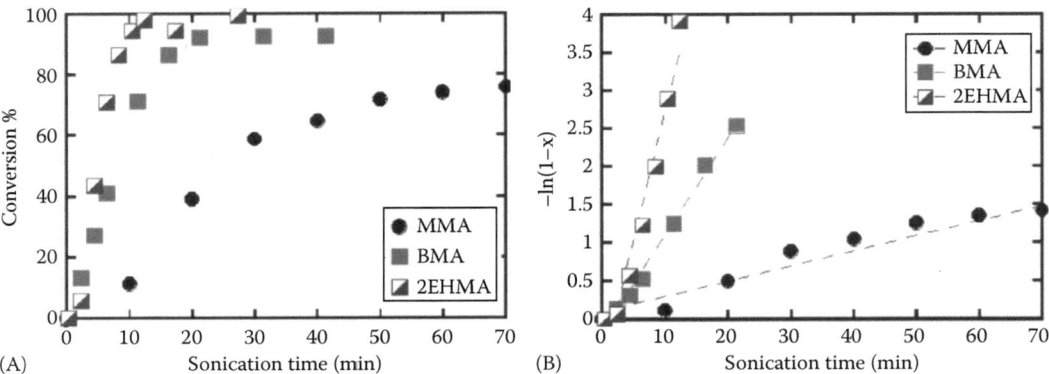

FIGURE 19.5 (A) Conversion % (±10%) of methacrylate monomers as a function of sonication time; (B) pseudo-first order, x, as the fractional conversion, as a function of sonication time for methacrylate monomers. (Modified from Teo, B.M., *Ultrason. Sonochem.*, 15(1), 89, 2008.)

Teo et al. (2008a, 2009b) reported on the microemulsion polymerization of n-butyl methacrylate at a range of acoustic frequencies using ionic and nonionic surfactants. The properties (particle size and molecular weight) of the latex depended on the concentration and type of surfactants used in their study. They concluded that the sonochemical microemulsion polymerization follows a continuous particle nucleation mechanism; however, no concrete conclusion was made in regard to the effect of acoustic frequency (Teo, 2010). Bradley et al. (2005) studied copolymerization of MMA and BA at different MMA:BA ratios using ultrasound irradiation. From their results from the analysis of the glass transition temperatures of the copolymers, they concluded that the microstructures of the copolymers were similar to those obtained in a conventional copolymerization system. This is indicative that ultrasound has no effect on the propagation step of the free radical polymerization process and it is useful as a source of free radicals to initiate free radical polymerization.

It has been shown by many authors (Biggs and Grieser, 1995; Cooper et al., 1996; Bradley and Grieser, 2002; Teo et al., 2008b) that the molecular weights of polymers prepared through sonochemical emulsion polymerization are generally higher than those prepared by conventional methods; therefore, there is a critical need to regulate the molecular weights of the polymers formed through ultrasound irradiation. For emulsion polymerization technology and many other industrial applications of polymers, it is crucial to have the ease of manufacturing and control of the polymer properties (molecular weight and glass transition temperature). By using a range of organic solvents, Teo et al. (Teo, 2010) were able to control the molecular weights of the polymers produced by ultrasound irradiation. Figure 19.6 shows the effect of increasing the volume percent of the organic solvent present in the miniemulsion system on the molecular weight of the polymer produced. This effect is due to chain transfer reactions. Under specific experimental conditions, the chain transfer reaction may result in the production of a new radical, which may then continue the kinetic chain by reinitiation.

SONOCHEMICAL SYNTHESIS OF POLYMER NANOCOMPOSITES

The fabrication of "functional" nanoparticles for practical use in a wide range of applications from medicine and biotechnology to electronics and catalysis is currently of considerable interest (Tirelli, 2006; Contreras-Cáceres et al., 2008). The encapsulation of "active" materials within a carrier polymer-based nanoparticle is seen as a significant and important innovation because polymeric nanocomposites have the potential to improve many existing technologies. One of the functions of encapsulation is protection through the isolation of the core from its external harsh environment (Erdem et al., 2000). For example, encapsulation protects probiotic bacteria from high-temperature food processing, from oxidation and moisture, from food and pharmaceutical industries, and

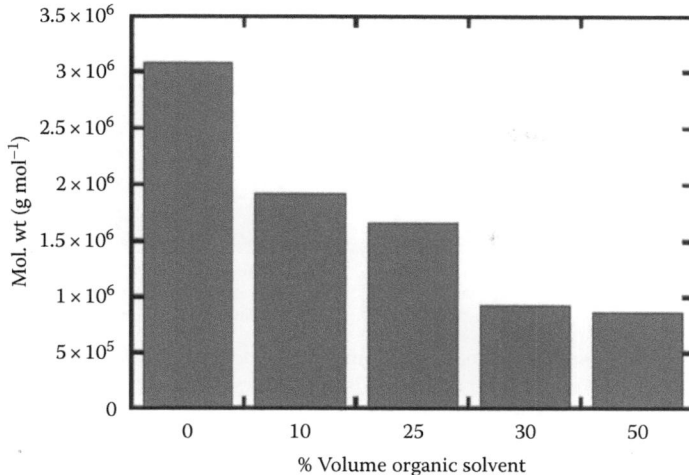

FIGURE 19.6 Effect of the amount of organic solvent on the molecular weight of polymer produced via ultrasound irradiation.

protects enzymes from being denatured by solvents. Encapsulation can also be used to deliver toxic materials such as pesticides and herbicides in agricultural and environmental applications. The second function of encapsulation is to enable controlled release in pharmaceutical and cosmetics applications. In the fragrance, flavor, and color trapping systems, it can control the time-release properties of these activities.

An intense area of investigation concerns the synthesis of nanocomposites via ultrasound irradiation methods. Nanocomposites are multiphase materials, whereby one phase, containing the filler, is dispersed in a second phase, known as the matrix. This results in the combination of individual properties of both the materials. Filler materials are often inorganic nanoparticles, whiskers, or fibers. The matrix can either be organic (polymer) or inorganic (ceramic or metal). The nanocomposites are often synthesized by simply mixing the filler and the matrix phases together to result in a heterogenous distribution. This is often achieved by the miniemulsion process. Miniemulsion polymerization allows for the convenient and direct incorporation of a large amount of hydrophobic materials, such as nanoparticles and pigments, by dispersion in the organic monomer phase (Espiard and Guyot, 1995; Ramirez and Landfester, 2003; Qiu et al., 2007). In order to successfully encapsulate hydrophobic materials within a latex particle, it is important to produce the appropriate latex particle size to allow encapsulation. In the first step of preparing nanocomposite materials, the material to be incorporated must be hydrophobic in nature and has to be dispersed in the monomer phase. Following this, miniemulsification of the monomer phase containing the hydrophobic materials into the water phase is performed by shearing the mixture with a homogenizer. In the next step, the monomer droplets containing the encapsulated materials are polymerized without changing their identity (Landfester et al., 1999). The main idea of the miniemulsion process is that particle formation proceeds by a droplet nucleation mechanism (whereby there is a 1:1 copy of the initial emulsion droplets to the corresponding polymer particles), such that a miniemulsion droplet can be treated as a "nanoreactor" and the encapsulation of preformed particles into polymer host particles is effective. This process is schematically illustrated in Figure 19.7.

In studies conducted by Wang et al. (Wang et al., 2001b; Xia et al., 2001), the incorporation of SiO_2, Al_2O_3, and TiO_2 into PBA by ultrasound-initiated emulsion polymerization was reported. They mentioned that ultrasound serves to provide better stirring and allows the breakdown of inorganic particles that have been agglomerated. This leads to better dispersion of the inorganic particles within the monomer droplets. The successful encapsulation of the inorganic particles was revealed by TEM and Fourier transform infrared (FTIR) spectroscopy. They concluded that the pH

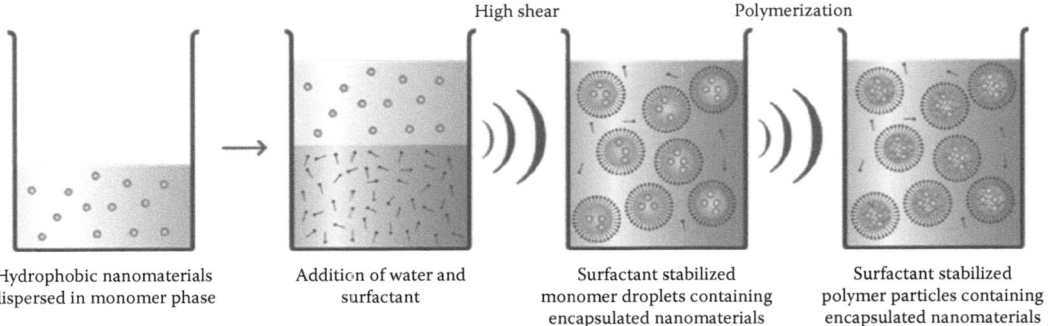

FIGURE 19.7 (See color insert.) Schematic diagram of the procedure for the encapsulation of hydrophobic nanomaterials and the 1:1 copy of monomer droplets to latex particles by the sonochemically driven miniemulsion polymerization pathway.

of the solution and the type of monomer, inorganic particles, and surfactants have a significant role in preparing such nanocomposites by the sonochemical method.

Bradley et al. (2003) described a one-step sonochemical method for incorporating fluorescent and phosphorescent molecules into latex particles. The monomer, MMA, containing either pyrene, 4-dicyanomethylene-2-methyl-6-(p-dimethylaminostyryl)-4H-pyran, or 1-bromo-naphthalene, was ultrasonically dispersed in an aqueous surfactant solution. Polymerization by ultrasound produced 60 nm latex particles. Figure 19.8 shows the fluorescence emission spectra of pyrene in the MMA emulsion and in the poly(MMA) latex particles. The absence of the excited state pyrene dimer emission (broad, featureless emission in the wavelength range 440–560 nm) in the latex particles indicates that pyrene molecules are embedded in the polymer matrix. The III/I peak ratio of the pyrene emission is approximately 2, indicating that there is minimal interaction between pyrene with water. From fluorescence lifetime studies, they showed that the incorporation of such molecules into latex particles can provide protection from environmental influences such as dissolved gases.

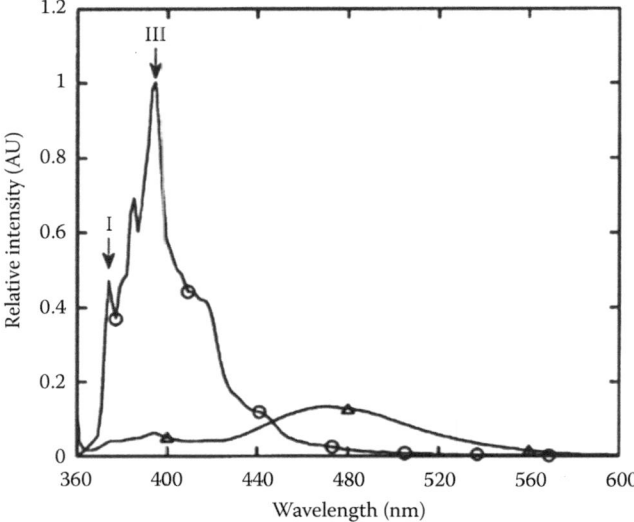

FIGURE 19.8 Fluorescence emission spectra of pyrene in an MMA emulsion (△) and in poly(MMA) latex particles (○). The 10 wt% monomer phase contained 2×10^{-2} M pyrene and was emulsified in a 25 mmol L^{-1} aqueous SDS solution. Excitation was at 350 nm. (Reprinted with permission from Bradley, M., Ashokkumar, M., and Grieser, F., Sonochemical production of fluorescent and phosphorescent latex particles, *J. Am. Chem. Soc.*, 125(2), 525–529. Copyright 2003 American Chemical Society.)

Applications of Ultrasound to Polymer Synthesis

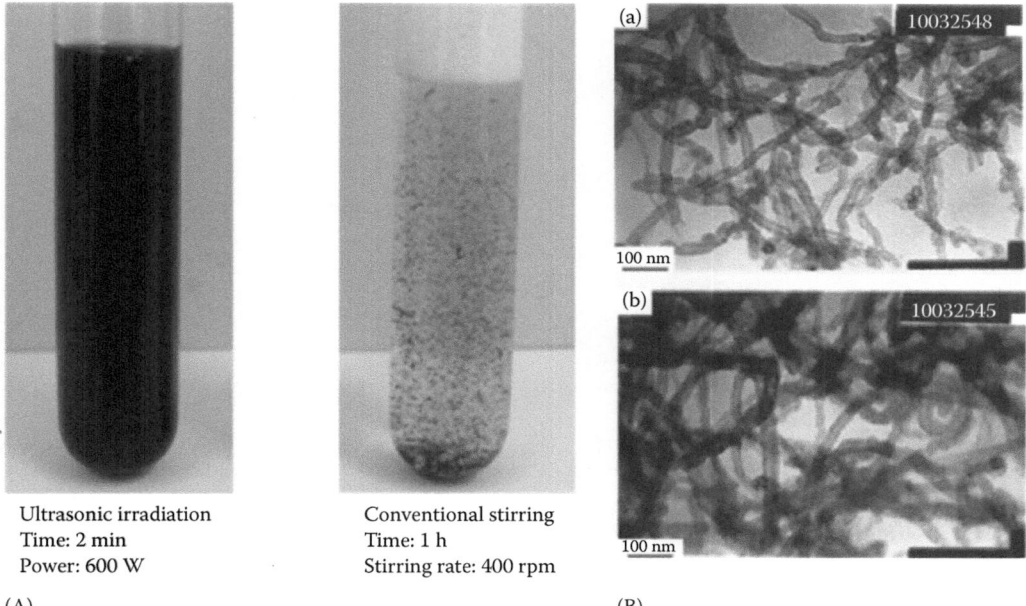

FIGURE 19.9 (A) Dispersion of 0.1 wt% MWCNTs in 0.1 wt% sodium lauryl sulphate (SLS) aqueous suspension; (B) TEM photographs of (a) raw MWCNTs and (b) PBA-encapsulated MWCNTs after Soxhlet extraction for 72 h with acetone. (Reprinted with permission from Xia, H., Wang, Q., and Qiu, G., Polymer-encapsulated carbon nanotubes prepared through ultrasonically initiated in situ emulsion polymerization, *Chem. Mater.*, 15(20), 3879–3886. Copyright 2003 American Chemical Society.)

A sonochemical approach in the fabrication of carbon nanotubes (CNTs) coated with polymer was adopted by Xia et al. (2003) showing that ultrasound irradiation provides excellent dispersion of the CNTs in aqueous solution as the aggregation of CNTs can be effectively broken down, as seen from Figure 19.9A. The in situ emulsion polymerization of MMA or BA to coat the CNTs was performed without the addition of any chemical initiators. TEM images of raw multiwalled carbon nanotubes (MWCNTs) and PBA-encapsulated MWCNTs after Soxhlet extraction are shown in Figure 19.9B; the encapsulated MWCNTs increased by 30 nm in diameter due to the polymer coating. The polymer-encapsulated MWCNTs were easily dispersed in a Nylon 6 matrix and the interfacial adhesion between the polymer coated MWCNTs and Nylon 6 was greatly improved. In another study conducted by Kim et al. (2007), MWCNTs and polystyrene composites were prepared via an in situ bulk polymerization under the application of ultrasound in the absence of chemical initiator. They found that ultrasound aids the dispersion of carbon nanotubes into the styrene monomer and that radicals for the polymerization reaction were generated from the decomposition of monomer by ultrasound cavitation.

The ultrasonic polymerization method has also been recently adopted to prepare magnetic-polymer nanocomposite materials (Teo et al., 2009a). These magnetic nanocomposites are reported to exhibit excellent colloidal stability and strong magnetic properties as shown in Figure 19.10. This synthetic procedure provides a "one-pot" composite material fabrication process that does not require the addition of chemical initiators and enables a clean and simply executed polymerization reaction. This method of producing nanocomposite materials, however, still has some limitations. For example, the distribution of magnetic nanoparticles between and within the latex beads was still rather heterogenous, as shown in Figure 19.11. This could be attributed to the interactions between the magnetic nanoparticles and the destabilization of the miniemulsion droplets (Landfester and Ramırez, 2003).

FIGURE 19.10 Photograph of the PBMA/Fe$_3$O$_4$ nanocomposite dispersion placed next to a strong permanent magnet that draws the fluid upward, as observed with regular magnetic fluids. (Reprinted with permission from Teo, B.M., Che, F., Hatton, T.A., Grieser, F., and Ashokkumar, M., Novel one-pot synthesis of magnetite latex nanoparticles by ultrasound irradiation, *Langmuir*, 25(5), 2593–2595. Copyright 2009 American Chemical Society.)

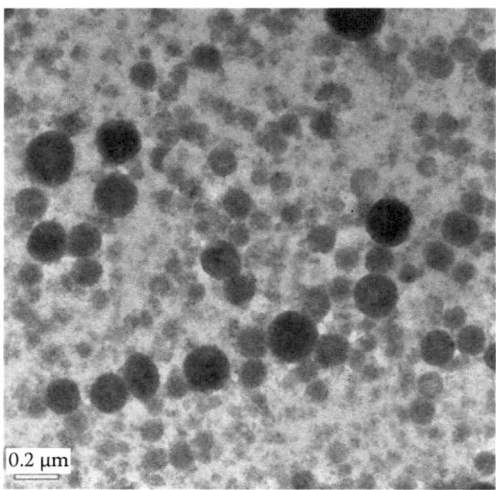

FIGURE 19.11 TEM image of the encapsulation of magnetite nanoparticles in poly(BMA) latex nanoparticles. (Reprinted with permission from Teo, B.M., Che, F., Hatton, T.A., Grieser, F., and Ashokkumar, M., Novel one-pot synthesis of magnetite latex nanoparticles by ultrasound irradiation, *Langmuir*, 25(5), 2593–2595. Copyright 2009 American Chemical Society.)

Gedanken et al. have prepared a range of polymeric nanocomposite materials via ultrasound irradiation (Wizel et al., 1998, 1999). In an early paper, they described two methods for encapsulating iron nanoparticles into poly(MA) by ultrasound irradiation. The first method described sonicating Fe(CO)$_5$ and distilled MA as neat liquids, whereas the second method described the two-step preparation of poly(MA)-coated amorphous iron nanoparticles. Comparing both the methods, they concluded that the second method gave a higher yield of Fe in the final product. There have been extensive studies in the area of sonochemical synthesis of nanocomposites summarized in a recent review by Gedanken (2007).

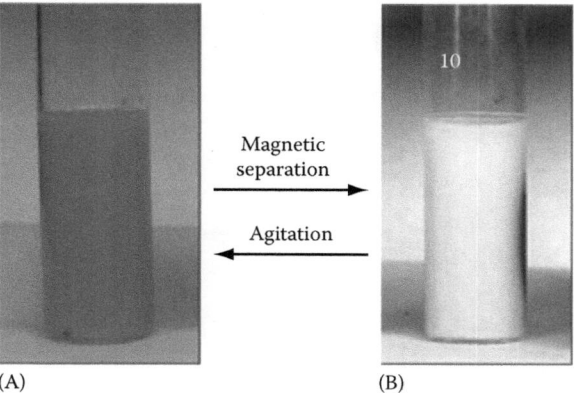

FIGURE 19.12 Magnetic separation and redispersion of a magnetic emulsion: (A) Without a magnetic field and (B) under a magnetic field. (Reprinted from *Ultrason. Sonochem.*, 14(1), Qiu, G.H., Wang, Q., Wang, C., Lau, W., and Guo, Y.L., Polystyrene/Fe_3O_4 magnetic emulsion and nanocomposite prepared by ultrasonically initiated miniemulsion polymerization, 55–61, Copyright (2007) with permission from Elsevier.)

Qiu et al. (2007) prepared PS/Fe_3O_4 magnetic nanocomposites through sonochemical miniemulsion polymerization. They reported that increasing the loading amount of Fe_3O_4 significantly increases the polymerization rate because the presence of Fe_3O_4 nanoparticles increases the number of radicals produced. They suggested that their PS/Fe_3O_4 magnetic particles could be separated from the magnetic emulsion by an external magnetic field resulting in a milky white dispersion as shown in Figure 19.12. This clearly indicates that the amount of Fe_3O_4 nanoparticles incorporated into the latex particles is very low.

Polymer/clay nanocomposites have also attracted intense research interest due to their enhancement of mechanical and thermal properties, as compared with the homopolymer. To achieve the optimum properties of such nanocomposites, many techniques have been developed and among these, the ultrasonic technique has proven to be an effective method for controlling the dispersion of clay in the polymer matrix (Kim et al., 2001; Delozier et al., 2002).

In other studies on the synthesis of polymeric nanocomposites, ZnO/poly(butyl methacrylate) (PBMA) and ZnO–PBMA/polyaniline (PANI) nanolatex composite particles of 50 nm were synthesized by a hydrothermal–sonochemical emulsion polymerization technique. The anticorrosive performance of the synthesized latex was evaluated on a steel substrate. The anticorrosive performance was compared for acid/salt/alkali-induced corrosion. It was found that the performance of the ZnO–PBMA coating for acid corrosion was superior to that of the ZnO–PBMA–PANI matrix (Sonawane et al., 2010).

Recently, asymmetric particles consisting of two hemispheres of different chemical composition have attracted much research attention owning to their possible use in a large variety of potential applications such as electronic displays (Takei and Shimizu, 1997; Perro et al., 2005), surfactants (Glaser et al., 2006; Walther et al., 2008), and biomedicine (Perro et al., 2005). Most methods for preparing such particles described in the literature are template assisted (Lattuada and Hatton, 2007; Qiang et al., 2008) or involve multiple steps in the procedure (Yin et al., 2001; Li et al., 2005) and mostly are conducted under nonambient preparation conditions. Hence, it is of interest to synthesize such particles by a simple and effective one-step procedure. The preparation of asymmetric nanocomposite doublets comprising PMMA/SiO_2 Janus particles through a one-step sonochemical miniemulsion polymerization method was demonstrated by Teo et al. (Teo, 2010). The formation of these doublets was realized by taking advantage of the phase separation between the growing PMMA particles and tetraethoxysilane (TEOS). Figure 19.13 shows the TEM images of the asymmetric particles prepared by sonochemical miniemulsion polymerization. Such particles can offer

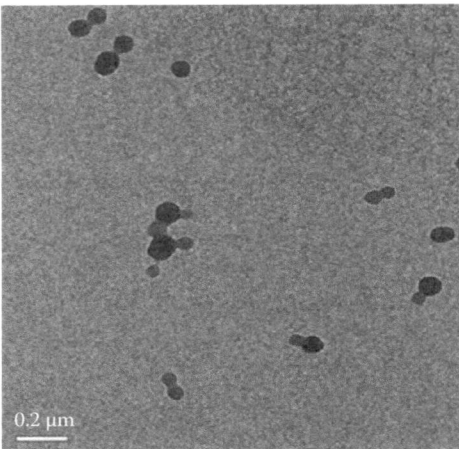 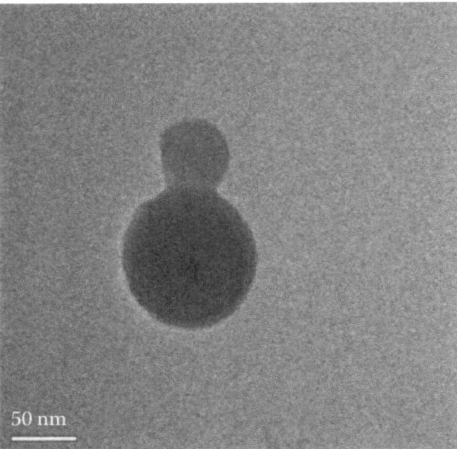

FIGURE 19.13 TEM images of asymmetric poly(MMA)-SiO$_2$ particles at different magnifications (left) scale bar represents 0.2 μm and (right) scare bar represents 50 nm.

FIGURE 19.14 Photograph of the interfacial behavior of three kinds of particles in a water–toluene dual phase system: (i) Poly(MMA), (ii) hydrophilic SiO$_2$ nanoparticles, and (iii) asymmetric poly(MMA)-SiO$_2$ nanocomposite particles.

the possibility of serving as Pickering emulsifiers due to their amphiphilic nature (Figure 19.14). This simple and straightforward method of preparing inorganic/polymer hybrid particles with asymmetric morphology can also be readily extended to preparing other nanocomposite materials by adjusting the experimental parameters such as the density and distribution of the coupling agent.

SONOCHEMICAL PREPARATION OF MICROSPHERES

Another interesting field of research concerning the use of ultrasound in polymeric material synthesis is the preparation of biopolymers, most notably, the synthesis of proteinaceous microspheres. These air- and oil-filled microspheres have a range of biomedical applications such as the use in targeted drug delivery, as echo contrast agents for sonography, as magnetic resonance imaging, and as microencapsulation of pharmaceuticals, neutraceuticals, and flavors (Langer, 1990; Webb et al., 1996; Lee et al., 2003). The first report on air-filled proteinaceous microspheres synthesized sonochemically was reported by Feinstein and coworkers (Lang et al., 1987) with their human serum albumin air-filled microspheres used as contrast agents in echosonography. In the synthesis of oil-filled proteinaceous microspheres of bovine serum albumin (BSA), Suslick and Grinstaff (1990) showed that 3 min sonication of a biphasic system comprising an aqueous solution of BSA and nonaqueous liquids was sufficient to form BSA protein microspheres containing the organic liquids. The proposed mechanism for the formation of these oil-filled microspheres is first due to emulsification by high-intensity ultrasound to disperse the nonaqueous phase into the aqueous protein solution. The authors ascertained that emulsification alone is not sufficient to form stable, long-lived microspheres and that a chemical process involving cross-linking of protein molecules through disulfide bond formation between cysteine residues was required. Based on chemical trapping experiments, superoxide (created during bubble collapse in the presence of oxygen) was identified as the cross-linking agent that renders the microspheres stable. The cross-linked shell of the microsphere was approximately ten protein molecules thick. In a later study by the same authors on sonochemical formation of air-filled microspheres, similar results were obtained (Grinstaff and Suslick, 1991). Although many studies have concluded the importance of cysteine and the formation of S–S bonds in the creation of microspheres (Suslick and Grinstaff, 1990; Grinstaff and Suslick, 1991; Suslick et al., 1994), it has been shown by Avivi and Gedanken (2002) that proteins that do not contain a thiol group can also be used to make microspheres. They applied this to making streptavidin microspheres and found that the microspheres were stable for many hours at room temperature and stable for at least a month at 4°C. The microspheres formed without cysteine residues were stabilized by intermolecular interactions such as hydrogen bonding, van der Waals forces, hydrophobic and electrostatic interactions. Their findings opened up a range of synthetic possibilities.

Since the first report of sonochemically synthesized proteinaceous microspheres, there has been a renewed interest in making such microspheres sonochemically (Shchukin and Mohwald, 2006; Cavalieri et al., 2008; Grinberg et al., 2009; Zhou et al., 2010). Gedanken and coworkers prepared magnetic proteinaceous microspheres that are useful for magnetic resonance imaging (Avivi et al., 2001). The microspheres composed of iron oxide–filled and coated BSA microspheres. More recently, Han et al. (Han et al., 2008; Radziuk et al., 2008) used the sonochemical method to make magnetic microspheres with chemically prepared magnetite embedded in polyelectrolyte multilayers and these magnetized microspheres can be guided by an external magnetic field. They have also demonstrated the influence of a range of ultrasound parameters on the size and size distribution of the microspheres. The selective targeting of drug-loaded protein microspheres to tumors has been demonstrated by many researchers. Suslick and coworkers (Toublan et al., 2006) reported on modified noncovalent, electrostatic layer-by-layer (LbL) protein microspheres that can selectively target protein microspheres to the integrin receptors that are overexpressed in several tumors. The protein microspheres are core-shell capsules consisting of a vegetable oil core and a BSA shell. Their research demonstrates the usefulness of these microspheres to targeted imaging and drug delivery systems. Cavalieri et al. (2008) successfully prepared stable lysozyme microbubbles using ultrasound-induced emulsification and cross-linking of chemically reduced lysozyme in aqueous solutions. Their lysozyme-coated microbubbles were stable for several months and also retained the enzymatic activity of lysozyme. LbL assembly of polyelectrolytes on these microbubbles to modify the surface properties of the microbubbles demonstrated the versatility of adsorbing potential drugs and/or biolabels for a range of therapeutic and diagnostic applications. In a later study, Zhou et al. (2010) demonstrated the ease of encapsulating a range of organic liquids into cross-linked lysozyme microspheres sonochemically. From their results, they found that the size and the stability of their microspheres were dependent on the nature of the encapsulated organic liquids, demonstrating the potential usefulness of the sonochemical method in a range of medical and food industries.

CONCLUDING REMARKS

Ultrasound-induced cavitation shows great promise as a relatively new technique for polymerization as discussed in this chapter. The recent development of this technique in polymer synthesis shows that sonochemistry accelerates polymerization reactions, allows for reactions to proceed to ambient temperatures without the need of additional chemical initiators, and the polymers synthesized have properties that cannot be achieved by conventional methods. The myriad of benefits that ultrasound-initiated events offers as an alternative to existing polymerization processes demonstrates that it should become a technology of choice for new and improved polymerization processes.

ACKNOWLEDGMENT

The authors would like to acknowledge the Albert Shimmins Postgraduate Writing up award granted by the Faculty of Science, University of Melbourne.

REFERENCES

Ashokkumar, M. and Grieser, F. 1999. Ultrasound assisted chemical processes. *Reviews in Chemical Engineering*, 15:41–83.

Avivi, S., Felner, I., Novik, I., and Gedanken, A. 2001. The preparation of magnetic proteinaceous microspheres using the sonochemical method. *Biochimica et Biophysica Acta—General Subjects*, 1527:123–129.

Avivi, S. and Gedanken, A. 2002. S–S bonds are not required for the sonochemical formation proteinaceous microspheres: The case of streptavidin. *Biochemical Journal*, 366:705–707.

Basedow, A.M. and Ebert, K.H. 1975. Mechanism of degradation of polymers in solution by ultrasound. *Macromolecular Chemistry and Physics*, 176:745–757.

Biggs, S. and Grieser, F. 1995. Preparation of polystyrene latex with ultrasonic initiation. *Macromolecules*, 28:4877–4882.

Bradley, M.A., Ashokkumar, M., and Grieser, F. 2003. Sonochemical production of fluorescent and phosphorescent latex particles. *Journal of American Chemical Society*, 125(2):525–529.

Bradley, M.A. and Grieser, F. 2002. Emulsion polymerization synthesis of cationic polymer latex in an ultrasonic field. *Journal of Colloid and Interface Science*, 251:78–84.

Bradley, M.A., Prescott, S.W., Schoonbrood, H.A.S., Landfester, K., and Grieser, F. 2005. Miniemulsion copolymerization of methyl methacrylate and butyl acrylate by ultrasonic initiation. *Macromolecules*, 38:6346–6351.

Brohult, S. 1937. Splitting of the haemocyanin molecule by ultra-sonic waves. *Nature*, 140:805.

Capek, I. 1996. On the kinetics of heterogeneous free radical crosslinking polymerization. *Journal of Dispersion Science and Technology*, 17:139–144.

Cass, P., Knower, W., Pereeia, E., Holmes, N.P., and Hughes, T. 2010. Preparation of hydrogels via ultrasonic polymerization. *Ultrasonics Sonochemistry*, 17:326–332.

Cavalieri, F., Ashokkumar, M., Grieser, F., and Caruso, F. 2008. Ultrasonic synthesis of stable, functional lysozyme microbubbles. *Langmuir*, 24:10078–10083.

Choi, H.J., Kim, C.A., Sohn, J.I., and Jhon, M.S. 2000. An exponential decay function for polymer degradation in turbulent drag reduction. *Polymer Degradation and Stability*, 69:341–346.

Chou, H.C. and Stoffer, J.O. 1999. Ultrasonically initiated free radical-catalyzed emulsion polymerization of methyl methacrylate (II): Radical generation process studies and kinetic data interpretation. *Journal of Applied Polymer Science*, 72:827–834.

Chun, M.S. and Park, O.O. 1994. On the intrinsic viscosity of anionic and nonionic rodlike polysaccharide solutions. *Macromolecular Chemistry and Physics*, 195:701–711.

Contreras-Cáceres, R., Sánchez-Iglesias, A., Karg, M., Pastoriza-Santos, I., Pérez-Juste, J., Pacifico, J., Hellweg, T., Fernández-Barbero, A., and Liz-Marzán, L.M. 2008. Encapsulation and growth of gold nanoparticles in thermoresponsive microgels. *Advanced Materials*, 20:1666–1670.

Cooper, G., Grieser, F., and Biggs, S. 1996. Butyl acrylate/vinyl acetate copolymer cater synthesis using ultrasound as an initiator. *Journal of Colloid and Interface Science*, 184:52–63.

Dar, Y.L., Farwaha, R., and Caneba, G.T. 2005. Free radical polymerization. In: *Encyclopedia of Chemical Processing*. Taylor & Francis, London, U.K.

Delozier, D.M., Orwoll, R.A., Cahoon, J.F. et al. 2002. Preparation and characterization of polyimide/organoclay nanocomposites. *Polymer*, 43:813–822.

Didenko, Y.T. and Suslick, K.S. 2005. Chemical aerosol flow synthesis of semiconductor nanoparticles. *Journal of American Chemical Society*, 127:12196–12197.

Doktycz, S.J. and Suslick, K.S. 1990. Interparticle collisions driven by ultrasound. *Science*, 247:1067–1069.

Eldik, R.V. and Hubbard, C.D. 1996. *Chemistry Under Extreme and Non-Classical Conditions*. John Wiley, New York.

El' Piner, I.E. 1964. *Ultrasound—Physical, Chemical, and Biological Effects*. Consultants Bureau Enterprises, New York.

Erdem, B., Sudol, D.E., Dimonie, V.L., and El-Aasser, M.S. 2000. Encapsulation of inorganic particles via miniemulsion polymerization. II. Preparation and characterization of styrene miniemulsion droplets containing TiO_2 particles. *Journal of Polymer Science: Part A: Polymer Chemistry*, 38:4431.

Espiard, P. and Guyot, A. 1995. Poly(ethyl acrylate) latexes encapsulating nanoparticles of silica: 2. Grafting process onto silica. *Polymer*, 36:4391.

Flory, P.J. 1937. The mechanism of vinyl polymerizations. *Journal of American Chemical Society*, 59:241–253.

Flosdorf, E.W. and Chambers, L.A. 1933. The chemical action of audible sound. *Journal of American Chemical Society*, 55:3051–3052.

Gedanken, A. 2007. Doping nanoparticles into polymers and ceramics using ultrasound radiation. *Ultrasonics Sonochemistry*, 14:418–430.

Gilbert, R.G. 1995. *Emulsion Polymerization: A Mechanistic Approach*. Academic Press, London, U.K.

Glaser, N., Adams, D.J., Boker, A., and Krausch, G. 2006. Janus particles at liquid–liquid interfaces. *Langmuir*, 22:5227.

Glyn, P.A.R., Van Der Hoff, B.M.E., and Reilly, P.M. 1972. A general model for prediction of molecular weight distributions of degraded polymers. Development and comparison with ultrasonic degradation experiments. *Journal of Macromolecular Science—Chemistry*, A6:1653–1664.

Gomes, V.G. 2005. Emulsion polymerization. In: *Encyclopedia of Chemical Processing*. Taylor & Francis, London, U.K.

Grinberg, O., Gedanken, A., Patra, C.R., Patra, S., Mukherjee, P., and Mukhopadhyay, D. 2009. Sonochemically prepared BSA microspheres containing Gemcitabine, and their potential application in renal cancer therapeutics. *Acta Biomaterialia*, 5:3031–3037.

Grinstaff, M.W. and Suslick, K.S. 1991. Air-filled proteinaceous microbubbles: Synthesis of an echo-contrast agent. *Proceedings of the National Academy of Sciences*, 88:7708–7710.

Guan, Z., Combes, J.R., Menceloglu, Y.Z., and Desimone, J.M. 1993. Homogeneous free radical polymerizations in supercritical carbon dioxide: 2. Thermal decomposition of 2, 2′ azobis (isobutyronitrile). *Macromolecules*, 26:2663–2669.

Han, Y.S., Radziuk, D., Shchukin, D., and Mohwald, H. 2008. Stability and size dependence of protein microspheres prepared by ultrasonication. *Journal of Materials Chemistry*, 18:5162–5166.

Hsieh, D.Y. and Plesset, M.S. 1961. Theory of rectified diffusion of mass into gas bubbles. *Journal of the Acoustical Society of America*, 33:206–215.

Kim, D.W., Blumstein, A., and Tripathy, S.K. 2001. Nanocomposite films derived from exfoliated functional aluminosilicate through electrostatic layer-by-layer assembly. *Chemistry of Materials*, 13:1916–1922.

Kim, S.T., Choi, H.J., and Hong, S.M. 2007. Bulk polymerized polystyrene in the presence of multiwalled carbon nanotubes. *Colloid and Polymer Science*, 285:593–598.

Kim, C.A., Choi, H.J., Kim, C.B., and Jhon, M.S. 1998. Drag reduction characteristics of polysaccharide xanthan gum. *Macromolecular Rapid Communications*, 19:419–422.

Kruus, P. 1991. Sonochemical initiation of polymerization. *Advances in Sonochemistry*, 2:1–22.

Kruus, P. and Patraboy, T.J. 1985. Initiation of polymerization with ultrasound in methyl methacrylate. *Journal of Physical Chemistry*, 89:3379–3384.

Landfester, K., Bechthold, N., Forster, S., and Antonietti, M. 1999. Evidence for the preservation of the particle identity in miniemulsion polymerization. *Macromolecular Rapid Communications*, 20:81.

Landfester, K. and Ramırez, L.P. 2003. Encapsulated magnetite particles for biomedical application. *Journal of Physics: Condensed Matter*, 15:1345–1361.

Lang, R.M., Borow, K.M., Neumann, A., Al-Sadir, J., and Feinstein, S.B. 1987. Effect of intracoronary injections of sonicated microbubbles on left ventricular contractility. *The American Journal of Cardiology*, 60:166–171.

Langer, R. 1990. New methods of drug delivery. *Science*, 249:1527–1533.

Lattuada, M. and Hatton, T.A. 2007. Preparation and controlled self-assembly of Janus magnetic nanoparticles. *Journal of American Chemical Society*, 129:12878–12889.

Lauterborn, W. and Vogel, A. 1984. Modern optical techniques in fluid mechanics. *Annual Review of Fluid Mechanics*, 16:223–244.

Lee, T.M., Oldenburg, A.L., Sitafalwalla, S., Marks, D.L., Luo, W., Toublan, F.J., Suslick, K.S., and Boppart, S.A. 2003. Engineered microsphere contrast agents for optical coherence tomography. *Optics Letters*, 28:1546–1548.

Leighton, T. 1994. *The Acoustic Bubble*. Academic Press, London, U.K.

Lepoint-Mullie, F., De Pauw, D., and Lepoint, T. 1996. Analysis of the new electrical model. *Ultrasonics Sonochemistry*, 3:73.

Li, Z.F., Lee, D.Y., Rubner, M.F., and Cohen, R.E. 2005. Layer-by-layer assembled Janus microcapsules. *Macromolecules*, 38:7876–7879.

Liao, Y., Wang, Q., Xia, H., Xu, X., Baxter, S.M., Slone, R.V., Wu, S., Swift, G., and Westmoreland, D.G. 2001. Ultrasonically initiated emulsion polymerization of methyl methacrylate. *Journal of Polymer Science: Part A: Polymer Chemistry*, 39:3356–3364.

Madras, G., Kumar, S., and Chattopadhyay, S. 2000. Continuous distribution kinetics for ultrasonic degradation of polymers. *Polymer Degradation and Stability*, 69:73–78.

Margulis, M.A. 1985. Sonoluminescence and sonochemical reactions in cavitation fields. A review. *Ultrasonics*, 23:157–169.

Margulis, M.A. 1995. *Sonochemistry and Cavitation*. Gordon Breach Publishers, Luxembourg, Belgium.

Morya, N.K., Iyer, P.K., and Moholkar, V.S. 2008. A physical insight into sonochemical emulsion polymerization with cavitation bubble dynamics. *Polymer*, 49:1910–1925.

Noltingk, B.E. and Neppiras, E.A. 1950. Cavitation produced by ultrasonics. *Proceedings of the Physical Society of London Section B*, 63:674–685.

Odell, J.A. and Keller, A. 1986. Flow-induced chain fracture of isolated linear macromolecules in solution. *Journal of Polymer Science Part B, Polymer Physics*, 24:1889–1916.

Odian, G. 2004. *Principles of Polymerization*. John Wiley & Sons, Inc., Hoboken, NJ.

Okudaira, G., Kamogawa, K., Sakai, T., Sakai, H., and Abe, M. 2003. Suspension polymerization of styrene monomer without emulsifier and initiator. *Journal of Oleo Science*, 52:167–170.

Ostroski, A.S. and Strambaugh, R.B. 1950. Emulsion polymerization with ultrasonic vibration. *Journal of Applied Physics*, 21:478–482.

Perro, A., Reculusa, S., Ravaine, S., Bourgeat-Lami, E., and Duguet, E. 2005. Design and synthesis of Janus micro- and nanoparticles. *Journal of Materials Chemistry*, 15:3745–3760.

Petrier, C., Micolle, M., Merlin, G., Luche, J.L., and Reverdy, G. 1992. Characteristics of pentachlorophenate degradation in aqueous solution by means of ultrasound. *Environmental Science & Technology*, 26:1639–1642.

Price, G.J. 1990. The use of ultrasound for the controlled degradation of polymer solutions. *Advances in Sonochemistry*, 1:231–287.

Price, G.J. 1996. Ultrasonically enhanced polymer synthesis. *Ultrasonics Sonochemistry*, 3:229–338.

Price, G.J., Hearn, M.P., Wallace, E.N.K., and Patel, A.M. 1996. Ultrasonically assisted synthesis and degradation of poly(dimethyl siloxane). *Polymer*, 37:2303–2308.

Price, G.J., Norris, D.J., and West, P.J. 1992. Polymerization of methyl methacrylate initiated by ultrasound. *Macromolecules*, 25:6447–6454.

Price, G.J. and Smith, P.F. 1993. Ultrasonic degradation of polymer solutions. 3. The effect of changing solvent and solution concentration. *European Polymer Journal*, 29:419–424.

Price, G.J., Smith, P.F., and West, P.J. 1991. Ultrasonically initiated polymerization of methyl methacrylate. *Ultrasonics*, 29:166–170.

Qiang, W.L., Wang, Y.L., He, P., Xu, H., Gu, H.C., and Shi, D.L. 2008. Synthesis of asymmetric inorganic/polymer nanocomposite particles via localized substrate surface modification and miniemulsion polymerization. *Langmuir*, 24:606–608.

Qiu, G.H., Wang, Q., Wang, C., Lau, W., and Guo, Y.L. 2007. Polystyrene/Fe_3O_4 magnetic emulsion and nanocomposite prepared by ultrasonically initiated miniemulsion polymerization. *Ultrasonics Sonochemistry*, 14:55–61.

Radziuk, D., Shchukin, D., and Mohwald, H. 2008. Sonochemical design of engineered gold–silver nanoparticles. *Journal of Physical Chemistry C*, 112:2462–2468.

Ramirez, L.P. and Landfester, K. 2003. Magnetic polystyrene nanoparticles with a high magnetite content obtained by miniemulsion processes. *Macromolecular Chemistry and Physics*, 204:22.

Rudin, A. 1999. *The Elements of Polymer Science and Engineering*. Academic Press, San Diego, CA.

Schmid, G. and Rommel, O. 1939. Disruptions of macro-molecules with ultrasounds. *Zeitschrift für Elektrochemie und Angewandte Physikalische Chemie*, 45:659–661.

Shchukin, D. and Mohwald, H. 2006. Sonochemical nanosynthesis at the engineered interface of a cavitation microbubble. *Physical Chemistry Chemical Physics*, 8:3496–3506.

Sohn, J.I., Kim, C.A., Choi, H.J., and Jhon, M.S. 2001. Drag-reduction effectiveness of xanthan gum in a rotating disk apparatus. *Carbohydrate Polymer*, 45:61–68.

Sonawane, S.H., Teo, B.M., Brotchie, A., Grieser, F., and Ashokkumar, M. 2010. Sonochemical synthesis of ZnO encapsulated functional nanolatex and its anticorrosive performance. *Industrial and Engineering Chemistry Research*, 49:2200–2205.

Strasberg, M. 1961. Rectified diffusion: Comments on a paper of Hsieh and Plesset. *Journal of the Acoustical Society of America*, 33:359.

Suslick, K.S. 1987. Sonochemistry of organometallic compounds. In: *High Energy Processes in Organometallic Chemistry*. American Chemical Society, Washington, DC.

Suslick, K.S. and Grinstaff, M.W. 1990. Protein microencapsulation of nonaqueous liquids. *Journal of American Chemical Society*, 112:7807–7809.

Suslick, K.S., Grinstaff, M.W., Kolbeck, K.J., and Wong, M. 1994. Characterization of sonochemically prepared proteinaceous microspheres. *Ultrasonics Sonochemistry*, 1:65–68.

Suslick, K.S., Hammerton, D.A., and Raymond, E.C. 1986. The sonochemical hot spot. *Journal of American Chemistry Society*, 108:5641–5642.

Takei, H. and Shimizu, N. 1997. Gradient sensitive microscopic probes prepared by gold evaporation and chemisorption on latex spheres. *Langmuir*, 13:1865–1868.

Teo, B.M. 2008. Ultrasound initiated miniemulsion polymerization of methacrylate monomers. *Ultrasonics Sonochemistry*, 15(1):89–94.

Teo, B.M. 2010. Ultrasonic polymer synthesis. In: *School of Chemistry*. The University of Melbourne, Melbourne, Australia.

Teo, B.M., Che, F., Hatton, T.A., Grieser, F., and Ashokkumar, M. 2009a. Novel one-pot synthesis of magnetite latex nanoparticles by ultrasound irradiation. *Langmuir*, 25(5):2593–2595.

Teo, B.M., Grieser, F., and Ashokkumar, M. 2008a. Microemulsion polymerizations via high-frequency ultrasound irradiation. *Journal of Physical Chemistry B*, 112:5265–5267.

Teo, B.M., Grieser, F., and Ashokkumar, M. 2009b. High intensity ultrasound initiated polymerization of butyl methacrylate in mini- and microemulsions. *Macromolecules*, 42:4479–4483.

Teo, B.M., Prescott, S.W., Ashokkumar, M., and Grieser, F. 2008b. Ultrasound initiated miniemulsion polymerization of methacrylate monomers. *Ultrasonics Sonochemistry*, 15:89–94.

Thompson, L.H. and Doraiswamy, L.K. 1999. Sonochemistry: Science and engineering. *Industrial and Engineering Chemistry Research*, 38:1215–1249.

Tirelli, N. 2006. (Bio)Responsive nanoparticles. *Current Opinion in Colloid & Interface Science*, 11:210–216.

Toublan, F.J.J., Boppart, S., and Suslick, K.S. 2006. Tumor targeting by surface-modified protein microspheres. *Journal of American Chemical Society*, 128:3472–3473.

Ugelstad, J., El-Aasser, M.S., and Vanderhoff, J.W. 1973. Emulsion polymerization: Initiation of polymerization in monomer droplets. *Journal of Polymer Science Part C: Polymer Letters*, 11:503–513.

Van Der Hoff, B.M.E. and Gall, C.E. 1977. A method for following changes in molecular weight distributions of polymers on degradation: Development and comparison with ultrasonic degradation experiments. *Journal of Macromolecular Science, Chemistry*, A11:1739–1758.

Walther, A., Hoffmann, M., and Müller, A.H.E. 2008. Emulsion polymerization using Janus particles as stabilizers. *Angewandte Chemie International Edition*, 47:711–714.

Wang, Q., Xia, H., Liao, Y., Xu, X., Baxter, S.M., Slone, V., Wu, S., Swift, G., and Westmoreland, D.G. 2001a. Ultrasonically initiated emulsion polymerization of n-butyl acrylate. *Polymer International*, 50:1252–1259.

Wang, Q., Xia, H.S., and Zhang, C.H. 2001b. Preparation of polymer/inorganic nanoparticles composites through ultrasonic irradiation. *Journal of Applied Polymer Science*, 80:1478–1488.

Webb, A.G., Wong, M., Kolbeck, K.J., Magin, R.L., Wilmes, L.J., and Suslick, K.S. 1996. Sonochemically produced fluorocarbon microspheres: A new class of magnetic resonance imaging agent. *Journal of Magnetic Resonance Imaging*, 6:675–683.

Wizel, S., Margel, S., Gedanken, A., Rojas, T.C., Fernandez, A., and Prozorov, R. 1999. The preparation of metal-polymer composite materials using ultrasound radiation: Part II. Differences in physical properties of cobalt-polymer and iron-polymer composites. *Journal of Materials Research*, 14:3913–3920.

Wizel, S., Prozorov, R., Cohen, Y., Aurbach, D., Margel, S., and Gedanken, A. 1998. The preparation of metal-polymer composite materials using ultrasound radiation. *Journal of Materials Research*, 13:211–216.

Xia, H., Wang, Q., Liao, Y., Xu, X., Baxter, S.M., Slone, R.V., Wu, S., Swift, G., and Westmoreland, D.G. 2002. Polymerization rate and mechanism of ultrasonically initiated emulsion polymerization of *n*-butyl acrylate. *Ultrasonics Sonochemistry*, 9:151–158.

Xia, H., Wang, Q., and Qiu, G. 2003. Polymer-encapsulated carbon nanotubes prepared through ultrasonically initiated in situ emulsion polymerization. *Chemistry of Materials*, 15(20):3879–3886.

Xia, H.S., Zhang, C.H., and Wang, Q. 2001. Study on ultrasonic induced encapsulating emulsion polymerization in the presence of nanoparticles. *Journal of Applied Polymer Science*, 80:1130–1139.

Yasui, K., Tuziuti, T., Sivakumar, M., and Iida, Y. 2004. Sonoluminescence. *Applied Spectroscopy Reviews*, 39:399–436.

Yin, Y.D., Lu, Y., and Xia, Y.N. 2001. A self-assembly approach to the formation of asymmetric dimers from monodispersed spherical colloids. *Journal of American Chemical Society*, 123:771–772.

Zhou, M., Leong, T.S.H., Melino, S., Cavalieri, F., Kentish, S., and Ashokkumar, M. 2010. Sonochemical synthesis of liquid-encapsulated lysozyme microspheres. *Ultrasonics Sonochemistry*, 17:333–337.

20 Mechanistic Aspects of Ultrasound-Enhanced Physical and Chemical Processes

Vijayanand S. Moholkar, Thirugnanasambandam Sivasankar, and Venkata Swamy Nalajala

CONTENTS

Introduction 502
Physical and Chemical Effects of Cavitation Bubbles 503
    Chemical Effect of Cavitation (Sonochemical Effect) 503
        Research in 1990s 503
        General Model of Storey and Szeri 504
        Contemporary Studies (Hoffmann and Coworkers and Lohse and Coworkers) 505
    Physical Effects of Cavitation 505
        Microstreaming 505
        Microturbulence 505
        Acoustic Waves (or Shock Waves) 506
        Microjets 506
Diffusion-Limited Model of Toegel and Lohse 507
    Numerical Solution 507
    Quantification of Radical Production 510
    Quantification of the Physical Effect of Cavitation (Generation of Convection) 510
        Microturbulence (or Microconvection) 510
        Shock Waves (or Acoustic Waves) 510
Mechanistic Aspects of Ultrasound-Enhanced Chemical Processes: Case Study
of Sonochemical Degradation of Recalcitrant Organic Pollutants 511
    Experimental Setup and Procedures 511
        Experimental Setup 511
        Analytical Procedure 512
    Results and Analysis of Degradation 512
        Preamble 512
        Experimental Results 513
    Simulation Results 514
    Discussion and Analysis 515
Mechanistic Aspects of Ultrasound-Enhanced Physical Processes: Case Study
of Sonocrystallization 522
    Experimental Procedure and Analysis 522
    Results 523
        Preamble 523
        Experimental and Simulation Results 524

Number of Nuclei, Nucleation Rate, and Growth Rate 526
Dominant Crystal Size and Span of CSD 528
Simulation Results 529
Discussion and Analysis 529
Concluding Views 531
References 531

INTRODUCTION

In the past few decades, we have observed development of several new technologies that offer sophisticated methods of introducing energy into the system to bring about a physical or chemical change. These methods are simpler, easier, cheaper, and safer, following the principles of "green chemistry," which aims at minimization of generation of hazardous waste, use of nontoxic precursors, reduced emissions, and discharges with milder operating conditions. One such technology is cavitation technology in which the physical and chemical transformations are brought about by ultrasound irradiation of the system. Cavitation is defined as nucleation, growth, and transient collapse of tiny gas bubbles driven by pressure fluctuations in the liquid. These fluctuations essentially create tension in the liquid, which is manifested in form of cavitation. Alternatively, intense local deposition of energy can also give rise to cavitation (Shah et al., 1999). In fact, the method of production of cavitation is the main criterion for distinguishing among different types of cavitation as follows:

1. Ultrasonic or acoustic cavitation is produced by pressure variation in a liquid due to the passage of an acoustic wave in the form of compressions and rarefactions.
2. Hydrodynamic cavitation is generated by pressure reduction in a flowing liquid (with concurrent rise in velocity head) through constrained geometries such as venture or orifice.
3. Optic cavitation is a result of the rupture of a liquid due to high-intensity light or laser.
4. Particle cavitation is produced by any type of elementary particle, such as proton, rupturing a liquid, as in a bubble chamber.

As far as cavitation for physical/chemical processing is concerned, only the acoustic and hydrodynamic modes are feasible. The other two modes of cavitation, viz., optic and particle, are mainly used for fundamental studies in cavitation bubble dynamics to generate cavitation under controlled conditions. In the past decades, voluminous literature has been published on applications of ultrasound and cavitation for intensification of numerous physical and chemical processes. These include wastewater treatment (i.e., degradation of various types of recalcitrant organic pollutants), synthesis of organic, inorganic, and biological nanomaterials, crystallization, food processing, extraction and leaching, wet textile treatments, biofuels applications such as synthesis of biodiesel, etc. (see state-of-the-art reviews and books by Suslick, 1988; Pandit and Moholkar, 1996; Ashokkumar and Grieser, 1999; Keil and Swamy, 1999; Shah et al., 1999; Thompson and Doraswamy, 1999; Adewuyi, 2001; Mason and Lorimer, 2002; Gogate and Pandit, 2004a,b). Although the potential and efficacy of ultrasound-enhanced processes has been clearly demonstrated, there has been little application on industrial scale. Several causes has led to this effect that includes low overall efficiency of ultrasonic processor (<20%), erosion of the ultrasound transducers, and directional sensitivity of the ultrasound field that results in nonuniform energy dissipation in the processor (EPRI Report, 1998). One more aspect of sonochemical processes that has posed hurdle in an effective scale-up of the process is the lack of proper understanding of the exact mechanism of the process, that is, the nature (either physical or chemical or both) of the role played by the ultrasound and cavitation in the process. Proper identification of the mechanism of the process will be of crucial importance for optimization of the energy dissipation in the process, which would help augmentation of the energy efficiency of the process.

In this chapter, we shall present an overview of our research attempts in the identification of the mechanism of a sonophysical (sonocrystallization) and sonochemical (degradation of recalcitrant organic pollutants) processes. The approach in both of these studies is similar, that is, coupling of experimental results with simulations of cavitation bubble dynamics using a mathematical model that takes into account essential physics and chemistry of cavitation bubbles. This approach is based on the fact that ultrasound manifests its physical and chemical effects through cavitation phenomenon. Concurrent analysis of the experimental and simulation results helps in determination of the physical mechanism of the system—as is demonstrated in subsequent sections.

The principal chemical effect of cavitation—popularly known as sonochemical effect—is generation of highly reactive radicals such as $H^\bullet$, $^\bullet OH$, $O^\bullet$, $HO_2^\bullet$ through transient collapse of cavitation bubbles, while the principal physical effect is generation of intense turbulence in the medium. The radicals generated by cavitation bubbles have several distinct effects on the reaction system such as (1) initiation of stubborn reactions with acceleration of their kinetics, (2) elimination of steps in synthesis allowing a "single pot" synthesis, (3) allowing use of cruder chemicals, and (4) complete switching of the reaction pathway. The physical effect of generation of convection in the system is mainly responsible for enhancement of heat and mass transfer characteristics of the system. These effects of are mainly beneficial in heterogeneous reaction system—either liquid–liquid or solid–liquid. In a liquid–liquid system, cavitation phenomenon occurring at the interface of the two liquids results in generation of fine emulsion, giving rise to enormous interfacial area between the liquids that markedly enhances mass transfer characteristics, and hence, chemical kinetics of the system. In a solid–liquid reaction system, such as catalytic reaction, the microemulsion generated by ultrasound and cavitation also enhances overall mass transfer characteristics, that is, transport of reactant species to catalyst surface and desorption of the product species. This results in continuous regeneration of the catalyst surface, which exposes the "active sites" for the fresh reactants. Catalyst poisoning also reduces due to the strong microconvection generated by cavitation bubbles. In the next section, we describe the physical and chemical effects of cavitation bubbles in greater details with underlying physical mechanism.

PHYSICAL AND CHEMICAL EFFECTS OF CAVITATION BUBBLES

CHEMICAL EFFECT OF CAVITATION (SONOCHEMICAL EFFECT)

A major problem in modeling of sonochemical processes is that direct quantitative measurement of the radial generation from transient collapse of cavitation bubbles is not possible. The extent of radical production depends on two variables: (1) amount of vapor entrapped in the bubble at the moment of transient collapse and (2) temperature peak reached in the bubble during collapse. The second variable has a greater impact on the production of radicals through dissociation of the vapor molecules. In addition, several other factors also influence the composition of the bubble contents. These are gas diffusion and rectification and the chemical reactions occurring among species produced from thermal dissociation of vapor molecules. Development of mathematical models for cavitation bubble dynamics including mass (solvent vapor as well as noncondensable gas) and heat transfer across bubble interface has been a matter of active research for past two decades. We present below a brief review of the same.

Research in 1990s

The problem of water vapor transport and entrapment in the cavitation bubble during radial motion, along with chemical reactions among the species generated, was treated with different perspectives by several authors in the decade of 1990s. The major contributions are made by Kamath et al. (1993), Prasad Naidu et al. (1994), Sochard et al. (1997), Yasui (1997a,b), Colussi et al. (1998), Gong and Hart (1998), Colussi and Hoffmann (1999), and Moss et al. (1999). We give below a summary of the approaches taken by these authors and the major findings of their studies.

Kamath et al. (1993) estimated the production of OH radicals by decoupling the bubble dynamics equation and the chemical kinetics. Prasad Naidu et al. (1994) modeled equilibrium production of various radicals using Rayleigh–Plesset equation for radial motion of bubble, coupled with Flynn's assumption (Flynn, 1964) that the bubble becomes a closed system during collapse when the partial pressure of gas becomes equal to the vapor pressure. The growth phase of the bubble was assumed to be isothermal, while the collapse phase, after bubble becoming closed system, was assumed to be adiabatic. Gong and Hart (1998) also took a similar approach of coupling bubble dynamics with the chemical kinetics, and explained some trends in sonochemistry. Sochard et al. (1997) modeled radical production in mildly forced bubbles taking into account nonequilibrium phase change and gas–vapor interdiffusion. Like earlier authors, Sochard et al. (1997) also assumed prevalence of equilibrium conditions inside the bubble. Moss et al. (1999) performed numerical simulations of the bubble motion keeping the amount of water vapor in the bubble constant and uniform during acoustic cycle, and without taking into account the chemical reactions. Principal conclusion of the study of Moss et al. (1999) was that inclusion of water vapor in the bubble interior leads to smaller adiabatic exponent that reduces the heating of the bubble, decreasing the final temperature attained therein. Yasui (1997a,b) accounted for the nonequilibrium phase change and chemical reactions in a collapsing air bubble. He used a Rayleigh–Plesset type equation for the bubble radius, with the water vapor transport by condensation—evaporation. Twenty-five reactions of radicals formed out of water vapor dissociation were also taken into account as per approach of Kamath et al. (1993). However, Yasui (1997a,b) assumed that transport of water vapor was condensation limited, that is, mass diffusion was assumed to be instantaneous, and not explicitly modeled. Primary outcome of Yasui's study was that some water vapor remains in the bubble even at the collapse. With intense heating of the bubble during the transient collapse, this water vapor undergoes endothermic dissociation, reducing the final temperature peak reached in the bubble.

General Model of Storey and Szeri

In landmark papers published in early last decade, Storey and Szeri (2000, 2001) presented a general treatment of the problem relaxing several assumptions made in earlier studies in 1990s. The Navier-Stokes equations for the gas mixture in the bubble were coupled to a scheme of 64 possible reactions among nine radical and molecular species generated from dissociation of water vapor, viz., H_2, O_2, H_2O, H, O, OH, HO_2, H_2O_2, and O_3. The transport properties (thermal and mass diffusion, viscosity) were calculated from the equations based on Chapman–Enskog theory and the equation of state was of Redlich–Kwong–Soave type. The rate of transport of water molecules was proportional to the difference between partial pressure of water in the bubble and the saturation pressure at interface. However, not all the water molecules that approach the surface stick to it, giving rise to nonequilibrium phase change. The fraction of water molecules that stick to the surface is the accommodation coefficient (σ_a). In other words, σ_a is a representative of the resistance to condensation at the interface during bubble collapse. The lower the value of σ_a, the greater the resistance and higher the amount of water vapor entrapped. Storey and Szeri (2000) used value of $\sigma_a = 0.4$, following Yasui (1997a) and Eames et al. (1997). The principal result of paper by Storey and Szeri (2000) was that water vapor transport in the bubble is a two-step process: diffusion to bubble wall and condensation. Thus, it is influenced by two timescales, viz., timescale of diffusion (t_{dif}) and timescale of condensation (t_{cond}), and their magnitudes relative to bubble dynamics (or oscillations) timescale, t_{osc}. In the earlier phases of bubble collapse, $t_{osc} \gg t_{dif}, t_{cond}$, which results in uniform bubble composition. As the bubble wall acceleration increases during collapse, the timescales for bubble dynamics and diffusion become equal. At this stage, rate of reduction of water vapor in the central region of bubble is lesser than at the bubble wall. With further acceleration of bubble wall, $t_{osc} \ll t_{dif}$, and the water vapor has insufficient time to diffuse to bubble wall, which results in nearly fixed distribution of water vapor in the bubble. Another mechanism that traps water vapor in bubble during collapse is the nonequilibrium phase change at bubble wall, as mentioned above. The timescale for the condensation varies inversely with σ_a. Qualitatively, when $t_{osc} \gg t_{cond}$, the condensation is in equilibrium with respect to the bubble motion. On the other hand, when $t_{cond} \gg t_{osc}$, no water vapor can escape bubble during collapse. Thus, the amount of water vapor trapped is sensitive to the value of σ_a.

The exact mechanism by which water vapor is trapped in the bubble is determined by the relative magnitudes of t_{osc}, t_{dif}, and t_{cond}. When the bubble dynamics timescale is smaller than either the diffusion or the condensation timescale, water vapor entrapment occurs. However, both mechanisms can contribute to the water vapor entrapment. Storey and Szeri (2000) showed that the condition $t_{osc} \ll t_{dif}$ is reached well before $t_{osc} \ll t_{cond}$. Thus, the water vapor trapping is diffusion limited.

Contemporary Studies (Hoffmann and Coworkers and Lohse and Coworkers)

Parallel studies conducted by Hoffmann and coworkers (Colussi et al., 1998; Colussi and Hoffmann, 1999) clearly show the effect of condensation timescale that varies inversely with the accommodation coefficient, as mentioned earlier. In their first paper, Hoffmann and coworkers used a very low value of $\sigma_a = 0.001$, without accounting for the finite rate of mass diffusion. With such unrealistically low value of σ_a, the condition $t_{osc} \ll t_{cond}$ reaches well before $t_{osc} \ll t_{dif}$. This makes the water vapor trapping condensation, rather than diffusion limited. In another study, Colussi and Hoffmann (1999) used a value of $\sigma_a = 0.3$, taking into account diffusive resistance generated in the bubble. With this value, the simulation results of Colussi and Hoffmann (1999) are in accordance with those of Storey and Szeri (2000) that the water vapor trapping is diffusion limited.

In view of the results of Storey and Szeri (2000) with full numerical simulations, Lohse group at University of Twente (Toegel et al., 2000; Toegel and Lohse, 2003) developed a diffusion-limited model using boundary layer approximation. This model has gained immense popularity in the sonochemistry community due to its simple structure, yet account for essential physics and chemistry of the cavitation bubble. The existence of boundary layer near bubble wall even at moment of extremely rapid collapse has been confirmed in the studies of Kwak and coworkers (Kwak and Yang, 1995; Kwak and Na, 1996, 1997) who showed that the spatial variation of temperature in the bubble was negligible except at the bubble wall. Other authors (Fujikawa and Akamatsu, 1980; Kamath et al., 1993) have also justified existence of the boundary layer. For the validation of the simple model, Toegel et al. (2000) have compared their results with those of Storey and Szeri (2000), finding an excellent qualitative and quantitative agreement. This is another confirmation that the water vapor transport is diffusion, rather than condensation, limited.

PHYSICAL EFFECTS OF CAVITATION

The physical effects of ultrasound and cavitation are severalfold. These effects are mainly responsible for generating strong convection in the bulk liquid medium through several mechanisms, as described below.

Microstreaming

The propagation of ultrasound waves through the liquid medium creates small-amplitude oscillatory motion of fluid elements around a mean position. This phenomenon is called microstreaming (Leighton, 1994). The velocity of the microstreaming is given as $v = P_A/\rho C$ where P_A is the pressure amplitude of ultrasound wave, ρ is the density of the medium, and C is the velocity of sound in the medium. For a typical pressure amplitude of 1.2 bar in water (with $\rho = 1000\,\text{kg m}^{-3}$ and $C = 1500\,\text{m s}^{-1}$), $v = 0.08\,\text{m s}^{-1}$. The oscillatory liquid motion is hindered near solid surfaces, and the direction of fluid movement becomes parallel to the solid boundary.

Microturbulence

Radial motion of cavitation bubble induces high-velocity oscillatory motion of the fluid in its vicinity. This is called microturbulence or microconvection (Moholkar et al., 2004). This phenomenon is explained as follows: during the expansion phase of radial motion, the fluid is displaced away from the bubble center. During the collapse phase, the liquid is pulled toward the bubble as it fills in the vacuum created in liquid with size reduction of the bubble. The mean velocity of the microturbulence depends on the amplitude of the oscillation of the bubble. For small-amplitude motion, the velocity of microturbulence is also small. However, if the amplitude of bubble oscillations is large (typically more than two times the original size), the ensuing collapse is transient, and the bubble velocity reaches (or even exceeds) the sonic velocity (1500 m s^{-1} in water) during final moments of collapse. Accordingly, the velocity of the microturbulence and convection generated is also quite intense. It should, however, be noted that phenomenon of microturbulence is restricted only in the region in close vicinity of the bubble. The velocity of the microturbulence diminishes very rapidly away from the bubble.

Acoustic Waves (or Shock Waves)

As mentioned above, during the compression phase of radial motion, the fluid elements in the vicinity of the bubble wall spherically converge toward bubble wall. For a gas bubble (containing noncondensable gas such as air), the adiabatic compression results in rapid rise of the pressure inside the bubble. At the point of minimum radius (or maximum compression), the bubble wall comes to a sudden halt and rebounces with high velocity. At this instance, the converging fluid elements are reflected back from the bubble interface (Willard, 1953). This reflection creates a high-pressure shock wave that propagates through the medium (Neppiras, 1980; Lauterborn and Hentschel, 1985, 1986; Blake et al., 1997; Ohl et al., 1998; Pecha and Gompf, 2000). The pressure magnitude of this wave is estimated, with numerical simulations, in the range of 50–100 bar.

Microjets

During radial motion driven by ultrasound wave, the cavitation bubble maintains spherical geometry as long as the motion of liquid in vicinity of it is symmetric and uniform, and thus, there are no pressure gradients. If the bubble is located close to a phase boundary (either solid–liquid or gas–liquid or liquid–liquid), the motion of liquid in its vicinity is hindered, resulting in development of pressure gradients around it. This nonuniformity of pressure results in the loss of spherical geometry of the bubble. Numerous authors have investigated this phenomenon with different approach, either numerical and/or experimental in past three decades (Benjamin and Ellis, 1966; Plesset and Chapman, 1971; Blake et al., 1986, 1987; Vogel et al., 1989; Phillip and Lauterborn, 1998). During the asymmetric radial motion, the portion of the bubble exposed to higher pressure collapses faster than the rest of the bubble, which gives rise to the formation of a high-speed liquid jet. However, the direction of this jet depends on the characteristics of the solid boundary (Blake et al., 1986, 1987). Rigid boundaries (e.g., metal surfaces) are characterized by the boundary condition of $\nabla \cdot \phi = 0$, where ϕ is velocity potential at the boundary, while a free (or pressure release boundary, e.g., gas–liquid interface) is characterized by boundary condition of $\phi = 0$. For a rigid boundary, the microjet is directed toward the rigid boundary, while for a free boundary, the microjet is directed away from the boundary. The velocity of these high-speed jets have been estimated (with help of high-speed photography) in the range of 120–150 m s^{-1} (Ohl et al., 1995; Phillip and Lauterborn, 1998). In case of rigid boundaries, these jets can cause severe damage at the point of impact and can erode the surface. These jets are also responsible for particle size reduction. In order to induce pressure gradient leading to asymmetric bubble collapse, the size of the solid boundary needs to be sufficiently large. If the size of the solids present in the vicinity of the bubble is of the same order as the maximum radius attained during radial motion, the pressure field surrounding the bubble is not disrupted by presence of the solid particles. Therefore, solid boundaries (or solid particles) with typical dimension of ≤150 μm cannot induce asymmetric bubble collapse nor microjet formation (Doktycz and Suslick, 1990).

In the next section, we give a brief account of the diffusion-limited model of Toegel and Lohse, which we have extensively used in our research on mechanistic aspects of ultrasound-enhanced physical, chemical, and biological processes.

DIFFUSION-LIMITED MODEL OF TOEGEL AND LOHSE

The essential equations and thermodynamic data of this model are given in Tables 20.1 and 20.2. The main components of the model for radial motion of cavitation bubble are as follows:

1. Equation for the radial motion of the bubble.
2. Equation for the diffusive flux of water vapor and pollutant vapor across bubble wall. In these equations, the binary diffusion coefficients for solvent vapor and dissolved solute are determined using Chapman–Enskog theory using Lennard-Jones 12–6 potential at the bulk temperature of the liquid medium. The overall diffusion coefficient in ternary mixture (e.g., N_2–O_2–solvent in case of air bubbles) or quaternary mixture (e.g., N_2–O_2–solvent–solute, where solute is sufficiently volatile to vaporize into the bubble) has been determined using Blanc's law. The diffusive penetration depth has been estimated using dimensional analysis.
3. Equation for heat conduction through bubble wall. The thermal conductivity of the bubble content (mixture of gas + solvent and/or solute vapor) is determined using Chapman–Enskog theory using Lennard-Jones 12–6 potential at the bulk temperature of the liquid medium. In this case, the thermal penetration depth is estimated using dimensional analysis.
4. Overall energy balance treating the cavitation bubble as an open system.

This model ignores the diffusion of gases across bubble wall (or the rectified diffusion) as the timescale for the diffusion of gases (which is of the order of few milliseconds) is much higher than the timescale for the radial motion of bubble (which is of the order of few microseconds). Data presented in Table 20.2 is for two solvents, viz., water and methanol.

NUMERICAL SOLUTION

The set of five ordinary differential equations given in Table 20.1 can be solved using Runge-Kutta fourth to fifth order adaptive step size method (Press et al., 1992). During the radial motion, the cavitation bubble may collapse at the instance of maximum compression, and its contents are released into the bulk medium. The word "collapse" essentially means fragmentation of the cavitation bubble. In a multibubble system, the radial motion of cavitation bubble is rather unstable and the fragmentation of the bubble can occur at the first compression after an initial expansion (Storey and Szeri, 2000). In view of this, the condition for the bubble collapse is taken to be first compression during radial motion. An alternate criterion for bubble collapse has been proposed by Mahulkar et al. (2008) on the basis of *material volume limitation*. According to this criterion, a bubble is assumed to collapse when the minimum bubble volume during radial motion becomes equal to the total hard core volume of the molecules in the bubble. Typically, a bubble with initial radius of 10 µm contains ~10^{11} molecules (of either gas or vapor) at the time of collapse (Sivasankar and Moholkar, 2009). For an average molecular diameter of 3.5×10^{-10} m (or 0.35 Å), the total hard core volume of the gas/vapor molecules would be 2.245×10^{-18} m^3, which would be equivalent to a sphere with radius 1.624 µm. Mahulkar et al. (2008) have applied this criterion for oscillation/collapse of steam bubbles (containing no noncondensable gas) injected in the liquid medium. This criterion, however, does not apply for gas bubbles, since the bubble never reaches so small size during radial motion. Typically, a compression ratio of 3.33 is seen for initial bubble size of 10 µm, corresponding to minimum bubble radius ~3 µm. For bubbles containing noncondensable gas, the pressure surge or pressure force inside the bubble during adiabatic compression resists the inertial forces outside the bubble causing compression, and thus, the bubble comes to a halt in its radial motion much earlier than the limiting condition proposed by Mahulkar et al. (2008) is reached. Moreover, the pressure force inside the bubble makes bubble "rebound" during

TABLE 20.1
Summarization of the Bubble Dynamics Formulation

Variable	Equation	Initial Condition
1. Radius of the bubble (R)	$\left(1-\dfrac{dR/dt}{c}\right) R \dfrac{d^2R}{dt^2} + \dfrac{3}{2}\left(1-\dfrac{dR/dt}{3c}\right)\left(\dfrac{dR}{dt}\right)^2$ $= \dfrac{1}{\rho_L}\left(1+\dfrac{dR/dt}{c}\right)(P_i - P_t) + \dfrac{R}{\rho_L c}\dfrac{dP_i}{dt} - 4\nu\dfrac{dR/dt}{R} - \dfrac{2\sigma}{\rho_L R}$	At $t = 0$ $R = R_0$ and $dR/dt = 0$
2. Bubble wall velocity (dR/dt)	Internal pressure in the bubble: $P_i = \dfrac{N_{tot}(t) kT}{[4\pi(R^3(t) - h^3)/3]}$ Pressure in bulk liquid medium: $P_t = P_0 - P_A \sin(2\pi f t)$	
3. Number of solvent (water or methanol) molecules in the bubble (N_{SL})	$\dfrac{dN_{SL}}{dt} = 4\pi R^2 D_{SL} \left.\dfrac{\partial C_{SL}}{\partial r}\right|_{r=R} \approx 4\pi R^2 D_{SL}\left(\dfrac{C_{SL,R} - C_{SL}}{l_{diff}}\right)$ $\dfrac{dN_{ST}}{dt} = 4\pi R^2 D_{ST} \left.\dfrac{\partial C_{ST}}{\partial r}\right|_{r=R} \approx 4\pi R^2 D_{ST}\left(\dfrac{C_{ST,R} - C_{ST}}{l_{diff}}\right)$	At $t = 0$ $N_{SL} = 0$ and $N_{ST} = 0$
4. Number of solute molecules in the bubble (N_{ST})	Instantaneous diffusive penetration depths: $l_{diff} = \min\left(\sqrt{\dfrac{RD_{SL}}{|dR/dt|}}, \dfrac{R}{\pi}\right)$; $l_{diff} = \min\left(\sqrt{\dfrac{RD_{ST}}{|dR/dt|}}, \dfrac{R}{\pi}\right)$	
5. Heat transfer through bubble (Q)	$\dfrac{dQ}{dt} = 4\pi R^2 \lambda \left.\dfrac{\partial T}{\partial r}\right|_{r=R} \approx 4\pi R^2 \lambda \left(\dfrac{T_0 - T}{l_{th}}\right)$ Thermal diffusion length: $l_{th} = \min\left(\sqrt{\dfrac{R\kappa}{|dR/dt|}}, \dfrac{R}{\pi}\right)$	At $t = 0$ $Q = 0$
6. Temperature of the bubble (T)	$C_{V,mix} dT/dt = dQ/dt - P_i dV/dt + (h_{SL} - U_{SL})dN_{SL}/dt$ Mixture heat capacity: $C_{V,mix} = \sum C_{V,i} N_i$ Molecular properties of solvent: Enthalpy: $h_{SL} = (f_{SL}+2)kT_0/2$ Internal energy of solvent: $U_{SL} = N_{SL}kT\left(\dfrac{f_{SL}}{2} + \sum_{i=1}^{3}\dfrac{\theta_i/T}{\exp(\theta_i/T)-1}\right)$ Heat capacity of monatomic species (i): $C_V = 3kN_i/2$ Heat capacity of di/tri and polyatomic species (e.g., $i = N_2/O_2/H_2O$): $C_{V,i} = N_i k\left(f_i/2 + \sum((\theta_i/T)^2 \exp(\theta_i/T)/(\exp(\theta_i/T)-1)^2)\right)$	At $t = 0$ $T = T_0$

Source: From *Ultrason. Sonochem.*, 16, Sivasankar, T. and Moholkar, V.S., Physical insights into the sonochemical degradation of recalcitrant organic pollutants with cavitation bubble dynamics, 769–781, Copyright (2009), with permission from Elsevier.

radial motion. In a steam bubble, however, the condition of bubble volume reaching hard core volume of molecules is likely to attain as the vapor molecules condense with pressure surge and merge with the bulk liquid medium. As a result, the pressure inside the bubble is not likely to reach very high values sufficient to have the "rebound" of the bubble. The bubble may undergo continuous compression with condensation of vapor inside it and ultimately disappear in the

TABLE 20.2
Thermodynamic Properties of Various Species

Species	Degrees of Freedom (Translational + Rotational) (f_j)	Lennard-Jones Force Constants σ (10^{-10} m)	ε/k (K)	Characteristic Vibrational Temperatures θ (K)
N_2	5	3.68	92	3350
O_2	5	3.43	113	2273
H_2O	6	2.65	380	2295, 5255, 5400
Ar	3	3.42	124	—
CH_3OH	15	3.626	481.8	500.59, 1674.41, 1708.94, 1854.22, 2169.26, 2356.26, 2376.4, 2392.22, 4581.62, 4649.23, 4752.8, 5923.74

Source: From *Ultrason. Sonochem.*, 16, Sivasankar, T. and Moholkar, V.S., Physical insights into the sonochemical degradation of recalcitrant organic pollutants with cavitation bubble dynamics, 769–781, Copyright (2009), with permission from Elsevier.

Notations: R, radius of the bubble; dR/dt, bubble wall velocity; c, velocity of sound in bulk liquid medium; ρ_L, density of the liquid; ν, kinematic viscosity of liquid; σ, surface tension of liquid; λ, thermal conductivity of bubble contents; κ, thermal diffusivity of bubble contents; θ, characteristic vibrational temperature(s) of the species; N_{SL}, number of solvent molecules in the bubble; t, time, D_{SL}, diffusion coefficient of solvent vapor; C_{SL}, concentration of solvent molecules in the bubble; $C_{SL,R}$, concentration of solvent molecules at the bubble wall or gas–liquid interface; Q, heat conducted across bubble wall; T, temperature of the bubble contents; T_o, ambient (or bulk liquid medium) temperature; k, Boltzmann constant; N_{Ar}, number of Ar molecules in the bubble; f_j, translational and rotational degrees of freedom; $C_{V,i}$, heat capacity at constant volume; N_{tot}, total number of molecules (gas + vapor) in the bubble; h, van der Waal's hard core radius; P_o, ambient (bulk) pressure in liquid; P_A, pressure amplitude of ultrasound wave; f, frequency of ultrasound wave.

medium. Various physical parameters required for numerical solution of the bubble dynamics model are determined as follows:

Acoustic frequency (f): This parameter depends on the type of sonicator used for experiments. Most of the commercial processors have frequency of 20 kHz.

Acoustic pressure amplitude (P_A): This parameter is determined using calorimetric measurements. Greater details of this method have been given by Sivasankar et al. (2007). Due to the attenuation in the medium, the actual pressure amplitude sensed by the cavitation bubble located away from the probe tip is lesser than that at the tip. A direct measurement of the local pressure amplitude in the vicinity of the cavitation bubble is rather difficult, requiring needle hydrophones. Nonetheless, one can assume 10%–15% attenuation of ultrasound waves as a *representative* value. The attenuation also depends on the bubble population in the medium. For a saturated or gassy medium, attenuation is higher than a degassed or unsaturated medium. For liquid media degassed to very low levels of dissolved gas, attenuation effect can be ignored.

Vapor pressure of solvent and solute: This parameter is easily calculated with Antoine's correlations at the temperature of the bulk medium. However, continuous monitoring of the solvent temperature is essential. If the temperature variation during sonication is small (typically <±2°C), it can be ignored. Some solutes such as NaCl reduce vapor pressure of solvent, and this effect needs to be properly accounted for with proper correlations available in literature.

Initial (or equilibrium) bubble radius (R_o): The bulk liquid medium contains cavitation nuclei, which are tiny gas pockets trapped in the crevices of reactor or gas bubbles already suspended in the medium. Depending on the conditions of the medium, these nuclei may have a wide or narrow size range. It is rather difficult to estimate the exact size distribution of these cavitation nuclei. However, for an unsaturated medium, the size range of these nuclei is expected to be smaller than for the saturated medium. For simulations, one can use a representative size (which could be median of the size distribution) of $R_o = 5\,\mu m$ for an unsaturated medium and $R_o = 10\,\mu m$ for the saturated medium.

QUANTIFICATION OF RADICAL PRODUCTION

At the instance of maximum compression during radial motion, the temperature inside the bubble reaches extreme (~5000 K). In addition, due to an extremely small bubble volume, the concentration of the species is quite high. Consequently, the rates of various reactions occurring in the bubble are very high. Therefore, thermodynamic equilibrium is likely to prevail in the bubble (Brenner et al., 2002; Krishnan et al., 2006). The equilibrium composition of the different species resulting from dissociation of solvent vapor and gas molecules can be estimated using technique of Gibbs energy minimization (Eriksson, 1975). A more rigorous approach in this regard would be to include various radical reactions in the mass balance equations along with heats of these reactions included in the energy balance (Yasui, 1997a; Storey and Szeri, 2000; Toegel and Lohse, 2003). Endothermicity of some of the radical reactions (e.g., $H_2O \leftrightarrows H^{\bullet} + {}^{\bullet}OH$) lowers the peak temperature reached during transient bubble collapse. However, addition of this feature in the model would change only the final quantitative answers, with trends remaining essentially unaltered.

QUANTIFICATION OF THE PHYSICAL EFFECT OF CAVITATION (GENERATION OF CONVECTION)

From the numerical solution of the bubble dynamics model, one can quantify the two components of convection generated due to cavitation bubble dynamics as follows:

Microturbulence (or Microconvection)

The radial motion of the bubble gives rise to an oscillatory velocity field in the close vicinity of the bubble, which we term as microturbulence (or microconvection). The velocity in the bulk liquid (V_{turb}) at a distance r from the bubble center is (Leighton, 1994):

$$V_{turb}(r,t) = \frac{R^2}{r^2}\left(\frac{dR}{dt}\right) \tag{20.1}$$

It should be noted, however, that microconvection described by Equation 20.1 is different from acoustic streaming, microstreaming, and ultrasound oscillatory velocity.

Shock Waves (or Acoustic Waves)

During radial motion, the bubble wall comes to a sudden halt at the instance of minimum radius. At this moment, the converging liquid elements toward bubble wall are reflected back, giving rise to shock waves (or acoustic waves). The amplitude of the acoustic wave (P_{AW}) radiated by the cavitation bubbles is (Grossman et al., 1997):

$$P_{AW}(r,t) = \frac{\rho}{4\pi r}\frac{d^2 V_b}{dt^2} = \rho\frac{R}{r}\left[2\left(\frac{dR}{dt}\right)^2 + R\frac{d^2R}{dt^2}\right] \tag{20.2}$$

where
  V_b is the volume of the bubble
  ρ is the density of the liquid

Typical value of r for the simulations can be taken as 1 mm. Equations 20.1 and 20.2 indicate that magnitudes of both P_{AW} and V_{turb} vary inversely with r. Thus, the intensity of the convection induced by the cavitation bubble is rather "local," that is, it is the highest in the close vicinity of the bubble and diminishes very rapidly away from it.

MECHANISTIC ASPECTS OF ULTRASOUND-ENHANCED CHEMICAL PROCESSES: CASE STUDY OF SONOCHEMICAL DEGRADATION OF RECALCITRANT ORGANIC POLLUTANTS

As noted earlier, the principal physical phenomenon underlying chemical effects of ultrasound is production of highly reactive radicals. However, depending on the nature of the reactant, two other mechanisms, which are consequences of the thermal effects of cavitation bubbles are also feasible. These are (1) thermal dissociation of reactants inside the bubble (for volatile compounds) and (2) thermal decomposition in the thin shell of liquid surrounding the bubble at the bubble–liquid interface (for nonvolatile compounds). As far as the radical-induced reactions are concerned, there are two possible locations: in the bulk liquid or in the thin shell of liquid surrounding the bubble (or the bubble–liquid interface). Exact manifestation of these mechanisms depending on the nature and type of reactants and other factors such as presence of radical scavenging/conserving species in the medium was evident from a comprehensive study on degradation of various types of pollutants conducted in our group (Sivasankar and Moholkar, 2009). We present below summary of this work.

The two major chemical pathways or mechanisms for the sonolytic degradation of a pollutant are (1) pyrolysis or thermal decomposition of the pollutant molecules entrapped inside the bubble and (2) hydroxylation, that is, the attack of $O^{\bullet}$, $^{\bullet}OH$, and $HO_2^{\bullet}$ radicals produced by the cavitation bubble leading to hydroxylated products. Hua et al. (1995) have proposed a third possible mechanism for the sonochemical degradation of organic pollutants, which is hydrolysis of the organic pollutants in the transient supercritical water packets formed in the close vicinity of cavitation bubble during transient collapse. Out of these pathways, the major pathway contributing to the overall degradation depends on the nature and physicochemical properties of the organic compound. Five model pollutants commonly found in industrial wastewater discharge were chosen in this study, viz., phenol (Ph), chlorobenzene (CB), nitrobenzene (NB), p-nitrophenol (PNP), and 2,4-dichlorophenol (2,4-DCP). For the experiments, we adopted techniques that caused variation in the nature of the cavitation phenomenon occurring in the medium. The trends in the degradation observed with these experimental conditions have been correlated to the simulations of the radial motion of cavitation bubbles.

EXPERIMENTAL SETUP AND PROCEDURES

Experimental Setup

Sonication of synthetically prepared aqueous solutions of pollutants was carried out in a jacketed glass reactor using a probe type ultrasonic processor operating at a frequency of 20 kHz (Sonics and Materials, Inc., Model VCX 500) in 15–5 min off pulse mode. A schematic of the experimental setup is shown in Figure 20.1. For a power output of 100 W, the ultrasound probe produced an acoustic wave with 150 kPa amplitude. For bubbling of different gases through the reaction medium during sonication, a glass sparger was used. Various techniques or experimental conditions applied during sonication of the solution of the pollutant are as follows: (1) variation in the initial concentration of the pollutant, (2) variation in the saturation level (or dissolved gas content) of the solution, (3) addition of salt (NaCl) to the solution of pollutant (either saturated or unsaturated), (4) sparging of gases such as Ar, N_2, O_2, and air through the pollutant solution

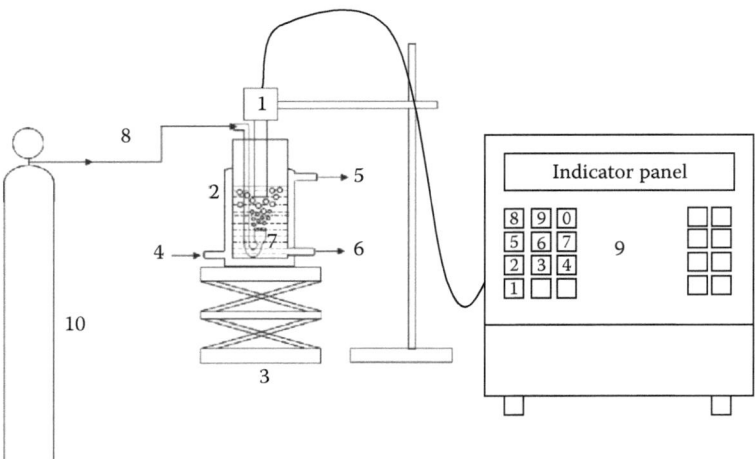

FIGURE 20.1 Schematic of experimental setup (Legends: 1—Ultrasound horn; 2—Jacketed glass reactor; 3—Laboratory jack; 4—Cooling water inlet; 5—Cooling water outlet; 6—Sample port; 7—Aerator; 8—Gas inlet; 9—Control unit of ultrasonic processor; and 10—Gas cylinder] (Reproduced from *Ultrason. Sonochem.*, 16, Sivasankar, T. and Moholkar, V.S., Physical insights into the sonochemical degradation of recalcitrant organic pollutants with cavitation bubble dynamics, 769–781, Copyright (2009), with permission from Elsevier.)

during sonication, and (5) addition of $FeSO_4 \cdot 7H_2O$ to the pollutant solution (either saturated or unsaturated). For each pollutant, only those techniques that were sufficient to establish the dominant physical mechanism of the sonochemical degradation were applied. For experiments with unsaturated (or degassed) medium, the dissolved oxygen (DO) content of the liquid medium was reduced by vacuumization.

Analytical Procedure

The extent of sonochemical degradation of Ph, NB, CB, and DCP was monitored with HPLC. Sonochemical degradation of PNP was analyzed using UV–Vis spectrophotometer. No analysis of the intermediate products of degradation was done, as these have been extensively studied in previous literature. Only the rate of disappearance of original pollutant was monitored and the analysis has been made on that basis. This approach is justified on the basis of principal aim of this study, that is, discernment of the physical (and *not* chemical) mechanism of the degradation of organic pollutants. Experimental techniques adopted in this study vary only the cavitation characteristics in the medium, without changing the degradation chemistry. Therefore, the intermediates and final products of the sonochemical degradation of the pollutants are expected to be same as reported in previous literature.

RESULTS AND ANALYSIS OF DEGRADATION

Preamble

Due to extremely high unstability and reactivity, reaction zone of the radicals generated from cavitation bubbles is restricted to only a small region in the vicinity of point of bubble collapse. The hydroxylation reaction would occur only if a pollutant molecule is present in the reaction zone. For dilute solutions, the probability of interaction between pollutant molecules and •OH radicals becomes an important factor influencing the extent of degradation of the pollutant. The extent of degradation through pyrolysis route is determined by evaporation of pollutant in the bubble that

TABLE 20.3
Physicochemical Properties of Various Organic Pollutants

Properties	Ph	ClBz	PNP	NB	2,4-DCP
Water solubility at 25°C (ppm)	83,000	500	16,000	1900	4500
Water–octanol partition coefficient (log K_{OW})	1.46	2.84	1.91	1.86	3.06
Vapor pressure at 25°C (Pa)	46	1598	0.0032	33.69	25.84

Source: From *Ultrason. Sonochem.*, 16, Sivasankar, T. and Moholkar, V.S., Physical insights into the sonochemical degradation of recalcitrant organic pollutants with cavitation bubble dynamics, 769–781, Copyright (2009), with permission from Elsevier.

Abbreviations: Ph, phenol; ClBz, chlorobenzene; PNP, *p*-nitrophenol; NB, nitrobenzene; 2,4-DCP, 2,4-dichlorophenol.

directly varies with partial pressure of pollutant at the bubble–bulk interface. Another factor of relevance is the hydrophobicity of the pollutant. Due to repulsive interactions between organic pollutant and water molecules, the pollutant molecules are "driven" toward the bubble interface with hydrophobic character. This results in enhanced concentration of pollutant molecules in the bubble–bulk interfacial region, which affects the degradation of the pollutant via both hydroxylation and pyrolysis routes. The physical properties of the five pollutants are shown in Table 20.3, which could be considered as benzene derivatives with grafting of $-OH$, $-NO_2$, and $-Cl$ groups. These groups impart different characteristics to the compounds. $-NO_2$ and $-Cl$ groups impart strong hydrophobic character, while $-OH$ group makes the compound hydrophilic. On the basis of physical properties, one can characterize various pollutants on *relative* basis as follows: (1) Ph: high solubility, high hydrophilicity, and nonvolatile; (2) CB: low solubility, high hydrophobicity, and volatile; (3) PNP: high solubility, moderate hydrophilicity, and nonvolatile; (4) NB: moderate solubility, moderate hydrophobicity, and nonvolatile; and (5) 2,4-DCP: low solubility, high hydrophobicity, and nonvolatile.

Experimental Results

The percentage degradation of five pollutants under different conditions is depicted in Table 20.4. The salient features of the experimental results are identified as follows:

Initial pollutant concentration: The extent of degradation of Ph and CB in saturated medium increased with increasing initial concentration.

Saturation level (or dissolved gas content) of the medium: The degradation of NB and 2,4-DCP increased with reduced saturation level (or degassing) of the medium, while degradation of PNP was higher for saturated medium.

Gas bubbling (or sparging) through the medium: For both Ph and 2,4-DCP, bubbling of N_2 through the medium resulted in least degradation. However, the overall trend for four gases in case of Ph was $O_2 > Air > Ar > N_2$, while for 2,4-DCP it was $Air > Ar > O_2 > N_2$.

Salt addition: Degradation of Ph showed marked enhancement with salt addition. Degradation of NB (in both saturated and unsaturated medium) and 2,4-DCP (in saturated medium) did not alter much, while degradation of PNP reduced slightly in both saturated and unsaturated medium.

Addition of $FeSO_4 \cdot 7H_2O$: For NB and 2,4-DCP, addition of $FeSO_4 \cdot 7H_2O$ resulted in slight improvement in degradation. On the contrary, PNP degradation showed marked (~100%) enhancement with $FeSO_4$ addition.

TABLE 20.4
Experimental Results on Degradation of Various Pollutants with Different Experimental Techniques

Experimental Parameters		\multicolumn{5}{c}{Degradation of Pollutants (%)}				
		Ph	ClBz	NB[b]	PNP[a]	2,4-DCP[b]
Initial concentration	50 ppm (saturated medium)	1.90 ± 0.14	65.86 ± 5.01	—	—	—
	100 ppm (saturated medium)	2.09 ± 0.08	65.59 ± 3.64	—	—	—
Dissolved gas content	Saturated medium	—	—	18.43 ± 0.61	4.22 ± 0.45	5.66 ± 0.66
	Unsaturated medium	—	—	31.80 ± 0.37	2.94 ± 0.25	8.68 ± 0.30
Salt addition (4% NaCl)	Saturated medium	6.23 ± 0.35 (50 ppm)	70.22 ± 6.58 (50 ppm)	20.31 ± 1.63	3.51 ± 0.04	4.98 ± 0.72
		5.25 ± 0.13 (100 ppm)	70.09 ± 2.21 (100 ppm)			
	Unsaturated medium	—	—	28.23 ± 5.24	2.79 ± 0.36	—
Gas bubbling (pure solution)	Nitrogen	1.39 ± 0.28	—	—	—	5.66 ± 0.20
	Argon	2.20 ± 0.49	—	—	—	6.92 ± 0.43
	Oxygen	4.09 ± 0.51	—	—	—	6.86 ± 0.90
	Air	2.53 ± 0.34	—	—	—	9.25 ± 0.04
$FeSO_4 \cdot 7H_2O$ addition (0.5 mM)	Saturated medium	—	—	22.94 ± 3.46	11.07 ± 0.25	5.73 ± 0.36
	Unsaturated medium	—	—	30.61 ± 2.70	4.31 ± 0.13	—

Source: From *Ultrason. Sonochem.*, 16, Sivasankar, T. and Moholkar, V.S., Physical insights into the sonochemical degradation of recalcitrant organic pollutants with cavitation bubble dynamics, 769–781, Copyright (2009), with permission from Elsevier.

[a] For all experiments, the initial concentration of PNP was 10 ppm.
[b] For all experiments, the initial concentration of NB and 2,4-DCP was 100 ppm.

SIMULATION RESULTS

Figures 20.2 and 20.3 show representative simulations of 5 μm air bubble and 10 μm argon bubble in dilute solution of nonvolatile pollutants such as 2,4-DCP. The evaporation of the pollutant into the bubble has been ignored in these simulations. Figure 20.4 shows simulation of 10 μm air bubble in CB solution of 50 ppm concentration, with evaporation of the pollutant in the bubble being taken into account. The summary of entire simulation results is given in Tables 20.5 through 20.7. The trend in production of •OH radical responsible for hydroxylation reaction varies as: Ar > Air > O_2 > N_2. Ar bubbles produce the highest number of radicals among all four gases. This is a consequence of the highest temperature peak reached at the collapse due to monatomic nature of Ar. The intensity of collapse of air, N_2, and O_2 bubbles is more-or-less the same, as indicated by the temperature peaks attained at the collapse of these bubbles. However, the equilibrium composition of the bubble varies due to scavenging of radicals by O_2 and N_2 molecules present in the bubble. For air bubble, the N_2 scavenges the H•, O•, and •OH radical to produce various species such as NO, N_2O, NO_2, HNO, and HNO_2. However, O_2 reacts with these species to regenerate O• and •OH radicals. O• radicals also react with H• and •OH radicals to produce $HO_2^•$ radicals. For O_2 bubbles, the yield of H_2O_2 and O_3 is much higher than other bubbles due to extensive scavenging of O•, H•, and •OH radicals by the O_2 molecules. This results in loss of oxidation potential. The overall yield of radicals is thus reduced. Production of radicals by N_2 bubbles is the least. This is a consequence of the extensive scavenging of the radical species by N_2 molecules. Moreover, there is no regeneration of O• and •OH radicals by O_2, as in case of air bubbles. The overall effect is massive loss of radicals that reduces the oxidation potential of N_2 bubbles.

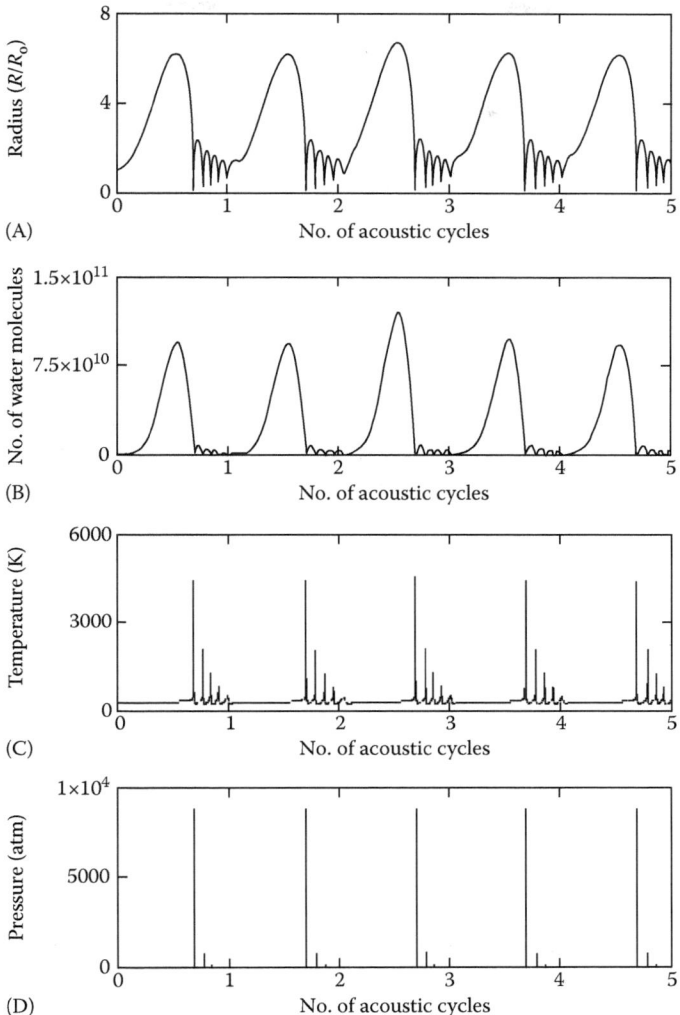

FIGURE 20.2 Simulation of radial motion of 5 μm air bubble in water. Time variation of (A) normalized bubble radius (R/R_o), (B) number of water molecules in the bubble, (C) temperature of the bubble, and (D) pressure inside the bubble. (Reproduced from *Ultrason. Sonochem.*, 16, Sivasankar, T. and Moholkar, V.S., Physical insights into the sonochemical degradation of recalcitrant organic pollutants with cavitation bubble dynamics, 769–781, Copyright (2009), with permission from Elsevier.)

DISCUSSION AND ANALYSIS

Before analyzing the experimental and simulation results and correlating them, we briefly outline as how the various experimental parameters used in the degradation studies alter characteristics of the cavitation phenomena in the system, and hence, the extent of degradation of the pollutants.

1. Rise in the initial concentration of pollutant increases in the probability of radical–pollutant interaction, and second, rise in the interfacial concentration of the pollutant due to which the extent of evaporation of the pollutant into the cavitation bubble increases.
2. Dissolved gas concentration affects the extent of radical production in the medium by altering the intensity of the transient collapse of the bubble via rectified diffusion. In a saturated medium, rectified diffusion of dissolved gas in the bubble during radial motion

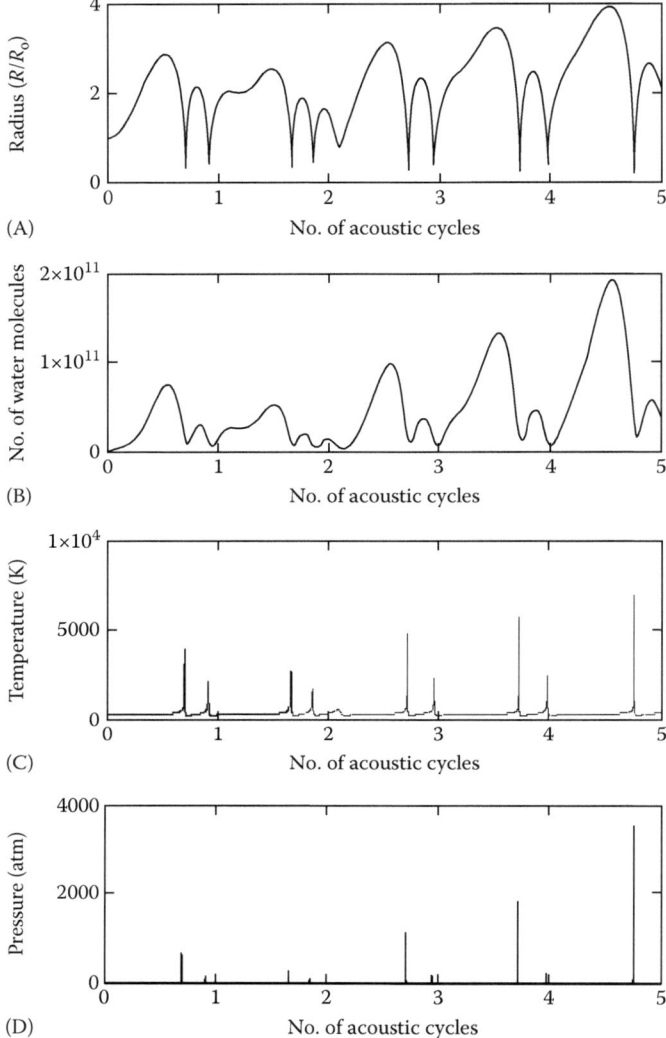

FIGURE 20.3 Simulation of radial motion of 10 μm argon bubble in water. Time variation of (A) normalized bubble radius (R/R_o), (B) number of water molecules in the bubble, (C) temperature of the bubble, and (D) pressure inside the bubble. (Reproduced from *Ultrason. Sonochem.*, 16, Sivasankar, T. and Moholkar, V.S., Physical insights into the sonochemical degradation of recalcitrant organic pollutants with cavitation bubble dynamics, 769–781, Copyright (2009), with permission from Elsevier.)

makes equilibrium radius of the bubble grow. This gas cushions the transient collapse of the bubble, reducing the intensity of collapse and rate of •OH production. In an unsaturated medium, gas inside the bubble dissolves into the medium during oscillations with consequent rise in the intensity of collapse and radical production.

3. Salt addition to the medium increases the ionic strength of the medium. This enhances the hydrophobic repulsive interactions between pollutant and water molecules. Due to this, the pollutant molecules are "pushed" toward the bubble interface causing increased concentration of pollutant molecules. Consequence of this is twofold: boosting of the probability of radical–pollutant interaction and greater evaporation of

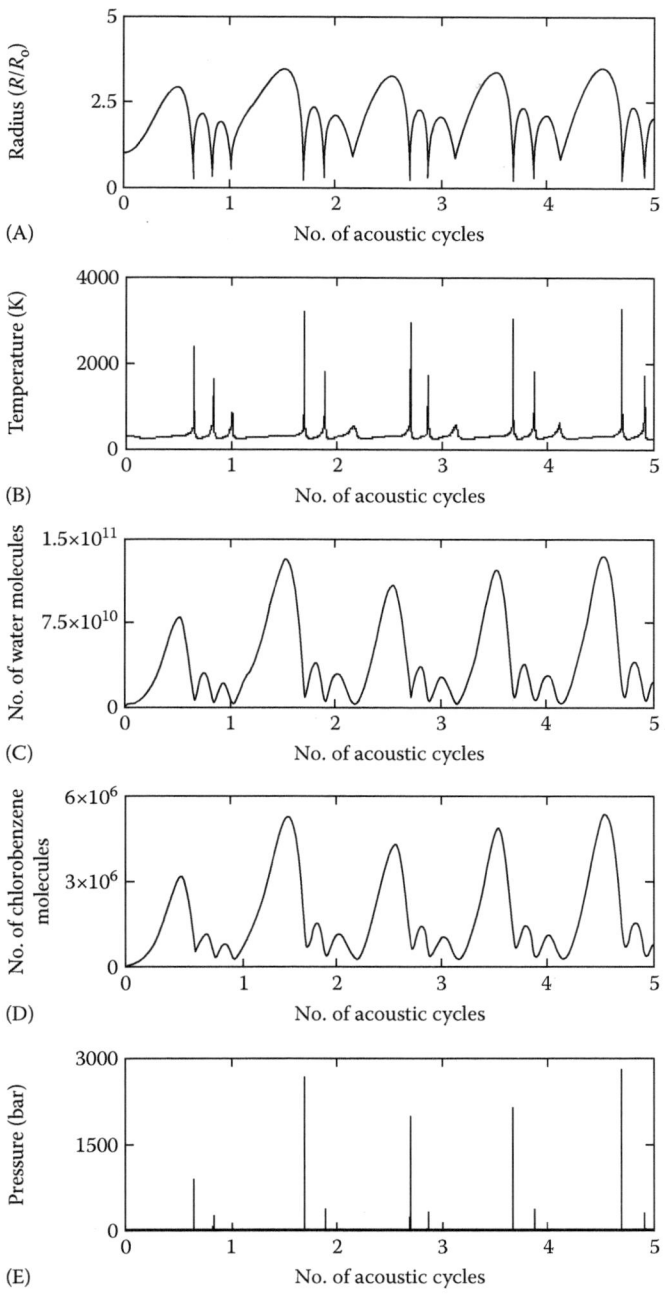

FIGURE 20.4 Simulation of the radial motion of 10 μm air bubble in 50 ppm aqueous solution of chlorobenzene (evaporation of solute in the bubble is taken into account). Time variation of (A) normalized bubble radius (R/R_o) (B) temperature in the bubble (C) number of water molecules in the bubble, (D) number of chlorobenzene molecules in the bubble, and (E) pressure inside the bubble. (Reproduced from *Ultrason. Sonochem.*, 16, Sivasankar, T. and Moholkar, V.S., Physical insights into the sonochemical degradation of recalcitrant organic pollutants with cavitation bubble dynamics, 769–781, Copyright (2009), with permission from Elsevier.)

TABLE 20.5
Summary of the Simulation Results for Ar, N_2, and O_2 Bubbles in Aqueous Solutions of Ph, NB, PNP, and 2,4-DCP

	Parameters for Simulations		
	Argon bubble	Nitrogen bubble	Oxygen bubble
	$R_o = 10\,\mu m$	$R_o = 10\,\mu m$	$R_o = 10\,\mu m$

Conditions at the First Collapse of the Bubble

	Argon bubble	Nitrogen bubble	Oxygen bubble
	$T_{max} = 3937\,K$	$T_{max} = 2397\,K$	$T_{max} = 2303\,K$
	$P_{max} = 628.7$ bar	$P_{max} = 1407$ bar	$P_{max} = 1539$ bar
	$N_{Ar} = 1.178 \times 10^{+11}$	$N_{N_2} = 1.178 \times 10^{+11}$	$N_{O_2} = 1.178 \times 10^{+11}$
	$N_{WT} = 1.51 \times 10^{+10}$	$N_{WT} = 1.48 \times 10^{+10}$	$N_{WT} = 1.47 \times 10^{+10}$

Equilibrium Composition of Bubble Contents at Transient Collapse

Species			
H_2O	6.587×10^{-1}	1.109×10^{-1}	1.099×10^{-1}
H_2	1.314×10^{-1}	9.211×10^{-4}	6.251×10^{-6}
OH	1.145×10^{-1}	2.838×10^{-4}	1.809×10^{-3}
H	3.888×10^{-2}	1.262×10^{-5}	6.232×10^{-7}
O_2	3.686×10^{-2}	1.279×10^{-4}	8.879×10^{-1}
O	1.899×10^{-2}	2.557×10^{-6}	1.206×10^{-4}
HO_2	5.463×10^{-4}	4.626×10^{-7}	2.778×10^{-4}
H_2O_2	7.150×10^{-5}	2.613×10^{-7}	1.808×10^{-5}
O_3	3.655×10^{-7}	0	5.395×10^{-6}
N_2	0	8.873×10^{-1}	0
NO	0	5.244×10^{-4}	0
N_2O	0	1.045×10^{-6}	0
N_{OH}	1.728×10^9	3.763×10^7	2.397×10^8

Source: From *Ultrason. Sonochem.*, 16, Sivasankar, T. and Moholkar, V.S., Physical insights into the sonochemical degradation of recalcitrant organic pollutants with cavitation bubble dynamics, 769–781, Copyright (2009), with permission from Elsevier.

Notations: T_{max}, temperature peak reached in the bubble at the time of first collapse; P_{max}, pressure peak reached in the bubble at the time of first collapse; N_{WT}, number of water molecules trapped in the bubble at the instance of first collapse; N_{O_2}, number of oxygen molecules in the bubble (for air and oxygen bubbles); N_{OH}, number of ·OH radicals produced per cavitation bubble.

pollutant into the bubble due to rise in the partial pressure of the pollutant at the bubble interface. Thus, both hydroxylation and pyrolysis routes of degradation are benefited by salt addition.

4. The technique of gas bubbling or sparging is used to seed external cavitation nuclei in the medium made up of desired gas. Different cavitation behavior displayed by monatomic and diatomic gases is the underlying principle for this technique. The peak temperature reached at the transient collapse of bubbles of monatomic gas such as Ar is higher than diatomic gases like N_2, air, and O_2, by virtue of low heat capacity of the former. On the other hand, presence of O_2 molecules in the air and O_2 bubbles gives rise to scavenging or conservation of radicals, that is, generation of new radical species due to reaction of oxygen with radical species released from cavitation bubbles. The phenomenon of scavenging/conservation influences the composition of the bubble at transient collapse. Due

TABLE 20.6
Summary of the Simulation Results for Variation in the Saturation Level of the Medium for NB, PNP, and 2,4-DCP Solutions

	Parameters for Simulations	
	Air bubble $R_o = 5\,\mu m$	Air bubble $R_o = 10\,\mu m$
	Conditions at First Compression of the Bubble	
	$T_{max} = 4426\,K$	$T_{max} = 2478\,K$
	$P_{max} = 8812\,bar$	$P_{max} = 1044\,bar$
	$N_{N_2} = 1.271 \times 10^{+10}$	$N_{N_2} = 9.305 \times 10^{+10}$
	$N_{O_2} = 3.379 \times 10^{+9}$	$N_{O_2} = 2.474 \times 10^{+10}$
	$N_{WT} = 2.70 \times 10^{+9}$	$N_{WT} = 1.09 \times 10^{+10}$
Species	**Equilibrium Composition of Bubble Contents at Transient Collapse**	
N_2	7.085×10^{-1}	7.122×10^{-1}
O_2	1.335×10^{-1}	1.805×10^{-1}
NO	1.188×10^{-1}	2.047×10^{-2}
H_2O	9.797×10^{-3}	8.379×10^{-2}
OH	1.242×10^{-2}	2.126×10^{-3}
O	1.214×10^{-2}	1.697×10^{-4}
NO_2	2.182×10^{-3}	5.093×10^{-4}
HO_2	7.135×10^{-4}	1.054×10^{-4}
H_2	3.419×10^{-4}	3.254×10^{-5}
N_2O	5.862×10^{-4}	3.151×10^{-5}
H	6.106×10^{-4}	4.006×10^{-6}
HNO_2	2.053×10^{-4}	4.659×10^{-5}
HNO	1.048×10^{-4}	1.651×10^{-6}
H_2O_2	1.972×10^{-5}	7.629×10^{-6}
O_3	2.914×10^{-5}	6.983×10^{-7}
N	6.444×10^{-5}	0
N_{OH}	2.334×10^{8}	2.736×10^{8}

Source: From *Ultrason. Sonochem.*, 16, Sivasankar, T. and Moholkar, V.S., Physical insights into the sonochemical degradation of recalcitrant organic pollutants with cavitation bubble dynamics, 769–781, Copyright (2009), with permission from Elsevier.

Note: Air bubble of 10 μm size represents saturated medium, while air bubble of 5 μm size represents unsaturated medium.

to dissolution of O_2 in liquid medium during sparging of air or O_2, the concentration of DO is maintained at saturation level. Scavenging/conservation induced by DO in the bulk liquid medium increases penetration depth or the "reaction zone" of radicals from location of bubble collapse. This has a favorable effect on the probability of radical–pollutant interaction.

5. Addition of Fe^{2+} ions in the medium gives rise to Fenton's reagent type action. In case of dilute solutions of pollutants, the probability of radical–pollutant interaction is low, as a result of which a large fraction of radicals undergo recombination. This is a loss of

TABLE 20.7
Summary of the Simulation Results for Air Bubbles in Chlorobenzene Solution

Species	Parameters for Simulation	
	50 ppm Solution	100 ppm Solution
	Conditions at the first compression of the bubble	
	$T_{max} = 2383\,K$	$T_{max} = 2383\,K$
	$P_{max} = 888.2\,bar$	$P_{max} = 888.2\,bar$
	$N_{cb} = 8.1 \times 10^5$	$N_{cb} = 8.1 \times 10^5$
	$N_w = 1.0502 \times 10^{10}$	$N_w = 1.0502 \times 10^{10}$
	Equilibrium Composition of Bubble Contents at Transient Collapse	
O_2	6.994×10^{-1}	6.994×10^{-1}
H_2O	2.955×10^{-1}	2.955×10^{-1}
OH	4.262×10^{-3}	4.262×10^{-3}
O	2.207×10^{-4}	2.207×10^{-4}
HOO	4.127×10^{-4}	4.127×10^{-4}
CO_2	1.375×10^{-4}	1.375×10^{-4}
H_2	3.879×10^{-5}	3.879×10^{-5}
H_2O_2	3.963×10^{-5}	3.963×10^{-5}
HCl	2.113×10^{-5}	2.113×10^{-5}
O_3	3.702×10^{-6}	3.702×10^{-6}
H	3.049×10^{-6}	3.049×10^{-6}
Cl	1.182×10^{-6}	1.182×10^{-6}
Cl_2	1.972×10^{-10}	1.972×10^{-10}

Source: From *Ultrason. Sonochem.*, 16, Sivasankar, T. and Moholkar, V.S., Physical insights into the sonochemical degradation of recalcitrant organic pollutants with cavitation bubble dynamics, 769–781, Copyright (2009), with permission from Elsevier.

Note: In these simulations, the evaporation of the pollutant into the cavitation bubble has been taken into account.

oxidation potential. The predominant radical species generated by transient collapse of bubble is •OH, and hence, the principal product formed out of radical recombination is H_2O_2. Fe^{2+} reacts with H_2O_2 to generate •OH radicals according to following reactions:

$$\bullet OH + \bullet OH \longrightarrow H_2O_2 \quad (R.20.1)$$

$$H_2O_2 + Fe^{2+} \longrightarrow Fe^{3+} + OH^- + \bullet OH \quad (R.20.2)$$

$$Fe^{3+} + H_2O_2 \longrightarrow Fe^{2+} + HO_2^\bullet + H^+ \quad (R.20.3)$$

$$Fe^{3+} + HO_2^\bullet \longrightarrow Fe^{2+} + O_2 + H^+ \quad (R.20.4)$$

The above reactions not only revert the oxidation potential loss but also help in deeper penetration of the radical action from the location of bubble collapse, which increases the probability of radical–pollutant interaction.

Based on above arguments and the physicochemical characteristics of the pollutant, physical mechanism of the degradation of each pollutant is established as follows:

1. Due to nonvolatile nature, Ph is not expected to evaporate into the bubble, and hence, the degradation mechanism of Ph is expected to be hydroxylation. The highest degradation of Ph in presence of O_2 indicates that the degradation occurs in the bulk medium. This is further corroborated by enhancement in degradation observed with salt addition and increasing initial concentration.
2. Due to volatile and hydrophobic nature, CB is expected to evaporate into the bubble and undergo pyrolytic decomposition resulting in very rapid degradation. CB has the highest degradation rate among all pollutants, which is one order of magnitude higher than any other pollutant. Marginal effect of salt addition on the degradation rate is attributed to hydrophobic nature of CB, as a result of which its concentration in the bubble–bulk interfacial region is near or at saturation. The extent of degradation increases with initial concentration, and this is attributed to higher evaporation of CB into the bubble at higher initial concentrations.
3. Due to nonvolatile nature of NB, it is not expected to evaporate into the bubble, and moreover, due to moderate hydrophobicity, its concentration in the interfacial region is higher than in the bulk. Significant rise in degradation of NB with unsaturation of the medium indicates hydroxylation as the predominant mechanism of degradation. Neither salt addition nor $FeSO_4$ addition creates any significant change in degradation indicates that hydrophobic interactions and radical scavenging is unimportant for NB degradation. This is attributed to the hydrophobicity of NB, due to which the concentration of NB at bubble interface is already near or at saturation, which does not change with salt addition. As a result, most of the degradation occurs in the bubble–bulk interfacial region, where probability of radical–pollutant interaction is at its maximum. Hence, addition of radical conserver like Fe^{2+} also does not make a significant difference to the extent of degradation.
4. PNP has nonvolatile and moderately hydrophilic character. However, the chemistry of its degradation is different from other pollutants (Kotronarou et al., 1991). The degradation commences with thermal decomposition (cleavage of C–N bond) in the boundary layer surrounding the bubble that gets heated up during collapse, followed by hydroxylation in the bulk medium. Second, very low initial concentration (10 ppm) makes the factor of radical scavenging crucially important in the overall degradation. Reduction in the degradation of PNP with degassing of the medium is a likely consequence of reduction in the thickness of thermal boundary layer surrounding the bubble, where thermal decomposition prior to hydroxylation occurs. Reduction in the degradation with salt addition is attributed to reduction in the DO content, due to which the scavenging action reduces. The importance of the scavenging action in the bulk medium is further corroborated by sharp rise in the degradation with the addition of Fe^{2+}. All of these results point at hydroxylation in the bulk medium as the predominant degradation mechanism.
5. The physicochemical nature of 2,4-DCP is essentially nonvolatile and hydrophobic. Sharp rise in degradation with unsaturation of the medium indicates hydroxylation at the bubble–bulk interfacial region as the predominant degradation pathway. Moreover, due to low solubility (4500 ppm), the concentration at the bubble–bulk interface is small, although the interfacial region is expected to be near or at saturation due to hydrophobic nature of the pollutant. These factors render scavenging action in the interfacial region a dominant factor in degradation. These arguments are further endorsed by the highest degradation observed in the presence of air. Relatively, low degradation rates with Fe^{2+} addition in comparison to air or O_2 sparging indicate that scavenging action in the bulk medium is unimportant. A straightforward interpretation of these results is that predominant region of degradation is the bubble–bulk interfacial region. The above inferences are summarized in Table 20.8 that lists the predominant mechanism and the location of degradation for the five pollutants.

TABLE 20.8
Predominant Mechanism and Location of Degradation of Various Organic Pollutants

Pollutant	Mechanism of Degradation	Location of Degradation
Phenol	Hydroxylation	Bulk medium
Chlorobenzene	Pyrolytic decomposition	Inside the cavitation bubble
Nitrobenzene	Hydroxylation	Bubble–bulk interfacial region
p-Nitrophenol	Thermal decomposition followed by hydroxylation	Bubble–bulk interfacial region (thermal decomposition) Bulk medium (hydroxylation)
2,4-Dichlorophenol	Hydroxylation	Bubble–bulk interfacial region

Source: From *Ultrason. Sonochem.*, 16, Sivasankar, T. and Moholkar, V.S., Physical insights into the sonochemical degradation of recalcitrant organic pollutants with cavitation bubble dynamics, 769–781, Copyright (2009), with permission from Elsevier.

The above case study of sonochemical degradation of pollutants is a vivid example of the mechanistic aspects of chemical processes induced by ultrasound and cavitation. The results clearly indicate as how chemical (radical production) and physical (heating of boundary layer surrounding the bubble) effects of cavitation have different manifestations of the reaction system. The above study is also an excellent example as how response of the chemical system to cavitation is governed by physicochemical nature of reactants. The above case study forms a general framework that could be extended to any other sonochemical system to deduce its physical mechanism.

MECHANISTIC ASPECTS OF ULTRASOUND-ENHANCED PHYSICAL PROCESSES: CASE STUDY OF SONOCRYSTALLIZATION

The driving force for crystallization is supersaturation of solute. Although supersaturation can be generated by various means, the most common methods are thermal swing (or temperature variation) and addition of an antisolvent. On a microscopic scale, the principal physical mechanisms underlying crystallization are nucleation and crystal growth. The macroscopic manifestation of these mechanisms is the size distribution of crystals and predominant crystal size resulting from the process. The rate of crystal growth depends on the rate of arrival of crystal unit at the crystal surface, which is governed by intensity of local convection in the vicinity of the crystal. Ultrasound is known to influence the crystallization system in several ways as follows: (1) reduction in induction time for crystallization, (2) reduction in amount of antisolvent required for crystallization, (3) narrowing down crystal size distribution (CSD) with simultaneous reduction in dominant crystal size, and (4) change in crystal geometry (in polymorphic systems). We have addressed the issue of discernment of physical mechanism of sonocrystallization. KCl–methanol–water has been chosen as model crystallization system, in which crystals of KCl are precipitated out of aqueous solution with methanol as antisolvent. Our study attempted to deduce as how microturbulence (or microconvection) and shock waves generated by cavitation bubbles influence nucleation and crystal growth, and as how this influence is manifested in the characteristics of CSD.

Experimental Procedure and Analysis

The experimental setup was similar to that used in experiments on degradation of pollutants. Crystallization under different conditions was carried out in a beaker (vol. 500 mL) made of borosilicate glass placed on a magnetic stirrer plate. Two kinds of probes, fabricated from high-grade titanium alloy, with diameters 12.5 (1/2 in.) and 25 mm (1 in.) were used for sonication. Pressure amplitude of ultrasound waves, as determined using calorimetric technique, was 1.5 bar. In order to maintain the temperature of the crystallization mixture constant, cooling water was circulated in the jacket around the glass beaker. In some experiments, argon gas was sparged through the crystallization mixture simultaneously with sonication. In some experiments, the crystallization magma was stirred using a magnetic needle. Crystallization was achieved by addition of aqueous KCl solution to methanol at a specific flow rate. Two specific initial concentrations of KCl were used in the experiments, viz., 200 and 300 g L^{-1}. The experiments were divided into four categories as follows:

Category 1: Addition of aq. KCl solution to mechanically stirred methanol
Category 2: Addition of aq. KCl solution to stagnant methanol
Category 3: Addition of aq. KCl solution to methanol with simultaneous sonication (without mechanical stirring)
Category 4: Addition of aq. KCl solution to methanol with simultaneous sonication and sparging of argon (without mechanical stirring of the solution)

In each of the above categories, two experimental parameters were varied, viz., volume ratio of aqueous solution to methanol (1:10 and 1:20) and diameter of the ultrasound probe (1/2 and 1 in.). The total time of addition of aq. KCl solution to methanol was 20 min, which was decided on the basis of mixing times for different power densities and vessel dimensions in ultrasonic processors reported by Kumar et al. (2006). Permutation–combination of these experimental parameters resulted in four experimental sets in first and second categories and eight experimental sets in third and fourth experimental categories. Table 20.9 gives the details of experimental conditions in each of these sets.

The crystal crop was characterized for size distribution and crystal habit. The size distribution of the crystal crop was measured using a laser particle size analyzer. Crystal habit was determined with SEM analysis of the dried samples of the KCl crystals.

Results

Preamble

The impact of ultrasonic irradiation on KCl crystallization is seen mainly in two characteristics of CSD, that is, dominant crystal size (or median) and span of the CSD. Factors governing these characteristics of CSD are (1) number of nuclei and (2) growth rate of the nuclei. With rise in nuclei population in crystallization magma, amount of precipitating material received by a single nucleus reduces, as same material is distributed among more nuclei. Consequently, the dominant crystal size of the CSD reduces. On the contrary, with larger growth rate, span of the CSD will reduce as precipitating crystal blocks are distributed uniformly among all nuclei. This distribution, however, depends on the level of convection. The higher the convection, the higher the mass transfer coefficient, and hence, the higher the growth rate.

Given these facts, we now try to guess as what would be the effect of two components of convection, viz., microturbulence and shock waves, created by cavitation bubbles. Microturbulence (or microconvection) has continuous nature with low-to-moderate velocities; however, it is highly localized. Therefore, its contribution to the "bulk convection" in the crystallization mixture is limited. The microturbulence is likely to affect the growth rate by proper distribution of the precipitating crystal blocks among the nuclei. Rise in number density of bubbles in the medium with sparging of a gas is expected to increase the intensity of bulk convection. On the other hand, shock waves are high-pressure amplitude entities with discrete or intermittent nature. These waves cause heavy drifting of molecules or particles in the medium with high velocities. These molecules and particles also collide with each other with great force during the drift. Consequently, these waves are

TABLE 20.9
Experimental Conditions in Various Categories and Sets

Category	Set	Concentration (g L⁻¹)	Volume Ratio	Sonicator Probe Diameter (mm)
Category 1 (crystallization in vigorously stirred methanol)	1A	300	1:20	N.A.
	1B	200	1:20	N.A.
	1C	300	1:10	N.A.
	1D	200	1:10	N.A.
Category 2 (crystallization in stagnant methanol)	2A	300	1:20	N.A.
	2B	200	1:20	N.A.
	2C	300	1:10	N.A.
	2D	200	1:10	N.A.
Category 3 (crystallization in sonicated methanol)	3A	300	1:20	12.5
	3B	200	1:20	12.5
	3C	300	1:20	25
	3D	200	1:20	25
	3E	300	1:10	12.5
	3F	200	1:10	12.5
	3G	300	1:10	25
	3H	200	1:10	25
Category 4 (Crystallization in sonicated methanol with argon sparging)	4A	300	1:20	12.5
	4B	200	1:20	12.5
	4C	300	1:20	25
	4D	200	1:20	25
	4E	300	1:10	12.5
	4F	200	1:10	12.5
	4G	300	1:10	25
	4H	200	1:10	25

Source: From *Ultrason. Sonochem.*, Nalajala, V.S. and Moholkar, V.S., Investigations in the physical mechanism of sonocrystallization, doi:10.1016/j.ultsonch.2010.06.016, Copyright (2010), with permission from Elsevier.

Note: N.A., not applicable.

likely to create clusters of molecules (overcoming the barrier of rise in surface and volume free energy), which could later form embryo and nuclei. Shock waves can also cause disruption of large (>200 μm) crystals into smaller ones, which could grow separately. Thus, the overall influence of shock waves is expected to be in terms of rise in the nucleation rate in the medium.

Experimental and Simulation Results

Representative CSDs in four categories of experiments are shown in Figure 20.5. The entire data obtained from CSD measurements is given in Table 20.10. The SEM micrographs of KCl crystals in different categories of experiments are shown in Figure 20.6. Analysis of the CSD data with mixed suspension mixed product removal (MSMPR) model gave following important parameters: (1) number of nuclei (n_o), (2) nucleation rate (B), (3) growth rate of crystals (G), (4) dominant crystal size or the mass median diameter (d_D), and (5) span or width of the CSD. The entire calculated data is presented in Table 20.11. The following trends are evident from analysis of this data.

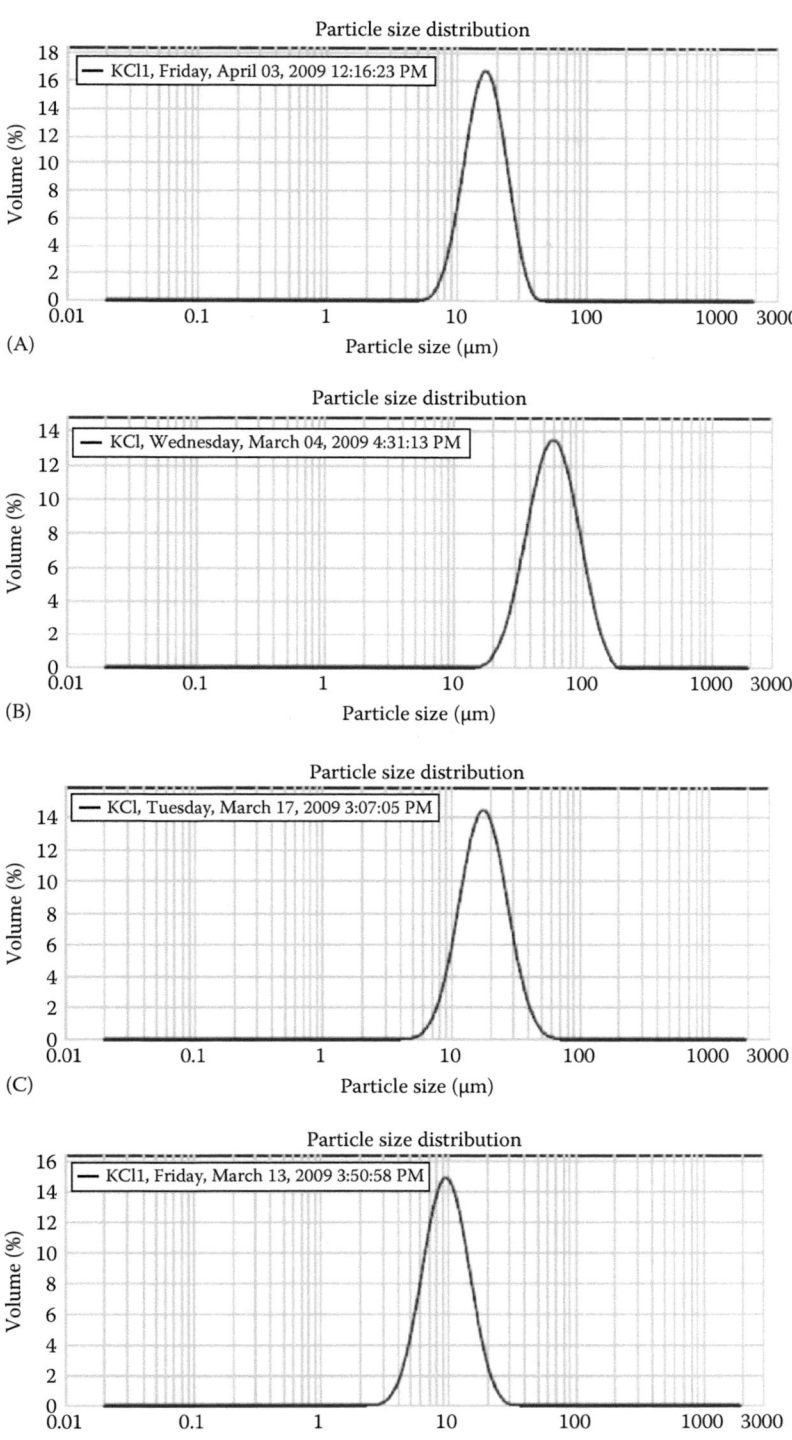

FIGURE 20.5 Representative results of measurement of crystal size distribution (CSD). (A) set 1A, (B) set 2A, (C) set 3B, and (D) set 4C. (Reproduced from *Ultrason. Sonochem.*, Nalajala, V.S. and Moholkar, V.S., Investigations in the physical mechanism of sonocrystallization. doi:10.1016/j.ultsonch.2010.06.016, Copyright (2010), with permission from Elsevier.)

TABLE 20.10
Results of the CSD Measurements

Category of Experiment	V_{FC} (−)	V_{TC} (m³)	Particle Size Distribution					
			$d(0.1)$ (µm)	$V_{FC,d(0.1)}$ (−)	$d(0.5)$ (µm)	$V_{FC,d(0.5)}$ (−)	$d(0.9)$ (µm)	$V_{FC,d(0.9)}$ (−)
1A	1.29E−02	9.03E−08	10.41	7.75	16.50	15.07	25.71	8.83
1B	1.18E−02	1.27E−07	12.15	8.00	18.53	16.38	28.21	7.73
1C	3.01E−02	2.11E−07	12.84	6.24	19.86	15.67	30.48	5.99
1D	2.43E−02	1.70E−07	12.03	7.75	18.80	15.54	29.29	8.28
2A	1.19E−01	8.31E−07	33.25	5.33	58.78	12.08	102.11	6.88
2B	5.35E−02	3.75E−07	13.45	2.64	40.95	8.22	82.83	5.33
2C	5.03E−02	3.52E−07	20.70	2.83	48.78	9.89	90.66	5.06
2D	3.47E−02	2.43E−07	16.77	3.63	33.48	10.77	61.39	5.01
3A	5.99E−02	4.19E−07	14.61	5.11	25.60	12.53	43.58	6.68
3B	3.38E−02	2.37E−07	10.40	6.53	17.72	13.02	30.05	4.79
3C	2.69E−02	1.88E−07	6.96	5.37	12.98	11.17	23.62	4.98
3D	1.27E−02	8.89E−08	5.56	4.18	11.27	10.07	20.77	5.77
3E	2.82E−02	1.97E−07	12.67	6.67	19.49	15.62	29.82	9.25
3F	2.15E−02	1.51E−07	10.24	7.55	16.76	13.96	27.53	5.91
3G	2.81E−02	1.97E−07	8.14	4.28	17.33	9.16	36.22	4.30
3H	1.72E−02	1.20E−07	6.06	5.41	11.28	11.13	21.02	4.87
4A	3.06E−02	2.14E−07	10.93	5.73	18.46	13.37	30.72	8.18
4B	2.85E−02	2.00E−07	9.45	7.72	16.55	12.60	28.45	6.16
4C	1.53E−02	1.07E−07	5.68	5.18	9.53	13.41	15.91	8.70
4D	4.00E−03	2.80E−08	4.77	3.85	9.97	10.20	18.58	5.88
4E	2.71E−02	1.90E−07	11.78	7.86	18.84	14.81	29.81	8.53
4F	2.31E−02	1.62E−07	9.68	5.74	15.82	14.12	25.88	7.58
4G	2.10E−02	1.47E−07	9.43	5.81	16.05	13.14	27.03	5.47
4H	7.80E−03	5.46E−08	5.22	5.52	9.80	11.01	17.73	7.20

Source: From *Ultrason. Sonochem.*, Nalajala, V.S. and Moholkar, V.S., Investigations in the physical mechanism of sonocrystallization, doi:10.1016/j.ultsonch.2010.06.016, Copyright (2010), with permission from Elsevier.

Notations: V_{FC}, total volume fraction of crystals in the crystallization mixture; V_{TC}, total volume of crystals in crystallization magma; $d(0.1)$, crystal size below which 10% of total crystals by weight lie; $d(0.5)$, crystal size below which 50% of total crystals by weight lie; $d(0.9)$, crystal size below which 90% of total crystals by weight lie; $V_{FC,d(0.1)}$, volume fraction of crystals with size $d(0.1)$; $V_{FC,d(0.5)}$, volume fraction of crystals with size $d(0.5)$; $V_{FC,d(0.9)}$, volume fraction of crystals with size $d(0.9)$.

Number of Nuclei, Nucleation Rate, and Growth Rate

1. Among all four categories of experiments, smallest number of nuclei (n_o) is seen for category 2. The growth rate (G) is the highest among all categories. Comparison of the values of these parameters among various sets in this category reveals little variation.
2. Values of n_o for category 1 are at least one order of magnitude higher than category 2. The growth rate (G) is slightly smaller but growth rate constant is higher than in category 2. Similar to category 2, values of n_o, B, and G show little variation among four sets in this category.
3. For category 3 experiments as well, nucleation showed marked rise—at least by an order of magnitude as compared to category 2. In addition, nucleation rate also showed a rise of about 40%–60%, as compared to category 1. However, growth rate showed reduction by at least an order of magnitude or higher than category 1. Comparing among various sets in this category, higher nuclei population and nucleation rate are seen for sets 3C and 3D and 3G and 3H, where sonicator probe of higher diameter (1 in.) was used.

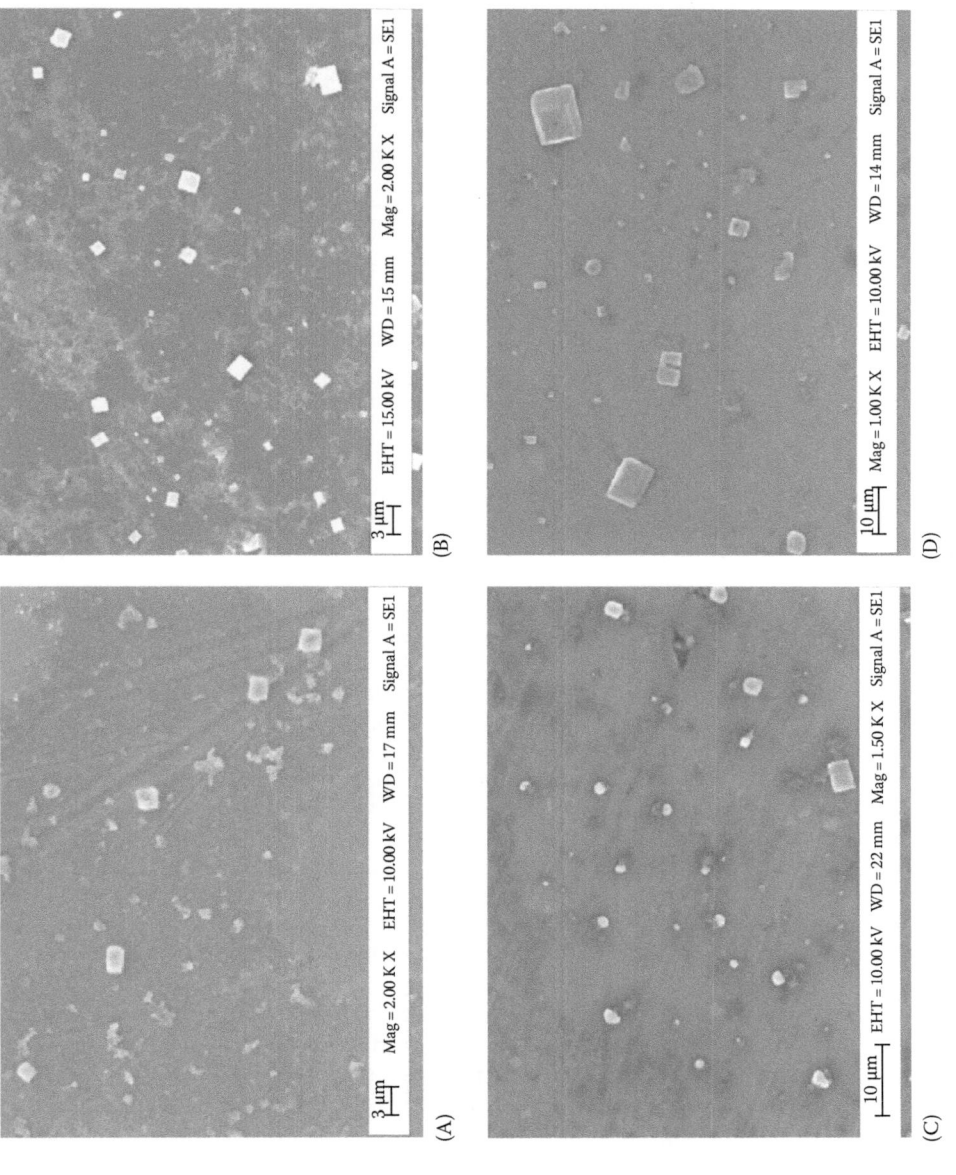

FIGURE 20.6 Representative SEM pictures of KCl crystals obtained in different experimental categories. (A) set 1A, (B) set 2D, (C) set 3C, and (D) set 4B. (Reproduced from *Ultrason. Sonochem.*, Nalajala, V.S., and Moholkar, V.S., Investigations in the physical mechanism of sonocrystallization, doi:10.1016/j.ultsonch.2010.06.016., Copyright (2010), with permission from Elsevier.)

TABLE 20.11
MSMPR Model Parameters for CSD

Category of Experiment	ln n_o (–)	n_o (m^{-4})	m (–)	G (ms^{-1})	B (m^{-3}s^{-1})	$d_{D,T}$ (m)	$d_{D,E}$ (m)	Span (–)
1A	17.55	4.18E+07	−17,1490	1.02E−08	0.43	1.75E−05	1.65E−05	0.93
1B	17.22	4.95E+07	−16,3423	1.07E−08	0.53	1.84E−05	1.85E−05	0.87
1C	17.89	5.85E+07	−154,354	1.14E−08	0.67	1.94E−05	1.99E−05	0.89
1D	17.84	5.61E+07	−153,951	1.14E−08	0.64	1.95E−05	1.88E−05	0.92
2A	15.63	6.16E+06	−45,819	3.83E−08	0.24	6.55E−05	5.88E−05	1.17
2B	16.00	8.90E+06	−67,839	2.59E−08	0.23	4.42E−05	4.09E−05	1.69
2C	15.62	6.06E+06	−55,442	3.16E−08	0.19	5.41E−05	4.88E−05	1.43
2D	15.95	8.44E+06	−81,310	2.16E−08	0.18	3.69E−05	3.35E−05	1.33
3A	17.73	5.01E+07	−105,860	1.66E−08	0.83	2.83E−05	2.56E−05	1.13
3B	18.48	1.06E+08	−181,104	9.69E−08	1.03	1.66E−05	1.77E−05	1.11
3C	18.88	1.59E+08	−226,874	7.73E−09	1.23	1.32E−05	1.30E−05	1.28
3D	18.27	8.64E+07	−240,060	7.31E−08	0.63	1.25E−05	1.13E−05	1.35
3E	17.56	4.24E+07	−134,208	1.31E−08	0.55	2.24E−05	1.95E−05	0.88
3F	18.25	8.44E+07	−188,966	9.28E−09	0.78	1.59E−05	1.68E−05	1.03
3G	17.84	5.59E+07	−158,969	1.10E−08	0.62	1.89E−05	1.73E−05	1.62
3H	18.84	1.53E+08	−258,992	6.77E−09	1.03	1.16E−05	1.13E−05	1.33
4A	17.73	5.01E+07	−141,195	1.24E−08	0.62	2.12E−05	1.85E−05	1.07
4B	18.54	1.13E+08	−186,978	9.38E−09	1.06	1.60E−05	1.65E−05	1.15
4C	18.84	1.51E+08	−257,107	6.82E−09	1.03	1.17E−05	9.53E−06	1.07
4D	17.44	3.75E+07	−266,464	6.58E−09	0.25	1.13E−05	9.97E−06	1.39
4E	17.93	6.14E+07	−152,214	1.15E−08	0.71	1.97E−05	1.88E−05	0.96
4F	17.96	6.30E+07	−169,163	1.04E−08	0.65	1.77E−05	1.58E−05	1.02
4G	18.08	7.08E+07	−186,905	9.39E−09	0.66	1.61E−05	1.61E−05	1.10
4H	18.31	8.96E+07	−272,495	6.44E−09	0.58	1.10E−05	9.80E−06	1.28

Source: From *Ultrason. Sonochem.*, Nalajala, V.S. and Moholkar, V.S., Investigations in the physical mechanism of sonocrystallization, doi:10.1016/j.ultsonch.2010.06.016, Copyright (2010), with permission from Elsevier.

Notation: n_o, population density of nuclei in the crystallization magma; m = quantity $(-1/Gt_r)$ where G is the growth rate of crystals and t_r is the time of crystallization; G, growth rate of the crystals; B, nucleation rate; $d_{D,T}$, dominant crystal size in the crystallization magma determined theoretically; $d_{D,E}$, dominant crystal size in the crystallization magma (determined with experimental measurement of CSD).

4. For category 4, nuclei population and nucleation rate were markedly higher, but the growth rate was smaller as in category 1. However, these values are comparable to those observed in various experimental sets in category 3.
5. Comparing among all experimental sets in any particular category, variation of the volume ratio of KCl solution to methanol does not show any remarkable effect on number of nuclei, rate of nucleation, and growth rate. We attribute this result to relative values of time of mixing and time of crystallization. As the latter was sufficient higher than former, all aliquots of KCl solution were thoroughly mixed in the magma, thus nullifying the effect of volume ratio.

Dominant Crystal Size and Span of CSD

1. Among all categories, the highest values of $d_{D,E}$ are seen for category 2. Same holds true for the span of CSD.
2. $d_{D,E}$ values for category 1 experiments are much smaller than category 2. Even the span of CSD gets lowered significantly. However, unlike category 2, little variation is seen in the values of $d_{D,E}$ and span of CSD among four sets of experiments in this category.

3. For category 3, values of $d_{D,E}$ are about 30%–50% smaller than category 1. However, it is noteworthy that the span of CSD is higher. A notable aspect is that large variation is seen in the $d_{D,E}$ values obtained in eight experimental sets in this category. For experimental sets where sonicator probe with 1 in. diameter was used (3C, D, and H), $d_{D,E}$ values are much smaller than other sets. The remarkable feature of CSD in this category is that $d_{D,E}$ is smaller but span of crystallization is greater as compared to the mechanically stirred system (category 1).
4. Smallest $d_{D,E}$ values were seen in category 4. Alike category 3, experimental sets employing sonicator probe of 1 in. diameters (viz., sets 4C, D, H) yielded smaller $d_{D,E}$ values. The span of crystallization also shows marginal reduction.
5. Parity of experimental ($d_{D,E}$) and theoretical ($d_{D,T}$) dominant crystal sizes also shows interesting features. In category 1, $d_{D,E}$ and $d_{D,T}$ match very closely (~99%), while in categories 3 and 4, match between $d_{D,E}$ and $d_{D,T}$ is somewhat smaller (~94%).

Simulation Results

The summary of the results of simulations of air and argon bubbles is given in Tables 20.12 and 20.13. It could be inferred that physical and chemical effects of dynamics of both bubbles are almost similar. Both of these bubbles create mild microturbulence (or microconvection) with velocities $\sim V_{turb} = 2\,cm\,s^{-1}$. However, the acoustic (or shock) wave emitted by both kinds of bubbles has relatively much high pressure amplitude $\sim P_{AW} = 11$ bar. The collapse conditions, that is, temperature and pressure peaks obtained at the transient collapse are also similar. Moreover, the extent of entrapment of methanol vapor molecules is also practically the same in both bubbles. The dissociation of entrapped methanol molecules at transient collapse is seen to yield mainly molecular (and not radical) species.

Discussion and Analysis

Correlation of the experimental and simulation results brings in following explanation for the trends in parameters n_o, B, G, $d_{D,E}$, and span of CSD.

1. In category 2, due to absence of any gross movement of bulk liquid, the movement of KCl crystal blocks was also restricted, which resulted in lower population of nuclei and nucleation rate as compared to other three categories. The precipitating material was distributed among smaller number of nuclei. Moreover, this distribution was likely to be nonuniform due to absence of convection in the medium. As a result, both $d_{D,E}$ and span of CSD are high.

TABLE 20.12
Results of Simulations of Cavitation Bubble Dynamics: Conditions at the Transient Collapse of the Cavitation Bubble

	Bubble Type	
Parameter	Argon (Category 4)	Air (Category 3)
T_{max} (K)	810	800
P_{max} (bar)	1455	1470
N_{MeOH}	3.88E+011	3.16E+011
V_{turb} (m s^{-1})	0.02	0.02
P_{AW} (bar)	11.32	11.72

Source: From *Ultrason. Sonochem.*, Nalajala, V.S. and Moholkar, V.S., Investigations in the physical mechanism of sonocrystallization, doi:10.1016/j.ultsonch.2010.06.016, Copyright (2010), with permission from Elsevier.

TABLE 20.13
Results of Simulations of Cavitation Bubble Dynamics: Equilibrium Composition of the Cavitation Bubble Contents in Methanol at Collapse Conditions

Species	Air Bubble	Argon Bubble
	\multicolumn{2}{c}{Equilibrium Composition (Mole Fraction)}	
CH_4	7.32E–01	7.21E–01
H_2O	2.28E–01	2.28E–01
H_2	2.06E–02	2.07E–02
CO_2	1.82E–02	1.72E–02
C_2H_6	3.16E–04	8.16E–05
CO	4.11E–04	7.33E–05
CH_3OH	1.55E–07	1.63E–07
C_2H_4	1.24E–07	1.32E–07
CH_3COOH	2.87E–08	1.99E–08
HCOOH	1.39E–08	1.56E–08
H_2CO	2.46E–08	1.85E–08

Source: From *Ultrason. Sonochem.,* Nalajala, V.S. and Moholkar, V.S., Investigations in the physical mechanism of sonocrystallization, doi:10.1016/j.ultsonch.2010.06.016, Copyright (2010), with permission from Elsevier.

Notation: T_{max}, temperature peak reached in the bubble at transient collapse; P_{max}, pressure peak reached in the bubble at transient collapse; N_{MeOH}, number of methanol molecules trapped in the bubble at transient first collapse; V_{turb}, microturbulence velocity generated by the cavitation bubbles (calculated as the average of the positive velocity, that is, directed away from the bubble center and negative velocity, that is, directed toward the bubble center); P_{AW}, pressure amplitude of the acoustic or shock wave emitted by the bubble.

2. In category 1, intense agitation generated high level of bulk convection in the medium, which is expected to be volumetrically uniform. As a result, population density of nuclei, rate of nucleation, and growth rate increased significantly. Rise in nuclei population reduced $d_{D,E}$. High convection also assisted uniform distribution of precipitating material among nuclei causing uniform growth of all crystals, and thus, span of the CSD also shows reduction.

3. Sharp rise in the nuclei population and rate of nucleation in category 3 is attributed to the shock waves generated by cavitation bubbles. However, due to their discrete or intermittent nature, the shock waves are not likely to contribute to the growth of the nuclei. The velocity of microconvection, however, is quite low (~2 cm s^{-1}) and localized. Thus, the overall bulk convection in category 3 is expected to be much lower than category 1. This results in reduction in overall growth rate of the crystals. The net outcome of all these factors is that $d_{D,E}$ for category 3 is lesser, yet span of the CSD is higher as compared to category 1.

The use of sonicator probe of higher diameter (1 in.) results in increase in the "active" insonated zone or volume in methanol, causing rise in cavitation intensity in the medium. A direct consequence of this is rise in population of nuclei and the nucleation rate, causing reduction in $d_{D,E}$. However, the rise in the bulk convection level is not expected to be proportionate, and hence, span of the CSD does not show a concomitant trend.

4. In category 4 experiments, contribution of cavitation bubbles to the convection (and hence rate of nucleation and nuclei population in the medium) is similar to that in category 3 experiments. However, as number density of cavitation bubbles is higher (due to external

seeding of nuclei), the overall cavitation activity in the medium is expected to be higher than category 3. Second, sparging of gas through methanol also induces bulk movement of liquid. Thus, the bulk convection levels are greater than in category 3. The overall consequence is that the nucleation rate is higher as compared to category 3, and these nuclei grow relatively more uniformly. Consequently, the values of $d_{D,E}$ and span of CSD are smaller than those in category 3.

The above case study is again a vivid example of the mechanistic aspects of a physical process induced and accelerated by ultrasound and cavitation. The principal physical effect of ultrasound and cavitation is generation of strong localized convection in the system. The two components of the convection, viz., microturbulence (or microconvection) and shock waves, have different impact on the crystallization process due to their nature (viz., discrete vs. continuous). The shock waves are expected to increase the nucleation rate in the system, and the microtubulence governs the growth of these nuclei. However, effect of the former is found to be more marked than the latter. This is evident from the result that nucleation rate shows an order of magnitude rise with sonication, while the growth rate (and hence the dominant crystal size) reduces with sonication as compared to the mechanically agitated system, where uniform velocity field is expected to prevail over the volume of the crystallization mixture. The ultimate manifestation of these effects is that as compared against the CSD of mechanically agitated crystallization system, the dominant crystal size of the CSD of sonocrystallization systems is smaller but the span of CSD is larger.

CONCLUDING VIEWS

In this chapter, we have attempted to describe the mechanistic aspects of the physical and chemical processes that are benefited by ultrasound irradiation with help of two case studies, viz., sonochemical degradation of pollutants with different physicochemical characteristics and sonocrystallization. As note earlier, ultrasound manifests its effects through phenomena of cavitation bubble dynamics. Our studies have shed light on nature or mechanism of the action of bubbles on the physical and chemical process. The methodologies adopted in the case studies presented above could form a general framework that can be extended to any other physical and chemical system. Establishment of the exact physical mechanism of a sonophysical or sonochemical process is vital to effective design and scale-up of the process. We hope that the framework resulting from our studies would fulfill this discrepancy.

REFERENCES

Adewuyi, Y. G. 2001. Sonochemistry: Environmental science and engineering application. *Industrial and Engineering Chemistry Research,* 40:4681–4715.

Ashokkumar, M. and Grieser, F. 1999. Ultrasound assisted chemical processes. *Reviews in Chemical Engineering,* 15(1):41–83.

Benjamin, T. B. and Ellis, A. T. 1966. The collapse of cavitation bubbles and the pressure thereby produced against solid boundaries. *Philosophical Transactions of Royal Society of London Series A,* 260:221–240.

Blake, J. R., Hooton, M. C., Robinson, P. B., and Tong, R. P. 1997. Collapsing cavities, toroidal bubbles and jet impact. *Philosophical Transactions of Royal Society of London Series A,* 355:537–550.

Blake, J. R., Taib, B. B., and Doherty, G. 1986. Transient cavities near boundaries. Part 1. Rigid boundary. *Journal of Fluid Mechanics,* 170:479–497.

Blake, J. R., Taib, B. B., and Doherty, G. 1987. Transient cavities near boundaries. Part 2. Free surface. *Journal of Fluid Mechanics,* 181:197–212.

Brenner, M., Hilgenfeldt, S., and Lohse, D. 2002. Single-bubble sonoluminescence. *Reviews of Modern Physics,* 74:425–484.

Colussi, A. J. and Hoffmann, M. R. 1999. Vapor supersaturation in collapsing bubbles: Relevance to mechanisms of sonochemistry and sonoluminescence. *Journal of Physical Chemistry A,* 103:11336–11339.

Colussi, A. J., Weavers, L. K., and Hoffmann, M. R. 1998. Chemical bubble dynamics and quantitative sonochemistry. *Journal of Physical Chemistry A,* 102(35):6927–6934.

Doktycz, S. J. and Suslick, K. S. 1990. Interparticle collisions driven by ultrasound. *Science,* 247:1067–1069.
Eames, I. W., Marr, N. J., and Sabir, H. 1997. The evaporation coefficient of water: A review. *International Journal of Heat and Mass Transfer,* 40:2963–2973.
EPRI Report. April 1998. Ultrasonic Chemistry—A Survey and Energy Assessment. Electric Power Research Institute, No. TR-109974, Palo Alto, CA, USA.
Eriksson, G. 1975. Thermodynamic studies of high temperature equilibria—XII: SOLGAMIX, a computer program for calculation of equilibrium composition in multiphase systems. *Chemica Scripta,* 8:100–103.
Flynn, H. G. 1964. Physics of acoustic cavitation in liquids. In *Physical Acoustics,* W. P. Mason (Ed.), pp. 57–172. New York: Academic Press.
Fujikawa, S. and Akamatsu, T. 1980. Effects of the non-equilibrium condensation of vapor on the pressure wave produced by the collapse of a bubble in a liquid. *Journal of Fluid Mechanics,* 97:481–512.
Gogate, P. R. and Pandit, A. B. 2004a. A review of imperative technologies for wastewater treatment I: Oxidation technologies at ambient conditions. *Advances in Environmental Research,* 8:501–551.
Gogate, P. R. and Pandit, A. B. 2004b. A review of imperative technologies for wastewater treatment II: Hybrid methods. *Advances in Environmental Research,* 8:553–597.
Gong, C. and Hart, D. P. 1998. Ultrasound induced cavitation and sonochemical yields. *Journal of Acoustical Society of America,* 104:2675–2682.
Grossmann, S., Hilgenfeldt, S., Zomack, M., and Lohse, D. 1997. Sound radiation of 3 MHz driven gas bubbles. *Journal of Acoustical Society of America,* 102:1223–1227.
Hua, I., Hochemer, R. H., and Hoffmann, M. R. 1995. Sonochemical degradation of *p*-nitrophenol in a parallel plate near field acoustic processor. *Environmental Science Technology,* 29:2790–2796.
Kamath, V., Prosperetti, A., and Egolfopoulos, F. N. 1993. A theoretical study of sonoluminescence. *Journal of Acoustical Society of America,* 94:248–260.
Keil, F. J. and Swamy, K. M. 1999. Reactors for sonochemical engineering—Present status. *Reviews in Chemical Engineering,* 15(2):85–155.
Kotronarou, A., Mills, G., and Hoffmann, M. R. 1991. Ultrasonic irradiation of p-nitrophenol in aqueous solution. *Journal of Physical Chemistry,* 95(9):3630–3638.
Krishnan, S. J., Dwivedi, P., and Moholkar, V. S. 2006. Numerical investigation into the chemistry induced by hydrodynamic cavitation. *Industrial and Engineering Chemistry Research,* 45:1493–1504.
Kumar, A., Kumaresan, T., Pandit, A. B., and Joshi, J. B. 2006. Characterization of flow phenomena induced by ultrasonic horn. *Chemical Engineering Science,* 61:7410–7420.
Kwak, H.-Y. and Na, J.-H. 1996. Hydrodynamic solutions for sonoluminescing gas bubble. *Physical Review Letters,* 77:4454–4457.
Kwak, H. Y. and Na, J. H. 1997. Physical processes for single bubble sonoluminescence. *Journal of Physical Society of Japan,* 66:3074–3083.
Kwak, H. Y. and Yang, H. 1995. An aspect of sonoluminescence from hydrodynamic theory. *Journal of Physical Society of Japan,* 64:1980–1992.
Lauterborn, W. and Hentschel, W. 1985. Cavitation bubble dynamics studied by high speed photography and holography. Part 1. *Ultrasonics,* 23:260–267.
Lauterborn, W. and Hentschel, W. 1986. Cavitation bubble dynamics studied by high speed photography and holography. Part 2. *Ultrasonics,* 24:59–64.
Leighton, T. G. 1994. *The Acoustic Bubble.* San Diego: Academic Press.
Mahulkar, A. V., Bapat, P. S., Pandit, A. B., and Lewis, F. M. 2008. Steam bubble cavitation. *AIChE Journal,* 54(7):1711–1724.
Moholkar, V. S., Warmoeskerken, M. M. C. G., Ohl, C. D., and Prosperetti, A. 2004. The mechanism of mass transfer enhancement in textile with ultrasound. *AIChE Journal,* 50:58–64.
Mason, T. J. and Lorimer, J. P. 2002. *Applied Sonochemistry: The Uses of Power Ultrasound in Chemistry and Processing.* Coventry: Wiley-VCH.
Moss, W. C., Young, D. A., Harte, J. A., Levalin, J. L., Rozsnyai, B. F., Zimmerman, G. B., and Zimmerman, I. H. 1999. Computed optical emissions from sonoluminescing bubbles. *Physical Review E,* 59:2986–2992.
Nalajala, V. S. and Moholkar, V. S. 2011. Investigations in the physical mechanism of sonocrystallization. *Ultrasonics Sonochemistry,* 18:345–355.
Neppiras, E. A. 1980. Acoustic cavitation. *Physics Reports,* 61:159–251.
Ohl, C. D., Lindau, O., and Lauterborn, W. 1998. Luminescence from spherically and aspherically collapsing laser induced bubbles. *Physical Review Letters,* 80:393–396.
Ohl, C. D., Phillip, A., and Lauterborn, W. 1995. Cavitation bubble collapse studied at 20 million frames per second. *Annual Physik,* 4:26–34.

Pandit, A. B. and Moholkar, V. S. 1996. Harness cavitation to improve processing. *Chemical Engineering Progress,* 96:57–69.

Pecha, R. and Gompf, B. 2000. Microimplosions: Cavitation collapse and shock wave emission on a nanosecond time scale. *Physical Review Letters,* 84:1328–1330.

Phillip, A. and Lauterborn, W. 1998. Cavitation erosion by single laser produced bubbles. *Journal of Fluid Mechanics,* 361:75–116.

Plesset, M. S. and Chapman, R. B. 1971. Collapse of an initially spherical cavity in the neighborhood of a solid boundary. *Journal of Fluid Mechanics,* 47:283–290.

Prasad Naidu, D. V., Rajan, R., Kumar, R., Gandhi, K. S., Arakeri, V. H., and Chandrasekaran, S. 1994. Modeling of a batch sonochemical reactor. *Chemical Engineering Science,* 49(6):877–888.

Press, W. H., Teukolsky, S. A., Flannery, B. P., and Vetterling, W. T. 1992. *Numerical Recipes* (2nd Edn.). New York: Cambridge University Press.

Shah, Y. T., Pandit, A. B., and Moholkar, V. S. 1999. *Cavitation Reaction Engineering.* New York: Plenum Press.

Sivasankar, T. and Moholkar, V. S. 2009. Physical insights into the sonochemical degradation of recalcitrant organic pollutants with cavitation bubble dynamics. *Ultrasonics Sonochemistry,* 16:769–781.

Sivasankar, T., Paunikar, A. W., and Moholkar, V. S. 2007. Mechanistic approach to enhancement of the yield of a sonochemical reaction. *AIChE Journal,* 53(5):1132–1143.

Sochard, S., Wilhelm, A. M., and Delmas, H. 1997. Modeling of free radicals production in a collapsing gas–vapor bubble. *Ultrasonics Sonochemistry,* 4:77–84.

Storey, B. D. and Szeri, A. J. 2000. Water vapor, sonoluminescence and sonochemistry. *Proceedings of Royal Society of London Series A,* 456:1685–1709.

Storey, B. D. and Szeri, A. J. 2001. A reduced model of cavitation physics for use in sonochemistry. *Proceedings of Royal Society of London Series A,* 457:1685–1700.

Suslick, K. S. 1988. *Ultrasound: Its Physical, Chemical and Biological Effects.* New York: VCH.

Thompson, L. H. and Doraiswamy, L. K. 1999. Sonochemistry: Science and engineering. *Industrial and Engineering Chemistry Research,* 38:1215–1249.

Toegel, R., Gompf, B., Pecha, R., and Lohse, D. 2000. Does water vapor prevent upscaling sonoluminescence? *Physical Review Letters,* 85:3165–3168.

Toegel, R. and Lohse, D. 2003. Phase diagrams for sonoluminescing bubbles: A comparison between experiment and theory. *Journal of Chemical Physics,* 118(4):1863–1875.

Vogel, A., Lauterborn, W., and Timm, R. 1989. Optical and acoustic investigations of the dynamics of laser induced cavitation bubbles near a solid boundary. *Journal of Fluid Mechanics,* 206:299–338.

Willard, G. W. 1953. Ultrasonically induced cavitation in water: A step-by-step process. *Journal of the Acoustical Society of America,* 25:669.

Yasui, K. 1997a. Alternative model for single-bubble sonoluminescence. *Physical Review E,* 56:6750–6760.

Yasui, K. 1997b. Chemical reactions in a sonoluminescing bubble. *Journal of Physical Society of Japan,* 66:2911–2920.

21 Ultrasound-Assisted Industrial Synthesis and Processes

*Cezar Augusto Bizzi, Edson Irineu Müller,
Érico Marlon de Moraes Flores, Fábio Andrei Duarte,
Mauro Korn, Matheus Augusto Gonçalves Nunes,
Paola de Azevedo Mello, and Valderi Luiz Dressler*

CONTENTS

Introduction 535
Food Processing and Preservation 537
    Effect on Living Cells 537
    Drying and Evaporation 539
    Emulsification and Mixing 540
    Crystallization and Freezing 541
    Other Applications 542
Applications in Biotechnology 542
    Microbial Cell Disruption 543
    Gene Transfer 543
    Biosensors 545
Synthesis 546
Textile Industry 549
Extraction 551
Sonocrystallization 553
Metal and Plastic Welding 562
Crude Oil Industry 563
Other Processes: Filtration, Separation, and Cleaning 565
Environmental Remediation Using Ultrasound 567
    Water and Wastewater Treatment 567
    Treatment of Biological and Chemical Contaminants in Water 568
    Soil and Sediments Remediation 571
        Soil Washing 571
        Air 572
Final Considerations 572
References 573

INTRODUCTION

Ultrasound (US) has been used since its discovery at the beginning of twentieth century for a variety of purposes in diverse areas such as chemical synthesis, medicine, and engineering processing. Besides considering its wide-ranging use and the recent developments, it is also necessary to consider US as a relatively new scientific field once there are increasing developments in the area of comprehension of ultrasonic effect and its characteristics. Once the great potential of US has been recognized for a variety of industrial chemical processes, new developments for large-scale processing have been proposed. In fact, acoustic cavitation is the key point for sonochemical processes, and the ability to increase or limit this effect is one of the main challenges for industrial applications of US. In spite of the fact that US has been extensively explored in laboratory scale, consolidated industrial applications are not so numerous (Thompson and Doraiswamy, 1999; Mason, 2000, 2003; Cravotto and Cintas, 2006). In addition, it is important to point out that when an industrial application is in progress it is generally not available in literature due to technological challenges and the market value for process developments. The optimization and design of sonochemical apparatus for large-scale applications are continuously in advancement. There is an extensive research for the development of new systems and large-scale applications have been developed using different designs with several transducers. In spite of the extensive effort on this issue, there are only relatively few examples of US applications in industrial process. The main causes lie in the complicated design of efficient reactors making difficult to reproduce the same conditions developed in laboratory scale (Cintas et al., 2010).

Table 21.1 summarizes the main areas related to US applications: (i) organic and inorganic chemistry/synthesis, as for modification and synthesis of polymers; preparation of organometallic compounds (Vajnhandl and Le Marechal, 2005); preparation of activated metals by reduction of metal salts; generation of activated metals; precipitation of metal oxides, catalysts (Cr, Mn, Co) (Vajnhandl and Le Marechal, 2005), crystallization (Mason, 2000), sonoelectrochemical synthesis (Mason, 2003), and preparation of nanomaterials (Mason, 2003); (ii) food technology, emulsification, mixing, blending, extraction, crystallization, foam destruction, degassing of liquids, particle aerosol precipitation, oxidation processes, influencing enzyme activity, sterilization, and ultrasonic airborne drying (temperature-sensitive powders–foodstuffs) (Mason, 2003; Vajnhandl and Le Marechal, 2005); (iii) environmental applications including sonolysis of organic pollutants in water and degradation of contaminants (Mason, 2000; Mason, 2003; Vajnhandl and Le Marechal, 2005); (iv) pharmaceutical industry, as for biological cell disintegration (extraction of active antigens for making vaccines and in the industry of lipids enzymes and viruses) and ultrasonic airborne drying (temperature-sensitive powders–pharmaceuticals) (Vajnhandl and Le Marechal, 2005); (v) medicine imaging, as nuclear magnetic resonance, diagnostic, enzyme activation, therapeutic, induction of thermolysis in tissues (cancer treatment), cleaning, atomization in inhalation therapy, air humidification (Mason, 2000; Vajnhandl and Le Marechal, 2005); and (vi) others such as welding (plastic and metal) (Mason, 2000), cutting (Mason, 2000), soldering, cleaning (Mason, 2000); deburring, erosion test, degassing of melts, spray pyrolysis, treatment of solid surfaces, dispersion, preparation of colloids, fuel atomization, electroless plating, wetting and impregnation, sieving, filtration

TABLE 21.1
Industrial Uses of US

Industrial Area	Examples of Applications
Biology	Homogenization and rupture of cell walls to content release
Engineering	Ultrasound-assisted drilling, grinding and cutting (especially with applications in glass and ceramic materials), and welding (plastic and metals)
Environmental	Water remediation and wastes treatment assisted by ultrasound
Geology	Eco ranging at sea (SONAR)
General industry	Dispersion of pigments and solids, cleaning and degreasing by immersion, filtration and metal casting
Dentistry	Cleaning and drilling of teeth
Medicine	Ultrasonic imaging and physiotherapy

Source: Adapted from Mason, T.J. and Lorimer, J.P., *Applied Sonochemistry: Uses of Power Ultrasound in Chemistry and Processing*, 2002, Wiley-VCH Verlag GmbH, Weinheim, Germany.

(Mason, 2000) and micromanipulation, extraction (Mason, 2000), and lithography (Mason, 2000; Mason, 2003; Vajnhandl and Le Marechal, 2005).

In the following sections, the main uses of US in industrial processes available in the literature from 1990 to 2010 are presented. Applications in medicine and dentistry as well as SONAR technology are not presented.

FOOD PROCESSING AND PRESERVATION

Due to the recent developments in US generation systems, as well as the increased understanding of cavitation phenomena, interest has increased in recent years related to the use of power US as an alternative for food processing and preservation. Equipment working at frequencies higher than 100 kHz present applications in stimulation of living cell activity, surface cleaning of food ultrasonically assisted extraction, crystallization, emulsification, filtration, drying, and freezing process as well as tenderization of meat. High-energy US (frequency ranging from 18 to 100 kHz) has been applied for degassing of liquid foods, for the induction of oxidation/reduction reactions, for extraction of enzymes and proteins, for enzyme inactivation, and for induction of nucleation for crystallization. Additionally, high-energy US combined with heat and/or pressure has been applied to inactivate thermal-resistant enzymes. Besides, data exist regarding the impact of US on the inactivation of microorganisms in conjunction with chemical antimicrobials, with heat alone or combined with moderate pressure (Knorr et al., 2004; Feng et al., 2008).

Some studies have been performed to elucidate the effect of US on the mechanism for biomolecule and cell inactivation. The most acceptable purpose was related to the collapse of transient bubbles in the liquid medium and therefore by mechanical and chemical effects of cavitation (Knorr et al., 2004; Feng et al., 2008; Jiranek et al., 2008). The formation of shock waves can cause cell envelope rupture facilitating the sonochemical reactions (Feng et al., 2008). Among the effects caused by US energy, the bactericidal properties could be mainly related to chemical effects including also the formation of OH^- and H^+ species and hydrogen peroxide (McClements, 1995).

Ultrasound energy has been used in food industry, mainly for processing and preservation. It could be observed during the process of dairy ingredients in order to improve functional properties by applying a high-intensity and low-frequency US. Several experimental parameters have been optimized for processing large volumes of whey- and casein-based dairy products in pilot-scale ultrasonic reactors. In this case, samples were sonicated at 20 kHz employing a radial ultrasonic transducer of 1 or 4 kW, with flow rates ranging from 0.2 to 6 L min^{-1}. Back pressure was maintained between ambient and 2 bar. The sonication of solutions up to 2.4 min led to a significant reduction in the viscosity of solution containing 18%–54% of solids. The viscosity of aqueous dairy ingredients treated with US was reduced from 6% to 50%, depending on acoustic power and contact time applied. A notable improvement in the gel strength of sonicated and heat-coagulated dairy systems was also observed. When sonication was combined with a previous heating treatment (80°C for 1 min or 85°C for 30 s), the thermal stability of the dairy ingredients containing whey proteins was significantly improved. On the whole, the sonication procedures described for processing dairy ingredients may be used to improve process efficiency, improve throughput, and produce value-adding ingredients in the dairy industry without changing the sample gelling properties and heat stability (Zisu et al., 2010).

Effect on Living Cells

In order to minimize the bacterial load of a product, conventional methods of inactivation usually involve thermal treatment, which often results in the formation of undesirable flavors and loss of nutrients. Modern preservation techniques have been developed for food industries in order to increase the safety of food products, thereby eliminating microbial activity with significant reduction of required heat. Cavitation caused by changes in pressure created by the ultrasonic waves is responsible for killing the bacteria in food products treated by US energy. The US effects on microorganisms are mainly the disruption of cell membrane, localized heating, and free radicals production. In combination with heat, the US energy can accelerate the rate of sterilization of foods, thus decreasing both the duration and intensity of thermal treatment and consequently possible food changes. The minimization of flavor loss, especially in sweet juices, and better homogeneity and savings of energy could be mentioned among the main advantages of US over heat pasteurization (Piyasena et al., 2003). In addition, it is possible to obtain different effects on living cells by applying US energy in diverse work conditions (Mason et al., 1996).

There is a number of applications using a low power US, which represents significantly milder cavitational effects (stable or noninertial cavitation) (Gogate and Kabadi, 2009). In this case, it is possible to increase the production of food products through the enhancement of efficiency of whole cells without disrupting the cell walls (Mason et al., 1996).

US energy has been applied for yogurt fermentation. The experiments have been performed using a US generator (working at 90, 255, and 450 W power at 20 kHz) with a 13-mm-diameter probe under exposure times from 1 to 10 min. A reduction of the total fermentation time by 0.5 h was obtained and increased water holding capacity and viscosity were observed using inoculated milk sample (150 g) blended with 30 g of yogurt starter using US treatment at 15°C. The increase in the US amplitude level before inoculation could mean a significant improvement of water holding capacity, viscosity, and syneresis effect (Wu et al., 2001). The syneresis characterized by the collection of whey on the yogurt surface has the same effect that leads the curd production in the milk processing due to the sudden removal of hydrophilic micropeptides.

A high-intensity US in a continuous-flow mode can be applied in order to inactivate bacteria in raw cow milk. In this case, sonicator has been equipped with a tip of 12.7 mm diameter and worked at a fixed frequency of 20 kHz. The mechanical vibration was transmitted to a titanium alloy disruptor horn. Higher output intensity level implies higher wave amplitude (maximum 120 μm), which could result in more intense cavitation action and greater cell disruption. The maximum available ultrasonic power delivered to the horn was 150 W. Using a flow rate from 0.66 to 3.0 L h^{-1} and temperature from 76.1°C to 33.1°C, respectively, US can be used to preserve milk samples. The ultrasonic-based procedure was more energy consuming and the necessary time was the same in comparison to the conventional heating procedure. However, some advantages could be mentioned in relation to the conventional techniques, such as the wall temperature which was equal to or lower than the liquid temperature, and could result in less deposition of proteins and minerals than in a continuous-flow heat exchanger. The milk homogenization by US with an additional increase of cheese yield and a size reduction of the fat globule was observed when the working temperature was close to 62°C. These results make the continuous-flow ultrasonic milk treatment a promising preservation technique, especially under mild conditions, resulting in a comparable microbiological quality in relation to the conventional treatment (Villamiel and Jong, 2000).

The inactivation and potential subsequent growth of microorganisms in orange juice was performed by using US energy. Process was carried out in a semi-industrial unit working in batch (static) or in continuous-flow (dynamic) conditions at changeable temperature. System could operate at low frequency (23 kHz and power level of 150, 300, and 600 W) and/or at high frequency (500 kHz with power level of 120 and 240 W). Ultrasound was applied in all cases with operation at 60% of maximum power and using a stainless steel high-amplitude horn. It was observed that limited level of microbial inactivation was obtained by selecting the batch ultrasonic treatment to 500 kHz/240 W for 15 min. Stabilization of microbial growth was observed in the substrate following 14 days of storage at both refrigeration (5°C) and higher (12°C) temperatures after batch US treatment. However, after continuous US treatment at flow rate of 3000 L h^{-1} negligible reduction of microbial counts was obtained which could be related with the microbial resistance caused by juice pulp. During the experimental procedure, it was not observed as a negative effect on orange juice color by the proposed US treatments. The proposed procedure could reduce the required processing temperatures, the processing times, or both pasteurization processes of liquid foods to achieve the same microbial lethality values that could be obtained with conventional heating methods. One of the main disadvantages observed during US-assisted process was the higher total energy input required in comparison to the conventional heating (Valero et al., 2007).

Recent regulations from the U.S. Food and Drug Administration have required processors to achieve a 5-log reduction in the numbers of the most resistant pathogens in their finished products. Despite conventional thermal processing of fruit juices remaining the most widely adopted technology for shelf-life extension and preservation of fruit juice, power US alone, or in combination with pressure or heat, is capable of achieving a desired 5-log reduction in foodborne pathogens related to fruit juices. However, consumer demand for nutritious foods, which are minimally and naturally processed, has led to an interest in nonthermal technologies (Tiwari et al., 2009a). The application of power US has been reported to be effective against foodborne pathogens found in orange juice (Valero et al., 2007), guava juice (Cheng et al., 2007), apple cider, and milk (D'Amico et al., 2006).

The nutritional quality of orange juice is primarily related to the ascorbic acid content (Zerdin et al., 2003). However, ascorbic acid is thermolabile and highly sensitive to various processing conditions, such as US treatment (Korn et al., 2002). The mechanism of vitamin C degradation follows aerobic and/or anaerobic pathways and depends on several processing conditions (Vieira et al., 2000). Ascorbic acid degradation kinetics of sonicated orange juice (acoustic energy densities ranging from 0.30 to 0.81 W mL^{-1} and treatment times of 2–10 min) were evaluated during storage and compared to orange juices thermally pasteurized. The sonication treatment led to enhancement of ascorbic acid retention in orange juice during storage (Tiwari et al., 2009b).

Sonication of tomato juices with ultrasonic processor at a frequency of 20 kHz and power of 1500 W for 2–10 min and pulse of 5 s was used to evaluate yeast inactivation. Hunter color values, pH, Brix, titratable acidity, and ascorbic acid concentration were also evaluated. No significant differences in pH, Brix, or titratable acidity were observed. However, sonication was found to have a significant effect on juice color, ascorbic acid content, and yeast inactivation. The results presented in this study indicate that power US treatment has potential for the inactivation of key spoilage microorganism relevant to tomato juice processing (Adekunte et al., 2010).

DRYING AND EVAPORATION

A crescent demand on dehydrated food brings the necessity for the development of new processes. Conventional dehydration methods are mainly based on hot-air drying, which has a widespread use but can generate a deterioration quality of the final product. It could represent an undesired flavor, color composition, vitamin degradation, and the loss of essential amino acids. In this way, US may be considered a promising choice once it permits the removal of moisture content of solids without producing a liquid phase change and can be applied to heat-sensitive food materials (Fuente-Blanco et al., 2006). These effects were observed because cavitation phenomena can strongly affect the degree of heat transfer. No cavitation occurs close to the boiling point of a liquid and acoustic streaming is the main factor related to the heat transfer rates, whereas at lower temperatures the effect of ultrasonic vibration is observed through violent motion of cavitation bubbles (Kim et al., 2004). In addition, when a high-intensity ultrasonic wave is directed to the solid material to be dried, the pressure changes in solids and this can be assimilated in the same way to a sponge squeezed and repeatedly released (Mulet et al., 2003). As a result, microscopic channels suitable for fluid movement could be created by reducing the diffusion boundary layer and increasing the convective mass transfer in foodstuff (Mulet et al., 2003; Fuente-Blanco et al., 2006). Moreover, ultrasonically enhanced drying can be carried out at lower temperatures than conventional methodology which reduce the probability of oxidation or degradation in food materials (Mason et al., 1996).

Some food products (e.g., fruits and vegetables) are sensitive to heat, leading to structural changes after thermal dehydration. Alternatively, a high-power US with rectangular plate transducer, working at 20 kHz, and power capacity of about 100 W has been used for food samples dehydration. Drying process can also be done in a chamber with an ultrasonic transducer located at the

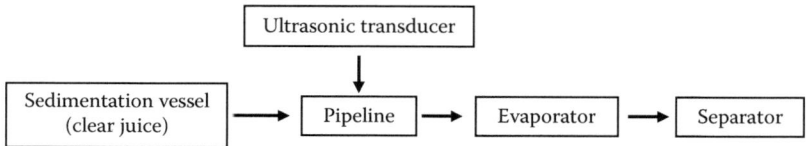

FIGURE 21.1 Flow chart of the structure and installation position of ultrasonic equipment at a sugar factory. (Adapted from Hu, A. et al., *Ultrason. Sonochem.*, 13, 329, 2006.)

upper part. The US is applied in combination with hot air to accelerate drying at low temperature, thereby preserving the integrity of the food products. The results obtained under sonication up to 100 W showed a direct increase of drying effect (water loss) with the acoustic power (Fuente-Blanco et al., 2006).

Additionally, US can present many effects resulting in improvements on evaporation process. The use of US was evaluated in industrial application for scale control in sugar industry. The device was made up of ultrasonic generator (20 kHz and 2000 W) linked to ultrasonic transducers, which are positioned on the outside of the bottom of pipeline. This system was installed between juice sedimentation vessel and the evaporator (Figure 21.1). The results indicated that not only the viscosity of sugar solution was reduced but also the heat transfer coefficient and evaporation intensity of evaporation system were improved by 42.4% and 15.2%, respectively. The scale was removed remarkably with no significant effects on white sugar quality. In addition, the time of cleaning an evaporator was reduced between 38% and 75% of the time without US and there was no significant influence of US on the final quality of white sugar (Hu et al., 2006).

Emulsification and Mixing

Emulsions are present in a wide variety of food industry products such as the production of margarine, tomato sauce, mayonnaise, and other similar blended products. Emulsion is a dispersion of two immiscible liquids, one of which is dispersed into the other as fine droplets or particles (Mason and Lorimer, 2002; Cucheval and Chow, 2008; Gaikwad and Pandit, 2008). Contrary to mechanical process, which can consume high power and provide little control of the droplet size distribution, US is also an alternative method to produce stable emulsion (Canselier et al., 2002). When the interface of two immiscible liquids is ultrasonically irradiated, an emulsion can be quickly obtained (Krishnan et al., 1981). In this way, tiny droplets of one liquid could be scattered into the other liquid, which constitutes the continuous phase. In addition, US can provide excess energy for the new interface formation, making it possible to obtain emulsions even in the absence of surfactant (Mason et al., 1996; Gogate and Kabadi, 2009). For any intensity above the cavitation threshold, there is a corresponding maximum concentration of emulsion (percentage of the dispersed phase holdup), which remains relatively stable. This limiting concentration of emulsion increases by increasing US intensity. The advantages of US to prepare emulsions include lower energy consumption, use of less or no surfactant, and production of a more homogeneous emulsion in comparison to the mechanical process (Gogate and Kabadi, 2009). Moreover, such emulsions are often more stable than those conventionally produced (Mason et al., 1996).

The efficiency of ultrasonic emulsification generally depends on the irradiation time, irradiation power, oil/water ratio, and physicochemical properties of the hydrophobic phase. Some recommendations for effective US effects in the emulsification must be considered (Mason and Lorimer, 2002; Gogate and Kabadi, 2009): (i) using US the droplet size is much smaller when compared to mechanical agitation under the same conditions, which makes more stable emulsions prepared by sonication (Gogate and Kabadi, 2009); (ii) the occurrence of conditions for transient cavitation is necessary to raise the efficiency of US energy (Cucheval and Chow, 2008); (iii) in general, an increase in the irradiation time decreases the droplet size of dispersed phase (Cucheval and Chow, 2008;

Gaikwad and Pandit, 2008); (iv) beyond a certain time of operation where an equilibrium condition in droplet size is achieved, further use of US energy is ineffective (Gogate and Kabadi, 2009); (v) increasing the ultrasonic irradiation power the fraction of volume of the dispersed phase is also increased, while the droplet size of the dispersed phase is decreased (Gaikwad and Pandit, 2008); (vi) smaller droplet size is generally observed for liquids with low viscosity, possibly attributed to higher cavitational effects in these systems (Gogate and Kabadi, 2009); and (vii) the use of US generally requires the addition of surfactants in order to obtain the same droplet size (Mason et al., 1996; Cucheval and Chow, 2008; Gogate and Kabadi, 2009).

In some cases, an optimal power input might exist, beyond which coalescence and cavitational bubble cloud formation can restrict its performance. The effect of US on sunflower oil emulsion and quality after US emulsification and processing was studied (Chemat et al., 2004). US treatment was performed at 20 and 45 kHz with a titanium horn. In spite of the fact that US-assisted food processing could be considered as a potential alternative to conventional processes, a metallic and rancid odor for emulsions and oil treated by US was observed.

CRYSTALLIZATION AND FREEZING

Crystallization is a process used in many industries, which presents a nucleation of solid crystals from a number of liquids influenced by US energy. It is important to mention that uncontrolled process of nucleation and subsequent crystallization can occur randomly and, as a consequence, the quality of the final product could be committed. In this way, high-intensity US can be used as an additional processing tool to modify the crystallization behavior of different systems (sonocrystallization) and therefore obtain the desirable physicochemical characteristics (Martini et al., 2008). US complies with a number of roles in the initiation of seeding and subsequent crystal formation and growth that can be attributed to the presence of shock waves in the solution, which improves the probability of collision of a molecule and a molecular aggregate (Luque de Castro and Capote, 2007a). It also has a secondary property which is beneficial in such processing applications, namely, the cleaning action of the cavitation efficiency stops the incrustation of crystals on cooling elements and thereby ensures continuous efficient heat transfer (Mason et al., 1996). The US treatment can also reduce the precipitation time (Gogate and Kabadi, 2009) and the texture of food products will be affected by the size of undissolved sugar crystals dispersed in the material. Crystallite size affects the rate of dissolution of sugar in food preparation. Normal crystallization of sugar from concentrated sucrose solutions leads to large, uneven-sized crystals, which can be broken down by subsequent sonication (Mason et al., 1996).

Effects of the application of ultrasonic power on the crystallization behavior of tripalmitoylglycerol (PPP) and cocoa butter was examined in terms of rate of nucleation and polymorphic control. Sonocrystallization was carried out during 15 s using a US equipment operating with a probe under 20 kHz at 100 W of power. High-purity PPP (>99%) and low-purity PPP (>80%) samples were employed to mimic real fat systems. For both PPP samples, the application of ultrasonic power accelerated the rate of nucleation as measured by induction time for the occurrence of crystals and by the number of crystals nucleated. As for cocoa butter, sonication for a short period accelerated the crystallization. Results indicated that US irradiation could be an efficient tool for controlling polymorphic crystallization of fats (Higaki et al., 2001).

A very important field related to crystallization in food industry is the formation of ice crystals during the freezing of water. Freezing is considered an unsteady state heat transfer process. The food loses heat through a surface in a cold environment and, as a consequence, heat transfer will occur in the bulk. It takes place by a combination of conduction (solids) and convection (liquids) processes (Sigfusson et al., 2004). The problem arises because the small ice crystals which are formed initially inside of the cellular material of the food continue to grow (Luque de Castro and Capote, 2007a). As these crystals increase in size they break some of the cell walls, leading to a partial destruction in the structure of the material (Mason et al., 1996). While the melting temperature of ice is constant at a given pressure, the nucleation of water to form ice can occur within a temperature range called as "supercooling range," depending on some physicochemical properties of the sample such as volume, purity, and gas content (Luque de Castro and Capote, 2007a). It represents a considerable delay between the initial crystallization and complete freezing at which point the temperature of the whole material can fall. Under the influence of US a much more rapid and even seeding occurs and this leads to a much shorter delay time. In addition, since there are a greater number of seeds, the final size of the ice crystals is smaller and cell damage is reduced (Mason et al., 1996). These effects may be due to the small bubbles produced by high-intensity ultrasonic waves that can act as nuclei, or because fluctuations in the pressure and temperature associated with the ultrasonic wave disturb the equilibrium between solid and liquid phases. By controlling the intensity, duration, and frequency of the ultrasonic waves, it may be possible to modify the size and concentration of the crystals produced (McClements, 1995).

OTHER APPLICATIONS

Soybean oil hydrogenation rate with copper or nickel catalyst was increased by using ultrasonic energy (Moulton Sr. et al., 1987) in a continuous system. Commercially refined and bleached soybean oils and Nysel® catalyst were mixed and filtered through a micro filter, and pumped through heat exchangers at 1 L h^{-1} up to the ultrasonic processing cell. Hydrogen pressure was controlled by a backpressure regulating valve in the vent line. Hydrogenated product was separated from hydrogen excess and withdrawn continuously. Ultrasonic power was supplied at 550 W and 20 kHz. Results showed that hydrogenation rate can be increased almost 20 times using the processing cell at 233°C and 112 psig.

Airborne ultrasonic technology has been proposed as a clean and commercial alternative to conventional methods for fermentation systems, carbonated beverages, and such processes where product quality or yield is adversely affected by foaming (Villamiel et al., 2000; Gallego-Juárez, 2007; Soria and Villamiel, 2010). Destruction and breaking of foams by US-assisted defoamers is mainly attributed to partial vacuum on the foam bubble surface created by high acoustic pressure, foam bubbles resonance which creates interstitial friction causing bubble coalescence, cavitation, and acoustic streaming (Mason et al., 2005). The main advantages of the use of US as an efficient additional step for food processing can be cited as the reduction or elimination of antifoam chemicals, the reduction of wastage in bottling production lines, and the increasing production throughput.

APPLICATIONS IN BIOTECHNOLOGY

The developing field of biotechnology constantly attracts new methods for further improvement of bioprocess performance (Rokhina et al., 2009). The process industry demands that operations must be performed in the most efficient way with respect to product quality, energy or time, or in terms of process economy. Novel technologies are constantly being required to reduce the total processing cost while maintaining or enhancing product quality in an environmentally benign manner (Gogate, 2008). Therefore, based on principles of green engineering and science, US has been used in several biotechnology applications such as cell disruption to release intracellular enzymes and organelles (Rokhina et al., 2009)—one of the first applications of power US for disruption of cell wall (Mason and Lorimer, 2002). Processes involving cell disruption are dependent on the ability to disrupt the membrane of a cell to release without damage the contents of this disrupted cell. Maximum cell disruption is obtained in a zone close to the US probe tip and the biological cells must be kept at this point for sufficient time allowing disruption to take place. Therefore, a delicate committed condition must be found between the power of the probe and the disruption rate since power US, which is associated to cavitational collapse energy and bulk heating effect, can denature the contents of the cell once released (Mason and Lorimer, 2002). In this way, ultrasonic methods have considerable potential for the introduction of macromolecules into cells (Liu et al., 2006), for microbial cell disruption (Gogate and Kabadi, 2009) and, additionally, to enhance the biosensor performance (Rokhina et al., 2009).

Microbial Cell Disruption

An efficient large-scale process for cell disruption can be considered the key factor in the economical industrial production of microbial components. High-speed shaker, bead mills, and high-pressure homogenizers are some of different strategies and methods usually employed for cell disruption (Figure 21.2). However, the typical efficiency for energy conversion of the previously discussed methods is in the range of 5%–10% (Geciova et al., 2002). The remaining energy is dissipated as heat, which needs to be efficiently removed to assure the integrity of these delicate bio-products. In order to improve the efficiency of cell disruption process, intense interest has been developed in the last decade in newer techniques, including acoustic and hydrodynamic cavitation (Harrison, 1991; Geciova et al., 2002; Gogate and Kabadi, 2009).

Sonication is one of the most widely used laboratory disruption methods. Its mechanism is associated with the cavitation phenomena. The cavitation bubble implosion can produce shock waves, which lead to viscous dissipative eddies resulting in the development of shear stress in the medium. Shear force creating eddies larger than the cells are more likely to move the cells rather than disrupt them, whereas eddies smaller than the cells are capable of generating disruptive shear stresses. Thus, larger cells can experience more disruptive eddies than smaller cells. Increasing the US power is possible to shift the size distribution toward smaller eddies, which in turn can increase the number of disruptive eddies acting on the cells resulting in greater disruption (Geciova et al., 2002).

Cell disruption process can proceed by complete breakage of the individual cells in certain devices, releasing all intracellular enzymes, or it can be shear driven, where only the cell wall can be disrupted. It was investigated during the release of invertase enzyme by the disruption of *Saccharomyces cerevisiae* cells using US, high pressure, or hydrodynamic cavitation. Using ultrasonic-induced cavitation (ultrasonic probe of 20 kHz at 600 W of power) the rate of release of invertase enzyme was constant and independent of the sonication time (15 or 20 min) under 10°C. In these conditions, the sonication time was varied linearly with enzyme concentrations. However, strong cavitation produced by US results in complete breakage of the cells, which represents a reduced degree of separation (low selectivity) and high energy consumption (Balasundaram and Pandit, 2001).

Gene Transfer

Recently, ultrasonic methods have been considered as a potential tool for the introduction of macromolecules into cells. All the transfer techniques can be divided into two broad categories: viral (retroviruses and adenoviroses) and nonviral (lipofection, electroporation, particle bombardment, microinjection, and sonoporation). In order to improve the efficiency and to make the procedure of nonviral gene transfer simpler, a US method was evaluated for genetic transformation (Liu et al., 2006).

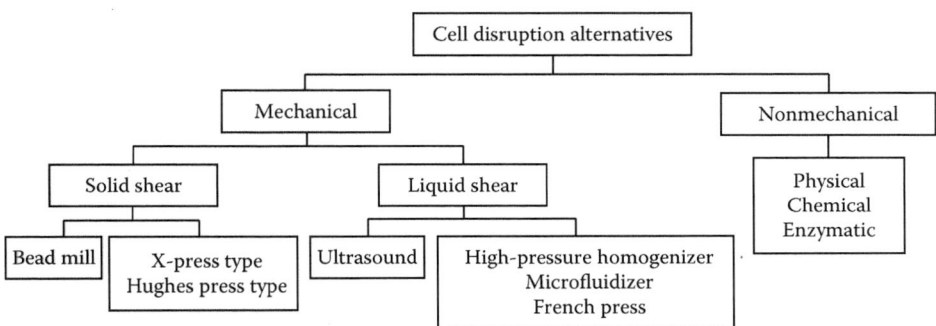

FIGURE 21.2 Methods for microbial cell disruption. (Adapted from Geciova, J. et al., *Int. Dairy J.*, 12, 541, 2002.)

The interaction of the broad US waves with cavitation bubbles resulted in mechanical effects produced on nearby cells. The main effect is cell lysis but sublethal membrane damages can also occur. In the last case, which is known as sonoporation effect, the membrane is transiently permeable to large molecules (Miller et al., 1999; Ohta et al., 2008).

It has been shown that sonication can alter the transient permeability of plasma through the cell membrane to facilitate uptake process (Liu et al., 2006). Besides, US as well as other mechanical methods is often more versatile, once it is based on disruption of cell membrane and is less dependent on cell type (Wyber et al., 1997; Liu et al., 2006; Gogate and Kabadi, 2009). The potential applications of US in biotechnology result from thermal and mechanical effects. The first one could be observed as a consequence of ultrasonic wave propagation in medium. The US energy could be partially absorbed by the medium resulting in temperature increase. On the other hand, acoustic cavitation and mass transfer enhancement are considered the main mechanical effects. Cavitation is considered a major mechanism for causing alterations to biological tissues, especially increased membrane permeability. In addition, US at low-intensity levels enhances the movement of the liquid medium, favoring mass transfer and reaction rates in both multiphase and homogeneous systems (Liu et al., 2006).

In order to well understand the acoustic permeabilization, two possibilities could be considered as the most important mechanism. Firstly, the violent collapse of cavitation bubbles can generate high-pressure and high-temperature shock waves, which could potentially cause localized rupture of the plasma lemma and lead to the uptake of exogenous solutes, followed by the reestablishment of membrane integrity. The second possibility originates from the electrochemical model predicting the existence of a critical hydrostatic pressure, at which the intrinsic membrane potential is sufficiently high to induce mechanical breakdown of the membrane. As a consequence, it is possible that the high oscillating pressure generated by the US field and the high-pressure shock waves could produce such high hydrostatic pressures causing reversible membrane breakdown. These two possibilities are said to be closely related and may simultaneously occur (Gogate and Kabadi, 2009).

Mild US irradiation has been proved an efficient method for transfection in animal cells and tissues in vitro and in vivo. In order to study the transfection efficiency for plasmid DNA in vitro, the US waves (118 MHz) were generated by piezoceramic, air-backed disc transducer with a diameter of 10 cm, mounted with its radiating face parallel to the side wall of the tank (Figure 21.3). The amplitude pressure was varied from 0.1 to 5 MPa and the total sonication time ranged from 10 s to 10 min. US waves were focused by a lens and coupled into a water tank due to the many tissues present in vivo properties similar to water with respect to US propagation. The spatial peak intensity in the focus was approximately 33 W cm^{-2}. The same equipment and water tank could be used for the in vivo experiments. Thus, the results obtained showed that sinusoidal-focused US can significantly enhance the transfection of plasmid DNA in the dunning prostate tumor (prostate carcinoma cells R3327-AT1) in vitro and, mainly, in vivo. The average number of positive cells in sonicated tumors was approximately 10-fold higher than that in non-sonicated controls, resulting in an in vivo

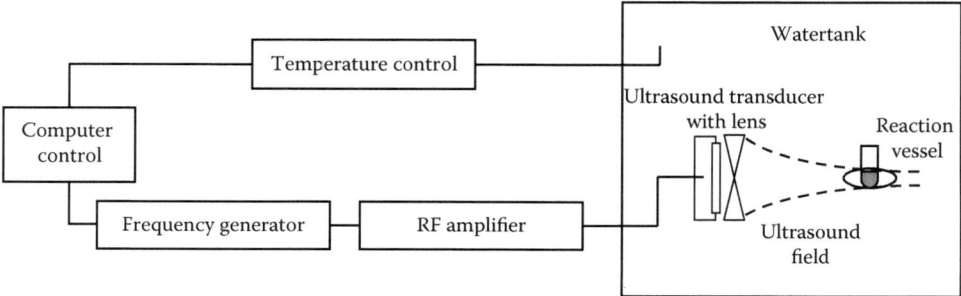

FIGURE 21.3 Sonication device scheme. (Adapted from Huber, P.E. and Pfisterer, P., *Gene Therapy*, 7, 1516, 2000.)

transfection efficiency of 5% after intratumoral DNA injection and US exposure after direct plasmid injection without marked side effects (Huber and Pfisterer, 2000).

Gene transfer by sonication employs the same simple procedure for the plant material to be transformed. The protoplasts, suspension cells, or small pieces of tissue are suspended in a few milliliters of sonication medium in a microcentrifuge tube. Plasmid DNA (and possibly carrier DNA) is then added, and after rapid mixing the samples are ready for sonication. The pulses of US are delivered by ordinary machines used for homogenization of various tissues. The microtip is immersed 2 to 3 mm into the suspension and pulses of selected intensity and duration are delivered. The cells are finally transferred to fresh growth medium (Liu et al., 2006). The plasmid DNA was introduced efficiently into sugar beet and tobacco protoplasts by a brief exposure (500–900 ms at 30–70 W of electric power) to 20 kHz US at 0.65–1.6 W cm^{-2} of acoustic power. Successful transformation was evidenced by transient expression of the introduced gene for chloramphenicol acetyltransferase (CAT). Optimal transient expression was obtained at rather high plasmid DNA concentrations (80–110 mg L^{-1}) and sucrose concentrations (21%–28%) in the sonication medium. In sugar beet protoplasts transient expression was reported to be 7–15-fold times higher than the expression obtained by electroporation (Joersbo and Brunstedt, 1990).

As previously discussed, the potential of US to introduce macromolecules into cells has been evaluated. A 20 kHz US was applied to the yeast cells suspension. For this reason, a probe-type sonicator (3 mm) operating at 20 kHz and 2 W of power was placed 3 mm below the surface of sample (1 μg of plasmid DNA) at 4°C. The efficiency of DNA delivery was scored as the number of cells transformed. Using yeast as a model system, it had established that transfer of DNA was optimal using short sonication time (30 s). DNA delivery into cells was dependent on the related effects of cavitation phenomena. Thus, cellular damage could occur by tensile forces on the cells during microstreaming as a result of stable cavitation, by the liquid jets or shock waves following unstable cavitation, and by the free radicals that may be produced during the final stages of bubble growth and collapse. In this way, under controlled conditions US was considered as an effective way of delivering plasmid DNA into cells (Wyber et al., 1997).

Biosensors

Biosensors have been used for the detection and characterization of bacteria without the need for a culture step (Zourob et al., 2005), for specific and sensitive detection of nucleic acid–based detection compared to immunological-based detection (Howkes et al., 2004), and in food quality for the detection of several types of compounds of interest such as carbohydrates, alcohols, phenols, carboxylic acids, amino acids, biogenic amines, heterocyclic, inorganic, and additive or contaminants compounds (El Kaoutit et al., 2007). US can enhance the biosensor performance. The main influence on the behavior of particles in aqueous suspension in a US standing wave could be the direct radiation force and acoustic streaming. The first one, which drives suspended particles toward and concentrates then in acoustic pressure node planes, has been applied to rapidly transfer cells in small-scale analytical separators. Furthermore, acoustic streaming has been employed for mixing small analytical samples. Both forces working together can enhance the rate of the reaction between suspended mixture cells and retroviruses (Kuznetsova and Coakley, 2007). In biosensors, a biological unit (e.g., enzyme or antibody), which is typically immobilized on the surface of the transducer (electrode), interacts with the analyte (which contains, e.g., a target bacterium or xenobiotic) and causes a change in a measurable property within the local environment near the transducer surface, thus converting a biochemical process into a measurable electronic signal (Kuznetsova and Coakley, 2007; Rokhina et al., 2009). The biosensor response could be mainly influenced by factors such as the mass-transport kinetics of analytes and products, as well as loading of the sensing molecule. In order to improve the sensitivity and efficiency of biosensors, US irradiation can be used to aggregate and drive (Howkes et al., 2004; Zourob et al., 2005; Kuznetsova and Coakley, 2007) microorganisms toward antibody-coated sensors using US standing waves (USW). Alternatively,

US could be used to facilitate enzyme immobilization on sonogel–carbon transducer electrodes, which are frequently used to detect, for example, xenobiotics and organophosphorus pesticides in waste streams (El Kaoutit et al., 2007, 2008; Rokhina et al., 2009).

The detection and characterization of bacteria without the need for a culture step is an important goal for microbiology. Rapid variant-specific sensing will facilitate new approaches to medical diagnosis, protection against bioterrorism, and food processing. In this way, a system was developed to detect bacteria using optical metal-clad leaky waveguide (MCLW) sensor with USW. A half-wavelength US reflector (1 mm glass slides) was used in order to obtain a node in the region of the reflector, at frequencies of approximately 3 MHz. The performance of a MCLW sensor for the detection of bacteria has been increased (>100 fold) by using USWs to drive bacteria onto the sensor surface. By forming the USW nodes at or within the surface of the MCLW, the diffusion-limited capture rate has been replaced by fast movement. The application of US for 3 min gave a detection limit for bacterial spores of 10^3 spores mL^{-1}. It represents an enhancement in the capture of bacteria spores from a flowing sample onto an immunocoated optical leaky waveguide surface by providing an attraction force that moves the bacteria spores directly across the lines of flow (Zourob et al., 2005).

SYNTHESIS

Acoustic energy in chemical processes has been used as a general explanation for the term sonochemistry. An increasing number of applications of US in synthesis, in the past few years, have made sonochemistry an attractive area of research. The main interest is related to the range of power US (20–50 kHz), which is known for enhancing chemical processes by providing enough energy to affect reactivity (Mason and Lorimer, 2002). The usefulness of sonochemical synthesis as a synthetic tool resides in its versatility. Cavitation generates sufficient kinetic energy to break chemical bonds. In addition, the application of US in chemical reaction can completely change the distribution of substrates or even to allow the formation of new products (Cravotto and Cintas, 2007a). The effects of US on reactivity can be thought in terms of the reaction types, as involving metal or solid surfaces, powders or particulate matter, emulsification or homogeneous reactions. The main benefits that can be pointed out in synthesis using US are the increase of reaction rate or at least the use of less forcing conditions, the reduction in induction periods, the reduction in reagents amount, less reaction steps can be performed and, in some situations, the reaction can follow an alternative pathway (Mason, 2003). Based on the advantages of US in synthesis it has been applied in many types of systems, as organic reactions in homogeneous or heterogeneous mixtures, polymerization, enzymatic synthesis, organometallic reactions, and mixtures involving metal activation and phase transfer catalysis and related reactions (Mason, 1997; Thompson and Doraiswamy, 1999; Cravotto and Cintas, 2006). However, in spite of industrial applications and large-scale systems being a promising idea, they are still a challenge that must be overcome.

Transesterification of vegetable oils for biodiesel production is among the common chemical reactions of industrial concern, mainly in the last years due to the increasing interest in the rational use of biomass and environmental requirements (Cintas et al., 2010; Vyas et al., 2010). Biodiesel is a renewable fuel obtained from vegetable oils and animal fats. The synthesis of biodiesel can be done using homogeneous (alkaline or acidic) catalysts or also heterogeneous catalysts via a transesterification reaction. Slow reaction kinetics and poor mass transfer are the main problems related to the biodiesel synthesis. The mechanical energy necessary for mixing and the activation energy required for starting the transesterification reaction can be provided by applying US to the mixture (Vyas et al., 2010). It was proved that low-frequency US can efficiently improve base-catalyzed transesterification (Scheme 21.1), allowing economical advantages, such as time saving and less energy consumption, over the classical process, being a valuable toll for the fatty acids transesterification (Stavarache et al., 2005). In this laboratory scale procedure, a reduction of four times in the reaction time was obtained using 6:1 molar ratio of alcohol/oil. The amount of catalyst was two

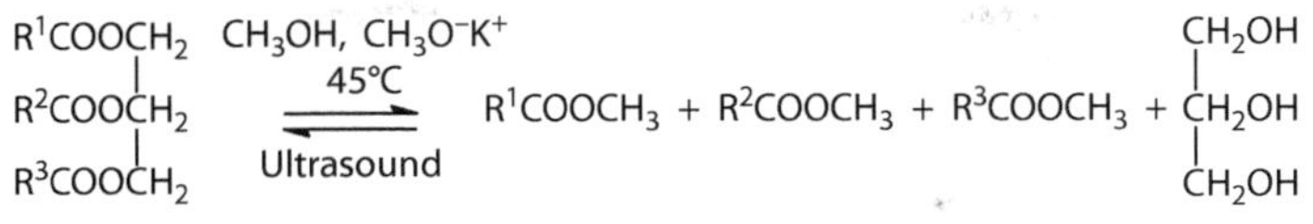

$$\begin{array}{c}R^1COOCH_2 \\ | \\ R^2COOCH_2 \\ | \\ R^3COOCH_2\end{array} \quad \underset{\text{Ultrasound}}{\overset{CH_3OH,\ CH_3O^-K^+\ 45°C}{\rightleftharpoons}} \quad R^1COOCH_3 + R^2COOCH_3 + R^3COOCH_3 + \begin{array}{c}CH_2OH \\ | \\ CH_2OH \\ | \\ CH_2OH\end{array}$$

SCHEME 21.1 Ultrasound assisted base-catalyzed transesterification of triglyceride with methanol. (From Mason, T.J. and Lorimer, J.P., Applied Sonochemistry: Uses of Power Ultrasound in Chemistry and Processing, Wiley-VCH Verlag GmbH, Weinheim, Germany.)

to three times lower with US at 28 and 40 kHz at room temperature. As the industrial production of biodiesel is performed at 60°C, the possibility of performing synthesis at room temperature can be pointed out as the main advantage of US-assisted procedure in terms of energy consumption (Stavarache et al., 2005). US-assisted transesterification was studied in laboratory scale for soybean (Jianbing et al., 2006; Georgogiani et al., 2007; Cintas et al., 2010), frying and fish oil (Armenta et al., 2007), and triolein (Hahn et al., 2009), and all procedures allowed yields higher than 88% by applying US with baths or probes for 10–60 min, at 25°C or 60°C (Vyas et al., 2010).

A continuous process driven by US was developed for vegetable oil transesterification using a 45 kHz ultrasonic transducer with a total power of 600 W (Figure 21.4a) (Stavarache et al., 2007). The system is composed of two reservoirs (2.62 and 6.35 L), ultrasonic reactors, pumps, and a separation unit. The flow rate was set to allow a resident time of 10, 20, or 30 min. At 20 min of resident time in a large reactor (processing 19 L), the conversion was about 90% (Stavarache et al., 2007). A new pilot flow reactor has been recently proposed for high-intensity US irradiation for the synthesis of biodiesel (Cintas et al., 2010). The flow reactor consists of three 21.5 kHz transducers in the bottom of the chamber cemented to a titanium alloy plate (100 × 325 × 0.9 mm). Transducers are high efficiency pre-stressed piezoelectric (PZT) rings compressed between two blocks. The system allows frequencies from 17 to 45 kHz. Optimal efficiency was observed at 21.5 kHz and power up to 900 W, corresponding to a mean value of 3 W cm^{-2} at the emitting surface. A peristaltic pump (30 W) is used to circulate the reactional mixture (oil, methanol, and sodium methoxide) contained in a cylindrical tank for 5 L through the sonication compartment. The cylindrical tank is thermostatted by a silicone flow (temperature range from 0°C to 90°C). Molar proportion of alcohol to oil is 6:1 and catalyst (MeONa) amount is 0.15% wt to the oil. The sonication chamber is a 0.5 L gastight

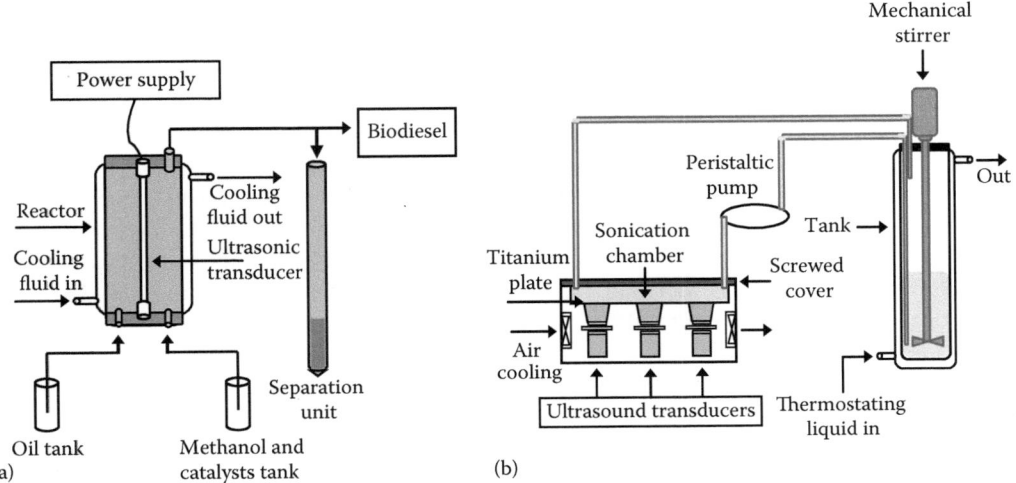

FIGURE 21.4 Detailed schemes for ultrasound-assisted biodiesel production: (a) continuous process for transesterification (Adapted from Stavarache, C. et al., *Ultrason. Sonochem.*, 12, 367, 2007.); and (b) pilot flow reactor for biodiesel synthesis. (Adapted from Cintas, P. et al., *Ultrason. Sonochem.*, 17, 985, 2010.)

tank completely full of circulating liquid by the peristaltic pump from the cylindrical tank. Reaction was studied in the range from 25 to 100 mL min^{-1} that corresponds to about 5–20 min of residence time. Optimal flow rate was observed at 55 mL min^{-1} (9 min for residence time) and in this flow rate the transesterification was studied at 500, 600, and 700 W. At 700 W, the temperature reached about 59°C and this temperature is not considered suitable for the process once it is close to the boiling point of methanol, reducing the cavitation and, as a consequence, the efficiency of reaction. In this way, 600 W power (reaching up to 50°C) power allows the best power/conversion ratio and a total conversion of soybean oil to methyl esters is obtained after 1 h of flow. It is believed that cavitation has the main effect over the mass transfer rates ensuring a uniform reactants distribution. The energy consumption under the best conditions is 0.28 kWh L^{-1}. Figure 21.4b shows a scheme of a pilot flow reactor for biodiesel synthesis assisted by US (Cintas et al., 2010). In addition to the laboratory and pilot process, an ultrasonic biodiesel processing unit is actually available for transesterification allowing lower reaction time and less catalyst for the synthesis (online document, available at http://www.hielscher.com, accessed in March 2011).

US-assisted synthesis has also been applied in the materials science as well as in nanotechnology (Mason, 2003). In summary, it is possible to point out that sonochemical methods related to materials science and nanotechnology are superior to the other techniques mainly for preparation of amorphous products, insertion of nanomaterial into mesoporous materials, and for deposition of nanoparticles on polymeric and ceramic surfaces. In metal oxides preparation using US, glass former materials are not required and additionally amorphous products are obtained in nanometer size (Mason, 2003). Nanosized catalysts are deposited as a smooth layer on the mesopores walls without blocking them by using ultrasonic waves, resulting in materials of better properties (Mason, 2003). When using US, a smoothed homogeneous coating layer of nanoparticles is formed on the surfaces of ceramics and polymeric materials by making chemical bonds or interactions with the substrate and cannot be removed by washing (Gedanken, 2004). Materials in nanometer size often exhibit distinct properties mainly due to the electronic structures with a high density of states but not continuous bands. A variety of synthetic methods can be used for preparation of nanostructured materials, including gas phase techniques (molten metal evaporation, flash vacuum thermal, and laser pyrolysis decomposition of volatile organometallics), liquid phase methods (reduction of metal halides with various strong reductants, colloidal techniques with controlled nucleation), and mixed phase approaches (synthesis of conventional heterogeneous catalysts on oxide supports, metal atom vapor deposition into cryogenic liquids, explosive shock synthesis). The choice of the appropriate synthetic route can determine the properties and suitability of the nanomaterial because its physical and chemical properties are strongly dependent on the synthesis conditions (Bang and Suslick, 2010). Various forms of nanostructured materials, including metals, alloys, oxides, sulfides, carbides, and nanostructured supported catalysts, can be obtained with a simple modification in reaction conditions. The sonochemical decomposition of volatile organic precursors and the enhanced mass transport due to the shock waves are important for the synthesis of nanocomposites (Bang and Suslick, 2010). Regarding many different methods for the fabrication of nanomaterials, the use of power US has been considered advantageous due to the capability of producing nanomaterials in amorphous state, shorter reaction times, insertion of nanoparticles into the pores of mesoporous materials, and the synthesis of inorganic fullerenes at room temperature (Mason, 2003). In particular, sonochemical reactions can result in nanosized products which can be amorphous in case of volatile products or crystallines when the solutes are nonvolatile. The morphology of the products can also be different, being spherical or similar, nanotubes, nanorods, fullerenes, and hollowed spheres (Gedanken, 2004).

Electrochemical synthesis using US is also an attractive field for sonochemistry. Large scale use of electrochemistry in industry is known as electroplating, but the main problem related to this kind of process is the high consumption of electrical power. Electrochemical reactions are essentially based on the transfer of ions to and/or from the electrode surface. The jet effect from cavitational collapse can assist the diffusion to and from the electrode. Additional beneficial effects when using

US include enhanced mass transport, changes in adsorption phenomena and surface effects, diminished electrode fouling, controlled reaction mechanisms and product distribution, increased yields, increased limiting current, and lessened cell power requirements. It is important to point out that many developments in the field of sonoelectrochemistry are reported practically in laboratory scale (Thompson and Doraiswamy, 1999; Mason and Lorimer, 2002; Mason, 2003).

TEXTILE INDUSTRY

Textiles and their end products represent the largest industry and the conventional method for textile dyeing involves mainly submitting the material under agitation to an aqueous solution containing the dye during a period of time, and the object is transport/diffuse dyes or chemical into the fiber (Vouters et al., 2004, Moore and Ausley, 2005). In this process sonochemical-forced impregnation can improve dyeing. The rate of penetration of the dye into the fabric can be increased by US (20 kHz, 5 W cm^{-2}) offering a commercial attractive option for reducing the immersion times and saving the quantities of dye used (Xie et al., 1999; Mason and Lorimer, 2002). Besides the process acceleration and the same or best results obtained by the conventional techniques, US allows process acceleration under lower temperature and lower chemical concentration. In this way, US-assisted textile dyeing has been developed for the textile industry, for dyeing and related processes as preparation of sizes, emulsions, dye dispersions and thickeners for print paste, as well as the treatment of process residues (Vajnhandl and Le Marechal, 2005).

The effects on dye dispersion quality, dye solubility in water, and dye uptake for textile materials were studied at high and low frequency, using natural and synthetic fibers. For example, for leather treatment, localized temperature raise and swelling effect due to US may improve diffusion dye/leather (Sivakumar and Rao, 2003). However, in spite of US-assisted wet textile process developments in lab-scale since its first report in 1941, it has not been implemented on an industrial scale up to now (Sivakumar et al., 2009a). In summary, improvements observed in US-assisted dyeing process are generally attributed to cavitation phenomena and there is a continuous effort to change it to industrial scale (Vouters et al., 2004; Vajnhandl and Le Marechal, 2005).

Low-frequency US field (26 kHz) was investigated for reduction of particle size of the dispersive dye, C.I. Dispersive Red 60 (Lee and Kim, 2001). It was observed that the volume of small particles increased whereas large particles relatively decreased in volume. The volume of particle size in the range from 65 to 72 μm decreased from 14% to 0% after 1 h of US and this reduction in the particle size enhanced the dye fixation. The diffusion and permeability of C.I. Direct Red 81 through a cellophane film was increased under 20 kHz US (Thakore et al., 1990). The dyeing process of silk using cationic, acid or metal-complex dyes at low temperatures, assisted by 26 kHz US (600 W) was evaluated and the results for dye uptake were compared with those obtained by conventional process. The dyeing under US showed an increase in dye uptake for all classes of dyes at lower temperatures (45°C and 50°C) and a reduction in dyeing time (15 min) was obtained in comparison to the conventional process (85°C for 60 min). In addition, no damage in fiber by cavitation was observed (Shukla and Mathur, 1995). The dyeing process of acrylic fibers using C.I. Astrazon Basic Red 5BL 200% was thoroughly investigated at 38.5 kHz and 100–500 W in a 5.75 L ultrasonic bath. The improvement of the US dyeing process was attributed to the acoustic cavitation phenomenon, which could lead to dispersion (breaking up of micelles and high molecular weight aggregates into uniform dispersions in the dye bath), degassing (expulsion of dissolved or entrapped gas or air molecules from acrylic fiber into liquid and removal by cavitation which facilitates the contact between dye and fiber), and diffusion (accelerating the rate of dye diffusion inside the fiber by piercing the insulating layer covering the acrylic fiber and accelerating the interaction or chemical reaction between the C.I. Astrazon Basic Red 5BL 200% dye and acrylic fiber) (Kamel et al., 2010).

In addition to the dyeing process, washing operations are performed to remove natural materials or impurities from the fiber surface. The suitability of using US also in this process has been

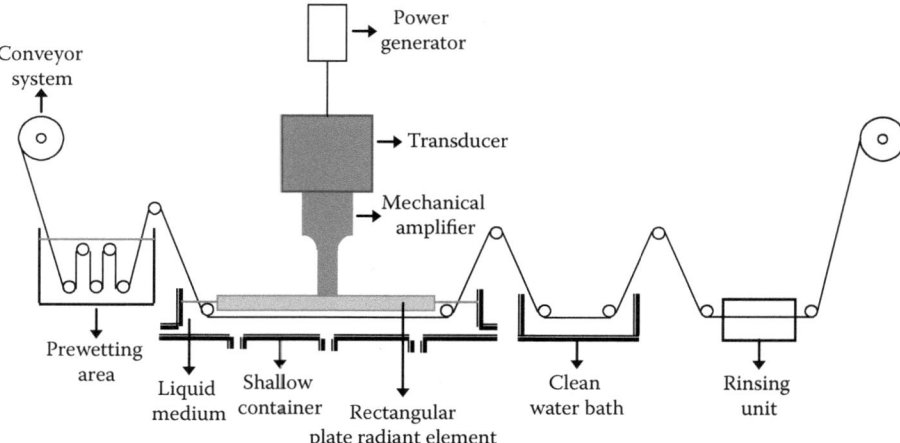

FIGURE 21.5 Schematic embodiment for continuous ultrasonic washing of textile. (Adapted from Gallego-Juarez, J.A. et al., *Ultrason. Sonochem.*, 17, 234, 2010.)

reported (Vouters et al., 2004; Sivakumar et al., 2010). A semi-industrial process for textile washing based on the application of the ultrasonic energy to textiles for washing using special vibrating plates was recently reported (Gallego-Juarez et al., 2010) which was based on a process for continuous textile washing previously patented (Figure 21.5) (Gallego-Juarez et al., 2001). The washing solution was 1.75 g L^{-1} of sodium dodecylbenzenesulphonate (SDBS) and 3.5 g L^{-1} of sodium triphosphate (STP). Textiles were submerged in the liquid layer of a few millimeters thickness and conveyed in a flat format through the radiator by a roller-type system. The plate radiator was designed with a grooved profile on its backside by removing mass from all the central area of the plate between its two nodal lines. The grooved-plate radiator was made in titanium alloy with a radiating surface of 110 cm^2, and it has a 600 W of power capacity allowing reaching an acoustic intensity up to 4.5 W cm^2 at 21 kHz. It was observed that the cleaning performance at 2 cm s^{-1} (measured through the reflectance) increases almost linearly with the intensity of US power up to a value in which saturation starts. The washing performance proved to be much higher than that obtained using a conventional washing machine even at relatively moderate acoustic intensities. The energy consumption for the ultrasonic washing process was considered very low, about 0.1 kWh kg^{-1} of textile (Gallego-Juarez et al., 2010).

US application in finishing process was also evaluated, for de-oiling of polyamide and de-sizing of cotton (Vouters et al., 2004). A dynamic device based on the principle of a jigger machine (a widespread textile finishing machine, constituted with a vat filled with water where two rolls above the vat allow making several passages in the bath) was developed. The vat capacity was 15 L and a plunge sonotrode of 6 cm in diameter operating at 24 kHz and up to 600 W was adapted for sonication. Textile pieces of 20–25 m long and 10 cm wide were used, with a fabric speed of 10 m min^{-1}. Ultrasonic power was 40 W L^{-1}. De-oiling process was evaluated for polyamide using a nonionic detergent and sodium hydroxide. Better results were obtained using temperatures in the range of 60°C–70°C. The residual grease content was always less than that obtained by the process without US, and chemicals, water, and energy saving was about 30%, 20%, and 40%, respectively. Cotton de-sizing was performed at 80°C for 30 min, using α-amylase. Using US, the manufacturing time for de-sizing was 30% saved. The final quality of fiber was better and a positive effect on the removal of sizing was obtained with US, without damages to the fibers. An additional evaluation related to the costs showed a 10% increase for US-assisted finishing in comparison to the conventional process, but it was accepted by finishers once final quality was improved.

EXTRACTION

Classical extraction procedures can be improved by the use of mechanical stirring for increasing the diffusion rate and the surface of solvent-material contact. However, in view of the unique characteristics of US related to relatively high mass transfer and cell disruption, it has been widely used to improve extraction processes in several application fields (Vinatoru et al., 1999; Luque de Castro and Capote, 2007a). These effects are much stronger at low frequencies (18–40 kHz) and practically negligible at 400–800 kHz (Cravotto et al., 2008). In this sense, US energy has been used for extraction of oils (Cravotto et al., 2008), phenols and (Japón-Luján et al., 2006, 2008) dyes (Sivakumar et al., 2009b), and processes enhancement (Cravotto et al., 2008; Riera et al., 2010), among others.

In a general way, ultrasonic system could be applied to extraction with organic solvent (Cravotto et al., 2008) and how it could be straightforwardly performed can be observed in Figure 21.6.

The particular effects produced by ultrasonic energy can be used to enhance other extraction procedures (Cravotto et al., 2008; Sivakumar et al., 2009b; Riera et al., 2010). Considering food processing, supercritical CO_2 has been considered a suitable solvent for extraction from vegetables. In spite of some important characteristics of supercritical CO_2 (i.e., nontoxic, recyclable, low cost, relatively inert, and non-flammable), this type of extraction can be affected by the slow kinetics of the process. In this case, as mechanical stirring is difficult to be applied, the use of US has demonstrated important benefits as a consequence of the mechanical effects produced in the supercritical environment (Riera et al., 2004, 2010; Balachandran et al., 2006; Hu et al., 2007). In this way, a system based on US-assisted supercritical fluid extraction (USFE) was developed for oil extraction from almonds and a vegetable product ("cocoa cake"). This system is composed of two units: supercritical fluid extraction (SFE) unit and high-power US (HPU) unit. Figure 21.7 shows a scheme of USFE system. For the extraction process, about 1.5 kg of sample is placed into a high-pressure extraction vessel (5 L capacity). The ultrasonic system is operated with power and frequency of 85 W and 19 kHz, respectively. For almond oil extraction, an improvement up to 90% can be obtained with pressure, temperature, and CO_2 flow rate set at 280 bar, 45°C, and

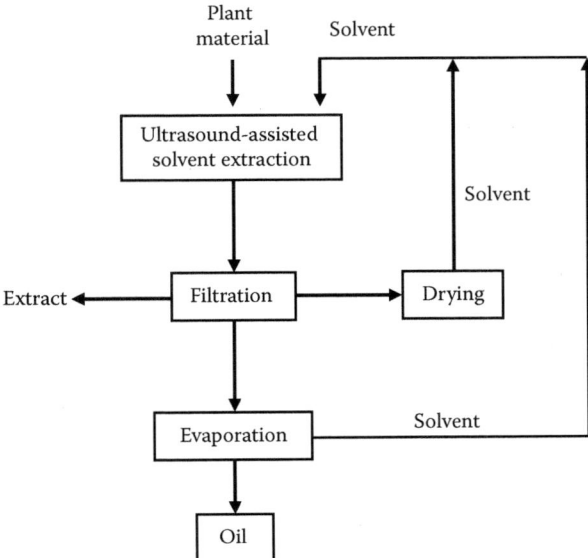

FIGURE 21.6 General operation unit scheme for organic solvent extraction assisted by ultrasound. (Adapted from Vinatoru, M. et al., Ultrasonically assisted extraction of bioactive principles from plants and their constituents, in Mason, T.J., ed., *Advances in Sonochemistry*, Vol. 5, JAI Press, London, U.K., 1999, pp. 209–247, 1999.)

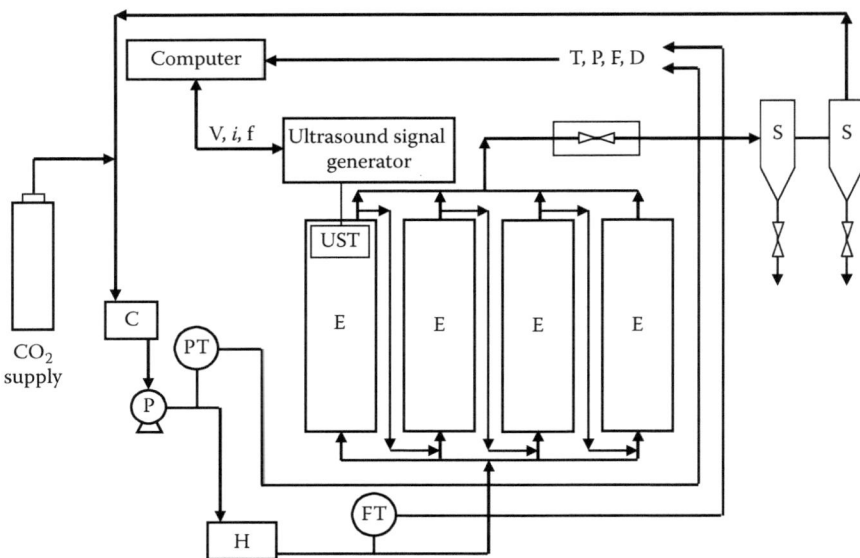

FIGURE 21.7 Scheme of the SFE pilot-plant assisted by ultrasound. Units: extractors (E), separators (S), cooler (C), high pressure pump (P), heater (H), pressure meter (PT), flow meter (FT), ultrasonic transducer (UST). Electrical parameters: voltage (V), current (i), frequency (f). Extraction parameters: temperature (T), pressure (P), CO_2 flow rate (F), density (D). (Adapted from Riera, E. et al., *Ultrasonics*, 50, 306, 2010.)

12.5 kg h^{-1}, respectively, after 3.5 h of extraction. On the other hand, the maximum improvement was 43% for "cocoa cake" oil extraction even increasing pressure and temperature. Even considering the high extraction, the behavior of the system showed high stability and good performance during the trials (Riera et al., 2010).

Reactors with similar design based on USFE have been developed for extraction of oil, pungent compounds, coixenolide, and lutein esters from almonds, ginger, adlay seed, and marigold, respectively. For all processes, operation frequency and temperature were set at 20 kHz and 40°C–55°C, respectively (Riera et al., 2004; Balachandran et al., 2006; Hu et al., 2007; Gao et al., 2009).

A new process based on US-assisted extraction (UAE), microwave-assisted extraction (MAE), or combination of both was developed for oil extraction from soybean germ and seaweed (Cravotto et al., 2008). In comparison with conventional extraction methods (static extraction at room temperature and Soxhlet), it is possible to reduce extraction times up to tenfold and increasing the yield by 50%–500%. This enhancement was observed using a cavitating tube (Cravotto et al., 2005; Cravotto and Cintas, 2007b) (19 kHz, 65 W) and double sonication employing an additional immersion horn (25 kHz, 60 W) at 45°C for 30 min. Using these conditions, yields up to 17.9% and 25.9% for soybean germ and seaweed, respectively, could be obtained.

Leaching of coloring matter from plant materials is an important application field of UAE processes. It is important to point out that synthetic dyes can be toxic after chronic exposure and the dyeing process is generally inefficient. In this sense, a process for both extraction of dye from natural resources (beetroot) and application in the substrate (leather) was recently developed according to environmental friendly purposes (Sivakumar et al., 2009b). In the specific case of beetroot, the coloring matter is strongly bound with plant cell membranes, hindering mass transfer. In a batch process recently developed (Sivakumar et al., 2009b) 1 g of sample and 50 mL 1:1 methanol–water were taken in the extraction vessel. For maximum dye extraction, US (power of 80 W, frequency of 20 kHz and operation in pulse mode of 1 s) was applied during 3 h at 45°C. Compared to magnetic stirring in the same conditions, an enhancement of 8% in the yield of colorant was observed. A similar process was developed for extraction of medicinal tincture from sage (*Salvia officinalis L.*) (Valachovic et al., 2001). For this process, an ultrasonic probe operating at 20 kHz (600 W) was placed in the top center of the extractor. The process was carried out using 5 kg of sample mass in a reactor with a capacity of 56 L, with solvent (65%, m/m, ethanol) flow rate set at 665 mL min^{-1}.

Oil treatment/extraction with US is required to avoid changes in the organoleptic characteristics, mainly in view of the low rate of temperature increase. For example (Japón-Luján et al., 2008), edible oils (olive, sunflower, and soya) were enriched with phenols in olive leaves (i.e., oleuropein, verbascoside, apigenin-7-glucoside, and luteolin-7-glucoside). For extraction step, milled leaves (1 g) were placed into an extraction chamber of the dynamic approach which was assembled and filled with oil impelled by the peristaltic pump. The oil was circulated through the solid sample for 20 min under ultrasonic irradiation (duty cycle 0.5 s, amplitude 50%, 20 kHz, 225 W with the probe in contact with the top surface of the extraction cell). The temperature of extraction was controlled and maintained constant at 25°C. Using the oil flow rate at 7 mL min^{-1}, it was possible to enrich olive oil with up to 14 mg L^{-1} oleuropein and around 2 mg L^{-1} apigenin-7-glucoside, luteolin-7-glucoside, and verbascoside. In addition, phenols originally present in the olive oil (hydroxytyrosol, apigenin, and luteolin) were not degraded after application of US and extracted phenols remained stable after enrichment (Japón-Luján, Janeiro, and Luque De Castro, 2008). In other processes (Virot et al., 2010), an experimental pilot study was carried out for extraction of polyphenols from apple pomace. For this process, a 30 L extraction tank consisting of a quadruple US transducer at 25 kHz (4 × 200 W) was used. After 45 min of extraction, polyphenol yields can be increased up to 20% compared to the conventional processes using 1:1 (v/v) ethanol as solvent for extraction with 15% (w/v) of solid/liquid ratio in the same time.

US-assisted extraction has been used to assist the solvent extraction of bioactive, thermolabile, or unstable compounds from herbs with reduction of solvents, temperature, and extraction time (Boonkird et al., 2008). In a pilot plant (Boonkird et al., 2008), a reactor with 20 L of capacity was used for extraction of capsaicinoids from chili pepper with yield about 76%. In this case, the operational conditions of US such as frequency and power were set at 26 kHz and 1080 W, respectively for extraction of 3 kg of chili pepper with 15 L of 95% (v/v) ethanol at 45°C. Compared to maceration, this UAE system provides a significant reduction in temperature and time required for extraction.

In summary, the industrial use of US results in significant intensification over conventional techniques such as maceration and soxhlet extraction. Additionally, the use of US can lead to a significant reduction in the solvent amount required for the process (Gogate and Kabadi, 2009).

SONOCRYSTALLIZATION

US energy can be used as auxiliary for crystallization of organic and inorganic compounds. Crystallization is a thermal separation, and it is generally used as a separation process that yields a solid product from a melt, a solution, or a vapor. As for all thermal separations, nonequilibrium conditions are required as a driving force for the crystallization process to occur. Evaporation of solvent and/or temperature reduction are the most common procedures employed to establish the nonequilibrium conditions. In principle, pressure can also be used to force the nonequilibrium condition. However, this parameter is usually not used for crystallization, mainly in industrial processes. In addition, parameters such as chemical reactions (reactive crystallization), change of dielectric constant/ionic strength of a solution (salting-out), and crystallization induced by a change in solvent composition (drowning-out) can be used for crystallization (Mamun et al., 2006; Ulrich and Jones, 2007).

The main feature distinguishing crystallization from other thermal separation processes is the fact that crystallization leads to a solid product. Considering the same material, crystallization is a higher selective process and operates at much lower temperatures than separation by distillation. In view of the low energy consumption and the high capability to obtain high-purity products, crystallization is increasingly used in different industrial processes as in chemical, pharmaceutical, and petrochemical industries. The key factors in the design of any thermal separation process are the thermodynamics and the kinetics of the system under consideration (Bernardo and Giulietti, 2009). The first one defines the limits of what can be achieved. The kinetics defines the timescale of the crystallization and therefore the size of the equipment required. Considering industrial applications of crystallization, it is fundamental to know phase diagrams and solubilities of the materials to be separated before to project any industrial process (Westhoff et al., 2002). In crystallization, the nucleation and crystal growth are of utmost importance. Both of these phenomena are dependent on a large number of variables that in some cases may be ill defined (Luque de Castro and Capote, 2007a).

Given a sufficient driving force, that is supersaturation or supercooling in the cases of solutions or melt, respectively, a liquid-to-solid phase transformation initiates with the initial formation of clusters and ordered collections of the crystallization species. These clusters, or nuclei, are the precursors to the crystals that will be formed. In order to grow into a detectable crystal, these nuclei have to reach a certain critical size. The critical nucleus size is governed by the excess free energy (Gibbs energy) of the nucleus, which is given by the sum of surface excess free energy and the volume excess free energy of the particle. Plotting ΔG versus the radius of the particle, the critical radius is a measure of the critical nucleus size and is determined by the maximum in the excess of free energy. Below the critical radius re-dissolution of the nucleus is energetically favorable, and above the critical radius growth is more favorable since it leads to a reduction of the excess of free energy.

Different nucleation mechanisms exist and the earlier discussion applies to primary homogeneous nucleation, which will occur if the system is free of impurities, mechanically undisturbed, and when the thermodynamic nonequilibrium is at or beyond the metastable limit. In this case, a metastable zone limit denotes the composition of the solution at a given set of conditions where the solutions become labile and at which point spontaneous nucleation must occur. However, under actual circumstances there are always impurities in a solution or melt, such as by-products from synthesis, or particulate impurities such as dust or particles resulting from abrasion from the equipment. Mechanical disturbances result from agitation of the solution or vibrations. Realistic situation is therefore the case of primary heterogeneous nucleation, which occurs at a much lower supersaturation and where impurities or rough vessel walls function as nuclei (Ulrich and Jones, 2007).

The mechanisms of nucleation are classified as primary and secondary (Figure 21.8), the secondary mechanism being the most frequently observed. This type of nucleation requires the presence of crystals of the material to be crystallized and occurs at much lower supersaturation than primary nucleation. As a rule, secondary nuclei are formed by the removal of structured assemblies from the surface of the crystals. Basically, three mechanisms lead to the secondary nuclei formation: initial breeding, where the nuclei results from simply placing crystals into a supersaturated solution

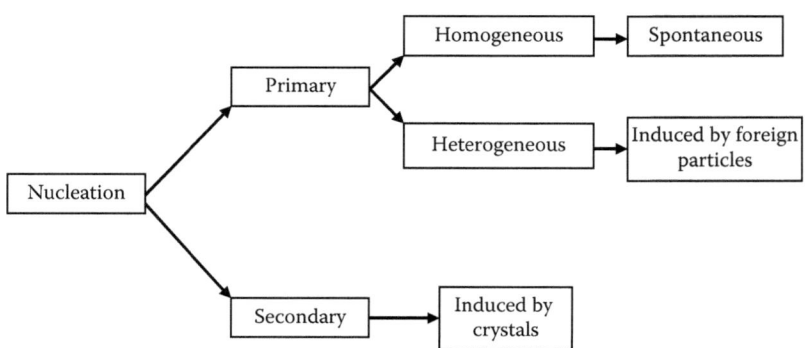

FIGURE 21.8 Classification of nucleation mechanisms. (Adapted from Westhoff, G.M. et al., *J. Cryst. Growth*, 237–239, 2136, 2002.)

or supercooled melt via the washing off of dust particles from the surface of the crystals; collision breeding, where the nuclei results from fragments of existing crystals which are broken off due to mechanical impact on crystal faces due to crystal–crystal, crystal–wall, or crystal–stirrer collisions; and fluid shear, where nuclei results from clusters or outgrowths being forced from the solid–liquid boundary layer due to shear forces resulting from liquid motion. A prerequisite for this behavior is that the growing crystal already has a size larger than the critical nucleus. Collision breeding is the most frequently observed and dominant secondary nucleation mechanism, at least in the majority of industrial crystallization (Westhoff et al., 2002).

Two aspects are considered in secondary nucleation. The positive aspect results from the fact that without secondary nucleation as a permanent source of new crystal nuclei, a continuous crystallizer with continuous crystal withdrawal would rapidly experience a lack of growing crystals. The negative aspect is that many more secondary nuclei are produced in an uncontrolled process than are required for crystallization. In this case, a very fine crystalline product is usually produced. However, it can be avoided by reducing the power input into the crystallizer. More important than the magnitude of the power input itself is the means by which the power is brought into the equipment. For nucleation and crystal grown, power is required homogenize the suspension and to transport the suspension. The power input is usually supplied to the crystallizer by pumps and impellers. Secondary nucleation rates can be controlled via diameter and tip speed of the impeller blade. Lower tip speeds and larger diameters result in lower secondary nucleation rates. However, when these measures are not sufficient to reduce the amount of fine particles, other measures may have to be taken to control the product. In other cases small particles will agglomerate. In the other cases a fine trap has to be introduced to the crystallizer. The trap consists of an additional loop within the crystallizer in which the conditions are such that all particles smaller than a given size passing though this loop are redissolved (Ulrich and Jones, 2007).

Primary nucleation is difficult to control and unreliable, as it will not always occur at precisely the same supersaturation. As a consequence, reproducibility of product quality cannot be guaranteed and the performance of the crystallizer will vary. Assuming that nuclei starts to grow at high supersaturation, liquid inclusions or dendritic growth are likely, as well as massive formation of small particles that tend to agglomerate. All of these phenomena lead to poor product purity and quality. Moreover, this may lead to strong tendency for "caking" in storage. If nucleation commences at too low a supersaturation, the ensuing crystal growth may be slow and will produce a problem with respect to production time or crystallizer size (Ulrich and Jones, 2007).

Therefore, to produce higher quality crystals in a reproducible manner, secondary nucleation by seeding is the preferred method of inducing crystallization. To achieve a good formation of crystals, it is always important to introduce the seed crystals at the same supersaturation condition. The optimum growth rate in the crystallizer should be established under conditions where the concentration in the solution remains at the center of the metastable zone. Maintaining constant growth rates requires good control of the supersaturation and one has to take into account the constantly increasing crystal surface area, which is ultimately responsible for the reduction in supersaturation (Ulrich and Jones, 2007).

With respect to seeding, the surface area of the seeds added has to be large enough to avoid any additional nucleation during the process is started or while it is in its early stages. Otherwise, a bimodal crystal size distribution could occur. On the other hand, the seeds should be small enough and low enough in number to produce the desired amount of product in the process. Choosing a large seed crystal size leads to a low yield; too many seed crystals lead to a small product crystal size. Moreover, an appropriate seeding, a good control of the supersaturation, and an optimized agitation provide the basis for a reproducible and controlled crystallization process in terms of number, as in size distribution of crystals in the final product.

An appropriate crystal growth only occurs if there is a driving force as a result of nonequilibrium thermodynamic conditions. The crystal growth rate is a physical property of a given material and depends on the temperature, pressure, composition of the mother liquor, supersaturation condition, fluid flow conditions, history of the crystals, nature of the surfaces of crystals, and the presence or absence of additives (impurities) in the mother liquor. Crystal growth rates are of key importance since they determine the retention time and therefore the size of the crystallizer.

Three physical phenomena occur in crystallization process: transport of the material to be crystallized from the bulk solution to the vicinity of the crystal surface, transfer of the material from the solution boundary layer to the solid state, and dissipation of the health of crystallization liberated at the point of growth. Crystal growth rates depend on temperature, supersaturation, and the fluid flow in the vicinity of the crystal surface. In general, the greater the crystal growth velocity, the higher the temperature and the higher the supersaturation. The influence of the fluid flow is more complex and in general a limit exists above which an increase in the flow velocity has no effect on the growth velocity. This limit coincides with a minimum thickness of the solid–liquid boundary layer that determines the mass transfer. The nature of the crystal surface also plays an important role in the growth rate. The perfection of the surface can influence the growth velocity, either accelerating or suppressing crystal growth. Furthermore, the liquid and its composition have a great influence on crystal growth (Ulrich and Jones, 2007).

Good yields and high product purity are desirable. These characteristics are achieved by maintaining a constant supersaturation throughout the crystallization process. In the case of batch processes, it requires a permanent adjustment of the driving force. In general, and mainly in the industrial process, solution crystallization should be best operated at the center of the metastable zone. Close to the upper limit of the metastable zone, a small fluctuation in the process variable is sufficient to induce nucleation which results in a reduction of the crystal size distribution and facilitate oscillations with respect to the output of crystal size distributions. On the other hand, if a process is operated close to the solubility limit, the crystallization process can be decreased.

Crystals have a well-defined three-dimensional order. Crystal structure is important even in industrial mass crystallization as the internal arrangement of the building blocks defines the macroscopically observable shape of the crystal and thus influences post-crystallization processing, that is, solid–liquid separation, drying, and packaging. Depending on the crystallization conditions, a given material may crystallize in different forms. The coexistence of chemically identical crystal forms is known as polymorphism. Polymorphism is important in the design of any crystallization process since different polymorphs will exhibit different physical properties. This aspect is critical in industries where product properties such as density, solubility, dissolution rate, bioavailability, color, and hardness are important. An example where these proprieties are important is in pharmaceutical products. The dissolution of the drug can influence the dose administrated to a patient: a wrong polymorph of a given drug may lead to inefficacy of treatment due to the low dissolution rate and, on the other side, the drug can be lethal if the dissolution is too fast (Dennehy, 2003).

Crystal shape or habit is another important physical characteristic of a crystal. This characteristic is determined by the crystal structure or the order of the individual building blocks. In some cases, crystals of the same chemical species with the same crystal structure can exhibit different habits. Different habits of the crystals can be a result of the growth rates of individual crystal faces that are subject to supersaturation effects and the influence of solvent and impurity molecules. For example, crystals of the same material grown from different solvents may have different habit. In general, a fast growth due to high supersaturation can lead to dendritic growth and results in high branched and fragile crystals. Fast growth can be economically interesting but can reduce the quality of the product.

The presence of impurities in the solution can also influence the crystal habit. Changes in crystal shape due to the presence of impurities can have negative and positive effects. Changes in filterability, drying, or the handling of the dry solids are negative effects, mainly if the crystals have acicular or tubular habit. However, the presence of impurities can be beneficial to tailor the shape of the crystals. In summary, different strategies can be used to influence the crystal habit (Ulrich and Jones, 2007).

Crystal size, usually defined as mean diameter of a size distribution, is an important requirement for some products. Flow properties, color, dissolution rates, agglomeration tendencies, mixing properties, and filterability are the factors that depend on the crystal size. However, the importance of mean size distribution of crystal depends on the application. In general, larger crystals are desirable but drugs, dyes, and pigments are typical examples where smaller particle size is of utmost importance.

Although crystallization is one of the oldest and economically most important separation and purification technologies in chemical industry, the design of crystallization processes still poses many challenges despite its wide application. Design is complicated compared to processes delivering a liquid product, because besides purity also properties like shape, polymorphic form, and size distribution have to be taken into account. These properties strongly influence the costs of the downstream solids processing train (e.g., filter area, drying time, and recrystallization steps) and also the added value of the final crystalline product (e.g., "caking" behavior, flowability, bioavailability, and color intensity) (Lakerveld et al., 2009). Thus many efforts have been made to improve the processes of crystallization not only with respect to the design of crystallizers, but also methods to induce crystallization.

The condition of supersaturation or supercooling alone is not a sufficient cause for a system to begin to crystallize. Before crystals can develop there must exist in the solution a number of minute solid bodies, embryos, nuclei, or seeds, which act as centers of crystallization. Nucleation may occur spontaneously or it may be induced artificially. It is not always possible, however, to decide whether a system has nucleated by itself or whether it has done so under the influence of some external stimulus. Nucleation can often be induced by agitation, mechanical shock, friction, and extreme pressures within solutions and melts. Relating to driving forces, nucleation of solid crystals from a number of liquids and melts can be influenced by US (Chow et al., 2003).

Crystallization in the presence of US, or sonocrystallization, exhibits a number of features specific to the ultrasonic wave. For most materials, these include: (a) a faster primary nucleation which is fairly uniform throughout the sonicated volume, (b) a relatively easy nucleation in materials which are usually difficult to nucleate otherwise, (c) the initiation of secondary nucleation, and (d) the production of smaller and purer crystals, which are more uniform in size (McCausland et al., 2001; Chow et al., 2003, 2005; Nalajala and Moholkar, 2010).

Applying US not only induces nucleation but also increases reproducibility. However, the precise mechanisms for US action on crystallization are not completely understood. US can induce primary nucleation in nominally particle-free solutions and at much lower supersaturation levels than would otherwise be the case. Another effect of US on nucleation is shortening the induction time between the establishment of supersaturation and the onset of nucleation and crystallization (Miyasaka et al., 2006; Luque de Castro and Capote, 2007b; Martini et al., 2008; Lakerveld et al., 2009; Wohlgemuth et al., 2009). In addition to the regions of extreme excitation, temperature and pressure created by bubble collapse, and concomitant release of shock waves, other postulates suggest that (a) subsequent rapid local cooling rates, calculated at 10^7–10^{10} K s^{-1}, play a significant role in increasing supersaturation; (b) localized pressure increase reduces the crystallization temperature; (c) the cavitation events allow the excitation energy barriers associated with nucleation to be overcome, in which case it should be possible to correlate the number of cavitation and nucleation events in a quantitative way; and (d) the application of US could induce modification of the wetting angle changing the crystallization behavior. Since all possible mechanisms appear simultaneously in the case of ultrasonic irradiation, it is not possible to analyze the accurate mechanism of nuclei formation (Luque de Castro and Capote, 2007b; Wohlgemuth et al., 2009). Moreover, US has been shown to significantly influence the reduction of agglomeration of crystals under given conditions. Agglomeration is not a desired condition in crystallization. Three US effects may contribute to this phenomenon. Thus, the shock wave, which is caused by cavitation, can decrease the contact between crystals to an extent precluding their bonding together. Also, some agglomeration invariably occurs at the nucleation stage. Nuclei possess a high surface area to volume ratio. It results in a high surface tension which nuclei tend to lower by adhering to one another. The surface tension decreases as crystals grow larger and become more stable, which hinder agglomeration. Finally, the excellent mixing conditions created by US also reduce agglomeration through control of the local nucleus population (Luque de Castro and Capote, 2007b).

US can affect the crystallization process by physical (facilitating the mixing and homogenization) and chemical (radical formation through cavitation) effects altering the induction time, supersaturation concentration, and metastable zone width (MZW). The influence is also a function of the particular medium to which this energy is applied (Luque de Castro and Capote, 2007b). Induction time, which is the time elapsed between the creation of supersaturation and the appearance of crystals, is significantly reduced by applying US. However, the induction time depends on the particular medium and working conditions. Therefore, contradictory results of induction time have been obtained, mainly for highly supersaturated solutions. For example, in the anti-solvent crystallization of roxithromycin using acetone–water mixture, the induction time decreases as supersaturation increases independently of the application of US. However, US significantly reduces the induction time, particularly at low levels of supersaturations (Guo et al., 2005). In general, the number of produced crystals initially increased when insonation time was increased, but decreased upon further increasing the insonation time. After this maximum, probably local heat effects of the probe caused dissolution while no new nuclei were produced due to the depletion of supersaturation (Kim et al., 2003; Guo et al., 2005). Therefore, the control of induction time is an important parameter when US is used to induce nucleation which depends on the desired final product. In general, higher induction times can increase the attrition of particles, reducing its size.

The MZW can also be reduced by the application of US (Guo et al., 2005). US decreases the apparent order of the primary nucleation rate and increases the rate of appearance of the solid. Seemingly, US modifies the mechanism of nucleation itself as its presence strongly reduces the apparent order of nucleation (Luque de Castro and Capote, 2007b). The decrease of supersaturation limit by US application has been attributed to its raising of the nucleation temperature. Thus, during nucleation, the cooling rate remains roughly constant; under silent conditions, however, a temperature rise is observed. After nucleation, the cooling rate decreases as the US power is raised.

Two opposing effects are involved: cooling is decelerated by the crystallization heat, but heat exchange is improved. US can induce nucleation under the conditions where spontaneous primary nucleation cannot occur in its absence, thus avoiding seeding and hence the introduction of foreign particles into the solution (Luque de Castro and Capote, 2007b). For example, in pharmaceutical industry, where seeding is not a good option since external engage in the crystallizers leads to difficulties in sterile processes and generates additional regulatory efforts. In this case, the use of US for nucleation can be a good choice (Kordylla et al., 2008).

US parameters such as frequency, intensity, power, horn tip design and horn immersion depth, volume of solution, and time of sonication have influence on the nucleation and crystallization process. In general, the use of low-frequency US (from 10 to 30 kHz) does not affect substantially the shape, mean size, and size distribution of the crystals. A possible cause is that wavelength is much larger than the size of the nuclei and crystals, and hence the effects of the insonation are similar (Li et al., 2003; Luque de Castro and Capote, 2007b). On the other hand, severe effects were observed in the crystallization of metallic glass by applying high-frequency US (Ichitsubo et al., 2004).

Increasing the US intensity and diameter of the horn tip increases the crystallization rate. These two effects physically contribute to the liquid flow patterns in the reaction vessel. An increase in US intensity is expected to result in heavier flow, while one in horn tip diameter should lead to more uniform flow patterns. From these observations, it can be concluded that the effect of cavitation known as "microstreaming" contributes little to crystallization, which is more markedly affected by "macrostreaming" (Nishida, 2004).

US power has a notable effect on crystallization. In general, there is a threshold for US power application. Increasing the US power up to a certain value increases the particle size. Above this value the particle size of the crystals decreases. Therefore, the particle size of the crystal can be controlled through the US power applied (Li-yun et al., 2005). Up to threshold limit, raising the US power produces shorter and thicker crystals. This fact can be attributed to mass transfer in the mixture being effectively accelerated and the driving force of crystallization increased as a result. With large kinetic energies and speeds, the solute molecules will have an increased opportunity to collide with each other, penetrate the stagnant film, and hence insert themselves into the crystal lattice more uniformly and easily. As the shape of the crystal depends on the growth rate at each face of the crystal, one may assume that the speed of insonated molecules is fast enough for them to approach each side of the crystal to compensate partly for differences in growth rate on each side in conventional crystallization, where diffusion control may occur. Therefore, one can expect a crystal insonated with a larger energy to be shorter and thicker (Li et al., 2003). In a general way, sonocrystallization provides a method for obtaining small crystals similar to supercritical fluid micronization, but with lower equipment costs and the ability to operate under ambient conditions (Luque de Castro and Capote, 2007b).

With an ultrasonic homogenizer, the flow pattern of the liquid depends on the distance from the horn tip. Since flow pattern (mixing) is the physical effect of US irradiation, any change in the flow pattern due to horn immersion may affect the crystallization rate. There is a specific horn immersion depth for each US device and irradiated medium which must be established experimentally on a case-by-case basis (Nishida, 2004; Luque de Castro and Capote, 2007b).

The volume of the solution to be irradiated by US exerts influence on mean crystal size. Increasing the volume of solution increases the mean crystal size. One explanation for this behavior is that a fixed US wave in a larger container produces weaker penetrating and reflecting waves, consequently vibration and cavitation at some point in the liquid are lower. This results in fewer nuclei, and hence in larger crystals being formed. Also, increased liquid volumes provide larger free spaces for crystals to reduce collision and abrasion with each other (Amara et al., 2001; Luque de Castro and Capote, 2007b).

In relation to US irradiation time, at short times the solution is not blended and precipitation occurs uniformly and little crystals are formed; longer times produce more crystals where the size decreases under continuous sonication.

Increasing the US irradiation time gives rise to the following sequence: at short times, the US wave fails to blend the solution and precipitant uniformly, so little precipitate is obtained after insonation; longer times produce apparent crystals, the size of which decreases under continuous sonication (Li et al., 2003). These results demonstrated that it is possible to "tailor" a crystal size distribution between the extreme cases of a short burst of US to nucleate at lower levels of supersaturation and allow growth to large crystal, and the production of small crystal via continuous (or perhaps a longer single burst) US application throughout the duration of the process, which can facilitate prolific nucleation at higher levels of supersaturation at the expense of some crystal growth. Pulsed or intermittent application of US can give intermediate effects. Therefore, the optimum time of US application needs to be determined experimentally (Luque de Castro and Capote, 2007b).

Whether US irradiation affects the characteristics of the crystals formed seemingly depends on the particular system. Some authors reported that US has not a significant effect on crystallization (Nishida, 2004). However, other works have clearly demonstrated the influence of US. For example, in the antisolvent crystallization of roxithromycin in an acetone–water mixture, the crystals exhibit a hexagonal and rhombus shape in the absence and presence of US, respectively (Guo et al., 2005). In the particular case of crystallization processes induced by the addition of an antisolvent, where high supersaturation levels may be produced very rapidly, the application of US reduces not only the induction times of nucleation but also the spread of variability in induction time at a given level of supersaturation (McLausland et al., 2001; McLausland and Cains, 2002). For a number of molecules it has been shown that significantly less antisolvent can be used in conjunction with US to induce crystallization in a controlled manner (Luque de Castro and Capote, 2007b).

The influence of US irradiation on crystal habit and mean size distribution was also shown for the crystallization of several substances as adipic acid (Wohlgemuth et al., 2010), dextrose monohydrate (Devarakonda et al., 2004), spectinomycin hydrochloride (Li et al., 2003), polystyrene (Mamun et al., 2006), hydroxyapatite (Li-yun et al., 2005) potash alum (Amara et al., 2001), disodium hydrogen phosphate dodecahydrate (Miyasaka et al., 2006), and ammonium sulphate (Virone et al., 2006). The effect of US has been attributed to an increased or decreased growth rate of some crystal faces under the influence of hot spots, which can alter the crystal lattice. On the other hand, abrasion may have some effect on the crystal habit.

Usually, small molecules with high levels of saturation, high nucleation rates, along with concomitant poorly controlled crystallization, leads crystals with undesirable needle habit. However, when a solution is treated with US at much lower levels of supersaturation, a high desired habit is produced. Besides, careful sonication allows the particle size to be controlled (Chow et al., 2003, 2005; Ruecroft et al., 2005).

Controlled application of US to a polymorphic system at the right level of supersaturation can help in isolating the ground-state polymorph (the most thermodynamically favored and less soluble) or one near the ground state. This availability to induce the formation of a given polymorph under US action is of paramount importance in the pharmaceutical industry (Dennehy, 2003; Kim and Kiang, 2003; Li et al., 2003, 2006; Bernardo and Giulietti, 2009).

Sonocrystallization also avoids the problems involved in intentional seeding, very common in industrial crystallization process. The effects of intentional seeding include narrowing of the MZW, shortening of induction times, and control of particle size distribution. In a batch process, seeds have to be added at precisely the correct time during the development of the supersaturation profile. In the case of sonication, very small seed crystals are generated which offers all the advantages of conventional seeding without many of the drawbacks such as handling, actual physical size of the seeds, and high quality of the seed mainly in pharmaceutical industry. Applying US is to easily control the exact point of nucleation and the degree of number of nuclei generated (Luque de Castro and Capote, 2007b).

The effects of US on crystal growth do not appear to be as remarkable as those on nucleation and arise largely from enhanced bulk-phase mass transfer. The mechanical disturbances created by both cavitation and ultrasonic streaming alter the fluid dynamics and increase bulk-phase mass transfer of solute to the surface of the growing crystal. The surface nucleation and integration effects at the crystal surface determine, however, the growth rate of each individual face and, hence, the habit of the crystal. Theoretical studies suggest that the effects of US on crystal growth rate depend on the magnitude of the supersaturation driven force. At low supersaturation, with growth velocities at the crystal faces around 10^{-10} m s^{-1}, the application of US doubles the growth rate, while at higher supersaturation with growth velocity around 10^{-7} m s^{-1} there appeared to be no effect. The US effect is explained by the hypothesis that, at low supersaturation, the quantity of available growth units in the vicinity of the crystal surface is small. Under these conditions, bulk-phase mass transfer becomes rate limiting in supplying growth units to the crystal surface, and its ultrasonic enhancement will enhance the growth rate (Luque de Castro and Capote, 2007b).

Most of the sonocrystallization applications have been carried out in laboratory scale at milligram to gram amounts using either high-intensity US probes or bath systems. In general, good results are shown using this scale level. However, in a similar way as for some applications it seems that there are still some difficulties when sonocrystallization will be applied in large scale. The main difficulty seems the construction of large-scale ultrasonic reactors, especially for use with high-intensity US probes. To circumvent these difficulties, uniform fields of ultrasonic energy density above the cavitational threshold can be created using multi-transducer systems. So, the volume of the cavitating region can be extended. Therefore, several transducers must be properly distributed around the reactor in order to avoid problems such as erosion of the reactor walls due to the effect of US (Ruecroft et al., 2005). These type of reactors can be used at kilogram scale or higher.

In summary, all processes should be examined on a case-by-case basis. There are also potential applications for power US technology that are scaled down from the conventional laboratory up to microgram scale. The applications consist mainly for microscale mixing, and where only very small quantities of material will be available. The size of the US devices limits the scale at which they may be operated. In addition, they also suffer further disadvantages regarding the distribution of the energy intensity that they deliver into the reactor. A bath will deliver nonhomogeneous acoustic fields throughout the medium with maximum amplitude at multiples of the half-wavelength of sound. The nonhomogeneity of the acoustic field means that extreme care must be taken in relation to depth and position where transducers are mounted in the crystallizer. In addition, the higher power levels will be at points closest to the base and will dissipate with increasing distance from the transducer. Despite some difficulties, several works illustrate the utility of US and how these systems can be used to improve crystallization processes and become new tools for the process chemist. A scheme of a relatively simple flow system used for crystallization is shown in Figure 21.9 (Dennehy, 2003; Kim and Kiang, 2003; Luque de Castro and Capote, 2007b).

Probe systems deliver very high intensity at the tip of the probe, but the energy density is concentrated in the axial direction, away from the tip, and falls away rapidly with distance according to an inverse square law. Indeed, it is not possible to transmit an intense cavitation field more than 2–5 cm beyond the end of the probe, nor is it a suitable means to transmit the acoustic energy into large process volume, thus making it difficult to scale-up. Probe systems only work effectively if operated in geometry where most of the working liquid is constrained within the longitudinal high-intensity region or where the liquid is stirred vigorously. In addition, these systems suffer from erosion and particle release at the tip surface, they may also be subject to cavitational blocking, and the large transducer displacement increases stress on the material of construction, resulting in possible failure. Another aspect of scale-up is that it is effect specific. Different applications depend on different effects arising from US. So, large-scale ultrasonic equipment have been developed in many different ways, which may be classified as (1) probes or other small area devices delivering very high local US intensities, in a flow cell or large volume, and (2) opposing parallel transducers arranged around a duct, through which the process solution or suspension flows (Dennehy, 2003; Ruecroft et al., 2005).

US is used both for crystallization of organic and inorganic molecules. There are several reports dealing with this issue and describing the main features of the use of US in this process. However, several works are based on fundamental studies using model molecules. Two applications focusing on the use of US in crystallization processes are described in the following.

Crystallization is a critical operation in the manufacture of active pharmaceutical ingredients (API) (Kim and Kiang, 2003; Ambrus, 2010). The development of a chemical process for the API final step may face many challenges related to crystallization and particle characteristics. Commonly encountered challenges include purity, yield, oiling-out/amorphism, polymorph control,

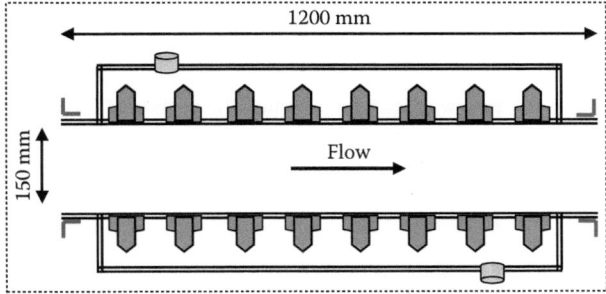

FIGURE 21.9 Scheme of a tubular flow system used for sonocrystallization. The black "points" along the tube corresponds to transducers arranged around the tube wall (40 transducers of 50 W each). (Adapted from Ruecroft, G. et al., *Organ. Process Res. Dev.*, 9, 923, 2005.)

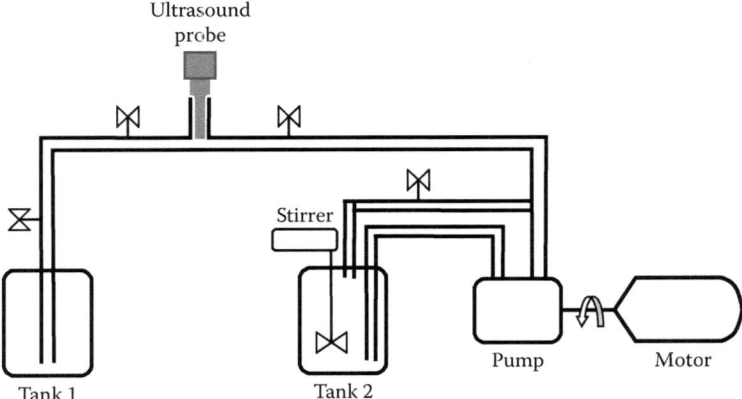

FIGURE 21.10 Schematic of a flow apparatus used to study sonocrystallization. (Adapted from Devarakonda, S. et al., *Cryst. Growth Des.*, 4, 687, 2004.)

and compound stability during processing. Other issues related to particle formation are particle size and distribution, crystal habit, filterability, crystal attrition or agglomeration, and the bulk powder properties related to formulation, such as flow ability, bulk density, and compatibility. Other important considerations are the scalability and reproducibility/robustness of the process to yield consistent product. Many of the problems can be addressed by the development of crystallization protocols with careful process control and optimized process conditions, such as solvent/antisolvent choice, temperature, agitation, and seeding. The other issues related to particle and bulk properties may be addressed by the development of specialized crystallization techniques, using an approach to "engineer" particles during crystal formation to manipulate the particle size or habit through the control of nucleation and growth mechanism (Kim and Kiang, 2003).

In this sense, the US can be applied in a flow crystallization system, which was designed to mimic the industrial process, in which dextrose monohydrate is in a flow environment before it finally passes into the batch crystallizer where it is slowly cooled (Devarakonda et al., 2004). The proposal of this investigation was focused on the impact of US on the seed size, lump breakage, and the induced nucleation of the dextrose monohydrate. The main effects of the US irradiation in the flow system are the seed size, and the low exposure/residence time. US can be used to break up undesirable dextrose lumps that form in the process pipes.

A saturated (or supersaturated) solution of dextrose is pumped through a flow system (Figure 21.10) where the US energy is applied, before the introduction into the crystallization vessel. US energy was supplied to the system via an ultrasonic probe, which is mounted on a flow pipe (approximately 12.5 mm internal diameter). The US system was operated at 20 kHz. The effect of US on the seed size of dextrose monohydrate was evaluated by pumping a saturated solution (50.44% dextrose by weight at 24.4°C with seeds) through the system at a fixed flow rate. US power was set at 80%. The flow system can also be used to study the effect of US on breakage of dextrose monohydrate lumps added to an unseeded saturated dextrose solution. The solution was pumped through the system at different residence times, while applying US energy set at 60%. After application of US, the solution is collected in the crystallizer. Using this system, the effect of US energy on the nucleation and subsequent crystallization of dextrose was investigated for saturated and supersaturated dextrose solutions (Devarakonda et al., 2004).

METAL AND PLASTIC WELDING

An interesting industrial application field of power US is welding of plastics or metals via specific heating at the junction between the pieces of material (Mason, 2003) and the development of new transducers design for welding devices is extremely important. Nowadays, frequencies about

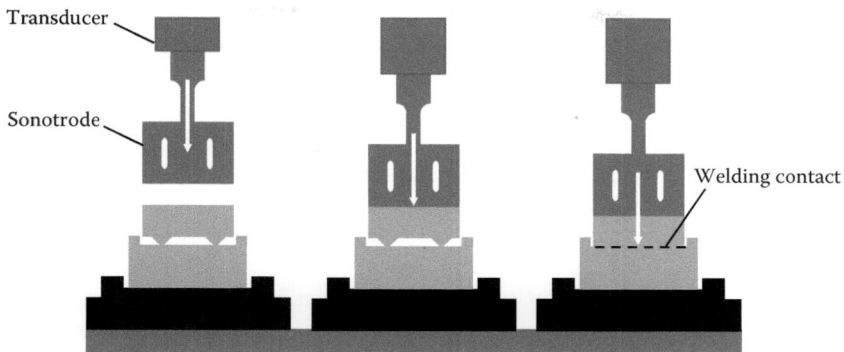

FIGURE 21.11 Scheme of plastic ultrasonic welding process. (Adapted from Truckenmüller, R. et al., *Sens. Actuat. A*, 132, 385, 2006.)

120–130 kHz have been frequently used for wire bonding. However, a new trend is related to the use of frequencies about 230 kHz. This higher frequency of the transducers results in a shorter bonding time, increasing the speed of the wire bonders. In addition, at higher frequencies, the same bond efficiency can be achieved at lower bonding temperatures (Pirrini, 2001).

Taking into account the plastic welding, it is necessary that any two plastic materials to be welded together should be chemically compatible. If they are not compatible, there can be no chemical bonds, even when they melt at the same temperature. Several factors which can affect the weldability of plastic parts including hygroscopicity, mold release agents, lubricants, plasticizers, fillers, flame retardants, regrind, pigments, and resin grades. The high quality of welding is dependent on a large number of factors such as the amplitude of the vibrations of the working and of the waveguide, frequency of mechanical vibrations, welding time, and static pressure caused by the clamping force of the waveguide on the welded parts (Gutnik et al., 2002). Moreover, the joint design is critical in achieving optimum assembly results. The joint design of a particular part depends on factors such as type of plastic, part geometry, and the requirements of the weld (Ensminger, 2009). Taking into account several uses of high-performance polymers, it is widely applied in the field of pharmaceutics, biotechnology, and life sciences. In this way, US energy can be used for welding of plastic parts. Figure 21.11 shows a scheme of the plastic ultrasonic welding process (Truckenmüller et al., 2006a,b).

Ultrasonic welding of metals or alloys, as well as plastic welding, is widely applied in industrial processes. In most of the cases, metal (or alloy) welding is performed using welding machines with frequency and power operating at 15 kHz and 1200–2400 W, respectively. However, the time required for welding is generally not higher than 5 s (Imai and Matsuoka, 2005; Ishikuro and Matsuoka, 2005; Watanabe et al., 2009). Aluminum welding is one typical application which is difficult by normal methods due to its tenacious oxide surface. By using ultrasonic metal welding—a form of low-temperature welding—the layer of oxide can be easily broken up and adsorbed within metal surrounding the weld. Welding by lateral vibration movement is readily achieved using this technique without the formation of brittle intermetallic compounds (Mason and Lorimer, 2002). Ultrasonic welding (metal and plastic welding) allows a very delicate joining of components, which makes use of US in welding an increasing area for US developments.

CRUDE OIL INDUSTRY

US-assisted oxidative desulfurization (UAOD) has received increasing interest in the past few years due to strict regulations which require that sulfur compounds have to be removed completely from fuels (online document, available at www.epa.gov, 2000; Babich and Moulijn, 2003;

Song, 2003; Wu and Ondruschka, 2010). Hydrodesulfurization (HDS) is the current industrial method to remove sulfur from crude oil, and it is a hydrotreatment process based on the use of hydrogen to break up the bonds of sulfur-containing compounds resulting in hydrogen sulfide and hydrocarbons. Nevertheless, this process requires high temperature (about 400°C), high hydrogen pressure (up to 100 atm) combined to the use of metallic catalysts (usually CoMo and NiMo-type), large reactors, and excessive residence time, resulting in higher operating costs. In addition, the conventional HDS process is efficient for mercaptans, thioethers, sulfides, and disulfides removal, but it has shown some limitations regarding the treatment of aromatic sulfur compounds such as thiophene, benzothiophene (BT), and dibenzothiophene (DBT). In this sense, several nonconventional methods have been developed to overcome the limitations of HDS, such as extraction with ionic-liquid, selective adsorption, electrochemical oxidation, biodesulfurization, and oxidative desulfurization (ODS) that can also be combined with US (Babich and Moulijn, 2003; Mello et al., 2009; Wu and Ondruschka, 2010). ODS has been considered a promising method for deep desulfurization technology because it can be carried out under mild conditions, such as relatively low temperature, pressure, and cost of operation when it is compared with HDS.

ODS is based on the conversion of sulfur compounds to its corresponding sulfoxides/sulfones which can be easily removed from oil due to its increased polarity. Several oxidizing systems have been studied in a laboratory scale, generally with the use of hydrogen peroxide and/or different combinations, mainly with a peroxyacid generated in situ by the reaction of hydrogen peroxide and an appropriate carboxylic acid (Babich and Moulijn, 2003; Ukkirapandian et al., 2008; Wu and Ondruschka, 2010). US has shown to improve the process due to cavitation that results in extreme local temperature and pressure with drastic liquid jets that arise from violent collapse of each bubble. US improvement for desulfurization can be mainly obtained from the better surface chemistry due to the enhancement in micro-mixing with changes in the reaction kinetics (Wu and Ondruschka, 2010). Sulfur compounds oxidation can be performed at the interface or in the bulk phase requiring good dispersion of the solvent and fuel phase that can also be improved by the very fine droplets created by the ultrasonic pulse (Deshpande et al., 2005).

In a recently reported scale-up study showing the efficiency of UAOD procedure (Wu and Ondruschka, 2010), a high sulfur removal of MGO (92%) was obtained. System was based on a single sonoreactor at a treatment rate of 12.5 lb h^{-1}. Two sonoreactors were connected in parallel to reach high sulfur reduction under a treatment rate of 25 lb h^{-1} which represents approximately 2 barrels per day. This sonoreactor demonstrated the feasibility of large-scale operation even in a relatively small installation under ambient environmental condition. The total processing costs were estimated as 5.49 cent gallon^{-1} excluding the cost for solvent extraction.

Regarding the use of US to oil treatment, SulphCo has patented and commercialized the Sonocracking™ technology, as shown in Figure 21.12. This technology is based on the use of high-power ultrasonics—the application of high-energy and high-frequency sound waves—in conjunction with catalyst and oxidant regimes to alter the molecular structure of crude oil fractions (i.e., petroleum products), crude and natural gas condensates, and crude oil. The Sonocracking™ process is designed to enhance the quality of petroleum products, condensates, and crude oil by oxidizing sulfur species,

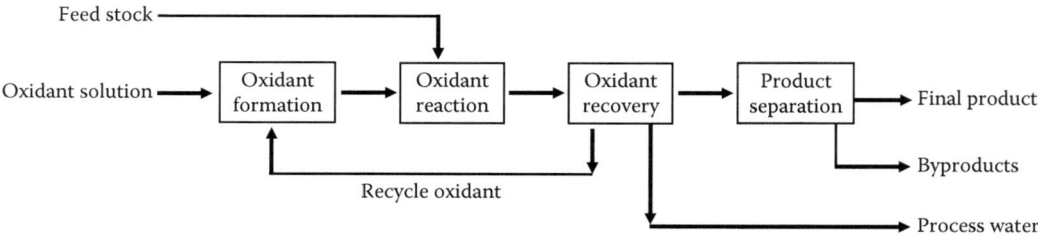

FIGURE 21.12 Scheme of industrial Sonocracking™ used for sulfur removal from crude oil fractions. (Adapted from www.sulphco.com)

enabling those oxidized sulfur compounds to be processed and removed using commercially available techniques such as adsorption and extraction (online document, available at www.sulphco.com, accessed in July 2010).

OTHER PROCESSES: FILTRATION, SEPARATION, AND CLEANING

There are industries with expertise in a wider range of activities involving the use of US, as in phase separation (Mason, 2003). In this way, the separation of fine particles from gases or liquids is a topic of permanent industrial attention (Sarabia et al., 2000). In addition, the use of US for separation of food components has also been recently investigated (Feng et al., 2008). However, ultrasonic standing waves have the ability, for example, to move cells in suspension into bands in a standing wave field separated by a half-wave acoustic wavelength (Sarabia et al., 2000; Stack et al., 2005). A pilot-scale process was developed to evaluate the suitability of US to coagulate grease from wastewater. US at $0.15\,W\,cm^2$ of transducer surface area, and $4.3\,mW\,mL^{-1}$ of treatment liquid was introduced (Stack et al., 2005). It was observed that sonoelectrocoagulation is capable of removing up to 100% of soap and grease from wool scouring effluent, allowing achieving compliance with regulatory limits of $250\,mg\,L^{-1}$ for effluents with pretreatment grease-plus-soap levels in the typically encountered range of 2000–$3000\,mg\,L^{-1}$.

An important application field of US for separation is related to the improvement of efficiency and/or selectivity of the flotation process (Cilek and Ozgen, 2009; Ozkan and Kuyumcu, 2007). In this sense, an investigation regarding the effect of US in the froth on performances of the pulp phase and froth phase in flotation of a complex sulfide ore was carried out (Cilek and Ozgen, 2009). US was applied during different stages of flotation, such as prior to or during flotation, using a flotation machine with a 2L cell, containing an ultrasonic probe (20 kHz and 200 W) located in the froth phase. Results showed that the use of US in the froth phase resulted in significant improvement of a complex sulfide ore flotation at intermediate and high level airflow rates. However, no significant differences in separation performance were obtained from the flotation with and without US at low airflow rates. In all cases, water recoveries were increased. In addition, to increase the effective pulp volume in a flotation cell by using the shallow froths, the quality of product can also be increased by the use of US in the froth phase, since the efficiency of the froth depth is increased by the use of US in the froth.

The separation of solids from liquids or isolation from original liquor is common to many industries. Generally, this step is performed with membranes of various sorts which have been employed in processes ranging from the simple filter pad through semipermeable osmotic type membranes to those which are used on a size-exclusion principle for the purification of polymeric materials. However, fouling is one of the main problems in membrane filtration and, as a consequence, there will always be the need to either replace filters or stop the operation and clean them on a regular basis. In industrial applications, this step is time- and money-consuming (Mason and Lorimer, 2002). The level of membrane fouling is dependent on the feed suspension properties (particle size, particle concentration, pH, ionic strength), membrane properties (hydrophobicity, charge, pore size), and hydrodynamics (cross-flow velocity transmembrane pressure) (Kyllönen et al., 2005). Since an efficient filter cleaning and improved filtration are required, the use of additional forces such as US has gained increasing attention in recent years once it is related to the increase in the flux primarily by breaking the cake layer and by decreasing the solute concentration at the membrane surface (Kobayashi et al., 1999; Lamminen et al., 2004; Muthukumaran et al., 2004).

The application of ultrasonic piezoceramic transducers operated in frequencies of 20, 40, or 200 kHz were evaluated in the enhancement of cross-flow membrane filtration using real industrial wastewaters from the paper industry (Kyllönena et al., 2006). The filters used in the cross-flow experiments were alumina-based ceramic membranes with mean pore size of 0.12, 0.19, 0.25, and $0.75\,\mu m$. The power intensity needed during filtration was so high that the membranes eroded gradually at some spots of the membrane surface. It was discovered that the ultrasonic field produced by the used transducers was uneven in pressurized conditions. On the other hand, the US treatment at atmospheric pressure during intermission pauses in filtration turned out to be an efficient and, at the same time, a gentle method to membrane cleaning. An input power of 120 W (power intensity of 1.1 W cm^{-2}) for a few seconds was enough for cleaning and the flow improvement was significant when 5 s of ultrasonic treatment was applied to the filters during a short pause in filtration. The US irradiation was carried out using either a frequency of 27 or 200 kHz at an input power of 200 W (1.8 W cm^{-2} or 4.1 W cm^{-2}) and the influence of ultrasonic irradiation was more pronounced when 27 kHz was used.

Although it is clear that US effectively cleans membrane surfaces and maintains high water fluxes, the mechanisms involved in cleaning surfaces by US are still largely unknown. According to the main mechanisms suggested for particle removal by US, this type of decontamination is effective due to the acoustic streaming, microstreaming, microstreamers, and microjets. Other possible ultrasonics mechanisms of surface cleaning include vibrations, chemical interactions with radicals, and shock waves resulting from the collapse of bubbles (Kuehn et al., 1996; Lamminen et al., 2004). The particular advantage of ultrasonic cleaning in this context is that it can reach crevices that are not easily reached by conventional cleaning methods, and objects for cleaning can range from large crates used for food packaging and transportation to delicate surgical implements such as endoscopes.

The overall process of most ultrasonic cleaning systems is heated stainless steel tank which contains a number of ultrasonic transducer and a detergent/sterilizing solution. The items to be cleaned are passed through the tank and subjected to the US as they pass in front of the transducer (Mason and Lorimer, 2002).

The cleaning effects of US are related to the microjets resulting from collapsing bubbles at a solid boundary and this feature was used in a laundry process as a promising technique to intensify the mass transfer (Warmoeskerken et al., 2002). The release of sodium chloride from the cotton test swatches was monitored by conductivity measurements of the bulk fluid in an ultrasonic bath working at 33 kHz. It has been found that the rinsing of salt from textiles can be speeded up with a factor of 6 when compared to a conventional washing procedure. In the same way, the combination of US and enzymatic treatment for noncellulosic component removal from cotton and consequent improvement of absorbent fiber capacity was reported (Yachmenev et al., 2001). Therefore, introducing magnetostrictive transducer working at 16 and 20 kHz and ultrasonic energy of 3 W cm^{-2} in the reaction chamber during enzymatic scouring of cotton fabric significantly improved pectinase efficiency. It was pointed out that the sonication of pectinase processing solutions did not impair the complex structure of the enzyme molecules but significantly improved their performance, and this combination could help overcome the longer processing time compared to conventional alkaline scouring.

Power US can also be used in the processing of minerals in order to clean their surfaces of oxidation products and fine coatings. In this sense, the use of US to remove impurities from the surfaces of silica sand for further glass manufacturing is reported (Du et al., 2010; Farmer et al., 2000a,b). A 12.7 mm tip diameter horn with 550 W and a cup horn with 63.5 mm diameter and 330 W of nominal power were used. Both processors worked at 20 kHz. Silica sand typically assaying from 0.025% to 0.30% Fe_2O_3 was subjected to sonication in water, dilute sulfuric acid, sodium hydroxide, and sodium carbonate solutions to determine whether a lower iron (less than 0.015% Fe_2O_3) product, suitable for tableware production, could be obtained. The results showed a reduction from 0.025% to 0.012% Fe_2O_3 using 1 min of sonication and sodium carbonate as washing solution.

The estimated capital and operating costs of commercially available ultrasonic processing units are comparable to that of wet high-intensity magnetic separation. Since the equipment available or required modifications will perform as expected, sonication can provide another mechanism not only for mineral processing, but for surface cleaning as a whole. In addition, it would be more environmentally friendly than the acid leaching process currently in use to reduce iron contamination on silica grains.

ENVIRONMENTAL REMEDIATION USING ULTRASOUND

A polluted ecosystem can result in agricultural productivity losses, contamination of water, and human and animal illness through direct ingestion of dust and the consumption of foods which have been grown on contaminated land (Adewuyi, 2001). With regard to the regulations for environmental remediation and its stricter limits, improved methods of both preventing and reducing pollution are required. In this way, the environmental treatment based on cavitation effects for the destruction of biological and chemical pollutants in water is of great interest and has received a considerable attention (Adewuyi, 2001; Mason et al., 2003; Gil et al., 2008; Collings et al., 2010; Cravotto et al., 2010; Sostaric and Weavers, 2010). However, ultrasonic energy devoted to environmental remediation is not restricted to these topics and it has also been used to remove airborne contaminants, surface cleaning and decontamination, washing of soils, sewage treatment, and to break down toxic compounds in water and soil (Mason and Lorimer, 2002).

WATER AND WASTEWATER TREATMENT

Water plays an essential role in supporting all life. In this sense, a lot of cases devoted to the destruction of biological and chemical pollutants are reported in aqueous systems. Among the disinfecting methods for water cleaning, the use of chlorine and its compounds have been known for many years (Mir et al., 1997; Phull et al., 1997; Ince et al., 2001). In addition, the mineralization of pollutants by ultrasonic irradiation or by coupling US with other free energy sources (e.g., UV radiation) or chemical oxidation using H_2O_2, ozone, or Fe(II) compounds are also found in literature as attractive approaches for water and wastewater remediation (Adewuyi, 2005b; Blume and Neis, 2005; Chand et al., 2007).

The main effects of power US in a liquid medium are thought to be the result of the effects of cavitation that generates forces with dramatic influence on biological systems. Therefore, US irradiation can actually be considered a combination of chemical reactions using the formed radicals and physical effects that are associated with an increase in temperature due to the local turbulence, acoustic streaming, and bubbles cavities generation (Bougrier et al., 2005). The high local temperatures and pressures provided by US would be able to disintegrate biological cells and denature enzymes. In addition, the imploding bubble produces high shear forces and liquid jets in the solvent that may also have sufficient energy to physically damage the cell wall. Stable oscillating refers to bubbles that oscillate in a regular fashion during many acoustic cycles and induce microstreaming in the surrounding liquid which can also induce stress in any microbiological species present. It is important to point out that US parameters, such as frequency, intensity, power, and irradiation time, can determine any potentially damaging effects on biological molecules. These molecular effects of US on biological cells are summarized in Table 21.2 (Mann and Krull, 2004; Rokhina et al., 2009).

Apart from the various factors dependent on the cavitation phenomena which affect the efficiency of disruption and inactivation of microorganisms, some factors like the size of the cell, shape of the cell, stage of development, and microbiological species have also been reported to affect the extent of disinfection (Scherba et al., 1991; Thacker, 1973).

Although the sonochemical effects have been observed many decades ago and despite the recent advances of sonochemistry, the mechanisms of homogeneous and heterogeneous sonochemistry are not fully understood. Concerning the thought of expensive power consumption, scale-up reactors design and handling difficulty have changed as a result of recently developed applications in synthesis and pollution control that have prompted interest in industrial scale operation (Adewuyi, 2005b; Mason et al., 2003; Mason and Lorimer, 2002; Suslick, 1990).

In spite of chlorination to be broadly used as an effective way for water disinfection, it has attracted some criticism due to the secondary effects since some species of bacteria produce colonies and spores, which can agglomerate in spherical clusters. In this way, the biocide can destroy microorganisms on the surface of such clusters but often leaves the innermost bacteria intact (Mason, 2007). For this reason, the need for an efficient microbial cell disruption operation is required to provide

TABLE 21.2
Molecular Effects of US on Biological Cells

Changes	Description
Physical changes	
Temperature effect	The extent of cell damage depends on the absorbed energy, the maximum temperature achieved, and the exposure duration (cell damage can include partial and full lysis)
Cavitation	Changes to ultrastructures within cells
	Altered enzyme stability
	Cellular effects caused by altered growth properties, which could lead to cell lysis
	Nucleus rupture and release of DNA
	Breakage of extracellular polymer substances
Chemical changes	
Radical generation induced by cavitation	Formation and release of compounds such as nitric acid, nitrous acid, and hydrogen peroxide due to interaction of formed radicals with the cell
	Decreased cellular stability
Stress-induced changes	
Acoustic streaming	Enhanced mass transport inside and outside the cell due to altered membrane permeability
	Alteration of cell surface charge
	Rupture of cell membranes

Sources: Adapted from Mann, T.L. and Krull, U.J., *Biosens. Bioelectron.*, 20, 945, 2004; Rokhina, E.V. et al., *Trends Biotechnol.*, 27, 298, 2009.

ultimate bacterial destruction in infected waters (Geciova et al., 2002; Ince et al., 2001). Studies involving the use of sonochemistry for microorganisms destruction have been of considerable interest since the 1920s when Harvey and Loomis (Harvey and Loomis, 1929) reported the reduction in light emission from a sewage suspension of rod-shaped *Bacillus fisheri* caused by sonication at 375 kHz under temperature-controlled conditions. In spite of the fact that this work has shown a significant effect of power US for killing bacteria and solutions sterilization, the method was not considered with commercial importance due to the relatively high cost of the process. After that, Hughes and Nyborg (Hughes and Nyborg, 1962) reported a work targeted to explain the mechanism of US interaction with microbial cells and they found an association with cavitation, localized heating, and free radical formation. In order to improve the efficiency of cell disruption, new techniques using acoustic cavitation have received great attention and improvements.

TREATMENT OF BIOLOGICAL AND CHEMICAL CONTAMINANTS IN WATER

The increase of chemical activity by ultrasonic waves is an emerging technology for water and wastewater remediation, owing to the advantages of high energy induced by extreme conditions achieved during the collapse of cavitation bubbles. This sonochemical condition provides pollutants destruction either directly via activating thermal decomposition reactions, or indirectly by the production of free radicals, such as of hydroxyl radical, in advanced oxidation process (AOP) (Ince et al., 2001). In this way, the effects of US alone and combined with chlorine upon the destruction of *Escherichia coli* in water samples was reported by Phull et al. (1997). According to this work, approximately 80% of the microorganism content in water samples was destroyed or inactivated after 15 min of sonication with an ultrasonic probe working at 20 kHz and 50 W of nominal power.

When combined with chlorine treatment, US enabled the reduction of chlorine required for disinfection. In addition, at higher frequencies such as 800 kHz the disinfection of water was more efficient than at lower frequencies (20 kHz).

In addition, the cavitation phenomena in combination with other oxidation process for hazardous environmental sample treatment have been reviewed (Gogate, 2008; Khanal et al., 2007; Mahamuni and Adewuyi, 2010; Pilli et al., 2010). In this sense, the synergistic effect due to the application of US in the presence of ozone (Lesko et al., 2006), advanced Fenton process (AFP) (Namkung et al., 2008), photocatalysts (Chen and Smirniotis, 2002), and enzymes (Entezari and Petrier, 2005) is reported.

In view of chemical reduction in water treatment and to reduce overall water consumption, non-chemical treatments are increasingly using US for wastewater treatment or potable water disinfection (Broekman et al., 2010), once diverse microorganisms have been stressed by US in aqueous medium. Unless controlled, microbiological growth in cooling water systems can lead to diminish operational efficiency and increasing maintenance costs and system downtime. In this sense, a patented water treatment system (Sonoxide®, http://ppd.herc.com/index.asp) assisted by US (Figure 21.13) was evaluated for planktonic and sessile bacteria in water as well as to prevent biofilm formation (Cordemans et al., 2003; Swinnen et al., 2008).

The system works by passing an air/water mixture or air-induced microbubbles through the Sonoxide chamber where bacterial cells are exposed to low-power and high-frequency ultrasonic energy. The ultrasonic emitters are in direct contact with the fluid to be treated and operate about 1.5–2.0 MHz and below 10 W cm^{-2} of power. The residence time in the ultrasonic chamber is about 1.5 s and the system size is based on the water quality (e.g., COD, total suspended solids) and system dynamics (e.g., volume, effluent water loss rates, and average retention time). The chamber can vary in size, number of US emitters (six to eight), and flow processing capabilities from 2 up to 70 m^3 per hour. Some examples of this US water treatment system were published (Broekman et al., 2010) for cooling water tower from university, dairy industry, chemical industry and sintering furnace, water from emulsion treatment, in glass washing system in glass manufacturing, and for water from automotive paint shop process. Results showed that the Sonoxide ultrasonic treatment technology enables control of bacteria, algae, and biofilms throughout an entire industrial system, allowing for reduction and sometimes elimination of chemical biocides.

In biological wastewater treatment, large amounts of biosolids (sewage sludge) are produced. Since the sludge is highly susceptible to degradation, it has to be stabilized by anaerobic digestion in order to enable an environmentally safe utilization and disposal. Due to the rate-limiting step of biological sludge hydrolysis, the anaerobic degradation is a slow process. In order to perform the reduction of digestion time and improvement of the biodegradability of organic biosolids mass, the impact of ultrasonic disintegration on subsequent anaerobic sludge digestion was investigated (Nickel and Neis, 2007). The experiments were performed in a vessel with 12 piezoceramic flat transducers fixed at each of the four sidewalls. The reactor volume was 1.3 L and the ultrasonic

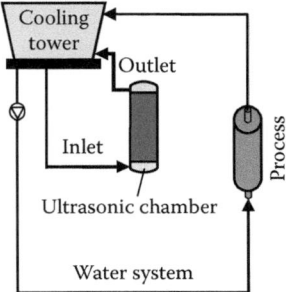

FIGURE 21.13 Sonoxide® ultrasonic water treatment system. (Adapted from Broekman, S. et al., *Ultrason. Sonochem.*, 17, 1041, 2010.)

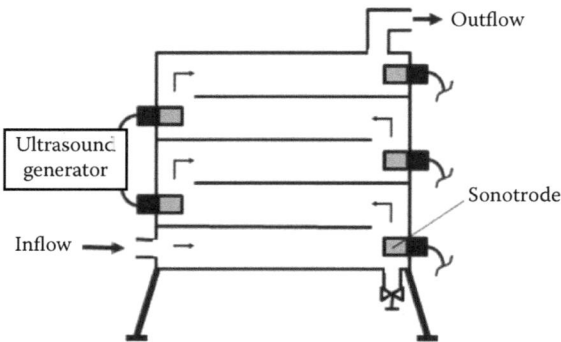

FIGURE 21.14 Ultrasound reactor for sludge treatment. (Adapted from Nickel, K. and Neis, U., *Ultrason. Sonochem.*, 14, 450, 2007.)

source worked at 31 kHz and acoustic intensities varying within 5 and 18 W cm^{-2}. It was observed that US resulted in a significant improvement in the overall process due to the increased volatile solid degradation rate (about 40%), increased biogas production, and a reduction from 60% to 52% of the nondegradable organic matter that exists in each type of biosolids. In the same work, a full-scale US reactor system, as shown in Figure 21.14, was also proposed. The reactor has a volume of 29 L and it is equipped with five 20 kHz sonotrodes each one supplied by a 2 kW generator. The work intensity range can be adjusted from 25 to 50 W cm^{-2}. In this device, the sludge is pumped up flow through the reactor channels, preventing accumulation of gas bubbles produced by degassing of the sludge water phase.

In recent years, US has been extensively used as an advanced oxidation process for wastewater remediation, owing to the sonochemical production of hydroxyl radicals in aqueous medium and the subsequent oxidation of pollutants (Mahamuni and Adewuyi, 2010). Moreover, US has been studied for the wastewater treatment of various chemical pollutants such as aromatic compounds, chlorinated aliphatic compounds, explosives, herbicides and pesticides, organic dyes, organic and inorganic gaseous pollutants, organic sulfur compounds, alcohols and oxygenates, pharmaceuticals, and personal care products (Adewuyi, 2001). However, due to the inefficient conversion of energy in producing ultrasonic cavitation and possible difficulties for the scale-up, no industrial installation for wastewater treatment has been even reported in the literature (Mahamuni and Adewuyi, 2010).

As a particular case, the presence of residual pharmaceuticals in the environment and in aquatic systems constitutes a serious environmental problem as these compounds (i) are extremely resistant to biological degradation processes and usually escape intact or as recalcitrant metabolites from conventional treatment plants; (ii) may impose serious toxic effects to humans and other living organisms; and (iii) are present at minute concentrations, thus requiring more sophisticated and laborious analytical tools for their accurate determination. Therefore, it is not surprising that research has recently been directed toward the application of nonbiological processes for the destruction of pharmaceuticals in water with emphasis on AOPs (Klavarioti et al., 2009).

In spite of the use of industrial ultrasonic processors for wastewater treatment being a prominent technology, it requires some initial experiments in laboratory scale. Regarding pharmaceutical wastes degradation using US, the degradation of diclofenac (DCF) by US irradiation at 20 kHz and 750 W in aqueous solutions was evaluated (Naddeo et al., 2010). Among the operating conditions, initial substrate concentration, applied power, liquid bulk temperature, solution pH, and the type of sparged gas on degradation and H_2O_2 formation were studied. The DCF conversion was enhanced at increased applied power densities and liquid bulk temperatures, acidic conditions, and in the presence of dissolved air or oxygen. The reaction rate increased with increasing DCF concentration in the range 2.5–5 mg L^{-1} but it remains constant in the range of 40–80 mg L^{-1}, indicating different kinetic regimes. Sanchez-Prado et al. (Sanchez-Prado et al., 2008) reported the sonochemical degradation of triclosan in various environmental samples (seawater, urban runoff, and domestic wastewater) as well as in model solutions (pure and saline water) with a horn-type sonicator operating at 80 Hz and nominal power of 135 W. In all cases, complete conversion was achieved at about 120 min. However, the rate of degradation is evidently affected by the matrix components present in the samples. In addition, sonochemical degradation was not accompanied by the formation of toxic metabolites which commonly appear as by-products of triclosan natural attenuation.

SOIL AND SEDIMENTS REMEDIATION

Many waterways are located in or close to industrial and urban areas, and sediments from surface runoffs are often contaminated with organic and inorganic contaminants. The soil contamination can arise from a number of potential sources, such as ash incinerators, nuclear plants, residual pollution from industrial sites, or the retention of pesticides used in agriculture. The difficulty in finding suitable disposal sites for contaminated dredged sediment and electroplating sludge is an example of the development of economic feasible techniques aiming to maximize the recovery and recycling of soil contaminants. Heavy metal contamination is a common problem at many hazardous waste sites. Once in the soil matrix, these metals are absorbed making remediation difficult. Unlike many organic pollutants that can be eliminated or reduced by chemical oxidation techniques or microbial activity, heavy metals will not be degraded (Meegoda and Perera, 2001; Li et al., 2010).

In spite of organic pollutants being, in most cases, more susceptible to conventional treatments, the determination of compounds such as pesticides, aliphatic and aromatic hydrocarbons in soils, and sediments requires complex operation of sample preparation, mainly due to the difficulty of quantitatively leaching the analyte from the solid sample. Because in some occasions the interactions established between analytes and solid matrix are very strong, the traditional methodologies based on Soxhlet extraction do not provide enough energy to release the analytes rapidly, thus requiring very long extraction times (8–48 h) (Caballo-López and Luque de Castro, 2003; Richter et al., 2006).

There are two ways in which acoustic energy can enhance soil washing. Methods are predominantly mechanical and involve a combination of abrasion to remove superficial impurities and improved solvent leaching of contaminants from the particles (Mason et al., 2004).

Soil Washing

Since the soil pollutants have a preferable trend to adsorb onto very small particles of soil such as silt, clay, and humic matter, which themselves tend to be attached to coarser sand and gravel particles, the soil washing is used to dislodge and separate these fine particles from the bulk soil.

A comparison of pollutants extraction by abrasive action was made between sonicated and conventional shaking (Newman et al., 1997). Granular pieces of brick impregnated with copper oxide were used as model for contaminated soil. The apparatus consisted of a stainless steel vibrating tray with a 20 kHz magnetostrictive ultrasonic transducer at the base of tray. The conventional condition was performed in the same way, but the used tray was attached to a sieve shaker. The analysis of the brick particles after 30 min sonication revealed an average reduction in copper content from 51 to 31 µg g^{-1}, representing a reduction of 40%, while the conventional shaking afforded only 6% reduction, resulting in residual copper content of 48 µg g^{-1}.

A US system coupled to a vacuum pressure device was used to facilitate the removal of chromium in dredged sediments from New York/New Jersey harbor (Meegoda and Perera, 2001). Full factorial experimental designs were performed to evaluate the treatment process for coarse and fine fractions of sediments. An ultrasonic probe type working at 1500 W and 20 kHz was used and the variables for evaluation were power, soil-to-water ratio, vacuum pressure, and dwell time. For coarse treatment, 92% of chromium removal was achieved with 1200 W power, 1:15 soil-to-water ratio, 15 psi vacuum pressure, and 15 min for the dwell time. The application of ultrasonic treatment to fine sediments resulted in an 83% as maximum chromium removal when factor levels were at 1200 W power, 1:50 soil-to-water ratio, and 90 min of dwell time. Though reasonable removal efficiencies were obtained for silt fraction (60%–80%), chromium removal of the clay fraction was very low (lower than 25%), and this technology was not effective when applied to the clay fraction.

Air

The inhalation of airborne particulate matter has been considered as a serious problem for public health. Fine particles originated in the emissions associated with carbon-fired power plants, cement factories, chemicals industries, and diesel-powered vehicles have increasingly become the focus of stricter government regulations. The ideal solution to the problem is to stop these emissions at source but current filters and electrostatic precipitators have problems in coping with the smallest particles. It has been shown that airborne acoustic energy in the ultrasonic frequency range can be used to precipitate suspended particles (aerosol or smoke). The use of power US is generally associated to the liquid medium, which can drive a range of reactions and processes. However, the removal of fine particles from gases is not so common due to some difficulties related to the use of US in gaseous systems (Sarabia et al., 2000; Riera et al., 2006).

The first drawback is the higher attenuation of acoustic waves in the propagation of sound through the air when compared with that through liquid medium. Another drawback is the transfer of acoustic energy generated in air into a liquid or solid material which is inefficient due to the incompatibilities between acoustic impedances of gases and solids or liquids. For these reasons, the US must be very powerful to be applied to the air treatment. An alternative approach to dust and mist suppression is the use of acoustic standing waves. When a sonic standing wave is set up in air, the suspended particles will migrate to the nodes of the sound wave. This phenomenon has been used in some applications like smoke particle removal, which remain suspended in air for a considerable period in view of its relatively light weight. The crossing of smoke particles through a standing wave field will increase the collision of particles in nodal points of field, resulting in the formation of larger fragments which will become large enough to fall to the floor of a deposition chamber (Mason and Lorimer, 2002).

The reduction of particle emissions in coal combustion fumes has been performed in a semi-industrial pilot plant (Gallego-Juarez et al., 1999). The investigation of particle agglomeration into ultrasonic chamber was driven by four high-power directional acoustic transducers working at 10 and/or 20 kHz, followed by an electrostatic precipitator. A fluidized bed coal combustor was used as fume generator with fume flow rates up to 2000 m^3 h^{-1}, gas temperature about 150°C, and particle concentrations from 1 to 5 g m^{-3}. A reduction of 40% in particle emission was achieved with the acoustic filter.

Regarding the air pollution caused by fine particle generated in diesel combustion, acoustic agglomeration of submicron particles in the 0.01–1 μm range was performed in a pilot-scale plant with a 97 kW diesel engine. It used an ultrasonic agglomeration chamber set at 20 kHz, a dilution system, a nozzle atomizer, and an aerosol sampling and measuring station (Riera et al., 2003). The influence of humidity on the agglomeration and precipitation of particles as well as the effect of US generated by a linear array of four high-power stepped-plate transducers was investigated. A small reduction in the number concentration of particles at the outlet of the chamber (*ca.* 25%) was observed. In the other way, by increasing the humidity (0.06 kg_{water} kg_{gas}^{-1}) the agglomeration rate was raised up to 56%.

FINAL CONSIDERATIONS

Energy efficiency and scale-up remains the major challenges related to industrial applications of sonochemistry. Whereas a number of sonochemical apparatus for sonochemistry are commercially available for laboratory studies, large-scale equipments remain relatively uncommon (Bang and Suslick, 2010). The problem of scale-up for sonochemical applications is commonly reported but it is in fact not as simple as the use of bigger versions of the equipment used in the laboratory scale. The volumes to be treated will be much larger than the ones used for laboratory studies, and the type of the process will govern the choice for reactor design. In this way, some processes can be more suitable to low-intensity sonication (e.g., using a bath-type reactor) whereas others may need higher-intensity irradiation (e.g., using a probe-based system) (Mason and Lorimer, 2002). In addition, while the production of US from electrical power can be extremely efficient the US coupling into useful cavitation remains a relatively low yield process (Bang and Suslick, 2010).

Besides uncommon large-scale application of US, some instrumentation to industrial application could be obtained from different companies. In this way, it could be pointed out that the industrial uses of US to liquid processing in the dispersing, blending, cleaning, cell disruption, sample preparation, homogenization, emulsification, and atomization, have been performed. In addition, it is possible to obtain equipment and accessories that work with large volumes operation at 20 kHz and power from 500 to 1500 W. Furthermore, US can be used in the plastic assembly industry, where equipment for plastic welding are available working with power ranging from 400 to 800 W (40 kHz), 1200 to 3300 W (20 kHz), and 3500 to 4500 W (15 kHz). Besides equipment to weld under vibration, spin and hot plate, and instrument to stack and tooling can be also found. In addition, it is possible to obtain instrumentation to metal welding that works in the similar frequency and power as previously mentioned, used to spot welder, wire splice, seam weld, tube sealer, and also tooling, as well as equipment to industrial cleaning to medical, optical, and wire materials (online document, available at www.sonicsandmaterials.com, accessed in March 2011; www.bransonultrasonics.com, accessed in March 2011).

As has been observed from the earlier discussion, a large amount of equipment in the industrial application of US are actually available. In addition, new instrumentation could be shortly produced as a result of recently scientific research in this novel area which employs the US energy. These facts, associated with the acoustic cavitation phenomenon comprehension, make the US a promising area for efficient, fast, clean, and economical industrial processes.

REFERENCES

Adekunte, A. O., Tiwari, B. K., Cullen, P. J., Scannell, A. G. M., and O'Donnell, C. P. 2010. Effect of sonication on colour, ascorbic acid and yeast inactivation in tomato juice. *Food Chemistry*, 122: 500–507.

Adewuyi, Y. G. 2001. Sonochemistry: Environmental science and engineering applications. *Industrial and Engineering Chemistry Research*, 40: 4681–4715.

Adewuyi, Y. G. 2005a. Sonochemistry in environmental remediation. 2. Heterogeneous sonophotocatalytic oxidation processes for the treatment of pollutants in water. *Environmental Science and Technology*, 39: 8557–8570.

Adewuyi, Y. G. 2005b. Sonochemistry in environmental remediation. 1. Combinative and hybrid sonophotochemical oxidation processes for the treatment of pollutants in water. *Environmental Science and Technology*, 39: 3409–3420.

Amara, N., Ratsimba, B., Wilhelm, A. M., and Selmmas, H. 2001. Crystallization of potash alum: Effect of power ultrasound. *Ultrasonics Sonochemistry*, 8: 265–270.

Ambrus, R., Amirzadi, N. N., Sipos, P., and Szabó-Révész, P. 2010. Effect of sonocrystallization on the habit and structure of Gemfibrozil crystals. *Chemical Engineering Technology*, 33: 827–832.

Armenta, R. E., Vinatoru, M., Burja, A. M., Kralovec, J. A., and Barrow, C. J. 2007. Transesterification of fish oil to produce fatty acid ethyl esters using ultrasonic energy. *Journal of American Oil Chemist's Society*, 84: 1045–1052.

Babich, I. V. and Molijn, J. A. 2003. Science and technology of novel processes for deep desulfurization of oil refinery streams: A review. *Fuel*, 82: 607–631.

Balachandran, S., Kentish, S. E., Mawson, R. et al. 2006. Ultrasonic enhancement of the supercritical extraction from ginger. *Ultrasonics Sonochemistry*, 13: 471–479.

Balasundaram, B. and Pandit, A. B. 2001. Selective release of invertase by hydrodynamic cavitation. *Biochemical Engineering Journal*, 8: 251–256.

Bang, J. H. and Suslick, K. S. 2010. Applications of ultrasound to the synthesis of nanostructured materials. *Advanced Materials*, 22: 1039–1059.

Bernardo, A. and Giulietti, M. 2009. Modeling of crystal growth and nucleation rates for pentaerythritol batch crystallization. *Chemical Engineering Research and Design*, doi:10.1016/j.cherd.2009.07.019

Blume, T. and Neis, U. 2005. Improving chlorine disinfection of wastewater by ultrasound application. *Water Science and Technology*, 52: 139–144.

Boonkird, S., Phisalaphong, C., and Phisalaphong, M. 2008. Ultrasound–assisted extraction of capsaicinoids from *Capsicum frutescens* on a lab- and pilot-plant scale. *Ultrasonics Sonochemistry*, 15: 1075–1079.

Broekman, S., Pohlmann, O., Beardwood, E. S., and Meulenaer, E. C. 2010. Ultrasonic treatment for microbiological control of water systems. *Ultrasonics Sonochemistry*, 17: 1041–1048.

Bougrier, C., Carrère, H., and Delgenès, J. P. 2005. Solubilisation of waste-activated sludge by ultrasonic treatment, *Chemical Engineering Journal*, 106: 163–169.

Caballo-López, A. and Luque de Castro, M. D. 2003. Continuous ultrasound-assisted leaching of phenoxyacid herbicides in soil and sediment with in-situ sample treatment. *Chromatographia*, 58: 257–262.

Canselier, J. P., Delmas, H., Wilhelm, A. M. and Abismail, B. 2002. Ultrasound emulsification – an overview. *Journal of Dispersion Science and Technology*, 23:333–349.

Chand, R., Bremner, D. H., Namkung, K. C., Collier, P. J., and Gogate, P. R. 2007. Water disinfection using the novel approach of ozone and a liquid whistle reactor. *Biochemical Engineering Journal*, 35: 357–364.

Chemat, F., Grondin, I., Sing, A. S. C., and Smadja, J. 2004. Deterioration of edible oils during food processing by ultrasound. *Ultrasonics Sonochemistry*, 11: 13–15.

Chen, Y. C. and Smirniotis, P. 2002. Enhancement of photocatalytic degradation of phenol and chlorophenols by ultrasound. *Industrial and Engineering Chemistry Research*, 41: 5958–5965.

Cheng, L. H., Soh, C. Y., Liew, S. C., and The, F. F. 2007. Effect of sonication and carbonation on guava juice quality. *Food chemistry*. 104: 1396–1401.

Chow, R., Blindt, R., Chivers, R., and Povey, M. 2003. The sonocrystallisation of ice in sucrose solutions: Primary and secondary nucleation. *Ultrasonics*, 41: 595–604.

Chow, R., Blindt, R., Chivers, R., and Povey, M. 2005. A study on the primary and secondary nucleation of ice by power ultrasound. *Ultrasonics*, 43: 227–230.

Cilek, E. C. and Ozgen, S. 2009. Effect of ultrasound on separation selectivity and efficiency of flotation. *Minerals Engineering*, 22: 1209–1217.

Cintas, P., Mantegna, S., Gaudino, E. C., and Cravotto, G. 2010. A new pilot flow reactor for high-intensity ultrasound irradiation. Application to the synthesis of biodiesel. *Ultrasonics Sonochemistry*, 17: 985–989.

Collings, A. F., Gwan, P. B., and Sosa-Pintos, A. P. 2010. Large scale environmental applications of high power ultrasound. *Ultrasonics Sonochemistry*, 17: 1049–1053.

Cordemans, E. M., Hannecart, B., Lepeltier, M. F., and Canivet, Y. 2003. Method and device for treating a liquid medium, US Patents 6,540,922 B1.

Cravotto, G., Binello, A., Di Carlo, S., Orio, L., Wu, Z. L., and Ondruschka, B. 2010. Oxidative degradation of chlorophenol derivatives promoted by microwaves or power ultrasound: A mechanism investigation. *Environmental Science Pollutants Research*, 17: 674–687.

Cravotto, G., Boffa, L., Mantegna, S. et al. 2008. Improved extraction of vegetable oils under high-intensity ultrasound and/or microwaves. *Ultrasonics Sonochemistry*, 15: 898–902.

Cravotto, G. and Cintas, P. 2006. Power ultrasound in organic synthesis: Moving to cavitational chemistry from academia to innovative and large-scale applications. *Chemical Society Reviews*, 35: 180–196.

Cravotto, G. and Cintas, P. 2007a. Forcing and controlling chemical reactions with ultrasound. *Angewandte Chemie International Edition*, 46: 5476–5478.

Cravotto, G. and Cintas, P. 2007b. The combined use of microwaves and ultrasound: Improved tools in process chemistry and organic synthesis. *Chemistry—A European Journal*, 13: 1902–1909.

Cravotto, G., Omiccioli, G., and Stevanato, L. 2005. An improved sonochemical reactor. *Ultrasonics Sonochemistry*, 12: 213–217.

Cucheval, A. and Chow, R. C. Y. 2008. A study on the emulsification of oil by power ultrasound. *Ultrasonics Sonochemistry*, 15: 916–920.

D'Amico, D. J., Silk, T. M., Wu, J. R., and Guo, M. R. 2006. Inactivation of microorganisms in milk and apple cider treated with ultrasound. *Journal of Food Protection*, 69: 556–563.

Dennehy, R. D. 2003. Particle engineering using power ultrasound. *Organic Process Research and Development*, 7: 1002–1006.

Devarakonda, S., Evans, J. M. B., and Myerson, A. S. 2004. Impact of ultrasonic energy on the flow crystallization of dextrose monohydrate. *Crystal Growth and Design*, 4: 687–690.

Deshpande, A., Bassi, A., and Prakash, A. 2005. Ultrasound-assisted, base-catalyzed oxidation of 4,6-dimethyldibenzothiophene in a biphasic diesel-acetonitrile system. *Energy Fuels*, 19: 28–34.

Du, F., Li, J., Li, X., and Zhang, Z. 2011. Improvement of iron removal from silica sand using ultrasound-assisted oxalic acid. *Ultrasonics Sonochemistry*, 18: 389–393.

ElKaoutit, M., Naranjo-Rodriguez, I., Temsamani, K. R., De La Vega, M. D., and Cisneros, J. L. H. H. 2007. Dual laccase-tyrosinase based sonogel-carbon biosensor for monitoring polyphenols in beers. *Journal of Agricultural and Food Chemistry*, 55: 8011–8018.

ElKaoutit, M., Naranjo-Rodriguez, I., Temsamani, K. R., Domínguez, M., and Cisneros, J. L. H. H. 2008. Investigation of biosensor signal bioamplification: Comparison of direct electrochemistry phenomena of individual Laccase, and actual Laccase-Tyrosinase copper enzymes, at a sonogel-carbon electrode. *Talanta*, 75: 1348–1355.

Ensminger, D. 2009. Properties of materials. In: *Ultrasonics: Data, Equations, and Their Practical Uses*, eds. Ensminger, D. and Stulen, F. B., CRC Press, Boca Raton, FL, pp. 285–322.

Entezari, M. H. and Petrier, C. 2005. A combination of ultrasound and oxidative enzyme: Sono-enzyme degradation of phenols in a mixture. *Ultrasonics Sonochemistry*, 12: 283–288.

Farmer, A. D., Collings, A. F., and Jameson, G. J. 2000a. Effect of ultrasound on surface cleaning of silica particles. *International Journal of Mineral Processing*, 60: 101–113.

Farmer, A. D., Collings, A. F., and Jameson, G. J. 2000b. The application of power ultrasound to the surface cleaning of silica and heavy mineral sands. *Ultrasonics Sonochemistry*, 7: 243–247.

Feng, H., Yang, W., and Hielscher, T. 2008. Power ultrasound. *Food Science and Technology International*, 14: 433–436.

Fuente-Blanco, S., Sarabia, E. R. F., Acosta-Aparicio, V. M., Blanco-Blanco, A., and Gallego-Juárez, J. A. 2006. Food drying by Power ultrasound. *Ultrasonics*, 44: e523–e527.

Gaikwad, S. G. and Pandit, A. B. 2008. Ultrasound emulsification: Effect of ultrasonic and physicochemical properties on dispersed phase volume and droplet size. *Ultrasonic Sonochemistry*, 15: 554–563.

Gallego-Juarez, J. A., Corral, G. R., Parga, G. N. V., Martinez, F. V., and van der Vlist, P. 2001. Process and device for continuous ultrasonic washing of textile. *United States Patent*, US 6,266,836 B1.

Gallego-Juarez, J. A., De Sarabia, E. R., Rodriguez-Corral, G., Hoffmann, T. L., and Gálvez-Moraleda, J. C. 1999. Application of acoustic agglomeration to reduce fine particle emissions from coal combustion plants. *Environmental Science and Technology*, 33: 3843–3849.

Gallego-Juarez, J. A., Riera, E., Acosta, V., Rodrígues, G., and Blanco, A. 2010. Ultrasonic system for continuous washing of textiles in liquid layers. *Ultrasonics Sonochemistry*, 17: 234–238.

Gallego-Juarez, J. A., Riera, E., de la Fuente, S., Rodriguez-Corral, G., Acosta-Aparicio, V. M., and Blanco, A. 2007. Application of high-power ultrasound for dehydration of vegetables: processes and devices. *Drying Technology*, 25: 1893–1901.

Gao, Y., Nagy, B., Liu, X. et al. 2009. Supercritical CO2 extraction of lutein esters from marigold (*Tagetes erecta L.*) enhanced by ultrasound. *The Journal of Supercritical Fluids*, 49: 345–350.

Geciova, J., Bury, D., and Jelen, P. 2002. Methods for disruption of microbial cells for potential use in the dairy industry—A review. *International Dairy Journal*, 12: 541–553.

Gedanken, A. 2004. Using sonochemistry for the fabrication of nanomaterials. *Ultrasonics Sonochemistry*, 11: 47–55.

Georgogiani, K. G., Kontominas, M. G., Tegou, E., Avlonitis, D., and Gergis, V. 2007. Biodiesel production: Reaction and processes parameters of alkali-catalyzed transesterification of waste frying oils. *Energy Fuels*, 21: 3023–3027.

Gil, S., Lavilla, I., and Bendicho, B. 2008. Mercury removal from contaminated water by ultrasound-promoted reduction/vaporization in a microscale reactor. *Ultrasonics Sonochemistry*, 15: 212–216.

Gogate, P. R. 2008. Treatment of wastewater streams containing phenolic compounds using hybrid techniques based on cavitation: A review of current status and the way forward. *Ultrasonics Sonochemistry*, 15: 1–15.

Gogate, P. R. and Kabadi, A. M. 2009. A review of applications of cavitation in biochemical engineering/biotechnology. *Biochemical Engineering Journal*, 44: 60–72.

Guo, Z., Zhang, M., Li, H., Wang, J., and Kougoulos, E. 2005. Effect of ultrasound on anti-solvent crystallization process. *Journal of Crystal Growth*, 273: 555–563.

Gutnik, V. G., Gorbach, N. V., and Dashkov, A. V. 2002. Some characteristics of ultrasonic welding polymers. *Fibre Chemistry*, 34: 426–432.

Hahn, H. D., Dong, N. T., Okitsu, K., Nishimura, R., and Maeda, Y. 2009. Biodiesel production through transesterification of triolein with various alcohols in an ultrasonic field. *Renewable Energy*, 34: 766–768.

Harrison, S. T. L. 1991. Bacterial cell disruption: A key unit operation in the recovery of intracellular products. *Biotechnology Advances*, 9: 217–240.

Harvey, E. N. and Loomis, A. L. 1929. The destruction of luminous bacteria by high frequency sound waves. *Journal of Bacteriology*, 17: 373–376.

Higaki, K., Ueno, S., Koyano, T., and Sato, K. 2001. Effects of ultrasonic irradiation on crystallization behavior of tripalmitoylglycerol and cocoa butter. *Journal of American Oil Chemists Society*, 78: 513–518.

Howkes, J. J., Long, M. J., Coakley, W. T., and McDonnel, M. B. 2004. Ultrasonic deposition of cells on a surface. *Biosensors and Bioelectronic*, 19: 1021–1028.

Hu, A., Zhao, S., Liang, H. et al. 2007. Ultrasound assisted supercritical fluid extraction of oil and coixenolide from adlay seed. *Ultrasonics Sonochemistry*, 14: 219–224.

Hu, A., Zheng, J., and Qiu, T. 2006. Industrial experiments for the application of ultrasound on scale control in the Chinese sugar industry. *Ultrasonics Sonochemistry*, 13: 329–333.

Huber, P. E. and Pfisterer, P. 2000. In vitro and in vivo transfection of plasmid DNA in the Dunning prostate tumor R3327-AT1 is enhanced by focused ultrasound. *Gene Therapy*, 7: 1516–1525.

Hughes, D. E. and Nyborg, W. L. 1962. Cell disruption by ultrasound. *Science*, 138: 108–114.

Ichitsubo, T., Matsubara, E., Kai, S., and Hirao, M. 2004. Ultrasound-induced crystallization around the glass transition temperature for $Pd_{40}Ni_{40}P_{20}$ metallic glass. *Acta Materialia*, 52: 423–429.

Imai, H. and Matsuoka, S. 2005. Finding the optimum parameters for ultrasonic welding of aluminum alloys. *JSME International Journal Series A*, 48: 311–316.

Ince, N. H., Tezcanli, G., Belen, R. K., and Apikyan, I. G. 2001. Ultrasound as a catalyzer of aqueous reaction systems: The state of the art and environmental applications. *Applied Catalysis B: Environmental*, 29: 167–176.

Ishikuro, T. and Matsuoka, S. 2005. Ultrasonic welding of thin alumina and aluminum using inserts. *JSME International Journal Series A*, 48: 317–321.

Japón-Luján, R., Janeiro, P., and Luque De Castro, M. D. 2008. Solid-liquid transfer of biophenols from olive leaves for the enrichment of edible oils by a dynamic ultrasound-assisted approach. *Journal of Agricultural and Food Chemistry*, 56: 7231–7235.

Japón-Luján, R., Luque-Rodríguez, J. M., and Luque de Castro, M. D. 2006. Dynamic ultrasound-assisted extraction of oleuropein and related biophenols from olive leaves. *Journal of Chromatography A*, 1108: 76–82.

Jianbing, J., Jianli, W., Yongchao, L., Yunliang, Y., and Zichao, X. 2006. Preparation of biodiesel with the help of ultrasonic and hydrodynamic cavitation. *Ultrasonics*, 44: 411–414.

Jiranek, V., Grbin, P., Yap, A., Barnes, M., and Bates, D. 2008. High power ultrasound as a novel tool offering new opportunities for managing wine microbiology. *Biotechnology Letters*, 30: 1–6.

Joersbo, M. and Brunstedt, J. 1990. Direct gene-transfer to plant protoplasts by mild sonication. *Biomedical and Life Sciences*, 9: 207–210.

Kamel, M. M., Helmy, H. M., Mashaly, H. M., and Kafafy, H. H. 2010. Ultrasound assisted dyeing: Dyeing of acrylic fabrics C. I. Astrazon Basic Red 5BL 200%. *Ultrasonics Sonochemistry*, 17: 92–97.

Khanal, S. K., Grewell, D., Sung, S., and Van Leeuwen, J., 2007. Ultrasound applications in wastewater sludge pretreatment: A review. *Critical Reviews in Environmental Science and Technology*, 37: 277–313.

Kim, H. Y., Kim, Y. G., and Kang, B. H. 2004. Enhancement of natural convection and pool boiling heat transfer via ultrasonic vibration. *International Journal of Heat and Mass Transfer*, 47: 2831–2840.

Kim, S., Wei, C., and Kiang, S. 2003. Crystallization process development of an active pharmaceutical ingredient and particle engineering via the use of ultrasonics and temperature cycling. *Organic Process Research and Development*, 7: 997–1001.

Klavarioti, M., Mantzavinos, D., and Kassinos, D. 2009. Removal of residual pharmaceuticals from aqueous systems by advanced oxidation processes. *Environment International*, 35: 402–417.

Knorr, D., Zenker, M., Heinz, V., and Lee, D. U. 2004. Applications and potential of ultrasonic in food processing. *Trends in Food Science and Technology*, 15: 261–266.

Kobayashi, T., Chai, X., and Fujii, N. 1999. Ultrasound enhanced cross–flow membrane filtration. *Separation and Purification Technology*, 17: 31–40.

Kordylla, A., Koch, S., Tumakaka, F., and Schembecker, G. 2008. Towards an optimized crystallization with ultrasound: Effect of solvent properties and ultrasonic process parameters. *Journal of Crystal Growth*, 310: 4177–4184.

Korn, M., Primo, P. M., and Sousa, C. S. 2002. Influence of ultrasonic waves on phosphate determination by the molybdenum blue method. *Microchemical Journal*, 73: 273–277.

Krishnan, R. S., Venkatasubramanian, V. S., and Rajagopal, E. S. 1981. Studies on ultrasonic emulsification. *Journal of Colloid Science*, 16: 41–48.

Kuehn, T. H., Kittelson, D. B., Wu, Y., and Gouk, R. 1996. Particle removal from semiconductor wafers by megasonic cleaning, *Journal Aerosol Science*, 27: S427–428.

Kuznetsova, L. A. and Coakley, W. T. 2007. Applications of ultrasound streaming and radiation force in biosensors. *Biosensors and Bioelectronics*, 22: 1567–1577.

Kyllönen, H. M., Pirkonen, P., and Nyström, M. 2005. Membrane filtration enhanced by ultrasound: A review. *Desalination*, 181: 319–335.

Kyllönen, H., Pirkonen, P., Nyström, M., Nuortila-Jokinen, J., and Grönroos, A. 2006. Experimental aspects of ultrasonically enhanced cross-flow membrane filtration of industrial wastewater. *Ultrasonics Sonochemistry*, 13: 295–302.

Lakerveld, R., Kramer, H. J. M., Jansens, P. J., and Grievinkb, J. 2009. The application of a task-based concept for the design of innovative industrial crystallizers. *Computers and Chemical Engineering*, 33: 1692–1700.

Lamminen, M. O., Walker, H. W., and Weavers, L. K. 2004. Mechanisms and factors influencing the ultrasonic cleaning of particle-fouled ceramic membranes. *Journal of Membrane Science*, 237: 213–223.

Lee, K. W. and Kim, J. P. 2001. Effect of ultrasound on disperse dye particle size. *Textile Research Journal*, 71: 395–398.

Lesko, T., Colussi, A. J., and Hoffmann, M. R. 2006. Sonochemical decomposition of phenol: Evidence for a synergistic effect of ozone and ultrasound for the elimination of total organic carbon from water. *Environmental Science and Technology*, 40: 6818–6823.

Li, H., Li, H., Guo, Z., and Liu, Y. 2006. The application of power ultrasound to reaction crystallization. *Ultrasonics Sonochemistry*, 13: 359–363.

Li, H., Wang, J., Bao, Y., Guo, Z., and Zhang, M. 2003. Rapid sonocrystallization in the salting-out process. *Journal of Crystal Growth*, 247: 192–198.

Li, C., Xie, F., Ma, Y., et al. 2010. Multiple heavy metals extraction and recovery from hazardous electroplating sludge waste via ultrasonically enhanced two-stage acid leaching, *Journal of Hazardous Materials*, 178: 823–833.

Liu, Y., Yang, H., and Sakanishi, A. 2006. Ultrasound: Mechanical gene transfer into plant cells by sonoporation. *Biotechnology Advances*, 24: 1–16.

Li-yun, C., Chuan-bo, Z., and Jian-feng, H. 2005. Influence of temperature, [Ca^{2+}], Ca/P ratio and ultrasonic power on the crystallinity and morphology of hydroxyapatite nanoparticles prepared with a novel ultrasonic precipitation method. *Materials Letters*, 59: 1902–1906.

Luque de Castro, M. D. and Capote, F. P. 2007a. *Analytical Applications of Ultrasound*. Amsterdam, The Netherlands: Elsevier.

Luque de Castro, M. D. and Capote, F. P. 2007b. Ultrasound-assisted crystallization (sonocrystallization). *Ultrasonics Sonochemistry*, 14: 717–724.

Mahamuni, N. N. and Adewuyi, Y. G. 2010. Advanced oxidation processes (AOPs) involving ultrasound for waste water treatment: A review with emphasis on cost estimation. *Ultrasonics Sonochemistry*, 17: 990–1003.

Mamun, A., Umemoto, S., Ishihara, N., and Okui, N. 2006. Influence of thermal history on primary nucleation and crystal growth rates of isotactic polystyrene. *Polymer*, 47: 5531–5537.

Mann, T. L. and Krull, U. J. 2004. The application of ultrasound as a rapid method to provide DNA fragments suitable for detection by DNA biosensors. *Biosensors and Bioelectronics*, 20: 945–955.

Martini, S., Suzuki, A. H., and Hartel, R. W. 2008. Effect of high intensity ultrasound on crystallization behavior of anhydrous milk fat. *Journal of American Oil Chemists Society*, 85: 621–628.

Mason, T. J. 1997. Ultrasound in synthetic organic chemistry. *Chemical Society Reviews*, 26: 443–451.

Mason, T. J. 2000. Large scale sonochemical processing: Aspiration and actuality. *Ultrasonics Sonochemistry*, 7: 145–149.

Mason, T. J. 2003. Sonochemistry and sonoprocessing: The link, the trends and (probably) the future. *Ultrasonics Sonochemistry*, 10: 175–179.

Mason, T. J. 2007. Sonochemistry and the environment—Providing a "green" link between chemistry, physics and engineering. *Ultrasonics Sonochemistry*, 14: 476–483.

Mason, T. J., Collings, A., and Sumel, A. 2004. Sonic and ultrasonic removal of chemical contaminants from soil in the laboratory and on a large scale. *Ultrasonics Sonochemistry*, 11: 205–210.

Mason, T., Riera, E., Vercet, A., et al. 2005. Application of ultrasound. In *Emerging technologies for food processing*, Ed. D. W. Sun 323–350. California: Elsevier Academic press.

Mason, T. J., Joyce, E., Phull, S. S., and Lorimer, J. P. 2003. Potential uses of ultrasound in the biological decontamination of water. *Ultrasonics Sonochemistry*, 10: 319–323.

Mason, T. J. and Lorimer, J. P. 2002. *Applied Sonochemistry: Uses of Power Ultrasound in Chemistry and Processing*. Weinheim, Germany: Wiley-VCH Verlag GmbH.

Mason, T. J., Paniwnyk, L., and Lorimer, J. P. 1996. The uses of ultrasound in food technology. *Ultrasonics Sonochemistry*, 3: S253–260.

Mello, P. A., Duarte, F. A., Nunes, M. A. G., et al. 2009. Ultrasound-assisted oxidative process for sulfur removal from petroleum product feedstock. *Ultrasonics Sonochemistry*, 16: 732–736.

McCausland, L. J. and Cains, P. W. 2003. Ultrasound to make crystals. *Chemistry and Industry*, 5: 15.

McCausland, L. J., Cains, P. W., and Martins, P. D. 2001. Use the power of sonocrystallization for improved properties. *Chemical Engineering Prog*ress, 97: 56–61.

McClements, D. J., 1995. Advances in application of ultrasound in food analysis and processing. *Trends in Food Science and Technology*, 6: 293–299.

Meegoda, J. N. and Perera, R. 2001. Ultrasound to decontaminate heavy metals in dredged sediments. *Journal of Hazardous Materials*, 85: 73–89.

Miller, D. L., Bao, S., Gies, R. A., and Thrall, B. D. 1999. Ultrasonic enhancement of gene transfection in murine melanoma tumors. *Ultrasound in Medicine and Biology*, 25: 1425–1430.

Mir, J., Morato, J., and Ribas, F. 1997. Resistance to chlorine of freshwater bacterial strains. *Journal of Applied Microbiology*, 82: 7–18.

Miyasaka, E., Ebihara, S., and Hirasawa, I. 2006. Investigation of primary nucleation phenomena of acetylsalicylic acid crystals induced by ultrasonic irradiation-ultrasonic energy needed to activate primary nucleation. *Journal of Crystal Growth*, 295: 97–101.

Miyasaka, E., Takai, M., Hidaka, H., Kakimoto, Y., and Hirasawa, I. 2006. Effect of ultrasonic irradiation on nucleation phenomena in a $Na_2HPO_4 \cdot 12H_2O$ melt being used as a heat storage material. *Ultrasonics Sonochemistry*, 13: 308–312.

Moore, S. B. and Ausley, L. W. 2005. System thinking and green chemistry on the textile industry: Concepts, technologies and benefits. *Journal of Cleaner Production*, 12: 585–601.

Moulton Sr., K. J., Koritala, S., Warner, K., and Frankel, E. N. 1987. Continuous ultrasonic hydrogenation of soybean oil. II. Operating conditions and oil quality. *Journal of American Oil Chemists' Society*, 64: 542–547.

Mulet, A., Cárcel, J. A., Sanjuán, N., and Bon, J. 2003. New food drying techniques—Use of ultrasound. *Food Science and Technology International*, 9: 215–221.

Muthukumaran, S., Yang, K., Seuren, A., et al. 2004. The use of ultrasonic cleaning for ultrafiltration membranes in the dairy industry. *Separation and Purification Technology*, 39: 99–107.

Naddeo, V., Belgiorno, V., Kassinos, D., Mantzavinos, D., and Meric, S. 2010. Ultrasonic degradation, mineralization and detoxification of diclofenac in water: Optimization of operating parameters. *Ultrasonics Sonochemistry*, 17: 179–185.

Nalajala, V. S. and Moholkar, V. S. 2011. Investigations in the physical mechanism of sonocrystallization. *Ultrasonics Sonochemistry*, 18: 345–355.

Namkung, K. C., Burgess, A. E., Bremner, D. H., and Staines, H. 2008. Advanced Fenton processing of aqueous phenol solutions: A continuous system study including sonication effects. *Ultrasonics Sonochemistry*, 15: 171–176.

Newman, A. P., Lorimer, J. P., Mason, T. J., and Hutt, K. R. 1997. An investigation into the ultrasonic treatment of polluted solids. *Ultrasonics Sonochemistry*, 4: 153–156.

Nickel, K. and Neis, U. 2007. Ultrasonic disintegration of biosolids for improved biodegradation. *Ultrasonics Sonochemistry*, 14: 450–455.

Nishida, I. 2004. Precipitation of calcium carbonate by ultrasonic irradiation. *Ultrasonics Sonochemistry*, 11: 423–428.

Ohta, S., Suzuki, K., Ogino, Y., et al. 2008. Gene transduction by sonoporation. *Development, Growth and Differentiation*, 50: 517–520.

Ozkan, S. G. and Kuyumcu, H. Z. 2007. Design of a flotation cell equipped with ultrasound transducers to enhance coal flotation. *Ultrasonics Sonochemistry*, 14: 639–645.

Phull, S. S., Newman, A. P., Lorimer, J. P., Pollet, B., and Mason, T. J. 1997. The development and evaluation of ultrasound in the biocidal treatment of water, *Ultrasonics Sonochemistry*, 4: 157–164.

Pilli, S., Bhunia, P., Yan, S., LeBlanc, R. J., Tyagi, R. D., and Surampalli, R. Y. 2011. Ultrasonic pretreatment of sludge: A review. *Ultrasonics Sonochemistry*, 18: 1–18.

Pirrini, L. 2001. Design of advanced ultrasonic transducers for welding devices. *IEE Transactions on Ultrasonics, Ferroelectrics, and Frequency Control*, 48: 1632–1639.

Piyasena, P., Mohareb, E., and McKellar, R. C. 2003. Inactivation of microbes using ultrasound: A review, *International Journal of Food Microbiology*, 87: 207–216.

Richter, P., Jiménez, M., Salazar, R., and Maricán, A. 2006. Ultrasound–assisted pressurized solvent extraction for aliphatic and polycyclic aromatic hydrocarbons from soils. *Journal of Chromatography A*, 1132: 15–20.

Riera, E., Blanco, A., García, J. et al. 2010. High-power ultrasonic system for the enhancement of mass transfer in supercritical CO2 extraction processes. *Ultrasonics*, 50: 306–309.

Riera, E., Elvira, L., González, I., Rodríguez, J. J., Muñoz, R., and Dorronsoro, J. L. 2003. Investigation of the influence of humidity on the ultrasonic agglomeration of submicron particles in diesel exhaust. *Ultrasonics*, 41: 277–281.

Riera, E., Gallego-Juarez, J. A., and Mason, T. J. 2006. Airborne ultrasound for the precipitation of smokes and powders and the destruction of foams. *Ultrasonics Sonochemistry*, 13: 107–116.

Riera, E., Golás, Y., Blanco, A. et al. 2004. Mass transfer in supercritical fluids extraction by means of power ultrasound. *Ultrasonics Sonochemistry*, 11: 241–244.

Rokhina, E. V., Lens, P., and Virkutyte, J. 2009. Low-frequency ultrasound in biotechnology: State of the art. *Trends in Biotechnology*, 27: 298–306.

Ruecroft, G., Hipkiss, D., Ly, T., Maxted, N., and Cains, P. W. 2005. Sonocrystallization: The use of ultrasound for improved industrial crystallization. *Organic Process Research and Development*, 9: 923–932.

Sanchez-Prado, L., Barro, R., Garcia-Jares, C. et al. 2008. Sonochemical degradation of triclosan in water and wastewater. *Ultrasonics Sonochemistry*, 15: 689–694.

Sarabia, E. R. F., Gallego-Juárez, J. A., Rodríguez-Corral, G. et al. 2000. Application of high-power ultrasound to enhance fluid/solid particle separation processes. *Ultrasonics*, 38: 642–646.

Scherba, G., Weigel, R. M., and O'Brien, W. D. 1991. Quantitative assessment of the germicidal efficacy of ultrasonic energy. *Applied Environmental Microbiology*, 57: 2079–2084.

Shukla, S. R. and Mathur, M. R. 1995. Low temperature ultrasonic dyeing of silk. *Journal of the Society of Dyers and Colourists*, 111: 342–345.

Sigfusson, H., Ziegler, G. R., and Coupland, J. N. 2004. Ultrasonic monitoring of food freezing. *Journal of Food Engineering*, 62: 263–269.

Sivakumar, V., Anna, J. L., Vijayeeswarri, J. et al. 2009b. Ultrasound assisted enhancement in natural dye extraction from beetroot for industrial applications and natural dyeing of leather. *Ultrasonics Sonochemistry*, 16: 782–789.

Sivakumar, V. and Rao, P. G. 2003. Studies on the use of power ultrasound in leather dyeing. *Ultrasonics Sonochemistry*, 10: 85–94.

Sivakumar, V., Swaminathan, G., Rao, P. G., Muralidharan, C., Mandal, A. B., and Ramasami, T. 2010. Use of ultrasound in leather processing industry: Effect of sonication on substrate and substances—New insights. *Ultrasonics Sonochemistry*, 17: 1054–1059.

Sivakumar, V., Swaminathan, G., Rao, P. G., and Ramasami, T. 2009a. Sono-leather technology with ultrasound: A boon for unit operations in leather processing—Review of our research work at Central Leather Research Institute (CLRI), India. *Ultrasonics Sonochemistry*, 16: 116–119.

Song, C. 2003. An overview of new approaches to deep desulfurization for ultra-clean gasoline, diesel fuel and jet fuel. *Catalysis Today*, 86: 211–263.

Sonoxide® is a registered trade mark of Ashland, Inc. http://ppd.herc.com/index.asp, accessed in March 2011.

Soria, A. C. and Villamiel, M. 2010. Effect of ultrasound on the technological properties and bioactivity of food: A review. *Trends in Food Science and Technology*, 21: 323–331.

Sostaric, J. S. and Weavers, L. K. 2010. Advancement of high power ultrasound technology for the destruction of surface active waterborne contaminants. *Ultrasonics Sonochemistry*, 17: 1021–1026.

Stack, L. J., Carney, P. A., Malone, H. B. et al. 2005. Factors influencing the ultrasonic separation of oil-in-water emulsions. *Ultrasonics Sonochemistry*, 12: 153–160.

Stavarache, C., Vinatoru, M., Maeda, Y., and Bandow, H. 2007. Ultrasonically driven continuous process for vegetable oil transesterification. *Ultrasonics Sonochemistry*, 14: 413–417.

Stavarache, C., Vinatoru, M., Nishimura, R., and Maeda, Y. 2005. Fatty acids methyl ester from vegetable oil by means of ultrasonic energy. *Ultrasonics Sonochemistry*, 12: 367–372.

Suslick, K. S. 1990. Sonochemistry. *Science*, 247: 1439–1455.

Swinnen, M., Meulenaer, E. D. C., Hannecart, B. O., and Beardwood, E. S. 2008. Device and process for treating cutting fluids using ultrasound. US Patents 7,404,906 B2.

Thacker, J. 1973. An approach to the mechanism of killing cells in suspension by ultrasound. *Biochemica et Biophysica Acta*, 304: 240–248.

Thakore. K. A., Smith, C. B., Hite, D., and Carlough, M. 1990. The effects of ultrasound and the diffusion coefficient of C. I. Direct Red 81 in cellulose. *Textile Chemist and Colorist*, 22: 21–22.

Thompson, L. H. and Doraiswamy, L. K. 1999. Sonochemistry: Science and engineering. *Industrial and Engineering Chemistry Research*, 38: 1215–1249.

Tiwari, B. K., O'Donnell, C. P., and Cullen, P. J. 2009a. Effect of non thermal processing technologies on the anthocyanin content of fruit juices. *Trends in Food Science and Technology,* 20: 137–145.

Tiwari, B. K., O'Donnell, C. P., Muthukumarappan, K., and Cullen, P. J. 2009b. Ascorbic acid degradation kinetics of sonicated orange juice during storage and comparison with thermally pasteurised juice. *LWT—Food Science and Technology* 42: 700–704.

Truckenmüller, R., Ahrens, R., Cheng, Y. et al. 2006a. An ultrasonic welding based process for building up a new class of inert fluidic microsensors and actuators from polymers. *Sensors and Actuators A*, 132: 385–392.

Truckenmüller, R., Cheng, Y., Ahrens, R. et al. 2006b. Micro ultrasonic welding: Joining of chemically inert polymer microparts for single material fluidic components and systems. *Microsystem Technologies*, 12: 1027–1029.

United States Environmental Protection Agency, Regulatory announcement: Heavy-duty engine and vehicle standards and highway diesel fuel sulfur control requirements, December, 2000. http://www.epa.gov, accessed in March 2011.

Ukkirapandian, V., Sadasivam, V., and Sivasankar, B. 2008. Oxidation of dibenzothiophene and desulphurization of diesel. *Petroleum Science and Technology*, 73: 423–435.

Ulrich, J. and Jones, M. J. 2007. Heat and mass transfer operations—Crystallization. Chemical engineering and chemical process technology. In: *Encyclopedia of Life Support Systems (EOLSS), Developed under the Auspices of the UNESCO*, eds. Pohorecki, R., Bridgwater, J., Molzahn, M., and Gani, R., Eolss Publishers, Oxford, U.K. http://www.eolss.net, accessed in March 2011.

Vajnhandl, S. and Le Marechal, A. M. 2005. Ultrasound in textile dyeing and the decolouration/mineralization of textile dyes. *Dyes and Pigments*, 65: 89–101.

Valachovic, P., Pechova, A., and Mason, T. J. 2001. Towards the industrial production of medicinal tincture by ultrasound assisted. *Ultrasonics Sonochemistry*, 8: 111–117.

Valero, M., Recrosio, N., Saura, D. et al. 2007. Effects of ultrasonic treatments in orange juice processing. *Journal of Food Engineering*, 80: 509–516.

Vieira, M. C., Teixeira, A. A., and Silva, C. L. M. 2000. Mathematical modeling of the thermal degradation kinetics of vitamin C in cupuaçu (*Theobroma grandiflorum*) nectar. *Journal of Food Engineering*, 43: 1–7.

Villamiel, M. and Jong, P. 2000. Inactivation of *Pseudomas fluorescens* and *Streptococcus thermophilus* in Trypticase® Soy Broth and total bacteria in milk by continuous-flow ultrasonic treatment and conventional heating. *Journal of Food Engineering*, 45: 171–179.

Villamiel, M., Verdumen, R., and Jong, P. 2000. Degassing of milk by high-intensity ultrasound. *Milchwissenschaft*, 55: 123–125.

Vinatoru, M., Toma, M., and Mason, T. J. 1999. Ultrasonically assisted extraction of bioactive principles from plants and their constituents. In: *Advances in Sonochemistry*, ed. Mason, T. J., JAI Press, London, U.K., pp. 209–247.

Virone, C., Kramer, H. J. M., van Rosmalen, G. M., Stoop, A. H., and Bakker, T. W. 2006. Primary nucleation induced by ultrasonic cavitation. *Journal of Crystal Growth*, 294: 9–15.

Virot, M., Tomao, V., Le Bourvellec, C. et al. 2010. Towards the industrial production of antioxidants from food processing by-products with ultrasound-assisted extraction. *Ultrasonics Sonochemistry*, 17: 1066–1074.

Vouters, M., Rumeau, P., Tierce, P., and Costes, S. 2004. Ultrasounds: An industrial solution to optimize costs, environmental requests and quality for textile finishing. *Ultrasonics Sonochemistry*, 11: 33–38.

Vyas, A. P., Verma, J. L., and Subrahmanyam, N. 2010. A review on FAME production process. *Fuel*, 89: 1–9.

Warmoeskerken, M. M. C. G., van der Vlist, P., Moholkar, V. S., and Nierstrasz, V. A. 2002. Laundry process intensification by ultrasound. *Colloids and Surfaces A: Physicochemical Engineering Aspects*, 210: 277–285.

Watanabe, T., Sakuyama, H., and Yanagisawa, A. 2009. *Journal of Materials Processing Technology*, 209: 5475–5480.

Westhoff, G. M., Butler, B. K., Kramer, H. J. M., and Jansens, P. J. 2002. Growth behaviour of crystals formed by primary nucleation on different crystalliser scales. *Journal of Crystal Growth*, 237–239: 2136–2141.

Wohlgemuth, K., Kordylla, A., Ruether, F., and Schembecker, G. 2009. Experimental study of the effect of bubbles on nucleation during batch cooling crystallization. *Chemical Engineering Science*, 64: 4155–4163.

Wohlgemuth, K., Ruether, F., and Schembecker, G. 2010. Sonocrystallization and crystallization with gassing of adipic acid. *Chemical Engineering Science*, 65: 1016–1027.

Wu, H., Hulbert, G. J., and Mount, J. R. 2001. Effects of ultrasound on milk homogenization and fermentation with yogurt starter. *Innovative Food Science and Emerging Technologies*, 1: 211–218.

Wu, Z. and Ondruschka, B. 2010. Ultrasound assisted oxidative desulfurization of liquid fuels and its industrial application. *Ultrasonics Sonochemistry*, 17: 1027–1032.

Wyber, J. A., Andrews, J., and D'Emanuele, A. 1997. The use of sonication for efficient delivery of plasmid DNA into cells. *Pharmaceutical Research*, 14: 750–756.

Xie, J. P., Lorimer, J. P., Walton, D. J., Mason, T. J., and Attenburrow, G. E. 1999. *Journal of the American Leather and Chemical Association*, 94: 146–157.

Yachmenev, V. G., Bertoniere, N. R., and Blancha, E. J. 2001. Effect of sonication on cotton preparation with alkaline pectinase. *Textile Research Journal*, 71: 527–533.

Zerdin, K., Rooney, M. L., and Vermue, J. 2003. The vitamin C content of range juice packed in an oxygen scavenger material. *Food Chemistry*, 82: 387–395.

Zisu, B., Bhaskaracharya, R., Kentish, S., and Ashokkumar, M. 2010. Ultrasonic processing of dairy systems in large scale reactors. *Ultrasonics sonochemistry*, 17: 1075–1081.

Zourob, M., Hawkes, J. J., Coakley, W. T. et al. 2005. Optimal leaky waveguide sensor for detection of bacteria with ultrasound attractor force. *Analytical Chemistry*, 77: 6133–6168.

22 Development of Sonochemical Reactor

Keiji Yasuda and Shinobu Koda

CONTENTS

Introduction 581
Quantification of Sonochemical Intensity 582
    Calorimetry 582
    Chemical Dosimetry 584
    Cavitation Noise 585
Reactor Design 585
Frequency of Sonication 586
Enhancement of Sonochemical Effects 589
    Superposition of Sonochemical Fields 589
    Pulsed Ultrasound 590
    Liquid Flow 591
    Particle Addition 592
Large-Size Sonochemical Reactor 592
    Effect of Liquid Height 592
    Large-Scale Sonochemical Reactor 594
References 595

INTRODUCTION

In pure liquids and solutions, generation and collapse of bubbles caused by ultrasound with a frequency range from 20 kHz to several megahertz result in extraordinarily high local temperature and pressure called "hot spot." Thus, as these extreme circumstances can be easily generated in pure liquids and solutions by ultrasound, the phenomenon is very attractive to physicists and chemists. From the 1990s, the ultrasound studies extended to practical application on various fields, that is, organic synthesis, environmental chemistry, material processing, and chemical engineering processes.

Cavitation caused by ultrasound leads to physical and chemical effects. The physical effects are related to microstreaming and mixing, which accelerate cleaning, extraction, polymer degradation, and other processes. The chemical effects are attributed to the production of OH and H radicals that can generate or influence some chemical reactions. It is well known that the intensity of these physical and chemical effects depends on various factors, such as the physical properties of media, emitted frequency, and intensity. Therefore, it is not easy to reproduce and compare the experimental results reported by different researchers. This might slow down the progress of industrial application in sonochemistry.

In order to obtain reliable data under sonication, it is necessary to operate the sonochemical reactors under constant ultrasonic intensity. The ultrasonic intensity is also a valuable parameter for the study of acoustic physical properties, and many methods have already been established for its measurement. Thus, the ultrasonic intensity is often measured in sonochemistry by checking one of the following characteristics: the acoustic pressure, vibrational amplitude, radiation pressure, and heat dissipated into solution. Moreover, as the cavitational intensity for different sonochemical reactors used at the same input electric power is also influenced by the shape of the reactors, the height of liquid from a transducer, and so on, one has to know the ultrasonic energy consumed in solution. Unfortunately, some workers report only the input electric power. For chemists and chemical engineers who are not familiar with dealing with the electrical part of ultrasonic devices, a simple method involving processes that can be either physical or chemical in character is desirable to quantify cavitational activity. Several chemical dosimetries have been proposed, but unfortunately, a universal method is not yet defined.

In addition to the reproducibility of the experimental results, we also need to develop large-scale sonochemical reactors for the practical applications. For example, using the sonochemical reactor with 1 m^3 in volume for 1 h sonication, we can only treat 24 m^3 of wastewater per day. Alternatively, by employing a large-scale and high-performance sonochemical reactor, the amount of wastewater can be substantially increased. Many workers have challenged to design various high-performance sonochemical reactors for laboratory and large-scale work. However, it is hard to compare the reactors for lack of the assessment criterions.

This chapter devotes to quantification of cavitation effects by using calorimetry and chemical dosimetry and their application to the assessment of sonochemical reactors. In addition, here we review various methods proposed for the enhancement of sonochemical reactions.

QUANTIFICATION OF SONOCHEMICAL INTENSITY

The ultrasonic power or intensity (Berlan and Mason, 1996) has been considered as one of the factors with importance for the quantification of sonochemical intensity. The measurement of sound pressure with a hydrophone is useful to determine the distribution of the ultrasonic intensity in a reaction vessel. However, it is not suitable to specify the mean ultrasonic power dissipated into the reaction system because it is revealed that the sound field is not distributed uniformly. Calorimetry is often used to specify the ultrasonic power dissipated into solutions (Contamine et al., 1995; Kimura et al., 1996). Many investigators have examined the correlation between generated thermal energy and sonochemical effects. Several chemical reactions (chemical dosimetry for the quantification of sonochemical intensity) have also been proposed. Other methods such as measurements of cavitation noise, cavitation bubbles, and sonochemical luminescence have also been used, but these methods are mainly feasible for the estimation of partial area of sonochemical reaction in the sonochemical reactor.

CALORIMETRY

The ultrasonic power dissipated into a liquid is calculated by the following equation:

$$E_{US} = \frac{dT}{dt} C_p M \tag{22.1}$$

where
C_p is the heat capacity of liquid (e.g., water: 4.2 J g^{-1})
M is the mass of water (g)

The value of (dT/dt) is the temperature rise per second. The initial temperature rise is measured at room temperature by using a thermocouple immersed in liquid.

Rotoarinoro et al. (1995) used water and toluene, and discussed the relationship between the electric power at transducer and the calorimetry for a sonochemical reactor with 20 kHz horn transducer. For same electric power value, the ultrasonic power in water was close to that in toluene.

Development of Sonochemical Reactor

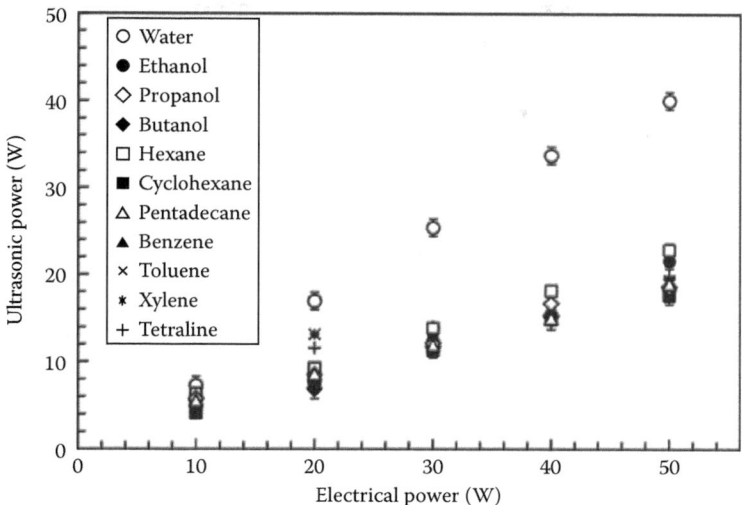

FIGURE 22.1 Influence of electrical power on ultrasonic power dissipated into sonicated medium for water and organic solvents.

Löning et al. (2002) examined by calorimetry the energy transformation from the electric power at transducer for several liquids by using a reactor with a 20 kHz horn transducer. They reported that more energy was consumed in high-viscous media than in low viscous liquids. Kuijpers et al. (2002) have investigated the energy efficiency of the ultrasound-induced radical formation from methyl methacrylate for a 20 kHz horn transducer. From the energy yields estimated for the formation of radicals due to the ultrasonic waves, it has been revealed that the highest energy conversion appeared at low temperatures and at low amplitudes.

Toma et al. (2011) measured the ultrasonic powers in water and 10 pure organic solvents by performing calorimetry for a cylindrically shaped sonochemical reactor with a bottom mounted vibrating plate at 500 kHz. Figure 22.1 shows the influence of electrical power on the ultrasonic power dissipated into the sonicated medium for water and organic solvents. The slopes are 0.82 for water and about 0.3–0.4 for organic solvents and indicating that the energy base efficiency for solvents is half from the efficiency of water. For some organic solvents investigated, an intense ultrasonic atomization was described. Based on these findings, the authors considered that the energy balance equation has to comprise an energy term for ultrasonic atomization as follows:

$$E_{electrical} = E_{dissipated} + E_{atomization} + E_{transducer} + E_{lost} \tag{22.2}$$

where
  $E_{electrical}$ is the electrical power applied at transducer
  $E_{dissipated}$ is the ultrasonic power accommodated into the sonicated liquid
  $E_{atomization}$ is the power lost through the liquid atomization
  $E_{transducer}$ is the heat stored at transducer
  E_{lost} stands for the part that is more difficult to be quantified such as the heat lost to the environment, liquid vaporization (apart from atomization), or ultrasonic wave attenuation

In the case of a large sonochemical reactor equipped with a transducer much smaller in diameter that the cross-sectional area of the reactor, the measured value of initial temperature rise strongly depends on the place of thermocouple in reactor. Sometimes, several thermocouples are set in the sonochemical reactor and ultrasonic power is estimated from the average value of initial temperature rises at different positions (Asakura et al., 2008b). In some situations, in order to measure the average value of initial temperature rises for large sonochemical reactors, the measurement of temperature rise under slow agitation may be useful.

CHEMICAL DOSIMETRY

Chemical dosimetry gives us the sonochemical efficiency in a whole reaction volume. Several chemical dosimetry methods have also been proposed, as shown in Table 22.1. The Wiessler reaction has been used as a popular chemical dosimeter (Weissler et al., 1950; Kimura et al., 1996). The iodine dosimetry method is based on the fact that sonication of water containing CCl_4 produces molecular chlorine, which reacts quickly with iodide ions in solution to liberate molecular iodine. However, the sample requires CCl_4, and in recent years some researchers avoid this method for environmental-protection reason.

The potassium iodide (KI) method has been used as a popular chemical dosimeter (Hart and Henglein, 1985; Koda et al., 2003). When ultrasound is irradiated into an aqueous KI solution, I^- ions are oxidized to give I_2 by OH radical. When excess I^- ions are present in solution, I_2 reacts with the excess I^- ion to form I_3^- ion. The concentration of I_3^- ion is measured by using a UV spectrometer at 355 nm (molar absorbance coefficient = 26,303 $dm^3\,mol^{-1}\,cm^{-1}$). The light absorbance measurement of sonochemical reaction is very simple and widely acceptable. Fricke reaction, an established method in radiation chemistry, is often used in sonochemistry, too (Fricke and Hart, 1935; Jana and Chatterjee, 1995; Nomura et al., 1996; Mark et al., 1998). When the ultrasound is generated into Fricke solution, the Fe^{2+} ions in the solution are oxidized to give Fe^{3+} ions by OH radicals. The concentration of Fe^{3+} ion is measured by using an ultraviolet-visible (UV) spectrometer at 304 nm (molar absorbance coefficient = 2,197 $dm^3\,mol^{-1}\,cm^{-1}$). Since the Fricke solution must be prepared with $Fe(NH_4)_2^{2-}(SO_4)_2 6H_2O$, H_2SO_4, and NaCl, there are safety concerns related to the handling of strong acids by the researchers.

In the fluorescence method, the sample under ultrasonic irradiation is the aqueous solution of terephthalic acid. The terephthalate ions react with hydroxyl radicals to generate highly fluorescent 2-hydroxyterephthalate ions (Mason et al., 1994; Mark et al., 1998). This method has very high sensitivity (Iida et al., 2005), but the apparatus for the fluorescence measurement is not a very common

TABLE 22.1
Chemical Dosimeters

Method	Main Reaction	Reference
Weissler	$CCl_4 + H_2O \rightarrow Cl_2 + CO + 2HCl$	Kimura et al. (1996), Weissler et al. (1950)
	$2KI + Cl_2 \rightarrow I_2 + 2KCl$	
KI	$2I^- + 2OH^{\cdot} \rightarrow I_2 + 2OH^-\ \ I_2 + I^- \leftrightarrow I_3^-$	Hart and Henglein (1985), Koda et al. (2003)
Fricke	$Fe^{2+} + OH^{\cdot} \rightarrow Fe^{3+} + OH^-$	Fricke and Hart (1935), Mark et al. (1998), Jana and Chatterjee (1995), Nomura et al. (1996)
Fluorescence	Terephthalate anion + $OH^{\cdot} \rightarrow$ 2-hydroxyterephthalate anion	Mason et al. (1994), Mark et al. (1998), Iida et al. (2005)
HNO_3	$N^{\cdot} + O^{\cdot} \rightarrow NO$	Koda et al. (1996)
	$NO + O^{\cdot} \rightarrow NO_2$	
	$NO_2 + OH^{\cdot} \rightarrow HNO_3$	
H_2O_2	$2OH^{\cdot} \rightarrow H_2O_2$	Sato et al. (2000)
Phenolphthalein	HIn^- (alkaline) $\leftrightarrow H^+ + In^{2-}$ (acidity)	Rong et al. (2001)
Porphyrin	Porphyrin + ultrasound $\rightarrow$ decomposition	Nomura et al. (1996), Kojima et al. (1998)
Rodamine B	Rodamine B + ultrasound $\rightarrow$ decomposition	Sivakumar and Pandit (2001)
Luminol	3-Aminophthalhydrazide + $2OH^{\cdot} \rightarrow$ 3-aminophthal acid dianion + N_2 + hv	Renaudin et al. (1994), Hatanaka et al. (2001), Price et al. (2010)

analytical instrument. Sonochemical productions of HNO_3 (Koda et al., 1996), H_2O_2 (Sato et al., 2000) in water are also used as a chemical dosimeter. However, the concentration measurements of HNO_3 and H_2O_2 are not simple operations. Phenolphthalein is a very popular pH indicator, and its transition interval is between pH = 8.3 and pH = 10. The pH value of the alkaline solution of phenolphthalein under sonication will decrease due to production of nitric and nitrous acids. Thus, the color of the aqueous solution of phenolphthalein will change from red to colorless. Therefore, the fading time of aqueous phenolphthalein solution is a measure of the ultrasonic intensity in a reaction vessel (Rong et al., 2001). This method is very simple but less accurate. Decompositions of colored substance such as porphyrin derivatives (Nomura et al., 1996; Kojima et al., 1998) and rodamine B (Sivakumar and Pandit, 2001) are also used for chemical dosimeter. The concentration changes of these colored substances are easy to be measured by using a UV spectrometer. However, the constitutional formula of colored substance is normally complex and the chemical reactions pathway for decomposition of colored substance is difficult to analyze.

The luminol (3-aminophthalhydrazide) reacts with hydroxyl radicals, which is produced by sonolysis of water, to generate luminescence (Renaudin et al., 1994). The light intensity measured by using a photomultiplier reflects the sonochemical reaction. In order to estimate sonochemical efficiency in a whole reactor, one needs to have a transparent sonochemical reactor and to gather the sonochemical luminescence from the whole reactor to a photo-detector such as a photomultiplier (Hatanaka et al., 2001). The light-emitting area observed in the sonochemical reactor under dark condition indicates sonochemical reaction zone. Price et al. (2010) have visualized sonochemical luminescence from luminol solution in reactor by using a CCD camera for 23 kHz horn and 515 kHz plate transducer. They have reported that the sonochemical reaction fields produced by a 515 kHz plate transducer and a 23 kHz horn sonicator were significantly different from each other.

CAVITATION NOISE

The analysis of cavitation noise is used for the measurement of the active amount of cavities (Lauterborn et al., 1999). Frequency spectra of acoustic signal measured by using hydrophone showed some noises at a half basic frequency (sub harmonic) and a twice of basic frequency (super harmonic) due to the collapse of cavitation bubbles. The noise level is reflected by active cavities amount for reaction. Segebarth et al. (2002) reported the correlation between the sonochemical production of peroxides and the acoustical noise spectra in anionic surfactant solution. Zeqiri et al. (2003a,b, 2006) also found out the spatial distribution of cavitation activity generated within an ultrasonic cleaning vessel and showed good qualitative agreement with the spatial distribution of cavitation determined through monitoring the erosion of a thin sheet of aluminum foil. Hodnett and Zeqiri (1997) reviewed the hydrophone, the optical interferometry, and the thermal methods for the measurement of ultrasonic intensity.

REACTOR DESIGN

The typical apparatus for sonochemistry consists of sonochemical reactor, temperature control unit, ultrasonic generator, and power amplifier. The cells of the sonochemical reactors can be made of acryl resin or glass, with or without a stainless steel frame, or stainless steel. A piezoelectric ceramics or magnetostrictive transducer is incorporated into the reactor and is used as a sound source. For the low frequency, the horn-type and Langevine-type transducers are used. For higher frequency, PZT (lead zirconate titanate) is used by mounting it at the bottom of the vessel or attaching the holder that incorporates PZT transducers. In addition to these component parts of reactor, the shape of reactor vessel, the attachment of transducer, and the irradiation method are taken into account to design a sonochemical reactor.

Figure 22.2 shows the cylindrical and rectangular sonochemical reactors. The former has a great advantage for theoretical analysis of sound fields, since its typical acoustic source is a circular disc. Beaker-scale reactors belong to a cylindrical sonochemical reactor. On the other hand, rectangular

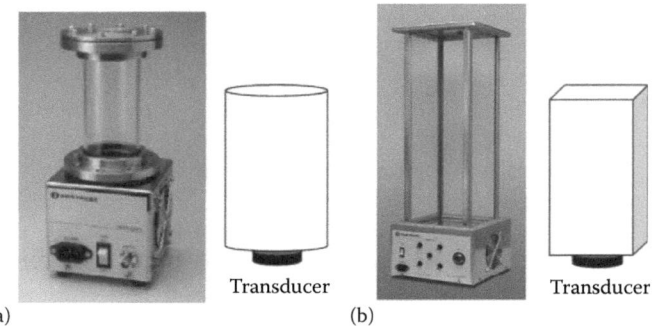

(a)   Transducer   (b)   Transducer

FIGURE 22.2 (a) Cylindrical and (b) rectangular sonoreactors.

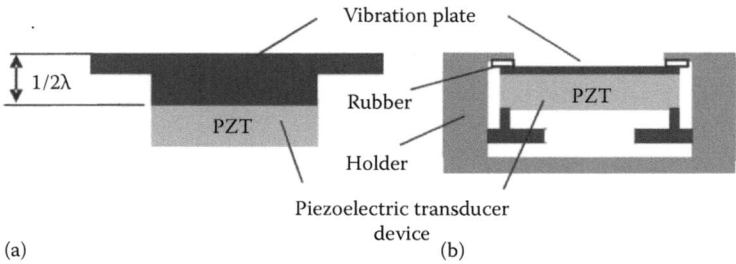

FIGURE 22.3 Types of transducer unit: (a) fixed unit—where the PZT transducer is mounted beneath the vibrating plate and (b) exchangeable unit assembly—where the PZT transducer and the vibrating plate are assembled and fitted with rubber into a holder.

sonochemical reactors are useful for practical application, since the rectangular sonochemical reactor is easy to be designed at a large scale. In addition, the rectangular sonochemical reactor is useful to visualize the liquid flow in the vessel. The hexagonal type of reactors has been designed to enhance the cavitation effects (Gogate et al., 2003).

The transducer unit for higher frequency has the vibrating plate mounted directly on transducer, as shown in Figure 22.3a, and fixes at the bottom of sonochemical reactors. A vibrating plate of the thickness of $(n\lambda/2)$ or much less than $(\lambda/10)$ is recommended for transmitting effectively the sound wave into liquid. Where the λ is sound wavelength in vibrating plate and n is integer. The other type of transducer unit consists of transducer, vibration plate, holder, and rubber, as shown in Figure 22.3b. This type of transducer unit is easily exchangeable and it facilitates the maintenance of reactors for industrial application.

In organic synthesis experiments, the reaction vessels are set in a water bath and sonication is performed through water. In this case, it is very important to know the acoustic intensity delivered into the reaction system, because a part of the ultrasonic energy dissipates into the water bath. To reduce experimental errors, the position of sonochemical reactors cells must be fixed.

Ultrasonic irradiation can be performed by using more than one acoustic source in order to enhance cavitation effects. The case of sonochemical reactor with multi-transducers will be described in the "Superposition of sonochemical fields" and "Large scale of sonoreactor" sections.

FREQUENCY OF SONICATION

Frequency is one of the important consideration factors in sonochemistry, since cavitation behavior depends strongly on the frequency. To our knowledge, Busnel and Picard (1952) and Busnel et al. (1953) reported first that the frequency dependence of the yield of I_2 in KI solution at different input electric powers has the maximum around 300 kHz. These results indicate that the sonochemical

Development of Sonochemical Reactor

effect caused by the sonolysis of water is most efficient at 300–500 kHz. Mark et al. (1998) have examined the frequency dependence of sonochemical effects by fluorescence spectroscopy using terephthalate solution. They also reported a strong frequency dependence of sonochemical reactions.

Koda et al. (2003) reinvestigated the frequency dependence of three typical sonochemical reactions (Fricke reaction, KI oxidation, and decomposition of porphyrin derivatives) and proposed the sonochemical efficiency given by Equation 22.3 as follows:

$$\text{SE} = \frac{m}{E_{US}} = \frac{C}{P_{US} t / V} \tag{22.3}$$

where
  m is number of reacted molecules
  E_{US} is ultrasound energy
  P_{US} is ultrasound power measured by calorimetry
  t is ultrasonic irradiation time
  V denotes solution volume, respectively

The value of C is the I_3^- or Fe^{3+} ion concentration in solution. Asakura et al. (2007, 2008a,b) and Kojima et al. (2010) have used the sonochemical efficiency value to estimate the performance of large sonochemical reactors.

Figures 22.4a through c demonstrates the frequency dependence of Fe^{3+} yield in Fricke solution, I_3^- yield in KI solution, and decomposition ratio of 5-, 10-, 15-, 20-Tetrakis (4-sulfotophenyl) porphyrin (hereafter abbreviated as TPPS) in solution, respectively. All results are divided by

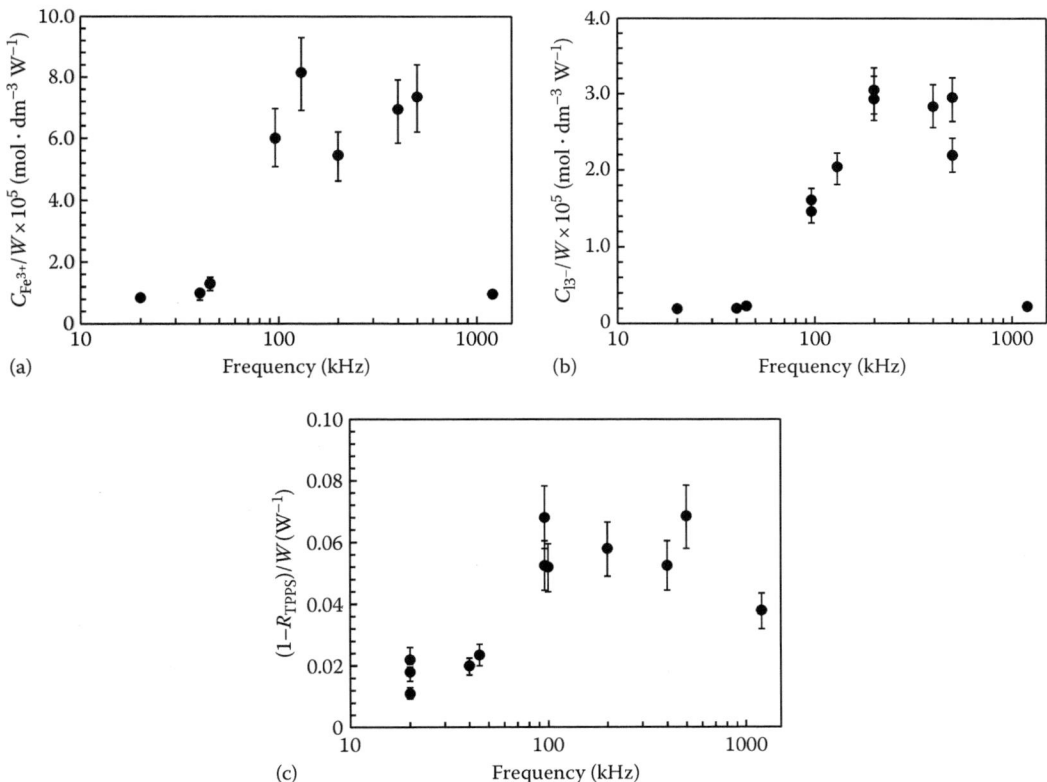

FIGURE 22.4 Frequency dependence of chemical effects per unit power. (a) Fricke reaction, (b) KI oxidation, and (c) decomposition of TPPS.

the ultrasonic power, which is measured by calorimetry and reflects ultrasonic power delivered into solution. The value of $(1 - R_{TPPS})$ is the decomposition conversion of TPPS. The frequency dependence of $(1 - R_{TPPS})$ is slightly different from those of KI oxidation and Fricke reaction. The decomposition ratios of TPPS around 300 kHz are about four times larger than those in the lower frequency region, while product yields of Fe^{3+} and I_3^- around 300 kHz are about 10 times larger than those below 100 kHz, where they seem to be independent of the frequency. Above 500 kHz, the sonochemical effects steeply decrease. At 1.2 MHz, the sonochemical effects are the same order of those obtained below 100 kHz. These results indicate that the mechanism of TPPS decomposition is not the same as those of KI oxidation and Fricke reaction induced by the sonolysis of H_2O. For TPPS decomposition, the direct sonolysis of TPPS at the interface of cavity may occur in addition to the sonolysis of H_2O.

The observed dependence of the chemical effects on the frequency can be attributed to several factors: temperature inside a bubble, cavitation threshold, bubble population, lifetimes of bubbles, etc. As the sonication frequency increases, the maximum temperature attained in the collapse of a bubble decreases according to bubble dynamics (Hung and Hoffmann, 1999). The cavitation threshold increases with frequency, which brings narrower active region to generate bubbles in a nonuniform sound field (Kojima et al., 2001). These factors act to reduce the chemical effects. On the other hand, the bubble population in a standing-wave field increases because of a shorter interval of active regions corresponding to the decrease of wavelength at a higher frequency. The lifetime of bubbles becomes shorter with increasing frequency (Leighton, 1997). This means that more radicals formed in a bubble can escape from the bubble to react with any other material present in the medium. These are all factors that lead to increased chemical effects. Other factors such as bubble cloud (Esche, 1952) and the number of vaporized molecules inside a bubble (Lauterborn and Holzfuss, 1986; Petrier et al., 1992; Yasui 2002) may also be responsible.

In radiation chemistry, an energy-specific yield, known as G-value, is defined as $G(-M)$ or $G(+P)$, with the number of molecules disappeared (M) or the number of molecules formed (P) per radiation energy of 100 eV absorbed. Therefore, Mark et al. (1998) have proposed a similar G-value in sonochemistry as the product yields expressed in terms of mol J^{-1}. Mason et al. have defined the sonochemical yield (SY) (Berlan and Mason, 1991) or the fluorescence yield (F/D) (Mason et al., 1994) to evaluate the sonochemical effects. The SY value indicates the measured effect divided by the input power and has units of $mol \cdot W^{-1} \cdot h^{-1}$ or $mol \cdot J^{-1}$ while the sonochemical efficiency (SE value) as indicated by Equation 22.3 is evaluated by the concentration of the compounds divided by the ultrasonic energy density $(mol \cdot dm^{-3})/(J \cdot dm^{-3})$. The F/D is the fluorescence intensity produced per unit ultrasound dosage and has unit of J^{-1}. The ultrasonic dosage is measured by calorimetry and for this reason the F/D value is similar to the SE value. However, they reported only the F/D value in the narrow frequency range from 20 to 60 kHz. The absorbance measurement of the sonochemical reactions is very simple and widely acceptable. The SE values for KI oxidation and Fricke reaction are summarized in Table 22.2 (Koda et al., 2003). These results show that the sonochemical effects have the maximum in the frequency range from 200 to 500 kHz.

The frequency dependence of physical effects is not yet clearly definite experimentally. The physical effects are expected to depend on the frequency, since the bubble size changes with the frequency. Acoustic degradation of the polymers in solution is mainly caused by shear force generated by the relative motion of the solvent along the polymer chains during the bubble collapse, that is, mechanical or physical effects. The polymer degradation rate under sonication changes with sonicated frequency. Koda et al. (2011) indicated that ultrasonic degradation of a water-soluble polymer in aqueous solution is more effective at 500 kHz frequency without a radical scavenger. Grönroos et al. (2001) revealed that the most extensive degradation of polyvinyl alcohol in aqueous solutions takes place at the lowest frequency of 23, 40, and 900 kHz. Our recent work has demonstrated that the degradation rate decreases as the frequency is increased. It is reasonable to recognize that the physical effects depend on the frequency, but the frequency-dependent behavior of the physical effects needs more investigation.

TABLE 22.2
Frequency Dependence of Sonochemical Efficiency (SE Value) for KI Oxidation and Fricke Reaction

	kHz	SE_{KI} (mol J^{-1})	SE_{Fricke} (mol J^{-1})
Shiga 20k	20	$0.60 \pm 0.02 \times 10^{-10}$	$2.3 \pm 0.1 \times 10^{-10}$
Shiga 40k	40	0.60 ± 0.02	2.8 ± 0.1
Aist 45k	45	0.67 ± 0.06	3.7 ± 0.1
Aist 96k	96	4.5 ± 0.2	
Nagoya 96k	96	4.1 ± 0.2	16.8 ± 1.0
Nagoya 130k	130	5.6 ± 0.4	22.6 ± 1.6
Shiga 200k	200	8.3 ± 0.6	15.2 ± 0.9
Toyama 400k	400	7.8 ± 0.2	19.3 ± 1.2
Nagoya 500k	500	7.1 ± 0.2	20.3 ± 1.2
Toyama 1200k	1200	0.64 ± 0.3	2.6 ± 0.1

ENHANCEMENT OF SONOCHEMICAL EFFECTS

Superposition of Sonochemical Fields

The ultrasonic irradiation with dual frequency, obtained by the superposition of two ultrasonic fields generated by two transducers, enhances sonoluminescence intensity and sonochemical reaction. Iernetti et al. (1997) observed that the sonoluminescence intensity of ultrasound at 700 kHz was increased by the addition of an ultrasonic radiation at 20 kHz. Ciuti et al. (2000) superposed the ultrasonic field at 27.2 kHz on that at 730 kHz and the sonoluminescence intensity increased by two or three orders of magnitude. Feng et al. (2002) reported that triple frequency irradiation at 28 kHz, 1 MHz, and 1.87 MHz enhanced the sonochemical reaction of KI and terephthalate ion. Zhao et al. (2002) studied the fluorescence enhancement of aqueous solution of terephthalate ion under orthogonal sonication of 28 kHz and 1.7 MHz frequency ultrasonic waves. Suzuki et al. (2004) examined the effect of superposition of two ultrasonic fields on the KI solution sonochemistry. In this work, one frequency was generated by a 20 kHz horn-type transducer and another frequency was generated by a plate-type transducer with operating frequencies in the range from 24.1 kHz to 1.17 MHz. In this case, the enhancement of the sonochemical effect for dual frequency irradiation was evident below 100 kHz. The frequency spectrum of acoustic noise indicated that the enhancement effect came from an increase in the number of cavitation bubbles. Yasuda et al. (2007) used an ultrasonic dual frequency reactor with a bottom and the side-wall transducer and investigated the fluorescence intensity of terephthalate ion in the frequency range from 176 to 635 kHz. The sonochemical reaction fields were visualized by using the sonochemical luminescence of luminol solution. Figure 22.5 shows the effect of ultrasonic frequency at side transducer on the ratio of fluorescence intensity for dual frequency to the sum value of fluorescence intensities for single frequency. The values of I_1 and I_2 indicate the fluorescence intensities for the single frequency irradiation from the bottom transducer and the side-wall transducer, respectively. The value of I_{1+2} indicates the fluorescence intensity for dual frequency. The effective electric power per transducer is 60 W. The fluorescence intensity ratio of dual frequency to the single frequency had maximum value when the frequency of the bottom transducer was comparable in magnitude to that of the wall sided. The photographs of sonochemical luminescence for dual frequency indicated that the sonochemical reaction fields extended in the reactor volume and became more intensive around the reactor center.

During the several years, many possible mechanisms have been proposed regarding the dual irradiation. The enhancement of sonochemical effects has been associated with the decrease of

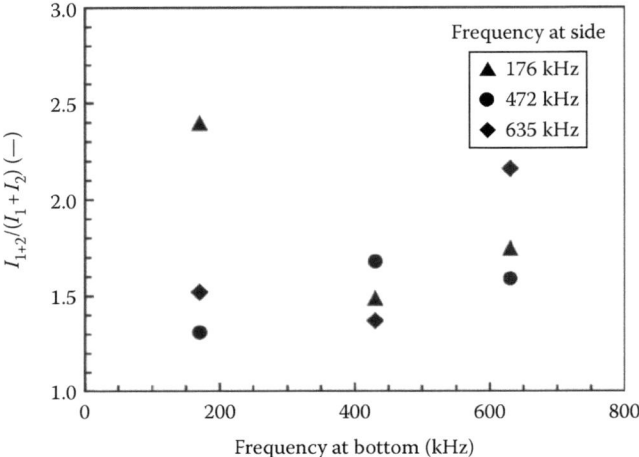

FIGURE 22.5 Effect of ultrasonic frequency at side transducer on the ratio of fluorescence intensity for dual frequency to sum value of fluorescence intensities for single frequencies. The effective electric power per one transducer is 60 W.

cavitation threshold (Iernetti et al., 1997; Ciuti et al., 2000), the production of cavity nuclei (Ciuti et al., 2000; Feng et al., 2002; Zhao et al., 2002), the destruction of bubble clusters (Ciuti et al., 2000), the enhancement of Bjerknes forces (Feng et al., 2002; Zhao et al., 2002), the nonlinear interaction (Feng et al., 2002; Zhao et al., 2002), the enhancement of mass transfer (Feng et al., 2002), the expansion of standing-wave region (Yasuda et al., 2007), and so on. The enhancement of sonochemical effects by using ultrasonic superposition has been reported by many researchers, but the mechanism remains an open question.

Pulsed Ultrasound

When ultrasound wave is generated with a certain pulsed mode, some cavitational effects may occur to a greater degree than when the same acoustic signal is applied in continuous wave. Many papers have been published until now and the works before the 1990s were put together by Leighton (1997). He clarified aspects opened by many papers, including the "activity" of the system, the recycling periodic cavities through resonance, the survival of unstabilized nuclei, the transient excitation, and the bubble migration. Readers interested in these works should refer to his book.

Mitome and Hatanaka (2002) used sonochemical reactors at 130.0 and 43.7 kHz and investigated changes in sonoluminescence intensity from distilled water under various experimental conditions. They used power-modulated pulsed ultrasound. A pulsing operation at a constant input energy enhances sonoluminescence intensity at lower power levels because of the higher amplitude of ultrasound. At higher power levels, the quenching effect due to excessive sound pressure appears and the pulsing operation is not effective. The pulsing operation is more effective at 130.0 than at 43.7 kHz, which corresponds well to the quenching mechanism based on the clustering of cavitation bubbles due to the secondary Bjerknes forces. Casadonte et al. (2005) explored the effects of pulse waveform and frequency on the oxidation of potassium iodide and the degradation of acid orange at 500 kHz. They also used power-modulated pulsed ultrasound. The square-shaped pulse generated more power than the triangle- or sine-shaped ones. Furthermore, an oxidation rate increased by a factor of 3 as compared with a continuous irradiation was observed under conditions of equivalent acoustic input power. Sonochemical reaction fields visualized by luminol sonoluminescence were increased under the pulsed conditions. Yang et al. (2005) decomposed the nonvolatile surfactants sodium 4-octylbenzene sulfonate and sodium dodecylbenzenesulfonate

Development of Sonochemical Reactor

at 354 kHz. They attributed the enhanced degradation of surfactants by pulsed ultrasound to the accumulation of surfactants on cavitation bubble surfaces.

Tuziuti et al. (2008) have studied the effect of liquid volumes on sonochemical-reaction efficiency at 152 kHz. They reported that the sonochemical efficiency was increased by pulsed ultrasound and the magnitude of the sonochemical effect enhancement strongly depends on liquid height. This behavior was related to both the residual pressure amplitude during the pulse-off time and the spatial enlargement of active reaction sites.

Liquid Flow

Liquid flow also enhances sonochemical reaction rate. Yasuda et al. (1999a) reported that the decomposition conversion of 5-, 10-, 15-, 20-Tetrakis (4-sulfotophenyl) porphyrin was enhanced by liquid mixing for a sonochemical reactor with plate-type transducer at 22.8 kHz and a stirring device. Hatanaka et al. (2006) have investigated the effects of ultrasonic frequency (20–131 kHz) and power on the intensity of the sonochemical luminescence of luminol, when liquid mixing was applied. The authors reported that the intensity became high when both values of ultrasonic frequency and power were low or they were high.

Kojima et al. (2010) used a rectangular sonochemical reactor at a frequency of 490 kHz with a stirrer to examine the effect of liquid flow on the sonochemical efficiency and the area of chemical luminescence in the reactor. Figure 22.6 illustrates the results of the sonochemical efficiency under different rotational speed of the stirrer. The plot shows clearly that the sonochemical efficiency increases with the rotational speed of the stirrer. The sonochemical efficiency at 350 rpm is nearly twice as high as that without mixing. They also observed the area of luminescence and the cavitation bubbles as a function of rotational speed. The photographs of luminol experiments demonstrated that the area of the luminescence increased with the increase in the rotational speed. The newly expanded region of chemical luminescence as a result of the increase in rotational speed corresponded approximately to the zone where newly trapped bubbles were observed. They considered that the formation of the steady standing wave in liquid under sonication was favorable for the increase of the number of active bubbles that contribute to the sonochemical reactions in the liquid (Asakura et al., 2008a). Hence, in order to enhance the sonochemical reaction rate, it is important to extend the field of the steady standing wave. However, large amplitude standing wave causes the aggregation of bubbles and thus, a reduction of the bubbles activity. As a result, the sonochemical efficiency decreases with increasing the electric power. On the contrary, the mechanical flow

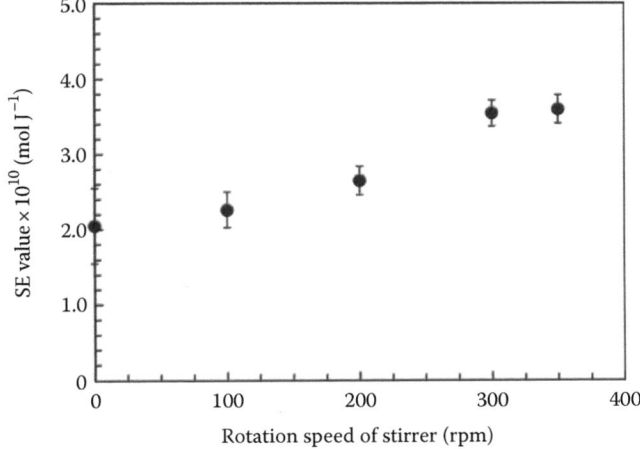

FIGURE 22.6 Sonochemical efficiency under different rotational speeds of stirrer.

generated by the rotational speed of stirrer is responsible for the increase of the horizontal flow rate, which prevents the active bubbles from aggregating in the standing wave. In addition, the stirrer supplies the acoustic field with reactant and cavitation nuclei required for the formation of the active bubbles. An increased liquid velocity might contribute to the increase of the sonochemical yield. Therefore, the combination of stirring with sonication enhances the sonochemical reaction rate.

PARTICLE ADDITION

The particle addition has a potential to enhance the yield in the sonochemical reaction. Yasuda et al. (1999b) used a sonochemical reactor with plate-type transducer at 22.8 kHz and silica or α-alumina particles in 30 μm in diameter. They found that the sonochemical reaction rate of 5-, 10-, 15-, 20-Tetrakis (4-sulfotophenyl) porphyrin increased by the addition of each particle type. Sekiguchi and Saita (2001) also observed that the ultrasonic decomposition of chlorobenzene in water for a horn transducer at 20 kHz was enhanced by the addition of α-alumina particle. Keck et al. (2002) investigated the influence of quartz particles in aqueous on the chemical effects of ultrasound. The formation rate of hydrogen peroxide shows a maximum value at 206 kHz in the frequency range from 68 to 1028 kHz when the diameter and the concentration of quartz particles was 3–5 μm and 4–8 g L^{-1}, respectively.

The particle cracks create cavitation nuclei for bubbles (Crum, 1979; Marschall et al., 2003). Tuziuti et al. (2005) investigated the correlation of acoustic cavitation noise and the I_3^- yield in solution by particle addition. They showed that the addition of alumina particles with an appropriate amount and size increased the yield of sonochemical reaction. The acoustic noise due to cavitation bubbles was increased by the particle addition. Tuziuti et al. (2006) also found that the sonochemical luminescence intensity and the I_3^- yield were enhanced by the addition of micrometer-sized air bubbles. The ultrasonic frequency was 141 kHz and bubble diameter 5–50 μm. They considered that the tiny bubbles added into the sonicated liquid are directly connected to an increase in the number of collapsing bubbles active for sonochemical reaction.

LARGE-SIZE SONOCHEMICAL REACTOR

In order to extend the industrial application of sonochemistry, an efficiently scale-up of the sonochemical reactors is required. An important issue related to the scale-up of the sonochemical reactors is the elucidation of the effect of liquid height or irradiation volume on sonochemical reactions.

EFFECT OF LIQUID HEIGHT

Iernetti (1971) compared the cavitation thresholds of distilled water at 700 kHz for variable sample volumes from 10^{-3} to 10^3 cm^3 and reported that the cavitation threshold increased with the volume. Renaudin et al. (1994) investigated the intensity of sonochemical luminescence at 500 kHz for liquid heights ranging from 25 to 75 mm. The chemiluminescence intensity and sample temperature clearly decrease when the liquid height increases. Kojima et al. (1998) examined the decomposition rate of 5-, 10-, 15-, 20-Tetrakis (4-sulphotophenyl) porphyrin in solutions at 126, 500, and 1000 kHz for liquid heights ranging from 20 to 70 mm. Their results also revealed that at laboratory scale, the effect of ultrasound on sonochemical reaction decreased with an increase in the liquid height.

Recently, Asakura et al. (2008a) have used cylindrical reactor 70 mm in diameter. The sonochemical efficiency of the cylindrical sonochemical reactor has been investigated as a function of frequency and liquid height when the irradiation frequencies were 45, 129, 231, and 490 kHz. The liquid height varied from 10 to 700 mm. The sonochemical efficiency of the cylindrical reactor was evaluated by KI dosimetry and calorimetry. Figure 22.7 shows the variation in the sonochemical efficiency value with the liquid height for different frequencies. It was found that the sonochemical efficiency value

Development of Sonochemical Reactor 593

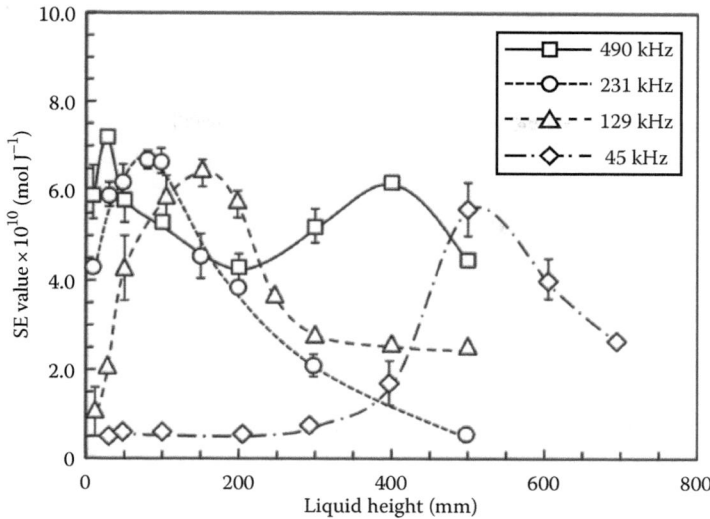

FIGURE 22.7 Variation of sonochemical efficiency value with liquid height for different frequencies.

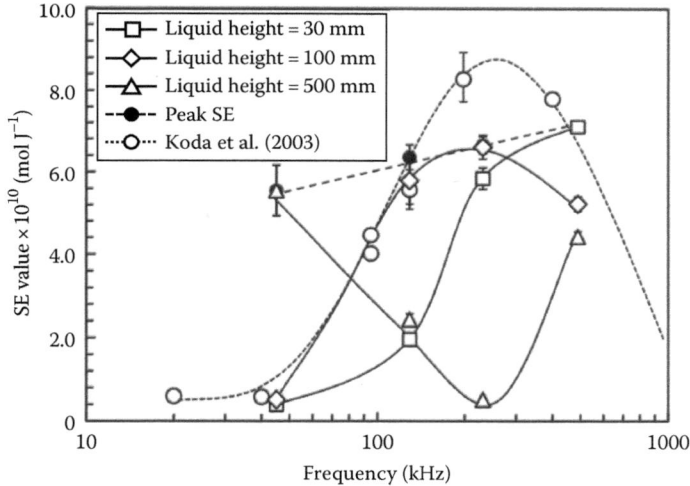

FIGURE 22.8 Variation of sonochemical efficiency value with frequency at liquid heights of 30, 100, and 500 mm; results reported by Koda et al. (2003) are also shown for comparison.

for each frequency depends on the liquid height. Figure 22.8 shows the variation in the sonochemical efficiency value with frequency at liquid heights of 30, 100, and 500 mm; results reported by Koda et al. (2003) are also shown for comparison. Although the experimental conditions such as types of reactors and irradiation methods are different, the frequency dependence of the sonochemical efficiency value for a liquid height of 100 mm is very similar in both these studies. For a liquid height of 30 mm, the sonochemical efficiency value at 45 kHz is very low and the peak of the sonochemical efficiency value appears at a high frequency. For a liquid height of 500 mm, the sonochemical efficiency value decreases with an increase in the frequency to about 200 kHz; above 200 kHz, the sonochemical efficiency value tends to increase. It should be emphasized that for a frequency of 45 kHz, the sonochemical efficiency value at a liquid height of 500 mm is approximately eight times the sonochemical efficiency value evaluated by Koda et al. (2003) at the laboratory scale.

LARGE-SCALE SONOCHEMICAL REACTOR

The studies of chemical effects have been mostly conducted by using laboratory-scale sonochemical reactors, whose volume was less than 1 dm³. In order to put the sonochemical processes of wastewater treatment and material processing in practical use, it is necessary to design and develop a large sonochemical reactor on a pilot scale (more than 100 dm³) or an industrial scale (more than 1 m³). Several researchers have conducted studies on large sonochemical reactors.

The large-scale sonochemical reactors have been developed for low frequency at 20–50 kHz. Mason (1992) reviewed various types (cleaning bath, submersible transducer assembly, probe system, and flow system) of large-scale sonochemical reactors. Gogate et al. (2003) developed hexagonal column sonochemical reactor. There were three types of transducers working at 20, 30, and 50 kHz attached at the each side wall of the reactor and the sample volume was 7.5 dm³. Asakura et al. (2007) used large-scale sonochemical reactor at 44.3 kHz with the inner dimensions of 350 mm × 350 mm × 850 mm (height). Thirteen transducers in the apparatus were fixed with a stainless steel plate and located at the reactor bottom. They investigated the effect of liquid height on the sonochemical reaction in the range of liquid height from 87 to 592 mm, which was equivalent to the sample volume of about 18–80 dm³. The sonochemical efficiency value was about six times higher than that of the laboratory-scale sonochemical reactor below 0.2 dm³.

Some large-scale sonochemical reactors for high frequency at 200–600 kHz have been reported. Gonze et al. (1998) have developed the cylindrical reactor at 500 kHz with 300 mm in diameter. The reactor has three transducers at the bottom and has a capacity of 28 dm³. They also developed many types of sonochemical reactors at 500 kHz and compared the sonochemical reaction performance. Destaillats et al. (2001) have used four transducers at 612 kHz and developed the sonochemical reactor with the volume of 6 dm³. They have reported that the decomposition reaction rate of dichloromethane in water exceeded those for the small-scale reactors by factors from 3 to 7.

Asakura et al. (2008b) developed a 500 kHz large-scale sonochemical reactor having the inner dimensions of 508 mm × 508 mm × 672 mm (height) with 12 transducers attached at the bottom. With this reactor, they investigated the effect of liquid height on the sonochemical reaction in the range of liquid height from 300 to 500 mm for sample volume of about 77–131 dm³. Figure 22.9 shows the effect of liquid height on sonochemical efficiency value. The closed circles in Figure 22.9 were data for the cylindrical reactor at 500 kHz (Asakura et al., 2008a). Both data are close

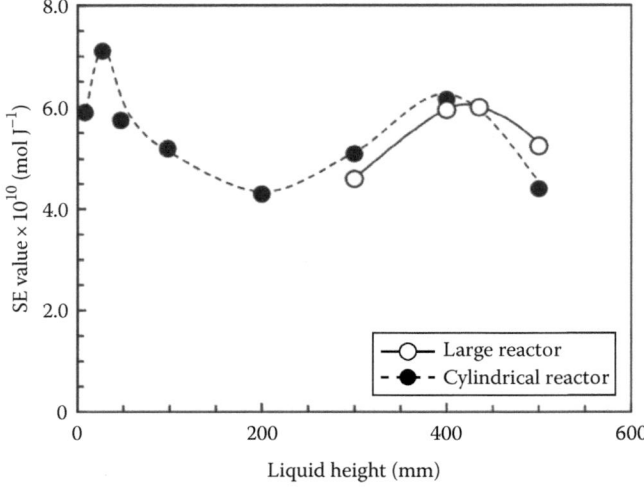

FIGURE 22.9 Effect of liquid height on sonochemical efficiency value at 500 kHz.

and have maximum value of sonochemical efficiency around 400 mm. This is because the ultrasound propagation at 500 kHz has high directivity and the sonochemical reaction fields spread only little in a direction perpendicular to ultrasonic propagation.

REFERENCES

Asakura, Y., Ishio, K., Kojima, Y. et al. 2007. Effect of liquid height on sonochemical reaction in a large-scale reactor of a rectangular parallelepiped using low frequency ultrasound. *Journal of Chemical Engineering of Japan*, 40: 1088–1092.

Asakura, Y., Nishida, T., Matsuoka, T., and Koda, S. 2008a. Effects of ultrasonic frequency and liquid height on sonochemical efficiency of large-scale sonochemical reactors. *Ultrasonics Sonochemistry*, 15: 244–250.

Asakura, Y., Yasuda, K., Kato, D., Kojima, Y., and Koda, S. 2008b. Development of a large sonochemical reactor at a high frequency. *Chemical Engineering Journal*, 139: 339–343.

Berlan, J. and Mason, T.J. 1991. Sonochemistry: From research laboratories to industrial plants. *Ultrasonics*, 30: 203–212.

Berlan, J. and Mason, T.J. 1996. Dosimetry for power ultrasound and sonochemistry. In *Advanced Sonochemistry*, Vol. 4, Mason T.J. (Ed.), pp. 1–73. Greenwich, U.K.: JAI Press Inc.

Busnel, R. and Picard, D. 1952. Rapport entre la longeur d'onde et l'oxydation de l'iodure de potassium par les ultrasons. *C. R. Académie des Sciences*, 235: 1217–1220.

Busnel, R.-G., Picard, D., and Bouzigues, H. 1953. Rapport entre la longeur d'onde et l'oxydation de l'iodure de potassium par les ultrasons. *Journal de Chimie Physique*, 50: 97–101.

Casadonte, D.J., Flores, M., and Petrier, C. 2005. Enhancing sonochemical activity in aqueous media using power-modulated pulsed ultrasound: An initial study. *Ultrasonics Sonochemistry*, 12: 147–152.

Ciuti, P., Dezhkunov, N.V., Francescutto, A., Kulak, A.I., and Iernetti, G. 2000. Cavitation activity stimulation by low frequency field pulses. *Ultrasonics Sonochemistry*, 7: 213–216.

Contamine, F.R., Willhelm, A.M., Berlan, J., and Delmas, H. 1995. Power measurement in sonochemistry. *Ultrasonics Sonochemistry*, 2: S43–S47.

Curm, L.A. 1979. Tensile strength of water. *Nature*, 278: 148–149.

Destaillats, H., Lesko, T.M., Knowlton, M., Wallace, H., and Hoffmann, M.R. 2001. Scale-up of sonochemical reactors for water treatment. *Industrial Engineering and Chemical Research*, 40: 3855–3860.

Esche, R. 1952. Untersuchungen der schwingungskavitation in flussigkeiten. *Acustica*, 2AB: 208–218.

Feng, R., Zhao, Y., Zhu, C., and Mason, T.J. 2002. Enhancement of ultrasonic cavitation yield by multi-frequency sonication. *Ultrasonics Sonochemistry*, 9: 231–236.

Fricke, H. and Hart, E.J. 1935. The oxidation of Fe^{++} to Fe^{+++} by the irradiation with X-rays of solutions of ferrous sulfate in sulfuric acid. *Journal of Chemical Physics*, 30: 60–61.

Gogate, P.R., Mujumdar, S., and Pandit, A.B. 2003. Large-scale sonochemical reactors for process intensification: Design and experimental validation. *Journal of Chemical Technology and Biotechnology*, 78: 685–693.

Gonze, E., Gonthier, Y., Boldo, P., and Bernis, A. 1998. Standing waves in a high frequency sonoreactor: Visualization and effects. *Chemical Engineering Science*, 53: 523–532.

Grönroos, A., Pirkonen, P., Heikkinen, J., Ihalainen, J., Mursunen, H., and Sekki, H. 2001. Ultrasonic depolymerization of aqueous polyvinyl alcohol. *Ultrasonics Sonochemistry*, 8: 259–264.

Hart, E.J. and Henglein, A. 1985. Free radical free atom reactions in the sonolysis of aqueous iodide and formate solutions. *Journal of Physical Chemistry*, 89: 4342–4347.

Hatanaka, S., Mitome, H., Yasui, K., and Hayashi, S. 2006. Multibubble sonoluminescence enhancement by fluid flow. *Ultrasonics*, 44: e435–e438.

Hatanaka, S., Tuziuti, T., Kozuka, T., and Mitome, H. 2001. Dependence of sonoluminescence intensity on the geometrical configuration of a reactor cell. *IEEE Transactions on Ultrasonics Ferroelectrics and Frequency Control*, 48: 28–36.

Hodnett, M. and Zeqiri, B. 1997. A strategy for the development and standardization of measurement methods for high power/cavitating ultrasonic fields: Review of high power field measurement techniques. *Ultrasonics Sonochemistry*, 4: 273–288.

Hung, H.M. and Hoffmann, M.R. 1999. Kinetics and mechanism of the sonolytic degradation of chlorinated hydrocarbons: Frequency effects. *Journal of Physical Chemistry A*, 103: 2734–2739.

Iernetti, G. 1971. Cavitation threshold dependence on volume. *Acustica*, 24: 191–196.

Iernetti, G., Ciuti, P., Dezhkunov, N.V., Reali, M., Francescutto, A., and Johri, G.K. 1997. Enhancement of high-frequency acoustic cavitation effects by a low-frequency stimulation. *Ultrasonics Sonochemistry*, 4: 263–268.

Iida, Y., Yasui, K., Tuziuti, T., and Sivakumar, M. 2005. Sonochemistry and its dosimetry. *Microchemical Journal*, 80: 159–164.

Jana, A.K. and Chatterjee, S.N. 1995. Estimation of hydroxyl free radicals produced by ultrasound in Fricke solution used as a chemical dosimeter. *Ultrasonics Sonochemistry*, 2: S87–S91.

Keck, A., Gilbert, E., and Köster, R. 2002. Influence of particles on sonochemical reactions in aqueous solutions. *Ultrasonics*, 40: 661–665.

Kimura, T., Sakamoto, T., Leveque, J.-M. et al. 1996. Standardization of ultrasonic power for sonochemical reaction. *Ultrasonics Sonochemistry*, 3: S157–S161.

Koda, S., Amano, T., and Nomura, H. 1996. Copolymerization of sodium styrene sulphonate and vinylpyrrolidone under ultrasonic irradiation. *Ultrasonics Sonochemistry*, 3: S91–S95.

Koda, S., Kimura, T., Kond, T., and Mitome, H. 2003. A standard method to calibrate sonochemical efficiency of an individual reaction system. *Ultrasonics Sonochemistry*, 10: 149–156.

Koda, S., Taguchi, K., and Futamura, K. 2010. Effects of frequency and a radical scavenger on ultrasonic degradation of water-soluble polymers, *Ultrasonics Sonochemistry*, 18: 276–281.

Kojima, Y., Asakura, Y., Sugiyama, G., and Koda, S. 2010. The effects of acoustic flow and mechanical flow on the sonochemical efficiency in a rectangular sonochemical reactor. *Ultrasonics Sonochemistry*, 17: 978–984.

Kojima, Y., Koda, S., and Nomura, H. 1998. Effects of sample volume and frequency on ultrasonic power in solutions on sonication. *Japanese Journal of Applied Physics*, 37: 2992–2995.

Kojima, Y., Koda, S., and Nomura, H. 2001. Effect of ultrasonic frequency on polymerization of styrene under sonication. *Ultrasonics Sonochemistry*, 8: 75–79.

Kuijpers, M.W.A., Kemmere, M.F., and Keurentjes, J.T.F. 2002. Calorimetric study of the energy efficiency for ultrasound-induced radical formation. *Ultrasonics*, 40: 675–678.

Lauterborn, W. and Holzfuss, J. 1986. Evidence for a low-dimensional strange attractor in acoustic turbulence. *Physical Letters A*, 115: 369–372.

Lauterborn, W., Kurz, T., Mettin, R., and Ohl, C.D. 1999. Experimental and theoretical bubble dynamics. In *Advances in Chemical Physics*, Vol. 110, Prigogine I. and Rice S.A. (Ed.), pp. 295–380. New York: John Wiley & Sons, Inc.

Leighton, T.G. 1997. *The Acoustic Bubbles*. San Diego, CA: Academic Press.

Löning, J.M., Horst, C., and Hoffmann, U. 2002. Investigations on the energy conversion in sonochemical processes. *Ultrasonics Sonochemistry*, 9: 169–179.

Mark, G., Tauber, A., Laupert, R., Schuchmann, H.-P. et al. 1998. OH-radical formation by ultrasound in aqueous solution—Part II: Terephthalate and Fricke dosimetry and the influence of various conditions on the sonolytic yield. *Ultrasonics Sonochemistry*, 5: 41–52.

Marschall, H.B., Mørch, K.A., Keller, A.P., and Kjeldsen, M. 2003. Cavitation inception by almost spherical solid particles in water. *Physics of Fluids*, 15: 545–553.

Mason, T.J. 1992. Industrial sonochemistry: Potential and practicality. *Ultrasonics Sonochemistry*, 1: 196.

Mason, T.J., Lorimer, J.P., Bates, D.M., and Zhao, Y. 1994. Dosimetry in sonochemistry: The use of aqueous terephthalate ion as a fluorescence monitor. *Ultrasonics Sonochemistry*, 1: S91–S95.

Mitome, H. and Hatanaka, S. 2002. Optimization of a sonochemical reactor using a pulsing operation. *Ultrasonics*, 40: 683–687.

Nomura, H., Koda, S., Yasuda, K., and Kojima, Y. 1996. Quantification of ultrasonic intensity based on the decomposition reaction of porphyrin. *Ultrasonics Sonochemistry*, 3: S153–S156.

Petrier, C., Jeunet, A., Luche, J.L., and Reverdyt, G. 1992. Unexpected frequency effects on the rate of oxidative processes induced by ultrasound. *Journal of American Chemical Society*, 114: 3148–3150.

Price, G.J., Harris, N.K., and Stewart, A.J. 2010. Direct observation of cavitation fields at 23 and 515 kHz. *Ultrasonics Sonochemistry*, 17: 30–33.

Ratoarinoro, C.F., Wilhelm, A.M., Berlan, J., and Delmas, H. 1995. Power measurement in sonochemistry. *Ultrasonics Sonochemistry*, 2: S43–S47.

Renaudin, V., Gondrexon, N., Boldo, P., Pétrier, C., Bernis, A., and Gonthier, Y. 1994. Method for determining the chemically active zones in a high-frequency ultrasonic reactor. *Ultrasonics Sonochemistry*, 1: S81–S85.

Rong, L., Kojima, Y., Koda, S., and Nomura, H. 2001. Simple quantification of ultrasonic intensity using aqueous solution of phenolphthalein. *Ultrasonics Sonochemistry*, 8: 11–15.

Sato, M., Itoh, H., and Fujii, T. 2000. Frequency dependence of H_2O_2 generation from distilled water. *Ultrasonics*, 38: 312–315.

Segebarth, N., Eulaerts, O., Reisse, J., Crum, L.A., and Matula, T.J. 2002. Correlation between acoustic cavitation noise, bubble population, and sonochemistry. *Journal of Physical Chemistry B*, 106: 9181–9190.

Sekiguchi, H. and Saita, Y. 2001. Effect of alumina particles on sonolysis degradation of chlorobenzene in aqueous solution. *Journal of Chemical Engineering of Japan*, 34: 1045–1048.

Sivakumar, M. and Pandit, A.B. 2001. Ultrasound enhanced degradation of Rhodamine B: Optimization with power density. *Ultrasonics Sonochemistry*, 8: 233–240.

Suzuki, T., Yasui, K., Yasuda, K. et al. 2004. Effect of dual frequency on sonochemical reaction rates. *Research of Chemical Intermediates*, 30: 703–711.

Toma, M., Fukutomi, S., Asakura, Y., and Koda, S. 2010. A calorimetric study of energy conversion efficiency of a sonochemical reactor at 500 kHz for organic solvents. *Ultrasonics Sonochemistry*, 18: 197–208.

Tuziuti, T., Yasui, K., Kozuka, T., Towata, A., and Iida, Y. 2006. Enhancement of sonochemical reaction rate by addition of micrometer-sized air bubbles. *Journal of Physical Chemistry A*, 110: 10720–10724.

Tuziuti, T., Yasui, K., Lee, J., Kozuka, T., Towata, A., and Iida, Y. 2008. Mechanism of enhancement of sonochemical-reaction efficiency by pulsed ultrasound. *Journal of Physical Chemistry A*, 112: 4875–4878.

Tuziuti, T., Yasui, K., Sivakumar, M., Iida, Y., and Miyoshi, N. 2005. Correlation between acoustic cavitation noise and yield enhancement of sonochemical reaction by particle addition. *Journal of Physical Chemistry A*, 109: 4869–4872.

Weissler, A., Cooper, H., and Snyder, S. 1950. Chemical effect of ultrasonic waves: Oxidation of potassium iodide solution by carbon tetrachloride. *Journal of the American Chemical Society*, 72: 1769–1775.

Yang, L., Rathman, J.F., and Weavers, L.K. 2005. Degradation of alkylbenzene sulfonate surfactants by pulsed ultrasound. *Journal of Physical Chemistry B*, 109: 16203–16209.

Yasuda, K., Tachi, M., Bando, Y., and Nakamura, M. 1999a. Effect of liquid mixing on performance of porphyrin decomposition by ultrasonic irradiation. *Journal of Chemical Engineering of Japan*, 32: 347–349.

Yasuda, K., Tanigawara, R., Tachi, M., Bando, Y., and Nakamura, M. 1999. Effect of particle properties on decomposition of organic material by ultrasonic irradiation. In *Proceedings of the Fifth International Symposium on Separation Technology between Korea and Japan*, Seoul, pp. 1079–1082.

Yasuda, K., Torii, T., Yasui, K. et al. 2007. Enhancement of sonochemical reaction of terephthalate ion by superposition of ultrasonic fields of various frequencies. *Ultrasonics Sonochemistry*, 14: 699–704.

Yasui, K. 2002. Effect of volatile solutes on sonoluminescence. *Journal of Chemical Physics*, 116: 2945–2954.

Zeqiri, B., Gelat, P.N., Hodnett, M., and Lee, N.D. 2003a. A novel sensor for monitoring acoustic cavitaion. Part I: Concept, theory, and prototype development. *IEEE Transactions on Ultrasonics Ferroelectrics and Frequency Control*, 50: 1342–1350.

Zeqiri, B., Hodnett, M., and Carroll, A.J. 2006. Studies of a novel sensor for assessing the spatial distribution of cavitation activity within ultrasonic cleaning vessels. *Ultrasonics*, 44: 73–82.

Zeqiri, B., Lee, N.D., Hodnett, M., and Gelat, P.N. 2003b. A novel sensor for monitoring acoustic cavitation. Part II: Prototype performance evaluation. *IEEE Transactions on Ultrasonics Ferroelectrics and Frequency Control*, 50: 1351–1362.

Zhao, Y., Zhu, C., Feng, R., Xu, J., and Wang, Y. 2002. Fluorescence enhancement of the aqueous solution of terephthalate ion after bi-frequency sonication. *Ultrasonics Sonochemistry*, 9: 241–243.

23 Ultrasound for Better Reactor Design:
How Chemical Engineering Tools Can Help Sonoreactor Characterization and Scale-Up

Jean-Yves Hihn, Marie-Laure Doche, Audrey Mandroyan, Loic Hallez, and Bruno G. Pollet

CONTENTS

Reactor Design: A Crucial Step 599
    Heterogeneous Distribution of the Acoustic Field 600
    Electrode Presence Is "Intrusive" 604
Direct Quantification of Ultrasound Effects by Electrochemistry 605
    The Well-Known Electrodiffusional Method or What Happens during an Electrochemical Measurement for Mass-Transfer? 606
    Reduce the Parameters Influence by Modeling 609
A New Tool: The Equivalent Velocity 613
    Application to the Measurement of Acoustic Energy Distribution 615
    Use for Cavitation Quantification 617
Physical Signification for the Dimensionless Number 618
Toward the "Ideal" Sonoreactor? 620
References 621

REACTOR DESIGN: A CRUCIAL STEP

"Aspirations for sonochemical processing are those which all production managers would welcome: faster reactions, better conversions, improved and perhaps new products." This sentence is taken from the last paragraph of a review by T.J. Mason (2000) in a paper concerning sonochemical processing, and it remains highly reliable!

    The enormous potential of sonochemical reactors in a wide variety of processes for chemical and allied industries is intact, whereas not exploited to date. A few fields use ultrasound in sonoreactors: They concern cleaning and decontamination, extraction and impregnation, crystallization and precipitation, and, to a greater or lesser extent, electrochemistry. In the majority of cases, some examples of large-scale use of sonoreactors are still valid, but generally, design is based on "intuition," and the results quoted in yield are impossible to predict. Indeed, rare are the tentative to design in depth a sonoreactor (Hihn et al., 2000; Soon et al., 2006; Viennet et al., 2009). For many authors, the necessity to take care of the scale-up aspects is acknowledged, but most of the time this only concerns the cavitation activity and intensity, using solutions based on bubble dynamics equations

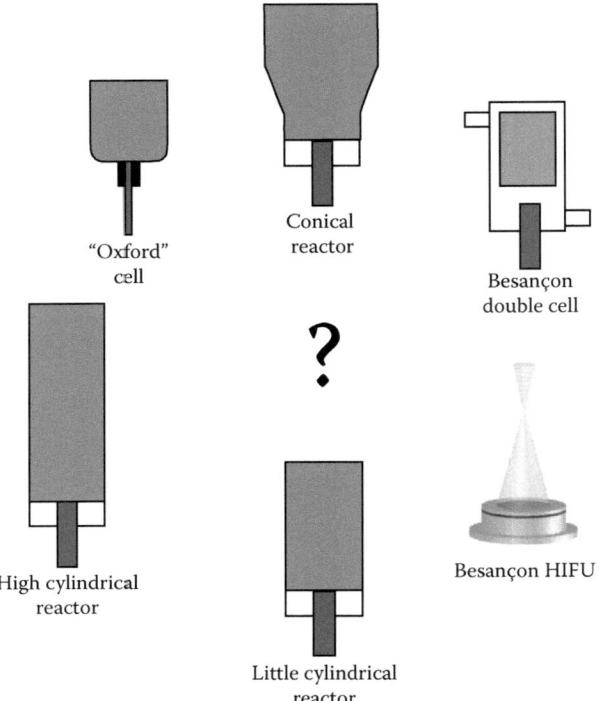

FIGURE 23.1 Huge varieties of the "best reactors."

as well as experimentation with different reactor types and reactions. Design correlations for collapse pressure and its relation to cavitational yield should assist designers in choice of the operating parameters for a desired cavitational effect (Gogate and Pandit, 2004). In the meantime, it cannot be dissociated from the techniques useful for good understanding of cavitational activity distribution (Sutkar and Pandit, 2009). Cavitation is the phenomenon with the most important effect for intensification of physical and chemical processing. However, even after a complete study of dynamic behavior of cavitation, this specificity creates problems in proposing reliable design. Therefore, operating strategies are needed. Surprisingly, while the physics of the phenomenon is well covered and recommendation for optimum reactor parameters and design still exists (Sutkar and Pandit, 2009), a more global approach is needed, especially in the case of sonoelectrochemistry. There are many examples of large-scale processes assisted by ultrasound. Electrochemical reactions are often complex, but it can be considered essentially as mass transfer or ion transfer to or from the electrode surface that is highly sensitive to asymmetric bubble collapse. Nevertheless, even in this restricted community, a large choice of reactors are still available (Figure 23.1) and are mostly dedicated to the laboratory using them for the first time.

HETEROGENEOUS DISTRIBUTION OF THE ACOUSTIC FIELD

The main problem in ultrasonic reactor design is heterogeneous distribution of the acoustic field. That is why many processes assisted by ultrasound experiments, successful on a laboratory scale, but are hard to transfer to industrial processes. For example, if ultrasound is known to induce a mass-transfer increase as well as an electrode surface activation, localization of the electrode is so crucial in sonoelectroreactors that major contradictions have been observed between the results from two different universities.

Therefore, knowledge of distribution of these active zones is very useful for reactor design, particularly in the case of these sonoelectrochemical applications, where reactions take place at the interface. Several methods have been developed for this determination such as aluminum erosion foil (Laborde et al., 1998), thermoprobes (Fry and Fry, 1954; Martin and Law; 1980; Boldo et al., 2004), and sonoluminescence (Gondrexon et al., 1995). Some authors use an interesting method for hydrodynamic and mass-transfer characterization, adapted to ultrasound effects: electrochemical mass-transfer measurement. This method is combined with a mechanical effect of cavitation and its principle is derived from the Nernst equation. Under diffusion-controlled conditions, the intensity of the limiting current is proportional to the mass-transfer coefficient of the active species at the surface of the working electrode. In practical terms, it consists of cycling voltammetry using the well known quasi-reversible redox couple $Fe(CN)_6^{3-}/Fe(CN)_6^{4-}$ (Trabelsi et al., 1996; Ligier et al., 2001; Viennet et al., 2001). Thanks to those records, quantitative information is obtained along a radial and axial distribution in the sonoreactor: This will be discussed in greater detail later on.

Moreover, visualization techniques provide decisive pictures for global comprehension of sonoreactor behavior, and therefore for their proper use. Burdin et al. have compared two laser techniques for the characterization of the acoustic cavitation cloud (Burdin et al., 1999) and Chouvellon et al. (Chouvellon et al., 2000) have used the tomography technique to study the velocity field in an ultrasonic reactor operating at 500 kHz. Our group has dedicated many papers to highlighting active zone distribution in an ultrasonic reactor (Viennet et al., 2004; Mandroyan et al., 2009a,b). The laser tomography technique is useful for visualizing the active zones with a high degree of accuracy: It consists of the use of a monochromatic incident laser light, which is diffracted by cavitation bubbles. Thus, it is possible to observe the zones of high bubble density, corresponding to the most active areas.

A large-scale reactor was especially designed for those experiments. It consists of a polymethyl methacrylate (PMMA) cylinder (optic quality) 450 mm in diameter and 500 mm high. Large dimensions were chosen because of the need of high bending curvature to solve optical problems, and to minimize the perturbations linked to the stirring effects induced by ultrasound irradiation. The transducers used here consist of a probe (Sonics & Materials, Danbury USA), with a 25 mm titanium horn as a radiating face, and a high-frequency transducer with a 56 mm glass radiating face (developed by the Laboratory of Molecular and Environmental Chemistry, University of Savoie, Chambéry, France). Both were fixed at the reactor bottom.

A laser argon beam (green line 534 nm) is converted to a thin horizontal or vertical static plane sheet of light, generated by the combination of a cylindrical optical component (which spreads the light in one direction) with a spherical component (which enables focusing Figure 23.2). Images are

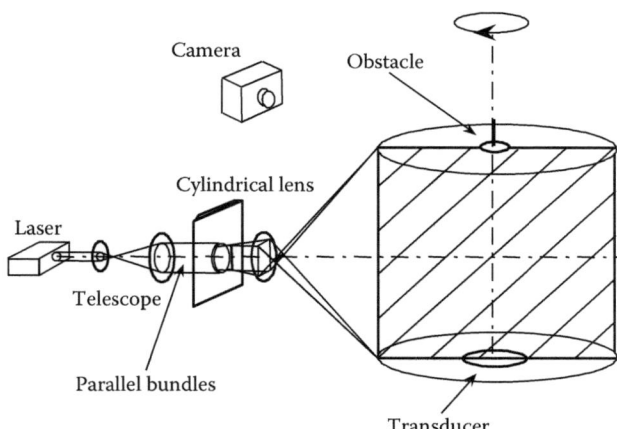

FIGURE 23.2 Equipment for laser visualizations. (From *Ultrason. Sonochem.*, 11, Viennet, R., Ligier, V., Hihn, J.-Y., Bereiziat, D., Nika, P., and Doche, M.-L., Visualisation and electrochemical determination of the actives zones in an ultrasonic reactor using 20 and 500 kHz frequencies, 125–129, Copyright (2004), with permission from Elsevier.)

FIGURE 23.3 (See color insert.) Laser visualization at 20 kHz. (From *Ultrason. Sonochem.*, 11, Viennet, R., Ligier, V., Hihn, J.-Y., Bereiziat, D., Nika, P., and Doche, M.-L., Visualisation and electrochemical determination of the actives zones in an ultrasonic reactor using 20 and 500 kHz frequencies, 125–129, Copyright (2004), with permission from Elsevier.)

collected by a camera, or by a video recording, to allow subsequent data processing, for example, for the determination of speed vectors within the liquid.

Figure 23.3 shows the results obtained by a static vertical sheet of light illuminating the zones close to the transducer. The bottom of the reactor vessel can be observed, with the flange adapter used to fix the transducer. A very high luminous intensity appears, near the transducer, which corresponds to a zone of maximum activity. As the cavitation bubbles are produced in zones with strong pressure variations, this result was foreseeable. Energy distribution is also focused in the middle of the reactor, because of the focusing effect of the cylindrical reactor.

It is easy to understand that incorrect positioning of a working electrode, for example, a few centimeters from the reactor axis, will give completely different results. This can lead us to think that in one case, ultrasound provides excellent effects, whereas in another case, it can be considered as inefficient. Similarly, locating the electrode in different positions all along the reactor axis will yield large discrepancies in their quantitative response. This was already a feeling shared by experimenters as most sonoelectrochemical applications for low frequencies are implemented close to the transducer.

When the zone close to the transducer is illuminated by a static horizontal sheet of light, it can be observed that the cavitation activity is uniformly distributed according to a radial direction from the ultrasound generator to the bulk of the most actives zones (Figure 23.4). Moreover, due to the large wavelength (75 mm) and the small transducer diameter, the length of the Fresnel zone is very short and the observations concern the far field zone (Fraunhofer zone). This homogeneous distribution is for the most part confirmed by other experiments. In the case of accelerated corrosion tests studies, homogeneous distributions of corrosion products were observed, confirming homogeneous ultrasound effects (Doche et al., 2001, 2003; Ligier et al., 2001).

If we look beyond the immediate vicinity of the transducer, a vertical sheet of light shows regular distribution of the more luminous zones in the reactor as a whole. These zones correspond to zones of strong pressure variations and are distributed with equal spaces from the transducer to the interface (Figure 23.5).

It is important to note that active zones become more luminous if the resonant case is reached, that is, fitting the transducer–reflector distance to an odd multiple of the quarter of the wavelength. This resonant case leads to standing waves of great pressure amplitude and the number of luminous zones corresponds to the number of zones of maximum pressure amplitude foreseeable. Finally, the light intensity of these zones decreases when the distance between interface and transducer increases, showing a decay in cavitation activity due to acoustic absorption. This confirms the sonoluminescence of Luminol, which showed the same global pattern (Renaudin et al., 1994;

FIGURE 23.4 (**See color insert.**) Horizontal cutting close to the transducer. (From *Ultrason. Sonochem.*, 11, Viennet, R., Ligier, V., Hihn, J.-Y., Bereiziat, D., Nika, P., and Doche, M.-L., Visualisation and electrochemical determination of the actives zones in an ultrasonic reactor using 20 and 500 kHz frequencies, 125–129, Copyright (2004), with permission from Elsevier.)

FIGURE 23.5 (**See color insert.**) Vertical light sheet in the entire reactor volume. (From *Ultrason. Sonochem.*, 11, Viennet, R., Ligier, V., Hihn, J.-Y., Bereiziat, D., Nika, P., and Doche, M.-L., Visualisation and electrochemical determination of the actives zones in an ultrasonic reactor using 20 and 500 kHz frequencies, 125–129, Copyright (2004), with permission from Elsevier.)

Laborde et al., 1998; Boldo et al., 2004; Hallez et al., 2010). A systematic study, with a variety of parameters such as transducer diameters and ultrasonic power input can be found in Besançon's works (Mandroyan et al., 2009a).

At higher frequencies, that is, 500 kHz, an organized distribution of active zones once again appears. The distance between two luminous zones is smaller than in the case of 20 kHz frequency due to the reduced wavelength (3 mm). The transducer-interface distance is less significant, and the resonant system is achieved in all cases. The wavelength is small enough to allow adaptation of liquid acoustic impedance.

Figure 23.6a shows a laser light sheet in the entire volume. Radial distribution of luminous intensity is not homogeneous; luminous zones are visible around a dark central circular zone. Figure 23.6b shows the pictures obtained with a horizontal light sheet and confirms that activity is lowest in the central area. Once again, this constitutes a real "trap" for the scientist during his experimental work, as an electrode placed in this kind of zone will present low acoustic activity, and can thus lead to the wrong conclusion that ultrasound has a negligible effect. This distribution of the cavitation activity is typical of energy distribution in the near field (Fresnel field—27 cm in our operating conditions). Closer to the reflector, this distribution is greatly modified. These results comply with

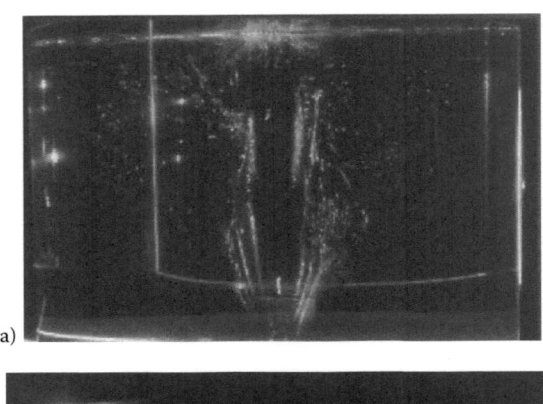

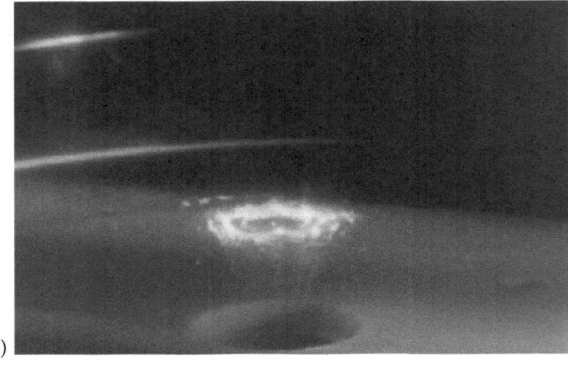

FIGURE 23.6 (**See color insert.**) (a) Visualization in the entire reactor volume; (b) radial distribution close to the transducer. (From *Ultrason. Sonochem.*, 11, Viennet, R., Ligier, V., Hihn, J.-Y., Bereiziat, D., Nika, P., and Doche, M.-L., Visualisation and electrochemical determination of the actives zones in an ultrasonic reactor using 20 and 500 kHz frequencies, 125–129, Copyright (2004), with permission from Elsevier.)

observations in the literature reporting that mass-transfer coefficients are maximum around a central zone, which presents weaker coefficients (Trabelsi et al., 1996).

In the presence of a reflector, even before the end of the Fresnel zone, ultrasound energy distribution is completely modified. A cone with bright lines spaced out by 0.75 mm (quarter of the wavelength) is produced throughout the volume. These observations confirm other results obtained in the literature, where authors obtain their best results close to the water/air interface at high frequency (Pétrier et al., 1992; Trabelsi et al., 1996; Viennet et al., 2004). By placing a horizontal laser light sheet near the target, the concentric circles referred to earlier appear throughout the volume. This result is very important and once again confirms the interest of this zone for chemical or electrochemical applications (Figure 23.7).

Electrode Presence Is "Intrusive"

Sonoelectrochemical experiments differ from sonochemical ones by the introduction of electrodes in the sonicated vessel. While ultrasound is known to markedly enhance electrochemical processes, this has always required the presence of an electrode in the acoustic field, which could greatly disturb the cavitation activity as well as the acoustic distribution itself. Consequently, electrode presence had to be taken into account and its influence evaluated. A systematic study was carried out in the Besançon group at low frequency (Mandroyan et al., 2009a,b).

In a sonoelectrochemical reactor, the working electrode is of prime importance and is usually positioned in front of the transducer, as per the results described earlier, even if it disturbs ultrasonic wave propagation. To reproduce this situation, the same experimental setup as described in Figure 23.2

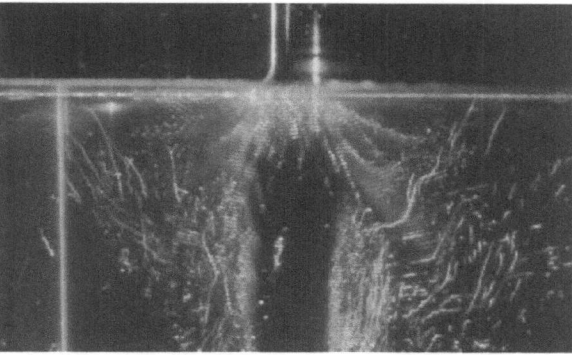

FIGURE 23.7 (**See color insert.**) Details close to the water/air interface at high frequency. (From *Ultrason. Sonochem.*, 11, Viennet, R., Ligier, V., Hihn, J.-Y., Bereiziat, D., Nika, P., and Doche, M.-L., Visualisation and electrochemical determination of the actives zones in an ultrasonic reactor using 20 and 500 kHz frequencies, 125–129, Copyright (2004), with permission from Elsevier.)

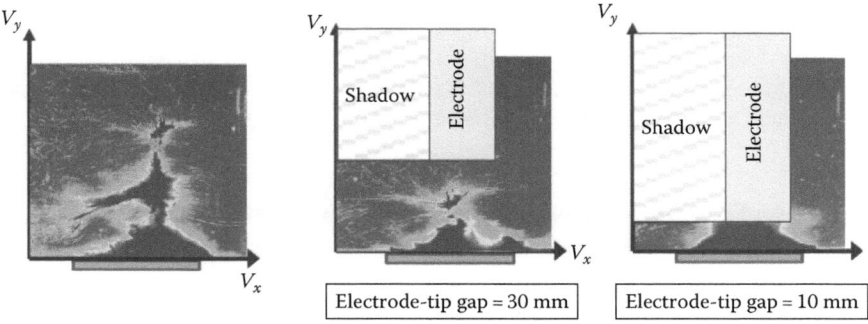

FIGURE 23.8 (**See color insert.**) Laser visualization of ultrasonic actives zones in electrode presence.

was used while introducing an unpolarized electrode in the reactor axis. Three different electrode/horn distances have been investigated, with different horn diameters as well as different ultrasonic power inputs for low frequencies (20 and 40 kHz). For all the situations tested, luminous intensity is barely modified by electrode presence, as long as the electrode is kept further than 30 mm away. Intensity increases weakly with the transmitted power indicating that bubble density in the studied area remains within the same range of magnitude. A completely different behavior is observed when the electrode is kept closer to the horn, as if the activity were "trapped" by the electrode with cavitation activity focusing itself on the obstacle. This is even more marked from 10 to 2 mm, where the activity appears to be confined in a "tube" between the horn and the electrode (Figure 23.8).

These results fully comply with the usual observations made by "sonoelectrochemists." Consequently, despite the apparent decrease in activity (average luminosity is lowest at a 2 mm electrode–horn distance), the closest location is the more frequently used.

Nevertheless, such intuition does not govern scientific practice, and the need for a characterization tool is confirmed. The choice of the reactor most appropriate for a precise situation requires additional investigations.

DIRECT QUANTIFICATION OF ULTRASOUND EFFECTS BY ELECTROCHEMISTRY

While it is commonly accepted that sonoelectrochemistry concerns enhancement of electrochemical processes, the investigation approaches used vary. The general problem of power ultrasound applied to surfaces is first to control and to characterize the disturbances they generate on these surfaces. It is only then that we will be able to concern ourselves reliably with the very important applications that they allow. At every step in the studies involving surface and ultrasonic irradiation, electrochemistry can be of vital help in understanding the phenomena. In this case, electrochemistry will be taken as a tool for ultrasound investigation. If we detail an electrochemical reaction, three types of phenomena appear: Mass transfer (including convection and electromigration), adsorption, and electronic transfer. Indeed, ultrasound may influence each step, a fact that can be highlighted by specific electrochemical investigations. This is something classic in the case of mass transfer, and the use of a redox couple (e.g., $Fe(CN)_6^{3-}/Fe(CN)_6^{4-}$) is common. Rarer are the studies that concern the phases of adsorption or of electronic transfer. For the adsorption case, very specific techniques such as impedance spectroscopy have to be carried out to show ultrasound influence. Electronic transfer was the subject of some kinetic studies, and the heterogeneous rate constant can be sensitive to ultrasonic irradiation. Consequently, the electrochemical reduction of silver thiosulfate shows a dependence of the standard heterogeneous rate constant up to 10-fold compared to the silent condition (Pollet et al., 2005). All these works increase knowledge of the phenomena involved in ultrasound effects. Concerning mass-transfer measurements, a large amount of information can be obtained. First, systematic measurement is an excellent tool for evaluating the power applied to a surface. For all applications involving a surface, it is necessary to have a criterion for energy dispersion evaluation, such as calorimetric power measurement in the case of homogeneous reactors. In certain conditions including a chemical engineering approach, electrochemistry can be a good solution. Second, the specific implosion mode of the cavitation bubble is in the bulk of the phenomena. Close to a surface, the cavitation bubbles imploded in an asymmetric way, leading to the formation of micro-jets and disturbances of the double layer. Nevertheless, even these methods are quite easy to set up: Data interpretation is not trivial and some discussions are necessary.

THE WELL-KNOWN ELECTRODIFFUSIONAL METHOD OR WHAT HAPPENS DURING AN ELECTROCHEMICAL MEASUREMENT FOR MASS-TRANSFER?

The first possibility consists of reducing the influence of various operation parameters in new criterion calculation. For example, it is possible to eliminate the influence of electroactive surface and species concentration without too many assumptions, by calculating in turn a current density and then a mass-transfer coefficient. The basis of this method is to record voltammograms of a reversible couple on an inert stationary electrode. The principle is to impose a kinetic energy controlled by diffusion, thanks to low electroactive species concentrations: to reach a limiting current density while increasing the potential, directly proportional to the mass transfer. This method is so frequently used that it is easy to forget its scientific background. Mass transfer is the displacement of ionic species and its overall behavior follows the Nernst–Planck equation:

$$J = -\underbrace{\sum_k D_k \cdot \frac{\partial C_k}{\partial x}}_{\substack{\text{Contribution} \\ \text{of diffusion} \\ \text{(1st Fick law)}}} - \underbrace{\left(\frac{F}{RT} \cdot \frac{\partial \phi}{\partial x}\right) \sum_k Z_k D_k C_k}_{\substack{\text{Contribution} \\ \text{of migration under} \\ \text{electrical field}}} + \underbrace{\sum_k C_k V_f}_{\substack{\text{Contribution} \\ \text{of forced} \\ \text{convection induced} \\ \text{by flow velocity}}}$$

J specific molecular flow in mol s^{-1} cm^{-2}

Due to systematic use of background salt, electrical migration of ion in the electric field can be ignored, so the ionic transfer results only of the contribution of convection and molecular diffusion. The resulting specific molecular flow is written as follows: (mol ions s^{-1} cm^{-2}):

$$\vec{J}_O = -D_O \vec{\text{grade}}\, C_O + C_O \vec{V}_O$$

where
- $\bar{J}_O$ is the flux of O species
- D_O is the diffusion coefficient derived from Stokes–Einstein equation in a solvent of dynamic viscosity η
- C_O is the concentration in species O
- $\vec{V}_O$ is the liquid velocity

The electron transfer through the interface yields the average flow of ions reacting at the electrode surface. Considering a Faradic yield equal to 1, where S is the electrode surface and n the number of electrons exchanged, the limiting current density will be j:

$$\left|\bar{J}_O\right| = \frac{I}{S n_O F} = \frac{\bar{j}}{n_O F}$$

$$\vec{J}_O\big|_{x=0} = \left|\bar{J}_O\right| = -D_O \left(\frac{\partial C_O}{\partial x}\right)_{x=0}$$

where
- I is the current
- F is the Faradic number

Then, directly to the electrode, one may write in a normal direction to the wall because electrolyte velocity is equal to zero at the electrode surface. The concentration profiles of the species O in close vicinity of the electrode assume the following appearance (Figure 23.9) in accordance with Fick's second law equations in unsteady-state regime.

If current density increases, there is an impoverishment in the electrode vicinity, and the concentration profiles bend until a limiting profile $C_O(x)$ reaching $C_O(0) = 0$ (at the electrode surface). The absolute value of the current density is described as follows:

$$j = +nF\left|\bar{J}_O\right| = +nFD_O \left(\frac{\partial C_O}{\partial x}\right)_{x=0}$$

The limiting concentration profile is defined by the hydrodynamic conditions and the limiting current density j_{Lim}. It is then possible to express the specific limiting transfer flow by a classical law linking the transfer flow to a potential difference (transfer driving force) multiplied by a

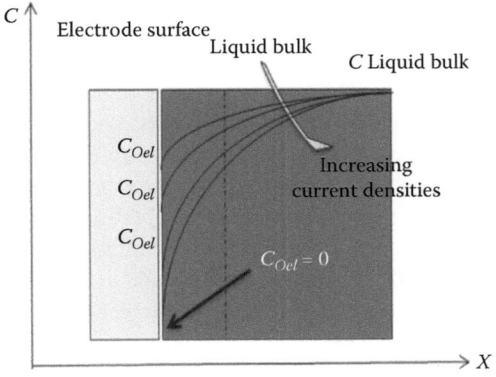

FIGURE 23.9 Concentration profiles close to the electrode.

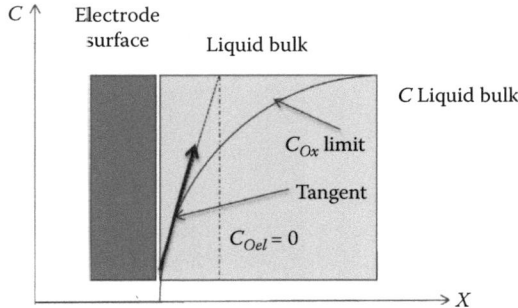

FIGURE 23.10 Limiting concentration profile close to the electrode.

conductance. This yields the following expression for the molecular flow where k_d is a constant of proportionality. Focusing on the limiting profile $C_O(x)$, it can be noted that the limiting flow is determined by the slope at $x = 0$ and is merged with its tangent (Figure 23.10).

$$\left|\overline{J}_O\right|_{Lim} = \left|\overline{J}_O\right|_{max} = k_d(C_{Os} - C_{Oe}) = k_d(C_{Os} - 0)$$

Finally, current potential polarization curves for the Fe^{III}/Fe^{II} reversible couple were recorded under steady state conditions for a low scan rate (typically $2\,mV\,s^{-1}$). The voltamograms exhibited typical sigmoidal current responses yielding signal plateaus at mass transfer–limited potentials (Hagan and Coury, 1994; Cooper and Coury, 1998). The mass transfer–limited currents under ultrasonic irradiation could be decomposed into a steady-state- and time-dependent components (oscillating around the average plateau current value). In our conditions, only steady-state current values were considered. The mass transfer–limited current was thus determined by a statistical study at the plateau.

$$j = nF\,k_d\,(C_{Os} - C_{Oe}) \implies \boxed{k_d = \frac{j_{Lim}}{n \times F \times C}}$$

This proportionality coefficient k_d points to the mass-transfer coefficient for diffusion/convection between the electrolyte and the electrode. It represents a conductance, with the dimension of a velocity ($m\,s^{-1}$) that does not refer to any model or hypothesis for its determination. It can be thus used in all situations, and its physical significance is precisely determined as k_d. Indeed, the distribution of the local coefficient is not uniform when limit layers are produced on the electrodes. In the same manner, k_d may be time dependent during transient condition, and a distinction had to be made between instantaneous transfer (e.g., in the case of a cavitation event) and average transfer during a time interval. Another application is that the mass transfer determined by the help of a quasi-reversible redox system is valuable for various other electrochemical species, so it is possible to predict limiting current density in another situation (e.g., in the case of plating).

For sonoreactor design, "translation" of limiting current density into a mass-transfer coefficient value reduces the influence of various operating parameters, and makes it possible to compare two experiments in the same experimental setup. In the following example, the problem was to know how to separate the contribution of stirring induced by ultrasound from its specific contribution to electropolymerization of poly(3,4-ethylenedioxythiophene) (PEDOT) (Et Taouil et al., 2011).

Mass-transfer experiments were also carried out under ultrasound irradiation (500 kHz, 25 W). The average mass transport–limited current density was found to be equal to $2.2\,mA\,cm^{-2}$. In order to work at the same stirring level both in silent conditions and under ultrasound, a calibration is made by measuring the mass-transfer coefficients at different rotating speeds (Figure 23.11). The comparison between the different curves showed that the rotation value implying the same transport-limited current density as 25 W ultrasound is 950 rotations per minute (rpm). The corresponding mass-transfer

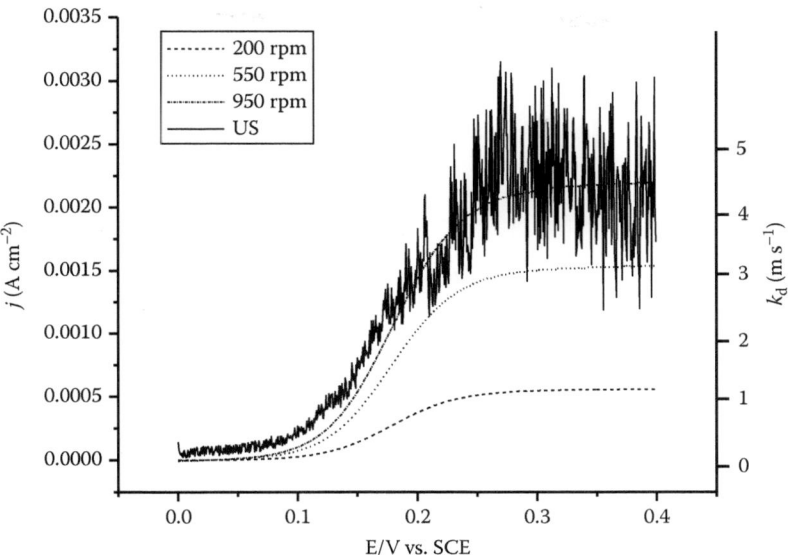

FIGURE 23.11 $Fe(CN)_6^{4-}/Fe(CN)_6^{3-}$ mass transfer curves on platinum at 2 mV s^{-1} scan rate. (From *Ultrason. Sonochem.*, 18, Et Taouil, A., Lallemand, F., Hihn, J.-Y., Melot, J.-M., Blondeau-Patissier, V., and Lakard, B., Doping properties of PEDOT films electrosynthesized under high frequency ultrasound irradiation, 140–148, Copyright (2011), with permission from Elsevier.)

coefficient is equal to 4.6×10^{-5} m s^{-1}. After that, PEDOT films were elaborated on fluorine-doped tin oxide (FTO) substrates within 100 s in silent conditions, under ultrasound, or with an electrode rotating at 950 rpm (equivalent stirring as in ultrasound presence as calibrated by mass-transfer measurement) (Figure 23.12). The decrease in current density in silent condition is due to the growth of the PEDOT film at the electrode surface, the conductivity of which is lower than that of the substrate. The chronoamperometry curve obtained under sonication exhibits a very interesting effect of high-frequency ultrasound. Indeed, when ultrasound is used, current density is much higher than in silent conditions, thus enhancing 3,4-ethylenedioxythiophene (EDOT) electropolymerization.

A deposit also made under the same potentiostatic conditions on a 950 rpm rotating electrode (without sonication) can be considered in similar agitation conditions (same mass-transfer coefficient k_d) as when ultrasound is used. Results present a clear increase in current density, showing a dependence of electrodeposition on the stirring conditions. Nevertheless, when comparing the two curves (US 25W and RDE recorded for the same $k_d = 4.6 \times 10^{-5}$ m s^{-1}), it is obvious that electrosynthesis current density is higher under ultrasonication: Ultrasound does not only have a stirring effect at the interface but other phenomena enhancing electropolymerization are to be considered (Et Taouil et al., 2011).

REDUCE THE PARAMETERS INFLUENCE BY MODELING

By extension, when using additional assumptions to linearize the motionless diffusion layer at the interface, the mass-transfer coefficient could be expressed with physical significance, and the Sherwood number appeared as a new criterion. Indeed, the complexity of turbulent systems require the description of the combination of the molecular mass transfer by diffusion and the mass transfer by convection in a single two-zone model (Figure 23.13).

The concentration profile is replaced by an equivalent simplified profile, linear with a slope equal to the tangent at the real profile in $x = 0$. In the thus-defined expression, only molecular diffusion takes place. Beyond this limit, the liquid is considered as perfectly mixed. Then k_d, which was a simple coefficient of proportionality, will assume a new signification. The limiting current becomes

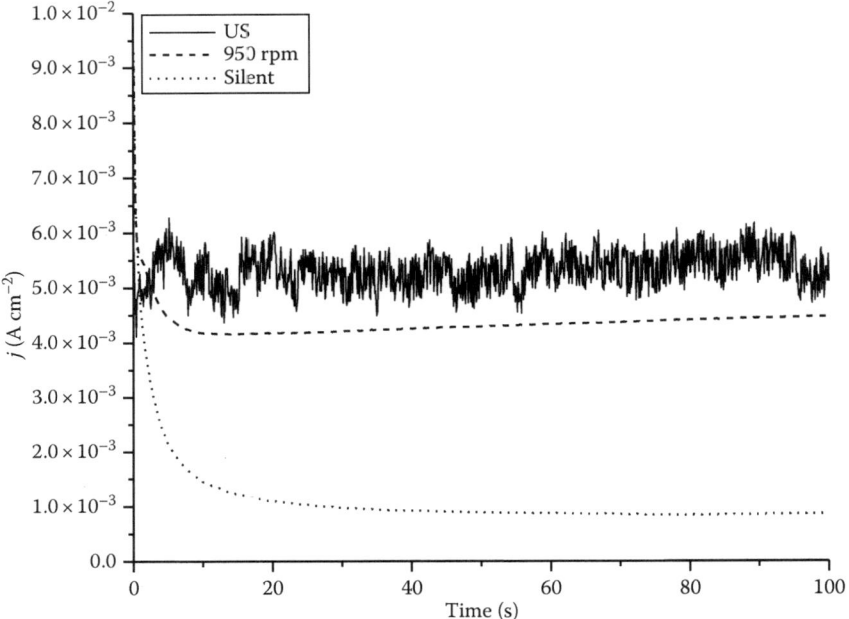

FIGURE 23.12 Chronoamperometry curve of PEDOT electrosynthesis in aqueous solution (5.10−3 M 3,4-ethylenedioxythiophene [EDOT] + 0.1 M LiClO$_4$) (1 V/SCE for 100 s). (From *Ultrason. Sonochem.*, 18, Et Taouil, A., Lallemand, F., Hihn, J.-Y., Melot, J.-M., Blondeau-Patissier, V., and Lakard, B., Doping properties of PEDOT films electrosynthesized under high frequency ultrasound irradiation, 140–148, Copyright (2011), with permission from Elsevier.)

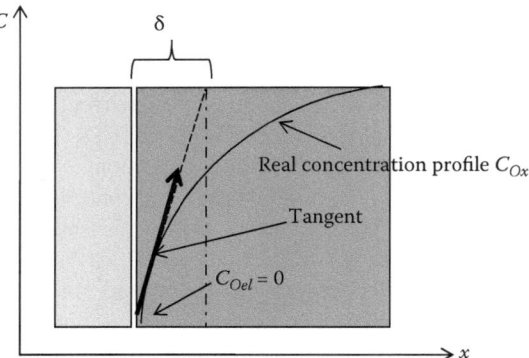

FIGURE 23.13 Two-zone model for mass transfer.

$$j_{\text{Lim}} = -nF D_O \left(\frac{\partial CO}{\partial x}\right)_{x=0} = +nF D_O \frac{(C_{Os} - C_{Oel})}{\delta}$$

and the well-known dimensionless Sherwood number appears in the expression

$$Sh = \frac{k_d d_h}{D_O} = \frac{d_h}{\delta}$$

Therefore, by using the simple and clear film model, it is possible to link the dimensionless number of Sherwood, which describes the mass transfer in many situations (Delmas and Couderc, 1981) to a ratio between a characteristic dimension d_H and the film thickness $\delta = D_O/k_d$. By this mean, the influence of a large number of operating parameters is reduced, and comparison between different experiments in different laboratories or experimental setups is easier. The raw recording is a current (A). The result can be written in terms of current density (removing the electrode area influence), then in term of mass-transfer coefficient k_d (removing the influence of the number of electrons exchanged and of electroactive species concentration), and finally in the Sherwood criterion (removing the influence of electrode size d_H and electrolyte nature through the diffusion coefficient).

The immediate application to sonoreactor design and use is to be able to compare different situations involving the use of ultrasound, allowing different teams of researchers to compare their results accurately. In the case of the setup of accelerated corrosion tests, Ligier et al. (2001) records systematically "ultrasonic conditions" in a single criterion that consists of a Sherwood number. Therefore, it is possible to reproduce the corrosion test in another setup, in the conditions by keeping the same Sherwood number.

As acoustic energy is not uniform in a sonoreactor, it is possible to map by moving systematically an electrode and by recording the mass-transfer coefficient (Faid et al., 1998). If these kinds of results are expressed in Sherwood number, their impact is increased because comparison between other experimental devices is facilitated. For example, to evaluate the corrosion rate of zinc exposed to ultrasonic irradiation in various electrolytes, the setup has been characterized by systematic records of the Sherwood number all along the sonoreactor axis (Figure 23.14) (Doche et al., 2003).

The Sherwood number decreases when the electrode is moved from the ultrasonic source. However, as from a distance equal to 60 mm, this decrease is not continuous and the curve presents a pseudo-sigmoidal shape: maximum stirring points alternating with minimum ones and separated by a distance of $\lambda/4$ (λ is the wavelength). This confirms the establishment of stationary waves due to a total reflection by water–air interface, and the same trend is confirmed irrespective of ultrasound frequency. Then, once quantitative information has been acquired on stirring level, it is possible to discuss the mechanisms. As the zinc corrosion mechanism in aerated electrolytes is reported to be controlled by the limiting step of oxygen diffusion, the average limiting oxygen diffusion current was taken to be the corrosion rate at the electrode. Results are plotted in Figure 23.15.

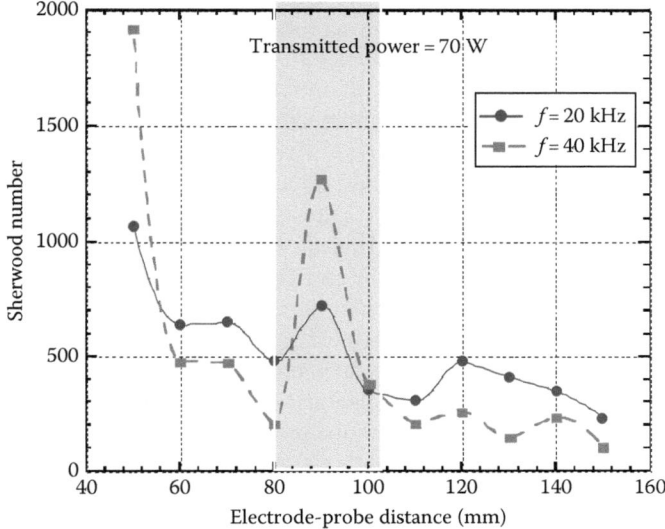

FIGURE 23.14 Comparison between Sherwood numbers as a function of electrode location in a sonoreactor at 20 and 40 kHz. (From *Ultrason. Sonochem.*, 10, Doche, M.-L., Hihn, J.-Y., Mandroyan, A., Viennet, R., and Touyeras, F., Influence of ultrasound power and frequency upon corrosion kinetics of zinc in saline media, 357–362, Copyright (2003), with permission from Elsevier.)

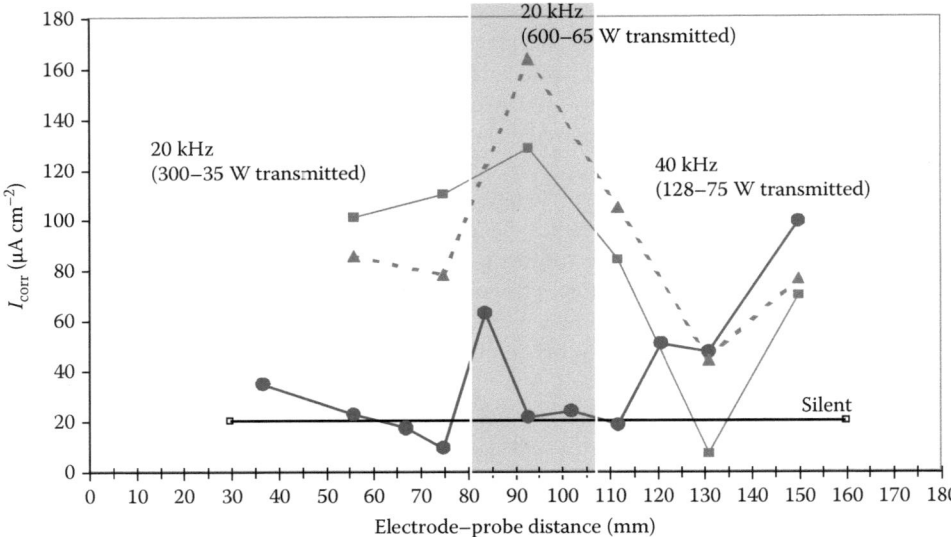

FIGURE 23.15 Comparison between zinc corrosion rates as a function of electrode location in a sonoreactor at 20 and 40 kHz. (From *Ultrason. Sonochem.*, 10, Doche, M.-L., Hihn, J.-Y., Mandroyan, A., Viennet, R., and Touyeras, F., Influence of ultrasound power and frequency upon corrosion kinetics of zinc in saline media, 357–362, Copyright (2003), with permission from Elsevier.)

It can be seen that corrosion current fluctuates according to electrode position and reaches a maximum value (seven to eight times higher than in silent conditions) for a distance of nearly 94 mm (nearly $\lambda + \lambda/4$). These results are consistent with mass-transfer measurements especially for the lowest frequency (20 kHz), when the larger wavelength reduces the error on the electrode location. However, there is some discrepancy, particularly when the electrode is set close to the electrolyte surface, which corresponds to a low stirring point but a high corrosion rate. The liquid surface (deformable interface) probably disturbs the wave system by acting as a reflector.

A final example of Sherwood use concerns the case of hydrodynamic measurement in the Besançon double cell, where studies of sonoelectrochemistry in various solvents are presented. Information obtained is rather different: Raw measurements of current densities are directly related to mass transport—reduction into dimensionless criteria highlighted the particular behavior of room-temperature ionic liquid (Costa et al., 2008). Electrochemical tests were carried out under ultrasonic irradiation for different ultrasound powers, different frequencies, three electrolytes, and the optimum 1.5 bar found previously for coolant overpressure (Figure 23.16). It is important to note that all inputs are expressed in terms of transmitted power, that is, measured during the experiments by calorimetry. Since this fact is taken into account, changes may not simply be ascribed to wave absorption.

Irrespective of operating conditions, ultrasound significantly improves the mass-transfer phenomenon. All recorded voltammograms exhibit a typical sigmoïdal current response yielding signal plateaus, where the oscillations show evidence of cavitational events. The current densities obtained in acetonitrile are higher than in water, but those obtained in (BuMIm) (Tf$_2$N) drop drastically. This can be easily explained by the evolution of Fe^{III}/Fe^{II} diffusion coefficients, which change inversely with viscosity values: Water 1×10^{-3} Pa s, acetonitrile 0.47×10^{-3} Pa s and ionic liquid 55.7×10^{-3} Pa s.

The results were then plotted in Sherwood numbers (Figure 23.17).

The dimensionless numbers are acknowledged as eliminating the electrolyte nature by taking the diffusion coefficients into account. As expected, the values in water and acetonitrile are comparable. It would appear, therefore, that stirring induced by ultrasound is identical in these two liquids for the same transmitted power inputs. Surprisingly, the values of Sherwood numbers are nearly five times higher in (BuMIm) (Tf$_2$N) than in the other liquids, despite the use of the adequate

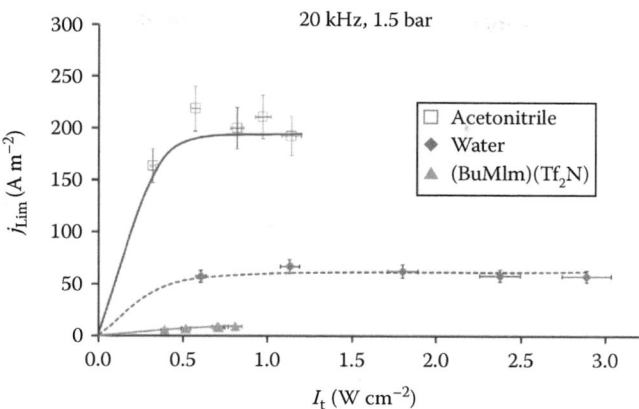

FIGURE 23.16 Mass transport–limited current density (j_{Lim}) versus acoustic intensity (I_t). Results obtained at 20 kHz for (□) acetonitrile, (♦) water, and (▲) (BuMIm) (Tf_2N) with a constant overpressure of 1.5 bar (Costa, C., Hihn, J.-Y., Rebetez, M., Doche, M.-L., Bisel, I., and Moisy, P., 2008, Transport-limited current and microsonoreactor characterization at 3 low frequencies in the presence of water, acetonitrile and imidazolium-based ionic liquids, *Phys. Chem. Chem. Phys.*, 10, 2149–2158, by permission of The Royal Society of Chemistry.)

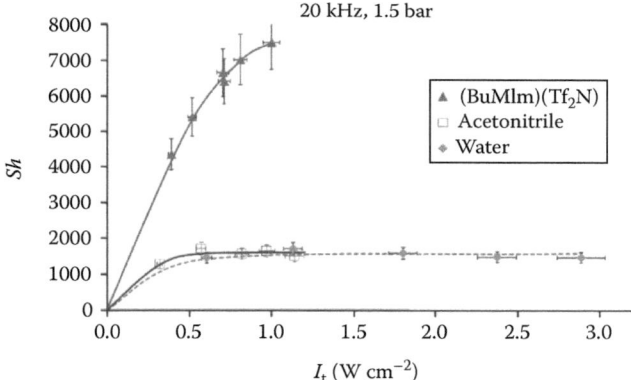

FIGURE 23.17 Sherwood numbers (*Sh*) versus acoustic intensity (I_t). Results obtained at 20 kHz for (□) acetonitrile, (♦) water, and (▲) (BuMIm) (Tf_2N) with a constant overpressure of 1.5 bar. (Costa, C., Hihn, J.-Y., Rebetez, M., Doche, M.-L., Bisel, I., and Moisy, P., 2008, Transport-limited current and microsonoreactor characterization at 3 low frequencies in the presence of water, acetonitrile and imidazolium-based ionic liquids, *Phys. Chem. Chem. Phys.*, 10, 2149–2158, by permission of The Royal Society of Chemistry.)

dimensionless number. This fact leads us to think that ultrasonic stirring close to the electrode is far more efficient in the presence of ionic liquid than for the other solvents. This is not trivial because although acetonitrile did not follow a Newtonian behavior, water, and (BuMIm) (Tf_2N) did!

This issue, highlighted by the choice of plotting the results in Sherwood numbers was the starting point of a complementary study concerning the particular behavior of ionic liquid exposed to ultrasonic irradiation (Costa et al., 2010).

A NEW TOOL: THE EQUIVALENT VELOCITY

Another possibility is the determination of the so-called equivalent fluid velocity as a characteristic of sonoreactors in an attempt to qualify and quantify all phenomena induced by ultrasound at the electrode surface (ultrasonic wind and cavitation activity) (Pollet et al., 2007;

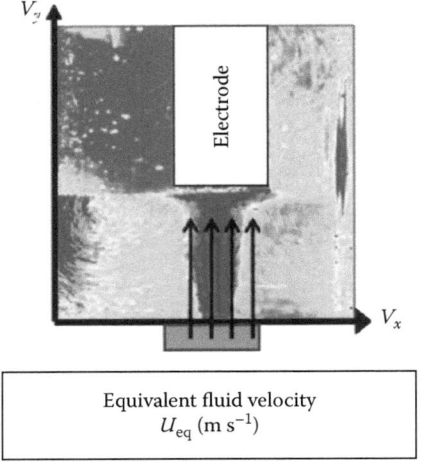

FIGURE 23.18 (See color insert.) Equivalent fluid velocity—coordinate system.

Mandroyan et al., 2010). By analogy between an electrode under ultrasonic irradiation and pure hydrodynamic flow, it is possible to deduce from mass balance equations an "equivalent" flow velocity necessary to obtain the same electrochemical signal under silent conditions as that obtained under sonication. This means that a unique adjustable parameter can be determined as a characteristic of sonoreactors.

Using the coordinate system shown in Figure 23.18, the general hydrodynamic equations in a laminar flow over a flat and plane plate and the mass transport of electroactive species to the disc electrode under steady-state conditions can be considered as per the Navier–Stokes equations. Assuming that the electrode is uniformly accessible, and that the Levich hypotheses can be employed, it can be written that

- The laminar flow over the flat and plane plate is the result of a flow normal to the electrode surface at large distances
- The flow velocity is equivalent over the entire electrode surface
- The electroactive species are consumed uniformly on its surface

In other words, the boundary conditions of hydrodynamics are as follows:

- If $y = 0$ then V_x (velocity component of the flow normal to the electrode) = V_y (velocity component of the flow perpendicular to the electrode) = 0
- If $y = \delta$ (limiting hydrodynamic layer thickness) and $y = \infty$, then $V_x = \bar{V}_\infty$ where $\bar{V}_\infty$ is the average infinite velocity, that is, at large distances from the disc
- If $\partial^2 V_x / \partial y^2 \to 0$ then the mass balance can be simplified as

$$V_x \cdot \frac{\partial V_x}{\partial x} + V_y \cdot \frac{\partial V_y}{\partial y} = 0$$

- If the concentration gradient versus surface length is equal to zero $\partial C/\partial x = 0$, then it becomes

$$\frac{\partial^2 C}{\partial y^2} = \frac{V_y}{D} \cdot \frac{\partial C}{\partial y}$$

Therefore, the velocity on the y-axis is given by

$$V_y = \left(\frac{1.33}{16}\right) \cdot \overline{V}_\infty^{3/2} \cdot \upsilon^{-1/2} \cdot x^{-3/2} \cdot y^2$$

where
  υ is the kinematic viscosity
  $\overline{V}_\infty$ is the average infinite velocity

This parameter can then be defined as the equivalent flow velocity U (m s^{-1}), at a large distance from the disc (outside the limiting hydrodynamic layer).

For all electrochemical measurements, the limiting current density is given by

$$j_{Lim} = n \cdot F \cdot D \cdot \left(\frac{\partial C}{\partial y}\right)_{y=0}$$

Combining this equation with that resulting from the mass balance, it appears

$$j_{Lim} = 0.34 \cdot n \cdot F \cdot D^{2/3} \cdot \upsilon^{-1/6} \cdot x^{-1/2} \cdot C^* \cdot U^{1/2}$$

To obtain an average limiting current density over the whole electrode surface, this equation is integrated in terms of x from 0 to r (electrode radius).

The equivalent flow velocity becomes

$$U = \frac{1}{(0.45 \cdot n \cdot F \cdot C^*)^2} \cdot D^{-4/3} \cdot \upsilon^{1/3} \cdot r \cdot i_{Lim}^2$$

It can be seen that for a given electrode geometry, at constant temperature, electrolyte concentration, etc., the only adjustable parameter is the equivalent flow velocity U. Using the observed experimental limiting current densities, it is possible to calculate the equivalent flow velocities (U) at various ultrasonic intensities (I) and electrode–horn distances (d) (Pollet et al., 2007).

APPLICATION TO THE MEASUREMENT OF ACOUSTIC ENERGY DISTRIBUTION

In specific applications of surface cleaning and electrochemistry, which consist of processes implanting surface irradiation by ultrasound, design of large-scale devices requires us to understand acoustic field distribution together with its quantification. In this specific case, which constitutes the first application of the "equivalent" flow velocity, it is interesting to use the systematic determination of U_{eq} throughout electrochemical measurements versus various operating parameters (powers, electrode–horn distances, reactor geometry, frequencies, etc.) (Figure 23.19).

Such an investigation was carried out in the Besançon group. (Mandroyan, 2010). Equivalent flow velocities were measured at a variety of distances between the working electrode and the transducer surface. For various parameters and for both 20 and 40 kHz frequencies, variation in "equivalent" flow velocity versus distance exhibits an exponential decrease shape. Velocities are included between 2 and 120 m s^{-1}. On almost all curves (an example is shown in Figure 23.20), a first peak is observable for an electrode-to-horn separation of 5 mm and sometimes, for greater distances, that is, a 20 mm electrode-to-horn separation, a second peak is visible. In this case, the distance between the two peaks is equal to a quarter of a wavelength (1.8 cm at 20 kHz).

As representation of equivalent flow velocity versus the distance between the working electrode and the transducer has the same shape irrespective of the experimental conditions, a numerical model was proposed to fit our curves and to identify some parameters by taking into account the characteristics of the ultrasonic wave (absorption coefficient, rate of cavitation bubbles, and acoustic power).

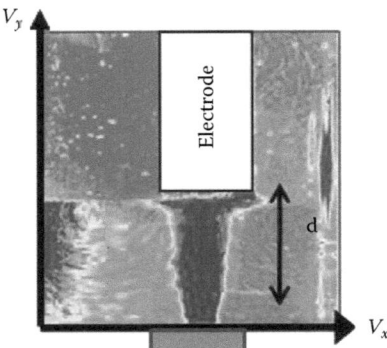

FIGURE 23.19 (See color insert.) Determination of the "equivalent flow" in the zone close to the transducer.

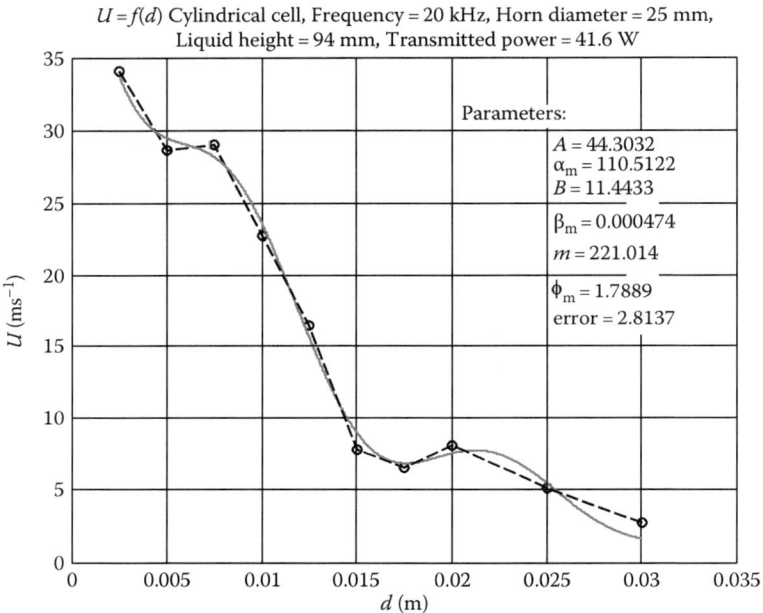

FIGURE 23.20 "Equivalent velocities" versus electrode to horn distance. Dashed curve: experimental points, continuous line: model simulation. (From *Ultrason. Sonochem.*, 17, Mandroyan, A., Hihn, J.-Y., Doche, M.-L., and Pothier, J.-M., A predictive model obtained by identification for the ultrasonic «equivalent» flow velocity at surface vicinity, 965–977, Copyright (2010), with permission from Elsevier.)

$$U = A \exp(-\alpha x) + B \exp\left(-\frac{x^2}{\beta}\right) \cdot \sin^2\left(dx + N\frac{\pi}{2} + \frac{\pi}{4}\right)$$

Attenuation in homogeneous media
Attenuation in presence of cavitation
Spatial variation of U
Adjustable phase shift

$$I = \frac{1}{2} \cdot \frac{P_A^2}{\rho.c} \quad avec \quad P_A = P_{A0} \sin\left(-\frac{\omega}{c} x\right)$$

Ultrasound for Better Reactor Design

Nevertheless, the flicked behavior of the ultrasonic processes in the vicinity of the electrode as well as bubble presence, which induce nonlinearities in wave propagation, imposes a new approach based on parameter identification by methods currently used in chemical engineering. Then the parameters of the model proposed by Mandroyan et al. (2010) and their best values are used to simulate it for comparison with the experimental points, as shown in Figure 23.20.

USE FOR CAVITATION QUANTIFICATION

Another possibility is the comparison between the equivalent flow velocities and the real velocities measured by particle image velocimetry (PIV) in sonoreactors. The laser tomography technique used in the earlier paragraphs (Figure 23.21) allowed us to visualize cavitation bubbles induced by ultrasonic wave propagation. This technique gives some information about the acoustic activity, but does not allow fluid motions to be quantified. PIV is a flow velocity measurement technique, which is extensively used to study fluid dynamics. It consists in determining particle displacement over time using a double-pulsed laser. In sonicated reactors, few papers report successfully on the measurement of acoustic streaming velocity vector fields produced by high-frequency probes. At low frequencies when cavitation is more intense, bubbles interfere with the tracers necessary for PIV. This problem can be overcome by seeding the inside reactor solution with fluorescent tracers and filtering the picture. A laser light sheet illuminates a plane in the flow, and the positions of particles in that plane are recorded using a digital camera. A fraction of a second later, another laser pulse illuminates the same plane, creating a second particle image (Mandroyan et al., 2009b).

Results consist of a map of the average velocity vector field (Figure 23.22). An intense axial flow marked by a high density of vertical vectors just above the transducer is clearly visible. Some vectors diagonally directed against the transducer reveal the presence of a recirculation flow near the reactor walls. The electrode acts as an obstacle to flow propagation, leading to the emergence of three different flow regions, as illustrated in Figure 23.22. A flow, visible on both sides of the electrode, corresponds to a tangential flow as it is similar to a boundary flow parallel to the electrode surface. Different velocities can be calculated as shown in Figure 23.22—V_{Axial} in the central flow, V_R in the recirculation, and V_T in the tangential flow—and expressed in terms of minimum, maximum, and average velocities.

$U_{\text{equivalent}}$ measured by electrochemistry is up to 10 times higher than real flow velocity, which can be explained by the fact that ultrasonic wind accounts only for ($\approx 10\%$), whereas the main contribution is asymmetric cavitation (micro-jets). Indeed, the acoustic streaming directed along the wave propagation direction is one of the consequences of the viscous forces in the liquid and corresponds

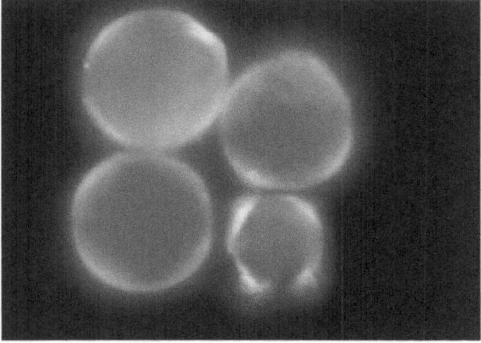

FIGURE 23.21 **(See color insert.)** Fluorescent tracers. (From *Ultrason. Sonochem.*, 16, Mandroyan, A., Doche, M.-L., Hihn, J.-Y., Viennet, R., Bailly, Y., and Simonin, L., Modification of the ultrasound induced activity by the presence of an electrode in a sono-reactor working at two low frequencies (20 and 40 kHz). Part II: Mapping flow velocities by particle image velocimetry (PIV), 97–104, Copyright (2009b), with permission from Elsevier.)

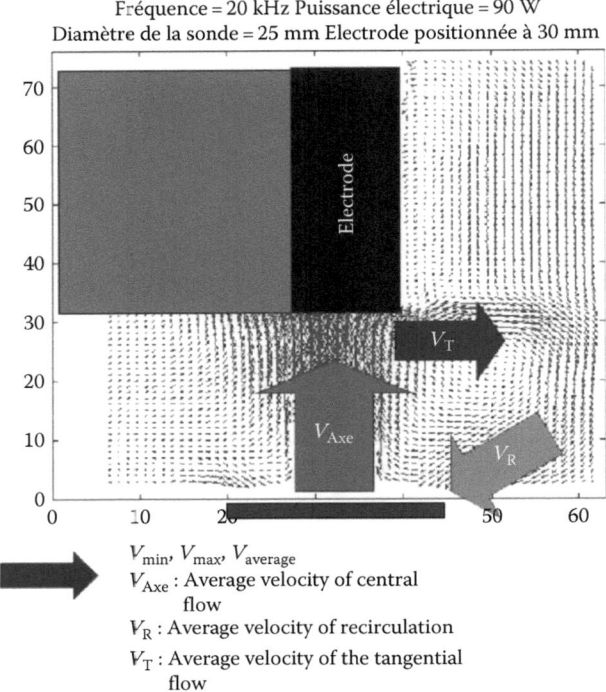

FIGURE 23.22 Velocity vector field in a low-frequency sonoreactor.

to macrostreaming, which contributes partially to an increase in mass transfer at the electrode. This can be quantified by the V_{Axial} measurement (around 0.2 m s^{-1}) (Mandroyan et al., 2009b). When these values are compared to the equivalent velocities obtained in the same operating conditions (2 m s^{-1}) (Mandroyan et al., 2010), a major difference is highlighted due to the contribution of the microstreamings induced by cavitation bubble movements. In point of fact, their implosion is also responsible for diffusion and migration mass transfers around the bubbles as well as another contribution to the increase in mass transfer at the electrode.

As a conclusion, thanks to the use of the notion of equivalent flow velocity, a very first quantitative estimation of the contribution of cavitation faced with the convective movement is measurable. Similar observations have been made in the Besançon group during high–intensity, focused ultrasound sonoreactor characterization. By comparing theoretical flow velocities obtained by Eckart equation simulation to equivalent flow velocities obtained by electrochemical measurements, it is possible to localize and quantify the cavitation zones close to the focal (Hallez, 2009).

PHYSICAL SIGNIFICATION FOR THE DIMENSIONLESS NUMBER

The last possibility is to give a final significance to the dimensionless number, such as in the case of the Reynolds expression. The Reynolds number is a unique criterion whose values determine the changes in fluid flow regimes irrespective of fluid nature or geometrical considerations (ratio of inertial strength over viscosity). In our case, $Re = Ur/\upsilon$, where Re is the Reynolds number (dimensionless), U is the flow velocity (m s^{-1}), r is the electrode radius (m), and υ is the kinematic viscosity (m^2 s^{-1}). In the chosen sonoreactor example, used for electrochemical applications at low frequencies, the notion of an equivalent fluid velocity is necessary to obtain a Reynolds value. The Sherwood number can then be expressed as a function of the Reynolds number and the Schmidt number in classical chemical engineering correlation.

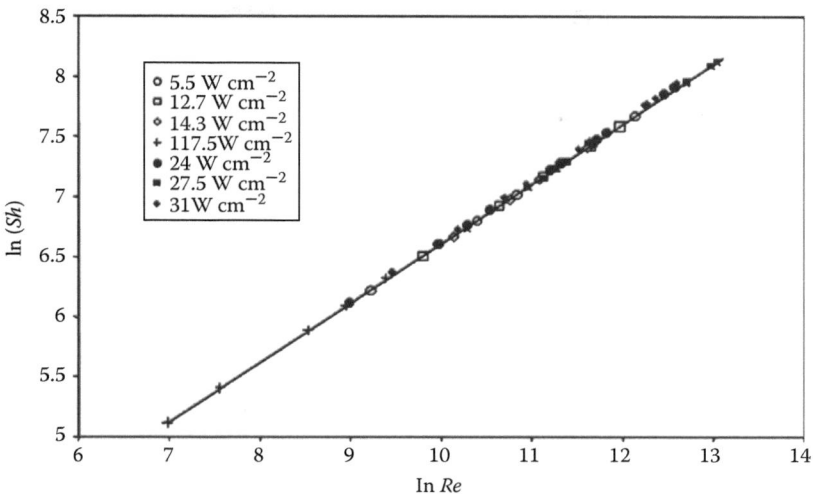

FIGURE 23.23 ln *Sh* versus ln *Re* at (298 ± 1) K—(Coventry cell 20 kHz, sodium hydroxide 1.0 mol dm^{-3}, ○ 5.5 W cm^{-2}, □ 12.7 W cm^{-2}, ◊ 14.3 W cm^{-2})—(Coventry cell 20 kHz, potassium chloride 1.0 mol dm^{-3}, + 117.5 W cm^{-2})—(Besançon cylindrical cell 40 kHz, sodium hydroxide 1.0 mol dm^{-3}, ● 24 W cm^{-2}, ■ 27.5 W cm^{-2}, ◆ 31 W cm^{-2}). (Pollet, B.G., Hihn, J.-Y., Doche, M.-L., Lorimer, J.P., Mandroyan, A., and Mason, T.J., 2007, Transport limited currents close to an ultrasonic horn equivalent flow velocity determination. *J. Electrochem. Soc.*, 154, E131–E138, by permission of The Electrochemical Society.)

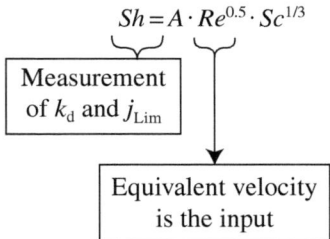

In this expression, *Sc* is the Schmidt number $Sc = \upsilon/D$ (dimensionless) and *Sh* is the Sherwood number, deduced from the experimentation $Sh = i_{Lim} r/n \cdot FD \cdot C^*$ (dimensionless). A very good correlation appears when the Sherwood number is plotted as a function of the Reynolds number at various ultrasonic frequencies and intensities, electrode–horn distances, electroactive species concentrations, background electrolytes, and cell geometries. It corresponds to a linear relationship between Sherwood and Reynolds numbers with a slope of 0.5, and intercepts of 1.6072 (for potassium chloride background electrolyte) and 1.6274 (for sodium hydroxide background electrolyte), respectively (Pollet et al., 2007) (Figure 23.23).

Such behavior was expected because the equivalent fluid velocity determination and the Sherwood numbers are derived from the same measurements. Nevertheless, the interest in such a representation consists in grouping the results in standard correlation as shown earlier to balance the different dimensionless numbers. After corrections by the respective values of the Schmidt number as a function of background electrolytes, the value of *A* is equal to 0.5, the Reynolds exponent is worth 1/2, and the Schmidt number about 1/3. This tallies with the results found in the literature for an electrode placed at a "normal" to a convective flow (*A* = 0.646, *Re* exponent = 1/2, *Sc* exponent = 1/3 Chin et al., or *A* = 0.532, *Re* exponent = 1/2, *Sc* exponent = 1/3 (Pletcher and Walsh, 1993).

This approach can be extended to other sonoreactors in the absence of electrochemistry, such as in the case of ultrafiltration assisted by ultrasound. In a filtration setup specially designed for high-frequency ultrasound use, the results are grouped in a single parameter to allow comparison with

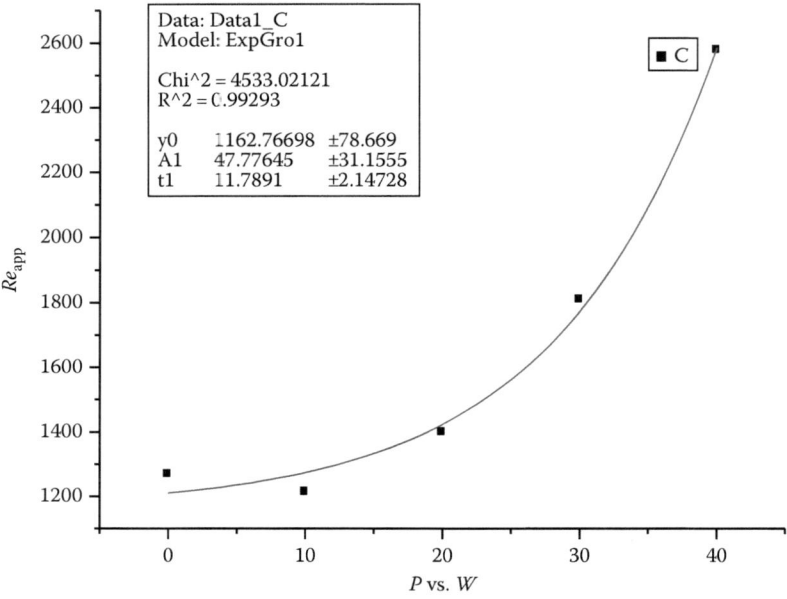

FIGURE 23.24 "Apparent Reynolds" versus ultrasonic power during ultrafiltration.

dead-end filtration assisted by ultrasound. (Simon et al., 2000). In this case, $Re = Udh/\upsilon$, where Re is the Reynolds number (dimensionless), U is the flow velocity (m s^{-1}), dh is the hydraulic diameter (m), and υ is the kinematics viscosity (m^2 s^{-1}).

For fixed operating conditions (i.e., a hydraulic diameter of 4.48×10^{-2} m and a water flow velocity of $U = 0.0283$ m s^{-1}), it is possible to determine the mass-transfer coefficient K by measurements on the permeate (Simon et al., 2000). This mass transfer can then be quoted in a Sherwood number, and the classical expression given earlier yields an "apparent Reynolds" number.

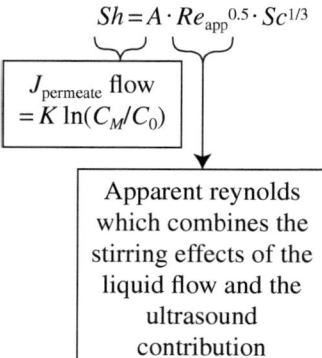

By identification, it is also possible to calculate the value of the apparent Reynolds number, which groups the contribution of ultrasound as well as flow circulation through the filtration module, as shown in Figure 23.24.

TOWARD THE "IDEAL" SONOREACTOR?

As for industrial situations, many reactors can be used successfully in various processes, with a design adapted by the people in charge of production. This is the case too in sonochemistry, where the increase in specific knowledge allows both researchers and engineers to ensure efficient design.

Nevertheless, if cavitational intensity and fluid circulation are frequently take into account, the setup of design laws derived directly from classical engineering tools for scale-up and design will be the key step for the forthcoming generation of sonoreactors.

REFERENCES

Boldo, P., Renaudin, V., Gondrexon, N., and Chouvellon, M. 2004. Enhancement of the knowledge on the ultrasonic reactor behaviour by an interdisciplinary approach. *Ultrasonics Sonochemistry*, 11: 27–32.

Burdin, F., Tsochatzidis, N.A., Guiraud, P., Wilhelm, A.M., and Delmas, H. 1999. Characterisation of the acoustic cavitation cloud by two laser techniques. *Ultrasonics Sonochemistry*, 6: 43–51.

Chouvellon, M., Largillier, A., Fournel, T., Boldo, P., and Gonthier, Y. 2000. Velocity study in an ultrasonic reactor. *Ultrasonics Sonochemistry*, 7: 207–211.

Cooper, E.L. and Coury, L.A. 1998. Mass transport in sonovoltammetry with evidence of hydrodynamic modulation from ultrasound. *Journal of Electrochemical Society*, 145(6): 1994–1999.

Costa, C., Doche, M.-L., Hihn, J.-Y., Bisel, I., Moisy, P., and Lévêque, J.M. 2010. Hydrodynamic sono-voltammetry of ferrocene in [Tf2N]-based ionic liquid media. *Ultrasonics*, 50: 323–328.

Costa, C., Hihn, J.-Y., Rebetez, M., Doche, M.-L., Bisel, I., and Moisy, P. 2008. Transport-limited current and microsonoreactor characterization at 3 low frequencies in the presence of water, acetonitrile and imidazolium–based ionic liquids. *Physical Chemistry and Chemical Physics*, 10: 2149–2158.

Delmas, H. and Couderc, J.P. 1981. Overall mass-transfer from a sphere in liquid-phase—Influence of the Schmidt number. *Letters in Heat and Mass Transfer*, 8: 271–280.

Doche, M.-L., Hihn, J.-Y., Mandroyan, A., Viennet, R., and Touyeras, F. 2003. Influence of ultrasound power and frequency upon corrosion kinetics of zinc in saline media. *Ultrasonics Sonochemistry*, 10: 357–362.

Doche, M.-L., Hihn, J.-Y., Touyeras, F., Lorimer, J.P., Mason, T.J., and Plattes, M. 2001. Electrochemical behaviour of zinc in 20 kHz sonicated NaOH electrolytes. *Ultrasonics Sonochemistry*, 8: 291–298.

Et Taouil, A., Lallemand, F., Hihn, J.-Y., Melot, J.-M., Blondeau-Patissier, V., and Lakard, B. 2011. Doping properties of PEDOT films electrosynthesized under high frequency ultrasound irradiation. *Ultrasonics Sonochemistry*, 18: 140–148.

Faid, F., Contamine, F., Wilhelm, A.M., and Delmas, H. 1998. Comparison of ultrasound effects in different reactors at 20 kHz. *Ultrasonics Sonochemistry*, 5: 119–124.

Fry, W.J. and Fry, R.B. 1954. Determination of absolute sound levels and acoustic absorption coefficients by thermocouples probes—Theory. *Journal of the Acoustical Society of America*, 26: 294–310.

Gogate, P.R. and Pandit, A.B. 2004. Sonochemical reactors: Scale up aspects. *Ultrasonics Sonochemistry*, 11: 105–117.

Gondrexon, N., Renaudin, V., Clément, M., Boldo, P., Pétrier, C., Gonthier, Y., and Bernis, A. 1995. Hétérogénéités au sein d'un réacteur parfaitement agité. Le réacteur ultrasonore *Récents Progrès en Génie des Procédés*, 9(41): 71–76.

Hagan, C.R.S. and Coury, L.A. 1994. Comparison of hydrodynamic voltammetry implemented by sonication to rotating-disk electrode. *Analytical Chemistry*, 66: 399–405.

Hallez, L. 2009. Les HIFU en sonochimie, irradiation ultrasonore de polymers, PhD thesis, University of Franche-Comté, Besançon, France.

Hallez, L., Touyeras, F., Hihn, J.-Y., and Klima, J. 2007. Energetic balance in an ultrasonic reactor using focused or flat high frequency transducers. *Ultrasonics Sonochemistry*, 14: 739–749.

Hallez, L., Touyeras, F., Hihn, J.-Y., Klima, J., Guey, J.-L., Spajer, M., and Baillo, Y. 2010. Characterization of HIFU transducers designed for sonochemistry application: Cavitation distribution and quantification. *Ultrasonics*, 50: 310–317.

Hihn, J.-Y., Bereiziat, D., Doche, M.-L., Chaillet, P., Lorimer, J.P., Mason, T.J., and Pollet, B. 2000. Double-structured ultrasonic high frequency reactor using an optimised slant bottom. *Ultrasonics Sonochemistry*, 7: 201–205.

Laborde, J.-L., Bouyer, C., Caltagirone, J.-P., and Gerard, A. 1998. Acoustic cavitation field prediction at low and high frequency ultrasounds. *Ultrasonics*, 36: 581–587.

Lakard, S., Hallez, L., Hihn, J.-Y., Touyeras, F., Fievet, P., and Cravotto, G. 2011. Influence of high-frequency ultrasounds on filtration treatments. *Ultrasonics Sonochemistry* (to be submitted).

Ligier, V., Hihn, J.-Y., Wéry, M., and Tachez, M. 2001. The effects of 20 kHz and 500 kHz ultrasound on the corrosion of zinc precoated steels in [Cl$^-$][SO$_4^{2-}$][HCO$_3^-$][H$_2$O$_2$] electrolytes. *Journal of Applied Electrochemistry*, 31: 213–222.

Mandroyan, A., Doche, M.-L., Hihn, J.-Y., Viennet, R., Bailly, Y., and Simonin, L. 2009b. Modification of the ultrasound induced activity by the presence of an electrode in a sono-reactor working at two low frequencies (20 and 40 kHz). Part II: Mapping flow velocities by particle image velocimetry (PIV). *Ultrasonics Sonochemistry*, 16: 97–104.

Mandroyan, A., Hihn, J.-Y., Doche, M.-L., and Pothier, J.M. 2010. A predictive model obtained by identification for the ultrasonic "equivalent" flow velocity at surface vicinity. *Ultrasonics Sonochemistry*, 17: 965–977.

Mandroyan, A., Viennet, R., Bailly, Y., Doche, M.-L., and Hihn, J.-Y. 2009a. Modification of the ultrasound induced activity by the presence of an electrode in a sonoreactor working at two low frequencies (20 and 40 kHz). Part I: Active zone visualization by laser tomography. *Ultrasonics Sonochemistry*, 16: 88–96.

Martin, C.J. and Law, A.N.R. 1980. The use of thermistor probes to measure energy distribution in ultrasound fields. *Ultrasonics*, 18: 127–133.

Mason, T.J. 2000. Large scale sonochemical processing: Aspiration and actuality. *Ultrasonics Sonochemistry*, 7: 145–149.

Petrier, C., Jeunet, A., Luche, J.-L., and Reverdy, G. 1992. Unexpected frequency-effects on the rate of oxidative processes induced by ultrasound. *Journal of the American Chemical Society*, 114: 3148–3150.

Pletcher, D. and Walsh, F.C. 1993. *Industrial Electrochemistry*, Blackie Academic and Professional, Glasgow, U.K.

Pollet, B.G., Hihn, J.-Y., Doche, M.-L., Lorimer, J.P., Mandroyan, A., and Mason, T.J. 2007. Transport limited currents close to an ultrasonic horn equivalent flow velocity determination. *Journal of the Electrochemical Society*, 154: E131–E138.

Pollet, B.G., Lorimer, J.P., Hihn, J.-Y., Touyeras, F., Mason, T.J., and Walton, D.J. 2005. Electrochemical study of silver thiosulphate reduction in the absence and presence of ultrasound. *Ultrasonics Sonochemistry*, 12: 7–11.

Renaudin, V., Gondrexon, N., Boldo, P., Pétrier, C., Bernis, A., and Gonthier, Y. 1994. Method for determining the chemically active zones in a high-frequency ultrasonic reactor. *Ultrasonics Sonochemistry*, 1: S81–S85.

Simon, A., Penpenic, L., Gondrexon, N., Taha, S., and Dorange, G. 2000. A comparative study between classical stirred and ultrasonically-assisted dead-end ultrafiltration. *Ultrasonics Sonochemistry*, 7: 183–186.

Soong, Y., Gamwo, I.K., Romanov, V., Dilmore, R.M., and Hedges, S.W. 2006. Design of an adapter for ultrasound transducer under high operating temperatures and pressures in a slurry bubble column reactor. *Chemical Engineering Research and Design*, 84: 133–138.

Sutkar, V.S. and Pandit, P.R. 2009. Design aspects of sonochemical reactors: Techniques for understanding cavitational activity distribution and effect of operating parameters. *Chemical Engineering Journal*, 155: 26–36.

Trabelsi, F., Ait-Lyazidi, H., Berlan, J., Fabre, P.-L., Delmas, H., and Wilhem, A.M. 1996. Electrochemical determination of the active zones in a high-frequency ultrasonic reactor. *Ultrasonics Sonochemistry*, 3: S125–S130.

Viennet, R., Doche, M.-L., Hihn, J.-Y., Mandroyan, A., and Cancel, N. 2001. Etude d'un réacteur ultrasonore fonctionnant à 20 et 40 kHz—Répartition de l'activité acoustique. *Récents Progrès en Génie des Procédés*, 15: 345–352.

Viennet, R., Hihn, J.-Y., Jeannot, M., and Berriet, R. 2009. Study of ultrasound transmission through an immersed glass plate in view of sonochemical reactor design optimization. *Advances in Acoustics and Vibration*, Article ID: 512839: 9.

Viennet, R., Ligier, V., Hihn, J.-Y., Bereiziat, D., Nika, P., and Doche, M.-L. 2004. Visualisation and electrochemical determination of the actives zones in an ultrasonic reactor using 20 and 500 kHz frequencies. *Ultrasonics Sonochemistry*, 11: 125–129.

24 Sonoelectrochemistry: From Theory to Applications

Bruno G. Pollet and Jean-Yves Hihn

CONTENTS

Introduction to Sonoelectrochemistry 623
    Effect of Power Ultrasound upon Chemical Reactions 625
    Ultrasonic Equipment Used in Sonoelectrochemistry 626
        Ultrasonic Baths and Ultrasonic Probes 626
        Other Sonoelectrochemical Cells 629
Use of Ultrasound in Electrochemistry 632
    Ultrasound Effect on Electrode-Kinetics 632
    Sono-Electrodeposition and Sono-Electroplating 638
    Sonoelectrochemistry and Corrosion 639
    Sono-Electro-Organic Chemistry 641
    Sono-Electropolymerization 642
    Sono-Electrochemiluminescence 643
    Environmental Sono-Electrochemistry 643
    Sono-Electroanalytical Chemistry 644
    Sonoelectrochemical Production of Nanomaterials 644
    Sonoelectrochemical Production of Fuel Cell Materials 645
        Effect of Aqueous Solutions 646
        Effect of Surfactants and Polymers 646
    Sonoelectrochemistry in "Exotic" Solvents 649
Conclusions 654
Acknowledgment 654
References 654

INTRODUCTION TO SONOELECTROCHEMISTRY

The use of ultrasound on electrochemical systems or sonoelectrochemistry was first observed by Moriguchi (1934) as early as 1934, and since then, it has continued to be an active and exciting research area. In the 1930s, Schmid and Ehert (1937a) also studied the ultrasonic effects on the passivity of metals and the generation of gases by electrolysis (Schmid and Ehert, 1937b). Nearly 30 years later, Kolb and Nyborg (1956) demonstrated the movement of liquid induced by ultrasound, known as acoustic streaming, and in the same year Penn et al. (1959) studied the effect of ultrasound on concentration gradient in the electrolyte and at the electrode surface. In the mid-1960s, for the first time, Bard (1963) showed that ultrasound caused an increase in the mass transport of electroactive species from the bulk solution to the electrode surface in controlled potential coulometry experiments. Since then, extensive work has been carried out in which high power ultrasound (20–100 kHz) was applied to various electrochemical processes leading to several industrial applications and many publications over a wide range of subject areas such as electrodeposition, electroplating, electrochemical dissolution, corrosion testing, nanotechnology, and fuel cell technology.

For over 70 years, nearly a thousand papers have been written on the subject with many original work, general reviews (Yegnaraman and Bharathi, 1992; Walton and Phull, 1996; Compton et al., 1997a; Pollet and Phull, 2001; Walton, 2002; Compton et al., 2003; Brett, 2008; Klima, 2010; Gonzales-Garcia et al., 2010) in sonoelectrochemistry (with the first one from Mason et al. (1990), including the effects of power ultrasound on special media), organic sonoelectrosynthesis (Walton and Mason, 1998; Cognet et al., 2000), sonoelectroanalysis (Wadhawan et al., 2001; Banks and Compton, 2003), sonoelectrochemical production of nanomaterials (Saez and Mason, 2009), and recently the sonoelectrochemical production of fuel cell materials (Pollet, 2010).

In all these papers and reviews, it was clearly shown that the effects of high-intensity ultrasonic irradiation on electrochemical processes lead to both chemical and physical effects, for example, mass-transport enhancement, surface cleaning, and radical formation. Many workers have also investigated the distribution of ultrasonic waves or energy in various electrochemical reactors operating in the lower ultrasonic frequency range (20–100 kHz) and at high ultrasonic powers. Several methods for such determination have been proposed, for example, aluminum foil erosion (Figure 24.1), sonoluminescence, calorimetric methods, chemical dosimetry, and laser-sheet visualization.

In sonoelectrochemistry, ultrasound is known for its capacity to promote, especially, heterogeneous reactions mainly through extremely increased mass transport, interfacial cleaning (Figure 24.1), and thermal effects. In addition, homogeneous chemical reactions have been reported to be affected and the generation of highly radical species [e.g., the production of H• and OH• radicals by sonolysis (see previous chapters)] in intense sound fields is an important aspect for the use of ultrasound. For example, in the detoxification of environmentally harmful wastes [containing heavy metals, hydrocarbons (PAH's) and chlorinated compounds (PCB's)], ultrasound has been found to be extremely beneficial.

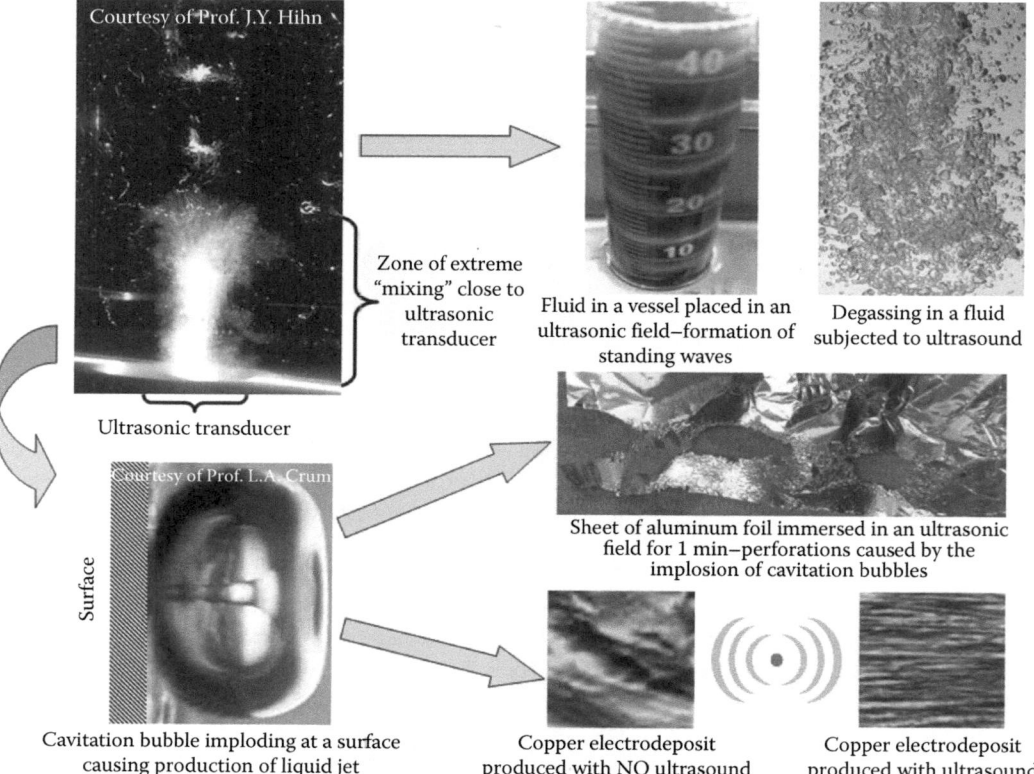

FIGURE 24.1 The effects of ultrasound in a liquid media.

Sonoelectrochemistry

The vast variety of ultrasonically induced effects observed in sonoelectrochemistry may be ascribed to the generation, pulsation, and collapse of cavitation bubbles in the electrolyte medium near the electrode surface (Figure 24.1). This ultrasonic cavitation occurs at low-to-high ultrasonic intensities (ultrasonic power *per* tip or transducer or cell base area). A pulsating cavitation bubble close to the electrode surface generates microstreaming. When the cavitation bubble reaches a resonant size, it collapses asymmetrically leading to the formation of high velocity jet of liquid toward the surface. This phenomenon leads to a thinning of the diffusion layer and can improve the overall mass transfer and hence the reaction rates.

Since most of the observed effects of ultrasound in electrochemical processes are thought to be due to the cavitation effect together with micro-streaming (Figure 24.1), the application of ultrasound is known to be very beneficial in the electrochemical industry. This has led to investigations into mass-transport, electron-transfer processes and electrode surface adsorption. This chapter deals with the aspect of electrochemistry combined with ultrasound and explains the various electrochemical phenomena occurring at the electrode surface when a potential is applied across it. For this purpose, electrode kinetic and mass-transport parameters will be defined. This chapter will also outline the theory, principles, and applications of sonoelectrochemistry in various branches of chemistry and how coupling ultrasound with electrochemistry could be used to improve electrochemical processes, enhance detection limits in the electroanalysis of toxic samples, and produce nanomaterials.

EFFECT OF POWER ULTRASOUND UPON CHEMICAL REACTIONS

Over the past few years, the use of power ultrasound has found wide applications in the chemical and processing industries where it is used to enhance both synthetic and catalytic processes and to generate new products. This area of research has been termed sonochemistry, which mainly concerns reactions involving a liquid leading to an increase in reaction rates, product yields and erosion of surfaces. However, the main reason for most of the observed effects of ultrasound on surfaces and chemical reactions is recognized as being due to "cavitation" effect, which occurs as a secondary effect when an ultrasonic wave passes through a liquid medium. Cavitation is a phenomenon where microbubbles are formed which tend to implode and collapse violently in the liquid leading to the formation of high velocity jets of liquid. Indeed, ultrasound consists of alternating compression and rarefaction cycles (see previous chapters). During rarefaction cycles, negative pressures developed by the high power ultrasound are strong enough to overcome the intermolecular forces binding the fluid. The succeeding compression cycles can cause the microbubbles to collapse almost instantaneously with the release of a large amount of energy. Bubble formation is a three-step process consisting of nucleation, bubble growth, and collapse of gas vapor–filled bubbles in a liquid phase. These bubbles transform the low energy density of a sound field into a high energy density by absorbing energy from the sound waves over one or several cycles and releasing it during very short intervals. Cavitation phenomenon is known to cause erosion, emulsification, molecular degradation, sonoluminescence, and sonochemical enhancements of reactivity solely attributed to the collapse of cavitation bubbles. It is now well accepted in the field that the cavitation bubble collapse leads to near adiabatic heating of the vapor that is inside the bubble, creating the so-called "hot-spot" in the fluid, where

1. High temperatures (ca. 5000 K) and high pressures (ca. 2000 atm) are generated with cooling rates of 10^{9-10} K s^{-1} when the collapsing of cavitation bubbles are observed. Here, water vapor is "pyrolyzed" into hydrogen radicals (H•) and hydroxyl radicals (OH•), known as water sonolysis.
2. The interfacial region between the cavitation bubbles and the bulk solution is paramount. The temperature is lower in the interior of the bubbles than the exterior, but high enough for thermal decomposition of the solutes to take place with greater local hydroxyl radical concentrations in this region.
3. The reactions of solute molecules with hydrogen atoms and hydroxyl radicals occur in the bulk solution at ambient temperature.

Ultrasonic Equipment Used in Sonoelectrochemistry

Ultrasonic Baths and Ultrasonic Probes

They are several forms of ultrasonic equipments commercially available including cleaning baths, probes systems, submersible transducers, whistle and tube reactors. In the laboratory there are two methods of generating acoustic wave in any liquid medium. They are (1) the ultrasonic probe system (Figure 24.2a) and (2) the ultrasonic cleaning bath (Figure 24.2b). Sonoelectrochemical experiments

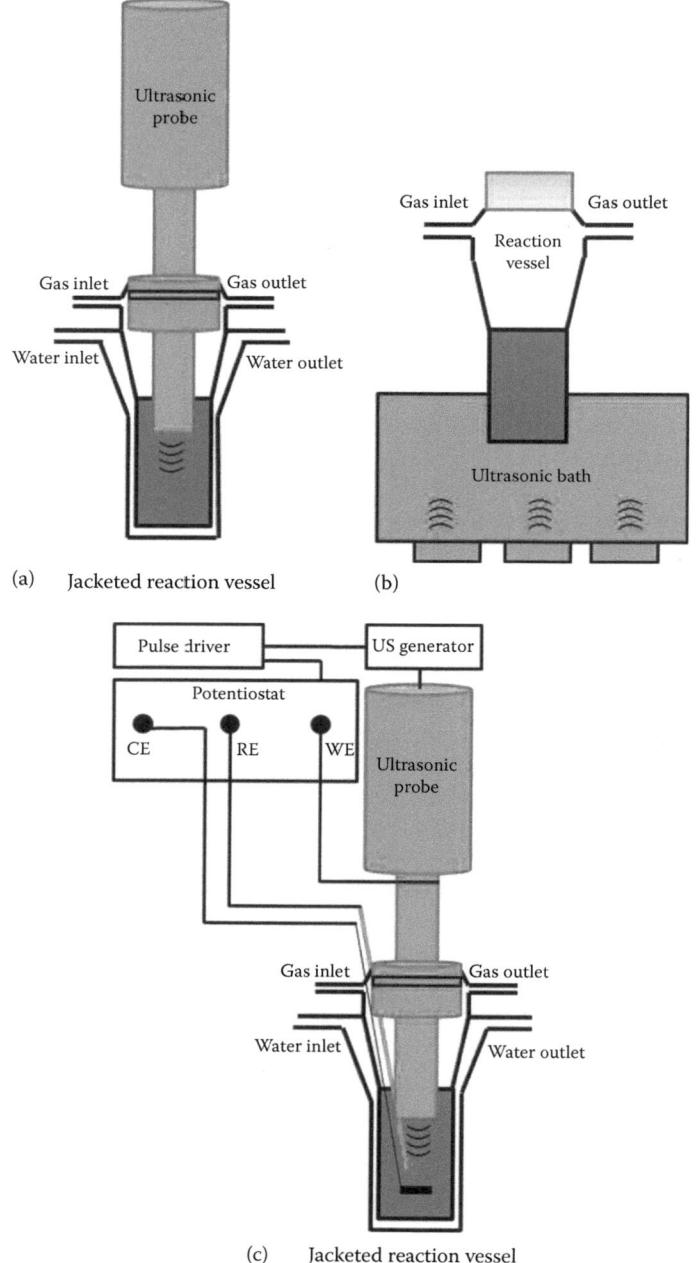

FIGURE 24.2 Sonochemical (a) ultrasonic probe system; (b) ultrasonic bath system and sonoelectrochemical (c) ultrasonic probe used as the cathode.

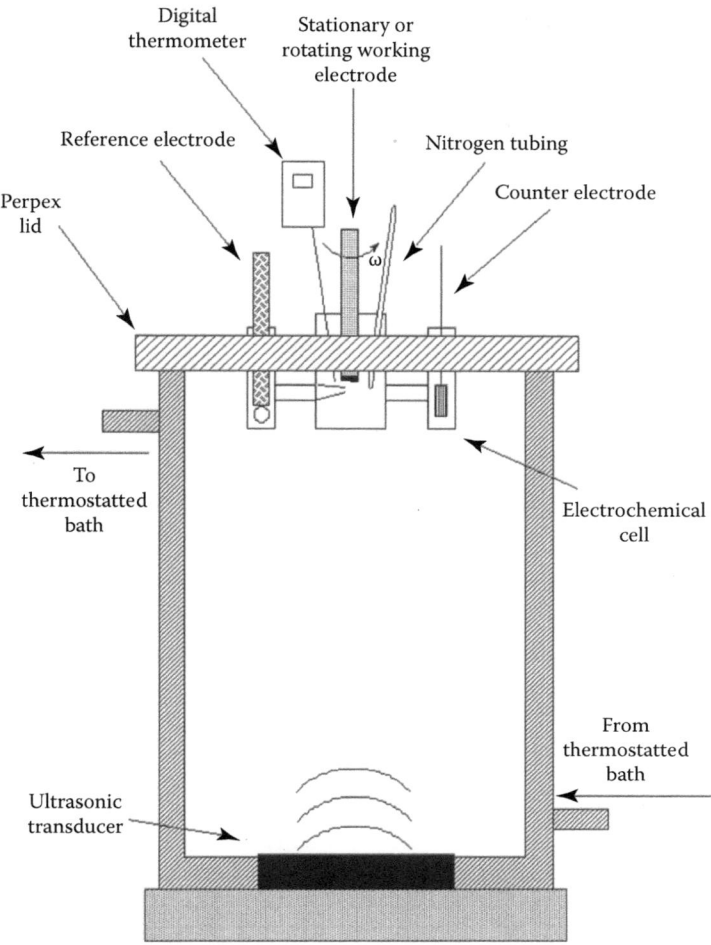

FIGURE 24.3 A 500 kHz ultrasonic bath arrangement used for sonoelectrochemical experiments.

are commonly performed using an ultrasonic bath or a probe arrangement. These are described in detail below.

1. *Bath arrangements*: The 40 and 500 kHz Ultrasonic Arrangements-For the 40 kHz (e.g., Kerry Ultrasonic Bath, Figure 24.2b) and 500 kHz (e.g., Undatim Ultrasonic Bath, Figure 24.3) ultrasonic bath arrangements, the sonoelectrochemical experiments are performed using a three-electrode electrochemical cell (see below). All three-electrode electrochemical cells employed have flat bottoms to maximize energy transfer. For example, energy is radiated vertically as sound waves from the base of the bath and through the glass walls of the cell into the electrochemical reaction itself. Thus it is more effective to employ a flat base for the cell allowing a greater transfer of ultrasonic energy. The base area of these cells is of importance as it allowed deduction of the ultrasonic intensity (power *per* tip or transducer or cell base area).

   The 800 kHz ultrasonic arrangement: For the 800 kHz (e.g., K.W. Meinhardt Ultrasonic Bath, Figure 24.4) ultrasonic bath arrangement, sonoelectrochemical experiments are performed using the ultrasonic bath as the electrochemical cell with the three electrodes placed directly into the cell. As previously, the reference is separated (approximately 1 mm) from

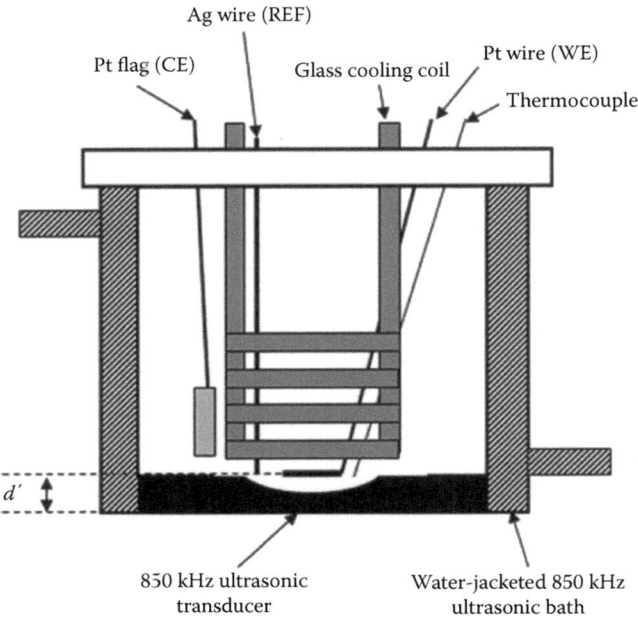

FIGURE 24.4 A 800 kHz ultrasonic bath arrangement used for sonoelectrochemical experiments.

the working electrode surface in order to minimize any rI problems. For this type of cell geometry setup, a coil or a mesh counter electrode was used. The concave transducer area is 12.64 cm².

2. *Ultrasonic probe arrangements*: Electrochemical cells used in sonoelectrochemistry—Four specially designed sonoelectrochemical cells are commonly used in sonoelectrochemistry. The use of these different cells enables different sample volumes and geometric configuration to be employed.

The three-electrode voltammetric compartment cell—Most of the sonoelectrochemical experiments are performed in a three-compartment cell (Figure 24.5). The working, counter, and reference electrodes are placed in the central and side compartments respectively. The total capacity of the cell is approximately 100 cm³ with the working electrode compartment cell being approximately 60 cm³. This cell can only be used in an ultrasonic bath.

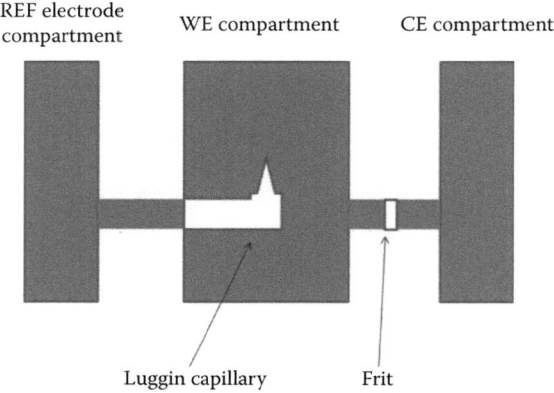

FIGURE 24.5 A three-compartment cell used in electrochemical and sonoelectrochemical experiments.

Other Sonoelectrochemical Cells

Oxford Cell

This sonoelectrochemical cell was originally designed by Compton et al. (1995) (Figure 24.6) and is used to investigate the effect of the ultrasonic horn–working electrode separation, cell geometry, and sonoelectrochemical systems. It consists of a 250 cm^3 one-compartment pyrex cell with an aperture at the bottom for electrode or probe insertion. An insulated cooling stainless steel coil linked to a thermostatically controlled bath is inserted into the electrochemical cell for cooling purposes. The ultrasonic probe and the working electrode are either inserted at the bottom of the cell or placed at the top of the cell in order that their surfaces are always parallel to each other. This geometry is known as the "face-on" geometry (Compton et al., 1994, 1995, 1997a, 2003).

Pollet Cell (The SonoEcoCell)

This new sonoelectrochemical cell was developed by Pollet et al. (1999, 2000) with a maximum volume of 1.2 dm^3 (Figure 24.7) for the heavy metals sonoelectrodeposition work undertook at Coventry University in the late 1990s. The cell consists of a stainless steel cylinder with three apertures, two on opposing sides of the wall and the third at the base thereby allowing access of three ultrasonic probes. An outer copper cooling jacket is fitted through which thermostatted water from

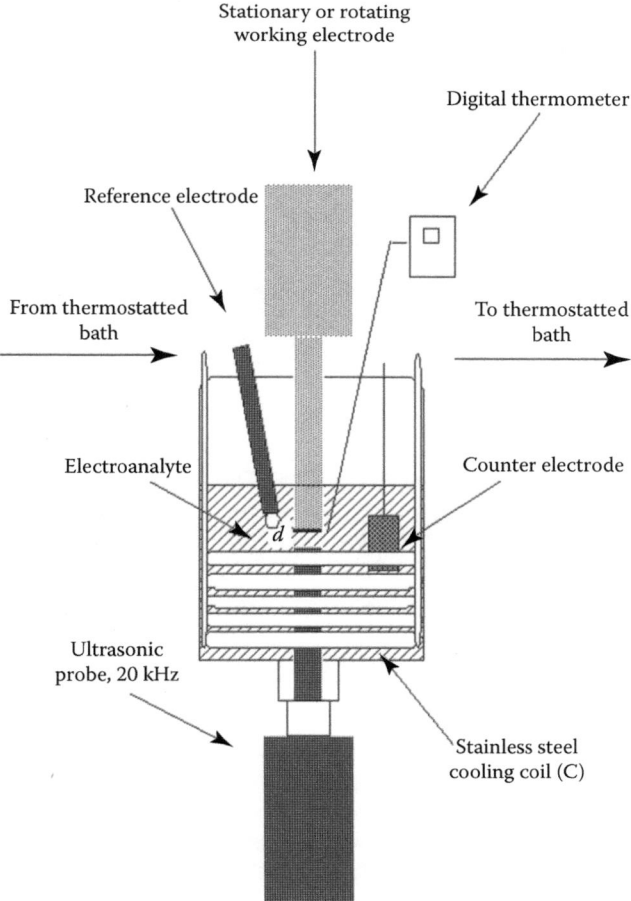

FIGURE 24.6 The Oxford cell arrangement with the exception of the ultrasonic source (sonic and materials VC50 20 kHz probe) at the bottom.

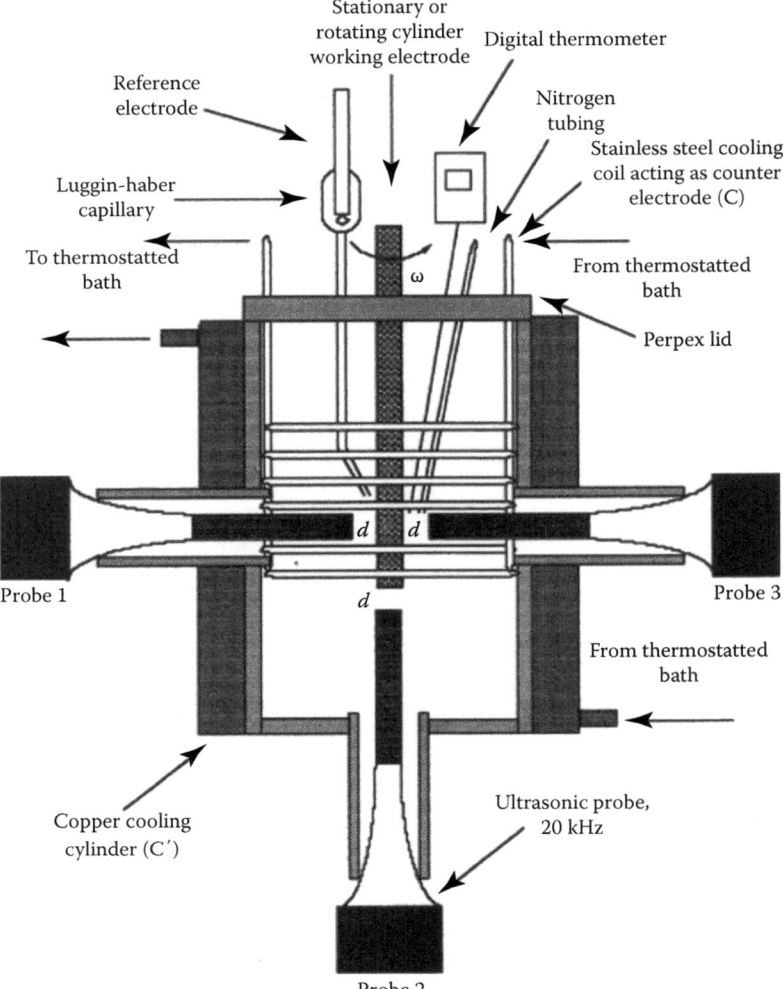

FIGURE 24.7 The Pollet cell—*SonoEcoCell* fitted with three ultrasonic probes (Vibra-Cell VC600, 20 kHz) for the electrolysis of metals from environmentally toxic samples.

the thermostatically controlled bath flowed to maintain the electrolyte at fixed temperature. The inner wall of the cylinder and the base is covered with Teflon spray. The lid consists of perspex disc with holes for the working, counter, and reference electrodes, nitrogen tubing, K-type thermocouple, and sample injection if required.

Besançon Cell (The Double Wall Cell)

This new sonoelectrochemical cell was developed by Hihn (Costa et al., 2008) for use in aqueous and "exotic" solvents at various coolant pressures (Figure 24.8). This jacketed cooling "microsonoreactor" is based on a particular design consisting of offsetting the ultrasonic probe out of the reaction volume (inner cell, $V_{ic} = 7\,cm^3$) in order to avoid any possible contamination and to ensure perfect electric insulation from the ultrasonic probe, which is usually made of Ti alloy (Ti-6Al-4V) (Figure 24.8). In order to obtain maximum ultrasonic effect, that is, ultrasonic intensity within the inner reaction volume, it is necessary to increase the coolant pressure up to 1.5 bar. Figure 24.9 shows the linear relationship between the ultrasonic (20 kHz) power and the coolant pressure and shows that it is possible to obtain "acceptable" ultrasonic powers to undertake sonoelectrochemical experiments.

Sonoelectrochemistry

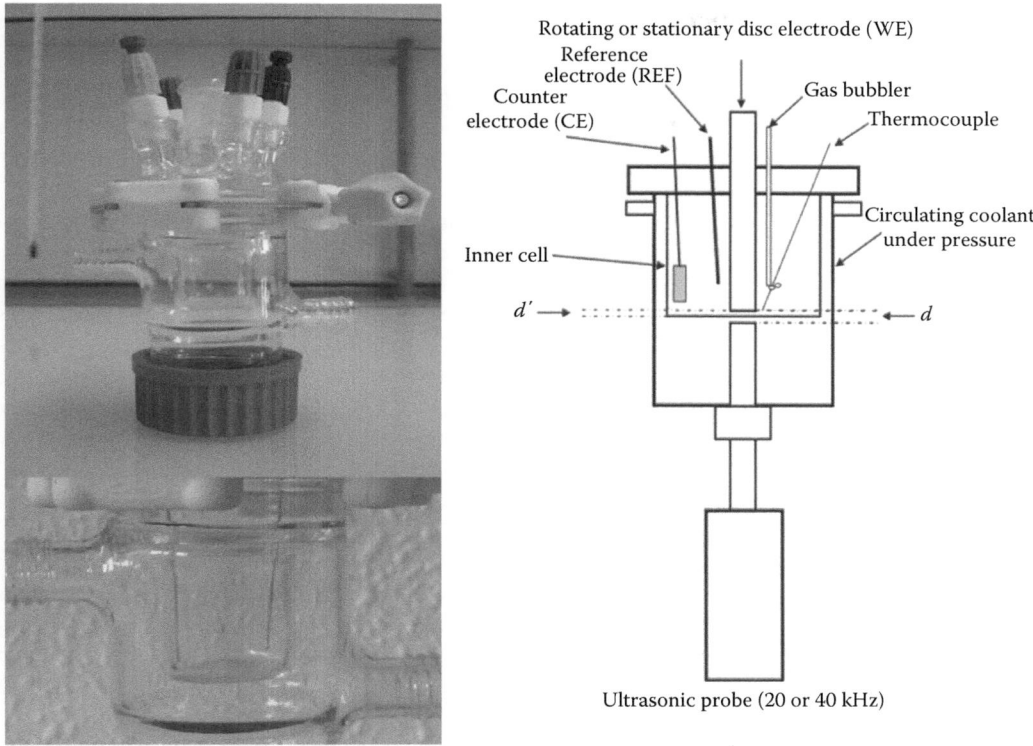

FIGURE 24.8 The Besançon cell—the double-structured cell used in sonoelectrochemistry of metals in RTILs and DESs.

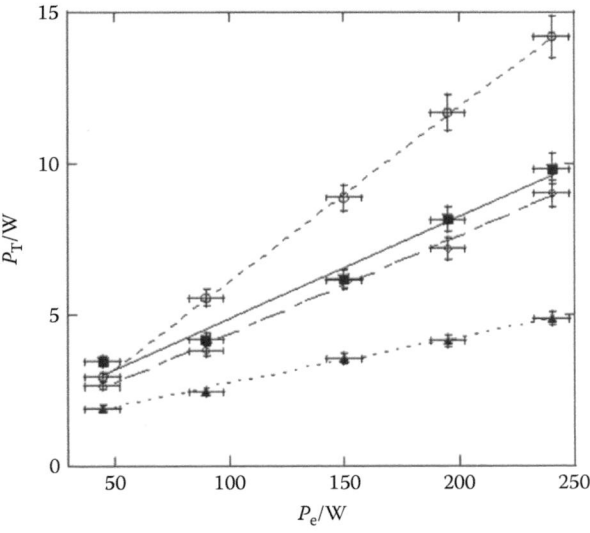

FIGURE 24.9 Plot of ultrasonic power (P_T) vs. electric output (P_e) for various overpressures (▲ 0 bar, ◊ 0.5 bar, ■ 1 bar and □ 1.5 bars) for 7 ml of argon-saturated water at 20 kHz (Costa et al., 2008).

USE OF ULTRASOUND IN ELECTROCHEMISTRY

The use of ultrasound in electrochemistry was initially developed and used in the metallurgical industry primarily for surface cleaning prior to electrodeposition. Recent studies have demonstrated that simultaneous ultrasonic irradiation of electrochemical systems can alter limiting parameters and significantly improve electrochemical reactions and techniques. These improvements include: enhanced diffusion processes, increased yields and current efficiencies, lower overpotentials, suppressed electrode fouling, and alteration of reaction mechanisms. There may be different origins for a variety of these effects, but one well-characterized effect of ultrasonic irradiation is the generation and subsequent collapse of cavitation bubbles in an electrochemical cell, which may be important both within the electrolyte medium and near to the electrode surface. The electrode surface causes asymmetrical collapse of a bubble, which in turn leads to the formation of a high velocity jet of liquid being directed toward the surface. This jetting, together with microstreaming, is thought to lead to the thinning of the mass transfer boundary layer at the electrode. This improves the overall mass transfer of the system and, as a consequence, the electrode reaction rates. This is experimentally observed by an increase in current and a modification in potential.

Many of the observed effects in sonoelectrochemistry may be explained by the enhancement of mass transport in diffusion-controlled processes. The extensive work of Coury and coworkers (Hagan et al., 1994; Compton et al., 1994; Lorimer et al., 1996; Pollet, 1998) were probably the first "modern" examples investigating mass transfer phenomena under sonication. Power ultrasound is known to decrease the diffusion layer thickness (δ) thereby giving substantial increase in limiting current (I_{lim}) attributed due to effects of cavitation and/or micro and macro-streaming. It is known in the field that both cavitational and acoustic streaming effects contribute significantly to the increase in observed experimental currents.

The experimental decrease in the diffusion layer thickness (<1 μm) is also known to be due to asymmetrical collapse of cavitation bubbles at the electrode surface leading to the formation of high velocity jets of liquid (duration of ca. 0.5–0.7 μs) being directed toward its surface. This jetting, together with acoustic streaming, is thought to lead to random punctuation and disruption of the mass transfer boundary layer at the electrode surface at close electrode-to-horn separations. Birkin and Silva-Martinez (1997) also showed that the nature of the solvent is paramount in assigning limiting currents. More recently, Pollet and Hihn (Pollet et al., 2007) showed, with aid of mathematical models based on mass-balance equations, that a Levich-like equation relating the limiting current density, the square root of ultrasonic intensity, and the inverse square root of the electrode-horn distance, may be generated for ultrasonic frequencies of 20 and 40 kHz allowing the generation of an "equivalent" flow velocity under sonication, an important and useful parameter in chemical engineering. They also observed that jets of liquid could reach the electrode surface with velocity above 50–200 m s^{-1}.

ULTRASOUND EFFECT ON ELECTRODE-KINETICS

Several workers have shown that ultrasound enhances mass-transport processes (Inazu et al., 1993; Zhang and Coury 1993; Hagan et al., 1994; Klima et al., 1994; Marken et al., 1995; Eklund et al., 1996; del campo et al., 1999; González-García et al., 2010). Compton et al. (1995, 1997a,b, 2003) have observed sigmoidal shape voltammograms and named these new shapes "sonovoltammograms" for quasi-reversible systems (Figure 24.10) (Costa et al., 2010). They concluded that sonovoltammetry is closely related to hydrodyamics methods, for example, polarography, dropping mercury electrode (DPE), rotating disc electrode (RDE), wall-jet electrode (WJE) and showed that Equation 24.1 (Compton et al., 1997a), which relates the limiting current (I_{lim}) to the diffusion layer thickness (δ), still applies in the presence of ultrasound:

$$I_{lim} = \frac{nFAD_0C^*}{\delta} \qquad (24.1)$$

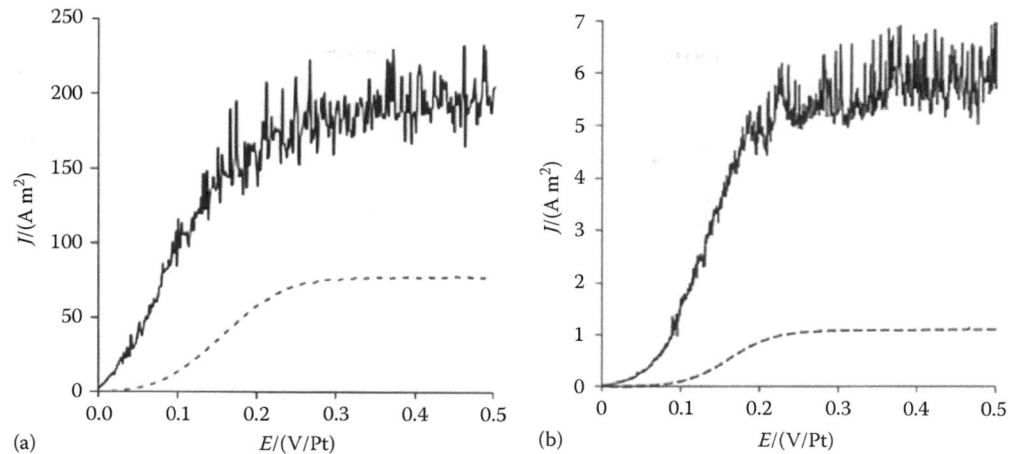

FIGURE 24.10 Anodic voltammograms of Fc(3 mM) at 293 K on a platinum disc electrode. (a) Acetonitrile, (b) [Omim] [Tf$_2$N]: silent conditions + rotating disc electrode at 5000 rpm (dot lines); Under ultrasound (f = 40 kHz; It = 1.1 W.cm^{-2}) (Straight line). (From *Ultrason. Sonochem.*, 18, Taouil, A. Et, Lallemand, F., Hihn, J.Y., Melot, J.M., Blondeau-Patissier, V. and Lakard, B., Doping properties of PEDOT films electrosynthesized under high frequency ultrasound irradiation, 140–148, Copyright (2011), with permission from Elsevier.)

where
I_{lim} is the limiting current
n is the number of electrons transferred during the electrochemical process
F is the Faraday constant
A is the electrode area
D_0 is the diffusion coefficient of the electroactive species
C^* is the bulk concentration of the electroactive species
δ is the diffusion layer thickness

In contrast to any conventional hydrodynamics responses (e.g., S-shaped voltammograms by RDE), sonovoltammograms usually exhibit "noisy" responses or a series of individual current-pulse responses in the mass-transport region of the voltammograms (Figure 24.10) corresponding to an intense transient cavitation in the vicinity of the electrode surface caused by individual collapse of cavitation bubbles. In viscous electrolytes, stirring ability of ultrasound is considerable.

Compton et al. (2003) also demonstrated that under sonication, millielectrodes (i.e., electrodes with radii larger than 1 mm) behave as microelectrodes (i.e., radii smaller than 1 mm) (Compton et al., 2003). They found that limiting currents obtained with a microelectrode in the absence of ultrasound were comparable with those obtained with a millielectrode in the presence of ultrasound. However, controversy exists in identifying if the steady-state voltammogram obtained in the presence of ultrasound is due only to cavitation effects, or to micro-jetting and acoustic streaming, or to a combination of the three.

An early report by Pollet and Hihn (Pollet et al., 2007), showed that, for a "face-on geometry" in the presence of ultrasound, the limiting current (I_{lim}) is inversely proportional to the horn-electrode distance (d), the electrode geometry (r_e), and is proportional to the ultrasonic intensity (Ψ) (at constant temperature, concentration, etc.) according to Equation 24.2 also known as the *Pollet equation*:

$$I_{lim} = 0.84 nFAD_0^{2/3} n^{-1/6} r_e^{-0.5} d^{-0.5} \Psi^{0.5} C^* \qquad (24.2)$$

where

I_{lim} is the limiting current
n is the number of electrons transferred during the electrochemical process
F is the Faraday constant
A is the electrode area
D_0 is the diffusion coefficient of the electroactive species
d is the horn-electrode distance
r_e is the electrode radius
ν is the kinematic viscosity
C^* is the bulk concentration of the electroactive species

They showed that the electrode-ultrasonic horn positioning is an important factor in assigning hydrodynamics parameters such as the limiting current (I_{lim}), the mass transport coefficient (m_o), the limiting solution velocity (U_{lim}), the Sherwood (Sh), and the Reynolds (Re) numbers.

Pollet et al. (2003) also showed that the geometry of the sonoelectrochemical set up affects mass-transport phenomenon. For example, Pollet investigated the effect of changing the position of an ultrasonic horn tip (i.e., vertical and horizontal) and a cylinder electrode (CE) on limiting currents [for the reduction of Ag(I) in sodium thiosulfate], in an attempt to find the optimum position required for maximum sonoelectrochemical effect. The importance of the ultrasonic intensity, the electrode-horn distance and positioning (angle) in assigning limiting currents was also investigated. For the cylinder electrode (CE) placed at an angle of 45° with respect to the ultrasonic horn, it was suggested that the 50% increase in limiting current for the "face-on" geometry is caused by an approximately 50% decrease in diffusion layer thickness for the "face-on" geometry compared to the "angular" geometry due to the difference in the sonicated areas for both geometries.

Figure 24.11c shows the limiting currents plotted against the square root of the emitted ultrasonic intensity ($\Psi^{1/2}$) for both the "face-on" and "angular" geometry. This figure shows linear relationships between the limiting current and $\Psi^{1/2}$ for the two electrode configurations. This observation suggests, as for the disc electrode (DE), that limiting currents obtained on a sonicated (CE) are also related to $\Psi^{1/2}$. In other words, whatever the positioning of the electrode, a square root dependence of ultrasonic intensity relationship is obtained for the two geometries employed. The figure also shows that, for an identical ultrasonic intensity transmitted to the total volume, the limiting current obtained for the "angular" geometry is much lower than that obtained for the "face-on" geometry. For example, at 30 W cm^{-2} power output, the limiting current obtained for the "face-on" geometry (I_{lim} = 150.4 mA) is 1.5 times higher than that obtained for the "angular" geometry (I_{lim} = 101.5 mA). Furthermore, in our conditions, this would correspond to an equivalent "effective rotation speed" of 15,000 and 8,500 rpm for the "face-on" and "angular" geometries respectively at maximum power (30 W cm^{-2}). These observations indicate that the positioning of the electrode with respect to the source of ultrasound is an important parameter in assigning ultrasonic hydrodynamics.

To account for the difference between the "face-on" and "angular" geometries, there are three possible explanations, which are detailed below. Please note that the following subscripts are used for the ultrasonic intensity (Ψ): x and y are spacial components, 0 indicates the ultrasonic intensity at $x = 0$ or/and $y = 0$, f.o. the "face-on" geometry, and ang. the "angular" geometry.

1. *Influence of the incident flow*: The first possible explanation is to consider the two geometries and resolve the flows into the component parts (see Figure 24.11a and b). For example, it is assumed that the total ultrasonic intensity, Ψ is given by Equation 24.3

$$\Psi = \Psi_x + \Psi_y \tag{24.3}$$

Sonoelectrochemistry

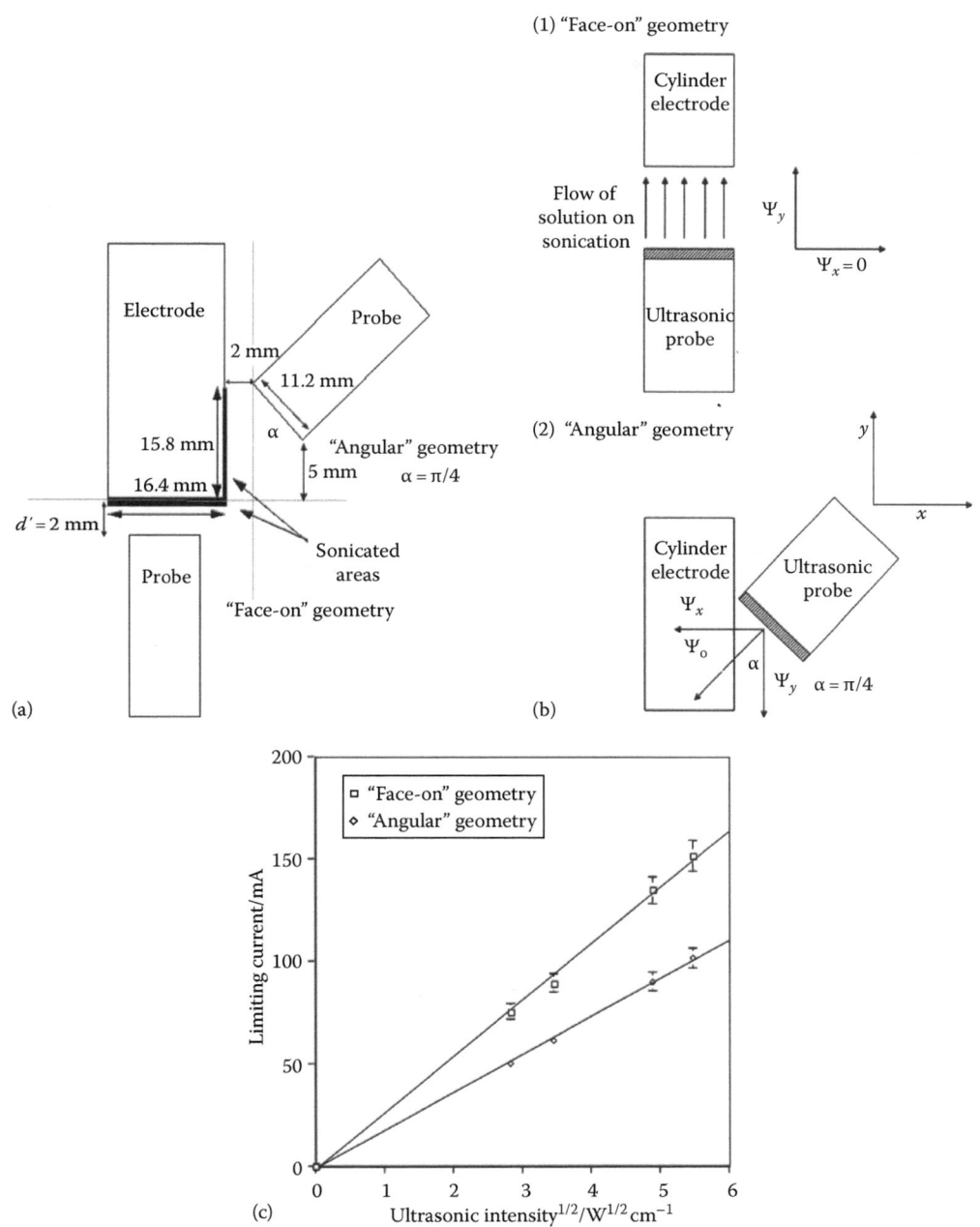

FIGURE 24.11 (a) Two geometries employed during this study—the "face-on" and "angular" geometries. The figure shows the two sonicated areas for both geometries; (b) Co-ordinate system used to describe the "face-on" geometry and "angular" geometry in terms of ultrasonic intensity; (c) Limiting currents plotted against the square root of ultrasonic intensity (20 kHz) for the reduction of silver (4 g dm^{-3}) on a sonicated cylinder electrode at (298 ± 1) K for the "face-on" and "angular" geometries.

It has also been demonstrated that the ultrasonic intensity within a fluid depends on distance d travelled thought that fluid with an absorption coefficient, α, according to Equation 24.4:

$$\Psi = \Psi_o \exp(-2ad') \tag{24.4}$$

Using $8.6 \times 10^{-8}\,\text{cm}^{-1}$ for the absorption coefficient of water at 298 K and at 20 kHz, Equation 24.4 leads to $\Psi = \Psi_o$ (incident ultrasonic intensity) in the range 2–10 mm electrode-probe distances. Thus, if the acoustic absorption is neglected, Equation 24.4 may be reduced to Equation 24.5:

$$\Psi_o = \Psi_{o,x} + \Psi_{o,y} \tag{24.5}$$

For the "face-on" geometry (see Figure 24.11), the ultrasonic intensity components are in the x and y directions, that is, $\Psi_{o,x,f.o.}$ (=0) and $\Psi_{o,y,f.o.}$ (=Ψ_o), respectively.

For the "angular" geometry (see Figure 24.11b), the ultrasonic intensity components are in the x and y directions, that is, $\Psi_{o,x,ang.} = \Psi_o \sin(45°) = 0.707\Psi_o$ and $\Psi_{o,y,s.o.} = \Psi_o \cos(45°) = 0.707\Psi_o$, respectively.

The ratio of the ultrasonic intensity on the y-axis for the two geometries gives Equation 24.6

$$\frac{\Psi_{o,y,f.o.}}{\Psi_{o,y,ang}} = \frac{\Psi_o}{0.707\Psi_o} = 1.41 \tag{24.6}$$

Assuming that the limiting current (I_{lim}) is proportional to the square root of the ultrasonic intensity ($\Psi_o^{1/2}$) (see Figure 24.11c), Equation 24.6 may be reduced to Equation 24.7:

$$\frac{I_{lim,f.o.}}{I_{lim,ang}} = \sqrt{1.41} = 1.18 \tag{24.7}$$

While the value of 1.18 can account in part for some of the observed 1.5 value for $I_{lim,f.o.}/I_{lim,ang.}$, it cannot account for it all and again suggests that the ultrasonic probe positioning is crucial for the determination of limiting currents.

For both cases, "face on" and "angular" geometries, the current density is variable over the total electrode area and therefore simple mathematical treatments and equations have obvious limitations. It is interesting to note that the electrode area in both cases may be regarded as two areas (1) a high current density area and (2) a low current density area. Ideally, it would be useful to treat the resulting system using Equation 24.6 containing two terms at high current density and low current density. However, low current density area is difficult to evaluate because of the complexity of the cylinder electrode and the "angular" geometries.

2. *Influence of hydrodynamics on the diffusion layer thickness*: This difference in limiting current between the two geometries may be attributed to a greater thinning of the diffusion layer in the case of the "face-on" geometry where electroactive species reaches all points of the electrode uniformly. In the "angular" geometry, the diffusion layer is nonuniform and roughly resembles that of a channel electrode. Further, it is interesting to note that the sonicated areas for the two geometries are different (see Figure 24.11a). It was deduced that the sonicated area for the "face-on" geometry is $2.13\,\text{cm}^2$ (CE base area) and for the "angular" geometry is $2.65\,\text{cm}^2$ (ellipsoid area). In order to compare diffusion layer thicknesses for both "face-on" and "angular" geometries, a hydrodynamic study can be made as follows:

For the "face-on" geometry, the limiting current is given by Equation 24.8

$$I_{lim,f.o.} = \frac{nFA_{f.o.}DC^*}{\delta_{f.o.}} \tag{24.8}$$

and for the "angular" geometry, the limiting current is given by Equation 24.9

$$I_{\text{lim,ang}} = \frac{nFA_{\text{ang}}DC^*}{\delta_{\text{ang}}} \tag{24.9}$$

provided that the non-sonicated areas of the electrode for both the "face-on" and "angular" geometries are not considered.

The ratio of the limiting currents for the two geometries gives Equation 24.10

$$\frac{I_{\text{lim,f.o.}}}{I_{\text{ang}}} = \frac{A_{\text{f.o.}}\delta_{\text{ang}}}{A_{\text{ang}}\delta_{\text{f.o.}}} \tag{24.10}$$

[Assuming that the diffusion coefficients (D) are similar for both geometries.]

Inserting experimental values of limiting current density and surface area for both geometries in Equation 24.10, a ratio of diffusion layer thicknesses at 30 W cm^{-2} power output may be deduced from Equation 24.11:

$$\frac{\delta_{\text{ang}}}{\delta_{\text{f.o.}}} = \frac{I_{\text{lim,f.o.}}A_{\text{ang}}}{I_{\text{lim,ang}}.A_{\text{f.o}}} = \frac{150.4 \times 2.65}{101.5 \times 2.13} = 1.84 \tag{24.11}$$

Thus, the overall "average" diffusion layer thickness is 1.8 times thinner for a "face-on" geometry than a "angular" geometry under identical conditions.

3. *Influence of localized temperature variations on diffusion coefficients*: The observed difference in limiting current between the two geometries may also be attributed to a change in diffusion coefficients due to localized temperature variations. Assuming that the diffusion coefficient for the "face-on" ($D_{0,\text{f.o.}}$) and the "angular" ($D_{0,\text{ang}}$) geometries are different and assuming that the diffusion layer for the "face-on" geometry is "hotter" than that for the "angular" geometry [e.g., 400 K for the "face-on" geometry [value deduced in Lorimer et al. (1998)] compared to 298 K for the "angular" geometry], it is possible to deduce the ratio of the diffusion coefficients of silver for the two geometries at maximum power (30 W cm^{-2}) (Equation 24.12)

$$\frac{D_{0,\text{f.o.}}}{D_{0,\text{ang}}} = \frac{I_{\text{lim,f.o}}}{I_{\text{lim,ang}}} = \frac{6.17 \times 10^{-5}}{1.64 \times 10^{-5}} = 3.76 \tag{24.12}$$

Since, in our conditions, the ratio of the limiting currents for the two geometries for 30 W cm^{-2} power output, which is given by Equation 24.13 is,

$$\frac{I_{\text{lim,f.o.}}}{I_{\text{lim,ang}}} = 1.5 \tag{24.13}$$

This suggestion may be dismissed. [The diffusion coefficient of silver at 400 K was deduced by assuming linearity of the plot of ln (D) vs. $1/T$. The equation of the line was in the form of ln (D) = $-1553/T - 5.8$. Values of D were obtained from the rotating disc experiments at 298, 310.5, and 323 K.] However, it is possible to deduce an "apparent" temperature by deducing an "apparent" diffusion coefficient for the "face-on" geometry. Assuming that the ratio of the limiting currents for both geometries is true (Equation 24.13), a value of the "apparent" diffusion coefficient may be deduced by employing Equation 24.12:

$$D_{\text{f.o}} = \frac{I_{\text{lim,f.o.}}}{I_{\text{lim,ang.}}} D_{\text{ang}} = 1.5 \times 1.62 \times 10^{-5} = 2.43 \times 10^{-5} \text{ cm}^2 \text{ s}^{-1} \tag{24.14}$$

[value of $D = 1.62 \times 10^{-5}$ cm^2 s^{-1} at 298 K].

Thus, by assuming linearity between ln (D) vs. $1/T$, the "apparent" temperature was deduced to be approximately 350 K. This result suggests that if the 1.5-increase in limiting current for the "face-on" geometry compared to the "angular" geometry was solely due to an increase in diffusion coefficient for the "face-on" geometry, this would lead to a bulk temperature of approximately 350 K, a value, which was never achieved under sonication at 298 K. However, it is interesting to note that this result compares favorably with the temperature of the diffusion layer for the "face-on" geometry, which is 400 K [value deduced in Lorimer et al. (1996, 1998)].

Apart from the remarkable effects of ultrasound on mass-transport processes, it has been shown that sonication alters the electrode potential. Several workers such as Moriguchi (1934), Lorimer et al. (1998), and Pollet et al. (2002) showed that the overpotential of hydrogen evolution on gold and platinum electrodes decreases in a sonicated environment. This finding was later observed by the Compton group on other metals a few years ago (Hyde and Compton, 2002). Pollet (1998) speculated that this decrease in overpotential was due to both a reduction in concentration gradients and in nucleation overpotential at the electrode surface and in the removal of adhered hydrogen bubbles and adsorbed materials. Even now, some controversy still surrounds the effect of ultrasound upon electrode kinetic parameters such as the *formal* (E^o) and half-wave ($E_{1/2}$) potentials and apparent heterogeneous rate constant (k_o), which are related to equilibrium potentials.

However, a recent detailed study by Pollet et al. (1998) showed that ultrasound affects heterogeneous electron-transfer kinetic parameters of a typical quasi-reversible system provided the electrode receives a strict cleaning procedure (for surface reproducibility) prior to any sonoelectrochemical experiments. It was shown that the half-wave potential ($E_{1/2}$) shifts cathodically and the apparent heterogeneous rate constant (k_o) increases with increasing ultrasonic intensity. Also, it was shown these parameters appear to be little affected by the frequency of simultaneous ultrasonic irradiation in the range 20–800 kHz, and is not influenced by choice of ultrasonic bath or probe as sonic source, provided measurements are made at constant ultrasonic intensity.

All these aspects at low and high ultrasonic frequencies have been reviewed in 2003 by Compton et al. (2003) and references therein. The authors concluded that the electrochemical processes in the presence of high-frequency ultrasound are governed by processes (microjetting and micromixing) considerably different from those that are important at lower frequencies (acoustic streaming).

SONO-ELECTRODEPOSITION AND SONO-ELECTROPLATING

Research has been extensively performed in the field of electrodeposition and electroplating. Several workers such as Kochergin and Vyaselva (1966) and Walker (1993) first reported the beneficial effect of ultrasound in metal deposition and plating. They found that one of the main advantages of ultrasound is the improvement of the deposition rate, the quality of the electrodeposit, and the electrode surface cleanliness.

Other workers (Kochergin and Vyaselva, 1966) in the electroplating industry have also shown that ultrasound offers several advantages when applied to electrodeposition. These advantages are the following: (1) increase in thickness of the electrodeposit, (2) increase in cathodic efficiency and improved porosity (3) increase in hardness and, (4) decrease in concentration overpotential.

1. *Increased electrodeposit thickness*: Ultrasound reduces the thickness of the diffusion layer as it produces surface cavitation, microjetting, and acoustic streaming. These three effects are known to assist the mass transfer of electroactive species from the bulk solution to the electrode surface. Thus, under these conditions, more electroactive species are depleted at the electrode surface which, in electrodeposition terms, means an increase in coating thickness.
2. *Increased cathodic efficiency and improved porosity*: Cathodic efficiency is an important parameter in the electroplating industry since it indicates the percentage of current used

for the electrodeposition of the metal. It is common that a cathodic efficiency of less than 100% is obtained. This is due either to the electrolysis of the background electrolyte (e.g., production of hydrogen) or the formation of a film at the electrode surface (e.g., metal oxide). It has been shown (Walker (1993)) that ultrasound is capable of removing hydrogen bubbles or oxide films present at the electrode surface, which may have shielded local sites required for electrodeposition thereby giving increased efficiency. Apart from an increase in cathodic efficiency, the porosity of the electrodeposit is improved in the presence of ultrasound Walker (1993). This is also due to the removal of these hydrogen bubbles within the electrodeposit by the collapse of cavitation bubbles and the intense acoustic streaming at the electrode surface.

3. *Increased hardness*: It has been demonstrated (Walker, 1993) that ultrasound increases the hardness of the electrodeposit due to the formation of small grain size. It was also shown (Walker, 1993) that in the presence of ultrasound the electrodeposit is more compact. This leads to an increase in dislocation density, that is, the electrodeposit is denser.

4. *Decreased concentration overpotential*: Concentration overpotential is an important factor for electrochemical reactions. This is due to concentration changes in the near vicinity of the electrode surface. It has been shown (Walker, 1993) that ultrasound reduces concentration overpotential, that is, decreases the solution resistance. This is due to an increase in mass-transport of metal ions from the bulk solution to the electrode surface.

Apart from the electrolytic deposition and plating of metals under insonation, many workers have explored the use of ultrasound in electroless metal plating on nonconductive substrates. For example, Touyeras et al. (2001, 2005) studied the electroless copper coating of epoxy plates in the presence of ultrasound (530 kHz). They found that ultrasound can increase the plating rates and practical adhesion up to 30% with an obvious decrease in internal stress. They also observed that there was some change of plating mechanism induced by sonication.

The same researchers also studied the electroless of Sn and Pd plating on nonconductive substrates under ultrasound at 530 kHz, whereby the ultrasonic irradiation was applied to the activation and to the plating steps. Effects were measured by following the final copper thickness obtained in 1 h of plating time, easily correlated to the average plating rate. They observed that ultrasound has a strong influence on the plating rates enhancement, and assumptions were made that this increase could be linked to the catalyst cleaning, which was confirmed by XPS measurements.

Recently, Touyeras et al. (2005) investigated the effects of ultrasound on electroless copper coating on both nonconductive and metallic substrates. In these experiments, the ultrasonic irradiation was both applied during activation (surface preparation for the electroless coating) and during plating steps in both cases. Several parameters were monitored, such as the plating rates, practical adhesion, hardness, internal stress vs. acoustic frequencies and powers. Optimum conditions for irradiation time, frequency, and power were determined for each step and it was clearly shown that ultrasound affects the electrodeposit properties. Coating mechanisms were put forward to explain enhancement of electrodeposit properties such as: increase in catalyst specific area, stirring dependence, surface energy evolution, hydrogen desorption, and structure of coating.

SONOELECTROCHEMISTRY AND CORROSION

Power ultrasound operating at 20 kHz has been used to investigate the influence of ultrasonic irradiation on several metals immersed in corrosive environments such as sodium chloride saline solutions (Morais and Brett, 2002; Gouveia-Caridade and Brett, 2004). The metals included three types of steel [carbon steel, chromium steel, and a high speed steel (containing several alloying elements besides chromium] as well as pure and commercial (copper-containing) aluminum. These metallic systems were selected in order to evaluate the influence of alloying elements as well as that of the main metallic elements, Al and Fe. In these experiments, the ultrasonic probe was placed directly above the metal surface and the effect of varying the probe-electrode distances ("face-on" geometry) and ultrasonic power was investigated. Three electrochemical techniques were used to evaluate the corrosion process and its rate—measurements of open circuit potentials (OCPs), polarization curves, and electrochemical impedance spectroscopy, before during and after insonation in the presence of dissolved oxygen. The corroded surfaces were examined by SEM after ultrasonic irradiation. For both types of aluminum—pure and commercial—which easily form a protective oxide layer, it was found that sonication led to the mechanical removal of the oxide layer and formation of oxychloro complexes; the oxide layer was then reformed with essentially the same characteristics as previously, independent of any pitting corrosion.

However, the steels revealed a different behavior, owing to the fact that significant pitting corrosion occurred in chloride solution since the oxide layer is not as robust as in the case of Al. Insonation led to an enhancement of pitting and the corrosion process, which was limited by the access of the cathodic reagent, in this case, oxygen. It was also found that the carbon steel had a higher corrosion resistance than in *silent* conditions, while that of the chromium and high speed steels exhibited a lower corrosion resistance. However, in all steels the irradiation led to permanent effects.

These results were generated in different accelerated corrosion tests by the Besançon group in various situations. For example, the corrosion resistance of new zinc coatings is commonly evaluated by long-term atmospheric exposure or by accelerated corrosion tests. This is particularly difficult when exposed to marine, rural, and urban and industrial atmospheres. If accelerated corrosion tests performed in specially designed electrolytes give satisfactory results as far as the nature of the zinc corrosion end-products and the corrosion sequences involved are concerned, the behavior of the surface is not acceptable (Ligier et al., 1999).

In the presence of electrolytes, zinc end-products grow in fissures generated on the zinc surface and then cover the whole zinc surface whereas on atmospheric zinc corrosion shows that atmospheric zinc corrosion end-products grow on the zinc surface without any fissures. Atmospheric metal corrosion can be considered as the result of chemical action by atmospheric pollutants and the more physical action of surface erosion by natural precipitation. This mechanical action must be added to the abrasive action of solid particles hitting the metal surfaces due to wind action. When dealing with electrolytes, an important way of inducing mechanical and even chemical changes is the use of an ultrasound source. Ligier et al used two ultrasound frequencies, 20 and 500 kHz, and showed a great improvement in term of corrosion form as well as corrosion rates (Ligier et al., 2001). On another hand, accelerated corrosion tests using low frequency ultrasound have been performed to simulate corrosion conditions in the cold end of an automotive exhaust system [middle range 1.4512 (AISI 409) stainless steel]. The conventional tests "Dip Dry test (DDT)" was compared to a similar test, but including an additional external stress thanks to an ultrasonic transducer. This new ultrasonic test (so called UST) appears to reduce the diagnostic time of the corrosion test by the combined action of the chemical corrosion process and of mechanical degradations. Both corrosion tests have been performed in two different media in order to simulate internal corrosion due to exhaust gas condensate and external cosmetic corrosion, greatly enhanced by road salt during winter. It respectively concerns a synthetic gas condensate, the composition of which is derived from what is obtained from motor gasoline combustion and NaCl solution. In both electrolytes the stainless steel suffers from pitting corrosion. As expected, the use of ultrasound allows pits growth to be achieved from the beginning of the exposure time, so that the maximum pit depth recorded after 180 h of immersion is twice than with the classical dip dry test. It seems that it does not modify the pit initiation mechanism but only increases growth kinetics (Doche et al., 2006).

SONO-ELECTRO-ORGANIC CHEMISTRY

Ultrasound in electro-organic chemistry or sonoelectro-organic chemistry has been less widely investigated than either the ultrasonic effect on electrode-kinetics or the sonoelectrodeposition of metals, with only a few reports prior to the 1990s. However, there has been an increasing upsurge of interest, and this new subject area has become an exciting and promising topic of sonoelectrochemistry (Luche, 1993; Walton, 1991).

In the mid-1980s, electro-organic systems such as the electrosynthesis of organoselenium and organotellurium were shown to be enhanced by sonication [Walton and Phull (1996) and references therein]. Ultrasound was found to offer several advantages: (1) no sacrificial cathode of elemental selenium or tellurium was required and (2) the production of the corresponding selenium and tellurium anions were obtainable.

Other workers (Gautheron et al., 1985; Klima et al., 1995; Gautheron et al., 1986) showed that the production of organosilanes and germanes, as polymeric systems, by electroreduction of the appropriate halospecies at a reactive metal cathode in aprotic media were increased by simultaneous ultrasonic irradiation. They also found that ultrasound enhances adsorption phenomena in some dissolving-metal chemical reactions.

Ultrasound has also been employed in electro-organic systems to facilitate an expected reaction and also to alter the course of an electro-organic reaction. For example, the Kolbé reaction (Torii, 1985; Vassiliev and Grinberg, 1991; Eberson and Utley, 1996), that is, the electro-oxidation of cyclohexane carboxylate (to form bicyclohexyl, cyclohexane, cyclohexene, methocyclohexane, methyl cyclohexanoate, and cyclohexanol) (Scheme 24.1) has been extensively studied in electro-organic chemistry (Eberson and Utley, 1996).

Under intense sonication, it was found that the product yield of cyclohexene and bicyclohexyl increases by eightfold and decreases by sevenfold, respectively, compared with silent conditions. In other words, for the Kolbé reaction, it was found (Torii, 1985) that ultrasound favors a two-electron mechanism (observed as an enhancement in alkene formation) compared with the one-electron transfer mechanism under silent conditions. Similar observations were found in the electroreduction of aryl compounds to amino derivatives (Vassiliev and Grinberg, 1991), for example, the reduction of nitrobenzene into p-amino phenol.

Several workers such as Luche (1993) in the field of sonoelectro-organic chemistry have shown that ultrasound improves current efficiencies and product yields, eliminates electrode fouling, and activates the electrode surface. Luche (1993) showed that these enhancements are mainly due to improved mass transfer of electroactive species, the formation of radical species, and improved "single electron transfer" (SET).

However, it should be emphasized that these improved processes may only be due to intense stirring and increase in bulk temperature, but not purely to sonication. Thus, comparison of the validity of a sonoelectrochemical process with that of a conventional electrochemical process and the stirring and thermostatic conditions for the silent process must be studied. In the absence of such information, the effect of ultrasound could be overemphasized and would lead to the false conclusion that the overall effect is due to ultrasound, whereas it originates partially from a stirring and a bulk temperature effect.

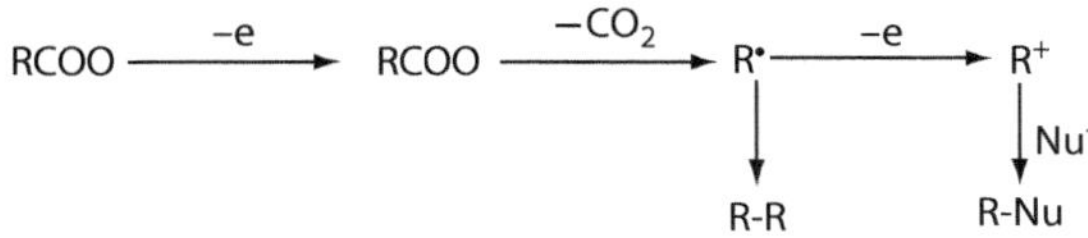

$$RCOO^- \xrightarrow{-e} RCOO^{\cdot} \xrightarrow{-CO_2} R^{\cdot} \xrightarrow{-e} R^+$$
$$R^{\cdot} \rightarrow R\text{-}R$$
$$R^+ \xrightarrow{Nu^-} R\text{-}Nu$$

SCHEME 24.1

SONO-ELECTROPOLYMERIZATION

In the late 1920s, Wood and Loomis (1927) reported that ultrasound increases the emulsification of oil and water. This observation prompted many investigations into the possible mechanisms for emulsification (Skotheim, 1986). Ultrasound has been found to increase the emulsion of polymerizations. Other workers [Walton and Phull (1996) and references therein] showed that, using styrene, a surfactant and an initiator (e.g., potassium persulfate), sonication offers several advantages in the styrene polymerization: a reduced induction period, a faster and better emulsion, and an increased oxidation rate. They are several possible explanations for the increase in oxidation rate: (1) H• and OH• radicals produced by sonolysis may recombine to form hydrogen peroxide and thus oxidize impurities, which could have acted as inhibitors, (2) ultrasonic degassing may remove absorbed oxygen, (3) the break down of the initiator may be accelerated by ultrasound, and (4) an increased mixing of the components producing a finer emulsion.

Ultrasound has also been shown to enhance the degradation of polystyrene in various solvents, for example, benzene [Walton and Phull (1996)] due to primarily to cavitation effects. For example, (1) the ultrasonic degradation of polystyrene has been observed to increase with decreasing ultrasonic frequency, which is attributed to the lower frequency providing a longer time for bubble growth and collapse, (2) the ultrasonic degradation of polyacrylic acid has been observed to decrease on the addition of a volatile compound (e.g., ether), which is attributed to a lower cavitational pressure due to an increase in vapor pressure, and (3) the ultrasonic degradation of polystyrene has been observed to be greater in deaerated solutions, which is attributed to a lowering of the cavitation threshold and an increase of cavitation bubbles.

In the mid-1980s, many workers (Osawa et al., 1987, 1992; Walton, 1991, Walton et al., 1992) studied the effect of ultrasound upon electroinitiated chain polymerizations. It was found (Walton, 1991) that ultrasound minimizes and, in some cases, removes the thin film of polymer blocking the electrode surface that impedes further electrochemical reactions.

By analogy with metal electrodeposition, the use of ultrasound in the preparation of conducting polymers (often called "organic metals"), for example, polypyrrole, polythiophene, or their derivatives was extensively used in the past 10 years. Since the discovery of conducting polymers in 1977, they have found a wide range of applications. Their very interesting properties enable them to be used as potential materials in a variety of areas, for example, field effect transistors, light-emitting diodes, solar cells, electrochromic devices, and protection of metals and biosensors.

Recently the sonoelectrofabrication and sonoelectrodeposition of conducting polymers has been shown to: (1) alter the effective reactivity ratio of the monomers, (2) produce an improvement in the percentage conversion versus the polymerization time characteristics, (3) improve the quality of the electrodeposited materials, (4) improve the mechanical characteristic of the electrodeposit, for example, flexibility and toughness (Osawa et al., 1987), (5) give higher yields and better quality films (Osawa et al., 1992), and (6) give a better polymer conductivity (e.g., $>100\,S\,cm^{-1}$).

Et Taouil et al. (2010) developed for the first time a novel masking technique against polymer deposition based on high intensity focused ultrasound (HIFU) irradiation to test a variety of background salts such as sodium salicylate (SS). SS was found to be the most effective electrolytic medium for pyrrole sonoelectropolymerization on copper as it led to a very efficient passivating oxide layer preventing copper dissolution, while enabling polymer formation independently from ultrasound. They observed that focused ultrasound increases copper dissolution in sodium oxalate electrolyte while preventing polypyrrole deposition, and they proposed this type of masking technique as an interesting and promising alternative to lithography as it offers advantages such as ease to carry out and allowing chemical waste reduction.

The same researchers also studied the use of high-frequency ultrasound (500 kHz, 25 W) for the electropolymerization of 3,4-ethylenedioxythiophene (EDOT) in aqueous medium in order to investigate its effects on conducting polymer properties. They showed that (1) mass transfer enhancement induced by sonication improves electropolymerization and (2) stirring effect is not the only phenomenon induced by ultrasound during electrodeposition. PEDOT films fabricated under ultrasonication presented increased doping levels revealed by x-ray photoelectron spectroscopy (XPS) analysis, especially in the case of thick films due to better incorporation of counter ions within the polymer matrix caused by mass transport improvement under ultrasound and possibly film heating by wave absorption for the highest thicknesses. They also observed that a dilation of the film under ultrasound led to an increase in film thickness with a refining of the surface structure (Et Taouil et al., 2010).

SONO-ELECTROCHEMILUMINESCENCE

It is well-recognized that when a sonic wave passes through an aqueous solution, luminescence is produced—*sonoluminescence* and is due to high-energy cavitation bubbles imploding within the aqueous media. Since the 1960s, sonoluminescence has been extensively studied and there are various reviews on the topic (Frenzel and Schultes, 1934; Walton and Phull, 1996 and references therein). Recently, many workers have investigated the effect of ultrasound upon electrochemiluminescent systems (Faulkner and Bard, 1976; Crum, 1994).

Electrochemiluminescence (ECL) is the production of light from an electrolytic system and is known to be a useful phenomenon, especially in analytical chemistry. For example in immunoassays, in which an ECL compound (e.g., ruthenium bypyridine, $Ru(bpy)_3^{2+}$) is attached to an antibody or an antigen and the change in ECL is monitored using a photomultiplier tube (PMT) [Walton and Phull (1996) and references therein].

It was shown that coupling ultrasound to the ECL of ruthenium bipyridine; luminol (3-aminophthalhydrazide); 1,5-diphenyl-3-styryl pyrazoline; rubrene; 9,10-diphenylanthracene; 9,10-dimethylanthracene; perylene, or the electro-oxidation of phenylacetate in acetonitrile solution offer various advantages [Walton and Phull (1996) and references therein]: (1) no fouling is formed at the electrode, (2) a greatly enhanced ECL emission is observed due to improved diffusion near the electrode, (3) a longer ECL emission over a fixed period of time is obtained, and (4) the detection of very low concentrations of analyte (e.g., antibody) can be easily achieved.

ENVIRONMENTAL SONO-ELECTROCHEMISTRY

Sonoelectrochemistry has also been proposed for the treatment of toxic wastes since it offers several advantages. For example, it has been shown [Walton and Phull (1996) and references therein] that the removal of phenol from industrial effluents by electrochemical oxidation is accelerated in the presence of ultrasound. Eighty percent oxidation of phenol to maleic acid was achieved when ultrasound was applied to a solution containing $100 g\ dm^{-3}$ phenol and $2 g\ dm^{-3}$ sodium chloride in an electrolyzer compared with 50% of the phenol being oxidized in the absence of ultrasound. An early report by Naffrechoux et al. (2000) showed that the sonochemical and sonoelectrochemical destruction of aromatic compounds in water samples is obtained at low frequency ultrasound. They also showed that hydroxyl radicals appear to be the main active reagent that reacts with the organic compound and the organic compound oxidation may be enhanced by combining various traditional techniques (e.g., O_3/H_2O_2, UV/H_2O_2, ultrasound/O_3, and UV/H_2O_2/ultrasound).

Sonoelectrochemistry can also be employed in the disinfection of sewage and potable (drinking) water. For example, in the water industry, chlorine disinfection has proved to be successful in eradicating water-borne diseases (e.g., *Cryptosporidium* and *E. coli*). Chlorine is often produced onsite by electrolyzing hydrochloric acid and is used for the disinfection of environmentally toxic effluents. It has been found (Walton and Phull, 1996) that in electrolyzing 22% hydrochloric acid, approximately 59% of chlorine was evolved in the presence of ultrasound when compared to 1% in the absence of ultrasound. Thus, in this case, sonoelectrochemical waste treatment may prevent waste of primary energy sources and environmental pollution.

Recent studies (Pollet, 1998; Pollet et al., 2000) in the sonoelectrochemical recovery of heavy metals, such as silver from photographic waste effluents, using the Pollet cell showed that ultrasound offers several advantages: (1) high current efficiencies are obtainable, for example, >99%, (2) low level of silver in "fix" solutions after treatment are also obtainable, for example, <1 ppm, (3) high level of silver in "fix" solutions may be treated—ca. 5000 ppm, (4) treated "fix" solutions may be reused, (5) no refining costs of the electrodeposited silver are required (6) treated "fix" solutions meet environmental regulations (permissible level set between 1 and 10 ppm), (7) "tailing" operation can be employed for further desilvering and (8) the sonoelectrochemical process is cost and time effective (2.2 h meet the permissible level of 1 ppm of silver) compared with traditional electrolytic processes (classical electrolyses [7 h], steel wool cartridges, and ion-exchange systems).

SONO-ELECTROANALYTICAL CHEMISTRY

The coupling of power ultrasound with conventional electroanalytical techniques, for example, anodic stripping voltammetry (ASV) and polarography is fairly a new subject area—sonoelectroanalytical chemistry or sonoelectroanalysis (Saterlay and Compton, 2000). This new technique offers great advantages over classical electroanalytical techniques such as: improved reaction rates, analysis times, increased analytical efficiencies, electrode life span, sensitivities and detection limits, reliability, analytical diversity, suppression of electrode fouling, and elimination of sample preparation.

Compton et al. (Saterlay and Compton, 2000; Compton et al., 2003) showed for the first time that power ultrasound can be used in conjunction with ASV—sono-ASV to overcome electrode passivation and mass-transport limitations in turn allowing sensitive electroanalyses to be performed in a range of complicated matrices, for example, fuels, effluents, foodstuffs, and blood. They also showed that copper in beer and in blood, lead in petrol and in river bed sediment, vanadium in aqueous media, and manganese in tea can be comfortably detected at ppb (parts-per-billion) levels by sono-ASV with no pre-sample preparation and treatment required. They demonstrated that sono-ASV is relatively inexpensive and readily available and offers an excellent alternative to classical analytical techniques, such as atomic spectroscopy techniques (AAS, AES and ICP-MS), which are costly to purchase (ca. £10K–£50K) and to maintain (ca. £5K per year). These techniques are also restricted by sample preparation and analysis times (usually a microwave acid digestion is required for metal analyses).

SONOELECTROCHEMICAL PRODUCTION OF NANOMATERIALS

Recently, an upsurge of interests has been observed in sonoelectrochemical synthesis of nanomaterials (below 200 nm) such as pure metals (Ag, Co, Al, Pd, Pt, Zn, Ni, W, Mg and Au), binary and ternary metallic alloys (FeCo, CdSe, PbTe, PbSe, Bi_2Se_3, and MoS_2 and Fe–Co–Ni), metal oxides (Cu_2O and MgO) and conducting polymers [polyaniline (PANI), polythiophene, polypyrrole, and poly(methylaniline] in various solvents (aqueous solutions, acids, alcohol, THF DBDG, and NTA), surfactants (CTAB, PVP, and PVA) and polymers (PEO disulfide, MPEO and PVP) (Saez and Mason, 2009).

Although, there are a various range of methods of producing nano-sized materials such as thermal decomposition, physical and thermal evaporation, laser ablation, laser-assisted catalytic growth (LCG), vapor–liquid–solid growth (VLS), ultrahigh-vacuum (UHV), ion implantation, biochemical, electrochemical, sonochemical, radiolysis, chemical reduction/oxidation and sol–gel; most of these techniques tend to be expensive and time-consuming.

An alternative method, which is both simple and cost-effective, is the use of Sonoelectrochemistry involving either depositing the electroactive species under continuous electrical current and ultrasound or producing nanosize materials at various currents and ultrasonic pulses (a few m s) at a vibrating electrode. Reisse et al. (1994) were the first to use the *pulsed* sonoelectrochemical technique to produce nano-metallics such as copper, in which the ultrasonic horn was used as the working electrode (in this case the cathode) in a three-electrode configuration (Figures 24.1c and 24.12a).

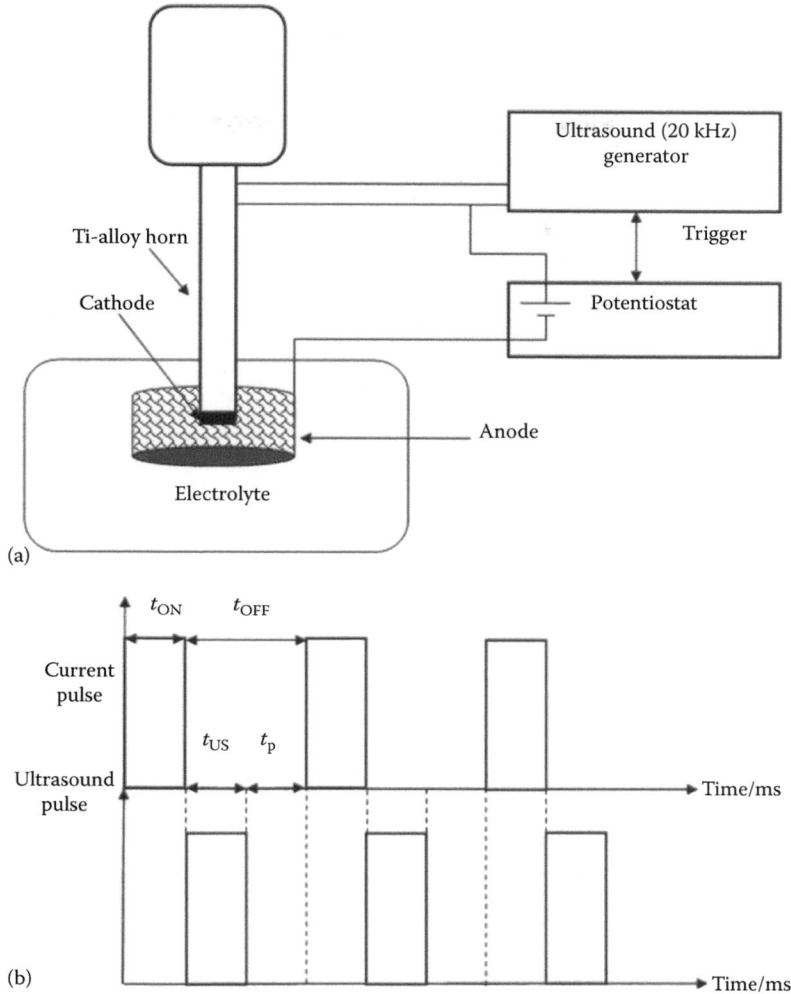

FIGURE 24.12 (a) Pulsed sonoelectrochemical set up and (b) time management.

This *sonoelectrode* was subjected to ultrasonic pulses, which were each followed by short applied current pulses. They showed that using this *time management* (Figure 24.12b) and during cavitation, a jet of liquid penetrates inside the cavitation bubble perpendicular to the *sonoelectrode* surface and the resulting impact is responsible for dislodging any nanopowder materials, which had been electrochemically deposited on the surface.

Recently, Haas and Gedanken (2006) showed that it is possible to produce spherical copper nanoparticles with a diameter range of 25–60 nm stabilized in CTAB and PVP (Figure 24.13) by applying a range of current density between 55 and 100 mA cm^{-2}. Reaction mechanisms between copper ions, PVP, and CTAB to form complexes were proposed.

Here we invite the reader to refer to the latest review by Saez and Mason (2009) and references therein describing all these aspects of the synthesis of nanomaterials by sonoelectrochemistry.

SONOELECTROCHEMICAL PRODUCTION OF FUEL CELL MATERIALS

There are a few studies reporting the use of the sonoelectrochemical method to produce noble metals (Pollet, 2010) and electrocatalysts for proton exchange membrane fuel cell (PEMFC) and direct methanol fuel cell (DMFC).

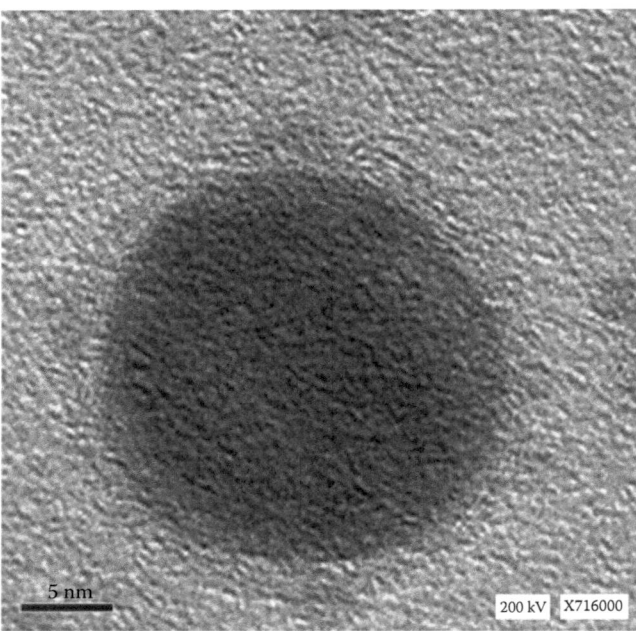

FIGURE 24.13 HR-TEM images of a copper nanoparticle produced by pulse sonoelectrochemistry in a CTAB solution. (With permission from Haas, I. and Gedanken, A., Sonoelectrochemistry of Cu^{2+} in the presence of cetyltrimethylammonium bromide: Obtaining CuBr instead of copper, *Chem. Mater.*, 18, 1184–1189, 2006. Copyright 2006 American Chemical Society.)

Effect of Aqueous Solutions

Recently, Zin et al. (2009) produced platinum nanoparticles from aqueous chloroplatinic solutions by the pulse sonoelectrochemical (20 kHz, up to 118 W cm^{-2}) method on titanium alloy electrodes in the absence of any surfactants, alcohols, and polymers by producing short applied current pulses triggered and followed immediately by ultrasonic pulses at the working electrode (in this case the cathode). The time management sequence employed was as follows:

1. A short current pulse of $|i| = 50$ mA cm^{-2} was applied to the *sonoelectrode*, whereby the titanium horn acted as an electrode only (t_{ON}); the time of this phase typically varied between 0.3 and 0.5 s.
2. Immediately after the electrochemical pulse was turned off, an ultrasonic pulse was sent to the *sonoelectrode* and here it acted only as a vibrating ultrasonic horn (t_{US}); this second phase lasted no more than 0.5 s.
3. A rest time, t_p, followed the two previous phases (this was useful to restore the initial electrolyte conditions close to the *sonoelectrode*).

They showed that Pt mean grain size ranging from 11 to 15 nm was produced and globular clusters had a mean size ranging between 100 and 200 nm, which in turn aggregated and built complex structures.

Effect of Surfactants and Polymers

Shen et al. (2008) produced uniform spherical three-dimensional dendritic Pt Nanostructures (DPNs) with an average dimension of 2.5 ± 0.5 nm, at room temperature sonoelectrochemically (20 kHz; 20 W; pulse width of the current = 1.0 s, resting time of the current pulse = 0.5 s; and duration of the ultrasonic pulse = 0.3 s) by using mixtures of hexachloroplatinum(IV) acid hydrate in various surfactants [PVP, poly(ethylene glycol)$_{20}$–poly(propylene glycol)$_{20}$–poly(ethyleneglycol)$_{20}$ (P123), SDS and poly-diallyl–dimethyl ammonium chloride (PDDA)] at current densities up to 40 mA cm^{-2}. They showed that at low current densities the reduction rate of Pt is slower, the number of nuclei is small and the rate of growth is faster than that of nucleation, which lead to large nanoparticles (5–7 nm). As the current density increases, the reduction rate of Pt^{4+} increases and the nucleation rate is faster than that of the growth leading to the generation of more nuclei and the formation of smaller primary Pt nanoparticles. They also showed that stabilized Pt nanoparticles do not aggregate in PVP due to stronger bonds between the Pt precursors with the C=O groups of the PVP. In other words PVP adheres to the nanoparticles through a charge-transfer interaction between the pyrrolidone rings and Pt atoms. The DPNs showed improved electrocatalytic activity towards MOR due to monodisperse Pt nanoparticles and improved porosity structure leading to large effective surface area. They proposed a mechanism whereby Pt ions are reduced by the electrical current and formed Pt primary nanoparticles, which are subsequently dislodged by the vibrating electrode. The primary nanoparticles, then in solution, spontaneously assemble together and form small spherical DPNs. They showed that ultrasound leads to small primary nanoparticles in favor of the crystallite reorganization and growth of a stable near-single crystal.

Shen et al. (2009) showed that by using the sonoelectrochemical (20 kHz; 20W; pulse on time of the current = 0.5 s; pulse off time of the current = 0.5 s, duration of the ultrasonic pulse = 0.3 s) method, it is possible to realize the morphology-controlled synthesis of palladium nanostructures [spherical (SNP), multitwinned (MTP) and spongelike (SSP)] at room temperature in the presence of various surfactants and polymers [cetyltrimethylammoniumbromide (CTAB), PVP, and PDDA] (Figure 24.14).

They showed that the size and shape of the Pd nanostructures may be controlled by varying the current density and the precursor solution pH value. Furthermore, the electrocatalytic activities of the produced spongelike Pd nanostructures for direct alcohol oxidation in alkaline media showed higher electrochemical active surface than other Pd nanostructures. Qui et al. (2003) also synthesized highly dispersed spherical Pd nanoparticles of a dendritic superstructure in the presence of CTAB by the pulse sonoelectrochemical (20 kHz) method at room temperature and a reaction time above 2.5 h. They explained that the dendritic-structured Pd nanoparticles had a treelike structure and agreed with the diffusion-limited aggregation (DLA) model, involving cluster formation by the adhesion of particles to a selected seed on contact and allowing the particle to diffuse and stick to the growing structure. They stipulated that it is possible that Pd particles reach the anode and grow into a dendritic structure. They concluded that the shape and size of spherical nanocrystalline Pd may be controlled by varying the current density (8–13 mA cm^{-2}), the interval between continuous ultrasonic pulses, ultrasonic intensity (20–120 W cm^{-2}), and the CTAB concentration. For example, they observed that the shape of the nanoparticles appeared irregular and agglomerated below 20 and 120 W cm^{-2}.

For the sonoelectrochemical production of nano-size metals, a mechanism has been proposed where metallic ions are reduced by a short current pulse to produce metallic nanoparticles on the *sonoelectrode* surface, which are then dislodged by the ultrasonic pulse. The metallic nanoparticles in solution tend to spontaneously assemble together whereby under insonation, the Ostwald ripening process is accelerated, leading to smaller primary nanoparticles (Shen et al., 2008, 2009). Table 24.1 shows a summary of the sonoelectrochemical production of non-precious, noble mono-metallics at various ultrasonic frequencies and powers in several surfactants. It is evident from the table that PEMFC and DMFC electrocatalyst nanoparticles can be produced sonoelectrochemically whereby the choice of surfactants plays an important role on the final nanoparticle size. For example, it is possible to fabricate Pt nanoparticles down to 2.5 nm in a solution containing chloroplatinic salt in PVP, P123, SDS, and PDDA at 20 kHz (20W) provided that the galvanostatic current and the ultrasound are pulsed at fixed short times, for example, pulse width of the current = 1.0 s; resting time of the current pulse = 0.5 s; and duration of the ultrasonic pulse = 0.3 s.

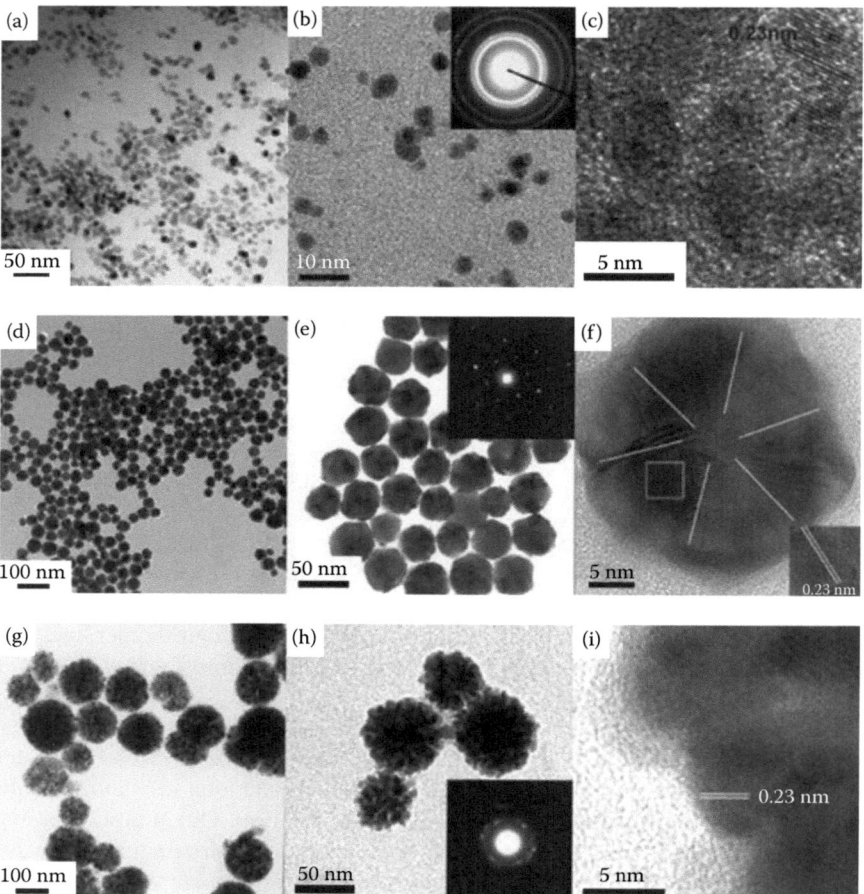

FIGURE 24.14 TEM and HR-TEM images of the Pd nanostructures prepared in different solutions. (a through c) Spherical nanoparticles (SNPs) prepared in CTAB solution. (d through f) Multi-twinned particles (MTPs) prepared in PVP solution. (g through i) Spherical spongelike particles (SSPs) prepared in PDDA solution. The insets in parts b, e, and h are SAED patterns of the corresponding Pd nanostructures. The inset in part f is the enlarge image of the square zone area. The white lines in part f indicate the twin boundaries separating the different adjacent facts. (With permission from Shen, Q., Min, Q., Shi, J., Jiang, L., Zhang, J.-R., Hou, W. and Zhu, J.-J., Morphology–controlled synthesis of palladium nanostructures by sonoelectrochemical method and their application in direct alcohol oxidation, *J. Phys. Chem.*, 113, 1267–1273, 2009. Copyright 2009 American Chemical Society.)

There are only few papers dealing with the use of sonochemical and sonoelectrochemical techniques to produce fuel cell electrodes; however, this section is added in this chapter as it merits some attention.

There is a strong interest in the electrodeposition of Co/CeO_2 composites for protection of cathodes in MCFC (Molten Carbonate Fuel Cell) and recently, Chr Argirusis et al. (2006) electrodeposited Co/CeO_2 and Ni/CeO_2 composites from electrolytes containing suspended nanoparticles of gadolinia doped ceria in the presence of ultrasound (20 kHz; 29 W cm^{-2}). They showed that sonication improves the co-deposition of ceria nanoparticles with Ni and Co, but no MCFC tests were carried out to investigate the performance of these new materials.

Currently there are different methods for mixing and milling the carbon support, the carbon-support electrocatalyst, and the ionomer or proton exchange membrane (PEM) either by intense

TABLE 24.1
Sonoelectrochemical Production of Nanosize Mono-Metallic Electrocatalysts at Various Ultrasonic Frequencies and Powers in Several Surfactants

Noble Metals	Ultrasonic Frequency (kHz)	Ultrasonic Power	Surfactant	Solvent	Particle Size (nm)	Authors
Pt	20	20 W	PVP, P123, SDS, PDDA	—	2.5 ± 0.5	Shen et al. (2008)
Pt	20	Up to 118 W cm^{-2}	—	—	<200	Zin et al. (2009)
Pd	20	—	PVP, PDDA, CTAB	—	<200	Shen et al. (2009)
Pd	20	20–120 W cm^{-2}	CTAB	—	5–10 (depending on current densities and CTAB concentrations)	Qui et al. (2003)

magnetic stirring, ultrasonic agitation, or ball milling to produce a PEMFC and DMFC catalyst "ink". This ink is either applied to (1) the gas diffusion layer (GDL) to form a gas diffusion electrode (GDE) or a catalyst coated substrate (CCS) or (2) the PEM to form a catalyst coated membrane (CCM). The CCSs are usually prepared by spreading, spraying, deposition, ionomer impregnation, electrodeposition (continuous and pulsed), and sputtering whereas the CCMs are fabricated by impregnation reduction, evaporation deposition, sputtering, dry spraying and decaling. For further detail about the above methods, the reader is invited to refer to the excellent reviews on analysis of PEM fuel cell design and manufacturing by Mehta and Cooper (2003), and Litster and McLean (2004) on PEM fuel cell electrodes.

For the first time, Pollet (2009) showed that the sonoelectrochemical method can be used for preparing PEMFC electrodes whereby platinum loaded on Nafion-bonded carbon anodes were prepared in K_2PtCl_4 aqueous solutions by galvanostatic pulse electrodeposition in the absence and presence of power ultrasound (20 kHz). Pollet found that PEMFC electrodes prepared sonoelectrochemically showed better performance compared to those prepared by (1) galvanostatic pulse method only (i.e., silent conditions) and (2) conventional method (Figure 24.15).

Here we invite the reader to refer to the latest review by Pollet (2010) of the use of ultrasound for the fabrication of fuel cell materials describing the ultrasonic, sonochemical, and sonoelectrochemical production of nano-size binary and ternary electrocatalysts and other important fuel cell materials such as the catalyst layer, the electrocatalyst non-carbonaceous/carbonaceous supports, and the membrane (Pollet, 2010).

SONOELECTROCHEMISTRY IN "EXOTIC" SOLVENTS

Many electrochemical processes such as the electrodeposition of metals in aqueous solvents are limited by reactions that control the potential window, that is, those involving the oxidation and reduction of the solvent. Non-aqueous solvents that include acetonitrile, DMF, DMSO, THF, methylene chloride, and propylene carbonate are considered suitable as they exhibit a potential window that is 1.5–2.5 times wider than that of aqueous acid electrolytes. A wide range of salts can be used for aqueous electrolyte solutions; however, salts for non-aqueous electrolyte solutions are more restricted and typically consist of large cations (e.g., tetraalkylammonium cations) and anions (e.g., hexafluorophosphate, tetrafluoroborate, and perchlorate) to ensure full dissociation.

In the last decade, room temperature ionic liquids (RTILs) have attracted considerable interest for use as non-aqueous solvents (Endres and Schweizer, 2000; Wassersheid and Welton, 2003; Endres,

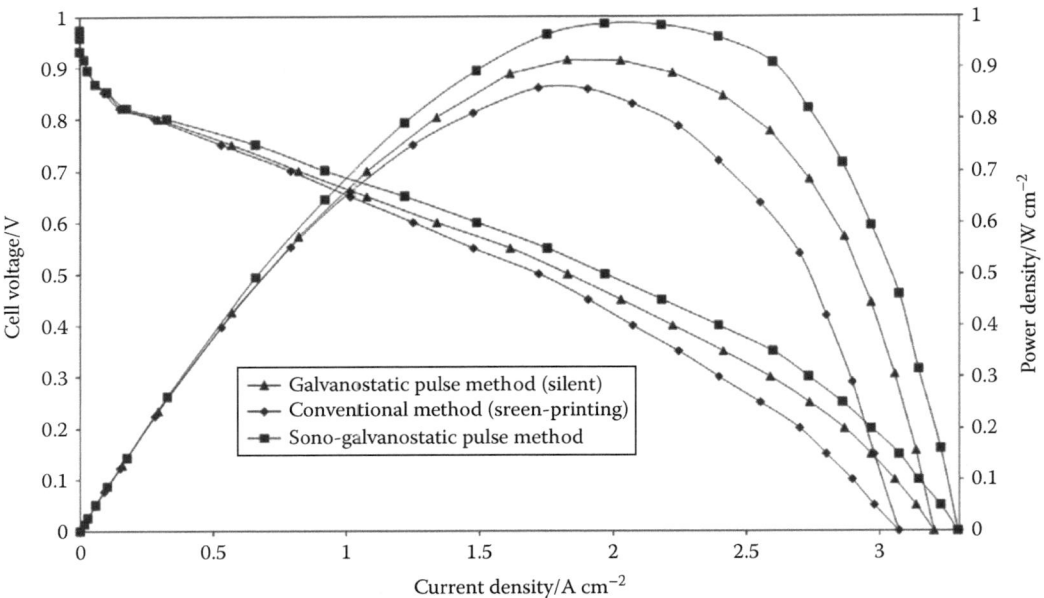

FIGURE 24.15 Comparison of MEA performance between anodes prepared by: (a) the galvanostatic pulse method in the absence of ultrasound [▲], (b) the sono-galvanostatic pulse method (20 kHz, 20 W cm^{-2}) [■] and (c) conventional method (0.30 mg Pt cm^{-2} electrodes) [♦]. The fuel cell testing parameters were H_2/O_2 (1.5/2 stoics), 70°C, and 1 atm.. (From *Electrochem. Commun.*, 11, Pollet, B.G., A novel method for preparing PEMFC electrodes by the ultrasonic and sonoelectrochemical techniques, 1445–1448, Copyright (2009), with permission from Elsevier.)

2004; Buzzeo et al., 2004; Galinski et al., 2006). As their name implies, they are compounds made up of ions rather than molecular species, by combinations of organic and/or inorganic cations and anions. Three recent reviews by Buzzeo et al. (2004), Endres (2004), and Lewandowski and coworkers (Galinski et al., 2006) cover the fundamental aspects of electrochemistry and electrodeposition from ionic liquids based upon the imidiazolium cation and $(F_3CSO_2)_2N^-$, BF_4^-, and PF_6^- discrete anions. These ionic liquids present many advantages such as (1) wide electrochemical windows of up to 4 V, which allows the electrodeposition of a number of metals that cannot be achieved in aqueous solutions, (2) relatively high conductivities and (3) low viscosities.

For example, the electrodeposition of copper in RTILs has been widely studied (Endres, 2004). It was found that the reduction of Cu^{2+} to metallic Cu occurs in two one-electron steps: in the first step Cu^+ is formed and in the second step the metal is deposited. Endres and Schweizer (2000) and Endres (2004) have shown, using in situ STM, that the bulk deposition of copper from acidic chloroaluminate liquids on gold is preceded by three processes. Furthermore, the electrode potential for the redox Cu(I)/Cu(II) is more positive than the surface oxidation of gold in the liquid. The electrodeposition of copper in chloraluminate ionic liquids, a basic 1-ethyl-3-methylimidazolium tetrafluoroborate and a Lewis acidic $ZnCl_2$-1-ethyl-3-methylimidazolium chloride RTIL has been recently investigated (Endres and Schweizer 2000; Endres, 2004).

However, the use of this type of ionic liquids have several disadvantages, such as toxicity and cost making them somewhat impractical for larger industrial applications such as metal electroplating and electrodeposition.

For the first time, Costa et al. (2008) showed that it is possible to sonoelectrodeposit metals in RTILs and use RTILs to study transport limited currents in a specially designed microsonoreactor (Besancon cell, see above) and characterization at three low frequencies in presence of water-, acetonitrile-, and imidazolium-based ionic liquids ([BuMIm][$(CF_3SO_2)_2N$]). They

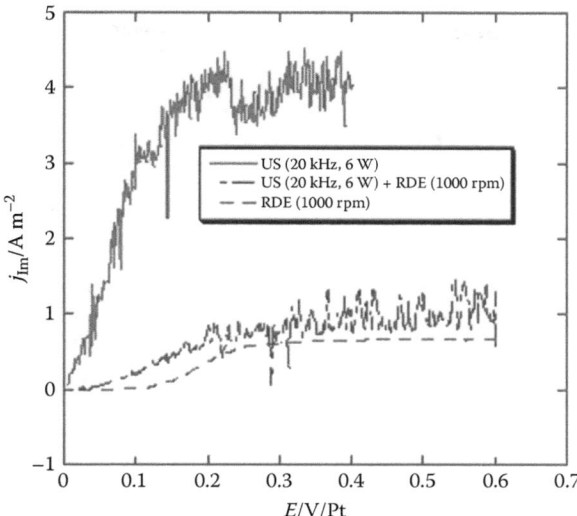

FIGURE 24.16 RDE (1000 rpm) and sono voltammograms (6 W, 20 kHz) of a quasi-reversible couple in [BuMIm][(CF$_3$SO$_2$)$_2$N] using the Besançon cell.

found that mass transfer enhancement was observed and was particularly high, characterized by an average Sherwood number of 6500 while the value obtained with an electrode rotating at 4500 rpm was only 1200. Figure 24.16 shows that there is a substantial increase in the current density in the presence of ultrasound (more than five times the level obtained at 1000 rpm for only 6 W transmitted to 7 mL). Oscillations in the presence of ultrasound suggest that cavitation occurs with an important drop in current density by combining both stirring modes (rpm + US) (Costa et al., 2010).

An alternative to the above nonaqueous solvents are Deep Eutectic Solvents (DES), which are a type of ionic solvent with special properties composed of a mixture forming an eutectic with a melting point much lower than either of the individual components (Abbott et al., 2003, 2004, 2005a,b, 2007). Compared to ionic liquids, they share many characteristics, but are only ionic mixtures and not ionic compounds, and are known to be less-toxic, air and moisture stable, biodegradable, and economically viable to large-scale processes (Abbott et al., 2003). DESs are either urea or ethylene glycol and choline chloride based ionic compounds (Abbott et al., 2003). Abbott et al. (2003) were the first to produce "type 3" eutectic-based ionic solvents: $R_1R_2R_3R_4N^+$. RZ and use simply amides ($Z = CONH_2$), acids ($Z = COOH$) and alcohols ($Z = OH$) as complexing agents, thus making the ionic liquids more versatile. They have shown that DESs can be successfully employed in electropolishing (Abbott et al., 2004), electroplating, and metal oxide processing (Abbott et al., 2005a). Recently, they have also shown that choline chloride (as the quaternary ammonium salt) and either urea or ethylene glycol (as hydrogen bond donors) based DES can be employed in the electrodeposition of zinc, tin, and zinc–tin alloys, but the electrochemical reactions are rather slow due to the high viscosity of the electrolytes. This, in turn, reduces the mass transfer of the electro-analyte due to low diffusion coefficients (Abbott et al., 2007).

Pollet et al. (2008) showed that deposition of copper in both water and DES is greatly affected by ultrasound at the two frequencies of 20 and 850 kHz employed. Limiting current densities were obtained in both solvents under sonication at 20 and 850 kHz and a 10-fold and 5-fold increase in currents in aqueous potassium chloride and glyceline 200 compared to silent conditions was observed respectively. The difference in viscosity of water (containing KCl) and glyceline 200 was found to be a crucial parameter in the evaluation of limiting current densities (Figure 24.17).

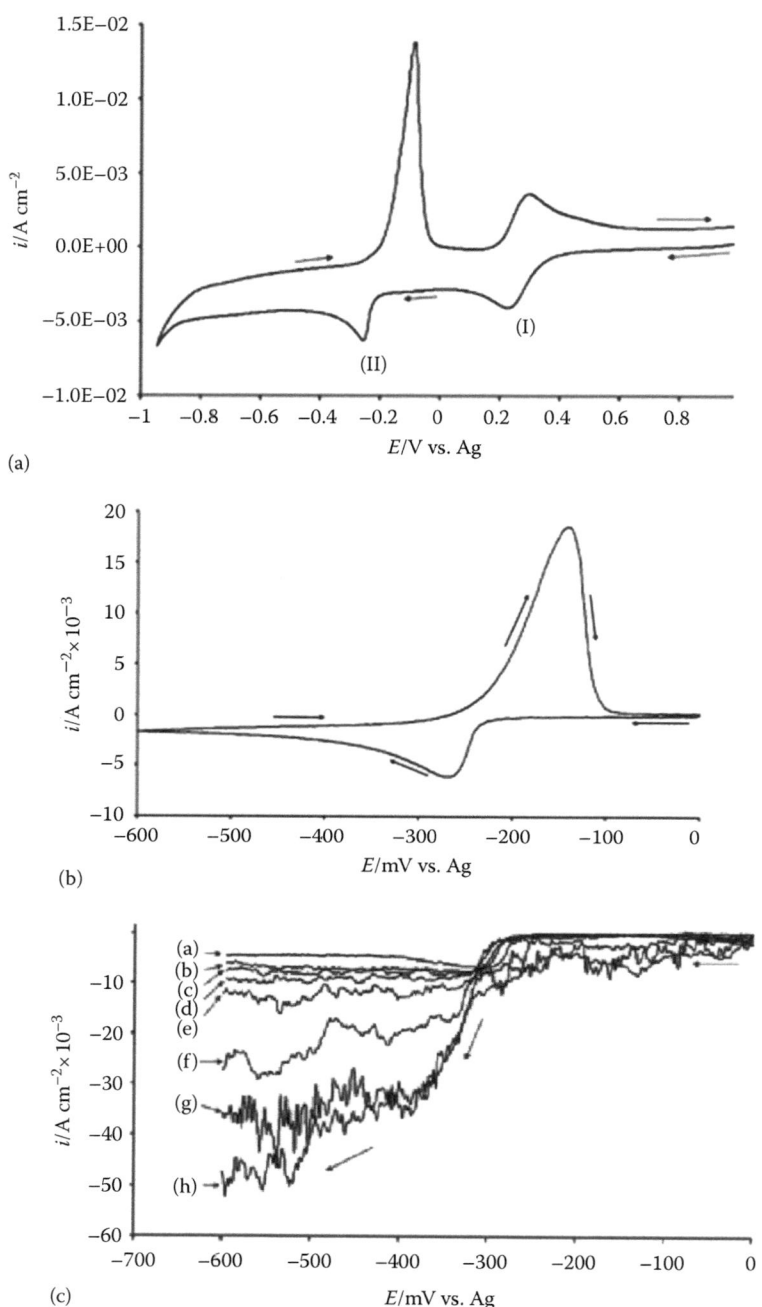

FIGURE 24.17 (a) Complete cyclic voltammogram of 2 g L^{-1} of CuCl$_2$ in 1.0 mol dm^{-3} KCl on Pt wire at a scan rate of 0.2 V s^{-1} and at (313 ± 1)K in the potential range [−1.0; + 1.0 V vs. Ag] under silent conditions. (b) Cyclic voltammogram of 2 g L^{-1} of CuCl$_2$ in 1.0 mol dm^{-3} KCl on Pt wire at a scan rate of 0.2 V s^{-1} and at 313 ± 1K in the potential range for copper deposition and dissolution under silent conditions. (c) Series of sono-voltammograms of 2 g L^{-1} of CuCl$_2$ in 1.0 mol dm^{-3} KCl on Pt wire at 20 and 850 kHz, at a scan rate of 0.2 V s^{-1}, at (313 ± 1) K and at various ultrasonic intensities: (a) 1.2 W cm^{-2} (850 kHz); (b) 1.7 W cm^{-2} (850 kHz); (c) 3.2 W cm^{-2} (850 kHz); (d) 4.1 W cm^{-2} (850 kHz); (e) 6.0 W cm^{-2} (20 kHz); (f) 29 W cm^{-2} (20 kHz); (g) 56 W cm^{-2} (20 kHz) and (h) 118 W cm^{-2} (20 kHz) in the potential range of copper deposition and dissolution.

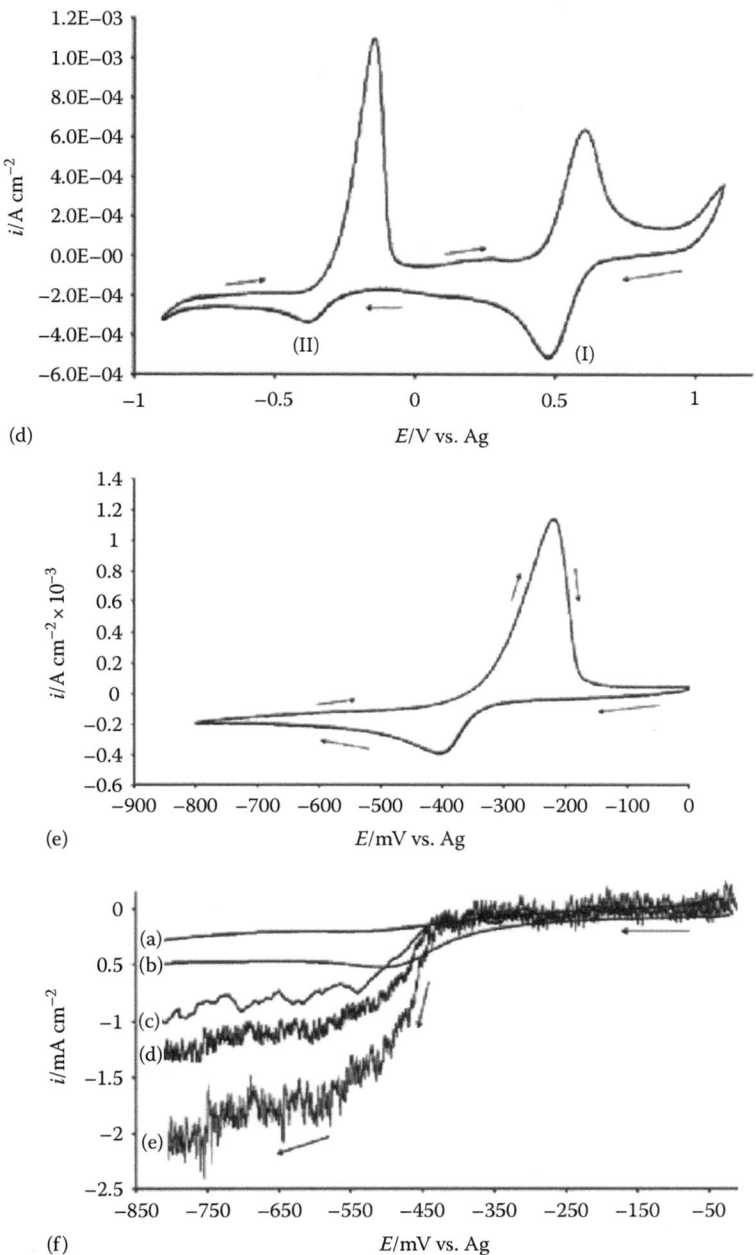

FIGURE 24.17 (continued) (d) Complete cyclic voltammogram of 2 g L^{-1} of CuCl$_2$ in glyceline 200 on Pt wire at a scan rate of 0.2 V s^{-1} and at (313 ± 1) K in the absence of ultrasound in the potential range [−1.0; + 1.1 V vs. Ag]. (e) Cyclic voltammogram of 2 g L^{-1} of CuCl$_2$ in glyceline 200 on Pt wire at a scan rate of 0.2 Vs^{-1} and at (313 ± 1) K in the absence of ultrasound in the potential range copper deposition and dissolution [0; −0.8 V vs. Ag]. (f) Linear sweep voltammograms of 2 g L^{-1} of CuCl$_2$ in glyceline 200 in the presence of ultrasound—(a) 5.8 W cm^{-2} (850 kHz), (b) 8.0 W cm^{-2} (20 kHz), (c) 41 W cm^{-2} (20 kHz), (d) 77 W cm^{-2} (20 kHz), and (e) 20 kHz (161 W cm^{-2}) at (313 ± 1) K and at a scan rate of 0.2 V s^{-1} in the potential range of copper deposition and dissolution.

CONCLUSIONS

Sonoelectrochemistry is not a new subject area, for example, sonoelectroanalytical and sonoelectrosynthesis experiments were performed nearly 80 years ago. The well-established effect of ultrasound in electrochemistry is the diminution of the diffusion layer (and thus the increase in limiting currents), the decrease in overall overpotential, and the modification in electrode surface (erosion) due to acoustic streaming together with microstreaming and the implosion of high-energy cavitation bubbles. Thus, these effects offer several benefits: an increase in mass-transport phenomena, an increase in reaction rates, an improvement in sensitivity and selectivity, greater product ratios and efficiencies, less electrolytic cell power requirements, a diminution of electrode fouling, an improvement in product coating, an acceleration in corrosion and electrode dissolution, and an enhancement in the production of microemulsions.

It is anticipated that the next few years will see the deployment of ultrasound in a variety of electroanalytical, electrosynthetic processes, fabrication of novel nanomaterials, and fuel cell materials.

ACKNOWLEDGMENT

The authors would like to thank Daniel Symes for working on the references section.

REFERENCES

Abbott, A.P., Boothby, D., Capper, G., Davies, D.L., and Rasheed, R.K. 2004. Deep eutectic solvents formed between choline chloride and carboxylic acids. *Journal of the American Chemical Society*, 126:9142–9147.

Abbott, A.P., Capper, G., Davies, D.L., Rasheed, R.K., and Shikotra, P. 2005a. Selective extraction of metals from mixed oxide matrices using choline based ionic liquids. *Inorganic Chemistry*, 44:6497–6499.

Abbott, A.P., Capper, G., Davies, D.L., Rasheed, R.K., and Tambyrajah, V. 2003. Novel solvent properties of choline chloride/urea mixtures. *Chemical Communications*, 1:70–71.

Abbott, A.P., Capper, G., McKenzie, K.J., and Ryder, K.S. 2007. Electrodeposition of zinc tin alloys from deep eutectic solvents based on choline chloride. *Journal of Electroanalytical Chemistry*, 599:288–294.

Abbott, A.P., Capper, G., Swain, B.G., and Wheeler, D.A. 2005b. Electropolishing of stainless steel in an ionic liquid. *Transactions of the Institute of Metal Finishing*, 83:51–53.

Argirusis, C., Matic, S., and Schneider, O. 2006. An EQCM study of ultrasonically assisted electrodeposition of Co/CeO_2 and Ni/CeO_2 composites for fuel cell applications. *Physica Status Solidi*, 205:2400–2404.

Banks, C.E. and Compton, R.G. 2003. Ultrasonically enhanced voltammetric analysis and applications: An overview. *Electroanalysis*, 15:329–346.

Bard, A. 1963. Polarographic methods in analytical chemistry. *Analytical Chemistry*, 35:1125–1128.

Birkin, P.R. and Silva-Martinez, S. 1997. A study on the effect of ultrasound on electrochemical phenomena. *Ultrasonic Sonochemistry*, 4:121–122.

Brett, C. 2008. Sonoelectrochemistry. In: Antonio Arnau Vives (Ed.)., *Piezoelectric Transducer and Applications*. Berlin/Heidelberg, Germany: Springer-Verlag, Chap. 15, pp. 399–411.

Buzzeo, M.C., Evans, R.G., and Compton, R.G. 2004. Non-haloaluminate room-temperature ionic liquids in electrochemistry. *Chemical Physics and Physical Chemistry*, 5:1106–1120.

Cognet, P., Wilhem, A.M., Delmas, H., Lyazidi, H.A., and Fabre, P.L. 2000. Ultrasound in organic electrosynthesis. *Ultrasonic Sonochemistry*, 7:163–167.

Compton, R.G., Eklund, J.C., and Marken, F. 1997a. Sonoelectrochemical processes: A review. *Electroanalysis*, 9:509–522.

Compton, R.G., Eklund, J.C., Marken, F., Rebitt, T.O., Akkermans, R.P., and Waller, D.N. 1997b. Dual activation: Coupling ultrasound to electrochemistry—A overview. *Electrochimica Acta*, 42:2919–2927.

Compton, R.G., Eklund, J.C., and Page, S.D. 1995. Sonovoltammetry—Heterogeneous electron transfer processes with coupled ultrasonically induced chemical reaction—The sono-EC reaction. *Journal of Physical Chemistry*, 99:4211–4214.

Compton, R.G., Eklund, J.C., Page, S.D., Sanders, G.H.W., and Booth, J. 1994. Voltammetry in the presence of ultrasound. Sonovoltammetry and surface effects. *Journal of Physical Chemistry*, 98:12410–12414.

Compton, R.G., Hardcastle, J.L., and del Campo, J. 2003. Sonoelectrochemistry, physical aspects. In: Bard-Stratmann (Ed.)., *Encyclopedia of Electrochemistry*, Pat Unwin (Ed.)., *Instrumentation and Electrochemical Chemistry*. Vol. 3, Wiley-VCH, Weinheim pp. 312–327.

Costa, C., Doche, M.L., Hihn, J.Y., Bisel, I., Moisy, P., and Lévêque, J.M. 2010. Hydrodynamic sono-voltammetry of ferrocene in [Tf2N](-)based ionic liquid media. *Ultrasonics*, 50:323–328.

Costa, C., Hihn, J.Y., Rebetez, M., Doche, M.L., Bisel, I., and Moisy, P. 2008. Transport-limited current and microsonoreactor characterization at 3 low frequencies in the presence of water, acetonitrile and imidazolium-based ionic liquids. *Physical Chemistry Chemical Physics*, 10:2149–2158.

Crum, L.A. and Roy, R.A., 1994. Sonoluminescence, *Science*, 266(no. 5183): 233–234.

del Campo, F.J., Coles, B.A., Marken, F., Compton, R.G., and Cordemans, E. 1999. High frequency sonoelectrochemical processes: Mass transport, thermal and surface effects induced by cavitation in a 500 kHz reactor. *Ultrasonic Sonochemistry*, 6:189–197.

Doche, M.L., Hihn, J.Y., Mandroyan, A., Maurice, C., Hervieux, O., and Roizard, X. 2006. A novel accelerated corrosion test for exhaust systems by mean of power ultrasound. *Corrosion Science*, 48:4080–4093.

Eberson, L. and Utley, J.H.P. 1996. Organic electrochemistry.

Eklund, J.C., Marken, F., Waller, D.N., and Compton, R.G. 1996. Voltammetry in the presence of ultrasound—A novel sono-electrode geometry. *Electrochimica Acta*, 41:1541–1547.

Endres, F. 2004. Ionic liquids: Promising solvents for electrochemistry. *Zeitschrift fur Physikalische Chemie*, 218:255–283.

Endres, F. and Schweizer, A. 2000. The electrodeposition of copper on Au(111) and on HOPG from the 66/34 mol% aluminium chloride/1-butyl-3-methylimidazolium chloride room temperature molten salt: An EC-STM study. *Physical Chemistry Chemical Physics*, 2:55.

Faulkner, L.R. and Bard, A.J. 1976. *Creation and Detection of the Excited State*, W.R. Ware (Ed.). New York: Marcel Dekker.

Frenzel, H. and Schultes, Z. 1934. Lumineszenz im ultraschallbeschickten Wasser. *Physical Chemistry*, B27:421–424.

Galinski, M., Lewandowski, A., and Stepniak, I. 2006. Ionic liquids as electrolytes. *Electrochemica Acta*, 51:5567–5580.

Gautheron, B., Tainturier, G., and Degrand, C. 1985. Ultrasound-induced electrochemical synthesis of the anions Se_2^{2-}, Se^{2-}, Te_2^{2-} and Te^{2-}. *Journal of the American Chemical Society*, 109: 5579–5581.

Gautheron, B., Tainturier, G., and Degrand, C. 1986. Electrochemical synthesis and properties of [.eta.5–C5H4R]2TiSe5 (R = H, Me, Me2CH). *Organometallics*, 5:942–946.

González-García, J., Esclapez, M.D., Bonete, P., Hernández, Y.V., Garretón, L.G., and Sáez, V. 2010. Current topics on sonoelectrochemistry. *Ultrasonics*, 50:318–322.

Gouveia-Caridade, C. and Brett, C.M.A. 2004. Review on Corrosion Protection of Materials, 23(2):14–19.

Haas, I. and Gedanken, A. 2006. Sonoelectrochemistry of Cu^{2+} in the presence of cetyltrimethylammonium bromide: Obtaining CuBr instead of copper. *Chemical Materials*, 18:1184–1189.

Hagan, C.R.S. and Coury, L.A. 1994. Comparison of hydrodynamic voltammetry implemented by sonication to a rotating disk electrode. *Analytical Chemistry*, 66:399–341.

Hyde, M.E. and Compton, R.G. 2002. How ultrasound influences the electrodeposition of metals. *Journal of Electroanalytical Chemistry*, 531:19–24.

Inazu, K., Nagata, Y., and Maeda, Y. 1993. Decomposition of chlorinated hydrocarbons in Aqueous solutions by ultrasonic irradiation, *Chemistry Letters*, 22(1):57–59.

Klima, J. 2011. Application of ultrasound in electrochemistry. An overview of mechanisms and design of experimental arrangement. *Ultrasonics*, 51(2):202–209.

Klima, J., Bernard, C., and Degrand, C. 1994. Sonoelectrochemistry: Effects of ultrasound on voltammetric measurements at a solid electrode. *Journal of Electroanalytical Chemistry*, 367:297–300.

Klima, J., Bernard, C., and Degrand, C. 1995. Sonoelectrochemistry: Transient cavitation in acetonitrile in the neighbourhood of a polarized electrode, *Journal of Electroanalytical Chemistry*, 399: 147–155.

Kochergin, S.M. and Vyaselva, G.Y. 1966. *Electrodeposition of Metals in Ultrasonic Fields*. New York: Consultants Bureau, p. 19.

Kolb, J. and Nyborg, W. 1956. Small-scale acoustic streaming in liquids. *Journal of the Acoustical Society of America*, 28:1237–1242.

Ligier, V., Hihn, J.Y., Wéry, M., and Tachez, M. 2001. The effects of 20 kHz and 500 kHz ultrasound on the corrosion of zinc precoated steels in [Cl^-] [SO_4^{2-}] [HCO_3^-] [H_2O_2] electrolytes. *Journal of Applied Electrochemistry*, 31:213–222.

Ligier, V., Wéry, M., Hihn, J.Y., Faucheu, J., and Tachez, M. 1999. Formation of the main atmospheric zinc end products: $NaZn_4Cl(OH)_6SO_4$, $6H_2O$, $Zn_4SO_4(OH)_6$, $5H_2O$, and $Zn_4Cl_2(OH)_4SO_4$, $5H_2O$ in $[Cl^-]$ $[SO_4^{2-}]$ $[HCO^{3-}]$ $[H_2O_2]$ electrolytes. *Corrosion Sciences*, 41:1139–1164.

Litster, S. and McLean, G. 2004. PEM fuel cell electrodes. *Journal of Power Sources*, 130:61–76.

Lorimer, J.P., Pollet, B., Phull, S.S., Mason, T.J., and Walton, D.J. 1998. Disc and cylindrical electrode with ultrasound. *Electrochimica Acta*, 43:449–455.

Lorimer, J.P., Pollet, B., Phull, S.S., Mason, T.J., Walton, D.J., and Geissler, U. 1996. The effect of ultrasonic frequency and intensity upon limiting currents at rotating disc and stationary electrodes. *Electrochimica Acta*, 41:2737–2741.

Luche, J.L. 1993. Sonochemistry from experiment to theoretical considerations. In T.J. Mason (Ed.), *Advances in Sonochemistry*, Vol. 3. London, U.K.: JAI Press.

Marken, F., Eklund, J.C., and Compton, R.G. 1995. Voltammetry in the presence of ultrasound: Can ultrasound modify heterogeneous electron transfer kinetics?. *Journal of Electroanalytical Chemistry*, 395:335–339.

Mason, T.J., Lorimer, J.P., and Walton, D.J. 1990. Sonoelectrochemistry. *Ultrasonics*, 28:333–337.

Mehta, V. and Cooper, J.S. 2003. Review and analysis of PEM fuel cell design and manufacturing. *Journal of Power Sources*, 114:32–53.

Morais, N.L.P.A. and Brett, C.M.A. 2002. Influence of power ultrasound on the corrosion of aluminium and high speed steel. *Journal of Applied Electrochemistry*, 32:653–660.

Moriguchi, N. 1934. *Journal of Chemical Society Japan*, 55:349.

Naffrechoux, E., Cahnoux, S., Petrier, C., and Suptil, J. 2000. Sonochemical and photochemical oxidation of organic matter. *Ultrasonic Sonochemistry*, 7:255–259.

Osawa, S., Ito, M., Tanaka, K., and Kuwano, J. 1987. Electrochemical polymerization of thiophene under ultrasonic field. *Synthetic Metals*, 18:145–150.

Osawa, S., Ito, M., Tanaka, K., and Kuwano, J. 1992. The application of ultrasonic waves to the electrochemical polymerization of thiophene. *Journal of Polymer Science—Part B Polymer Physics*, 30:19–24.

Penn, R., Yeager, E., and Hovorka, F. 1959. Effect of ultrasonic waves on concentration gradients. *Journal of the Acoustical Society of America*, 31:1372–1376.

Pollet, B.G. 1998. The effect of ultrasound upon electrochemical processes, PhD thesis. Coventry University.

Pollet, B.G. 2009. A novel method for preparing PEMFC electrodes by the ultrasonic and sonoelectrochemical techniques. *Electrochemical Communication*, 11:1445–1448.

Pollet, B.G. 2010. Review: The use of ultrasound for the fabrication of fuel cell materials. *International Journal of Hydrogen Energy*, 35(21):11986–12004.

Pollet, B.G., Hihn, J.Y., Doche, M.L., Lorimer, J.P., Mandroyan, A., and Mason, T.J. 2007. Transport limited current close to an ultrasonic horn: Equivalent flow velocity determination. *Journal of Electrochemical Society*, 154:E131–138.

Pollet, B.G., Hihn, J.Y., Lorimer, J.P., Phull, S.S., Mason, T.J., and Walton, D.J. 2002. The effect of ultrasound upon the oxidation of thiosulphate on stainless steel and platinum electrodes. *Ultrasonic Sonochemistry*, 9:267–274.

Pollet, B.G., Hihn, J.Y., and Mason, T.J. 2008. Sono-electrodeposition (20 and 850 kHz) of copper in aqueous and deep eutectic solvents. *Electrochemica Acta*, 53:4248–4256.

Pollet, B., Lorimer, J.P., Phull, S.S., and Hihn, J.Y. 2000. Sonoelectrochemical recovery of silver from photographic processing solutions. *Ultrasonic Sonochemistry*, 7:69–76.

Pollet, B.G., Lorimer, J.P., Phull, S.S., Mason, T.J., and Hihn, J.Y. 2003. A novel angular geometry for the sonochemical silver recovery process at cylinder electrodes. *Ultrasonic Sonochemistry*, 10:217–222.

Pollet, B., Lorimer, J.P., Phull, S.S., Mason, T.J., Walton, D.J., Hihn, J.Y., Ligier, V., and Wéry, M. 1999. The effect of ultrasonic frequency and intensity upon electrode kinetic parameters for the $Ag(S_2O_3)_2^{3-}/Ag$ redox couple. *Journal of Applied Electrochemistry*, 29:1359–1366.

Pollet, B.G. and Phull, S.S. 2001. Sonoelectrochemistry—Theory, principles, and applications. *Recent Research Developments in Electrochemistry*, 4:55–78.

Qiu, X.F., Xu, J.Z., Zhu, J.M., Zhu, J.J., Xu, S., and Chen, H.Y. 2003. Controllable synthesis of palladium nanoparticles via a simple sonoelectrochemical method. *Journal of Materials Research*, 18:1399–1404.

Reisse, J., Francois, H., Vandercammen, J., Fabre, O., Kirsh-De Mesmaeker, A., Maershalk, C., and Delplancke, J.L. 1994. Sonoelectrochemistry in aqueous electrolyte: A new type of sonoelectroreactor. *Electrochimica Acta*, 39:37–39.

Sáez, V. and Mason, T.J. 2009. Sonoelectrochemical synthesis of nanoparticles. *Molecules*, 14:4284–4299.

Saterlay, A.J. and Compton, R.G. 2000. Sonoelectroanalysis: An overview. *Fresenius Journal of Analytical Chemistry*, 367:308–313.

Schmid, G. and Ehret, L. 1937a. Beeinflussung der Metallpassivität durch Ultraschall. *Zeitschrift Fur Elektrochemie*, 43(6):408–415.
Schmid, G. and Ehret, L. 1937b. Beeinflussung der Elektrolytischen Abscheidungspotentiale von Gasen durch Ultraschall. *Zeitschrift Fur Elektrochemie*, 48(8):597–608.
Shen, Q., Jiang, L., Zhang, H., Min, Q., Hou, W., and Zhu, J.J. 2008. Three-dimensional dendritic Pt nanostructures: Sonoelectrochemical synthesis and electrochemical applications. *Journal of Physical Chemistry*, 112:16385–16392.
Shen, Q., Min, Q., Shi, J., Jiang, L., Zhang, J.R., Hou, W., and Zhu, J.J. 2009. Morphology-controlled synthesis of Palladium nanostructures by sonoelectrochemical method and their application in direct alcohol oxidation. *Journal of Physical Chemistry*, 113:1267–1273.
Skotheim, T.A. 1986. *The Handbook of Conducting Polymers*. Vol. 1 and 2, New York: Marcel Dekker.
Taouil, A., Lallemand, F., Hallez, L., and Hihn, J.Y. 2010. Electropolymerization of pyrrole on oxidizable metal under high frequency ultrasound irradiation. Application of focused beam to a selective masking technique. *Electrochimica Acta*, 55(28):9137–9145.
Taouil, A., Et, Lallemand, F., Hihn, J.Y., Melot, J.M., Blondeau-Patissier, V., and Lakard, B. 2011. Doping properties of PEDOT films electrosynthesized under high frequency ultrasound irradiation. *Ultrasonic Sonochemistry*, 18:140–148.
Torii, S. 1985. *Electro-Organic Syntheses, Part 1: Oxidations*. Tokyo: Kodansha.
Touyeras, F., Hihn, J.Y., Bourgoin, X., Jacques, B., Hallez, L., and Branger, V. 2005. Effects of ultrasonic irradiation on the properties of coatings obtained by electroless plating and electroplating. *Ultrasonics Sonochemistry*, 12:13–19.
Touyeras, F., Hihn, J.Y., Delalande, S., Viennet, R., and Doche, M.L. Ultrasound influence on the activation step before electroless coating. *Ultrasonics Sonochemistry*, 10:363–368.
Touyeras, F., Hihn, J.Y., Doche, M.L., and Roizard, X. 2001. Electroless copper coating of epoxide plates in an ultrasonic field. *Ultrasonics Sonochemistry*, 8:285–290.
Vassiliev, Y.B. and Grinberg, V.A. 1991. Adsorption kinetics of electrode processes and the mechanism of Kolbe electrosynthesis: Part I. Adsorption of carboxylic acids and the nature of the particles chemisorbed on platinum electrodes. *Journal of Electroanalytical Chemistry*, 283:359–378.
Wadhawan, J.D., Marken, F., and Compton, R.G. 2001. Biphasic sonoelectrochemistry. A review. *Pure and Applied Chemistry*, 73:1947–1955.
Walker, R. 1993. The effect of ultrasound on electrodeposition and electroplating. In T.J. Mason (Ed.)., *Advances in Sonochemistry*, Vol. 3. Cirencester: JAI Press.
Walton, D.J. 1991. *Electronic Materials—From Silicon to Organics*, L.S. Miller and J.B. Mullin (Ed.). New York: Plenum Press.
Walton, D.J. 2002. Sonoelectrochemistry—The application of ultrasound to electrochemical systems. *Arkovic*, 3:198–218.
Walton, D.J., Hall, C.E., and Chyla, A. 1992. Characterization of poly(pyrroles) by cyclic voltammetry. *The Analyst*, 117:1305–1308.
Walton, D.J. and Mason, T.J. 1998. Organic sonoelectrochemistry. *Synthetic Organic Sonochemistry*, 4:263–300.
Walton, D.J. and Phull, S.S. 1996. Sonoelectrochemistry. *Advances in Sonochemistry*, 4:205–284.
Wassersheid, P. and Welton, T. 2003. *Ionic Liquids in Synthesis*. Weinheim, Germany: Wiley-VCH Verlag.
Wood, R.W. and Loomis, A.L. 1927. The physical and biological effects of high intensity high-frequency sound waves. *Philosophical Magazine*, 4:417.
Yegnaraman, V. and Bharathi, S. 1992. Sonoelectrochemistry—An emerging area. *Electrochemistry*, 8:84–85.
Zhang, H. and Coury Jr., L.A., 1993. Effects of high-intensity ultrasound on glassy carbon electrodes. *Analytical Chemistry*, 65:1552–1556.
Zin, V., Pollet, B.G., and Dabalà, M. 2009. Sonoelectrochemical (20 kHz) production of platinum nanoparticles from aqueous solutions. *Electrochimica Acta*, 54:7201–7206.

25 Combined Ultrasound–Microwave Technologies

Pedro Cintas, Giancarlo Cravotto, and Antonio Canals

CONTENTS

Introduction 659
Chemical Analysis 659
Natural Product Extraction 662
Synthesis and Catalysis 666
Preparation of Nanomaterials 668
Acknowledgments 670
References 670

INTRODUCTION

Power ultrasound (US) and microwave (MW) dielectric heating are among the most simple, inexpensive, and effective tools in applied chemistry. These green techniques dramatically enhance heat and mass transfer, inducing faster and more selective chemical transformations. Sonochemistry has a longer history than microwave chemistry; however, in the last two decades, the latter has found a growing number of relevant applications. While popular wisdom just associates MW with superior heating and US with efficient agitation, these techniques are capable of doing much more and this potential provides additional impulse to their expansion in synthesis and processing. The reproducibility of these techniques, however, still requires further attention, especially since some design parameters are sometimes overlooked. Surprisingly, the additive and even synergic effects that have been observed in combined US/MW irradiations have overcome all expectations (Cravotto and Cintas, 2007). Maeda and Amemiya first described the surprising synergic effects in sono- and chemiluminescence experiments carried out under simultaneous US/MW irradiations (Maeda and Amemiya, 1995). Although sequential or simultaneous irradiation with US and MW sources entails technical and safety considerations, their coupling can easily be performed on a lab scale. It is hoped that such equipment will be commercially available in the near future at a reasonable price.

This chapter is divided into four sections dealing with the applications in chemical analysis, natural product extraction, synthesis and catalysis, and preparation of nanomaterials.

CHEMICAL ANALYSIS

Much wider attention has been given to the applications of MW dielectric heating and US energy in analytical chemistry than in other chemical disciplines (Kingston and Jassie, 1988; Zlotorzynski, 1995; Kingston and Haswell, 1997; Mason and Lorimer, 2002; Luque-García and Luque de Castro, 2003a,b; Priego-Capote and Luque de Castro, 2004; Luque de Castro and Priego-Capote, 2007; Capelo-Martínez, 2009). Among these applications, we can find MW and US spectroscopy (Townes and Schawlow, 1975; Migliori and Sarrao, 1997; Colorado State University, Department of Physics, 2010). They have, however, been mainly used for sample preparation (i.e., digestion, extraction, dissolution, etc.) (Kingston and Jassie, 1988; Kingston and Haswell, 1997; Luque-García and

Luque de Castro, 2003a,b; Priego-Capote and Luque de Castro, 2004; Luque de Castro and Priego-Capote, 2007; Capelo-Martínez, 2009). MW and US have been used to speed up and increase the efficiency of sample preparation, save energy, and reduce the amount of reagents and the risk of contamination. MW radiation has been recently used to improve the single drop microextraction of chlorobenzenes from water samples (Vidal et al., 2007) and to speed up the cloud point extraction of Rh, Pd, and Pt from pharmaceutical products (Simitchiev et al., 2008). US energy has also been used to increase the efficiency and velocity of microextraction techniques (Regueiro et al., 2008). However, few applications have been published to date on the combination effects of both types of energies on sample preparation (Lagha et al., 1999; Chemat et al., 2004; Canals et al., 2006; Domini et al., 2009) since simultaneous irradiation with MW and US entails technical and safety considerations (Cravotto and Cintas, 2007). In all cases, a simultaneous US/MW irradiation occurred; however, the Canals' system (Canals et al., 2006; Domini et al., 2009) was the only one that could irradiate the sample in a direct way. This means that the US probe was placed in direct contact with the reactive mixture inside the MW cavity. This strongly improved the sonochemical efficiency.

In 1999, Chemat et al. (Lagha et al., 1999) published the pioneering application of simultaneous US/MW irradiation used for the digestion and dissolution of the refractory mineral material Co_3O_4 (mineral model) in nitric acid as well as for the determination of copper in olive oil (food product model). The authors used a Prolabo Maxidigest 350 monomode MW oven (max power 300 W, Figure 25.1). The reacting mixture was placed in a borosilicate open vessel (20–150 mL) joint with the US cup horn (Branson Sonifier 250, diameter of the transducer's tip 18 mm) by means of decaline, introduced into a double jacket (200 mL). This liquid was chosen because of its low viscosity, which enables good US propagation meanwhile is MW transparent. The US probe was placed distant from the electromagnetic field in order to avoid interactions and short circuits. MW digestion and classical convective–conductive protocols were used for comparison. From the kinetic study it was concluded that the dissolution rate with the combined US/MW system was two and six times faster than the MW only and classical heating systems (total metal dissolution in only 1, 3, and 6 h), respectively. This acceleration was due to the synergistic combination of both types of energy: MW heating created hot spots at the surface of the metal enhancing the dissolution's temperature and rate, whereas the acoustic cavitation eroded the metal oxide and enhanced the mass transfer for the digested metal. The average MW and US power required for total solid dissolution was 150 W for both energy sources. Under optimized conditions (i.e., MW power: 180 W for 27 min and 290 W for 3 min; US power: 150 W for 30 min) all the Cu in the olive oil was digested in only 30 min, whereas the MW alone or the conventional method took 45 and 60 min, respectively. The efficiency, especially in heterogeneous (solid/liquid) media, of the combined MW and US irradiations was well proved in this study.

Later, Chemat et al. (2004) used the same manifold (Figure 25.1) to study the digestion of edible oils for copper analysis and food products for total Kjeldahl nitrogen analysis. A MW digestion method and a classical convective-conductive standard method AFNOR were used for comparison. The amount of nitrogen recovered was comparable in the three digestion procedures; however, US/MW Kjeldahl digestion method was complete within 10 min, whereas the digestion times were 30 min and 3 h for the MW alone digestion and the classical standard method, respectively. Therefore, the US/MW Kjeldahl digestion method was more efficient for viscous liquid and solid products. This improvement was observed even though the samples were not directly subjected to US, which detracted from its optimal efficiency. When all the copper of the edible oils was analyzed [digestion conditions as Lagha et al. (1999)], the digestion time values were 30, 45, and 60 min for US/MW, MW, and classical method, respectively.

Canals et al. (2006) have patented the combination of both energies for analytical sample preparation (i.e., digestion, dissolution, and extraction) purposes. A scheme of the combined US/MW system is shown in Figure 25.2 while Figure 25.3 displays an image of the prototype. It is based on a CEM Discover MW oven operating at 2.45 GHz with a nominal power of 300 W (5). The US system (6) is a Dr. Hielscher UP 200 S (200 W power; 24 kHz frequency) equipped with a glass horn

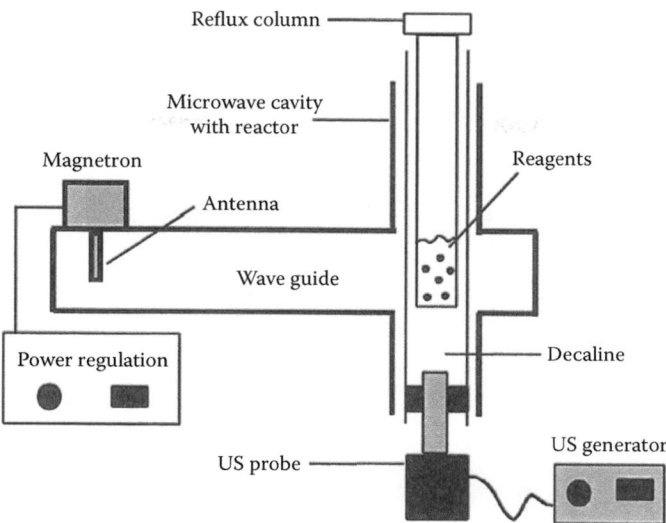

FIGURE 25.1 Schematic drawing of the single mode microwave system combined with ultrasound. (Reprinted from *Ultrason. Sonochem.*, 11, Chemat, S., Lagha, A., Amar, H.A., and Chemat, F., Ultrasound-assisted microwave digestion, 5–8, Copyright (2004), with permission from Elsevier.)

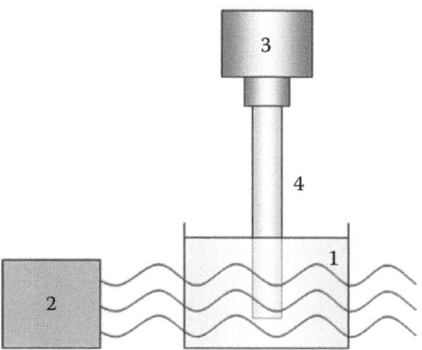

FIGURE 25.2 Scheme of direct and simultaneous US/MW irradiation system. (1) Reaction mixture (sample + reagents or liquid phase immiscible with sample), (2) MW radiation source, (3) US energy source, and (4) horn (sonotrode) (manufactured with an MW transparent material).

of 12 mm diameter (4). This new design of the US probe (4) is in direct contact with the reactive mixture (1). This combined digestion system is easy to handle, safe, and efficient. The sample vessel is operated at atmospheric pressure and must be chemically inert and transparent to MW. The system enables working simultaneously or separately with both energies by simply disconnecting one generator (2 or 3). For safety reasons and in order to eliminate MW leakage through the hole used to introduce the sonotrode, a metallic cylinder (4 cm diameter; 5 cm high) is used. This cylinder acts as a cut-off filter for MW.

Chemical oxygen demand (COD) was measured to assess the applicability of this combined system. A classical convective-conductive heating system, three MW systems (one closed and two open), and one US-assisted digestion systems were used for comparison (Domini et al. 2006). It is well known that pyridine is a difficult compound for COD determination; therefore, this analyte was used as a model. Table 25.1 shows the results obtained on COD determination with all the digestion systems evaluated. The new US/MW system supplies a recovery value of 75% in only 1 min, whereas the recovery values obtained with the remaining digestion systems were 12%, 27%, 21%,

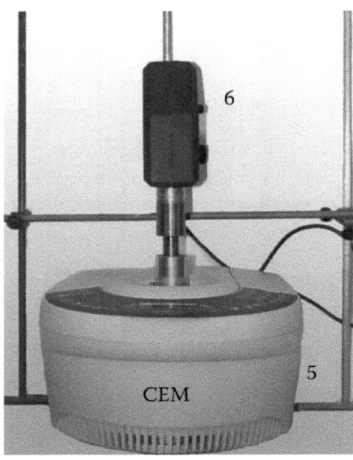

FIGURE 25.3 Photography of simultaneous and direct US/MW irradiation prototype. (5) MW unit; (6) US unit.

52%, and 46% for classical, closed MW, open MW system no. 1 (Star system 2, CEM), US, and open MW system no. 2 (Discover, CEM) digestion methods, respectively.

A recent publication (Domini et al., 2009) described the application of combined US/MW technologies in Kjeldahl method for total nitrogen determination in real food samples. Multivariate analysis has been used to optimize the most influential variables for the US/MW digestion using tryptophan as the model substance. The optimum conditions were H_2SO_4 volume, 10 mL; H_2O_2 volume, 5 mL, weight of sample, 0.05 g; MW power, 500 W; US power, 50 W; and digestion time, 7 min. A modification of the classical Kjeldahl method and a US-assisted digestion method were used for comparison. Digestion under simultaneous US/MW irradiation was carried out in a closed vessel placed in a professional multimode oven (Figure 25.4a) (Mycrosynth, Milestone®, BG, Italy) equipped with a US probe with a Pyrex® horn (Figure 25.4b) (frequency 21.4 kHz, tip diameter 17 mm). The setup was developed in the laboratory of Cravotto at the University of Torino in collaboration with Danacamerinisas (TO, Italy). The same US probe (Figure 25.4b) was also used in the digestion experiments carried out under US alone. The significant reduction in digestion time (30 and 25 min for classical Kjeldahl method and US-assisted digestion method, respectively) and in consumption of reagents evidence that simultaneous and direct US/MW irradiation is a powerful and promising tool for low-pressure preparation (i.e., digestion, dissolution, and extraction) of solid and liquid samples.

NATURAL PRODUCT EXTRACTION

Although in recent years several applications of combined US/MW irradiation for plant extraction have appeared in the literature, the great potential of this hybrid technique has not been adequately exploited as yet. US could dramatically improve the extraction of a target component mainly through the phenomenon of cavitation. The mechanical ultrasonic effect promotes the release of soluble compounds from the plant body by disrupting cell walls, enhancing mass transfer, and facilitating solvent access to the cell content. Meanwhile, MW heats the whole sample very quickly and volumetrically, inducing the migration of dissolved ions. The simultaneous irradiation of both energy sources increases solvent penetration into the matrix, facilitates analyte solvation, and usually increases the solubility of target compounds.

The combination of US-assisted extraction (UAE) and MW-assisted extraction (MAE) has been successfully employed by Cravotto and Cintas as complementary techniques in the extraction of oils from vegetable sources, viz., soybean germ and a cultivated marine microalgae rich in docosahexaenoic acid (DHA) (Cravotto et al., 2008a). This was achieved by inserting a nonmetallic horn in a professional

TABLE 25.1
Chemical Oxygen Demand (COD) Values and Precision Obtained with Different Digestion Systems (Canals et al., 2006)[a]

Compound	Theoretical Value ThCOD (mg L^{-1})	Classical Semimicro Method[b]			Closed Microwaves[b]			Open Microwaves No. 1 (Star System 2, CEM)[b]			Ultrasounds[b]			Open Microwaves No. 2 (Discover, CEM)[c]			Simultaneous–Directly Irradiated with US/MW[c]		
		COD (mg L^{-1})	RSD (%)	R[d] (%)	COD (mg L^{-1})	RSD (%)	R[d] (%)	COD (mg L^{-1})	RSD (%)	R[d] (%)	COD (mg L^{-1})	RSD (%)	R[d] (%)	COD (mg L^{-1})	RSD (%)	R[d] (%)	COD (mg L^{-1})	RSD (%)	R[d] (%)
Pyridine	99.05	12.1	4.2	12	26.6	18.9	27	20.8	11.8	21	51.4	6.0	52	45.6	13.3	46	74.2	5.2	75

[a] Number of replicates: $n = 5$.
[b] Irradiation time: classical—2 h; closed MW—4 min; open MW—4 min; US—1 min (Domini et al., 2006).
[c] Irradiation time: 1 min.
[d] Recovery.

(a)

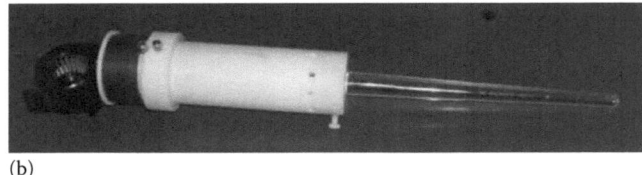

(b)

FIGURE 25.4 (a) Simultaneous and direct US/MW irradiation in multimode MW oven (Milestone®) equipped with a pyrex® horn developed at University of Torino; (b) US horn made of pyrex®. (Reprinted from *Ultrason. Sonochem.*, 16, Domini, C., Vidal, L., Cravotto, G., and Canals, A., A simultaneous, direct microwave/ultrasound-assisted digestion procedure for the determination of total Kjeldahl nitrogen, 564–569, Copyright (2009), with permission from Elsevier.)

MW oven. If on one hand, the double simultaneous irradiation can bring additive or even synergic effects on the extraction phenomenon of vegetal matrices, on the other, the nonmetallic horns can only be used at moderate power. Pyrex, quartz, or Peek® (Cravotto and Cintas, 2007) can be safely used in the range of 30–90 W, above which the intrinsic structure of the material can be irreversibly damaged.

Lianfu and Zelong applied a US transducer with fixed power and frequency (50 W, 40 kHz) to the bottom of an MW cavity. Although no technical information was given on this equipment coupling, they described its efficient application to the extraction of lycopene from tomato paste (Lianfu and Zelong, 2008). This US/MW-promoted extraction performed in open vessel was compared with a UAE method. The former required less time (6 vs. 29 min) and gave a better yield (~8% more).

A similar set-up was used by Wang et al. to extract inulin and polyphenols from burdock root, which is a popular vegetable in Japan. The simultaneous US/MW irradiation was performed on 5 g burdock root powder suspended in 75 mL distilled water in a 250 mL Erlenmeyer flask with a condenser. Apart from a much shorter extraction time, the authors claimed an improvement in the extract's functional and swelling properties (Lou et al., 2009).

Other Chinese researchers combined US and MW technologies in a sequential way to extract and dry isoflavonoids from *Pueraria lobata*, namely, UAE for cell disruption and extraction, and then fast MW-vacuum drying to obtain the final product (Hu et al., 2008). The authors showed that the US-based method followed by MW drying under vacuum increases extraction efficiency while preserving the pharmacological properties of the active compounds.

Among the several techniques used to extract the volatile components in cucumbers (vacuum distillation-freeze concentration, solid-phase microextraction), combined US/MW has also been described (Beb et al., 2008).

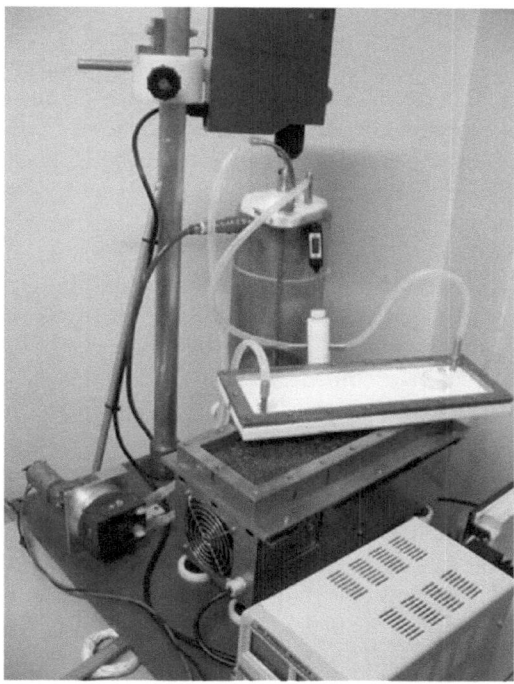

FIGURE 25.5 A prototype of pilot US-flow reactor for plant extraction.

Although lab-scale simultaneous US/MW irradiation for plant extraction seems to be very advantageous, its scaling up is hardly feasible. As already demonstrated for UAE (Cravotto and Binello, 2010) (Figure 25.5), the combined US/MW flow reactors that have already been exploited for water decontamination (Cravotto et al., 2005) seem to be the most promising tools for the scale-up of such applications (Cravotto et al., 2007a). Combined UAE–MAE extraction can be easily achieved in a sequential way, where a peristaltic pump circulates the plant powder suspension: first through the sonication chamber (Figure 25.5) and then through an MW cavity (Figure 25.6).

FIGURE 25.6 A modified MW domestic oven incorporating a flow unit for plant extraction.

SYNTHESIS AND CATALYSIS

The combined use of MW and US irradiations constitutes a very promising innovation in the field of heterogeneous catalysis. Additional effects are to be expected when dielectric heating is associated with the large amount of energy released in cavitational collapse, causing particle fragmentation and molecular excitation. Several examples of combined US/MW irradiation applied to synthesis and catalysis have appeared in the literature (Cravotto and Cintas, 2006, 2007). The first reports were published by Peng and Song regarding the hydrazinolysis of esters in solvent-free conditions (Peng and Song, 2001), a straightforward ether synthesis via the Williamson reaction in the absence of phase-transfer catalysts (Peng and Song, 2002), and the preparation of 3-aryl acrylic acids through a Knoevenagel-Doebner reaction in aqueous media (Peng and Song, 2003). They employed a simple, home-made, yet efficient device by inserting a detachable horn (the material of which was not specified) into a modified domestic oven. Potentially hazardous MW leakage was prevented by a copper mesh screen tightly fastened onto the horn. A few years later they also reported the synthesis of β-aminoketones via an aqueous Mannich reaction (Peng et al., 2005) and the flash synthesis of 4H-pyrano[2,3-c] pyrazoles in aqueous media (Peng et al., 2006).

In an alternative, less efficient set-up that avoided subjecting the horn to the high-frequency field, a low-viscosity apolar liquid (decalin), exposed to US outside the MW oven, conveys the radiation of the former through a double-jacketed pyrex vessel to the reacting mixture placed in the oven itself (Figure 25.1). This device already described previously (Chemat et al., 1996, 1997) was used for the esterification of stearic acid with butanol under heterogeneous catalysis, reporting improved results in comparison with the MW-promoted reaction.

Several synthetic applications of the combined US/MW technique have been reported by Cravotto and associates. This approach was described as a "new synergy in green organic synthesis" in particular when applied to the preparation and use of room-temperature ionic liquids (ILs) as solvents (Lévêque and Cravotto, 2006). With a series of synthetically useful aryl–aryl couplings, catalyzed by Pd/C or Pd (II) acetate, the combined US/MW technique gave better yields and shorter reaction times than individual irradiations with either US or MW (Cravotto et al., 2005) (Scheme 25.1 and Table 25.2). Suzuki homo- and cross-couplings of several arylboronic acids and aryl halides were strongly improved using the patented US/MW flow reactor depicted in Figure 25.7 (Buffa et al., 2004). Even electron-deficient aryl chlorides reacted using palladium (II) acetate as catalyst and required neither phosphine ligands nor phase-transfer catalysts.

Cravotto et al. showed that Heck reactions could conveniently be carried out under simultaneous US/MW irradiation to afford products in high yields while using very low ligandless catalyst loads (Scheme 25.2) (Palmisano et al., 2007a). With styrene, electron-poor aryl chlorides such as 4-chloroacetophenone and 4-chloronitrobenzene gave good yields after 1 h in the presence of 0.25 mol% $Pd(OAc)_2$ and a co-catalyst (Wilkinson 0.005 mol% or CuBr 4.0 mol%) or 2.0–3.0 mol% Pd/C. In most cases MW heating gave comparable results (although yields were 5%–20% lower) in somewhat longer times, whereas under conventional heating acceptable yields were only achieved after 18 h.

US/MW-assisted Suzuki and Heck reactions with low catalyst loads have been performed on poorly reactive 3,5-dichloro-2-pyrazinones. Due to the optimal heat and mass transfer, simultaneous US/MW irradiation strongly improved the kinetics and yields of these chemical modifications (Garella et al., 2010).

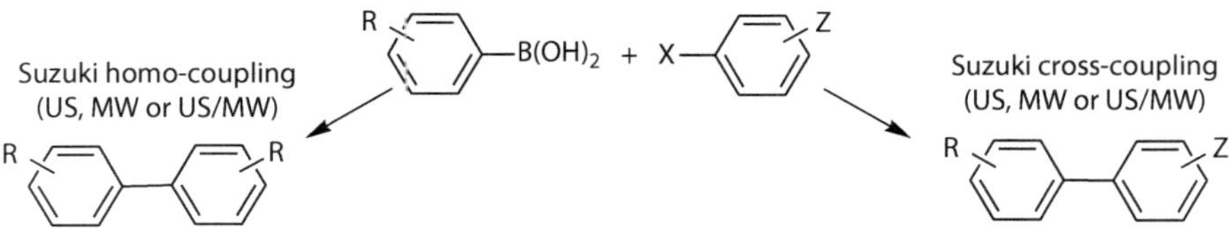

SCHEME 25.1 Suzuki reactions under US and MW, alone or combined.

TABLE 25.2
Comparison of Yields of Suzuki Reactions under US and MW, Alone or Combined

Aryl Halide	Boronic Acid	US Yield (%)	MW Yield (%)	US/MW Yield (%)
3-bromoanisole	phenylboronic	54	64	88
2-iodothiophene	phenylboronic	40	37	59
4-chloro-nitrobenzene	phenylboronic	22	30	57
—	thianthrene-1-boronic	48	55	69
—	4-*t*-butylboronic	68	74	86

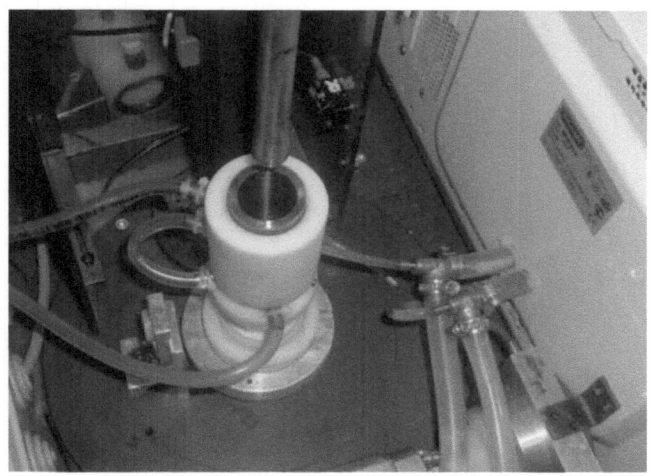

FIGURE 25.7 A loop US/MW flow reactor.

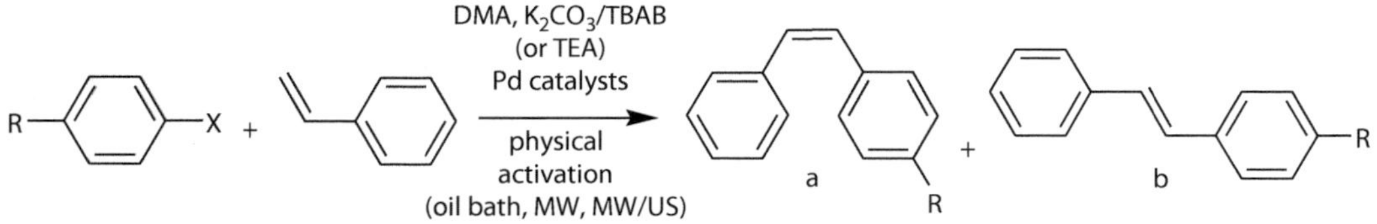

X = I or Br;	R = OCH_3	(*Z*)-4-methoxystilbene (1a),	(*E*)-4-methoxystilbene (1b)
X = Br or Cl;	R = $COCH_3$	(*Z*)-4-acetylstilbene (2a),	(*E*)-4-acetylstilbene (2b)
X = Cl;	R = NO_2	(*Z*)-4-nitrostilbene (3a),	(*E*)-4-nitrostilbene (3b)

SCHEME 25.2 Heck couplings using very low ligandless catalyst loads under MW, or combined US/MW irradiation.

The use of combined US/MW irradiation promotes redox reactions, such as the preparation of azo and azoxy compounds by selective reduction of nitroarenes using zinc powder and ammonium chloride in DMF/water (Cravotto et al., 2006) or by oxidation of primary aromatic amines with $KMnO_4$ in DMF (Wu et al., 2008). The technique was successfully applied to a catalyst-free synthetic procedure for the regioselective opening of epoxides by *N*-nucleophiles in water (Palmisano et al., 2007b) and to the one-pot synthesis of second-generation ILs cutting down reaction times and improving yields (Cravotto et al., 2007b). Analogous results were even achieved in the alkylation of

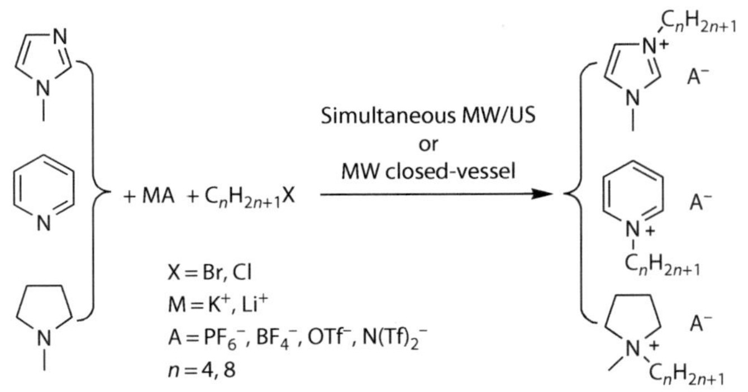

SCHEME 25.3 Solvent-free one-pot synthesis of ionic liquids under simultaneous US/MW irradiation.

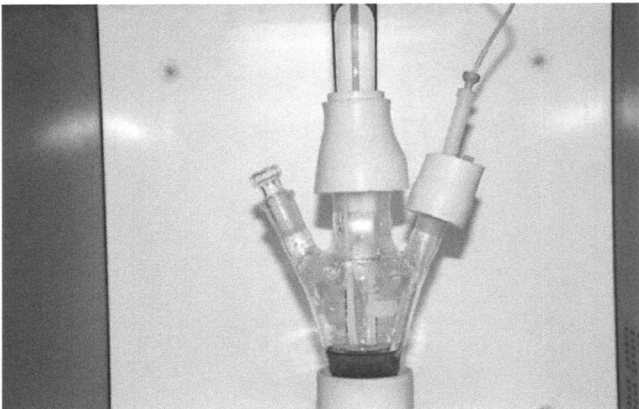

FIGURE 25.8 Click reaction catalyzed by metallic copper under US/MW irradiation.

N-heterocycles with poorly reactive but inexpensive alkyl chlorides that usually require long reaction times and drastic conditions (Cravotto et al., 2008b) (Scheme 25.3).

Collina and associates described a US/MW-assisted procedure for the preparation of α,β-unsaturated carboxylic acids and esters from aldehydes via tandem Wittig olefination and hydrolysis (Rossi et al., 2009).

The 1,3-dipolar cycloaddition of azides and alkynes has become the model for click reactions. US and MW irradiations, alone or combined, have been used to accelerate this reaction especially under heterogeneous catalysis, for example, charcoal-supported Cu(II) or Cu(I) catalysts as employed by Cintas and associates (Cintas et al., 2007). One of the most attractive procedures used to carry out click reactions exploits metallic copper as a solid catalyst (Cravotto et al., 2010). This is a green, sustainable method, which is extremely fast when performed under combined US/MW irradiation (Figure 25.8) (Cintas et al., 2010).

PREPARATION OF NANOMATERIALS

US and MW irradiations used separately have been widely used for the preparation of nanomaterials as documented by several reviews (Kerner et al., 2001; Gedanken, 2004; Dahl et al., 2007). Both techniques contribute to greener nanosyntheses, within shorter reaction times, reduced energy consumption, and better product yields. Such remarkable improvements provide a unique platform for the growth of novel nanostructures (Nemamcha et al., 2006). Noteworthy is the possibility to scale-up the production of nanoparticles using these methodologies for industrial applications

Combined Ultrasound–Microwave Technologies

FIGURE 25.9 Field emission scanning electron microscopy (FESEM) of CdS nanoflowers prepared by US/MW irradiation at 413 K for 45 min by reacting $CdCl_2$, sulfur powder, and another sulfur source [(a) and (b): $CS(NH_2)_2$; (c) and (d): C_2H_5NS] taken at different resolutions. The inset shows isolated CdS nanoflower clusters. (Reprinted from *Ultrason. Sonochem.*, 15, Tai, G. and Guo, W., Sonochemistry-assisted microwave synthesis and optical study of single-crystalline CdS nanoflowers, 350–356, Copyright (2008), with permission from Elsevier.)

(Gerbec et al., 2005). Many studies have explored the effects of temperature, solvent, reductant, and reaction times on nanoparticle formation. Although MW and US irradiations have proven to be useful tools in synthesizing nanomaterials, they have some limitations. These are limited thickness penetration and, in some cases, the poor experimental repeatability (Nikolai, 2002).

In recent years a few reports have documented the favorable combination of US and MW for the preparation of nanomaterials. Ishikawa et al. described the preparation of platinum nanoparticles in a heterogeneous solid–liquid system by simultaneous US and MW irradiations. The average size of the Pt nanoparticles resulted in fine particles with an excellent homogeneous distribution (Ishikawa et al., 2008). Tai and Guo developed a US/MW-assisted procedure for the controlled synthesis of cadmium sulfide 3D nanostructures (flower-like hexagonal nanopyramids and/or nanoplates) of high purity (Figure 25.9). This efficient method was fast, high-yield, seedless, template-free, and environmentally friendly (Tai and Guo, 2008).

Shen studied the US/MW combination for a rapid synthesis of Pb(OH)Br nanowires (Shen, 2009). The ionic liquid 1-butyl-3-methylimidazolium bromide ([BMIM]Br) was employed both as a reactant and as structure-directing agent in this reaction. Combined US/MW irradiation at 50 W in each case greatly reduced the reaction time (10 min) and significantly increased the product yield (45%) in comparison with conventional heating (24 h reaction time, 23% yield). Combined US/MW irradiation also changed the resultant wires from 20–30 μm in diameter and 2–3 mm in length to 80–800 nm and 50–100 μm, respectively. A further increase in the power of the combined irradiation (from 50 to 250 W) led to a dramatic reduction in the reaction time (80 s) with a slightly increased yield up to *ca.* 48%.

Ag-doped CdS nanoparticles have been successfully synthesized by Guo et al. under combined US/MW irradiation (Figure 25.10). These results show that this synthetic technique can be applied as the method of choice for the production of noble metal-doped semiconductor nanostructures (Ma et al., 2010).

We can conclude that combined US/MW irradiation will surely find a wide range of practical applications in nanoscience and nanotechnology.

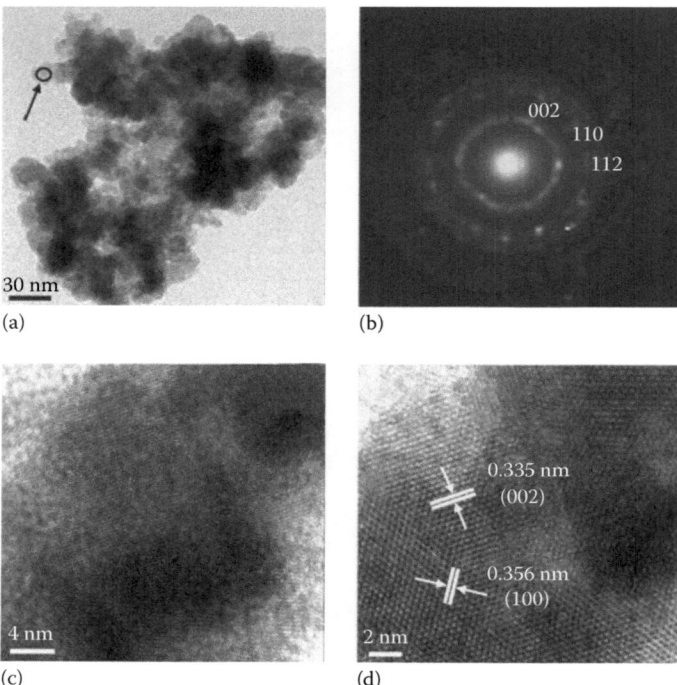

FIGURE 25.10 Transmission electron microscopy (TEM) images of Ag-doped (5%) CdS nanoparticles. (a) TEM image of the sample showing spherical particles with an approximate size of 15 nm; (b) electron diffraction pattern recorded on the nanoparticles (wurtzite phase); (c) and (d) high-resolution TEM images of the as-prepared sample. (Reprinted from *Ultrason. Sonochem.*, 17, Ma, J., Tai, G., and Guo, W., Ultrasound-assisted microwave preparation of Ag-doped CdS nanoparticles, 534–540, Copyright (2010), with permission from Elsevier.)

ACKNOWLEDGMENTS

Financial support from the Spanish Ministry of Science and Innovation (Grants CTQ2008-06730-C02-01 and MAT2009-14695-C04-01), the Regional Government of Valencia (Spain) (ACOMP/2009/144), and the Italian Ministry of Education, University and Research (PRIN 2008) is gratefully acknowledged.

REFERENCES

Beb, H.-X., Lu, D.-D., Bian, N.-S., and Wu, M.-F. 2008. Studies on the chemical constituents of cucumber. *Tianran Chanwu Yanjiu Yu Kaifa (Natural Product Research & Development, in Chinese)*, 20: 388–394.

Buffa, C., Cravotto, G., Dal Lago, G., and Omiccioli, G. 2004. Combined microwave-ultrasound reactors for chemical synthesis. Italian patent (2005), IT 2004-VR167 20041022, *Chem. Abs.*, 152: 195062.

Canals, A., Hidalgo, M., Domini, C.E., and Cravotto, G. 2006. Method and apparatus for direct irradiation of a liquid or solid sample with microwave or ultrasound irradiation simultaneously, consecutively or alternating. Spanish patent (2008), ES 2304839 A1 20081016, *Chem. Abs.*, 150: 578426.

Capelo-Martínez, J.L. 2009. *Ultrasound in Chemistry: Analytical Applications*. Weinheim: Wiley-VCH.

Chemat, S., Lagha, A., Amar, H.A., and Chemat, F. 2004. Ultrasound assisted microwave digestion. *Ultrasonics Sonochemistry*, 11: 5–8.

Chemat, F., Poux, M., Di Martino, J.L., and Berlan, J. 1996. An original microwave-ultrasound combined reactor suitable for organic synthesis: Applications to pyrolysis and esterifications. *Journal of Microwave Power and Electromagnetic Engineering*, 31: 19–22.

Chemat, F., Poux, M., and Galema, S.A. 1997. Esterification of stearic acid by isomeric forms of butanol in a microwave oven under homogeneous and heterogeneous reaction conditions. *Journal of the Chemical Society, Perkin Transactions*, 2: 2371–2374.

Cintas, P., Barge, A., Tagliapietra, T., Boffa, L., and Cravotto, G. 2010. Alkyne-azide click reaction catalyzed by metallic copper under ultrasound. *Nature Protocols*, 5: 607–616.

Cintas, P., Martina, K., Robaldo, B. et al. 2007. Improved protocols for microwave-assisted Cu(I)-catalyzed Huisgen 1,3-dipolar cycloadditions. *Collection of Czechoslovak Chemical Communications*, 72: 1014–1024.

Colorado State University, Department of Physics. 2010. Ultrasonic spectroscopy. http://www.physics.colostate.edu/groups/ultrasound/resonant–ultrasound.htm (accessed June 2010).

Cravotto, G., Beggiato, M., Palmisano, G., Lévêque, J.M., and Bonrath, W. 2005. High-intensity ultrasound and microwave, alone or combined, promote Pd/C-catalyzed aryl–aryl couplings. *Tetrahedron Letters*, 46: 2267–2271.

Cravotto, G. and Binello, A. 2010. Innovative techniques and equipments for flavours extraction. Application and effectiveness of ultrasound and microwaves. *Household and Personal Care Today*, 1: 32–34.

Cravotto, G., Boffa, L., Bia, M., Bonrath, W., and Heropoulos, G. 2006. An easy access to aromatic azo compounds under US/MW irradiation. *Synlett*, 2605–2608.

Cravotto, G., Boffa, L., Calcio Gaudino, E. et al. 2008b. Preparation of second generation ionic liquids by efficient solvent-free alkylation of *N*-heterocycles with chloroalkanes. *Molecules*, 13: 149–156.

Cravotto, G., Boffa, L., Lévêque, J.M., Estager, J., and Bonrath, W. 2007b. A speedy one-pot synthesis of ionic liquids under microwave/ultrasound irradiation. *Australian Journal of Chemistry*, 60: 946–950.

Cravotto, G., Boffa, L., Mantegna, S. et al. 2008a. Improved extraction of natural matrices under high-intensity ultrasound and microwave, alone or combined. *Ultrasonics Sonochemistry*, 15: 898–902.

Cravotto, G. and Cintas, P. 2006. Power ultrasound in organic synthesis: Moving cavitational chemistry from academia to innovative and large-scale applications. *Chemical Society Reviews*, 35: 180–196.

Cravotto, G. and Cintas, P. 2007. The combined use of microwaves and ultrasound: New tools in process chemistry and organic synthesis. *Chemistry—A European Journal*, 13: 1902–1909.

Cravotto, G., Di Carlo, S., Curini, M., Tumiatti, V., and Roggero, C. 2007a. A new flow reactor for the treatment of polluted water with microwave and ultrasound. *Journal of Chemical Technology and Biotechnology*, 82: 205–208.

Cravotto, G., Fokin, V.V., Garella, D., Binello, A., and Barge, A. 2010. Huisgen 1,3-dipolar cycloaddition catalyzed by metallic copper under ultrasound. *Journal of Combinatorial Chemistry*, 12: 13–15.

Cravotto, G., Tumiatti, V., and Roggero, C. 2005. Process for the degradation and/or detoxification of chemical and biological pollutants. International Patent WO2005IB03028.

Dahl, J.A., Maddux, B.L.S., and Hutchison, J.E. 2007. Toward greener nanosynthesis. *Chemical Reviews*, 107: 2228–2269.

Domini, C.E., Hidalgo, M., Marken, F., and Canals, A. 2006. Comparison of three optimized digestion methods for rapid determination of chemical oxygen demand: Closed microwaves, open microwaves and ultrasound irradiation. *Analytica Chimica Acta*, 561: 210–217.

Domini, C., Vidal, L., Cravotto, G., and Canals, A. 2009. A simultaneous, direct microwave/ultrasound-assisted digestion procedure for the determination of total Kjeldahl nitrogen. *Ultrasonics Sonochemistry*, 16: 564–569.

Garella, D., Tagliapietra, S., Pravinchandra Mehta, V., Van der Eycken, E., and Cravotto, G. 2010. Straightforward functionalization of 3,5-dichloro-2-pyrazinones under simultaneous microwave and ultrasound irradiation. *Synthesis*, 136–140.

Gedanken, A. 2004. Using sonochemistry for the fabrication of nanomaterials. *Ultrasonics Sonochemistry*, 11: 47–55.

Gerbec, J.A., Magana, D., Washington, A., and Strouse, G.F. 2005. Microwave-enhanced reaction rates for nanoparticle synthesis. *Journal of the American Chemical Society*, 127: 15791–15800.

Hu, Y., Wang, T., Wang, M.-X. et al. 2008. Extraction of isoflavonoids from *Pueraria* by combining ultrasound with microwave vacuum. *Chemical Engineering Process*, 47: 2256–2261.

Ishikawa, D., Hayashi, Y., and Takizawa, H.J. 2008. Preparation of platinum nanoparticles in heterogeneous solid–liquid system by ultrasound and microwave irradiation. *Journal of Nanoscience and Nanotechnology*, 8: 4482–4487.

Kerner, R., Palchik, O., and Gedanken, A. 2001. Sonochemical and microwave-assisted preparations of PbTe and PbSe. A comparative study. *Chemistry of Materials*, 13: 1413–1419.

Kingston, H.M. and Haswell, S.J. 1997. *Microwave-Enhanced Chemistry: Fundamentals, Sample Preparation, and Applications*. Washington, DC: American Chemical Society.

Kingston, H.M. and Jassie, L.B. 1988. *Introduction to Microwave Sample Preparation: Theory and Practice.* Washington, DC: American Chemical Society.

Lagha, A., Chemat, S., Bartels, P.V., and Chemat, F. 1999. Microwave-ultrasound combined reactor suitable for atmospheric sample preparation procedure of biological and chemical products. *Analusis,* 27: 452–457.

Lévêque, J.M. and Cravotto, G. 2006. Microwaves, power ultrasound and ionic liquids. A new synergy in green organic synthesis. *Chimia,* 60: 313–320.

Lianfu, Z. and Zelong, L. 2008. Optimization and comparison of ultrasound/microwave assisted extraction (UMAE) and ultrasonic assisted extraction (UAE) of lycopene from tomatoes. *Ultrasonics Sonochemistry,* 15: 731–737.

Lou, Z., Wang, H., Wang, D., and Zhang, Y. 2009. Preparation of inulin and phenols-rich dietary fibre powder from burdock root. *Carbohydrate Polymers,* 78: 666–671.

Luque de Castro, M.D. and Priego-Capote, F. 2007. *Analytical Applications of Ultrasound.* Amsterdam, the Netherlands: Elsevier.

Luque-García, J.L. and Luque de Castro, M.D. 2003a. Ultrasound: A powerful tool for leaching. *Trac—Trends in Analytical Chemistry,* 22: 41–47.

Luque-García, J.L. and Luque de Castro, M.D. 2003b. Where is microwave-based analytical equipment for solid sample pre-treatment going? *Trac—Trends in Analytical Chemistry,* 22: 90–98.

Ma, J., Tai, G., and Guo, W. 2010. Ultrasound-assisted microwave preparation of Ag-doped CdS nanoparticles. *Ultrasonics Sonochemistry,* 17: 534–540.

Maeda, M. and Amemiya, H. 1995. Chemical effects under simultaneous irradiation by microwaves and ultrasound. *New Journal of Chemistry,* 19: 1023–1028.

Mason, T.J. and Lorimer, J.P. 2002. *Applied Sonochemistry: Uses of Power Ultrasound in Chemistry and Processing.* New York: Wiley-VCH.

Migliori, A. and Sarrao, J.L. 1997. *Resonant Ultrasound Spectroscopy: Applications to Physics, Materials Measurements, and Nondestructive Evaluation.* New York: Wiley-Interscience.

Nemamcha, A., Rehspringer, J.L., and Khatmi, D. 2006. Synthesis of palladium nanoparticles by sonochemical reduction of palladium (II) nitrate in aqueous solution. *Journal of Physical Chemistry B,* 110: 383–387.

Nikolai, K. 2002. Microwave-assisted reactions in organic synthesis—Are there any nonthermal microwave effects? *Angewandte Chemie International Edition,* 41: 1863–1866.

Palmisano, G., Bonrath, W., Boffa, L. et al. 2007a. Heck reactions with very low ligandless catalyst loads accelerated by microwaves or simultaneous microwaves/ultrasound irradiation. *Advanced Synthesis and Catalysis,* 349: 2338–2344.

Palmisano, G., Tagliapietra, S., Barge, A. et al. 2007b. Efficient regioselective opening of epoxides by nucleophiles in water under simultaneous ultrasound/microwave irradiation. *Synlett,* 2041–2044.

Peng, Y., Dou, R., Song, G., and Jiang, J. 2005. Dramatically accelerated synthesis of β-aminoketones via aqueous Mannich reaction under combined microwave and ultrasound irradiation. *Synlett,* 2245–2247.

Peng, Y. and Song, G. 2001. Simultaneous microwave and ultrasound irradiation: A rapid synthesis of hydrazides. *Green Chemistry,* 3: 302–304.

Peng, Y. and Song, G. 2002. Combined microwave and ultrasound assisted Williamson ether synthesis in the absence of phase-transfer catalysts. *Green Chemistry,* 4: 349–351.

Peng, Y. and Song, G. 2003. Combined microwave and ultrasound accelerated Knoevenagel-Doebner reaction in aqueous media: A green route to 3-aryl acrylic acids. *Green Chemistry,* 5: 704–706.

Peng, Y., Song, G., and Dou, R. 2006. Surface cleaning under combined microwave and ultrasound irradiation: Flash synthesis of 4H-pyrano[2,3-c]pyrazoles in aqueous media. *Green Chemistry,* 8: 573–575.

Priego-Capote, F. and Luque de Castro, M.D. 2004. Analytical uses of ultrasound I. Sample preparation. *Trac—Trends in Analytical Chemistry,* 23: 644–653.

Regueiro, J., Llompart, M., García-Jares, C., García-Monteagudo, J.C., and Cela, R. 2008. Ultrasound-assisted emulsification-microextraction of emergent contaminants and pesticides in environmental waters. *Journal of Chromatography A,* 1190: 27–38.

Rossi, D., Urbano, M., Baraglia, A. et al. 2009. Polymer-assisted solution-phase synthesis under combined ultrasound and microwave irradiation: Preparation of α,β-unsaturated esters and carboxylic acids, key intermediates of novel sigma ligands. *Synthetic Communications,* 39: 3254–3262.

Shen, X.F. 2009. Combining microwave and ultrasound irradiation for rapid synthesis of nanowires: A case study on Pb(OH)Br. *Journal of Chemical Technology and Biotechnology,* 84: 1811–1817.

Simitchiev, K., Stefanova, V., Kmetov, V. et al. 2008. Microwave-assisted cloud point extraction of Rh, Pd and Pt with 2-mercaptobenzothiazole as preconcentration procedure prior to ICP-MS analysis of pharmaceutical products. *Journal of Analytical Atomic Spectrometry,* 23: 717–726.

Tai, G. and Guo, W. 2008. Sonochemistry-assisted microwave synthesis and optical study of single-crystalline CdS nanoflowers. *Ultrasonics Sonochemistry,* 15: 350–356.

Townes, C.H. and Schawlow, A.L. 1975. *Microwave Spectroscopy* (2nd edn). Mineola, NY: Dover Publications.

Vidal, L., Domini, C.E., Grané, N., Psillakis, E., and Canals, A. 2007. Microwave-assisted headspace single-drop microextraction of chlorobenzenes from water samples. *Analytica Chimica Acta,* 592: 9–15.

Wu, Z.-L., Ondruschka, B., Cravotto, G., Garella, D., and Asgari, J. 2008. Oxidation of primary aromatic amines under irradiation with ultrasound and/or microwaves. *Synthetic Communications,* 38: 2619–2624.

Zlotorzynski, A. 1995. The application of microwave radiation to analytical and environmental chemistry. *Critical Reviews in Analytical Chemistry,* 25: 43–76.

26 Integrating Ultrasound with Other Green Technologies:
Toward Sustainable Chemistry

Julien Estager

CONTENTS

From Sustainability to Green Chemistry and Ultrasound 675
Green Chemistry: How Ultrasound Can Join This Noble Cause 677
    Greater Yields and Chemical Rates for Greener Reaction 678
    Water as a Clean Reactive Medium 680
    The Best Solvent Is No Solvent 681
Ultrasound: A Boon for Heterogeneous Reactions 682
Coupling Ultrasound with Other Green Technologies: One Step Further
Toward Green Chemistry 684
    Coupling Ultrasound and Microwave: The Best of Two Waves? 684
    Ultrasound and Supercritical Fluids: A Super Extracting Agent? 684
Sonoelectrochemistry: Electrons in the Sonic Waves 686
    Ultrasound and Ionic Liquids: Sonicating the Liquid Salts 687
Ultrasound and Photochemistry: Some Light in the Waves 689
Other Green Methodologies Involving Ultrasound 690
Conclusion: The Next Step 691
Acknowledgments 692
References 692

FROM SUSTAINABILITY TO GREEN CHEMISTRY AND ULTRASOUND

Considering the global interest in the last Copenhagen summit (2009 United Nations Climate Change Conference, Copenhagen, Denmark, December 7–18, 2009), one can assume that sustainability is now considered to be an urgent matter to deal with, at least as a political issue. In fact, this term has been defined more than 20 years ago by the United Nations, following the work of the Brundtland commission, as "meeting the needs of the present generation without compromising the future generations to meet their own needs" (Brundtland, 1987). Furthermore, green chemistry has been defined as *the design of chemical products and process to reduce or eliminate the use or generation of hazardous substances* (Anastas and Warner, 1998). Nowadays, green chemistry has become one of the prominent branches of chemistry, as demonstrated, for example, with the creation of well-renowned U.S. Presidential Green Chemistry Awards in 1995 or release of the first volume of *green chemistry* by the Royal Society of Chemistry in 1999. In order to help chemists to move toward sustainable development, Anastas and Warner proposed 12 rules, which may be considered as a toolbox for "green chemists" (Anastas and Warner, 1998). Sometimes referred to as Twelve Principles of green chemistry, they are summarized in Figure 26.1.

Prevention: it is better to prevent waste than to treat or clean up waste after it is formed

Atom economy: synthetic methods should be designed to maximize the incorporation of all material used in the process into the final product

Less hazardous chemical synthesis: whenever practicable, synthetic methodologies should be designed to use and generate substances that pose little or no toxicity to human health and the environment

Designing safer chemicals: chemical products should be designed to preserve efficacy of the function while reducing toxicity

Safer solvents and auxiliaries: the use of auxiliary substances (i.e. solvents, separation agents, etc.) should be made unnecessary whenever possible and, when used, innocuous

Design for energy efficiency: energy requirement of chemical processes should be recognized for their environmental and economic impact and should be minimized. If possible, synthetic methods should be conducted at ambient pressure and temperature

Use of renewable feedstock: a raw material or feedstock should be renewable rather than depleting whenever technically and economically practicable

Reduce derivatives: unnecessary derivatisation (use of blocking groups, protection/deprotection, temporary modification of physical or chemical processes) should be minimized or avoided if possible because such steps require additional reagents and can generate waste

Catalysis: catalytic reagents (as selective as possible) are superior to stoichiometric reagents

Design for degradation: chemical products should be designed so that at the end of their function they break down into innocuous degradation products and do not persist in the environment

Real time analysis for pollution prevention: analytical methodologies need to be further developed to allow real-time, in-process monitoring and control prior to the formation of hazardous substances

Inherently safer chemistry for accident prevention: substances and the form of a substance used in a chemical process should be chosen to minimize the potential for chemical accidents, including releases, explosions and fires

FIGURE 26.1 The 12 principles of green chemistry.

Sheldon et al. (2007) elegantly summarized these different principles by defining green chemistry as "efficiently utilise (preferably renewable) raw materials, eliminate waste and avoid the use of toxic and hazardous reagents and solvents in the manufacture and application of chemical products" (Sheldon et al., 2007).

Subsequently, in order to quantify the "greenness" of a process, different values were invented; as it was pointed out by Lord Kelvin in the nineteenth century, *to measure is to know*. Thus, Barry Trost developed the concept of atom economy that concerns the green efficiency of a chemical reaction (Trost, 1991), while Roger Sheldon introduced the more industrial Environmental factor (E-factor), the mass of waste (in kg) divided by the mass of product obtained (Sheldon, 1994). Depending on the industries concerned, the E-factor can reach dramatic values as shown in Table 26.1 (Sheldon, 2008). Surprisingly, this factor showed that the pharmaceutical industry, which is traditionally perceived as "clean," generates huge amounts

TABLE 26.1
Examples of E-Factor for Different Industries

Industrial Segment	Volume/Ton Per Annum	E-Factor
Bulk chemistry	10^4–10^6	<1–5
Fine chemical industry	10^2–10^4	5 → 50
Pharmaceutical industry	10–10^3	25 → 100

of waste per mass of valuable product, while petrochemistry (bulk chemistry) emerges as material-efficient branch of chemical industry.

In all these definitions, chemistry is always linked to process, and if these good intentions should make it to the real world, they must remain close to the industry. Quite humorously, Poliakoff et al. emphasized this by replacing the 12 principles by 24 others which acronym is IMPROVEMENTS PRODUCTIVELY (Tang et al., 2008). In fact, various industrial processes already follow the trends of green chemistry. For example, Eastman developed an enzymatic esterification process that led to a dramatic decrease of by-product amounts, and significant energy savings (*Office of pollution prevention and toxics, The Presidential Green Chemistry Challenge Awards Program, Summary of 2009 Awards, Entries and Recipents*, U.S. Environmental Protection Agency, Washington, DC, EPA 744K09001, 2009). In 2006, Merck developed a new synthetic pathway for Sitagliptin, a medicine for type II diabetes treatment, which reduces the number of derivatizations required. On the industrial scale, this modification can imply a great reduction in the quantity of solvent used and, simultaneously, a great decrease of the process operational cost (*Office of pollution prevention and toxics, The Presidential Green Chemistry Challenge Awards Program, Summary of 2006 Awards, Entries and Recipents*, U.S. Environmental Protection Agency, Washington, DC, EPA 744R06003, 2006).

As pointed out by Tucker (2010), practicing green chemistry requires more an evolution of our way to consider a problem than an expansion of our knowledge. A lot of tools that may help to reach a more sustainable chemistry already exist—the challenge is to use the already available technology with another objective—to have a green process. It is our belief that ultrasound must be a part of this evolution.

As it has been extensively explained in the previous chapters of this present book, chemistry can make a great profit by taking advantage of the specific effects of sonication. Since the pioneering work of Woods and Loomis (1927), who discovered the effect of ultrasound on chemical reaction rate, intensive studies led to the creation of a new term in chemistry—sonochemistry. Driven by the formation, growth, and collapse of micrometric bubbles, sonochemistry can force reactions that would be impossible in conventional conditions. Even if this interpretation is still arguable, it is largely admitted in the scientific community that the sonication of a medium leads to the formation of bubbles—called cavitation bubble—that can implode very violently creating a hot spot where the temperature can reach 5200 K and a pressure of 1000 bar (Suslick et al., 1990). These extreme conditions can lead to different effects that may be of chemical or physical nature. Chemical effects of ultrasound involve generally the homolytic breaking of chemical bonds and consequently the formation of radicals, for example, H$^\bullet$ and $^\bullet$OH in the case of water (Ince et al., 2001). Physical effects of ultrasound include reduction of particle size during nanoparticle synthesis and processing, microemulsion formation, enhancement of the mass transfer, or dispersion of solids in a medium. The objective of this chapter is to explain how these effects can be beneficent for green chemistry.

Firstly, it will be shown how sonochemistry can be in itself a great help for green chemistry. Then, we will focus on the coupling of ultrasound with other "green" techniques and different examples of their applications will be given.

GREEN CHEMISTRY: HOW ULTRASOUND CAN JOIN THIS NOBLE CAUSE

Even if sonochemistry is quite an old field, its green potential has been only investigated since the end of 1990s, when green chemistry really started to arise (Cintas and Luche, 1999). In fact, the correlation seems nowadays quite obvious considering that, on many occasions, sonication leads to an improvement in terms of yield or selectivity. Indeed, such improvements are classically obtained by adding chemicals into the medium—increasing the potential for waste and non eco-friendly posttreatment—whereas ultrasound is cleaner because of its purely physical nature. Ultrasound clearly reaches Sheldon's objective of "efficiently utilise (preferably renewable) raw materials, eliminate waste and avoid the use of toxic and hazardous reagents and solvents" (Sheldon et al., 2007) since it can, for example,

1. Change the course of a reaction without adding extra chemicals, for example, in the case of the famous sonochemical switching described by Ando et al. (1984)
2. Improve chemical rate (saving time and energy) and selectivity (reduction of waste)

Different examples will be described in this chapter and this first part will show how ultrasound can be a real benefit for green chemistry.

GREATER YIELDS AND CHEMICAL RATES FOR GREENER REACTION

In literature, there are many examples where the use of ultrasound improved a reaction, in term of yield, rate, or both (Luche, 1998). However, very few of them clearly explain where this beneficial effect is coming from. In an attempt to rationalize these observations, Jean–Louis Luche defined a set of three empirical rules to distinguish between the different effects that may be induced by sonication (Luche, 1993). Chemical reactions have been separated into three main categories.

Type I In a homogeneous medium, the chemical effect of ultrasound can be rationalized by considering that single electron transfers (SET) are increased by sonication. Metal complexes will undergo a ligand–metal cleavage that produces coordinatively unsaturated species. This type of reaction is sometimes called "true sonochemistry," since ultrasound takes part in the chemical mechanism.

Type II In a heterogeneous system, the mechanical effects of the ultrasound can affect greatly the kinetics and the yields of organic reaction. In the case of solid–liquid interface, reduction of particle size has to be taken into consideration and in the case of liquid–liquid system, mass transfer and microemulsion phenomena should be considered. This type of reaction is sometimes referred to as false sonochemistry.

Type III In a heterogeneous system, the reaction can follow a "SET" pathway. This kind of reaction is affected by both the chemical and physical effects of ultrasound and it is consequently difficult to determine the effective influence of the ultrasound. These are sometimes called ambivalent reactions. Ando's sonochemical switching is probably the most famous example of a Type III reaction (Ando et al., 1984).

Considering these observations from a green chemistry perspective, an improvement of yield, selectivity, or rate is beneficial whatever its reason. An increase of yield means a decrease of the E-factor, an increase of selectivity may imply less of a generally environmentally costly work-up (use of solvent for extraction, recrystallisation) and an increase of the kinetics generally leads to a decrease in terms of energy consumption as the reaction time becomes shorter. The potential of sonochemistry in this aspect is really large, and there is still plenty of space for development—as shown comprehensively in this book.

To quote two examples amongst many, Blasco–Jimenez et al. (2009) described recently an ultrasound-assisted alkylation of imidazole via a Michael addition catalyzed by 3-aminopropyl-trimethylsilane grafted on MCM-41/niobosilicate (Scheme 26.1).

As a solid support was used, some physical effect due to ultrasound may be expected, and an increase of conversion has been observed with different catalysts as shown in Figure 26.2.

In 2008, Bhor et al. (2008) described the coupling of naphthols catalyzed by an iron catalyst. Referring to a mechanistic study on the coupling of phenol done by Toda et al. (1989), a radical pathway has been suggested, as described in Scheme 26.2.

SCHEME 26.1 Ultrasound-assisted alkylation of imidazole via Michael addition.

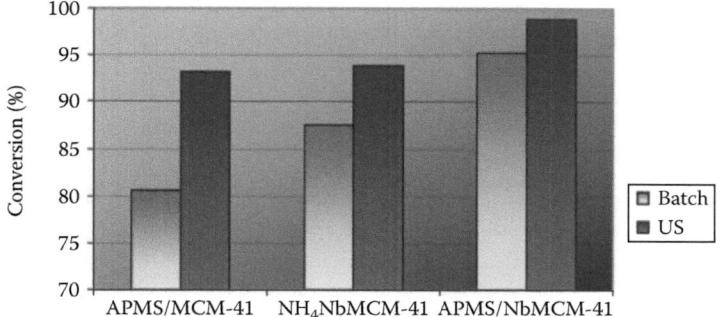

FIGURE 26.2 Alkylation of imidazole with ethylacrylate after 3 h at 60°C. (Reproduced from *Catal. Today*, 142, Blasco–Jiménez, D., Lopez–Peinado, A.J., Martin–Aranda, R.M., Ziolek, M., and Sobczak I., Sonocatalysis in solvent-free conditions: An efficient eco-friendly methodology to prepare N-alkylimidazoles using amino-grafted NbMCM-41, 283–287, Copyright (2009), with permission from Elsevier.)

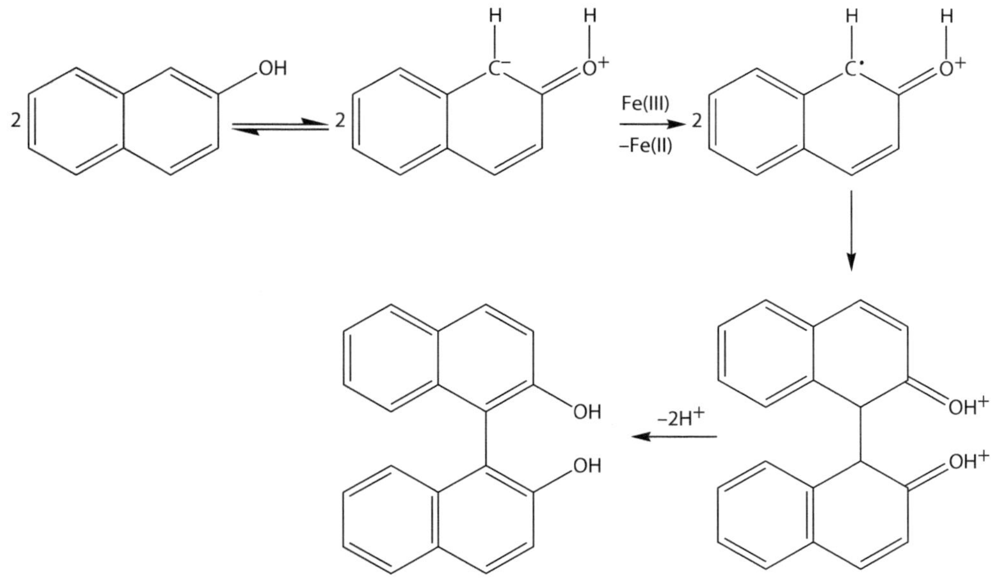

SCHEME 26.2 Radical pathway for the homo-coupling reaction of 2-naphthol.

In each case, the type of reaction is different, as the Michael addition is a type II reaction whereas the homocoupling of naphtols is a type III one. However, in both cases, they can be considered as an improvement in terms of green chemistry. Thus, organic sonochemistry can be considered as a green science in itself. As pointed out by Tucker (2010), looking to an already well-established technology with a new perspective can lead to great achievements.

WATER AS A CLEAN REACTIVE MEDIUM

The use of volatile organic solvents in the chemical industry is one of the major obstacles in the path of green chemistry. Indeed, they are usually used in large excess in order to solubilize the different compounds, often harmful or toxic, and usually volatile, increasing the risk of release and/or accident. The use of benign solvents, especially water, has been proposed as an alternative and widely investigated for a long time (Breslow, 2007). Because of its very specific interaction with water or aqueous solutions, sono-assisted reactions in this medium have been widely studied and opened a great opportunity in terms of sustainable chemistry. It is now well known that the ultrasonic irradiation of water results in the formation of very reactive radical species due to the homolytic cleavage of H_2O. Moreover, the homolytic cleavage of oxygen gas in the presence of water during the sonication leads to the formation of different oxidants as described in the below set of Equations 26.1 (Hamdaoui and Naffrechoux, 2008).

$$H_2O \rightarrow H^{\bullet} + HO^{\bullet}$$

$$O_2 \rightarrow 2O$$

$$O + H_2O \rightarrow 2HO^{\bullet}$$

$$H^{\bullet} + HO^{\bullet} \rightarrow H_2O$$

$$2HO^{\bullet} \rightarrow O + H_2O$$

$$H^{\bullet} + O_2 \rightarrow HOO^{\bullet}$$

$$2HO^{\bullet} \rightarrow H_2O_2$$

$$2HOO^{\bullet} \rightarrow H_2O_2$$

EQUATION 26.1 Species generated through sonication of water in presence of oxygen.

The presence of these very reactive species, coupled with the possible pyrolysis using the energy released by the collapsing cavitation bubble, can be very useful, for example, for the degradation of organic compounds. This methodology is an alternative chemical-free destructive method for the degradation of organic matter, often termed "aquasonolysis."

4-Chlorophenol decomposition is maybe the most studied example of aquasonolysis. This compound is released in the environment as a by-product in many industrial processes and has been labeled as a "priority pollutant" by the EPA and the European Union. It has an upper concentration limit of 0.5 mg/L for supplied water, and consequently it must be constantly monitored in aquatic environment (Rodriguez et al., 2000). In a search of suitable degradation method, Hao et al. (2004) described the aquasonolysis of 4-chlorophenol at 1.7 MHz in the absence of air. In these conditions, a degradation of the pollutant proceeds through the cleavage of the aromatic ring. Various studies have been performed to optimize the experimental conditions of this process and, for example, it has been shown that the rate constant decreases when the initial concentration of pollutant increases (Jiang et al., 2006). Because of their capacity to penetrate the cavitation bubble, volatile organic compounds (VOCs) are particularly vulnerable to sonolysis, often through a pyrolysis mechanism. This particularity is more crucial from the green chemistry viewpoint, since almost one half of the

Integrating Ultrasound with Other Green Technologies

$$H_2O +))) \rightleftharpoons HO^\bullet + {}^\bullet H$$

$$HO^\bullet \longrightarrow 1/2\ H_2O_2$$

$$H^\bullet \longrightarrow 1/2\ H_2$$

$$HO^\bullet + phenol \longrightarrow Products$$

SCHEME 26.3 Mechanism for degradation of phenol via aquasonolysis.

189 air pollutants regulated by the Clean Air Act Amendment in 1990 are VOCs. As an example of an excellent ultrasound application, one may quote the work of Peters (2001), who described degradation of a range of chlorinated VOC at high frequency. In most of the cases, the chlorinated amount of VOCs is reduced by one order of magnitude within a couple of hours. In another paper, Inazu et al. (1993) showed that this decomposition leads mainly to the formation of chloride, hydrogen, carbon monoxide, and carbon dioxide.

Oxidation of phenol has also been intensively studied, and Pétrier et al. (1994) observed that the kinetics of degradation is more rapid at a frequency of 487 kHz than at 20 kHz, highlighting the influence of the radical anion in the process. The oxidation mechanism of phenols has been described by Serpone et al. (1992) and is shown in Scheme 26.3. In this mechanism, as in many other aquasonolyses, ultrasound acts as an initiator for the generation of radical ion.

These different phenomena of degradation have been known for a long time (e.g., Schmitt et al., 1928, described the sono-oxidation of various compounds in 1928) and are nowadays used with an impressively wide range of chemicals. This area has been reviewed in a very useful publication by Adewuyi (2001). In order to make one step further toward Green Synthesis, one should now consider the nature and toxicity of the products of aquasonolysis, especially when considering their industrialization, or pursue a way to reach total mineralization. Combining different existing techniques may be a solution to achieve this ultimate goal, and different examples of such pursuits will be provided further in this chapter.

THE BEST SOLVENT IS NO SOLVENT

Benign solvents, such as water, can be used to carry out experiments in an environmentally friendly way. However, another way to reach this green objective would be to eliminate the solvent completely from the process. By increasing the reactant concentration, this methodology can increase the reaction rate and is widely used, for example, in microwave chemistry. On the other hand, solvents play a major role in directing the reaction via solvent effects and aid to bring the solubilized reactants into contact, which in turn leads to higher yields and kinetics. Using ultrasound for solventless reactions can remove the problem of solubility by taking advantage from the physical effect of ultrasound such as micromixing or enhanced mass transfer.

Various recent publications about solvent-free synthesis under sonication described protocols employing solid catalysts. For example, Chtourou et al. (2010) described recently the synthesis of *trans*-chalcones using acidic clays as catalyst (Scheme 26.4).

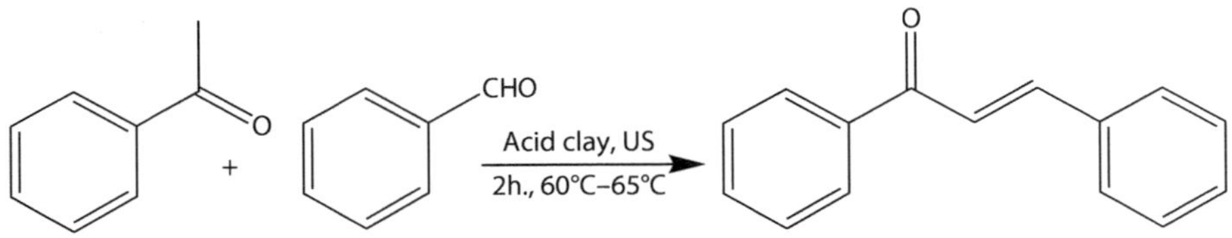

SCHEME 26.4 Synthesis of chalcones under solvent-free conditions.

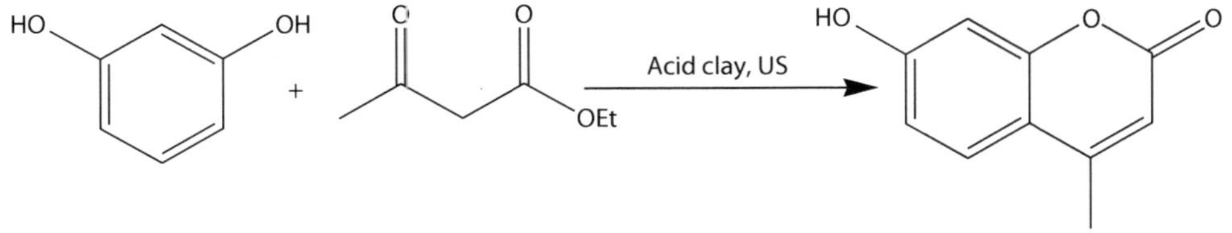

SCHEME 26.5 Pechman condensation under solvent-free conditions.

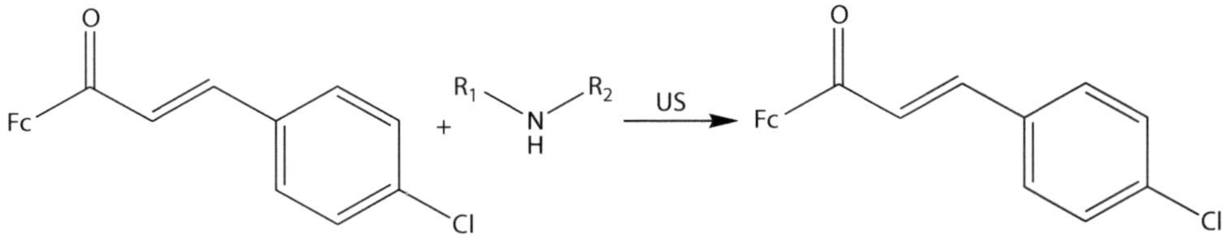

SCHEME 26.6 Solvent-free addition of amine to ferrocenylenones.

Under these particular conditions, it is very difficult to determine the real effect of ultrasound, because the enhancement due to sonication can be explained by efficient mixing and also by the reduction of the particle size of the solid. The same applies, for instance, to the synthesis of coumarines from resorcinol and β-ketoesters by Pechmann condensation (Gutierrez-Sanchez et al., 2009), as described in Scheme 26.5.

This methodology has also been studied for the synthesis of 2H-chromen-2-ones and once again, a solid catalyst (copper perchlorate) has been used, making further interpretation difficult (Puri et al., 2009). Solvent-free reactions can also be performed without solid catalysts, for example, the Michael addition of secondary amines on ferrocenylenones (Yang et al., 2005), illustrated in Scheme 26.6.

Other reactions can be carried out using the same idea, for example, the Hantzsch reaction (Wang et al., 2008) or synthesis of pyrazolones (Mojtahedi et al., 2008), but most of the time the role of the ultrasound is not explained in the publication. Still, an increase in term of yield and/or kinetics is generally observed compared to silent conditions, which proves the usefulness of ultrasound.

Despite these significant improvements, solvent-free reactions under ultrasound should always be considered cautiously, especially when the final product is a solid. It is vital to bear in mind that solidification of the medium can lead to the destruction of the ultrasonic apparatus (sonic probes, ceramics) due to the reflection of the wave. In this case, a solvent has to be used in order to avoid precipitation of the medium.

ULTRASOUND: A BOON FOR HETEROGENEOUS REACTIONS

As pointed out previously, sonochemistry is especially useful in the case of type II reactions, that is, multiphasic reactions. In many cases, ultrasound can greatly enhance both yields and rates through its physical effects. In the case of liquid–liquid systems, the phenomena of micromixing, mass transfer, and microemulsion can force a reaction to occur. The usual way to perform chemical reactions in liquid–liquid systems is to use a phase-transfer catalyst (PTC) (Makosza, 1975). This methodology, usually involving salts with tetraalkylammonium cations, can be considered as a part of green chemistry, since it implies a reduction of the amount of organic solvent (which is in this case replaced by water) and requires only catalytic amounts of recoverable PTC (Makosza, 2000). However, even if—considering Anastas' 12 principles—the use of catalyst is one of the

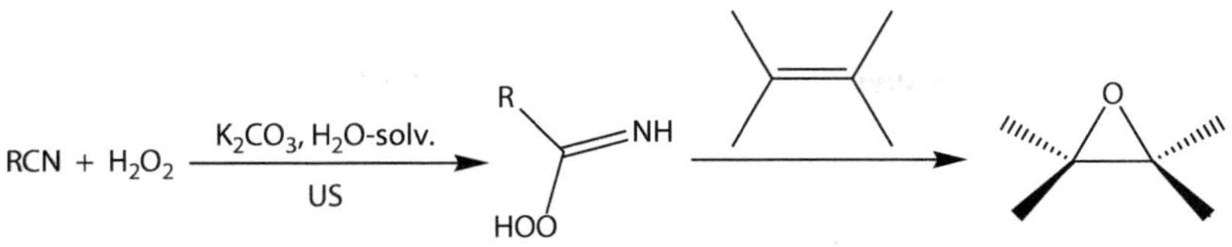

$$RCN + H_2O_2 \xrightarrow[US]{K_2CO_3,\ H_2O\text{-solv.}} \underset{HOO}{\overset{R}{\underset{|}{C}}}=NH \longrightarrow \text{epoxide}$$

SCHEME 26.7 Epoxidation of alkenes using alkylcyanide and hydrogen peroxide.

major requirements of green chemistry, the best catalyst of all remains no catalyst. In this perspective, ultrasound can be considered as a "physical PTC," and many examples of such heterogeneous reactions are described in the literature. For example, the epoxidation of alkenes by H_2O_2–RCN is possible in an aqueous medium as described in Scheme 26.7 (Braghiroli et al., 2006).

Physical effects of ultrasound can also be used for solid–liquid heterogeneous systems, and in this case the main effect of ultrasound lies in reducing the size of the solid particles, and therefore increasing their surface area. The use of solid catalysts is highly recommended in terms of green chemistry, because they are easier to handle and usually do not have vapor pressure, thus reducing the risk of gas release or explosion. Unfortunately, they usually exhibit lower reactivity, so the use of ultrasound can be a very interesting approach to overcome this problem. Solid–liquid systems are by far the most studied in sonochemistry. For example, ultrasound can be useful in the elimination reaction over the solid base catalysts, as in the case of the β-elimination for β-bromoacetals using solid KOH, described by Diez–Barra et al. (1992) (Scheme 26.8).

In this study, the authors justified the use of PTC as a necessity for KOH to react efficiently. However, subsequently, it has been proven that ultrasound can efficiently replace this catalyst. The same ultrasonic effect is widely used for reactions involving metals, for example, in reduction (Peng et al., 2005), cyclisation (Ranu and Mandal, 2006; Zhang et al., 2008), or coupling reactions (Deshmukh et al., 2001; Cravotto et al., 2005a,b). In the case of triphasic liquid–liquid–solid systems, the combination of micro-emulsion and reduction of particle size can lead to very interesting results, as found for the one-pot synthesis of some ionic liquids (Scheme 26.9) (Estager et al., 2007b)

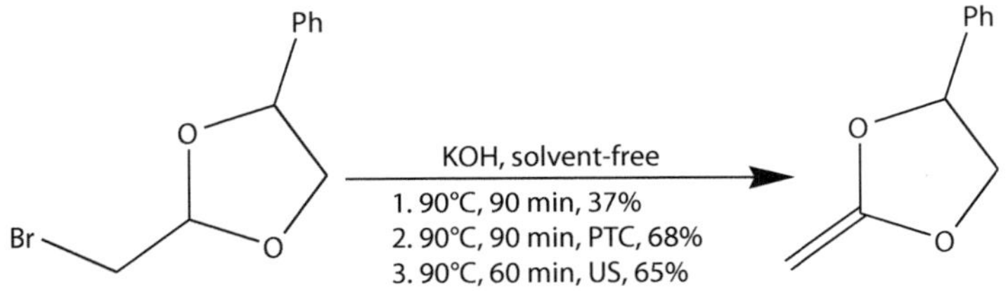

KOH, solvent-free
1. 90°C, 90 min, 37%
2. 90°C, 90 min, PTC, 68%
3. 90°C, 60 min, US, 65%

SCHEME 26.8 B-elimination for β-bromoacetals.

Imidazole + RBr + KPF_6 $\xrightarrow{\text{Solvent-free, US}}$ 1-alkyl-3-methylimidazolium PF_6^-

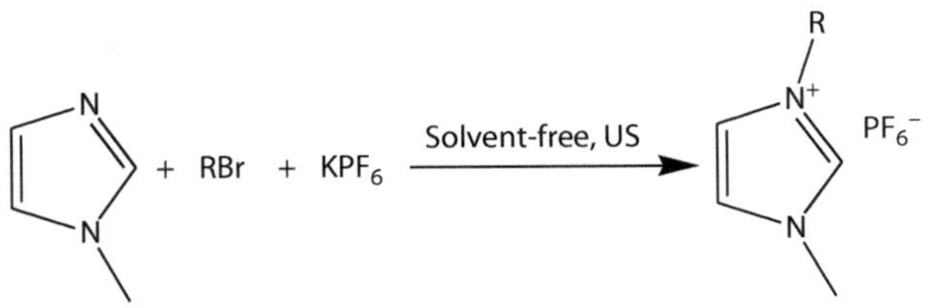

SCHEME 26.9 One-pot synthesis of ionic liquids, illustrated by the example of synthesis of 1-alkyl-3-methylimidazolium hexafluorophosphate.

COUPLING ULTRASOUND WITH OTHER GREEN TECHNOLOGIES: ONE STEP FURTHER TOWARD GREEN CHEMISTRY

As it shown previously, there are numerous links between sonochemistry and green chemistry, and by slightly changing our approach, it is possible to develop really interesting sustainable processes involving ultrasound. An even more fascinating way toward green sonochemistry is to combine ultrasound with other techniques, known to be valuable in terms of sustainable chemistry, in search of new synergetic effects. This chapter describes different coupled techniques described in the scientific literature.

Coupling Ultrasound and Microwave: The Best of Two Waves?

Since the pioneering work from Gedye et al. (1986), an intensive work has been carried out by different groups to explain how microwave can have such a powerful influence on yields and kinetics. Some studies about microwave chemistry (Loupy, 2006) and its relation to green chemistry (Strauss and Varma, 2006) can provide a more comprehensive understanding of this technique and its applications. In a nutshell, microwave irradiation of polar compounds leads to a very quick heating of the system—hundreds of degrees Celsius within minutes. Specific effects are also suspected to occur, but they still remain disputable (Perreux and Loupy, 2002). Concerning the coupling of microwave and ultrasound, the most important to note is that there can hardly be any direct interactions between the two waves, due to the frequency difference (the usual microwave frequency being 2.45 GHz). Consequently, one can expect to maintain all the advantages of both techniques. Following the work from Maeda and Amemiya (1995), various studies have been performed on the subject, as described elsewhere within this book. From the perspective of green chemistry, coupling of the microwave and ultrasound can lead to very fast reaction rates, and reducing the reaction time is one of the major steps toward energy efficiency. As a point in case, Cravotto et al. (2005a,b) studied Suzuki coupling and showed that dual irradiation gives better results than microwave or ultrasound alone. The same protocol can be performed for the synthesis of 3-aryl acrylic acids, as described in Scheme 26.10 (Peng and Song, 2003).

Also for this solvent-free Knoevenagel–Doebner reaction, the combination of both irradiations gives better results than microwave or ultrasonic irradiations alone. Coupling the very efficient heating of microwave with the efficient mass transfer induced by sonication enables, in this case, a dramatic reduction of the reaction time from 7 h in conventional conditions to 65 s. Moreover, such coupling can be also useful for the extraction of natural compounds, for which microwave and ultrasound are efficient methods (Filgueiras et al., 2000). As an example, the extraction of vegetable oils under double irradiation has been studied, and interesting results have been obtained (despite the fact that some other methods were more efficient) (Cravotto et al., 2008). Even if this domain is still quite limited (Cravotto and Cintas, 2007), the various effects promoted by simultaneous use of these two irradiations make this field exceptionally promising.

Ultrasound and Supercritical Fluids: A Super Extracting Agent?

A supercritical fluid (SCF) is defined as a substance at a temperature and pressure above its critical point. For example, supercritical CO_2 is obtained by heating CO_2 above 304.1 K at a pressure above 72.8 atm, whereas supercritical water requires much harsher conditions (647.1 K and

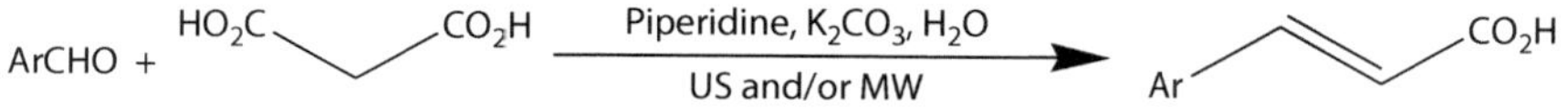

$$ArCHO + HO_2C\text{-}CH\text{=}CH\text{-}CO_2H \xrightarrow[\text{US and/or MW}]{\text{Piperidine, }K_2CO_3\text{, }H_2O} Ar\text{-}CH\text{=}CH\text{-}CO_2H$$

SCHEME 26.10 One-pot synthesis of ionic liquids, illustrated by the example of synthesis of 1-alkyl-3-methylimidazolium hexafluorophosphate.

217.8 atm.). Their applications in chemistry have been widely studied (Jessop and Leitner, 1999) and their interest in green chemistry, based on the low toxicities in their non-supercritical state, is now unambiguous among the scientific community (Tanchoux and Leitner, 2002). The different techniques involving supercritical fluids take advantage from the specific physicochemical properties of this state of the matter, possessing properties of both gas and liquid, such as low viscosity, high diffusivity (like gases), and also a good ability to dissolve matter (like liquids). Considering these properties, supercritical fluids, especially supercritical CO_2, are very good solvents for extraction processes. The most famous process including supercritical CO_2 is probably the decaffeination of coffee beans via the Zosel process (Zosel, 1977). Reaching supercritical conditions for water requires significant amounts of energy, and scH_2O itself is extremely corrosive. Therefore, as opposed to "standard" water, it can hardly be described as a "green" solvent.

Using sonication for extraction, and combining it with a supercritical CO_2 as a green extracting solvent, can be a very valuable process considering green chemistry principles. For example, Riera et al. (2004) described the extraction of oil from almonds by supercritical CO_2 and with aid of sonication at 20 kHz and 50 W, using a power piezoelectric transducer of the Langevin type. A scheme of the extractor used is shown on Figure 26.3.

An interesting study by the same group shows the evolution of the characteristic impedance of CO_2 in sub- and supercritical conditions. Coupling of ultrasound and supercritical CO_2 can lead to an increase of both rate and yield of extraction. Hu et al. (2007) used the same methodology to extract oil and coixenolide from adlay seeds, both compounds being beneficial for health. By using the technology discussed earlier, they observed an increase of the extraction yields even for softer experimental conditions, as presented in Table 26.2.

In attempt to extract pungent compounds from ginger, Balachandran et al. (2006) also used a combination of the two methods, and suggested an explanation for the increased rate and yield they observed. Reportedly, such increase can hardly be explained by either abrasive effects or by the creation of turbulence, but can more likely be assigned to the increase of mass transport due to the cellular damage. In other words, the ultrasound destroys the cell wall, while the supercritical CO_2

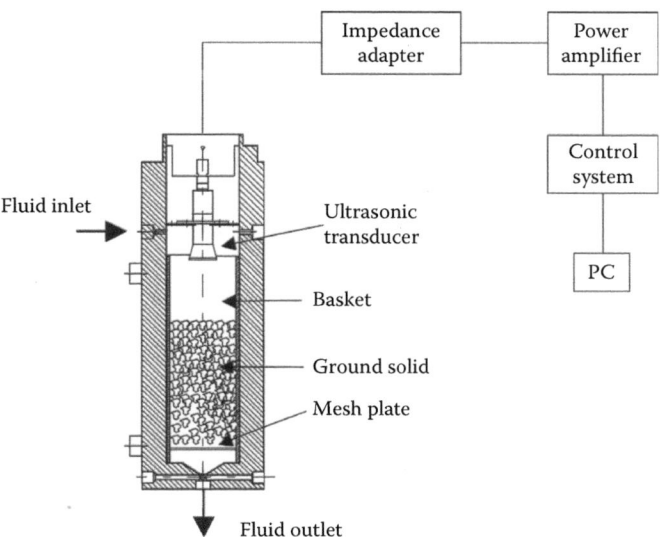

FIGURE 26.3 Scheme of the extractor used by Riera et al. (2004) for oil extraction from almonds. (Reproduced from *Ultrason. Sonochem.*, 11, Riera, E., Golas, Y., Blanco, A., Gallego, J.A., Blasco, M., and Mulet, A., Mass transfer enhancement in supercritical fluids extraction by means of power ultrasound, 241–244, Copyright (2004), with permission from Elsevier.)

TABLE 26.2
Extraction of Oil and Coixenolide from Adley Seeds

Compound	Extraction	T/°C	P/MPa	t/h	Flow$_{CO_2}$/(L/h)	(%)$_{ext.}$
Adlay oil	SCF	45	25	4.0	3.5	84.95
Adlay oil	SCF +)))	40	20	3.5	3.0	96.36
Coixenolide	SCF	45	25	4.0	3.5	84.72
Coixenolide	SCF −)))	40	20	3.5	3.0	96.55

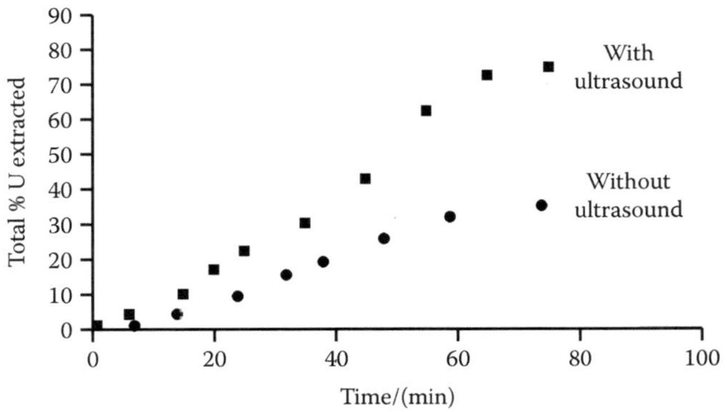

FIGURE 26.4 Supercritical extraction of UO_3 by CO_2 at 60°C and 150 atm. (Trofimov, T.I., Samsonov, M.D., Lee, S.C., Smart, N.G., and Wai, C.M.: Ultrasound enhancement of dissolution kinetics of uranium oxides in supercritical carbon dioxide. *J. Chem. Technol. Biotechnol.* 2001. 76. 1223–1226. Copyright Wiley-VCH Verlag GmbH & Co. KGaA. Reproduced with permission.)

extracts the oil. Furthermore, ultrasound can be used to solubilize uranium oxides into supercritical CO_2 containing 4,4-trifluoro-1-(2-thienyl)-1,3-butanedione (HTTA) and tributylphosphate, as shown in Figure 26.4 (Trofimov et al., 2001).

The authors suggest that ultrasound helps to destroy the complex $UO_2(TTA)_2H_2O$ on the oxide surface, therefore facilitating the rest of the process. UO_2 powder can also be dissolved in supercritical CO_2 using CO_2-soluble tri-*n*-butylphosphate-HNO_3-H_2O complex as an extracting agent (Enokida et al., 2002). In this case, the use of ultrasound increases the efficiency of the process by one order of magnitude. Other compounds can also be extracted with this method, for example, diesel oil from diesel-contaminated soil (Park et al., 2008).

SONOELECTROCHEMISTRY: ELECTRONS IN THE SONIC WAVES

By nature, electrochemistry can be related to green chemistry since it proposes to use electrons to drive a reaction instead of a chemical or a catalyst, which, of course, improves the atom economy of the process. The concept of sonoelectrochemistry will be explained more comprehensively elsewhere within this book, but to depict it simply, the cavitation phenomenon is able to prevent passivation of the electrode, by continuous removal of the material that may be electrodeposited on its surface. Moreover, the other effects of ultrasound can be also useful; for example, for the electrochemical oxidation of carboxylic acids in a biphasic system (Kolbe coupling) (Wadhawan et al., 2001) where good yields and high current efficiency were be obtained. As another example, Del Campo et al. (2001) proved that an electron solvated in ammonia at 60°C can be generated in presence of ultrasound, and subsequently used for Birch reductions (Del Campo et al., 2001). Thus, ultrasound can be used once again to improve even more an already green method.

Ultrasound and Ionic Liquids: Sonicating the Liquid Salts

The field of ionic liquids has experienced an incredible development during the last decade and is nowadays well established in green chemistry. These compounds are by definition salts with melting points below 100°C. As totally ionic species, they are generally composed of a bulky organic cation and an organic or inorganic anion. They have been used for a wide range of applications, such as organic chemistry, inorganic chemistry, and electrochemistry (Wasserscheid and Welton, 2008), and are now used in different industrial processes (Plechkova and Seddon, 2008). Their green potential arises from their specific physicochemical properties, mainly their very low vapor pressure that reduces the risks of air pollution and accidental release. They are also considered as nonflammable (in most cases), and are often used in recyclable processes. The very low vapor pressure of ionic liquids can be very useful for sonochemistry, helping to avoid nebulisation phenomena or reduce the tendency of the solvent to cavitation (Flanningan et al., 2005).

Coupling of ionic liquids and ultrasound have been used in different areas of chemistry. First of all, different publications described the synthesis of ionic liquids using ultrasound (Lévêque et al., 2007). As the metathesis step requires the use of a salt, such as potassium hexafluorophosphate or lithium *bis*(trifluoromethane)sulfonimide, it is believed that sonication reduces particles sizes leading to increased reactivity. The degradation of ionic liquids has also been achieved under ultrasound, with hydrogen peroxide and acetic acid in aqueous solution, as described in Scheme 26.11 (Li et al., 2007).

Ionic liquids have been used during various sono-assisted organic reactions in the recent years. For example, Heravi (2009) described the synthesis of quinoline in butylimidazolium tetrafluoroborate as described in Scheme 26.12.

Here the ionic liquid plays the role of acidic catalyst for this tandem addition/annulation reaction of *o*-amino-aryl ketones and α-methyleneketones, and the medium has been recycled twice without any loss of acidity. The same catalyst has also been used for the sono-assisted synthesis of 1,8-dioxo-octahydroxanthene derivatives (Venkatesan et al., 2008) or for the synthesis of 3,4-dihydropyrimidin-2-(1H)-ones *via* the Biginelli reaction (Gholap et al., 2004). Ionic liquids without these acidic properties have also been studied; for instance, for the acetalization of alcohol (Gholap et al., 2003), in which the ionic liquid enabled an easy work-up, or for the benzoin condensation in which

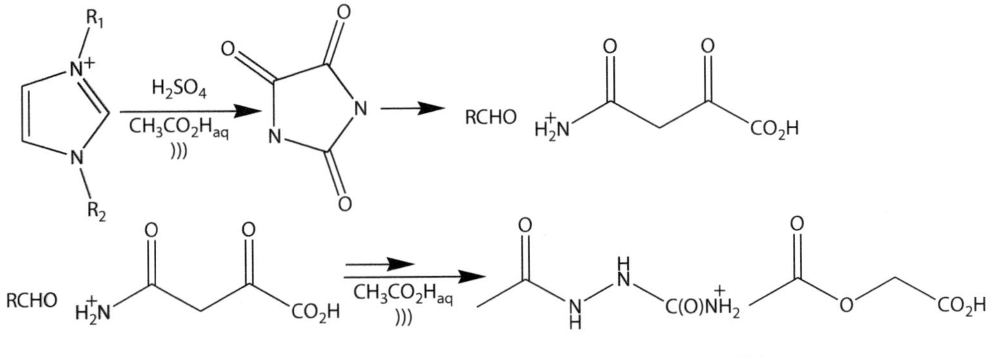

SCHEME 26.11 Sono-degradation of an imidazolium cation in oxidative conditions.

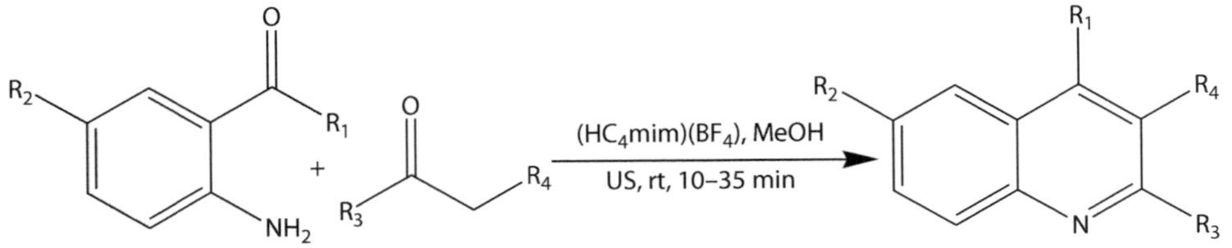

SCHEME 26.12 Synthesis of quinoline derivatives under ultrasound.

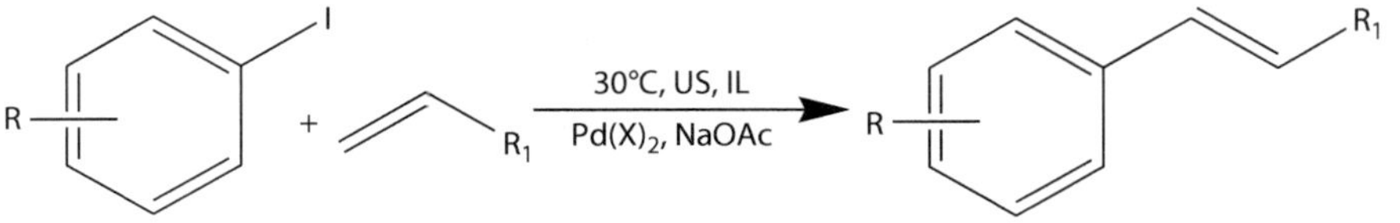

SCHEME 26.13 Heck reaction under sonication in imidazolium based ionic liquids.

the imidazolium cation acted as a platform for the reaction via its C2 position (Estager et al., 2007a). The labile properties of the hydrogen in C2 position of the imidazolium-based ionic liquids have also been employed for organometallic catalysis such as the sono-assisted Heck reaction (Scheme 26.13) (Deshmukh et al., 2001).

In the above process, no reaction occurred in molecular solvent, the influence of the ionic liquid is unquestionable. This phenomenon has been explained via the isolation and characterization of the in situ-formed complex, described in Scheme 26.14.

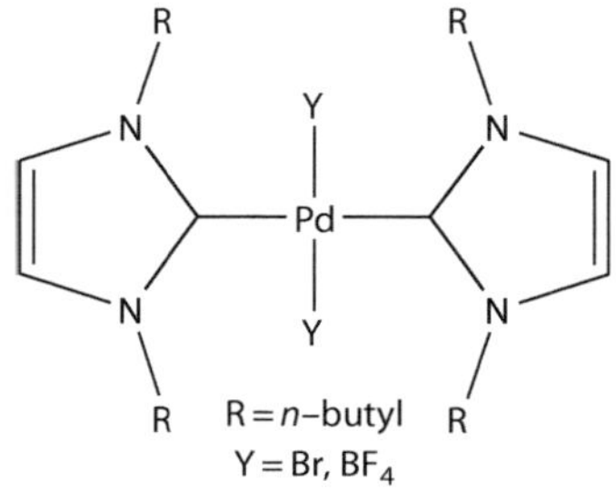

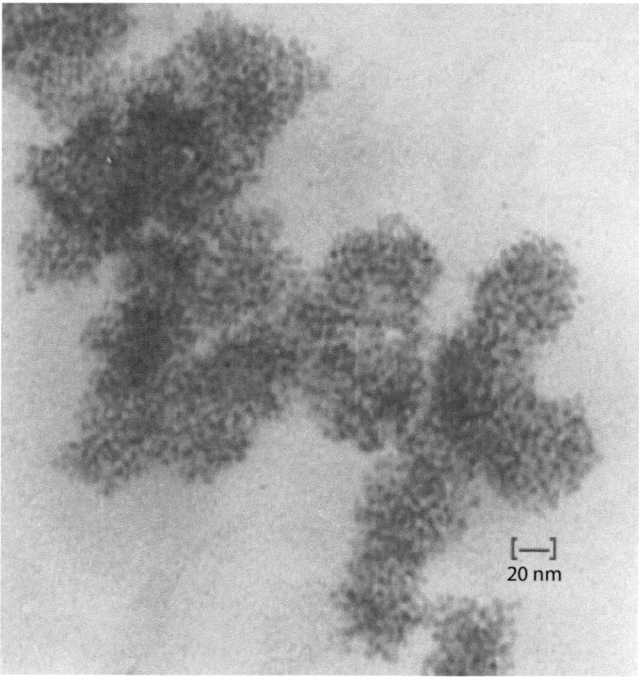

SCHEME 26.14 Pd-biscarbene formed during sono-assisted Heck reaction in imidazolium based ionic liquids and TEM image of these palladium clusters. (From Deshmukh, R.R., Rajagopal, R., and Srinivasan, K.V., Ultrasound promoted C–C bond formation: Heck reaction at ambient conditions in room temperature ionic liquids, *Chem. Commun.*, 1544–1545, 2001. Reproduced by permission from The Royal Society of Chemistry.)

Suzuki coupling has also been performed by the same team, using the same idea, and it was demonstrated that the formation of the Pd–biscarbene complex enabled the reaction to occur (Rajagopal et al., 2002). Coupling ionic liquids and ultrasound is of a great interest in the field of nanochemistry. Nanostructures are objects that have at least one dimension within nanoscale—that is, 1–100 nm. Such structures may exhibit very interesting properties in catalysis, microelectronics, or photonics because of their large surface area and thus the large number of surface atoms, or the 3D confinement of electrons. Sonication is now a well-established technique for the synthesis of such compounds, since cavitation enables the thermal decomposition of organometallic precursors leading to the formation of nanosized particles, but also the removal of surface contaminants that enable to form more uniform clusters (Dhas and Suslick, 2005). Even if this subject has been described comprehensively elsewhere in this book, we will briefly describe the sono-assisted synthesis of HgS and PbS nanoparticles (Zhu et al., 2000) to demonstrate their "green" nature. In this study, 1-decanethiol and ethylenediamine have been used not only to activate the elementary sulfur via the formation of a complex, but also to avoid aggregation phenomena that lead to the formation of larger particles. Ionic liquids can be an alternative solution to avoid aggregation of nanoparticles because of their surfactant-like structure. Moreover, the organized structure of ionic liquids can act as a template for porous nanomaterials, whereas their high charge and polarization can stabilize the nanostructures electrostatically and sterically (Zhou, 2005). Thus, Alammar et al. (2009) described recently, a templateless synthesis of CuO nanorods from copper(II) acetate hydrate using ultrasound in 1-butyl-3-methylimidazolium *bis*(trifluoromethylsulfonyl)imide as the solvent (Alammar et al., 2009). The roles of both ionic liquids and ultrasound can be crucial as it was shown recently for the sonoassisted synthesis of ZnO nanoparticles in 1-hexyl-3-methylimidazolium bistriflimide, where it is thought that the ionic liquid is involved in the mechanism due to the high polarization of the $[C_6mim]^+$ cation (Goharshadi et al., 2009). Ultrasound and ionic liquids have also been used to synthesize organized soft material consisting of imidazolium ionic liquids and single-wall carbon nanotubes, making use of the interaction of the imidazolium ring and the π electron on the surface of the nanotubes (Figure 26.5), leading to gels with fascinating properties (Fukushima and Aida, 2007).

Finally, it is important to notice that there are some really interesting synergies between ionic liquids and ultrasound. For example, ionic liquids can be designed to tune their solubility of different compounds, what makes them really useful for extraction and/or solubilization of materials. In both cases, the physical effects of ultrasound—enhanced mass transfer and microemulsion formation for extraction, and reduction of particles' size for solubilization—can play a major role in these processes. As an example, ultrasound and ionic liquids have been studied for the solubilization of cellulose in 1-allyl-3-methylimidazolium chloride or in $[C_4mim][Cl]$ (Mikkola et al., 2007).

ULTRASOUND AND PHOTOCHEMISTRY: SOME LIGHT IN THE WAVES

Photosonochemistry is not a particularly new topic, since Toy and Stringham (1984) have described the sonolysis of methyl disulfide and hexafluorobutadiene as early as 1984. Since then, different studies have been carried out in order to determine any possible interactions between these two process, for instance in synthesis (Toma et al., 2001). Photochemistry enables the formation of HO• radicals, and can be used with ultrasound, especially for the degradation of pollutants. For example, Hamdaoui and Naffrechoux (2008) have recently described the photosonochemical degradation of *p*-chlorophenol (4-CP) in water. In this study, the role of the hydroxide radical has been proven unambiguously by the use of *tert*-butyl alcohol, but it is interesting to note that the photosonochemical rate of degradation is higher than the additive rate of photochemical and ultrasonic degradations. In fact, it is vital to note that even if sonochemistry and ultrasound can both induce the decomposition via HO•, some more specific effects may occur—such as the direct electron transfer from the organic compounds to semiconductor oxide in the case of photolysis or thermal pyrolysis in the

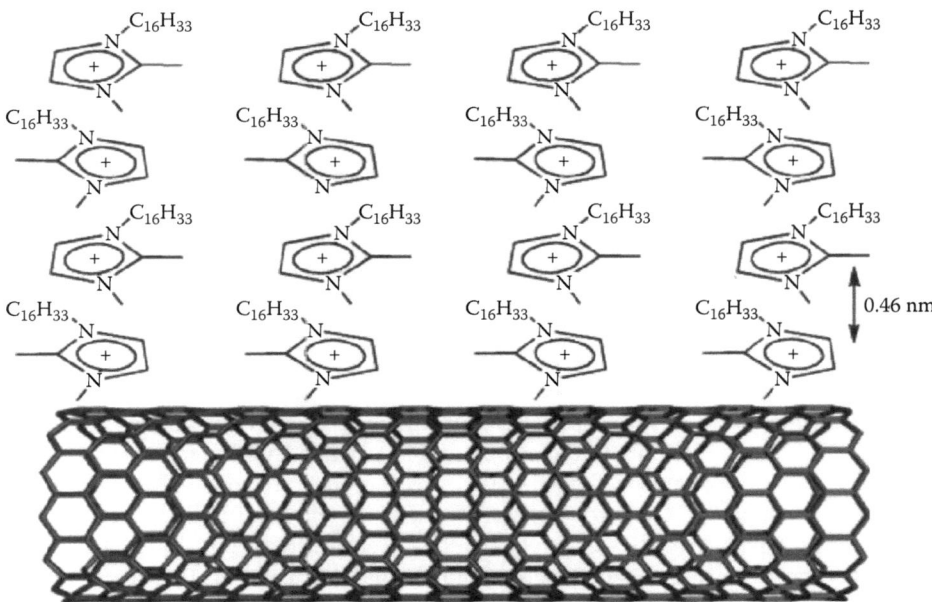

FIGURE 26.5 Organization of single-wall nanotubes/imidazolium ionic liquid medium. (Fukushima, T. and Aida, T.: Ionic liquids for soft functional materials with carbon nanotubes. *Chem. Eur. J.*, 13. 5048–5058. Copyright Wiley-VCH Verlag GmbH & Co. KGaA. Reproduced with permission.)

case of ultrasound (Augugliaro et al., 2006). In the case of *p*-chlorophenol, the authors suggested a triple mechanism to explain this observation; firstly, *p*-chlorophenol can be degradated via a classical sonolysis mechanism in water; secondly, a specific photodegradation mechanism can take place involving the formation of the radical cation CP⁺ as intermediate; thirdly, ozonolysis can occur via the photoformation of ozone in the air and its efficient sono-assisted mixing with the aqueous solution. Photosonochemical degradation has been widely studied, for example, the sono-photolysis of Naphtol blue (Stock et al., 2000) or the destruction of salicylic acid using TiO_2 on zeolites as photocatalysts (Reddy et al., 2003).

Even if it was extensively studied in the past (Toma et al., 2001), organic photosonochemistry seems to have been set aside recently, despite remaining examples such as oxidation of substituted 1,4-dihydropyridines (Memarian and Abdoli-Senejani, 2008). Other systems can also take benefits from this double irradiation. For instance, Harada (2001) described a water splitting phenomenon using TiO_2 powder under a Xe lamp as photocatalyst. In this study, he described a mechanism involving the two different irradiations, which is described in the set of Equations 26.2.

$$H_2O \rightarrow \tfrac{1}{2}H_2O + \tfrac{1}{2}H_2O_2(\;)))\;)$$

$$H_2O_2 \rightarrow \tfrac{1}{2}O_2 + H_2O_2(h\nu)$$

EQUATION 26.2 Water spitting using photosonocatalysis.

OTHER GREEN METHODOLOGIES INVOLVING ULTRASOUND

Along all the classical methods that have been linked to sonochemistry, catalysis is probably the most studied, and various reviews can be found in the literature (Ragaini and Bianchi, 1998; Toukoniity et al., 2008). Especially emphasized are benefits from the physical effects of ultrasound in

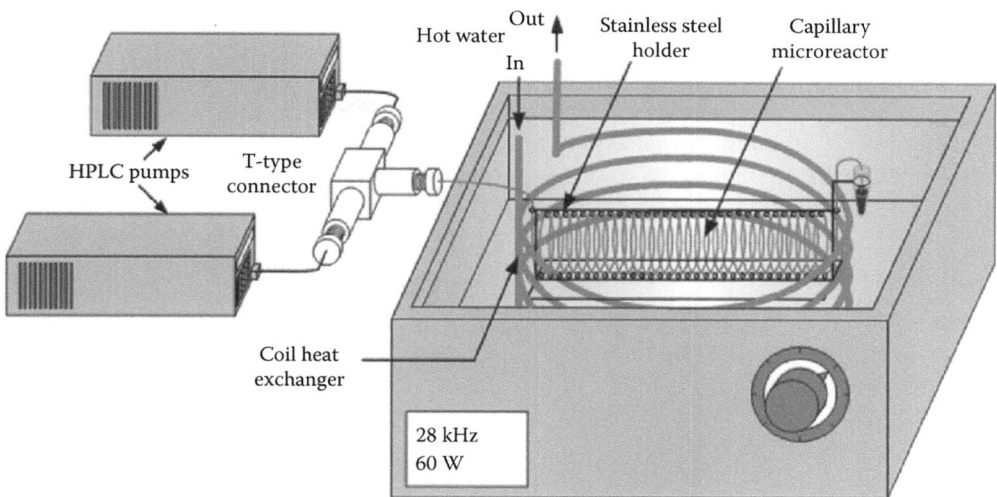

FIGURE 26.6 Scheme of micro-sonoreactor. (Reproduced from *Chem. Eng. Process.*, 48, Aljbour, S., Yamada, H., and Tagawa, T., Ultrasound-assisted phase transfer catalysis in a capillary microreactor, 1167–1172, Copyright (2009), with permission from Elsevier.)

heterogeneous catalysis. This field is too large to be comprehensively described in this chapter, but a recent article using an enzyme as a catalyst is particularly interesting (Lee et al., 2008). The authors investigated the transesterification of sugar in a lipase-catalyzed synthesis in 1-butyl-3-methylimidazolium triflate. Using ultrasound improved the enzyme activity without degrading it, leading to very interesting results. Enzymatic chemistry is interesting from the green chemistry viewpoint, because enzymes are generally extremely efficient and selective catalysts, and operate under very mild conditions, obviously close to the biological ones. Even if some other examples are available in the literature, such as the resolution of racemic 1,2-azidoalcohols using lipase Amaro PS (Brenelli and Fernandes, 2003), or the biodegradation of distillery wastewater using an enzyme cellulase (Sangave and Pandit, 2006), the enzymatic sonochemistry remains quite limited but full of potential.

Another possibility to reach Green sonochemistry is probably the design and characterization of new sonoreactors. A great effort has been made for scaling-up sonochemical protocols, and some recent publications describe the design of micro-sonoreactors in which mass transfer is particularly efficient (Costa et al., 2008). The use of microreactors usually implies better yields; thanks to very efficient mass transfer and lower energy consumption due to good heat transfer. It is therefore very interesting not only for green chemistry, but also for process intensification as such. This methodology has been recently investigated by Aljbour et al. (2009) for the liquid–liquid phase transfer reaction between benzyl chloride and sodium sulfide. A capillary microreactor immersed in a sonic bath was used in this case, as illustrated in Figure 26.6 (Aljbour et al., 2009).

CONCLUSION: THE NEXT STEP

The main objective of this chapter was to prove that there is an immense potential for sonochemistry within the green chemistry field. As the various phenomena induced by cavitation are well established and accepted by the scientific community, sonochemists have a great goal ahead—to work toward a more sustainable development of chemistry. As it has been demonstrated in this chapter, a lot of the tools already exist and can be found in patents or open-literature, and probably more are still to be invented. However, this challenging field is not as simple as it may seem, requiring different scientists, from sonochemists to physicists, to work together to achieve further progress. For example, who would have thought of introducing an ultrasonic horn into a microwave reactor, in the times when all of the horns were made of metal? Moreover, the biggest challenge of all is maybe for chemical engineers who will have to work on the scale-up of these new methodologies, as green chemistry has to offer COST EFFECTIVE and INDUSTRIALLY APPLICABLE solutions, if we do not want it to stay confined within the research laboratories. As it has been shown comprehensively in this book, there are various industrial syntheses and processes involving ultrasound. Green sonochemistry has to follow the same path to make a step further, as it was demonstrated very recently in the case of combined microwave–ultrasound (Leonelly and Mason, 2010).

ACKNOWLEDGMENTS

The author would like to thank Kenneth R. Seddon for his support, and also Malgorzata Swadzba-Kwasny and Markus Fanselow from the QUILL Research Centre, Belfast, for their kind help and constructive comments.

REFERENCES

Adewuyi, Y.G. 2001. Sonochemistry: Environmental science and engineering applications. *Industrial Engineering and Chemical Research*, 40: 4681–4715.
Alammar, T., Birkner, A., and Mudring, A.V. 2009. Ultrasound-assisted synthesis of CuO nanorods in a neat room-temperature ionic liquid. *European Journal of Inorganic Chemistry*, 19: 2765–2768.
Aljbour, S., Yamada, H., and Tagawa, T. 2009. Ultrasound-assisted phase transfer catalysis in a capillary microreactor. *Chemical Engineering Processing*, 48: 1167–1172.
Anastas, P.T. and Warner, J.C. 1998. *Green Chemistry: Theory and Practice*, New York: Oxford University Press.
Ando, T., Sumi, S., Kawate, T., Ichibana J., and Hanafusa, T. 1984. Sonochemical switching of reaction pathways in solid-liquid two-phase reactions. *Journal of the Chemical Society and Chemical Communications*, 45: 439–440.
Augugliaro, V., Litter, M., Palmisano, L., and Soria, J. 2006. The combination of heterogeneous photocatalysis with chemical and physical operations: A tool for improving the photoprocess performance. *Journal of Photochemistry and Photobiology C: Photochemistry Reviews*, 7: 127–144.
Balachandran, S., Kentish, S.E., Mawson, R., and Ashokkumar, M. 2006. Ultrasonic enhancement of the supercritical extraction from ginger. *Ultrasonics Sonochemistry*, 13: 471–479.
Bhor, M.D., Nandurkar, N.S., Bhanushali, M.J., and Bhanage, B.M. 2008. Ultrasound promoted selective synthesis of 1,1′-binaphtyls catalyzed by Fe impregnated pillared Montmorillonite K10 in presence of TBHP as oxidant. *Ultrasonics Sonochemistry*, 15: 195–202.
Blasco-Jiménez, D., Lopez-Peinado, A.J., Martin-Aranda, R.M., Ziolek, M., and Sobczak, I. 2009. Sonocatalysis in solvent-free conditions: An efficient eco-friendly methodology to prepare N-alkylimidazoles using amino-grafted NbMCM-41. *Catalysis Today*, 142: 283–287.
Braghiroli, F.L., Barboza, S.C.S., and Serra, A.A. 2006. Sonochemical epoxidation of cyclohexene in RCN-H_2O_2. *Ultrasonics Sonochemistry*, 13: 443–445.
Brenelli, E.C.S. and Fernandes, J.L.N. 2003. Stereoselective acylations of 1,2-azidoalcohols with vinyl acetate, catalyzed by lipase Amano PS. *Tetrahedron Asymmetry*, 14: 1255–1259.
Breslow, R. 2007. A fifty-year perspective on chemistry in water. In *Organic Reactions in Water*, Lindstroem U.M. (Ed.), pp. 1–28, Oxford, U.K.: Blackwell Publishing Ltd.
Brundtland, C.G. 1987. *Our Common Future*, The World Commission on Environmental Development, Oxford: Oxford University Press.
Chtourou, M., Abdelhédi, R., Frikha, M.H., and Trabelski, M. 2010. Solvent free synthesis of 1,3-diaryl-2-propenones catalyzed by commercial acid-clays under ultrasound irradiation. *Ultrasonics Sonochemistry*, 17: 246–249.
Cintas, P. and Luche, J.L. 1999. Green chemistry, *Green Chemistry*, 1: 115–125.
Costa, C., Hihn, J.Y., Rebetez, M., Doche, M.L., Bisel, I., and Moisy, P. 2008. Transport-limited current and microsonoreactor characterization at 3 low frequencies in the presence of water, acetonitrile and imidazolium-based ionic liquids. *Physical Chemistry and Chemical Physics*, 10: 2149–2158.
Cravotto, G., Boffa, L., Mantegna, S., Perega, P., Avogadro, M., and Cintas, P. 2008. Improved extraction of vegetable oils under high-intensity ultrasound and/or microwaves. *Ultrasonics Sonochemistry*, 15: 898–902.

Cravotto, G., Beggiato, B., Penani, A. et al. 2005a. High-intensity ultrasound and microwave, alone or combined, promote Pd/C-catalyzed aryl-aryl couplings. *Tetrahedron Letters*, 46: 2267–2271.

Cravotto, G. and Cintas, P. 2007. The combined use of microwave and ultrasound: Improved tools in process chemistry and organic synthesis. *Chemical European Journal*, 13: 1902–1909.

Cravotto, G., Palmisano, G., Tollari, S., Nano, G.M., and Penoni, A. 2005b. The Suzuki homocoupling under high-intensity ultrasound. *Ultrasonics Sonochemistry*, 12: 91–94.

Del Campo, F., Neudeck, A., Compton, R., Marken, F., Bull, S., and Davies, S. 2001. Sonoelectrochemistry at platinum and boron-doped diamond electrodes: Achieving 'fast mass transport' for 'slow diffusers'. *Journal of Electroanalytical Chemistry*, 507: 144–151.

Deshmukh, R.R., Rajagopal, R., and Srinivasan, K.V. 2001. Ultrasound promoted C-C bond formation: Heck reaction at ambient conditions in room temperature ionic liquids. *Chemical Communications*, 17: 1544–1545.

Dhas, N.A. and Suslick, K.S. 2005. Sonochemical preparation of hollow nanospheres and hollow nanocrystals. *Journal of American Chemical Society*, 127: 2368–2369.

Diez-Barra, F., De la Hoz, A., Diaz-Ortiz, A., and Prieto, R. 1992. Ultrasound and phase-transfer catalysis without solvent in elimination reactions: Synthesis of cyclic ketene acetals. *Synletters*, 11: 893–894.

Enokida, Y., Abd El-Fatah, S., and Wai, C.M. 2002. Ultrasound-enhanced dissolution of UO_2 in supercritical CO_2-soluble extractant for dissolution of uranium dioxide. *Industrial Engineering and Chemical Research*, 41: 2282–2286.

Estager, J., Lévêque, J.M., Cravotto, G., Boffa, L., Bonrath, W., and Draye, M. 2007a. One-pot and solventless synthesis of ionic liquids under ultrasonic irradiation. *Synletters*, 13: 2065–2068.

Estager, J., Lévêque, J.M., Turgis, R., and Draye, M. 2007b. Neat benzoin condensation in recyclable room-temperature ionic liquids under ultrasonic activation. *Tetrahedron Letters*, 48: 755–759.

Filgueiras, A.V., Capelo, J.L., Lavilla I., and Bendicho, C. 2000. Comparison of ultrasound-assisted extraction and microwave-assisted digestion for determination of magnesium, manganese and zinc in plant samples by flame atomic absorption spectrometry. *Talanta*, 53: 433–441.

Flanningan, D.J., Hopkins, S.D., and Suslick, K.S. 2005. Sonochemistry and sonoluminescence in ionic liquids, molten salts, and concentrated electrolyte solutions. *Journal of Organometallic Chemistry*, 690: 3513–3517.

Fukushima, T. and Aida, T. 2007. Ionic liquids for soft functional materials with carbon nanotubes. *Chemical European Journal*, 13: 5048–5058.

Gedye, R., Smith, F., Westaway, K., Ali, H., Baldisera, L., Laberge, L., and Roussel, J. 1986. The use of microwave ovens for rapid organic synthesis. *Tetrahedron Letters*, 27: 279–282.

Gholap, A.R., Venkatesan, K., Daniel, T., Lahoti, R.J., and Srinivasan, K.V. 2003. Ultrasound promoted acetylation of alcohols in room temperature ionic liquid under ambient conditions. *Green Chemistry*, 5: 693–696.

Gholap, A.R., Venkatesan, K., Daniel, T., Lahoti, R.J., and Srinivasan, K.V. 2004. Ionic liquid promoted novel and efficient one pot synthesis of 3,4-dihydropyrimidin-2(1H)-ones at ambient temperature under ultrasound irradiation. *Green Chemistry*, 6: 147–150.

Goharshadi, E.K., Ding, Y., Jorabchi, M.N., and Nancarrow, P. 2009. Ultrasound-assisted green synthesis of nanocrystalline ZnO in the ionic liquid [hmim][NTf$_2$]. *Ultrasonics Sonochemistry*, 16: 120–123.

Gutierrez-Sanchez, C., Calvino-Castilda, V., Perez-Mayoral, E. et al. 2009. Coumarins preparation by Pechmann under ultrasound irradiation. Synthesis of Hymecromone as insecticide intermediate. *Catalysis Letters*, 182: 318–322.

Hamdaoui, O. and Naffrechoux, E. 2008. Sonochemical and photosonochemical degradation of 4-chlorophenol in aqueous media. *Ultrasonics Sonochemistry*, 15: 981–987.

Hao, H., Chen, Y., Wu, M., Wang, H., Yin, Y., and Lü, Z. 2004. Sonochemistry of degrading p-chlorophenol in water by high frequency ultrasound. *Ultrasonics Sonochemistry*, 11: 43–46.

Harada, H. 2001. Sonophotocatalytic decomposition of water using TiO_2 photocatalyst. *Ultrasonics Sonochemistry*, 8: 55–58.

Heravi, M.R.P. 2009. An efficient synthesis of quinolines derivatives promoted by a room temperature ionic liquid at ambient conditions under ultrasound irradiation via tandel addition/annulations reaction of O-aminoaryl ketones with alpha-methylene ketones. *Ultrasonics Sonochemistry*, 16: 361–366.

Hu, A.J., Zhao, S.V., Liang, H., Qiu, T.Q., and Chen, G. 2007. Ultrasound assisted supercritical fluid extraction of oil and coixenolide from adley seed. *Ultrasonics Sonochemistry*, 14: 219–224.

Inazu, K., Nagata, Y., and Maeda, Y. 1993. Decomposition of chlorinated hydrocarbons in aqueous solution by ultrasonic irradiation. *Chemical Letters*, 57–60.

Ince, N.H., Tezcanli, G., Belen, R.K., and Apikyan, I.G. 2001. Ultrasound as a catalyzer of aqueous reaction systems: The state of the art and environmental applications. *Applied Catalysis B: Environment*, 29: 167–176.

Jessop, P. and Leitner, W. 1999. *Chemical Synthesis Using Supercritical Fluids*, Weinheim, Germany: Wiley-VCH.

Jiang, Y., Pétrier, C., and Waite, D. 2006. Sonolysis of 4-chlorophenol in aqueous solution: Effects of substrate concentration, aqueous temperature and ultrasonic frequency. *Ultrasonics Sonochemistry*, 13: 415–422.

Lee, S.H., Nguyen, H.M., Koo, Y.M., and Ha, S.H. 2008. Ultrasound-enhanced lipase activity in the synthesis of sugar ester using ionic liquids. *Process Biochemistry*, 43: 1009–1012.

Leonelly, C. and Mason, T.J., 2010. Microwave and ultrasonic processing: Now a realistic option for industry. *Chemical Engineering and Process*, 49: 885–900.

Lévêque, J.M., Estager, J., Draye, M., Cravotto, G., Boffa, L., and Bonrath, W. 2007. Synthesis of ionic liquids using non conventional activation methods. An overview. *Monatshefte für Chemie*, 138: 1103–1113.

Li, X., Zhao, J., Li, Q., Wang, L., and Tsang, S.C. 2007. Ultrasonic chemical oxidative degradations of 1,3-dialkyimidazolium ionic liquids and their mechanistric elucidations. *Dalton Transactions*, 19: 1875–1880.

Loupy, A. 2006. *Microwave in Organic Synthesis*, 2nd edn., Weinheim, Germany: Wiley-VCH.

Luche, J.L. 1993. Sonochemistry from experiments to theoretical considerations. In *Advances in Sonochemistry*, Mason T.J. (Ed.), pp. 85–124, London, U.K.: JAI Press.

Luche, J.L. 1998. *Synthetic Organic Sonochemistry*, New York: Plenum Press.

Maeda, M. and Amemiya, H. 1995. Chemical effects under simultaneous irradiation by microwaves and ultrasound. *New Journal of Chemistry*, 19: 1023–1028.

Makosza, M. 1975. Two-phase reactions in the chemistry of carbanions and halocarbenes-useful tool in organic synthesis. *Pure and Applied Chemistry*, 43: 439–462.

Makosza, M. 2000. Phase-transfer catalyst. A general green methodology in organic synthesis. *Pure and Applied Chemistry*, 72: 1399–1403.

Memarian, H.R. and Abdoli-Senejani, M. 2008. Ultrasound-assisted photochemical oxidation of unsymmetrically substituted 1,4-dihydropyridines. *Ultrasonics Sonochemistry*, 15: 110–114.

Mikkola, J.P., Kirilin, A., Tuuf, J.C. et al. 2007. Ultrasound enhancement of cellulose processing in ionic liquids: From dissolution towards functionalization. *Green Chemistry*, 9: 1229–1237.

Mojtahedi, M.M., Javadpour, M., and Abaee, M.S. 2008. Convenient ultrasound synthesis of substituted pyrazolones under solvent-free conditions. *Ultrasonics Sonochemistry*, 15: 228–232.

Park, I.B., Son, Y., Song, I.S., Kim, J., and Khim, J. 2008. Remediation of diesel-contaminated soil using supercritical carbon dioxide and ultrasound. *Japanese Journal of Applied Physics*, 47: 4314–4316.

Peng, Y. and Song, G. 2003. Combined microwave and ultrasound accelerated Knoevenagel-Doebner reaction in aqueous media: A green route to 3-aryl acrylic acids, *Green Chemistry*, 5: 704–706.

Peng, Y., Zhong, W., and Song, G. 2005. Efficient and mild room temperature reduction of benzophenones under ultrasound irradiation. *Ultrasonics Sonochemistry*, 12: 169–172.

Perreux, L. and Loupy, A. 2002. Nonthermal effects of microwave in organic synthesis. In *Microwave in Organic Synthesis*, Loupy, A. (Ed.), pp. 61–114, Weinheim, Germany: Wiley-VCH.

Peters, D. 2001. Sonolytic degradation of volatile pollutants in natural groundwater: Conclusions from a model study. *Ultrasonics Sonochemistry*, 8: 221–226.

Pétrier, C., Lamy, M.F., Francony, A. et al. 1994. Sonochemical degradation of phenol in dilute aqueous solutions: Comparison of the reaction rates at 20 and 487 kHz. *Journal of Physical Chemistry*, 98: 10514–10520.

Plechkova, N.V. and Seddon, K.R. 2008. Applications of ionic liquids in the chemical industry. *Chemical Society Reviews*, 37: 123–150.

Puri, S., Kaur, B., Parmar, A., and Kumar, H. 2009. Ultrasound-promoted greener synthesis of 2H-chromen-2-ones catalyzed by copper perchlorate in solventless media. *Ultrasonics Sonochemistry*, 16: 705–707.

Ragaini, V. and Bianchi, C.L. 1998. Sonochemical catalytic reactions. In *Synthetic Organic Sonochemistry*, Luche, J.L. (Ed.), pp. 235–261, New York: Plenum Press.

Rajagopal, R., Jarikote, D.V., and Srinivasan, K.V. 2002. Ultrasound promoted Suzuki cross-coupling reactions in ionic liquid at ambient conditions. *Chemical Communications*, 616–617.

Ranu, R.C. and Mandal, T. 2006. Indium(I) iodide as a radical initiator: Intramolecular cyclization of functionalized bromo-alkynes to substituted tetrahydrofurans. *Tetrahedron Letters*, 47: 2859–2861.

Reddy, E.P., Daydov, L., and Smirniotis, P. 2003. TiO_2-loaded zeolites and mesoporous materials in the sono-photocatalytic decomposition of aqueous organic pollutants: The role of the support. *Applied Catalysis B: Environment*, 42: 1–11.

Riera, E., Golas, Y., Blanco, A., Gallego, J.A., Blasco, M., and Mulet, A. 2004. Mass transfer enhancement in supercritical fluids extraction by means of power ultrasound. *Ultrasonics Sonochemistry*, 11: 241–244.

Rodriguez, I., Llompart, M.P., and Cela, R. 2000. Solid-phase extraction of phenols. *Journal of Chromatography*, 885: 291–304.

Sangave, P.C. and Pandit, A.B. 2006. Ultrasound and enzyme assisted biodegradation of distillery waste water. *Journal of Environmental Management*, 80: 36–46.

Schmitt, F.O., Johnson, C.H., and Olson, A.R. 1928. Oxidations promoted by ultrasonic radiation. *Journal of American Chemical Society*, 51: 370–375.

Serpone, N., Terzian, R., Colarusso, P., Minero, C., Pelizetti, E., and Hidaka, H. 1992. Sonochemical oxidation of phenol and three of its intermediate products in aqueous media: Pyrocatechol, hydroquinone, and benzoquinone. Kinetic and mechanistic aspects. *Research on Chemical Intermediates*, 18: 183–202.

Sheldon, R.A. 1994. Consider the environmental quotient. *CHEMTECH*, 24: 38–47.

Sheldon, R.A. 2008. E factors, green chemistry and catalysis: An odyssey. *Chemical Communications*, 29: 3352–3365.

Sheldon, R.A., Arends, I.W.C.E., and Hanefeld, U. 2007. *Green Chemistry and Catalysis*, Weinheim, Germany: Wiley-VCH.

Stock, N.L., Peller, J., Vinodgopal, K., and Kamat, P.V. 2000. Combinative sonolysis and photocatalysis for textile dye degradation. *Environmental Science and Technology*, 34: 1747–1750.

Strauss, C.R. and Varma, R.S. 2006 Microwaves in green and sustainable chemistry. *Topics in Current Chemistry*, 266: 199–231.

Suslick, K.S., Hammerton, D.A., and Cline, R.E. Jr. 1990. Sonochemical hot spot. *Journal of American Chemical Society*, 108: 5641–5642.

Tanchoux, N. and Leitner, W. 2002. Supercritical carbon dioxide as an environmentally benign reaction medium for chemical syntheses. In *Handbook of Green Chemistry and Technologies*, Clark, J., and Macquarrie, D. (Eds.), pp. 482–501, Oxford, U.K.: Blackwell Science Ltd.

Tang, S.Y., Bourne, R.A., Smith, R.L., and Poliakoff, M. 2008. The 24 principles of green engineering and green chemistry: "Improvements productively". *Green Chemistry*, 7: 268–269.

Toda, F., Tanaka, K., and Iwata, J. 1989. Oxidative coupling reactions of phenols with iron(III) chloride in the solid state. *Journal of Organic Chemistry*, 54: 3007–3009.

Toma, S., Gaplovsky, A., and Luche, J.L. 2001. The effect of ultrasound on photochemical reactions. *Ultrasonics Sonochemistry*, 8: 201–207.

Toukoniity, B., Mikkola, J.P., Murzin, D.Y., and Salmi, T. 2008. Utilization of electromagnetic and acoustic irradiation in enhancing heterogeneous catalytic reactions. *Trends in Chemical Engineering*, 11: 1–37.

Toy, M.S. and Stringham, R.S. 1984. Ultrasonic photolysis of methyl disulfide and hexafluorobutadiene. *Journal of Fluorine Chemistry*, 25: 213–218.

Trofimov, T.I., Samsonov, M.D., Lee, S.C., Smart, N.G., and Wai, C.M. 2001. Ultrasound enhancement of dissolution kinetics of uranium oxides in supercritical carbon dioxide. *Journal of Chemical Technology and Biotechnology*, 76: 1223–1226.

Trost, B.M. 1991. The atom economy: A search for synthetic efficiency. *Science*, 254: 1471–1477.

Tucker, J.L. 2010. Green chemistry: Cresting a summit towards sustainability. *Organic Process Research and Development*, 14: 328–331.

Venkatesan, K., Pujari, S.S., Lahoti, R.J., and Srinivasan, K.V. 2008. An efficient synthesis of 1,8-dioxo-octahydro-xanthene derivatives promoted by a room temperature ionic liquid at ambient conditions under ultrasound irradiation. *Ultrasonics Sonochemistry*, 15: 548–553.

Wadhawan, J., Del Campo, F., Compton, R. et al. 2001. Emulsion electrosynthesis in the presence of power ultrasound. Biphasic Kolbe coupling processes at platinum and boron-doped diamond electrodes. *Journal of Electroanalytical Chemistry*, 507: 135–143.

Wang, S.X., Li, Z.Y., Zhang, J.C., and Li, J.T. 2008. The solvent-free synthesis of 1,4-dihydropyridines under ultrasound irradiation without catalyst. *Ultrasonics Sonochemistry*, 15: 677–680.

Wasserscheid, P. and Welton, T. 2008. *Ionic Liquids in Synthesis*, 2nd edn., Weinheim, Germany: Wiley-VCH.

Woods, R. and Loomis, A. 1927. The physical and biological effects of high frequency sound waves of great intensity. *Philosophia Magazine*, 4: 414–436.

Yang, J.M., Ji, S.J., Gu, D.G., Shen, Z.L., and Wang, S.Y. 2005. Ultrasound-irradiated addition of ferrocenylenones under solvent-free and catalyst-free conditions at room temperature. *Journal of Organometallic Chemistry*, 690: 2989–2995.

Zhang, Z.H., Li, J.J., and Li, T.S. 2008. Ultrasound-assisted synthesis of pyrroles catalyzed in zirconium chloride in solvent-free conditions. *Ultrasonics Sonochemistry*, 15: 673–676.

Zhou, Y. 2005. Recent advances in ionic liquids for synthesis of inorganic nanomaterials. *Current Nanoscience*, 1: 35–42.

Zhu, J., Liu, S., Palchnik, O., Koltypin, Y., and Gedanken, A. 2000. A novel sonochemical method for the preparation of nanophasic sulfides: Synthesis of HgS and PbS nanoparticles. *Journal of Solid State Chemistry*, 153: 342–348.

Zosel, K. US Patent Application No 784,744, May 19, 1977.

Index

A

ABS, *see* Acrylonitrile butadiene styrene
Acoustic cavitation and sonochemistry
    bubble growth and collapse, 476–477
    bubble radius-time curve, 476
    chemical effects stemming, 476–477
    physical effects, 477
    rectified diffusion, 476
Acoustic frequency, 509
Acoustic pressure amplitude, 509
Acoustic waves
    diffusion-limited model, 510–511
    physical effect of cavitation bubbles, 506
Acrylonitrile butadiene styrene (ABS), 43–44
Active pharmaceutical ingredients (API), 561
Advanced oxidation process (AOP), 568
Anaerobic digestion
    merits and demerits, 327–328
    oxidation–reduction reaction, 326
    stages of, 326–327
Atom efficiency (AE), 11

B

Barbier reaction, 295
Baylis–Hillman (BH) reaction, 285–286
Benign solvents, 680–681
Besançon cell, 630–631
Best available technologies (BAT), 15
BH reaction, *see* Baylis–Hillman reaction
Biginelli reaction, 233–235
Biodegradation, 408
Biosensors, 545–546
Bioventing, 408
Bubble radius, 509

C

Carbon efficiency (CE), 11
Cavitation
    definition, 502
    heterogeneous sonochemistry, 266
    homogeneous sonochemistry, 265
    liquid–liquid system, 503
    microbubbles, 265
    solid–liquid reaction system, 503
    sonocatalysis, 266
    types of, 502
Cavitation bubbles
    chemical effects
        contemporary studies, 505
        research in 1990s, 503–504
        Storey and Szeri model, 504–505
    physical effects
        acoustic waves, 506
        microjets, 506
        microstreaming, 505
        microturbulence, 505–506
        quantification, 510
    thermal effects, 511
Cefuroxime acetil (CA) nanoparticles, 89–90
Chemical aerosol flow synthesis, 120
Chemical dosimetry, 584–585
Chemical oxygen demand (COD), 661, 663
4-Chlorophenol decomposition, 680
Clean Air Act Amendment, 1990, 681
C-M bonds complex
    azine derivatives, 199–200
    copper and nickel atoms, 191–192
    divinylzinc complex, 184–185
    electrochemical study, 188
    ferrocene, 187–188
    Fischer carbene complex, 185
    Grignard reagent, 184
    halogen-containing metal complexes, 196–197
    LiPc particles, 194
    metalated peptide complexes, 201
    N-containing carboxylates, 198–199
    N-containing metal complexes, 189
    nicotinic and isonicotinic acid coordination mode, 197–198
    non-substituted metal phthalocyanines, 190–191
    N-,O-,S-containing metal complex, 202
    nucleophilic aromatic substitution, 188–189
    O-containing metal complexes, 191–195
    oxygen-reducing catalyst, 189
    PcM formation mechanism, 190–191
    phenylethyne cobalt hexacarbonyl complex, 185–186
    phthalocyaninate formation, 192–193
    porphyrin application, 189–190
    porphyrin preparatiom, 189
    Schiff base complex, 197
    S-containing metal complexes, 195–196
    S,O-containing metal complexes, 201–202
    SWCNTs sonochemical route, 187–188
    Te-containing metal complexes, 196
    Znq_2 nanorods formation, 200–201
    Znq_2 product TEM image, 200
COD, *see* Chemical oxygen demand
Condensation reaction
    chalcone synthesis, 267–268
    Claisen–Schmidt condensation, 267
    malononitrile, 266
    nitromethane and aromatic aldehyde, 268–269
    pyridine-catalyzed Knoevenagel condensation, 266–267

Contaminated soil treatment, persistent organic pollutant
　biological technology, 408
　coupling effect, 416
　desorption, 411
　electrokinetic process, 416
　environmental chemical cycle, 407
　ex situ and in situ treatment, 408–409
　physicochemical technology, 408
　soil and sediment pollutant, 407
　ultrasonication
　　activated carbon amendment, 415
　　advanced oxidative soil remediation, 413–414
　　electrokinetic remediation, 414–415
　　soil-flushing, 412–413
　　surfactant-aided soil-washing, 415
　ultrasonic energy, 412
　ultrasound and sonochemistry
　　application, 409
　　cavitation bubble collapse, 409–410
　　hot-spot model, 410
　　sonochemical process, 411
Cross-coupling reactions, 272–273
Crude oil industry, 563–565
Crystallization, 541, 556–557
Crystal size distribution (CSD)
　characteristics of, 523
　dominant crystal size and span of, 528–529
　MSMPR model, 528
　representative results of measurement, 525–526
Cycloaddition reaction
　cyclopentadiene and dienophiles, 284
　Diels–Alder reaction, 282
　furano diene derivative, 283
　mechanistic survey, 284
　one-pot process, 282–283
Cyclocondensation reaction
　aryl-14-H-dibenzo[a,j]xanthene synthesis, 241
　2,3-bis(4-hydroxyphenyl)indole synthesis, 239
　3-carboxycoumarine synthesis, 236–237
　2,4-diarylthiazole synthesis, 241, 243
　5,5-disubstituted hyndantoin synthesis, 239
　environmentally friendly sonocatalysis, 241–242
　ferrocenyl substituted 3-cyanopyridine derivative synthesis, 237–238
　heterocyclic pyrimidine synthesis, 239–240
　4H-pyrano[2,3-c]pyrazole synthesis, 236–237
　oxindole and 3-oxo-tetrahydroisoquinoline derivative synthesis, 237–238
　pyrazole and isoxazole synthesis, 241
　pyrazolone synthesis, 236
　pyridazinone derivative synthesis, 241–242
　pyrido[2,3-d]pyrimidine derivative synthesis, 243–244
　4(3H)-quanazolinone synthesis, 241, 243
　quinoxaline derivative synthesis, 243
　Spiro [indole-3,5'-[1,3] oxathiolane synthesis, 240
　2,4,5-triarylimidazole derivative synthesis, 243
　2,4,5-trisubstituted imidazole synthesis, 236
Cylindrical sonoreactors, 585–586

D

Degradation of diclofenac (DCF), 570
Dendritic Pt Nanostructures (DPNs), 646
De-oiling process, 550

Diels–Alder cycloaddition, 36
Diels–Alder reaction, 282
Diffusion-limited model
　bubble dynamics equations, 508
　components of, 507
　microturbulence, 510
　numerical solution, 507–510
　radical production quantification, 510
　shock waves, 510–511
　thermodynamic properties, 509
Dimensionless number
　Reynolds number, 618
　Schmidt number, 619
Directivity distribution, ultrasound waves
　acoustic pressure, 135
　apparent attenuation, 136
　directivity parameter, 135
　elastic/viscoelastic forces, 134
　experimental system, 136–137
　mechanical oscillator, 134
　planar wave attenuation, 136
　three-dimensional field, half-cut potato, 136–137
　true attenuation losses, 136
　ultrasonic wave transducer, 134
Double wall cell, see Besançon cell
Drying process, 539–540
Dry weight (DW) content, 151
Dyeing process, 549–550

E

E-factor for industries, 676
Electrical discharge theory, 476
Electrochemical metallization
　electroless copper plating
　　advantages, 52
　　colloidal palladium-tin catalyst bath, 51–52
　　dispersive properties, 53
　　mass transfer coefficients, 51
　　ultrasonic agitation, 52–53
　　ultrasonic frequency ranges, 51
　electroless nickel plating, 50–51
　electroplating
　　acoustic field, 46
　　cathodic current efficiency, 47
　　composite coating composition and hardness, 48
　　dispersive properties, 48
　　fume suppressant, 49
　　megasonics, 47
　　microjetting, 47
　　mist reduction, chrome plating electrolyte, 49
　　RDE, 47–48
　　sonication, 49
　liquid electrolyte, 46
Electrochemistry, ultrasound applications
　aqueous solutions, 646
　electrode-kinetics
　　angular geometry, 634–636
　　anodic voltammograms, 632–633
　　diffusion layer thickness, 636–637
　　face-on geometry, 634–636
　　incident flow, 634–636
　　localized temperature variations, 637
　　Pollet equation, 633–634

Index 699

sonoelectrochemical effect, 634
sonovoltammograms, 632
ultrasonic intensity, 638
environmental sonoelectrochemistry, 643–644
exotic solvents, 649–653
fuel cell materials, 645–646
mathematical models, 632
nanomaterials, sonoelectrochemical synthesis, 644–645
sono-electroanalytical chemistry, 644
sono-electrochemiluminescence, 643
sonoelectrochemistry and corrosion, 639–640
sono-electrodeposition and sono-electroplating, 638–639
sono-electro-organic chemistry, 641
sono-electropolymerization, 642–643
surfactants and polymers effect
   dendritic-structured Pd nanoparticles, 647
   nanosize metals, 647
   nanosize mono-metallic electrocatalysts, 647, 649
   Pt nanoparticles, 646–647
   sonoelectrochemical method, 649
ultrasonic irradiation, 632
Emulsification, 540–541
Emulsion polymerization
  compartmentalization, 480
  compartmentalized pseudo-bulk system, 480
  critical micelle concentration, 479
  pseudo-bulk system, 480
  theory of, 479–480
  zero-one system, 480
Environmental sono-electrochemistry, 643–644
Epoxidation, 247
Equivalent velocity tool
  acoustic energy distribution applications, 615–617
  boundary conditions of hydrodynamics, 614
  cavitation quantification, 617–618
  coordinate system, 614
  Levich hypotheses, 614
  Navier–Stokes equations, 614
Esterification, 225
Etherification, 225–226
Evaporation process, 540

F

Fick's second law equations, 607
Field emission scanning electron microscopy (FESEM), 669
Fine particle fraction (FPF), 86
Food control measurements, 174–175
Food-processing applications
  bioactive compounds extraction, 170–172
  extraction process, 168–170
  microbial and enzyme inactivation, 165–168
  oil and protein extraction, 172–173
  in separation experiments, 173–174
Food technology, ultrasound, 164
Free radical polymerization
  initiation process, 477–478
  propagation stage, 478
  termination stage, 478–479
Fricke reaction, 587
Friedel–Crafts product, 34

Fruit and vegetable quality evaluation
  acoustic parameters, 132
  continuous-touch systems, 140
  cut half fruit specimens measurements
   directional wave propagation, 145
   pressure waves measurement, 141
   single-touch system, 141, 144
   time-domain waveform, 144
   two-dimensional field, half-cut melon, 144–145
  food-processing industry, 130
  human sensory perceptions, 130
  nondestructive/noninvasive techniques, 130
  piezoelectric element, 131
  portable systems
   fixed-load system operating, 147–148
   variable-load system operating, 148–150
  primary attenuation mechanism, 131
  pulse-echo mode, 131–132
  quality assessment (*see* Quality assessment, fruits and vegetables)
  quality-related parameters
   mechanical parameters, 139
   physicochemical indices, 139–140
   textural attributes, 138
  single-touch systems, 140
  through-transmission mode, 131–132
  tissue segments measurements
   beam-focusing elements, 141
   destructive tests, 140
   pre-and postharvest processes, 141–143
   single-touch system, 141
  ultrasound hardware, 138
  ultrasound wave propagation, fresh agricultural tissue attributes, 132
   directivity distribution (*see* Directivity distribution, ultrasound waves)
   wave attenuation and amplitude, 133–134
   wave modes and velocities, 133
  whole fruits measurements
   beam-focusing elements, 146
   continuous-touch systems, 145–146
   nondestructive evaluation, 145
   output pulse amplitude, 146
   penetrometer, 147
   surface absorption, 146
Fuel cell materials, 645–646

G

Gibbs energy, 554
Green chemistry
  anthropogenic wastes
   cost-efficiency, 6
   definition of chemicals and life, 4
   hazard and vulnerability, 5
   LCA, 5–6
   mass, 5
   permanent renewable and nonrenewable sources, 5
   toxic chemicals, 4
   toxicity analysis, 7
   types, 6
  chemical wastes
   carbon monoxide, 8–9
   hydrogen cyanide, 7

hydrogen fluoride, 8
ibuprofen, 8
monomer acrylonitrile, 7–8
PAN, 7
solvent-free syntheses, 9
synthetic chemistry metrics (*see* Synthetic chemistry metrics)
synthetic process, 9
chemistry curse, causes
  Agent Orange, 2–3
  Bhopal disaster, 3
  negative associations, 1
  2,3,7,8-tetrachlordibenzo-*p*-dioxine, 3
  1,2,4,5-tetrachlorobenzene, 3
  thalidomide, 3–4
  1,1,1-trichloro-2,2-di(4-chlorophenyl)ethane, 2
  World War I and II, 2
coupling reactions, 684
methodology and methods, 17–18
principles and goals, 16–17
principles of, 676
sustainable development
  Agenda 21, 13
  BAT, 15
  environmental protection and conservation, 12
  Laws of Sustainability, 14
  "Our Common Future," WCED, 12
  pollution prevention, 15
  Pollution Prevention Act, 14
  REACH, 15
  UNFCCC, 13
  World3 model, 12
  WSSD, 13
ultrasound contribution
  solvents, 681–682
  water as clean reactive medium, 680–681
  yields and chemical rates, 678–679
VOCs, 680–681

H

Heck reaction, 294
Henry's law, 377–378
Heterocycles synthesis, 274–276
Heterogeneous catalysis
  advanced Fenton process, 434–435
  biphasic solid-liquid medium, 419
  EDTA effect, 427–428
  FAZA, 434
  functional and biochemical change, 428–429
  hybrid method
    agro-industrial effluent, 439–440
    photoassisted process, 436–438
    P25 TiO_2 catalyst, 439
    solar photocatalysis, 440–441
    sonophoto-Fenton process, 441
    synergistic effect, 439
    2,4,6-trichlorophenol sonophotochemical degradation, 440
    UV and US irradiation, 440
  pollutant sonodegradation, 421–426
  radical oxidation and recombination reactions, 434
  solid particles and US waves interaction, 420–421
  sonocatalytic system, hydrogen peroxide, 430–433
  sono-enzyme peroxide degradation system, 436
  sono-Fenton degradation process, 429
  sonophoto-Fenton process, 419–420
  ultrasound irradiation, 419
  US/TiO_2 system, 420–421
  ZVI, 427–428
Heterogeneous reactions, 682–683
Heterogeneous sonochemistry, 266
High-intensity ultrasound, 164
Homogeneous sonochemistry, 265
Hot-spot model, 410
Hotspot theory, 476–477
Hydrodesulfurization (HDS), 564
Hydrodynamic cavitation, 502

I

Incineration, 408
Industrial environmental factor, 676
Inhalation drug delivery system
  aerodynamic particle size, 82
  agglomeration, 82–83
  amorphous particles, 86
  budesonide, 84
  cohesive–adhesive balance, 82
  DPIs, 81–82, 87
  drug molecules, 75
  fluticasone propionate and salmeterol xinafoate, 87
  FPF, 86
  HFA–pMDI formulation, 84
  in vitro aerosolization, 84
  lactose, 87–89
  pulmonary drug delivery, 81
  salbutamol sulphate, 84–86
  SAXS technique, 83–84
  scanning electron micrograph, NaCl particles, 83–84
  SD-SCSS, 86–87
  spray drying, 83
  supersaturation, 85
  US irradiation, 83
Initiator efficiency, 478
Ionic liquids
  one-pot synthesis of, 683–684
  and ultrasound, 687

K

Keller–Miksis equation, 487
Khand reaction, 185–186
KI oxidation, 587
Knoevenagel condensation, 233

L

Large-scale sonochemical reactors, 594–595
Large-size sonochemical reactor
  liquid height effect, 592–593
  sonochemical efficiency, 593–594
Life-cycle assessment (LCA), 5–6
Liquid flow effects, 591–592
Liquid height effect, 592–593
Liquid-liquid extraction (LLE), 349
Loop US/MW flow reactor, 667
Low-intensity ultrasound, 164

Index

M

Mannich and Baylis–Hillman reaction, 285–286
Mannich reaction, 270–272
Mass transport-limited current density *vs.* acoustic intensity, 612–613
Metal and plastic welding, 562–563
Michael additions, 269
Microbial cell disruption, 543
Microjets phenomenon, 506
Microorganisms, ultrasound effects, 165
Microstreaming phenomenon, 505
Microturbulence, 505–506
Microwave–ultrasound combined technologies, 684
    chemical analysis, 659–662
    COD, 661, 663
    nanomaterials preparation, 668
        FESEM, 669
        TEM, 669–670
    natural product extraction, 662, 664–665
    synthesis and catalysis
        click reaction catalysis, 668
        loop US/MW flow reactor, 667
        solvent-free one-pot synthesis, 668
        Suzuki reactions, 666
Milestone®, 664
Mixed suspension mixed product removal (MSMPR) model, 528
Multicomponent reaction (MCR)
    1-amidoalkyl-2-naphthol synthesis, 231–232
    2-amino-3,5-dicarbonitrile-6-thio-pyridine synthesis, 232–233
    1,8-dioxo-octahydro-xanthene synthesis, 231–232
    fused heterocyclic pyrimidine synthesis, 230
    β-indolyl ketone synthesis, 232
    one-pot three-component Mannich-type reaction, 230
    one-pot three-component synthesis, 229
    thiazine synthesis, 231
    three-component Mannich reaction, 228–229
    Ugi four-component reaction, 228–229
    ultrasound-promoted regioselective synthesis, 229–230

N

N-acylation, 228
N-alkylation, 227
Natural organic matter degradation
    aromaticity, 386–388
    chemical nature, 383
    ^{13}C NMR, 386
    environmental media NOM concentration, 382–383
    H_2O_2 formation, 384–385
    HPSEC spectra, 386, 389
    humic substance properties, 383
    hydrophobicity, 386
    implication, 390
    molecular weight, 388–389
    shearing degradation, 384
    sonochemical degradation, 383–384
    TOC reduction, 385
    total acidity, 389–390
    UV/Vis absorbance, 385–386

Navier–Stokes equations, 614
Nernst–Planck equation, 606–607
New chemical entity (NCE), 97
Nitrogen quaternization, 36
Nucleation mechanism, 554–555
Nucleophilic substitution reaction, 250–251

O

Office of Pollution Prevention and Toxics (OPPT), 14
Oil extraction technologies, 172–173, 552–553
Optic cavitation, 502
Organic chemistry
    acoustic cavitation, 214
    C-alkylation/acylation, 226–227
    cavitation, 265–266
    C-C bond formation reaction
        allylation reaction, 219–220
        Baylis–Hillman adduct synthesis, 219
        bis(indolyl)methane synthesis, 216–217, 221
        Claisen–Schmidt condensation, 219–220
        complex pharmaceuticals, 214
        Diels–Alder cyclization reaction, 214–215
        Heck reaction, 216–217
        Michael reaction, 218–219
        pinacol coupling reaction, 217–218
        regioselective alkylation, 215–216
        solid-liquid phase transfer catalysis, 214–215
        Sonogashira reaction, 219–220
        Suzuki homocoupling reaction, 218
        Suzuki reaction, 215–216
        ultrasound irradiation technique, 218–219
    C-N bond formation reaction
        2-amino-1,4-naphthoquinoline synthesis, 224
        aromatic nucleophilic substitution reaction, 222
        5-aryl-1,3-diphenylpyrazole synthesis, 221
        aryl hydrazone synthesis, 223
        ferrocenyl enone addition, 224
        glycoluril synthesis, 221
        oxime synthesis, 222
        sonochemical nitration, 222
        ultrasound-mediated one-pot synthesis, 223
    C-O bond formation reaction, 224–225
    condensation reaction
        chalcone synthesis, 267–268
        Claisen–Schmidt condensation, 267
        malononitrile, 266
        nitromethane and aromatic aldehyde, 268–269
        pyridine-catalyzed Knoevenagel condensation, 266–267
    cross-coupling reaction, 272–273
    cyclocondensation reaction
        aryl-14-H-dibenzo[a,j]xanthene synthesis, 241
        2,3-bis(4-hydroxyphenyl)indole synthesis, 239
        3-carboxycoumarine synthesis, 236–237
        2,4-diarylthiazole synthesis, 241, 243
        5,5-disubstituted hyndantoin synthesis, 239
        environmentally friendly sonocatalysis, 241–242
        ferrocenyl substituted 3-cyanopyridine derivative synthesis, 237–238
        heterocyclic pyrimidine synthesis, 239–240
        4H-pyrano[2,3-c]pyrazole synthesis, 236–237
        oxindole and 3-oxo-tetrahydroisoquinoline derivative synthesis, 237–238

pyrazole and isoxazole synthesis, 241
pyrazolone synthesis, 236
pyridazinone derivative synthesis, 241–242
pyrido[2,3-d]pyrimidine derivative synthesis, 243–244
4(3H)-quanazolinone synthesis, 241, 243
quinoxaline derivative synthesis, 243
Spiro [indole-3,5′-[1,3] oxathiolane synthesis, 240
2,4,5-triarylimidazole derivative synthesis, 243
2,4,5-trisubstituted imidazole synthesis, 236
environmental concern, 277
epoxidation, 247
esterification, 225
etherification, 225–226
green techniques, 263
heterocycles synthesis, 274–276
imine preparation, 276–277
Mannich reaction, 270–272
MCR
    1-amidoalkyl-2-naphthol synthesis, 231–232
    2-amino-3,5-dicarbonitrile-6-thio-pyridine synthesis, 232–233
    1,8-dioxo-octahydro-xanthene synthesis, 231–232
    fused heterocyclic pyrimidine synthesis, 230
    β-indolyl ketone synthesis, 232
    one-pot three-component Mannich-type reaction, 230
    one-pot three-component synthesis, 229
    thiazine synthesis, 231
    three-component Mannich reaction, 228–229
    Ugi four-component reaction, 228–229
    ultrasound-promoted regioselective synthesis, 229–230
Michael addition, 269
miscellaneous reaction, 252–253
N-acylation, 228
N-alkylation, 227
name reaction
    Biginelli reaction, 233–235
    Curtius rearrangement, 235
    Knoevenagel condensation, 233
    Ullmann reaction, 235
nucleophilic substitution reaction, 250–251
O-acylation, 227
oxidation, 245–247
oxime deprotection, 251
piezoelectric materials, 264
preparation of ionic liquids, 251–252
reduction
    benzylacetamide synthesis, 247–248
    mandelic acid synthesis, 248
    4-methyl-oxazole-5-carboxylic acid amide dehydration, 249
    methyl trans 3-[2,4,6-trimethyl phenyl]-isoxazoline synthesis, 248
    trifluoromethyl ketone hydrogenation, 249
    α, β-unsaturated γ-dicarbonyl compound, 250
ring opening reaction, 244–245
S-acylation, 227
sonochemical switching, 263
sulfonation, 228
ultrasonic cleaning bath, 264–265
ultrasound range diagram, 264

Organic pollutant
    adsorption process, 448
    air sample analysis
        different extraction technique, 366–367
        matrix effects, 365–366
        PAH, extraction efficiency, 366
        PCB, extraction efficiency, 365–366
        PUF plug, 364–365
        ultrasonication, 362–363
        ultrasonic extraction optimization, 364
    analysis of, 347
    bioremediation, 447
    bubble formation, growth, and collapse, 449
    bubble parameters, 449–450
    catalyst, 464
    chromatographic analysis, 348–349
    cleanup procedure, 349
    concurrent analysis, 465
    dye degradation, 455, 462–463
    energy transfer efficiency, 454
    green remediation technology, 470
    laboratory-based US-assisted remediation, 455
    liquid-liquid extraction, 349
    LLE and SPE method, 346
    operating pressure and temperature, 453
    PCB and PAH, 346–347
    polychlorinated biphenyls, 345
    polycyclic aromatic hydrocarbons, 345–346
    Rayleigh–Plesset equation, 451
    reaction zone, 449–450
    reagents and solvents, 348
    real-water, soil, and air samples, 351
    shake-flask extraction, 350
    soil analysis
        design matrix, 356, 358, 360–361
        extraction solvent, 361–362
        fortification level, PAH recovery, 362
        fortified real soil sample extraction efficiency, 362–363
        optimum extraction procedure, 358
        recovery experiment, 360
        soxhlet and shake-flask extraction, 358–359
        spiked soil, PCB recovery, 358–359
        USE applicability, 356, 358
    solid phase extraction, 349–350
    sonication system, 469–470
    soxhlet extraction, 347, 350
    synergistic effect, 464–465
    ultrasonication, 348
    ultrasonic frequency and intensity, 452–453
    ultrasonic reactor, 453–454
    ultrasonic solvent extraction, 350
    ultrasound-assisted degradation
        benefits, 455
        dye, 465–466
        hormone, 469
        pesticides, insecticides, and herbicides, 467–468
        pharmaceuticals, 468–469
        remediation, 455–461
    US transducer, 454
    water analysis
        design matrix, 352
        emulsification and mass-transfer phenomena, 351

extraction efficiency, 351, 353–354
extraction time, 355
fortification level, PAH recovery, 355–356
fortified real water sample extraction efficiency, 356–357
solvent volume, 352–353
spiked distilled water, PCB recovery, 353
USE procedure validation, 353
zero valent iron, 448
Organometallic chemistry
  Barbier reaction, 295
  catalytic applications, 203–204
  coordination polymers, 202–203
  direct C-M bonds complex
    azine derivatives, 199–200
    copper and nickel atoms, 191–192
    divinylzinc complex, 184–185
    electrochemical study, 188
    ferrocene, 187–188
    Fischer carbene complex, 185
    Grignard reagent, 184
    halogen-containing metal complexes, 196–197
    LiPc particles, 194
    metalated peptide complexes, 201
    N-containing carboxylates, 198–199
    N-containing metal complexes, 189
    nicotinic and isonicotinic acid coordination mode, 197–198
    non-substituted metal phthalocyanines, 190–191
    N-,O-,S-containing metal complex, 202
    nucleophilic aromatic substitution, 188–189
    O-containing metal complexes, 191–195
    oxygen-reducing catalyst, 189
    PcM formation mechanism, 190–191
    phenylethyne cobalt hexacarbonyl complex, 185–186
    phthalocyaninate formation, 192–193
    porphyrin application, 189–190
    porphyrin preparatiom, 189
    Schiff base complex, 197
    S-containing metal complexes, 195–196
    S,O-containing metal complexes, 201–202
    SWCNTs sonochemical route, 187–188
    Te-containing metal complexes, 196
    Znq_2 nanorods formation, 200–201
    Znq_2 product TEM image, 200
  five-membered rings, 297–299
  Heck reaction, 294
  indium-mediated procedure, 292
  metal-mediated alkylation, 295–296
  nanoparticle synthesis, 204–205
  σ-and π-organometallics, 183
  organotllurides, 293
  palladium-catalyzed coupling process, 292
  Reformatsky reaction, 291
  six-membered rings, 299–301
  SONITEK, 183–184
  Sonogashira reaction, 294–295
  spiro heterocycles, 301–302
  three-membered rings, 296–297
  UO_2 dissolution, 204–205
  US-mediated Suzuki reaction, 292
Oxford cell, 629
Oxidative desulfurization (ODS), 564

P

Particle addition effects, 592
Particle cavitation, 502
Particle rounding technology, 79
Perfluorooctane sulfonate (PFOS)
  half-life, 379
  sonochemical degradation, 379–380
  ultrasound, 380
Perfluorooctanoate (PFOA)
  half-life, 379
  sonochemical degradation, 379–380
  ultrasound, 380
Pharmatose, 88
Photosonochemistry, 689–690
Phytoremediation, 408
Plastic ultrasonic welding process, 563
Pollet cell (SonoEcoCell), 629–630
Pollet equation, 633–634
Polyacrylonitrile (PAN), 7
Poly(3,4-ethylenedioxythiophene) (PEDOT) films, 608
Poly(D,L-lactide-coglycolide)/hydroxyapatite (PLGA/Hap) core-shell nanospheres, 60
Polymer synthesis
  acoustic cavitation and sonochemistry
    bubble growth and collapse, 476–477
    bubble radius-time curve, 476
    chemical effects stemming, 476–477
    physical effects, 477
    rectified diffusion, 476
  acoustic cavitation effect, 475
  emulsion polymerization
    compartmentalization, 480
    compartmentalized pseudo-bulk system, 480
    critical micelle concentration, 479
    pseudo-bulk system, 480
    theory of, 479–480
    zero-one system, 480
  free radical polymerization
    initiation process, 477–478
    propagation stage, 478
    termination stage, 478–479
  microspheres, 494–495
  miniemulsion polymerization, 480–481
  polymer nanocomposites
    asymmetric poly(MMA)-SiO_2 particle, 493–494
    encapsulation, 488–489
    magnetic separation and redispersion, 493
    miniemulsion polymerization pathway, 489–490
    MWCNT, 491
    one-step sonochemical method, 490
    PBMA/Fe_3O_4 nanocomposite dispersion, 491–492
    poly(BMA) latex nanoparticle, 491–492
    pyrene fluorescence emission spectra, 490
    synthesis of, 493
    water-toluene dual phase system, 493–494
  sonochemical polymerization, heterogenous system
    horn-type sonicator, 485
    Keller–Miksis equation, 487
    methacrylate monomer, 487–488
    miniemulsion polymerization pathway, 485–486
    MMA, 486–487
    monomer droplet growth, 483–484
    n-butyl methacrylate, 488

poly(styrene) latex particle synthesis, 484
polymer molecular weight, 488–439
styrene, 484
sonochemical polymerization, homogenous system, 483
ultrasonic depolymerization, 481–482
ultrasonic polymer synthesis, 482–483
Polymer technology
cavitation effect, 62
depolymerization and synthesis, 59–60
hydrogel, 61
ibuprofen, 61–62
isotactic polypropylene and b-isotactic polypropylene, 61
PLGA/Hap core-shell nanospheres, 60
polydispersity-index variation, 60
sol-gel sheaths, 61
Primary nucleation, 555
Protein extraction technologies, 172–173
Pulsed laser ablation method, 115
Pulsed sonoelectrochemistry, 112–113
Pulsed ultrasound effects, 590–591
Pyrex®, 664

Q

Quality assessment, fruits and vegetables
apple, 154
avocado
attenuation $vs.$ DW content, 151–152
attenuation $vs.$ firmness and linear regressions, 153
continuous-touch ultrasonic systems, 152
directional decay rate, 150
firmness, wave attenuation, and velocity, 152–153
low-temperature storage, 153
nonlinear regression, 152
physicochemical changes, 151
wave velocity measurements, 150
mango
continuous-touch ultrasonic systems, 154
nondestructive determination, 156
parabolic expressions, averaged firmness values, 155
polynomial expression, averaged sugar contents, 155
ultrasonic attenuation, 154, 156
olives, 156–157
physicochemical changes, 150
plum, 157
potato, 157–158
tomato, 158

R

Rayleigh–Plesset (RP) equation, 451
Reactor design
acoustic field
active zones, 602–603
characterization, 601
Fresnel zone, 602
laser argon beam, 601–602
luminous intensity, 603
ultrasonic reactor design, 600
dimensionless number
Reynolds number, 618
Schmidt number, 619

direct quantification, ultrasound effects
chronoamperometry curve, 609–610
dimensionless numbers, 612–613
electrochemical processes, 605
electronic transfer, 606
mass-transfer coefficient, 609
mass transport–limited current density $vs.$ acoustic intensity, 612–613
PEDOT electrosynthesis, 609–610
Sherwood number $vs.$ acoustic intensity, 611–613
two-zone model, 609–610
well-known electrodiffusional method, 606–609
zinc corrosion mechanism, 611–612
equivalent velocity tool
acoustic energy distribution applications, 615–617
boundary conditions of hydrodynamics, 614
cavitation quantification, 617–618
coordinate system, 614
Levich hypotheses, 614
Navier–Stokes equations, 614
sonochemical processing, 599
sonoelectrochemical reactor, 604–605
sonoreactor, 620–621
Rectangular sonoreactors, 585–586
Reformatsky reaction, 291
Registration, Evaluation, Authorization and Restriction of Chemicals (REACH), 15
Reynolds number, 618
Ring opening reaction, 244–245
Rotating disk electrode (RDE), 47–48

S

S-acylation, 227
Salbutamol sulphate (SS), 84–86
Scanning electron micrographs (SEM), 45
Schiff base complex, 197
Schmidt number, 619
Secondary nucleation, 555
Shake-flask extraction, 350
Sherwood number $vs.$ acoustic intensity, 611–613
Shock waves
diffusion-limited model, 510–511
physical effect of cavitation bubbles, 506
Sludge treatment, 570
Sodium chloride aerosols, 100
Soil and sediments remediation, 571
Soil vapor extraction, 408
Soil washing, 571–572
Solar photocatalysis, 440–441
Solid phase extraction (SPE), 349–350
Sonicslurry®, 79
Sonocatalysis, 266
Sonochemical effect, *see* Cavitation bubbles, chemical effects
Sonochemical polymerization
heterogenous system
horn-type sonicator, 485
Keller–Miksis equation, 487
methacrylate monomer, 487–488
miniemulsion polymerization pathway, 485–486
MMA, 486–487
monomer droplet growth, 483–484
n-butyl methacrylate, 488

poly(styrene) latex particle synthesis, 484
polymer molecular weight, 488–489
styrene, 484
homogenous system, 483
Sonochemical reactor
characteristics, 582
chemical effects, 581
hot spot, 581
large-size sonochemical reactor
liquid height effect, 592–593
sonochemical efficiency, 593–594
liquid flow effects, 591–592
particle addition effects, 592
physical effects, 581
pulsed ultrasound effects, 590–591
reactor design, 585–586
sonication frequency
acoustic degradation, 588
chemical effects, 587
G-value, 588
terephthalate solution, 587
TPPS, 587–588
sonochemical intensity quantification
calorimetry, 582–584
cavitation noise, 585
chemical dosimetry, 584–585
superposition of sonochemical fields, 589–590
ultrasonic intensity, 581–582
Sonochemical switching, 281
Sonochemistry
bubble formation, liquids, 26–27
definition, 23
high-energy cavities
bubble dynamics, 27
hot spot, 27–28
ligand substitution, volatile metal carbonyls, 28
sonication, 28
sonoluminescence, 28–30
pharmaceutical sciences
cell therapy, 101
chemotherapy, 101
drug synthesis, 98–99
NCE, 97
photo-acoustic evaluation, pharmaceutical tablets, 100
polymerization and depolymerization, 99
sonocrystallization, aerosol formation, 100
TDD, 100–101
ultrasonic driven powder transport system, 97–98
ultrasonic extraction, 99
ultrasonic irradiation of toxic effluent, 101–102
sonochemical activation
anti-Arrhenius effect, 35
cavitation bubbles, homogeneous medium, 30–31
cycloadditions, 36
divergent pathways, 34–35
experimental parameters, 32–33
frequency effects, 33
Friedel–Crafts substitution, 34
heterogeneous liquid–liquid system, 31
hydrogenation and dehydrogenation reactions, 36
hydrogen radicals, 32
hydroxystannation, alkenes, 35
liquid–solid interface, 31

MBSL spectra, 30–31
microbubbles, 30
multicomponent reaction, 36
oxidation, 31–32
radical species formation and recombination, 32
reaction pathways, 34
single-and multi-bubble cavitation, 29
sonochemical switching, 34
sonolysis, 30
temporal pressure evolution, 33
ultrasound-accelerated synthesis, ionic liquids, 36–37
ultrasound-promoted coupling, 36–37
sound spectrum, 24–25
true sonochemistry, 281
and ultrasound, 409
ultrasound-assisted chemistry, 23
Sonocracking™ process, 564–565
Sonocrystallization
ultrasound-assisted industrial synthesis
agglomeration, 557–558
applications, 560–561
crystal growth rate, 555–556
crystal structure, 556
enhanced bulk-phase mass transfer, 560
gibbs energy, 554
growth rate, 560
microstreaming, 558
MZW, 558
nucleation mechanisms, 554–555
organic and inorganic compounds, 553
polymorphism, 556
primary nucleation, 555
probe systems, 561
saturated solution, 562
secondary nucleation, 555
supersaturation/supercooling condition, 557
thermal separation process, 554
tubular flow system, 561
ultrasonic homogenizer, 559
US irradiation affects, 559–560
ultrasound-enhanced physical and chemical processes
dominant crystal size, 529–531
experimental and simulation results, 524–529
experimental conditions, 524
experimental procedure and analysis, 522–523
simulation results, 529
Sono-electroanalytical chemistry, 644
Sonoelectrochemical cell
Besançon cell (double wall cell), 630–631
Oxford cell, 629
Pollet cell (SonoEcoCell), 629–630
Sonoelectrochemical reactor, 604–605
Sonoelectrochemical synthesis, nanoparticles
alloy nanopowders, 58
conducting polymer nanoparticles, 59
conductive polymer nanoparticles, 55
experimental conditions, 55–56
massive nucleation, 54
metallic nanopowders, 55, 57–58
nanopowder production, 53–54
pulse distribution, 54
pulsed sonoelectrochemistry method, 53

semiconductor nanopowders, 58–59
three-electrode configuration, 54
Sono-electrochemiluminescence, 643
Sonoelectrochemistry
    acoustic streaming, 623
    alloy nanopowders, 113
    cavitation bubbles, 625
    conducting polymer nanoparticles, 113
    and corrosion, 639–640
    current density, 114
    electrochemical current pulse time, 114
    electrochemical synthesis, 111
    electrochemical systems, 623
    fuel cell materials, 624
    homogeneous chemical reactions, 624
    metallic nanopowders, 112–113
    nucleation process, 111
    power ultrasound effect, 625
    semiconductor nanopowders, 113
    sonotrode/sonoelectrode, 111–112
    stabilizer, 114–115
    synthesis cell temperature, 114
    ultrasonic equipments
        sonoelectrochemical cell, 629–631
        ultrasonic baths and ultrasonic probes, 626–628
    ultrasound applications
        aqueous solutions, 646
        corrosion and sonoelectrochemistry, 639–640
        electrode-kinetics, 632–638
        environmental sono-electrochemistry, 643–644
        exotic solvents, 649–653
        fuel cell materials, 645–646
        mathematical models, 632
        nanomaterials, sonoelectrochemical synthesis, 644–645
        surfactants and polymers effect, 646–649
        ultrasonic irradiation, 632
    ultrasound pulse intensity, 114
    ultrasound pulse time, 114
Sono-electrodeposition, 638–639
Sono-electro-organic chemistry, 641
Sono-electroplating, 638–639
Sono-electropolymerization, 642–643
Sonogashira reaction, 294–295
Sonoreactor, 620–621
Soxhlet extraction, 347, 350
SPE, see Solid phase extraction
Spherical spongelike particles (SSPs), 647–648
Spray dried sonocrystallized salbutamole sulphate (SD-SCSS), 86
Stokes–Einstein equation, 607
Sulfonation, 228
Supercritical fluid extraction (SFE), 551
Suzuki coupling, 688
Suzuki reaction, 292–293
Synthetic chemistry metrics
    atom efficiency, 11
    carbon efficiency, 11
    conversion X, 9–10
    EcoScale, 11–12
    effective mass yield, 10
    environmental factor, 10
    one-step chemical synthetic process, 9
    reaction mass efficiency, 11
    yield of target product, 10
Synthetic organic chemistry
    carbon-heteroatom (C-X) bond formation reaction, 307–308
    cavitation, 281
    chemical disposals and energy consumptions, 281
    condensation reaction, 289–291
    cycloaddition reaction
        cyclopentadiene and dienophiles, 284
        Diels–Alder reaction, 282
        furano diene derivative, 283
        mechanistic survey, 284
        one-pot process, 282–283
    multicomponent reaction, 285–286
    nucleophilic addition reaction, 287–289
    organometallic chemistry
        Barbier reaction, 295
        five-membered rings, 297–299
        Heck reaction, 294
        indium-mediated procedure, 292
        metal-mediated alkylation, 295–296
        organotllurides, 293
        palladium-catalyzed coupling process, 292
        Reformatsky reaction, 291
        six-membered rings, 299–301
        Sonogashira reaction, 294–295
        spiro heterocycles, 301–302
        three-membered rings, 296–297
        US-mediated Suzuki reaction, 292
    oxidation and reduction reaction
        aldehydes and ketone coupling, 304–305
        Baeyer–Villiger oxidation, 303–304
        cinnamaldehyde hydrogenation, 304
        RuI_3-mediated degradation, 303
    protection and deprotection procedures, 305–307
    true sonochemistry, 281

T

TDD, see Transdermal drug delivery
TEM, see Transmission electron microscopy
5,10,15,20-Tetraquis(*p*-hydroxyphenyl)porphyrin (TPPOH), 189
Textile dye degradation
    Acid Green 20, 63
    Acid Orange 7, 62, 64
    Congo Red, 63
    decolorization, 62–63
    Fenton reagent process treatments, 62
    H_2O_2-Fe_3O_4 system, 64
    hydrophobic enrichment, 65
    power density effect, 62–63
    Reactive Orange 16, 64
    Rhodamine B, 63–64
Thermal desorption, 408
Transdermal drug delivery (TDD), 100–101
Transmission electron microscopy (TEM), 669–670
Triphenylphosphine oxide (TPPO), 101
True sonochemistry, 281
Tubular flow system, 561
Two-zone model, 609–610
Type I homogeneous medium, 678
Type II and III heterogeneous system, 678

Index

U

Ullmann reaction, 235
Ultrasonication
    activated carbon amendment, 415
    advanced oxidative soil remediation, 413–414
    electrokinetic remediation, 414–415
    soil-flushing, 412–413
    surfactant-aided soil-washing, 415
Ultrasonic cavitation, 502
Ultrasonic reactor, 453–454
Ultrasonic solvent extraction (USE), 350
Ultrasonic spray pyrolysis (USP)
    macroporous materials, polymer-based nanocomposites, 118
    mesostructured nanomaterials, organic–inorganic hybrid nanocomposites, 117
    mesostructured nanomaterials, organic–inorganic hybrid nanocomposites, 117
    nanosized structures, metal salt-based nanocomposites, 118–119
    nanosized structures, silica-based nanocomposites, 118
    nanostructured semiconductors, chemical aerosols, 119–120
    phase isolation, 116
    sacrificial materials, 117
    thermal decomposition, solid/liquid particles, 116
Ultrasonic water treatment system, 569
Ultrasound and materials science
    acoustic cavitation, 42
    definition of ultrasound, 41–42
    electrochemical metallization (*see* Electrochemical metallization)
    environmental protection
        aerobic and anaerobic sewage treatment, 62
        aromatic hydrocarbons treatment, 65–66
        physical and chemical affects, 62
        sewage sludge treatment, 66
        surface decontamination, 66–67
        textile dye degradation (*see* Textile dye degradation)
    heterogeneous solid-liquid system, 42
    polymer technology
        cavitation effect, 62
        depolymerization and synthesis, 59–60
        hydrogel, 61
        ibuprofen, 61–62
        isotactic polypropylene and b-isotactic polypropylene, 61
        PLGA/Hap core-shell nanospheres, 60
        polydispersity-index variation, 60
        sol-gel sheaths, 61
    pulse echo technique, 41
    sonochemical surface modification
        ABS *vs.* etching time, 43–44
        adhesion, 42–43
        atomic percentage oxygen, 45–46
        definition, 42
        desmear process, 42–43
        lead zirconium titanate, 43
        Noryl (polyphenylene/polyester) material, 45–46
        printed electronics, 42
        SEM, 45
        ultrasonic frequency, weight loss, 46
        weight loss, ceramic material, 43–44
        wet manufacturing techniques, 43
        XPS measurements, 44
    sonoelectrochemical synthesis, nanoparticles
        alloy nanopowders, 58
        conducting polymer nanoparticles, 59
        conductive polymer nanoparticles, 55
        experimental conditions, 55–56
        massive nucleation, 54
        metallic nanopowders, 55, 57–58
        nanopowder production, 53–54
        pulse distribution, 54
        pulsed sonoelectrochemistry method, 53
        semiconductor nanopowders, 58–59
        three-electrode configuration, 54
    therapeutic ultrasound, 67
Ultrasound and sonication
    application of, 329–330
    cavitation, 328–329
    merits and demerits, 329
    parameter monitoring, 332–333
    sludge disintegration and control parameter, 330–332
Ultrasound-assisted anaerobic digestion
    anaerobic digestion
        merits and demerits, 327–328
        oxidation–reduction reaction, 326
        stages of, 326–327
    enhanced biogas production, 336–337
    innovative pretreatment, 325
    sludge
        dewaterability, 332–333
        management and/or conditioning technique, 323–324
        minimization, 326
        stabilization, 324
        treatment, 323
    sonicated sludge, 332, 334–335
    thermal pretreatment, 323–324
    ultrasound and sonication
        application of, 329–330
        cavitation, 328–329
        merits and demerits, 329
        parameter monitoring, 332–333
        sludge disintegration and control parameter, 330–332
    ultrasound technology and anaerobic digestion, 325
Ultrasound-assisted degradation
    benefits, 455
    dye, 465–466
    hormone, 469
    pesticides, insecticides, and herbicides, 467–468
    pharmaceuticals, 468–469
    remediation, 455–461
Ultrasound-assisted industrial synthesis
    air, 572
    biological and chemical contaminants
        biological wastewater treatment, 569–570
        industrial ultrasonic processors, 570–571
        sludge treatment, 570
        ultrasonic chamber, 569
        ultrasonic water treatment system, 569
        ultrasonic waves, 568
    biosensors, 545–546
    crude oil industry, 563–565

extraction processes, 551
filtration, 565–566
food processing and preservation
   applications, 542
   crystallization and freezing, 541–542
   drying and evaporation, 539–540
   emulsification and mixing, 540–541
   living cells, 537–539
gene transfer
   plasmid DNA, 545
   sonoporation effect, 544
industrial application, 573
metal and plastic welding, 562–563
microbial cell disruption, 543
oil extraction, 552–553
separation and cleaning, 565–566
soil and sediments remediation, 571
soil washing, 571–572
sonocrystallization
   agglomeration, 557–558
   applications, 560–561
   crystal growth rate, 555–556
   crystal structure, 556
   enhanced bulk-phase mass transfer, 560
   Gibbs energy, 554
   growth rate, 560
   microstreaming, 558
   MZW, 558
   nucleation mechanisms, 554–555
   organic and inorganic compounds, 553
   polymorphism, 556
   primary nucleation, 555
   probe systems, 561
   saturated solution, 562
   secondary nucleation, 555
   supersaturation/supercooling condition, 557
   thermal separation process, 554
   tubular flow system, 561
   ultrasonic homogenizer, 559
   US irradiation affects, 559–560
synthesis
   acoustic energy, 546
   biodiesel, 546–547
   electroplating, 548–549
   nanostructured materials, 548
   sonochemical, 546
   vegetable oil, 547–548
textile industry
   de-oiling process, 550
   dyeing process, 549–550
   ultrasonic washing process, 550
USFE system, 551
water and wastewater treatment, 567–568
Ultrasound-assisted particle engineering
   cavitation and acoustic streaming, 76
   crystallization from solution
      solution atomization and sonication, 77–78
      sonocrystallization, 76–77
   drug micronization, 75
   inhalation drug delivery system
      aerodynamic particle size, 82
      agglomeration, 82–83
      amorphous particles, 86
      budesonide, 84

      cohesive–adhesive balance, 82
      DPIs, 81–82, 87
      drug molecules, 75
      fluticasone propionate and salmeterol xinafoate, 87
      FPF, 86
      HFA–pMDI formulation, 84
      in vitro aerosolization, 84
      lactose, 87–89
      pulmonary drug delivery, 81
      salbutamol sulphate, 84–86
      SAXS technique, 83–84
      scanning electron micrograph, NaCl particles, 83–84
      SD-SCSS, 86–87
      spray drying, 83
      supersaturation, 85
      US irradiation, 83
   limitations, 93–94
   melt sonocrystallization, 79–80, 91
   particle design with enhanced dissolution rate
      caffeine/maleic acid, 92
      CA nanoparticles, 89–90
      cocrystal, 92
      gembibrozil, 92
      ibuprofen, 91
      insonation, 91
      MSC celecoxib, 91
      sonoprecipitation, 89
      supersaturation, 89
   scale-up, 93
   slurry/suspension, 78–79
   small-scale applications, 76
   ultrasound-assisted polymer extrusion, 80–81
Ultrasound-assisted synthesis, nanomaterials
   cavitation and nebulization, 106, 116
   nanostructured material synthesis, 107, 117
   sol–gel technique, metal oxide nanostructures, 108–109
   sonochemical decomposition, metal chalcogenide synthesis, 109–110
   sonochemical reduction, 107–108
   sonoelectrochemistry
      alloy nanopowders, 113
      conducting polymer nanoparticles, 113
      current density, 114
      electrochemical current pulse time, 114
      electrochemical synthesis, 111
      metallic nanopowders, 112–113
      nucleation process, 111
      semiconductor nanopowders, 113
      sonotrode/sonoelectrode, 111–112
      stabilizer, 114–115
      synthesis cell temperature, 114
      ultrasound pulse intensity, 114
      ultrasound pulse time, 114
   ultrasound-assisted laser ablation, 115–116
   ultrasound-induced deposition, 110–111
   USP (see Ultrasonic spray pyrolysis)
Ultrasound effects, enzymes, 166
Ultrasound-enhanced physical and chemical processes
   sonochemical degradation
      analytical procedure, 512
      characteristics of cavitation phenomena, 515–520
      experimental results, 513–514
      experimental setup, 511–512

Index

physical mechanism of degradation, 520–521
simulation results, 514–515
sonocrystallization
dominant crystal size, 529–531
experimental and simulation results, 524–529
experimental conditions, 524
experimental procedure and analysis, 522–523
simulation results, 529
Ultrasound-mediated amorphous to crystalline transition (UMAX), 78–79
Ultrasound wave propagation, fresh agricultural tissue
attributes, 132
directivity distribution
acoustic pressure, 135
apparent attenuation, 136
directivity parameter, 135
elastic/viscoelastic forces, 134
experimental system, 136–137
mechanical oscillator, 134
planar wave attenuation, 136
three-dimensional field, half-cut potato, 136–137
true attenuation losses, 136
ultrasonic wave transducer, 134
wave attenuation and amplitude, 133–134
wave modes and velocities, 133
UMAX, *see* Ultrasound-mediated amorphous to crystalline transition
United Nations Environment Programme (UNEP), 408
United Nations Framework Convention on Climate Change (UNFCCC), 13
US-assisted oxidative desulfurization (UAOD), 563
US-assisted supercritical fluid extraction (USFE), 551
USP, *see* Ultrasonic spray pyrolysis

V

Volatile organic compounds (VOCs), 680–681

W

Water and wastewater treatment
anthropogenic contaminant
bisphenol A degradation, 381
levodopa, 380
MTBE degradation, 381–382
PFOS and PFOA half-life, 379
PFOS and PFOA sonochemical degradation, 379–380
ultrasound, 380
ultrasound-mediated pollutant degradation, 379
cavitational bubble, 374
contaminant property
hydrophobicity, 378–379
volatility, 377–378
Fenton's reagent, 375
filtration process
membrane filtration and membrane fouling, 392–393
membrane fouling control, 396–397
membrane integration, 395–396
solution chemistry effect, 397–398
ultrasonic control mechanism, 393–395
NOM degradation
aromaticity, 386–388
chemical nature, 383
^{13}C NMR, 386
environmental media NOM concentration, 382–383
H_2O_2 formation, 384–385
HPSEC spectra, 386, 389
humic substance properties, 383
hydrophobicity, 386
implication, 390
molecular weight, 388–389
shearing degradation, 384
sonochemical degradation, 383–384
TOC reduction, 385
total acidity, 389–390
UV/Vis absorbance, 385–386
OH radical degradation, 375
sonophysical effect, 375
temperature profile, 374
ultrasonic factors
frequency, 376
power, 376
pulsed or continued sonication, 377
ultrasonic technology, 373–374
ultrasound and disinfection, 390–392
Water, clean reactive medium, 680–681
Well-known electrodiffusional method
Fick's second law equations, 607
limiting concentration profile, 608
mass transfer experiments, 606, 608–609
Nernst–Planck equation, 606–607
PEDOT films, 609
Stokes–Einstein equation, 607
World Commission on Environment and Development (WCED), 12
World Summit on Sustainable Development (WSSD), 13

X

X-ray photoelectron microscopy (XPS) measurements, 44

Z

Zero valent iron (ZVI), 427–428